STRUCTURAL STEEL DESIGN

FOURTH EDITION

STRUCTURAL STEEL DESIGN

FOURTH EDITION

ABI AGHAYERE

JASON VIGIL

MERCURY LEARNING AND INFORMATION

Boston, Massachusetts

Mercury Learning and Information
121 High Street, 3rd Floor
Boston, MA 02110
info@merclearning.com

A. Aghayere and J. Vigil. *Structural Steel Design, Fourth Edition.*
ISBN: 978-1-50152-330-4

Library of Congress Control Number: 2024948824

242526321 This book is printed on acid-free paper in the United States of America.

To my dearly beloved wife and forever bride, Josie Osigbemhe Aghayere, for your selfless love and encouragement all through the years, and…

*To my beloved children - Osa, Ito, Odosa, and Eghosa,
and to my adorable grandchildren - Zeke, Micah, and Archer, and….*

*To my dearly beloved mother of blessed memory,
Regina Ekeneza-Obasogie (1937–2014), for your selfless love and dedication,
and for instilling in me the value of education and hard work.*

Contents

Preface

The knowledge and expertise required for the design of steel-framed structures is essential to architectural, civil, and structural engineers, as well as to students intending to pursue careers in the field of structural design and construction. This textbook provides the fundamentals of structural steel design from a structural systems framework that involves not only the isolated design of individual structural steel elements or components, but the design of these elements within the context of the entire structural system. This enables the reader to see the connection between each structural element and how and where they fit within the entire structural system, in addition to engendering a big picture view of structural steel design.

We strive to bridge the divide between engineering education and professional practice by exposing the reader to structural engineering principles and theory as well as practical applications of the design methods, so that what is learned in this text is directly applicable and relevant to professional design practice. We go beyond standard textbook-style exercises to include real-world-type structural engineering exercises, together with a project-based approach that instills the fundamentals of structural behavior and design, and enables the student to gain an appreciation of the design constraints, and learn from the iterative nature of the structural design process.

The intent is that by the time the reader has worked through all the chapters of the text and the corresponding design project tasks related to each chapter, they will have completed the design of an entire steel structure, enabling them to see the big picture and thus reinforcing the connections between the design of individual structural steel elements and the design of a complete structural system.

The American Institute for Steel Construction encourages the exposure of students to the design of structural steel elements within a realistic building context. We present practical details, real-world examples, and end-of-chapter exercises that not only provide the reader with an essential background on structural steel design, but also provide subject material that closely mirrors the details and examples found in professional engineering practice, thus making the relevance of the material studied to engineering practice more apparent to the student. Many of the examples and the end-of-chapter exercises presented are characteristic of the types of technical exercises that the reader will encounter in professional engineering practice.

This approach of balancing theory and practical applications will enhance the learning of structural steel design and engender design creativity in the reader.

INTENDED AUDIENCE

This text is ideal for students in any undergraduate course in structural steel design and will sufficiently prepare them to apply the fundamentals of structural steel design to typical projects that they might come across in professional engineering practice. It is suitable for students in civil engineering, architectural engineering, structural engineering, and construction engineering programs, as well as architecture programs and civil engineering technology, construction engineering technology, and construction management programs. This text can also serve as a good resource and reference for practicing structural engineers, and as a useful tool for the mentorship of entry-level civil/structural engineers in industry. It is also a helpful reference and study guide for the Fundamentals of Engineering (FE), the Professional Engineering (PE), and the Structural Engineering (SE) exams. This text covers the course content for the Steel Design I course and a majority of the course content for the Steel Design II course in the curriculum for basic education of a structural engineer proposed by the National Council for Structural Engineering Associations (NCSEA).

UNIQUE FEATURES OF THIS TEXT

We blend the "how's" and "why's" of structural steel design into a balanced structural steel design text that bridges the gap between theory and engineering practice.

We present numerous realistic details and diagrams to help illustrate the structural steel design process, similar to what is done in professional engineering practice, while at the same time discussing the theory behind the design equations. In Chapter 14, we introduce two student design projects to expose the reader to the important aspects of a real-world structural steel building project and a pedestrian steel bridge design project. Several other unique features of this text are listed below:

1. The use of real-world structural plans and details, and practical real-world examples, including practical information on structural drawings.
2. An introduction to the techniques for laying out floor and roof framing, and for sizing floor and roof decks for gravity loads, as well as how to lay out and distribute lateral forces to the the lateral force resisting systems in buildings. General rules of thumb for selecting beam spacing, depth and thickness of deck, and the direction of framing for the beams and girders are discussed.
3. A discussion of other rules of thumb for sizing of structural steel to allow for the quick selection of common structural members (e.g., open-web steel joists, beams, and columns).
4. The calculation of gravity and lateral loads in accordance with the ASCE 7 provisions is included.
5. The design of column base plates and anchor rods for axial loads, shear, uplift, and moments.
6. Step-by-step design of moment frames with a design example, as well as the design of moment connections.
7. An introduction to floor vibration analysis and design based on AISC Design Guide No. 11.
8. A chapter on practical considerations helps to reinforce the connection between structural element/member design and structural system design in practice.
9. The behavior, analysis and design of horizontal diaphragms, chords, and collectors (or drag struts) is discussed. A discussion of practical details that ensures adequate and identifiable lateral load path from the roof and floor diaphragms to the lateral force resisting systems is presented.

10. An introduction to the flexure analogy method for the analysis of torsion in steel members is discussed, and practical real-world examples are provided.
11. An introduction to the strengthening and rehabilitation of steel structures, and to performance-based design.
12. Coverage of other important topics, including beam copes and beam web holes and their reinforcing; X-braces using tension rods, clevises, and turnbuckles; stability bracing of beams and columns; beam design for net wind uplift loads; and introduction to coatings for structural steel.
13. The impact of differential axial shortening of columns in tall steel buildings is discussed.
14. The axial compression load capacity and the bending moment capacity of WT and double angle sections, which are frequently used as the top and bottom chords of floor and roof trusses, are discussed. The axial compression load capacity of a plane and flanged cruciform section is also discussed.
15. The vertical and lateral deflection criteria for beams and girders, and the lateral vibration criteria for tall buildings are discussed.
16. Sample structural steel plans, bracing elevations, moment connection details, and details of the connection of collectors or drag struts to the lateral force resisting system, and sample roof and floor plan notes are presented to acquaint the reader with the information that should be included in a structural drawing package for a typical project.

NEW TO THE FOURTH EDITION

- Content has been updated to correspond to the latest AISC Specification for Structural Steel Buildings (ANSI/AISC 360-22) and the 16th edition of the *AISC Steel Construction Manual.*
- Load calculations have been revised to conform to the ASCE 7-22 load standard, and examples involving uplift forces at the columns of the lateral force resisting systems are presented. Calculation of additional vertical forces (compression or tension) on the columns of the lateral force resisting systems (LFRS) caused by the overturning moments from lateral wind and seismic loads. Discussion of vehicular impact loads on exposed columns and parking garage edge barriers.
- Composite columns—embedded steel columns and concrete-filled tubes and box section—subject to axial compression loads and flexure are covered.
- Roof and floor deck selection examples using both LRFD and ASD methods are presented, including discussion of deck fireproofing methods, and the impact of fire resistance rating requirements on roof and floor deck sizes.
- Expanded discussion of residual stresses and brittle fracture, and the AISC Code requirements for heavy shapes (i.e., members thicker than 2 inches) to prevent brittle fracture.
- Torsional buckling of steel members with lateral bracing that is offset from the shear center (new AISC Sections E4d and E4e); and coverage of torsional bracings.
- Other topics covered include ductility requirements for shear studs and where the requirements can be neglected; construction joint requirements in large composite floor decks, the various categories of Architecturally Exposed Structural Steel (AESS), and web sidesway buckling and web compression buckling; discussions of structural integrity requirements, and prevention of progressive or disproportionate collapse; and discussion of fire protection

requirements and the various methods of fireproofing steel framing; tension rods in entrance canopies, and expanded discussions of the instantaneous center method for eccentrically loaded welded connections.

- Additional end-of-the-chapter exercises that mimic steel design exercises from professional practice have been added.
- Additional real-world examples of structural steel framing, structural details, bracing elevations, and moment connection details, and photos are included.
- Additional topics include wind loads on parapets, seismic analysis of podium structures, and the requirements for foundation ties in seismic design.

INSTRUCTOR'S RESOURCES

Upon adoption, an online Instructor's Manual is available to qualified instructors for download by contacting the publisher at *info@merclearning.com*.

Acknowledgments

An endeavor of this magnitude is not possible without the support of countless individuals, so we would like to express our gratitude to our students and other readers who through the years have provided feedback on the content of the previous three editions of this text. The first author would like to acknowledge the Late Robert Halsall, founding president of Halsall Associates Ltd. (now WSP), for providing me the opportunity to acquire the training and experience that have been the solid foundation for my professional career. We are also grateful to our colleagues in the consulting engineering industry and in academia who have provided feedback to us over the years.

We thank our families for their enduring support and encouragement. We would also like to thank the reviewers of this text and would welcome and appreciate any comments or questions regarding the contents of the book.

Abi Aghayere, Avondale, PA
Jason Vigil, Rochester, NY
March 2025

CHAPTER

1

Introduction to Steel Structures

1-1 INTRODUCTION

The primary purpose of this book is to present the design procedures for steel-framed buildings and structures using both the limit states design approach and the allowable strength design method in a practical, concise, and easy-to-follow format that includes the design of all the major structural steel elements. In the United States, both the load and resistance factor design (LRFD) method and the allowable strength design (ASD) method are prescribed in the latest specification for the design of steel-framed buildings from the American Institute of Steel Construction, AISC 360-22; this specification, known as "a consensus document" of the AISC Specification Committee, follows a unified approach where the nominal strength, R_n, is the same for both the LRFD and ASD methods. Although the latest AISC manual follows the unified format, with the LRFD requirements presented side-by-side with the ASD requirements, only one of these two design methods is usually taught in detail at most colleges. For the design of steel bridges in the United States, the prescribed design method for bridges receiving federal funding is the LRFD method, and most undergraduate programs in civil engineering, architectural engineering, structural engineering, and civil engineering technology in the United States offer at least one course in structural steel design using the LRFD method [1, 2]. In addition, many countries, including Australia, Canada, France, New Zealand, and the United Kingdom (among many others), have long adopted a limit states design approach like the LRFD method for the design of steel buildings, and most of the research in steel structures uses the limit states design approach. A comparative study of the cost differences between allowable stress design and LRFD methods for steel high-rise building structures indicated a cost savings of up to 6.9% in favor of the LRFD method [3]. Though the trend is toward the use of the LRFD method, especially for tall buildings and complex structures and bridges, the ASD method is still in use in some segments of the structural steel industry in the United States, especially for the design of steel connections. Consequently, for many chapters and sections in this text, we have adopted a dual format, and provided discussions and example exercises for both the LRFD and the ASD methods.

The unique feature of this text is a balanced approach that blends theory with practical applications that include the use of realistic structural plans and details in the examples and end-of-chapter exercises, the discussion of structural loads and structural steel component design within the context of the entire structural building system, and the discussion of other pertinent topics that are essential to the design of real-world structural steel buildings in practice. The blending or inseparability of theory and engineering practice [4] guides our approach in this text as we endeavor to go beyond component design to a holistic structural systems-based design while emphasizing structural behavior and enhancing the students' ability to design realistic structural members and systems. The importance of a practice-oriented approach in civil engineering education has been highlighted and advocated by Roessett and Yao [5] and by the Carnegie Foundation for the Advancement of Teaching. The Carnegie Study, "Educating Engineers: Designing for the Future of the Field" [6], also stresses the need to teach for professional practice in undergraduate engineering education. The American Institute for Steel Construction has developed a web-based teaching tool for structural steel design that incorporates realistic structural drawings, the calculation of structural loads, and structural steel component design within the context of an entire building design case study [7].

The intended audiences for this book are students taking a first or second course in structural steel design, structural engineers, architects, and other design and construction professionals seeking a structural steel design text that effectively blends theory with practical real-world applications. The book will be well suited for a steel design course with or without a design project. The reader is assumed to have a working knowledge of statics, mechanics of materials or applied mechanics, and some structural analysis. We recommend that the reader have the *Steel Construction Manual,* 16th ed. (AISC 325-23) [8] available.

1-2 THE MANUFACTURE OF STRUCTURAL STEEL

The three most commonly used construction materials around the world are structural steel, concrete, and wood. Structural steel, an alloy of mostly iron and carbon, is manufactured by one of two methods: the electric arc furnace (EAF) method and the basic oxygen furnace (BOF) method. Almost all of the structural steel in buildings can be recycled at the end of the life cycle of the building, and most of the structural steel shapes produced in the United States today are from recycled scrap steel using the electric arc furnace (EAF) process, with approximately 95% of the steel used in structural shapes in the United States obtained from recycled steel scrap material [9]. The recycling process of structural steel does not lead to any loss of physical properties, and structural steel shapes produced in the United States contain "an average of 93%" of recycled steel scrap." The old recycled steel scraps — some of which are from "cars, refrigerators, decommissioned buildings," and other steel materials (see aisc.org/sustainability)—are melted in an electric arc furnace after sorting and removing undesirable elements such as radioactive materials. Different types of scrap steel are used, depending on their chemical composition and density. The scraps are fed into the EAF, where the temperatures reach up to 3000°F. The scraps of steel are melted into molten steel and drained off at the bottom of the furnace, while the slag by-product and other impurities rise to the surface from where they can be skimmed off and removed. The resulting slag by-product is used as aggregate in road construction [9]. The molten steel is then transported to another furnace, where the steel is heated again and undergoes de-oxidation and desulfurization, and the steel is further refined by the addition of any desired chemical alloys. The carbon content is continuously monitored during this heating process until the desired carbon content of the molten steel is achieved. Various chemical elements, such as manganese, vanadium, copper, nickel, and others, can be added to produce the desired chemical composition or ASTM specification of the molten steel. The amount of carbon and alloys in steel impacts the steel properties such as yield strength, ductility, resistance to corrosion, toughness, resistance to brittle fracture, and the weldability of the structural steel. The higher the carbon content, the higher the strength and hardness, but the lower the ductility and toughness. A sample of the molten steel is taken periodically during the manufacturing process and tested to determine its chemical composition.

After this process, the molten steel is cast into long "beam blanks." They are cooled, cut to lengths and then stored until they are ready for rolling. During the rolling process, the beam blanks are placed in furnaces, reheated and melted at temperatures of up to 2400°F. The molten steel is then passed through a set of rollers (multiple times for some of the rollers) to form the desired standard steel shapes (e.g., wide flange or channel shapes). Samples of the standard steel shapes are also cut periodically, and the section and material properties (e.g., tensile yield stress) are measured at the plant in accordance with the ASTM specification and as part of the quality control process [9, 10]. After this process, the structural shapes are cooled and cut to lengths of up to 80 ft. to prepare them for transport to the steel fabricators, where the steel will be further cut to lengths specific to a given project and then detailed to receive other connecting steel members [9, 10].

In the BOF method for the manufacture of virgin steel [11], the three basic raw materials are iron ore, coal, and limestone, which are all mined. The process starts with the conversion of the coal into coke by grinding the coal and baking it in ovens in the absence of air for several hours. This process prevents the burning of the coal (since there is no oxygen) and facilitates the removal of impurities and gases from the coal. The resulting coke from the coal-baking process has approximately 90% carbon content (much higher than raw coal) with fewer impurities. The coke is cooled with water and used both as a source of fuel and as a reactant in the extraction of iron from the iron ore.

The three basic ingredients—iron ore, coke (obtained from the coal-baking process), and limestone — are fed into a blast furnace — a large vessel where the reaction of the coke with the preheated air produces heat and carbon monoxide that helps to break down the iron ore, producing pure iron and carbon dioxide. A by-product known as slag is also formed from the reaction of the limestone with the impurities in the iron ore [11]. The pig or molten iron is drained from the bottom of the blast furnace, desulfurized, and then fed into the basic oxygen furnace (BOF). The pig iron from the blast furnace has a carbon content of approximately 4% and is converted to steel in the basic oxygen furnace (BOF) by reducing the carbon content to between 0.4% and 1.5%. The oxygen in the BOF oxidizes the carbon and other impurities in the pig iron and melts the steel, producing slag and other impurities, which float to the top of the denser molten steel. The chemical composition of the molten steel is monitored, and after the desired composition is achieved, the molten steel is tapped. Alloys can be added at this stage, as needed. The molten steel is then cast into thick slabs of steel. These thick slabs can then be melted again in an electric arc furnace (EAF) — without losing its strength — and the molten steel is passed through a series of rollers to shape the steel into different hot-rolled structural steel shapes, such as wide flange shapes or channels [11]. The EAF method of steel production is less energy intensive than the BOF method. The EAF method acknowledges "the fact that steel is a cradle-to-cradle material and not a cradle-to-grave" material because it can be "recycled an infinite number of times without loss of strength, durability or flexibility." Terms like "circular life-cycle" have been used to describe the EAF method versus "extractive" for the BOF method. The BOF process requires very expensive equipment and is capital intensive.

1-3 PROPERTIES OF STRUCTURAL STEEL

The forerunners to structural steel were cast iron (a brittle material) and wrought iron (a more ductile material than cast iron); these were widely used in building and bridge structures until the mid-nineteenth century. Their low tensile strength and low ductility due to high carbon content led to bridge failures [12, 13]. The high carbon content in cast iron makes welding to cast iron difficult, so renovations to existing cast iron structures usually involve the use of bolted connections with "drilled or tapped holes (i.e., holes with threads)." In 1856, steel was first manufactured in the United States, and since then it has been used in the construction of many buildings and bridge structures. Some notable examples of buildings constructed mainly of structural steel include the 1450-ft.-tall, 110-story Willis Tower (formerly known as the Sears Tower) in Chicago, and the 1474-ft.-tall Taipei 101 building in Taiwan, with 101 floors. Structural steel is an alloy of iron and carbon and is manufactured in various standard shapes and sizes by steel rolling mills; it has a unit weight of 490 lb./ft.^{3}, a modulus of elasticity of 29,000 ksi, and a Poisson's ratio of approximately 0.30. The carbon content of commonly used structural steel varies from approximately 0.15% to 0.30% by weight, with the iron content as high as 95% [14]. The higher the carbon content, the higher the yield stress, but the lower the ductility and weldability. Higher-carbon steels are also more brittle. Structural steel is widely used in the United States for the construction of different types of building structures, from low-rise industrial, institutional, and office buildings to high-rise office and residential buildings. Figure 1-1a shows the progression of a multi-story steel-framed building under construction.

INTECH

FIGURE 1-1a A multi-story steel-framed building under construction.

Structural steel is also used in the construction of bridges, communication towers, amusement park structures, stadiums, and other structures. Steel offers competitive advantages because of its high strength-to-weight ratio. Some of the advantages of structural steel as a building material are as follows:

1. Structural steel has a high strength-to-weight ratio.
2. The properties of structural steel are uniform and homogeneous and highly predictable.
3. It has high ductility, thus providing adequate warning of any impending collapse.
4. Structural steel is sustainable as it can be recycled without any diminished material properties or loss of strength. In fact, a very high percentage of the structural steel used in many modern structures is made from recycled steel.
5. Steel structures are easier and speedier to fabricate and erect, compared to concrete or light-frame wood structures.
6. The erection of steel structures is not as affected by weather compared to other building materials, enabling steel erection to take place even in the coldest of climates. Figure 1-1d shows steel beams being erected during the construction of a multi-story steel-framed building.

7. It is very adaptable to existing conditions and relatively easier to retrofit existing steel structures because of the relative ease of connecting the new framing members to existing structural steel members.
8. Structural steel beams and girders are not susceptible to long-term effects such as creep, unlike concrete and wood structural systems that support gravity loads.
9. Steel-framed buildings are more amenable to adaptive reuse and repurposing where major structural design changes have to be accommodated.
10. Structural steel is the material of choice for the retrofitting or renovations of existing steel or concrete structures; for example, an existing building undergoing adaptive reuse that requires an existing concrete or steel column to be cut and removed to create a larger open space. Structural steel girders or trusses are typically used on both sides of the concrete or steel column that is to be removed.

Some of the disadvantages of structural steel as a building material include the following:

1. Steel is susceptible to corrosion, and it must be protected by galvanizing or by coating with zinc-rich paint, especially structures exposed to weather or moisture, although corrosion-resistant steels are also available. Consequently, maintenance costs could be high compared to other structural materials.
2. Steel is adversely affected by high temperatures and therefore often needs to be protected from fire by fireproofing. "The strength of steel remains essentially unchanged until about 600°F. The steel retains about 50% of its strength at 1100°F [and] loses all of its capacity when it melts at about 2700°F. However, for design purposes, it is usually assumed that all capacity is lost at about 2200°F" [8]. Examples of sprayed-on fireproofing for the fire protection of structural steel are shown in Figures 1-2b and 1-2c.

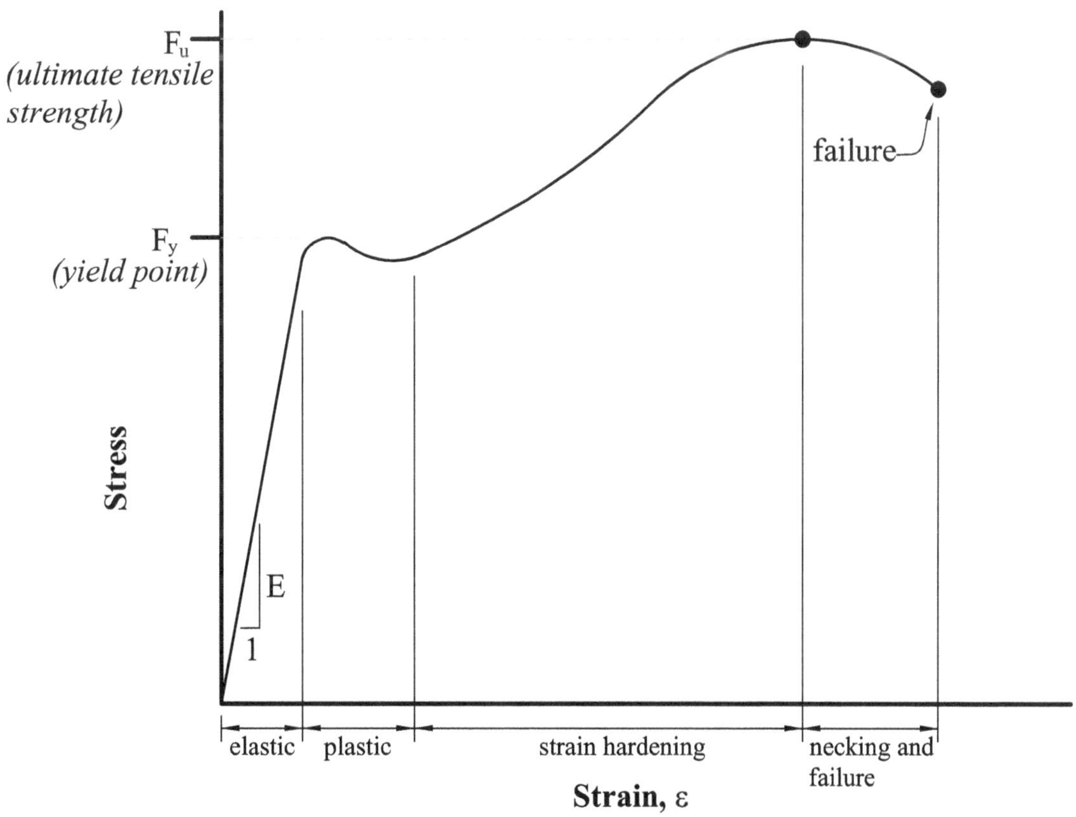

b) Steel with a defined yield point

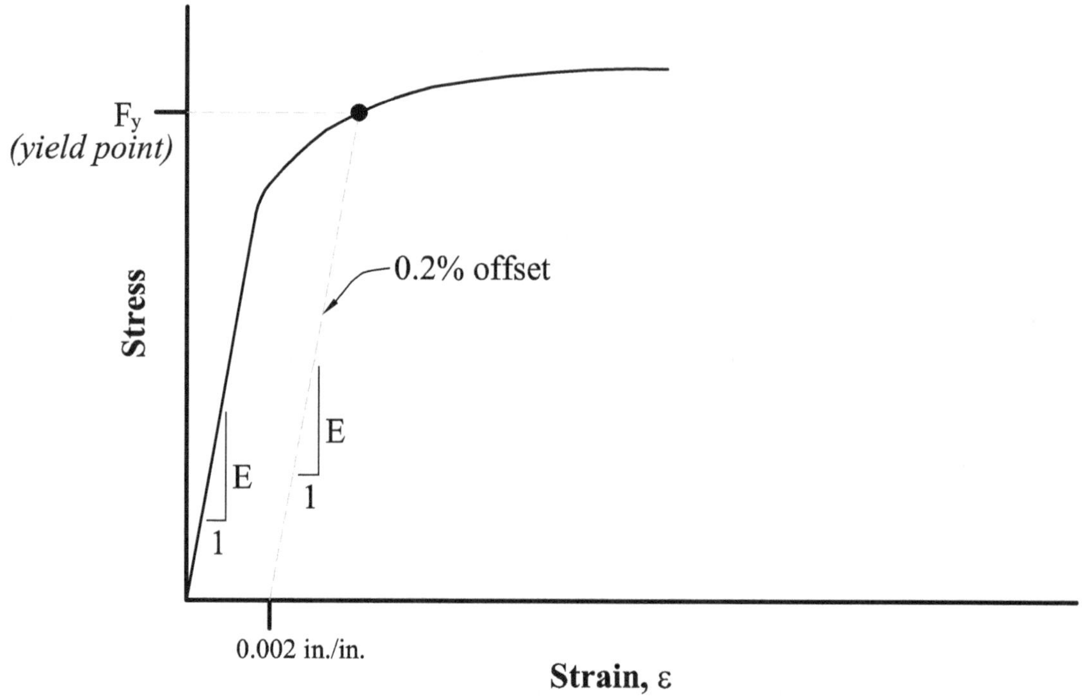

c) Steel without a defined yield point

FIGURE 1-1b,c Typical stress-strain diagram

FIGURE 1-1d Erection of structural steel beams.

3. Depending on the types of structural details used and the thickness of the steel member, structural steel may be susceptible to brittle fracture due to the presence of stress concentrations from residual stresses arising from differential cooling after a steel member is heated from, say, welding, and to fatigue failure from cyclic or repeated loadings causing stress reversals in the members and connections.

The two most important properties of structural steel used in structural design are the tensile yield stress, F_y, and the ultimate tensile strength, F_u. These properties are determined by a tensile test that involves subjecting a steel specimen to tensile loading and measuring the load and axial elongation of the specimen until failure. The tensile stress is computed as the applied tension load divided by the original cross-sectional area of the specimen, and the strain is the elongation divided by the original length of the specimen. A typical stress-strain curve for structural steel is shown in Figure 1-1b; it consists of a linear elastic region with a maximum stress equal to the yield stress, a plastic region in which the stress remains relatively constant at the yield stress as the strain increases, and a strain hardening region, the peak of which determines the ultimate tensile strength, F_u. The Young's modulus, E, is the slope of the linear elastic or straight-line region of the stress-strain curve. The ability of structural steel to sustain large deformations under constant load without fracture is called ductility, an important structural property that distinguishes structural steel from other commonly used building materials such as concrete and wood. The longer the flat horizontal segment or plastic region of the stress-strain curve, the higher the ductility of the steel (see Figure 1-1b).

For high-strength steels where the steel stress-strain curve has no defined yield point, the yield stress is determined using the 0.2% offset method (see Figure 1-1c), defined as the point where a line drawn with a slope E passing through the 0.2% elongation point (or 0.002 strain) on the horizontal axis intersects the stress-strain curve. It should be noted that high-strength steels have much less ductility than mild steel. For structural design, the stress-strain diagram for structural steel is usually idealized as shown in Figure 1-2.

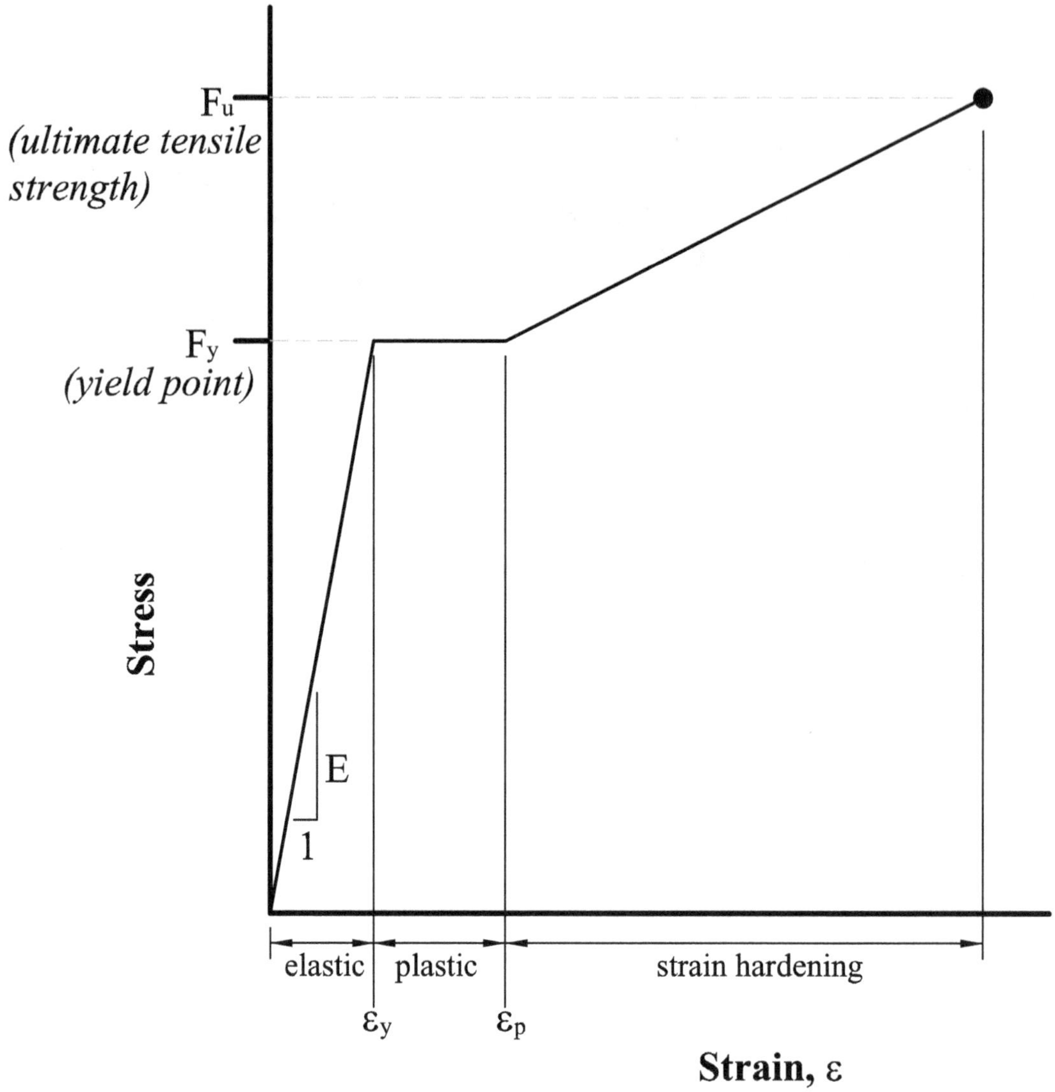

FIGURE 1-2 Idealized stress-strain diagram

The behavior of structural steel discussed previously occurs at normal temperatures, usually between –30°F and +120°F [12]. At very low temperatures, steel becomes more brittle and thus possesses lower ductility, and it loses strength when subjected to elevated temperatures. At a temperature of approximately 1300°F, both the strength and stiffness of steel are approximately 20% of its strength and stiffness at normal temperatures [15]. Because of the adverse effect of high temperatures on steel strength and behavior, structural steel used in building construction is often fireproofed by spray-applying cementitious materials or fibers directly onto the steel member (see Figures 1-2b and 1-2c) or by enclosing the steel members within plaster, concrete, gypsum wall board, or masonry enclosures. Intumescent coatings and ceramic wool wraps are also used as fire protection for steel members. Fire ratings for various building occupancies—specified in terms of the time in hours it would take a structural assembly to completely lose its strength in a fire event—are specified in the International Building Code (IBC) [16]. The topic of fire protection of structural steel will be discussed in more detail in Chapter 14 of this text.

FIGURE 1-2b Sprayed-on fireproofing of beams, girders, and columns.

FIGURE 1-2c Sprayed-on fireproofing of columns and diagonal bracing.

TABLE 1-1 Tensile Yield Strength and Ultimate Tensile Strength of Commonly Used Structural Steel.

Steel type	ASTM Specification of structural steel	F_y (ksi)	F_u (ksi)
Carbon steel	A36	36	58–80
	A53 Grade B	35	60
	A500 Grade B	46	58
	A500 Grade C	50	62
	A529	50	65-100
		55	70-100
High-strength, low-alloy	A913	50	65
		60	75
		65	80
		70	90
	A992	80	95
		50–65	65
High-strength, low-alloy	A572	42	60
		50	65
		55	70
		60	75
		66	80
Corrosion-resistant, high-strength, low-alloy	A588	50	70
	A847	50	70

Adapted from *AISCM* Table 2-4
where
F_u = Ultimate tensile strength, ksi, and
F_y = Yield stress, ksi.

Structural steel is specified using the American Society for Testing and Materials (ASTM) specifications, which define its properties. Prior to the 1960s, structural steel used in building construction in the United States was made mainly from ASTM A7 grade, with a minimum specified yield stress of 33 ksi. The Steel Construction Manual of the American Institute of Steel Construction (*AISCM*) lists the different types of structural steel used in steel building construction (see *AISCM* Tables 2-4 and 2-5), the applicable ASTM standard specification, and the corresponding tensile yield strength and ultimate tensile strength properties; these are shown in Table 1-1 for commonly used structural steels for building design and construction in the United States.

As indicated in Tables 2-4 and 2-5 of the 16th edition AISC Steel Manual (*AISCM*) [8], ASTM A992 Grade 50 (F_y = 50 ksi and F_u = 65 ksi) is the preferred high-strength steel material specification used for the main structural members (i.e., wide flange or W-shapes, Channels, C- and MC-shapes) in building construction in the United States, while ASTM A572 Grade 50 steel (F_y = 50 ksi and F_u = 65 ksi) is now typically the preferred material specification for M-shapes, HP-shapes, angles (L-shapes), and plates up to 4 inches thick. Plates over 4 to 8 inches thick are available in ASTM A36 (i.e., F_y = 36 ksi and F_u = 68 ksi).

Where structural steel with a higher yield stress is desired for the main structural steel members (e.g., wide flange or W-shapes), ASTM A913 (Grades 60, 65, 70, or 80), or ASTM A572 (Grades 55, 60, or 65), or ASTM A529 (Grade 55) may be specified [8, 17]. ASTM A913 Grades 65 and 70 steels were used in the award-winning 54-story 150 North Riverside high-rise office building in Chicago to achieve reductions in the weight of the steel members, and hence reduced costs. In this building, the four-story transfer diagonal struts used to support the multi-story exterior columns above the 4th floor level consisted of a series of W36 × 925 diagonal steel members, currently the "heaviest hot-rolled steel sections in the world" [18].

Hollow structural sections (HSS) are cold-formed carbon steel sections made by cold-bending flat steel plates into the desired tube steel shape and then welding the ends of the plate together. Large HSS sections may be formed by welding two half pieces together. The more commonly used hollow structural sections (HSS), with shapes that can be square, rectangular, or round, are commonly specified as ASTM A500 Grade C (with yield strengths of 50 ksi and ultimate tensile strength of 62 ksi). These are cold-formed carbon steel sections. Round structural pipes that have a lower yield strength than HSS are produced in ASTM A53 Grade B with a yield strength of 35 ksi and ultimate strength of 60 ksi. It is preferable and more common in design practice to use round HSS instead of pipe sections because HSS members have higher strengths than pipe sections. Although ASTM A500 Grade C is the preferred specification for rectangular and round HSS, other material specifications that are now available for HSS include ASTM A1085 Grade 50 (F_y = 50 ksi and F_u = 65 ksi), a carbon steel; also available for rectangular HSS is ASTM A1065 Grade 50 (F_y = 50 ksi and F_u = 60 ksi), a high-strength low alloy steel more commonly used for large HSS sections. The advantages of ASTM A1085 specification over ASTM A500 are as follows:

1. It provides tighter tolerances for the HSS wall thickness and corner radius, resulting in more efficient designs.
2. It is more suitable for dynamic loading and for use in bridge structures due to its material toughness.
3. It has a maximum limit of 70 ksi for the yield strength (compared to a higher variability in the strengths of ASTM A500), which results in a lower overstrength factor and therefore a more efficient seismic design.

It is envisaged that in the future, ASTM A1085 will gradually replace ASTM A500 as the more preferred materials specification for HSS [17, 19]. For round or rectangular HSS requiring protection against atmospheric corrosion, ASTM A847 (F_y = 50 ksi and F_u = 70 ksi), a corrosion-resistant high-strength low alloy steel, can be specified.

Where resistance to corrosion is desired, as in the case of exposed steel members, ASTM A588 steel (F_y = 50 ksi and F_u = 70 ksi), a corrosion-resistant high-strength low alloy steel, can be specified. This "weathering" steel provides protection from corrosion through the formation, over time, of a thin oxide coating on the surface of the structural member when exposed to the atmosphere. The use of this steel obviates the need for painting, and it is frequently used in bridge structures. However, for many building structures where corrosion protection is required, this is more commonly provided by coating the structural steel members with zinc-rich paint or by hot-dip galvanizing of the structural member, a process where the structural member is coated with zinc by dipping the entire member into a molten zinc bath at a temperature of approximately 850°F. The length and shape of the member to be galvanized are often limited by the size of the zinc bath used in the galvanizing process.

In general, it is recommended at the beginning of any project to confirm with the local steel fabricators the types (ASTM specifications) and grades of steel that are readily available to avoid paying a premium cost for structural steel that would have to be placed on special order if not locally available [17].

Another property of structural steel that is of interest to the structural engineer is the coefficient of thermal expansion, which has an average value of 6.5×10^{-6} in./in. per °F with variations in temperatures of up to 100°F [8]. This property is used to calculate the expected expansion and contraction of a steel member or structure and is useful in determining the required size of expansion joints in building structures or the magnitude of forces that will be induced in a structure, if the thermal expansion or contraction is restrained. For enclosed heated and air-conditioned buildings, it is common practice to assume a maximum temperature range of between 50°F and 70°F[55]. However, because buildings are usually unheated and unenclosed during construction, the maximum temperature change may exceed these values, and the structural engineer would be wise to consider these increased temperature changes — which would vary depending on the location of the building — in the design and detailing of the structural members [20]. Structurally, expansion joints in buildings are usually achieved either using a line of double columns, that is, a column line on both sides of the expansion joint, or using low-friction sliding bearings that are supported off a bracket connected to the columns on one side of the expansion joint (see Figures 1-3a and 1-3b). The designer should ensure that the sliding bearing details allow for the anticipated lateral movement and rotation of the beam or girder; faulty details that do not allow expansion, or contraction, or rotation, or sliding bearings that are unintentionally restrained because of the buildup of debris could result in large unintended lateral forces being transmitted to the structure, which could result in structural failure. The unanticipated restraint of an expansion joint connection was the cause of the loading dock slab collapse at the Pittsburgh Convention Center in 2007 [21].

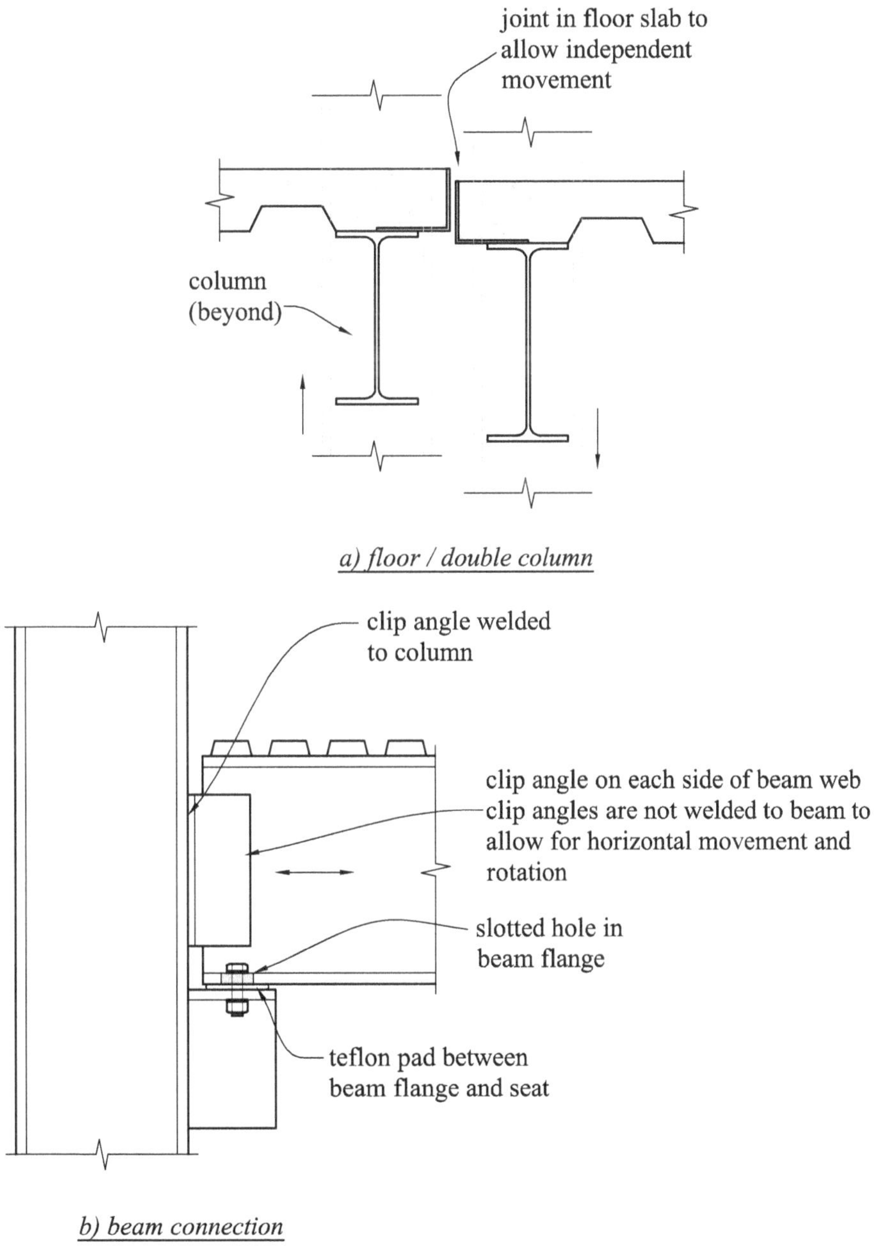

FIGURE 1-3a,b Expansion joint details

Residual Stresses

Due to the different rates of cooling in a structural steel member during the final stages of the manufacturing process (i.e., after the rolling process), initial self-equilibrating non-uniform longitudinal or axial stresses will exist in the member prior to any loads being applied. These preexisting non-uniform and self-equilibrating longitudinal stresses that are caused by the different cooling rates of the different fibers of the steel section are called residual stresses. The fibers of the steel section that are the first to cool (for example, the tips of wide flanges and the middle section of the web and the thinner elements) will develop compressive residual stresses as they restrain the fibers that are the last to cool from shrinking or undergoing axial shortening; for example, the interior of the wide flanges near the web-flange intersection, which are the last to cool will develop tension residual stresses. The tension and compressive residual stresses are in equilibrium. Figure 1-4a shows the longitudinal residual stresses in a rolled wide flange section [22, 23] caused by uneven cooling after the steel is rolled and the resulting restraint against distortion, which results in compression stresses at the flange tips and tension stresses at the web-flange intersection. In addition to the rolling process in the manufacture of structural steel, other processes that could result in residual stresses include:

- Cold bending or straightening,
- Thermal cutting of structural members,
- Drilling of holes in structural members, and
- The heat generated from the welding of structural members and the uneven cooling of the elements that make up the member.

Figure 1-4b shows the residual stresses in a plate during welding or thermal cutting, and resulting from differential cooling after welding or flame cutting the plate section.

The residual stresses are usually in internal equilibrium in a structural steel section, and, though they may result in the premature attainment of inelastic action under loading, they have no impact on the plastic moment or tension capacity (i.e., the strength) of a steel member due to the ductility of structural steel. Residual stresses do, however, affect the load-deformation relationship and the stiffness of a structural member, and residual stresses also cause fracture when excessive tension residual stresses occur in combination with other conditions [22]. The impact of residual stresses is most significant for axially loaded members, such as columns, because it causes a reduction in the modulus of elasticity, which decreases from the linear elastic modulus (E) to the tangent modulus (E_T), thus resulting in a reduction in the Euler buckling load [23, 24, 25]. Residual stresses also reduce the moment capacity of flexural members that are susceptible to local or lateral torsional buckling. A residual stress value of 30% of the yield stress, F_y (i.e., 15 ksi for 50 ksi steel), is assumed in the AISC Specification when calculating the moment capacity of non-compact beams (i.e., beams subject to local flange or web buckling) and beams subject to lateral torsional buckling (see Chapter 6).

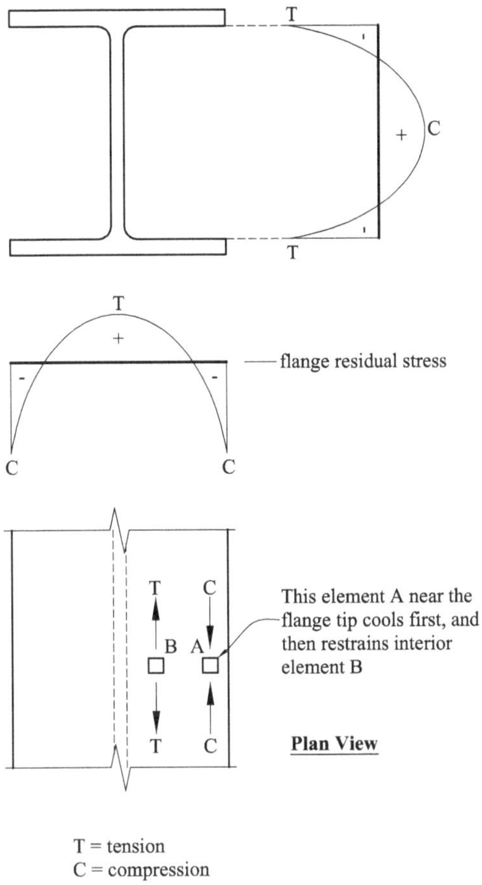

FIGURE 1-4a Residual stresses in a rolled wide flange section

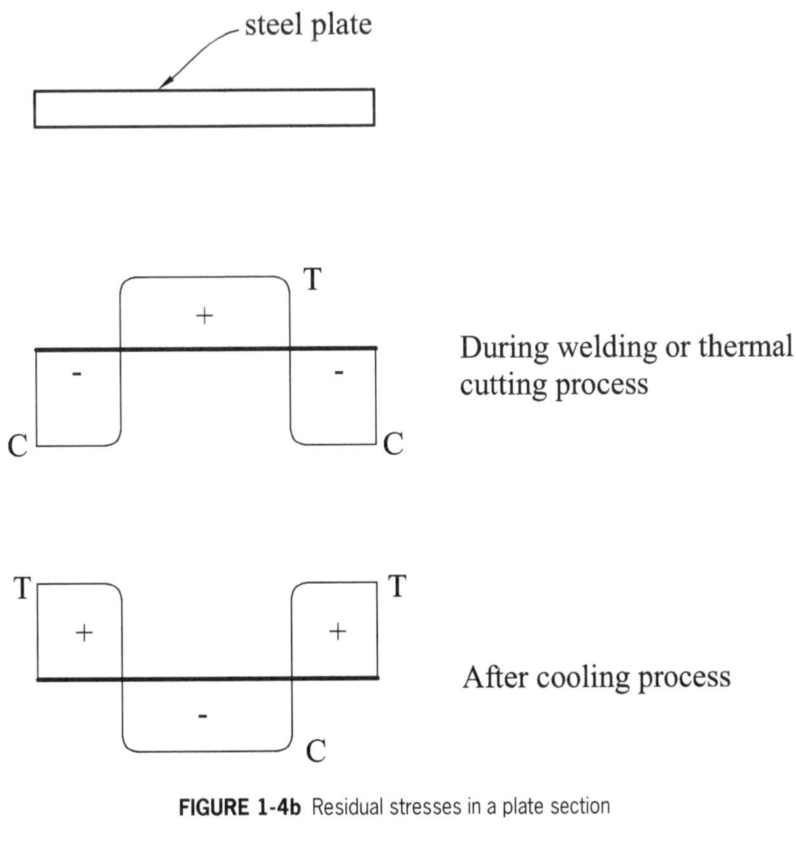

FIGURE 1-4b Residual stresses in a plate section

Brittle Fracture and Fatigue

A brittle fracture is the sudden failure of a structural steel member due to tensile stresses that cause a cleavage of the member, and it occurs without warning. Brittle fracture results from low ductility and poor fracture toughness of the structural member or connection. Other factors affecting brittle fracture include size or thickness of the steel member (members thicker than 2 inches have a propensity for brittle fracture), the presence of geometric discontinuities in a steel member resulting in stress concentrations (e.g., notches and weld access holes that are not ground smooth), rate of application of load, and temperature. The lower the temperature of the steel, the lower the ductility and toughness of the steel member. The importance of brittle fracture and the need for the structural engineer to seriously consider this failure mode in the design of steel structures was brought to the fore in December 2018 with the fracture cracks that were discovered in the 4-in. thick bottom flange of the welded steel transfer plate girders at the Transbay (Salesforce) Transit Center in San Francisco, CA [47, 53]. The AISC Specifications A3.1.1d and A3.1.1e specifies that for steel members that are thicker than 2 inches, "the [Charpy V-Notch] impact test shall meet a minimum average value of 20 ft-Ibf absorbed energy at a maximum temperature of +70°F (+21°C)." Note that some structural members may be subjected to low ambient temperatures, much lower than 70°F especially during construction or if the steel is exposed during service; therefore, the effect of low temperatures on the fracture toughness should be considered in the specification for the structural steel, because the Charpy energy at lower temperatures could be much lower that the required 20-ft-Ibf. The thicker the member, the greater the imperative to investigate the potential for brittle fractures using the Charpy V-Notch (CVN) Impact test, which is an indirect measure of the fracture toughness of the steel material. The CVN requirements for a particular structural element must be indicated on the contract documents.

The process of drilling holes in a thick flange or plate can cause microcracks to form that can lead to brittle fracture because the drilling process generates high temperatures at the hole locations, while the areas further away from the drilled hole location are cool and acts as a heat sink and therefore restrains the areas around the hole which cools rapidly – forming martensite ("a hard brittle phase of steel"), leading to cracking at the hole location, especially if the edges of the holes have geometric discontinuities or reentrant corners that are have not been ground smooth to the "bright metal" [8, 47, 53]. Brittle fracture can also occur due to welding of thick steel members because of the resulting high temperatures in the weld; the thick steel member becomes a heat sink for the weld, which leads to the undesirable rapid cooling of the weld, resulting in shrinkage and microcracking of the weld due to the restraint provided by the cooler parts of the thick steel member. Several ways for mitigating brittle fracture include:

- Preheating of the steel before cutting or drilling of holes, or before welding. This helps to slow the cooling rate of the steel or weld, thus minimizing the formation of a hard microstructure; preheating also helps to dissipate the hydrogen from the welding process, thus reducing the propensity of the weld to become hard and brittle, preventing hydrogen cracking of the weld.
- Controlling the cooling rate of the steel or weld by wrapping the member with thermal blankets. Fast cooling of steel is undesirable as it leads to the formation of a hard and brittle microstructure with a propensity for cracking and brittle fracture [23].
- Avoiding reentrant corners or jagged edges around the edges of drilled or cut holes by grinding the edges of the holes smooth.

Brittle fractures can also occur when steel is subjected to extremely low temperatures, which results in low Charpy energy values, an indicator of low fracture toughness. Based on the lessons learned from the Transbay (Salesforce) Transit Center girder fractures, the following are recommended for plates thicker than 2 inches (i.e., heavy shapes) [53]:

- The CVN test should be carried out not just at the outer fibers of the thick plate, but also at the center of thickness of the plate and at ¼-point and ¾ point of the thickness. This will help prevent the interior of the plate thickness from developing low toughness.
- The required Charpy energy at the lowest ambient temperature that the structural member will be subjected to during its life cycle, should not be less than 20 ft-Ib.

Whereas *brittle* fracture of a steel member results from a few applications or even a single application of loading, *fatigue* failure, on the other hand, occurs due to repeated applications of loading to a structure over time, starting with a small fatigue crack. Note that residual stresses resulting from the temperature rise in a welded steel member, if the member or weld cools rapidly, can also be the source of the initiation of a fatigue crack [51]. An example of a structure where the members and connections are subject to fatigue loading are the space trusses shown in Figure 1-5a that support aerial cable cars at one of Canada's National Parks. Another iconic structure that is subject to fatigue loading is the London Eye, located on the banks of the River Thames in London, UK (see Figure 1-5b). It consists of a cantilevered 3-D truss wheel structure supported by two 3-D braced frame towers in the river and laterally supported by an A-frame and six backstay cables.

Members repeatedly loaded, primarily in tension, are more susceptible to developing fatigue cracks. Fatigue strength reduction in steel bridges can occur due to fatigue cracks caused by the grit blasting of steel bridges to ensure corrosion protection coating adherence because the grit blasting results in geometric discontinuities or changes in geometry that become the initiation point for the crack [51]. The behavior of steel in fatigue is measured in terms of the stress range,

FIGURE 1-5a Aerial cable cars supported by steel space trusses at Banff National Park in Alberta, Canada (photo by Abi Aghayere)

FIGURE 1-5b The London Eye (photo by Abi Aghayere)

S, and the number of cycles of live loading and unloading, N, resulting in a typical S-N curve. Only the live loads are critical to fatigue failures; the dead loads are not. The stress range is the difference between the maximum and minimum stresses in a member or connection during one load cycle [24, 26]. The greater the number of cycles of loading and unloading, N, the smaller the allowable stress range, S, that a structural element or connection can resist.

The fatigue strength of a steel structure is usually determined at service or unfactored load levels; it is a function of the stress category, which in turn is greatly dependent on the connection details used in the structure (see Table A-3.1 in Appendix 3 of *AISCM*) [8]. It is also a function of the stress range in the member, as previously indicated. The stress range is usually higher when the steel member and connections are subjected to stress reversals (e.g., tension stress in one cycle and compression stress in another cycle) due to live loads. When a member or connection is subjected to only compression stresses, fatigue crack propagation cannot occur, so fatigue does not need to be considered for such members. The lower the stress range, the higher the fatigue strength of the member or connection. The calculated stress range due to live loads should not exceed the design or allowable stress range calculated from the equations in Section 3.3 and the fatigue/stress category factors from Table A-3.1 of Appendix 3 in ANSI/AISC 360-22, Specification for Structural Steel Buildings [8]. The fatigue resistance of steel members and connections must be considered when the number of live load application cycles exceeds 20,000 cycles.

In most buildings, fatigue is not an issue and therefore is not normally considered in the design of structural members in building structures, except for crane runway girders and members supporting machinery. Fatigue is a major consideration in the design of steel bridges and industrial crane support structures where the maximum number of live load cycles may exceed 2 million cycles of loading and unloading during the life of the structure [24, 27].

1-4 STRUCTURAL STEEL SHAPES AND ASTM SPECIFICATION

The general requirements for the mechanical properties, cross-sectional dimensions, chemical composition, and standard mill practice for rolled structural steel shapes, bars, and plates are specified in the ASTM A6 specification. Part 1 of the AISC Manual (*AISCM*) summarizes the information from ASTM A6. For hollow structural sections (HSS) and structural steel pipes, the ASTM A500 and ASTM A53 specifications, respectively, apply. Table 1-2 shows standard structural steel shapes and the corresponding preferred material specifications. The ASTM A6 specification prescribes the permissible maximum percentages of alloy elements such as carbon, manganese, chromium, nickel, copper, molybdenum, and vanadium in structural steel to ensure adequate weldability and resistance to corrosion and brittle fracture. In the specification, the percentage by weight of each of these chemical elements is combined to produce an equivalent percentage carbon content that is called the carbon equivalent (CE) [22, 23, 24]. Note that there are several carbon equivalent (CE) equations in the literature. The CE equation gives a "rough idea" of the weldability of the steel. Table 1-3 shows the major chemical elements in structural steel and their effects on the steel properties, as well as the advantages and disadvantages of each element [22, 23, 24]. The carbon equivalent is useful in determining the weldability of older steels in the repair or retrofitting of existing or historical structures where the structural drawings and specifications may not be available; it is used to determine what, if any, special precautions are necessary for welding to these steels to prevent brittle fractures and cracking of the welds. The higher the CE value, the lower the allowable cooling rate [22] and therefore, the lower the weldability of the steel. To ensure good weldability, the carbon equivalent, as calculated from equation (1-1), should be no greater than 0.5% [14, 23, 24].

Precautionary measures for steels with higher carbon equivalents include preheating the steel and using low-hydrogen welding electrodes. Alternatively, bolted connections could be used in lieu of welding.

TABLE 1-2 Structural steel shapes and corresponding preferred material specification (see AISCM Tables 2-4 and 2-5).

Structural steel shapes	**ASTM Specification**	**Min F_y (ksi)**	**Min F_u (ksi)**
W-shape	A992[†] A913[*]	50–65 50–80	65 65-95
M- and S-shapes, HP and angles (L)	A572	50	65
Channels (C- and MC-shapes)	A992[†]	50	65
Plates – up to 8 in. thick	A36	36	58–80
Plates - up to 4 in. thick	A572	50	65
Miscellaneous plates and steel where 50 ksi yield strength is not necessary	A36	36	58
Steel pipe	A53 Grade B ^	35	60
Round HSS	A500 Grade C ^	50	62
Square and rectangular HSS	A500 Grade C	50	562

[*] A913 is a low-alloy, high-strength steel.
[†] Preferred material specification for the different shapes.
^ Round HSS (A500) is preferred and stronger than steel pipes (A53 Grade B)

The equivalent carbon content or carbon equivalent [22, 23, 24] is given as

$$CE = C + \frac{(Cu + Ni)}{15} + \frac{(Cr + Mo + V)}{5} + \frac{(Mn + Si)}{6} \leq 0.5\%, \tag{1-1}$$

where

C = Percentage carbon content by weight,
Cr = Percentage chromium content by weight,
Cu = Percentage copper content by weight,
Mn = Percentage manganese content by weight,
Mo = Percentage molybdenum by weight,
Ni = Percentage nickel content by weight,
V = Percentage vanadium content by weight,
Si = Percentage silicon by weight.

Table 1-3 Advantages and disadvantages of alloy chemical elements used in structural steel.

Chemical element	Major advantages	Disadvantages
Carbon (C)	Increases the strength of steel.	Too much carbon reduces the ductility and weldability of steel.
Copper (Cu)	When added in small quantities, it increases the corrosion resistance of carbon steel as well as the strength of steel.	Too much copper reduces the weldability of steel.
Vanadium (V)	Increases the strength and fracture toughness of steel and does not nega-tively impact the notch toughness and weldability of steel.	
Nickel (Ni)	Increases the strength and the corro-sion resistance of steel. Increases frac-ture toughness.	Reduces weldability
Molybdenum (Mo)	Increases the strength of steel. In-creases corrosion resistance.	Decreases the notch toughness of steel.
Chromium (Cr)	Increases the corrosion resistance of steel when combined with copper and increases the strength of steel. It is a major alloy chemical used in stainless steel.	
Columbium (Cb)	Increases the strength of steel when used in small quantities.	Greatly reduces the notch toughness of steel.
Manganese (Mn)	Increases the strength and notch toughness of steel. Improves ductili-ty.	Reduces weldability.
Silicon (Si)	Used for deoxidizing hot steel during the steel-making process and helps to improve the toughness of the steel.	Reduces weldability.
Other alloy elements found in very small quantities include nitrogen; those elements permitted only in very small quantities in-clude phosphorus and sulfur.		Phosphorous and sulfur lowers the ductility of steel.

EXAMPLE 1-1

Expansion Joints in Steel Buildings and the Effect of Temperature Changes on Restrained Girders

a. A steel building is 600 ft. long, with expansion joints provided every 200 ft. If the maximum anticipated temperature change is 50°F, determine the size of the expansion joint that should be provided. Assume the coefficient of thermal expansion, α, for structural steel is 6.5×10^{-6} in./in./°F.

b. If the longitudinal movement of a W24x94 girder in this building is inadvertently restrained at both ends of the girder, determine the magnitude of the axial force in kips induced in the girder and its connections due to a 50°F temperature change during construction.

Solution

a. From the mechanics of materials, the anticipated longitudinal expansion or contraction due to temperature change is $\delta = \alpha \ell \Delta T$

Where,

α = coefficient of thermal expansion = 6.5×10^{-6} in./in. per °F for steel

$\Delta T = T_{high} - T_{low}$ = maximum change in temperature

ℓ = original length of the structural member

$\delta = \alpha \ell \Delta T = (6.5 \times 10^{-6}$ in./in. per °F) (200 ft. × 12 in./ft.) (50°F) = 0.78 in.

This is the total change in length for the 200-ft.-long segment of the building. Assuming one-half of this expansion or contraction takes place at each end of the 200-ft-long segment, the expansion or contraction at each end would be ½ (0.78 in.) = 0.39 in.

Since the portion of the building on both sides of the expansion joints can expand at the same rate simultaneously, the total required minimum width of the expansion joint = 2 (0.39 in.) = 0.78 in.

Therefore, use a 1-in.-wide expansion joint.

It should be noted that the nonstructural elements in the building, such as the exterior cladding (e.g., brick wall) and backup walls (e.g., block wall), as well as the interior partition walls, must be adequately detailed to accommodate the anticipated expansion and contraction.

b. Axial force in the restrained W24 x94 girder due to a 50°F temperature change or differential during construction:

Cross-sectional area of the girder, A = 27.7 in.^{2}

The modulus of elasticity, E = 29,000 ksi

From the mechanics of materials, the axial deformation of the girder is $\delta = \dfrac{P\ell}{AE} = \alpha \ell \Delta T$

Rearranging the preceding equation, the axial force in the girder is obtained as

$$P = \frac{\alpha \ell \Delta T \, AE}{\ell} = \alpha \, \Delta T \, AE$$

$$= (6.5x10^{-6}\ \text{in./in./}{}^{\circ}F)(50^{\circ} F)(27.7\,\text{in}^2)(29{,}000\,\text{ksi}) = 261\ \text{kips}$$

The ratio of the change in length of this girder to its original length is

$$\frac{\delta}{\ell} = \frac{\alpha \ell \Delta T}{\ell} = \alpha \, \Delta T = (6.5x10^{-6}\ \text{in./in./}{}^{\circ}F)(50^{\circ} F) = 0.000325 \ or \ 0.0325\%$$

Even though the change in length of the girder caused by the temperature change is only 0.0325% of the original length of the girder, it is obvious that the axial force induced in the girder is quite large if the ends of the girder are restrained, such that it cannot be ignored. Restrained end conditions, as illustrated in this example, can occur during construction when exposed beams or girders are supported by and attached to concrete basement walls, and the beams or girders in turn provide lateral support to the top of the basement walls. Breakout of chunks of the concrete in the basement wall below the beam or girder can occur if the beams or girders are subjected to significant temperature fluctuations from daytime highs to nighttime lows during construction because the building at this time is not yet enclosed. One of the authors has investigated a failure at the end supports of exposed beams during construction caused by temperature changes where the beams at one end were anchored through a bearing plate to a concrete basement wall. The axial force induced in the beam due to temperature changes was large enough to cause a breakout of a chunk of the concrete basement wall below the beam bearing plates. The following steps could have been taken to prevent the breakout in the concrete basement wall:

- Provide temporary lateral bracing to the top of the concrete basement walls.
- Sit the steel beams on the embedded bearing plates without attaching them to the plates to temporarily allow the beams to expand or contract until the building is enclosed.
- After the building is enclosed, and thus the temperature differential is greatly reduced, weld the steel beams to the beam bearing plates and fill the beam pocket with non-shrink grout.
- Remove the temporary lateral bracing for the concrete basement walls.

EXAMPLE 1-2

Carbon Equivalent and Weldability of Steel

A steel floor girder in an existing building needs to be strengthened by welding a structural member to its bottom flange. The grade of steel for the existing girder is unknown, and to determine its weldability, material testing has revealed the following percentages by weight of the alloy chemicals in the girder:

$$C = 0.25\%$$
$$Cr = 0.15\%$$
$$Cu = 0.25\%$$
$$Mn = 0.45\%$$
$$Mo = 0.12\%$$
$$Ni = 0.30\%$$
$$V = 0.12\%$$
$$Si = 0.20\%$$

Calculate the carbon equivalent and determine if this steel is weldable.

Solution

Using equation (1-1), the carbon equivalent, CE, is calculated as

$$\text{CE} = 0.25\% + \frac{(0.25\% + 0.30\%)}{15} + \frac{(0.15\% + 0.12\% + 0.12\%)}{5} + \frac{(0.45\% + 0.20\%)}{6}$$

$$= 0.47\% < 0.5\%.$$

Therefore, the steel in the girder is weldable.

However, because of the high carbon equivalent, precautionary measures such as using low-hydrogen welding electrodes and preheating of the member are recommended. Since this is an existing structure, the effect of preheating the member should be thoroughly investigated so as not to create a fire hazard. If preheating of the member is not feasible, alternative strengthening approaches that preclude welding may need to be investigated.

Steel Shapes

The two main types of steel shapes are:

- **Rolled steel shapes** — These are standardized hot-rolled shapes with cross-sectional dimensions and properties specified in the American Institute for Testing and Materials specification (ASTM A6) and are also presented in Part 1 of the *AISCM* [8]. Components of *rolled* steel shapes with flange thickness thicker than 2 inches are classified in Sections A3.1.1d of AISC 360-22, Specifications for Structural Steel Buildings, as rolled *heavy shapes*.

- **Built-up shapes** — Where standardized structural shapes cannot be used (e.g., where the load to be supported exceeds the capacity of the sections listed in the *AISCM*), built-up shapes could be made from plate stock. Examples include welded plate girders, plane and flanged cruciform shapes, and box girders. Components of *built-up* shapes with plates thicker than 2 inches are classified in Sections A3.1.1e of AISC 360-22, Specifications for Structural Steel Buildings, as built-up *heavy shapes*. For heavy rolled and built-up heavy shapes, Sections A3.1.1d and A3.1.1e require the Charpy V-notch (CVN) impact test (i.e., the notch toughness test) per ASTM A6 that measures the energy absorption ability of a steel material when subjected to an impact load in the presence of a V-notch and provides an indirect measure of the fracture toughness of the steel member [53].

Standard hot-rolled shapes are most commonly used for building construction, while built-up shapes are commonly used in bridge construction. Note that built-up shapes, such as welded plate girders, are also used in building construction as transfer girders to support multistory columns above an atrium or an auditorium,,m or above a roadway or a railway to create a large column-free area, resulting in large, concentrated loads that need to be supported. Welded plate girders will be covered in Chapter 13.

Examples of the rolled standard shapes listed in Part 1 of the *AISCM* include the following:

Wide-Flanged Sections: W-Shapes and M-Shapes

Wide-flanged sections are I-shapes (also known as W-shapes) that are commonly used as beams or columns in steel-framed structures (see Figure 1-6). They are also sometimes used as the top and bottom chord members of trusses and as diagonal braces in braced frames. The inner and outer flange surfaces of W-shapes are parallel; M-shapes are similar to W-shapes, but they are not as readily available or widely used as W-shapes, and their sizes are also limited. The listed M-shapes in *AISCM* have a maximum depth of 12.5 in. and a maximum flange width of 3.75 in.

A W14 × 90, for example, implies an I-shaped member with a nominal depth of 14 in. and a self-weight of 90 lb./ft. Similarly, an M12 × 10 indicates a miscellaneous shape with a nominal depth of 12 in. and a self-weight of 10 lb./ft. It should be noted that because of the variations from mill to mill in the fillet sizes used in the production of wide flange shapes and also the wear and tear on the rollers during the steel shape production process, the decimal k-dimensions (k_{des}) specified for these shapes in Part 1 of the *AISCM* should be used for design, while the fractional k-dimensions (k_{det}) are to be used for detailing.

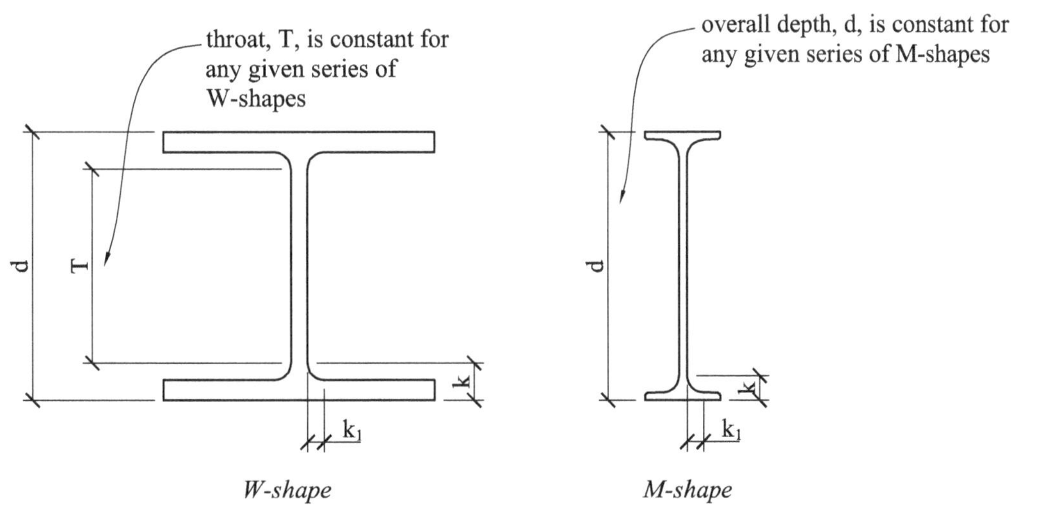

FIGURE 1-6 W- and M- shapes

S-Shapes

S-shapes (see Figure 1-7), also known as American Standard beams, are similar to I-shapes except that the inside flange surfaces are sloped. The inside face of the flanges usually has a slope of 2:12, with the flange thickness varying from a smaller value at the flange tips to a larger flange thickness closest to the web of the beam. S-shapes were the first rolled beam sections in the United States; they are no longer used for typical roof or floor framing but are only frequently used as monorail and crane support beams. An S12 × 35 implies a member with a 12-in. actual depth and a self-weight of 35 lb./ft. length of the member.

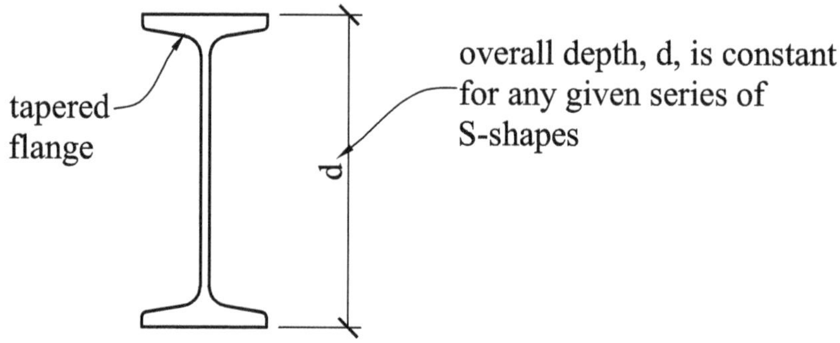

FIGURE 1-7 S-shapes

HP-Shapes

HP-shapes (see Figure 1-8) are similar to W-shapes and are commonly used as H-piles in bearing pile foundations. These H-piles, which can be as long as 100 ft. or more, are used to support the superstructure loads, and they are typically socketed and grouted into the bedrock to resist the uplift loads on the structure. HP-shapes have thicker flanges and webs compared to W-sections, and the nominal depth of these sections is usually approximately equal to the flange width, with the flange and web thicknesses approximately equal. The thick webs help to resist the high impact loads that these sections are subjected to during pile driving operations.

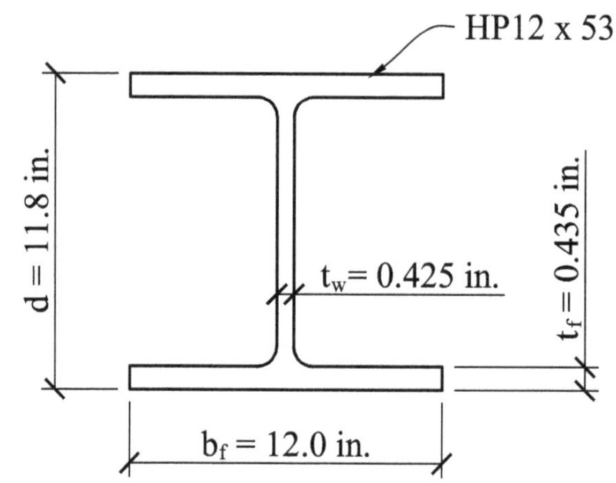

FIGURE 1-8 HP shapes

Channel or C- and MC-Shapes

Channels are C-shaped members with the inside faces of the channel flanges tapered from a minimum thickness at the flange tip to a maximum thickness at the channel web (see

Figure 1-9). They are commonly used as beams to support light loads, such as in catwalks and as stair stringers, and they are also used as girts to provide lateral support to metal cladding, and used to frame the edges of roof openings. C-shapes are American Standard channels, while MC-shapes are miscellaneous channels. A C12 × 30 member implies a C-shape with an actual depth of 12 in. and a weight of 30 lb./ft., while an MC 12 × 35 member implies a miscellaneous channel with an actual depth of 12 in. and a self-weight of 35 lb./ft. Both C12 and MC12 sections are recommended for the stringers of steel stairs because of their wide flanges that allow enough space for adequate fillet weld connection of the handrail posts to the top surface of the top flange of the channel [22].

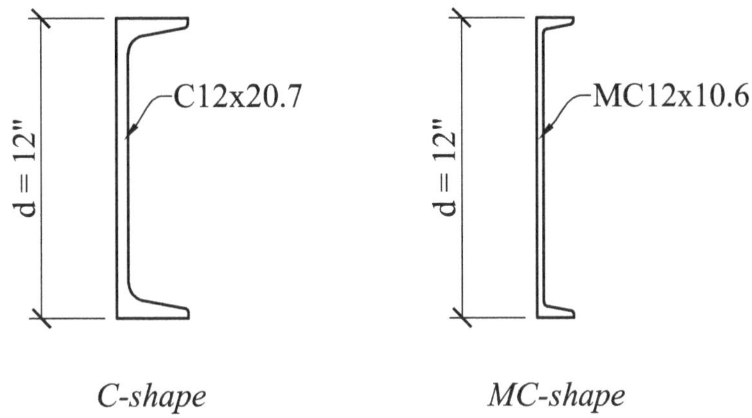

FIGURE 1-9 C- and MC- shapes

Angle (L) Shapes

Angles (see Figure 1-10) are L-shaped members with legs of equal or unequal lengths, and they are used as lintels to support brick cladding and block wall cladding above door and window openings and as web members in trusses. They are also used as X-braces, chevron braces, or knee-braces in braced frames (see Figure 1-10b) where they could occur as single angles or as double angles placed back-to-back. Double angles are frequently used for the end connections for beams and girders. An angle with the designation L4 × 3 × $^1/_4$ implies a member with a long leg length of 4 in., a short leg length of 3 in., and a thickness of $^1/_4$ in. While all other rolled sections have two orthogonal axes of bending (i.e., *x–x* and *y–y*), single angles have three axes about which the member could bend or buckle (i.e., *x–x*, *y–y*, and *z–z*).

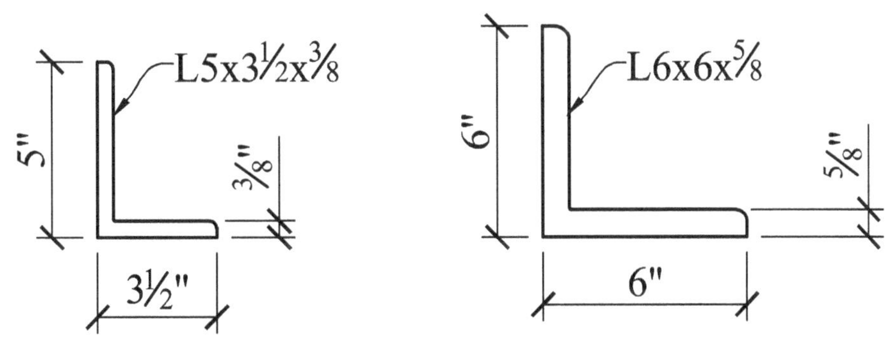

FIGURE 1-10 Angle shapes

FIGURE 1-10b Knee braced frame with double angles

Structural Tees—WT-, MT-, and ST-Shapes

Structural tees (see Figure 1-11) are made by cutting in half a wide flange section (i.e., an I- or W-shape), or M-shape, or S-shape. For example, if a W14 × 90 is cut in half, the resulting shapes will be WT 7 × 45, where the nominal depth is 7 in. and the self-weight of each piece is 45 lb./ft. WT-shapes are commonly used as brace members and as top and bottom chords of trusses. They are also used to strengthen existing steel beams where a higher moment capacity is required. Similarly, ST- and MT-shapes are made from S- and M-shapes, respectively.

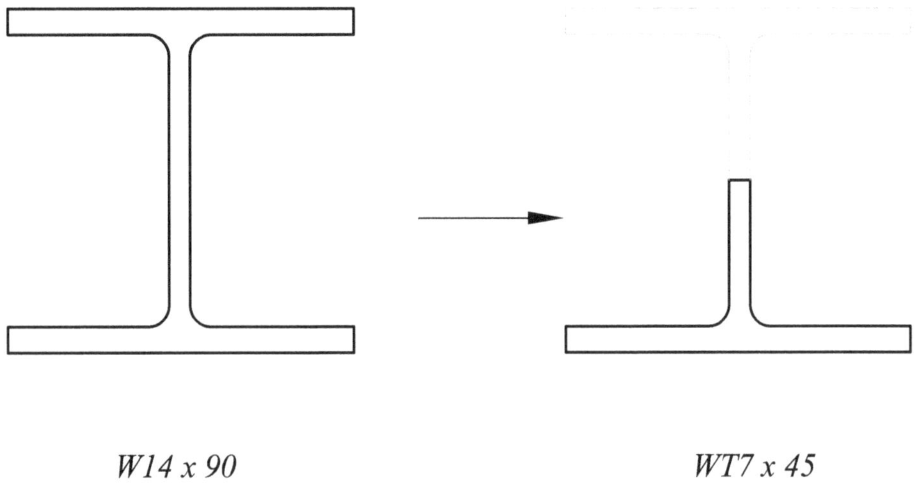

FIGURE 1-11 WT-shape

Plates and Bars

Plates and bars (see Figure 1-12) are flat stock members that are used as stiffeners, gusset plates, and X-braced members. They are also used to strengthen existing steel beams and as supporting members in built-up steel lintels. Plates are also used in plane cruciform columns. There is very little structural difference between bars and plates, and although historically, flat stock with widths not exceeding 8 in. were generally referred to as bars, while flat stock with widths greater than 8 in. were referred to as plates, it is now common practice to refer to flat stock universally as plates. As an example, a PL 6 × ½ implies a 6-in.-wide by ½-in.-thick plate. In practice, plate widths are usually specified in ½-in. increments, while thicknesses are specified in $^1/_8$-in. increments. The practical minimum thickness for plates is ¼-in., with a practical minimum width of 3 in. to accommodate the required minimum bolt edge distances.

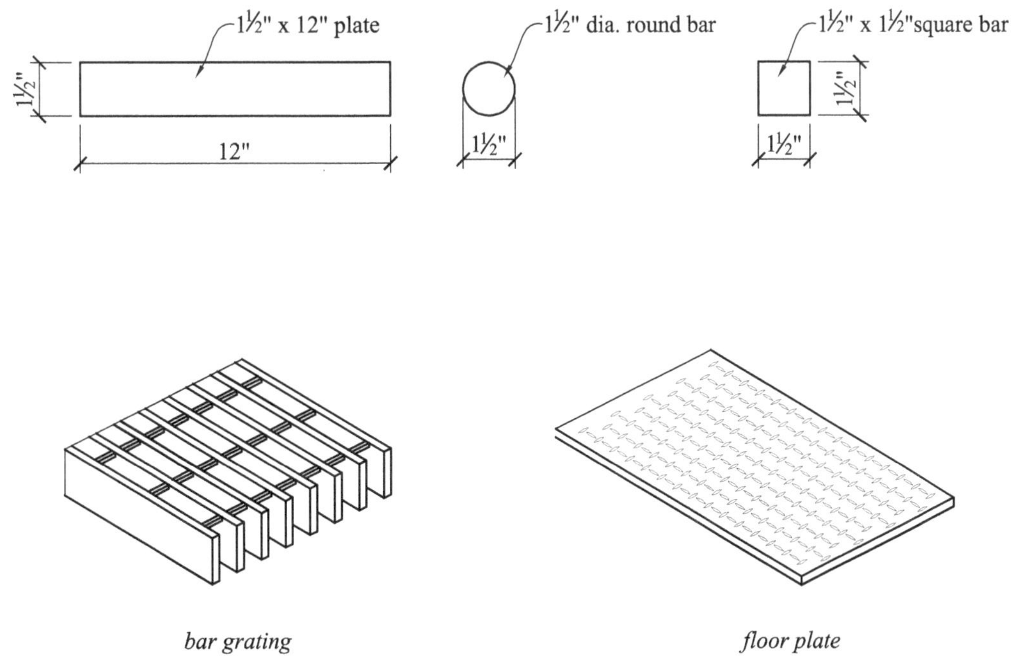

FIGURE 1-12 Plates and bars

Hollow Structural Sections (HSS)

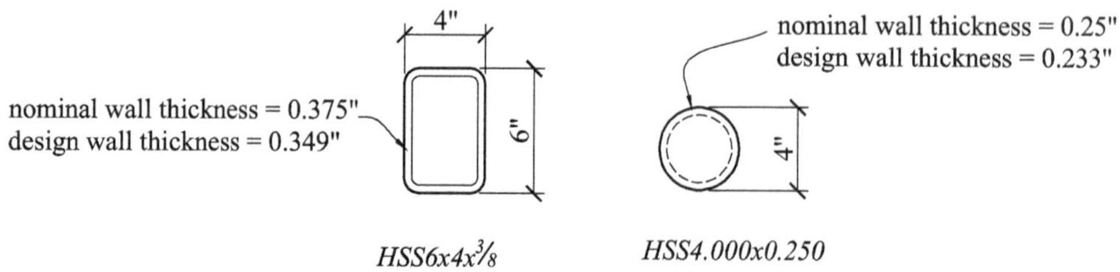

FIGURE 1-13a Rectangular and round HSS

FIGURE 1-13b Square HSS column

All the shapes discussed above are made from hot-rolled steel sections, whereas hollow structural sections (HSS) are welded cold-formed carbon steel made by cold-bending a flat piece of carbon steel into rectangular, square, or round tubular shapes and then welding the ends together (see Figures 1-13a and 1-13b). HSS are commonly used as columns, lintel beams, struts, girts, hangers, lateral bracings, and braced-frame members in building structures; they are also used in bridge structures. HSS members, because of their closed shape, have better resistance to lateral-torsional buckling and torsion than open sections such as wide flange sections (i.e., W-shapes) or channels. Therefore, HSS members are frequently used as lintel beams spanning large openings, especially where the eccentricity of the supported cladding loads may result in large torsional moments in the HSS lintel beam. HSS members, because of their excellent torsional properties, are also used for beams that are curved in plan where they are subjected to torsional moments from the gravity loads acting on the curved beam. Examples of HSS members are as follows:

- HSS 6 × 4 × ¼ (long side vertical—LSV) implies a rectangular hollow structural section with outside wall dimensions of 6 in. in the vertical direction and 4 in. in the horizontal direction and a nominal wall thickness of ¼ in., except at the rounded corners. Note that rectangular HSS whose vertical and horizontal dimensions are even numbers are more generally readily available [22]. Rectangular HSS with long side vertical (LSV) are typically used as stringers in monumental steel stairs (e.g., HSS 12 × 6), and they are also used as lintels to support block walls and brick cladding loads because of their excellent torsional as well as flexural properties. Rectangular and square HSS members are used as columns, diagonal braces, knee braces, lintels and girts in building structures (see Figure 1-13c)
- HSS 6.625 × 0.28 implies a round hollow structural section with an outside wall diameter of 6.625 in. and a nominal uniform wall thickness of 0.28 in. It is recommended to specify round HSS sections that have similar cross-sectional dimensions as standard steel pipe sections because such round HSS are generally more readily available in stock and do not require a special order [22].

Jumbo size HSS members up to 22 in. square with 1 in. maximum nominal wall thickness, rectangular HSS up to 34 in. × 10 in. with 1-in. maximum nominal wall thickness, and round HSS up to 28 in. diameter with maximum nominal wall thickness of 1-in. are now available and produced domestically. Round and square HSS members have aesthetic appeal, and they are very efficient structural members since they have no weak axis and are therefore commonly used as exposed columns in buildings, and as roof trusses (see Figure 1-13d). For HSS members and box sections produced according to ASTM specification A1065 or A1085, the design or actual wall thickness is equal to the nominal wall thickness (i.e., $t = t_{nominal}$), and the maximum yield stress for A1085 is 70 ksi [52]. For HSS produced to other ASTM specifications, the design wall thickness is 0.93 × the nominal wall thickness (i.e., $t = 0.93t_{nominal}$).

Structural Pipes

Structural pipes are round structural tubes similar to HSS members (see Figure 1-13a) that are sometimes used as columns and as struts for bracing basement excavations. They are also used as posts or columns for lightly loaded structures such as mezzanines and in residential buildings to support the ground floor girder. They are available in three strength categories: standard (Std), extra strong (X-strong), and double-extra strong (XX-strong). The bending moment capacity and the axial compression load capacity of pipe sections are tabulated in Tables 3-15 and 4-6, respectively, of the *AISCM*. Pipe sections are designated with the word "pipe," followed by the nominal diameter, and then the letter or word representing the strength category (e.g., Pipe 4 Std.; Pipe 4x-strong; Pipe 4xx-strong). The designation "Pipe 4 Std." represents a nominal 4-in. standard pipe, while "Pipe 4x-strong" represents a 4-in. extra-strong pipe, and "Pipe 4xx-strong" represents a 4-in. double-extra-strong pipe. The maximum outside diameter of available steel pipes is 26 inches, and the maximum available nominal wall thickness is 1.0 inch (see *AISCM* Table 1-14).

FIGURE 1-13c HSS members used as columns, lintels or girts, diagonal braces and knee braces in building structures (photo by Abi Aghayere)

FIGURE 1-13d 3-D roof trusses with round HSS members (photo by Abi Aghayere)

Built-Up Sections

Built-up sections (see Figure 1-14) include welded plate girders and other sections built from plates and standard rolled sections (e.g., W-section with plate welded to the bottom flange; plane and flanged cruciform sections). Cruciform columns are used where adequate moment capacity is desired about both orthogonal axes. Flanged cruciform columns also have the added advantage that plumbing and other service pipes can be hidden out of sight within the cruciform shape. Plate girders are used to support heavy loads where the listed standard rolled steel sections are inadequate to support the loads. Built-up sections can also be used as lintels and as reinforcement for existing beams and columns. Other built-up shapes include double angles (e.g., 2L $5 \times 5 \times {}^{1}/_{2}$) and double channels (e.g., 2C 12×25) placed back-to-back in contact with each other or separated by spacers, and W- and S-shapes with cap channels that are used to increase the bending capacity of W- and S-shapes about their weaker (*y–y*) axis.

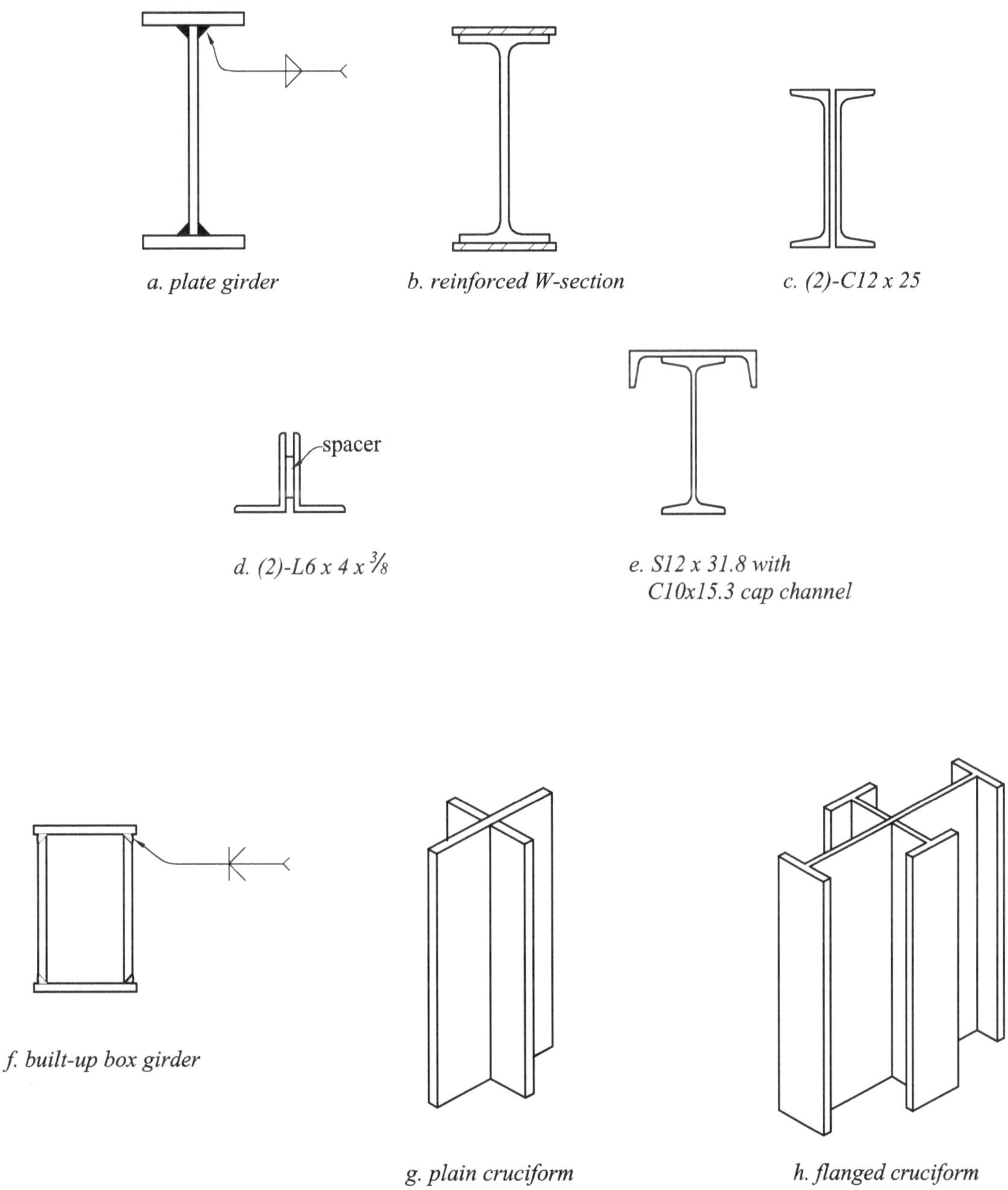

FIGURE 1-14 Built-up sections

Elastic and Plastic Section Modulus for Built-up Sections

While the section properties of standard rolled steel sections are listed in Part 1 of the *AISCM*, the section properties for built-up sections, such as the elastic section modulus, S, and the plastic section modulus, Z, have to be calculated using the principles from mechanics of materials. The elastic section modulus is calculated assuming a triangular (elastic) stress distribution (see Figure 1-15) with the maximum stress occurring in the extreme tension and compression fibers and with zero stress at the elastic neutral axis (ENA). The location of the elastic neutral axis (ENA) is equal to the centroid of the cross-section. For symmetrical sections, the ENA will be at the mid-depth of the section. The sum of the moments of the resultant elastic compression and tension forces about the ENA is equal to the yield moment, M_y; the yield moment capacity about the strong axis is,

$$M_y = F_y S_x = C_e \ell_{ae},$$

Where,

ℓ_{ae} = the distance between the locations of the resultant elastic compression force, C_e, and the resultant elastic tension force, T_e (the elastic stress distribution is similar to that shown in Figure 1-15). The values of these parameters will depend on the geometry of the section.

The elastic section modulus is,

$$S_x = \frac{C_e \ell_{ae}}{F_y}.$$

For illustration purposes, we will now consider a solid rectangular section with a width, b, and a depth, d, the resultant elastic compression force will be,

$$C_e = \frac{1}{2}(b)\left(\frac{d}{2}\right)F_y = \frac{bd}{4}F_y \text{ and}$$

$$\ell_{ae} = d - \frac{1}{3}\left(\frac{d}{2}\right) - \frac{1}{3}\left(\frac{d}{2}\right) = \frac{2}{3}d,$$

from which we obtain the following elastic section modulus for a solid rectangular section, a familiar equation from the mechanics of materials, as follows:

$$S_x = \frac{C_e \ell_{ae}}{F_y} = \frac{\left(\frac{bd}{4}\right)F_y\left(\frac{2}{3}\right)(d)}{F_y} = \frac{bd^2}{6}$$

Note that the elastic section modulus, S, is equal to the moment of inertia of the section divided by the distance between the elastic neutral axis (ENA) and the extreme fiber of the section; for a solid rectangular section, this would mean that the elastic section modulus can be calculated as

$$S_x = \frac{I_{xx}}{y_{top}} = \frac{I_{xx}}{y_{bottom}} = \frac{\frac{bd^3}{12}}{\frac{d}{2}} = \frac{bd^2}{6},$$ which is similar to the equation derived previously.

The location of the plastic neutral axis (PNA) is obtained from the equilibrium of the tension and compression forces on the section. For a homogeneous steel section with the same steel grade, the PNA divides the section into equal areas in tension and compression. Note that for symmetrical homogeneous sections, the location of the PNA and the ENA coincide with the centroid of the section. In calculating the plastic section modulus about the strong axis, Z_x, it should be noted that the critical section will be fully yielded across the section with a maximum stress of F_y in tension and in compression (the plastic stress distribution is similar to that shown in Figure 1-15). With the cross-section fully plastic with a yield stress of F_y, the sum of the moments of the resultant compression and tension forces about the PNA gives the nominal plastic moment capacity about the strong axis. This is:

$$M_p = F_y Z_x = C_p \ell_{ap},$$

Where,

ℓ_{ap} = the distance between the locations of the resultant plastic compression force, C_p, and the resultant plastic tension force, T_p (see Figure 1-15).

The plastic section modulus is,

$$Z_x = \frac{C_p \ell_{ap}}{F_y} = \frac{(A_c F_y)\ell_{ap}}{F_y} = A_c \ell_{ap},$$

Where,

A_c = the area of the compression zone of the plastic section, which for a rectangular homogeneous section is one-half the gross area of the section or $\frac{bd}{2}$.

For a homogeneous solid rectangular section of width, b, and depth, d, the resultant plastic compression force is,

$$C_p = A_c F_y = \frac{bd}{2} F_y$$

and, $\ell_{ap} = d - \frac{1}{2}\left(\frac{d}{2}\right) - \frac{1}{2}\left(\frac{d}{2}\right) = \frac{d}{2},$

from which we can obtain the plastic section modulus for a solid rectangular section,

$Z_x = \frac{C_p \ell_{ap}}{F_y} = \frac{\left(\frac{bd}{2} F_y\right)\left(d/2\right)}{F_y} = \frac{bd^2}{4}$, another familiar equation for rectangular sections from the mechanics of materials.

A similar approach to the preceding discussion can be used to determine the elastic and plastic section moduli about the strong axis for any section. The corresponding elastic and plastic section moduli for double-symmetric built-up I-shape plate girders can be derived as (see Figure 1-15):

$$S_x = 2t_w\left(\frac{d}{2} - t_f\right)\left[\frac{1}{2}\left(\frac{d-2t_f}{d}\right)\right]\left[\frac{2}{3}\left(\frac{d}{2} - t_f\right)\right] + 2b_f t_f\left[\frac{1}{2}\left(1 + \frac{d-2t_f}{d}\right)\right]\left(\frac{d-t_f}{2}\right)$$

$$Z_x = 2b_f t_f\left(\frac{d}{2} - \frac{t_f}{2}\right) + t_w\left(\frac{d}{2} - t_f\right)\left(\frac{d}{2} - t_f\right)$$

$$= \frac{b_f d^2}{4} - \frac{(b_f - t_w)(d - 2t_f)^2}{4}$$

Note that the equations derived above will be conservative for rolled steel shapes because they neglect the fillets at the intersection of the web and flanges of those shapes. The plastic and elastic section moduli as well as other sectional properties of commercially available rolled steel shapes are listed in Part 1 of the AISC Manual. The shape factor (Z_x/S_x) is defined as the ratio of the plastic section modulus to the elastic section modulus. It is a measure of how effectively the member is utilized in bending [29]. The larger the shape factor, the less effective the member is utilized in bending (e.g., the shape factor for a rectangular section = 1.5). A large shape factor means that a large portion of the member is still below the yield stress limit, F_y, when the outer fibers reach the yield stress. Sections with low shape factors (e.g., wide flange or I-shaped sections) are very effective in resisting bending about their strong axis because the stress in a large portion of the cross-section is close to the yield stress, F_y, when the outer fibers of the flange reach yield. Thus, the difference between the yield moment, M_y, and the plastic moment, M_p, is smaller for I-shapes or wide flange sections (W-shapes) than for solid rectangular shapes.

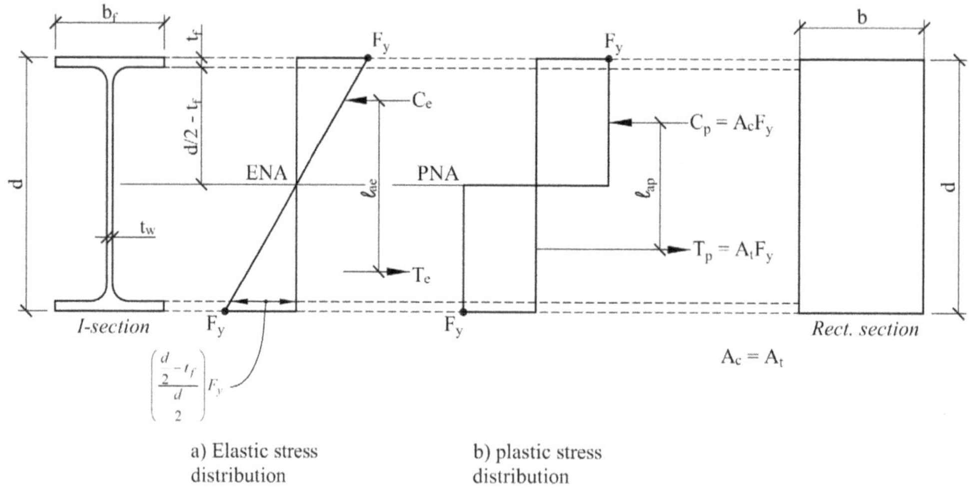

FIGURE 1-15 Elastic and plastic stress distributions for an I-shape and a solid rectangular shape

1-5 BASIC STRUCTURAL STEEL ELEMENTS

Due to the loads imposed on a structure, the structural elements may be subjected to one or more of the following load effects: tension, compression, flexure or bending, and shear. We will now discuss the basic structural steel elements and members that are used to resist gravity loads and lateral loads in steel-framed buildings, as shown in Figures 1-16 and 1-17. The typical floor-to-floor height of steel buildings is approximately 10 ft. for residential buildings and approximately 13 ft. or more for office buildings. The depth of the ceiling plenum space for office buildings (i.e., the space that contains the structural steel framing, the HVAC ducts, sprinkler systems, lighting, and ceiling) is typically approximately 4 ft.

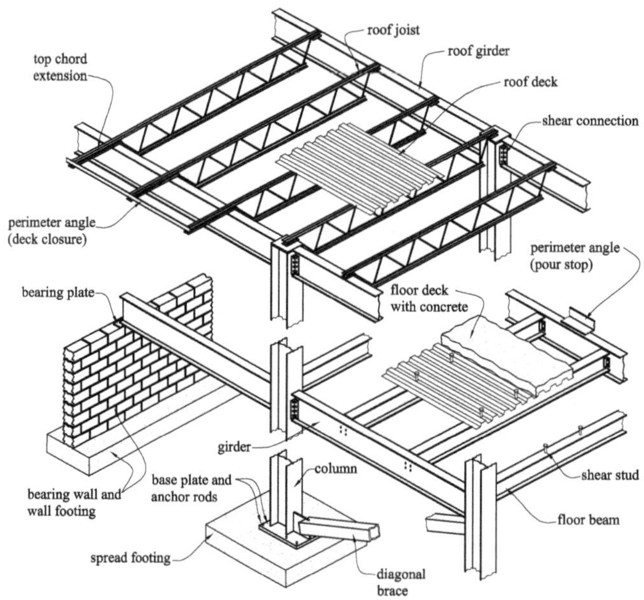

FIGURE 1-16 Typical steel building—basic structural elements (3D)

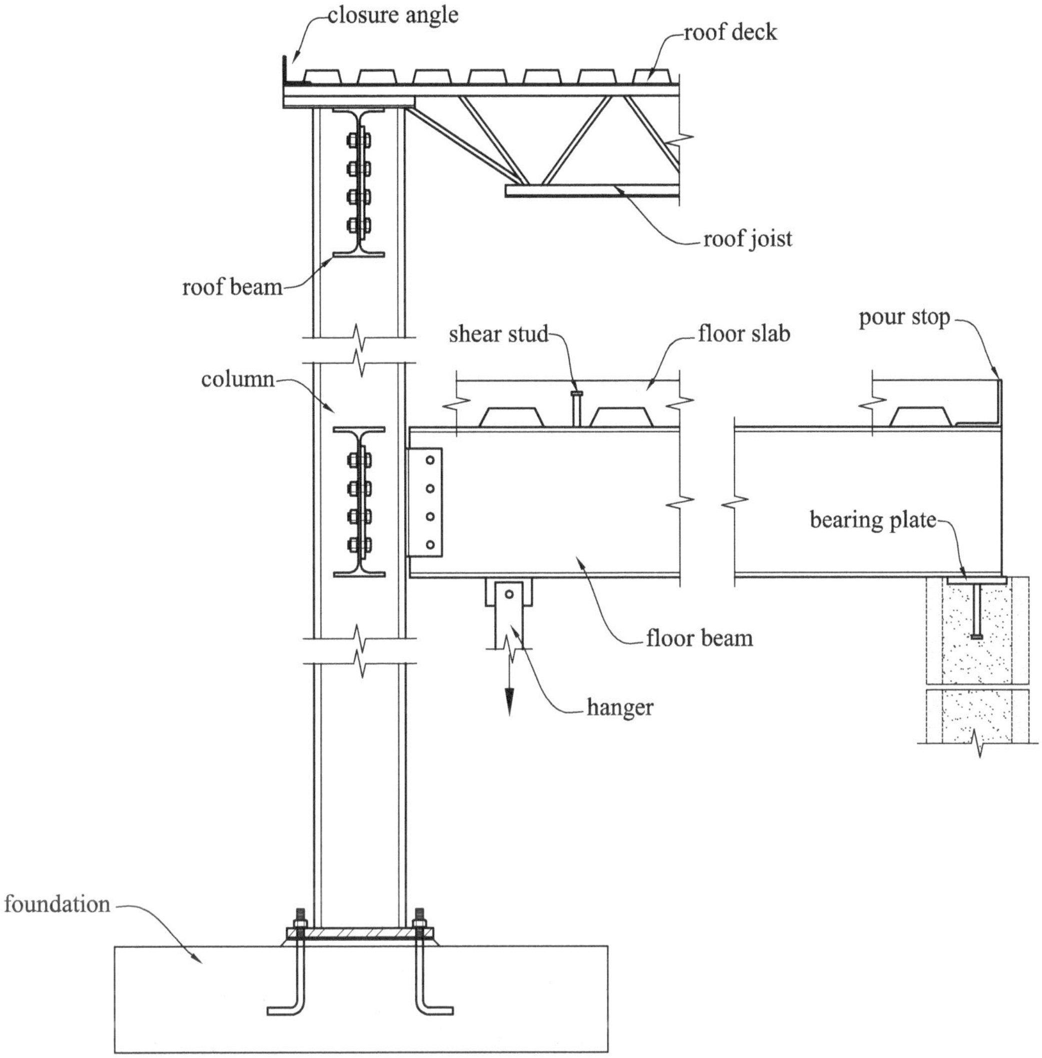

FIGURE 1-17 Typical steel building cross-section—basic structural elements

Beams and Girders

- The infill beams or joists support the floor or roof deck directly and span between the girders. The roof or floor deck usually spans in one direction between the roof or floor infill beams.
- The girders support the infill beams and span between the columns. While the beams along the column lines are usually connected to the web of the columns, girders are typically connected to the column flanges since the girders support heavier reactions than the typical infill beams. Thus, the girder reaction eccentricity at the columns is resisted by the bending of the column about its stronger axis.

Columns

These are vertical members that support axial compression loads only. They are sometimes referred to as struts when they are used in the horizontal position (as in bracings for soil excavations) or as diagonal struts to resist axial compression loads from discontinued columns. In real-world structures, structural members are rarely subjected to pure compression loads alone since the members cannot be fabricated perfectly straight, and we cannot assure that the line of application of the axial loads will line up perfectly with the centroidal axis of the column.

Beam–Columns

Beam-columns are members that support axial tension or axial compression loads in addition to bending moments. In practice, typical building columns usually act as beam-columns due to the eccentricity of the beam and girder reactions relative to the column centroidal axis.

Hangers

Hangers are vertical members that support axial tension loads in addition to any bending moment from the eccentricity of the axial tension load.

The reader should refer to Figures 1-16 and 1-17, and the other structural elements are discussed in greater detail later in this chapter.

1-6 TYPES OF STRUCTURAL SYSTEMS IN STEEL BUILDINGS

The common types of structural systems (i.e., a combination of several structural elements or members) used in steel building structures include trusses, moment or "rigid" frames, and braced frames, or a combination of these systems. Trusses are used predominantly to resist gravity loads, whereas braced frames and moment resisting frames are used to resist lateral loads.

Reinforced concrete core walls can also be used as the lateral force resisting system (LFRS) in steel-framed buildings to resist lateral loads, in lieu of steel braced frames or moment frames. A recently completed 25-story 392-ft tall steel-framed office building with a 250 ft × 110 ft footprint and located in a high seismic zone was laterally braced with special ductile-reinforced concrete coupled core walls with a footprint of 94 ft × 33 ft and a maximum wall thickness of 30 inches, and with a maximum concrete compressive strength of 8,000 psi. The special ductile reinforced concrete shear walls were interconnected across door openings and corridors with ductile-reinforced concrete coupling beams [54].

The concrete-filled composite steel plate shear wall, also known as "Speedcore," is a relatively new type of composite shear wall that has so far been used in at least two high-rise buildings in the West Coast [30]. In this shear wall system, invented by Ron Klemencic of Magnuson Klemencic Associates (MKA), the concrete acts like a sandwich material between parallel steel plates. The steel plates are connected with steel tie rods, and composite action between the steel plates and the concrete infill is ensured using headed steel studs.

Trusses

The typical truss profiles shown in Figures 1-18a through 1-18f consist of top and bottom chord members. The vertical and diagonal members are called web members. While the top and bottom chords are usually continuous members, the web members are connected to the top and bottom chords using bolted or welded connections. Trusses may occur as roof framing members over large spans or as transfer floor trusses supporting gravity loads from discontinued columns or shearwalls from upper levels. These transfer trusses, which could be as deep as one story or even two stories, support gravity loads and typically span between columns or girders at both ends of the truss. In practice, the top and bottom chords of floor or roof trusses are usually built as continuous members and not as pin-jointed members, as is frequently assumed in a typical structural analysis of trusses. Thus, the forces in the top and bottom chords will be a combination of moments and axial loads. When trusses are used in a vertical configuration, they essentially behave like braced frames since the vertical column members will be fabricated as continuous members, at least between the column splices.

Horizontal outrigger steel trusses are sometimes used to tie the concrete core walls or shear walls in tall buildings to the perimeter columns to create interaction between the concrete core walls and the perimeter columns in resisting the lateral loads, and thus reducing lateral drift in the building [45]. Outriggers are typically needed when the ratio of the core height to the core width at its base is greater than approximately 6. These outriggers, which are usually at least a full story high, may also occur as heavy "hat" or "cap" trusses supported on the central concrete core walls on the top of the building. Full-story-high intermediate outrigger trusses may also be used in addition to the "hat" or "cap" trusses. The steel outrigger trusses help to engage the perimeter columns in providing the moment restraint to the core walls and thus stiffening the core walls and reducing lateral drift as well as the moment at the base of the core walls. Without the outrigger trusses, the concrete core walls under lateral loads behave as a vertical cantilever fixed at its base; however, with the outrigger trusses and the engaged perimeter columns, the concrete core wall behaves as a vertical flexural member fixed at its base with some moment restraint provided at the intermediate outrigger locations (i.e., the core wall rotation is restrained at the outriggers), but the core wall is allowed to move laterally. This rotational restraint of the concrete core walls by the outriggers and the perimeter columns leads to a reduction in the moment at the base of the core walls and a reduction in the lateral translation and lateral acceleration of the building. The stiffer the flexural rigidity of the outrigger truss, the more effective the interaction between the core walls and the perimeter columns will be. For very tall buildings, large forces are generated at the connection of the outrigger trusses to the concrete core walls, making the connection of the steel trusses to the concrete core walls critical.

Trusses are also used to span over theater stages to support the floor or roof framing as well as the rigging loads from the stage operable walls or the curtains that are hung from the stage rigging and hanging grid, and the sound equipment, lighting, and props. Floors or roofs with long cantilevers can be framed effectively with deep trusses. Heavy steel trusses are also used to support tall buildings over existing rail lines or subways, and these steel trusses may be supported directly on pile caps. "Hat" trusses on top of tall buildings have also been used to support hanging perimeter columns where the building footprint is restricted and narrower at the lower levels (see Figure 1-18f).

Figure 1-18g shows a building entrance canopy roof framing with Vierendeel trusses or frames consisting of top and bottom chords and only vertical web members (i.e., no diagonal web members in the truss), with moment connections between the vertical web members and the top and bottom chord members. Vierendeel frames are advantageous over traditional trusses with diagonal web members where an unobstructed view of the exterior is desired. Furthermore, for buildings where the exterior bays on several floors or stories cantilever beyond the ground floor footprint of the building, vertical Vierendeel cantilevered frames can be used to achieve multi-level cantilevered floors in lieu of using individual cantilevered beams/girders with moment connections at each floor/roof level.

Figure 1-18h shows a truss cantilevering over a roadway, with the cantilevered span of the truss consisting of a regular truss configuration with diagonal and vertical web members, while the back span is a Vierendeel truss (i.e., a truss with no diagonal, but only vertical, web members). Trusses are also very effective as pedestrian bridges spanning over roadways as shown in Figure 1-18i, where the top and bottom chords of the truss are made from wide-flange steel members, and the diagonal and vertical web members consists of hollow structural sections (HSS).

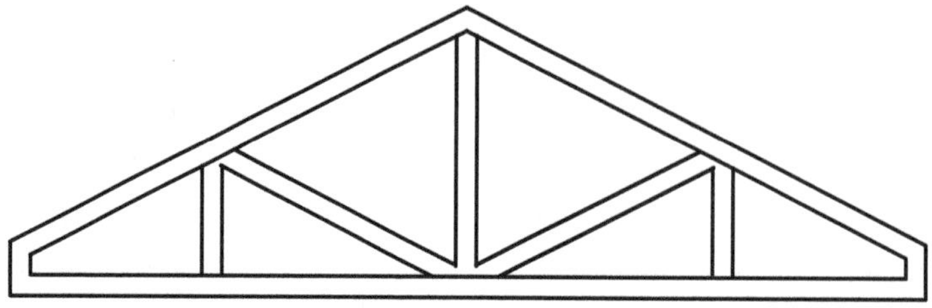

a) Howe truss

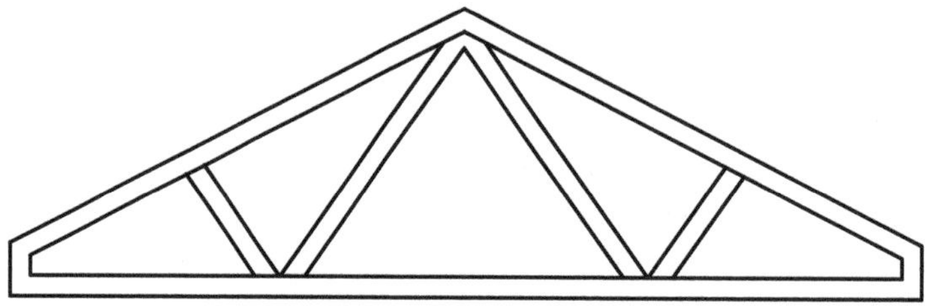

b) Simple Fink or Warren Truss

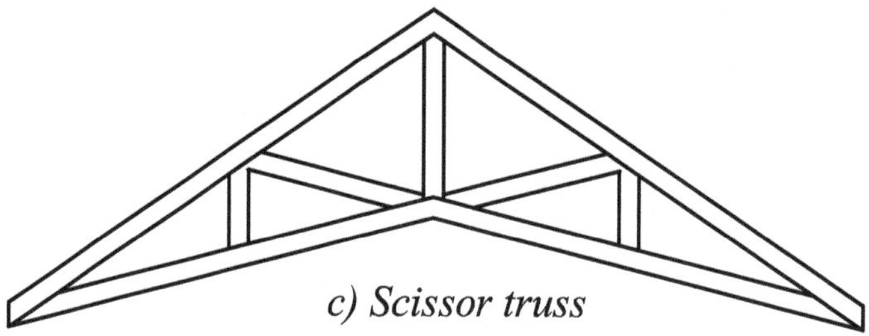

c) Scissor truss

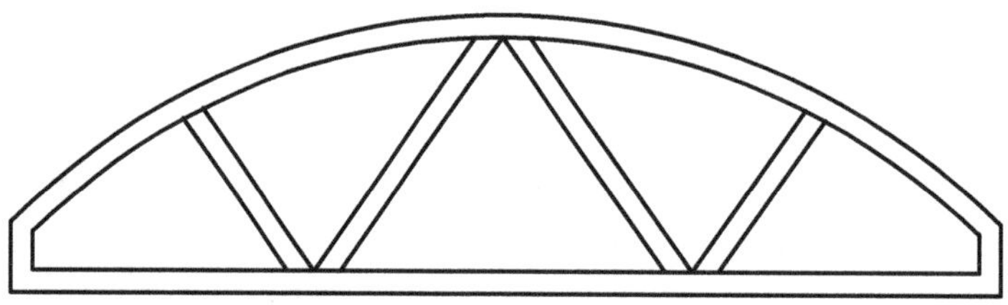

d) Bowstring Truss

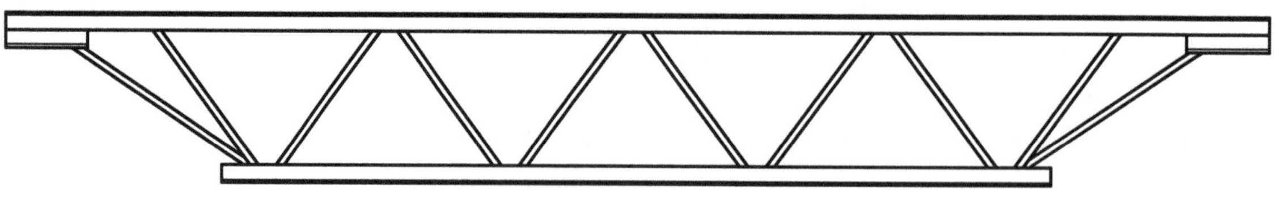

e) Pre-engineered roof or floor truss

FIGURE 1-18 a TO e Typical truss profiles

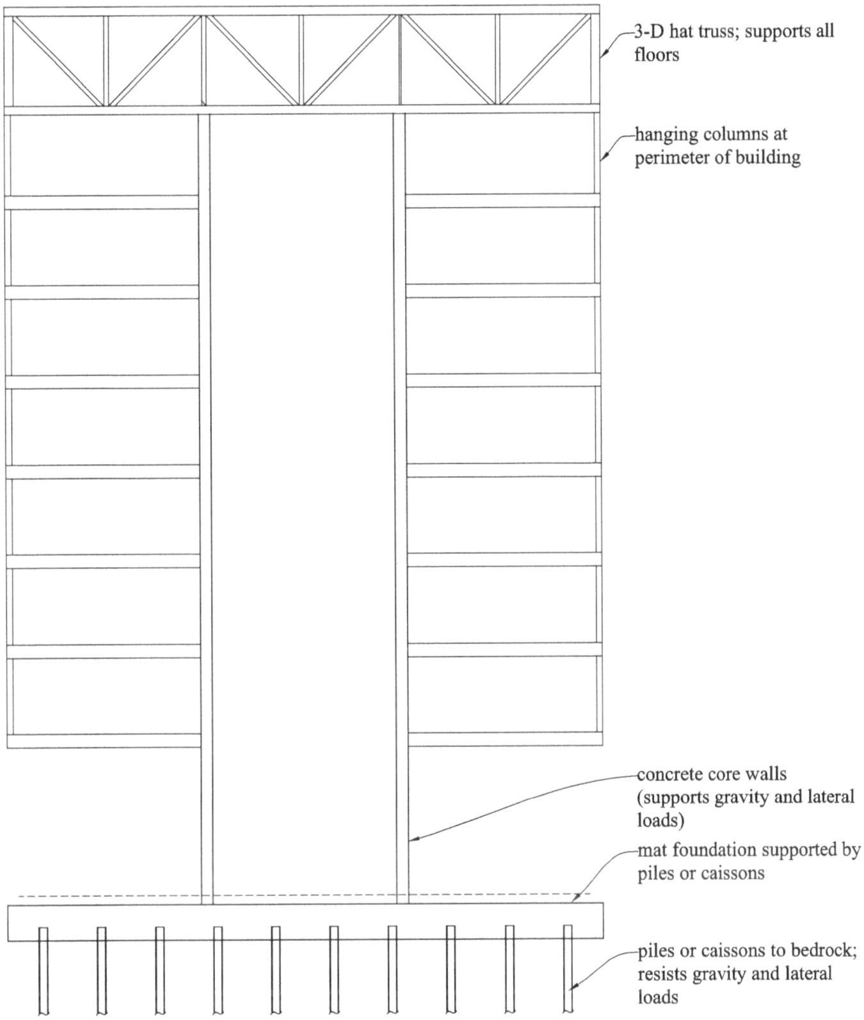

FIGURE 1-18f Typical hat truss

FIGURE 1-18g Vierendeel truss entrance canopy roof framing (photo by Abi Aghayere)

FIGURE 1-18h Cantilevered truss over a roadway

FIGURE 1-18i Pedestrian truss bridges spanning over a roadway in New York City

Frames

Frames are structural steel systems used to resist lateral wind or seismic loads in buildings. The two main types of building frames are moment-resisting frames and braced frames.

Moment-Resisting Frames: Moment-resisting frames resist lateral loads through the bending of the beams or girders and the columns. The connections between the beams or girders and the columns are designed and detailed as shown in Figure 1-19 to resist moments due to gravity and lateral loads. Note that moment-resisting frames are more laterally flexible than braced frames or shear walls. The moment connections used in moment resisting frames will be covered in Chapter 11.

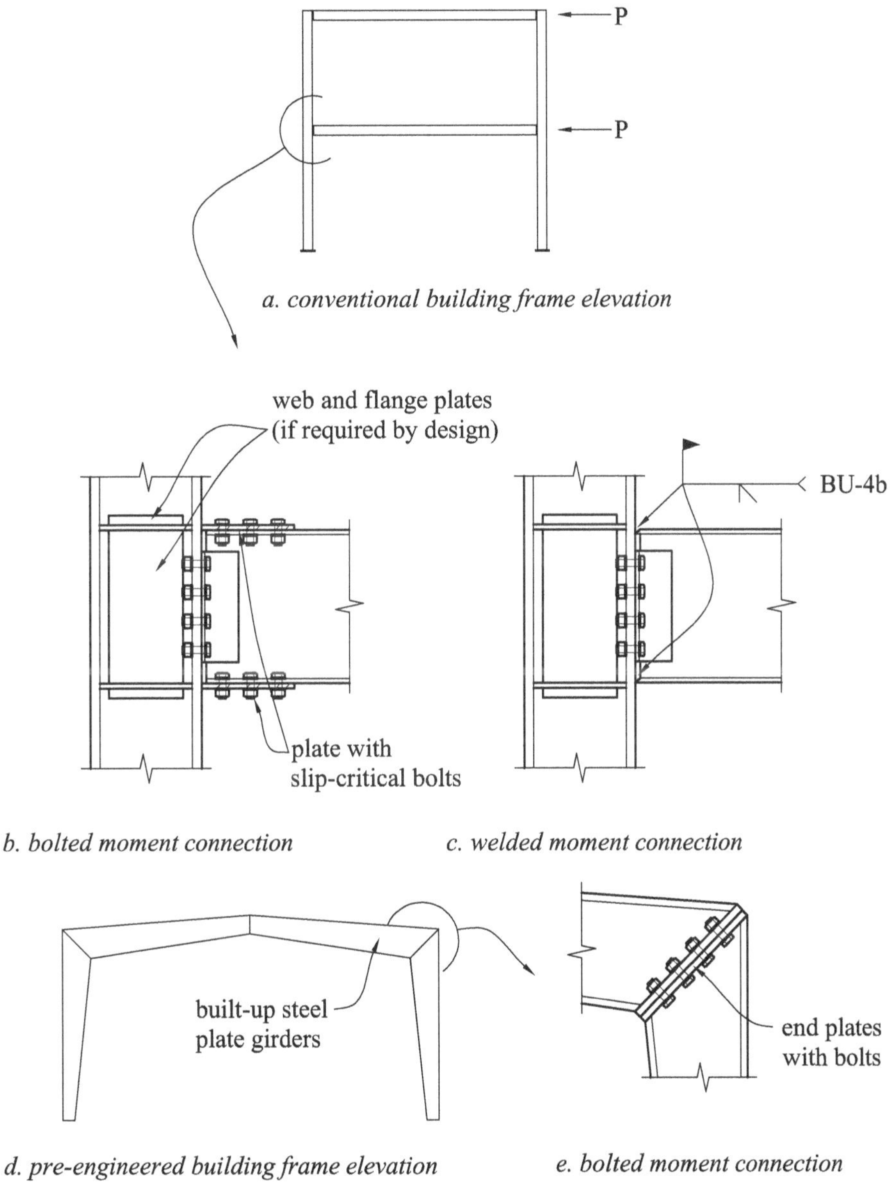

FIGURE 1-19 Typical moment resisting frames

Braced Frames: Braced frames (see Figure 1-20) resist lateral loads through axial compression and/or axial tension in the diagonal members. Examples include X-braced frames, diagonal braced frames, Chevron- or inverted-V braced frames, and knee-braced frames. These frames are usually more rigid than a typical moment frame and exhibit smaller lateral deflections. Figures 1-21a, 1-21b, and 1-21c show different types of braced frames used in steel buildings.

Another type of braced frame used for structures with tall story heights is the multi-tiered braced frame (MTBF), which consists of two or more stacked levels of braced frames within each story, with each braced frame panel connected to the ones above and below with horizontal struts [48]. The braced frame within each vertical panel could be X-braces, Chevron or inverted V-braces, or single diagonal braces. For structural efficiency, the columns in MTBF are oriented in such a way that the weak (y-y) axis bending of the column occurs in the plane of the braced frame, and any out-of-plane bending of the columns occurs about their strong (x-x) axis.

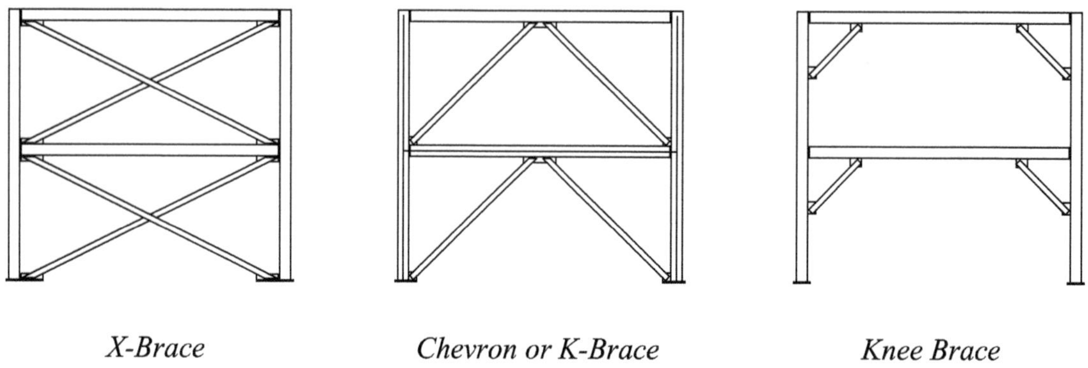

FIGURE 1-20 Typical braced frames

FIGURE 1-21a Braced frame and columns in a multi-story steel building under construction (photo courte-sy of Obinna Otti).

FIGURE 1-21b Braced frames: Plate and steel rod X-bracings in a steel building under construction (photo by Abi Aghayere).

FIGURE 1-21c Braced frames: Single diagonal and Chevron braces in steel buildings under construction (photo by Abi Aghayere).

1-7 BUILDING CODES, DESIGN SPECIFICATIONS, AND THE STEEL CONSTRUCTION MANUAL

The stakeholders in a typical building project include the owner, the architect, the structural, civil/site, geotechnical, mechanical, electrical, and plumbing engineers, the general contractor, the sub-contractor and steel fabricator/erector who convert the engineers' designs in the form of contract drawings and specifications into a real-life structure, and the state and city building officials who are responsible for granting the building permit before construction and the certificate of occupancy ("C of O") after construction. The city officials also carry out inspections during construction to ensure that it conforms to the city's building code.

Building construction in the United States and in many parts of the world is regulated using building codes that prescribe a consensus set of minimum requirements that will ensure public health and safety and also include economic considerations. A code consists of standards and specifications (or recommended practice) and covers all aspects of design, construction, and functions of buildings, including occupancy and fire-related issues, but it only becomes a legal document within any jurisdiction after it is adopted by the legislative body in that jurisdiction. Once adopted by a jurisdiction, the code becomes the legal binding document for building construction in that locality, and the design and construction professionals are bound by the minimum set of requirements specified in the code. The International Building Code (IBC 2024) [16], published by the International Code Council (ICC), has replaced the former model codes—the Uniform Building Code (UBC), Building Officials and Code Administrators (BOCA), National Building Code (NBC), and the Standard Building Code (SBC)—and is the basis for most local and state building codes in the United States [24]. The IBC is a consensus document developed by many stakeholders, and a new edition is published every three years. The IBC references the ASCE 7 Load Standard [31] for the calculation of structural loads, and this load standard, which is reviewed every few years, is the basis for the calculations of the structural loads in this text (see Chapters 2 and 3). The steel material section of the IBC references the AISC specifications as the applicable specification for the design of steel members in building structures. The ICC also publishes the *International Existing Building Code* for the renovation of existing structures, as well as the *International Residential Code* for the design of one- and two-family dwellings.

The premier technical specifying and trade organization in the United States for the fabricated structural steel construction industry is the American Institute of Steel Construction (AISC). This nonprofit organization publishes and produces several technical manuals, design guides, and specifications related to the design and construction of steel buildings, such as the Steel Construction Manual or *AISC Manual* (AISC 325-23)—hereafter referred to as the "*AISCM*." The *AISCM* includes the AISC 360-22, Specification for the Design of Steel Buildings, and the properties of standard steel shapes and sizes [8]. This manual, with its first edition published in 1927 and now in its 16th edition, consists of 18 chapters as listed below and provides the dimensions and properties of several standardized structural shapes, as well as several design aids, some of which will be used later in this text. The *AISCM* includes design provisions for both the load and resistance factor design (LRFD) method and the allowable strength design (ASD) method. The *AISCM* consists of the following chapters:

Part 1: Dimensions and Properties
Part 2: General Design Considerations
Part 3: Design of Flexural Members

Part 4: Design of Compression Members
Part 5: Design of Tension Members
Part 6: Design of Members Subject to Combined Forces
Part 7: Design Considerations for Bolts
Part 8: Design Considerations for Welds
Part 9: Design of Connecting Elements
Part 10: Design of Simple Shear Connections
Part 11: Design of Moment Connections
Part 12: Design of Simple Connections for Combined Forces
Part 13: Design of Bracing Connections and Truss Connections
Part 14: Design of Beam Bearing Plates, Column Base Plates, Anchor Rods, and Column Splices
Part 15: Design of Hanger Connections, Bracket Plates, and Crane-Rail Connections
Part 16: Specifications and Codes
Section 16.1 includes the AISC Specifications (AISC 360-22)—Chapters A through N; Appendices 1 through 8; and the Commentary on the Specifications—which provides research and background information and the limitations to the design equations presented in the specifications.
Section 16.2 contains the "Specifications for Structural Joints Using High-Strength Bolts," prepared by the Research Council on Structural Connections (RCSC);
Section 16.3 contains the "Code of Standard Practice for Steel Buildings and Bridges" (AISC 303-22). Both the AISC Specifications and the Code of Standard Practice are accredited by the American National Standards Institute (ANSI).
Part 17: Miscellaneous Data and Mathematical Information (e.g., weights of common building materials)
Part 18: Symbols and Index

The AISC Specifications (i.e., Part 16.1), which form the basis for the design of structural steel buildings in the United States, were first adopted in 1923 and are divided into fourteen main chapters (i.e., Chapters A through N) as follows:

Chapter A: General provisions - Table A3.1 lists the commonly used ASTM standard designations and permitted grades and strengths that are approved for use for hot-rolled shapes, hollow structural sections, plates, and bars under the AISC 360-22 Specification. AISC Specification Section A3.1b allows other materials to be used "for specific applications when the suitability of the material is determined to be acceptable by the engineer of record (EOR)." [20]

Chapter B: Design Requirements (e.g., loads and load combinations; LRFD and ASD requirements; diaphragms and collectors; structural integrity; fire protection; simple and moment connections; corrosion protection for structural steel; member properties, etc.).
Chapter C: Design for stability
Chapter D: Design of members for tension
Chapter E: Design of members for compression
Chapter F: Design of members for flexure
Chapter G: Design of members for shear
Chapter H: Design of members in combined forces and torsion (e.g., axial compression plus bending and axial tension plus bending, etc.)
Chapter I: Design of composite members
Chapter J: Design of connections
Chapter K: Additional requirements for HSS and Box-section connections
Chapter L: Design for serviceability

Chapter M: Fabrication and erection of structural steel (e.g., fabrication requirements—e.g., dimensional tolerances, thermal cutting, etc.; erection requirements; shop and erection drawings, shop painting, etc.)
Chapter N: Quality control and quality assurance

For each type of steel member and connection in Chapters D through K, the AISC specification identifies the possible failure modes and the limit states that should be considered for design. The required strength using ASD load combinations (sometimes called the allowable strength) or the design strength (LRFD) are a function of the nominal strength, R_n, which is the same for both design methodologies. The eight (8) appendices, which are also a mandatory part of the specification, are listed as follows:

Appendix 1: Design by advanced analysis—elastic and inelastic analyses
Appendix 2: Design of filled composite members (high strength)
Appendix 3: Fatigue
Appendix 4: Structural design for fire conditions
Appendix 5: Evaluation of existing structures
Appendix 6: Member stability bracing—column, beam, and beam-column bracing
Appendix 7: Alternative methods of design for stability—Effective length and first-order analysis methods
Appendix 8: Approximate analysis—Approximate second-order elastic analysis with B_1 and B_2 factors

The appendices are followed by the "Commentary on the Specification for Structural Steel Buildings," which consists of discussions and commentary on the fourteen chapters and eight appendices. In general, the *decimal values* in the tabulated dimensions of structural steel sections from Part 1 of the *AISCM* are used in design calculations, while the *fractional values* are used for detailing purposes only by the steel detailers. The AISC web site at *www.aisc.org* contains a lot of useful information and resources related to steel design and construction, including the *Modern Steel Construction* magazine and the Steel Solutions Center, among others. *Modern Steel Construction* regularly publishes useful and interesting articles related to the practical design and construction of steel structures.

1-8 THE STRUCTURAL STEEL DESIGN AND CONSTRUCTION PROCESS

The design process for a structural steel building is iterative in nature and typically starts out with the schematic or conceptual design phase. With the schematic drawings developed by the architect for the owner of a building, the structural engineer carries out preliminary designs to determine the preliminary sizes of the members for each structural material and structural system (gravity and lateral) considered. This information is used to determine the most economical and efficient structural material and structural system for the building. This is sometimes called the design development phase. After the structural material and structural system to be used are selected, this is followed by the final design phase where the roof and floor framing members and the lateral force resisting systems (LFRS) are laid out and all the member sizes are proportioned to resist the applied loads. This process results in a set of construction or contract documents that include structural plans, sections, details, and contract specifications for the materials used on the project (includes requirements for materials to be used, material testing, quality control, etc). The construction documents are used to obtain the construction permit from the Authority Having Jurisdiction (AHJ). The final design phase is followed by the bidding phase, after which the construction drawings are issued for construction. The construction documents issued for construction must meet the requirements in AISC Specification Section A4.1. This is followed by the construction administration (CA) phase, which includes shop drawing production and review and approval by the architect and structural engineer of record (SER), and then followed by fabrication and erection of the building by the contractor. The shop drawing review process provides one final opportunity for the structural engineer of record (SER) to ensure that the fabrication drawings and details meet their design intent as presented in the contract documents. Once the shop drawings are approved, steel fabrication and erection can commence. The importance of proper fabrication and erection procedures and constructible details to the successful construction of a steel project cannot be overemphasized. Note that for some speculative building projects, construction may commence even before the construction documents are fully complete, unlike bridge projects that typically have the construction documents completed before construction begins.

In the United States and Canada, the design of simple connections (i.e., simple shear connections) is sometimes delegated by the structural engineer of record (EOR) to the general contractor and their steel fabricator (i.e., delegated design); the steel fabricator typically hires a third-party structural engineer to design these connections using the loads and reactions provided by the structural engineer of record (SER) on the structural drawings and/or specifications. The connection designs and the detail drawings of the connections are also submitted to the structural engineer of record for review and approval. In other cases, especially for the more complicated connections, such as moment connections, the SER may provide schematic or full connection designs directly to the fabricator. During the construction phase, although the structural engineer of record may visit the construction site occasionally, it is common practice for the owner to retain the services of a materials inspection firm — especially when special inspections are mandated by the code — to periodically inspect the fabrication and erection of the structural steel to ensure that the construction is carried out in accordance with the structural drawings and contract specifications.

1-9 GRAVITY AND LATERAL LOAD PATHS AND STRUCTURAL REDUNDANCY

Structural loads are resisted or supported by structural members, resulting in tension, compression, shear, or flexural stresses. The load path is defined as the route through which a gravity or lateral load is transmitted from its point of application on the structure to the foundation and the supporting soil or bedrock. It is important that the structural engineer develop an adequate continuous load path when designing a structure because any structural deficiency in the integrity of the load path could lead to collapse; gaps or discontinuities in the load path commonly result from inadequate connections between adjoining structural elements rather than the failure of an individual structural member. The typical path of a gravity load from its point of application on the structure to the foundation is as follows: The gravity load applied to the roof, floor deck, or slab is transmitted horizontally to the infill beams, which in turn distribute the load horizontally to the girders. The girders and the beams along the column lines transfer the load as vertical reactions to the columns, which in turn transmit the load safely to the foundation and to the ground or bedrock. Therefore, in essence, the flow of gravity load is as follows:

- Load applied to the Slab or Roof Deck → Beams → Girders → Columns → Foundations.

The gravity load path is illustrated in Figure 1-22.

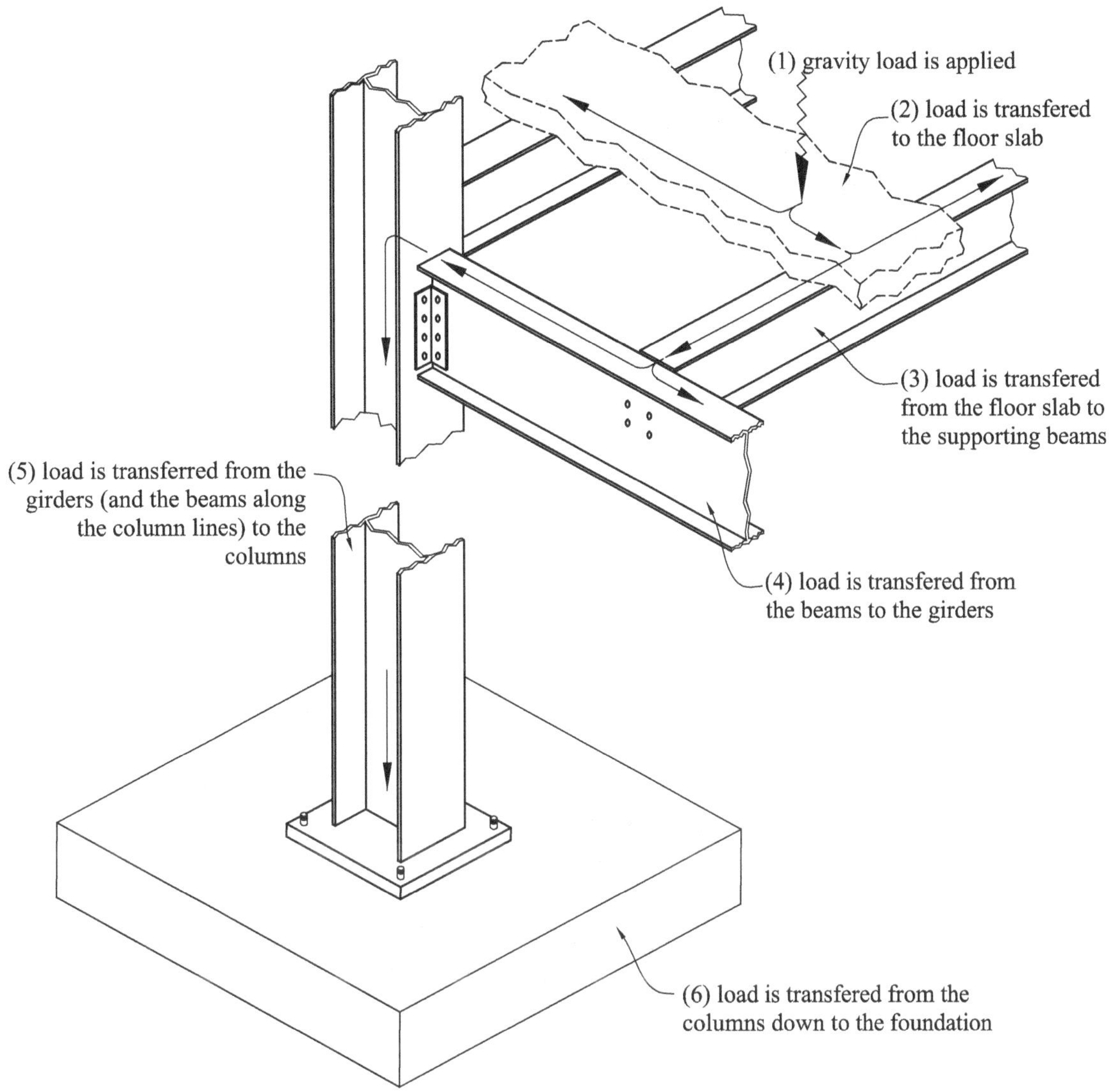

FIGURE 1-22 Gravity load path

For the *wind lateral load path*, the wind load is applied directly to the surface area of the vertical wall on the windward side of the structure, and this wall bends vertically and transfers the horizontal reactions to the horizontal roof or floor diaphragms (i.e., the composite deck and floor slabs). The horizontal diaphragms in turn transfer and distribute the lateral or horizontal wind forces to the vertical lateral force resisting system (LFRS) — (e.g., moment-resisting frames with welded or bolted beam-to-column connections, braced frames—e.g., X-braced, Chevron braced, or diagonal braced frames, and/or shear walls)—that have their longitudinal axis parallel to the lateral wind load, and these vertical lateral force resisting systems then transmit the lateral forces to the foundation and the soil or bedrock. Therefore, in essence, the lateral wind load path is as follows:

- Wind pressure on the windward vertical wall surface area → Roof or Floor Diaphragm → Vertical Lateral Force Resisting System → Foundations.

The lateral load path for wind loads is illustrated in Figure 1-23.

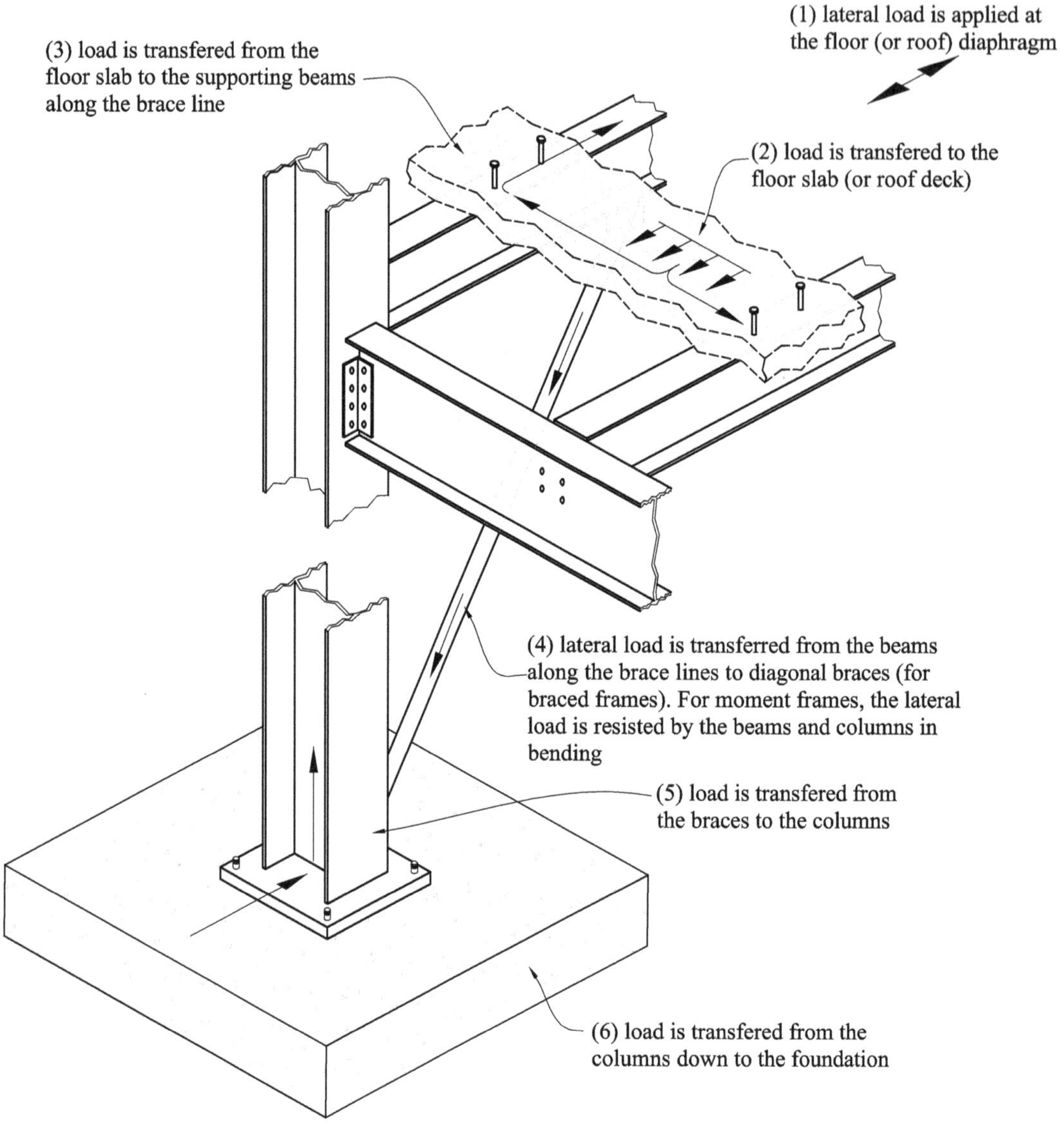

FIGURE 1-23 Lateral load path

The *seismic lateral load path* begins with ground motion or shaking of the structure from a seismic event, which results in inertial forces being applied to the building structure at its base (i.e., the base shear). The base shear is assumed to be distributed to each horizontal diaphragm level (i.e., the floors and the roof), and the lateral seismic forces are assumed to be concentrated at the floor and roof diaphragm levels; the diaphragms, in turn, transmit the lateral seismic forces to the vertical lateral force resisting systems (LFRS) — (e.g., moment-resisting frames with welded or bolted beam-to-column connections, braced frames — e.g., X-braced, Chevron braced or diagonal braced frames, and/or shear walls)—that have their longitudinal axis parallel to the seismic lateral load, and then the vertical LFRS transmits the lateral forces to the foundation and then to the soil or bedrock. In essence, the *seismic lateral load path* is as follows:

- Ground motion → Base shear → Diaphragm lateral loads → Vertical Lateral Force Resisting System (LFRS) → Foundations.

Structural redundancy — which is highly desirable in structural systems — is the ability of a structure to support loads through more than one load path, thus preventing progressive collapse when the only load path fails. In a redundant structure, there are alternate load paths available so that failure of one member does not lead to failure of the entire structure; thus, the structure is able to safely transmit the load to the foundation through the alternate load paths (see Chapter 14 for discussions on progressive or disproportionate collapse prevention).

1-10 ROOF AND FLOOR FRAMING LAYOUT, DECK SELECTION, AND FIREPROOFING

Once the gravity loads acting on a building are determined, the next step in the design process, before any of the structural members or elements can be designed, is to lay out the roof and floor framing and the lateral force-resisting systems. In this section, the criteria for the economical layout of roof and floor framing are presented. The self-weight of roof and floor framing (i.e., beams and girders) varies from approximately 5 psf to 10 psf. In calculating the allowable loads of roof and floor decks using proprietary deck load tables, the self-weight of the framing should be subtracted from the total roof or floor loads since the beams and girders support the deck and the loads acting on it. The proprietary deck load tables for some manufacturers are based on service loads.

In laying out roof or floor framing members (i.e., the deck, the beams, and the girders), the following criteria should be noted for constructability and economy:

- The infill beams or joists (supported by the girders) should be framed in the longer direction of the bay, while the girders should span in the shorter direction. Thus, the girder length should be less than or equal to the length of the infill beam or joist. This ensures optimum beam and girder sizes.
- To reduce labor cost and simplify the design, reduce the number of infill beams using optimum beam spacing. The fewer the infill beams, the fewer the beam-to-girder connections that must be fabricated and erected. It is also economical to group together member sizes; for example, rather than using a mix of beam sizes, it may be more economical to use the larger beam size, thus standardizing all the beam connections and simplifying the steel erection [32].
- Avoid skewed beam-to-girder or beam-to-column connections with small acute angles as much as possible because they are costly to erect and fabricate.
- The infill beams along the column lines should be connected to the web of the steel columns, while the girders should be connected to the column flanges because the girders support heavier loads and therefore have reactions that are, in general, greater than those of the infill beams. This arrangement ensures that the moments from the girder reactions are resisted by the column bending about its stronger axis (see Chapter 8).
- The span of the deck should be as close to the Steel Deck Institute (SDI) maximum allowable span [33] as possible to minimize the required number of infill beams or joists and therefore the number of connections. The minimum slab thickness on composite metal deck required to satisfy the Underwriters Laboratories Inc. Fire Resistance Rating requirements may be different than that required from the load requirements using the metal deck manufacturer's load tables; the larger slab thickness should be specified. As an example, to achieve a 2-hour fire resistance rating for an unprotected floor deck, options include a 3.25-inch thick light-weight concrete slab or a 4.5-inch normal-weight concrete slab on top of a 1.5-inch, 2-inch, or 3-inch deep metal floor deck, depending on the spacing of the infill beams. For a 3-hour fire resistance rating, the required thickness of the slab on top of the metal floor deck becomes 4-3/16-inch light-weight concrete or 5¼-inch normal-weight concrete.

If the underside of the composite deck is protected against fire with sprayed-on fiber or cementitious fireproofing, the required minimum slab thicknesses are 2.5 inches for both normal-weight and light-weight concrete.

- As much as possible, specify 22-gage or 20-gage decks because these are the most commonly available deck thicknesses. Other deck thicknesses, such as the 18-gage deck, can be obtained, but at a higher or premium cost.
- Decks are available in lengths of 30 ft. to 42 ft. and widths of 2 ft. to 3 ft., and these parameters should be considered in the design of the deck. It should be noted that deck lengths longer than 30 ft. are often too heavy for two construction workers to safely handle on site.
- Roof decks are readily available in 1½-in. and 3-in. depths, with gages ranging from 16-ga through 22-ga; however, the 22-ga wide-rib deck is more commonly used in practice.
- Floor decks are readily available in 1½-in., 2-in., and 3-in. depths, with gages ranging from 16-gage through 22-gage; however, the 20-gage and the 18-gage wide-rib decks are more commonly used. Floor decks can be composite or noncomposite. Composite floor decks have protrusions inside the deck ribs that engage the hardened concrete within the ribs to provide the composite action between the metal deck and the concrete in resisting the floor loads. Noncomposite floor decks do not have these protrusions and are called form decks. The form deck does not act compositely with the concrete slab in resisting gravity loads but acts only as formwork to support the wet weight of the concrete during construction. Therefore, the concrete slab alone must be reinforced (usually with welded wire fabric) to support the applied loads.
- Noncomposite floor deck (i.e., form deck) is available in depths ranging from $^{9}/_{16}$ in. to 3 in. and in gages ranging from 16-gage to 28-gage.
- Although 3-in.-deep deck costs more than 1½-in. deck, they can span longer distances and thus minimize the number of infill beams and connections used.
- To protect against corrosion, roof and floor decks could be painted or galvanized (G60 galvanized or G90 galvanized), but where spray-applied fire protection is to be applied to the deck, care should be taken in choosing a classified paint product to ensure adequate bonding of the fireproofing to the metal deck.
- To maximize the strength or capacity to support gravity loads, the deck should be selected to span over at least four beams, creating a deck that is continuous over three spans with its inherent redundancy (i.e., the so-called 3-span or triple-span deck). This 3-span deck configuration implies that the minimum length of deck that the contractor can use on site will be three times the spacing between the beams or joists, and since the number of beam or joist spacings may not necessarily be a multiple of three, the deck sheets will have to be overlapped.
- Note that the load capacities given in the referenced roof and floor deck load tables may be unfactored. In the roof deck load capacity tables, the unfactored *allowable total load* is tabulated, whereas in the floor deck load capacity tables, the *superimposed load capacity* is tabulated. This superimposed load capacity should be compared to the applied superimposed floor load, which is the total unfactored dead and live load on the floor deck less the self-weight of the concrete slab and the composite metal deck.

- The thickness of the concrete slab and the depth of the metal deck selected based on strength considerations should also be compared to the required thickness of slab and depth of deck that satisfies the required fire resistance rating of the structure (the larger thickness or depth governs), and thus determine if coating or other fire protection or fireproofing of the floor deck and steel framing are required.
- The fire resistance rating affects the depth of the concrete slab on top of the metal floor decks and the required ceiling system or other types of fire protection for the roof and floor deck. Sprayed-on fireproofing or gypsum wall board enclosures for steel beams, girders, and columns may be required. Unless an unprotected deck is specified by the architect for aesthetic appeal, the metal deck may need fire protection, and the typical types of fire protection for roof and floor decks include the following:
    - Gypsum board ceiling,
    - Exposed grid ceiling systems (*https://www.armstrongceilings.com/commercial/en/suspension-systems/prelude-xl-hrc-grid.html*)
    - Sprayed-on cementitious fireproofing at the underside of the deck (not usually preferred by architects)

  The use of lightweight concrete slabs on top of the metal floor deck is common for tall buildings to achieve the required fire resistance rating, and also to reduce the overall weight on the structure and therefore reduce the size of the floor framing, columns, and foundations. The fire-resistance rating (FRR) is "the time a building element, component, or assembly can confine a fire to a given area, perform a given structural function, or both as determined by testing." Fire resistance ratings in buildings vary from 1 hour to 4 hours, depending on the building occupancy (see IBC Table 601) [16]. As a result of the World Trade Center collapse, tall buildings that are more than 420 ft. in height now require an additional one-hour increase in the fire resistance rating of the structural components and assemblies. An example of a floor system with an unprotected deck, and with steel framing—beams, girders, and columns—fireproofed with a red intumescent paint system can be seen at *https://cjsarchitects.com/work/project:625-goodman/*.
- Note that in hybrid buildings, to achieve increased construction speed and lighter weight as well as enhance sustainability, cross-laminated timber (CLT) panels (typically five-ply) or glue-laminated timber (GLT) panels with an acoustic mat on top of the panels that provides sound absorption properties have been used for the roof and floors in lieu of a concrete slab on composite metal deck.

In designing any structure, the reader should strive for simplicity in the structural layout and details and not just focus solely on achieving the least weight for the steel members, because labor costs, which consist of steel fabrication and erection, are approximately twice the material costs [34]. Thus, the least weight may not necessarily always result in the least cost. In addition, the engineer needs to understand or have a feel for how the steel structure will be erected, and when in doubt, it is advisable to check with a local steel fabricator. The sizing of roof and floor decks and the layout of roof and floor framing members are illustrated in the following two examples.

EXAMPLE 1-3

Roof Framing Layout and Roof Deck Selection

A roof framing plan (see Figure 1-24) consists of 30-ft. by 40-ft. bays, supporting a total dead load of 30 psf and a roof snow load of 80 psf. Determine the layout of the roof deck and the size of the deck. Assume a self-weight of 5 psf for the steel framing. The metal deck is made from Grade 50 steel.

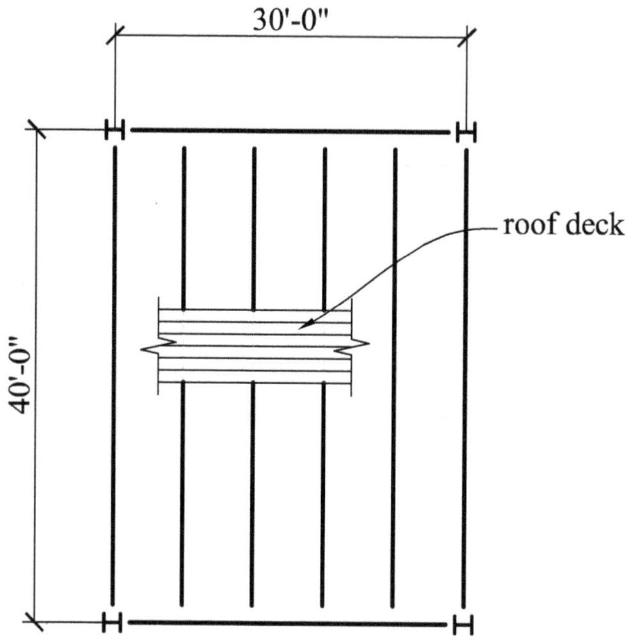

FIGURE 1-24 Roof framing layout for Example 1-3

Solution

The roof deck supports the total dead load less the weight of the steel framing.

Dead load on the roof deck alone = 30 psf - 5 psf (weight of steel framing) = 25 psf

Snow load on the roof = 80 psf

For our example, try 1.5B-36 × 20-ga galvanized roof deck (Grade 50 steel); the reader should refer to the *Vulcraft Steel Roof and Floor Deck Manual* [33] or similar deck manufacturers' catalogs.

Preferably, the deck should span continuously over at least **four** beams (i.e., a 3-span or triple-span deck). It should be noted that the deck span selected must be such that the shorter bay dimension (i.e., the distance between adjacent column supports in the shorter direction) is a multiple.

LRFD Solution

Applied total factored dead plus snow load on the roof deck = 1.2(25 psf) + 1.0(80 psf) = 110 psf

Applied total unfactored dead plus snow load on the roof deck = 25 psf + 80 psf = 105 psf

Try a 7-ft. 6-in. span (of which the 30-ft. bay dimension is a multiple), and this results in 3 equal spaces or deck spans per bay, which assures a triple-span deck.

From the Vulcraft Deck Load Table, for 1.5B-36 × 20 gage deck and by linear interpolation, the total load capacity, ϕW_n = (173 psf + 133 psf)/2 = 153 psf > 110 psf OK.

Therefore, the deck selected is adequate for strength under total factored loads.

Similarly, by linear interpolation, the allowable total unfactored load that causes a deflection of L/240 = (78 psf + 52 psf)/2 = 65 psf < 105 psf, Not good.

Therefore, the deck selected is *not* adequate for the deflection criteria under total load.

Therefore, choose a smaller deck span for which 30 ft. is also a multiple.

Try 1.5B-36 × 20 gage deck with a span of **6 ft. 0 in.**, resulting in five equal spaces or deck spans per bay, which assures a triple-span deck.

From the Vulcraft Deck Load Table, Total load capacity, ϕW_n = 235 psf > 110 psf OK.
Allowable total load that causes a deflection of L/240 = 124 psf > 105 psf OK.
The roof deck selected is adequate. Therefore, use **1.5B – 36 × 20-ga galvanized metal deck (Grade 50 steel) with infill beams @ 6' 0" o.c.**
Note that though a 22-gage deck meets the strength criteria, it is not adequate for the deflection criteria.

Note that the triple-span or 3-span deck requirement assumed in this example must be specified on the structural drawings or in the contract specifications. Recall that this requirement means the deck must cover at least three spans. The implication of the triple-span deck assumed is that the minimum length of roof deck supplied by the contractor to the site must be at least 18 ft long (i.e., 3 spans × 6 ft deck span). For strength considerations, the triple or 3-span deck is the strongest, followed by the 2-span deck, and then the 1-span deck.

For ease of transportation, maneuverability on site, and to assure a lighter weight and the ability for the deck to be lifted by hand, some designers sometimes specify the 2-span deck, especially for large deck spans. In some special situations, the single or 1-span deck may also be specified.

ASD Solution

Applied total unfactored dead plus snow load on the roof deck = 25 psf + 80 psf = 105 psf
Try a 7-ft. 6-in. span (of which the 30-ft. bay dimension is a multiple), and this results in 3 equal spaces or deck spans per bay, which assures a triple-span deck.
From the Vulcraft Deck Load Table, for 1.5B-36 × 20 gage deck and by linear interpolation, the allowable total load, W_n/Ω = (115 psf + 88 psf)/2 = 102 psf < 105 psf Not good.
Therefore, the deck selected is *not* adequate for strength under total loads.
Similarly, by linear interpolation, the allowable total unfactored dead plus snow load that causes a deflection of L/240 = (78 psf + 52 psf)/2 = 65 psf < 105 psf, Not good.
Therefore, the deck selected is *not* adequate for the deflection criteria under total load.

Therefore, choose a smaller deck span for which 30 ft. is also a multiple.
Try 1.5B-36 × 20 gage deck with a span of **6 ft. 0 in.**, resulting in five equal spaces or deck spans per bay, which assures a triple-span deck.
For the 6 ft. triple-span deck, the allowable total deck load, W_n/Ω = 156 psf > 105 psf, OK.
Allowable total load that causes a deflection of L/240 = 124 psf > 105 psf OK.
The roof deck selected is adequate. Therefore, use **1.5B – 36 × 20-ga galvanized metal deck (Grade 50 steel) with infill beams @ 6' 0" o.c.**
Note that though a 22-gage deck meets the strength criteria, it is not adequate for the deflection criteria.

EXAMPLE 1-4

Floor Framing Layout and Floor Deck Selection

A typical floor framing plan (see Figure 1-25) consists of 30-ft. by 40-ft. bays and supports a total floor dead load of 95 psf and a floor live load of 150 psf. Determine the layout of the floor framing and the size of a composite floor deck, assuming normal-weight concrete.

Assume a self-weight of 7 psf for the framing, and normal-weight concrete with a 28-day compressive strength of 3000 psi. The composite deck is made from Grade 50 steel.

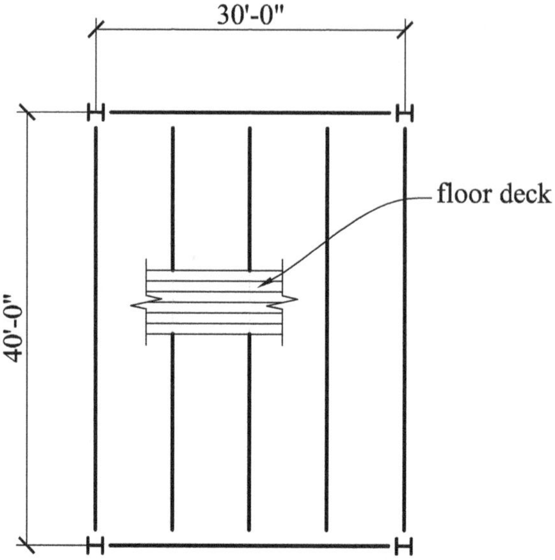

FIGURE 1-25 Floor framing layout for Example 1-4

Solution

LRFD Solution

Self-weight of steel framing = 7 psf

Since the roof deck does not support the weight of the steel framing,

Dead load on the floor deck alone = 95 psf – 7 psf = 88 psf

Floor live load = 150 psf

The total *factored* applied load on the floor deck = 1.2(88 psf) + 1.6(150 psf) = **346 psf**

The total *unfactored* applied load on the floor deck = 88 psf + 150 psf = **238 psf**

Try **2-in concrete slab on 3VLI-36 × 22-gage composite metal deck (grade 50 steel).**

The total slab depth = 2 in. + 3 in. = 5 in, and

Normal-weight concrete with 28-day compressive strength = 3000 psi

From the Vulcraft Deck Load Tables [33], for the **3-span or triple-span deck** condition, we find that the maximum unshored construction clear span = 11 ft. 0 in.

Self-weight of concrete and composite metal deck = 44 psf (see Vulcraft deck load tables).

The Superimposed *factored* applied load on the floor deck = 346 psf – 1.2(44 psf) = **293 psf**

The Superimposed *unfactored* applied load on the floor deck = 238 psf – 44 psf = **194 psf**

Note that the deck span to be selected must be less than or equal to 11 ft. 0 in. (i.e., the maximum allowable span for this deck without shoring during construction when it spans over a minimum of **three spans).** In addition, the shorter bay dimension of the floor (i.e., the 30 ft. dimension) must be a multiple of the selected deck span.

Try a 10-ft. 0-in. deck span (i.e., the spacing between the infill beams), of which the 30-ft. bay dimension is a multiple, and 10 ft. is less than 11 ft. 0 in., therefore OK.

From the Vulcraft Deck Load Tables [33], for a **3VLI-36 × 22-gage composite metal deck (Grade 50 steel)**, the *Superimposed* Design Load Capacity for the composite floor deck at 10-ft. span is:

ϕW_n = **360 psf** > 293 psf, OK.

For a maximum deck deflection of L/360, the allowable *unfactored Superimposed* design load is:

329 psf > 194 psf, OK.

Therefore, use **2-in concrete slab on 3VLI-36 × 22-gage galvanized composite metal deck (Grade 50 steel)**, with total slab depth = 2 in. + 3 in. = 5 in.

ASD Solution

Self-weight of steel framing = 7 psf

Since the roof deck does not support the weight of the steel framing,

Dead load on the floor deck alone = 95 psf – 7 psf = 88 psf

Floor live load = 150 psf

The total unfactored applied load on the floor deck = 88 psf + 150 psf = 238 psf

Try **2-in concrete slab on 3VLI-36 × 22-gage composite metal deck (grade 50 steel).**

The total slab depth = 2 in. + 3 in. = 5 in, and

Normal-weight concrete with 28-day compressive strength = 3000 psi

From the Vulcraft Deck Load Tables [33], for the **3-span or triple-span deck** condition, we find that the maximum unshored construction clear span = 11 ft. 0 in.

Self-weight of concrete and composite metal deck = 44 psf (see the Vulcraft deck load tables [33]).

Superimposed Applied Load on the floor deck = (The total *unfactored* dead load plus live load on the floor deck – the *unfactored* self-weight of the concrete and composite deck) = 238 psf – 44 psf = **194 psf**

Note that the deck span to be selected must be less than or equal to 11 ft. 0 in.—the maximum allowable span for this deck without shoring during construction when it spans over a minimum of **three spans.** In addition, the shorter bay dimension of the floor (i.e., the 30 ft. dimension) must be a multiple of the selected deck span.

Try a 10-ft. 0-in. deck span (i.e., the spacing between the infill beams), of which the 30-ft. bay dimension is a multiple, and 10 ft. is less than 11 ft. 0 in., therefore OK.

From the Vulcraft Deck Load Tables [33],

Superimposed Allowable Load on the composite floor deck at a 10-ft. span is:

W_n/Ω = **233 psf** > 194 psf, OK.

For a maximum deck deflection of L/360, the *unfactored Superimposed* Allowable load is: 233 psf > 194 psf, OK.

Therefore, use **2-in concrete slab on 3VLI-36 × 22-gage galvanized composite metal deck (Grade 50 steel)**, with total slab depth = 2 in. + 3 in. = 5 in.

Fire Resistance Rating and Fireproofing:

With the 2-in concrete slab on 3VLI-36 × 22-gage galvanized composite metal deck (Grade 50 steel), a 2-hour fire resistance rating (FRR) can be obtained using cementitious or sprayed fiber fireproofing on the underside of the composite metal deck (see the Vulcraft Composite Deck Underwriters' Laboratory (UL) Fire Rated Assemblies [33]). If an unprotected deck is desired, the 2-hour FRR can be obtained by increasing the concrete thickness above the top of the deck, and available options include the following:

- 3¼ in. Light Weight (LW) concrete slab on 3 in. composite metal deck, or
- 4½ in. Normal weight (NW) concrete slab on 3 in. composite metal deck

-11 SUSTAINABILITY IN THE DESIGN AND CONSTRUCTION OF STEEL STRUCTURES

Sustainability has been defined as "*meeting the needs of the present without compromising the capability of future generations to meet their own needs.*" In the United States, buildings account for approximately 40% of all energy usage [35, 36]. Consequently, there has been a growing trend toward sustainable design and construction of steel structures, where the focus is on minimizing negative environmental impact. The most common and popular rating system for the design of "green" buildings is the United States Green Building Council's (USGBC) Leadership in Energy and Environmental Design (LEED) certification system, introduced in 1998. This is a point-based building evaluation system that involves a checklist of the "quantifiable aspects of a project" [37, 38]. In the LEED 2009 rating system, the following levels of certification are possible using third-party verification: LEED certified, LEED Silver, LEED Gold, and LEED Platinum. Several structural as well as nonstructural issues are considered in calculating the LEED points for buildings. In many cases, the nonstructural issues, such as natural lighting, the type of paint used, the type of heating, ventilation, and air-conditioning (HVAC) systems, and the type of roofing membrane and system, play a greater role in the calculation of the LEED points than do the structural components. Building materials also have an impact on the environmental footprint of a building — from the energy required in the process of manufacture or production, plus the energy required to transport the structural material to the project site, and the energy required for the erection process. The environmental impact of building materials remains relatively flat throughout the life span of a building, whereas the impact of the non-structural building systems (such as HVAC, lighting, etc.) continues to increase over the life span of the building. However, as the nonstructural building systems become more energy efficient, the relative impact of the building materials on the environmental footprint of a building will increase. It is worthwhile for structural engineers to include sustainability principles in their designs, working collaboratively with all stakeholders — including fabricators, mechanical and electrical engineers — to arrive at the most efficient and optimal structure from the standpoint of design, fabrication, and erection. Using modern tools like building information modeling (BIM) to coordinate the various aspects of a structure, interferences and the resulting costly redesigns or changes during construction can be minimized [35, 36].

The current version of LEED (LEED v4.1) follows a performance-based approach in the areas of project design, operation, and maintenance that requires measurable results throughout the lifecycle of the project [38]. According to the USGBC web site (*https://www.usgbc.org/leed/v41*), "LEED helps buildings to focus on efficiency and leadership to deliver the triple bottom line returns of people, planet, and profit. LEED v4.1 raises the bar on building standards to address energy efficiency, water conservation, site selection, material selection, day lighting, and waste reduction."

The Structural Engineering Institute (SEI) of the American Society of Civil Engineers (ASCE) has published a guide titled "*Whole Building Life Cycle Assessment (WBLCA)—Reference Building Structure and Strategies*" [40, 41] that can aid the structural engineer in designing buildings that minimize environmental impact. The American Institute of Steel Construction (AISC) has developed Environmental Product Declarations (EPDs) for structural steel sections, plates, and hollow structural sections that provide information about the average environmental lifecycle impact of these products, including the impact from the manufacturing and fabrication processes. These EPDs are updated every five years and can be found on the AISC web page at *https://www.aisc.org/epd*. The State of California has already enacted legislation that mandates suppliers of construction materials for state-sponsored projects to report the EPDs for the products used, including structural steel.

Some suggestions for reducing lifecycle impacts that are within the purview of the structural engineer include using the following:

- The least amount of material;
- Low cement concrete;
- Structural details that minimize energy loss;

- Structural systems that are de-constructible; and
- Alternate structural systems that have a lower carbon footprint [42]; for example, using steel braced frames instead of moment-resisting frames for the lateral force resisting system may lead to a reduction in the carbon footprint [48].

Since each structural component has a carbon footprint because of the embodied energy involved in its production through its erection in a structure, their environmental impact on a project can be minimized by optimizing the member sizes and limiting the distance from the building site to the location where the material is sourced or produced.

The structural engineer can also contribute to operational energy savings on a project by avoiding structural details where the structural steel members penetrate the building envelope or by minimizing thermal bridging. Thermal bridging occurs when heat flows across a building envelope (i.e., across the building's exterior wall or roof or exposed ceiling) from a higher temperature space to a lower temperature space through a highly conductive structural steel member [41, 42]. Some examples of the conditions where thermal bridging occurs and therefore thermal breaks are needed—because of penetration of the building envelope by structural steel members — include the following [41, 42]:

- Balcony and overhang framing that extend from the temperature-controlled interior part of the building to the exposed exterior part of the building,
- Exposed exterior steel columns supporting a floor framing and penetrating an exterior ceiling or soffit,
- Parapets,
- Shelf angles,
- Rooftop posts or stub columns that are used to support rooftop units or equipment and are connected to the roof structural framing,
- Entrance canopy beams supported by interior girders or columns.

Thermal bridging or heat flow from a high temperature space to a lower temperature space across a thermal barrier (e.g., a building's exterior wall) causes condensation issues and energy loss from the inside of the building through the highly conductive steel material; thermal bridging effects can be mitigated using structural thermal break or low thermal conductivity materials as shims or infill between the exposed structural steel member and the protected interior structural steel framing, which are bolted together. The thermal break material prevents direct contact between the interior and exterior steel members, thus preventing energy loss while ensuring structural integrity. Examples of structural thermal break materials include fiber-reinforced plastic (FRP) shims and several other proprietary products [41, 42, 43, 44]. There are currently no code provisions in the United States addressing the design and installation of structural thermal breaks [44].

1-12 DESIGN CONSTRAINTS

There are several design constraints [50] that influence the design of structures and these include:

- **Safety** (i.e., strength)
- **Serviceability** (e.g., deflection, vibrations, cracking)
- **Durability** (e.g., corrosion resistance)
- **Economy** (i.e., optimal member sizes; use as few infill beam connections as possible)

- **Constructability** (i.e., buildable details)
- **Sustainability** (e.g., LEED ratings; minimizing embedded energy and reducing lifecycle impacts)
- **Resilience** (e.g., going beyond the minimum prescriptive code-level life safety requirements, and using performance-based design to ensure speedy recovery and business continuity after a hazard event)

References

[1] Albano, L. D. 2006. "Classroom Assessment and Redesign of an Undergraduate Steel Design Course: A Case Study." *ASCE Journal of Professional Issues in Engineering Education and Practice* 132, no. 4 (October): 306–311.

[2] Gomez-Rivas, A., and P. George. 2002. "Structural Analysis and Design: A Distinctive Engineering Technology Program." In *Proceedings of the 2002 American Society for Engineering Education Annual Conference and Exposition*, Montreal, Canada.

[3] Sarma, K. C., and A. Hojjat. 2005. "Comparative Study of Optimum Designs of Steel High-rise Building Structures Using Allowable Stress Design and Load and Resistance Factor Design Codes," *Practice Periodical on Structural Design and Construction*, February: 12–17.

[4] Hines, E. M. 2012. "Principles of Engineering Education – Part I." *STRUCTURE Magazine*, April: 60–61.

[5] Roesset, J. M., and J. T. P. Yao. 2001. "Suggested Topics for a Civil Engineering Curriculum." In *Proceedings of the 2001 American Society for Engineering Education Annual Conference and Exposition*, Albuquerque, NM.

[6] Sheppard, Sheri D., Kelly Macatangay, Anne Colby, and William M. Sullivan. 2008. *Educating Engineers: Designing for the Future of the Field*. Carnegie Foundation for the Advancement of Teaching. San Francisco: Jossey-Bass.

[7] Estrada, H. 2007. "Using the AISC Steel Building Case Study in a Structural Engineering Course Sequence." *Proceedings of the 2007 American Society for Engineering Education Annual Conference and Exposition*, Honolulu, HI.

[8] American Institute of Steel Construction. 2023. "*Steel Construction Manual,*" 16th ed. Chicago, IL: AISC.

[9] Mckee, B., and H. Timothy. 2007. "Structural Steel—How it's Done." *Modern Steel Construction*, August: 22-29.

[10] Weisenberger G. 2014. "Keep on Rolling – An Inside Look at a Modern High-tech Steelmaking Operation." *Modern Steel Construction*, February.

[11] Terpstra C., and C. Aumiller. 2015. "From Pellet to Plate – A Step-by-Step Look at the Production of Flat Steel." *Modern Steel Construction*, April.

[12] Spiegel, L., and G. Limbrunner. 2002. "*Applied Structural Steel Design*. 4th ed. Upper Saddle River, NJ: Prentice Hall.

[13] Heins, C. P. 1979. "*Design of Modern Steel Highway Bridges*." New York: John Wiley & Sons.

[14] Tamboli, A. R. 1997. *Steel Connections Design Handbook—LRFD Method*. New York: McGraw-Hill.

[15] Gewain, R. G., R. I. Nester, and A. Farid. 2003. "*Facts for Steel Buildings—Fire*." AISC, Chicago, IL: AISC.

[16] International Codes Council. 2024. "*International Building Code—2024.*" Falls Church, VA: ICC.

[17] Tavarez, J. 2018. "Are You Properly Specifying Materials". *Modern Steel Construction*, April.

[18] Weisenberger, G. 2017. "Above and Beyond." *Modern Steel Construction*, July.

[19] Olson, K. 2014. "*ASTM A1085: An Update to a Classic Material Specification,*" STRUCTURE Magazine, January.

[20] Fisher, J. M. 2005. "Expansion Joints: Where, When, and How." *Modern Steel Construction,* April: 25-28.

[21] Rosenblum, C. L. 2007. "*Probers Eye Expansion Joint in Pittsburgh Slab Mishap.*" Engineering News–Record (ENR), February. *https://www.enr.com/articles/32303-probers-eye-expansion-joint-in-pittsburgh-slab-mishap-2-8-2007?v=preview* (accessed August 16, 2024).

[22] Brockenbrough, R. L., and F. S. Merritt. 2011. "*Structural Steel Designer's Handbook,*" *Fifth Edition*. New York: McGraw Hill.

[23] Lay, M. G. 1982. "*Structural Steel Fundamentals—An Engineering and Metallurgical Primer.*" Australian Road Research Board.

[24] Geschwindner, L.F., R. O. Disque, and R. Bjorhovde. 1994. *Load and Resistance Factor Design of Steel Structures*. Prentice Hall.

[25] Geschwindner, L. F. 2012. *Unified Design of Steel Structures*. 2nd ed. John Wiley.

[26] Galambos, T.V., F. J. Lin, and B. G. Johnston. 1996. "*Basic Steel Design with LRFD.*" Prentice Hall.

[27] MacCrimmon, R. A. 2009. *Crane-Supporting Steel Structures*. Markham, Ontario, Canadian Institute of Steel Construction.

[28] Mamlouk, M. S., and J. P. Zaniewski. 2006. *Materials for Civil and Construction Engineers*. Upper Saddle River, NJ: Prentice Hall.

[29] Budynas, R. G. 1999. *Advanced Strength and Applied Stress Analysis*. 2nd ed. New York: McGraw-Hill.

[30] Broberg M., S. Shafaei, and J. Seo J. 2019. "Speedcore and Composite Plate Shear Walls." *Modern Steel Construction*, February.

[31] American Society of Civil Engineers. 2022. "*ASCE-7: Minimum Design Loads for Buildings and Other Structures.*" Reston, VA: ASCE.

[32] Ruby, J. and J. Matuska. "Structural Steel Economy: Revisiting the Assumptions." *Modern Steel Construction,* January.

[33] Vulcraft Steel Deck Product Catalog. 2020/2021. NUCOR. *https://vulcraft.com/products/deck* (accessed September 16, 2024).

[34] Carter, C. J., T. M. Murray, and W. A. Thornton. 2002. "Economy in Steel." *Modern Steel Construction*, April.

[35] Weisenberger, G. 2010. "Buying in." *Modern Steel Construction*, January.

[36] Weisenberger, G. 2011. "Sustainability and the Structural Engineer." *Practice Periodical on Structural Design and Construction*, November: 146-150.

[37] Farneth, S. 2007. "Sustaining the Past." *Green Source—The Magazine of Sustainable Design,* October: 25-27.

[38] USGBC. U.S Green Building Council. *https://www.usgbc.org/leed/v41/* (accessed June 4, 2024)

[39] Yang, F. 2018a. "Capturing Points for Whole Building LCA in LEED v4." *STRUCTURE Magazine*, April.

[40] Yang, F. 2018b. "Whole Building Life Cycle Assessment – Reference Building Structure and Strategies." Sustainability Committee of the Structural Engineering Institute (SEI), ASCE. Reston, Virginia: ASCE.

[41] Peterman, K.D., Wang Lizhong., M. D. Webster, J. A. D'Aloisio, and J. F. Hajjar. 2018. "Double Impact." *Modern Steel Construction*, September: 38-43.

[42] Hamel, S., and K. Peterman. 2019. "Thermal Breaks in Building Envelopes – Recent Research Findings." *STRUCTURE Magazine*, January: 28-29.

[43] VanGeem, M. 2018. "Challenges and Complexities of Proposal for Mitigating Thermal Bridges." *Concrete International*, September: 22–23.

[44] Weisenberger, G. 2018. "Breaking Up Is(n't) Hard to Do." *Modern Steel Construction*, September: 43–49.

[45] Roberts, E. J., S. Han, and A. C. Polimeni. 2018. "Supertall in Seoul." *ASCE Civil Engineering*, July/August.

[46] Horiuchi, C. and N. Wang. 2019. "Structural Design and Embodied Carbon." *STRUCTURE Magazine*, March: 44-45.

[47] Post, N. M. 2019. "Contractors Defend Work at Troubled Salesforce Transit Center." *Engineering News Record*, March 27. *https://www.enr.com/articles/46576-contractors-defend-work-at-troubled-salesforce-transit-center* (accessed June 23, 2019)

[48] Gannon, M. 2019. "A Story to Brace For." *STRUCTURE Magazine*, July: 12-14.

[49] Olson, K. 2021. "Well-Rounded." *Modern Steel Construction*, January: 16-19.

[50] Hajjar, J. F. 2024. "Leading the Way in Sustainability and Resilience in Structural Engineering." *STRUCTURE Magazine*, March: 9.

[51] Leishear, R. A. 2020. "Coating Preparations Reduce the Strength of Bridges." *STRUCTURE Magazine*, October: 10-12.

[52] Tavarez, J. 2022. "Are You Properly Specifying Materials?" *Modern Steel Construction*, June: 16-22.

[53] Engelhardt, M., J. Fisher, B. Kozy, T. Sabol, R. Shaw, W. Mohr, and S. Wolf. 2020. "Recommendations to Industry Arising from Investigation of Fractured Plate Girders and Associated Studies at the Transbay Transit Center, San Francisco." Report of the Metropolitan Transportation Commission, Peer Review Panel for the Transbay Transit Center, February 14: 25 pp. *https://mtc.ca.gov/sites/default/files/PRP_Final_Report.pdf* (last accessed June 21, 2024).

[54] Sarkisian, M., P. Lee, R. Garai, J. Zhou, A. Zha, and J. Chhabra. 2021. "95 State At City Creek, Salt Lake City, UT." *STRUCTURE Magazine*, November: 42-45.

[55] Brady, M. D. 2011. "Expansion Joint Considerations for Buildings" *Modern Steel Construction*, May:

Exercises

1-1. List three advantages and disadvantages of steel as a building material, and research the Internet for the three tallest steel building structures in the world, indicating the types of gravity and lateral load resisting systems used in these buildings.

1-2. List the various types of standard shapes available in the *AISCM*.

1-3. What are the smallest and the largest wide flanges or W-shapes listed in the *AISCM*?

1-4. Determine the self-weight, moment of inertia (I_x), and cross-sectional areas for the following hot-rolled standard sections:

W14 × 22

W21 × 44

HSS 6 × 6 × 0.5

L6 × 4 × ½

C12 × 30

WT 18 × 128

1-5. Determine the weight, area, and moment of inertia (I_x) of the built-up sections in Figure 1-26:

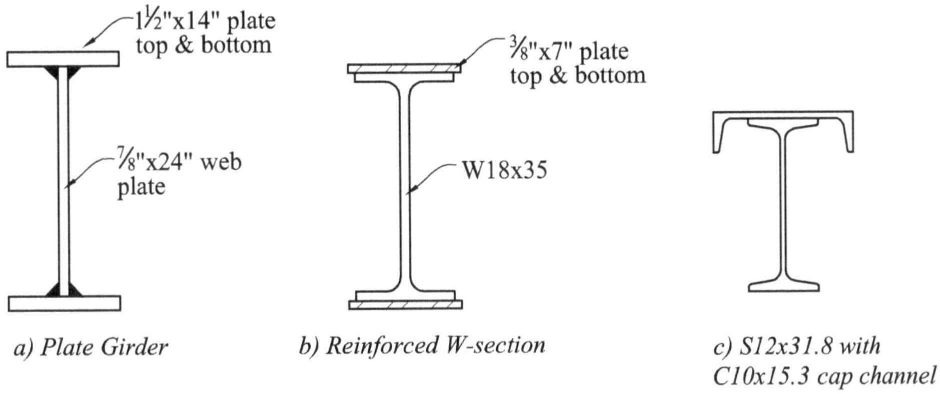

FIGURE 1-26 Compound shapes for Exercise 1-5

1-6. List the basic structural elements used in a steel building.

1-7. Determine the most economical layout of the roof framing (joists and girders) and the gage (thickness) of the roof deck for a building with a 25-ft. by 35-ft. typical bay size. The total roof dead load is 25 psf, and the snow load is 35 psf. Assume a 1½-in.-deep galvanized wide-rib roof deck and an estimated weight of roof framing of 6 psf.

1-8. Repeat Exercise 1-7 using a 3-in.-deep galvanized wide-rib roof deck.

1-9. Determine the most economical layout of the floor framing (beams and girders), the total depth of the floor slab, and the gage (thickness) of the floor deck for a building with a 30-ft. by 47-ft. typical bay size. The total floor dead load is 110 psf and the floor live load is 250 psf. Assume normal-weight concrete, a $1^1/_2$-in.-deep galvanized composite wide-rib floor deck, and an estimated weight of floor framing of 10 psf.

1-10. Repeat problem 1-9 using a 3-in.-deep galvanized composite wide-rib floor deck.

1-11. A steel floor girder in an existing building needs to be strengthened by welding a structural member to its bottom flange. The steel grade is unknown, but material testing has revealed the following percentages by weight of the following alloy chemical elements in the girder:

C = 0.16%

Cr = 0.10%

Cu = 0.20%

Mn = 0.8%

Mo = 0.15%

Ni = 0.25%

V = 0.06%

Si = 0.20%

Calculate the CE and determine the weldability of the structural steel.

1-12. A steel building is 900 ft. long, and it has been decided to provide expansion joints every 300 ft. If the maximum anticipated temperature change is 70°F,

a. determine the required minimum size of the expansion joint.
b. If the expansion joint along a W24 x55 girder line were to be inadvertently restrained from expanding or contracting, calculate the magnitude of the axial force in kips that will be exerted on the beam and its end connections due to this restraint.

1-13. For the steel roof framing plan shown in Figure 1-27, select an appropriate deck from a metal deck manufacturer's load table. The snow load is 32 psf, and the dead load, including the deck, is 27 psf (service loads). Provide the reference material for the deck selection from the manufacturer.

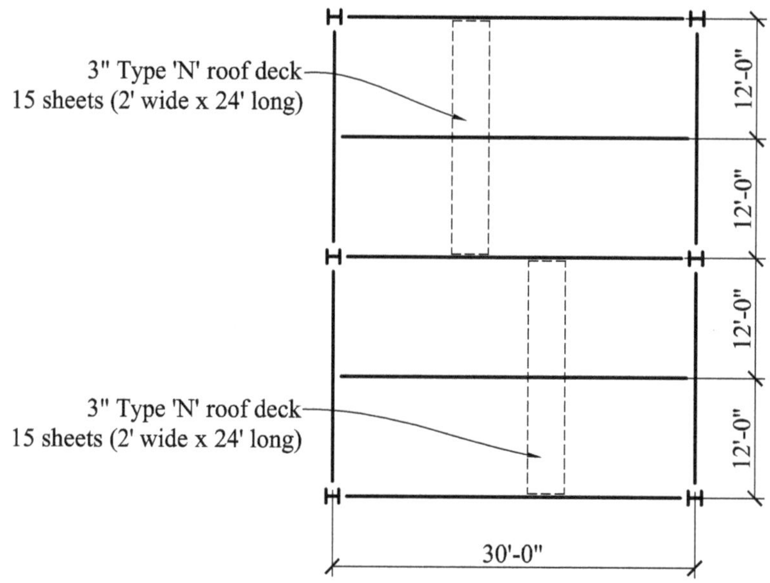

FIGURE 1-27 Roof Framing Plan for Exercise 1-13

1-14. The framing plan in Figure 1-28 shows a steel-framed platform supported at the four corners for gravity loads. Beams B-1 and B-2 framed into G-1 and G-1 are supported at each end by a hanger.

a. Select an appropriate deck from a manufacturer's load table. The service live load is 250 psf, and the service dead load is 20 psf plus the concrete and deck. Provide the reference material for the deck selection from the manufacturer.
b. Describe the complete load path for a point load on the middle of the deck.

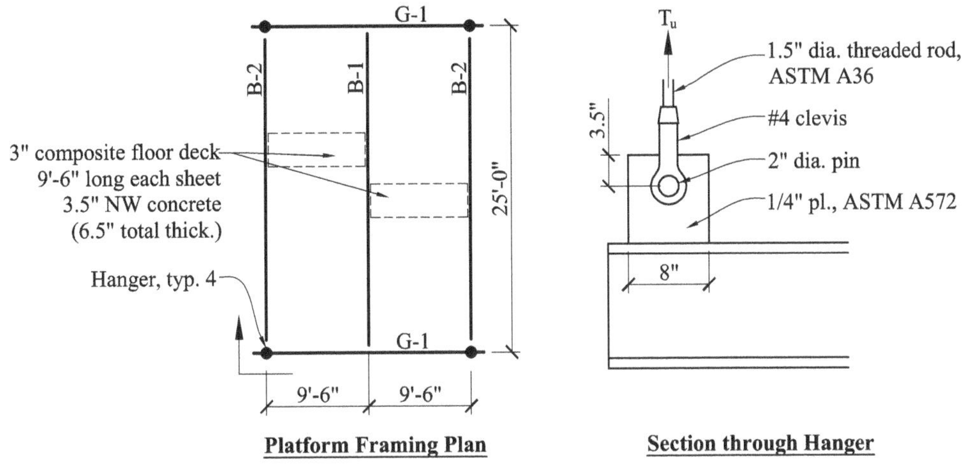

FIGURE 1-28 Platform framing plan and sections for Exercise 1-14

1-15. For the steel framing plans shown in Figure 1-29, select the one that is the most ideal and provide two reasons to support your answer.

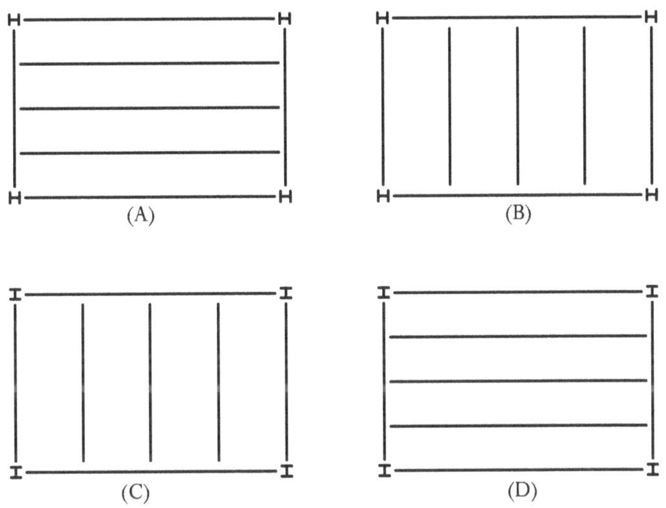

FIGURE 1-29 Steel framing plans for Exercise 1-15

1-16. Determine an appropriate floor deck thickness (gage) and beam spacing for the typical bay and floor deck detail shown in Figure 1-30. Note that the depth of the composite metal deck and the total depth of the concrete-on-metal deck are given. Assume only a 2-span condition exists. Total floor dead load = 80 psf (includes the weight of the floor slab and deck), and the live load is 175 psf. Neglect the weight of the framing in your calculations.

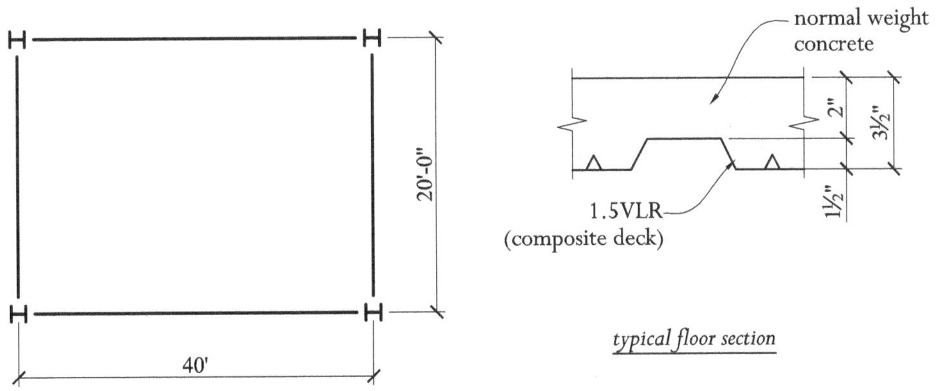

FIGURE 1-30 Typical bay and floor deck for Exercise 1-16

1-17. Determine the moment of inertia about the x–x axis and the weight per foot of the composite shapes in Figure 1-31.

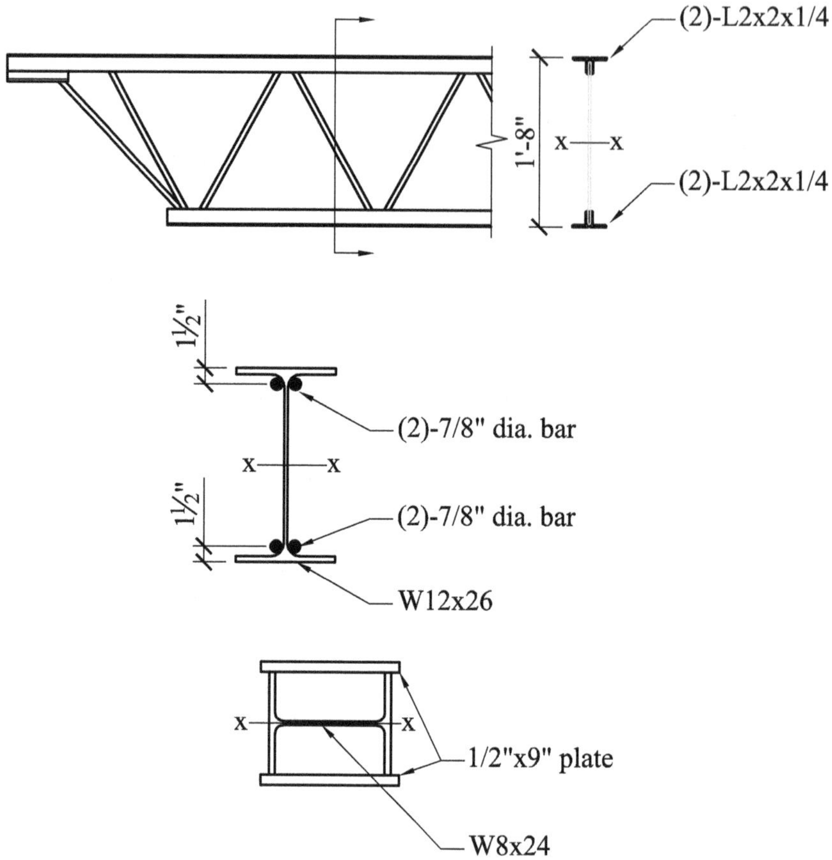

FIGURE 1-31 Composite shapes for Exercise 1-17

1-18. Considering a section through the rectangular hole shown in the beam in Figure 1-32, determine the area, weight per foot, and moment of inertia (I_x) for this shape and explain where a hole like this would occur in a steel structure.

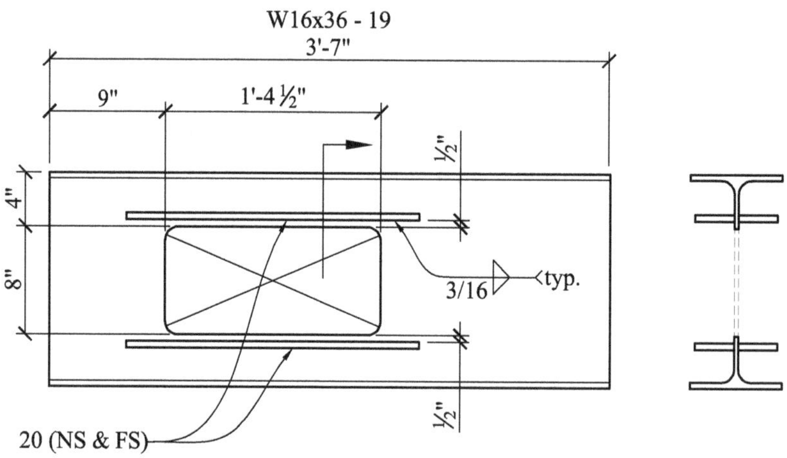

MARK	QTY	SIZE	LENGTH		WT.
19	1	W16x36	3	7	129
20	4	PL. ½x3	1	10½	38

FIGURE 1-32 Beam with an opening in the web for Exercise 1-18

1-19. Using the appropriate stress distribution and the equilibrium of forces and moments from statics and mechanics of materials, derive the equations for the elastic section modulus (S_{xx}) and the plastic section modulus (Z_{xx}) for a solid rectangular section with a width, b, and a depth, h. Calculate the Z_x/S_x ratio for a 13" × 82" deep rectangular section.

1-20. A built-up T-section has a flange that is 10 inches wide by 2 inches thick and a web that is 18 inches deep by 2 inches thick, giving a total depth of 20 inches for the T-section. Determine the following section properties:

a. The location of the elastic neutral axis (ENA), y_e, measured from the top of the T-section.
b. The moment of inertia, I_x, about the strong axis of the T-section.
c. The elastic section modulus, S_x, of the T-section.
d. The location of the plastic neutral axis (PNA), y_p, measured from the top of the T-section.
e. The plastic section modulus, Z_x.

1-21. Using the appropriate stress distribution and the equilibrium of forces and moments from statics and mechanics of materials, derive the equations (and calculate the values) for the elastic section modulus, S_x, and the plastic section modulus, Z_x, for the wide flange built-up section shown in Figure 1-33. Calculate the Z_x/S_x ratio for the section. Compare the Z_x/S_x ratio for this built-up wide flange section to that of the solid rectangular section in Exercise 1-19 and discuss your observations.

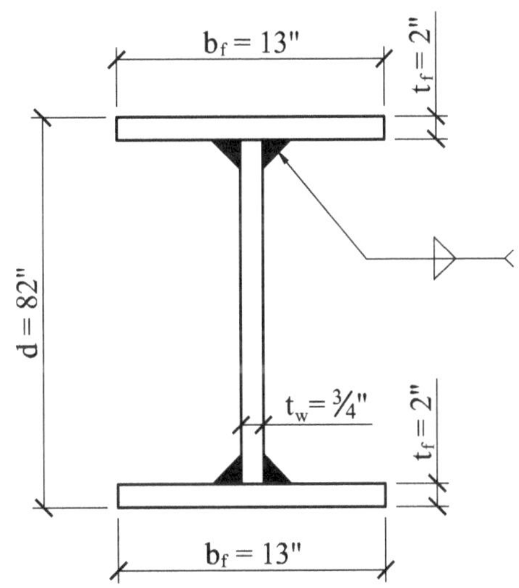

FIGURE 1-33 Built-up shape for Exercise 1-21

1-22. A 115 ft. × 115 ft. floor plan consists of 23 ft. × 23 ft. bay modules with the infill beams spaced at 11 ft. – 6 in. on center. Assume the dead load on the floor (excluding the steel framing self-weight) is 70 psf and a live load of 50 psf.

a. Using a lightweight concrete slab on top of an unprotected composite metal deck, determine the most economical concrete slab thickness, as well as the shallowest depth and gage of the composite metal deck, to satisfy both the strength requirements and a 2-hour fire resistance rating. Use a typical deck manufacturer's load table (LRFD or ASD) from the Internet.
b. Repeat Part (a) using normal-weight concrete and assuming a dead load of 80 psf (excluding the steel framing self-weight) and a 50 psf live load.
c. Repeat Part (a) assuming the infill beam spacing is reduced to 7 ft.-68 in. on center.

CHAPTER

2

Design Methods, Load Combinations, and Gravity Loads and Load Paths

2-1 INTRODUCTION TO DESIGN METHODS

The intent of structural design is to select structural systems, member sizes, and connections whose strength or capacity is greater than or equal to the effect of the applied loads or the demand (i.e., the ultimate limit state, ULS), and whose vertical and lateral deflections, floor vibrations, and lateral accelerations are within the allowable limits (i.e., the serviceability limit states, SLS). There are two main methods prescribed in the AISC Specification [1] for the design of steel structures: the load and resistance factor design (LRFD) method and the allowable strength design (ASD) method; however, Appendix 1 of the specification also allows the use of inelastic methods of design, such as the plastic design (PLD) method [1].

The LRFD requirements presented in the AISC 2022 specification for structural steel buildings are similar to the previous LRFD Specifications. The allowable *strength* design method in the AISC 2022 specification is similar to the allowable *stress* design method in previous specifications in the sense that both are carried out at the service load level. The difference between the allowable *stress* design and the allowable *strength* design is that the provisions for the allowable strength design method are given in terms of forces in the AISC 2022 specification, while the provisions for the allowable stress design method were given in terms of stresses in previous AISC specifications prior to 2010. It should be noted that in the current AISC specification, the design provisions for both the ASD and LRFD methods are based on ultimate limit state or strength design principles. In both methods, the design should be such that no applicable strength or serviceability limit states are exceeded. For the strength limit state (e.g., yielding, rupture, buckling, etc.) in the LRFD method, the goal is to not exceed the ultimate load capacity of the structural member or structural system under the factored design loads; for the ASD method, the goal is not to exceed the allowable load under the service design loads. For the serviceability limit state (e.g., deflections, lateral drift, vibrations, lateral acceleration, crack widths), the goal is for the structural member or structural system to remain functional and to serve its intended purpose under the service or unfactored loads. Both the ASD and LRFD methods as presented in the AISC Specifications (and indeed other specification-based prescriptive codes) provide a sense of whether an individual structural member or element is safe or not, but it does not provide any information on the expected level of damage in the structural member, nor does it provide any information on the system-level structural behavior or performance under a given hazard or hazard level. These prescriptive codes are based on performance measures that emphasize life safety (i.e., no loss of life), though the structure may be damaged beyond repair. The prescriptive codes provide defined quantitative values of load capacity and/or member size depending on the occupancy of the building and the type of structural system. These codes provide the minimum requirements to ensure life safety, but the owner of a structure may request a higher level of safety than provided in the prescriptive codes, though this would result in higher costs to the owner. The current Steel Construction Manual (*AISCM)* [1] presents a dual side-by-side approach—ASD and LRFD—for all the design aids and tables, with the same nominal or theoretical strength of the member (i.e., P_n, V_n, M_n, etc.) for both design methods. The three design methods for steel structures—LRFD, ASD, and the plastic design method—are discussed in this chapter, but only the ASD and LRFD methods are used in subsequent chapters, except that the plastic design method is further discussed in Appendix A.

Load and Resistance Factor Design Method

The load and resistance factor design (LRFD) method is a reliability or probabilistic-based limit state design approach that takes into account the uncertainties or statistical variations in the strength of a structural member and the loads acting on the member, and the type of failure (e.g., ductile versus brittle failure) using different load and resistance or strength reduction factors; in contrast, the ASD method is a deterministic design approach that accounts for uncertainties in the loads and strength, and the type of failure using only one uniform empirical factor of safety. The factor of safety in the ASD method does not account for the different statistical variabilities of the different loads acting on the structural member. The uncertainties in structural design arise from the statistical variability in material properties and member dimensions; the statistical variability in the loads acting on the structure; the variability in the strength or capacity of the structure or member; and the limitations of the structural model and methods of analysis [20].

In the LRFD method, the safety margin is realized using load factors (that are usually greater than 1.0) and resistance or strength reduction factors that are usually less than 1.0; these factors are determined from probabilistic analysis, based on a survey of the reliability indices inherent in existing buildings [2, 3], and a preselected reliability index that accounts for the risks, the consequences, and the modes of failure of the structural member. The load factors vary depending on the type of load because of the different degrees of uncertainty in predicting each load type; the strength reduction factors prescribed in the AISC specification also vary depending on the load effects, again to account for the varying uncertainties in predicting the strengths for the different load effects and the mode of failure. For example, dead loads are more easily predicted than live or wind loads; therefore, the maximum load factor for dead loads is generally less than that for live loads or wind loads. The load factors account for the possibility of overload in the structure.

In the ASD method, the safety margin for a structural member is realized by reducing the nominal resistance by a single factor of safety, which is independent of the type of applied loads but dependent on the mode of failure of the structural member. The fundamental difference between the ASD and the LRFD methods is that the ASD method uses only one safety factor to account for all the uncertainties in the loads on the structural member and the strength of the structural member, whereas the LRFD method uses different factors (i.e., the load factors) for the different load types and another factor (the resistance factor) for the structural strength [3, 24]. Thus, the LRFD method provides more uniform reliability and level of safety for all members in the structure, even for different loading conditions.

Probabilistic Framework of the Limit State Design Method

The intent of any structural design philosophy is to provide an acceptable level of safety. In the limit state design philosophy, due to the unpredictability of the loads acting on structures, the variability of material and section properties, and the uncertainties in our structural analysis and design methods coupled with the consequences of failure (e.g., failure of a building used for public assembly versus a farm shed) and the modes of failure (e.g., whether ductile or brittle failure), both the loads or load effect, Q, and the resistances or strength, R, are treated as independent random parameters [1-3, 24]. A frequency distribution of the load effect, Q, and the resistance, R, assumed to follow a normal or Gaussian type probability distribution (i.e., a bell curve), is depicted in Figure 2-1.

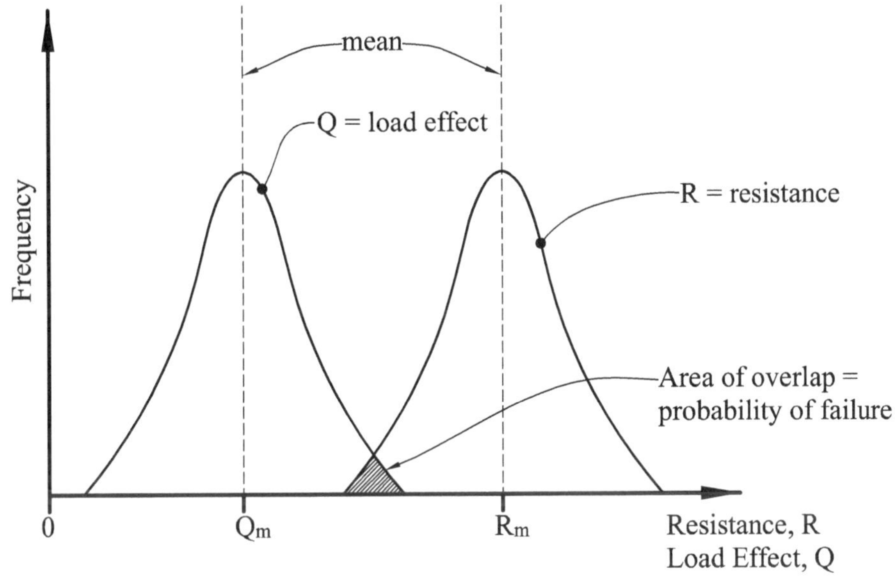

FIGURE 2-1 Frequency distribution of load effect, Q and Resistance, R

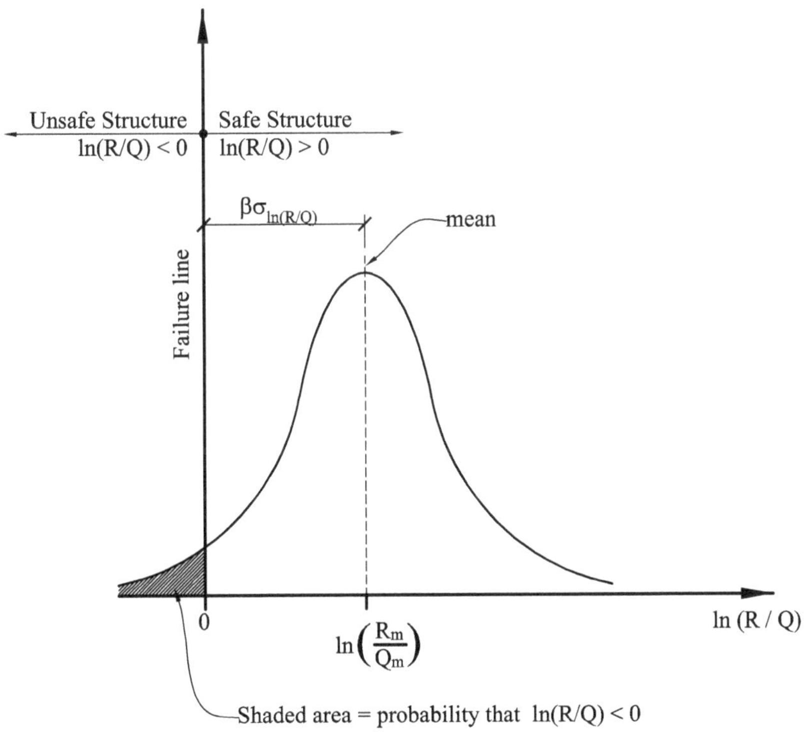

FIGURE 2-2 Reliability index

Whenever $R > Q$, there is a margin of safety against collapse or failure of the structure, and the structural behavior is satisfactory [24]. However, since R and Q are random variables, there is a possibility that in certain situations, the resistance, R, may be less than the load effect, Q, in which case the structural behavior will be unsatisfactory, and the structure will have reached a limit state where its performance is no longer satisfactory. The shaded area in Figure 2-1 represents the conditions when $R < Q$, and the area of this overlap represents the probability of failure. For any structure, we cannot be 100% certain that Q will never be less than R. The goal is to make the overlap area as small as possible so that the probability of failure of the structure is at an acceptably low, but finite, level [24]. Since R and Q are random variables, it is possible to estimate the means (R_m and Q_m) and the standard deviations or coefficients of variations, V_R and V_Q, of the variables that make up R and Q for typical structural elements and structural loadings. Figure 2-1 is replotted in Figure 2-2 with the horizontal axis as the natural logarithm of (R/Q) or $\ln(R/Q)$. In Figure 2-1, when $R\text{-}Q < 0$ or $R/Q < 1$, the structure is deemed unsafe; similarly, in Figure 2-2, when $\ln(R/Q) < \ln(1)$ (i.e., $\ln(R/Q) < 0$), the structure is deemed unsafe. Therefore, the shaded area in Figure 2-2 is the probability that failure will occur [24]. In deriving the LRFD method, a simple formula for the reliability index, β, which is a function of the mean values of R and Q but independent of their actual distributions, is given as follows [1, 22, 23, 24]:

$$\beta = \frac{\ln\left(\frac{R_m}{Q_m}\right)}{\sqrt{V_R^2 + V_Q^2}} \tag{2-1}$$

Where,

β = safety or reliability index
Q_m = mean value of the load effect, Q
R_m = mean value of the resistance or strength, R
V_Q = coefficient of variation of the load effect or demand, $Q = \dfrac{\sigma_Q}{Q_m}$
V_R = coefficient of variation of the resistance, strength, or capacity, $R = \dfrac{\sigma_R}{R_m}$
σ_Q = standard deviation of the load effect, Q
σ_R = standard deviation of the resistance or strength, R

$\sigma_{\ln(R/Q)} \approx \sqrt{V_R^2 + V_Q^2}$ in Figure 2-2

A higher value of the reliability index, β, indicates a higher margin of safety. There is variation in the β values obtained for different live load-to-dead load (L/D) ratios, for different tributary areas, and for the different structural elements in a structure. The higher the L/D ratio, the lower the safety or reliability index, β, since live load is more unpredictable or more variable than dead load. Conversely, the lower the L/D ratio, the higher the safety or reliability index. Furthermore, ductile structural elements do not fail suddenly, and therefore, they have a lower target reliability index, while structural elements that are susceptible to sudden failures (e.g., brittle fracture) would require a higher target reliability index (see ASCE 7 Table 1.3-1 [2]). The value of the target reliability, β, selected for bolted or welded connections in the LRFD method, varied between 4 and 5; on the other hand, for adequately braced compact rolled beams and tension members failing in yielding—both of which have a strength reduction factor, ϕ, of 0.9, a target β value of 2.4 to 3.1 is selected, depending on the live load-to-dead load ratio [1]. The higher β value selected for connections reflects the complexity in modeling their structural behavior, their mode of failure, and the relatively greater difficulty in installing connections [1]. In the LRFD approach, though the probability of failure is not directly computed for a given structure, the load and resistance factors were derived based on an acceptably low annual probability of failure, P_F (see ASCE 7-22 Table 1.3-1) or an acceptably high target reliability index, β. For example, for a Risk Category II building with ductile elements, the target reliability index, β, is 3.0 and the probability of failure, P_F, from ASCE 7-22 Table 1.3-1 is 3×10^{-5} per year, which "results in an approximate mean failure interval of 30,000 years" [39]. Therefore, structures that are designed to meet all the applicable code criteria for loads, strength, and serviceability have an acceptably low risk or low probability of failure, indicating that no structure is designed to be 100% failure-proof or perfectly safe; some reasonable level of risk is assumed in the design of any structure. ASCE 7 Table 1.3-1 [2] gives the target annual probability of failure and the target reliability indices as a function of the modes of failure of the structural element and the four risk categories for all load conditions, except earthquakes, tsunamis, or extraordinary events.

For a comprehensive discussion of the reliability-based design approach, the reader should refer to Ref. [3, 22, 24]. It should be noted that design and construction errors or carelessness and other human errors are not accounted for in the load and resistance factors or in the safety factor discussed previously in this section. Design and construction errors must be minimized through quality control processes by the engineer of record and the contractor.

As previously stated, the LRFD method uses a limit states design method; a limit state is the point at which a structure or structural member reaches its limit of usefulness. The basic LRFD limit state design equation requires that the design strength, ϕR_n, be greater than or equal to the sum of the factored loads or load effects (i.e., the demand). Mathematically, this can be written as

$$\phi R_n \geq Q_u, \tag{2-2}$$

where

R_n = Theoretical or nominal strength or resistance of the member determined using the AISC specifications,

Q_u = Required strength or sum of the factored loads or load effects (or the demand) using the LRFD load combinations = $\Sigma Q_i \gamma_i$. (e.g., for a structural member subjected to a floor dead load of D and a floor live load, L, Q_u = 1.2D + 1.6L)

Q_i = Service load or load effect,

γ_i = Load factor (usually greater than 1.0), and

ϕ = Resistance or strength reduction factor (usually less than 1.0).

Note that the service load (i.e., the unfactored or working load), Q_i, is the load applied to the structure or member during normal service conditions, while the factored or ultimate load, Q_u, is the load applied on the structure at the point of failure or at the ultimate limit state.

Allowable Strength Design (ASD) Method

In the ASD method, a member is selected so that the allowable strength is greater than or equal to the applied service load or load effect, or the required strength, R_a. The allowable strength is the nominal or theoretical strength divided by a safety factor that is only dependent on the limit state being considered; that is,

$$R_n / \Omega \geq R_a, \tag{2-3}$$

where

R_n/Ω = Allowable strength,

R_a = Required *allowable* strength, or applied service load or load effect determined using the ASD load combinations, and

Ω = Safety factor.

Note, as previously indicated, that the theoretical or nominal strength, R_n, is the same for both the LRFD and ASD methods. The safety factor is dependent on the mode of failure of the structural member or the limit state under consideration. As discussed previously, the allowable strength design method uses a single factor of safety to account for the uncertainties in the strength of the member and the applied loads on the member.

Relationship Between Safety Factor, Ω, *L/D* Ratio, and Strength Reduction Factor, Φ

For a structural member supporting a dead load, *D*, and a live load, *L*, the relationship between the factor of safety, the live load-to-dead load ratio (*L/D*), and the resistance factor, ϕ, can be derived as follows using equations (2-2) and (2-3), and noting that the factored load, $Q_u = 1.2D + 1.6L$:

$$\text{For ASD,} \quad \frac{R_n}{\Omega} \geq R_a = D + L$$

$$\text{For LRFD,} \quad \phi R_n \geq Q_u = 1.2D + 1.6L$$

Multiplying the ASD equation above by ϕ and rearranging, we obtain, in the limit,

$$\phi R_n = \phi\Omega(D + L)$$

Substituting this equation into the LRFD equation, we obtain

$$\phi\Omega(D + L) = 1.2D + 1.6L$$

or

$$\phi\Omega\left(1 + \frac{L}{D}\right) = 1.2 + 1.6\frac{L}{D}$$

The factor of safety, Ω, can then be written as,

$$\Omega = \left(\frac{1}{\phi}\right)\left(\frac{1.2 + 1.6\frac{L}{D}}{1 + \frac{L}{D}}\right) \tag{2-4}$$

TABLE 2-1 Factor of safety, Ω, versus *L/D* ratio

L/D	Factor of safety, Ω
0	$\frac{1.2}{\phi}$
1	$\frac{1.4}{\phi}$
2	$\frac{1.47}{\phi}$
3	$\frac{1.5}{\phi}$
4	$\frac{1.52}{\phi}$
5	$\frac{1.53}{\phi}$

The minimum required factor of safety for different *L/D* ratios is shown in Table 2-1. Note that as the *L/D* ratio increases, the variability of the load increases because of the greater uncertainty inherent in live loads compared to dead loads, and hence the minimum required factor of safety also increases.

This means that when the variability of the load is low, a lower factor of safety can be tolerated. Furthermore, the factor of safety, Ω, is inversely proportional to the strength reduction factor, ϕ.

The higher the strength reduction factor, the lower the required factor of safety. For instance, from the factor of safety equation (2-4), we find that a structural member with an *L/D* ratio of 3 and a strength reduction factor of 0.9 results in a required factor of safety of 1.67, whereas a member with an *L/D* ratio of 3 and a strength reduction factor of 0.75 results in a required factor of safety of 2.0.

The original LRFD specification was calibrated to the 1978 ASD specification at a live load-to-dead load (*L/D*) ratio of 3; this yields factors of safety, Ω, of 1.67 and 2.0 for $\phi = 0.9$ and $\phi = 0.75$, respectively [1, 24]. The probabilistic-based design methods such as the LRFD method produce more reliable and potentially more economical designs that have an "acceptably low probability of failure" compared to the ASD method. For unique structures, the resistance and load factors, ϕ and γ, specific to the project can be derived from reliability or probability-based analyses. Other examples of reliability or probabilistic-based prescriptive codes include the American Concrete Institute Code (ACI 318), the AASHTO Bridge Code, and the LRFD method for structural wood design in the National Design Specification (NDS) for Wood Structures.

Plastic Design Method (PLD)

Plastic design is an optional method in Appendix 1 of the AISC Specifications that can be used for the design of continuous steel beams and girders. In the plastic design method, the structure is assumed to fail after the formation of a plastic collapse mechanism due to the presence of plastic hinges. The load at which a collapse mechanism forms in a structure is called the collapse or ultimate load; the load and resistance factors used for plastic design are the same as those used in the LRFD method. The plastic analysis and design of continuous beams is presented in Appendix A of this text.

Performance-based Design (PBD) Method

The prescriptive methods previously discussed provide a set of design rules that are meant to satisfy the life safety goals by achieving a low probability of failure (see ASCE 7 Table 1-3.1 [2]) and meeting the serviceability requirements. However, these methods do not specify or indicate the expected damage levels or the specific performance levels. That is, the extent of the damage that might occur in a structure or structural member after a hazard event is not specified or indicated in the prescriptive codes.

A design method that is increasingly being used for the seismic design of tall buildings and is now permitted in the International Building Code (IBC) as an alternative design method is performance-based design (PBD). PBD is a design approach that focus on goals beyond life safety. It is based on defined performance objectives or goals that must be related to some design criteria or some parameter threshold, like lateral deflection, inter-story drift, and lateral acceleration limits; these performance objectives and corresponding damage levels are jointly determined by the project stakeholders, and they are dependent on the specific hazard and its intensity, as well as the occupancy and therefore the risk category of the structure. Other factors include the importance of the structure to the resilience and recovery of the community after a hazard event, the acceptable risk of casualties, and the financial loss that the owner of the structure and the community are willing to bear [21, 33, 34]. From the performance objectives, the design strategy to achieve each objective is developed. The expected system response is determined using sophisticated nonlinear structural analysis methods that take into account material and geometric nonlinearities. These performance objectives lead to a design strategy that is intended to result in an economical structure with better performance than the minimum design options specified in the prescriptive life-safety codes (e.g., ACI 318, AISC 360, NDS, AASHTO, etc.). The PBD method can also result in more design creativity and innovation—attributes that are thought by some to be hampered by the prescriptive or specification-based codes. In PBD, the engineer of record (EOR) for the structure must prove to a peer review committee and the code official either through analysis or testing that the PBD method will result in a structural system that meets the performance objectives and that is of equal or better performance than the prescriptive code-designed structure. PBD is currently mostly used in the seismic engineering of tall buildings, but its extension to other hazard events is only a matter of time. The typical performance or expected damage levels in PBD are categorized as follows [21]:

Level 1 or Mild Damage Level: After the hazard event, there is no structural damage and the structure is safe to occupy and is functional (i.e., immediate occupancy, IO). Injury levels are minor, and damage to the building contents is minimal.

Level 2 or Moderate Damage Level: The structure has some structural and non-structural damage, but it can be repaired and can be occupied and operational while the repairs are being carried out. Injury levels are moderate, and the likelihood of mass casualties is very low.

Level 3 or High Damage Level or Life Safety Level: There are both significant structural and non-structural damage, and repair is possible but will take some time and result in delays in the re-occupancy of the structure. There might be moderate levels of injuries; the likelihood of mass casualties is low. This performance level is closest to the life-safety level, the same minimum level provided in the prescriptive or specification-based codes, such as the AISC Specifications. For instance, this performance level for seismic hazards corresponds to the earthquake-resistant design of structures subjected to the design earthquake equal to two-thirds of the maximum considered earthquake ground motion, or (2/3)*MCE* (see Chapter 3).

Level 4 or Severe Damage Level or Collapse Prevention Level: There is severe and substantial structural and non-structural damage, though there is no structural collapse, but the structure is not repairable and will have to be demolished; and there are occupancy injuries and likelihood of death due to injuries from this "near collapse level." Note that the building may collapse due to seismic aftershocks. This performance level is the basis for the earthquake-resistant design of structures subjected to the maximum considered earthquake ground motion, or *MCE* (see Chapter 3).

Depending on the risk category of a building and the intensity of a hazard event, a performance or damage level may be specified. The designer would then relate the selected performance level to the analysis and design of the structural systems. For example, a moderate damage level (Level 2) could be specified for a risk category IV building under a very large or very rare earthquake event, or a "severe damage level" (level 4) might be specified for a risk category II building. The structure is then analyzed under this seismic hazard level using sophisticated non-linear analytical techniques like the finite element method to obtain the damage levels in the structural members. The goal is to ensure that the damage levels in the structural members do not exceed the specified damage level. It should be noted that currently, "except for structural systems assigned to ASCE 7 Risk Category IV, the role of individual buildings in fulfilling community resilience goals is seldom recognized by codes" [36]. In this era of artificial intelligence and looking forward to the future, some have argued that performance-based design is the structural engineering "profession's new hope." For more information on PBD, the reader should refer to Ref. [21, 33, and 34].

2-2 STRENGTH REDUCTION OR RESISTANCE FACTORS

The strength reduction or resistance factors (f) account for the variability of the material and section properties and the mode of failure; they are, in general, usually less than 1.0. These factors are specified for various limit states in the AISC specification and are shown in Table 2-2.

TABLE 2-2 Resistance Factors

Limit state	Resistance factor (ϕ)
Shear	1.0 or 0.9
Flexure	0.90
Compression	0.90
Tension (yielding)	0.90
Tension (rupture)	0.75

2-3 LOAD FACTORS AND LOAD COMBINATIONS

The individual structural loads acting on a building structure do not act in isolation but may act simultaneously with other loads on the structure. Load combinations are the possible permutations and intensity of different types of loads that can occur together on a structure at the same time. The building codes recognize that all structural loads may not act on the structure or structural member at the same time, and the maximum magnitude of the multiple load types acting on a structure may not occur at the same time. The load combinations or critical combination of loads to be used for design are prescribed in the ASCE 7 Load Standard [2] and in Section 1605 of the International Building Code (IBC) [4]. These load combinations include the overload factors for the LRFD method, which are usually greater than 1.0, and account for the possibility of overload of the structure. For LRFD load combinations including flood loads, F_a, or atmospheric ice loads, or self-straining forces, T, the reader should refer to Sections 2.3.2, 2.3.3, and 2.3.4, respectively, of the ASCE 7 standard. It should be noted that for most building structures, a number of the loads in the ASCE 7 load combinations will be zero - such as the load due to lateral earth pressure, H, the load due to self-straining forces, T, and the loads due to fluid pressure, F, resulting in more simplified load combination equations.

LRFD Load Combinations

The basic load combinations for LRFD (excluding fluid loads F and self-restraining force T, which will be zero for most building structures) are as follows:

1. $1.4D$
2. $1.2D + 1.6L + (0.5L_r \text{ or } 0.3S \text{ or } 0.5R)$
3. $1.2D + 1.6(L_r \text{ or } 1.0S \text{ or } 1.6R) + (f_1L \text{ or } 0.5W)$
4a. $1.2D + 1.0W + f_1L + (0.5L_r \text{ or } 0.3S \text{ or } 0.5R)$
4b. $1.2D + 1.0W_T + f_1L + (0.5L_r \text{ or } 0.5R)$
5a. $0.9D + 1.0W$ (D always counteracts or opposes W in this load combination)
5b. $0.9D + 1.0W_T$ (D always counteracts or opposes W_T in this load combination)
6. $1.2D + E_v + E_h + f_1L + 0.15S$
7. $(0.9D - E_v) + E_h$ (D always counteracts or opposes E_h in this load combination)

Where,

f_1 = 1 for areas of public assembly with live loads that exceed 100 psf and parking garages.
= 0.5 for all other live loads and floor live loads, $L_o \leq 100$ psf
D = Dead load
L = Floor live load
L_r = Roof live load

S = Snow load
R = Rain load
W = Wind load
W_T = Tornado load
E = Combined load effect due to *horizontal* and *vertical* earthquake-induced forces
E_h = Horizontal component of the earthquake
E_v = Vertical component of the earthquake

NOTE:

- The above LRFD load combinations can also be obtained from IBC Section 1605.2.
- For LRFD, where the effect of the load, H, due to lateral earth pressure, ground water pressure (hydrostatic pressure), or pressure of bulk materials adds to the primary variable load effect (L, L_r, S, R, W, or E), the load factor on H shall be taken as 1.6.
- For LRFD, where H counteracts or resists the primary variable load effect, the load factor for H should be set equal to 0.9 when the load, H, is permanent (e.g., lateral soil pressure) or a load factor of zero for all other conditions. (ASCE 7 Section 2.3.1). Note that dead loads are *not* primary variable loads or load effects.
- In load combination 7, the net acting downward load will oppose E_h, and this is how this will occur in a braced frame or moment frame column: The force, E_h, causes an overturning moment, which can be resolved into an upward (tension) and downward acting force-couple (recall that T = C = overturning moment divided by the distance between the force couple). As E_h tends to lift up the LFRS column at one end (i.e., the "T" force), it is pushing down at the other end (i.e., the "C" force). The "T" force is then resisted by the net downward load of $0.9D - E_v$.
  If the magnitude of the net downward load is smaller than the uplift tension force caused by the overturning moment, which was caused by the lateral seismic force, E_h, then there will be a net uplift force on the column; therefore, the column, the column base plate, the anchor rods, and the column foundation will all have to be designed or checked for this net uplift force. Note that other load combinations will result in a maximum downward acting load (i.e., compression), Pu, that the column must also be designed for.

ASD Load Combinations

When designing for strength under service load conditions, the ASD load combinations as given in the following equations should be used [1]:

1. D
2. $D + L$
3. $D + (L_r \text{ or } 0.7S \text{ or } R)$
4. $D + 0.75L + 0.75\,(L_r \text{ or } 0.7S \text{ or } R)$
5a. $D + (0.6W)$
5b. $D + (0.6W_T)$

6a. $D + 0.75(0.6W) + 0.75L + 0.75(L_r$ or $0.7S$ or $R)$
6b. $D + 0.75(0.6W_T) + 0.75L + 0.75(L_r$ or $0.7S$ or $R)$
7a. $0.6D + 0.6W$ (D always counteracts or opposes W in this load combination)
7b. $0.6D + 0.6W_T$ (D always counteracts or opposes W_T in this load combination)
8. $D + (0.7E_v) + (0.7E_h)$
9. $D + (0.525E_v) + (0.525E_h) + 0.75L + 0.1S$
10. $(0.6D - 0.7E_v) + (0.7E_h)$ (D always counteracts or opposes E_h in this load combination)

NOTE:

- The above ASD load combinations can also be obtained from IBC Section 1605.3, and they are also used to check structural members for serviceability limit states such as deflections and vibrations.
- For ASD, where the effect of the load, H, due to lateral earth pressure, ground water pressure, or pressure of bulk materials adds to the primary variable load effect (L, L_r, S, R, W, or E), the load factor on H in the ASD load combinations shall be taken as 1.0.
- For ASD, where H counteracts or resists the primary variable load effect, the load factor for H should be set equal to 0.6 when the load, H, is permanent (e.g., lateral soil pressure) or a load factor of zero for all other conditions. Note that dead loads are *not* primary variable loads or load effects.

Note that for the columns that are part of the lateral force resisting system, e.g., braced frames or moment frames, the overturning moments due to lateral wind load, W, and the lateral seismic load, E_h, will result in upward (i.e., tension) and downward (i.e., compression) forces in the columns and the column bases (see Figure 2-2b). For the direction of the lateral loads shown in Figure 2-2b, the columns AB, BC, and CD will be subjected to tension forces from the lateral wind and seismic loads, while columns EF, FG, and GH will be subjected to additional compression forces.

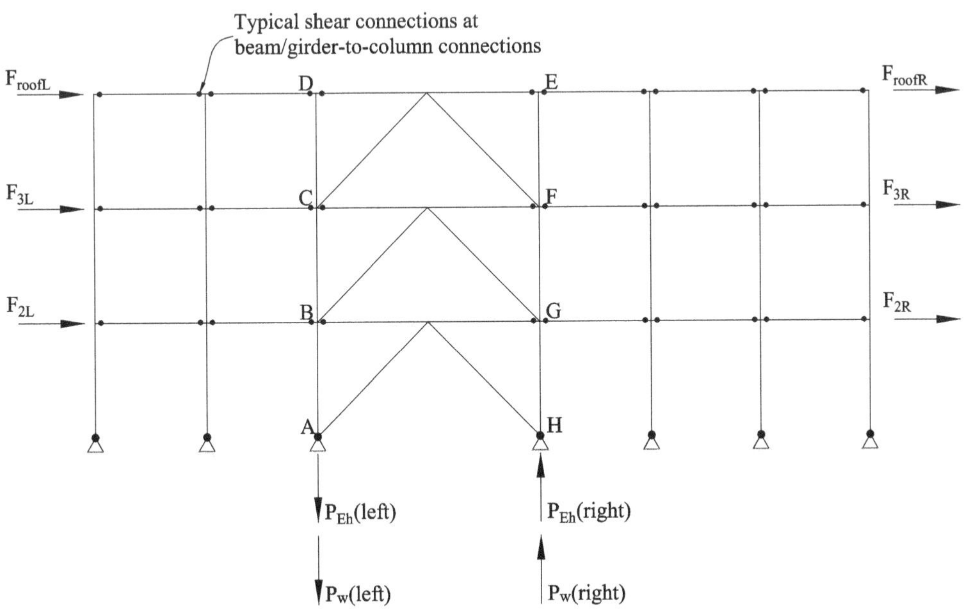

FIGURE 2-2b Additional Forces in the Columns of the LFRS

Examples of structures where hydrostatic pressures, H, may be critical include underground storage tanks or buildings located in areas with a high water table, where the lowest level or basement slab is below the water table and therefore subjected to hydrostatic buoyancy pressures due to the high water table; the stability of such structures should be checked to ensure that there is enough dead load to resist the upward hydrostatic pressures, H. Basement structural slab-on-ground below grade supported by piles or drilled shafts that are anchored to bedrock can be used to resist hydrostatic uplift pressures. Hydrostatic basement slabs as thick as 18 to 30 inches have been used in practice to resist hydrostatic uplift pressures of more than 1000 psf, depending on the height of the groundwater table above the underside of the hydrostatic slab and the lateral distance between the drilled shafts or piles. To resist the uplift forces, the structural basement slab-on-ground are anchored to the drilled caissons or piles, which are in turn socketed into the bedrock; the bottom of the slab and the sides of the basement walls are typically protected with waterproofing membranes to prevent water ingress into the building.

The applicable load combinations for the LRFD method involving uplift pressures will be $0.9D + 1.6H$ (see IBC Equation 16-6); the applicable load combination for the ASD method involving hydrostatic pressures is $0.6D + 1.0H$ (see IBC Equation 16-15) [4]. Note that the dead load, D, opposes the upward hydrostatic pressures, H, in the preceding load combinations. Therefore, to avoid flotation of the structure, the designer should ensure that $0.9D$ is greater than or equal to $1.6H$ when the LFRD method is used and $0.6D$ is greater than or equal to $1.0H$ when the ASD method is used.

In both the LRFD and ASD load combinations presented previously, downward loads are assumed to have a positive (+) sign, while upward loads have a negative (-) sign. For the LRFD method, load combinations 1 through 4, and 6 are used to maximize the downward loads, while load combinations 5 and 7 are used to maximize the uplift load or overturning effects. Therefore, in load combinations 5 and 7, the wind load, W, and the seismic load, E, take on *only negative or zero values*, while in all the other load combinations, W and E take on positive values.

For the ASD method, load combinations 1 through 6, and 8 and 9 are used to maximize the downward acting loads, while load combinations 7 and 10 are used to maximize the uplift load or overturning effects. Therefore, in load combinations 7 and 10, the wind load, W, and the seismic load, E, take on *only negative or zero values*, while in all the other load combinations, W and E take on positive values.

The seismic load effect, E, consists of a combination of the horizontal seismic load effect, E_h, and the vertical seismic load effect, E_v. The notations used in the load combinations presented previously are defined as follows:

E_h = ρQ_E = Horizontal component of the earthquake (see Chapter 3)

E_v = $0.2\, S_{DS}D$ = Vertical component of the earthquake (affects mostly columns and foundations)

E = Combined load effect due to *horizontal* and *vertical* earthquake-induced forces (see Chapter 3)

= $E_h + E_v = \rho Q_E + 0.2\, S_{DS}D$ in LRFD load combinations 6 and ASD load combinations 8 and 9.

= $E_h - E_v = \rho Q_E - 0.2\, S_{DS}D$ in LRFD load combination 7, and ASD load combination 10.

D = Dead load

Q_E = Horizontal earthquake load effect due to the base shear, V (i.e., forces, reactions, moments, shears, and so on caused by the horizontal seismic force)

= F_x (see Chapter 3)

S_{DS} = Design spectral response acceleration at short period (see Chapter 3)
H = Lateral soil pressure, hydrostatic pressures, and pressure of bulk materials
L = Floor live load
L_r = Roof live load
W = Wind load
W_T = Tornado load (see ASCE 7 Chapter 32)
S = Snow load
R = Rain load
ρ = Redundancy factor

The vertical seismic load effect, E_v (= 0.2 $S_{DS}D$), can be taken as zero when S_{DS} is less than or equal to 0.125. A redundancy factor, r, must be assigned to the seismic lateral force resisting system in both orthogonal directions of the building. This factor accounts for the presence or absence of multiple seismic lateral force-resisting load paths. The value of the redundancy factor is obtained from ASCE 7, Sections 12.3.4.1 and 12.3.4.2 as a function of the seismic design category (SDC), which will be discussed in Chapter 3, and the corresponding values of E_h are as follows:

Seismic Design Category, SDC	**Redundancy factor, ρ**	***Horizontal component of the seismic load effect, E_h***
B or C	1.0	E_h = (1.0)Q_E or (1.0)F_x
D, E, or F*	1.3	E_h = (1.3)Q_E or (1.3)F_x

The redundancy factor is permitted to be taken as 1.0 if the conditions in ASCE 7-22 Sections 12.3.4.2.1 and 12.3.4.2.2 are satisfied.

For the LRFD load combinations 6 and 7, the modified equations accounting for the redundancy factor are as follows:

6. $1.2D + 0.2S_{DS}D + 1.0\rho Q_E + f_1L + 0.15S$, or
$(1.2 + 0.2S_{DS})D + 1.0\rho Q_E + f_1L + 0.15S$

7. $(0.9 - 0.2S_{DS})D + 1.0\rho Q_E$

The determination of wind and seismic loads will be covered in Chapter 3.

Applicable Load Combinations for Design

All structural elements must be designed for the most critical of the load combinations presented in the previous section. Since most floor beams are usually only subjected to dead load, *D*, and floor live load, *L*, the most likely controlling load combinations for floor beams and girders will be LRFD load combinations 1 or 2 for limit states design and ASD load combinations 1 and 2 for allowable strength design and for checking serviceability (i.e., deflections and vibrations). To determine the most critical load combinations for roof beams and columns, all the applicable load combinations must be checked, but both the LRFD and ASD load combination 2 are more likely to control the design of most floor beams and girders. In Examples 2-1 through 2-9, the load effects, *H* and *F*, are assumed to be zero, as is the case for most building structures.

Special Seismic Load Combinations and the Overstrength Factor

For certain special structures and elements, the maximum seismic load effect, E_m, and the special load combinations specified in ASCE 7, Section 12.4.3 should be used. Some examples of special structures and elements for which the seismic force is amplified by the overstrength factor, Ω_o, include the following:

- **Drag strut** or **collector elements** (see ASCE 7, Section 12.10.2.1) in building structures (except for light-frame buildings). Drag struts will be discussed in Chapter 14.
- Structural elements supporting discontinuous systems, such as columns supporting discontinuous shear walls (see ASCE 7, Section 12.3.3.3).

The seismic load effect with overstrength factor for these special elements is given as

$$E_m = E_{mh} \pm E_v = \Omega_o Q_E \pm 0.2\, S_{DS}\, D,$$

where Ω_o is the overstrength factor from ASCE 7, Table 12.2-1, and the seismic load effect with overstrength, $E_{mh} = \Omega_o Q_E = \Omega_o F_x$.

For structural elements supporting discontinuous systems, such as columns supporting discontinuous shear walls or other lateral force resisting systems (LFRS), the overturning moment in the shear wall results in axial compression and tension forces in the supporting columns at the two ends of the discontinued shear wall. For structures with this type of vertical discontinuity, the use of the overstrength factor protects the gravity load resisting system (i.e., the columns supporting the shearwall) from overloads that may result from the shearwall having a higher strength (and therefore a higher overturning moment capacity) than it was designed for, resulting in higher overturning axial tension and compression forces on the supporting columns. For these special structures, the applicable LRFD load combinations are as follows:

6. $1.2D + E_v + E_{mh} + f_1 L + 0.15S = (1.2D + 0.2S_{DS}D) + \Omega_o Q_E + f_1 L + 0.15S$
7. $(0.9D - E_v) + E_{mh} = (0.9D - 0.2S_{DS}D) + \Omega_o Q_E$ (D always counteracts or opposes E_{mh} in this load combination)

EXAMPLE 2-1 (LRFD)

Load Combinations, Factored Loads, and Load Effects

A simply supported floor beam 20 ft. long is used to support service or working loads as follows:

$w_D = 2.5$ kips / ft.,

$w_L = 1.25$ kips / ft.

a. Calculate the required shear strength or factored shear, V_u.
b. Calculate the required moment capacity or factored moment, M_u.
c. Determine the required *nominal* moment strength.
d. Determine the required *nominal* shear strength.

Solution

For floor beams, only LRFD load combinations 1 and 2 need to be considered in calculating the factored loads.
The simply supported span of the beam, $\ell = 20$ft.

a. Factored loads:

The corresponding factored loads, w_u, are calculated as follows:

1. $w_u = 1.4D = 1.4\, w_D = 1.4(2.50 \text{ kip/ft.}) = 3.5 \text{ kip/ft.}$
2. $w_u = 1.2D + 1.6L = 1.2\, w_D + 1.6\, w_L = 1.2(2.5) + 1.6(1.25)$
   $= 5.0$ kips/ft. (governs)

$$\text{Maximum factored shear, } V_{u_{max}} = \frac{w_u \ell}{2} = \frac{(5)(20 \text{ ft.})}{2} = 50 \text{ kips}$$

b. $$\text{Maximum factored moment, } M_{u_{max}} = \frac{w_u \ell^2}{8} = \frac{(5)(20 \text{ ft.})^2}{8} = 250 \text{ ft.-kips}$$

Using the limit state design equation yields the following:

c. Required nominal moment strength, $M_n = M_u/\text{f} = 250/0.90 = 278$ ft.-kips.
d. Required nominal shear strength, $V_n = V_u/\text{f}_v = 50/1.0 = 50$ kips.

Note that f = 0.9 for shear for some steel sections (see Chapter 6).

EXAMPLE 2-2 (LRFD)

Load Combinations, Factored Loads, and Load Effects

Determine the required moment capacity, or factored moment, M_u, acting on a floor beam if the calculated service load moments acting on the beam are as follows: $M_D = 55$ ft.-kip, $M_L = 30$ ft.-kip. Calculate the required nominal moment strength, M_n.

Solution

1. $M_u = 1.4\, M_D = 1.4(55 \text{ ft.-kip}) = 77$ ft.-kips
2. $M_u = 1.2\, M_D + 1.6\, M_L = 1.2\ (55 \text{ ft.-kips}) + 1.6\ (30 \text{ ft.-kips})$
$= 114$ ft.-kips (governs)

$M_u = 114$ ft.-kips

The LRFD limit states the equation for flexure is

$$\phi M_n \geq M_u$$

where, $\phi = 0.9$ *for flexure*

$$\textit{Therefore, } M_n = \frac{M_u}{\phi} = \frac{114}{0.9} = 126.7\ ft. - kips$$

EXAMPLE 2-3 (LRFD)

Load Combinations, Factored Loads, and Load Effects

Determine the ultimate or factored load for a roof beam subjected to the following service loads:

Dead load = 35 psf

Snow load = 25 psf

Wind load = 15 psf upward

= 10 psf downward

Solution

$D = 35$ psf

$S = 25$ psf

$W = -15$ psf or 10 psf

The values for service load not given above are assumed to be zero; therefore,

Floor live load, $L = 0$

Rain live load, $R = 0$

Seismic load, $E = 0$ (i.e., $E_v = 0$ and $E_h = 0$)

Roof live load, $L_r = 0$

Since the live load, L, does not exceed 100 psf, and it is not an assembly occupancy or a parking garage, therefore, $f_1 = 0.5$.

Using the LRFD load combinations, and noting that only downward-acting loads should be substituted in load combinations 1 through 4, and 6 and upward wind or seismic loads in load combinations 5 and 7, the controlling factored load is calculated as follows:

1. $w_u = 1.4D = 1.4\,(35) = 49$ psf

2. $w_u = 1.2D + 1.6L + 0.5S$

$= 1.2\,(35) + 1.6\,(0) + 0.5\,(25) = 55$ psf

3a. $w_u = 1.2D + 1.6S + 0.5L$

$= 1.2\,(35) + 1.6\,(25) + 0.5\,(0) = 82$ psf

3b. $w_u = 1.2D + 1.6S + 0.5W$

$= 1.2\,(35) + 1.6\,(25) + 0.5\,(10) =$ **87 psf** (governs)

4. $w_u = 1.2D + 1.0W + 0.5L + 0.5S$

$= 1.2\,(35) + 1.0\,(10) + 0.5\,(0) + 0.5\,(25) = 64.5$ psf

5. $w_u = 0.9D + 1.0W$ (D *must* always oppose W in load combination 5)

$= 0.9\,(35) + 1.0\,(-15)$ (*upward wind load is taken as negative*)

$= 16.5$ psf (this implies a net downward load due to wind; i.e., there is no net uplift due to wind)

6. $w_u = 1.2D + 1.0E + 0.5L + 0.2S$

$= 1.2\,(35) + 1.0\,(0) + 0.5\,(0) + 0.2\,(25) = 47$ psf

7. $w_u = 0.9D - E_v + E_h$ (D *must* always oppose E in load combination 7)

$= 0.9\,(35) - (0) + (0)$

$= 31.5$ psf

In this example, load combinations 5 and 7 resulted in a net positive (or downward) load. However, load combinations 5 and 7 may sometimes result in net negative (or upward) factored loads that would also have to be considered in the design of the structural member.

- For strength calculations, the controlling factored load is $w_u =$ **87 psf.**

This load will be multiplied by the tributary width of the beam to convert it to Ib/ft. or kips/ft. load that will be used in the structural analysis of the beam. The concepts of tributary width and tributary area will be discussed later in this chapter. The controlling service loads will be calculated using the ASD load combinations in a subsequent example.

EXAMPLE 2-4 (LRFD)

Load Combinations, Factored Loads, and Load Effects

a. Determine the factored axial load or the required axial strength for columns that are part of a single bay braced frame (similar to the braced frame shown in Figure 2-2b) with the given service loads.

b. Calculate the required *nominal* axial compression strength, P_n, of the column.

The service axial loads on the column are as follows:

$P_D = 75$ kips (dead load)

$P_L = 150$ kips (floor live load)

$P_S = 50$ kips (snow load)

$P_W = \pm 100$ kips (wind load)

$P_{E_h} = \pm 50$ kips (seismic load)

Solution

a. Since the vertical component of the seismic load and S_{DS} are not given (see Chapter 3), neglect the vertical component of the seismic load, P_{E_v}, and consider only the horizontal component of the seismic load, P_{E_h}. Note that downward acting loads take on positive values, while upward loads take on negative values. The factored loads are calculated as follows:

1. $P_u = 1.4\,P_D = 1.4\,(75 \text{ kips}) = 105$ kips

2. $P_u = 1.2\,P_D + 1.6\,P_L + 0.5\,P_S$

$= 1.2\,(75) + 1.6\,(150) + 0.5\,(50) =$ **355 kips** (governs)

3a. $P_u = 1.2\,P_D + 1.6\,P_S + 0.5\,P_L$

$= 1.2\,(75) + 1.6\,(50) + 0.5\,(150) = 245$ kips

3b. $P_u = 1.2\,P_D + 1.6\,P_S + 0.5\,P_W$

$= 1.2\,(75) + 1.6\,(50) + 0.5\,(100) = 220$ kips

4. $P_u = 1.2\,P_D + 1.0\,P_W + 0.5\,P_L + 0.5\,P_S$

$= 1.2\,(75) + 1.0\,(100) + 0.5\,(150) + 0.5\,(50) = 290$ kips

Note that P_D must always oppose P_W and P_{E_v} in load combinations 5 and 7:

5. $P_u = 0.9\,P_D + 1.0\,P_W$

$= 0.9\,(75) + 1.0\,(-100) =$ **– 32.5 kips** (governs)

6. $P_u = 1.2\,P_D + P_{E_v} + P_{E_h} + 0.5\,P_L + 0.2\,P_S$

$= 1.2\,(75) + (0) + (50) + 0.5\,(150) + 0.2\,(50) = 225$ kips

7. $P_u = 0.9\,P_D - P_{E_v} + P_{E_h}$

$= 0.9\,(75) - (0) + 1.0\,(-50) = 17.5$ kips

- The factored axial *compression* load on the column is 355 kips.
- The factored axial *tension* force on the column is 32.5 kips.

The column, base plate, anchor bolts, and foundation will need to be designed for both the downward factored load of 355 kips and the factored net uplift, or tension, load of 32.5 kips.

b. Nominal axial compression strength of the column, $P_n = P_u/\phi = 355/0.90 = 394$ kips.

EXAMPLE 2-5 (LRFD)

Load Combinations—Factored Loads and Load Effects

Repeat Example 2-4, assuming the structure is to be used as a parking garage.

Solution

a. For parking garages or areas used for public assembly or areas with a floor live load, L, greater than 100 psf, the multiplier of the floor live load, f_1, in load combinations 3, 4, and 6 is 1.0. The factored loads are calculated as follows:

1. $P_u = 1.4\,P_D = 1.4\,(75\text{ kips}) = 105\text{ kips}$

2. $P_u = 1.2\,P_D + 1.6\,P_L + 0.5\,P_S$

$= 1.2\,(75) + 1.6\,(150) + 0.5\,(50) = 355\text{ kips}$

3a. $P_u = 1.2\,P_D + 1.6\,P_S + 1.0\,P_L$

$= 1.2\,(75) + 1.6\,(50) + 1.0\,(150) = 320\text{ kips}$

3b. $P_u = 1.2\,P_D + 1.6\,P_S + 0.5\,P_W$

$= 1.2\,(75) + 1.6\,(50) + 0.5\,(100) = 220\text{ kips}$

4. $P_u = 1.2\,P_D + 1.0\,P_W + 1.0\,P_L + 0.5\,P_S$

$= 1.2\,(75) + 1.0\,(100) + 1.0\,(150) + 0.5\,(50) =$ **365 kips** (governs)

Note that P_D must always oppose P_W and P_E in load combinations 5 and 7:

5. $P_u = 0.9\,P_D + 1.0\,P_W$

$= 0.9\,(75) + 1.0\,(-100) =$ **– 32.5 kips** (governs)

6. $P_u = 1.2\,P_D + P_{E_v} + P_{E_h} + 1.0\,P_L + 0.2\,P_S$

$= 1.2\,(75) + (50) + (0) + 1.0\,(150) + 0.2\,(50) = 300\text{ kips}$

7. $P_u = 0.9\,P_D - P_{E_v} + P_{E_h}$

$= 0.9\,(75 - 0) + 1.0\,(-50) = 17.5\text{ kips}$

- The factored axial *compression* load on the column is 365 kips.
- The factored axial *tension* force on the column is 32.5 kips.
- The column, base plate, anchor bolts, and foundation will need to be designed for both the downward factored load of 365 kips and the factored net uplift, or tension, load of 32.5 kips.

b. Nominal axial compression strength of the column, $P_n = P_u / \phi = 365 / 0.90 = 406$ kips.

The corresponding ASD load combination examples will now be presented:

EXAMPLE 2-6 (ASD)

ASD Load Combinations

A simply supported floor beam 20 ft. long is used to support service or working loads as follows:

$w_D = 2.5$ kips / ft.,
$w_L = 1.25$ kips / ft.

a. Calculate the ASD design shear, V.
b. Calculate the ASD design moment, M.
c. Determine the nominal moment strength, M_n, assuming a factor of safety for bending of 1.67.
d. Determine the nominal shear strength, V_n, assuming a factor of safety for shear of 2.0.

Solution

For floor beams, only ASD load combinations 1 and 2 need to be considered.
The simply supported span of the beam, ℓ = 20 ft.

a. Unfactored loads:

The corresponding unfactored loads, w, are calculated as follows:

1. $w = D = w_D$ = 2.50 kip/ft. = 3.5 kip/ft.
2. $w = D + L = w_D + w_L$ = 2.5 + 1.25 = 3.75 kips/ft. (governs)

Maximum unfactored shear, $V_{max} = \frac{w\ell}{2} = \frac{(3.75)(20\text{ ft.})}{2} = 37.5\text{ kips}$

b. Maximum unfactored moment, $M_{max} = \frac{w\ell^2}{8} = \frac{(3.75)(20\text{ ft.})^2}{8} = 187.5\text{ ft.-kips}$

The ASD design equation gives the allowable strength or applied service load or load effect, as $R_a = R_n/\Omega$, where Ω is the factor safety.

c. The allowable moment strength, $M_a = M_n/\Omega$; therefore, the nominal moment strength required, $M_n = \Omega M_a$ = (1.67)(187.5 ft-kip) = 313 ft.-kips.
d. The allowable shear strength, $V_a = V_n/\Omega$; therefore, the nominal shear strength required, $V_n = \Omega V_a$ = (2.0)(37.5 kips) = 75 kips.

Compare these required nominal strength values to those obtained for the LRFD method in Example 2-1. The ASD method gave higher values of the required nominal strengths, M_n and V_n (278 ft.-kips versus 313 ft.-kips and 50 kips versus 75 kips).

EXAMPLE 2-7 (ASD)

Determine the unfactored moment and the nominal moment capacity (assuming a factor of safety of 1.67) for a floor beam if the calculated service load moments acting on the beam are as follows: M_D = 55 ft.-kip, M_L = 30 ft.-kip.

Solution

Using the ASD load combinations, we obtain

1. $M = M_D$ = 55 ft.-kip
2. $M = M_D + M_L$ = 55 ft.-kips + 30 ft.-kips
   = 85 ft.-kips (governs)

The allowable moment strength, $M_a = M_n/W$; therefore, the nominal moment strength required, $M_n = WM_a$ = (1.67)(85 ft-kip) = 142 ft.-kips

Compare to the nominal moment strength, M_n, of 126.7 ft.-kips obtained for the LRFD method in Example 2-2.

EXAMPLE 2-8 (ASD)

Determine the ASD design load for a roof beam subjected to the following service loads:

Dead load = 35 psf

Snow load = 25 psf

Wind load = 15 psf upward

= 10 psf downward

Solution

$D = 35$ psf

$S = 25$ psf

$W = -15$ psf or 10 psf

The values for service loads not given above are assumed to be zero; therefore,

Floor live load, $L = 0$

Rain live load, $R = 0$

Seismic load, $E = 0$ (i.e., $E_v = 0$ and $E_h = 0$)

Roof live load, $L_r = 0$.

Since $S > L_r > R$, use $S = 25$ psf

Using the ASD load combinations and noting that only downward-acting loads should be sub*stituted in load combinations 1 through 6, and 8 and 9, and upward wind or seismic loads in* load combinations 7 and 10, the controlling factored load is calculated as follows:

1. D = 35 psf
2. D + L = 35 psf + 0 psf = 35 psf
3. D + (L_r or S or R) = 35 psf + 25 psf = **60 psf** (controls)
4. D + 0.75L + 0.75(L_r or S or R) = 35 psf + 0 psf + (0.75)(25 psf) = 53.8 psf
5. D + (0.6W) = 35 psf + 0.6(10 psf) = 41 psf
6. D + 0.75L + 0.75(0.6W) + 0.75(L_r or S or R) = 35 psf + 0.75(0 psf) + 0.75(0.6)(10 psf) + 0.75(25 psf) = 58.3 psf
7. 0.6D + 0.6W = 0.6(35 psf) + 0.6(−15 psf) = 12 psf (D always opposes or counteracts W in this load combination)
8. D + 0.7E_v + 0.7E_h = 35 psf + 0.7(0) + 0.7(0) = 35 psf
9. D + 0.525E_v + 0.525E_h + 0.75L + 0.75S = 35 psf + 0.525(0 psf) + 0.525 (0 psf) + 0.75(0 psf) + 0.75 (25 psf) = 53.8 psf
10. 0.6D − (0.7E_v) + (0.7E_h) = 0.6(35 psf) − 0.7(0 psf) + 0.7(0 psf) = 21 psf (D always opposes or counteracts E in this load combination)

Therefore, the roof beam will be designed for a total service or unfactored load of **60 psf**. This load will be multiplied by the tributary width of the beam to convert it to a load with units of Ib/ft. or kips/ft. that will be used in the structural analysis of the beam. The concept of tributary width and tributary area will be discussed later in this chapter.

EXAMPLE 2-9 (ASD)

a. Determine the ASD design axial load for a column in an office building with the given service loads.

b. Calculate the required *nominal* axial compression strength, P_n, of the column if the factor of safety is assumed to be 1.67.

The service axial loads on the column are as follows:
D = 75 kips (dead load)
L = 150 kips (floor live load)
S = 50 kips (snw load)
W = ±100 kips (wind load)
E_h = ±50 kips (seismic load)

Solution

a. Since the vertical component of the seismic load is not given and we are not given the value of the design spectral response acceleration at short period, S_{DS}, neglect the vertical component of the seismic load, P_{E_v}. Note that downward loads are assumed to be positive, while upward loads take on negative values. The ASD design loads are calculated as follows:

1. D = 75 kips
2. D + L = 75 kips + 150 kips = 225 kips
3. D + (L_r or S or R) = 75 kips + 50 kips = 125 kips
4. D + 0.75L + 0.75(L_r or S or R) = 75 kips + 0.75(150 kips) + 0.75(50 kips) = 225 kips
5. D + (0.6W) = 75 kips + 0.6(100 kips) = 135 kips
6. D + 0.75L + 0.75(0.6W) + 0.75(L_r or S or R) = 75 kips + 0.75(150 kips) + 0.75(0.6)(100 kips) + 0.75(50 kips) = **270 kips** (governs)
7. 0.6D + 0.6W = 0.6(75 kips) + 0.6(-100 kips) = **–15 kips** (D always opposes or counteracts W in this load combination)
8. D + $0.7E_v$ + $0.7E_h$ = 75 kips + 0.7(0 kips) + 0.7(50 kips) = 110 kips
9. D + $0.525E_v$ + $0.525E_h$ + 0.75L + 0.75S = 75 kips + 0.525(0 kips) + 0.525 (50 kips) + 0.75(150 kips) + 0.75 (50 kips) = 251.3 kips
10. 0.6D – ($0.7E_v$) + ($0.7E_h$) = 0.6(75 kips) – 0.7(0 kips) + 0.7(-50 kips) = 10 kips (D always opposes or counteracts E_v in this load combination)
    - The required ASD design axial *compression* strength of the column is **270 kips.**
    - The required ASD design axial *tension* strength of the column is **15 kips.**

The column, base plate, anchor bolts, and foundation will be designed for both the downward total unfactored axial load of 270 kips and the net uplift, or tension, load of 15 kips.

b. The nominal axial compression strength of the column, $P_n = P_a W$ = 270(1.67) = 451 kips.

Compare the nominal axial compression strength (451 kips) to the value of 394 kips obtained for the LRFD method in Example 2-4. Again, the ASD method yields a higher required nominal axial compression strength, P_n, than the LRFD method.

EXAMPLE 2-10 (ASD AND LRFD)

A buried rectangular tank with a horizontal cross-sectional area of 44 ft.^{2} weighs 1300 Ib. The density of the 2 ft. deep saturated soil above the top of the tank is specified by the geotechnical engineer as 110 pcf. If the depth of the tank is 4 ft. and the water table is assumed to be located at or near grade, calculate the net buoyancy uplift force on the tank using the LRFD method and the ASD method. How should the effect of the net buoyancy uplift forces on this tank be mitigated?

Solution

Depth of tank = 4 ft.

Depth of saturated soil above the top of the buried tank = 2 ft.

The water table is at or near the grade level; therefore, the total depth of water to the base of the tank is

h = 2 ft. + 4 ft. = 6 ft.

Density of water, γ_w = 62.4 pcf

Weight of tank = 1300 Ibs

Weight of saturated soil above the buried tank = (44 ft.^{2})(2 ft.)(110 pcf) = 9680 Ibs

Total dead load when the tank is empty, D = 1300 + 9680 = 10,980 Ibs

Hydrostatic or buoyancy pressure at the base of the tank = $\gamma_w h$ = (62.4 pcf)(6 ft.) = 374.4 psf

Total uplift force at the base of the tank, H = (374.4 psf) (44 ft.^{2}) = 16,474 Ibs

The downward load, D, is taken as positive while the upward force, H, is taken as negative.

LRFD method:

Substituting D = +10,980 Ibs and H = −16,474 Ibs into the appropriate LRFD load combination that addresses uplift or overturning (i.e., ASCE 7 LRFD load combination 5 with "H" included or IBC 2018 Equation 16-6), we obtain the load combination as follows:

$0.9D + 1.6H = (0.9)(10980 \text{ Ibs}) + (1.6)(-16474 \text{ Ibs}) = -16{,}477 \text{ Ibs.}$

Therefore, the factored net uplift force at the base of the tank = 16,477 Ibs.

ASD method:

Substituting D = +10,980 Ibs and H = −16,474 Ibs into the appropriate ASD load combination that addresses uplift or overturning (e.g., ASCE 7 ASD load combination 7 with "H" included or IBC 2018 Equation 16-15), we obtain the load combination as follows:

$0.6D + 1.0H = (0.6)(10980 \text{ Ibs}) + (1.0)(-16474 \text{ Ibs}) = -9{,}886 \text{ Ibs.}$

Therefore, the unfactored net uplift force at the base of the tank = 9,886 Ibs.

The *factored* net uplift force of 16,477 Ibs for the LRFD method (or the *unfactored* net uplift force of 9,886 Ibs for the ASD method) will need to be counteracted or resisted using hold-down anchors or weighing down the tank by attaching it to a thick concrete base slab that extends beyond the tank footprint to engage additional weight of soil. For an adequate design, the sum of the dead load from the empty tank plus the additional dead load engaged by the hold-down anchors (e.g., the weight of the concrete base slab and the additional weight of soil engaged) must be large enough so that $0.9D$ is greater than or equal to $1.6H$ for the LRFD method, or $0.6D$ is greater than or equal to $1.0H$ for the ASD method.

2-4 INTRODUCTION TO STRUCTURAL DESIGN LOADS

Structural loads are the forces applied to a structure (e.g., dead load, floor live load, roof live load, snow load, rain load, ice load, wind load, earthquake or seismic load, flood loads, earth, and hydrostatic pressure). For buildings, the magnitude of these loads are specified in the ASCE 7 Load Standard; in this standard, buildings are grouped into four different occupancy types, which are used to determine the risk category of the building (see ASCE 7 Table 1.5-1), which in turn are used to obtain the importance factor, I (see ASCE 7 Table 1.5-2) that is used for snow, seismic, or ice load calculations [2]. Maps of the 3-second gust wind speeds for Exposure Category C that can be used for calculating the factored design wind pressures are given in ASCE 7 Figures 26.5-1A, B, C, and D for the four different risk categories (RC). These wind speed maps give the ultimate strength design level wind speeds, which already incorporate the importance factor for wind; hence, no importance factors for wind are given in ASCE 7, Table 1.5-2. The importance factor is a measure of the consequences of failure of a building to public safety, and the higher the importance factor, the larger the snow, ice, or seismic loads on the structure. ASCE 7, Table 1.5-1 should be used in conjunction with ASCE 7 Table 1.5-2 to determine the importance factors for snow loads, ice loads, and seismic loads, respectively. The International Building Code (IBC) stipulates in Section 1603 that all structural design loads must be listed on the contract documents. This information is usually listed on the first structural drawing, including the General Notes and the Typical Structural Details.

Gravity Load Resisting Systems

The two main types of floor systems used to resist gravity loads in building structures consist of *one-way* and *two-way* load distribution systems. In one-way floor systems, the floor or roof load is distributed horizontally predominantly in one direction (i.e., the direction with the shorter span), whereas in two-way floor systems, the load is distributed horizontally in two orthogonal directions. For steel structures, the one-way load distribution system occurs much more frequently than the two-way system; consequently, only one-way systems are discussed further in this text. There are several *one-way* roof and floor systems used in steel buildings, and these systems support gravity loads in one-way action by virtue of their construction and because the bending strength of the floor or roof system in one direction is several times greater than the strength in the orthogonal direction. These types of one-way systems span in the stronger direction of the floor or roof panel regardless of the aspect ratio of the panel. Examples of one-way systems used in steel buildings include:

- Metal roof decks (used predominantly for roofs in steel buildings),
- Concrete-filled composite metal floor decks (used predominantly for floors in steel buildings) and concrete-filled non-composite metal deck or form deck, and
- Hollow-core precast concrete planks.
- Metal cladding spanning vertically between horizontal girts or spanning horizontally between wind columns in resisting wind pressure perpendicular to the face of the cladding.
- Cross-laminated timber (CLT) floor or roof panels on steel beams.

2-5 GRAVITY LOADS IN BUILDING STRUCTURES

The common types of gravity load that act on building structures are *roof dead load, floor dead load, roof live load, snow load,* and *floor live load,* and these are discussed in the following sections.

2-6 DEAD LOADS

Dead loads are permanent or nonmovable loads that act on a structure, and they include the weight of all materials that are permanently attached to the structure, including the self-weight of the structural member. Since the dimensions of structural members are known or can be determined, and the material density is also a known quantity, the dead loads, which are predominantly the self-weight of the structural member, can be determined with greater accuracy than any other type of load, and they are not as variable as live loads. Examples of items that would be classified as dead loads include floor finishes, partitions, mechanical and electrical (M&E) components, fireproofing, glazing, and cladding. The floor and roof dead loads are typically uniformly distributed gravity loads expressed in units of pounds per square foot (psf) or kips per square foot (ksf) of the horizontal projected plan area. For sloped members, the dead load, which is in units of psf of the sloped area, must be converted to units of psf of the horizontal projected plan area. In certain cases, concentrated dead loads, such as a heavy safe with a small footprint, may also have to be considered in the structural analysis and design. The typical checklists of the items to include for dead loads and the estimated weights for roof and floor dead load components in steel buildings are as follows:

Common Roof Dead Load Components in Steel Buildings

Framing (self-weight)	5 psf to 8 psf
Fireproofing	2 psf
Metal deck	2 psf (1½ in. deck) 3 psf (3 in. deck) 6 psf (7½ in. deck)
5 ply membrane roofing with gravel ballast (see Figure 2-2c)	6.5 psf
Rigid insulation	1.5 psf per inch of thickness
Plywood sheathing	0.4 psf per ⅛ in. thickness
Asphalt shingles	3.0 psf
Suspended ceiling	2.0 psf
Mechanical/electrical components*	5 psf to 10 psf

* Includes HVAC ducts, electrical and plumbing conduits (does not include weight of rooftop units)

FIGURE 2-2c Flat roof with gravel ballast, and primary and secondary drains

Common Floor Dead Load Components in Steel Buildings

Framing (self-weight)	6 psf to 12 psf
Fireproofing	2 psf
Composite metal deck	1 psf (9⁄16 in. deck) 2 psf (1½ in. deck) 3 psf (3 in. deck)
Concrete slab:	
lightweight	10 psf per inch of thickness
normal weight	12.5 psf per inch of thickness
Concrete on composite metal deck	see deck manufacturers catalog for weights
Floor finishes	
¼" ceramic tile	10 psf
1" slate	15 psf
Gypsum fill	6.0 psf per inch of thickness
⅞" hardwood	4.0 psf
Partitions	15 psf (minimum)
Suspended ceiling	2.0 psf
Mechanical/electrical components*	5 psf to 10 psf
Concrete ponding load	10% of the weight of the concrete slab on the metal deck [29].
Glass curtain wall façade	25 psf of vertical surface area

* Includes HVAC ducts, electrical and plumbing conduits (does not include weight of mechanical or electrical equip-ment on a floor)

The ASCE 7 Load Standard [2] specifies a minimum partition load of 15 psf, and partition loads need not be considered when the tabulated unreduced floor live load, L_o, is greater than 80 psf because of the low probability that partitions will be present in occupancies with higher live loads. Partition loads may be greater than the 15 psf minimum specified in the ASCE 7 Load Specifications; therefore, the actual partition load should be estimated for each project. It should be noted that partition loads are classified in ASCE 7 as live loads, but it is common design practice to treat partition loads as dead loads, and this approach is adopted in this text.

Another dead load that should be considered for steel-framed floors is the concrete ponding load. This is the dead load of the additional concrete that is needed to achieve a flat floor in a non-composite beam construction (see Chapter 6) due to the deflection of the non-composite beams and girders during the construction phase as the wet concrete is poured. A concrete ponding dead load of 10% of the self-weight of the concrete and metal deck slab system is suggested in Ref. [29] and adopted in this text. Some engineers use concrete ponding load values of up to 15%. The concrete ponding dead load can be neglected for steel framing where an upward camber is specified for the floor beams or if the beams are shored. For floor framing with non-composite beams, concrete ponding dead load should also be considered if the beams and girders are not cambered.

The total dead load can vary from approximately 25 psf to 35 psf for steel-framed roofs, and approximately 80 psf to 100 psf for steel-framed floors with composite deck.

Load Distribution: Tributary Widths and Tributary Areas

When a gravity load is applied to a structure, the load is distributed to various structural members as the load is transmitted from its point of application through the roof or floor system to the beams and girders, to the columns, and ultimately to the foundations and the soil or bedrock. This path through which the gravity load travels or is transmitted is called the gravity load path. There also needs to be a continuous lateral load path, and this will be discussed in Chapters 3 and 14. Continuous gravity and lateral load paths are very important considerations in the design of structures. A structural system with an inadequate load path is susceptible to collapse. The magnitude of the gravity load distributed to the beams, girders, or columns is dependent on the tributary width and tributary areas of the supporting beams and girders and the tributary areas of the columns.

The tributary widths and tributary areas are used to determine the distribution of floor and roof loads to the individual structural members. The tributary width (TW) of a beam is defined as the width of the floor slab or roof deck supported by the beam; this is equal to one-half the distance to the adjacent beam on the right-hand side plus one-half the distance to the adjacent beam on the left-hand side of the beam whose tributary width is being determined. The tributary width for a typical interior beam is calculated as

TW = ½ (Distance to adjacent beam on the right-hand side) + ½ (Distance to adjacent beam on the left-hand side).

For an edge, perimeter, or spandrel beam, the tributary width is calculated as

TW = ½ (Distance to adjacent beam) + the floor or roof edge distance measured from the longitudinal axis of the spandrel or edge beam.

The tributary area, A_T, of a beam, girder, or column is the floor or roof plan area supported by the structural member. The tributary area of a beam is obtained by multiplying the span of the beam (i.e., its length between supports) by its tributary width. The tributary area of a column is the plan area bounded by lines located at one-half the distance to all the adjacent columns surrounding the column whose tributary area is being calculated. For edge or corner columns, the tributary area will also include the area of the floor or roof extending from the longitudinal axes of the perimeter beams and girders framing into the column to the exterior edge of the floor or roof. The following points should be noted:

- Beams are usually subjected to uniformly distributed loads (UDL) from the roof deck or floor slab.
- Girders are usually subjected to concentrated loads due to the reactions from the beams. These concentrated loads or reactions from the beams have their own tributary areas.
- The tributary area of a girder is the sum of the tributary areas of all the concentrated loads acting on the girder.
- Perimeter or edge beams and girders—also known as spandrel beams and spandrel girders—support an additional uniform load due to the loads acting on the floor or roof area extending beyond the longitudinal axes of the beam or girder to the edge of the roof or floor. When the exterior wall cladding is an infill system where the weight of the exterior wall cladding is supported at each floor level, the spandrel beams and girders will also directly support the weight of the exterior cladding (i.e., the brick veneer or glazing and the masonry block or cold-formed steel stud back-up walls). For by-pass exterior wall systems, the cladding load is not supported by the spandrel beams and girders, but instead, the weight of the exterior cladding is supported on a concrete foundation wall, and the spandrel beam and girders only provide lateral support to the exterior wall cladding. The by-pass exterior wall system is permitted only in buildings not greater than three stories or 30 ft. in height [7].

EXAMPLE 2-11

Calculation of Tributary Width and Tributary Area

Using the floor framing plan shown in Figure 2-3, determine the following parameters:

a. Tributary width and tributary area of a typical interior beam,
b. Tributary width and tributary area of a typical spandrel or perimeter beam,
c. Tributary area of a typical interior girder,
d. Tributary area of a typical spandrel girder,
e. Tributary area of a typical interior column,
f. Tributary area of a typical corner column, and
g. Tributary area of a typical exterior column.

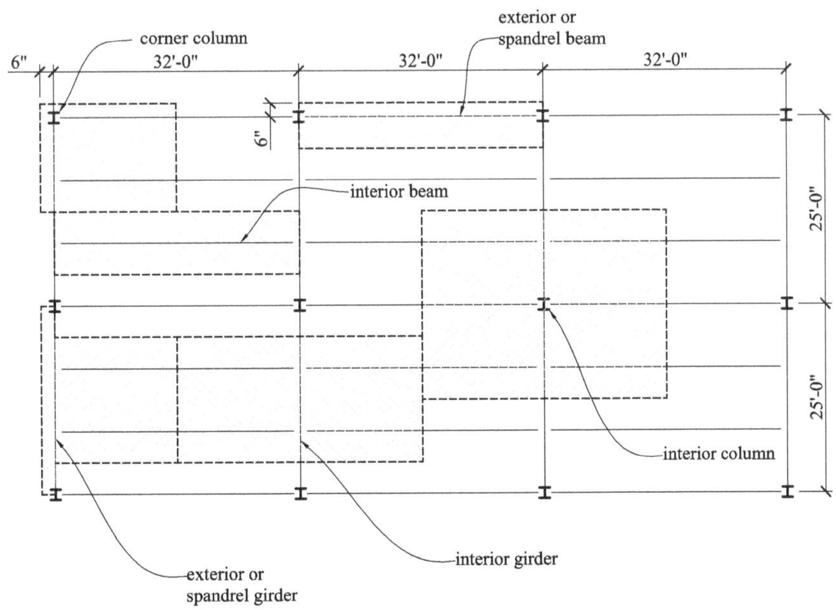

FIGURE 2-3 Tributary width and tributary areas

Solution

The tributary widths and areas are calculated in Table 2-3.

TABLE 2-3 Tributary Widths and Areas

Member	Tributary width (TW)	Tributary area (A_T)
Typical interior beam	25 ft./3 spaces = 8.33 ft.	(8.33 ft.) (32 ft.) = 267 ft.^{2}
Typical spandrel beam	(25 ft./3 spaces)/2 + 0.5 [ft.] edge distance = 4.67 ft.	(4.67 ft.) (32 ft.) = 150 ft.^{2}
Typical interior girder	—	$\left(\frac{267}{2}\text{ft.}^2\right)(4\,\text{beams}) = 534\text{ ft.}^2$
Typical spandrel girder	—	$(0.5\text{ft. edge dist})(25\text{ ft.}) + \left(\frac{267}{2}\text{ft.}^2\right)(2\text{ beams}) = 280\text{ ft.}^2$
Typical interior column	—	(32 ft.) (25 ft.) = 800 ft.^{2}
Typical corner column	—	(32 ft./2 + 0.5 ft. edge distance) × (25 ft./2 + 0.5 ft. edge distance) = 215 ft.^{2}
Typical exterior column (short side of building)	—	(32 ft./2 + 0.5 ft. edge distance) (25 ft.) = 413 ft.^{2}
Typical exterior column (long side of building)	—	(25 ft./2 + 0.5 ft. edge distance) (32 ft.) = 416 ft.^{2}

2-7 LIVE LOADS

In general, any gravity load that is not permanently attached to the structure is considered to be a live load. The three main types of live loads that act on building structures are *floor live loads*, *roof live loads*, and *snow loads*. Floor live loads, L, are the minimum live loads that are specified in ASCE 7, Table 4.3-1, or IBC, Table 1607.1; the magnitude depends on the use of the structure and the tributary area (A_T) supported by the structural member. Floor live loads are usually uniform loads expressed in units of pounds per square foot (psf) of horizontal plan area. In certain cases, the code also specifies alternate concentrated floor live loads (in lb. or kip) that need to be considered in design, but in most cases, the uniform loads govern the design of structural members.

Since live loads are transient loads and do not remain permanently on a structure, pattern or skipped live load scenarios should be investigated for continuous beams and girders and beams or girders with overhangs or cantilevers so as to obtain the maximum load effects (i.e., moments, shears, reactions, and deflections). To obtain the maximum cantilever moment and deflection for a beam with a cantilever, the overhanging or cantilevered span of the beam should be loaded with the full dead and live load, while the backspan is loaded with only the dead load. This loading scenario will also result in the maximum uplift reaction force in the far-end support of the backspan. The effect of human-induced vibrations (for example, jumping or aerobic exercises) should also be investigated for the overhanging span or cantilever.

The live loads that act on roof framing members are roof live load, $\boldsymbol{L_r}$, rain loads, R, and snow load, $\boldsymbol{S}$. From the load combinations presented in Section 2-3, it becomes apparent that roof live loads, rain loads, and snow loads *do not* occur on a structure at their maximum values simultaneously (i.e., at the same time). Roof live loads rarely govern the design of structural members in the higher snow regions of the United States, except where the roof is a special purpose roof used for promenades or as a roof garden. Rain loads may be critical for flat roofs with high parapets.

2-8 FLOOR LIVE LOADS

Floor live loads in building structures are occupancy loads that depend on the use of the structure. These loads are assumed to be uniform loads expressed in units of psf, and the values are specified in building codes such as the International Building Code (IBC) and the ASCE 7 Load Standard, first published in 1988; the precursor to ASCE 7 was ANSI A58.1—Minimum Design Loads for Buildings and Other Structures, first published in 1982. The floor live loads are determined from statistical analyses of many load surveys. The codes also specify alternate concentrated live loads for some occupancies because, for certain design situations, live load concentrations—as opposed to uniform live loads—may be more critical for design. The uniform floor live loads are most commonly used in practice; the concentrated live loads are used only in rare situations where, for example, punching shear may be critical, such as in slabs supporting vehicles (parking garage) or forklift loads, or in the design of stair treads where the concentrated loads, instead of the uniform loads, may control the design of the structural member. A selected listing of the recommended minimum live loads is presented in Table 2-4 [2, 4], with a more comprehensive listing provided in ASCE 7 Table 4.3-1 and ASCE 7 Table C4.3-1 [2]; particular attention should be paid to the footnotes below these tables. For example, in addition to the vertical live loads specified for stadiums and arenas, lateral loads of 24 lb. per linear ft. of seat applied in a direction parallel to each row of seats and 10 lb. per linear ft. of seat applied in a direction perpendicular to each row of seats are specified in the footnotes of ASCE 7 Table 4.3-1. This lateral load accounts for the action of spectators moving and swaying in unison. This lateral load is equivalent to approximately 12 psf of plan area assuming 2 ft. spacing between the seats; this same lateral load has recently been recommended for the design of exterior decks based on results of preliminary tests at Washington State University [19]. Note that the balcony live loads have changed from the previous editions of ASCE 7. Due to the lack of redundancy of cantilevered balcony structures, the live load on balconies is now 1.5 times the live load for the occupancy (see Table 2-4 or ASCE Table 4.3-1 or IBC Table 1607.1) that the balcony serves, but not exceeding 100 psf. Thus, for example, balconies in office buildings are designed for a live load of 75 psf (i.e., 1.5 × 50 psf). Some engineers design balcony framing for 100 psf because it lacks redundancy and they are frequently used as assembly areas. Note that balconies are also permanently exposed to the elements and thus susceptible to hidden deterioration from corrosion due to failure of the waterproofing system.

Some live loads that are not listed in the ASCE 7 tables include the live load for exhibition halls, which is commonly assumed as 350 psf due to the heavy materials and equipment that may be displayed on the floor of an exhibition hall. The bottom chords of roof or floor trusses spanning over an auditorium, theater, or exhibition hall may support rigging loads of 12 psf or higher, and in addition, these trusses may also support the dead and live loads from suspended catwalks.

For bridge structures, the two main types of live loads are the truck axle loads, which are concentrated loads, and the lane loads, which are a continuous uniform load plus a single concentrated load. The lane loads represent the effect of several small vehicles on a bridge (the continuous uniform load) plus a single truck represented by the single concentrated load.

2-9 FLOOR LIVE LOAD REDUCTION

To account for the low probability that floor structural elements with large tributary areas will have their entire tributary area loaded with the live load at one time, the IBC and ASCE 7 Load Specifications allow for floor live loads to be reduced, provided that certain conditions are satisfied. The reduced design live load of a floor, L in psf, is given as

$$L = L_o\left[0.25 + \frac{15}{\sqrt{K_{LL}A_T}}\right] \tag{2-5}$$

$\geq 0.50\ L_o$ for members supporting *one floor* (e.g., slabs, beams, and girders)
$\geq 0.40\ L_o$ for members supporting *two or more floors* (e.g., columns)

L_o = Unreduced design live load from Table 2-4 (or IBC, Table 1607.1, or ASCE 7, Table 4.3-1)

K_{LL} = Live load element factor, which is a ratio of the influence area, A_I, of the structural member (i.e., the floor area from which the member derives its load) to the tributary area of the member (see ASCE 7, Table 4.7-1 or IBC Table 1607.10.1). For instance, the influence area for a typical interior column is the total area of the floor bounded by all the adjacent columns. Thus, a typical interior column in a 30 ft. × 30 ft. grid will have a tributary area of 900 ft.^{2} and an influence area of 3600 ft.^{2} (i.e., 60 ft. × 60 ft.); therefore, K_{LL} = 3600/900 = 4. . Similarly, the influence area for a typical interior beam is the total area bounded by the beams to immediately to the right and to the left of the beam whose influence area is being calculated. Thus, a typical interior beam spaced at 10 ft. on centers and spanning 30 ft. will have a tributary area of 300 ft.^{2} (i.e., 10 ft. × 30 ft.) and an influence area of 600 ft.^{2} (i.e., 20 ft. × 30 ft.); therefore, K_{LL} = 600/300 = 2.

The K_{LL} values for different structural elements are given as follows:

= 4 (interior columns and exterior columns without cantilever slabs)
= 3 (edge columns with cantilever slab)
= 2 (corner columns with cantilever slabs, edge beams without cantilever slabs, interior beams)
= 1 (all other conditions not identified above)

A_T = Summation of all the applicable floor tributary areas, in ft.^{2}, supported by the structural member, excluding the roof tributary area

- For beams and girders (including continuous beams or girders), A_T is as defined in Section 2-6.
- For **one-way slabs**, A_T must be less than or equal to $1.5s^2$, where s is the slab span.
- For a member supporting more than one floor area in multistory buildings (e.g., multi-story columns), A_T will be the summation of all the applicable floor areas supported by that member. As an example, for a 10-story building supporting office occupancy on the second through 9th floors and mechanical occupancy on the 10th floor, the cumulative total axial live load on the ground floor column will be calculated as the sum of the column tributary areas of the second through 9th floors multiplied by the reduced office floor live load on the ground floor column based on this cumulative tributary area, plus the unreduced live load from the 10th floor mechanical occupancy multiplied by the column tributary area at the 10th floor level, plus the roof live load or the roof snow load multiplied by the

column tributary area at the roof level. This cumulative live load is added to the cumulative dead load (which is never reduced) to obtain the total unfactored cumulative axial load on the column. The total factored cumulative axial load on the column can similarly be calculated using the LRFD load combinations from Section 2-3.

The ASCE 7 Load Specification [2] ***does not*** permit floor live load reductions for floors satisfying any one of the following conditions:

- $K_{LL} A_T \leq 400$ ft.^{2}
  In this case, the tributary area is small enough so that there is a high probability that the whole of the tributary area will be loaded with the code-specified live load, L_o.
- Floor live load, $L_o > 100$ psf, due to the high probability of overloading.
  However, the floor live load may be reduced by 20% for members supporting two or more floors in non-assembly occupancies.
- Floors used for assembly occupancies such as auditoriums, stadiums, exhibition halls, theaters, etc., because of the high probability of overloading, especially during an emergency. Furthermore, floors used for mechanical equipment occupancy are not allowed to be reduced.
- Floors used for passenger car garage floors, except that the live load can be reduced by 20% for members supporting two or more floors.

The following should be noted regarding the tributary area, A_T, used in calculating the reduced floor live load:

1. The infill beams are usually supported by girders, which, in turn, are supported by columns, as indicated in our previous discussions on load paths. The tributary area, A_T, for beams is usually smaller than those for girders and columns, and, therefore, beams will have smaller floor live load reductions than girders or columns. The question arises as to which A_T to use for calculating the reduced live loads on the girders:
   - **a.** For the design of the beams, use the A_T of the beam to calculate the reduced live load that is used to calculate the beam moments, shears, and reactions. These load effects are used for the design of the beam and the beam-to-girder or beam-to-column connections. Note that, in design practice, because of the relatively small tributary areas for beams, it is common practice to neglect live load reduction for beams.
   - **b.** For the girders, recalculate the beam reactions since the tributary area supported by the girders is larger than that supported by the beams; for recalculating the reaction from the beams on the girder, use the tributary area, A_T, of the girder in calculating the live load reduction on the beam. Note that using the larger tributary area of the girder in recalculating the live load reduction in the infill beams will result in the infill beam reactions on the girders that are smaller than those calculated in Part (a). These smaller beam reactions are *only* used for the design of the girders.
2. For columns, A_T is the ***summation*** of the tributary areas of all the floors with *reducible live loads* at and above the level at which the column load is being determined, excluding the roof tributary areas, the mechanical occupancy tributary areas, and the assembly occupancy areas supported by the column. See Figure 2-3b for the definition of A_T or ΣA_T for multi-story columns.

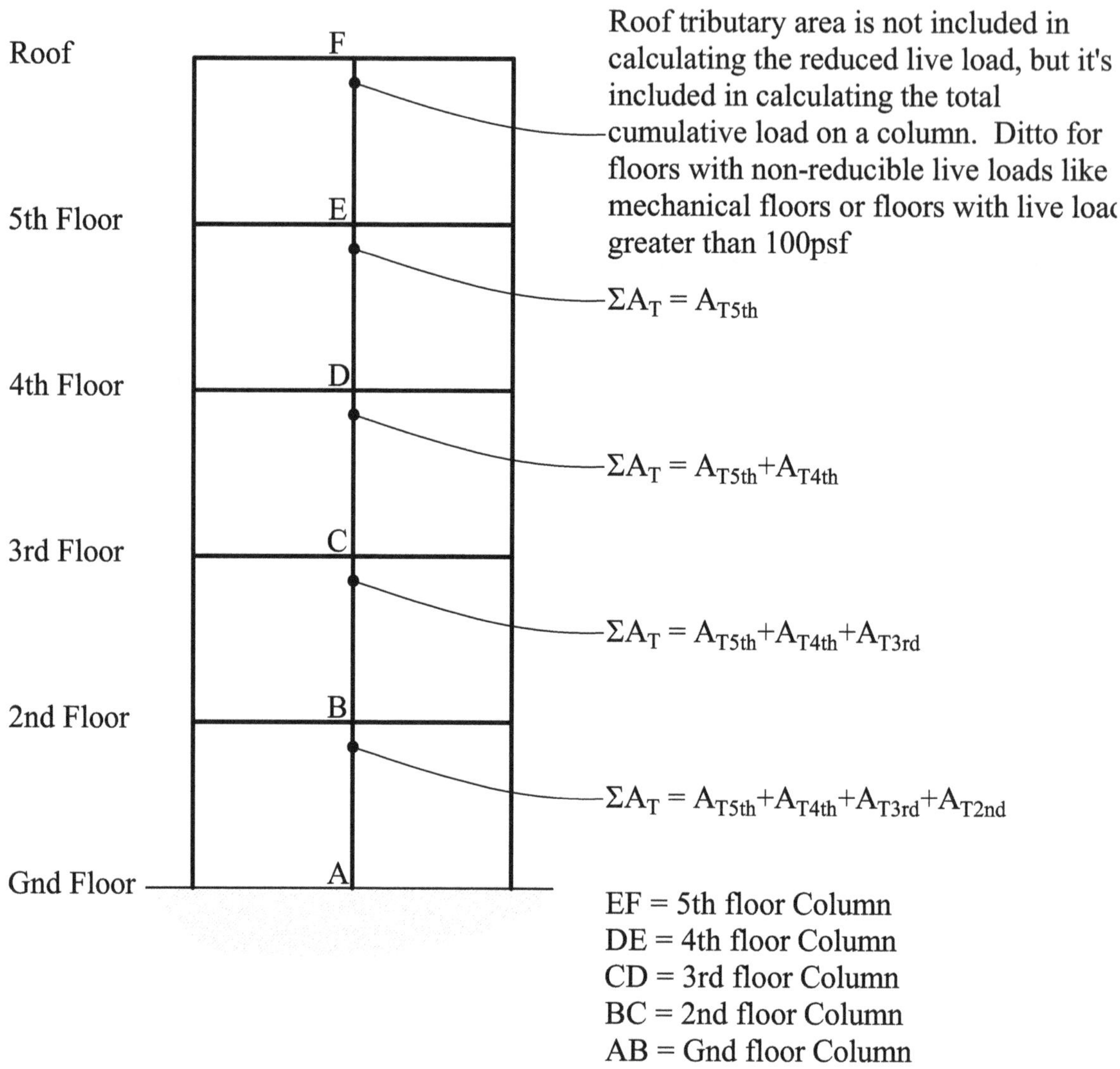

FIGURE 2-3b Definition of A_T or ΣA_T for multi-story columns

TABLE 2-4 Minimum uniformly distributed and concentrated floor live loads.

Minimum uniformly distributed and concentrated live loads		
Occupancy	**Uniform load (psf)**	**Concentrated load (lb.)**
Balconies Exterior One- and two-family residences only, and not exceeding 100 ft.^{2}	 100 60	 — —
Dining rooms and restaurants	100	—
Helipads	60	See ASCE 7
Libraries Corridors above first floor Reading rooms Stack rooms	 80 60 150	 1000 1000 1000

(continued)

Minimum uniformly distributed and concentrated live loads		
Occupancy	**Uniform load (psf)**	**Concentrated load (lb.)**
Office buildings		
Lobbies and first-floor corridors	100	2000
Offices	50	2000
Corridors above first floor	80	2000
Residential (one- and two-family dwellings)	40	—
Hotels and multifamily houses		
Private rooms and corridors serving them	40	—
Public rooms and corridors serving them	100	
Roofs		
Ordinary flat, pitched, and curved roofs	20	
Promenades	60	
Gardens or assembly	100	
Single panel point of truss bottom chord or at any point along a beam		2000
Schools		
Classrooms	40	1000
Corridors above first floor	80	1000
First-floor corridors	100	1000
Stairs and exit ways	100	300 lb. over an area of 4 in.^{2}
One- and two-family residences only	40	
Storage		
Light	125	
Heavy	250	
Stores		
Retail		
First floor	100	1000
Upper floors	75	1000
Wholesale	125	1000

-10 ROOF LIVE LOAD

Roof live loads, L_r, represent the weight of maintenance personnel and equipment on a roof, and this has a maximum value of 20 psf as indicated in ASCE 7 Table 4.3-1 and is reducible. Other types of roof live loads include the weight of moveable nonstructural elements such as planters or other decorative elements on a roof or the use of a roof for assembly purposes (the maximum live load is 100 psf). Like floor live loads, the unreduced roof live loads are also tabulated in the ASCE 7 Table 4.3-1. According to ASCE 7 Section 4.8.3, live loads on roofs that are used for special occupancies, such as roof gardens, can be reduced using equation (2-5).

A typical section through a tree planter in a roof garden includes the following components:

- The structural slab, which must be sloped or pitched for adequate drainage.
- A concrete topping slab layer on top of the structural slab.
- Waterproofing membrane on top of the topping slab.

- Root barrier layer that prevents tree roots from penetrating into the concrete topping and structural slabs.
- The drainage panel layer consists of lightweight aggregates.
- Filter fabric
- Wire mesh. Straps are attached to and looped around the wire mesh to anchor the tree down.
- The saturated soil medium can be as deep as 4 ft. and contains the tree root system.
- Mulch layer.

The live load for roof gardens is 100 psf, but the dead load, which consists of the weight of the components listed above (e.g., saturated soil, planters, and trees, etc.), can be as high as 600 psf or more.

Roofs that are used for assembly occupancies should be designed for a roof live load that is the same as that used for floors with the same occupancy and note that live loads for roofs that are used for assembly occupancies cannot be reduced.

However, roof live loads on ordinary flat, pitched, or curved roofs that represent the weight of maintenance personnel and equipment can be reduced as described in the following section.

Roof Live Load Reduction for Ordinary Flat, Pitched, and Curved Roofs

For ordinary flat, pitched, and curved roofs, the ASCE 7 Load Standard [2] allows the roof live load, L_o, to be reduced according to the following formulas (*note that for all other types of roofs, $L_r = L_o$*):

$$\text{Design roof live load, in psf, } L_r = L_0\, R_1\, R_2, \tag{2-6}$$

where,

$12 \text{ psf} \leq L_r \leq 20 \text{ psf}$, and

L_o = Roof live load from ASCE 7, Table 4.3-1 (typically 20 psf).

The roof live load for ordinary flat, pitched, and curved roofs is 20 psf.

The reduction factors, R_1 and R_2, in equation (2-6) are determined as follows:

$$\begin{aligned} R_1 &= 1.0 && \text{for } A_T \leq 200 \text{ ft.}^2 \\ R_1 &= 1.2 - 0.001A_T && \text{for } 200 \text{ ft.}^2 < A_T < 600 \text{ ft.}^2 \\ R_1 &= 0.6 && \text{for } A_T \geq 600 \text{ ft.}^2 \end{aligned}$$

$$\begin{aligned} R_2 &= 1.0 && \text{for F} \leq 4 \\ R_2 &= 1.2 - 0.05\text{F} && \text{for } 4 < \text{F} < 12 \\ R_2 &= 0.6 && \text{for F} \geq 12 \end{aligned}$$

F = Number of inches of rise per foot for a pitched or sloped roof

$(\text{e.g., F} = 3 \text{ for a roof with a 3-in-12 pitch or slope})$

= Rise-to-span ratio multiplied by 32 for an arch or dome roof

A_T = Tributary area in square feet (ft.^2)

For landscaped roofs, the weight of the landscaping material should be included in the dead load calculations and should be computed assuming the soil will be fully saturated.

Calculating the Gravity Loads on Beams, Girders, and Columns:
After obtaining the uniform dead load on a floor or roof in psf and the corresponding uniform live load in psf, the loads on the structural members are calculated as follows:

Beams:
After calculating a beam's tributary width, the unfactored uniform dead load on the beam in Ib/ft can be obtained by multiplying the tributary width by the psf uniform dead load. A similar calculation is carried out for the live loads. Use the LRFD load combinations to calculate the factored total load on the beam in pounds per linear foot (i.e., Ib/ft or plf), and then analyze the beam and determine its reactions, shears, and moments using structural analysis principles.

Girders:
The beam reactions become the loads on the girder that supports the beams. Note that for roof live loads, L_r, or in the case of floor live load reduction, where the live load on a structural element is dependent on the tributary area supported by that member, the beam live load reactions on the girder should be obtained from an analysis of the beams using the uniform roof live loads (or the reduced floor live loads in the case of floor live load reduction) corresponding to the tributary area of the girder. After obtaining the beam live load and dead load reactions on the girder, analyze the girders for those reactions and other loads directly supported by the girder, such as wall loads or hanging loads (if any), using the LRFD load combinations and structural analysis principles, and thus obtain the girder reactions, shears, and moments.

Columns in steel buildings:
There are two ways to calculate the axial load on a column:

1. **Using Tributary Area**: Calculate the factored psf load from the floor or roof using the LRFD load combinations, and then multiply the tributary area of the column by the factored psf load from the floor or roof to obtain the factored axial load on the column from that floor or roof level. For columns supporting multiple floors, see the column load run-down example in Example 2-13.
2. **Using Beam and Girder Reactions**: An alternate approach for calculating the factored axial load on a column from a floor or roof level is by summing up the factored reactions from the beams and girders framing into the column at that level.

EXAMPLE 2-12

Roof Live Load

For the ordinary flat roof framing of a one-story building shown in Figure 2-4, determine the design roof live load, L_r, for the following structural members neglecting snow or rain loads:

a. Typical interior beam,
b. Typical spandrel or perimeter beam,
c. Typical interior girder,
d. Typical spandrel girder, and
e. Typical interior column.
f. Assuming the roof dead load is 30 psf, calculate the design loads for all the five structural elements above using the LRFD load combinations from Section 2-3.

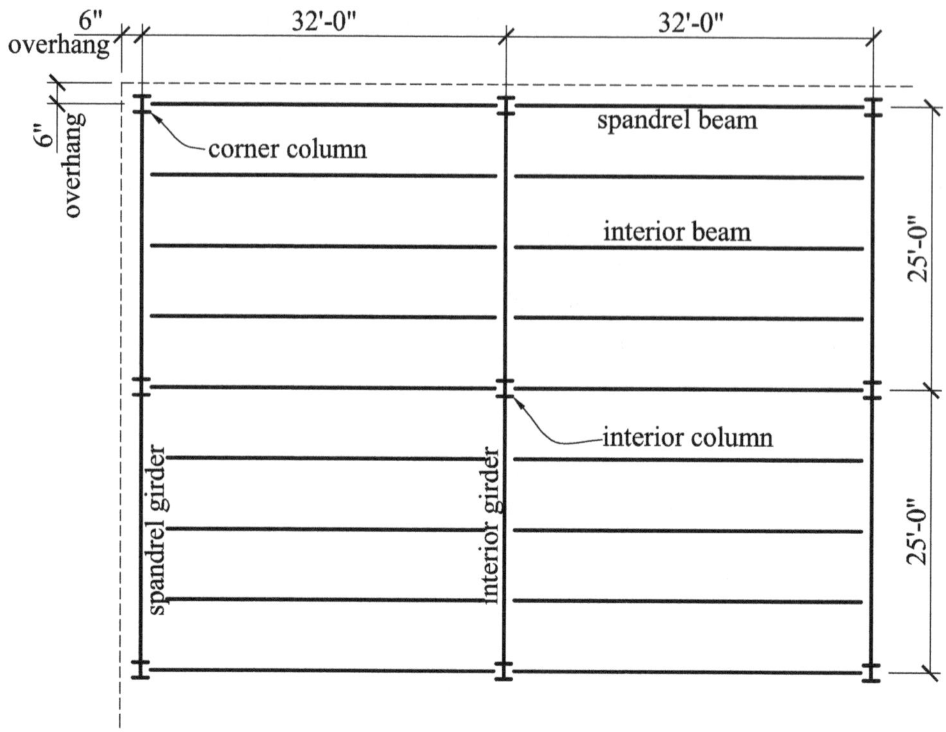

FIGURE 2-4 Roof framing for Example 2-12

Solution

The unreduced roof live load, L_o, is 20 psf from ASCE-7, Table 4.3-1, and the tributary width and tributary areas are calculated in Table 2-5.

TABLE 2-5 Tributary Widths and Areas

Member	Tributary width (TW)	Tributary area (A_T)
Typical interior beam	25 ft./4 spaces = 6.25 ft.	(6.25 ft.) (32 ft.) = 200 ft.^{2}
Typical spandrel beam	(25 ft./4 spaces)/2 + 0.5 ft. edge distance = 3.63 ft.	(3.63 ft.) (32 ft.) = 116 ft.^{2}
Typical interior girder	—	$\left(\frac{200}{2}\ \text{ft.}^2\right)(6\ \text{beams}) = 600\ \text{ft.}^2$
Typical spandrel girder	—	$(0.5\ \text{ft. edge dist})(25\ \text{ft.}) + \left(\frac{200}{2}\ \text{ft.}^2\right)(3\ \text{beams}) = 313\ \text{ft.}^2$
Typical interior column	—	(32 ft.) (25 ft.) = 800 ft.^{2}

a. Typical interior beam: Tributary area, A_T = 200 ft.^{2}, therefore, R_1 = 1.0
For a flat roof, F = 0, therefore, R_2 = 1.0
Using equation (2-6), the design roof live load, L_r = 20 (1.0) (1.0) = 20 psf

b. Typical spandrel or perimeter beam: Tributary area, A_T = 116 ft.^{2}, therefore, R_1 = 1.0
For a flat roof, F = 0, therefore, R_2 = 1.0
Design roof live load, L_r = 20 (1.0) (1.0) = 20 psf

c. Typical interior girder: Tributary area, A_T = 600 ft.^{2}, therefore, R_1 = 0.6
For a flat roof, F = 0; therefore, R_2 = 1.0
Design roof live load, L_r = 20 (0.6) (1.0) = 12 psf

d. Typical spandrel girder: Tributary area, A_T = 313 ft.^{2}, therefore, R_1 = 1.2 - 0.001 (313) = 0.89
For a flat roof, F = 0, therefore, R_2 = 1.0
Design roof live load, L_r = 20 (0.89) (1.0) = 17.8 psf

e. Typical interior column: Tributary area, A_T = 800 ft.^{2}, therefore, R_1 = 0.6
For a flat roof, F = 0, therefore, R_2 = 1.0
Design roof live load, L_r = 20 (0.6) (1.0) = 12 psf

f. Factored loads on the structural elements - Beams, Girders, and Columns:

Typical interior beam: total factored uniform load = [1.2(30 psf) + 1.6(20 psf)](6.25 ft) = 425 plf.
Typical spandrel or perimeter beam: total factored uniform load = [1.2(30 psf) + 1.6(20 psf)](3.63 ft) = 247 plf.

Typical interior girder:
The total uniform factored load in psf based on this girder tributary area = 1.2(30 psf) + 1.6(12 psf) = 55.2 psf.
The corresponding factored beam reaction on the girder due to this uniform load at each load point where the beam frames into the girder = (55.2 psf)(6.25 ft) x ½ (32 ft) = 5520 Ib. There are six (6) beam reactions supported by this girder (two reactions at each of the three load points). The total factored beam reaction supported by the typical interior girder at each load point will be 2 beams x 5520 Ib = 11,040 Ib.

Typical spandrel girder:
The total uniform factored load in psf based on this girder tributary area = 1.2(30 psf) + 1.6(17.8 psf) = 64.5 psf.
The corresponding factored beam reaction on the girder due to this uniform load at each load point where the beam frames into the girder = (64.5 psf)(6.25 ft) x ½ (32 ft) = 6450 Ib. Therefore, there will be three (3) beam reactions supported by this girder (one reaction at each of the three load points). The total factored beam reaction on the typical spandrel girder at each load point will be 6450 Ib.
In addition, there is a uniform load on the spandrel girder due to the 6 in. overhang = 64.5 psf (0.5 ft edge distance) = 32.3 plf.

Typical interior column:
The uniform factored load in psf based on the column tributary area = 1.2(30 psf) + 1.6(12 psf) = 55.2 psf.
The corresponding factored axial load on the typical interior column = (55.2 psf)(800 ft.^{2}) = 44, 160 = **44.2 kips.**

Alternatively, the axial load on the column can be obtained by summing up the beam and girder reactions that frame the typical interior column. Two typical interior girders and two typical interior beams frame this column. Calculate the beam and girder reactions using the 55.2 psf load above that is based on the column tributary area.

The corresponding factored interior beam reaction supported by the typical interior girder due to this uniform load at each load point where the beam frames into the girder = (55.2 psf)(6.25 ft) x ½ (32 ft) = 5520 Ib.
The corresponding factored interior girder reaction supported by the column due to the six beam reactions = (6)(5520)/2 = 16,560 Ib.

Since two typical interior girders and two typical interior beams frame into the typical interior column in Figure 2-4, the sum of all the interior beam and interior girder reactions = (2)(16,560 Ib) + (2)(5520 Ib) = 44,160 Ib = **44.2 kips**, which is the same column axial load obtained above using the column tributary area method.

EXAMPLE 2-13

Column Load Rundown (with and without Floor Live Load Reduction)

A three-story building, with columns that are spaced at 20 ft. in both orthogonal directions, is subjected to the roof and floor loads shown here. Using a tabular format, calculate the cumulative **factored** and unfactored axial loads on a typical interior column **with and without live load reduction**. Assume a minimum roof slope of $^1/_4$ in./ft. for drainage.

Roof Loads :

Dead load, $D_{\text{roof}} = 30$ psf

Snow load, $S = 48$ psf

Roof live load, L_r = per code

Second and Third Floor Loads :

Dead load, $D_{\text{floor}} = 110$ psf

Floor live load, $L = 40$ psf

Solution

At each level, the tributary area, A_T, supported by a typical interior column is 20 ft. × 20 ft. = 400 ft.^{2}

Roof Live Load, L_r:

For an ordinary flat roof, L_o = 20 psf (ASCE 7, Table 4.3-1).

From Section 2-10, the roof slope factor, F, is ¼; therefore, R_2 = 1.0.

Since the tributary area, A_T, of the column = 400 ft.^{2}, R_1 = 1.2 - 0.001 (400) = 0.8

Using equation (2-6), the design roof live load is

$$L_r = L_o R_1 R_2 = 20\ R_1 R_2 = 20(0.8)(1.0) = 16 \text{ psf},$$

Since 12 psf < L_r < 20 psf; therefore, L_r = 16 psf.

From the LRFD load combinations, the maximum factored roof live load, $1.6L_r$, is smaller than the factored snow load, $1.0S$ = 48 psf; therefore, the snow load, S, is more critical than the roof live load, L_r, in the applicable load combinations. The reduced or designed floor live loads (in psf) for the second and third floors are calculated using Table 2-6.

All other loads, such as W, H, T, F, R, and E, are zero for the roof and all the floors. The applicable *LRFD* load combinations from Section 2-3 will be used to calculate the factored column axial loads. Only load combinations 2 and 3 are pertinent for this column.

A close examination of these load combinations will confirm that the most critical factored load combinations (i.e., LRFD) are load combinations 2, ($1.2D + 1.6L + 0.3S$), and 3, ($1.2D + 1.0S + 0.5L$). A similar examination of the ASD load combinations reveals that load combinations

2, $(D + L)$, and 3, $(D + 0.75L + 0.75(0.7S))$, are the two most critical load combinations for the calculation of the unfactored axial load on the typical interior column for this building. We will now illustrate how some of the numbers in Table 2-7 were obtained for the case with floor live load reduction (i.e., the top three rows of Table 2-7).

Load combination 2 (LC2) - (1.2D + 1.6L + 0.3S):

For the case with *floor live load reduction*, the loads from load combination 2 or LC2 are now calculated:

For the *roof level*, the value of the factored load for LC2 is 1.2(30psf) + 0.3(48psf) = 50.4 psf.

For the *third floor* level, the value of the factored load for LC2 is 1.2(110psf) + 1.6(25psf) = 172 psf. (Note: 25 psf is the reduced floor live load at the third floor level).

For the *second floor* level, the value of the factored load for LC2 is 1.2(110 psf) + 1.6(21 psf) = 165.6 psf. (Note: 21 psf is the reduced floor live load at the second floor level).

The above psf loads are the values shown in the 6th column of Table 2-7

To obtain the axial load in kips for LC2, these factored uniform loads in psf are multiplied by the column tributary area as follows:

For the *roof level*, the factored axial load for LC2 = (50.4 psf)(400 ft^2) = 20.2 kips

For the *third floor level*, the factored axial load for LC2 = (172psf)(400 $ft.^2$) = 68.8 kips

For the *second floor level*, the factored axial load for LC2 = (165psf)(400 $ft.^2$) = 66 kips

The above are the left-hand values tabulated under the 8th column in Table 2-7.

Load combination 3 (LC3) – (1.2D + 0.5L + 1.0S):

For the case with *floor live load reduction*, we will now calculate the loads using load combination 3 or LC3:

For the *roof level*, the value of the factored load for LC3 is 1.2(30psf) + 1.0(48psf) = 84 psf.

For the *third floor level*, the value for LC3 is 1.2(110psf) + 0.5(25psf) = 144.5 psf.

For the *second floor level*, the value for LC3 is 1.2(110 psf) + 0.5(21psf) = 142.5 psf.

The above factored uniform loads are the values shown in the 7th column of Table 2-7

To get the axial load in kips for LC3, these psf loads are multiplied by the column tributary area as follows:

For the roof level, the factored axial load for LC3 = (84psf)(400 ft^2) = 33.6 kips.

For the third floor level, the factored axial load for LC3 = (144.5 psf)(400 ft.^{2}) = 57.8 kips.

For the second floor level, the factored axial load for LC3= (142.5 psf)(400 ft.^{2}) = 57 kips.

The above are the right-hand values tabulated under the 8th column in Table 2-7.

Using both LRFD combinations 2 (i.e., $1.2D + 1.6L + 0.3S$) and 3 ($1.2D + 1.0S + 0.5L$), the maximum factored column axial loads with and without floor live load reductions are calculated in Table 2-7. The corresponding values for the unfactored column axial loads with and without floor live load reductions are calculated in Table 2-8. Tables 2-7 and 2-8 are known in design practice as the **column load summation table** or the **column load rundown table**.

Therefore, the ground floor column will be designed for a cumulative reduced factored axial compression load of 152.6 kips; the corresponding factored axial load without floor live load reduction is 177 kips. The reduction in factored column axial load due to floor live load reduction is only 14% for this three-story building. Thus, the effect of floor live load reduction on columns and column footings is not as critical for low-rise buildings as it would be for taller buildings.

Although in the previous discussions and example we used the tributary area method for calculating the column loads, the column load can also be determined, as indicated earlier, by summing the reactions of all the beams and girders that frame into the column. This method will prove to be useful later when the columns must be designed for combined axial loads and bending moments due to the eccentricity of the beam and girder reactions relative to the centroid of the supporting columns.

TABLE 2-6 Reduced or Design Floor Live Load Calculation Table

Member	Levels supported	A_T (summation of floor tributary area)	K_{LL}	Unreduced floor live load, L_o (psf)	Live load reduction factor, $0.25 + 15/\sqrt{K_{LL}A_T}$	Design live load, (L or S)
Third-floor column (i.e., column below roof)	Roof only	Floor live load reduction NOT applicable to roofs	—	—	—	48 psf (snow load)
Second-floor column (i.e., column below third floor)	1 Floor + Roof	(1 Floor) (400 ft.^{2}) = 400 ft.^{2}	$K_{LL} = 4$; $K_{LL}A_T$ = (4)(400) = 1600 > 400 ft.^{2} Therefore, live load reduction is allowed	40 psf	$\left[0.25 + \frac{15}{\sqrt{1600}}\right]$ = 0.625	0.625 (40) = **25 psf** ≥ 0.50 L_o = *(0.5)(40psf) = 20 psf*
Ground or first-floor column (i.e., column below second floor)	2 Floors+ Roof	(2 Floors) (400 ft.^{2}) = 800 ft.^{2}	$K_{LL} = 4$; $K_{LL}A_T$ = (4)(800) = 3200 > 400 ft.^{2} Therefore, live load reduction is allowed	40 psf	$\left[0.25 + \frac{15}{\sqrt{3200}}\right]$ = 0.52	0.52 (40) =**20.8 psf** ≈ **21 psf** ≥ 0.40 L_o = (0.4)(40psf) = 16 psf

TABLE 2-7a Factored column load summation table (see also Tables 2-7b and 2-7c).

1 **Level**	2 **Tributary area, (ft.^{2}) A_T**	3 **Dead load, D (psf)**	4 **Live load, L_o (S or L_r or R on the roof) (psf)**	5 **Design live load floor: L Roof: S or L_r or R (psf)**	6 **Factored uniform load at each level, w_{u1} Roof: 1.2D + 0.3S Floor: 1.2D + 1.6L (psf) [LRFD Load Combo #2]**	7 **Factored uni-form load at each level, w_{u2} Roof: 1.2D + 1.0S Floor: 1.2D + 0.5L (psf) [LRFD Load Combo #3]**	8 **Factored column axial load, P, at each level, (A_T) (w_{u1}) or (A_T) (w_{u2}) (kips)**	9 **Cumulative factored axial load, ΣP LC2 (kips)**	10 **Cumulative factored axial load, ΣP LC3 (kips)**	11 **Maximum cumulative factored axial load, ΣP (kips)**
With Floor Live Load Reduction										
Roof	400	30	48	48	50.4	84	20.2 or 33.6	20.2	33.6	**33.6**
Third floor	400	110	40	25	172	144.5	68.8 or 57.8	89	91.4	**91.4**
Second floor	400	110	40	20.8	**165.6**	**142.5**	*See foot-note	153*	148*	152.6
Without Floor Live Load Reduction										
Roof	400	30	30	30	50.4	84	20.2 or 33.6	20.2	33.6	**33.6**
Third floor	400	110	40	40	196	152	78.4 or 60.8	98.6	94.4	**98.6**
Second floor	400	110	40	40	196	152	78.4 or 60.8	177	156	**177**

*For the column segment below the second floor (i.e., the ground floor column), the reduced floor live load at the *second floor level* applies to all the tributary areas of all the floors (with reducible live load) supported by this column (i.e., both the second and third floors in this case). Therefore, the total cumulative reduced factored axial loads in the ground floor column (i.e., the column below the *second floor level*) are calculated as follows:

ΣP_{LC2} = 50.4 psf (400 ft.^{2}) + **165.6 psf** (400 ft.^{2} + 400 ft.^{2}) = 152.6 kips, and
ΣP_{LC3} = 84 psf (400 ft.^{2}) + **142.5 psf** (400 ft.^{2} + 400 ft.^{2}) = 148 kips.
The maximum *factored* column loads (***with*** floor live load reduction) are as follows:
Third-story column (i.e., column below roof level) = 33.6 kips.
Second-story column (i.e., column below the third floor) = 91.4 kips.
First-story column (i.e., column below the second floor) = 152.6 kips.
The maximum factored axial column loads (***without*** floor live load reduction) are as follows:
Third-story column (i.e., column below roof level) = 33.6 kips.
Second-story column (i.e., column below the third floor) = 98.6 kips.
First-story column (i.e., column below the second floor) = 177 kips.

Table 2-7b (LRFD) — Live load reduction: Columns #6 and #7 of Table 2-7a Explained.

	Column #6: 1.2D + 0.5S + 1.6L	**Column #6: 1.2D + 1.6S + 0.5L**
Roof level D = 30 psf S = 30 psf TA (column) = 400 ft^2	**1.2D + 0.3S** = 51 psf ΣP_{roof} = 0.051 ksf × 400 = 20.4 kips	**1.2D + 1.0S** = 84 psf ΣP_{roof} = 0.084 ksf × 400 ft^2 = **33.6 kips**
3rd Floor level D = 110 psf L^ = 25 psf åTA (column) = 400 ft^2 (3rd flr only)	**1.2D + 1.6L^** = 172 psf ΣP_{3rd} = 20.4 kips + 0.172 ksf × 400 ft^2 = 89.2 kips	**1.2D + 0.5L^** = 144.5 psf ΣP_{3rd} = 33.6 kips + 0.1445 ksf × 400 ft^2 = **91.4 kips**
2nd Floor level D = 110 psf L* = 20.8 psf 21 psf ΣTA (column) = 400 ft^2 + 400 ft^2 = 800 ft^2 (2nd and 3rd floors)	**1.2D + 1.6L*** = 165.6 psf ΣP_{2nd} = 20.4 kips + 0.1656 ksf × 800 ft^2 = **153 kips**	**1.2D + 0.5L*** = 142.5 psf ΣP_{2nd} = 33.6 kips + 0.1425 ksf × 800 ft^2 = 148 kips

^ The reduced live load at the 3rd floor level applies ONLY to the 3rd floor.
* The reduced live load at the 2nd floor level applies to both the 2nd floor and 3rd floor!

In general, the reduced floor live load calculated at any floor level applies to that particular floor level plus ALL floor levels above—for which the floor live load is allowed to be reduced per ASCE 7 Load Standard.

Table 2-7c (LRFD) — No live load reduction: Columns #6 and #7 of Table 2-7a Explained.

	Column #6: 1.2D + 0.5S + 1.6L	**Column #6: 1.2D + 1.6S + 0.5L**
Roof level D = 30 psf S = 30 psf TA (column) = 400 ft^2	**1.2D + 0.3S** = 1.2(30) + 0.3(48) = 51 psf ΣP_{roof} = 0.051 ksf x 400 = 20.4 kips	**1.2D + 1.0S** = 1.2(30) + 1.0(48) = 84 psf ΣP_{roof} = 0.084 ksf x 400 ft^2 = **33.6 kips**
3rd Floor level D = 110 psf L = 40 psf ΣTA (column) = 400 ft^2 (3rd flr only)	**1.2D + 1.6L** = 1.2(110) + 1.6(40) =196 psf ΣP_{3rd} = 20.4 kips + 0.196 ksf x 400 ft^2 = 20.4 + 78.4 = **98.8 kips**	**1.2D + 0.5L** = 1.2(110) + 0.5(40) = 152 psf åP_{3rd} = 33.6 kips + 0.152 ksf x 400 ft^2 = = 33.6 + 60.8 = 94.4 kips
2nd Floor level D = 110 psf L = 40 psf ΣTA (column) = 400 ft^2 (2nd flr only)	**1.2D + 1.6L** = 1.2(110) + 1.6(40) =196 psf ΣP_{2nd} = 98.8 kips + 0.196 ksf x 400 ft^2 = **177 kips**	**1.2D + 0.5L** = 1.2(110) + 0.5(40) = 152 psf ΣP_{2nd} = 94.4 kips + 0.152 ksf x 400 ft^2 = 156 kips

Note:
When there is no live load reduction, the tributary area at each level supported by a column is calculated separately, and the tributary loads contributed from all levels above the column are summed up cumulatively to determine the total cumulative load on the column.

TABLE 2-8a Unfactored column load summation table (see also Tables 2-8b and 2-8c).

1 Level	2 Tributary area, (A_T) (ft.^{2})	3 Dead load, D (psf)	4 Live load, L_o (S or L_r or R on the roof) (psf)	5 Design live load roof: S or L_r or R; Floor: L (psf)	6 Unfactored total load at each level, w_{s1}; Roof: D; Floor: D + L(psf) [ASD Load Combo #2]	7 Unfactored total load at each level, w_{s2}; Roof: D + 0.75(0.7S); Floor: D + 0.75L (psf) [ASD Load Combo #4]	8 Unfactored column axial load at each level, $P = (A_T)(w_{s1})$ or $(A_T)(w_{s2})$ (kips)	9 Cumulative unfactored axial load, ΣP_{D+L} (kips)	10 Cumulative unfactored axial load, $\Sigma P_{D+0.75L+0.75(0.7S)}$ (kips)	11 Maximum cumulative unfactored axial load, ΣP (kips)
With Floor Live Load Reduction										
Roof	400	30	48	48	30	55.2	12 or 22.1	12	22.1	**25.4^ ^**
Third floor	400	110	40	25	135	128.8	54 or 51.5	66	73.6	**73.6**
Second floor	400	110	40	21	**131**	**125.6**	*See footnote	116.8*	122.6*	122.6
Without Floor Live Load Reduction										
Roof	400	30	30	48	30	55.2	12 or 22.1	12	22.1	**25.4^ ^**
Third floor	400	110	40	40	150	140	60 or 56	72	78.1	**78.1**
Second floor	400	110	40	40	150	140	60 or 56	132	134.1	**134.1**

*For the column segment below the second floor (i.e., the ground floor column), the reduced floor live load at the *second floor level* applies to all the tributary areas of all the floors (with reducible live loads) supported by the column (i.e., both the second and third floors in this case). Therefore, the total cumulative unfactored reduced axial loads in the ground floor column (i.e., the column below the *second floor level*) are calculated as follows:

ΣP_{D+L} = 30 psf (400 ft.^{2}) + **131 psf** (400 ft.^{2} + 400 ft.^{2}) = 116.8 kips, and
$\Sigma P_{D+0.75L+0.75(0.7S)}$ = 55.2 psf (400 ft.^{2}) + **125.6 psf** (400 ft.^{2} + 400 ft.^{2}) = 122.6 kips.
The maximum unfactored column loads (***with*** *floor live load reduction*) are as follows:
Third-story column (i.e., column below roof level) = 25.4 kips. (ASD load combination 3)
Second-story column (i.e., column below the third floor) = 73.6 kips.
First-story column (i.e., column below the second floor) = 122.6 kips.
The maximum unfactored axial column loads (***without*** *floor live load reduction*) are
^ ^ ^ ^ Third-story column (i.e., column below roof level) = (400)(30psf + (0.7)(48psf)) = 25.4 kips. (ASD load combination 3)
Second-story column (i.e., column below the third floor) = the higher of (22.1 + 56) or (12 + 60) = 78.1 kips.
First-story column (i.e., column below the second floor) = the higher of (78.1 + 56) or (72 + 60) = 134.1 kips

Table 2-8b (ASD)—Live load reduction: Columns #6 and #7 Table 2-8a Explained.

	Column #6: D + L	**Column #7: D + 0.75(0.7S) + 0.75L**	**Additional Load Combo: D + 0.7S**
Roof level D = 30 psf S = 30 psf TA (column) = 400 ft^2	**D** = 30 psf ΣP_{roof} = 0.030 ksf × 400 = 12 kips	**D + 0.75(0.7S)** = 30 + 0.75(0.7)(48) = 55.2 psf ΣP_{roof} = 0.0552 ksf × 400 ft^2 = 22.1 kips	**D + 0.7S** = 30 + 0.7(48) = 63.6 psf ΣP_{roof} = 0.0636 ksf × 400 ft^2 = **25.4 kips**
3rd Floor level D = 110 psf L^ = 25 psf ΣTA (column) = 400 ft^2 (3rd flr only)	**D + L^** = 110 + 25 = 135 psf ΣP_{3rd} = 12 kips + 0.135 ksf × 400 ft^2 = 66 kips	**D + 0.75L^** = 110 + 0.75(25) = 128.8 psf ΣP_{3rd} = 22.1 kips + 0.1288 ksf × 400 ft^2 = **73.6 kips**	**D** = 110 psf ΣP_{3rd} = 25.4 kips + 0.110 ksf x 400 ft^2 = 69.4 kips
2nd Floor level D = 110 psf L* = 20.8 psf » 21 psf ΣTA (column) = 400 ft^2 + 400 ft^2 = 800 ft^2 (2nd and 3rd floors)	**D + L^** = 110 + 21 = 131 psf ΣP_{2nd} = 12 kips + 0.131 ksf × 800 ft^2 = 116.8 kips	**D + 0.75L^** = 110 + 0.75(21) = 125.8 psf ΣP_{2nd} = 22.1 kips + 0.1258 ksf × 800 ft2 = **122.6 kips**	**D** = 110 psf ΣP_{2nd} = 25.4 kips + 0.110 ksf × 800 ft^2 = 113.4 kips

^ The reduced live load at the 3rd floor level applies ONLY to the 3rd floor.
* The reduced live load at the 2nd floor level applies to both the 2nd floor and 3rd floor!

In general, the reduced floor live load calculated at any floor level applies to that particular floor level plus ALL floor levels above—for which the floor live load is allowed to be reduced per ASCE 7 Load Standard.

Table 2-8c. (ASD) – No Live Load Reduction: Columns #6 and #7 Explained.

	Column #6: D + L	**Column #7: D + 0.75(0.7S) + 0.75L**	**Additional Load Combo: D + S**
Roof level D = 30 psf S = 30 psf TA (column) = 400 ft^2	**D** = 30 psf ΣP_{roof} = 0.030 ksf × 400 = 12 kips	**D + 0.75(0.7S)** = 30 + 0.75(0.7)(48) = 55.2 psf ΣP_{roof} = 0.0552 ksf × 400 ft^2 = 22.1 kips	**D + 0.7S** = 30 + 0.7(48) = 63.6 psf ΣP_{roof} = 0.0636 ksf × 400 ft^2 = **25.4 kips**
3rd Floor level D = 110 psf L = 40 psf ΣTA (column) = 400 ft^2 (3rd flr only)	**D + L** = 110 + 40 = 150 psf ΣP_{3rd} = 12 kips + 0.150 ksf × 400 ft^2 = 72 kips	**D + 0.75L** = 110 + 0.75(40) = 140 psf ΣP_{3rd} = 22.1 kips + 0.140 ksf × 400 ft^2 = **78.1 kips**	**D** = 110 psf ΣP_{3rd} = 25.4 kips + 0.110 ksf × 400 ft^2 = 69.4 kips
2nd Floor level D = 110 psf L = 40 psf ΣTA (column) = 400 ft^2 (2nd flr only)	**D + L** = 110 + 40 = 150 psf ΣP_{2nd} = 72 kips + 0.150 ksf × 400 ft^2 = 132 kips	**D + 0.75L^** = 110 + 0.75(40) = 140 psf ΣP_{2nd} = 78.1 kips + 0.140 ksf × 400 ft^2 = **134.1 kips**	**D** = 110 psf ΣP_{2nd} = 69.4 kips + 0.110 ksf × 400 ft^2 = 113.4 kips

Note:
When there is no live load reduction, the tributary area at each level supported by a column is calculated separately, and the tributary loads contributed from all levels above the column are summed up cumulatively to determine the total cumulative load on the column.

Additional Vertical Loads (Compression or Tension) on the Columns of LFRS due to Overturning Moments from Lateral Wind and Seismic Loads

Overturning moments on the lateral force resisting systems (LFRS) will result in additional axial loads on the columns at each level and at the base of the columns of the LFRS. Using the LRFD load combinations, the additional axial tension load and additional axial compression load can be calculated from the overturning moments by summing moments about one of the columns of the LFRS.

From the overturning moment taken about the ground floor level or the base of the column, the equivalent upward and downward vertical loads caused by lateral wind and seismic loads can be obtained. Similarly, using the overturning moment calculated at the 2nd floor level, the equivalent upward and downward vertical loads on the columns caused by lateral wind and seismic loads at the 2nd floor level can also be obtained. See the free body diagrams (FBD) in Figure 2-4b.

For illustration purposes, consider the two-story 30-ft. tall X-braced frame shown below, with 15 ft. floor-to-floor height; we will now proceed to calculate the additional factored axial tension and compression forces in the columns of the lateral force resisting system (LFRS) at the 2nd floor level and at the base of the columns due to the lateral loads, F_{roof} and F_2, using the LFRD load combinations as shown below (assuming $f_1 = 1$).

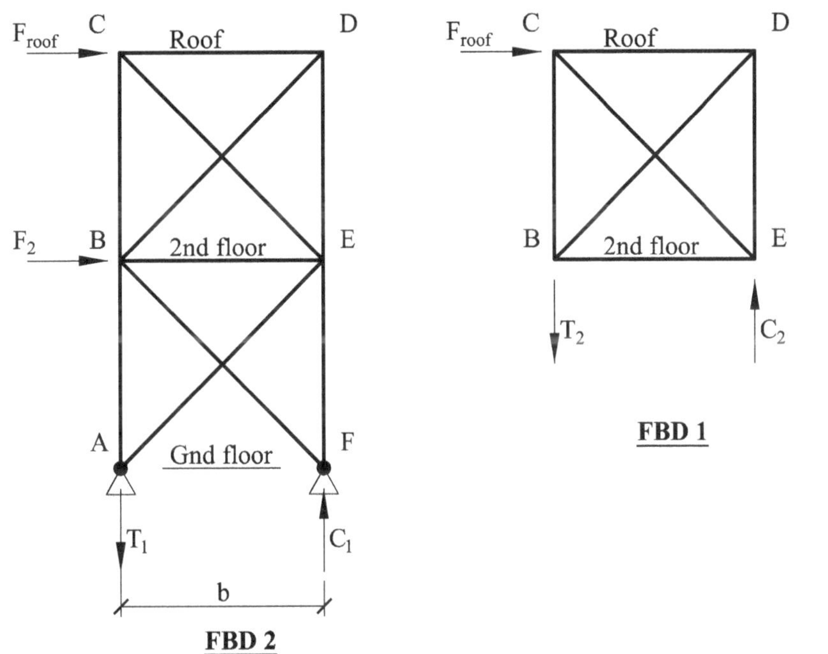

FIGURE 2-4b Vertical forces on columns and column bases from overturning moments due to the lateral forces on the LFRS.

The lateral forces in the two story example in Figure 2-4b, F_{roof} and $F_{2nd\ flr}$ (or F_2) can be caused by wind load W or by horizontal earthquake load, E_h. Note that in the load combinations, W or E_h will correspond to the T_1 and T_2 values, and the C_1 and C_2 values.

To calculate the tension and compression forces in LFRS columns and column bases caused by the lateral loads only, we consider the two free body diagrams as follows:

FBD 1:

The overturning moment at the 2nd floor level, $OM_{2ndflr}=15F_{roof}$

Summing moments about point E, the vertical tension force in the column is calculated as follows:

$$T_2 = -\frac{OM_{2ndflr}}{b} = -\frac{15F_{roof}}{b}$$

Summing moments about point B, the vertical compression force in the column is calculated as follows:

$$C_2 = +\frac{OM_{2ndflr}}{b} = +\frac{15F_{roof}}{b}$$

FBD 2:

The overturning moment at the ground floor level, $OM_{grdflr} = 30F_{roof} + 15F_{2nd\,flr}$

Summing moments about point F, the vertical tension force in the column is calculated as follows:

$$T_1 = -\frac{OM_{grdflr}}{b} = -\frac{(30F_{roof} + 15F_{2nd\,flr})}{b}$$

Summing moments about point A, the vertical compression force in the column is calculated as follows:

$$C_1 = +\frac{OM_{grdflr}}{b} = +\frac{(30F_{roof} + 15F_{2nd\,flr})}{b}$$

The LRFD load combinations to be considered are as follows:

Wind Load Combinations:

- **LC 4a: 1.2D + 1.0W + L + 0.3S**
  In this load combination, all downward acting loads like D, L, and S are *positive*, i.e., +ve, and the upward-acting reaction due to the overturning moment from the lateral wind loads is also *positive* (+ve) because it indicates a compression or downward-acting force in the column).
  For this load combination, the column axial loads at the ground floor and at the 2nd floor levels will be calculated, where W in the load combination is replaced by the compression reaction (which indicates a downward acting force in the column) caused by the overturning moments from lateral wind loads, as follows:
  - Ground Floor Column Axial Load: **1.2D + 1.0W + L + 0.3S = 1.2D + 1.0C_1 + L + 0.3S**
  - 2nd Floor Column Axial Load: **1.2D + 1.0W + L + 0.3S = 1.2D + 1.0C_2 + L + 0.3S**
- **LC 5a: 0.9D + 1.0W**
  In this load combination, all downward acting loads like D are *positive*, i.e., +ve, and the downward-acting reaction due to the overturning moment from the lateral wind loads is *negative* (-ve) because it indicates a tension or upward-acting force in the column.
  For this load combination, the column axial loads at the ground floor and at the 2nd floor levels will be calculated, where W in the load combination is replaced by the tension reaction (which indicates an upward acting force in the column) caused by the overturning moments from lateral wind loads, as follows:
  - Ground Floor Column Axial Load: **0.9D + 1.0W = 0.9D + 1.0T_1**
    (Note that T_1 as shown in FBD2 is negative. See the free body diagram in Figure 2-4b)

- 2nd Floor Column Axial Load: **0.9D +1.0W = 0.9D + 1.0T_2**
  (Note that T_2 as shown in FBD1 is negative. See the free body diagram in Figure 2-4b)

Seismic Load Combinations:

- **LC 6: 1.2D + E_v + E_h + L + 0.15S**
  In this load combination, all downward acting loads like D, L, and S are *positive* (+ve) and E_v = 0.2S_{DS}D is also *positive*; and the upward-acting reaction due to the overturning moment from the lateral seismic loads, E_h, is also *positive* (+ve) because it indicates a compression or downward-acting force in the column).
  For this load combination, the column axial loads at the ground floor and at the 2nd floor levels will be calculated, where E_h in the load combination is replaced by the compression reaction (which indicates a downward-acting force in the column) caused by the overturning moments from seismic loads, as follows:
  - Ground Floor Column Axial Load: **1.2D + E_v + E_h + L + 0.15S = 1.2D + 0.2S_{DS}D + C_1 + L + 0.15S** = **(1.2 + 0.2S_{DS})D + C_1 + L + 0.15S**
  - 2nd Floor Column Axial Load: **1.2D + E_v + E_h + L + 0.15S** = **(1.2 + 0.2S_{DS})D + C_2 + L + 0.15S**
- **LC 7: 0.9D – E_v + E_h**
  In this load combination, all downward acting loads like D are *positive*, and the downward-acting reaction due to the overturning moment from the seismic loads are *negative*). (E_v = 0.2S_{DS}D is negative, and the vertical reaction due to the overturning moment from the lateral seismic loads, E_h, is also *negative*, -ve).
  For this load combination, the column axial loads at the ground floor and at the 2nd floor levels will be calculated, where E_h in the load combination is replaced by the tension reaction (which indicates an upward-acting force in the column) caused by the overturning moments, as follows:
  - Ground Floor Column Axial Load: **0.9D - E_v + E_h** = **(0.9 - 0.2S_{DS})D + T_1**
    (Note that T_1 as shown in FBD2 is negative. See the free body diagram in Figure 2-4b)
  - 2nd Floor Column Axial Load: **0.9D - E_v + E_h** = **(0.9 - 0.2S_{DS})D + T_2**
    (Note that T_2 as shown in FBD1 is negative. See the free body diagram in Figure 2-4b)

After the net compression and net tension forces are obtained, the following design checks should be carried out:

1. Are the maximum factored downward axial loads (i.e., compressive loads) in the LFRS columns calculated in Load Combinations 4 and 6 larger than the previously calculated column compression load due to gravity loads only? If so, the columns have to be designed for the higher compression loads.
2. Given that the reactions T_1 and T_2 are acting downwards and therefore causes a tension or negative force in the columns (see the free body diagrams in Figure 2-4b), determine if there is a net tension force on the LFRS columns or a **net uplift force** at the base of the ground floor column from Load Combinations 5a and 7. If so, that could indicate the following:
   a. The column, its base plates, anchor rods, and the column foundation must be designed for any **net uplift** or net tension load.

b. If there is a **net uplift** at a LFRS column base, appropriate anchor rods that can resist uplift forces must be specified. See Figures 8-35 and 8-41 in Chapter 8 for the types of anchor rods that can be used at columns with net uplift load.

EXAMPLE 2-14

For the braced frame in Figure 2-4b, assuming a corner column with a tributary area of 247 ft^2 and the following loads, determine the maximum net factored uplift and the maximum compression load on the column of the X-braced frame using the LRFD load combinations (see Section 2-3). Assume the width of the braced frame, b = 24 ft, and f_1 = 1.0.

Wind Loads:

F_{roof} = 13 kips
F_2 = 22 kips

Seismic Loads:

F_{roof} = 16 kips
F_2 = 20 kips
S_{DS} = 0.18

Dead Loads:

Roof: 30 psf
2nd Floor: 85 psf

Roof Snow Load:

Roof: 40 psf

Floor Live Load:

2nd Floor: 50 psf

Solution

P_D (roof) = 30 psf (247 ft^2) = 7,410 Ib = 7.4 kips
P_D (2nd floor) = 85 psf (247 ft^2) = 20,995 Ib = 21 kips
P_L (2nd floor) = 50 psf (247 ft^2) = 12,350 Ib = 12.4 kips
P_S (roof) = 40 psf (247 ft^2) = 9,880 Ib = 9.9 kips

Wind Load Effects:

The overturning moment at the 2nd floor level due to wind loads only is

$$OM_{2ndflr} = 15F_{roof} = (15\,ft)(13\,kips) = 195\ ft.kips$$

$$T_2 = -\frac{OM_{2ndflr}}{b} = -\frac{195}{24} = -8.13\,kips$$

$$C_2 = +\frac{OM_{2ndflr}}{b} = +\frac{195}{24} = 8.13\,kips$$

The overturning moment at the ground floor level due to wind loads only is

$$OM_{grdflr} = 30F_{roof} + 15F_{2nd\,flr} = (30\,ft)(13\,kips) + (15\,ft)(22\,kips) = 720\ ft.kips$$

Summing moments about point F (i.e., the base of the column in tension), the vertical tension force in the column is calculated as follows:

$$T_1 = -\frac{OM_{grdflr}}{b} = -\frac{720}{24} = -30\,kips$$

Summing moments about point A (i.e., the base of the column in compression), the vertical compression force in the column is calculated as follows:

$$C_1 = +\frac{OM_{grd\,flr}}{b} = +\frac{720}{24} = 30\,kips$$

Load Combination 4a:

2nd Floor Column:
1.2D + 1.0W + L + 0.3S = 1.2(7.4 kips) + 1.0(8.13 kips) + 0 kips + 0.3(9.9 kips) = **20 kips** (net compression in the column)

Ground Floor Column Axial Load:

1.2D + 1.0W + L + 0.3S = 1.2(7.4 kips + 21 kips) + 1.0(30 kips) + 12.4 kips + 0.3(9.9 kips) = **79.5 kips** (net compression in the column)

Load Combination 5a:

2nd Floor Column Axial Load:
0.9D + W = 0.9(7.4 kips) - (8.13 kips) = **-1.5 kips** (net tension in column)

Ground Floor Column Axial Load:
0.9D + W = 0.9(7.4 kips + 21 kips) - (30 kips) = **-4.5 kips** (net tension in column and net uplift on column base plate)

Seismic Load Effects:

The overturning moment at the 2nd floor level due to seismic loads only is

$$OM_{2ndflr} = 15F_{roof} = (15\,ft)(16\,kips) = 240\ ft.kips$$

$$T_2 = -\frac{OM_{2ndflr}}{b} = -\frac{240}{24} = -10\,kips$$

$$C_2 = +\frac{OM_{2ndflr}}{b} = +\frac{240}{24} = 10\,kips$$

The overturning moment at the ground floor level due to seismic loads only is

$$OM_{grdflr} = 30F_{roof} + 15F_{2nd\,flr} = (30\,ft)(16\,kips) + (15\,ft)(20\,kips) = 780\ ft.kips$$

Summing moments about point F, the vertical tension force in the column is calculated as follows:

$$T_1 = -\frac{OM_{grdflr}}{b} = -\frac{780}{24} = -32.5\,kips$$

Summing moments about point A, the vertical compression force in the column is calculated as follows:

$$C_1 = +\frac{OM_{grdflr}}{b} = +\frac{780}{24} = 32.5\,kips$$

Load Combination 6:

2nd Floor Column Axial Load:
1.2D + E_v + E_h + L + 0.15S = 1.2D + 0.2S_{DS}D + E_h + L + 0.15S
= 1.2(7.4 kips) + (0.2)(0.18)(7.4 kips) + (10 kips) + 0 kips + 0.15(9.9 kips) = **20.6 kips** (net compression)

Ground Floor Column Axial Load:

$1.2D + E_v + E_h + L + 0.15S = 1.2D + 0.2S_{DS}D + E_h + L + 0.15S$
$= 1.2(7.4 \text{ kips} + 21 \text{ kips}) + (0.2)(0.18)(7.4 \text{ kips} + 21 \text{ kips}) + (32.5 \text{ kips}) + 12.4 \text{ kips} + 0.15(9.9 \text{ kips})$ = **81.5 kips** (net compression)

Load Combination 7:

2nd Floor Column Axial Load:

$0.9D - E_v + E_h = 0.9D - 0.2S_{DS}D + E_h = (0.9 - 0.2S_{DS})D + E_h$
$= [(0.9 - (0.2)(0.18)](7.4 \text{ kips}) - (10 \text{ kips})$ = **-3.6 kips** (net tension)

Ground Floor Column Axial Load:

$0.9D - E_v + E_h = 0.9D - 0.2S_{DS}D + E_h = (0.9 - 0.2S_{DS})D + E_h$
$= [(0.9 - (0.2)(0.18)](7.4 \text{ kips} + 21 \text{ kips}) - (32.5 \text{ kips})$ = **- 8.0 kips** (net tension in column and net uplift on column base plate)

The maximum net uplift force at the base of the column = **8 kips**
The maximum compression load at the base of the column = **81.5 kips**

The ground floor columns of the X-braced frame in Figure 2-4b would have to be designed for a **compression load of 81.5 kips**, while the column and its base/foundation will have to be designed for a **tension/net uplift load of 8 kips**. The reader should verify that the maximum compression load from Load Combinations 1 through 3 (i.e., gravity load combinations only) from Section 2-3 is **59 kips**, which is much less than the maximum compression load calculated above, considering the lateral loads. Therefore, the lateral loads on the LFRS in buildings (e.g., wind, seismic, and lateral pressure) can lead to an increase in compression forces in some of the columns of the LFRS, and net tension/uplift forces in other columns.

EXAMPLE 2-15

The typical frame in Figure 2-4c below is part of a series of building frames that cantilevers over an existing railroad track. Given the gravity and lateral wind loads acting on the typical frame, calculate the net uplift force at support A considering all the applicable LRFD load combinations. Assume $f_1 = 0.5$.

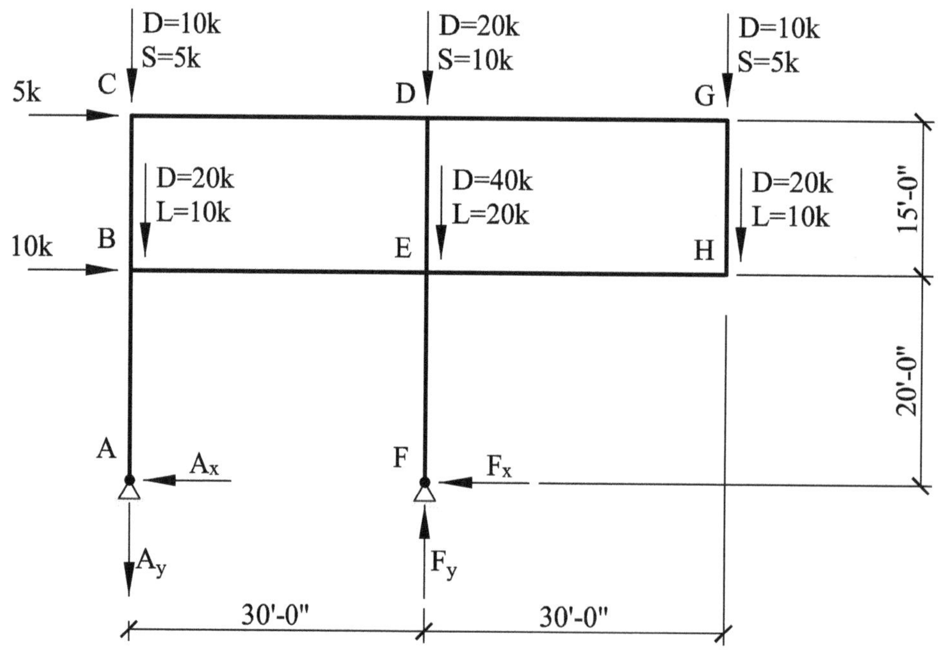

FIGURE 2-4c Cantilever frame

Solution

Note that A_x and F_x reactions cannot be determined from equilibrium equations alone, but we can at least calculate the A_y and F_y reactions by summing moments about point F. If A_y is assumed to act downwards as shown (i.e., a net uplift is assumed initially but will be checked later), there will be a tension force in the column and an uplift force that has to be resisted at the base of the column at A.

A positive value of A_y would indicate that the previous assumption is correct and a net uplift tension load does indeed exist. However, the negative value of A_y in Figure 2-4c would actually indicate that no net tension uplift exists in the column and the column is actually in compression.

Note that the reaction A_y will need to be calculated using each of the LFRD equations, and the largest value of A_y is used. The results will be presented in a tabular format.

Summation of forces about point F results in the following equilibrium equation:

$$\sum M_F = (-A_y)(30) + \gamma_{wind}(F_{B,wind})(20) + \gamma_{wind}(F_{C,wind})(15+20) - \sum\gamma_{gravity}(C_{gravity})(30) - \sum\gamma_{gravity}(B_{gravity})(30)$$
$$+\sum\gamma_{gravity}(G_{gravity})(30) + \sum\gamma_{gravity}(H_{gravity})(30) = 0$$

$$Therefore,\ A_y = \frac{1}{30}\begin{bmatrix}\gamma_{wind}(F_{B,wind})(20) + \gamma_{wind}(F_{C,wind})(15+20) - \sum\gamma_{gravity}(C_{gravity})(30) \\ -\sum\gamma_{gravity}(B_{gravity})(30) + \sum\gamma_{gravity}(G_{gravity})(30) + \sum\gamma_{gravity}(H_{gravity})(30)\end{bmatrix}$$

Where,

γ_{wind} = wind load factor in the LRFD load combination

$\gamma_{gravity}$ = the appropriate load factors for the gravity loads from the LRFD load combinations

Load combination	$F_{B,wind}$	$F_{C,wind}$	$C_{gravity}$	$B_{gravity}$	$G_{gravity}$	$H_{gravity}$	A_y
1.4D	0	0	1.4(10)	1.4(20)	1.4(10)	1.4(20)	0 kips
1.2D + 1.6L + 0.3S	0	0	1.2(10) + 1.6(0)+ 0.3(5)	1.2(20) + 1.6(10) + 0.3(0)	1.2(10) + 1.6(0) + 0.3(5)	1.2(20) + 1.6(10) + 0.3(0)	0 kips
1.2D + 1.0S + 0.5L	0	0	1.2(10) + 1.0(5)+ 0.5(0)	1.2(20) + 1.0(0) + 0.5(10)	1.2(10) + 1.0(5) + 0.5(0)	1.2(20) + 1.0(0) + 0.5(10)	0 kip
1.2D + 1.0S + 0.5W	0.5(10)	0.5(5)	1.2(10) + 1.0(5)	1.2(20) + 1.0(0)	1.2(10) + 1.0(5)	1.2(20) + 1.0(0)	6.25 kips (uplift
1.2D + 1.0W + 0.5L + 0.3S	1.0(10)	1.0(5)	1.2(10) + 0.5(0) + 0.3(5)	1.2(20) + 0.5(10) + 0.3(0)	1.2(10) + 0.5(0) + 0.3(5)	1.2(20) + 0.5(10) + 0.3(0)	**12.5 kips** (uplift)
0.9D + 1.0W	1.0(10)	1.0(5)	0.9(10)	0.9(20)	0.9(10)	0.9(20)	**12.5 kips** (uplift)

The maximum net uplift reaction at support A = **12.5 kips.**

2-11 SNOW LOAD

The risk-targeted design and site-specific ground snow load values are obtained from the ASCE Design Ground Snow Load Geodatabase or from the ASCE 7 Hazard Tool (*https://ascehazardtool.org*) [2]. These new ground snow load values in ASCE 7-22 [2] are at the ultimate strength level; hence, the new ASCE 7 load factor for snow is 1.0 as opposed to the value of 1.6 in the 2016 edition of ASCE 7. The uniform hazard 50-year mean recurrence interval (MRI) from previous editions of ASCE 7 has therefore been replaced with a strength level snow load that is tied to the reliability or safety levels prescribed in Table 1.3-1 of ASCE 7-22. These new snow load values result in a more uniform reliability or risk of failure across the United States [2, 39]; for a few areas of the United States, specific snow load case studies may be required in order to establish the ground snow loads. The value of the roof snow load depends on, but is usually lower than, the ground snow load, p_g, because of the increased effect of wind at the higher elevations. The roof snow load is also a function of the roof exposure (i.e., how much the roof is exposed to wind on all sides), the roof slope, the building occupancy, the temperature of the roof—whether it is heated or not, or whether it is a structure that is intentionally kept below freezing temperatures, such as freezer buildings or cold room—and the terrain conditions at the building site; however, the roof snow load is independent of the tributary area of a structural member. The steeper the roof slope, the smaller the roof snow load, because steep roofs are less likely to retain snow, and conversely, the flatter the roof slope, the larger the roof snow load. Flat roofs are more susceptible to failure from snow load if the drainage fails and refreezing of melted snow occurs. The accumulation of ice together with the snow increases the load on the roof considerably since the density of ice is several times that of snow. Several roof failures were reported in the Northeastern United States during the winter of 2010 because of the unexpected additional loads caused by ice accumulation [11, 12, 13].

Depending on the type of roof, the snow load can either be a uniformly balanced load or an unbalanced load. A balanced snow load is a uniform snow load distributed over the entire roof surface, while an unbalanced snow load is a partial uniform or nonuniform distribution of snow load over the roof surface or a portion of the roof surface.

The procedure for calculating snow loads on a roof surface is as follows:

1. Determine the ground snow load from the ASCE Design Ground Snow Load Geodatabase or from the ASCE 7 Hazard Tool (*https://ascehazardtool.org*).
2. Determine the snow exposure factor, C_e, from ASCE 7 Table 7.3-1 based on the terrain roughness category and the degree of exposure of the roof. Refer to ASCE 7 Section 26.7 for the description of the terrain categories (B, C, and D) that are needed in ASCE Table 7.3-1.
3. Determine the thermal factor, C_t, from ASCE 7 Tables 7.3-2 and 7.3.3. The thermal factor is a function of the ground snow load and the thermal resistance, or the resistance to heat flow through the roof assembly per unit area (i.e., R-value), and the thermal transmittance (U-value) of the roof. With increased levels or thickness of roof insulation being used in practice to meet the energy code requirements, resulting in higher R-values, less heat will flow through the roof to melt the snow layer, and hence there will be a larger amount of snow on the roof.

   The higher the ground snow load for a given R-value, the smaller the thermal factor. Furthermore, for a given ground snow load, the higher the R-value (i.e., the more insulated a roof is), the less likelihood that the snow will melt since the insulation restricts the heat flow to the roof, therefore the higher the thermal factor.
   Modern ventilated roofs, with the levels of insulation required to meet the energy code

requirements, essentially behave thermally like unheated roofs and have a C_t value of 1.2; heated structures with unventilated roofs have a C_t value ranging from 1.0 to 1.2 depending on the ground snow load (see ASCE 7 Table 7.3-3); freezer buildings or cold rooms have a C_t value of 1.3; and unheated and open-air buildings such as loading docks or sheds, bus shelters, or parking garages have a C_t value of 1.2. Coordination with the architect will be needed to ascertain the R-value for the roof that will be used to determine the correct thermal factor, C_t.

4. Determine the Risk Category of the building from ASCE 7 Table 1.5-1. The Risk Category (RC) is a measure of the consequences of failure of the building due to a hazard event. There are four risk categories (I, II, III, IV) for the hazard events of snow, ice loads, seismic, and wind loads. Note that occupancies with ballrooms or assembly occupancies each having 300-plus people capacity while the total number of occupants in the 300-plus occupant assembly spaces totaling more than 2500 persons will be assigned to the higher RC III rather than RC II; this would typically be applicable to hotel occupancies. Theaters with assembly occupancy spaces of 300-plus person capacity should be assigned to RC III.

5. Calculate the flat roof snow load, p_f, from equation 2-7:

$$\text{Flat roof snow load, } p_f = 0.7\ C_e\ C_t\ p_g\ (\text{psf}) \geq p_m \tag{2-7}$$

6. Determine the maximum minimum flat roof snow load, $P_{m,max}$ from ASCE 7 Table 7.3-4 based on the Risk Category:

Risk Category	$P_{m,max}$
I	25 psf
II	30 psf
III	35 psf
IV	40 psf

$$\text{If } p_g \leq p_{m,\max},\quad p_m = p_g\ (\text{psf}) \tag{2-8}$$

$$\text{If } p_g > p_{m,\max},\quad p_m = p_{m,\max}\ (\text{psf}) \tag{2-9}$$

The minimum snow load, P_m, should *not* be combined with sliding or drift snow load [11]. It is the balanced flat roof snow load, P_f, that should be combined with drift or sliding snow.

7. Determine the design snow load accounting for the sloped roof factor, C_s, which is a function of the roof slope and the temperature below the roof (see ASCE 7 Figure 7.4-1):

$$\text{Design sloped roof snow load, } P_s = C_s P_s\ (\text{psf}) \tag{2-10}$$

For non-slippery roofs in a heated building (i.e., a warm roof),

$$C_s = \frac{1}{40}(70-\theta) \leq 1.0 \quad \text{for roof slope } \theta \leq 70°;$$

$$C_s = 0 \quad \text{for roof slope } \theta \geq 70°$$

$$C_s = 1.0 \quad \text{for roof slope } \theta \leq 30°$$

For cold roofs and slippery roof surfaces, refer to ASCE 7 Figure 7.4-1.

8. Consider partial loading for continuous beams per ASCE 7 Section 7.5, where alternate span loading might create maximum loading conditions.

9. The unbalanced loading from the effects of wind must be considered per ASCE 7 Section 7.6.
10. Determine snowdrift loads on adjacent lower roofs per ASCE 7 Section 7.7.
11. Determine snowdrift loads at roof projections and parapet walls per ASCE 7, Section 7.8. Where the side length of a roof projection or rooftop equipment is smaller than 15 feet, snowdrift does not have to be considered.
12. Determine the effects of sliding snow from higher sloped roofs onto adjacent lower roofs per ASCE 7 Section 7.9.
13. Check the requirements for rain-on-snow surcharge per ASCE 7 Section 7.10. Roofs with a slope of less than *W*/50 (where *W* is the horizontal distance in feet from eave to ridge) in locations where $p_g \leq p_{m,max}$ but not zero must be designed for an additional rain-on-snow surcharge load of **8 psf**. Note that this additional load only applies to the balanced load case and does *not* need to be used for snowdrift, sliding snow, or unbalanced or partial snow load calculations.
14. The requirements for ponding instability should be checked using AISC Design Guide 40. Ponding is the additional load that results from rain-on-snow or melted snow water acting on the deflected shape of a flat roof, which, in turn, leads to increased deflection and therefore increased roof loading because of the additional space available (due to the deflected roof framing) for the snow and rain to accumulate.
15. Check existing adjacent lower roofs within 20 ft. of a new higher building for the effect of drifting or sliding snow (ASCE 7 Section 7.12).

The following should be noted with respect to using the "slippery surface" values in ASCE 7, Figure 7.4-1, to determine the roof slope factor, C_s:

- Slippery surface values can only be used where the roof surface is free of obstruction and sufficient space is available below the eaves to accept all the sliding snow. Examples of slippery surfaces include metal, slate, glass, and bituminous, rubber, and plastic membranes with a smooth surface.
- Membranes with imbedded aggregates, asphalt shingles, and wood shingles shall not be considered slippery surfaces.

EXAMPLE 2-16

Balanced Snow Load

An office building located in an area with a ground snow load of 85 psf has an essentially flat roof with a slope of ¼ in. per foot for drainage. Assuming a partially exposed heated roof and terrain category "C," calculate the design roof snow load using the ASCE 7 load standard and assuming the following:

a. Modern ventilated roof
b. Heated structure with unventilated roof with R-50

Solution

The ground snow load, p_g = 85 psf
Roof slope, q (for $^1/_4$ in./ft. of run for drainage) = 1.2 degrees
From ASCE 7, Table 7.3-1, exposure factor, C_e = 1.0 (partially exposed roof and terrain category *C*)

The building is in Risk Category II; therefore, from ASCE 7 Table 7.3-4,
$p_{m,max} = 30$ psf $< p_g$

For q = 1.2° and assuming a *heated roof* (see ASCE 7, Figure 7.4-1),

$$C_s = \frac{1}{40}(70-1.2) = 1.72 \leq 1.0, \textit{ therefore, } C_s = 1.0$$

Minimum snow load:

$$p_g = 85 \; psf > p_{m,\max} = 30 \; psf, \textit{ therefore } p_m = p_{m,\max} = 30 \; psf$$

Part (a)
For a modern ventilated roof, the thermal factor, C_t = 1.2 (similar to an unheated roof)

From equation (2-7), the flat roof snow load, $p_f = 0.7\ C_e\ C_t\ p_g$
= (0.7) (1.0) (1.2) (85) = **71.4 psf**
$> P_m$ = 30 psf

Using equation (2-10), the design roof snow load, $p_s = C_s\, p_f$ = (1.0) (71.4) = 71.4 psf

Part (b)
For heated structure with unventilated roof, thermal factor, C_t = 1.183 for R = 50 and p_g = 85 psf, by linear interpolation from ASCE 7-22 Table 7.3-3

From equation (2-7), the flat roof snow load, $p_f = 0.7\ C_e\ C_t\ p_g$
= (0.7) (1.0) (1.183) (85) = **70.4 psf**
$> P_m$ = 30 psf

Using equation (2-10), the design roof snow load, $p_s = C_s\, p_f$ = (1.0) (70.4) = 70.4 psf
The calculated snow load will need to be combined with the dead load and all the other applied loads using the load combinations given in Section 2-3.

Windward and Leeward Snowdrift

Snowdrift loads on the lower levels of multilevel roofs are caused by wind transporting snow from the higher roof and depositing it onto the lower roof or to balconies or canopies (i.e., snow deposited on the *leeward* side). It can also occur where the wind encounters roof obstructions, such as high roof walls, penthouses, high parapet walls, and mechanical rooftop units, and the snow is deposited on the *windward* side of the obstruction. Snow drift loads are the cause of "roughly 75% of all snow-related structural problems" [37]. The snow accumulation roof loads are sometimes not accounted for by structural designers or properly indicated on the structural drawings, resulting in structurally deficient buildings. The snowdrift loads are additional snow loads on the lower roof that are superimposed on the *balanced* flat roof snow load, P_f. The distribution of the snowdrift load is assumed to be triangular in shape (see Figure 2-5). The two kinds of snowdrift are windward and leeward drifts. *Windward* snowdrift occurs when the wind moves the snow from one area of a lower roof to another area on the same lower roof adjacent to a high wall or other sufficiently high obstruction. *Leeward* snowdrift occurs as the wind moves the snow from an upper roof and deposits the snow on a lower roof adjacent to the wall of the higher roof building. The length of roof used for the calculation of the snowdrift is the length of the roof area from which the drifting snow is derived (see Figure 2-5). The snow drift loading is based on a winter wind parameter.

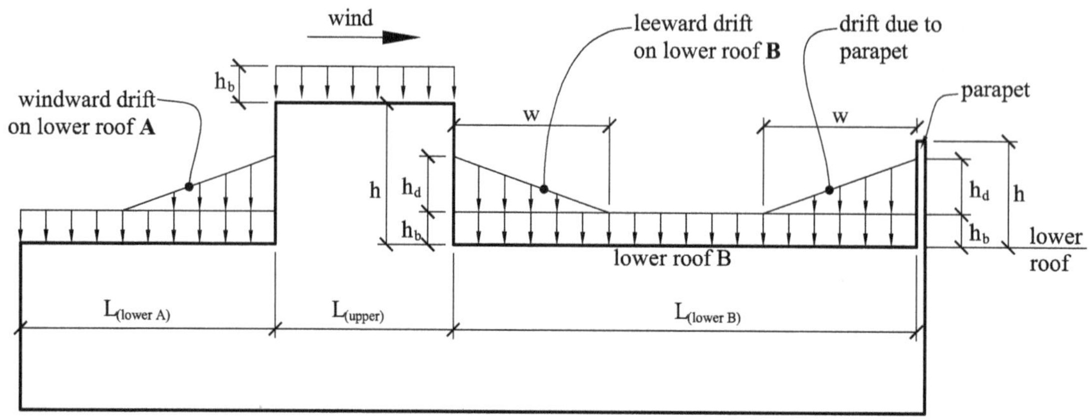

For the wind direction shown:

- $L_{lower\ A}$ is used to calculate windward drift on lower roof A
- L_{upper} is used to calculate leeward drift on lower roof B
- $L_{lower\ B}$ is used to calculate leeward drift on lower roof B
- $L_{lower\ B}$ is used to calculate windward drift at the parapet of roof B

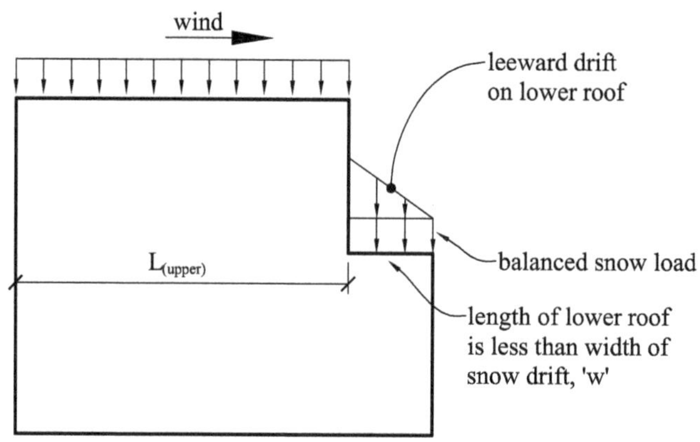

FIGURE 2-5 Snow drift diagrams

Sliding Snow Load

Where a higher pitched or gable roof is adjacent to a lower flat roof, there is a tendency for snow to slide from the higher roof onto the lower roof (see Figure 2-6). The ASCE 7 load standard assumes that only 40% of the snow load on the pitched roof will slide onto the lower roof because of the low probability that the snow load on both the pitched higher roof and the low roof will be at their maximum values at the same time when the sliding occurs. The magnitude of the snow that slides from a pitched higher roof (slippery roof with slopes greater than $^1/_4$ in./ft. or nonslippery roof with slopes greater than 2:12) onto a lower flat roof is assumed to be a uniform load, p_{SL}, distributed over a length of 15 ft. on the lower roof. This snow load is in addition to the flat roof balanced snow load, p_f, on the lower roof. However, sliding snow load should not be combined with or superimposed on drift snow load or unbalanced, partial, or rain-on-snow loads [2].

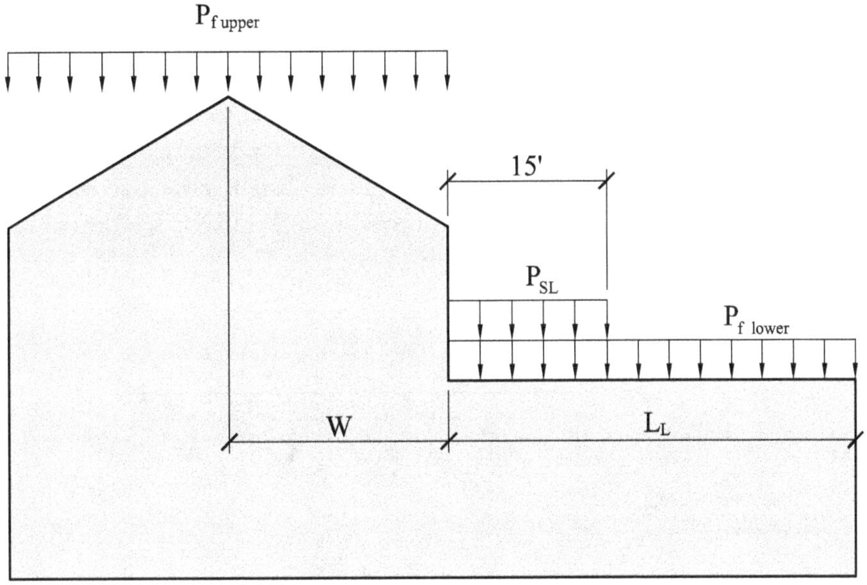

FIGURE 2-6 Sliding snow diagram

Windward Snowdrift at Roof Projections

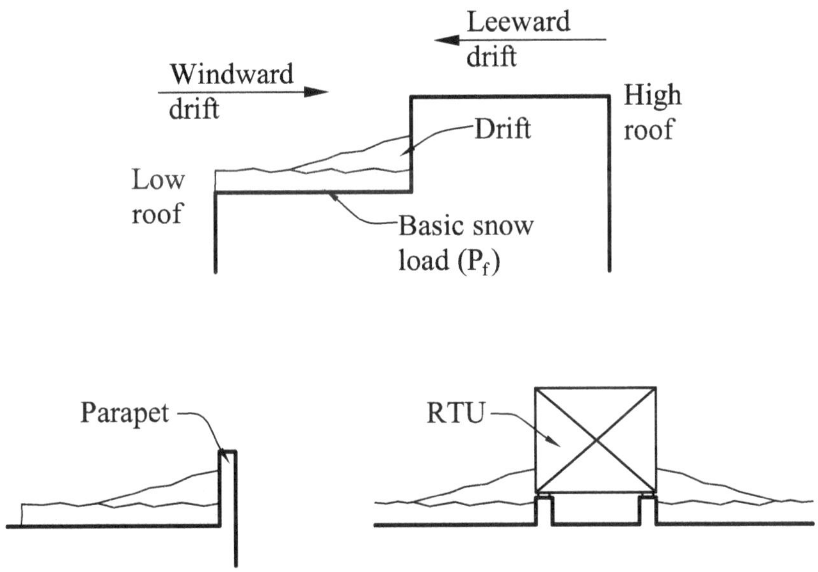

FIGURE 2-7 Snow drift due to roof projections

When drifting snow transported by wind is obstructed by a roof projection such as high parapets, signs, and rooftop equipment units (RTU), snow will tend to accumulate on both the windward and leeward sides of the roof-top units (see Figure 2-7). The snow drifts at parapet walls are windward snow drifts. For roof-top units with their bottoms raised at least 2 ft. above the balanced snow load, no snow drift is assumed. This windward snowdrift load can be neglected if the width of the roof projection perpendicular to the wind direction is less than 15 ft. [2]. Consequently, snowdrift loads do not need to be considered at rooftop equipment that has a plan size smaller than 15 ft. × 15 ft.

Partial Loading for Continuous and Cantilevered Roof Beams

In addition to designing continuous or cantilevered flat roof beams for the balanced design roof snow load, p_s, Section 7.5 of ASCE 7 prescribes a pattern of full and partial loading for the design of continuous beams. It involves applying one-half of the design snow load and the full design snow load in a checkered pattern that creates maximum load effects. The three different load cases specified in ASCE 7 are illustrated in Figure 2-8a.

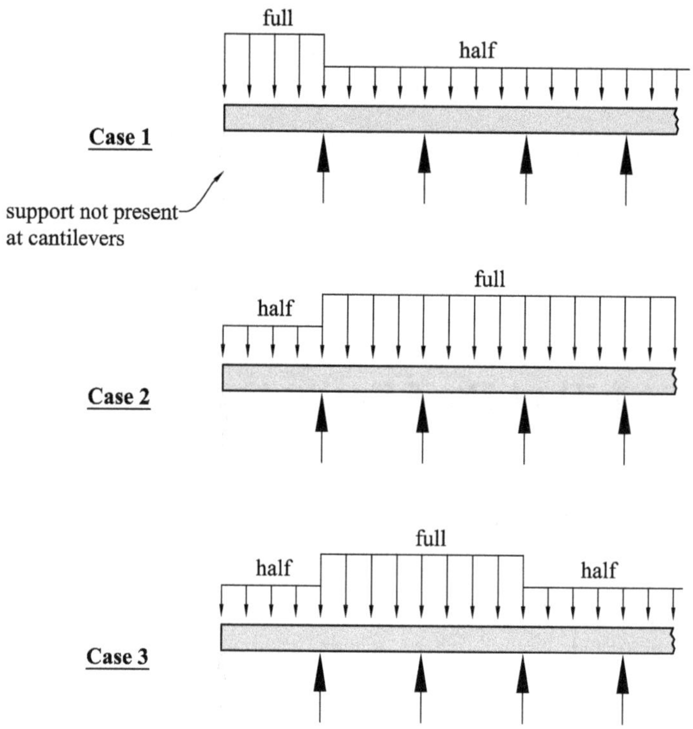

FIGURE 2-8a Partial loading diagram for continuous roof beams

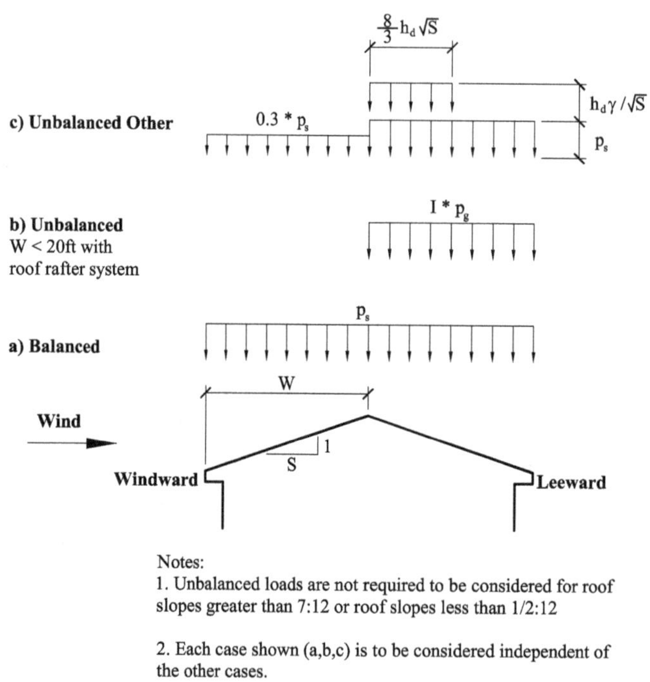

FIGURE 2-8b Balanced and unbalanced snow loads for hip and gable roofs

Snow Loads on Hip and Gable Roofs

In addition to designing hip and gable roofs for the balanced design roof snow load, p_s, Section 7.6 of ASCE 7 prescribes two unbalanced pattern loadings for the design of gable roofs as shown in Figure 2-8b. These unbalanced loads are actually caused by "across-the-ridge leeward snow drifts" [37]. These unbalanced loads may govern the design in certain situations and should be considered in the design, together with the balanced snow load.

Snow Loads on Other Types of Structural Systems

The snow load caused by ice dams and icicles on the eaves of overhanging roofs is covered in ASCE 7, Section 7.4.5. The snow load on such roofs is twice the unheated flat roof snow load (i.e., $2p_{f\ (unheated)}$) and should be applied as a uniformly distributed load on the roof overhang over a length of 5 ft. If the roof overhang is longer than 5 ft., the unheated snow load, $p_{f\ (unheated)}$, should be applied to the remainder of the roof overhang.

Snow Stockpiling on Parking Garage Roofs

For parking garages without snow chutes, the live load due to stockpiled snow from snow plowing operations on parking garage roofs during a snowstorm should be considered. Parking garage floors used for passenger vehicles are usually designed for a live load of 40 psf, but the garage roof is also designed for additional loads due to snow. Thus, the roof of parking garages is usually designed for a live load of 40 psf plus the flat roof snow load, p_f. This would amount to a live load on the parking garage roof of about 60 psf in Philadelphia and 75 psf in Rochester, New York. Additionally, some areas of the parking garage roof for garages without snow chutes are usually designated to accommodate stockpiled snow from snow plowing operations during a snowstorm. Those areas of the parking garage roof, which should be clearly identified on the structural drawings and marked on the finished structure, are subjected to far greater live loads than other areas of the roof or floors of the parking garage. The amount of snow load in the stockpiled snow area of a parking garage roof is equal to the flat roof snow load, p_f, in psf multiplied by the total parking garage roof area divided by the plan area of the designated snow stockpiling section. The density of compacted snow is 20 pcf and can be as high as 30 pcf when wet, while the density of frozen snow or ice is about 56 pcf. So, if a maximum height of 6 ft of stockpiled snow is specified in some designated areas on the roof framing plan of a parking garage, the resulting live load on the parking garage roofs for the areas with stockpiled snow could be as large as **336 psf** (i.e., 56 pcf x 6 ft). The density of ice is used instead of the density of snow (which is smaller) because of the potential of freezing the compacted stockpiled snow and the high probability of clogged roof drains. The calculated snow load is a heavy live load that must be considered in designing the parking garage roof framing when there are no snow chutes for snow disposal.

As a side note, the currently specified live load of 40 psf for parking garages translates to 6500 Ib vehicles on a 9 ft. x 18 ft. parking stall, but as the trend for electric vehicles (EVs) continues, the increased loads from EV vehicles, which are heavier than internal combustion vehicles, will translate to parking garage live loads that are higher than the current 40 psf. However, a recent study [42] found that the current 40 psf live load is adequate for the currently available electric vehicles. With heavier SUV EVs coming on stream, the applied garage live loads may be higher than the currently specified 40 psf live load.

More information on snow loads for other types of roofs, such as curved, folded plate, and dome roofs, and a detailed treatment of structural loads in general can be found in Ref. [18].

2-12 RAIN LOADS

Rain loads (see ASCE 7 Chapter 8) are applicable only to flat roofs with parapets since accumulation of rain will generally not occur on roofs without parapets or on pitched roofs. The higher the roof parapet, the greater the rain load. Building codes require that roofs with parapets have two independent drainage systems—primary and secondary (or overflow) drains at each drain location [2, 10]. See Figure 2-2c for a flat roof with primary and secondary drainage. The secondary drain must be located at least 2 in. above the main roof level where the primary drain is located; the secondary drain is the means for roof drainage if the primary drain is blocked. The design rain load, R, is calculated based on the assumption that the primary drainage system is blocked (see Figure 2-9). Thus, the total depth of water to be considered consists of the depth from the roof surface to the inlet of the secondary drainage (the static head) plus the depth of water that rises above the inlet of the secondary drainage due to the hydraulic head of the flowing water.

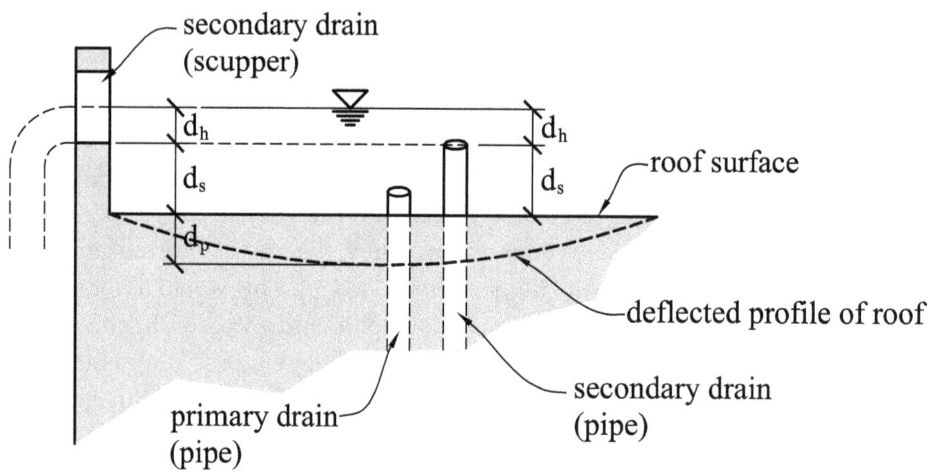

FIGURE 2-9 Primary and secondary roof drains

Roof drainage is a structural engineering, architectural, and mechanical engineering or plumbing issue; therefore, proper coordination is required among these disciplines to ensure adequate design. For flat roofs with susceptible bays as defined in ASCE 7, Section 8.4, ponding or the additional rain load due to the deflection of the flat roof framing must also be considered in the design. Some flat roof structures have failed due to the additional rain load accumulated because of the deflection of the flat roof framing.

Assuming the primary drainage is blocked, the rigid roof rain load, R, which is a function of the static head, the hydraulic head of water, and the drainage area, is given as (see ASCE 7 Equation (8.2-1):

$$R\,(\text{psf}) = (5.2\,psf\,/\,in.\,)(d_s + d_h + d_p), \tag{2-11}$$

Where

5.2 psf/in. = density of water (i.e., 62.4 pcf or 62.4psf/12-in)

d_s = Depth in *inches* from the undeflected roof surface to the inlet of the secondary drainage system (i.e., the *static head* of water),

d_h = Depth of water in *inches* above the inlet of the secondary drainage or above the top of the secondary drainage pipe (i.e., the *hydraulic head*) obtained from ASCE 7, Table C8.3-1. The hydraulic head is a function of the tributary roof area, A, drained by each drain, the size of the drainage system, and the flow rate, Q. Use ASCE 7 Tables C8.3-3 and C8.3-5 to determine the hydraulic head for rectangular and circular scuppers.

d_p = Ponding head = depth of water in *inches* due to deflections of the roof framing subjected to unfactored rain load and unfactored dead load. This accounts for the additional rain loads resulting from the deflection of the roof structural framing.

R = rain load in psf

Q = Flow rate out of a single drainage system, in gallons per minute = 0.0104 Ai (see ASCE 7 Equation C8.2-1 ASCE 7, Table C8.2-1)

A = Tributary roof area (in ft.^{2}) drained by a single secondary drain

Note: If there is a vertical wall next to a flat roof area, the drainage area of the secondary drains in the flat or low roof area should be increased by 50% of the vertical surface area of the walls to account for rainfall on the wall surface that will flow to the low roof (See the definition of A in ASCE 7-22 Section C8.1.2; this is also specified in Section 1106.4 of the International Plumbing Code (IPC))

i = design rainfall precipitation intensity (in inches/hr.) for 15-min duration rainfall event for the secondary drainage based on the precipitation intensity estimates corresponding to the design storm return period obtained from the National Oceanographic and Atmospheric Administration (NOAA) precipitation frequency data server at the following URL: *http://hdsc.nws.noaa.gov.*

The IBC Code Section 1603 stipulates that the rain load must be listed on the construction documents, in addition to other pertinent structural loads [32]. The design storm return periods are as shown below:

Risk Category	Design storm return period
I and II	100 years
III	200 years
IV	500 years

EXAMPLE 2-17

Rain Loads

The roof drainage plan for a risk category II building located in Avondale, Pennsylvania, with a 100-year, 15-minute duration precipitation intensity of 5.43 inch/hr. is shown in Figure 2-10.

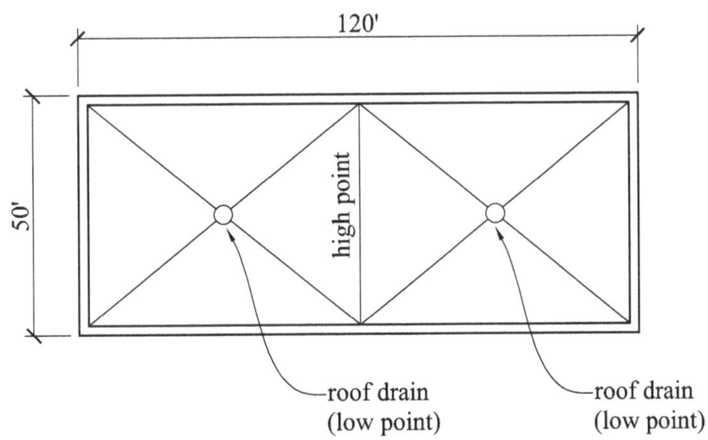

FIGURE 2-10 Roof drainage plan

a. Assuming a 4-in.-diameter secondary drainage pipe with an 8-in.-diameter overflow dam that is set at 3 in. above the roof surface, determine the design rain load, R.

b. Assuming a 6-in.-wide channel scupper secondary drainage system that is set at 3 in. above the roof surface, determine the design rain load, R.

c. Assuming a 24-in.-wide, 6-in.-high closed scupper secondary drainage system that is set at 3 in. above the roof surface, determine the design rain load, R.

Solution

a. drain size = **4-in.-diameter secondary drainage or outlet with** an 8-in.-diameter overflow dam

d_s = Depth in inches from the undeflected roof surface to the inlet of the secondary drainage system (i.e., *the static head of water*) = 3 in. (given)

A = Tributary area of roof drained by one secondary drain or outlet

$$= \frac{(120 \text{ ft.})(50 \text{ ft.})}{2 drains} = 3000 \text{ ft.}^2 \text{ per drain}$$

For RC II, the design storm return period = 100 years. For the 100-year design storm return period or average recurrence interval, the 15-minute duration design rainfall precipitation intensity, i = 5.43 in./hr. (see NOAA data server at *http://hdsc.nws.noaa.gov*)

$Q = 0.0104\,Ai$ = (0.0104) (3000 ft.^{2}) (5.43 in.) = 169 gallons/min.

Alternatively, Q can be calculated as follows:

$$Q = (3000 ft^2 / drain)\left(\frac{5.43\ in/hr}{12\ in/ft}\right) = 1357.5\,ft^3/hr$$

$$= (1357.5\,ft^3/hr)(7.4805\ gallons/ft^3)$$

$$= \left(\frac{1357.5\,ft^3}{60 \text{ min}}\right)(7.4805\ gallons/ft^3)$$

$$= 169\,gallons/\text{min}$$

Using **ASCE 7 Table C8.2-1** for a **4-in.-diameter secondary drainage** system with an 8-in.-diameter overflow dam and a flow rate, Q = 169 gallons/min., we find by linear interpolation from the values in the table that the depth of water in inches above the inlet of the secondary drainage system is

$$d_h = \frac{(2 \text{ in.} - 1.5 \text{ in.})}{(200 \text{ gallons / min} - 150 \text{ gallons / min})}(169 \text{ gallons / min} - 150 \text{ gallons / min}) + 1.5 \text{ in.}$$

$$= 1.69 \text{ in.}$$

Using equation (2-11) and neglecting the ponding depth, d_p, the rain load is

$$R\,(\text{psf}) = 5.2\,(d_s + d_h + d_p) = 5.2\,(3 \text{ in.} + 1.69 \text{ in.} + 0) = \mathbf{24.4\,psf}$$

$$(\text{assuming primary drainage is blocked})$$

b. 6-in.-wide channel scupper

d_s = Depth in inches from the undeflected roof surface to the inlet of the secondary drainage system (i.e., the static head of water) = 3 in. (given)

A = Area drained by one secondary drainage = 3000 ft.^{2}

i = design rainfall precipitation intensity = 5.43 in./hr.

Q = 0.0104 Ai = (0.0104) (3000 ft.^{2}) (5.43 in.) = 169 gallons/min.

Using ASCE 7, Table C8.2-3 for a **6-in.-wide channel scupper** secondary drainage with a flow rate, Q = 169 gallons/min., we find by linear interpolation from the values in the table that

d_h = Depth of water in inches above the inlet of the secondary drainage system

$$= \frac{(5\text{ in.} - 4\text{ in.})}{(194\text{ gallons / min} - 140\text{ gallons / min})}(169\text{ gallons/min} - 140\text{ gallons/min}) + 4\text{ in.}$$

$= 4.54$ in.

Using equation (2-11), the rain load is

$$R\ (\text{psf}) = 5.2\left(d_s + d_h + d_p\right) = 5.2\ (3\text{ in.} + 4.54\text{ in.} + 0) = \mathbf{39.2\ psf}$$

(assuming primary drainage is blocked)

c. 24-in.-wide, 6-in.-high closed scupper

d_s = Depth in inches from the undeflected roof surface to the inlet of the secondary drainage system (i.e., the static head of water) = 3 in.

A = Area drained by one secondary drainage = 3000 ft.^{2}

i = design rainfall precipitation intensity = 5.43 in./hr.

Q = 0.0104 Ai = (0.0104) (3000 ft.^{2}) (5.43 in.) = 169 gallons/min.

Using ASCE 7, Table C8.2-3 for a **24-in.-wide, 6-in.-high closed scupper** system with a flow rate, Q = 169 gallons/min., we find by linear interpolation from the values in the table that

d_h = Depth of water in inches above the inlet of the secondary drainage

$$= \frac{(2\text{ in.} - 1\text{ in.})}{(200\text{ gallons / min} - 72\text{ gallons / min})}(169169\text{ gallons/min} - 140\text{ gallons/min}) + 1\text{ in.}$$

$= 1.76$ in.

Using equation (2-11), the rain load is

$$R\ (\text{psf}) = 5.2\left(d_s + d_h + d_p\right) = 5.2\ (3\text{ in.} + 1.76\text{ in.} + 0) = \mathbf{24.8\ psf}$$

(assuming primary drainage is blocked)

2-13 ICE LOADS DUE TO FREEZING RAIN

The weight of ice formed on exposed structures such as towers, cable systems, and pipes in the northern parts of the United States due to freezing rain must be accounted for in the design of these structures (see ASCE 7 Chapter 10); it is quite common in these areas to see downed power lines and tree limbs due to the weight of accumulated ice loads. Figure 2-11 shows ice accumulation on exposed tree limbs in Rochester, New York, during an ice storm.

FIGURE 2-11 Ice accumulation on exposed tree limbs (Photo by Abi Aghayere)

Similar ice accumulations occur on exposed structural steel members. The weight of ice on these structures is usually added to the snow load on the structure.

The procedure for calculating the ice load on a structural element in accordance with ASCE 7 Chapter 10 is as follows:

1. Determine the Risk Category based on the occupancy of the structure from ASCE 7 Table 1.5-1.
2. Using the Risk Category of the structure from step 1, the risk-targeted atmospheric nominal ice thickness corresponding to the mean recurrence interval below, the concurrent wind speed, V_c, and the temperature concurrent with the ice thickness caused by freezing rain is obtained from ASCE 7 Figures 10.4-2 through 10.4-5, 10.5-1, 10.5-2, 10.6-1, and 10.6-2, or preferably from the ASCE 7 Hazard Tool (*https://ascehazardtool.org/*).

Risk Category	Mean Recurrence Interval (MRI)
I	250-years
II	500-years
III	1000-years
IV	1400-years

3. Determine the topographical factor, K_{zt}, from ASCE 7 Section 26.8

$$K_{zt} = (1 + K_1K_2K_3)^2 = 1.0 \text{ for flat land (per ASCE 7 Section 26.8} - 1), \tag{2-12}$$

where the multipliers K_1, K_2, and K_3 account for the wind speed-up effect for buildings located on a hill or escarpment and are determined from ASCE 7 Figure 26.8-1.

4. Determine the height factor, f_z, to account for the increase in ice thickness as the height above ground increases:

$$f_z = \left(\frac{z}{33}\right)^{0.10} \leq 1.4 \tag{2-13}$$

where z = Height of the structural member in feet above the ground.

5. The design uniform radial ice thickness in inches due to freezing rain is given as (see ASCE 7 equation 10.4-5):

$$t_d = t f_z (K_{zt})^{0.35} \qquad (2\text{-}14)$$

6. The weight of ice on exposed surfaces of structural shapes and prismatic members is calculated in accordance with ASCE 7, Section 10.4.1, as follows:

$$\text{Ice load, } D_i = \gamma_{ice} \pi t_d (D_c + t_d), \text{ in pounds per foot,} \qquad (2\text{-}15)$$

where

D_c = Characteristic dimension, shown for various shapes in ASCE 7 Figure 10.4-1, and

g_{ice} = Density of ice = **56 pcf** (minimum value per ASCE 7 Section 10.4.1).

For large 3-D objects, the volume of ice is calculated as follows:

$$\text{Flat plates: } V_i = (\lambda \pi t_d)(\text{Area of one side of the plate}) \qquad (2\text{-}16)$$

$\lambda = 0.8$ for vertical plates

$= 0.6$ for horizontal plates

$$\text{Domes or spheres with radius, r}: V_i = \pi t_d \times (\pi r^2) \qquad (2\text{-}17)$$

The ice load, D_i, must now be combined with the dead load and all other applicable loads in the load combinations from Section 2-3. The load combinations must be modified according to ASCE 7 Sections 2.3.3 and 2.4.3 when ice loads are considered.

7. Determine the dead load, D, of the structural member in pounds per foot.
8. Determine the snow load, S, on the ice-coated structural member.
9. Determine the live load, L, on the structural member (e.g., for a pipe transporting liquid, determine the live load due to the liquid carried in the pipe).
10. Determine the wind velocity pressure q_z for wind velocity V_c in accordance with ASCE 7 Section 26.10.
11. Determine the wind force coefficients, C_f, in accordance with ASCE 7 Sections 10.5 and 29.4.
12. Determine the gust effect factor from ASCE 7 Section 26.11.
13. Determine the wind-on-ice force, W_i, in accordance with ASCE 7 Chapter 29, considering the increased projected surface area of the ice-coated structural member that is exposed to wind (see ASCE 7 Chapter 10). The projected area will be increased by t_d all around the perimeter of the exposed structural member.
14. Calculate the maximum total load using the modified load combinations that include ice loading. The **LRFD** load combinations from Section 2-3 are modified per ASCE 7 Section 2.3.3 to include ice load as follows:

    2b. $1.2\,D + 1.6\,L + 0.2D_i + 0.3S$

    4c. $1.2D + L + D_i + W_i + 0.3S$

    4d. $1.2D + D_i$

    5c. $0.9D + D_i + W_i$

The **ASD** or service load combinations with ice load included are as follows:

1b. D + 0.7 D_i

2b. D + L + 0.7 D_i

3b. D + 0.7D_i + 0.7W_i + 0.7S

7c. 0.6D + 0.7D_i + 0.7W_i

EXAMPLE 2-18

Ice Loads (LRFD)

Determine the ice load, dead load, and live load on exposed 50-in.-diameter horizontal steel pipes for a chemical plant located in West Grove, Pennsylvania, in the United States. The pipes have a wall thickness of 1 in. and carry liquid with a density of 64 pcf. The top of the pipe is at an elevation of 100 ft. above the ground, and the site is flat. Assume that the snow load is 35 psf, the density of the ice is 56 pcf, the density of the steel is 490 pcf, and the wind load on the pipe due to the wind velocity, V_c, has been calculated to be 10 psf.

Solution

1. Determine the Risk Category of the structure (from ASCE 7 Table 1.5-1):
   Chemical plant = Risk Category IV building (from ASCE 7 Table 1.5-1).
2. Determine the nominal ice thickness, t, and the concurrent wind speed, V_c, from ASCE 7 Figure 10.4-2D or directly and preferably from the ASCE 7 Hazard Tool:
   Thickness (t) = 0.98 in.
3. Determine the topographical factor, K_{zt}:

$$K_{zt} = \left(1 + K_1\ K_2\ K_3\right)^2 = \mathbf{1.0} \quad \text{for flat land}$$

4. Determine the height factor, f_z:

$$f_z = \left(\frac{z}{33}\right)^{0.10} = \left(\frac{100}{33}\right)^{0.10}$$

$$= 1.12 < 1.4$$

$$\therefore\ f_z = 1.12,$$

   where z = Height of the pipe in feet above the ground = 100 ft.

5. The design uniform radial ice thickness in inches due to freezing rain is calculated as

$$t_d = \ t\, f_z\, \left(K_{zt}\right)^{0.35} = (0.98 \text{ in.})(1.12)(1.0)^{0.35} = 1.1 \text{ in.}$$

6. The weight of ice on exposed surfaces of a steel pipe is calculated in accordance with ASCE 7 Section 10.4.1 as follows:

$$\text{Ice load, } D_i = \gamma_{\text{ice}} \pi t_d \left(D_c + t_d\right)$$

$$= (56 \text{ pcf})(\pi)(1.1 \text{ in.}/12)(50 \text{ in.} + 1.1 \text{ in.})/12 = 69 \text{ lb./ft.},$$

   where
   D_c = Characteristic dimension, shown for various shapes in
   ASCE 7 Figure 10.4-1 = 50 in. for 50-in.-diameter pipe,
   γ_{ice} = Density of ice = 56 pcf (minimum value per ASCE 7 Section 10.4.1), and

7. Determine the dead load, D, of the structural member in pounds per linear foot:

$$\text{Dead load of pipe, } D = \left[\frac{\pi(50 \text{ in.})^2}{4} - \frac{\pi(50 \text{ in.} - 1 \text{ in.} - 1 \text{ in.})^2}{4}\right](490 \text{ pcf}/144)$$

$$= 524 \text{ lb./ft}$$

8. The snow load on the ice-coated pipe, $S = 35 \text{ psf } (D_c + t_d)$

$$= 35 \text{ psf}\left(\frac{50 \text{ in.}}{12} + \frac{1.1 \text{ in.}}{12}\right)$$

$$= 149 \text{ lb./ft.}$$

9. Determine the live load, L, of the pipe due to the liquid carried by the pipe:

$$\text{Live load in pipe, } L = \frac{\pi(50 \text{ in.} - 1 \text{ in.} - 1 \text{ in.})^2}{(4)(144)}(64 \text{ pcf}) = 805 \text{ lb./ft.}$$

10. Calculate the wind-on-ice load, W_i (the wind pressure has previously been calculated and given as 10 psf, which obviates the need for steps 10 through 12):

$$W_i = (\text{wind pressure})(D_c + t_d) = (10 \text{ psf})(50 \text{ in.} + 1.1 \text{ in.})/12 = 43 \text{ lb./ft.}$$

11. The ice load, D_i, must now be combined with the dead load and other applicable loads to determine the maximum total factored load on the pipe. The most critical limit states load combinations for downward-acting loads are calculated using the modified LRFD load combinations:

1. $1.4D = 1.4\,(524) = 734$ lb./ft.

2b. $1.2D + 1.6L + 0.2D_i + 0.3S$
$= 1.2\,(524) + 1.6\,(805) + 0.2\,(69) + 0.3\,(149) =$ **1976 lb./ft. (governs)**

3. $1.2D + 1.6\,(L_r \text{ or } 0.7S \text{ or } R) + (0.5L \text{ or } 0.8W)$
$= 1.2\,(524) + 1.6\,(0.7)(149) + 0.5\,(805) = 1198$ lb./ft.

4c. $1.2D + L + D_i + W_i + 0.3S$
$= 1.2\,(524) + 805 + 69 + 43 + 0.3\,(149) = 1591$ lb./ft.

4d. $1.2D + D_i$
$= 1.2(524) + 69 = 698$ lb./ft.

5c. $0.9D + D_i + W_i$
$= 0.9\,(524) + 69 + 43 = 584$ lb./ft.

Therefore, the controlling factored-downward load on the pipe for which the pipe will be designed is **1976 lb./ft.**

2-14 MISCELLANEOUS LOADS

The following miscellaneous loads will be discussed in this section of the chapter:

- **Fluid loads,**
- **Flood loads**—see ASCE 7 Sections 2.3.2 and 2.4.2.,
- **Self-straining loads** (e.g., temperature)—see ASCE 7, Sections 2.3.4 and 2.4.4.,
- **Lateral pressure due to soil, water, and bulk materials,**

- **Impact loads** (see Table 2-11),
- **Live loads from miscellaneous structural elements -** See Table 2-12 for loads on:
  - Handrails and balustrades, and
  - Vehicle edge barriers in parking structures
    - **Vehicular impact loads:** Note that several vehicle edge barrier failures have been reported in the literature, some leading to multiple fatalities. Some have recommended that vehicle edge barriers be designed for a force of 20 kips, which is equivalent to the force required to bring a 6,000 Ib vehicle traveling at 10 mph to a stop, with the barrier deflecting 1 ft [40]. This lateral force is obtained by equating the kinetic energy of the moving vehicle to the potential energy, that is:

$$\frac{1}{2}\left(\frac{6000\,Ib}{32.2\,ft./s^2}\right)\left[(10\,mph)\left(\frac{5280\,ft}{3600}\right)\right]^2 = (F)(1\,ft)$$

Thus, the lateral force, F = 20,000 Ib = 20 kips.
This recommended lateral force is much greater than the 6 kips currently specified in ASCE 7-22 [40]. Therefore, the design of vehicle barriers in parking structures should account for the heavier weight of electric vehicles that use parking facilities today. EVs will apply greater lateral forces than currently specified in ASCE 7 [40, 43].
For vehicular impact forces on bridge piers or building columns that are located close to major roads (e.g., see Figure 14-32b and 14-32c) and therefore susceptible to vehicular impact, the pier or column or column protection systems (such as bollards) should be designed for an equivalent static horizontal force of 2670 kN (or 600 kips) applied at a height of 5 ft. above the bottom of the pier or column or the column protection system, in accordance with the American Association of State Highway and Transportation Officials (AASHTO) Code [44]. See Chapter 14 for discussions on prevention of progressive or disproportionate collapse.

- **Construction live loads** (see Table 2-13 and Ref [5]).
Note that the basic LRFD and ASD load combinations from ASCE 7 presented in Section 2-3 of this text that include dead, live, snow, wind, and seismic loads have been derived from reliability analyses that are "*supported by an extensive statistical database*" [2]. The load combinations for other loads—including some of those in the following sections—are based more on judgment than on a rigorous engineering analysis, and hence they are not included in the basic load combinations in order to distinguish the load combinations that are based on engineering principles from those that are based on judgment [17].

Fluid Loads, *F*

The ASCE 7 load standard uses the symbol F to denote loads due to fluids with well-defined pressures and maximum heights. This load is *separate* and *distinct* from the soil or hydrostatic pressure load, H, or the flood load, F_a. Not much guidance is given in ASCE 7 regarding this load, but since it has the same load factor as the dead load, D, in the LRFD load combinations, that would indicate that this load pertains to the weight of fluids that may be stored in a building or structure (e.g., storage tanks).

TABLE 2-11 Impact Factors

Type of load or equipment supported by the structural member	Impact factor*
Elevator loads	See ASME A17
Elevator machinery	See ASME A17
Hoists for façade and building maintenance equipment	2.5
Light machinery or motor-driven units	1.2
Reciprocating machinery or power-driven units	1.5
Monorail cranes (powered)	1.25
Cab-operated or remotely operated bridge cranes (powered)	1.25
Pendant-operated bridge cranes (powered)	1.10
Bridge cranes or monorail cranes with hand-geared bridge, trolley, and hoist	1.0

* All equipment impact factors shall be increased where larger values are specified by the equipment manufacturer.

† Ref: ASCE 7 Sections 4.6.3, 4.6.4, and 4.9.3.

TABLE 2-12 Live loads on miscellaneous structural elements.

Structural element	Live load
Handrails and balustrades	50 lb./ft. applied at the top along the length of the handrail or balustrade in any direction, or a single concentrated load of 200 lb. applied in any direction at any point.
Grab bar	A single concentrated load of 250 lb. was applied in any direction at any point.
Vehicle barriers for *passenger cars*	6000 lb. of horizontal load applied in any direction to the barrier system applied over an area of 1 ft. x 1 ft., acting at a height of between 1.5 ft. and 2.25 ft above the floor or ramp surface.
Fixed ladders with rungs	A single concentrated load of 300 lb. applied at any point to produce maximum load effect plus additional 300 lb. of concentrated load for every 10 ft. of ladder height.
Rails of fixed ladders extending above floor or platform	Concentrated load of 100 lb. in any direction, at any height.
Ship ladders with treads instead of rungs	Live load similar to stairs (typically 100 psf (see ASCE 7, Table 4.3-1).
Anchorage for attachment of fall arrest equipment	Required unfactored live load = 3100 lb.

Adopted from ASCE-7, Section 4.5.

TABLE 2-13 Construction Live Loads

Type of construction	Construction live load (psf)
Very light duty construction	20
Light duty construction	25
Medium duty construction	50
Heavy duty construction	75

Flood Loads, F_a

This pertains to flood loads acting against a structure, but most building structures are not subjected to flood loads, except for buildings in coastal regions. The individual flood load components that make up F_a include hydrostatic, hydrodynamic, wave, and debris impact loads. The procedures for calculating these individual flood load components are covered in Chapter 5 of ASCE 7. For load combinations, including flood load, F_a, refer to ASCE 7, Section 2.3.2. Note that a high load factor of 2.0 is used on F_a in the flood load combinations in ASCE Section 2.3.2 because of the relative lack of information on flood loads. More information on how to combine the individual flood load components that make up the flood load, F_a, can be found in Ref. [15, 16]. In designing buildings for flooding in coastal regions (e.g., ocean-front buildings), the first habitable floor level must be located above the design flood elevation (DFE), which can be obtained from Ref. [28]. Below the DFE level, all walls are designed to withstand wind forces, but any non-structural wall must be designed to "break away" under flooding loads because the flood must be able to "flow through" the building below the DFE level. The lowest floor levels in these buildings are typically used as open parking areas [27, 28].

Self-Straining Force, T

Self-straining loads (e.g., temperature) arise due to the restraining of movement in a structure caused by expansion or contraction from temperature or moisture change, creep, or differential settlement. If these movements are unrestrained, the temperature force is practically zero. This is the case for most building structures, except for posttensioned members, where restrained shrinkage could lead to self-straining loads in the member. Some cladding systems, when subjected to expansion due to temperature effects, could develop sizable self-straining forces, but these could be alleviated by proper detailing of the cladding connections. An example of the effect of self-straining force due to temperature changes was previously discussed in Section 1-3 and in Example 1-1.

Lateral Pressure Due to Soil, Water, and Bulk Materials, H

Lateral and hydrostatic pressures of soil on retaining walls and the lateral pressures exerted by bulk solids against the walls of bins and silos are denoted in ASCE 7 by the symbol H. This notation is also used to denote the upward hydrostatic pressures on base slabs and foundation mats of buildings since these upward forces usually act simultaneously with the lateral hydrostatic or soil pressures, depending on the elevation of the water table. An example on the calculation of the net buoyancy uplift force on an underground storage tank was previously illustrated in Example 2-10.

Impact Loads

Impact loads are dynamic loads that are caused by the sudden application of a live load on a structure. This can be caused either by a load dropping onto a structure or a crane suddenly picking up a load, or from the wheels of a vehicle or a crane hitting bumps in a bridge or crane girder resulting in an increased vertical load on the structure; the resulting amplification of the static live load is called the impact factor. Structures subject to impact loads include bridges and industrial crane-supporting structures. Only live loads can cause impact; therefore, only the live load is amplified by the impact factors. Thus, the service live load for the structure is the static **live** load multiplied by an impact factor, which may range from 1.25 to greater than 2.0, depending on the cause of the impact and the elevation of the object relative to the structure. For structures subjected to impact loads from falling objects, the impact factor may be much larger than 2.0, depending on the height of the falling object above the structure. Examples of minimum impact factors specified in ASCE 7 are as shown in Table 2-11. For the calculation of the loads and the design of bridge crane girders and crane runway girders and supporting columns, the reader should refer to Ref. [31].

Extraordinary Event Loads

The ASCE 7 load standard provides load combinations for extraordinary events such as explosions, fires, and vehicular impacts, which have a low probability of occurrence (see ASCE 7, Section 2.5). The load combination for checking the strength and stability of the structure under these events is

$(0.9 \text{ or } 1.2)D + A_k + 0.5L + 0.15S$,

where A_k is the load effect arising from the extraordinary event.

After a damaging event occurs, the load combination for checking the residual load bearing capacity of the damaged member is

$(0.9 \text{ or } 1.2)D + 0.5L + 0.2\ (L_r \text{ or } 0.7S \text{ or } R)$.

The load factor of 0.9 is used for D when the dead load, D, counteracts the primary variable load, A_k, and a load factor of 1.2 is used for D when D is additive to the primary variable load, A_k.

For progressive collapse from the removal of a column in a building, the General Services Administration prescribes a floor load of 2(1.2D + 0.5L or 0.2S) at floor levels above the column that is removed and a floor load of 1.2D + 0.5L or 0.2S at floor levels at or below the removed column [45].

2-15 VERTICAL AND LATERAL DEFLECTION, AND LATERAL ACCELERATION CRITERIA

Vertical Deflection Limits

The limits on vertical deflections due to gravity loads are performance objectives intended to ensure occupant comfort and to ensure the functionality of, and prevent excessive cracking of, plaster ceilings, architectural partitions and finishes, and supported exterior cladding. These deflection limits are typically specified in terms of the joist, beam, or girder span, and the deflections are calculated based on elastic analysis of the structural member under service or unfactored loads. The service loads, instead of the factored loads, are used in the deflection calculations because, under the ultimate limit state, when failure is imminent, deflection of the structure is no longer important. The maximum allowable deflections recommended in IBC (Table 1604.3) are as follows [4]:

Maximum allowable *floor* deflection due to service live load $\leq \frac{L}{360}$,

Maximum allowable *floor* deflection due to service total dead plus live load $\leq \frac{L}{240}$,

Maximum allowable *roof* deflection due to service live load $\leq \frac{L}{360}$, and

Maximum allowable *roof* deflection due to service total dead plus live load $\leq \frac{L}{240}$,

Elevator machine room beam, static deflection $\leq \frac{L}{1666}$

Equipment support platforms, static deflection $\leq \frac{1}{2}$ in.

where

L = Simple span length of the flexural member.

The roof framing deflection limits presented above assume that the roof structural members support deflection-sensitive elements such as plastered ceilings, which would crack under excessive deflections. See Table 6-4 in Chapter 6 for more beam deflection limits. For members that directly support masonry wall partition or cladding and glazing (e.g., spandrel beams and girders, or interior beams and girders directly supporting a partition wall), the allowable total deflection due to the weight of the cladding or partition wall plus the superimposed dead load plus the live load should be limited to $\frac{L}{600}$ or 0.3 in., whichever is smaller, to reduce the likelihood of cracking of the masonry cladding or partition wall or glazing [7].

It should be noted that the most critical deflection is the deflection that occurs after the cladding is in place (i.e., the live load deflection) because as the masonry cladding or glazing is being installed, the beam or girder would deflect, and the cladding will tend to conform to the shape of the deflected member, and any curvature in a masonry wall cladding can be corrected at the mortar joints by the mason. For prefabricated members or composite steel beams and girders, a camber is sometimes specified to help control the total deflection. To avoid too much camber, it is common practice to specify a camber approximately equal to the dead load deflection of the structural member, but no less than ¾ inch. Support restraints should also be considered in determining the required camber, and this is discussed in Chapters 6 and 7.

For **operable partition walls or hanging partitions**, the typical structural framing is as shown in Figure 2-12. The required vertical deflection limit should be obtained from the operable partition wall manufacturer's data sheet. A typical maximum deflection limit for operable walls under superimposed loads is $^1/_8$ in. per 12 ft. of wall length, or a maximum vertical deflection limit of $L/1152$, where L is the span of the beam or girder supporting the operable wall, and the "1152" in the denominator is obtained from (12 ft.)(12 in./ft)/0.125 in. = 1152 [6]. The vertical deflection limits for structural framing supporting overhead cranes, which can be as low as L/700 to ensure adequate crane performance, should be confirmed with the crane supplier.

In all the deflection limits above, the length, L, for cantilevers should be taken as twice the length of the cantilever or overhang.

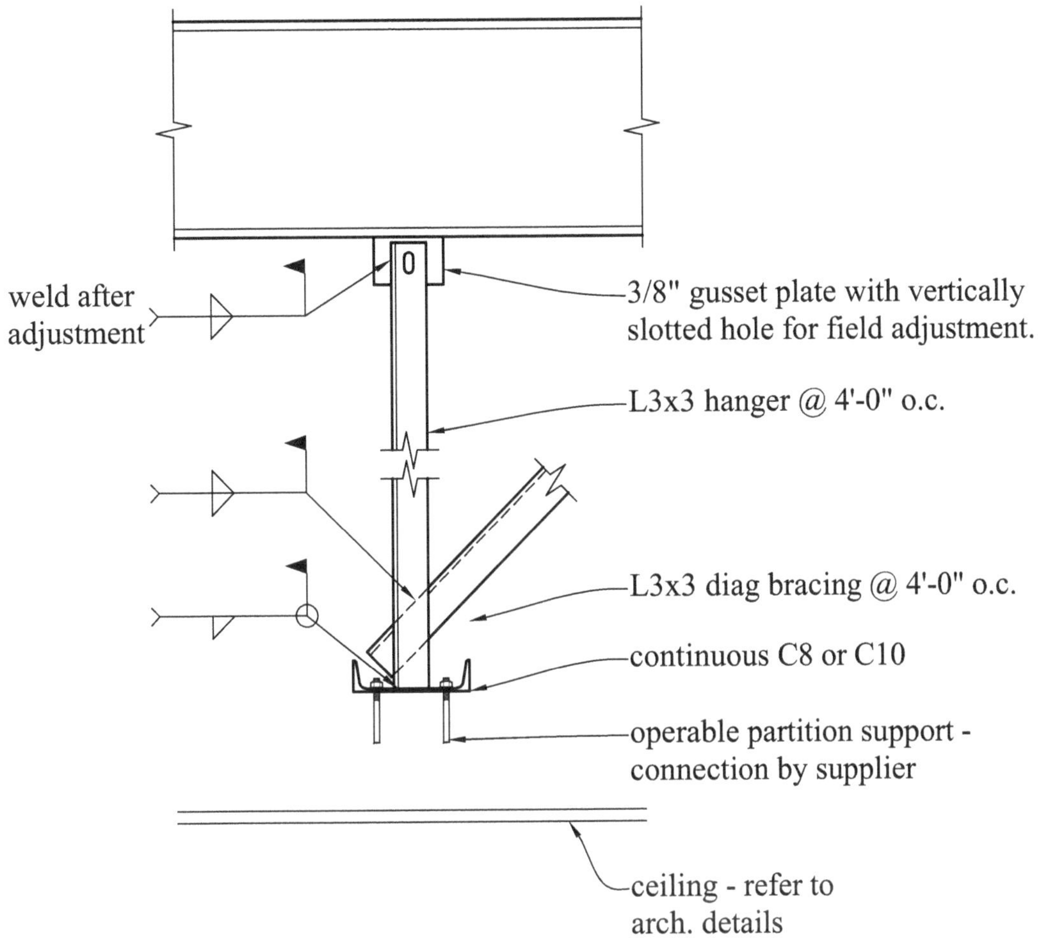

FIGURE 2-12 Typical operable or folding partition wall support detail

Lateral Deflection (Drift) Limits

The deflection limits discussed thus far are for vertical deflections due to gravity loads. There are also performance objectives for the lateral deflections in buildings. These are expressed as limits on the total ***lateral*** or horizontal deflection of steel buildings and the inter-story drift caused by wind or seismic loads. These performance objectives are intended to ensure occupant comfort, the proper functioning of elevators, the functionality of the interior partition walls and exterior cladding panels, and the prevention of cracking [14]. The wind load used for serviceability considerations is a matter of engineering judgment and should be decided in consultation with the client [2] and the exterior cladding supplier. The serviceability wind speed that could be used to check the lateral drift limits in buildings includes the unfactored 10-year mean return interval (MRI) wind speed, the 50-year MRI wind speed, or the 100-year MRI wind speed.

The unfactored 3-second gust wind speeds in Exposure C for the 10-year MRI, 25-year MRI, 50-year MRI, and 100-year MRI, respectively, can be obtained using the ASCE 7 Hazard Tool [2]. Note that the 700-year MRI or the 1700-year MRI or the 3000-year MRI 3-second gust wind speeds in Ref. [2] should not be used to check serviceability because these are the factored wind speeds and would be too conservative for serviceability considerations.

The overall drift and inter-story drift limits used by engineers in practice vary between 1/800 and 1/400 of the overall height of the building and 1/250 and 1/500 of the floor-to-floor height of the building under a 10-year wind, respectively. The Canadian Standards Association (CSA)

Steel Code indicates that the maximum total drift due to service wind loads be limited to 1/400 of the total building height and the inter-story drift be limited to 1/500 of the floor-to-floor height for buildings with brick cladding and 1/400 for all others [8]. It is common practice among some designers to use the preceding lateral drift limits for typical buildings under the 10-year MRI wind speed. [9]. Others have recommended using a maximum inter-story drift of 1/600 of the floor-to-floor height under a 20-year MRI wind speed or between 1/300 and 1/500 of the floor-to-floor height under a 50-year MRI wind speed [14]. The most commonly used inter-story and total lateral deflection (drift) limit for tall buildings appears to be 1/400 under a 50-year MRI wind speed [14, 41]. In the author's experience, an inter-story and total lateral deflection (drift) limit of 1/400 under a 10-year MRI wind speed will suffice for low- to mid-rise buildings. For more deflection-sensitive buildings (e.g., super tall and slender buildings), a total lateral deflection (drift) limit of 1/400 under a 100-year MRI 3-second gust wind speed (see Figure CC.2-4 of the ASCE 7 Commentary [2]) should be used. For example, the following stringent lateral deflection criteria were used for the serviceability design of a 55-story high-rise building [30]:

- Overall lateral displacement at the top of the building = $H/500$ under a 100-year MRI wind, where H is the total height of the building.
- Inter-story drift limit = $h/350$ under a 100-year MRI wind, where h is the floor-to-floor or story height.
- Inter-story drift limit of $h/50$ or $0.02h$ under strength-level seismic forces

The Petronas Towers in Kuala Lumpur, Malaysia, comprising 1483 ft. tall twin buildings of 88 stories, used a maximum total lateral deflection (drift) limit of $H/560$ under a 50-year MRI wind [14].

The lateral deflection limits prescribed in ASCE 7 for seismic loads are much higher than the lateral deflection limits commonly used by designers in practice for wind loads, and the seismic deflection limits are calculated using strength-level (i.e., factored) seismic forces because the design philosophy for earthquakes focuses on life safety and not on serviceability conditions. The seismic lateral deflection limits are given in ASCE 7, Table 12.12-1 [2].

Lateral Acceleration Limits

Satisfying the lateral deflection criteria for multi-story buildings may not be sufficient to assure the comfort of building occupants; therefore, lateral acceleration performance criteria are also required to ensure the comfort of tall building occupants during high wind events. These performance criteria, which are measured in terms of acceptable lateral accelerations, are meant to minimize motion or seasickness of occupants in tall, slender buildings during windstorm events. The lateral acceleration limits prescribed by the International Organization for Standardization (ISO) to ensure occupant comfort are 15 to 18 milli-g for residential buildings, 15 to 20 milli-g for hotels, and 20 to 25 milli-g for office buildings under a 10-year mean return period (MRI) wind event [25, 26, 41]. The lateral acceleration of a building is dependent on its lateral stiffness and its mass and damping. For tall, slender buildings, the acceleration cannot be controlled by stiffness alone, and since the mass or dead weight of the building is fixed, the only efficient way available to control the motion of the building is increasing the damping, which reduces the lateral acceleration of the building. The types of damping systems used in tall, slender buildings to minimize lateral accelerations due to wind forces include tuned liquid sloshing dampers (filled with water) and tuned mass dampers (TMD)—a heavy mass and spring system [25, 26]. These damping systems are usually installed at the highest floor level of the building, and they minimize the lateral accelerations from wind by opposing the building motion caused by wind [35].

Though they help to control the lateral vibrations in the building, these damping devices are heavy, and they increase the weight on the building, which has to be taken into account in calculating the seismic loads on the building; in addition, they occupy large floor areas that cannot be leased out for living occupancies, and therefore, these spaces will not be revenue-generating.

2-16 DEMAND-TO-CAPACITY RATIO

In general, for new designs, the demand, D (i.e., the applied loads or load effects), should be less than or equal to the design capacity, C; thus, the demand-to-capacity ratio, or D/C, for new designs should be less than or equal to 1.0.

For existing buildings, the International Building Code (IBC) Sections 502.4 and 502.5 state that:

"Any *existing gravity load-carrying* structural element for which an addition and its related alterations cause an increase in design dead, live, or snow load, including snow drift effects, of more than *5 percent* shall be replaced or altered as needed to carry the gravity loads required by the International Building Code (IBC) for new structures."

"Any *existing lateral load-carrying* structural element whose demand-to-capacity ratio with the addition considered is not more than *10 percent* greater than its demand-to-capacity ratio with the addition ignored shall be permitted to remain unaltered."

The above IBC requirements imply that for existing buildings, the Code allows the demand-to-capacity ratio, or D/C, to be up to a maximum of 1.05 for *gravity-load* supporting elements and 1.10 for *lateral-load* supporting elements; thus, no alteration is required for existing structural elements up to these maximum values. However, care should be taken not to apply these higher demand-to-capacity limits more than once in the life of the existing structure; in addition, the designer should also consider that the inherent assumption that "the current strength of a structure is equal to the original design strength" may not be valid due to environmental degradation or deterioration of the structural elements [38].

References

[1] American Institute of Steel Construction. 2023. *Steel Construction Manual*. 16th ed. Chicago, IL: AISC.

[2] American Society of Civil Engineers. *Minimum design loads for buildings and other structures (ASCE 7-22)*, ASCE, Reston, VA, 2022.

[3] Ellingwood, B.,V. Galambos, J. G. MacGregor, and C. A. Cornell. 2008. *Development of a Probability-Based Load Criterion for American National Standard A58*. NBS Special Publication 577. Washington, DC: U.S. Department of Commerce, National Bureau of Standards, June.

[4] International Codes Council. 2021. *International building code (IBC)—2021,* Falls Church, VA: ICC.

[5] American Society of Civil Engineers. 2015. *ASCE 37-14, Design Loads on Structures During Construction*, Reston, VA: ASCE.

[6] ASTM. 2001. *E 557-00, Standard Guide for the Installation of Operable Partitions*. July.

[7] ACI. 2005. *Building Code Requirement for Masonry Structures (ACI 530-05).* American Concrete Institute (ACI).

[8] CSA. 2006. *CAN/CSA-S16-01*. Canadian Standards Association, Ottawa Canada.

[9] Griffis, L. G. 1993. "Serviceability Limit States Under Wind Loads". *Engineering Journal*, First Quarter.

[10] International Codes Council. 2015. *International plumbing code—2015*, Falls Church, VA: ICC.

[11] O'Rourke, M. 2011. "Snow and Rain Provisions in ASCE 7-10". *STRUCTURE Magazine*, February.

[12] O'Rourke, M. 2010. *Snow Loads: Guide to the Snow Load Provisions of ASCE 7-10*. American Society of Civil Engineers (ASCE), Reston, VA.

[13] Knapschefer, J. 2011. "Hundreds of Roofs Buckle After Storms Slam Northeast". *Engineering News Record (ENR)*: 16.

[14] Smith, R. 2011. "Deflection Limits in Tall Buildings—Are They Useful?" *ASCE Structures Congress*: 515-527.

[15] FEMA. 2011. *Coastal Construction Manual (CCM), FEMA P-55*. Federal Emergency Management Administration, Washington, DC.

[16] Coulbourne, W. L. 2011. "ASCE 7-10 Changes to Flood Load Provisions". *ASCE Structures Congress*.

[17] McAllister, T. P., and B. R. Ellingwood. 2011. "Load Combination Requirements in ASCE Standard 7-10: New Developments". *ASCE Structures Congress*.

[18] Fanella, D. A. 2012. *Structural Loads—2012 IBC and ASCE/SEI 7-10*. International Code Council.

[19] Parsons, B. J., D. A. Bender, D. J. Dolan, and F. E. Woeste. 2014. "Lateral Loads Generated by Occupants on Exterior Decks". *STRUCTURE Magazine*, January.

[20] Wight, J. K. and J. G. MacGregor. 2009. *Reinforced Concrete – Mechanics and Design, 5th ed.* Upper Saddle River, New Jersey: Pearson Education.

[21] Taranath, B. S. 2017. *Tall Building Design – Steel, Concrete, and Composite Systems*. Boca Raton, Florida: CRC Press/Taylor & Francis Group.

[22] Nowak, A. S. and K. R. Collins. 2001. *Reliability of Structures*. New York: McGraw Hill.

[23] Ratay, R. T. 2010. *Forensic Structural Engineering Handbook. 2nd ed.* New York: McGraw Hill.

[24] Galambos, T. V. 1981. "Load and Resistance Factor Design". *Engineering Journal, American Institute for Steel Construction (AISC),* Third Quarter.

[25] Marcus S., H. Mena, and F. Yalniz. 2016. "56 Leonard". *STRUCTURE Magazine*, June: 28-30.

[26] Fields D., and R. Klemencic. 2018. "Stacked Frustrums Create Chicago's Newest Super-Tall Tower". *STRUCTURE Magazine*, June: 26-28.

[27] McCarthy, E. M. 2017. "New Clearwater – A Challenging Wind and Flood Design". *STRUCTURE Magazine*, June.

[28] Society of Civil Engineers. 2014. *ASCE 24-14. Flood Resistant Design and Construction*, Reston, Virginia: ASCE.

[29] Ruddy, J. L. 1986. "Ponding of Concrete Deck Floors". *AISC Engineering Journal,* Third Quarter.

[30] Poon, D., S. Pacitto, C. Kwitkin, and S. S. Shim. 2013. "Inclined to Succeed". *ASCE Civil Engineering*, May: 74-79.

[31] MacCrimmon, R. A. 2009. *Crane-Supporting Steel Structures*. Markham, Ontario: Canadian Institute of Steel Construction, Markham.

[32] O'Rouke, M. and A. Longabard. 2019. *"Do Structural Engineers Design for Rain Loads?" STRUCTURE Magazine,* April: 28-30.

[33] Dusenberry, D. O. 2019. "Performance-Based Design is the Future". *STRUCTURE Magazine*, February: 50.

[34] Poland, C. D. 2019. "Performance-Based Earthquake Design – Lessons Learned from a Building Code Option". *STRUCTURE Magazine*, March: 32-33.

[35] Reid, R. L. 2019. "Skinny Scrapers". *ASCE Civil Engineering*, July/August: 45-53.

[36] Ellingwood, B. R., J. W. van de Lindt, T. P. McAllister. 2020. "A New Challenge to the Practice of Structural Engineering". *STRUCTURE Magazine*, November: 28-31.

[37] O'Rouke, M., and C. Letchford. 2021. "A New and Unexpected Roof Snow Drift". *STRUCTURE Magazine*, August: 36-38.

[38] Wagner, K. 2021. "The Basics of the 5% Rule". *STRUCTURE Magazine*, March: 28-29.

[39] Maguire, M., B. Bean, J. Harris, A. Liel, and S. Russell. 2022. "Ground Snow Loads for ASCE 7-22." STRUCTURE Magazine, February: 16-18

[40] Gamble, W. L. 2022. "Parking Structure Failure and Incidents." *Concrete International*, October: 39–44.

[41] ACI Committee 375. 2016. "Perormance-Based Design of Buildings Subjected to Wind Loads". *Concrete International*, December: 37-41.

[42] Hurwitz, L. J., B. Fisher, and C. Schneemann. 2024. "Electric Vehicles and Parking Structures". *STRUCTURE Magazine*, July: 22-27.

[43] McDonald, J., J. Moody, M. Perkins, and M. Pabiel. 2024. "Engineering Infrastructure to Support Society Resiliency – Electric Vehicle Weights Endangering Parking Structures." *STRUCTURE Magazine*, January: 36-39.

[44] Chen, L., H. Wu, and T. Liu. 2021. "Vehicle Collision with Bridge Piers: A State-of-the-Art Review." *Advances in Structural Engineering*, 24(2): 385-400. DOI: 10.1177/1369433220953510

[45] General Services Administration. 2013. *GSA Alternate Path Analysis and Design Guidelines for Progressive Collapse Resistance*. October 24, Revision 1, January 28, 2016.

Exercises

2-1. Define the term "limit state." What is the difference between the ASD method and the LRFD method?

2-2. a. What are the reasons for using resistance factors in the LRFD method?

b. List the resistance factors for shear, bending, tension yielding, and tension fracture.

c. Explain why the load factor for dead loads (1.2) is lower than that for live loads (1.6).

2-3. a. Determine the factored axial load or the required axial strength, P_u, of a column in a braced frame in an office building with a regular roof configuration. The service axial loads on the column are as follows:

$$P_D = 200 \text{ kips } (\text{dead load})$$

$$P_L = 300 \text{ kips } (\text{floor live load})$$

$P_S = 150$ kips (snow load)

$P_W = \pm 60$ kips (downward and uplift load on column due to wind load)

$P_E = \pm 40$ kips (downward and uplift load on column due to seismic load)

b. Calculate the required *nominal* axial compression strength, P_n, of the column.

2-4. a. Determine the ultimate or factored load for a roof beam subjected to the following service loads:

Dead load = 29 psf (dead load)

Snow load = 35 psf (snow load)

Roof live load = 20 psf

Wind load = 25 psf upward

15 psf downward

b. Assuming a roof beam span of 30 ft. and a tributary width of 6 ft., determine the factored moment and shear.

2-5. List the floor live loads for the following occupancies:

- Library stack rooms,
- Classrooms,
- Heavy storage,
- Light manufacturing, and
- Offices.

2-6. Determine the tributary widths and tributary areas of the beams, girders, and columns in the roof framing plan shown in Figure 2-13.

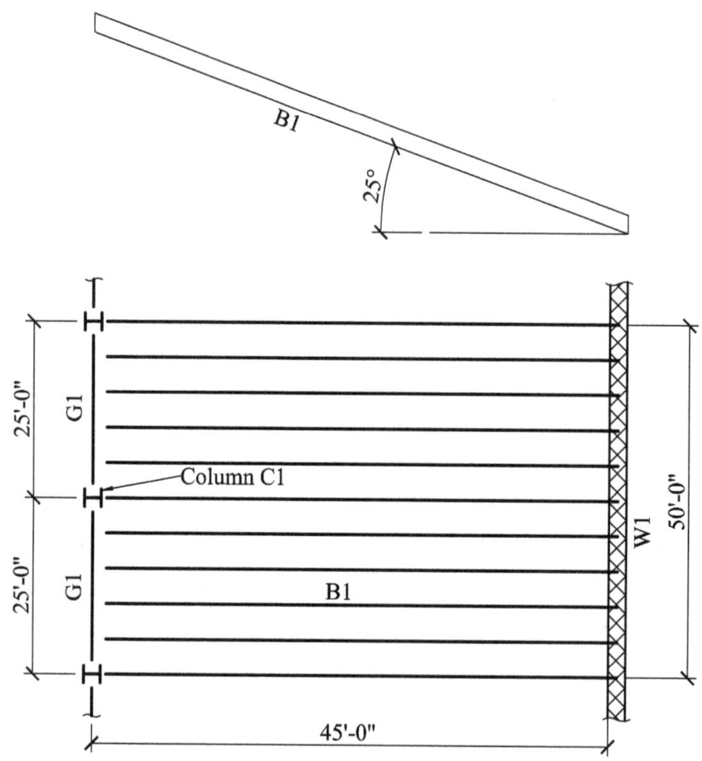

FIGURE 2-13 Roof framing plan for Exercise 2-6

Assuming a roof dead load of 30 psf and an essentially flat roof with a roof slope of $^1/_4$ in./ft. for drainage, determine the following loads using the ASCE 7 load combinations. Neglect the rain load, R, and assume the snow load, S, is zero:

a. Uniform dead and roof live loads on the typical roof beam14 in pounds per foot.

b. Concentrated dead and roof live load reactions on the typical roof girder in pounds.

c. Total factored axial load on the typical interior column in pounds.

d. Total factored axial load on the typical corner column in pounds.

2-7. A three-story building has columns spaced at 18 ft. in both orthogonal directions and is subjected to the roof and floor loads listed below. Using a column load summation table, calculate the cumulative axial loads on a typical interior column with and without live load reduction. Assume a roof slope of $^1/_4$ in./ft. per foot for drainage.

Roof Loads:

Dead load, $D_{\text{roof}} = 20$ psf

Snow load, $S = 40$ psf

Second- and Third-Floor Loads:

Dead load, $D_{\text{floor}} = 40$ psf

Floor live load, $L = 50$ psf

2-8. a. Determine the **dead load** (*with and without partitions*) in pounds per square foot of floor area for a steel building floor system with W24 × 55 beams (weighs 55 lb./ft.) spaced at 6 ft. 0 in. o.c. and W30 × 116 girders (weighs 116 lb./ft.) spaced at 35 ft. on centers. The floor deck is 3.5-in. normal weight concrete on 1.5 in. × 20 ga. composite steel deck.

- Include the weights of 1-in. light-weight floor finish, suspended acoustical tile ceiling, mechanical and electrical equipment (assume an industrial building), and *partitions*.
- Since the beam and girder sizes are known, you must calculate the actual weight, in pounds per square foot, of the beam and girder by dividing their weights in in. pounds per foot by their tributary widths.

b. Determine the dead loads in kips/ft. for a typical interior beam and a typical interior girder. Assume that the girder load is uniformly distributed.

c. If the floor system is to be used as a heavy manufacturing plant, determine the controlling factored loads in kips/ft. for the design of the *typical interior beam*.

d. Determine the factored concentrated loads on the *typical interior girder*.

- Use the LRFD load combinations.
- Note that *partition* loads need not be included in the dead load calculations when the floor live load is greater than 80 psf.

e. Determine the factored shear, V_u, and the factored moment, M_u, for a typical beam and a typical girder.

- Assume that the beams and girders are simply supported.
- The span of the beam is 35 ft. (i.e., the girder spacing).
- The span of the girder is 30 ft.

2-9. The building with the steel roof framing shown in Figure 2-14 is in Rochester, New York.

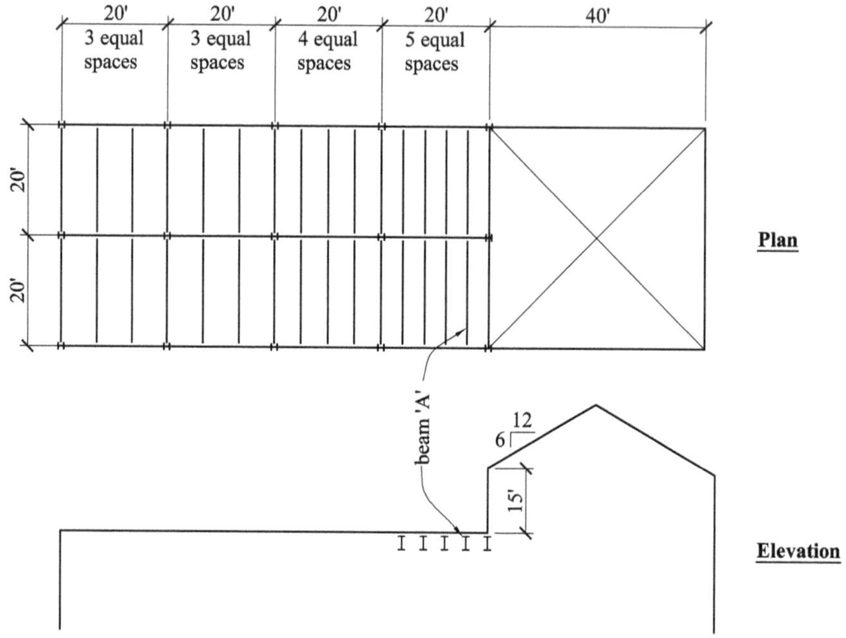

FIGURE 2-14 Roof plan and building elevation for Exercise 2-9

Assuming terrain category C and a partially exposed roof and a ground snow load of 35 psf, determine the following:

a. Balanced snow load on the *lower* roof, p_f.

b. Balanced snow load on the *upper* roof, p_f.

c. Design snow load on the *upper* roof, p_s.

2-10. An eight-story office building consists of columns located 30 ft. apart in both orthogonal directions. The roof and typical floor gravity loads are as follows:

Roof Loads:

Dead load = 80 psf

Snow load = 40 psf

Floor Loads:

Floor dead load = 120 psf

Floor live load = 50 psf

a. Using the column tributary area and a column load summation table, determine the total unfactored and factored vertical loads in a typical interior column in the first story, neglecting live load reduction.

b. Using the column tributary area and a column load summation table, determine the total unfactored and factored vertical loads in a typical interior column in the first story, considering live load reduction.

c. Develop an Excel spreadsheet to solve parts (a) and (b) and verify your results.

2-11. Use the following framing plan and floor section (see Figures 2-15a and 2-15b):

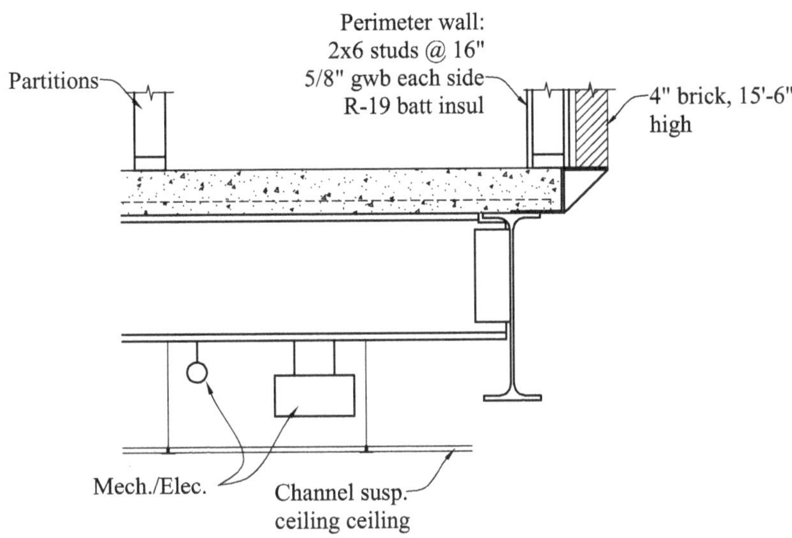

FIGURE 2-15a Floor section

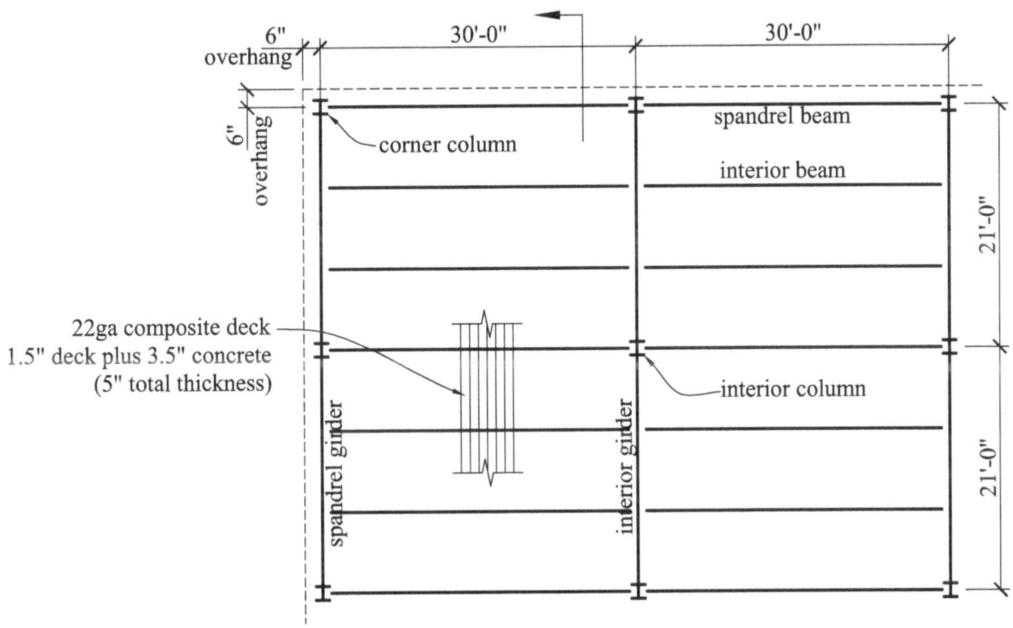

FIGURE 2-15b Floor framing plan

Framing members

Interior beam: W16 × 31

Spandrel beam: W21 × 50

Interior girder: W24 × 68

Floor deck: refer to the manufacturer's catalog based on the detail. Provide the reference.

Assume office occupancy, $L = 50$ psf

a. Determine the floor dead load in psf to the interior beam.
b. Determine the weight of the perimeter wall (brick and stud wall) in plf.
c. Determine the service dead and live loads to the spandrel and interior beams in plf.
d. Using the LRFD load combinations, determine the factored loads to the spandrel and interior beams in plf.

e. Determine the factored maximum moment and shear in the spandrel and interior beams.
f. Determine the factored loads to the interior girder.
g. Determine the factored maximum moment and shear in the interior girder.

2-12. Refer to the following details in Figure 2-16.

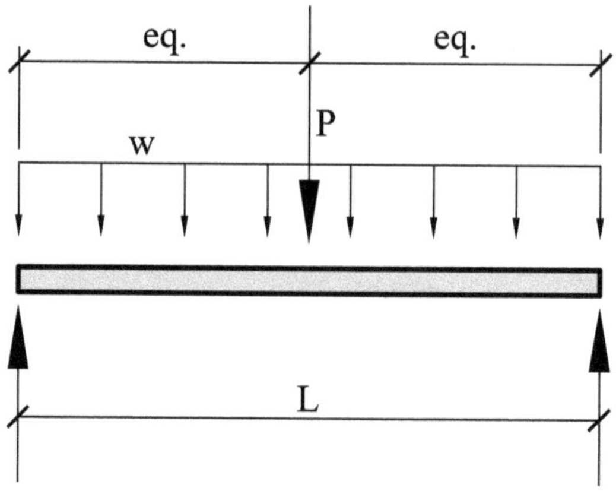

FIGURE 2-16 Beam loading for Exercise 2-12

Given Loads (Exercise 2-12):

Uniform Load, w	*Concentrated Load, P*
D = 500 plf	D = 11 k
L = 800 plf	S = 15 k
S = 600 plf	W = +12 k or −12 k
Beam length = 25 ft.	E = +8 k or − 8 k

Determine the following:

a. Describe a practical framing scenario where these loads could all occur as shown in Figure 2-16.
b. Determine the maximum moment for each individual load effect (D, L, S, W, and E).
c. Develop a spreadsheet to determine the worst-case bending moments for the code-required load combinations.

2-13. Refer to the following details in Figure 2-17.

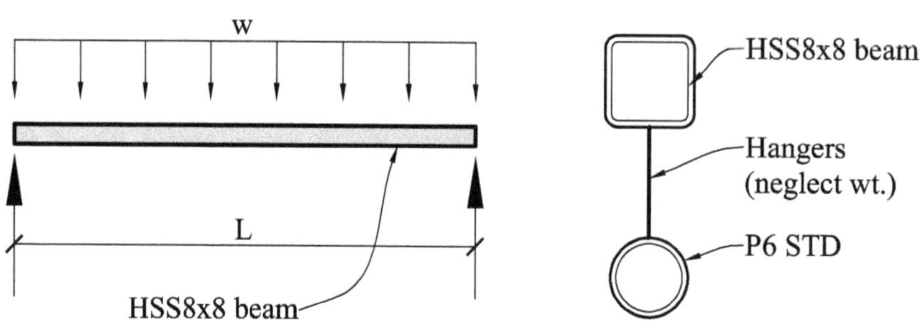

FIGURE 2-17 Beam loading for Exercise 2-13

Given:

Beam is HSS8 × 8 × ⅜, length = 28 ft.

Pipe 6 STD is hung from the beam and is full of water (assume load is uniformly distributed).

⅝" thick ice is around the HSS8 × 8 and P6

Find:

a. The uniform service load in plf for each load item (self-weight, ice, and water)
b. The maximum service bending moment in the beam.

2-14. Refer to the following details in Figure 2-18.

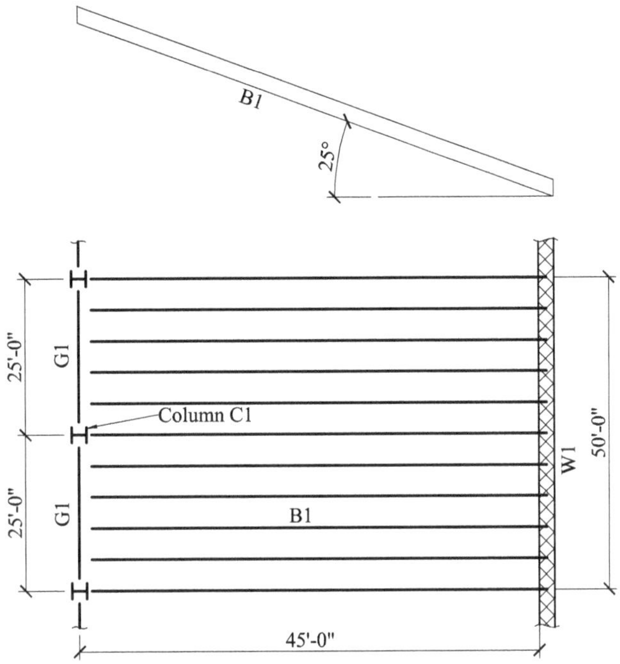

FIGURE 2-18 Roof framing plan for Exercise 2-14

Assuming a roof dead load of 25 psf and a 25° roof slope, determine the following using the ASCE 7-factored load combinations. Neglect the rain load R and assume that the snow load S is zero:

a. The tributary areas of B1, G1, C1, and W1.
b. The uniform dead and roof live load and the factored loads on B1 in plf.
c. The uniform dead and roof live load on G1 and the factored loads in plf. (assume G1 is uniformly loaded).
d. The total factored axial load on column C1 in kips
e. The total factored uniform load on W1 in plf (assume a tributary length of 50 ft.).

2-15. A three-story building has columns spaced at 25 ft. in both orthogonal directions and is subjected to the roof and floor loads shown below. Using a column load summation table, calculate the cumulative axial loads on a typical interior column. Develop this table using a spreadsheet.

Roof Loads:	second and third floor loads:
Dead, D = 20 psf Snow, S = 45 psf	Dead, D = 60 psf Live, L = 100 psf

NOTE: All other loads not shown are taken as zero.

2-16. Using the floor plan as shown in Figure 2-19, assume a floor live load, L_o = 60 psf. Determine the tributary areas and the design floor live load, L, in psf by applying a live load reduction, if applicable. Assume the slab overhangs the perimeter beams by 9".

B-1, interior beam

B-2, spandrel beam

G-1, interior girder (assume point loads)

C-1, interior column

C-2, corner column

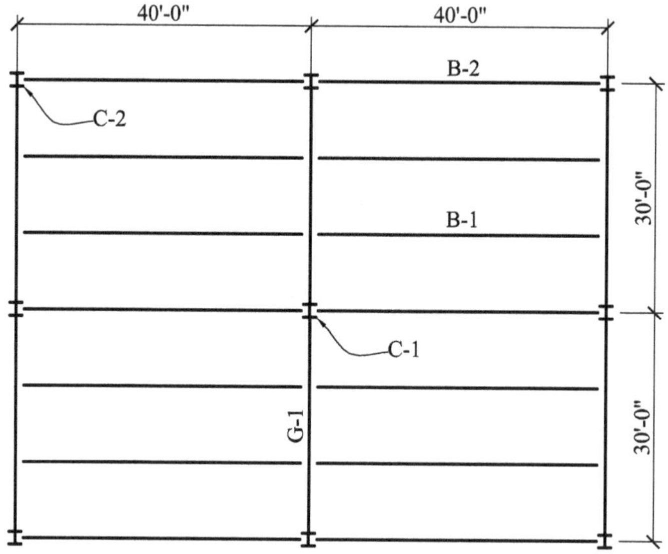

FIGURE 2-19 Roof framing plan for Exercise 2-16

2-17. An interior column in a multistory office building with a 50 psf live load supports a total cumulative floor tributary area of 7200 ft^2. Calculate the reduced floor live load (in **psf**) that will be used for the design of the bottom story column.

2-18. Calculate the flat roof snow load, P_f, in **psf** for a police station in an area where the ground snow load is 75 psf. The building is heated with a ventilated roof and has a flat roof that is partially exposed. Assume Terrain Category "C."

CHAPTER

3

Lateral Loads and Lateral Force-Resisting Systems

3-1 LATERAL LOADS ON BUILDINGS

The types of lateral loads that act on structures include wind loads, seismic loads, lateral earth pressure, and hydrostatic pressures. These lateral forces produce the following load effects in the structure:

- Overturning moment, resulting in uplift/tension forces and increased compression forces in the columns of the lateral force resisting systems (LFRS).
- Sliding or base shear.

In this text, only the two main types of lateral loads—wind and seismic loads—will be discussed.

Wind Loads—Cause and Effect

All exposed structures are acted upon by wind pressures (see Figure 3-1), and as the wind flows around a structure, the surface nearest to the wind direction (i.e., the windward face) is acted on by a positive wind pressure (i.e., the wind is "pushing" into the wall surface); the surface opposite the wind direction (i.e., the leeward face) is acted on by a negative wind pressure (i.e., suction, which implies that the wind is "pulling" away from the wall surface). The side walls of the building will be subjected to suction, while the roof will be mostly subjected to suction or uplift pressures, though there are areas of the roof that may be subjected to downward acting wind pressures. The wind pressures on structures are affected by atmospheric pressure effects (i.e., boundary layer effects or the drag on the wind flow near the ground surface) and aerodynamic effects (i.e., wind flow around structures that results in external and internal wind pressures and suction). Therefore, the factors affecting the calculation of wind forces on structures include:

- Wind speed
- Height of the structure
- Type of terrain at the site where the structure is located
- The influence of nearby or adjacent structures
- The shape of the structure
- The frequency of lateral vibration of the structure, which is related to the stiffness of the structure and the wind gust factor.

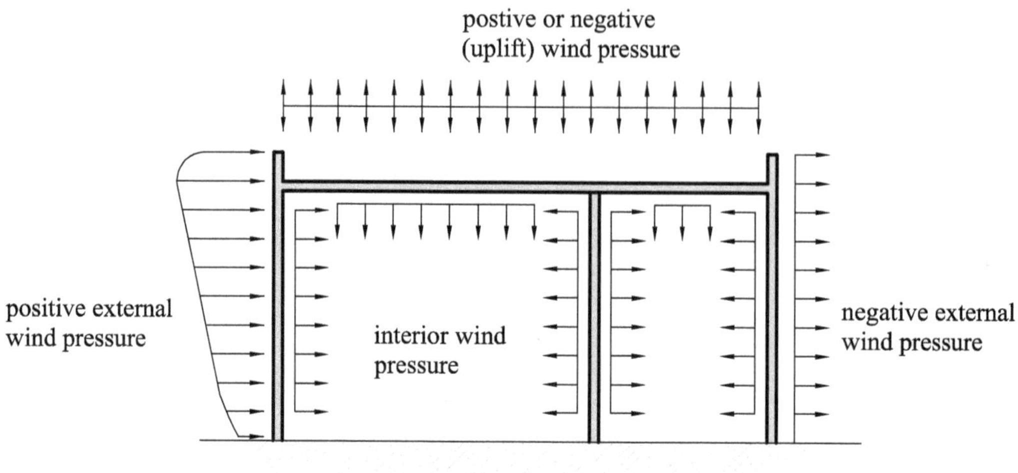

a) External and internal wind pressures (elevation of building)

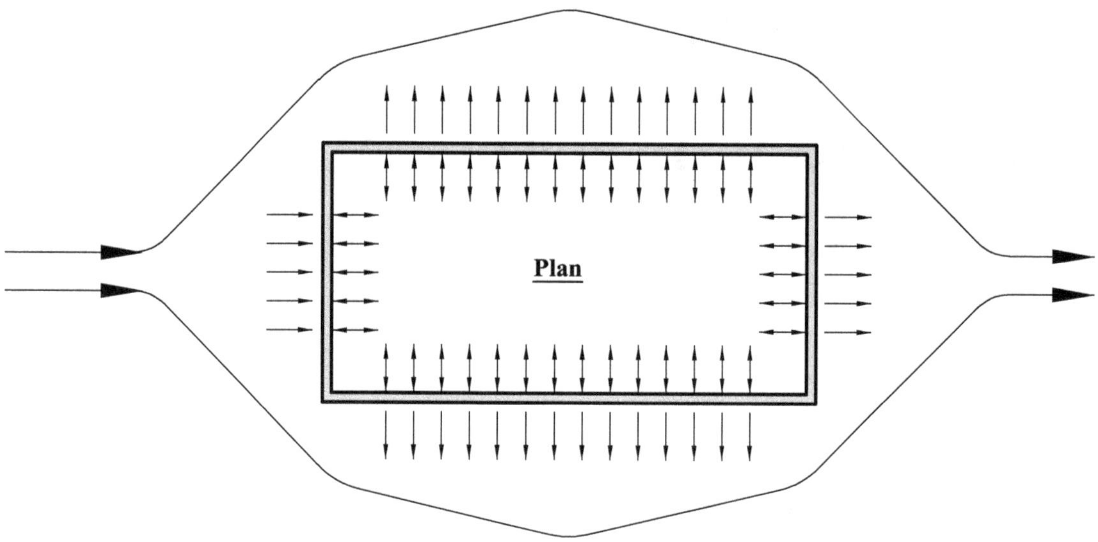

b) Air flow around a building (plan view)

FIGURE 3-1 Wind pressures on building surfaces

Wind is a dynamic force, but for simplicity and convenience, the lateral wind pressure is represented in ASCE 7 [1] by an equivalent static pressure, $P = qGC_p$, in psf, and this wind pressure is proportional to the velocity wind pressure, q, the wind gust response factor, G, and the shape-dependent pressure coefficients, C_p [1]. The velocity wind pressure, q, is a function of the wind speed and is represented by the Bernoulli equation as follows [1]:

$$q = \frac{1}{2} C \rho V^2 \tag{3-1}$$

Where,

C = a constant that accounts for the drag coefficient and the shape factor of the building,

ρ = the mass density of air = 0.002378 slugs/ft^3 or 0.002378 Ib-s^2/ft/ft^3.

Since the wind speed, V, is usually specified in miles per hour (mph), and the velocity wind pressure is usually expressed in psf, equation (3-1) can be rewritten with these appropriate units and the appropriate unit conversions to obtain the velocity pressure as follows:

$$q\ (\text{psf}) = \frac{1}{2}C(0.002378\text{Ib} \cdot \text{s}^2/\text{ft.}^4)\, V^2 \left(\frac{5280\,\text{ft}}{3600\ \text{s}}\right)^2 = 0.00256CV^2 \tag{3-2}$$

Where,
q = velocity pressure in Ib/ft^2 or psf
V = wind speed in mph

Thus, the velocity pressure, q (in psf), is proportional to the square of the 3-second gust wind speed (in mph) and is a function of the topography, the building height, the building risk category (RC), the building stiffness and hence its frequency, and the percentage of wall openings in the building. Using a form of equation (3-2), the velocity pressure in psf is given in the ASCE 7 load standard [1] as

$$q\ (\text{psf}) = 0.00256K_zK_{zt}K_eV^2 \tag{3-3}$$

Where,
K_z = velocity pressure exposure coefficient (see ASCE 7, Table 26.10-1)
K_{zt} = topographic factor (see ASCE 7, Figure 26.8-1)
K_e = ground elevation factor (see ASCE 7, Section 26.9 and Table 26.9-1). This factor accounts for the lower mass density of air as the height above sea level of the ground surface level increases. It leads to lower wind pressures for structures at locations with higher ground elevations (e.g., Denver, Colorado) [15].
V = 3-second gust wind speed in mph hat can be obtained from the following ASCE 7 figures or or from the ASCE 7 Hazard Tool:
ASCE 7, Figure 26.5-1A for Risk Category I buildings and structures
ASCE 7, Figure 26.5-1B for Risk Category II buildings and structures
ASCE 7, Figure 26.5-1C for Risk Category III buildings and structures
ASCE 7, Figure 26.5-1D for Risk Category IV buildings and structures

Note that these are ultimate or factored wind speeds, with the risk categories (RC) and the wind load factors and importance factors already built into the mapped 3-second wind speed values; therefore, the maximum load factor for wind in the ASCE 7 load combinations (see Chapter 2) is taken as 1.0.

The wind pressures act perpendicular to the building surfaces. The minimum exterior wind pressure is 16 psf per ASCE 7; per IBC Section 1607.14, the minimum interior horizontal wind pressure for the design of interior walls and partitions is 5 psf [6]. Note that the interior walls and partitions in open buildings (e.g., buildings with multiple large openings, such as loading dock overhead doors that may remain open in a wind event), may need to be designed for pressures higher than 5 psf because the interior walls may be subjected to increased pressures from the exterior wind pressures that are transmitted to the interior of the building through the large openings.

Buildings are also subjected to in-plane torsional moments due to wind, and these in-plane torsional moments must be calculated using the wind load cases presented in ASCE 7 Figure 27.3-8. For checking serviceability conditions under wind loads, such as lateral drift limits, a different set of unfactored wind velocities can be obtained using the ASCE hazard tool (*https://ascehazardtool.org*) or from the following ASCE 7 figures [1]:

ASCE 7, Figure CC.2-2 – 10-year MRI 3 s gust wind speed in mph
ASCE 7, Figure CC.2-3 – 25-year MRI 3 s gust wind speed in mph
ASCE 7, Figure CC.2-4 – 50-year MRI 3 s gust wind speed in mph
ASCE 7, Figure CC.2-5 – 100-year MRI 3 s gust wind speed in mph

For lateral deflection and lateral acceleration criteria, see also Section 2-15.

Special attention should be paid to canopy or open structures (e.g., gas station structures) because these structures are susceptible to large wind uplift pressures that may lift off the roof of the building, and note that the wind pressures on canopies or overhangs of buildings are usually much higher than the pressures at other locations of the building.

Seismic or Earthquake Loads—Cause and Effect

According to the Federal Emergency Management Administration (FEMA.gov), "earthquakes are some of the most destructive and unpredictable natural phenomena, causing deaths, injuries, and extensive property damage in populated areas." Earthquakes are caused by the relative movement of the tectonic plates in the earth's crust, and these movements, which occur suddenly, originate at planes of weaknesses in the earth's crust called faults (e.g., the San Andreas fault), causing a release of the stress that has been built up, resulting in a release of massive amounts of energy [2]. This energy causes ground motion, which results in the vibration of buildings and other structures. Although earthquake action causes motion in all directions, only the horizontal and vertical motions are of the most significance to buildings and bridges. Thus, the effect of earthquakes on a building is basically akin to an applied ground displacement at the base of the building, unlike wind forces, which are directly applied to the building.

The point at which the earthquake originates within the earth's crust is called the hypocenter, and the point on the earth's surface directly above the hypocenter is called the epicenter. The size or intensity of the earthquake is measured by the moment magnitude scale (MMS), which is related to the amount of energy released by the earthquake. The amount of damage from an earthquake is also dependent on the distance of the structures from the epicenter of the earthquake. The moment magnitudes of earthquakes vary from M1.0 to M9.9. "The moment magnitude provides an estimate of earthquake size that is valid over the complete range of magnitudes, a characteristic that was lacking in other magnitude scales" such as the Richter scale, which is no longer widely used [12b]. Another measure of the damage potential of an earthquake is the Modified Mercalli Intensity (MMI), which is a qualitative human assessment of the damage potential of an earthquake in contrast to the moment magnitude scale, which is a quantitative measurement [1, 12].

The majority of the world's earthquakes occur along the so-called "Ring of Fire" in the Pacific Rim, which includes countries such as Mexico, the Philippines, Japan, Chile, and States such as California and Alaska in the United States [25], but devastating earthquakes resulting in many thousands of deaths and untold infrastructure damage have occurred also in other places around the Globe, for example, in countries such as Turkey, Haiti, and Italy. Examples of previous earthquakes include the 1985 Mexico City earthquake with a moment magnitude of 8.0 that resulted in more than 10,000 deaths, and multiple earthquakes in series with a maximum moment magnitude of 7.8 hit Turkey and Syria in February 2023 and "killed more than 50,000 people and toppled more than 5600 buildings." In April 2024, a magnitude 4.8 earthquake hit the eastern United States, with the epicenter in Lebanon, New Jersey, but there were no reports of injuries or any significant damage to infrastructure [25].

In the United States, the design of buildings to resist earthquakes focuses mainly on life safety, but the damage from the design basis earthquake is not quantified or defined, and the affected building may be damaged beyond repair. For some signature structures, such as high-rise buildings in seismically active zones like California, performance-based design (PBD) can be used to achieve performance objectives that go above and beyond life safety and include resilience and business continuity and recovery. For example, the China seismic design code prescribes simple performance objectives for several levels of earthquake, as shown below [30]:

Earthquake Levels	Performance Objectives
Frequent earthquake with MRI = 50 years	"No damage"
Moderate earthquake with MRI = 475 years	"Minor damage, but repairable"
Severe earthquake with MRI = 2475 years	"No collapse"

Note that no building or structure is designed to be "earthquake proof"; there is always some probability of collapse, though small, due to the seismic event. Furthermore, the building code does not currently account for multiple earthquakes occurring in succession or in series; for example, if a building survives a design-level seismic event and is damaged, can the damaged building survive subsequent seismic aftershocks? "In the U.S., similar to Turkiye, the building code does not consider an explicit performance objective for multiple earthquakes in series. While the implied ability of a building to sustain aftershocks is often made, the quantity and intensity of those aftershocks and the explicit capacity of a damaged building to resist them are not quantified" [32]. In the current ASCE 7 Load Standard, "explicit design for aftershocks is not a requirement." [1, 26]. The resistance of as-damaged buildings to earthquake afterschocks or to multiple hazards in series has implications for the post-disaster assessment of damaged structures [32].

Seismic waves with long periods and therefore low frequency (f = 1/T) will affect high-rise buildings the most, whereas seismic waves with short periods and therefore higher frequency will tend to affect low-rise buildings the most.

Many major cities in the United States including San Francisco have an old building stock, and existing buildings that were built before the advent of the new seismic codes—that were grandfathered and have not been updated to the new codes—may be particularly vulnerable to earthquakes. These buildings possess low ductility and inadequate energy dissipating abilities. Vulnerable buildings include [23]:

- Buildings with unreinforced masonry shear walls and unreinforced masonry in-fill wall frames,
- Buildings with non-ductile reinforced concrete building frames, and
- Buildings with a soft or weak first story, typically caused by large exterior front wall openings or store fronts at the ground floor level.

Given the enormous loss of lives and the thousands of collapsed buildings that resulted from the 2023 Turkey-Syria earthquake, concerns have been raised about the vulnerability and the "severe earthquake risk" of old building stock in the United States, especially in San Francisco. As one structural engineer opined, "These buildings, when they get damaged, they can get damaged to such a degree that they lose their vertical load-bearing capacity and they collapse." [31].

The Force Induced at the Base of a Structure due to an Earthquake

The *theoretical elastic* dynamic force exerted on a structure by an earthquake is obtained from Newton's second law of motion as follows:

$$F = V = Ma = \left(\frac{W}{g}\right) a = W\left(\frac{a}{g}\right), \tag{3-4}$$

where

V = the force induced at the base of a structure due to the ground acceleration (i.e., the base shear)

M = Mass of structure,

a = ground acceleration of the structure induced by the earthquake,
g = Acceleration due to gravity,
a/g = Seismic coefficient, and
W = Weight of the structure.

The ASCE 7 Load Standard uses a modified version of equation (3-4) together with a design response spectrum to determine the seismic base shear on a structure. The code equation considers the damping (or internal friction of the material), structural, and foundation properties. The base shear, V, is then converted to some code-calculated static lateral force at each floor level and roof level of the building. It should be noted that the lateral forces measured in buildings during actual earthquake events are usually greater than the code-equivalent lateral forces (see Figure 3-2) [1, 20]. However, experience indicates that buildings that have been designed elastically to these code-equivalent static forces have always performed well during actual earthquakes. The reason for this is the ductility of building structures, that is, the ability of structures to dissipate seismic energy without failure through inelastic action, such as cracking and yielding. In general, for many structures, the inelastic lateral displacement is much greater than the elastic lateral displacement, as depicted in Figure 3-2.

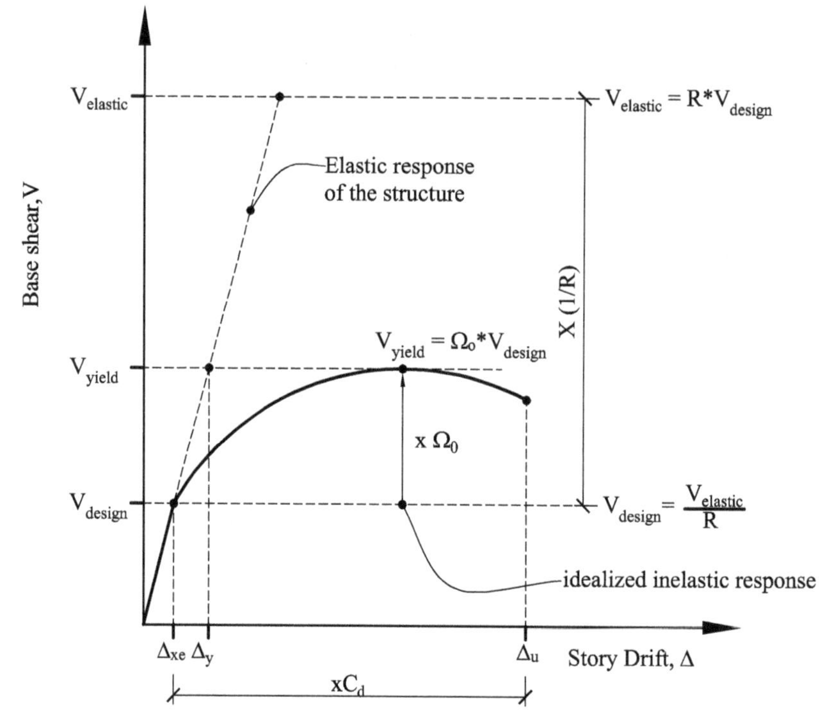

Δ_u = inelastic deflection used to calculate the design story drift, Δ, ($\Delta_u = C_d\Delta_{xe}$)

Δ_{xe} = maximum deflection under design seismic forces from elastic analysis

C_d = deflection amplification factor or fudge factor that amplifies the elastic deflection

R = Seismic response modification coefficient or fudge factor that reduces the elastic seismic force, $V_{elastic}$, to a strength level design force, V_{design}

Ω_o = overstrength factor (see Section 2-3)

FIGURE 3-2 Seismic inelastic force-deformation curve

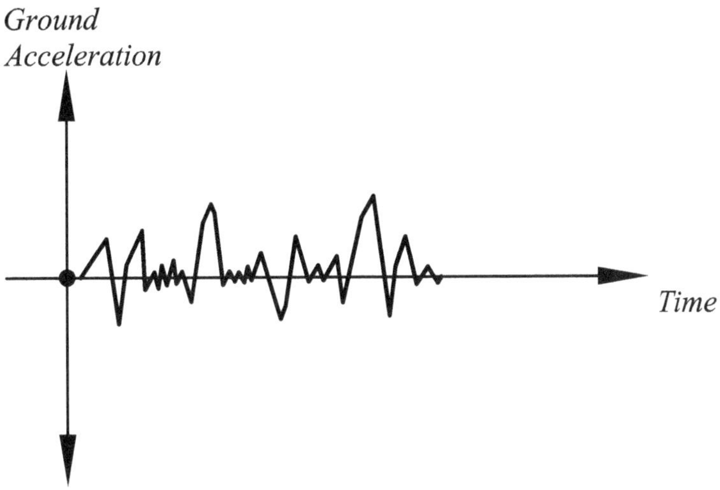

FIGURE 3-3 Acceleration-time plot

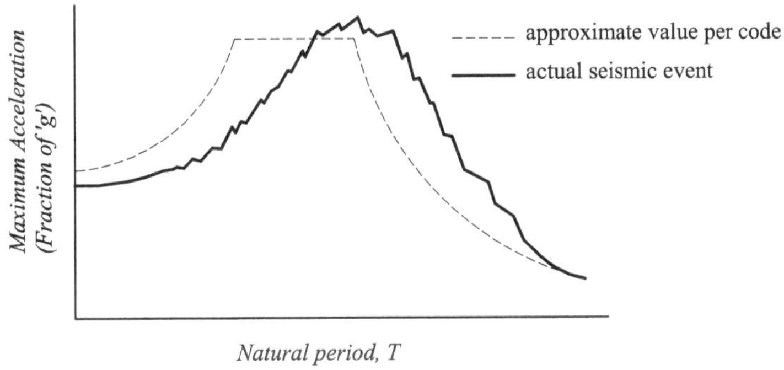

FIGURE 3-4 Acceleration Response Spectra

During an earthquake event, the induced ground acceleration of the structure, measured with an accelerograph, varies in an erratic or random manner, having low and high points as shown in the ground acceleration time history in Figure 3-3. This plot of the ground acceleration or motion during an earthquake is called an accelerogram [2]. A plot of the absolute maximum accelerations of different buildings with different periods, T, yields an acceleration response spectrum similar to that shown in Figure 3-4.

Ductility

In reality, unlike wind loads, "earthquakes act more like an applied displacement on the building [or structure] rather than a set of applied loads" [29]; therefore, ductility, which is the ability of a structural system or element to sustain large inelastic deformations and thus some structural damage under constant load without collapse, resulting in a "significant amount of energy dissipation," is of the utmost importance in designing structures for earthquakes. The horizontal length of the load-deformation curve of a structural member from first yield to when the member collapses (see Figure 3-2) is a measure of the ductility of the structural system; the higher the ratio of the deflection at the ultimate limit state to the deflection at first yield, the more ductile the structural member is. The more ductile a structure is, the better the seismic resistance and behavior of the structure. Ductility is usually achieved in practice by proper *detailing* of the structure and its connections as prescribed in the materials sections of ASCE 7. For seismic design using ASCE 7, system-level ductility is accounted for through the structural system response modification factor, R. This will be discussed later in this chapter.

Similarities and Differences between Wind and Seismic Forces

The similarities and differences between wind and seismic forces affect the design provisions for these forces in the ASCE 7 Load Standard [1]. These are summarized as follows:

- Both wind and seismic loads are dynamic in nature, but earthquakes are even more so than wind.
- During an earthquake, there are no direct external forces applied to the various levels of a structure, but instead, the lateral forces computed using the ASCE 7 Load Standard are only approximations of the internal forces that will arise from the ground motion or acceleration at the base of the structure and the inertial resistance of the structure to this motion. On the other hand, during a windstorm, direct external forces from the wind are applied to the structure.
- Seismic forces on a structure depend on structural and foundation properties and the dynamic properties of the earthquake. The softer the soil, the higher the earthquake forces on the structure. Wind forces, however, depend mainly on the shape and surface area of the structure that is exposed to wind.
- Because of the highly dynamic nature of earthquakes, compared with wind, safety is not necessarily ensured by using a stiffer structure for seismic resistance. In fact, the stiffer a structure is, the higher the seismic forces that the structure attracts. Therefore, in designing for seismic forces, the structural stiffness and the ductility of the structure are both equally important.
- The code-specified seismic forces are smaller than the actual elastic inertial forces induced on the structure during a seismic event; however, buildings have been known to perform well in earthquakes because of the ductility of these structures, which allows the structure to dissipate seismic energy through controlled structural damage but without collapse [3]. Therefore, the structure is designed for a reduced earthquake force that is much smaller than the elastic earthquake force that the structure will experience. This reduction in seismic force is represented by the structural system response modification factor, R (see Figure 3-2). Consequently, when designing for earthquake effects, it is not enough to design just for the code-specified seismic forces; the prescribed seismic detailing requirements in the materials section of the International Building Code (IBC) or Chapter 14 of the ASCE 7 or in References [9a] and [9b] also must be satisfied in order to ensure that adequate ductility is present in the structure to facilitate the inelastic action that justifies using the reduced seismic force indicated in Figure 3-2. On the other hand, when designing for wind forces, stiffness is a more important criterion, and ductility is not as important because of the lesser dynamic nature of wind. Therefore, we design for the full wind forces.
- To calculate seismic forces, the base shear is first determined, and this base shear is converted into equivalent static lateral forces at each floor or diaphragm level of the building using a linear or parabolic distribution based on the modal response of the structure. For wind forces, the design wind pressures are first calculated, followed by the calculation of the lateral forces at each level based on the vertical surface tributary area of each level of the building, and from this, the wind base shear is calculated.
- For wind loads, two sets of lateral forces are required for the design of a structure:
  - Lateral wind forces acting on the main wind force-resisting system (MWFRS) and

- Wind forces acting on smaller elements (i.e., with tributary areas not greater than 700 ft.^{2}) known as the components and cladding (C&C). Elements that are not part of the MWFRS should be designed for the larger C & C wind pressures.

- For seismic design, two sets of lateral forces are required for the design of a structure:
  - Lateral forces, F_x, on the vertical lateral force-resisting system (LFRS), and
  - Lateral forces, F_P, on the horizontal diaphragms (i.e., the roof and floors)—because of the different dynamic behavior of the horizontal diaphragms compared with that of the vertical LFRS during an earthquake event.
  - In addition, the seismic lateral forces on the parts and components (i.e., structural and nonstructural components) of the building also need to be calculated.

The comparison between seismic and wind loads is summarized in the table below:

SEISMIC	WIND
The seismic force, or base shear, is applied *indirectly* to the structure. In reality, the impact of an earthquake on a structure is more like a set of displacements applied to the structure instead of a set of applied loads.	The wind load is applied *directly* to the various levels of the structure.
In ASCE 7, the seismic base shear V is distributed to the various levels of the structure to model the internal forces that will occur in the building due to the ground motion at the base of the structure.	
Seismic forces depend on the properties of the structure and its foundation. The softer the soil, the higher the seismic forces.	The wind forces depend on the shape and surface area of the structure that is exposed to the wind.
In seismic design, *structural stiffness and ductility are both equally important*. Therefore, the design seismic force is the maximum elastic seismic force reduced by the R-factor, the measure of the ductility of the seismic lateral force-resisting system.	For wind design, *strength or stiffness is of more importance*. Therefore, the full wind load is used in the design without any reduction.
For seismic design, two sets of lateral forces are required for the design of the structure: – The lateral force, F_x, on the lateral force-resisting system (LFRS), and – The lateral force, F_P, acting on the horizontal diaphragms (i.e., the roof and floors). – In addition, the seismic lateral forces on parts and components that are permanently attached to the structure (e.g., architectural components and mechanical/electrical pipes) also need to be calculated.	For wind design, two sets of lateral forces are required: – The wind forces on the main wind force-resisting systems (MWFRS). These are used for the design of the lateral force-resisting systems (i.e., braced frames, moment frames, or shear walls). – The wind forces on the components and cladding (C & C). These are used for the design of the smaller elements in the structure to account for spikes in wind pressures.

3-2 LATERAL FORCE-RESISTING SYSTEMS IN STEEL BUILDINGS

The different types of lateral force-resisting systems (LFRS) commonly used in steel buildings to resist wind and seismic lateral loads are discussed in this section. Each of these LFRS may be used solely to resist the lateral force in both orthogonal directions in a building, or a mixed LFRS or a combination of these systems may also be used. In taller buildings (30 stories or higher), a mixed LFRS of moment-resisting frames and shear walls is an efficient system for resisting lateral forces in the same direction [3]. However, for low- and mid-rise buildings, it is typical in design practice to use only one type of LFRS in each orthogonal direction to resist the lateral forces. Depending on architectural considerations, the LFRS may be located internally within the building or on the exterior face of the building. When laying out the LFRS in buildings, soft stories should be avoided where possible. Per ASCE 7 Table 12.3-2, "Soft Story Irregularity is defined to exist where there is a story in which the lateral stiffness is less than 60-70% of that in the story above or, where there are at least three stories above, less than 70-80% of the average stiffness of the three stories above." For moment resisting frames, the columns should be stronger than the beam/girder-to-column connections, and the beam/girder-to-column connections should be stronger than the beams/girders, so that the hinges will form in the beams, and not in the connections or in the columns.

The lateral force distributed to each LFRS is a function of the in-plane rigidity of the roof and floor diaphragms (diaphragms can be classified as either flexible, rigid, or semi-rigid). The definitions of flexible and rigid diaphragms are given in ASCE 7, Section 12.3, and this topic is discussed in detail in Chapter 14. The different lateral force-resisting systems (LFRS) used to resist lateral wind and seismic loads are as follows (note that for wind loads, these LFRS are also called the main wind force-resisting systems, MWFRS):

- **Moment-resisting frames**
  - Fully restrained (FR) moment frames
  - Partially restrained (PR) moment frames
- **Braced frames**
  - X-braced frames
  - Chevron braced frames
  - Knee braced frames
  - Single diagonal braced frames
- **Shear walls**
  - Reinforced concrete shear walls
  - Masonry shear walls
  - Steel plate shear walls
  - Steel plate composite shear walls (Speedcore)

Fully Restrained Moment-Resisting Frames (i.e., Rigid Frames)

For frames with fully restrained moment connections, sometimes called "rigid" or moment frames (see Figure 3-5), the beams and girders are connected to the columns with moment-resisting connections, and the lateral load is resisted by the bending of the beams/girders and columns. For maximum efficiency in steel buildings, the columns in the moment-resisting frames in both orthogonal directions should be oriented so that they are subjected to bending about their strong axis, and the moment-resisting connection can be achieved by welding steel plates to the column flange and bolting or welding these plates to the beam/girder top and bottom flanges, or welding the beam/girder top and bottom flanges directly to the column flange. To support the shear and reaction from the gravity loads, the web of the steel beam/girder is connected to the column flange using shear connection plates (i.e., shear tabs) or double angles that are either welded or bolted to the beam web and to the column flange.

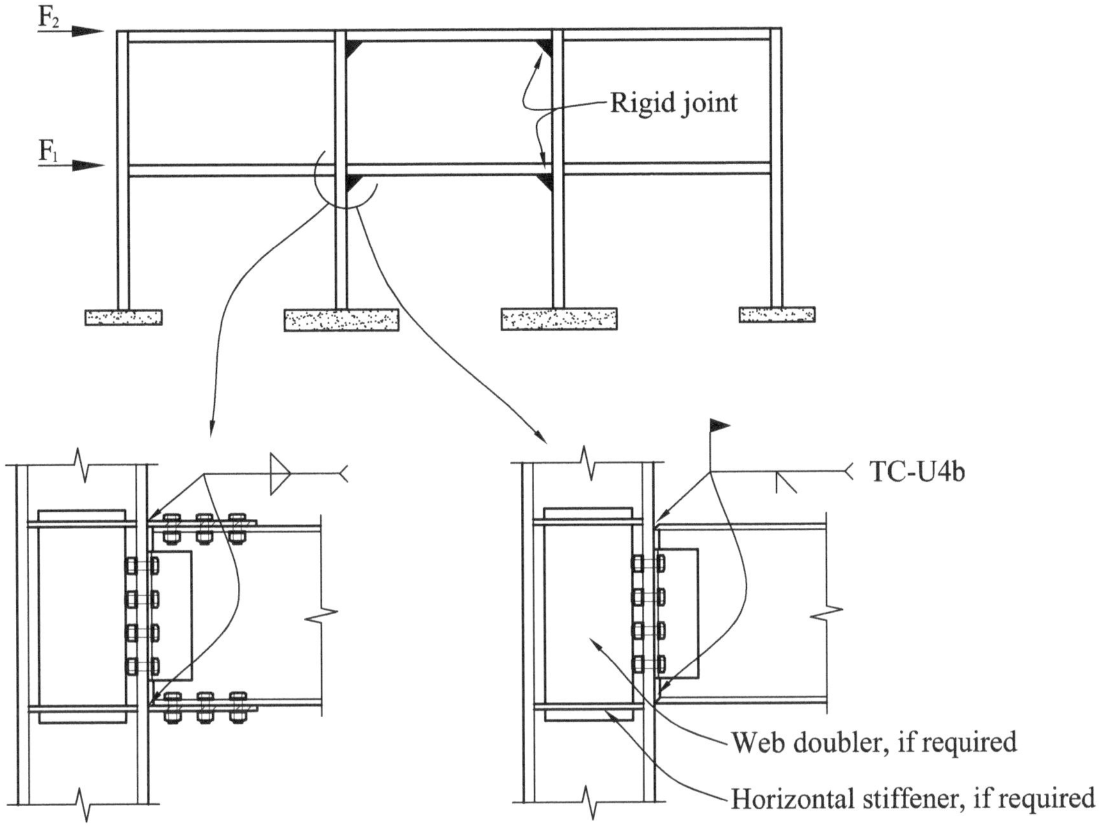

FIGURE 3-5 Steel moment-resisting frames

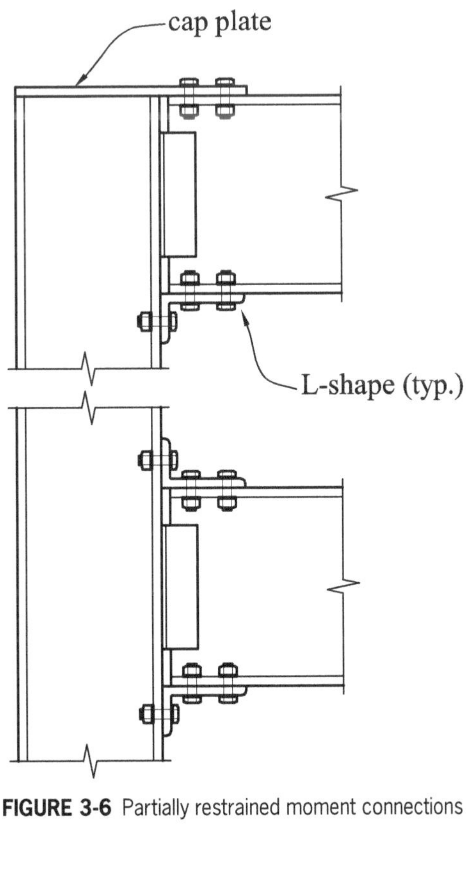

FIGURE 3-6 Partially restrained moment connections

Partially Restrained Moment Connections (Semirigid Frames)

For semirigid frames with partially restrained moment connections (see Figure 3-6), the rigidity of the beam-to-column connections is generally less than that of the fully restrained moment frame. In these connections, the flanges of the steel beam or girder are usually connected to the column flange with angles, and the beam or girder web is connected to the column flange using shear connection plates or angles. However, at the roof level, the top flange of the beam or girder is typically connected to the column with a steel plate that also acts as a cap plate for the column. The deflection of a partially restrained moment frame is generally higher than that of a fully restrained moment frame.

Braced Frames

For braced frames, the lateral load is resisted through axial tension and/or axial compression forces in the diagonal bracing members. The beam-to-column connections in braced frames are usually simple shear connections with no moment-resisting capacity. Examples of braced frames are shown in Figure 3-7; these include X-bracing, Chevron or inverted V-bracing, diagonal bracing, V-bracing, and knee bracing (see also Figures 1-21b and 1-21c). The X-bracing and V-bracing in braced frames offer the least flexibility for the location of doors, while chevron bracing, diagonal bracing, and knee bracing offer the most flexibility, allowing for door, window, and corridor openings between the diagonal braces, and these are usually the bracing systems preferred by architects. Note that knee-braced frames are a hybrid between braced frames and moment-resisting frames because they resist lateral loads by a combination of axial deformation in the knee braces and bending in the beams and columns. In lieu of moment frames, knee braces are commonly used in low-rise school buildings with continuous windows on the exterior walls. Depending on their energy dissipation capacity, braced frames can be classified as eccentric braced frames (EBF) or concentric braced frames (CBF). Concentrically braced frames include X-braced members and chevron braces where the diagonal braces meet at the same work point on the beam, thus there is no vertical shear in the beam from the axial forces in the diagonal braces due to lateral loads. Note that for gravity loads, the diagonals in the Chevron brace (see Figure 3-7b) provide vertical support to the beam or girder at their midspans. In eccentrically braced frames, on the other hand, the diagonal braces do not meet at the same work point on the beam, thus inducing shear forces in the beam between the diagonal brace locations. Another form of braced frames used in buildings with tall story heights is the multi-tiered braced frames with one or more braced panels within a single story with horizontal struts at the top and bottom of each panel. These braces, which may consist of plates (X-braces) or HSS or W-shapes for Chevron and V-braces, are sometimes hidden within interior walls or in the exterior walls where there are no windows; in some cases, the architect may choose to expose the diagonal braces for aesthetic purposes.

A more recent braced frame system used in steel buildings for seismic resistance are buckling-restrained braced frames (BRBF). In this system, the diagonal braces are restrained from lateral buckling and therefore can attain their cross-sectional axial material strengths in both tension and compression. The core steel brace can be a plate or cruciform-shaped axially loaded member that is enclosed in a tube shell, but the exterior tubing is not attached to the buckling-restrained brace member. The interior space between the inside wall of the tube steel and the BRB is filled with mortar, with a bond breaker between the mortar and the buckling restrained brace (BRB) to ensure that the BRB is freely able to deform axially without restraint from the mortar or the tube steel. Thus, the outer tube shell and the mortar provide lateral restraint against flexure and buckling to the axially loaded "core steel" of the BRB member but without contributing to its cross-sectional axial load capacity; thus, the BRB member can attain its yield strength in both tension and compression and is able to reach the strain hardening phase without buckling. These BRB members have a "full and essentially balanced hysteresis"—i.e., they possess identical behavior in both tension and compression over multiple cycles of loading and unloading—resulting in a much higher capacity to dissipate energy; therefore, they are very efficient in resisting seismic loads [17]. Steel buckling-restrained braced (BRB) frames can be used to reduce the earthquake or seismic lateral forces acting on a building because of the increased value of the seismic response coefficient (R = 8) compared to steel ordinary concentrically braced frames (R = 3¼).

Though the moment frame shown in Figure 3-5 and the braced frames shown in Figure 3-7 extend only over one bay (i.e., extends over two adjacent columns), braced frames or moment frames may extend over multiple bays (i.e., extend over more than 2 columns) of a building to help "spread out" the overturning moment and thus minimize the net uplift force at the bracing frame or moment frame columns. For example, each of the four moment frames in Example 3-1 (Figure 3-11) and Example 3-3 (Figure 3-17b) are 3-bay moment frames (i.e., each moment frame extends over three bays or over four columns).

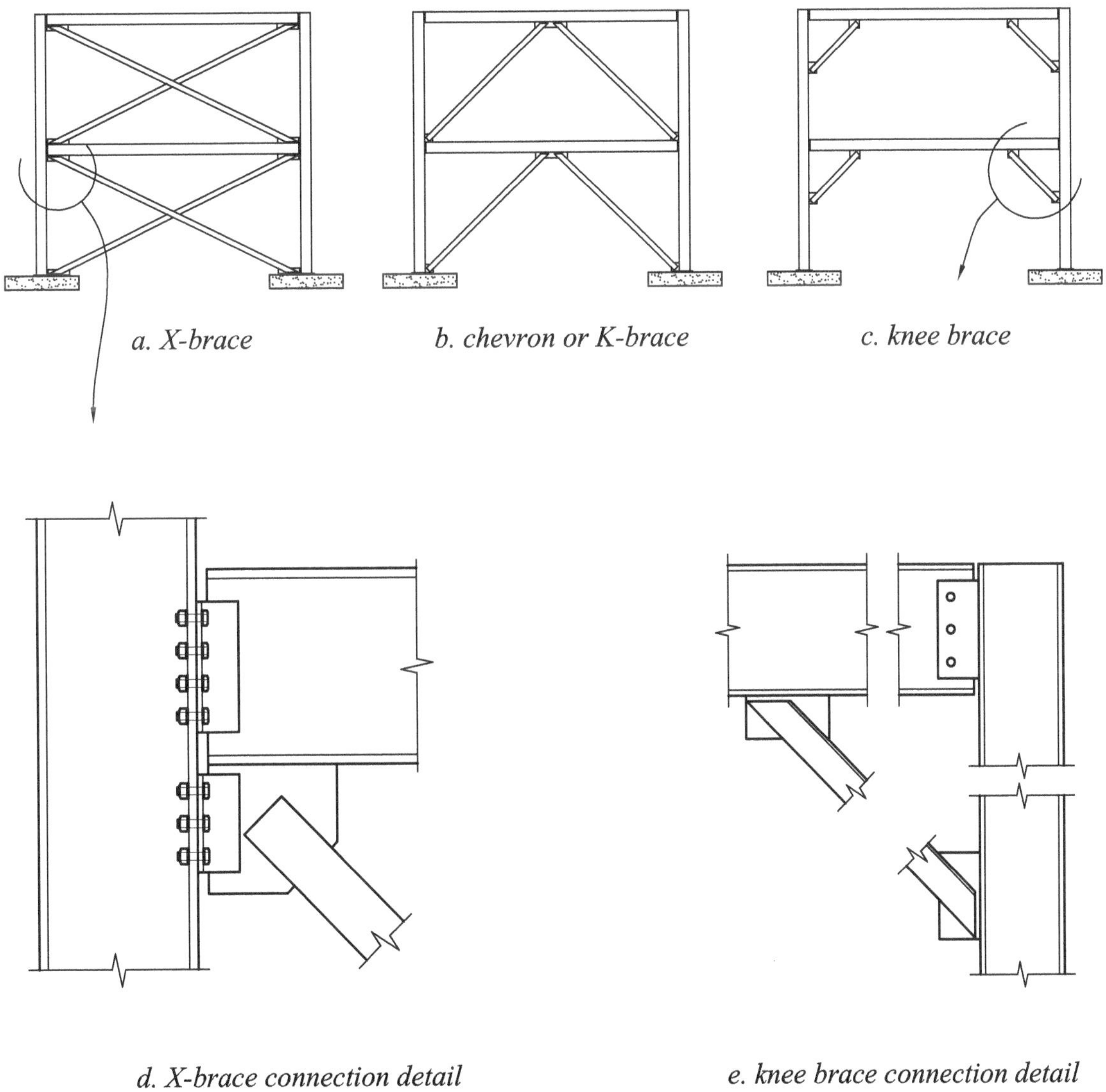

FIGURE 3-7 Braced frames

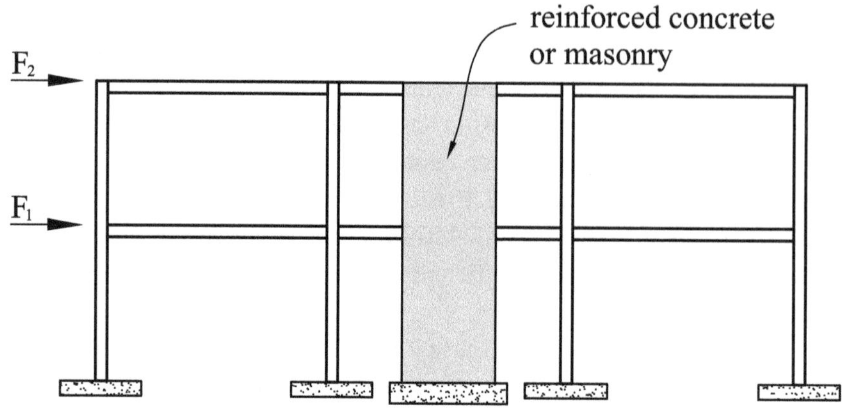

FIGURE 3-8a Reinforced concrete and masonry shear walls

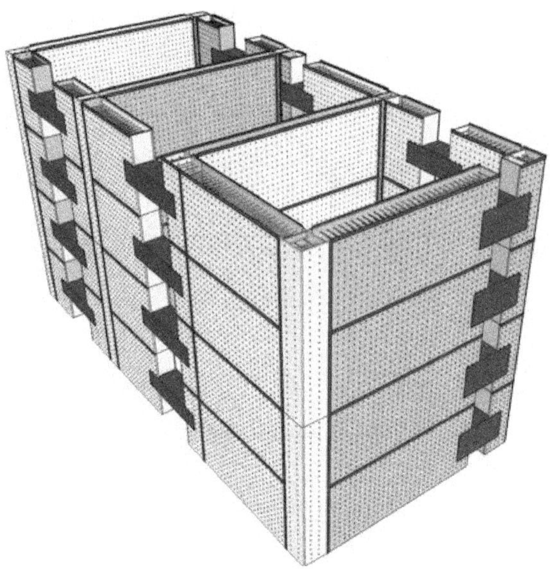

FIGURE 3-8b Steel Plate Composite Coupled Shear Wall Module - Speedcore (Photo courtesy of Magnusson Klemencic Associates)

FIGURE 3-8c Steel Plate Composite Coupled Shear Walls (Speedcore) under construction (Photo courtesy of Magnusson Klemencic Associates)

Shear Walls

Shear walls (see Figures 3-8a through 3-8c) are planar structural elements that act as vertical cantilevers fixed at their bases; they are usually constructed of concrete, masonry, plywood sheathing, or steel plates. They could be located internally within the building or on the exterior face of the building. The concrete or masonry walls around the stairs and elevator shafts are frequently used as shear walls. Shear walls are very efficient lateral-force-resisting elements. Concrete shear walls that are perforated by door, window, or corridor openings are called coupled shear walls, and these may be modeled approximately as moment-resisting frames. The portion of the wall above and below the door, or window, or corridor openings that connects one solid wall segment to an adjacent wall segment is called a coupling beam. The actual strength of a coupled shear wall lies somewhere between the strength of the wall moment frame and the strength of an unperforated or solid shear wall of the same overall dimensions. The coupling beams in coupled shear walls must be adequately reinforced and detailed for the shear and bending moment that they are subjected to.

Another type of shear wall system that has been used for at least two high-rise buildings is the composite plate shear walls that are concrete filled, also called "Speed Core" [18] (see Figures 3-8b and 3-8c). They consist of two longitudinal steel plates connected with steel cross ties at specified horizontal and vertical spacings and with shear studs welded on the interior face of the two plates that allow the concrete infill to interact with the steel plates to create the composite action of the sandwich system. The ends of the longitudinal steel plates are connected to transverse steel end plates that also have studs welded to their interior faces. Essentially, this "composite corewall system" is a rectangular steel box infilled with self-consolidating concrete (SCC) and without the need for vertical steel reinforcement (i.e., rebar) in the concrete because the steel plates provide the reinforcement. Shear studs spaced at regular intervals horizontally and vertically are welded to the inside face of the longitudinal and transverse steel plates. Similar to steel columns, the Speedcore composite shear wall system is spliced at a minimum of 4 ft above the floor level (see Chapter 8 for the column splice locations), and fireproofing is provided using a protective coating to the outer steel plates. This sandwich shear wall system obviates the need for concrete formwork since the steel plates act as the formwork for the concrete infill and has been predicted to cut down the superstructure construction time by up to 50% [19]. The design of the Speedcore shear wall system is covered in AISC Design Guide 37.

Although moment-resisting frames are the most flexible of all the lateral force-resisting systems discussed above, they provide the most architectural flexibility for the placement of windows, doors, and corridors, while shear walls and X-braced frames provide the least architectural flexibility. The distribution of lateral forces to the lateral force-resisting systems (LFRS) and the diaphragm elements (i.e., collectors or drag struts and chords) in building structures is discussed in Chapter 14.

3-3 WIND LOADS

The two parts of building structures for which wind loads are computed are the main wind force-resisting system (MWFRS) and the components and cladding (C&C). The MWFRS consists of the shear walls, braced frames, and moment-resisting frames that are parallel to the direction of the lateral wind force, including the roof and floor diaphragms and the chords and collectors (i.e., drag struts). The C&C are small individual structural components or members (with tributary areas not greater than 700 ft.^{2}) and the wind load acts perpendicular to these members. Examples of C&C include walls, stud walls, cladding, a roof deck fastener resisting a net uplift wind force, and a cladding fastener resisting a suction force. The wind pressures on C&C are usually higher than the wind pressures on the MWFRS because of local spikes in wind pressure that occur over small areas of the C&C. The C&C wind pressure is a function of the effective wind area, A_e, given as

$$\begin{aligned} A_e &= \text{Span of member} \times \text{Tributary width} \\ &\geq (\text{Span of member})^2/3 \end{aligned} \tag{3-5}$$

For cladding and deck fasteners, the *effective wind area*, A_e, shall not exceed the area that is tributary to each fastener. In calculating wind pressure, positive pressures are indicated by a force "pushing" into the wall or roof surface, while negative pressures are shown "pulling" away from the wall or roof surface. The minimum design wind pressure for MWFRS and C&C is 16 psf and is applied to the vertical projected area of the wall surfaces for the MWFRS and perpendicular to the wall or roof surface for C&C (see Figure 3-9 or ASCE 7, Figure C27.1-1).

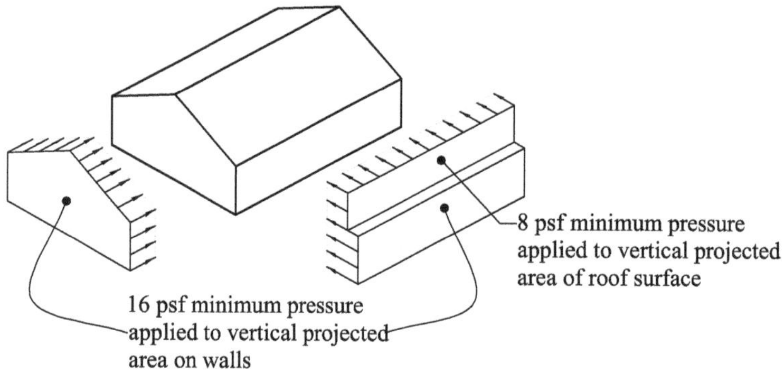

FIGURE 3-9 Minimum wind load diagram

3-4 CALCULATION OF WIND LOADS ON BUILDING STRUCTURES

The calculation of wind loads as well as tornado loads is covered in Chapters 26 through 32 of ASCE 7-22. The outline of the chapters is as follows:

Chapter 26: General requirements and parameters for wind loads

Chapter 27: Wind loads on the main wind force-resisting system (MWFRS)—Directional method

Chapter 28: Wind loads on the main wind force-resisting system (MWFRS) —Envelope method

Chapter 29: Wind loads on building appurtenances and other structures—MWFRS—Directional method

Chapter 30: Wind loads on components and cladding (C & C).
The importance of exterior cladding and their attachment to the building frame was evident in the devastation of oceanfront high-rise buildings in Acapulco, Mexico, due to Hurricane Otis on October 5, 2023. The exterior cladding was light and adequate to meet the seismic requirements, but not strong enough or adequately attached to the structural frame to resist the 165 mph winds from the hurricane [27].

Chapter 31: Wind tunnel procedure. This procedure is typically used for very tall and wind-sensitive buildings where *across-wind oscillations* occur simultaneously with *along-wind oscillations* [1], and for slender buildings with a height-to-width ratio greater than 5.0 [4]. The methods for wind load calculations presented previously only consider *along-wind motion* of the structure. The wind pressures on buildings with unique topographic features should also be determined using the wind tunnel method, and wind tunnel tests are also used to determine the effect of wind on pedestrians around the base of buildings.

Chapter 32: Tornado loads. Wind pressures due to tornados are presented in this section of ASCE 7-22.

In Chapter 26, the general requirements and parameters for determining wind loads are defined and discussed. The wind hazard maps for the different risk categories and the definitions of the wind load parameters, such as building enclosure categories—open, partially enclosed, partially open, and enclosed buildings—used to determine the internal pressures are provided in ASCE 7 Chapter 26. The basic factored wind speed maps, the topography factor, and the building exposure categories that are dependent on the terrain roughness are also provided in Chapter 26 of ASCE 7.

The directional method found in ASCE 7-22 Chapter 27 is used in this text for calculating wind loads on buildings. More detailed treatment of wind and seismic loads can be found in Ref. [11]. In the directional method, the applied wind load is a function of the external wind pressures on the windward, leeward, and side walls of the building. On the windward face, the wind pressure varies and increases with height, while on the leeward side, the wind pressure is constant throughout the height of the building, and with a leeward side wind pressure equivalent to the wind pressure at the mean roof height, h, of the building. This method is applicable to regularly shaped buildings of all heights, and the exterior pressure coefficients (C_P) used in the directional method were obtained from past wind tunnel testing on models of prototypical buildings. In the directional method, the information required to compute the lateral wind loads is as follows:

a. The factored wind speed is based on the risk category (RC) and the site location. Note that there is no importance factor for wind loads because the importance factor is embedded in the ultimate or factored wind speed values. The unfactored wind speed is used for lateral deflection calculations. The factored and unfactored wind speeds can be obtained for any location in the United States from the wind speed maps in ASCE 7-22 or by using the ASCE 7 hazard tool: *https://ascehazardtool.org*.

b. The external exposure category is based on the surface roughness category (B, C, or D). This is used to determine the velocity pressure coefficients, K_z and K_h.

c. The enclosure classification—i.e., open, partially open, partially enclosed, or enclosed—is used to determine the internal pressure coefficient. Many buildings can be classified as "enclosed," though the "partially enclosed" designation is more conservative and accounts better for future uses of the building.

d. The topographic factor, K_{zt}, takes on a value of 1.0 for flat land.

e. The wind gust factor, G, which accounts for wind turbulence (since the wind velocity varies with time), and is dependent on the lateral natural frequency of vibration of the building, f. The higher the natural frequency, the stiffer the structure is and the less dynamically sensitive to wind the structure is.

- For $f \geq 1$, the structure is classified as rigid, and the wind gust factor, G = 0.85.
- For $f < 1$, the structure is classified as flexible, and the wind gust factor, G, which will be greater than 0.85, is obtained from ASCE 7-22 Section 26.11.

f. The ground elevation factor, K_e, is an adjustment factor for the air density. The higher the ground elevation above sea level, the lower the air density, and hence the lower the ground elevation factor. A lower ground elevation factor means a lower velocity wind pressure (see equation (3-3)).

The following are assumed in the directional method:

a. The wind pressure acts perpendicular to the surface of the structure.
b. The windward wind pressure varies and increases with height.
c. The leeward wind pressure is obtained at the mean roof height and is uniform or constant over the height of the building.
d. The roof wind pressures are uniform.

In the directional method, a simple diaphragm building is also assumed, which means that the wind forces are assumed to be concentrated at the horizontal diaphragm levels, i.e., at the roof and the floor levels, and the lateral forces at each diaphragm level are then distributed to the vertical LFRS (i.e., the shear walls, braced frames, or moment-resisting frames) that are parallel to the wind direction [1]. Therefore, the net wind pressure on the building for the design of the main wind force-resisting system (MWFRS) is the sum of the magnitude of the windward pressure and the leeward wind pressure since the interior pressures cancel out.

The velocity pressure on the windward face is lower closer to the ground due to the drag forces that occur closer to the ground surface. Using equation (3-3), the general equations for the velocity wind pressure *at a height, z,* above the ground surface can be written as (ASCE 7 Section 26.10):

$$q_z = 0.00256 K_z K_{zt} K_e V^2,\ psf \tag{3-6}$$

The general equations for the velocity wind pressure *at the mean roof height, h,* above the ground surface is given as

$$q_h = 0.00256 K_h K_{zt} K_e V^2,\ psf \tag{3-7}$$

Where,

q_z = velocity wind pressure, in psf, at a height z, in ft, above the ground surface

q_h = velocity wind pressure, in psf, at the mean roof height, h.

V = the basic wind speed in miles per hour (mph) obtained from ASCE 7, Section 26.5. The basic wind speed corresponds to the 3-second gust speed at 33 ft above the ground surface for Exposure Category C.

K_z = Velocity pressure exposure coefficients at height z above the ground level (ASCE 7 Table 26.10-1), and it depends on the Exposure Category.

K_h = Velocity pressure exposure coefficients at the mean roof height, h, in feet above the ground level (ASCE 7 Table 26.10-1), and it depends on the Exposure Category

K_{zt} = topography factor (ASCE 7 Table 26.6-1); K_{zt} is equal to 1.0 for flat land.

K_e = ground elevation factor (ASCE 7 Table 26.9-1), and it is permitted and conservative to assume a value of 1.0 for all cases.

The general equation for the net design wind pressure on the wall surface of a building structure is given as

$$p_{design} = q_z K_d GC_p - q_h K_d (GC_{pi}),\ \text{psf} \tag{3-8}$$

For the *windward* wall (WW), the factored design wind pressure is

$$p_{WW} = q_z K_d GC_{p,WW} - q_h K_d (GC_{pi}),\ \text{psf} \tag{3-9}$$

For the *leeward* wall (LW), the factored design wind pressure is

$$p_{LW} = q_h K_d GC_{p,LW} - q_h K_d (GC_{pi}),\ \text{psf} \tag{3-10}$$

Where,

K_d = wind directionality factor from ASCE 7 Table 26.6-1. It has a value of 0.85 for the main wind force-resisting systems (MWFRS) and components and cladding (C&C).

The windward pressure "*pushes*" into the windward wall, and this is regarded as a positive pressure, whereas the leeward pressure "*pulls*" away from the leeward wall and therefore is regarded as a negative pressure or suction.

The net design pressure on the main wind force-resisting system (MWFRS) for the entire building structure is given as

$$\begin{aligned} p_{net} &= p_{WW} - p_{LW} \\ &= q_z K_d GC_{p,WW} - q_h K_d (GC_{pi}) - \left[q_h K_d GC_{p,LW} - q_h K_d (GC_{pi})\right] \\ &= q_z K_d GC_{p,WW} - q_h K_d (GC_{pi}) - q_h K_d GC_{p,LW} + q_h K_d (GC_{pi}) \qquad (3\text{-}11) \\ &= q_z K_d GC_{p,WW} - q_h K_d GC_{p,LW} \end{aligned}$$

Note that the terms containing the internal pressures cancel out in equation (3-11). Since the exterior wall pressure coefficients for the windward wall, $C_{p,WW}$, are *positive* and those for the leeward wall, $C_{p,LW}$, are *negative* (see ASCE 7, Figure 27.3-1), therefore, the net design wind pressure on the MWFRS for the entire building structure will be the **algebraic sum** of the magnitudes of the windward and leeward wind pressures. That is,

$$p_{design} = |p_{WW}| + |p_{LW}| = |q_z K_d GC_{p,WW}| + |q_h K_d GC_{p,LW}| \qquad (3\text{-}12)$$

Where,

G = wind gust-effect factor (see ASCE 7, Section 26.11) with a value of 0.85 for rigid buildings (i.e., buildings with a fundamental lateral frequency of vibration, $f \geq 1.0$ Hz). For flexible buildings (i.e., $f < 1$ Hz), the gust-effect factor is calculated using the equations in ASCE 7, Section 26.11. The wind gust-effect factor accounts for the effect of wind turbulence on the structure in the direction of the wind. It does not account for flutter, across-wind loading, vortex shedding, or any aerodynamic torsional vibration of the building [1].

f = natural frequency.

For wind load calculations, the natural lateral frequency of vibration of a structural steel moment-resisting frame building in Hz or cycles per second can be approximated as $f = 22.2/h^{0.8}$, where h is the mean roof height in feet. The natural lateral frequency of vibration for other lateral-force-resisting systems in Hz or cycles per second is approximated as $f = 75/h$, where h is the mean roof height in feet (ASCE 7-22, Section 26.11.3). See ASCE 7-22, Section 26.11.2.1 for the limitations on the approximate lateral frequency equations.

$C_{p,WW}$ = External pressure coefficient for the windward wall from ASCE 7, Figure 27.3-1.

$C_{p,LW}$ = External pressure coefficient for the leeward wall from ASCE 7, Figure 27.3-1. The external pressure coefficients for the leeward wall are dependent on the building aspect ratio or shape factor, L/B, where,
L is the plan dimension of the building *parallel* to the wind direction, and
B is the plan dimension *perpendicular* to the wind direction.

(GC_{pi}) = Internal wind pressure coefficient from ASCE 7 Section 26.13-1.
q_h and q_z are as defined previously.

The directional method of ASCE 7-22, Chapter 27 is the only wind load calculation method that will be used in this text because of its versatility, and this method will be illustrated in

Example 3-1. For other methods for calculating wind pressures, or for wind pressures on non-structural elements, see ASCE 7-22, Chapters 28 and 29.

The wind pressure distribution of the Directional Procedure of Chapter 27 is assumed to act perpendicular to the surface of the wall or roof, and the windward wall pressure varies with the height of the building (due to the lower pressure near the surface as a result of surface friction), while the leeward wall pressure measured at the mean roof height is uniform or constant over the height of the building; the roof pressures are also uniform (see Figures 3-10b and 3-10c).

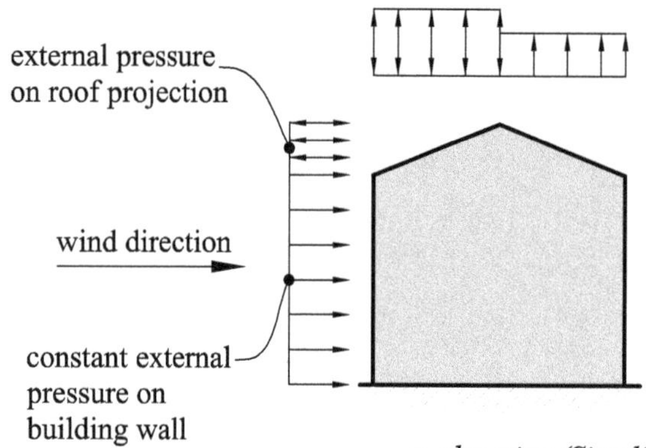

a. elevation (Simplified Method, ASCE 7 Ch. 28, Part 2)

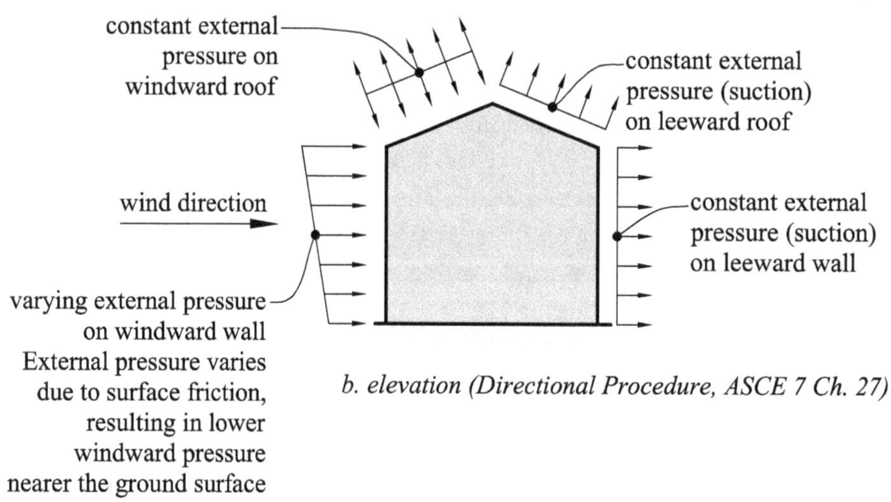

b. elevation (Directional Procedure, ASCE 7 Ch. 27)

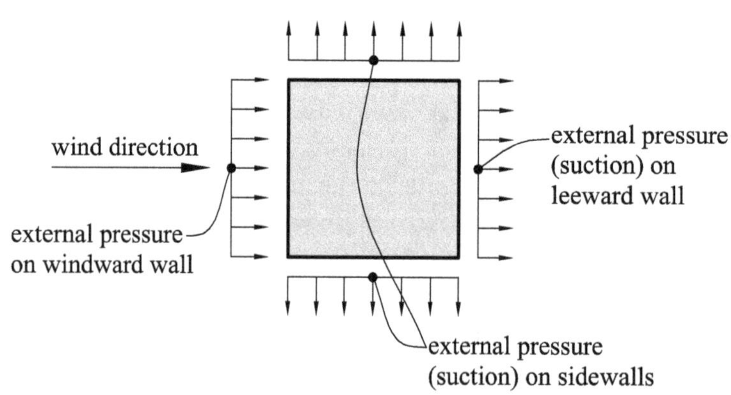

c. plan view of wind pressure

FIGURE 3-10a, b, & c External wind pressure distribution

Orthogonal Wind Load Effects and Wind Load Cases for the MWFRS (ASCE 7 Figure 27.3-8)

To account for orthogonal wind load effects, that is, the effect of wind loads acting simultaneously in two orthogonal directions or the effect of diagonal winds and the effect of in-plane torsion from wind loads, four wind load cases are prescribed in ASCE 7 Figure 27.3-8. For the main wind force-resisting system (MWFRS) in any building, the four wind load cases are as follows:

- **Wind Load Case 1:**
  Accounts for full design pressure on the windward ($p_{WW,X}$ or $p_{WW,Y}$) and leeward faces ($p_{LW,X}$ or $p_{LW,Y}$) of the building. This load case is calculated separately for the two orthogonal (X- and Y-) directions of the building, and the wind load is assumed to act *independently* (i.e., not simultaneously) in both directions *without* any in-plane torsional moment. That is:
  - Full-design wind pressures acting on the windward and leeward faces of the building simultaneously
  - The wind load is considered ***separately*** in the two orthogonal (i.e., X- and Y)-directions
- **Wind Load Case 2:**
  This load case accounts for 75% of the full design pressures on the windward and leeward faces of the building *plus* an in-plane torsional moment, M_T, about a vertical axis of the building due to the eccentricity of $0.15B_X$, where B_X is the plan dimension of the building perpendicular to the X-direction wind. This load case is calculated *separately* for the Y-direction wind using an eccentricity of $0.15B_Y$, where B_Y is the plan dimension of the building perpendicular to the Y-direction wind. For this load case, the wind loads in the X- and Y- directions are assumed to act *independently*.

  X-direction wind:

  $e_X = \pm 0.15B_X$

  Windward lateral pressure $= 0.75p_{WW,X}$

  Leeward lateral pressure $= 0.75p_{LW,X}$

  $M_T = 0.75(p_{WW,X} + p_{LW,X})(B_X)(e_X)$, ft-kips/ft. of tributary height.

  $= F_{x,X}(e_X)$, ft-kips/ft. of tributary height.

  Where, $F_{x,X} = 0.75(p_{WW,X} + p_{LW,X})(B_X)$, kips/ft. of tributary height.

  Y-direction wind (applied *independently* of the X-direction wind):

  $e_Y = \pm 0.15B_Y$

  Windward lateral pressure $= 0.75p_{WW,Y}$

  Leeward lateral pressure $= 0.75p_{LW,Y}$

  $M_T = 0.75(p_{WW,Y} + p_{LW,Y})(B_Y)(e_Y)$, ft-kips/ft. of tributary height.

  $= F_{x,Y}(e_Y)$, ft-kips/ft. of tributary height.

  Where, $F_{x,Y} = 0.75(p_{WW,Y} + p_{LW,Y})(B_Y)$, kips/ft. of tributary height.
- **Wind Load Case 3:**
  75% of the full design pressures (i.e., $0.75p_{WW}$ and $0.75p_{LW}$) is assumed to act *simultaneously* on the windward ($0.75p_{WW,X}$ or $0.75p_{WW,Y}$) and leeward faces ($0.75p_{LW,X}$ or $0.75p_{LW,Y}$) of the building in the two orthogonal (X- and Y-) directions. Thus, this is a combined load case that occurs *simultaneously* in both the X- and Y- directions but without any in-plane torsional moment. In summary,
  - 75% of the full design wind pressure acts on the windward and leeward faces of the building.

- The wind loads are considered ***concurrently*** in both orthogonal (i.e., X- and Y-)-directions.

- **Wind Load Case 4:**
This load case accounts for 56.3% of the full design pressures acting *simultaneously* on the windward and leeward faces of the building in both orthogonal (X- and Y-) directions, plus a combined in-plane torsional moment, M_T, about a vertical axis of the building due to the eccentricities of $0.15B_X$ and $0.15B_Y$, where B_X is the plan dimension of the building perpendicular to the X-direction wind and B_Y is the plan dimension of the building perpendicular to the Y-direction wind.

X-direction wind:

$e_X = \pm 0.15B_X$

Windward lateral pressure $= 0.563p_{WW,X}$

Leeward lateral pressure $= 0.563p_{LW,X}$

Y-direction wind (applied *simultaneously* with the X-direction wind load):

$e_Y = \pm 0.15B_Y$

Windward lateral pressure $= 0.563p_{WW,Y}$

Leeward lateral pressure $= 0.563p_{LW,Y}$

The in-plane torsional moment, M_T, due to the combined X- direction and Y-direction wind loads acting *simultaneously* is

$M_T = 0.563(p_{WW,X} + p_{LW,X})(B_X)(e_X) + 0.563(p_{WW,Y} + p_{LW,Y})(B_Y)(e_Y)$, ft.-kips/ft. of tributary height.

$= F_{x,X}(e_X) + F_{x,Y}(e_Y)$, ft.-kips/ft. of tributary height.

Where, $F_{x,X} = 0.563(p_{WW,X} + p_{LW,X})(B_X)$, kips/ft. of tributary height.

$F_{x,Y} = 0.563(p_{WW,Y} + p_{LW,Y})(B_Y)$, kips/ft. of tributary height

The parameters required in the directional method of ASCE 7 Chapter 27 are discussed in the following sections:

Exposure Categories

An exposure category is a measure of the terrain surface roughness and the degree of exposure of the building to the wind. The three main exposure categories are Exposures Categories B, C, and D. Exposure B is the most commonly occurring exposure category, although many engineers tend to specify exposure C for most buildings. An airport is an example of an Exposure C condition. The exposure category is determined by first establishing the type and extent of the ground surface roughness at the building site using Table 3-1, and then using Table 3-2 to determine the exposure category (see ASCE 7-22, Sections 26.7.2 and 26.7.3).

TABLE 3-1 Ground surface roughness categories

Ground surface roughness	Description
B	Urban and suburban, and mixed wooded areas with numerous spaced obstructions the size of a single-family dwelling or larger
C	Open terrain or open country with scattered obstruction is generally less than 30 ft. in height.
D	Flat, unobstructed areas and water surfaces; includes smooth mudflats, salt flats, and unbroken ice.

TABLE 3-2 Exposure Categories

Exposure category*	Description
B	Surface roughness B occurs in the *upwind*^ direction for a distance of $20h \geq 2600$ ft. relative to the building location, where h is the height of the building. Examples: City centers; suburban residential areas
C	Where exposure category B or D does not occur Examples: Open country; flat grass land
D	Exposure D occurs if any one of the following conditions is satisfied: 1. For a building with a mean roof height of h located on the shoreline or coastline, the length of the water body in front of the building, $S_{water} \geq$ greater of 5000 ft or 20h, where h is the mean roof height of the building. 2. For a building with a mean roof height of h located inland at a distance of S_{land} from the shoreline or coastline (with surface roughness B or C between the shoreline and the building), with the length of the water body beyond the shoreline, $S_{water} \geq$ greater of 5000 ft or $20h$, and $S_{land} \leq 20h$. 3. For a building with a mean roof height of h located inland at a distance of S_{land} from the shoreline or coastline (with surface roughness B or C between the shoreline and the building), with the length of the water body in front of the coastline, $S_{water} \geq$ greater of 5000 ft or $20h$, and $S_{land} \leq 600$ ft. Examples include Ocean-front and beach-front structures; structures in coastal areas (note: these structures would also have to be designed for flood loads).

* Where a building or structure is located in a transition zone between two exposure categories, use the exposure category that produces the higher wind loads. In the author's experience, the most widely used exposure category is Exposure Category C.

^ Upwind is the ground surface starting from the building or structure toward the direction opposite the direction of the wind. For example, if the wind is acting on a building from the west to the east, then the upward direction is west of the building while the downwind direction is east of the building.

Basic Wind Speed, *V*

The wind speeds are given in ASCE 7-22 for two load levels: the serviceability load level wind speed and the ultimate load level wind speed. The wind speeds at the serviceability load levels are used for lateral drift or lateral deflection calculations and are given for 10 year, 25 year, 50 year, and 100 year mean return intervals (MRI) in ASCE hazard tool (*https://ascehazardtool.org*). The ultimate wind speeds used for strength level or ultimate limit state design have higher mean return intervals. The basic wind speed is a 3-second gust wind speed in miles per hour based on [1]:

- 15% probability of exceedance in 50 years for risk category (RC) I buildings (mean return interval (MRI) of 300 years)
- 7% probability of exceedance in 50 years for risk category (RC) II buildings (mean return interval (MRI) of 700 years)
- 3% probability of exceedance in 50 years for risk category (RC) III buildings, (mean return interval (MRI) of 1700 years)
- 1.6% probability of exceedance in 50 years for risk category (RC) IV buildings (mean return interval (MRI) of 3000 years)

Wind speed increases as the height above the ground level increases because of the reduced drag effect of terrain surface roughness at higher elevations. See ASCE 7 Figures 26.5-1A through 26.5-1D for the ultimate or factored wind speed maps for the continental United States corresponding to the risk categories and MRI, or they can be obtained using the ASCE 7 Hazard Tool . Note that the load factor and the importance factor for wind have already been embedded within the ultimate wind speed maps; therefore, there are no importance factors given for wind loads in ASCE 7 Table 1.5-2.

For structures in the United States, the ASCE 7 Hazard Online tool (see *https://asce7hazardtool.online/*) can be used to more accurately determine the wind and seismic design parameters for a building. In fact, all the design parameters for seven different hazards, including wind, seismic, snow, ice, rain, flood, and tsunami, can be accurately determined using this online tool. For example, the basic factored wind speed, and the unfactored wind speed for 10-year, 25-year, 50-year, and 100-year MRI can be obtained using this ASCE 7 Hazard Tool.

Components and Cladding (C&C) Wind Pressures

The C&C wind pressures are typically calculated for the roofs and walls of a building. These pressures are used to design the cladding, the backup walls, and the attachment of the cladding to the backup wall; they are also used to design the roof and their attachment to the structure, and to design for net uplift wind loads on the roof framing. The C&C pressures are divided into two zones on the wall (Zone 4 – interior areas and Zone 5- edge areas), and three zones on the roof (Zone 1 – interior areas, Zone 2 – edge areas, and Zone 3 – corner areas). The external pressure coefficients (GC_p) for each of these zones are given in ASCE 7-22 Figures 30.3-1 for walls and Figure 30.3-2A for roofs in buildings with h not greater than 60 ft. For flat roof buildings taller than 60 ft., the external pressure coefficients (GC_p) are given in ASCE 7-22 Figure 30.4-1 for the roof and walls. The external pressure coefficient is a function of the tributary area or effective wind area of the structural member. The C&C factored design wind pressures for low rise buildings (i.e., h not greater than 60 ft.) for all walls and roofs is $p = q_h K_d[(GC_p) - (GC_{pi})]$. For buildings taller than 60 ft., the C&C factored design wind pressures for the windward wall is $p = q_z K_d(GC_p) - q_h K_d(GC_{pi})$, and the factored design wind pressures for the roof, leeward wall and sidewalls, is $p = q_h K_d[(GC_p) - (GC_{pi})]$.

EXAMPLE 3-1

Directional method for wind loads (MWFRS)

The typical floor plan and elevation for a six-story office building measuring 100 ft. × 100 ft. in plan and laterally braced with four 3-bay ordinary moment-resisting frames are shown in Figure 3-11. The building is a risk category (RC) II building located in an area with a 3-second factored wind speed of 115 mph. Determine the factored design wind forces on the main wind force-resisting systems (MWFRS) using the directional method in ASCE 7 Chapter 27. Assume that the building is *partially enclosed*, located on a flat terrain, and the roof is flat (except for the minimum roof slope for drainage), with no roof overhang.

Note: In Figure 3-11, the dark arrows are used to denote moment connections in the moment frames, with each moment frame extending over three bays (i.e., each moment frame includes four columns). There are two moment frames in each orthogonal direction.

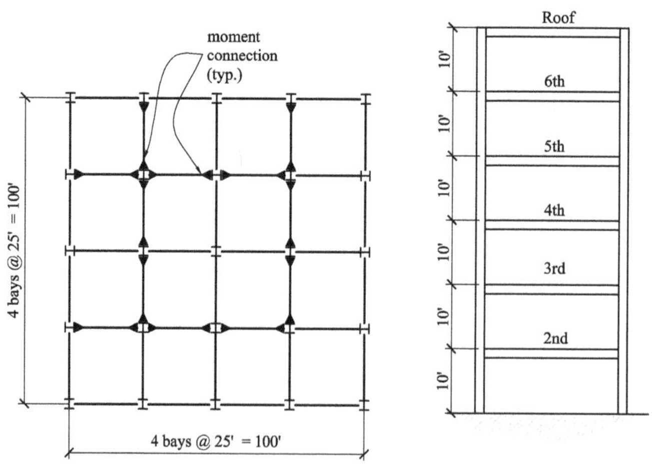

a) plan view *b) elevation*

FIGURE 3-11 Building plan and elevation for Example 3-1

Solution

The solution to this example using the directional procedure in Chapter 27 of the ASCE 7 Load Standard is now carried out in a step-by-step format.

1. **Risk category (RC) II**
   (Ref: ASCE 7 Table 1.5-1)
2. **Basic wind speed, V** = 115 mph
   (Use the ASCE 7 Hazard tool or ASCE 7 Figures 26.5-1A through D of ASCE 7)
3. **Determine wind load parameters**
   a. **Wind Directionality Factor** (main wind force-resisting system, MWFRS), K_d = 0.85
   (Ref: ASCE 7 Table 26.6-1)
   b. **Exposure Category** = Terrain category "C"
   (See Tables 3-1 and 3-2 or ASCE 7 Sections 26.7.3 and 26.7.2. Note that there are three Exposure Categories: B, C, & D)
   c. **Topographic factor, *Kzt*** = 1.0 assuming a flat land
   (See ASCE 7 Table 26.8-1)
   d. Calculate the lateral natural frequency, *f*, of the building to determine if the building is rigid or flexible. A rigid building possesses a lateral natural frequency of vibration, $f \geq 1.0$ Hz or cycles per second (ASCE 7 Section 26.2), and for rigid buildings, the gust-effect factor, G, is taken as 0.85 (ASCE 7, Section 26.11.1). A flexible structure possesses a lateral natural frequency of vibration, $f < 1$. The approximate lateral natural frequency of building structures can be estimated using ASCE 7 Section 26.11.3. For this 6-story building with steel moment-resisting frames, the approximate lateral natural frequency is calculated as $f = 22.2/(h^{0.8}) = 22.2/(60 \text{ ft})^{0.8} = 0.84$ Hz or cycles per second. Therefore, this building will be classified as flexible with respect to wind loads, and thus the gust-effect factor, *G*, will not be 0.85; it will have to be calculated using the cumbersome equations in ASCE 7 Sections 26.11.4 and 26.11.5. The gust-effect factor, *G*, using which the equations from ASCE 7 Section 26.11.3, is found to be **0.92**; the reader is encouraged to independently confirm this number.
   e. **Enclosure Classification** = Partially Enclosed
   The Enclosure Classification of "partially enclosed" was specified for this example. ASCE 7 Sections 26.12 and 26.2 provide the definitions for the Enclosure Classifications. The Enclosure classification is used to determine the internal pressure coefficients (GC_{Pi}) in step 3f below (see ASCE 7, Section 26.13-1), which depends on the amount of wall and roof openings in the building relative to the total surface area of the building envelope. To determine the Enclosure Classification, each wall at a time is assumed to be the windward wall.
   f. **Internal Pressure Coefficient**, GC_{pi} = +0.55 or -0.55 (for Partially Enclosed Building; See ASCE 7 Table 26.10-1). Note that the internal pressures do not impact the net design wind pressures and the design wind forces on the MWFRS for the entire building, because the internal pressures cancel out as depicted in equations (3-11) and (3-12), as well as in equations (3-11) and (3-12), as well as in ASCE 7 Figure 27.3-1 or ASCE 7 Figure 27.3-8. The internal pressure coefficient impacts the design of the individual exterior walls and the roof, and their anchorages. Since only the net lateral wind pressure and the wind load on the entire building are required in this example, therefore, the internal pressures are not useful or needed.
4. **Velocity Pressure Coefficients, K_z and K_h:** The velocity pressure coefficient, K_z, on the windward face of the building varies with height and is also a function of the External

Exposure Category (i.e., whether Exposure Category B, C, or D). The higher the height or level, the higher the value of K_z, because nearer the ground surface, the wind velocity is reduced by the surface roughness and obstructions. The velocity pressure coefficient, K_h, on the leeward face of the building is uniform for the full height of the building and is calculated at the mean roof height of the building. The K_z and K_h factors are obtained from ASCE 7, Table 26.10-1, and with the given Exposure Category "C," the values of the velocity pressure exposure coefficients, K_z and K_h, are tabulated in Table 3-3.

TABLE 3-3 Velocity pressure exposure coefficients for Example 3-1

Diaphragm Level	Height, z at each diaphragm level	K_z	K_h (i.e., at $z = h$)
Roof	60 ft	1.13	1.13
6th flr	50 ft	1.09	1.13
5th flr	40 ft	1.04	1.13
4th flr	30 ft	0.98	1.13
3rd flr	20 ft	0.90	1.13
2nd flr	10 ft	0.85	1.13

Ref: ASCE Table 26.10-1

5. Determine the velocity pressure, q_z and q_h, using equations (3-6) and (3-7) as follows:
   q_z (psf) $= 0.00256\ K_z\ K_{zt}\ K_e\ V^2 = 0.00256\ K_z\ (1.0)(1.0)(115\text{ mph})^2 = 33.88\ K_z$
   q_h (psf) $= 0.00256\ K_h\ K_{zt}\ K_e\ V^2 = 0.00256\ (1.13)\ (1.0)(1.0)(115\text{ mph})^2 = 38.26$
6. Determine the External Pressure Coefficient, C_p, from ASCE 7 Figure 27.3-1 for the walls:
   L = Length of building (plan dimension parallel to the wind direction) = 100 ft
   B = Width of building (plan dimension perpendicular to the wind direction) = 100 ft
   Therefore the shape factor, L/B = 100/100 = 1.0 (this will be used in ASCE 7 Table 27.3-1 to determine the external pressure coefficients for the leeward wall. For the windward wall, there is only one value for the external pressure coefficient, and it is valid for all shape factors, L/B).

   C_p (windward wall) $= C_{p,WW} = 0.8$ (for all values of the shape factor, L/B)
   C_p (leeward wall) $= C_{p,LW} = -0.5$ (i.e., for L/B = 1)

7. Calculate the wind pressures on each surface of the building using the diagrams in ASCE 7 Figure 27.3-1 (see also ASCE 7, Equation 27.3-1). For this exercise, we will focus only on the wall pressures on the **windward** face and the **leeward** face of the building. The external wind pressures, p, on the windward wall and the leeward wall are,
   $p_{windward\ wall} = p_{WW} = q_z\ K_d G C_{p,WW} = 33.88(0.85)\ K_z\ (0.92)(0.8) = 21.2\ K_z$ psf
   $p_{leeward\ wall} = p_{LW} = q_h\ K_d G C_{p,LW} = 38.26(0.85)(0.92)(-0.5) = -15$ psf
   where, $K_d = 0.85$
   Note that the external wind pressure on the windward wall varies with height. Some engineers use a linearly varying wind pressure distribution between floor levels instead of the simpler uniform distribution between floor levels used by the authors in Figure 3-12. The internal wind pressures for the windward and leeward walls are:

   $p_{internal} = q_h\ K_d\ (GC_{pi}) = (38.26)(0.85)\ (\pm 0.55) = \pm 17.9$ psf

   For this exercise, the internal pressures are not required since only the net wind pressures on the MWFRS of the entire building are required. As a result, the internal pressures will cancel out.

8. **Net lateral wind pressure on the MWFRS, $p = p_{WW} + |p_{LW}|$**
The net lateral wind pressure on the MWFRS is the algebraic sum of the magnitudes of the windward and leeward pressures. Note that this net wind pressure (see Table 3-4) varies with the height of the building as shown in Figure 3-13. Also, in calculating the net lateral wind pressure on the building, the internal pressures cancel out.

TABLE 3-4 Net Lateral Wind Pressure on MWFRS in Example 3-1

Diaphragm Level	Height at each diaphragm level	K_z	K_h	$p_{WW} = 21.2\,K_z$ (psf)	p_{LW} (psf) (see steps 5 and 7)	$\lvert p_{LW}\rvert$ (psf)	Net lateral wind pressure $p = p_{WW} + \lvert p_{LW}\rvert = q_z GC_{p,WW} + \lvert q_h GC_{p,LW}\rvert$ ≥ 16 psf minimum (psf)
Roof	60 ft	1.13	1.13	24	−15	15	24 + (15) = 39
6th flr	50 ft	1.09	1.13	23.1	−15	15	23.1 + (15) = 38.1
5th flr	40 ft	1.04	1.13	22.05	−15	15	37.05
4th flr	30 ft	0.98	1.13	20.8	−15	15	35.8
3rd flr	20 ft	0.90	1.13	19.08	−15	15	34.08
2nd flr	10 ft	0.85	1.13	18.02	−15	15	33.02

Note:
Negative (-ve) sign in the table above indicates wind suction (i.e., wind pulling away from the wall).
Positive (+ve) sign in the table above indicates wind pressure (i.e., wind pushing into the wall)
For this building, $p_{WW,X} = p_{WW,Y} = p_{WW}$ and $p_{LW,X} = p_{LW,Y} = p_{LW}$

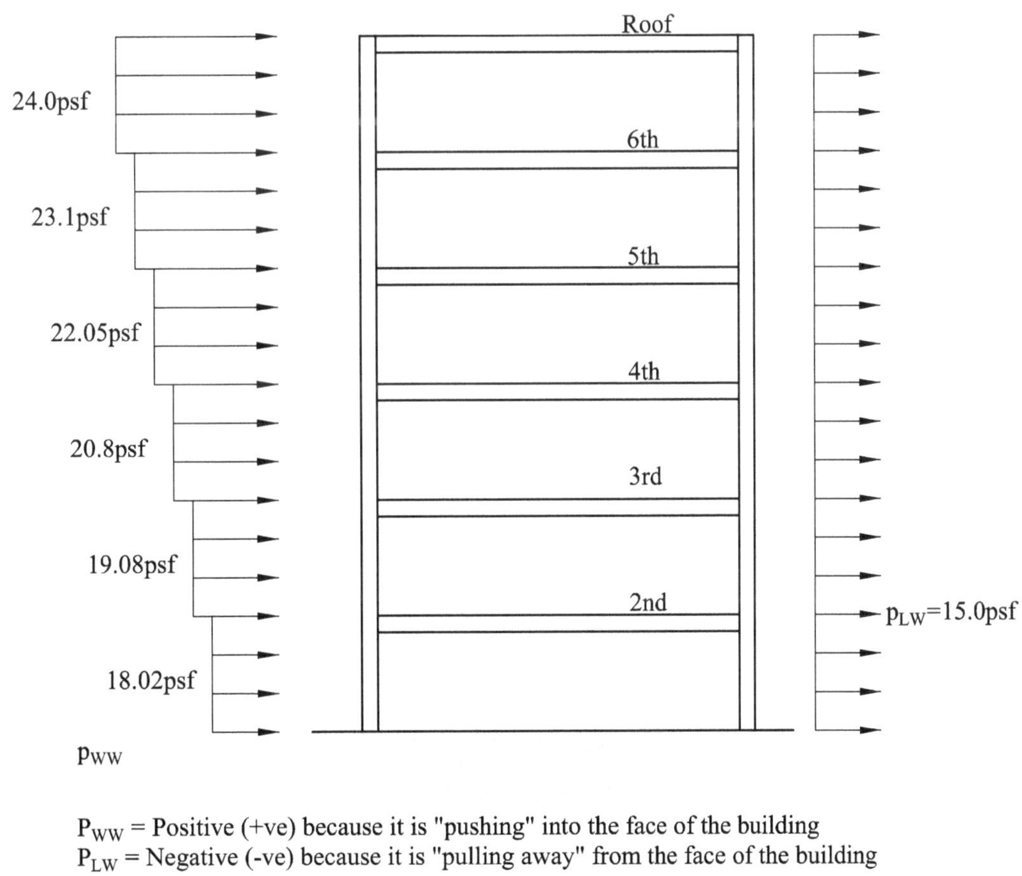

FIGURE 3-12 Windward and Leeward Wind Pressures on the Building (Load Case 1; see ASCE 7, Figure 27.3-8)

The windward and leeward wind pressures on the MWFRS for the entire building for load case 1 are shown in Figure 3-12, and the net design lateral wind pressures are shown in Figure 3-13. The summary tables for load cases 2 through 4 will be presented later.

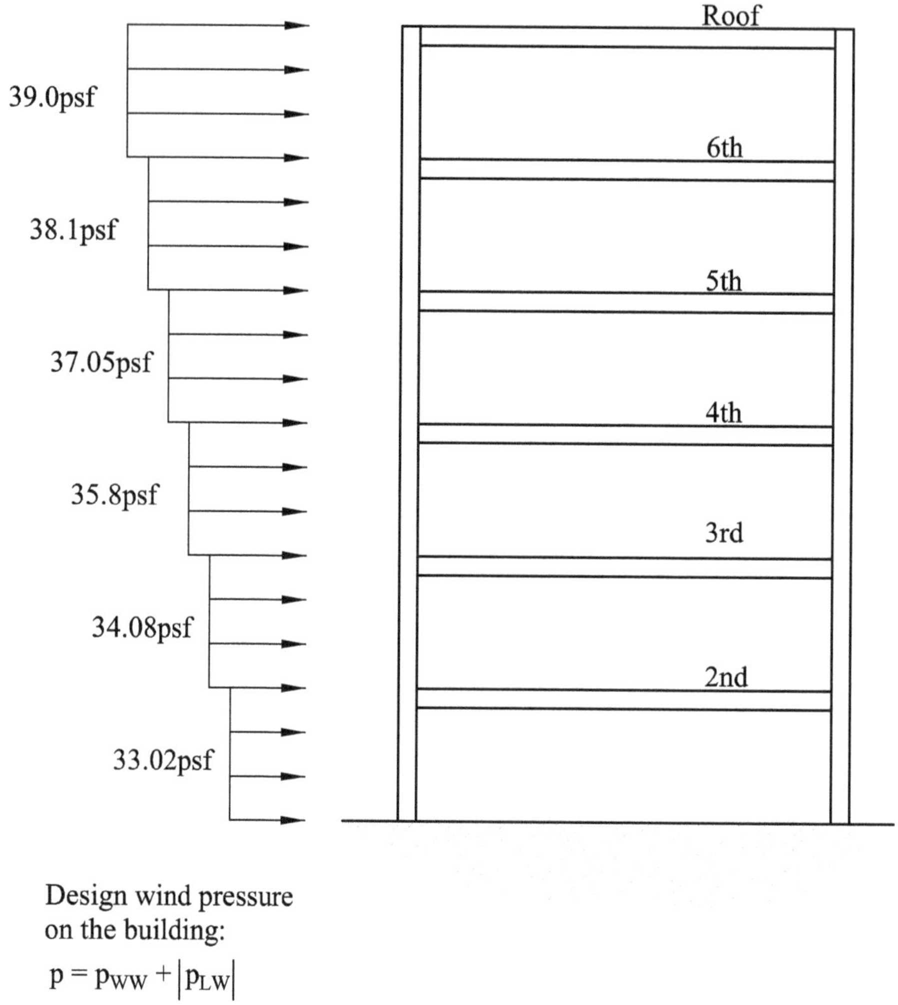

FIGURE 3-13 Net Design Wind Pressures on the Building (Load Case 1)

9. The total lateral force on the building at each level is calculated using the net lateral design wind pressures (see Table 3-5a and Figure 3-14).

   $F_x = [(p_U)(TH_U) + (p_L)(TH_L)]$(Width of building perpendicular to the wind direction)

   TH_U = the upper portion of the Tributary height **above** the floor level under consideration. Note: At the roof level, TH_U = height of parapet, or equal to zero where there is no parapet.

   TH_L = the lower portion of the Tributary height **below** the floor level under consideration

   p_U = uniform net lateral design wind pressure **above** the floor level under consideration

   p_L = uniform net lateral design wind pressure **below** the floor level under consideration

TABLE 3-5a Total Lateral Wind Force on the Building (Load Case 1)

Diaphragm Level	Height at each diaphragm level	$p = p_{WW} + \lvert p_{LW} \rvert$ ≥ 16 psf (minimum) (psf)	Tributary height (ft)	Total factored lateral force on the building at each diaphragm level, F_x (This is the "W" load in the load combinations in Section 2-3)
Roof	60 ft	39	$TH_U = 0$ $TH_L = 10/2 = 5$	39psf(100')(5') = **19.5 kips**
6th flr	50 ft	38.1	$TH_U = 10/2 = 5$ $TH_L = 10/2 = 5$	39psf(100')(5') + 38.1psf(100')(5') = **38.6 kips**
5th flr	40 ft	37.05	$TH_U = 10/2 = 5$ $TH_L = 10/2 = 5$	38.1psf(100')(5') + 37.05psf(100')(5') = **37.5 kips**
4th flr	30 ft	35.8	$TH_U = 10/2 = 5$ $TH_L = 10/2 = 5$	37.05psf(100')(5') + 35.8psf(100')(5') = **36.4 kips**
3rd flr	20 ft	34.08	$TH_U = 10/2 = 5$ $TH_L = 10/2 = 5$	35.8psf(100')(5') + 34.08psf(100')(5') = **35 kips**
2nd flr	10 ft	33.02	$TH_U = 10/2 = 5$ $TH_L = 10/2 = 5$	34.08psf(100')(5') + 33.02psf(100')(5') = **33.6 kips**
				Total base shear, V= 200.6 kips

Note that for load case 1, the X- and Y- direction forces will be the same since the building is symmetrical. The forces in Figure 3-14 are applied independently in the X-direction and in the Y-direction. For p_{WW} and p_{LW}, see Table 3-4.

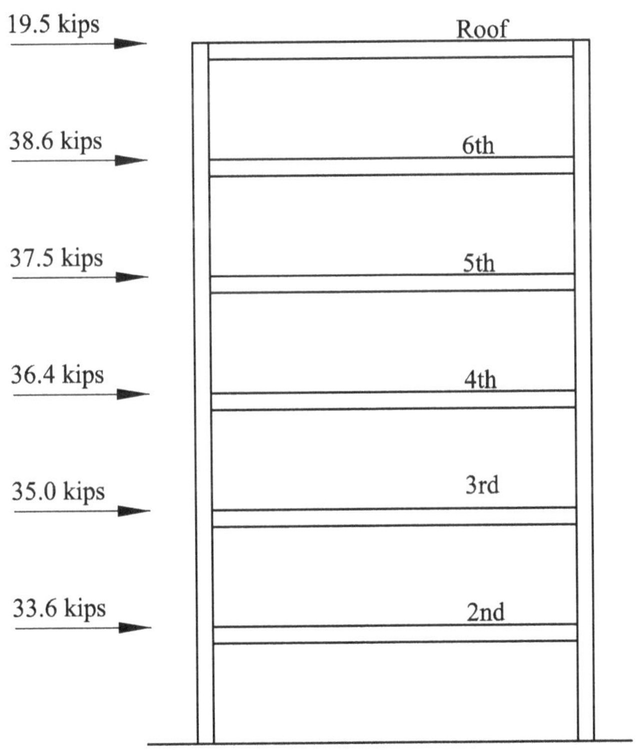

Notes: These lateral forces will be distributed to the LFRS in the building depending on the rigidity of the diaphragm and the rigidity of the LFRS

FIGURE 3-14 Design lateral forces on the building at each level

In Table 3-5a, the design lateral forces on the six-story building were calculated and shown in Figure 3-14 for wind load case 1. As previously discussed, ASCE 7 Figure 27.3-8 presents four wind load cases for the main wind force-resisting system (MWFRS) in a building that are required to be checked. We present in the tables below the wind loads for load cases 2 through 4.

Wind load case 2:

B_X = 100 ft.

B_Y = 100 ft.

$e_X = \pm 0.15 B_X = \pm 0.15(100 \text{ ft.}) = \pm 15$ ft.

Windward lateral pressure = $0.75 p_{WW,X}$

Leeward lateral pressure = $0.75 p_{LW,X}$

For the **X-direction** wind,

$M_T = 0.75(p_{WW,X} + p_{LW,X})(B_X)(e_X) = 0.75(p_{WW,X} + p_{LW,X})(100 \text{ ft.})(\pm 15 \text{ ft.})$ ft-kips/ft. of tributary height

$= F_{x,X}(e_X)$, ft-kips/ft. of tributary height

Due to the symmetry of this building, the X- forces and the in-plane torsional moments calculated for the X-direction wind load will be the same for the Y-direction wind load, but the X- and Y-direction forces/moments for this load case are applied *independently*.

Wind load case 3:

B_X = 100 ft.

B_Y = 100 ft.

$e_X = 0$

$e_Y = 0$

For the **X-direction** wind,

Windward lateral pressure = $0.75 p_{WW,X}$

Leeward lateral pressure = $0.75 p_{LW,X}$

For the **Y-direction** wind,

Windward lateral pressure = $0.75 p_{WW,Y}$

Leeward lateral pressure = $0.75 p_{LW,Y}$

The in-plane torsional moment due to wind is calculated at each diaphragm level as

$M_T = 0$ ft-kips/ft

Due to the symmetry of this building, the X- direction wind load and the in-plane torsional moments will be the same for the Y-direction, and note that the X- and Y-direction forces and moments in this load case are applied *simultaneously*.

Wind load case 4:

B_X = 100 ft.

B_Y = 100 ft.

$e_X = \pm 0.15 B_X = \pm 0.15(100 \text{ ft.}) = \pm 15$ ft.

$e_Y = \pm 0.15 B_Y = \pm 0.15(100 \text{ ft.}) = \pm 15$ ft.

For the **X-direction** wind,

Windward lateral pressure = $0.563 p_{WW,X}$

TABLE 3-5b Total Lateral Wind Force on the Building (Load Case 2)

Diaphragm Level	Height at each diaphragm	$0.75\,p_{WW}$ = (0.75) 21.2 K_z (psf)	$0.75\lvert p_{LW}\rvert$ (psf)	$p = 0.75p_{WW} + 0.75\lvert p_{LW}\rvert$ ≥ 16 psf (minimum) (psf)	Tributary height (ft)	Total <u>factored</u> lateral force on the building at each diaphragm level, $F_x = [(p_U)(TH_U) + (p_L)(TH_L)](B_X)$ (This is the "W" load in the load combinations in Section 2-3)	In-plane Torsional Moment, M_{Tx} *at each diaphragm level* $= F_x(e_X) = F_x(\pm 15\text{ ft.})$ (ft-kips)
Roof	60 ft	18	11.25	29.25	$TH_U = 0$ $TH_L = 5$	29.25psf(100')(5') = **14.63 kips**	±(14.63)(15ft.) = ±220
6th flr	50 ft	17.33	11.25	28.58	$TH_U = 5$ $TH_L = 5$	29.25psf(100')(5') + 28.58psf(100')(5') = **28.9 kips**	±(28.9)(15ft.) = ±434
5th flr	40 ft	16.54	11.25	27.79	$TH_U = 5$ $TH_L = 5$	28.58psf(100')(5') + 27.79psf(100')(5') = **28.1 kips**	±(28.1)(15ft.) = ±422
4th flr	30 ft	15.6	11.25	26.85	$TH_U = 5$ $TH_L = 5$	27.79psf(100')(5') + 26.85psf(100')(5') = **27.3 kips**	±(27.3)(15ft.) = ±410
3rd flr	20 ft	14.31	11.25	25.56	$TH_U = 5$ $TH_L = 5$	26.85psf(100')(5') + 25.56psf(100')(5') = **26.3 kips**	±(26.3)(15ft.) = ±395
2nd flr	10 ft	13.52	11.25	24.77	$TH_U = 5$ $TH_L = 5$	25.56psf(100')(5') + 24.77psf(100')(5') = **25.2 kips**	±(25.2)(15ft.) = ±378

TH_U = upper tributary height of the floor level and TH_L = lower tributary height of the floor level
For p_{WW} and p_{LW}, see Table 3-4.

TABLE 3-5c Total Lateral Wind Force on the Building (Load Case 3)

Diaphragm Level	Height at each diaphragm	X-direction			Y-direction			Trib height (ft)	Total <u>factored</u> lateral force on the building in the X-direction at each diaphragm level, $F_{x,X} = [(p_U)(TH_U) + (p_L)(TH_L)](B_X)]$ (This is the "W" load in the load combinations in Section 2-3)	Total <u>factored</u> lateral force on the building in the Y-direction at each diaphragm level, $F_{x,Y} = [(p_U)(TH_U) + (p_L)(TH_L)](B_Y)]$ (This is the "W" load in the load combinations in Section 2-3)	In-plane Torsional Moment, M_{Tx} *at each diaphragm level* $= F_{x,X}(e_X) + F_{x,Y}(e_Y) = F_{x,X}$ (0 ft.) + $F_{x,Y}$(0 ft.) (ft-kips)
		$0.75p_{WW,X}$ = (0.75) 21.2 K_z (psf)	$0.75\lvert p_{LW,X}\rvert$ (psf)	$p = 0.75p_{WW,X} + 0.75\lvert p_{LW,X}\rvert$ ≥ 16 psf (minimum) (psf)	$0.75p_{WW,Y}$ = (0.75) 21.2 K_z (psf)	$0.75\lvert p_{LW,Y}\rvert$ (psf)	$p = 0.75p_{WW,Y} + 0.75\lvert p_{LW,Y}\rvert$ ≥ 16 psf (minimum) (psf)				
Roof	60 ft	18	11.25	**29.25**	18	11.25	**29.25**	$TH_U = 0$ $TH_L = 5$	29.25psf(100')(5') = **14.63 kips**	29.25psf(100')(5') = **14.63 kips**	0
6th flr	50 ft	17.33	11.25	**28.58**	17.33	11.25	**28.58**	$TH_U = 5$ $TH_L = 5$	29.25psf(100')(5') + 28.58psf(100')(5') = **28.9 kips**	29.25psf(100')(5') + 28.58psf(100')(5') = **28.9 kips**	0
5th flr	40 ft	16.54	11.25	**27.79**	16.54	11.25	**27.79**	$TH_U = 5$ $TH_L = 5$	28.58psf(100')(5') + 27.79psf(100')(5') = **28.1 kips**	28.58psf(100')(5') + 27.79psf(100')(5') = **28.1 kips**	0
4th flr	30 ft	15.6	11.25	**26.85**	15.6	11.25	**26.85**	$TH_U = 5$ $TH_L = 5$	27.79psf(100')(5') + 26.85psf(100')(5') = **27.3 kips**	27.79psf(100')(5') + 26.85psf(100')(5') = **27.3 kips**	0
3rd flr	20 ft	14.31	11.25	**25.56**	14.31	11.25	**25.56**	$TH_U = 5$ $TH_L = 5$	26.85psf(100')(5') + 25.56psf(100')(5') = **26.3 kips**	26.85psf(100')(5') + 25.56psf(100')(5') = **26.3 kips**	0
2nd flr	10 ft	13.52	11.25	**24.77**	13.52	11.25	**24.77**	$TH_U = 5$ $TH_L = 5$	25.56psf(100')(5') + 24.77psf(100')(5') = **25.2 kips**	25.56psf(100')(5') + 24.77psf(100')(5') = **25.2 kips**	0

TH_U = upper tributary height of the floor level and TH_L = lower tributary height of the floor level
For p_{WW} and p_{LW}, see Table 3-4.

TABLE 3-5d Total Lateral Wind Force on the Building (Load Case 4)

Diaphragm Level	Height at each diaphragm	X-direction			Y-direction			Trib height (ft)	Total **factored** lateral force on the building in the X-direction at each diaphragm level, $F_{x,X}$ = $[(p_U)(TH_U) + (p_L)(TH_L)](B_X)$ (This is the "W" load in the load combinations in Section 2-3)	Total **factored** lateral force on the building in the Y-direction at each diaphragm level, $F_{x,Y}$ = $[(p_U)(TH_U) + (p_L)(TH_L)](B_Y)$ (This is the "W" load in the load combinations in Section 2-3)	In-plane Torsional Moment, M_{Tx} *at each diaphragm level* $= F_{x,X}(\pm 15\text{ ft.}) + F_{x,Y}(\pm 15\text{ ft.})$ (ft-kips)
		$0.563p_{WW,X} = (0.563)\ 21.2\ K_z$ (psf)	$0.563\lvert p_{LW,X}\rvert$ (psf)	$p = 0.563p_{WW,X} + 0.563\lvert p_{LW,X}\rvert$ ≥ 16 psf (minimum) (psf)	$0.563p_{WW,Y} = (0.563)\ 21.2\ K_z$ (psf)	$0.563\lvert p_{LW,Y}\rvert$ (psf)	$p = 0.563p_{WW,Y} + 0.563\lvert p_{LW,Y}\rvert$ ≥ 16 psf (minimum) (psf)				
Roof	60 ft	13.5	8.45	**21.95**	13.5	8.45	**21.95**	$TH_U = 0$ $TH_L = 5$	21.95psf(100')(5') = **11 kips**	21.95psf(100')(5') = **11 kips**	**±330**
6th flr	50 ft	13.0	8.45	**21.45**	13.0	8.45	**21.45**	$TH_U = 5$ $TH_L = 5$	21.95psf(100')(5') + 21.45psf(100')(5') = **21.7 kips**	21.95psf(100')(5') + 21.45psf(100')(5') = **21.7 kips**	**±651**
5th flr	40 ft	12.4	8.45	**20.85**	12.4	8.45	**20.85**	$TH_U = 5$ $TH_L = 5$	21.45psf(100')(5') + 20.85psf(100')(5') = **21.15 kips**	21.45psf(100')(5') + 20.85psf(100')(5') = **21.15 kips**	**±635**
4th flr	30 ft	11.7	8.45	**20.15**	11.7	8.45	**20.15**	$TH_U = 5$ $TH_L = 5$	20.85psf(100')(5') + 20.15psf(100')(5') = **20.5 kips**	20.85psf(100')(5') + 20.15psf(100')(5') = **20.5 kips**	**±615**
3rd flr	20 ft	10.74	8.45	**19.19**	10.74	8.45	**19.19**	$TH_U = 5$ $TH_L = 5$	20.15psf(100')(5') + 19.19psf(100')(5') = **19.67 kips**	20.15psf(100')(5') + 19.19psf(100')(5') = **19.67 kips**	**±590**
2nd flr	10 ft	10.15	8.45	**18.6**	10.15	8.45	**18.6**	$TH_U = 5$ $TH_L = 5$	19.19psf(100')(5') + 18.6psf(100')(5') = **18.9 kips**	19.19psf(100')(5') + 18.6psf(100')(5') = **18.9 kips**	**±567**

TH_U = upper tributary height of the floor level and TH_L = lower tributary height of the floor level
For p_{WW} and p_{LW}, see Table 3-4.

Leeward lateral pressure = $0.563p_{LW,X}$

For the **Y-direction** wind,

Windward lateral pressure = $0.563p_{WW,Y}$

Leeward lateral pressure = $0.563p_{LW,Y}$

The in-plane torsional moment due to wind is calculated at each diaphragm level as

$M_T = 0.563(p_{WW,X} + p_{LW,X})(B_X)(e_X) + 0.563(p_{WW,Y} + p_{LW,Y})(B_Y)(e_Y)$, ft-kips/ft. of tributary height

$= F_{x,X}(e_X) + F_{x,Y}(e_Y)$ ft-kips/ ft. of tributary height

Due to the symmetry of this building, the X- forces and the in-plane torsional moments will be the same for the Y-direction, and note that the X- and Y-direction forces and in-plane torsional moments in this load case are applied *simultaneously*.

Approximate Natural Frequency for Wind Load Calculations

The approximate natural or fundamental lateral frequency of structures used for calculating wind loads is given in ASCE 7, Section 26.11.3 for the different lateral force-resisting systems as follows:

Lateral Force-Resisting System	Approximate Fundamental Lateral Frequency of Vibration ^
Structural steel moment-resisting frames	$f = \frac{22.2}{h^{0.8}}$ *Hz or cycles per second*
Concrete moment-resisting frames	$f = \frac{43.5}{h^{0.9}}$ *Hz or cycles per second*
Structural steel and concrete buildings with *other* lateral force-resisting systems (e.g., shear walls, X-brace, Chevron brace, and single diagonal brace frames)	$f = \frac{75}{h}$ *Hz or cycles per second*

^ *h* is the mean roof height in feet in the above equations. See ASCE 7 Section 26.11.2.1 for the limitations on the approximate lateral frequency equations, where the maximum building height is limited to 300 ft. and the maximum height-to-minimum plan width ratio is 4.0.

Building Enclosure Classifications

There are four possible types of building enclosure classifications defined in ASCE 7 Section 26.2: **open, partially open, partially enclosed**, and **enclosed**. These classifications are used to determine the internal pressure coefficients (GC_{Pi}) from ASCE 7 Table 26.13-1 which depends on the amount of wall and roof openings in the building relative to the total building envelope surface area. ASCE 7 Sections 26.2 and 26.12 provide the definitions for the Enclosure Classification. To determine the Enclosure Classification, each wall at a time is considered as the windward wall - that is, the wall that receives the positive pressure with surface area, A_{os}. The most critical classification, which results in the highest wind pressures, is used. The corresponding internal pressure coefficients (GC_{pi}) are obtained from ASCE 7-22, Table 26.13-1.

The internal wind pressures are used to determine the net wind pressures on the individual walls and the roof. These internal pressures are not needed in the calculation of the net lateral wind pressure on the entire building since they cancel out as shown in equations (3-11) and (3-12).

3-5 EFFECT OF NET FACTORED UPLIFT LOADS ON ROOF BEAMS, JOISTS, AND ENTRANCE CANOPIES

The net factored uplift roof loads due to wind are normally calculated using the Components and Cladding (C&C) wind pressures. Typically, the C&C roof pressures are obtained assuming the smallest effective wind area, A_e, of 10 ft.^{2}, which is conservative for most structural members, except for members with very small tributary areas. For members with larger tributary areas, the C&C roof pressures can be calculated using a more realistic tributary area, A_e, of 100 ft.^{2}.

Beams that support entrance canopies are usually subjected to high net uplift wind pressures as well as gravity loads from roof dead loads, downward wind pressures, snow loads, and rain loads. One end of these beams is usually supported at the exterior wall of the building, while the exposed end of the beam is supported by cables or tension rods. For the gravity loads (i.e., dead and snow), the cable or tension rod is adequate to support the exterior or free end of the beam. However, when the canopy beam is subjected to a net wind uplift, the cable or tension rods will now be in compression, which they are incapable of resisting. To mitigate this situation, the canopy beams should be connected to their support at the building exterior wall with a moment connection so that they are capable of resisting the moment from the net wind uplift loads since the cables or tension rods are ineffective for these upward loads. Note that components and cladding (C & C) or structural elements with tributary areas greater than 700 ft.^{2} can be designed using the provisions for the MWFRS (see ASCE 7-22 Chapter 27 and Section 30.2.3). For structural elements having tributary areas not greater than 700 ft.^{2}, the larger C & C wind pressures from ASCE 7 Chapter 30 should be used for the design of those elements.

Example 3-2 Net Uplift Wind Loads

To illustrate the calculation of net uplift wind loads, assume that the dead, snow, and roof live loads on the roof framing of a building have already been determined as follows:

Dead load, D = 25 psf

SNow load, S = 56 psf

Roof live load, L_r = 20 psf (actual value depends on the tributary area of the member under consideration).

- Assume the wind pressures for C&C calculated using ASCE 7, Figure 30.5-1 are as follows:
  W = +16.0 psf (this positive or +ve wind load will be used in the LRFD load combinations 1 through 4)
  W = –41.8 psf (this negative or –ve wind load value will be used in the LRFD load combination 5 to determine if there is any net uplift force on the roof framing)
- Assume the roof framing consists of steel beams that span 20 ft. with a spacing of 5 ft. on centers.

Calculate the maximum loads, shear, and moments in a typical beam.

Solution

Summary of Roof Loads

The following is a summary of the loads acting on the roof of the building:

Dead load, D = 25 psf

SNow load, S = 56 psf

Maximum roof live load L_r = 20 psf

Since L_r < S and $0.5L_r$ < 0.3S; therefore, the snow load, S, governs and will be used in the load combinations.

Wind load, $W = +16.0$ psf and -41.8 psf

The load and resistance factor design (LRFD) load combinations from Chapter 2 will now be used to determine the governing or controlling load case for this steel-framed roof. The factored loads on the roof are calculated as follows:

LRFD Load Combinations

1. 1.4 (25 psf) = 35 psf
2. 1.2 (25 psf) + 1.6 (0) + 0.3 (56 psf) = 46.8psf
3. 1.2 (25 psf) + 1.0 (56 psf) + 0.5 (16.0 psf) = **94 psf** (controls for downward loads)
4. 1.2 (25 psf) + 1.0 (16 psf) + 0 + 0.3 (56 psf) = 62.8 psf
5. 0.9 (25 psf*) + 1.0 (–41.8 psf) = **–19.3 psf** (controls for net wind uplift load; load acts upward)

* This is the dead load assumed to be present when the wind acts on the building. The actual value may be less than the dead load used for calculating the maximum factored downward load in load combinations 1 through 4. This dead load value should be carefully determined so as not to create an unconservative design for wind uplift forces.

Load Effects in Roof Beams with Net Uplift Loads

The roof beams in this example will have to be designed for a downward-acting load of **94 psf** and a net factored uplift load of **-19.3 psf**. The girders will have to be designed for the corresponding beam reactions. Note that this beam must be checked for both downward and uplift loads. Most roof beams have their top flanges fully braced by roof decking or framing, but in designing beams for moments due to net uplift loads, the unbraced length of the compression flange of the beam (which is the bottom flange for uplift loads) will, in most cases, be equal to the full span or length of the beam. For steel beams, this could lead to a substantial reduction in strength that could make the uplift moments more critical than the moments caused by the factored downward loads.

Downward Loads and Load Effect:

With the roof beam span, L, of 20 ft. and a tributary width of 5 ft., the moments and reactions in the typical roof beam are calculated as follows:

Factored uniform downward load, w_u (downwards) = (94 psf)(5 ft.) = 470 lb./ft.

Maximum +ve moment, M_u^{+ve} = (470 lb./ft.)(20 ft.)^2/8 = 23,500 ft.-lb.

(**unbraced length, L_b = 0 ft.** since the compression flange of the beam is laterally braced by the roof deck)

Maximum downward load reaction, R_u^{+ve} = (470 lb./ft.)(20 ft.)/2 = 4700 lb.

Net Wind Uplift Load and Load Effect:

Net factored uniform uplift load, w_u (upwards) = (–19.3 psf)(5 ft.) = –96.5 lb./ft.

Maximum –ve moment due to the net wind uplift, M_u^{-ve} = (–96.5 lb./ft.)(20 ft.)^2/8 = –4825 ft.-lb.

(**unbraced length, L_b = 20 ft.**)

Maximum upward load reaction, R_u^{-ve} = (–96.5 lb./ft.)(20 ft.)/2 = –965 lb.

If *open-web steel joists* were used for the roof framing of this building instead of the infill steel beams, a net uplift wind pressure could lead to the collapse of these joists if the uplift load is not adequately considered in the design of the joists. When subjected to a net uplift wind load, the bottom chord of the open-web steel joists and the long diagonal members, which are typically in tension under gravity or downward loads (and, which in many cases, may have slenderness ratios of between 200 to 300), will be in compression due to the net wind uplift loads. The bottom chords of the joist may have unbraced lengths that far exceed the default values that may be assumed in standard structural analysis software. The actual unbraced length of the joist bottom chord is the smaller of the span of the joist or the spacing between joist bridgings. Under this condition, these bottom chord members may be inadequate to resist the resulting compression loads unless they have been designed for this stress or load reversal.

The combination of a light roof system and an inaccurately calculated net wind uplift load led to the collapse of a commercial warehouse roof in the Dallas area in 2001 [7]. In this warehouse structure, the net uplift due to wind loads caused stress reversals in the end diagonal web members and the bottom chord members of the roof joist, which had only been designed to resist axial tension forces resulting from the gravity or downward acting roof dead and live loads.

3-6 WIND LOADS ON PARAPETS, ROOF SCREEN WALLS, AND ROOFTOP EQUIPMENT

For buildings with parapets that extend above the roof, additional wind pressures on the parapet will add to the lateral wind load on the MWFRS. The combined net wind pressures on parapets, taking into account the wind pressure in front of and behind the parapet (see ASCE 7-22 Section 27.3.4), is given as:

$$P_p = q_p K_d \left(GC_{pn}\right)$$

P_p= Combined net pressure on the parapet caused by the combination of the net pressures from the front and back parapet surfaces, psf

q_p = Velocity pressure evaluated at the top of the parapet, psf
= $0.00256(K_z \textit{ at top of parapet }) K_{zt} K_e V^2$

(GC_{pn}) = Combined net pressure coefficient, which is +1.5 for windward parapet or −1.0 for leeward parapet. The "+1.5" is the coefficient for the pressure on the front or windward face of the parapet (i.e., the pressure is pushing toward the front face of the parapet wall), and the "−1.0" is the coefficient for the pressure or suction on the back surface or leeward face of the parapet (i.e., this suction indicates that the pressure is pulling away from the back face of the parapet wall).

K_d = the directionality factor = 0.85 for MWFRS from ASCE 7-22 Table 26.6-1

Therefore, the total wind pressure on the MWFRS due to the parapet is

$$P_p = q_p (0.85)\left[(1.5)-(-1.0)\right] = 2.125 q_p \ \ psf$$

Roof screen walls are typically used to conceal the rooftop equipment from view. In some cases, the roof screen walls are near the edges of the roof, and the wind pressures on these walls can be treated using the same approach for wind pressures on parapet walls. Roof screen walls (see Figure 3-15) typically consist of galvanized metal cladding supported by HSS girts that frame between vertically cantilevered posts made of hollow structural sections (HSS). For very tall roof screen walls, vertically cantilevered HSS posts will be inadequate to resist the bending from lateral wind loads and the corresponding lateral deflection. In this case, steel braced frames perpendicular to the cladding or screen, and spaced at regular intervals should be provided to laterally support the cladding. The braced frames are tied together with steel members parallel to the cladding to ensure laterally bracing parallel to the screen wall.

The wind pressures on MWFRS due to rooftop structures and equipment (see ASCE 7 Section 29.4.1) are

$$P_h = q_h K_d (GC_r)$$

Where,

q_h = Velocity pressure at mean roof height of the building;

Bh = Vertical surface area of the building perpendicular to the wind direction, ft.^{2}

= Width of the building perpendicular to the wind direction, B x mean roof height of the building, h.

A_f = Vertical projected area of the rooftop structure or equipment on a plane perpendicular to the wind direction, ft.^{2}

(GCr) = 1.9 for $A_f \leq 0.1$ Bh

(GCr) = 1.0 for A_f = Bh

Linear interpolation for (GCr) is permitted, that is,

$$1.0 \leq (GC_r) = 2 - \frac{A_f}{Bh} \leq 1.9$$

FIGURE 3-15 Roof screen walls in a steel-framed building under construction (concealing rooftop mechanical equipment)

3-7 DESIGNING FOR TORNADOES

A tornado is "a violently rotating column of air that touches the ground and is usually attached to the base of a thunderstorm." The wind pressures due to tornadoes are much higher than those for regular straight-line wind events, especially the wind uplift pressures on roofs. Tornadoes are classified into the following categories: EF0, EF1, EF2, EF3, EF4, and EF5, with the maximum 3-second peak gust wind speeds of 85 mph, 110 mph, 135 mph, 165 mph, and 200 mph, respectively; EF5 tornadoes have wind speeds greater than 200 mph. In areas where tornadoes are prevalent, the tornado wind speed maps and the methods presented in ASCE 7-22 Chapter 32 [1] should be used to determine the tornado wind pressures and loads on the MWFRS and the C & C. Most tornado winds are of EF2 intensity or less, and these are the ones covered in Chapter 32 of ASCE 7-22. Only safe rooms or storm shelters need to be designed for high-intensity tornadoes such as EF4 or EF5, and for these, the reader should refer to ICC 500—*Standard for the Design and Construction of Storm Shelters*.

3-8 CALCULATION OF SEISMIC LOADS

The seismic limit state is based upon system performance and not member performance, and considerable energy dissipation is assumed [1]. Seismic loads are calculated differently for the primary system (i.e., the lateral force-resisting system) than for parts and components such as architectural, mechanical, and electrical fixtures. In the United States, seismic design is based on target reliabilities or probabilities of failure specified in ASCE 7 Tables 1.3-2 and 1.3-3. This results in a maximum considered earthquake (MCE) with a 2% probability of exceedance in 50 years, equivalent to a mean recurrence interval of approximately 2500 years; and a design basis earthquake (DBE) with a 10% probability of exceedance in 50 years, equivalent to a mean recurrence interval of approximately 500 years. The seismic load calculations for the primary system are covered in ASCE 7, Chapter 12, while the seismic load calculations for parts and components (i.e., nonstructural components) are covered in ASCE 7, Chapter 13 [1].

The design basis or the performance objectives for seismic design using the ASCE 7 Load Standard are as follows (see ASCE 7 Sections C11.5 and C11.6 [1]):

1. **Under maximum considered earthquake ground motion, MCE_R:**
   The structure may be **severely damaged** but has a **low probability of collapse** (~10% probability of collapse for risk category (RC) I and II buildings, 5% for RC III buildings, and 2.5% for RC IV buildings) under a rare seismic event—that is, the risk-targeted **maximum considered earthquake ground motion, MCE_R.** See ASCE 7-22 Table 1.3-2.
2. **Under maximum design basis earthquake (DBE) ground motion, $\frac{2}{3}MCE_R$:**
   The structure has a lower probability of collapse, but there is **failure of the nonstructural components**, and a **very low probability of loss of life** under this moderate level earthquake ground motion of $\frac{2}{3}MCE_R$. According to the ASCE 7 Commentary (Section C11.8.3[1]), the factor of "2/3" was obtained from seismic code calibration studies, which indicated a "reserve capacity of more than 1.5 times relative to collapse" for most buildings under the design earthquake ground motion, hence the MCE_R ground motion multiplier of 1/1.5 or 2/3. Under this design earthquake ground motion intensity of $\frac{2}{3}MCE_R$, the corresponding probabilities of collapse are 2.5% for

risk category (RC) II buildings, 1% for RC III buildings, and 0.5% for RC IV buildings (See ASCE 7-22 Section C1.3.1.3).

3. RC IV structures under **maximum design earthquake ground motion, $\frac{2}{3}MCE_R$:** Structures that are essential to disaster response and recovery efforts (i.e., risk category (RC) IV structures) must remain functional during and after the design earthquake ground motion, $\frac{2}{3}MCE_R$ (see Table 3-6). Note that under the maximum considered earthquake ground motion, RC IV structures may not be functional, but life safety is a performance expectation.

With the preceding performance objectives, a risk category (RC) II structure achieves life safety under the design basis earthquake ground motion, $\frac{2}{3}MCE_R$; it achieves collapse prevention but sustains severe damage when the structure is subjected to the maximum considered earthquake ground motion, MCE_R.

In terms of the level of damage or devastation expected in a seismic event, the ASCE 7 Commentary (Section C1.3.1.3) indicates that for risk category (RC) II structures subjected to the MCE_R, total or partial collapse should be expected to occur in 10% of new buildings, and failure of *non-structural components* that could pose injury risk to human lives will occur in 25% of new buildings; similar values for risk category (RC) IV buildings are 2.5% and 9%, respectively [1]. The expected target reliability or the expected probability of failure for other risk categories for structural and non-structural members subjected to the MCE_R can be found in ASCE 7 Tables 1.3-2 and 1.3-3 [1].

Table 3-6 shows the relationship between the expected seismic performance levels, the risk category (RC), and the intensity and frequency of the earthquake [1, 22]. Note that many commercial structures are in risk category (RC) II, and the expected seismic performance from Table 3-6, which indicates "collapse prevention" with a 10% probability of collapse, means that most of the buildings will not collapse, but they may be badly damaged beyond repair, and as such, many of these structures may have to be demolished, resulting in huge economic losses and untold hardships for many citizens. The resilience of community infrastructure—including buildings and bridges—and business continuity after a hazard event such as an earthquake is a topic of immense interest [23], and one of the key questions *is:*

> *Are all stakeholders fully aware of, and able to survive, the risk of huge economic losses and untold financial hardships that would ensue due to the level of structural damage from these hazard events that is implied by the minimum provisions in the current prescriptive- or specifications-based codes and standards?*

It should be emphasized to building owners that the prescriptive- and specification-based codes provide only minimum requirements that are focused on life safety—this is "*akin to a car that crumples in a crash: the vehicle absorbs the impact to protect passengers, but it is totaled*" [28]. Higher performance levels may be sought through the use of performance-based design (see Chapter 2), though this will lead to an increase in the initial cost of the project. The Structural Engineering Institute (SEI) of the American Society of Civil Engineers (ASCE) is actively promoting the development and implementation of "performance-based codes and standards" and the incorporation of "resilience in the built environment," and these are two of the key initiatives of its "Vision for the Future of Structural Engineering."

Table 3-6 Expected seismic performance versus risk category and earthquake frequency

Earthquake Frequency and Magnitude	Risk Category (RC) I and II	Risk Category (RC) III	Risk Category (RC) IV
Frequent	Immediate occupancy (IO)	Between IO and OP	Operational (OP)
Design earthquake = $\frac{2}{3}MCE_R$ (Rare). The design earthquake is the basis of the equivalent lateral force (ELF) method.	Life safety (LS) (2.5% probability of collapse)	Between LS and IO (1% probability of collapse)	Immediate occupancy (IO). Note: IO implies that the structure must be functional after the hazard event, and its strength and stiffness must not be impaired. (0.5% probability of collapse)
Maximum considered earthquake = MCE_R (Very rare)	Collapse prevention (CP). (10% probability of collapse under MCE)	Between CP and LS. (5% probability of collapse under MCE)	Life safety (LS). (2.5% probability of collapse under MCE)

(Adapted from ASCE 7-22 Figure C11.5-1)

One of the important parameters needed to calculate the seismic loads on structures is the Seismic Design Category (SDC) covered in ASCE 7 Section 11.6. The seismic design category (SDC) affects the following:

- Drift limits under design earthquake ground motions.
- Design forces in collectors (ASCE 7 Sections 12.10.2 and 12.10.3.4).
- Foundation ttie requirements.
- Amplification of torsional moments in horizontal diaphragms.
- Seismic forces due to orthogonal effects.
- The redundancy factor.
- The structural height limitations (see ASCE 7 Table 12.2-1 and Section 12.2)
- Increase in seismic force due to structural irregularities and permitted/prohibited structural irregularities (see ASCE 7 Tables 12.3-1 and 12.3-2),
- The mandated special seismic detailing requirements and quality assurance plans. For example, structures in SDC A, B, or C do not require special detailing, whereas structures in SDC D, E, or F have special detailing requirements that are specified in ASCE 7 Chapter 14.

The SDC is dependent on the following parameters:

- The Site Class, which depends on the soil type at the project site.
- The mapped spectral acceleration parameter at short period, S_S, and the mapped spectral acceleration parameter at 1-s period, S_1.
- The short period and 1-s period response acceleration parameters, S_{DS} and S_{D1}, respectively, at the project site.
- The risk category (RC) (see ASCE 7, Tables 1.5-1 and 1.5-2, 11.6-1, and 11.6-2).

The six SDCs identified in ASCE 7 are A through F, and the four occupancy categories and their corresponding seismic importance factors are tabulated in Table 3-7 (see ASCE 7, Figure C1.5-1, Tables 1.5-1 and 1.5-2). Note that for structures in SDC A, B, or C, there are no special detailing requirements. There are special detailing requirements for structures in SDC D, E, or F. The step-by-step procedure for determining the seismic design category (SDC) is described in Table 3-8.

TABLE 3-7 Risk Category and Seismic Importance Factor

Risk Category (RC)	Type of Occupancy	Seismic Importance Factor, I_e
I	Buildings that represent a low risk to human life in the event of failure (e.g., buildings that are not always occu-pied). Less than 5 "lives at risk" inside and outside of the structure due to the failure of the structure (see ASCE 7 Figure C1.5-1) [1]. Examples include unoccupied storage facilities, barns, and sheds.	1.0
II	All buildings and other structures except those listed in risk categories I, III, and IV. Less than 500 "lives at risk" inside and outside of the structure due to the failure of the structure (see ASCE 7 Figure C1.5-1) [1].	1.0
III	Assembly buildings: Buildings used for assembly occupancies where there is a higher risk to the occupants if failure were to occur (e.g., theaters, elementary or grade schools, prisons). Less than 5000 "lives at risk" inside and outside of the structure due to the failure of the structure (see ASCE 7 Figure C1.5-1) [1].	1.25
IV	Essential facilities that need to be operational after a hazard event and facilities that manufacture or store hazardous chemicals, wastes or fuels e.g., police and fire stations, hospitals, aviation control towers, communication towers, emergency shelters, power-generating stations, water treatment plants, and national defense facilities. More than 5000 "lives at risk" inside and outside of the structure due to the failure of the structure (see ASCE 7 Figure C1.5-1) [1].	1.50

Table 3-8 Determination of the seismic design category (SDC)

Step	Short-period Ground Motion, S_s	Long-period ground Motion, S_1
1. Determine the risk-targeted maximum considered earthquake (MCE_R) spectral response acceleration parameters S_S, S_1, S_{MS}, and S_{M1} for the building location from the ASCE hazard tool (*https://ascehazardtool.org*).	At short period (0.2 sec.), S_s (site class B), given as a fraction or percentage of acceleration due to gravity, g.	At long period (1 sec.), S_1 (site class B), given as a fraction or percentage of acceleration due to gravity, g. Check if the notes in step 6 are applicable.
2. Determine site classification (usually specified by the geotechnical engineer) or obtained from ASCE 7, Chapter 20 (see Table 20.3-1).		
• If site class is F	Site-specific design is required.	Site-specific design is required.

(Continued)

Step	Short-period Ground Motion, S_s	Long-period ground Motion, S_1
• If data are available for shear wave velocity, standard penetration resistance (SPT), and undrained shear strength	Choose from site classes A through E.	Choose from site classes A through E.
• If no soil data is available	Use site class D.	Use site class D.
3. Calculate the design spectral response acceleration (percentage of g), which is two-thirds of the MCE_R spectral response acceleration parameter.	$S_{DS} = \frac{2}{3} S_{MS}$ (ASCE 7, equation 11.4-1)	$S_{D1} = \frac{2}{3} S_{M1}$ (ASCE 7, equation 11.4-2)
4. Determine the risk category (RC) of the structure from ASCE 7, Tables 1.5-1 and 1.5-2. (see Table 3-7)		
5. Determine the SDC*.	Use ASCE 7, Table 11.6-1*. for S_{DS}	Use ASCE 7, Table 11.6-2 for S_{D1}
6. Select the most severe SDC* (see ASCE 7, Section 11.6) from step 7.	Compare the two values of the SDC from step 5 and select the more severe SDC value. In addition, note the following per ASCE 7 Section 11.6: • For risk categories I, II, or III (see ASCE 7, Tables 1.5—1 and 1.5-2), with $S_1 \geq 0.75$g, SDC = E. • For Risk Category IV, with $S_1 \geq$ 0.75g, SDC = F.	

* When $S_1 < 0.75$, ASCE 7 Section 11.6 allows the SDC to be determined from ASCE 7 Table 11.6.1 alone, provided the following conditions are satisfied:

- In each orthogonal direction, the fundamental period of the structure, T_a, determined in equations 3-17, 3-18a, or 3-18b is less than 0.8 T_s, where $T_s = S_{D1}/S_{DS}$ (ASCE 7 Section 11.4.5).
- In each orthogonal direction, the fundamental period used to calculate the story drift is less than T_s.
- The seismic response coefficient is calculated using equations (3-16a) through (3-16f).

It is recommended, whenever possible, to aim for seismic design category (SDC) A, B, or C, but it should be noted that the SDC value for any building will depend largely on the risk-targeted MCE_R spectral response acceleration parameters, the soil conditions at the site, and the structural properties of the building. If SDC D, E, or F is obtained, this will trigger increased seismic lateral forces as well as special detailing requirements for the structure, and the reader should refer to Chapter 14 of the ASCE 7 Load Standard for additional detailing and special inspection requirements.

3-9 SEISMIC ANALYSIS OF BUILDINGS USING THE ASCE 7 LOAD STANDARD

The four methods available in the ASCE 7 Load Standard [1] for the seismic analysis of structures are as follows:

- Equivalent Lateral Force (ELF) method (ASCE 7 Section 12.8)
- Modal Response Spectrum Analysis method (ASCE 7 Section 12.9.1)

- Linear Response History Analysis (ASCE 7 Section 12.9.2)
- Non-linear Response History Analysis (ASCE 7 Chapter 16)

While the equivalent lateral force (ELF) method is a statics-based method, both the modal response spectrum method and the linear response history analysis are linear dynamic analyses using 3-D models. The fourth method—the nonlinear response history analysis method—also uses 3-D models and accounts for the nonlinear material and geometry of the structure, as well as the foundation-structure interaction. The structure is subjected to a suite of ground motions, and this method requires independent peer review. For a typical building, the ELF method will result in higher seismic base shear than the three other methods.

A minimum seismic lateral force at each level, equal to $0.01W_x$, can be used for structures in SDC A in lieu of the equivalent lateral force (ELF) method, where W_x is the portion of the total seismic dead load (see Section 3-9) tributary to or assigned to level x (see ASCE 7, Sections 1.4.2 and 11.7).

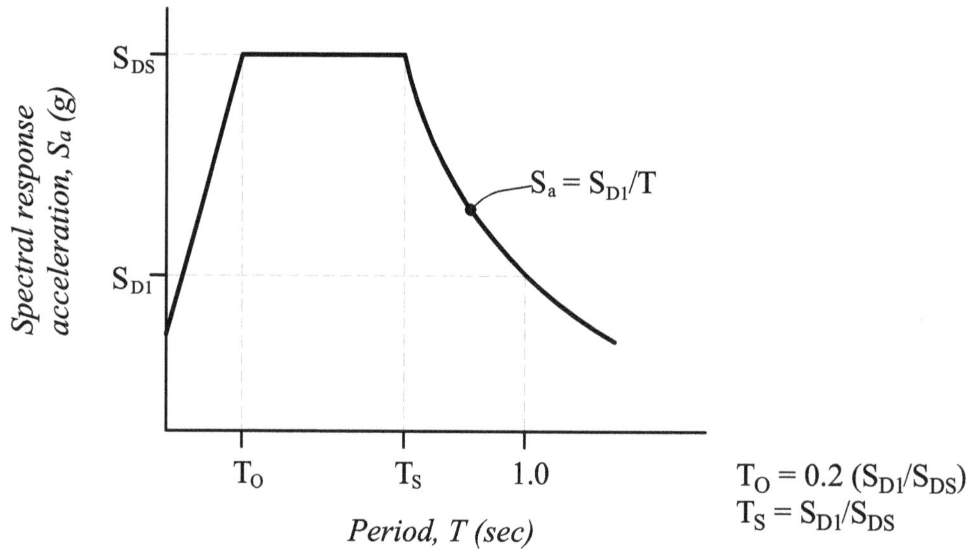

FIGURE 3-16 Design response spectra

A plot of the design response spectra from Figure 1615.1.4 of the 2003 International Building Code (IBC) that gives the maximum ground acceleration versus the period, T, for various buildings is shown in Figure 3-16. S_{D1} and S_{DS} are defined in Section 3-10. The two-period design response spectra specified in ASCE 7-22 Section 11.4.5.2 and Figure 11.4-1 consist of four zones with the corresponding spectral acceleration as follows:

Period of lateral vibration, T	Spectral acceleration, S_a
$< T_o$	$S_{DS}\left(0.4+0.6\dfrac{T}{T_o}\right)$
$T_o \le T \le T_s$	S_{DS}
$T_s < T \le T_L$	$\dfrac{S_{D1}}{T}$
$> T_L$	$\dfrac{S_{D1}T_L}{T^2}$

T_L = long-period transition period (see ASCE 7-22 Figures 22-14 through 22-17)

The response spectra can be used to determine the maximum ground acceleration for any building as a function of the period of the building. This ground acceleration can then be used to calculate the base shear and the seismic design forces on the building, and this is the basis of the equivalent lateral force (ELF) method of seismic analysis. The design response spectrum approach is appropriate for structures where the fundamental or first mode of lateral vibration dominates the lateral vibration behavior of the structure. However, for tall, slender structures where the higher modes of vibration also play a significant role in the lateral vibration of the building, the response spectrum approach is no longer applicable, and other more sophisticated methods of seismic analysis that also consider the contributions of the higher modes of vibration could be used, such as:

- Modal response spectrum analysis (MRSA) method,
- Linear response history analysis (LRHA), and
- Non-linear response history analysis (NLRHA) method, though this method is time consuming computationally and requires peer review, which increases the design cost.

In using the equivalent lateral force (ELF) method, the mapped risk-targeted maximum considered earthquake (MCE_R) spectral response acceleration parameters for short period (0.2-s), S_S, and long period (1.0-s), S_1, respectively, and the modified values, S_{MS} and S_{M1}, are required and can be obtained directly from the United States Geological Survey (USGS) Seismic Design Geodatabase [12a] for the applicable site class or from the ASCE hazard tool (*https://ascehazardtool.org*). These parameters are based on a damping of 5% of critical damping [1]. Other required parameters include the mapped long-period transition period, T_L, for the location of the structure, which can also be obtained from the ASCE hazard tool (*https://ascehazardtool.org*). Recall that the design basis earthquake (DBE) spectral accelerations are two-thirds of the maximum considered earthquake (MCE_R) spectral acceleration, S_1 and S_s, respectively. Alternatively, S_{MS} and S_{M1} can also be obtained from ASCE 7 Figures 22-1 through 22-8, and T_L can be obtained from ASCE 7 Figures 22-14 through 22-17. To use the ASCE 7 hazard tool (*https://ascehazardtool.org*), the location of the project, the risk category, and the site class are required as input.

3-10 EQUIVALENT LATERAL FORCE (ELF) METHOD

In the equivalent lateral force method (ELF), the dynamic seismic forces caused by the ground motion or acceleration are replaced by a set of equivalent static lateral forces at each diaphragm level of the structure (see ASCE 7, Section 12.8). The ELF is the most widely used method for seismic analysis and will be used in this text; the method allows a *linear elastic* analysis to be used. The ELF assigns more importance to the height of a floor or roof level from the base of the building compared to the lumped seismic weight at that level. It is a static linear elastic analysis method that considers predominantly the first mode shape (i.e., similar to a static cantilever beam). The restrictions in ASCE 7-16 to the use of the ELF procedure for certain heights and structural irregularity conditions have been removed in ASCE 7-22 (see ASCE 7-22 Section C12.6) [1]. The designer is responsible for selecting the best procedure that meets the needs of the building being designed. Note that 2D analysis is permitted for the ELF procedure for some cases, whereas the other seismic design methods—modal response spectrum analysis (MRSA) procedure and linear response history analysis (LRHA)—require 3D modeling. Note that any seismic design procedure that is based on non-linear dynamic analysis requires an independent design review (see ASCE 7 Section 16.1.4). The same goes for structures that use seismic base isolation and damping systems (ASCE 7 Sections 17.1 and 18.5).

The definitions of horizontal and vertical structural irregularities for seismic design are given in ASCE 7 Tables 12.3-1, 12.3-1a, and 12.3-2. Structures that have continuous load paths do not have the vertical or horizontal irregularities identified in ASCE 7 Chapter 12. As indicated previously, ASCE 7-22 Table 12.2-1 provides information about R, C_d, Ω_o, and structural system limitations, including structural height, as a function of the seismic design category (SDC).

For the ELF method, the factored seismic base shear (applied separately in each of the two orthogonal directions) is calculated as

$$V = C_s W, \tag{3-14}$$

Where,

C_s = the seismic response coefficient, calculated using one of two methods.

W = the total seismic dead load, which includes the following loads:

- Dead load on the diaphragm (i.e., the roof and floor levels), including its self-weight.
- Weight of the exterior cladding or wall loads, including parapets and roof screen walls.
- A minimum of 25% of the floor live load for warehouses and structures used for storage of goods, wares, or merchandise (public garages and open parking structures are excluded).
- Partition load or 10 psf, whichever is greater, but this only applies when an allowance for a partition load was included in the floor load calculations.
- Total operating weight of all permanent equipment, e.g., mechanical systems attached to the structure, air conditioners, any equipment attached to the structure. (*For mechanical room equipment, since most of this equipment is attached to concrete housekeeping pads, which are not usually attached to the floor to avoid transmission of equipment vibrations to the structure, the authors recommend taking the operating weight of this equipment as 50% of the mechanical room live load for schools and residential buildings (including the weight of the housekeeping pads) and 75% for mechanical rooms in industrial buildings.*)
- 15% of the flat roof or balanced snow load if the flat roof snow load, P_f, exceeds 45 psf. *A higher snow load results in a tendency for the bottom part of the accumulated snow to adhere or stick to the structure and thus contribute to the seismic load.*
- Weight of landscaping and other materials on roof gardens.
- Weight of fluids or bulk materials present in the building during normal use [1].

Therefore, assuming the 2nd floor is the first diaphragm above the base of the structure (i.e., the level at which the ground motion is transmitted to the structure), the total seismic dead load is

$$W = W_{2nd} + W_{3rd} + W_{4th} + \ldots + W_{roof} = \sum_{x=2nd}^{roof} W_x, \tag{3-15}$$

Where,

W_x = Seismic dead load tributary to the diaphragm at level x and lumped at level x.

The seismic response coefficient, C_s, is a measure of the ground acceleration relative to the acceleration due to gravity. The seismic response coefficient in equation (3-14) can be calculated by one of two methods (Method 1 or Method 2) from ASCE 7-22 Section 12.8.1.1. In this text, Method 2 will be used, and the seismic response coefficient is calculated from Equations (3-16a) through (3-16f) as follows:

$$C_s = \frac{S_{DS}}{\left(\frac{R}{I_e}\right)} \quad (3\text{-}16a)$$

$$\leq \frac{S_{D1}}{T\left(\frac{R}{I_e}\right)} \quad \text{for } T \leq T_L \quad (3\text{-}16b)$$

$$\leq \frac{S_{D1}T_L}{T^2\left(\frac{R}{I_e}\right)} \quad \text{for } T > T_L \quad (3\text{-}16c)$$

$$\geq \frac{0.5 S_1}{\left(\frac{R}{I_e}\right)} \quad \text{for buildings and structures with } S_1 \geq 0.60, \quad (3\text{-}16d)$$

$$\geq 0.044 S_{DS} I_e \quad (3\text{-}16e)$$

$$\geq 0.01 \quad (3\text{-}16f)$$

Where,

Equation (3-16a) is the seismic response coefficient in the constant acceleration region. Equation (3-16b) is for the constant velocity region, and Equation (3-16c) is for the constant displacement region.

I_e = Importance factor for earthquake from ASCE 7, Tables 1.5-1 and 1.5-2,
S_{DS} = Short-period design spectral response acceleration parameter (ASCE 7 Sections 11.4.4 or 11.4.7),
S_{D1} = 1-sec. period design spectral response acceleration parameter (ASCE 7 Sections 11.4.4 or 11.4.7),
S_1 = The mapped risk-targeted maximum considered earthquake (MCE_R) ground motion parameter for 1.0-s spectral response acceleration (ASCE 7, Figures 22-2 through 22-8),
T = Fundamental period of vibration of the structure,
T_L = Mapped long-period transition period (see ASCE 7, Figures 22-14 through 22-17, [1]), and
R = Structural system response modification factor, also known as the ductility factor (see ASCE 7, Table 12.2-1).

The **structural system response modification coefficient, *R*,** is a measure of the ductility of the seismic lateral force-resisting system, and it accounts for the inelastic behavior and the energy dissipation capacity of the lateral force-resisting system (see Figure 3-2). An R value of 1.0 corresponds to a purely linear elastic structure, and this would result in the highest seismic lateral force, but it would be highly uneconomical to design structures assuming linear elastic behavior. The use of higher R values that would reduce the base shear values to well below the elastic values is borne out of experience from previous earthquakes in which buildings have been known to resist earthquake forces significantly higher than the design base shear values because of inelastic behavior and the inherent redundancy in the building structure; this was previously discussed in Section 3-1 regarding Figure 3-2. A higher R-value reflects superior non-linear performance of the lateral force-resisting system (LFRS), with superior energy dissipation abilities. These systems have "a well-balanced and symmetric hysteretic behavior" when subjected to cyclic loading.

The seven basic lateral force-resisting systems prescribed in ASCE 7, Table 12.2-1, each with its own unique R-values, are described as follows:

A. ***Bearing Wall Systems:*** These are lateral force-resisting systems (LFRS) that *support both gravity and lateral loads* (e.g., concrete or masonry shear walls that support both gravity and lateral loads). Since the same wall system supports both gravity and lateral loads, it lacks redundancy, and therefore, the R values are smaller than other lateral force-resisting systems.

B. ***Building Frame Systems:*** In the building frame system, the gravity load is predominantly supported by the building frame consisting of the beams, girders, and columns, while the *lateral load is resisted by shear walls and/or braced frames*. Note that the diagonal braces in braced frames do not support the building gravity loads. The LFRS that support only lateral load are classified as building frame systems. It has better structural performance because the LFRS supports only a single type of load (i.e., lateral load); therefore, the R values are higher than those for bearing wall systems. Thus, building frame systems are more economical than bearing wall systems. Examples of building frame systems include braced frames (X-braced frames, Chevron braced frames, single diagonal braced frames, and multi-tiered braced frames), masonry shear walls in steel-framed buildings, steel plate shear walls and steel plate composite shear walls.

   It should be noted that concrete or masonry shear walls that do not support gravity loads may also be classified as building frame systems, and it is not necessary that the walls be isolated from the building frame. If a shear wall that is built integral with a building frame has confined columns at the ends of the wall and/or confined columns within the wall length and has a beam or girder immediately above and in the plane of the wall spanning between these columns to support the gravity loads, such a shear wall can be classified as a building frame system; the reason is that if the shear wall panel itself were removed, the gravity loads can still be supported by the end columns and/or the confined columns and the beam or girder spanning between these columns [5]. The greater a structural system's ability to dissipate seismic energy, the higher the R-value or system response modification coefficient. Individual steel and concrete composite plate shear walls (system B-14) have a system response modification coefficient, R of 6.5, but the steel and concrete coupled composite plate shear walls (system B-28) or the Speedcore shear wall system with its coupling beams (see Figures 3-8b and 3-8c) have an R value of 8 because of its greater ability to dissipate seismic energy.

   Note that for higher seismic regions, the use of ordinary concentric braced frames (OCBF) is not permitted because of their inherent unbalanced hysteretic behavior in tension and compression (i.e., the tension brace and the compression brace have different load-deformation characteristics). Once the compression brace buckles, only the tension brace is active unless and until the lateral load reverses direction. For high seismic regions, buckling restrained braced frames (BRBF) are preferred because of their balanced or symmetric hysteretic behavior. That is, under repeated cyclic lateral loading, both the tension diagonal brace and the compression diagonal brace have the same yield strength and the same load-deformation curve since the compression brace is restrained from buckling. In ASCE 7 Table 12.2-1, BRBF used in the dual lateral force-resisting systems (System D-13) has a system response modification coefficient, R, of 8 (indicating excellent ductility), whereas OCBF (System B-3) has an R value of $3\frac{1}{4}$.

C. ***Moment Resisting Frame Systems***

D. ***Dual Systems with Special Moment-Resisting Frames Capable of Resisting at least 25% of the Prescribed Seismic Forces:*** Shear wall or buckling restrained braced frames (BRBF) combined with special moment-resisting frames (SMRF) where the SMRF resist at least 25% of the prescribed seismic forces. These are systems with some redundancies and therefore, they have higher *R* values.

E. ***Dual Systems with <u>Intermediate</u> Moment-Resisting Frames Capable of Resisting at least 25% of the Prescribed Seismic Forces:*** Shear wall or buckling restrained braced frames (BRBF) combined with intermediate moment-resisting frames (IMRF), where the IMRF resists at least 25% of the prescribed seismic forces. These are systems with some redundancy, and therefore, they have higher *R* values.

F. ***Shear Wall-Frame Interactive System with*** ordinary reinforced concrete moment-resisting frames and ordinary reinforced concrete shear walls.

G. ***Cantilevered Column Systems:*** Structures where a large proportion of their total weight is concentrated at the top of the structure (e.g., water storage towers).

H. ***Steel Systems Not Specifically Detailed for Seismic Resistance, Excluding Cantilevered Column Systems (R = 3.0)***: This system is only permitted for SDC A, B, or C and is frequently chosen by design professionals in practice because it is the more economical option for structures with SDC A through C and obviates the need for any special steel detailing requirements. Steel systems with R = 3 are expected to remain "nominally elastic." Using *R* values greater than 3.0 triggers the special detailing requirements in the AISC Seismic Detailing Provisions for Steel Buildings (AISC 341-22) [9]. To avoid the increased costs associated with the seismic detailing requirements that are triggered whenever *R* exceeds 3.0, an *R*-value of 3.0 or less is recommended for design whenever possible, but this is not permitted in highly seismic regions or buildings with SDC D, E, or F. In fact, the lateral force-resisting systems for most buildings in low to moderate seismic regions of the United States have an R value of 3, requiring no ductile detailing [22].

In addition to the structural system response modification coefficient (i.e., *R*-values) provided in ASCE 7, Table 12.2-1, the following additional seismic design parameters and information are also provided in this table:

- **Deflection amplification factor, C_d.** This factor magnifies the elastic deflection, Δ_e, calculated from the linear elastic analysis to account for the inelastic action due to seismic effects (see Figure 3-2). Thus, the inelastic deflection, $\Delta_u = C_d\Delta_e$.
- **Overstrength factor, Ω_o.** This amplification factor (see ASCE 7 Sections 12.3.3.3 and 12.4.3) accounts for members (e.g., discontinued shear walls supported on columns) whose strength may be higher than the forces or loads for which they were designed. For instance, a discontinued shear wall supported by columns may have higher strength than it was designed for and therefore result in higher reaction forces on the supporting columns than the shear wall design called for. Therefore, the analysis and design of the columns and their connections to the discontinued shear wall above and the foundation below are to include the overstrength factor in the load combinations (see ASCE 7 Section 12.4.3). The foundation system used to resist overturning in cantilever column systems (e.g., water tank support structure) in SDC B through F must include the overstrength factor in the load combinations (see ASCE 7 Sections 12.2.5.2 and 12.4.3). Other types of structural elements for which the overstrength factor applies include drag struts (or collectors) and chords in horizontal diaphragms (see ASCE 7 Section

12.10.2.1). The overstrength factor also applies to diaphragm and LFRS connection elements (e.g., bolts, welds, gusset plates, columns, and anchor bolts at diagonal brace locations, which need to be stronger than the diagonal brace) in lateral force-resisting systems in order to avoid creating a weak link in the secondary elements along the seismic lateral load path [21]. In ASCE 7 Section 12.3.3.4, the diaphragm forces are required to be increased by 25% for diaphragms and their connections in structures in SDC D through F with certain types of horizontal or vertical irregularities, but this increase can be neglected if the overstrength factor is included in the seismic load calculations.

- Structural system limitations, including building height limits as a function of the seismic design category (SDC), and
- The sections in ASCE 7 where the material-specific seismic detailing requirements are specified.

Notes:

- The higher the period, T, of a structure, the smaller the seismic coefficient, C_s, calculated from equation (3-16), and therefore the smaller the seismic base shear on the structure. A high-period structure indicates a structure with low lateral frequency, and since lateral stiffness is proportional to the lateral frequency, this would indicate a relatively more flexible structure, which would therefore attract smaller seismic forces.
- The higher the R value (meaning a more ductile LFRS), the smaller the seismic coefficient, C_s, calculated from equation (3-16), and therefore the smaller the seismic force exerted on the structure.
- Where there is more than one type of LFRS in the same orthogonal direction in a structure, use the more stringent seismic design parameters from ASCE 7 Table 12.2-1 for that orthogonal direction; that is, use the *lowest* R value and the *highest* C_d and Ω_o values.

Fundamental Period of Lateral Vibration, *T* (for all types of buildings) for Seismic Analysis

The period of lateral vibration of a structure cannot be determined until the structure is designed and the structural properties are known, but the seismic forces cannot be determined without knowing the period of vibration. Consequently, for seismic analysis, the ASCE 7 Load Specification allows the use of approximate methods for estimating the period of lateral vibration of building structures. In reality, the period of vibration of a seismic lateral force resisting system is not constant, but will vary from a higher value when the seismic LFRS is in the elastic state to a lower value as it cracks or yields. The most commonly used equation for calculating the approximate fundamental period used for the seismic analysis for all types of buildings is given in ASCE 7 Section 12.8.2.1 as

$$T = T_a = C_t\left(h_n^{\,x}\right), \tag{3-17}$$

Where,

T = approximate fundamental period of the building

C_t and x are obtained from Table 3-9 (ASCE 7, Table 12.8-2) [1], and

h_n = Height (in feet) from the **base** of the building to the highest level (i.e., the roof), where the base of the building is the level where the horizontal seismic ground motions are imparted to the structure (see ASCE 7-22 Figures C11.2-2 through C11.2-9):

- For a building without basements, the base of the building is at the top of the slab-on-grade.
- For a building with a basement and stiff soil around the building over the depth of the basement, and with a level site, the base of the building will be at the ground floor level.
- For a building with a basement and soft soil around the building over the depth of the basement, and with a level site, the base of the building will be at the top of the lowest basement floor slab on grade.
- For a building with a basement and a sloping grade or grade elevation lower than the ground floor, the base of the building will be at the top of the lowest basement floor slab on grade.

TABLE 3-9 C_t values for various structural systems

Structural System	C_t	x
^Steel moment-resisting frames	0.028	0.8
^Concrete moment-resisting frames	0.016	0.9
Steel eccentrically braced frames (EBF)	0.03	0.75
Steel buckling-restrained Braced frames (BRBF)	0.03	0.75
All other structural systems (e.g., shear walls, regular braced frames)	0.02	0.75

^moment-resisting frames that resist 100% of the lateral seismic force

Fundamental Period of Vibration, *T* (for moment-resisting frames only)

An alternate equation for the approximate fundamental period of lateral vibration of building structures consisting of *concrete or steel moment-resisting frames* and valid only for structures not exceeding 12 stories above the base or the level at which the earthquake ground motion is transmitted to the structure, and with an average story or floor-to-floor height of at least 10 ft. is given as follows (see ASCE 7 Section 12.8.2.1):

$$T = T_a = 0.1\,N, \tag{3-18a}$$

where N is the number of stories above the base of the building.

Note that the number of stories will include the basement levels for buildings with underground levels.

Fundamental Period of Vibration, *T* (for concrete and masonry shear walls not exceeding 120 ft in height)

Another alternate approximate equation for the fundamental period of lateral vibration for seismic analysis and design that is applicable only *to concrete or masonry shear walls* not exceeding 120 ft in height is given as follows (see ASCE 7 Section 12.8.2.1):

$$T = T_a = \frac{0.0019}{\sqrt{C_w}} h_n \tag{3-18b}$$

where

$$C_w = \frac{100}{A_B} \sum_{i=1}^{x} \frac{A_i}{\left[1 + 0.83\left(\frac{h_n}{D_i}\right)^2\right]} \tag{3-18c}$$

A_B = Area of base of structure, ft.^{2}
A_i = Web area of shear wall i in ft.^{2}
D_i = Length of shear wall i in ft
h_n = Structural height or the height from the base of the structure to the highest level of the seismic lateral force-resisting system, in ft.
x = Number of shear walls in the building effective in resisting seismic lateral forces in the direction under consideration.

The higher the fundamental period of lateral vibration of a structure, T_a, the smaller the stiffness of the structure and therefore, the smaller the seismic force acting on the structure. If a dynamic structural analysis software is used to determine the fundamental period of lateral vibration of the structure, the calculated period, $T_{dynamic}$, is limited to a maximum value, T_{max}, in order to minimize the effect of the error in calculating the fundamental period when more elaborate methods of analysis are used. The ASCE 7 load specification sets an upper limit on the calculated period of vibration using the factor C_u, which varies between 1.4 and 1.7. The maximum period, T_{max}, that can be used to calculate the seismic forces is determined as follows:

$$T_{\text{dynamic}} \leq T_{\text{max}} = C_u \, T_a, \tag{3-19}$$

where
C_u = Factor that depends on S_{D1} and is obtained from ASCE 7, Table 12.8-1, and
T_a = Approximate period of vibration determined from the approximate equations presented previously.

Note that equation (3-17) is more commonly used in practice to calculate the period of lateral vibration for seismic analysis and design, and equation (3-19) is used only when the natural period of lateral vibration of the structure is determined from a dynamic structural analysis.

The inelastic action of structures during a seismic event, resulting from cracking and yielding of the members, causes an increase in the damping ratio and in the fundamental period of lateral vibration of the structure, which leads to reduced seismic forces. Note, however, that the linear elastic analysis procedure for seismic design in ASCE 7 uses a constant fundamental period of vibration, T, even though in reality, the period is not constant but continues to increase as the structural elements yield, and hinges are formed. This change in the fundamental period of lateral vibration, which results in a change in the stiffness of the structure, is considered in non-linear analysis techniques such as a pushover analysis.

3-11 VERTICAL DISTRIBUTION OF SEISMIC BASE SHEAR, *V*

Since most structures are multiple degrees-of-freedom systems with several modes of lateral vibration, the distribution of the seismic base shear is a combination of the contributions from all the significant modes of lateral vibration of the structure.

The force distribution to each level is a function of the seismic weight, W_x, at each level, the height or stiffness of the structure, and the predominant modes of lateral vibration. The exponent k in equation (3-21) (see Table 3-10) is an attempt to capture the contributions from the higher modes of lateral vibration. The higher modes result in higher lateral shears at the higher or upper levels of the building. When k is 1, the first or the fundamental mode of lateral vibration dominates, which implies a linear distribution of the seismic lateral forces over the height of the building.

TABLE 3-10 *k* values

Building Period, *T*, in seconds	*k*
≤ 0.5	1 *(no whiplash effect; and implies a linear distribution of the seismic lateral forces)*
$0.5 < T < 2.5$	$1 + 0.5\,(T - 0.5)$
≥ 2.5	2

The factored concentrated seismic lateral force at any diaphragm level x of the vertical LFRS is given as

$$F_x = C_{vx}\,V, \tag{3-20}$$

Where,
V = factored seismic base shear, and

$$C_{vx} = \frac{W_x h_x^k}{\sum_{i=1}^{n} W_i h_i^k}, \tag{3-21}$$

F_x = factored concentrated seismic lateral force at level x
W_x = the portion of the total gravity load of the building, W, that is tributary to level x (includes the weight of the floor or roof, plus the weight of the perimeter or interior walls tributary to diaphragm level x),
W_i = the portion of the total gravity load of the building, W, that is tributary to level i (includes the weight of the floor or roof, plus the weight of the perimeter or interior walls tributary to diaphragm level i),
h_i and h_x = Height (in feet) from the **base** of the building to level $\boldsymbol{i}$ or $\boldsymbol{x}$,
k = Exponent related to the building period (refer to Table 3-10).

Note: The height of a vertical wall that is tributary to a particular level, x, is one-half the distance between level x and level $x + 1$ plus one-half the distance between level x and level x - 1.

Level $\boldsymbol{i}$ = any diaphragm level in the building ($\boldsymbol{i}$ = 1 for the first elevated level above the base of the structure),
Level $\boldsymbol{x}$ = that diaphragm level under design consideration, and
Level $\boldsymbol{n}$ = uppermost level of the building.

For an accuracy check, note that the sum of all the concentrated lateral forces at each diaphragm level of the structure must sum up to the total seismic base shear, V, that is, $V = \sum_{i=1}^{n} F_i$

The intent of the lateral force distribution assumed in Equation (3-20) is to obtain the maximum overturning moment at the base of the building. Note that for structures in SDC A, it is permitted to calculate the seismic lateral force at each diaphragm level (ASCE 7 Sections 11.7 and 1.4.2) as:

$$F_x = 0.01 W_x$$

Orthogonal Seismic Load Effects

For structures in SDC A through C, the seismic load on the structure in the two orthogonal directions are considered independently (see ASCE 7, Sections 12.5.1.1, 12.5.2 and 12.5.3). However, for structures in SDC D through F, the orthogonal effects of seismic loads must be considered when certain structural irregularities are present (see ASCE 7, Sections 12.5.1.2 and 12,5.4); in that case, the seismic lateral load on a structure in SDC D through F consists of 100% of the seismic lateral load in one orthogonal direction plus 30% of the seismic lateral load in the perpendicular direction considered simultaneously (ASCE 7 Section 12.5.1.2).

Diaphragm Lateral Forces, F_{px}

The diaphragm lateral force is the maximum lateral force that can act on a diaphragm during an earthquake, regardless of what lateral forces are applied to the other floor or roof levels of the building. The lateral forces calculated in Equation (3-20) are based primarily on the first mode of vibration and are therefore not representative of the maximum lateral force in the horizontal diaphragms. There are three methods given in ASCE 7 Section 12.10 for calculating the seismic lateral forces in horizontal diaphragms.

The traditional method for the seismic design of horizontal diaphragms given in ASCE 7 Sections 12.10.1 and 12.10.2 is adopted in this text. Note that the traditional method is not permitted for precast concrete diaphragms in SDC C through F; for this type of diaphragm, the permitted alternative method is found in ASCE 7 Section 12.10.3. The third alternative method, found in ASCE 7 Section 12.10.4, is permitted only for one-story buildings with flexible diaphragms (i.e., wood structural panel or bare steel deck diaphragms without concrete topping) and rigid vertical lateral force-resisting systems (LFRS) such as concrete shear walls, precast concrete shear walls, masonry shear walls, steel concentrically braced frames, steel and concrete composite braced frames, or steel and concrete composite shear walls.

For multi-story buildings, the diaphragm design force, F_{px}, is usually larger than the story force, F_x or F_i from equation (3-20), because the response of the diaphragms at lower levels of multi-story buildings to the higher modes of lateral vibration may result in forces larger than those calculated for the lateral force-resisting system using equation (3-20).

The seismic lateral forces on the *floor and roof diaphragms* (see ASCE 7-22 Section 12.10.1.1) can be obtained from Equation (3-22) as follows:

$$F_{px} = \frac{\sum_{i=x}^{n} F_i}{\sum_{i=x}^{n} W_i} W_{px} \tag{3-22}$$

$$\geq 0.2 S_{DS} I_e W_{px}$$

$$\leq 0.4 S_{DS} I_e W_{px}$$

Where,

F_{px} = the diaphragm lateral force at level x

F_i = the design lateral force at level i from equation (3-20)

$\sum_{i=x}^{n} F_i$ = the story shear just below the diaphragm at level x

$\sum_{i=x}^{n} W_i$ = the sum of the seismic weights at level x and all the diaphragms above level x

W_{px} = the seismic weight tributary to the diaphragm at level x

W_i = the seismic weight tributary to level i

The minimum value of the diaphragm force, $0.2S_{DS}I_eW_{px}$, typically controls for high-rise buildings in high seismic regions. The larger of the design lateral force, F_i, and the diaphragm lateral forces, F_{px}, are used in the analysis and design of the roof and floor diaphragms and their chords and drag struts. Note that the diaphragm design force, F_{px}, is applied separately to each diaphragm since their maximum value does *not* occur simultaneously or in the same direction in all the diaphragms because of the difference in the mode shapes for the higher modes [16]. Note that alternative design provisions for determining the diaphragm lateral force, F_{px}, are given in ASCE 7-22, Sections 12.10.3 and 12.10.4. The diaphragm chords and drag struts or collectors should also be designed and detailed for these in-plane lateral forces in the diaphragms, and more information on these structural elements can be found in Chapter 14 of this text.

3-12 STORY DRIFT AND STRUCTURAL SEPARATION DUE TO SEISMIC LATERAL FORCES

The lateral displacements at each level of the seismic lateral force-resisting system due to the design earthquake can be calculated using elastic analysis. These displacements are amplified to account for inelastic action (see Figure 3-2). The lateral deflection at level x of the LFRS is calculated as

$$\delta_{DE} = \frac{C_d \delta_e}{I_e} + \delta_{di} \qquad (3\text{-}23)$$

δ_{DE} = inelastic lateral deflection at level x of the seismic lateral force-resisting system due to the design earthquake

δ_e = elastic lateral deflection at level x of the seismic lateral force-resisting system due to the design earthquake, taking into account the effects of accidental torsion and torsional amplification

C_d = deflection amplification factor for the LFRS from ASCE 7, Table 12.2-1

I_e= seismic importance factor from ASCE 7, Table 1.5-2

δ_{di} = elastic lateral deflection of the horizontal diaphragm at level x due to the design earthquake

The story drift is the difference between the lateral deflections, δ_x, at the centers of mass at the top and bottom of the story under consideration, and the allowable story drift, D_a, for different structural systems and different risk categories are tabulated in ASCE 7, Table 12.12-1. For most steel structures, the allowable story drift, D_a, are as follows:

- $0.020h_{sx}$ for risk category (RC) I and II;
- $0.015h_{sx}$ for risk category (RC) III; and
- $0.010h_{sx}$ for risk category (RC) IV,

Where,
h_{sx} = the story height below level x.

For moment-resisting frames in seismic design category (SDC) D, E, or F, the allowable story drift is $\frac{\Delta_a}{\rho}$ (see ASCE 7, Section 12.12.1.1), where ρ is the redundancy factor defined previously in Section 2-3. The more redundancy or alternate load path that a structure has, the lower the redundancy factor.

To avoid damaging contact (i.e., pounding) between two adjacent structures, the structures should be separated by a minimum distance, δ_{SS}, which is equal to the square root of the sum of squares of the maximum design earthquake lateral deflections of the two adjacent structures at their adjacent edges (see ASCE 7, Section 12.12.2). That is,

$$\delta_{SS} = \sqrt{(\delta_{DE1})^2 + (\delta_{DE2})^2} \tag{3-24}$$

Where,

δ_{SS} = minimum separation distance between two adjacent buildings
δ_{DE1} = design earthquake lateral displacement of building 1
δ_{DE2} = design earthquake lateral displacement of building 2

3-13 TWO-STAGE SEISMIC ANALYSIS FOR PODIUM STRUCTURES

Podium structures usually involve a multistory light-frame wood or mass timber upper structure built on top of one or more stories of concrete or steel podium building (see Figure 3-17a), and both the upper and lower (podium) portions of the structure are required to be separated for fire resistance. The podium or lower structure typically consists of retail or parking occupancy, so a 3-hr fire resistance is typically required for the podium "roof" slab that supports the upper light frame wood structure. ASCE 7-22 Section 12.2.3.2 allows a two-stage seismic analysis for such structures, treating both the upper and lower portions as two distinct structures, provided the following conditions are satisfied:

- **a.** The lateral stiffness of the stiffer lower portion (i.e., the podium structure) is at least 10 times the lateral stiffness of the flexible upper structure, and
- **b.** The period of lateral vibration of the entire structure (i.e., both the upper and lower portions) should not be greater than 1.1 times the period of the flexible upper portion.

The two-stage process involves the following steps:

1. Carry out a seismic analysis of the flexible upper portion of the structure using the top of the podium structure as the base of the upper structure and assuming the upper portion of the structure is fixed at its base (i.e., at the top level of the lower podium structure). The **equivalent lateral force (ELF)** method or the modal analysis method can be used to analyze the upper portion of the structure. Note the R_{upper} and ρ_{upper} for the upper structure.
2. Next, carry out the seismic analysis of the stiffer lower building (i.e., the podium structure) using the **equivalent lateral force (ELF) method**. In analyzing the lower portion, the reactions from the upper portion due to the horizontal earthquake, E_h, must be amplified by $\frac{\left(R/\rho\right)_{upper}}{\left(R/\rho\right)_{lower}} \geq 1.0$.

As an example, for a podium building where the LFRS for the upper portion consists of light frame (wood) walls sheathed with wood structural panels rated for shear resistance (i.e., R = 7 from ASCE 7-22 Table 12.2-1) and the lower or podium steel structure is not specifically detailed for seismic resistance (i.e., R = 3), the amplification factor for the reactions from the upper portion due to the horizontal earthquake, E_h, assuming $\rho = 1$ for both upper and lower portions, is equal to 7/3 or 2.33.

(Figure 3-17a continued)

FIGURE 3-17a 4-over-1 Podium structure under construction

3-14 FOUNDATION TIES IN SEISMIC DESIGN

To ensure that foundations behave as an integral unit, ASCE 7 Sections 12.13.7.2 and 12.13.8.2 require deep foundations, individual pile caps, caissons, or drilled piers in SDC C through F to be interconnected in two orthogonal directions with tension and compression ties (i.e., concrete grade beams with continuous top and bottom reinforcement capable of resisting a tension/compression axial force equivalent to $0.10S_{DS}D$,

Where,

D = larger of pile cap or column factored dead plus live load

Additionally, individual spread footings founded on soil defined as site class E or F in ASCE 7 Chapter 20 shall also be interconnected with ties in both orthogonal directions (ASCE 7 Section 12.13.8.2).

3-15 STRUCTURAL DETAILING REQUIREMENTS FOR SEISMIC DESIGN

The seismic limit state is based on the behavior of the structural system and not on the behavior of a structural member; it assumes considerable energy dissipation through repeated cycles of inelastic deformations because of the large demand force exerted on the structure during an earthquake and the huge cost that would be required to ensure linear elastic behavior (see Figure 3-2). Because of this premise, many material-specific detailing requirements are needed to ensure inelastic and ductile behavior of the structure. These detailing requirements as well as the requirements relating to construction quality assurance are specified in ASCE 7, Chapter 14 [1]. After the seismic design forces on a structure have been determined and the lateral load resisting systems have been designed for these forces, the structure must also be detailed to conform to the structural system requirements that are required by the seismic design category (SDC). Buildings in SDC A, B, or C do not generally require stringent detailing requirements for ductility. However, buildings in SDC D, E, or F require stringent detailing requirements for ductility for the seismic lateral force-resisting system and other components of the building. The reader should refer to ASCE 7, Chapter 14, for the specified design and detailing requirements as a function of the SDC and the construction material, as well as any special inspection requirements [1].

EXAMPLE 3-3

Seismic Lateral Forces in a Multistory Building—Equivalent Lateral Force Method

The typical floor plan and elevation of a six-story office building measuring 100 ft. × 100 ft. in plan and laterally braced with four 3-bay ordinary moment-resisting frames is shown in Figure 3-17b. Determine the seismic lateral forces on the building and on each moment frame, and on the diaphragms assuming the following design parameters:

Roof dead load = 24.5 psf
Snow load = 45.5 psf
Floor dead load = 75 psf (includes partition load)
Cladding (glazing) = 20 psf

The mapped risk-targeted maximum considered earthquake (MCE_R) spectral response acceleration at 0.2-s and 1.0-s periods, respectively, are $S_S = 0.25g$ and $S_1 = 0.072g$

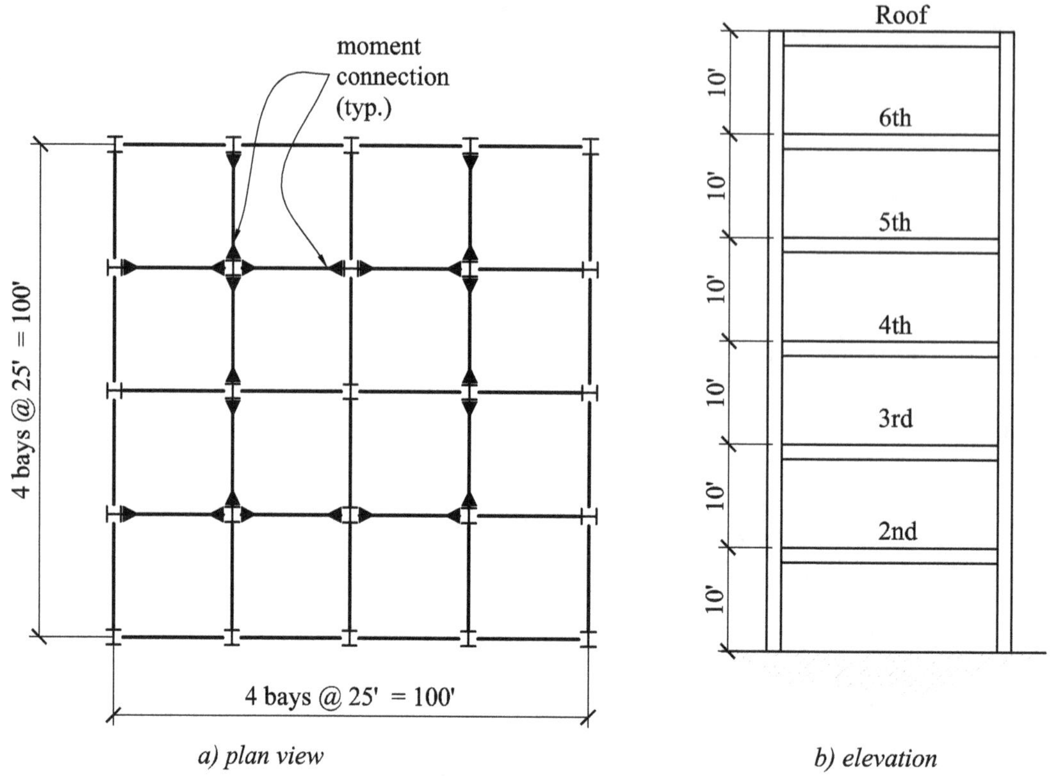

FIGURE 3-17b Building plan and elevation for Example 3-3

Solution

1. Determine the SDC (see Table 3-11.)

Table 3-11 Determining the seismic design category (SDC) for example 3-5

Step	Short-Period Ground Motion, S_s	Long-Period Ground Motion, S_1
1. Determine the risk-targeted maximum considered earthquake (MCE_R) spectral response acceleration parameters S_S, S_1, S_{MS}, and S_{M1} for the building location from the ASCE hazard tool (*https://ascehazardtool.org*).	$S_s = 0.25$ Use the fraction of g (i.e., use 0.25) in the calculations. $S_{MS} = 0.4$	$S_1 = 0.072 < 0.75$ Use the fraction of g (i.e., use 0.072) in the calculations. $S_{M1} = 0.17$
2. Determine site class (usually specified by the geotechnical engineer). • If site class is F	Do site-specific design.	Do site-specific design.
• If data are available for shear wave velocity, standard penetration resistance (SPT), and undrained shear strength	Choose from site classes A through E.	Choose from site classes A through E.
• If no soil data are available	**Use site class D.**	**Use site class D.**

3. Calculate the design spectral response acceleration (percentage of g).	$S_{DS} = \frac{2}{3} S_{MS}$ (ASCE 7, equation 11.4-1) $= \frac{2}{3}(0.4) = 0.27$	$S_{D1} = \frac{2}{3} S_{M1}$ (ASCE 7, equation 11.4-2) $= \frac{2}{3}(0.17) = 0.11$
4. Determine the risk category (RC) of the structure from ASCE 7, Tables 1.5-1 and 1.5-2 or Table 3-7.	Standard occupancy building $\Rightarrow I_e = 1.0$	Standard occupancy building $\Rightarrow I_e = 1.0$
5. Determine the SDC.	**SDC = B** (ASCE 7, Table 11.6-1 for S_{DS})	**SDC = B** (ASCE 7, Table 11.6-2 for S_{D1})
6. Choose most severe SDC (i.e., the higher SDC value).	Compare the second and third columns from the previous step $\Rightarrow$ **USE SDC = B**	

2. Select the method of seismic analysis.
From Table 3-11, the SDC is found to be B. The equivalent lateral force method (ELF) will be used for the seismic analysis of this building.

3. Calculate the seismic dead load at each level, W_i, and the total seismic dead load, W (see Table 3-12).

The flat roof snow load given for this building is $P_f = 45.5$ psf > 45 psf. Therefore, 15% of the flat roof snow load should be included in the seismic dead load calculations to account for the snow that may stick to the roof structure.

TABLE 3-12 Assigned seismic weights at each level of the building

Diaphragm Level	Height of Diaphragm from Base, h_i	Weight, *Wi* at each Diaphragm level
Roof	60 ft.	W_{roof} = (24.5 psf)(100 ft.)(100 ft.) + (15%)(45.5 psf)(100 ft.)(100 ft.) +(20 psf)(2)(100 ft. + 100 ft.)(10 ft./2) = **353 kips**
Sixth floor	50 ft.	W_6 = (75 psf)(100 ft.)(100 ft.) + (20 psf)(2)(100 ft. + 100 ft.)[(10 ft. +10 ft.)/2] = **830 kips**
Fifth floor	40 ft.	W_5 = (75 psf)(100 ft.)(100 ft.) +(20 psf)(2)(100 ft. + 100 ft.)[(10 ft. + 10 ft.)/2] = **830 kips**
Fourth floor	30 ft.	W_4 = (75 psf)(100 ft.)(100 ft.)+(20 psf)(2)(100 ft. + 100 ft.)[(10 ft. + 10 ft.)/2]= **830 kips**
Third floor	20 ft.	W_3 = (75 psf)(100 ft.)(100 ft.) + (20 psf)(2)(100 ft. + 100 ft.)[(10 ft. + 10 ft.)/2] = **830 kips**
Second floor	10 ft.	W_2 = (75 psf) (100 ft.)(100 ft.) + (20 psf)(2)(100 ft. + 100 ft.)[(10 ft. + 10 ft.)/2] = **830 kips**

Note: the total seismic weight, $W = \Sigma W_i = 4503$ kips.

4. Determine the seismic coefficient, C_s.

For this building, the lateral loads are resisted solely by the moment-resisting frames. Therefore, from ASCE 7, Table 12.2-1, we could select ordinary steel moment-resisting frames (system C-4) for which we obtain the following seismic design parameters for the system:

$R = 3.5,\ C_d = 3,$ and $\Omega_o = 3.$

The equivalent lateral force method is permitted in SDC A through C.

To use system C-4, certain detailing and system-specific requirements must be met, and the AISC 341 specification for seismic detailing of steel buildings [9a, 9b] must be used (see ASCE 7, Sections 12.2.5 and 14.1).

However, if a "steel system not specifically detailed for seismic resistance" or system "H" from ASCE 7-22, Table 12.2-1 is used (*whenever possible and for economic reasons, this is a highly recommended system for steel buildings.*), the less stringent AISC 360 specification [10] for structural steel buildings is permitted to be used. From a cost point of view for this building, we will adopt system "H" (i.e., steel systems not specifically detailed for seismic resistance), and the following seismic design parameters are obtained for the system from ASCE 7, Table 12.2-1:

$R = 3,\ C_d = 3,$ and $\Omega_o = 3.$ (*Note that structural steel system "H" is* <u>*only permitted in SDC A, B, or C*</u>).

The equivalent lateral force method is permitted for this building with SDC B.

$C_t = 0.028$ and $x = 0.8$ (from Table 3 − 9 for steel moment resisting frames)

h_n = Roof height = 60 ft.

$$T = T_a = \text{Approximate of the building} = C_t\left(h_n^{0.8}\right)$$

$$= 0.028 \times \left(60^{0.8}\right) = 0.74s < T_L = 6 \text{ s} \ \ (\text{see ASCE 7, Figure 22} - 12)$$

$S_1 = 0.072\text{g} < 0.6\text{g}$

From equations (3-16a) through (3-16f), the seismic response coefficient is calculated as

$$C_s = \frac{S_{DS}}{\left(\frac{R}{I_e}\right)} = \frac{0.27}{\left(\frac{3}{1.0}\right)} = 0.09$$

$$\leq \frac{S_{D1}}{T\left(\frac{R}{I_e}\right)} \text{ for } T \leq T_L = \frac{0.12}{0.74\left(\frac{3}{1.0}\right)} = \mathbf{0.054}$$

$$\geq 0.044 S_{DS} I_e = 0.044(0.27)(1.0) = 0.012$$

$$\geq 0.01$$

Therefore, $C_s = 0.054$

5. Calculate the seismic base shear, V.

   $V = C_s W = (0.054)(4{,}503 \text{ kips}) = 243 \text{ kips}$

   Since the same structural system is used in both orthogonal directions, the calculated base shear will be the same in both the N-S and E-W directions.

6. Determine the vertical distribution of the seismic base shear (i.e., determine the lateral force, F_x, at each diaphragm level).

   For $T = 0.74$ sec., and from Table 3 − 10, $k = 1 + 0.5(0.74 - 0.5) = 1.12$

The seismic lateral forces at each level of the building are calculated in Table 3-13.

Table 3-13 Seismic Lateral Force for Example 3-5

Diaphragm Level	Height of diaphragm from base, h_i	Dead Load at each diaphragm Level, W_i	$W_i(h_i)^k$	$Cvx = W_i(h_i)^k / \Sigma Wi(hi)^k$	$F_x = C_{vx}$ V	E_h	$E_v = 0.054W_i$	ΣE_v
Roof	60 ft.	353 kips	34,618	0.154	37 kips	37 kips	19 kips	19 kips
Sixth floor	50 ft.	830 kips	66,363	0.295	72 kips	72 kips	45 kips	64 kips
Fifth floor	40 ft.	830 kips	51,687	0.230	56 kips	56 kips	45 kips	109 kips
Fourth floor	30 ft.	830 kips	37,450	0.167	41 kips	41 kips	45 kips	154 kips
Third floor	20 ft.	830 kips	23,781	0.106	26 kips	26 kips	45 kips	199 kips
Second floor	10 ft.	830 kips	10,942	0.049	12 kips	12 kips	45 kips	244 kips
$\Sigma W_i(h_i)^k = 224{,}841$				$\Sigma F_i = 244$ kips ≈ 243 kips				

- The F_x forces calculated in Table 3-13 are the factored seismic lateral forces acting at each diaphragm level of the building in both the N–S and the E–W directions. That is, the horizontal component of the earthquake, $E_h = \rho Q_E = \rho F_x = (1.0)F_x = F_x$

  The redundancy factor, $\rho = 1.0$ for SDC B & C.

  The vertical component of the earthquake is,

  $E_v = 0.2S_{DS}D = (0.2)(0.27)D = 0.054D$

  The seismic force, $E = E_h \pm E_v = F_x \pm 0.054D$

  That is, the horizontal component of the earthquake is F_x, and the vertical component is $0.054D = 0.054W_i$

  Since the seismic design category (SDC) is "B," no orthogonal seismic load effects have to be considered. That is, the seismic lateral load will be considered separately or independently in the two orthogonal directions (see ASCE 7 Section 12.5.2).

- If the building has *rigid diaphragms* (as many steel buildings do), the F_x forces at each diaphragm level will be distributed to the lateral force resisting systems (LFRS) in each direction in proportion to the relative lateral stiffness of the moment-resisting frames that are parallel to the direction of the seismic lateral force.

- If the building has *flexible diaphragms,* the forces at each diaphragm level are distributed in proportion to the plan area of the building tributary to each moment frame.
  The different types of horizontal diaphragms—rigid, flexible, and semi-rigid—and the design of horizontal diaphragms for flexure and shear, and the distribution of lateral forces by the diaphragms to the LFRS in steel buildings are discussed in more detail in Chapter 14.

- The overturning moment for the entire building at the base of the building can be calculated using the F_x forces as follows:
  Overturning moment = (37 kips)(60 ft.) + (72 kips)(50 ft.) + (56 kips)(40 ft.) + (41 kips)(30 ft.) + (26 kips)(20 ft.) + (12 kips)(10 ft.) = 9930 ft.-kips.

Diaphragm Lateral Forces, F_{px}

For this building, the diaphragm lateral forces are calculated using the traditional method (see ASCE 7-22 Section 12.10.1.1) from equation (3-22), and the results are shown in Table 3-14.

$$F_{px} = \frac{\sum_{i=x}^{n} F_i}{\sum_{i=x}^{n} W_i} W_{px}$$

$$\geq 0.2 S_{DS} I_e W_{px} = (0.2)(0.27)(1.0)\, W_{px} = 0.054\, W_{px}$$

$$\leq 0.4 S_{DS} I_e W_{px} = (0.4)(0.27)(1.0)\, W_{px} = 0.108\, W_{px}$$

For the **roof level**, the lower and upper limits above are calculated to be

$$0.054\, W_{px} = (0.054)(353\,\text{kips}) = 19.1\,\text{kips} \quad \text{(lower limit)}$$

$$0.108\, W_{px} = (0.108)(353\,\text{kips}) = 38.2\,\text{kips} \quad \text{(upper limit)}$$

For each of the **other floor levels**, the lower and upper limits are calculated to be

$$0.054\, W_{px} = (0.054)(830\,\text{kips}) = 44.8\,\text{kips} \quad \text{(lower limit)}$$

$$0.108\, W_{px} = (0.108)(830\,\text{kips}) = 89.6\,\text{kips} \quad \text{(upper limit)}$$

Note that for this example, the diaphragm lateral force, F_{px}, is greater than or equal to the seismic lateral force at each level, F_x or F_i from Table 3-13 or F_i from Table 3-14. The diaphragm lateral load is assumed to be applied to the diaphragm as a uniform lateral load with a value of $\frac{F_{px}}{\ell}$, where ℓ is the length of the diaphragm perpendicular to the seismic lateral load. For the seismic lateral load acting in the X-direction, $\ell = B_x$; for the seismic lateral load acting in the Y-direction, $\ell = B_y$ (see Figure 14-8b). The diaphragm behaves as a deep horizontal beam spanning between the vertical lateral force-resisting systems that are parallel to the direction of the seismic lateral force. See Chapter 14 for the analysis, design, and detailing of horizontal diaphragms and their components.

TABLE 3-14 Diaphragm Lateral Force, F_{px}.

Diaphragm Level	Dead Load at each diaphragm Level, W_i	$\sum_{i=x}^{n} W_i$	W_{px}	F_i	$\sum_{i=x}^{n} F_i$	$F_{px\,(minimum)} = 0.2S_{DS}I_eW_{px}$	$F_{px\,(maximum)} = 0.4S_{DS}I_eW_{px}$	$F_{px} = \frac{\sum_{i=x}^{n} F_i}{\sum_{i=x}^{n} W_i} W_{px}$
Roof	353 kips	353	353 kips	37 kips	37 kips	19.1kips	38.2 kips	**37 kips**
Sixth floor	830 kips	1,183	830 kips	72 kips	109 kips	44.8 kips	89.6 kips	**76.5 kips**
Fifth floor	830 kips	2,013	830 kips	56 kips	165 kips	44.8 kips	89.6 kips	**68 kips**
Fourth floor	830 kips	2,843	830 kips	41 kips	206 kips	44.8 kips	89.6 kips	**60.1 kips**
Third floor	830 kips	3,673	830 kips	26 kips	232 kips	44.8 kips	89.6 kips	**52.4 kips**
Second floor	830 kips	4,503	830 kips	12 kips	244 kips	44.8 kips	89.6 kips	**45 kips**

References

[1] American Society of Civil Engineers. 2022. *ASCE 7: Minimum Design Loads for Buildings and Other Structures (Provisions and Commentary).* Reston, VA.

[2] Green, N. B. 1981. *Earthquake resistant building design and construction.* New York: Van Nostrand Reinhold.

[3] McNamara, R. J. 2005. "Some Current Trends in High Rise Structural Design". STRUCTURE Magazine, September: 19–23.

[4] Gamble, S. 2003. "Wind Tunnel Testing, a Breeze Through". STRUCTURE Magazine, November.

[5] Ghosh, S. K., and S. Dowty. 2007. "Code simple". *Structural Engineer,* February: 18.

[6] International Codes Council. 2024. *International Building Code (IBC)—2024,* ICC, Falls Church, VA: ICC.

[7] International Codes Council. 2006. *International Building Code (IBC)—2006.* Falls Church, VA: ICC 2006.

[8] Nelson, E. L., D. Ahuja, S. M. Verhulst, and E. Criste. 2007. The Source of the problem. *Civil Engineering, ASCE,* January: 50–55.

[9a] ANSI/AISC. 2022. Seismic Provisions for Structural Steel Buildings (ANSI/AISC 341-22), *American Institute for Steel Construction.*

[9b] American Institute of *Steel Construction.* 2022. *ANSI/AISC 358: Prequalified Connections for Special and Intermediate Steel Moment Frames for Seismic Applications,* Chicago, IL: AISC.

[10] American Institute of Steel Construction. 2023. *Steel construction manual,* 16th ed, Chicago, IL: AISC.

[11] Fanella, D. A. 2012. *Structural Loads—2012 IBC and ASCE/SEI 7-10.* Falls Church, VA: International Code Council.

[12a] USGS. 2024. *United States Geological Survey: https://www.usgs.gov/programs/earthquake-hazards.* Website accessed 07-22-2024

[12b] USGS. 2024. *United States Geological Survey:*Moment magnitude, Richter scale - what are the different magnitude scales, and why are there so many? | U.S. Geological Survey (usgs.gov). Website accessed 07-22-2024

[13] Brockenbrough, R. L. and F. S. Merritt. 2011. *Structural Steel Designer's Handbook. 5th* ed., New York: McGraw Hill.

[14] Williams, A. 2006. *Structural Design Examples. International Code Council (ICC),* Falls Church, VA: ICC.

[15] Scott, D. R. 2018. "ASCE 7-16 Wind Load Provisions". *STRUCTURE Magazine,* July.

[16] Kennedy, J. 2018. "Developing Diaphragm Analysis." *Modern Steel Construction,* May: 17-20.

[17] Marshall, J. D., B. Saxey, and Z. Xie. 2018. "Quantifying Inelastic Force and Deformation Demands on Buckling Restrained Braces and Structural System Response." *Engineering Journal,* Fourth Quarter: 209–229.

[18] Bhardway, S. R. and A. H. Varma. 2017. "Nuclear Option". *Modern Steel Construction,* November.

[19] Post, N. 2017. "Steel Core System Could Transform Office Tower Construction." *Engineering News Record (ENR),* September 27.

[20] SEAOC Seismology Committee. 2008. "A Brief Guide to Seismic Design Factors". *STRUCTURE Magazine*, September: 30-32.

[21] Heausler, T. F. 2015. "The Most Common Errors in Seismic Design." *STRUCTURE Magazine*, September: 30–32.

[22] Hines, E., and L. Fahnestock. 2016. "Ductility in Moderation." *Modern Steel Construction*, February: 32-37.

[23] Ghosh, S. K. 2019. "Is Seismic Design by U.S. Codes and Standards Deficient?," Part 1, *STRUCTURE Magazine*, July: 8-10.

[24] Ghosh, S. K. 2019. "Is Seismic Design by U.S. Codes and Standards Deficient?," Part 2, *STRUCTURE Magazine*, August: 18-20.

[25] Irfan, U. 2024. "We Know Where the Next Big Earthquakes will Happen – But Not When," *VOX*, April 5.

[26] Zimmerman, R., E. Sonmez, and R. Collins. 2023 "When the Ground Shook – Post-Disaster Observations of the Kahramanmaras Turkiye Earthquake Sequence – Part 2." *STRUCTURE Magazine*, October: 38–42.

[27] Bruneau, M. 2023. "Acapulco was Built to Withstand Earthquakes, But Not Hurricane Otis' Destructive Winds – How the Building Codes Failed This Resort City." *The Conversation*, November 7.

[28] Thompson, A. 2023. "How to Engineer Buildings That Withstand Earthquakes." *Scientific American,* February 22. *https://www.scientificamerican.com/article/how-to-engineer-buildings-that-withstand-earthquakes/*(accessed August 17, 2024)

[29] Eatherton, M. R., and M. A. Matthew. 2024. "Seismic Sensibility." *Modern Steel Construction*, August: 14-18.

[30] Poon, D. C. K., L. Hsiao, Y. Zhu, L. Joseph, S. Zuo, G. Fu, and O. Ihtiyar. 2011. "Non-Linear Time History Analysis for the Performance-Based Design of Shanghai Tower." *Structures Congress*.

[31] Ingram, D. "Unpublished List Reveals 3,400 San Francisco Buildings may be at Severe Earthquake Risk." *NBC News. https://www.nbcnews.com/science/science-news/san-francisco-concrete-building-list-sf-risk-earthquake-retrofit-cost-rcna84497* (accessed January 10, 2025).

Exercises

3-1. For a two-story office building 140 ft. × 140 ft. in plan and with a floor-to-floor height of 13 ft. located in your city, calculate the following wind loads, assuming an X-brace is

a. Average horizontal wind pressure in the transverse and longitudinal directions.

b. Total base shear in the transverse and longitudinal directions.

c. Force to each X-brace frame in the transverse and longitudinal directions.

Assume the building is enclosed and the exposure category is D.

3-2. For a two-story building 140 ft. × 140 ft. in plan with a floor-to-floor height of 13 ft., calculate the following seismic loads:

a. Seismic base shear and force at each level, assuming the minimum lateral force procedure.

b. Seismic base shear and force at each level, assuming the simplified procedure.

Assume a roof dead load of 25 psf, a floor dead load of 90 psf, and a flat roof snow load of 35 psf. Include the weight of the cladding (55 psf) around the perimeter of the building in the weight of the roof and floor levels. Use $S_{DS} = 0.27$, $S_{D1} = 0.12$, and $R = 3.0$.

3-3. A five-story office building, 80 ft. × 80 ft. in plan, with a floor-to-floor height of 12 ft. and an essentially flat roof, is laterally supported by 10-ft-long shear walls on each of the four faces of the building. The building is located in New York City (assume a 120-mph basic wind speed, exposure category C, and risk category I building).

a. For the MWFRS, determine the factored wind loads at each floor, the base shear, and the overturning moments at the base of the building using the ASCE 7 directional method.

b. Assuming two shear walls in each direction, determine the unfactored wind lateral force at each floor level for each shear wall, the base shear, and the overturning moment.

c. Repeat part b using unfactored wind loads.

3-4. Find the following for the building described in Exercise 3-3. Assume that the rigid diaphragms and the ordinary reinforced concrete shear walls support gravity, as well as lateral loads.

a. Determine the factored seismic lateral force at each level of the building in the N–S and E–W directions. (Neglect torsion.)

b. Calculate the factored seismic force at each level for a typical shear wall in the N–S and E–W directions.

c. Calculate the factored total seismic base shear for a typical shear wall in the N–S and E–W directions.

d. Calculate the factored seismic overturning moment at the base of a typical shear wall in the N–S and E–W directions.

e. Instead of having walls on each of the four faces of the building (i.e., two shear walls in each direction), the building has five ordinary concentric steel x-brace frames in a building frame system (located 20 ft. apart) in both the N–S and E–W directions. Recalculate the factored seismic force at each level of a typical interior X-brace frame, the factored base shear, and the overturning moment. Assume that the structural steel system is not specifically detailed for seismic resistance.

f. Recalculate the forces and moments in part e, assuming the building has flexible diaphragms.

Assume the following design parameters:

- Average dead load for each floor is 150 psf.
- Average dead load for roof is 30 psf; the balanced roof snow load, P_f, is 35 psf; and the ground snow load, P_g, is 50 psf.
- Average weighare bearing wall systems with ordinary reinforced concrete shear walls.
- Short-term spectral acceleration, $S_S = 0.25g$.
- 1-sec. spectral acceleration, $S_1 = 0.07g$.
- No geotechnical report is available.
- Neglect torsion.

3-5. The roof of a one-story, 100-ft. × 120-ft. warehouse with a story height of 20 ft. is framed with open-web steel joists and girders as shown in Figure 3-18. Assuming a roof dead load of 15 psf, determine the net factored wind uplift load on a typical interior joist. The building is located in Dallas, Texas. Assume exposure category C.

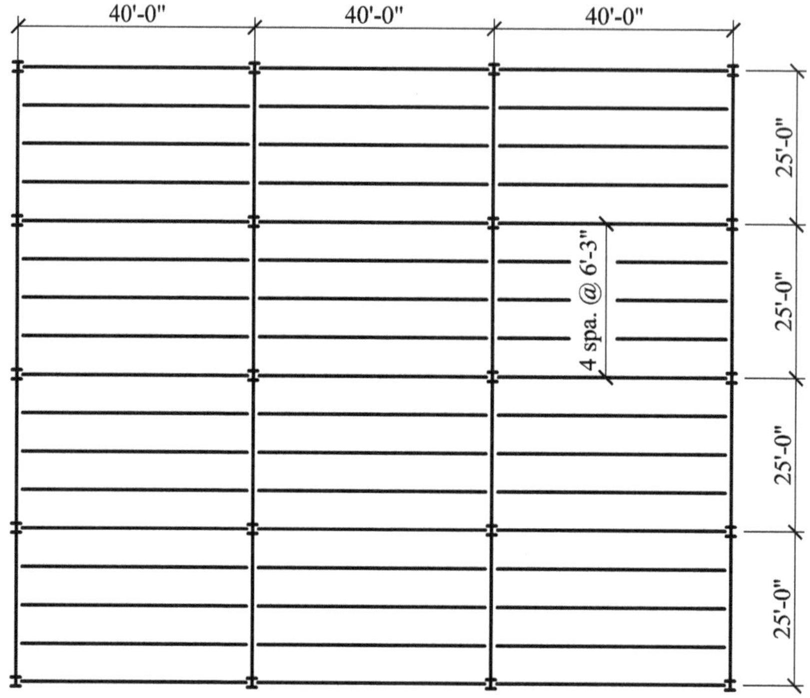

FIGURE 3-18 Warehouse roof framing plan for Exercise 3-5

3-6. A two-story steel structure, 36 ft. × 75 ft. in plan, is shown in Figure 3-19 with the following given information. The floor-to-floor height is 12 ft., and the building is enclosed and located in Rochester, New York, on a site with a category C exposure. Assuming the following additional design parameters, calculate the following:

Floor dead load = 100 psf

Roof dead load = 30 psf

Exterior walls = 10 psf

Snow load, $P_f = 40$ psf

Site class = D

Importance, $I_e = 1.0$

$S_S = 0.25\%$

$S_1 = 0.07\%$

$R = 3.0$

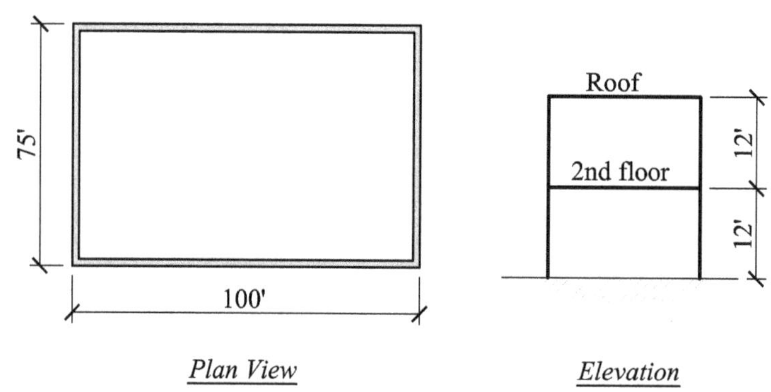

FIGURE 3-19 Building plan and elevation for Exercise 3-6

a. The total horizontal wind force on the MWFRS in both the transverse and longitudinal directions.

b. The gross vertical wind uplift pressures and the net vertical wind uplift pressures on the roof (MWFRS) in both the transverse and longitudinal directions.

c. The seismic base shear, V, in kips using the equivalent lateral force method.

d. The lateral seismic load at each level in kips.

3-7. Calculate the following for the building plan shown in Figure 3-20. The building is located in Orlando, FL. The exposure category is B, and the building has normal occupancy. Use the directional method for wind load calculations.

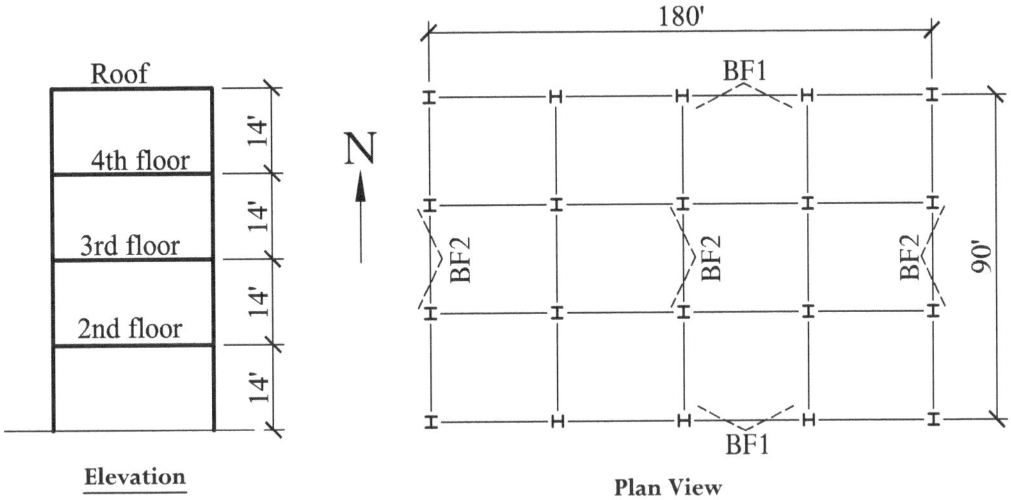

FIGURE 3-20 Building plan and elevation for Exercise 3-7

a. Table of values for P_{30}, end zone width "a," λ, and average wind pressure, "q." Consider horizontal wind loads in the transverse and longitudinal directions only (neglect vertical loads and components and cladding).

b. Calculate the wind force at each level in the transverse and longitudinal directions. Since the parapet height is not given, ignore the wind pressure on the exposed area of the parapet.

c. Draw elevations of Braces BF1 and BF2 showing the lateral wind forces at each level. Calculate the overturning moment and base shear in each of the typical braced frames, BF1 and BF2.

d. Describe the load path for the wind load on this structure. Assume the exterior walls are framed with full-height wall studs that frame between each level.

3-8. Find the following parameters for the building in Figure 3-21. Use service loads only.

a. Wind force at each level and the base shear in the transverse direction.

b. Uniform dead load to B-1 in plf.

c. Draw an FBD of the braced frame at B-1 showing the dead and wind loads.

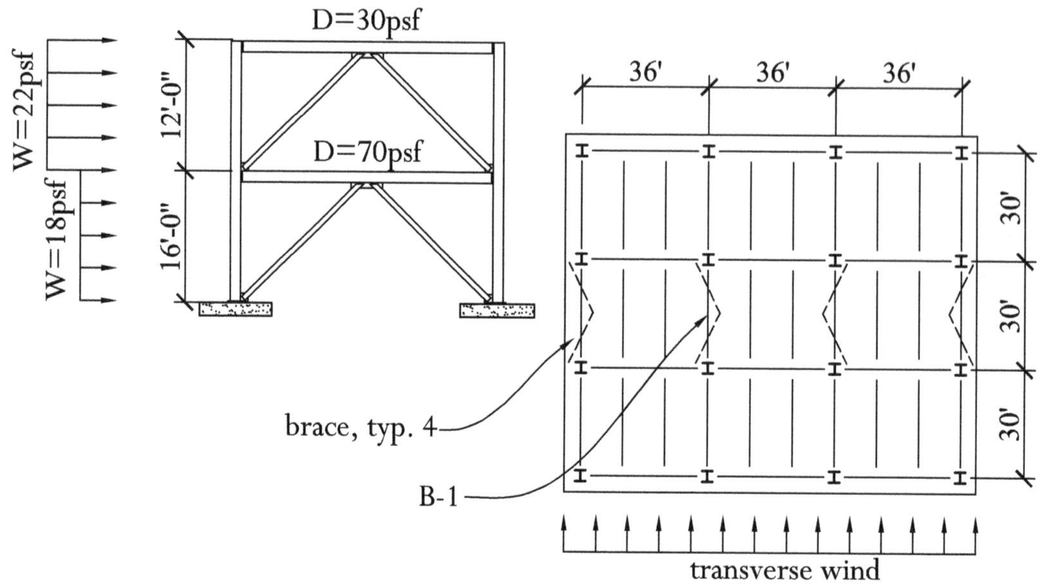

FIGURE 3-21 Building plan and elevation for Exercise 3-8

3-9. For the roof plan shown for a single-story building in Figure 3-22, describe the load path through each element shown if the roof is subjected to wind loads in the upward direction.

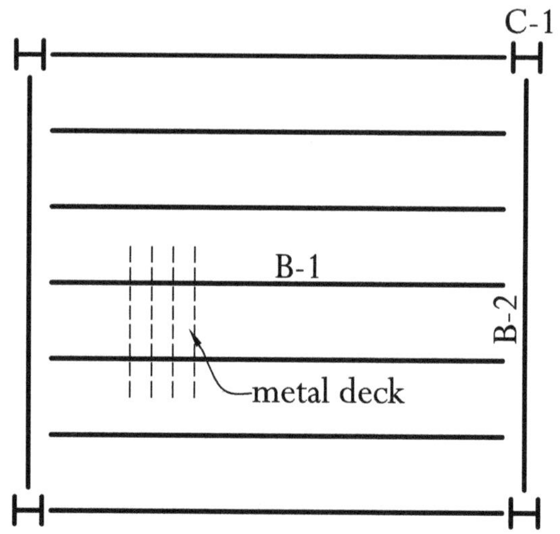

FIGURE 3-22 Roof framing for Exercise 3-9

3-10. Calculate the following for the building shown in Figure 3-23. The mapped risk-targeted maximum considered earthquake (MCE_R) spectral response acceleration at 0.2-s and 1.0-s periods, respectively, are: $S_S = 0.40g$ and $S_1 = 0.16g$. The dead load on each floor is 80 psf, and the dead load on the roof is 20 psf; the cladding weight is 20 psf. The flat roof snow load (P_f) is 40 psf. The lateral force-resisting system is classified as an "Intermediate Steel Moment Frame," and given the aspect ratio of the roof and floor diaphragms, assume a rigid diaphragm in distributing the lateral loads to the moment-resisting frames.

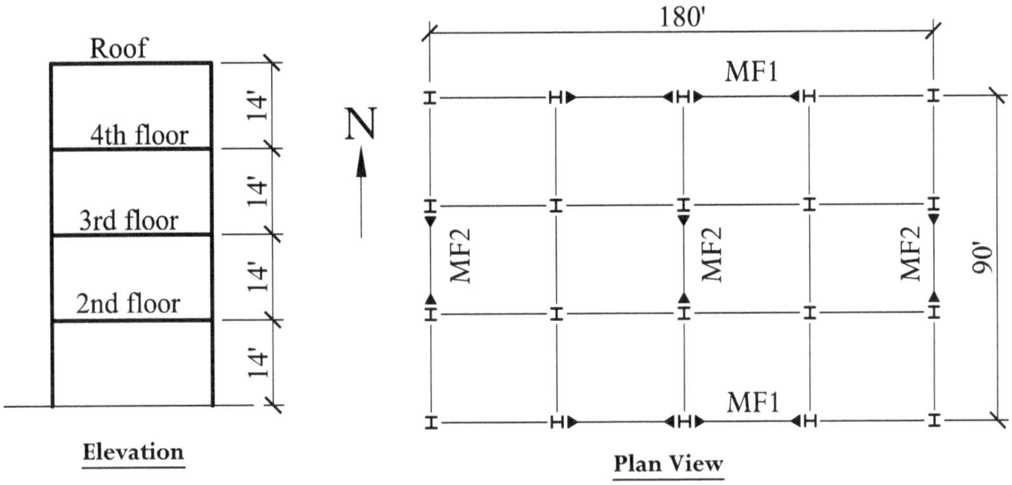

FIGURE 3-23 Building plan and elevation for Exercise 3-10

a. Determine the SDC. The soil site class is C, and the building has normal occupancy.

b. Determine the seismic base shear, V, and the force at each level using the equations for the "Equivalent Lateral Force" method. Assume $T < T_L$.

c. Determine the force at each level in each moment frame MF1 and MF2, and sketch a line diagram showing the elevation of each moment frame and the applied lateral seismic loads.

d. Plot the lateral shear force at each level of the building and discuss how and why the distribution of the story shear forces differs from a linear distribution.

3-11. For the four-story building in Exercise 3-10 (Figure 3-23), calculate the following wind loads using the Directional Method from Part 1 of ASCE 7 Chapter 27. The lateral force-resisting system is classified as an "Intermediate Steel Moment Frame," and given the aspect ratio of the roof and floor diaphragms, assume a rigid diaphragm in distributing the lateral loads to the moment-resisting frames. For each part below, sketch a line diagram showing the elevation of each moment frame and the applied lateral wind loads. Assume the basic wind speed is 115 mph, a partially enclosed classification, and terrain/exposure category C.

a. Determine the factored lateral wind force in kips in the East-West direction at each level of the building for each moment frame MF1.

b. Determine the factored lateral wind force in kips in the north-south direction at each level of the building for each moment frame MF2.

c. Compare the wind load results to the seismic load results obtained in Exercise 3-10. Briefly discuss your findings and conclusions as to which lateral loads—wind or seismic—will control the design of this building.

d. Compute the wind forces in kips at each level of each of the moment-resisting frames MF1 and MF2 that will be used in checking the lateral deflections of the moment-resisting frames.

3-12. A one-story building that will be used as an emergency shelter has the roof plan shown in Figure 3-24. The building is laterally braced with X-braces, as shown in the figure. Assume the following seismic design parameters and determine other parameters as needed:

- Total seismic dead load assigned to the roof, W = 500 kips
- $S_{DS} = 0.30g$
- $S_{D1} = 0.15g$
- $T_L = 6$ sec
- $R = 3$
- $C_t = 0.02$ and $x = 0.75$ for X-braces
- $S_1 < 0.6g$

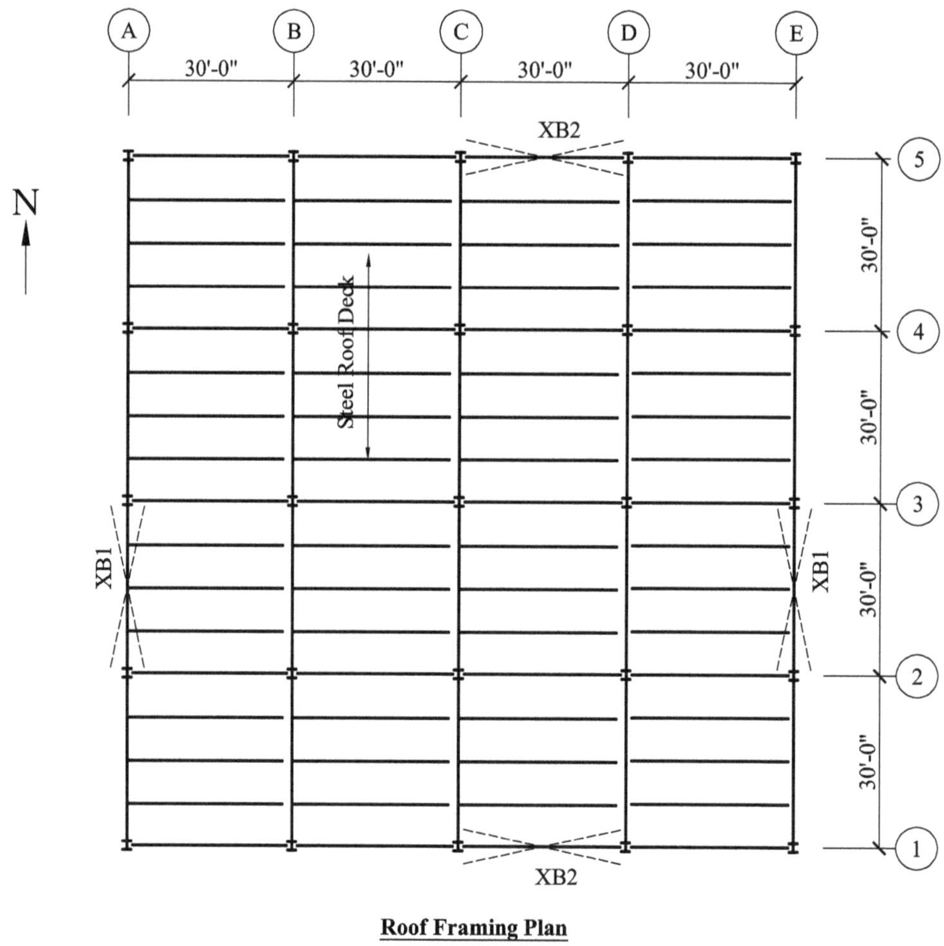

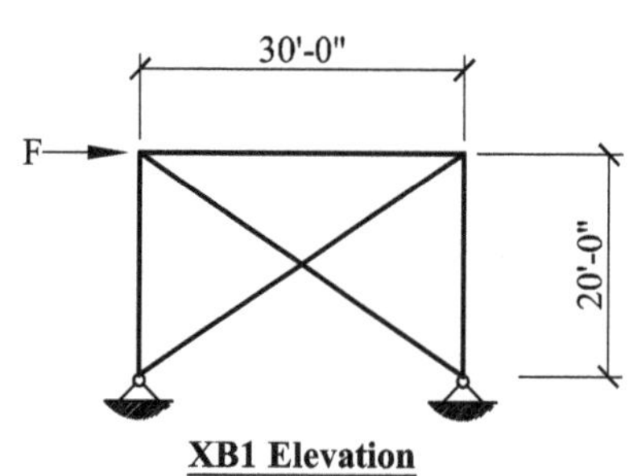

FIGURE 3-24 Building plan and elevation for Exercise 3-12

a. Using the Equivalent Lateral Force (ELF) method, calculate the factored seismic base shear, **V**, (in kip) on the building. Note that the seismic weight, W, concentrated at the roof level of the one-story building is given.

b. Calculate the factored force, F, (in kip) at the roof level of the X-braced frame, XB1, along grid line A. Assume a rigid diaphragm in distributing the lateral loads.

c. Using the factored lateral force acting at the roof level of the X-braced frame, XB1, along grid line A, calculate the factored axial tension force, **Pu**, in the diagonal X-brace, XB1.

3-13. Assume the wind load parameters for the building in Figure 3-24 are as follows:

- Building with lateral natural frequency, f > 1 Hz
- Risk category (RC) II
- Basic Wind Speed = 150 mph
- Exposure Category "C"
- Flat land
- Partially enclosed
- Assume L/B = 1.0

a. Calculate K_z and K_h at each level of the one-story building.

b. Calculate the velocity pressures, q_z, on the windward wall and q_h on the leeward wall in psf, for the main wind force-resisting systems (MWFRS). Assume K_e = 1.0.

c. Assuming L/B = 1.0, determine the external pressure coefficients for the windward wall and the leeward wall using the ASCE 7 table.

d. Calculate the design external wind pressure (in psf) on the windward wall, P_{WW}, and the design external wind pressure (in psf) on the leeward wall, P_{LW}. Illustrate with a diagram showing the wind pressures acting on the one-story building and the proper directions of the wind pressures, P_{WW} and P_{LW}.

e. Calculate the net lateral design wind pressure, *P*, in psf acting on the building. Illustrate with a diagram showing the NET factored wind pressure acting on the one-story building.

f. Using the factored net wind pressures from Part (e), calculate the factored concentrated lateral wind load acting at the roof level of X-brace XB1 along grid line A.

3-14. A building located in a hurricane zone in the United States is to be used as an emergency hurricane shelter. The seismic activity in the area is low.

a. What risk category (RC) should be assigned for the wind design of this building, and why?

b. What risk category (RC) should be assigned for the seismic design of this building, and why?

3-15. A 17-story highrise building located on an oceanfront site has a 95 ft x 225 ft footprint with the 225 ft plan dimension parallel to the N-S direction. The upper floor elevations relative to the ground floor elevation of 0 ft are as follows: 15 ft, 25 ft, 34 ft, 45 ft, 56 ft, 68 ft, 79 ft, 91 ft, 102 ft, 114 ft, 125 ft, 137 ft, 149 ft, 160 ft, 172 ft, 183 ft, and the roof elevation is at 195 ft elevation. Assume a factored wind speed of 175 mph. Using the ASCE 7 directional method,

a. Calculate the design wind pressures on the MWFRS of the building for the E-W wind.

b. Calculate the design wind pressures on the MWFRS of the building for the N-S wind.

3-16. A 7-story building with a 390 ft x 390 ft footprint is located on a site with Exposure Category C. The upper floor elevations relative to the ground floor elevation of 0 ft are as follows: 15 ft, 26.5 ft, 37 ft, 47.5 ft, 58 ft, and 68.5 ft, and the roof elevation is at 79 ft, with the top of the parapet at 82 ft elevation. Assume a factored wind speed of 200 mph.

Using the ASCE 7 directional method, calculate the design wind pressures on the MWFRS of the building, including the wind pressures on the 3 ft tall parapet.

3-17. Calculate the additional wind pressure in psf on the MWFRS due to a 3 ft tall parapet on top of the roof of a 100 ft × 100 ft × 30 ft tall building. Assume the following additional design parameters:

- V = 175 mph
- Building is enclosed and in Exposure Category D
- Rigid structure
- Flat land
- $K_e = 1.0$

CHAPTER

4

Tension Members

4-1 INTRODUCTION

Axially loaded members stressed in tension are used in steel structures in various forms; they occur as web and chord members in roof and floor trusses, and as hangers and sag rods, diagonal braces for lateral stability, lap splices, and in moment connections (see Figures 4-1a to 4-1h for some examples of tension members). They also occur as horizontal floor and roof members in resisting the horizontal tension at the top of sloped columns or diagonal struts, and as chords and drag struts or collectors in roof and floor diaphragms in resisting lateral loads (see Chapter 14).

Beams and columns are subject to buckling due to compression stresses (such as lateral-torsional buckling, Euler buckling, and local buckling) and must be checked for this failure mode, but tension members are not susceptible to lateral instability since compression stresses do not exist. The exception to this is the special case when the applied tension load is eccentric to the member in question, inducing an applied moment and therefore creating the possibility of lateral torsional buckling.

The basic design check required for a tension member is to provide enough cross-sectional area to resist the applied tensile force. One factor that has to be considered is the non-uniform cross-sectional area of tension members with bolted end connections. The cross-sectional area of the tension member is smaller at the bolt locations than elsewhere along its length due to the presence of the bolt holes. Members with concentric axial tension loads are rare in practice, and instead, tension members are often subjected to eccentric axial loading; for example, a single angle subjected to a concentric axial tension load may also be subjected to bending moment at its ends because of the eccentricity of its connection gusset plates at the end supports of the angle. Even though slenderness is not a direct design concern for tension members since they do not buckle, the AISC specification does recommend an upper limit on the slenderness ratio L_c/r for tension members. This upper L_c/r limit is equal to 300 for tension members and 200 for compression members, where L_c is the effective length, KL, of the member and r is the radius of gyration (see Chapter 5). This recommendation does not apply to rods or hangers in tension and is not a mandatory requirement.

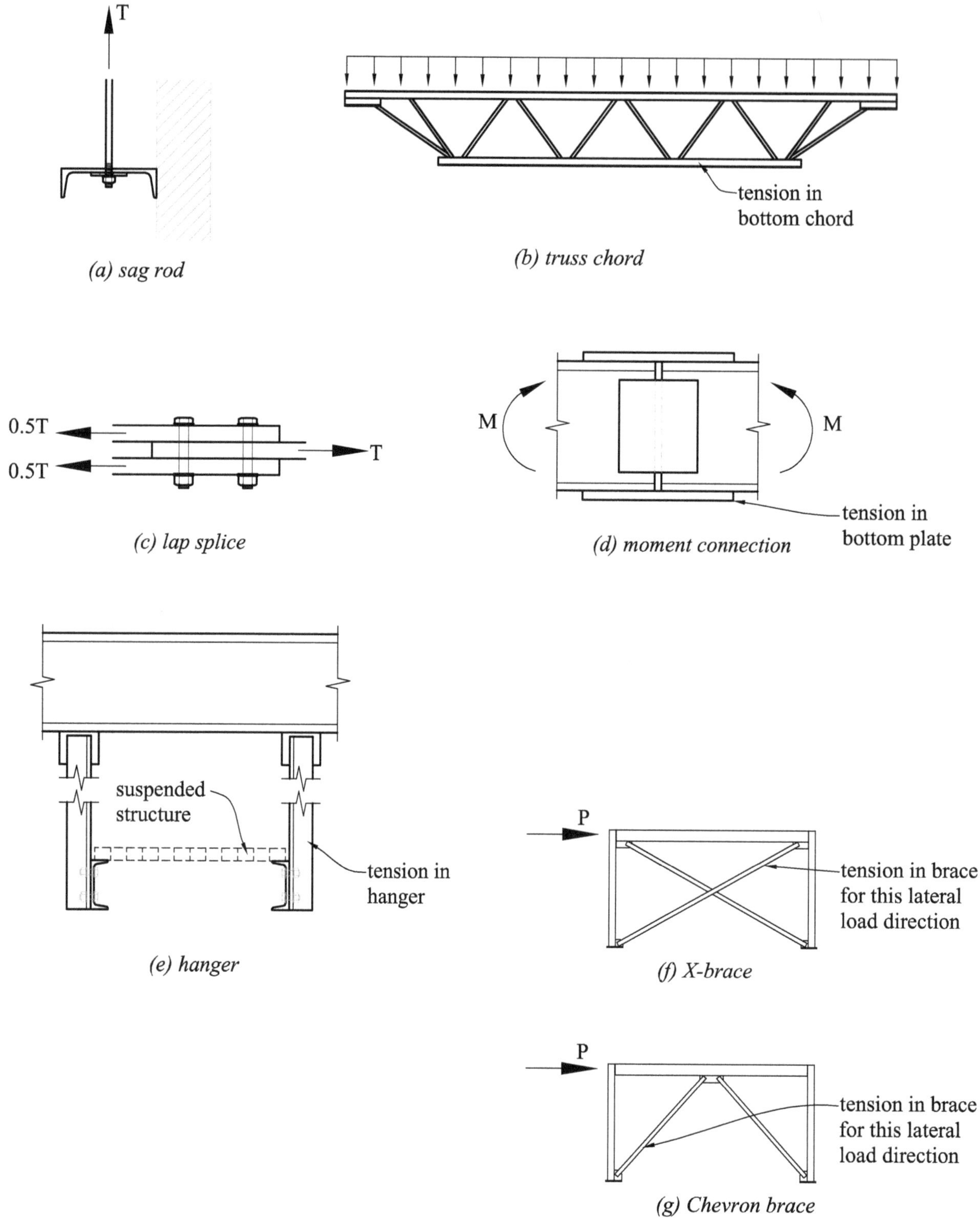

FIGURE 4-1 a-g Common tension members

FIGURE 4-1h Tension cables in the Arthur Ravenel Jr. Cable-stayed Bridge in Charleston, S.C. (built in 2005)

4-2 FAILURE MODES AND ANALYSIS OF TENSION MEMBERS

For members subjected to tension, the four failure modes are as follows:

1. Tensile yielding on the gross section
2. Tensile rupture on the net section
3. Block shear failure
4. Connection failure

The two most basic modes of failure are tensile yielding on the gross section and tensile rupture at the net section. Tensile yielding occurs when the stress on the *gross area* of the section is large enough to cause excessive deformation. Tensile rupture occurs when the stress on the *effective area* of the net cross-section is large enough to cause the member to fracture or rupture (perpendicular to the tension force), which usually occurs across a line of bolts where the tension member is weakest (see Section D of the AISC Specifications). The effective area is usually smaller than the gross cross-sectional area, and the reduction in the cross-sectional area occurs due to the non-uniformity of axial tension stresses (a phenomenon also known as shear lag) and due to the presence of bolt holes in bolted connections. The block shear failure mode will be discussed later in this chapter, and the failure of the bolted and welded connections at the ends of the tension member will be covered in Chapters 9 and 10, respectively. For tension members with bolted connections, the bolts will frequently be subjected to shear. For now, the bolt shear capacities, which will be derived later in Chapter 9, are given below:

- ¾" dia A325N bolt: $\phi R_n = \phi F_{nv} A_b = 0.75$ (54 ksi)p$(3/4)^2/4$ = **17.9 kips**
- 7/8" dia A325N bolt: $\phi R_n = \phi F_{nv} A_b = 0.75$ (54 ksi)p$(7/8)^2/4$ = **24.3 kips**
- 1" dia A325N bolt: $\phi R_n = \phi F_{nv} A_b = 0.75$ (54 ksi)p$(1)^2/4$ = **31.8 kips**

Tensile Yielding on the Gross Section:

Failure occurs in this mode by tensile yielding of the member at the gross cross-section, leading to elongation of the member. In the LRFD method, the ultimate tension capacity or design tensile strength, ϕP_n, for the failure mode of tensile yielding in the gross section is given as

$$\phi P_n = \phi F_y A_g, \tag{4-1a}$$

where

ϕ_t = 0.90,

P_n = Nominal tensile yield strength = $F_y A_g$

F_y = Minimum yield stress, and

A_g = Gross cross-sectional area of the tension member.

For the allowable strength design (ASD) method, the allowable tension load for the tensile yielding failure mode is given as

$$P_a = \frac{P_n}{\Omega_t} = \frac{F_y A_g}{\Omega_t} \tag{4-1b}$$

Where,

Ω_t = Factor of safety for tensile yielding = 1.67

Tensile Rupture in the Net Section:

In this failure mode, the tension member fails by tensile rupture at the net cross-section, leading to a complete severing of the member into two parts. For the LRFD method, the design tensile strength due to tensile rupture on the effective net area is given as

$$\phi_R P_n = \phi_R F_u A_e, \tag{4-2a}$$

where

ϕ_R = 0.75,

P_n= Nominal tensile strength in fracture or rupture = $F_u A_e$,

F_u = Minimum ultimate tensile strength, and

A_e = Effective cross-sectional area of the tension member.

For the allowable strength design (ASD) method, the allowable tension load due to tensile rupture is given as

$$P_a = \frac{P_n}{\Omega_R} = \frac{F_u A_e}{\Omega_R} \tag{4-2b}$$

Where,

Ω_R = Factor of safety for tensile rupture = 2.0

The design tensile strength of a tension member is the smaller of the two expressions indicated in equations (4-1a) and (4-2a) for the LRFD method and equations (4-1b) and (4-2b) for the ASD method. Note the lower strength reduction factor and higher safety factor for the tensile rupture failure mode—a brittle failure mode that occurs suddenly without warning, therefore indicating low ductility—compared to the corresponding values for the yielding failure mode—a mode of failure that gives ample warning of impending failure, therefore indicating comparatively higher ductility.

For the LRFD method, the limit state design equation requires that the design tensile strength, ϕP_n, be greater than or equal to the applied maximum factored load or load effect, P_u.

That is, $\phi P_n \geq P_u$.

In the ASD method, the allowable tension load, P_a, must be greater than or equal to the applied maximum unfactored applied tension load, P_s.

That is, $P_a \geq P_s$

The applied factored and unfactored loads are calculated using the LRFD and ASD load combinations presented in Chapter 2.

Effective Area, A_e:

The *gross cross-sectional area*, A_g, of a tension member is simply the total cross-sectional area of the member in question. The *effective net area*, A_e, of a tension member is a product of the net cross-sectional area, which accounts for the presence of bolt holes, if any, at the critical section, and the shear lag factor, U. Shear lag in tension members is a phenomenon that occurs due to the non-uniform axial tension stress distribution at the connections of the tension member because all of the elements of the tension member are not connected or attached to the supporting member or gusset plate. The axial tension stress is usually higher at the bolt or weld locations and lower further away from the connectors, thus causing stress concentrations at the tension member supports. One common example where shear lag occurs is when only one leg of an angle tension member is connected to gusset plates at its ends; the angle in this case will also be subjected to moments in addition to axial tension stress. Shear lag leads to a reduction in the strength of a tension member. Thus, the shear lag factor is a measure of the efficiency of the connections of the tension member; the higher the shear lag factor, U, the more uniform the axial tension stress is, and the smaller the impact of shear lag, indicating a higher efficiency of the tension member connection. The effective cross-sectional area, A_e, of a tension member is given as:

$$A_e = A_n U, \tag{4-3}$$

where

A_n = Net cross-sectional area of the tension member, and

U = Shear lag factor.

Note that for a tension member with welded connections, the net cross-sectional area is equal to the gross cross-sectional area, since there are no bolt holes (i.e., $A_n = A_g$ for tension members with welded connections).

The net area of the critical cross-section of a tension member with a group of bolts that lie on lines parallel and perpendicular to the axial load direction (see Figure 4-2) is equal to the difference between the gross cross-sectional area and the sum of the area of the bolt holes along the failure plane:

$$A_n = A_g - \sum A_{\text{holes}}, \tag{4-4}$$

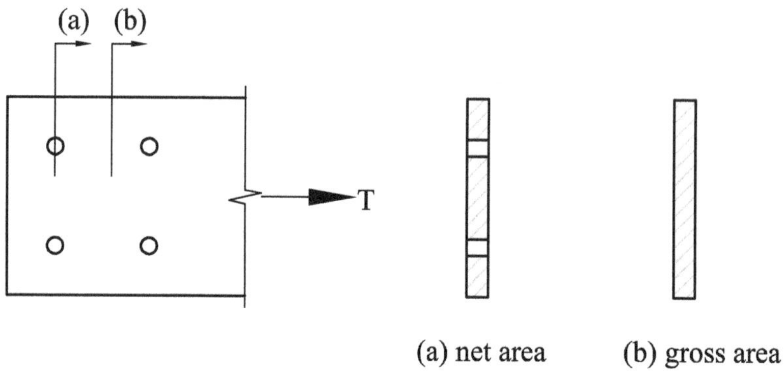

FIGURE 4-2 Tension member with in-line fasteners

where

$$\sum A_{\text{holes}} = n(d_b + {}^{1}\!/_{8})t \text{ for 7/8 in.-diameter bolts and smaller}$$

$$= n\left(d_b + \frac{3}{16}\right)t \text{ for 1-in.-diameter bolt and larger}$$

n = Number of bolt holes along the failure plane, perpendicular to the tension force,
d_b = Bolt diameter, and
t = Material thickness

Section B4.3b of the AISC specification indicates that when calculating the net area for shear and tension, an additional $^1/_{16}$ in. should be added to the nominal width of a bolt hole to account for the roughened edges that result from the bolt hole punching or drilling process. For standard bolt holes, the bolt hole diameter used for strength calculations would be the nominal standard bolt hole dimension given in *AISCM*, Table J3.3 plus $^1/_{16}$ in. (see *AISCM* Section B4.3b)

- For bolt diameters of $^7/_8$ in. or smaller, the nominal bolt hole dimension is $^1/_{16}$ in. larger all around than the bolt diameter for standard (std) holes; the actual bolt hole diameter used in the design calculations will be $d_b + {}^1/_{16}$ in. + $^1/_{16}$ in. or $d_b + {}^1/_8$ in. for bolted connections in tension.
- For bolt diameters of 1-in. or larger, the bolt hole diameter used in the design calculations will be $d_b + {}^3/_{16}$ in.

Though tension members typically have higher axial tension stresses at the bolt hole locations, it is assumed that all bolts in a tension connection will support an equal portion of the applied total load, and this is achievable because of the ductility of the steel member.

For tension members with more than one line of bolts parallel to the tension load, a staggered layout of bolts may be used to satisfy minimum bolt spacing requirements when the member width is limited, and they are also used to optimize the net cross-sectional area for the fracture or rupture failure modes. For tension members with a staggered bolt layout, several possible planes of fracture or rupture need to be investigated, and some of the possible failure paths will follow a zig-zag pattern. When the fracture failure plane passes straight through a line of bolt holes (i.e., the failure plane is perpendicular to the tension force as represented by line *ABDE* in Figure 4-3), then the net cross-sectional area is as noted in equation (4-4). For a failure plane where one or more segments of the failure plane are inclined at an angle less than 90-degrees relative to the axial tension force direction (e.g., line *ABCDE* in Figure 4-3), the following term is added to the net width of the member for each diagonal portion that is present along the failure plane:

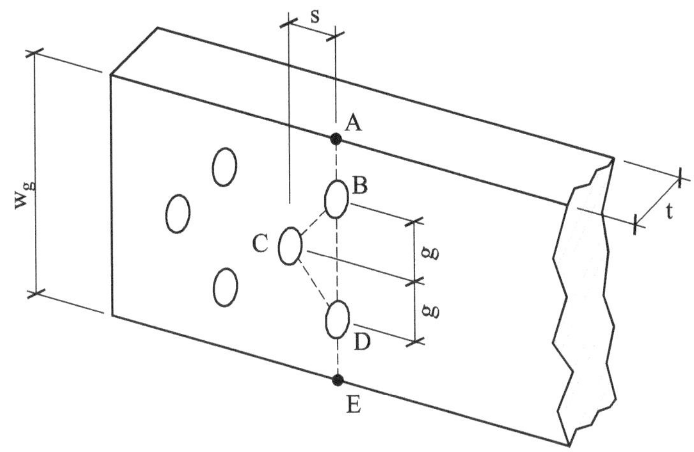

FIGURE 4-3 Tension member with diagonal fasteners

$$\frac{s^2}{4g}, \tag{4-5}$$

where
s = Longitudinal center-to-center spacing or pitch between two consecutive bolt holes, and
g = Transverse center-to-center spacing or gage between two consecutive bolt holes.

This modification accounts for the increase in strength due to the added cross-sectional area because of the diagonal or angled portion of the failure plane [1]. In Figure 4-3, note that the failure plane *ABCDE* has two diagonal failure planes: *BC* and *CD*. The expression for the net width then becomes

$$w_n = w_g - \Sigma d_h + \Sigma \frac{s^2}{4g}, \tag{4-6}$$

where
w_n = Net width,
w_g = Gross width, and
d_h = bolt hole diameter.

Multiplying equation (4-6) by the thickness of the member yields

$$w_n t = w_g t - \Sigma d_h t + \Sigma \frac{s^2}{4g} t. \tag{4-7}$$

Since $A_n = w_n t$ and $A_g = w_g t$, equation (4-7) can be rewritten as follows:

$$A_n = A_g - \Sigma d_h t + \sum \frac{s^2}{4g} t \tag{4-8}$$

As previously indicated, the shear lag factor (U in equation (4-3)) accounts for the nonuniform axial tensile stress distribution when some of the elements of a tension member are not directly connected to the supporting member or connection, and therefore not fully active in transferring the tension load to the supports - such as when only one leg of a single angle or only the web of a channel member, or only the web or flange of a WT section is welded or bolted to the supporting gusset plate member (see Figure 4-4).

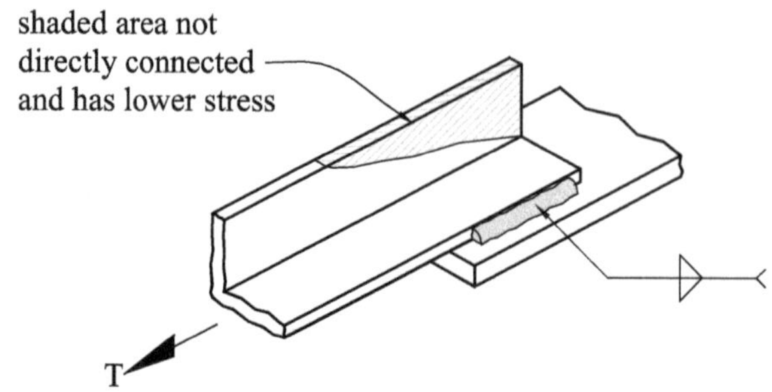

FIGURE 4-4 Shear lag effect

Table D3.1 of the *AISCM* gives the values of the shear lag factor, U, for various connection configurations. With the exception of plates and round hollow structural sections (HSS), members with a single concentric gusset plate and longitudinal welds, the shear lag factor can be calculated as follows:

$$U = 1 - \frac{\bar{x}}{\ell}, \tag{4-9}$$

Where,

$\bar{x}$ = Eccentricity of the connection or the distance between the centroid of the connected member and the connection plane

ℓ = Connection length measured parallel to the tension load (for bolts, it is the out-to-out distance between extreme bolts).

The variables $\bar{x}$ and ℓ in equation (4-9) are illustrated in Figure 4-5. The longer the connection length, the more uniform the axial tension stress distribution across the tension member would be, and therefore the smaller the effect of the shear lag, and hence the larger the shear lag factor, U.

The shear lag factor for plates and round HSS members with a single concentric gusset plate are included in Table 4-1. Also shown in Table 4-1 are alternate values of U for single angles and W, M, S, and HP shapes that may be used in lieu of equation (4-9).

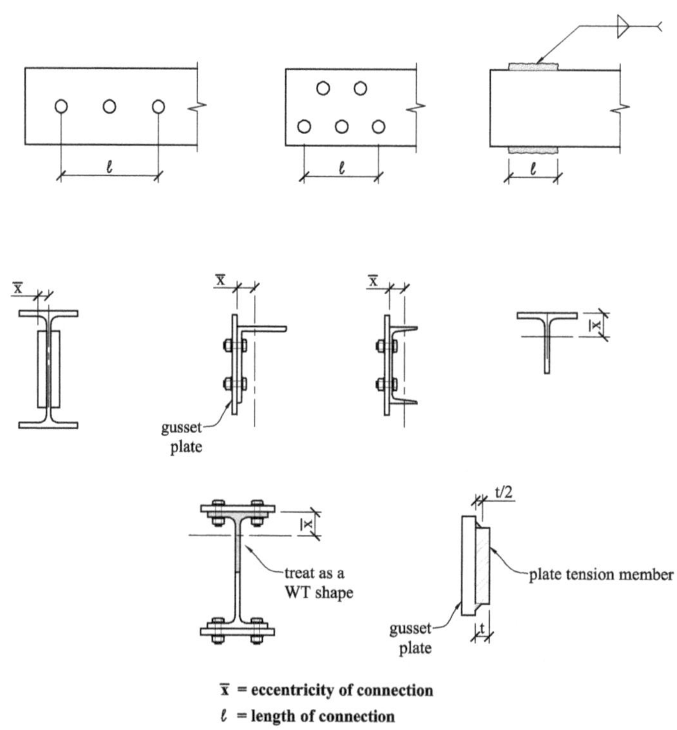

FIGURE 4-5 Determination of $\bar{x}$ and ℓ

Table 4-1 Shear lag factor for common tension member connections.

Tension member type	Description	Shear lag factor, U	Example
1a. All tension members where the axial tension load is transmitted directly to all of its component elements	All bolted*	$U = 1.0$	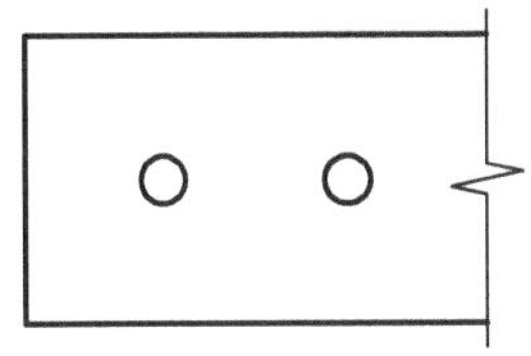
1b. All tension members where the axial tension load is transmitted directly to all of its component elements	All welded	$U = 1.0$	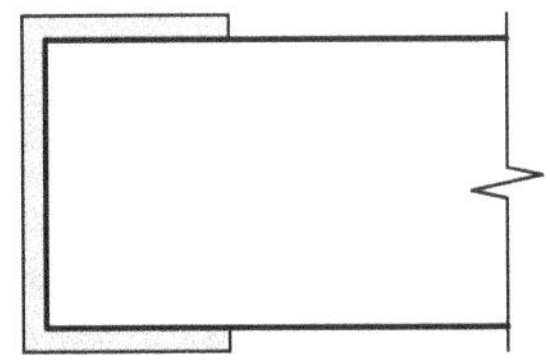
*2. All tension members with axial tension load transmitted to some but not all of the elements connected by bolts or a combination of long and transverse welds	Bolts or longitudinal plus transverse welds	$U = 1 - \frac{\bar{x}}{\ell}$ see Figure 4-5 for the definition of $\bar{x}$ and ℓ	
3. All tension members connected to some but not all the elements of the member	Transverse welds only	$U = 1.0$ A_n = area of directly connected element	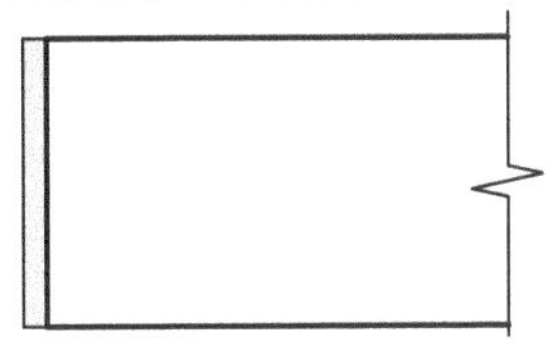
4. Plates, angles, channels with welds at heels, tees, and W-shapes	Longitudinal welds only	$U = \frac{3\ell^2}{3\ell^2 + w^2}\left(1 - \frac{\bar{x}}{\ell}\right)$ where $\ell = \frac{\ell_1 + \ell_2}{2}$ $\ell_1 \geq 4(\textit{fillet weld size})$ $\ell_2 \geq 4(\textit{fillet weld size})$ =	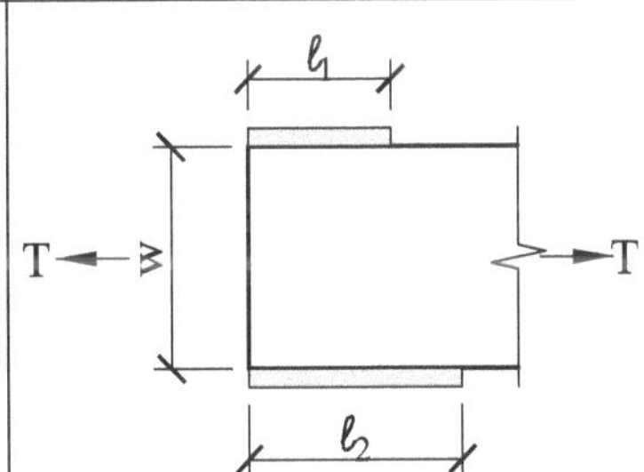
5a. Round HSS	Single concentric gusset plate through slots in HSS	$U = \left[1 + \left(\frac{\bar{x}}{\ell}\right)^{3.2}\right]^{-10}$ $\bar{x} = \frac{R\sin\theta}{\theta} - \frac{1}{2}t_p$ θ *in radians*	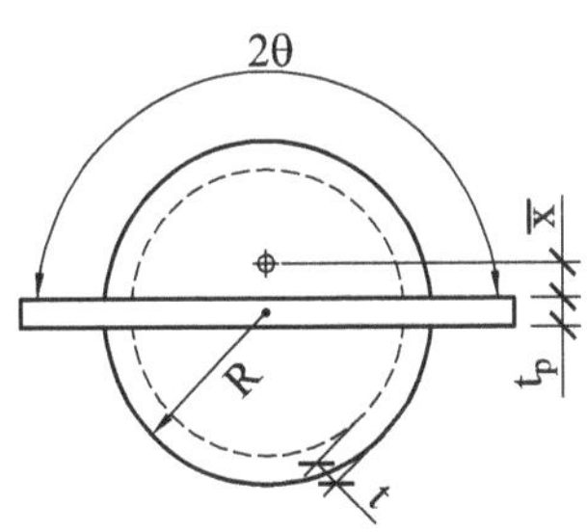
5b. Rectangular HSS	Single concentric gusset plate through slots in HSS	$U = 1 - \frac{\bar{x}}{\ell}$ $\bar{x} = b - \frac{2b^2 + tH - 2t^2}{2H + 4b - 4t}$	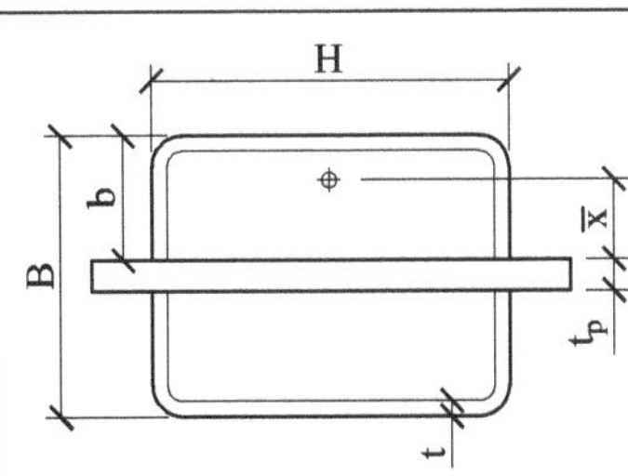

(Continued)

Tension member type	Description	Shear lag factor, U	Example
6. Rectangular HSS	Two-sided gusset plates	$U = \frac{BU_B + HU_H}{H + B}$ $U_B = \frac{3\ell^2}{3\ell^2 + B^2}$ $U_H = \frac{3\ell^2}{3\ell^2 + H^2}$	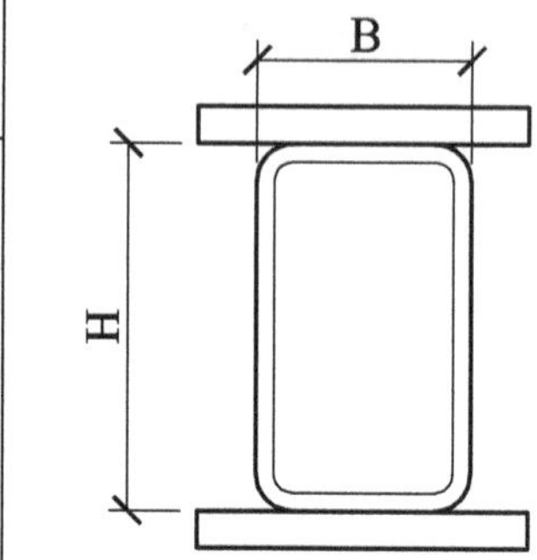
7. *W*, *M*, *S*, or HP, or Tees Cut from these shapes	Flange connected with three or more fasteners per line in the direction of the load	$b_f \geq \frac{2}{3}d,\ U = 0.90$ $b_f < \frac{2}{3}d,\ U = 0.85$	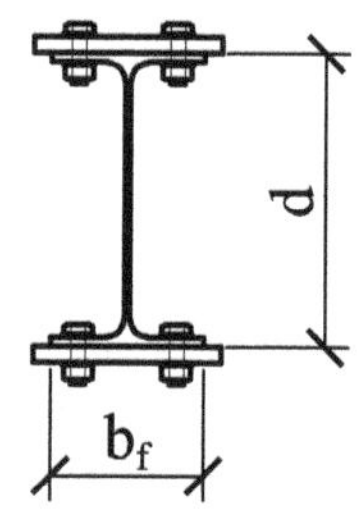
	Web connected with four or more fasteners per line in the direction of the load	$U = 0.70$	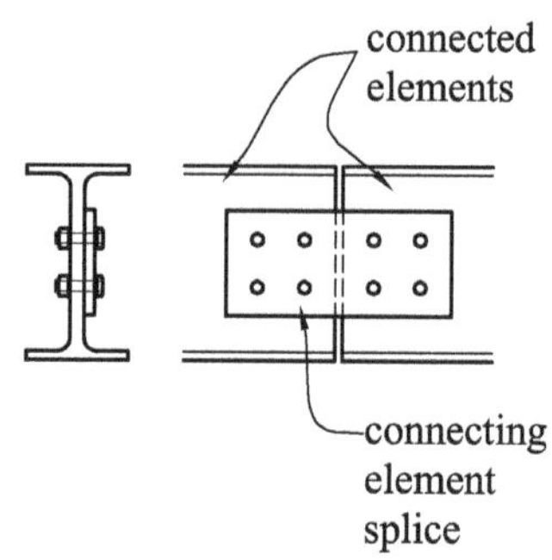
8. Single and double angles	Four or more fasteners per line in the direction of the load	$U = 0.80$	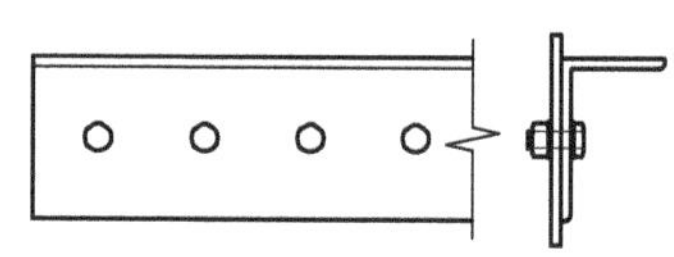
	Two or three fasteners per line in the direction of the load	$U = 0.60$	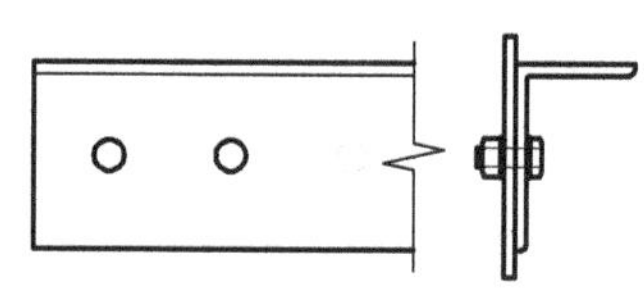

Source: Adapted from *AISCM*, Table D3.1 [1]. Copyright © American Institute of Steel Construction. Reprinted with permission. All rights reserved.

B = overall width of HSS, measured 90° to the plane of the connection, in.

D = outside diameter of round HSS

H = overall height of rectangular HSS, measured in the plane of the connection.

* For open sections, the shear lag factor U ≥ the ratio of the gross area of the connected elements or parts of the member to the gross area of the member. This provision does not apply to plates or HSS. (See AISC Specification Section D3).

EXAMPLE 4-1

U-Value for a Bolted Connection

a. For the bolted tension member shown in Figure 4-6, determine the shear lag factor, U; the net area, A_n; and the effective area, A_e.

b. For a double-angle tension member (2L 7 × 4 × ½) connected to a gusset plate with one line of 4 bolts with a typical spacing of 3 in. between adjacent bolts, determine which of the following configurations will yield a higher U value:

i. Using the double angle with *long legs back-to-back* (i.e., LLBB)

ii. Using the double angle with *short legs back-to-back* (i.e., SLBB)

Solution

a. From the section property tables in part 1 of the *AISCM*, we find that for an $L5 \times 5 \times {}^3\!/_8$,

$\bar{x} = 1.37$ in.

$A_g = 3.65 \text{ in.}^2$

Total length of the connection (i.e., distance between extreme bolts), $\ell = 9$ in.

Shear Lag Factor:

$$U = 1 - \frac{\bar{x}}{\ell}$$

$$= 1 - \frac{1.37 \text{ in.}}{9 \text{ in.}} = 0.848$$

Alternatively, $U = 0.80$ from Table 4-1. The larger value of $U = 0.848$ can be used.

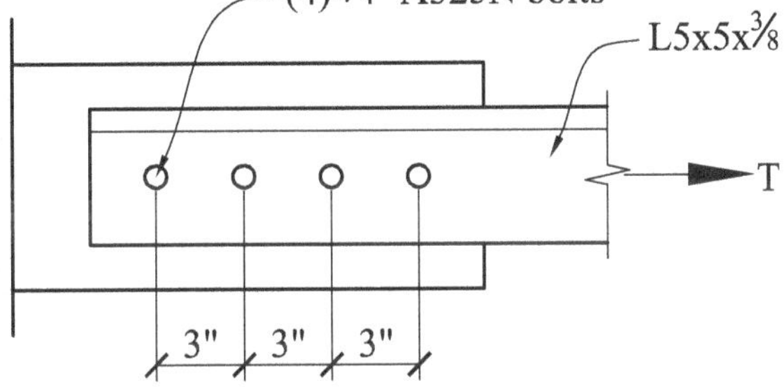

FIGURE 4-6 Detail for Example 4-1

Net Area of the Angle:

At any critical section where rupture may occur perpendicular to the tension load, there is only one bolt hole. Therefore, the net area is

$$A_n = A_g - \sum A_{\text{holes}}$$

$$= 3.65 - (1)\left(\frac{3}{4} + \frac{1}{8}\right)(0.375) = 3.32 \text{ in.}^2$$

Effective Area:

$$A_e = A_n U$$

$$= (3.32)(0.848) = 2.82 \text{ in.}^2$$

b. (i) From the section property tables in part 1 of the *AISCM*, we find that for an $L7 \times 4 \times ½$ (LLBB),

$\bar{x} = 0.91$ in.

Since there are 4 bolts, the total length of the connection (i.e., distance between extreme bolts) is,

$\ell = (4\text{-}1)(3 \text{ in.}) = 9 \text{ in.}$

Shear Lag Factor (L7 x4x ½ LLBB):

$$U = 1 - \frac{\bar{x}}{\ell}$$

$$= 1 - \frac{0.91 \text{ in.}}{9 \text{ in.}} = 0.9$$

(ii) From the section property tables in part 1 of the *AISCM*, we find that for an $L7 \times 4 \times$ ½ (SLBB),

$\bar{y} = 2.40$ in. This is equivalent to the $\bar{x}$ for the short leg

Since there are 4 bolts, the total length of the connection (i.e., distance between extreme bolts) is,

$\ell = (4\text{-}1)(3\text{in.}) = 9\text{in.}$

Shear Lag Factor (L7 x4x ½ SLBB):

$$U = 1 - \frac{\bar{x}}{\ell}$$

$$= 1 - \frac{2.40 \text{ in.}}{9 \text{ in.}} = 0.73$$

Therefore, the shear lag factor for LLBB is higher than that for SLBB, which means that the double-angle tension member is stronger when used with long legs back-to-back (LLBB).

EXAMPLE 4-2

U-Value for a Welded Connection

For the welded tension member shown in Figure 4-7, determine the shear lag factor, *U*; the net area, A_n; and the effective area, A_e.

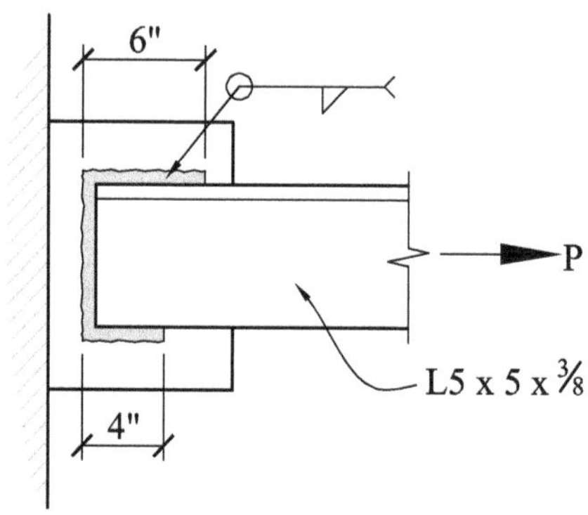

FIGURE 4-7 Detail for Example 4-2

Solution

From the section property tables in part 1 of the *AISCM*, we find that for an L5 × 5 × $^3/_8$,

$\bar{x} = 1.37$ in., and

$A_g = 3.65 \text{ in.}^2$

Shear Lag Factor:

The average length of the longitudinal welds, $\ell = \frac{6\text{in.} + 4\text{in.}}{2} = 5$ in.

$$U = 1 - \frac{\bar{x}}{\ell}$$

$$= 1 - \frac{1.37 \text{ in.}}{5 \text{ in.}} = 0.726$$

There is not an alternate value to use from Table 4-1, so $U = 0.726$.
Since there are no bolt holes, $A_n = A_g = 3.65 \text{ in.}^2$.

Effective Area:

$$A_e = A_n U$$

$$= (3.65)(0.726) = 2.65 \text{ in.}^2$$

EXAMPLE 4-3

Maximum Factored Load (LRFD) and Maximum Allowable Load (ASD) in a Tension Member

a. Determine the maximum factored load (LRFD) and the maximum allowable load (ASD) that can be applied in tension to the plate shown in Figure 4-8. The material is ASTM A36 steel and is welded on three sides to the gusset plate.

b. If the unfactored dead load on the tension member is 20 kips, what is the maximum unfactored floor live load in psf that can be supported by the tension member assuming it supports a tributary area of 200 ft.^{2}

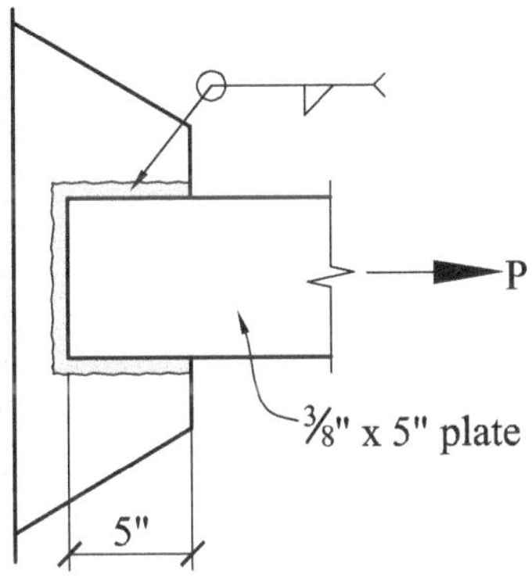

FIGURE 4-8 Detail for Example 4-3

Solution

From the *AISCM*, Table 2-4:

$$F_y = 36 \text{ ksi}$$

$$F_u = 58 \text{ ksi to } 80 \text{ ksi } (\text{use } F_u = 58 \text{ ksi}).$$

Gross and Effective Area:

$$A_g = (0.375 \text{ in.})(5 \text{ in.}) = 1.88 \text{ in.}^2$$

$$A_n = A_g \text{ (no bolt holes)}$$

$$U = 1.0 \text{ (Table 4-1, all-welded plate)}$$

$$A_e = A_n U$$

$$= (1.88 \text{ in.}^2)(1.0) = 1.88 \text{ in.}^2$$

LRFD Method:

a. From equation (4-1a), the nominal tensile strength based on tensile yielding on the gross area of the tension member is

$$P_n = F_y A_g = (36)(1.88 \text{ in.}^2) = 67.7 \text{ kips}$$

For the LRFD method, the design tensile strength based on yielding on the gross area of the tension member is

$$\phi_t P_n = (0.9)(67.7) = 60.8 \text{ kips (tensile yielding failure mode; LRFD method)}$$

From equation (4-2a), the nominal tensile strength due to fracture or rupture on the effective cross-sectional area is given as

$$P_n = F_u A_e = (58 \text{ ksi})(1.88 \text{ in.}^2) = 109 \text{ kips}$$

The design tensile strength due to fracture or rupture on the effective cross-sectional area is given as

$$\phi_R P_n = (0.75)(109 \text{ kips}) = 81.6 \text{ kips (tensile fracture or rupture failure mode; LRFD method)}$$

The smaller value of the design strength, ϕP_n, governs. That is, the design tensile strength of the tension member is 60.8 kips. Since the design equation for the LRFD method requires that the design tensile strength be greater than or equal to the maximum factored load, P_u, (i.e., $\phi P_n \geq P_u$); therefore, the maximum factored load that can be supported by the tension member according to the LRFD method is

$\boldsymbol{P_u}$ **= 60.8 kips**

ASD Method:

For the ASD method, the allowable tension load based on yielding on the gross area of the tension member is given in equation (4-1b) as

$$P_a = \frac{P_n}{\Omega_t} = \frac{F_y A_g}{\Omega_t}$$

$$= \frac{67.7}{1.67} = 40.5 \text{ kips (tensile yielding failure mode; ASD method)}$$

Where,

Ω_t = Factor of safety for tensile yielding = 1.67

Note that for each failure mode, the previously calculated nominal strength, P_n, in the LRFD method is equally applicable to the ASD method. For the allowable strength design (ASD) method, the allowable tension load due to tensile fracture or rupture on the effective cross-sectional area is given as

$$P_a = \frac{P_n}{\Omega_R} = \frac{F_u A_e}{\Omega_R} \tag{4-2b}$$

$$= \frac{109}{2.0} = 54.5 \text{ kips} \quad \text{(tensile fracture or rupture failure mode; ASD method)}$$

Where,

Ω_R = Factor of safety for tensile rupture = 2.0

The smaller value of the allowable load, P_a, governs. That is, the allowable load of the tension member is 40.5 kips. Since the design equation for the ASD method requires that the allowable load be greater than or equal to the maximum unfactored load, P_s, (i.e., $P_a \geq P_s$); therefore, the maximum total unfactored load or total service load that can be supported by the tension member according to the ASD method is

$\boldsymbol{P_s}$ **= 40.5 kips**

b. Maximum Unfactored Live Load with a dead load of 20 kips
LRFD:

$P_u = 1.2P_D + 1.6P_L$
Therefore, 60.8 kips = 1.2(20 kips) + $1.6P_L$
Solving the above equation gives P_L = 23 kips.

a. With a tributary area of 200 ft.^{2}, the maximum unfactored floor live load is
L = 23(1000)/200 = **115 psf**

ASD:
$P_s = P_D + P_L$
Therefore, 40.5 kips = 20 kips + P_L
Solving the above equation gives P_L = 20.5 kips.
With a tributary area of 200 ft.^{2}, the maximum unfactored floor live load that can be supported by the tension member, L = 20.5(1000)/200 = **103 psf**

EXAMPLE 4-4

Tension Member Analysis

Determine if the channel is adequate for the applied tension load shown in Figure 4-9. The channel is ASTM A36 steel and is connected with four $\frac{5}{8}$-in.-diameter bolts. The tension member is subjected to service dead and live loads of 28.5 kips and 25.5 kips, respectively. Neglect block shear.

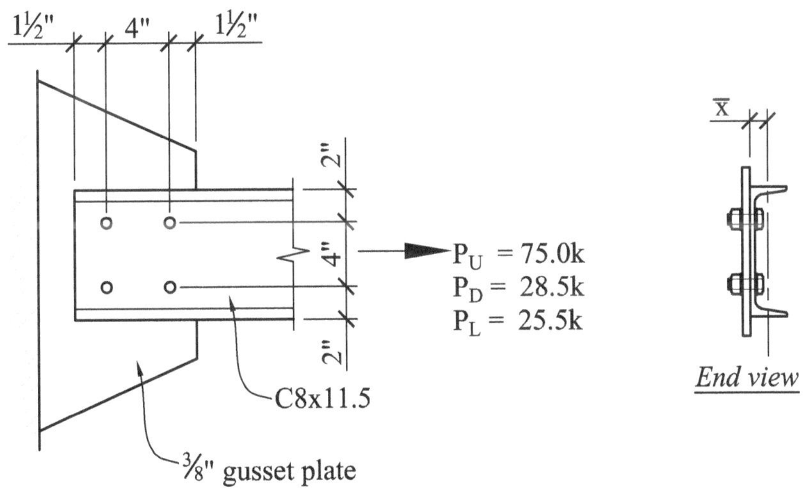

FIGURE 4-9 Detail for Example 4-4

Solution

From the *AISCM*, Table 1-5:

$$A_g = 3.37 \text{ in.}^2$$

$$\bar{x} = 0.572$$

$$t_w = 0.220 \text{ in.}$$

From the *AISCM* Table 2-3:

$$F_y = 36 \text{ ksi}$$

$$F_u = 58 \text{ ksi to } 80 \text{ ksi } (\text{use } F_u = 58 \text{ ksi})$$

Net Area of the Channel:

$$A_n = A_g - \sum A_{\text{holes}}$$

$$= 3.37 - \left[(2)\left(\frac{5}{8}+\frac{1}{8}\right)(0.220)\right] = 3.04 \text{ in.}^2$$

Effective Area of the Channel:

From Figure 4-5, the shear lag factor is

$$U = 1 - \frac{\bar{x}}{\ell}$$

$$= 1 - \frac{0.572 \text{ in.}}{4 \text{ in.}} = 0.857$$

$$A_e = A_n U$$

$$= (3.04)(0.857) = 2.61 \text{ in.}^2$$

LRFD Method:

P_u = 1.2(28.5) + 1.6(25.5) = 75 kips

From Section 4-2, the nominal tensile strength due to tensile yielding on the gross area is

$$P_n = F_y A_g = (36)(3.37) = 121.3 \text{ kips}$$

The design tensile strength due to yielding is obtained from Equation (4-1a) as

$$\phi_t P_n = (0.9)(121.3) = 109 \text{kips} > P_u = 75 \text{ kips. OK}$$

From Section 4-2, the nominal tensile strength due to fracture or rupture on the effective area is

$$P_n = F_u A_e = (58)(2.61 \text{ in.}^2) = 151.4 \text{ kips}$$

The design tensile strength due to fracture or rupture on the effective area is

$$\phi_R P_n = (0.75)(151.4) = 113.6 \text{ kips} > P_u = 75 \text{ kips. OK}$$

The design tensile strength of the tension member is the smaller of the two values calculated above. Thus, the design strength, $\phi P_n = 109 \text{ kips} > P_u = 75 \text{ kips}$. OK

ASD Method:

P_s = 28.5 + 25.5 = 54 kips

For the ASD method, the allowable tension load based on yielding on the gross area of the tension member is given in equation (4-1b) as

$$P_a = \frac{P_n}{\Omega_t} = \frac{F_y A_g}{\Omega_t}$$

$$= \frac{121.3}{1.67} = 72.6 kips$$ (tensile yielding failure mode; ASD method)

Where,

Ω_t = Factor of safety for tensile yielding = 1.67

Note that for each failure mode, the previously calculated nominal strength, P_n, in the LRFD method is equally applicable to the ASD method.

For the allowable strength design (ASD) method, the allowable tension load due to tensile fracture or rupture on the effective cross-sectional area is given as

$$P_a = \frac{P_n}{\Omega_R} = \frac{F_u A_e}{\Omega_R} \quad (4\text{-}2b)$$

$$= \frac{151.4}{2.0} = 75.7 \text{kips (tensile fracture or rupture failure mode; ASD method)}$$

Where,

Ω_R = Factor of safety for tensile rupture = 2.0

The smaller value of the allowable load, P_a, governs. That is, the allowable load of the tension member is 72.6 kips. Since the design equation for the ASD method requires that the allowable load, P_a, be greater than or equal to the maximum unfactored load, P_s, (i.e., $P_a \geq P_s$); therefore, the maximum unfactored or service load that can be supported by the tension member according to the ASD method is

P_a = 72.6 kips > P_s = 54 kips OK.

The tension member has been checked using both the LRFD and ASD methods and found to be adequate in both cases. Note that the block shear failure mode has not been checked in this example; this will be covered later in this section.

EXAMPLE 4-5

Tension Member Analysis with Staggered Bolts

Determine the maximum factored load that can be applied in tension to the angle shown in Figure 4-10. The angle is ASTM A36 steel and is connected with four $^3/_4$-in.-diameter bolts. Neglect block shear.

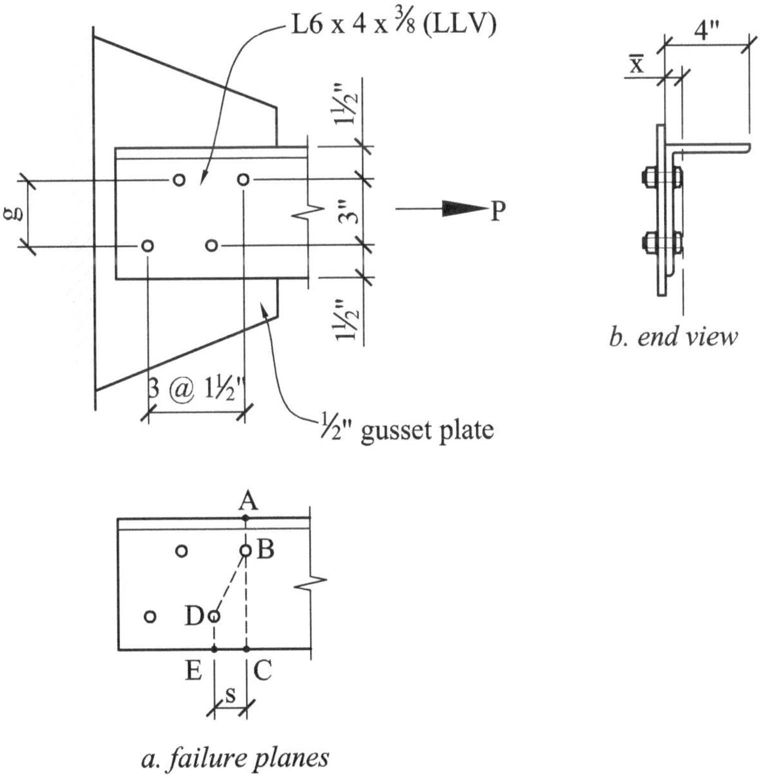

FIGURE 4-10 Detail for Example 4-5

Solution

From the *AISCM* Table 1-7:

A_g = 3.61 in.^{2}

$\bar{x}$ = 0.933 in.

t = 0.375 in.

From the *AISCM*, Table 2-3:

$F_y = 36$ ksi

$F_u = 58$ ksi to 80 ksi (use $F_u = 58$ ksi)

Net Area of the Angle:

$$A_n = A_g - \Sigma d_h t + \Sigma \frac{s^2}{4g} t$$

Failure Plane *ABC*:

$$A_n = 3.61 - \left[\left(\frac{3}{4} + \frac{1}{8}\right)(0.375)\right] + 0 = 3.28 \text{ in.}^2$$

Failure Plane *ABDE*:

s = 1.5 in and g = 3 in.

$$A_n = 3.61 - \left[(2)\left(\frac{3}{4} + \frac{1}{8}\right)(0.375)\right] + \left[\frac{(1.5)^2}{(4)(3)}(0.375)\right] = 3.02 \text{ in.}^2$$

The failure plane along *ABDE* controls since it has a smaller net cross-sectional area.

Effective Area of the Angle:

The shear lag factor, $U = 1 - \dfrac{\bar{x}}{\ell}$

$$= 1 - \frac{0.933}{(3)(1.5)} = 0.792$$

Alternatively, $U = 0.60$ from Table 4-1. The larger value is permitted to be used, so $U = 0.792$.

$$A_e = A_n U$$

$$= (3.02)(0.792) = 2.39 \text{ in.}^2$$

From equation (4-1a), the tensile strength based on tensile yielding on the gross area is

$$\phi_t P_n = \phi_t F_y A_g$$

$$= (0.9)(36)(3.61) = 116 \text{ kips}$$

From equation (4-2a), the tensile strength based on the tensile rupture on the effective area is

$$\phi_R P_n = \phi_R F_u A_e$$

$$= (0.75)(58)(2.39 \text{ in.}^2) = 104 \text{ kips}$$

The smaller value controls, therefore, the maximum factored tension load, $\boldsymbol{P_u}$ **= 104 kips.**

For the ASD method, the available strength based on tensile yielding on the gross area from equation (4-1b) is:

$$\frac{P_n}{\Omega_t} = \frac{F_y A_g}{\Omega_t} = \frac{(36)(3.61)}{1.67} = 77.8 \text{kips}$$

And the available tensile strength based on the tensile rupture on the effective area is:

$$\frac{P_n}{\Omega_R} = \frac{F_u A_e}{\Omega_R} = \frac{(58)(2.39)}{2.0} = 69.3 \text{kips}$$

The smaller value controls, therefore, the maximum allowable total unfactored tension load, $P_a = 69.3$ kips.

(Note: Block shear should also be checked, which is covered in the following section.)

Block Shear Failure Mode:

In previous sections we discussed the strength of members in pure tension, and the failure modes of tensile yielding at the gross cross-section and tensile rupture at the net cross-section were considered. In Chapters 9 and 10, we will consider additional failure modes due to bolt or weld connection failure. In addition to checking the tension member and the connected ends of tension members for tensile failure, there are certain tension member connection configurations where tensile failure occurs due to the tearing out of a segment of the tension member or the connecting element from the rest of the connection (see Figure 4-11). The failure planes usually occur along the centerlines of the bolt holes for bolted connections and along the outline of the welds for welded connections.

Failure in block shear can occur by a combination of shear yielding plus tensile rupture as depicted by the right-hand side of equations (4-11a) and (4-11b) or by a combination of shear rupture plus tensile rupture as depicted by the left-hand side of equations (4-11a) and (4-11b). Therefore, both the shear and tension failure planes contribute to the strength of the connection in block shear. From *AISCM* Section J4.2, the nominal strength based on *shear yielding* is

$$R_n = 0.6\, F_y A_{gv}, \tag{4-10a}$$

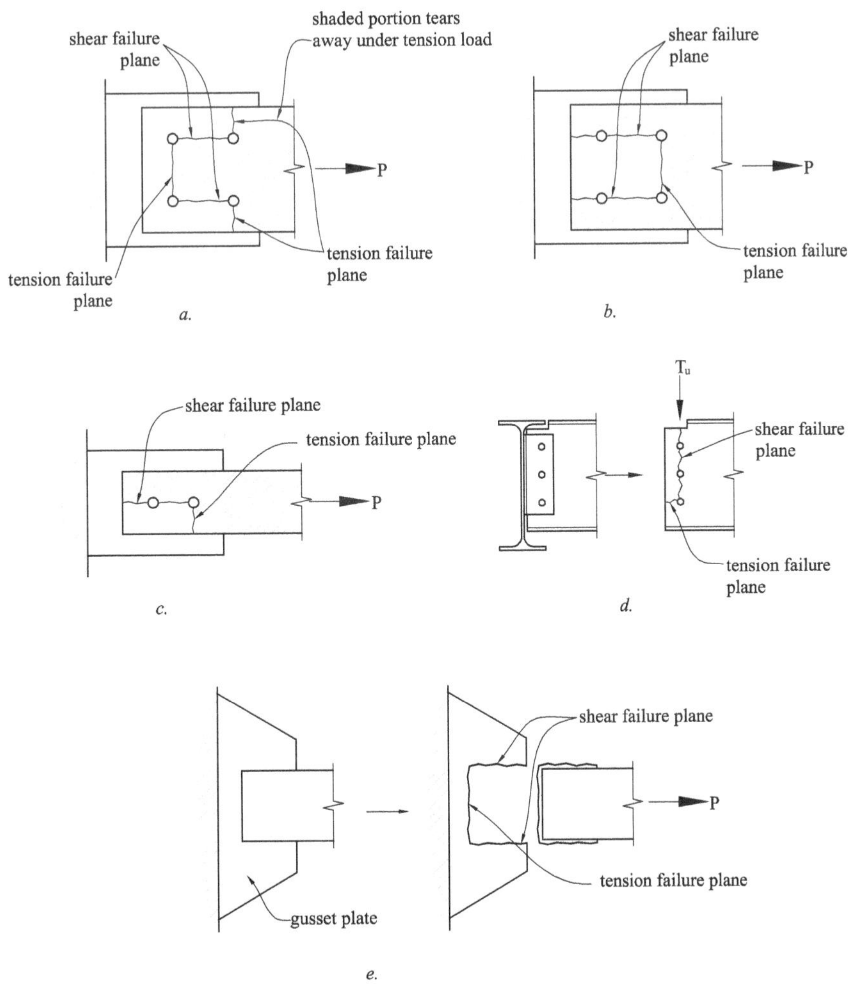

FIGURE 4-11 Block shear failure modes

and the nominal strength based on *shear rupture* is

$$R_n = 0.6\,F_u A_{nv}, \tag{4-10b}$$

Where,

A_{gv} = Gross area subject to shear, and

A_{nv} = Net area subject to shear (see equation (4-4)).

Combining the available tension and shear strengths gives the expressions for the design block shear strength and the available block shear strength as follows (see *AISCM* Section J4.3):

$$\phi P_n = \phi\left(0.6F_u A_{nv} + U_{bs} F_u A_{nt}\right) \le \phi\left(0.6F_y A_{gv} + U_{bs} F_u A_{nt}\right) \text{ for the LRFD method} \tag{4-11a}$$

$$\frac{P_n}{\Omega} = \frac{\left(0.6F_u A_{nv} + U_{bs} F_u A_{nt}\right)}{\Omega} \le \frac{\left(0.6F_y A_{gv} + U_{bs} F_u A_{nt}\right)}{\Omega} \text{ for the ASD method} \tag{4-11b}$$

where

ϕ = 0.75,

Ω = 2.0

F_u = Minimum ultimate tensile strength,

F_y = Minimum yield stress,

A_{gv} = Gross area subjected to shear,

A_{nt} = Net area subjected to tension (see equation (4-4)),

A_{nv} = Net area subjected to shear (see equation (4-4)), and

U_{bs} = 1.0 for uniform tension stress (applies to most tension connections except for beam-end connections with multiple rows of bolts. See *AISCM* Figure C-J4.2a)

= 0.50 for non-uniform tension stress distribution (e.g., beam-end connection with multiple rows of bolts. See *AISCM* Figure C-J4.2b).

$U_{bs}F_uA_{nt}$ = nominal strength based on tensile rupture

The U_{bs} term in equations (4-11a) and (4-11b) is a reduction factor that accounts for a non-uniform stress distribution. The Commentary Section C-J4.3 of the *AISCM* (*AISCM* Figure C-J4.2) gives examples of connections with uniform and nonuniform tension stress distribution, but for most cases that have a uniform stress distribution, U_{bs} = 1.0. One example with a nonuniform tension stress distribution where the U_{bs} = 0.50 is a coped beam connection with more than one vertical row of bolts in the web of the supported beam (see Figure C-J4.2b in the Commentary of the AISC Specifications [1]).

EXAMPLE 4-6

Tension Member with Block Shear

For the connection shown in Example 4-4, determine if the channel and gusset plate are adequate for the applied tension load considering block shear. Assume that the width of the plate is such that block shear along the failure plane shown in Figure 4-12 controls the design of the plate.

Solution

From the *AISCM* Table 2-3:

$$F_y = 36 \text{ ksi}$$

$$F_u = 58 \text{ ksi to } 80 \text{ ksi } \left(\text{use } F_u = 58 \text{ ksi}\right)$$

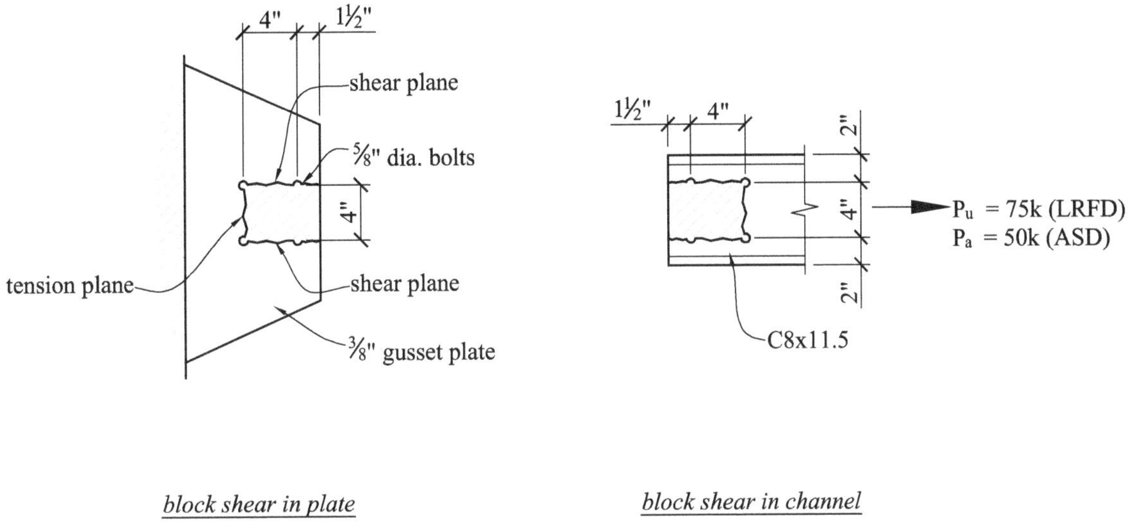

FIGURE 4-12 Detail for Example 4-6

Gusset Plate Dimensions:

Shear plane:

$$A_{gv} = (2\,shear\,planes)(4\,\text{in.} + 1.5\,\text{in.})(0.375\,\text{in.}) = 4.12\ \text{in.}^2$$

$$A_{nv} = A_{gv} - \sum A_{\text{holes}}$$

$$= 4.12 - \left[(2\ \text{sides})(1.5\,holes)\left(\frac{5}{8} + \frac{1}{8}\right)(0.375\,\text{in.})\right] = 3.28\ \text{in.}^2$$

Tension plane:

$$A_{nt} = A_{gt} - \sum A_{\text{holes}}$$

$$= \left[(4\,\text{in.})(0.375\,\text{in.})\right] - \left(\frac{1}{2}hole + \frac{1}{2}hole\right)\left[\left(\frac{5}{8} + \frac{1}{8}\right)(0.375\,\text{in.})\right] = 1.21\ \text{in.}^2$$

Channel Dimensions:

Shear plane:

$$A_{gv} = (2\,shear\,planes)(4\,\text{in.} + 1.5\,\text{in.})(0.220\,\text{in.}) = 2.42\ \text{in.}^2$$

$$A_{nv} = A_{gv} - \sum A_{\text{holes}}$$

$$= 2.42 - \left[(2\ shear\ planes)(1.5\,holes)\left(\frac{5}{8} + \frac{1}{8}\right)(0.220\,\text{in.})\right] = 1.92\ \text{in.}^2$$

Tension plane:

$$A_{nt} = A_{gt} - \sum A_{\text{holes}}$$

$$= \left[(4\,\text{in.})(0.220\,\text{in.})\right] - \left(\frac{1}{2}hole + \frac{1}{2}hole\right)\left[\left(\frac{5}{8} + \frac{1}{8}\right)(0.220\,\text{in.})\right] = 0.715\ \text{in.}^2$$

$$U_{bs} = 1.0\ (\text{tension stress is uniform})$$

LRFD:

The design block shear strength is found from equations (4-11a) for the LRFD method:

$$\phi P_n = \phi(0.6F_u A_{nv} + U_{bs} F_u A_{nt}) \le \phi(0.6F_y A_{gv} + U_{bs} F_u A_{nt})$$

For the ***Gusset Plate***, the tensile strength in block shear is

$$\phi P_n = 0.75[(0.6)(58)(3.28) + (1.0)(58)(1.21)]$$

$$\le 0.75[(0.6)(36)(4.12) + (1.0)(58)(1.21)]$$

$$= 138 \text{ kips} \le 119 \text{ kips.}$$

The smaller value controls, so the design tension strength of the gusset plate in block shear is 119 kips, which is greater than the applied factored load of P_u = 75 kips.

For the ***Channel*** (t_w = 0.22 in.), the tensile strength in block shear is

$$\phi P_n = 0.75[(0.60)(58)(1.92) + (1.0)(58)(0.715)]$$

$$\le 0.75[(0.60)(36)(2.42) + (1.0)(58)(0.715)]$$

$$= 81.2 \text{ kips} \le 70.3 \text{ kips.}$$

The smaller value controls, so the design tension strength of the channel in block shear is 70.3 kips, which is less than the applied factored load of P_u = 75 kips. Hence the channel is not adequate in block shear.

ASD:

The allowable block shear strength is found from equations (4-11b) for the ASD method:

$$\frac{P_n}{\Omega} = \frac{(0.6F_u A_{nv} + U_{bs} F_u A_{nt})}{\Omega} \le \frac{(0.6F_y A_{gv} + U_{bs} F_u A_{nt})}{\Omega}$$

For the ***Gusset Plate***, the allowable tensile strength in block shear is

$$\frac{P_n}{\Omega} = \frac{(0.6)(58)(3.28) + (1.0)(58)(1.21)}{2.0} \le \frac{(0.6)(36)(4.12) + (1.0)(58)(1.21)}{2.0}$$

$$= 92.1 \text{ kips} \le 79.6 \text{ kips.}$$

The smaller value controls, so the available strength of the gusset plate in block shear is 79.6 kips, which is greater than the applied load of P_a = 50 kips.

For the ***Channel*** (t_w = 0.22 in.), the allowable tensile strength in block shear is

$$\frac{P_n}{\Omega} = \frac{(0.6)(58)(1.92) + (1.0)(58)(0.715)}{2.0} \le \frac{(0.6)(36)(2.42) + (1.0)(58)(0.715)}{2.0}$$

$$= 54.1 \text{ kips} \le 46.9 \text{ kips.}$$

The smaller value controls, so the allowable tensile strength of the channel in block shear is 46.9 kips, which is less than the applied total service or unfactored load of P_a = 50 kips. Hence the channel is not adequate in block shear.

(Note: The bolts also should be checked for shear and bearing, but bolt strength is covered later in Chapter 9.)

EXAMPLE 4-7

Block Shear in a Gusset Plate (LRFD Method)

For the HSS-to-gusset plate connection shown in Figure 4-13, determine the length, ℓ, required to support the applied tension load, considering the strength of the gusset plate only. The plate is ASTM A529, Grade 50 steel with a specified F_u = 70 ksi. Is the gusset plate thickness adequate?

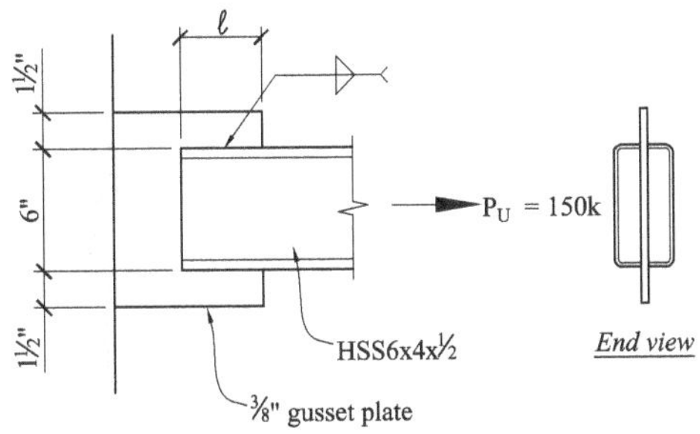

FIGURE 4-13 Details for Example 4-7

Solution

From the *AISCM*, Table 2-4:

$$F_y = 50 \text{ ksi}$$

$$F_u = 70 \text{ ksi}$$

From equation (4-1a), the tensile strength due to tensile yielding on the gross area of the plate is

$$\phi P_n = \phi F_y A_g$$

$$= (0.9)(50)(1.5 + 6 + 1.5)(0.375) = 151 \text{ kips} > 150 \text{ kips. OK}$$

From equation (4-2a), the tensile strength due to fracture or rupture at the net cross-sectional area is

$$\phi P_n = \phi F_u A_e = \phi F_u U A_n = \phi F_u \left[UA_g\right], \quad (\textit{where, } A_n = A_g \text{ since there are no bolt holes})$$

$$= (0.75)(70\,\text{ksi})\left[(1.0)(1.5\,\text{in.} + 6\,\text{in.} + 1.5\,\text{in.})(0.375\,\text{in.})\right] = 177 \text{ kips} > 150 \text{ kips. OK}$$

(Note: Initially, assume the shear lag factor, $U = 1.0$; this will be checked later.)

The design block shear strength is found from equation (4-11a) as follows:

$$\phi P_n = \phi(0.6F_u A_{nv} + U_{bs} F_u A_{nt}) \le \phi(0.6F_y A_{gv} + U_{bs} F_u A_{nt})$$

$$U_{bs} = 1.0 \text{ (tension stress is uniform)}$$

Since $A_{nv} = A_{gv}$, the right-hand side of the equation will control; and for block shear, the failure mode in the gusset plate will coincide with the outline of the HSS similar to Figure 4-11e, consisting of a tension plane that is 6 in. wide, plus two shear planes, each with a length, ℓ.

Therefore, $\phi P_n = \phi(0.6F_y A_{gv} + U_{bs} F_u A_{nt})$

$$150 = (0.75)\left[(0.6)(50)(A_{gv}) + (1.0)(70)(6)(0.375)\right]$$

Therefore, $A_{gv} = 1.42 \text{ in.}^2$ and since there are two shear planes,

$$\ell_{\min} = \frac{1.42}{(2)(0.375)} = 1.89 \text{ in.}$$

The minimum length of engagement is 1.89 in. in order to develop adequate strength in the gusset plate. The minimum length, ℓ, of the slot should be equal to or greater than the height of the connecting HSS member, or 6 in. $(\ell \geq H)$. Therefore, the minimum length, ℓ, is 6 in.

$$B = 4\,\text{in.}$$

$$t_p = 0.375\,\text{in.}$$

$$t = 0.5\,\text{in.}$$

$$b = \left(\frac{4\,\text{in.} - 0.375\,\text{in.}}{2}\right) = 1.813\,\text{in.}$$

$$H = 6\,\text{in.}$$

From Table 4-1, the shear lag factor for the HSS member is

$$\bar{x} = b - \frac{2b^2 + tH - 2t^2}{2H + 4b - 4t}$$

$$= 1.813 - \frac{2(1.813)^2 + (0.5)(6) - 2(0.5)^2}{2(6) + 4(1.813) - 4(0.5)} = 1.29\,in.$$

$$U = 1 - \frac{\bar{x}}{\ell}$$

$$= 1 - \frac{1.29}{6} = 0.785$$

This actual value of the shear lag factor is then used to recalculate the strength of the gusset plate member in tensile fracture or rupture.

$$\phi P_n = \phi F_u A_e = \phi F_u \left[U A_g\right]$$

$$= (0.75)(70\text{ ksi})\left[(0.785)(1.5\text{in.} + 6\text{ in.} + 1.5\text{ in.})(0.375\text{in.})\right] = 139\text{ kips} < 150\text{ kips. Not Good}$$

Therefore, increase the gusset plate thickness, t_p, from 3/8" or 0.375 in. to 0.5 in. With this revised gusset plate thickness, $\bar{x}$ and U are reclaculated as 1.24 and 0.793, respectively. The tension capacity in fracture with a ½ in. thick gusset plate is recalculated as

$$\phi P_n = (0.75)(70\text{ ksi})\left[(0.793)(1.5\text{in.} + 6\text{ in.} + 1.5\text{ in.})(0.5\text{ in.})\right] = 187\text{ kips} > 150\text{ kips.} \quad \text{O.K.}$$

EXAMPLE 4-8

Hanger in Tension with Block Shear

The W12 × 53 tension member shown in Figure 4-14 has two rows of three 1-in.-diameter A325N bolts in each flange. Assuming ASTM A572 steel and considering the strength of the W12 × 53 and the gusset plate,

1. Determine the design tension strength of the W12 × 53,
2. Determine the service dead load that can be supported if there is no live load,
3. If a service dead load P_D = 100 kips is applied, what is the maximum service live load, P_L, that can be supported?

Solution

1. For A572 steel, F_y = 50 ksi, F_u = 65 ksi (*AISCM* Table 2-4).

For 1-in.-diameter bolts, $d_{\text{hole}} = 1 \text{ in.} + \frac{3}{16} \text{in.} = 1.19 \text{in.}$

From part 1 of the AISCM, for W12 × 53,

$A_g = 15.6 \text{ in.}^2$

$b_f = 10 \text{ in.}$

$t_f = 0.575 \text{ in.}$

$d = 12.1 \text{ in.}$

WT 6 × 26.5 results from cutting a W12 × 53 into two halves across the web.

$\bar{x} = 1.02 \text{ in.}$ (see *AISCM* Table 1-8, and use the value for WT 6x26.5)

$A_n = A_g - \Sigma A_{\text{holes}}$

$= 15.6 - (4 \text{ holes})(1.19)(0.575) = 12.9 \text{ in.}^2$

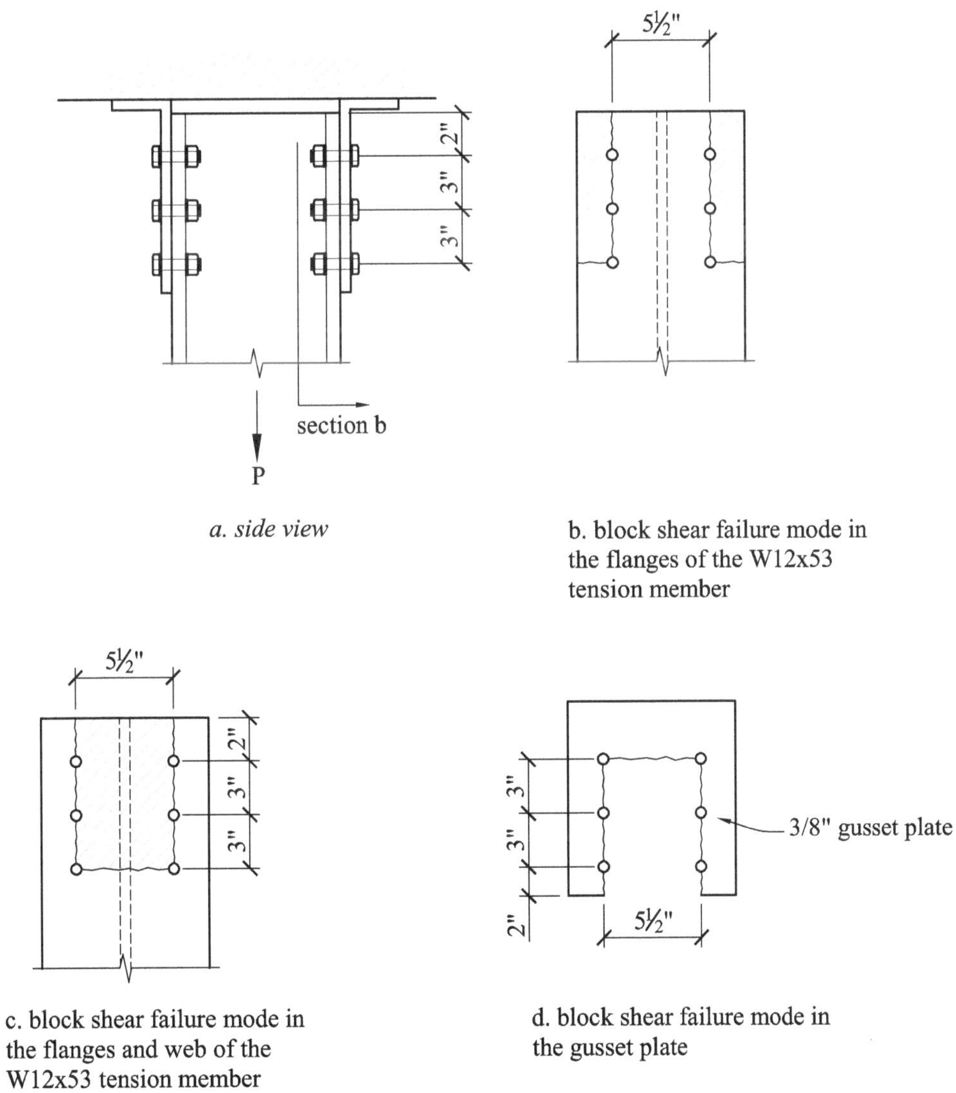

FIGURE 4-14 Detail for Example 4-8

Note: At the net cross-section, there are two (2) bolt holes in the top and bottom flanges for a total of four bolt (4) holes.

From Table 4-1, we obtain the shear lag factor, $U = 0.90$, since

$$b_f = 10 \text{ in} > \frac{2}{3}d = \frac{2}{3}(12.1) = 8.07 \text{ in.}$$

Using the calculation method, the shear lag factor can be determined as follows:

$$U = 1 - \frac{\bar{x}}{\ell} = 1 - \frac{1.02}{6} = 0.83$$

ℓ = centerline distance between outermost or extreme bolts = 3 in. + 3 in. = 6 in.

The AISC Code allows the larger value of $U = 0.9$ to be used.

$$A_e = UA_n = (0.9)(12.9) = 11.6 \text{ in.}^2$$

The design tension strengths are calculated as follows:

a. Yielding failure mode: $\phi P_n = 0.9A_gF_y = (0.9)(50)(15.6) = \mathbf{702\ kips}$

b. Tensile rupture failure mode: $\phi P_n = 0.75F_uA_e = (0.75)(65)(11.6) = \mathbf{566\ kips}$

c. Block shear failure mode of **W12x53** (Figure 4-15):

$$A_{gv} = (2\,flanges)(2\,shear\,planes)(2 \text{ in.} + 3 \text{ in.} + 3 \text{ in.})(0.575 \text{ in.}) = 18.4 \text{ in.}^2$$

$$A_{gt} = (2\,flanges)(2\,tension\,planes)(2.25 \text{ in.})(0.575 \text{ in.}) = 5.18 \text{ in.}^2$$

$$A_{nv} = A_{gv} - \sum A_{\text{holes}}$$

$$= 18.4 - (2\,flanges)(2\,shear\,planes)(2.5\,holes)(1.19 \text{ in.})(0.575 \text{ in.}) = 11.56 \text{ in.}^2$$

$$A_{nt} = A_{gt} - \sum A_{\text{holes}}$$

$$= 5.18 - (2\,flanges)(2\,tension\,planes)\left(\frac{1}{2}hole\right)(1.19 \text{in.})(0.575 \text{ in.}) = 3.81 \text{ in.}^2$$

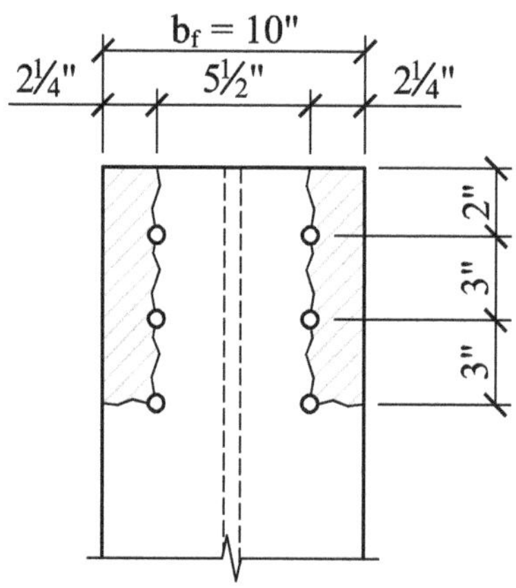

FIGURE 4-15 Block shear in W12 × 53

The design block shear strength of the W12x53 is found from equation (4-11a) as follows with $U_{bs} = 1.0$:

$$\phi P_n = \phi(0.6F_uA_{nv} + U_{bs}F_uA_{nt}) \le \phi(0.6F_yA_{gv} + U_{bs}F_u\,A_{nt})$$

$$\phi P_n = (0.75)[(0.6)(65)(11.56) + (1.0)(65)(3.81)]$$

$$\le (0.75)[(0.6)(50)(18.4) + (1.0)(65)(3.81)]$$

524 kips $\le$ 600 kips

$\phi P_n = \mathbf{524\ kips}$ (block shear capacity of W12x53)

Check the block shear failure mode of the 3/8" thick ***Gusset Plate***:

$$A_{gv} = (2\,plates)(2\,shear\,planes)(3\text{ in.} + 3\text{ in.} + 2\text{in. })(0.375\text{ in.}) = 12\text{ in.}^2$$

$$A_{gt} = (2\,plates)(5.5\text{ in.})(0.375\text{ in.}) = 4.13\text{ in.}^2$$

$$A_{nv} = A_{gv} - \Sigma A_{\text{holes}}$$

$$= 12 - (2\ plates)(2\,shear\,planes)(2.5\,holes)(1.19\text{ in.})(0.375\text{ in.}) = 7.54\text{ in.}^2$$

$$A_{nt} = A_{gt} - \Sigma A_{\text{holes}}$$

$$= 4.13 - (2\,flanges)\left(\frac{1}{2}hole + \frac{1}{2}hole\right)(1.19\text{ in.})(0.375\text{ in.}) = 3.24\text{ in.}^2$$

Recall that the bolt hole diameter is 1.19 in. for a 1 in.-diameter bolt.

The design block shear strength of the gusset plate is found from equation (4-11a) as follows with $U_{bs} = 1.0$:

$$\phi P_n = \phi(0.6F_u A_{nv} + U_{bs} F_u A_{nt}) \le \phi(0.6F_y A_{gv} + U_{bs} F_u\ A_{nt})$$

$$\phi P_n = (0.75)[(0.6)(65)(7.54) + (1.0)(65)(3.24)]$$

$$\le (0.75)[(0.6)(50)(12) + (1.0)(65)(3.24)]$$

378 kips $\le$ 428 kips

Therefore, the block shear capacity of the ***Gusset Plate*** = **378 kips.**

The overall tension capacity of the hanger and its connection will be the smallest of the tension capacities of the W12 × 53 hanger in yielding, fracture, and block shear, and the block shear capacity of the Gusset Plate. Therefore, for this hanger and its connection, the tension capacity, $\phi P_n = 378$ kips.

Thus, the tension capacity of this hanger is limited by the block shear capacity of the 3/8 in. gusset plate. If the gusset plate thickness is increased to at least the thickness of the flange of the W12x53 hanger (i.e., t = 0.575 in.), the block shear capacity of the gusset plate will now be higher than that of the W12 × 53 tension member. In this case, the tension capacity of the hanger and its connection will be limited by the tension capacity of the W12 × 53 hanger and not by the block shear capacity of the gusset plate.

Summary (assuming gusset plate is as thick as the flange of the W12 × 53 hanger):

Tensile yielding failure mode : $\phi P_n = 702$ kips

Tensile rupture failure mode : $\phi P_n = 566$ kips

Block shear failure mode : $\phi P_n = \mathbf{524\ kips}$

The smallest of the three values governs. Therefore, the design strength, ϕP_n = **524 kips**

2. Since there is no live load, that means $P_L = 0$; the two load combinations to be considered are:

$P_u = 1.4\ P_D$ (controls)

$P_u = 1.2\ P_D + 1.6\ P_L = 1.2\ P_D + 1.6\ (0) = 1.2\ P_D$

Using the limit states design equation gives

$P_u = 1.4\ P_D \le \phi P_n = 524$ kips

$$1.4\ P_D \le 524 \text{ kips}$$

Therefore, we obtain P_D = 374 kips (the maximum unfactored dead load that can be supported).

3. Given P_D = 100 kips, determine the live load, P_L, that can be safely supported:
   The limit state design equation is

$$P_u = 1.2\ P_D + 1.6\ P_L \le \phi P_n = 524 \text{ kips}$$

$$(1.2)(100) + 1.6\ P_L \le 524 \text{ kips}$$

Therefore, P_L = 253 kips (maximum unfactored live load that can be supported).

4-3 DESIGN OF TENSION MEMBERS

The design of tension members requires consideration of additional failure modes at the connection points not specifically addressed in this chapter but will be addressed later in the text. For tension members with welded connections, the design strength of the welds in shear and tension must be considered. For tension members with bolted connections, the design strength of the bolts in shear, tension, and bearing must be considered. Load eccentricity effects at the connection points also must be considered. In this section, we will consider the design strength in tension of the actual member only. The reader should refer to Chapters 9 and 10 for the design of bolted and welded connections.

Tension members need to have enough gross cross-sectional area for strength in tensile yielding and enough effective area for strength in tensile rupture. Recall that the effective area accounts for shear lag effects.

In addition to having enough design strength in tensile yielding and tensile rupture, the block shear failure mode at the connected ends also needs to be checked. In some cases, there might be more than one block shear failure mode. The design strength of the tension member is the smallest of the strength of all the possible failure modes, including yielding, fracture, block shear, and connection failure. The strength of bolts in shear, tension, and bearing will be covered in Chapter 9, and the strength of welds in shear will be discussed in Chapter 10.

Examples of X-braced frames with diagonal tension members in tension are shown in Figures 4-15a through 4-15c. Figures 4-15d and 4-15e consist of diagonal members that resist both tension and compression axial forces. Figure 4-15f shows the details at the connection of HSS X-braces to a gusset or knife plate.

Slenderness effects should also be considered for tension-only diagonal braces. The AISC specification recommends a slenderness limit L_c/r of 300 to prevent flapping, flutter, or sagging of the member, but this is not a mandatory requirement, and it does not apply to tension hangers. If this slenderness limit cannot be met, the *tension-only diagonal braces* that consist of plates or small diameter tension rods or light angles can be pretensioned to prevent sag or vibration. Heavier diagonal brace members such as HSS or wide flange sections are not susceptible to much sag, and it would be difficult to stretch these members. The amount of this pretension force can vary between 5% and 10% of the applied tension force, and the pretension force will reduce the design strength of the member accordingly by increasing the forces used for design. One way to induce this pretension is to fabricate the member shorter than its true length or by "inducing draw." The AISC specification suggests the pretension or "draw" values in Table 4-2 for tension-only diagonal bracing members.

TABLE 4-2 Recommended pretension values for slender tension-only diagonal brace members.

Length of tension member, L	Pretension or "Draw" member length deduction
$L \leq 10$ ft.	0 in.
10 ft. $< L \leq 20$ ft.	$\frac{1}{16}$ in.
20 ft. $< L \leq 35$ ft.	$\frac{1}{8}$ in.
$L > 35$ ft.	$\frac{3}{16}$ in.

FIGURE 4-15a X-braced frames in an industrial building with diagonal members in tension

FIGURE 4-15b HSS X-bracing with gusset or knife plates in HSS slots

FIGURE 4-15c Exposed X-braced frame in an industrial building

FIGURE 4-15d Wide flange X-bracing, functioning also as gravity load-carrying members

FIGURE 4-15e HSS diagonal bracings

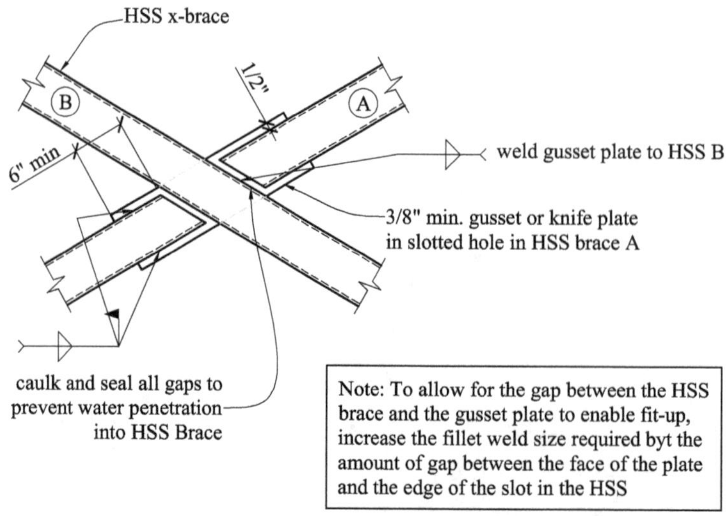

FIGURE 4-15f HSS X-braces at gusset or knife plate connection.

When members are fabricated shorter than their actual lengths in accordance with Table 4-2, they will be drawn up in the field in order to reduce the amount of flap or sag in the member. However, it can be shown that for a given cross-sectional area of the tension member, the increase in tension force in the member can be as high as 50% or more when using the values in Table 4-2, which is not desirable. The authors have found that an increase of 5% to 10% in the applied tension load has proven to be effective in practice and they would recommend using these lower values. The authors further recommend that the designer consult with the steel fabricator on any given project to determine the proper amount of draw that might be appropriate based on the actual fabrication and erection procedures.

Figure 4-16 shows a possible pretension detail where the tension-only diagonal brace member is intentionally fabricated shorter than its actual length so that once the connection is tightened to its final position, any sag or vibration of the hanger is prevented.

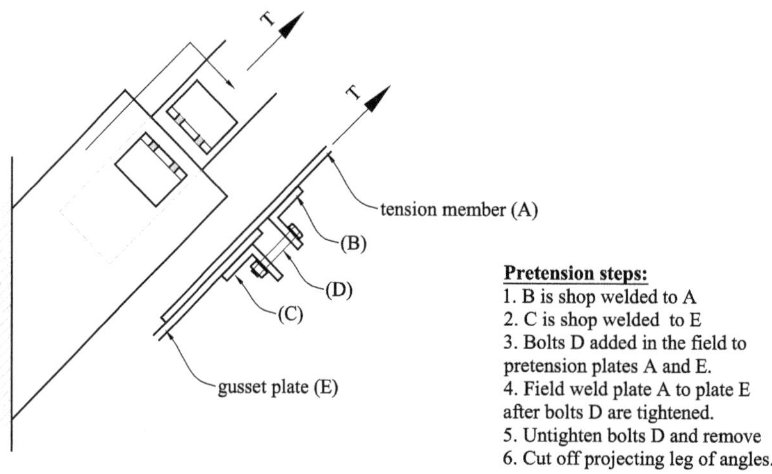

FIGURE 4-16 Pre-tensioned connection (for plate X-bracings only)

The process for the design of a tension member can be summarized as follows:

1. Calculate the maximum factored and unfactored tension loads, respectively, on the member, P_u and P_s, and determine the minimum gross area from the tensile yielding failure mode equation from Section 4-2:

$$A_g \geq \frac{P_u}{0.9F_y} \text{ (for LRFD)} \tag{4-12a}$$

$$A_g \geq \frac{\Omega_t P_s}{F_y} = \frac{1.67P_s}{F_y} = \frac{P_s}{0.6F_y} \text{ (for ASD)} \tag{4-12b}$$

2. Determine the minimum net area from the tensile rupture failure mode equation:

$$A_n \geq \frac{P_u}{0.75F_u U} \text{ (for LRFD)} \tag{4-13a}$$

$$A_n \geq \frac{\Omega_t P_s}{F_u U} = \frac{2P_s}{F_u U} = \frac{P_s}{(0.5F_u)U} \text{ (for ASD)} \tag{4-13b}$$

where the net area is found from equation (4-4):

$$A_n = A_g - \Sigma A_{\text{holes}}$$

Therefore, the required gross area for the tensile rupture failure mode is

$$A_g \geq \frac{P_u}{0.75F_u U} + \Sigma A_{\text{holes}} \text{ (for LRFD)} \tag{4-14a}$$

$$A_g \geq \frac{2P_s}{F_u U} + \Sigma A_{\text{holes}} \text{ (for ASD)} \tag{4-14b}$$

3. Use the larger A_g value from equations (4-12a) and (4-14a) for the LRFD method and (4-12b) and (4-14b) for the ASD method; select a trial member size based on the larger value of A_g.
4. For tension members, AISC specification Section D1 *suggests* that the slenderness ratio $L_c/r_{\text{min}} = KL/r_{\text{min}}$ should be less than 300 to prevent flapping or flutter of the member.
   Where
   K = Effective length factor (usually assumed to be 1.0 for tension members),
   L = Unbraced length of the tension member, and
   r_{min} = *Smallest* radius of gyration of the member.
   The smallest radius of gyration for rolled sections can be obtained from Part 1 of the *AISCM*. For other sections, such as plates, the radius of gyration can be calculated from

$$r_{\text{min}} = \sqrt{\frac{I_{\text{min}}}{A_g}} > \frac{L}{300}, \tag{4-15}$$

   where I_{min} is the smallest moment of inertia.
   If equation (4-15) cannot be satisfied (i.e., the member is too slender), the member should be pretensioned. Allow for 5% to 10% pretension force in the design of the member; that is, increase P_u (LRFD) in equations (4-12a, 4-13a, and 4-14a) and P_s (ASD) in equations (4-12b, 4-13b, and 4-14b) by 5% to 10%. See Figure 4-16 for a pretension detail.
5. **LRFD**: For the LRFD method, use equation (4-11a) to determine the block shear capacity of the selected tension member.
   If ϕP_n(block shear) is greater than or equal to P_u, the member is adequate.
   If ϕP_n(block shear) is less than P_u, increase the member size and repeat step 5 until ϕP_n(block shear) $\geq P_u$.
   ASD: For the ASD method, use equation (4-11b) to determine the allowable tensile strength in block shear for the selected tension member.
   If $\frac{P_n}{\Omega}$ (block shear) is greater than or equal to P_s, the member is adequate.
   If $\frac{P_n}{\Omega}$ (block shear) is less than P_s, increase the member size and repeat step 5 until $\frac{P_n}{\Omega}$ (block shear) $\geq P_s$.

EXAMPLE 4-9

Design of a Tension Member (LRFD Method)

Using the LRFD method, design the X-brace in the first story of the building shown in Figure 4-17, which is subjected to wind loads. Use a steel plate that conforms to ASTM A36 and assume a single row of bolts.

Note: Some X-brace configurations have slender diagonal members such that they can only support loads in tension (e.g., X-braced frames with plate and angle diagonal members). In Figure 4-17, all the diagonal brace members are assumed to support only axial tension forces because they are too slender to support axial compression forces. Therefore, for the load direction show in Figure 4-17, only the shaded diagonal members support lateral loads. However, when the lateral loads change direction, the unshaded diagonal members will now be in tension.

The lateral loads shown in Figure 4-17 are the loads acting on *each* X-braced frame. The wind loads acting on the entire building must be distributed to the various braced frames in the building that are parallel to the direction of the lateral load. If the horizontal diaphragm (i.e., the roof and floors) is assumed to be *rigid* (i.e., with a high in-plane stiffness), the lateral load is distributed in proportion to the stiffness of each braced frame. If the diaphragm is *flexible*, the lateral load is distributed in proportion to the tributary area of each braced frame. See Chapter 14 for further discussions on lateral load distribution.

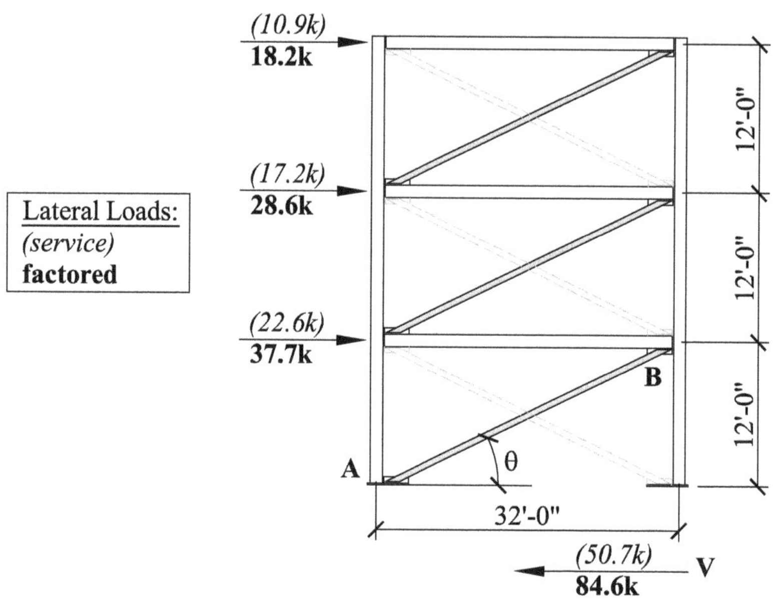

FIGURE 4-17 Detail for Example 4-9

Solution

The service load factor for wind is 0.6 (from the ASCE 7 ASD load combinations presented in Chapter 2).

Loads to Each Level:

Factored Lateral Loads:	Service Lateral Loads:
$P_r = 18.2$ kips	$(0.6)(18.2) = 10.9$ kips
$P_3 = 28.6$ kips	$(0.6)(28.6) = 17.2$ kips
$P_2 = 37.7$ kips	$(0.6)(37.7) = 22.6$ kips

V_u (base shear) $= 18.2 + 28.6 + 37.7 = 84.5$ kips

(For use in the ASD method, the unfactored base shear is

V_s (unfactored base shear) $= 10.9 + 17.2 + 22.6 = 50.7$ kips

$$\theta = \tan^{-1}\left(\frac{12 \text{ ft.}}{32 \text{ ft.}}\right) = 20.6°$$

$$T_{AB} = \frac{V_u}{\cos\theta} = \frac{84.5}{\cos 20.6} = 90.3 \text{ kips for LRFD}$$

For ASD, $T_{AB} = \frac{V_s}{\cos\theta} = \frac{50.7}{\cos 20.6} = 54.2$ kips

Bolts and welds will be covered in Chapters 9 and 10, but for now, assume the shear strength of A325N bolts in single shear to be as follows:

- $\phi R_n = 17.9$ kips for $\frac{3}{4}$-in.-diameter bolt
- $\phi R_n = 24.3$ kips for $\frac{7}{8}$-in.-diameter bolt

1. The required gross area, $A_g \geq \frac{P_u}{0.9F_y}$ using the LRFD method

   (Allow for additional 5% to 10% pretension. Assume 7.5% pretension.)

   $$A_g \geq \frac{(90.3)(1.075)}{0.9(36)} = 2.99 \text{ in.}^2 \text{ (LRFD)}$$

   Note: If the ASD method were to be used, the gross area required due to tensile yielding failure would be

   $$A_g \geq \frac{1.67P_s}{F_y} = \frac{(1.67)(54.2)(1.075)}{(36)} = 2.70 \text{ in.}^2 \text{ (ASD)}$$

   We will now proceed with using only the LRFD method for this example. The reader is encouraged to also work through the example using the ASD method.
   Try $\frac{1}{2}$-in. × 6-in. plate, $A_g = (0.5)(6) = 3.0$ in.^{2} > A_g required (for tensile yielding failure mode) = 2.99 in.^{2} OK.

2. Number of bolts:

   Five $\frac{3}{4}$-in.-diameter bolt: $\phi R_n = (5)(17.9) = 89.5 \text{ kips} < 90.3 \text{ kips}$

   Five $\frac{7}{8}$-in.-diameter bolt: $\phi R_n = (5)(24.3) = 112.5 \text{ kips} > 90.3 \text{ kips}$

   Six $\frac{3}{4}$-in.-diameter bolt: $\phi R_n = (6)(17.9) = 107.4 \text{ kips} > 90.3 \text{ kips}$

   Therefore, try five $\frac{7}{8}$-in.-diameter bolts in a single row parallel to the tension load.
   The shear lag factor, U, is 1.0 for plates connected with bolts (from Table 4-1), indicating that there is no shear lag in this case and the tension stress distribution is uniform across the tension member. For tensile fracture or rupture failure mode, the required gross area using the LRFD method is

   $$A_g \geq \frac{P_u}{0.75F_uU} + \sum A_{\text{holes}}$$

   $$A_g \geq \frac{(90.3)(1.075)}{(0.75)(58)(1.0)} + (1 \text{ hole})\left(\frac{7}{8} + \frac{1}{8}\right)(0.5 \text{in.}) = 2.73 \text{ in.}^2 < 3.0 \text{ in.}^2 \text{ OK}$$

3. From step 1, the required gross cross-sectional area, A_g = 2.99 in.^{2}, and from step 2, A_g = 2.73 in.^{2}, both are less than the actual gross area of the trial member size, which is 3.0 in.^{2}.

4. Check the slenderness ratio:

$$L = \sqrt{(12)^2 + (32)^2} = 34.2 \text{ ft.}$$

$$I_{\min} = \frac{bh^3}{12} = \frac{(6)(0.5)^3}{12} = 0.0625 \text{ in.}^2$$

$$r_{\min} = \sqrt{\frac{I_{\min}}{A_g}} > \frac{L}{300}$$

$$r_{\min} = \sqrt{\frac{0.0625}{3}} \geq \frac{(34.2)(12)}{300}$$

$= 0.144 < 1.37$; therefore, pretension of diagonal brace is required.

The slenderness limit is exceeded, and the assumption that the X-brace diagonal tension member is pretensioned is justified.

5. Check the block shear strength.

The bolt spacing and configuration shown in Figure 4-18 will be used.

Shear planes:

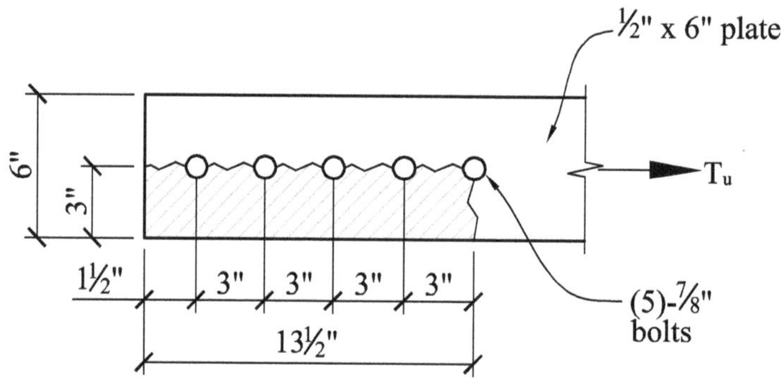

FIGURE 4-18 Block shear failure mode in $^1/_2$" × 6" plate tension menber

$$A_{gv} = (13.5 \text{ in.})(0.5 \text{ in.}) = 6.75 \text{ in.}^2$$

$$A_{nv} = A_{gv} - \Sigma A_{\text{holes}}$$

$$= 6.75 - (4.5\ holes)\left(\frac{7}{8} + \frac{1}{8}\right)(0.5 \text{ in.}) = 4.5 \text{ in.}^2$$

Tension planes:

$$A_{gt} = \left(\frac{6 \text{ in.}}{2}\right)(0.5 \text{ in.}) = 1.5 \text{ in.}^2$$

$$A_{nt} = A_{gt} - \Sigma A_{\text{holes}}$$

$$= 1.5 - (0.5 \text{ holes})\left(\frac{7}{8} + \frac{1}{8}\right)(0.5) = 1.25 \text{ in.}^2$$

The design tension strength for the block shear failure mode is

$$\phi P_n = \phi(0.60F_u A_{nv} + U_{bs} F_u A_{nt}) \le \phi(0.60F_y A_{gv} + U_{bs} F_u A_{nt})$$

$$\phi P_n = 0.75[(0.60)(58)(4.5) + (1.0)(58)(1.25)]$$

$$\le 0.75\ [(0.60)(36)(6.75) + (1.0)(58)(1.25)]$$

171 kips $\le$ 163 kips

Therefore, $\phi P_n = 163$ kips (block shear capacity) $> T_u = 90.3$ kips OK

Use a 6-in. by $\frac{1}{2}$-in. plate with five (5) $\frac{7}{8}$-in.-diameter A325N bolts.

EXAMPLE 4-10

Design of a Single-Angle Tension Member (LRFD Method)

Using the LRFD method, design a tension member given the following parameters:

- Service loads: P_D = 40 kips, P_L = 66 kips;
- Single angle required;
- Unbraced length, L = 20 ft.;
- ASTM A36 steel; and
- Two lines of four $\frac{3}{4}$-in.-diameter bolts.

Solution

$$P_u = 1.4\ P_D = (1.4)(40) = 56 \text{ kips}$$

$$P_u = 1.2\ P_D + 1.6\ P_L = [(1.2)(40)] + [(1.6)(66)] = 154 \text{ kips (controls)}$$

1. $A_g \ge \dfrac{P_u}{0.9F_y}$ $\left(\text{assume slenderness ratio, } \dfrac{K\ell}{r} < 300\right)$

$$A_g \ge \frac{(154)}{(0.9)(36)} = 4.75 \text{ in.}^2$$

2. Shear lag factor, U, is 0.80 for single angles (from Table 4-1). Alternatively, U may be calculated using $\ell = (3)(3 \text{ in.}) = 9$ in. (three spaces between bolt holes at 3 in. on centers), but an angle size would have to be assumed.

$$A_g \ge \frac{P_u}{0.75F_u U} + \Sigma A_{holes}$$

$$A_g \ge \frac{(154)}{(0.75)(58)(0.80)} + (2 \text{ holes})\left(\frac{3}{4} + \frac{1}{8}\right)(t)$$

$A_{g\ required} = 4.43 + 1.75t$, where t is the thickness of the angle

$$r_{min} = \frac{L}{300} = \frac{(20)(12)}{300} = 0.80 \text{ in.}$$

Summary of angle selection

t	A_g Required		Selected Angle	r_z
	A_g (Step 1)	A_g (Step 2)		
$\frac{1}{4}$ in.	4.75 in.^{2}	4.86 in.^{2}	None worked	–
$\frac{5}{16}$ in.	4.75 in.^{2}	4.97 in.^{2}	None worked	–
$\frac{3}{8}$ in.	4.75 in.^{2}	5.08 in.^{2}	None worked	–
$\frac{7}{16}$ in.	4.75 in.^{2}	5.19 in.^{2}	L8 × 6 × $\frac{7}{16}$ (wt. = 20.2 lb./ft.)	1.31 in.
$\frac{1}{2}$ in.	4.75 in.^{2}	5.30 in.^{2}	L6 × 6 × $\frac{1}{2}$ (wt. = 19.6 lb./ft.)	1.18 in.
			L8 × 4 × $\frac{1}{2}$ (wt. = 19.6 lb./ft.)	0.863 in.

3. Select L8 × 4 × $\frac{1}{2}$ because of its lighter weight; also, it would have greater block shear capacity than the L6 × 6 × $\frac{1}{2}$ (same weight).
4. The minimum slenderness ratio calculated = 0.8 in.;
   r_{zz} for L 8 × 4 × ½ = 1.18 in. > 0.8 in. OK.
5. Check block shear capacity. The spacing of the bolts will have to be assumed. See Figure 4-19 for the assumed bolt layout.

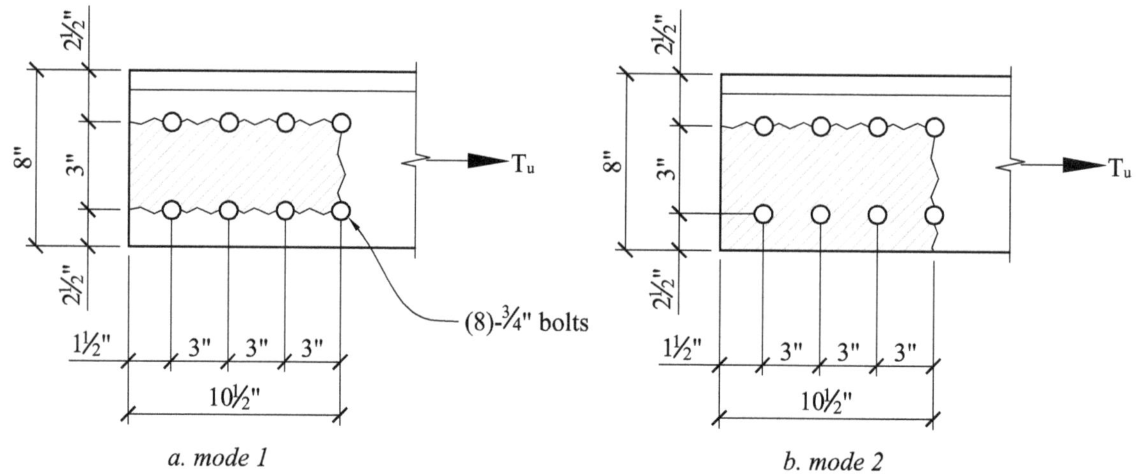

FIGURE 4-19 Block shear failure mode in L8 × 4 × ½ tension menber

Block Shear Failure Mode 1:

Shear plane:

$$A_{gv} = (2\,shear\ planes)(10.5)\ (0.5) = 10.5\ \text{in.}^2$$

$$A_{nv} = A_{gv} - \Sigma A_{\text{holes}}$$

$$= 10.5 - (2\ shear\ planes)(3.5\ \text{holes})\left(\frac{3}{4}+\frac{1}{8}\right)\ (0.5\ \text{in.}) = 7.43\ \text{in.}^2$$

Tension plane:

$$A_{gt} = (3\ \text{in.})(0.5\ \text{in.}) = 1.5\ \text{in.}^2$$

$$A_{gt} = (3\ \text{in.})(0.5\ \text{in.})$$

$$= 1.5 - (2)(0.5\ hole)\left(\frac{3}{4}+\frac{1}{8}\right)(0.5\ \text{in.})\ (0.5\ \text{in.}) = 1.06\ \text{in.}^2$$

The design tension strength for the block shear failure mode is

$$\phi P_n = \phi(0.60F_u A_{nv} + U_{bs} F_u A_{nt}) \le \phi(0.60F_y A_{gv} + U_{bs} F_u A_{nt})$$

$$= 0.75[(0.60)(58)(7.43) + (1.0)(58)(1.06)]$$

$$\le 0.75[(0.60)(36)(10.5) + (1.0)(58)(1.06)]$$

240 kips $\le$ 216 kips

Therefore, $\phi P_n = 216$ kips (mode #1 block shear capacity) $> T_u = 154$ kips OK

Block Shear Failure Mode 2:

Shear plane:

$$A_{gv} = (10.5 \text{ in.})(0.5 \text{ in.}) = 5.25 \text{ in.}^2$$

$$A_{nv} = A_{gv} - \Sigma A_{\text{holes}}$$

$$= 5.25 - (3.5 \textit{ holes})\left(\frac{3}{4} + \frac{1}{8}\right)(0.5 \text{ in.}) = 3.72 \text{ in.}^2$$

Tension plane:

$$A_{gt} = (3 \text{ in.} + 2.5 \text{ in.})(0.5 \text{ in.}) = 2.75 \text{ in.}^2$$

$$A_{nt} = A_{gt} - \Sigma A_{\text{holes}}$$

$$= 2.75 - (1.5 \textit{ holes})\left(\frac{3}{4} + \frac{1}{8}\right)(0.5 \text{ in.}) = 2.09 \text{ in.}^2$$

The design tension strength for the block shear failure mode is

$$\phi P_n = \phi(0.60F_u A_{nv} + U_{bs} F_u A_{nt}) \le \phi(0.60F_y A_{gv} + U_{bs} F_u A_{nt})$$

$$= 0.75[(0.60)(58)(3.72) + (1.0)(58)(2.09)]$$

$$\le 0.75[(0.60)(36)(5.25) + (1.0)(58)(2.09)]$$

= 188 kips $\le$ 175 kips

Therefore, $\phi P_n = 175$ kips (*mode #2 block shear capacity*) $> T_u = 154$ kips OK

The block shear capacity is the smaller of the block shear capacities for block shear failure modes 1 and 2. Therefore, $\phi P_n = 175$ kips.

Select an $L8 \times 4 \times \frac{1}{2}$ with two lines of four (4) $\frac{3}{4}$-in.-diameter bolts.

4-4 TENSION RODS

Threaded rods with circular cross sections are commonly used in a variety of structural applications. Depending on the structural application, tension rods might be referred to as hanger rods or sag rods. Hangers are tension members that are hung from one member to support other members. Sag rods are often provided to prevent a member from deflecting (or sagging) under its own weight, as is the case with girts on the exterior of a building (see Figure 4-1a). Tension rods are also commonly used as diagonal bracing in combination with a clevis and turnbuckle to resist lateral wind and seismic loads. ASTM A36 is the preferred material specification for threaded rods.

There are two basic types of threaded rods, and the more commonly used type is a rod where the nominal diameter is greater than the root diameter at the threaded section (i.e., D > K). The tensile capacity is based on the available cross-sectional area at the root, where the threaded portion of the rod is the thinnest (see Figure 4-20). The second type of threaded rod is one with an upset end, where the root diameter equals the diameter of the rod (i.e., K = D). Upset end tension rods are not commonly used because their fabrication process can be cost-prohibitive.

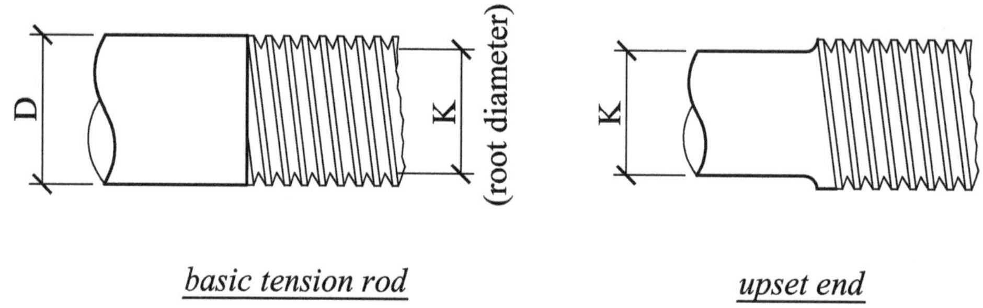

FIGURE 4-20 Basic tension rod and upset end tension rod

As stated previously, the slenderness ratio (L_c/r) of tension members should be no more than 300, but this requirement is not mandatory, and it does not apply to hangers in tension. The AISC specification does not limit the size of tension rods, but the practical minimum diameter of the rod should not be less than $^5/_8$ in. since smaller diameter rods are more susceptible to damage during construction.

The design strength of tension rods is the same as for bolts in tension (see Chapter 9). The design tensile strength and the allowable tensile strength of a tension rod, respectively, are given as follows (see Section J3.7 of the AISC specification):

$$\phi R_n = \phi F_n A_b, \text{ for LRFD} \tag{4-16a}$$

$$\frac{R_n}{\Omega} = \frac{F_n A_b}{\Omega}, \text{ for ASD} \tag{4-16b}$$

where

$\phi = 0.75$,

$\Omega = 2.0$

R_n = Nominal tensile strength of a tension rod = $F_n A_b$

F_n = Nominal tension stress from *AISCM* Table J3.2, and

A_b = Nominal unthreaded body area = $\dfrac{\pi d_b^2}{4}$

d_b = unthreaded diameter of the tension rod (same as dimension "D" in Figure 4-20)

From the *AISCM* Table J3.2,

$$F_n = 0.75 F_u \tag{4-17}$$

The 0.75 factor in equation (4-17) accounts for the difference between the nominal unthreaded body diameter (D) and the diameter of the threaded rod at the root (K), where the stress is most critical. Combining equations (4-16) and (4-17) yields

$$\phi R_n = \phi 0.75 F_u A_b, \text{ for LRFD} \tag{4-18a}$$

$$\frac{R_n}{\Omega} = \frac{0.75 F_u A_b}{\Omega}, \text{ for ASD} \tag{4-18b}$$

The F_u term in the above equations is the minimum tensile stress of the threaded rod. There are several acceptable grades of threaded rods available (*AISCM* Section A3.4 and *AISCM* Table 2-6), the most common of which are summarized in Table 4-3.

When tension rods are used as diagonal bracing, they are commonly used in combination with a clevis at the ends and possibly a turnbuckle to act as a splice for the tension rod (see Figure 4-21). Clevises and turnbuckles are generally manufactured in accordance with AISI C-1035, but the load capacities are based on testing done by various manufacturers (e.g., see Tables 1 and 2 in Ref. [6]). The different grades of threaded rods are given in Table 4-3.

TABLE 4-3 Grades of Threaded Rods

Material specification		Diameter range, in.	F_y, ksi	F_u, ksi
ASTM A36		Up to 15	36	58 to 80
ASTM A193 Gr. B7 *(corrosion-resistant)*		4 to 7	75	100
		2.5 to 4	95	115
		2.5 and under	105	125
ASTM F1554	Grade 36	0.25 to 4	36	58 to 80
	Grade 55	0.25 to 4	55	75 to 95
	Grade 105	0.25 to 3	105	125 to 150

Clevises are designated by a number (No. 2 through No. 8) and have a corresponding maximum diameter associated with each number designation. Each clevis also has a corresponding maximum pin diameter associated with each number designation [6]. The pins are also proprietary and are generally designed to have a capacity equal to or greater than that of the clevis, provided that the diameter of the pin is at least 125% of the diameter of the threaded rod.

Turnbuckles are designated by the diameter of the connecting threaded rod. Alternatively, sleeve nuts can be used in lieu of turnbuckles. Sleeve nuts develop the full capacity of the tensile strength of the threaded rod, provided that the threaded rod has the proper thread engagement. Sleeve nuts are generally manufactured to conform to AISI C-1035. Turnbuckles are usually preferred over sleeve nuts from a cost standpoint. Figure 4-21 shows the previously mentioned connecting elements for tension rods. For cast or forged steel products such as pins, turnbuckles, clevises, and sleeve nuts, the reader should refer to the manufacturer's catalog for load capacity information (e.g., see Tables 1 and 2 in Ref. [6]). Note that the gusset plate that the pins connect to needs to be checked and designed for shear, tension, and bearing.

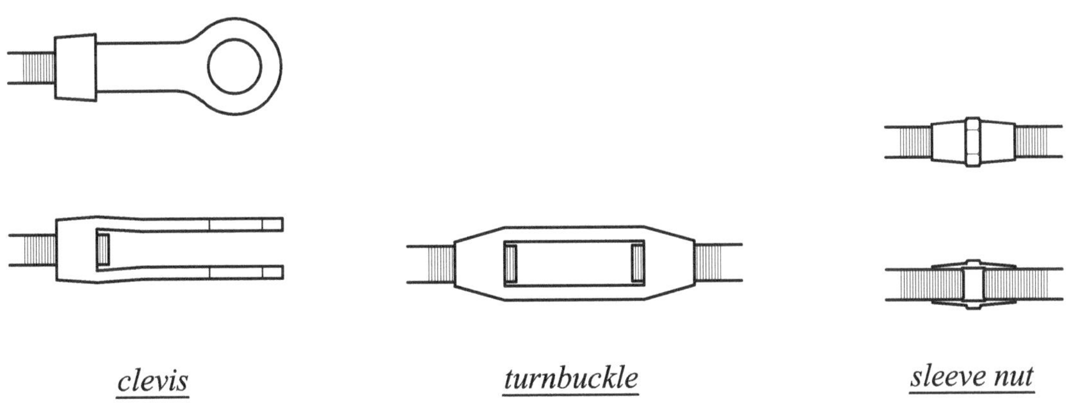

FIGURE 4-21 Connectors for tension rods

Tension members connected with a single pin, as is the case with threaded tension rods used as diagonal bracing, are subject to the failure modes covered in Section D5 of the AISC specification. Pin-connected members differ from a single-bolt connection in that deformation is **not** permitted in the gusset plate at the pin hole so that the pins can rotate freely. For a pin-connected member, the gusset plate needs to be checked for the following failure modes:

- Tensile rupture of the gusset plate on the net area (Figure 4-22a),
- Shear rupture of the gusset plate on the effective area (Figure 4-22b), and
- Bearing on the gusset plate over the projected area of the pin (Figure 4-22c).

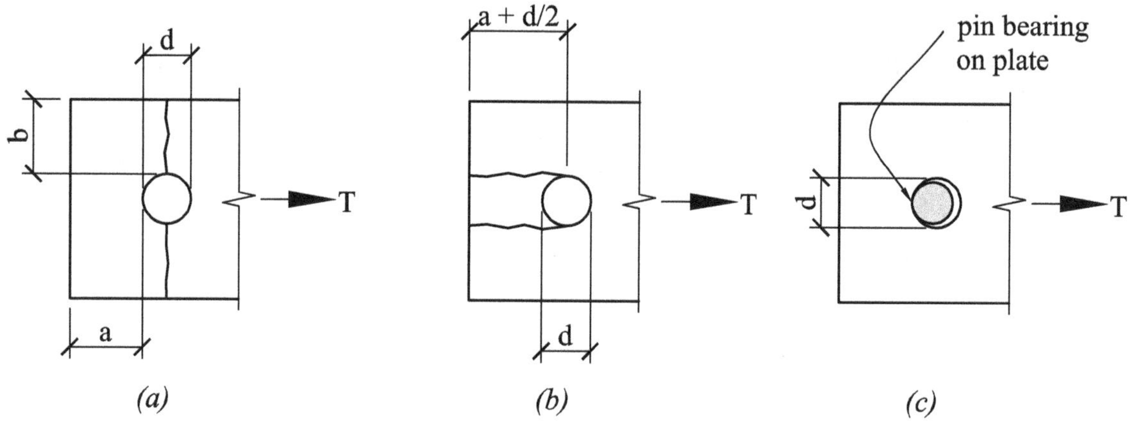

FIGURE 4-22 Failure modes for pin-connected members

The expression for *tensile rupture* on the net effective area is given as

$$\phi P_n = \phi 2tb_e F_u; \text{ for LRFD} \tag{4-19a}$$

$$\frac{P_n}{\Omega} = \frac{2tb_e F_u}{\Omega}; \text{ for ASD} \tag{4-19b}$$

The expression for *shear rupture* on the effective area is given as (see AISC Specification Section D5):

$$\phi P_n = \phi 0.6 C_r F_u A_{sf}, \text{ for LRFD} \tag{4-20a}$$

$$\frac{P_n}{\Omega} = \frac{0.6 C_r F_u A_{sf}}{\Omega}, \text{ for ASD} \tag{4-20b}$$

The expression for bearing on the projected area of pins in reamed, drilled, or bored holes, including **through-bolts** or **through-rods** in unstiffened box sections or HSS members, is given as (see *AISC* Specification Section J7):

$$\phi P_n = \phi 1.8 F_y A_{pb}, \text{ for LRFD} \tag{4-21a}$$

$$\frac{P_n}{\Omega} = \frac{1.8 F_y A_{pb}}{\Omega}, \text{ for ASD} \tag{4-21b}$$

Where,

$\phi = 0.75$,

$\Omega = 2.0$

C_r = reduction factor for shear rupture on pin-connected members is a function of d_h–d

	$d_h - d \le \frac{1}{32}$ *in.*	$\frac{1}{32}$ *in.* $< d_h - d \le \frac{1}{16}$ *in.*
C_r =	1.0	0.95

$A_{sf} = 2t\left(a + \frac{d}{2}\right)$, in.^{2} = shear area on the failure plane

a = Shortest distance from the edge of the pinhole to the member edge measured parallel to the direction of the force,

b_e = $2t + 0.63$ in. $\leq b$,

b = Distance from the edge of the hole to the plate edge perpendicular to the direction of the force,

d = Pin diameter = p (the pin diameter is noted as p in Ref [6]),

d_h = *diameter of pin hole*

t = Plate thickness,

A_{pb} = Projected bearing area (i.e., $A_{pb} = dt = pt$),

A_{sf} = Area on the shear failure path,

F_y = Minimum yield stress, and

F_u = Minimum tensile stress.

For pin-connected members, there are also dimensional requirements for the gusset plate that must be satisfied in accordance with AISC Specification Section D5.2, and these are indicated below and are shown in Figure 4-23. Note that the edges of the gusset plate can be cut 45°, provided that the distance from the edge of the bolt hole to the cut is not less than the primary edge distance.

$$a \geq 1.33 b_e \tag{4-22}$$

$$w \geq 2b_e + d \tag{4-23}$$

$$c \geq a \tag{4-24}$$

For pins with diameters less than 3 inches, the diameter of the pin hole should not be more than 1/32-inch greater than the pin diameter; for pins with diameters of 3 inches or greater, the diameter of the pin hole should not be more than 1/16-inch greater than the pin diameter.

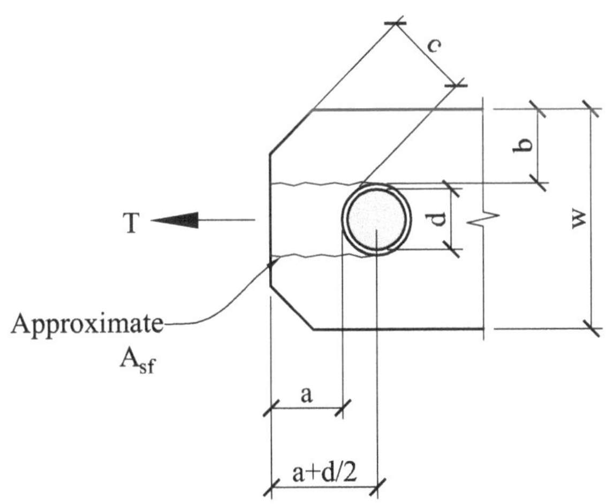

FIGURE 4-23 Dimensional requirements for pin-connected members

EXAMPLE 4-11

Tension Rod Design (LRFD method)

1. For the braced frame shown in Figure 4-24, design the threaded rod X-bracing to resist the applied lateral load shown. The threaded rod conforms to ASTM A36 and assumes $d_{pin} = 1\frac{3}{4}$-in. = 1.75 in.
2. Determine if the gusset plate connection is adequate. Assume the plate is ASTM A36 steel.

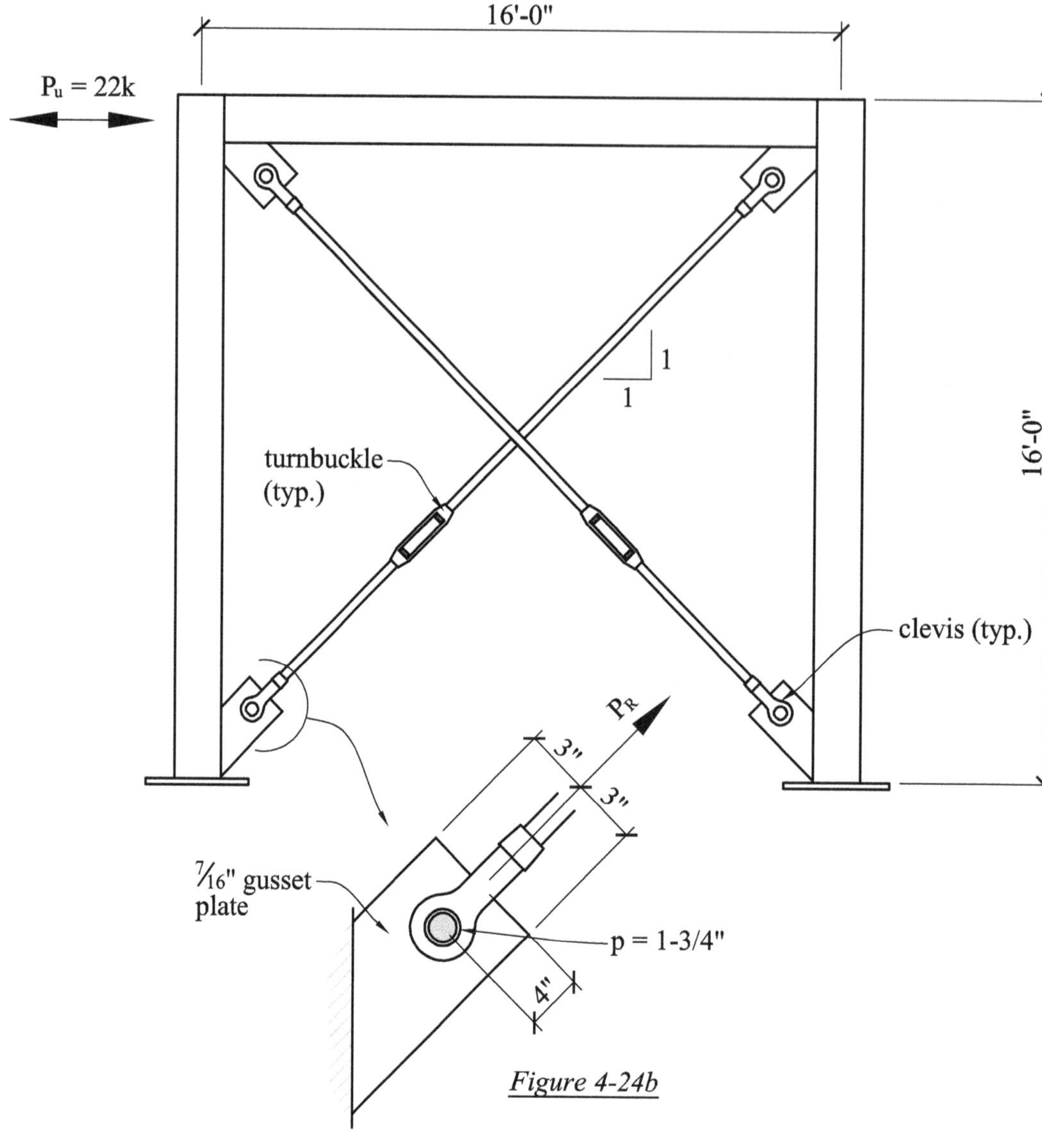

FIGURE 4-24 Detail for Example 4-11

Solution

Load to each tension rod: *(Note: Only one tension rod is engaged when the lateral load is applied since the threaded rod is too slender to support axial compression loads.)* We will assume that the slenderness ratio, L_c/r, is greater than 300 and therefore requires pretensioning; a pretension force of 10% of the applied factored load on the tension member is assumed.

$$P_{u,rod} = \frac{(22\,\text{kips})(1.10)}{\cos 45} = 34.3\text{ kips}$$ (factored load on the tension rod and connectors, including a 10% pretension force)

Unfactored tension load $P_s = 0.6P_u$ (using the ASD load combinations from Section 2-3)
$= 0.6(34.3\text{ kips}) = 21\text{ kips}$

From a forged steel product manufacturer's catalog (see Ref. [6], for an unfactored rod tension load of 21 kips, the maximum pin diameter (using AISC factor of safety of 3 from Table 1 of Ref. [6] and noting the rated capacity of 25 kips which is greater than the unfactored load demand of 21 kips) is
p = 1¾-in., and the corresponding clevis is Clevis No. 3 (see Table 1 of Ref. [6])

The turnbuckle corresponding to Clevis No. 3 should be used.
From Table 2 of Ref [6], the threaded rod diameters that can be used with pin diameter, p = 1¾ in. and Clevis No. 3 are: 7/8 in., 1-in., 1-1/8in., or 1¼ in.
Next, select the threaded rod diameter from those above that can resist the factored tension load on the diagonal brace.
From the *AISCM* Table 2-6, the threaded rod yield and ultimate strengths are
F_y = 36 ksi
$F_u = 58$ ksi to 80 ksi (use $F_u = 58$ ksi)

1. **Check the 1¼-in. threaded rod:**

$$A_b = \frac{\pi(1.25)^2}{4} = 1.23 \text{ in.}^2 \ (A_b \textit{ is also tabulated in AISCM} \text{ Table 7-2})$$

$$\phi R_n = \phi 0.75 F_u A_b$$

$$= (0.75)(0.75)(58)(1.23) = 40.1 \text{ kips} > 34.3 \text{ kips}$$

Use a $1\frac{1}{4}$-in. threaded rod with a No. 3 clevis and a turnbuckle.

2. **Check tensile rupture on the net effective area:**

The pin diameter, $p = d = 1.75$ *in.*

The pin hole diameter = p + 1/8 = 1.75 + 1/8 = 1.875 in.
The distance from the edge of the bolt hole to the plate edge, perpendicular to the direction of the applied force is

$$b = \frac{(3+3)-(1.875)}{2} = 2.06 \text{ in.}$$

$$b_e = 2t + 0.63 \text{ in.} \le b$$

$$= (2)(7/16) + 0.63 = 1.5 \text{ in.} \le b = 2.06 \text{ in.},$$

Therefore, use $b_e = 1.5$ in.

$$\phi P_n = \phi 2 t b_e F_u$$

$$= (0.75)(2)(7/16)(1.5)(58) = 57.1 \text{ kips} > 34.3 \text{ kips OK}$$

3. **Check shear rupture on the effective area of the gusset plate:**

The shortest distance from the edge of the pin hole to the edge of the gusset plate, parallel to the direction of the force, is

$$a = 4 - \frac{1.75 + \frac{1}{8}}{2} = 3.06 \text{ in.}$$

$$A_{sf} = 2t\left(a + \frac{d}{2}\right)$$

$$= (2)(7/16)\left(3.06 + \frac{1.75}{2}\right) = 3.44 \text{ in.}^2$$

Assume $d_h - d \le \frac{1}{32}$ *in.*, therefore C_r = 1.0

$\phi P_n = \phi 0.6 C_r F_u A_{sf}$

$= (0.75)(0.6)(1.0)(58)(3.44) = 89.9 \text{ kips} > 34.3 \text{ kips, OK}$

4. **Check bearing on the projected area of the pin:**

$\phi P_n = \phi 1.8 F_y A_{pb} = \phi 1.8 F_y pt$

$= (0.75)(1.8)(36)(1.75)(7/16) = 37.2 \text{ kips} > 34.3 \text{ kips OK}$

5. **Check the dimensional requirements for the gusset plate:**

$a \geq 1.33 b_e$

$a = 3.06 \text{ in.} \geq 1.33(1.5) = 2 \text{ in.}, \quad OK$

$w \geq 2b_e + d$

$w = (3+3) = 6 \text{ in.} \geq (2)(1.5) + 1.75 = 4.75 \text{ in.} \quad \text{OK}$

$c \geq a$, is not applicable.

Therefore, the 7/16-in. gusset plate is adequate.

Entrance Canopy Structures

There are certain unique issues with beams that support entrance canopy structures (see Figure 4-41) due to net wind uplift loads; the loads for which canopy-support beams should be designed are as follows:

1. Deign the canopy beams for these downward loads: 1.2D + (1.0S or 1.6R or 1.6Lr). The beam reactions will be resisted by the support at A and by the tension rod at B.
2. Design the canopy beam and their supports for the net wind uplift loads: 0.9D+1.0W. This net uplift wind load may result in a compression force in the diagonal tension rod, which the tension rod is incapable of resisting.

   Where, W = upward wind load = -ve

   D = dead load = +ve
3. To mitigate this issue, Beam AB should be connected to support A with a moment-resisting connection; therefore, Beam AB should be designed as a cantilever to resist the net wind uplift load. The diagonal tension rod is active and is in tension for downward loads (i.e., dead, snow or roof live load or rain, and downward-acting wind loads), but it will be inactive and in compression for net wind uplift loads.

References

[1] American Institute of Steel Construction. 2002. *Steel Design Guide Series 17: High-Strength Bolts—A Primer for Structural Engineers*. Chicago, IL: AISC.

[2] American Institute of Steel Construction. 2006. *Steel Design Guide Series 21: Welded Connections—A Primer for Structural Engineers*. Chicago, IL: AISC.

[3] American Institute of Steel Construction. 2023. *Steel Construction Manual*. 16th ed. Chicago, IL: AISC.

[4] ASCE. 2022. *ASCE 7-22: Minimum Design Loads for Buildings and Other Structures*. Reston, VA: ASCE.

[5] International Codes Council. 2024. *International Building Code—2024*. Falls Church, VA: ICC.

[6] Industrial Hardware and Specialties Incorporated. *https://www.industrialhardware.com/pub/media/knowledge_base/wire_rope_chain_fittings/clevis/clevis_chart-min.pdf* Accessed January 9, 2025

Exercises

4-1. Determine the tensile capacity of the $\frac{1}{4}$-in. × 6-in. plate shown in Figure 4-25. The joint is made with $\frac{5}{8}$-in.-diameter bolts; the plate is ASTM A992, grade 50.

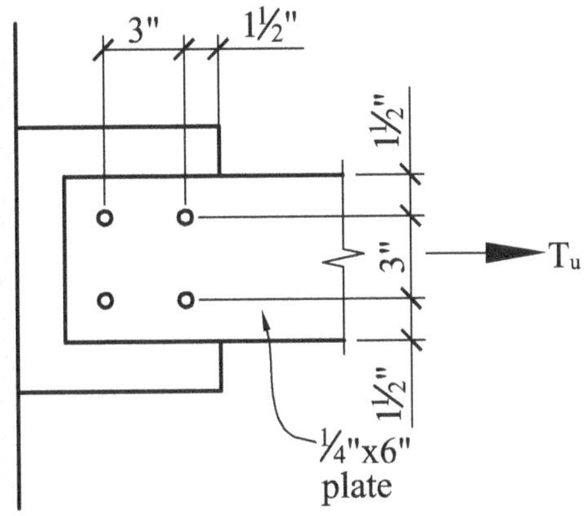

FIGURE 4-25 Detail for Exercise 4-1

4-2. Determine the following for the joint shown in Figure 4-26, assuming the bolts are $\frac{3}{4}$-in.-diameter and the angle and plate is ASTM A36.

a. Tensile capacity of the angle, and
b. Required gusset plate thickness to develop the tensile capacity of the angle determined in part a.

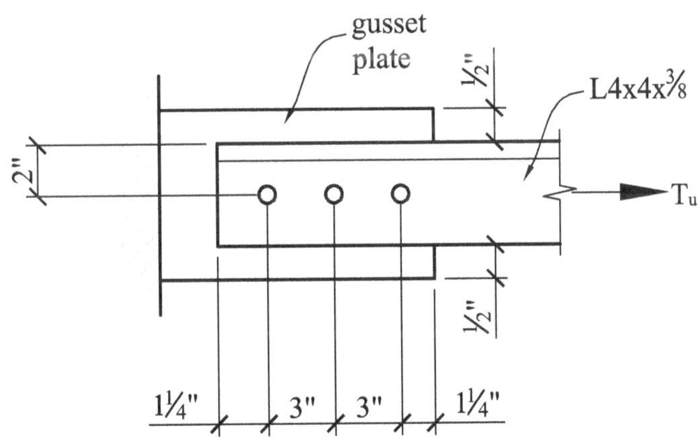

FIGURE 4-26 Detail for Exercise 4-2

4-3. Determine the maximum factored tensile force that can be applied based on the following:

a. Capacity of the angle in tension only
b. Capacity of the angle in block shear (see failure diagram in Figure 4-27).
The angle is ASTM A572, Grade 50.

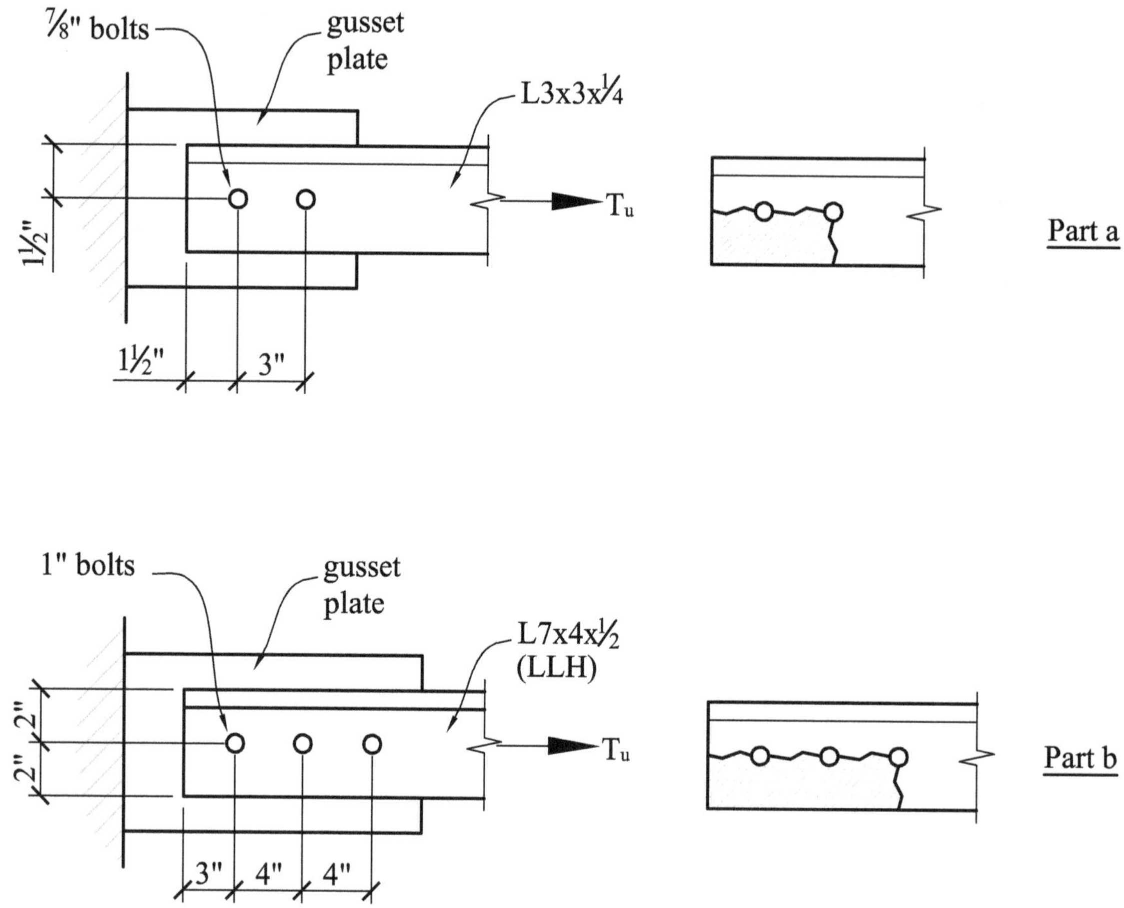

FIGURE 4-27 Details for Exercise 4-3

4-4. Determine the following using ASTM A36 steel (Figure 4-28):

a. Maximum tension force T_u that can be applied based on the tensile capacity of the angles.
b. Maximum tension force T_u that can be applied based on block shear.
c. Based on (a) and (b), if the service dead load is T_D = 50 kips, determine the maximum service live load that can be applied. Consider the load combination 1.2D + 1.6L only.

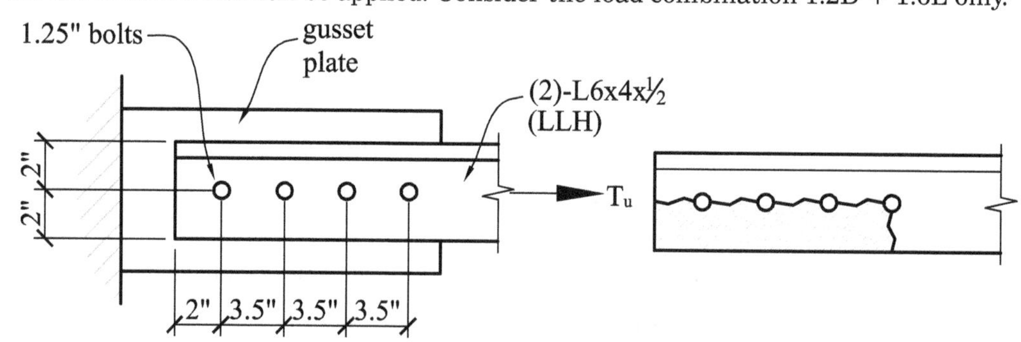

FIGURE 4-28 Details for Exercise 4-4

4-5. Determine the design strength of the $\frac{3}{8}$" × 4" plate shown in Figure 4-29, assuming the steel is ASTM A36.

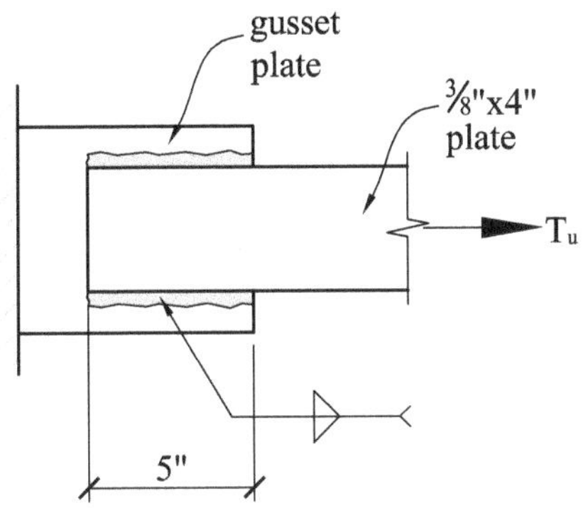

FIGURE 4-29 Detail for Exercise 4-5

4-6. Determine the design strength of the connection shown in Figure 4-30, assuming the steel is ASTM A529, grade 50. The plates are $^3/_8$ in. thick. Neglect the strength of the bolts.

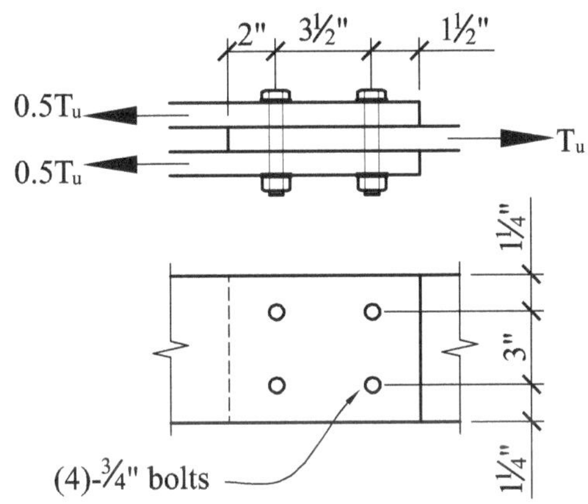

FIGURE 4-30 Detail for Exercise 4-6

4-7. Determine the net area of the members shown in Figure 4-31.

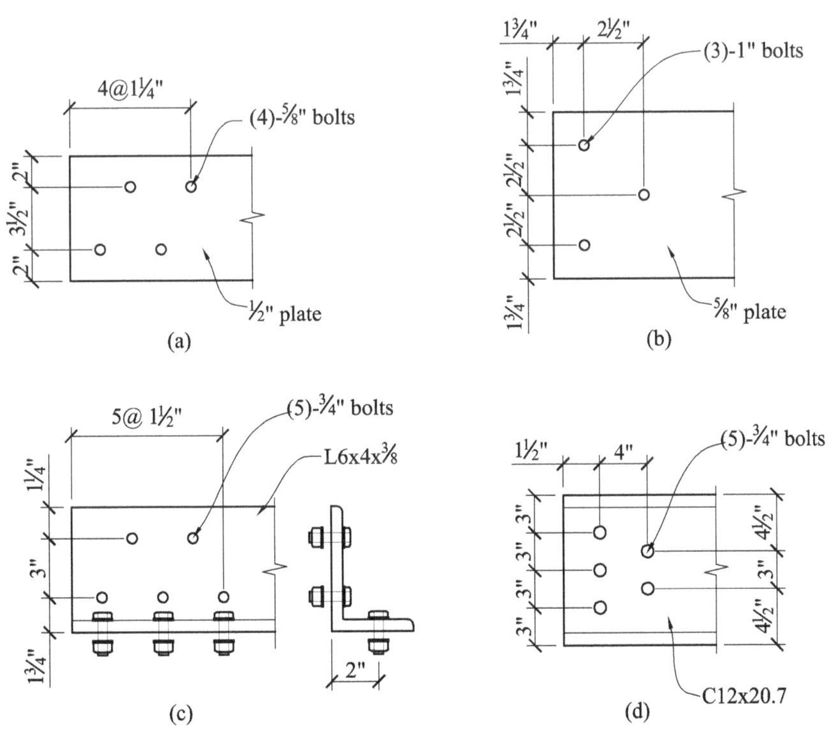

FIGURE 4-31 Detail for Exercise 4-7

4-8. For the two-story braced frame shown, design the following, assuming the threaded rod and the gusset plate conform to ASTM A36 and the clevises and turnbuckles conform to AISI C-1035:

a. Clevis, turnbuckle, and threaded rod at each level, and
b. Gusset plate, assuming a $^3/_8$-in. thickness.

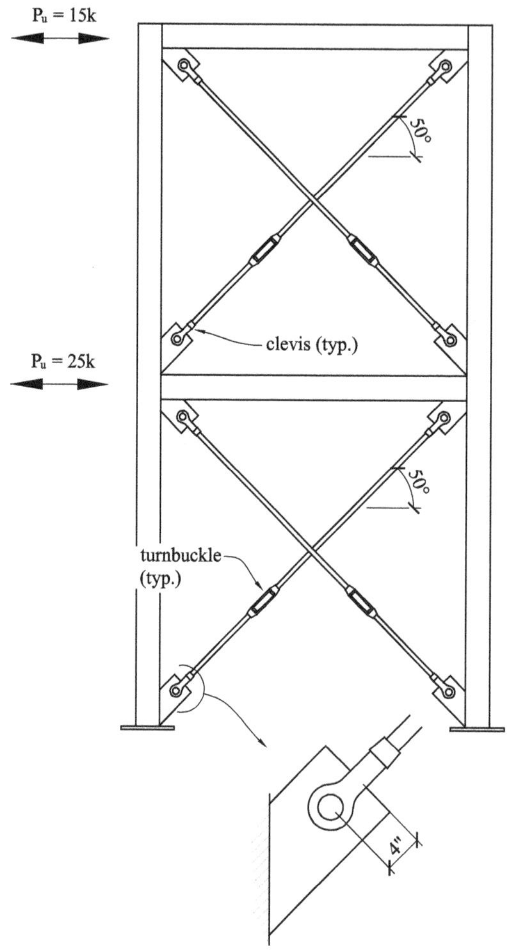

FIGURE 4-32 Detail for Exercise 4-8

4-9. Using the lateral loads and geometry shown in exercise 4-8, complete the following, assuming the tension member is an L4 × 4 × $^3/_8$, ASTM A36, and connected as shown in Figure 4-33:

a. Determine if the L4 × 4 × $^3/_8$ tension member is adequate.
b. Design the gusset plate for the full tension capacity of the angle.

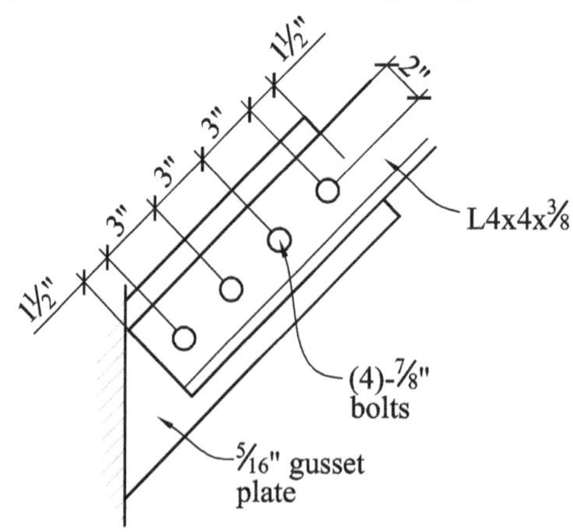

FIGURE 4-33 Detail for Exercise 4-9

4-10. Determine the maximum factored tensile force that can be applied as shown in Figure 4-34 and based on the following:

a. Capacity of the angle in tension only, and
b. Capacity of the angle in block shear.
The angle is ASTM A992, grade 50.

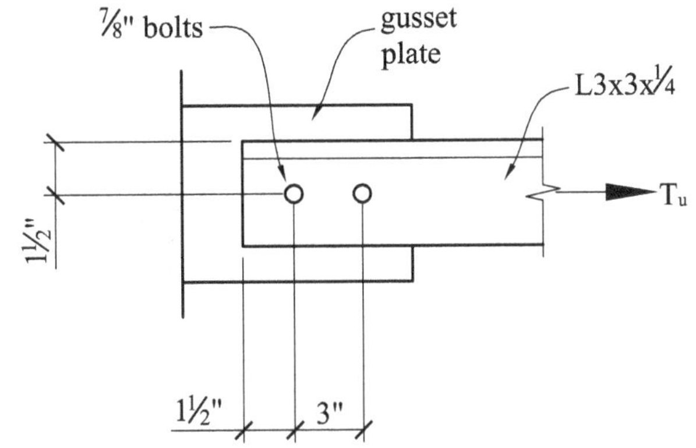

FIGURE 4-34 Detail for Exercise 4-10

4-11. For the truss shown in Figure 4-35, determine if member *CD* is adequate for the service loads shown. The steel is ASTM A36 and assumes ¾-in.-diameter bolts (in double shear). Ignore the strength of the gusset plate.

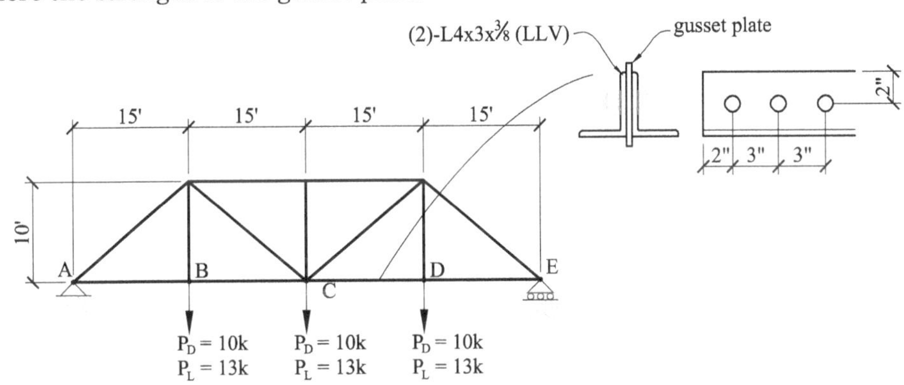

FIGURE 4-35 Detail for Exercise 4-11

4-12. For the canopy support details shown in Figure 4-36

a. Determine the required threaded rod size for member *AB*, assuming ASTM A36 steel for the threaded rod.
b. Determine an appropriate clevis and gusset plate size and thickness, assuming the plate at point *A* is 6 in. wide. Use ASTM A36 steel for the plate. The clevis conforms to AISI C-1035.

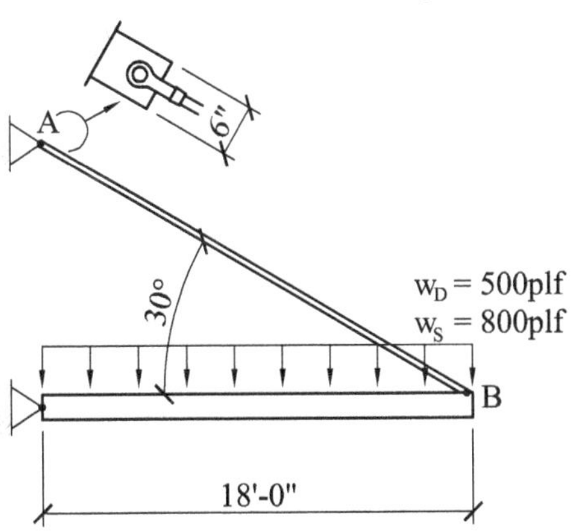

FIGURE 4-36 Detail for Exercise 4-12

4-13. For the tension rod connection shown in Figure 4-37:

a. Find the minimum required diameter of the threaded rod.
b. Determine if the gusset plate is adequate and check the plate dimensions.

The plate is ASTM A36, and the threaded rod is ASTM F1554, grade 55.

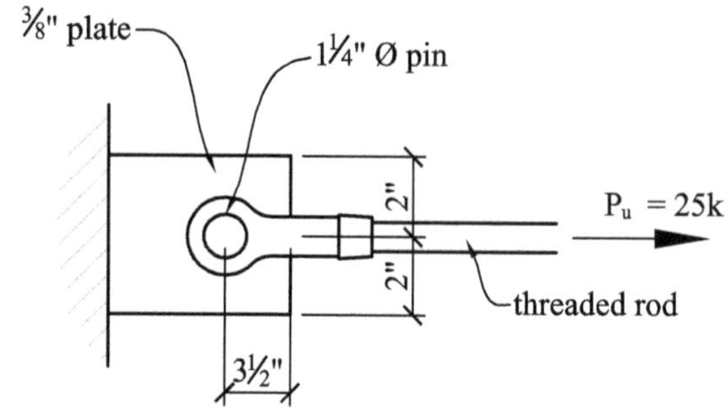

FIGURE 4-37 Detail for Exercise 4-13

4-14. The framing plan in Figure 4-38 shows a steel-framed platform supported at the four corners for gravity loads. Beams B-1 and B-2 frame into G-1, and G-1 is supported at each end by a hanger.

Assuming the total service loads are 150 psf live and 100 psf dead, calculate that maximum factored tension, T_u, and determine if the threaded rod, clevis, and plate are adequate for the applied tension load.

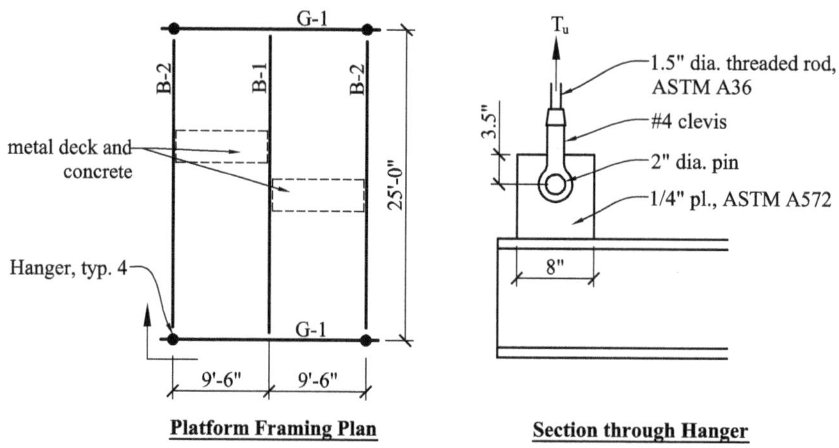

FIGURE 4-38 Details for Exercise 4-14

4-15. Develop a spreadsheet with the following minimum parameters and submit results for a Pipe 4 STD (Figure 4-39):

P_u = Axial tension load
B = Round HSS size (outside diameter)
F_{yB} = Yield strength for the HSS; *use ASTM A53 for this problem*
F_{yp} = Yield strength for the plate; *use ASTM A36 for this problem*
l = Engagement length; *use 5" for this problem*
y = Edge distance; *use 3" for this problem*
t = Plate thickness; *use 1/4" for this problem*

The spreadsheet should be used to check/design the following:

a. The U factor and the maximum axial load that can be applied based on this
b. The maximum axial load that can be applied based on block shear in the plat
c. The maximum axial load that can be applied based on gross tension only on the plate

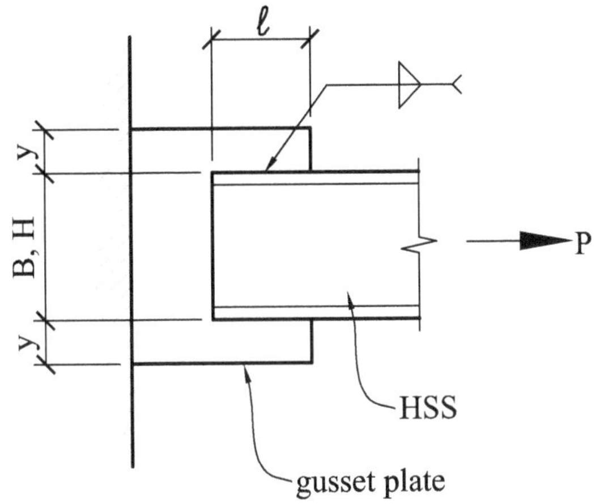

FIGURE 4-39 Details for Exercise 4-15

4-16. Develop a spreadsheet with the following minimum parameters and submit results for an L4 × 4× ½ (see Figure 4-40):

T_u = Axial tension load

Angle parameters: x-bar, t, x (leg length)

F_{yB} = Yield strength for the L; *use ASTM A572 for this problem*

F_{yp} = Yield strength for the plate; *use ASTM A36 for this problem*

d_b = Bolt diameter; *use 1" for this problem*

N = Number of bolts, *2 minimum; use 4 for this problem*

S = Bolt spacing; *use 3" for this problem*

y = Edge distance; *use 2" for this problem*

t = Plate thickness; *use 3/8" for this problem*

Use the spreadsheet to check/design the following:

a. The U-factor and the maximum axial load that can be applied based on this,
b. The maximum axial load that can be applied based on net tension only across the plate,
c. The maximum axial load that can be applied based on block shear in the angle,
d. The number of bolts required so that the joint achieves at least 85% of the gross tension strength of the angle.

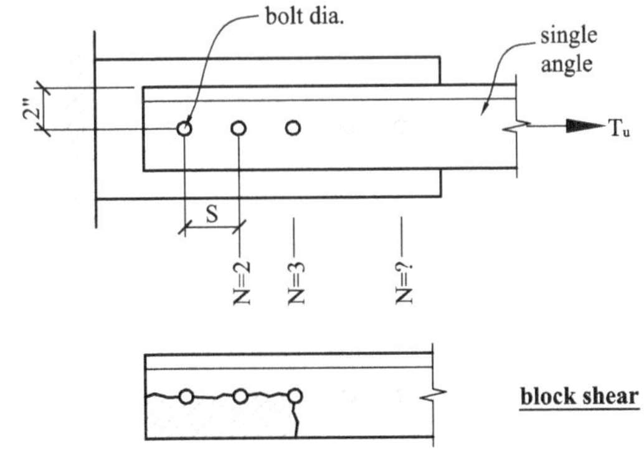

FIGURE 4-40 Details for Exercise 4-16

4-17. Refer to the framing plan and section shown in Figure 4-41. Typical member B1 is supported at the left end at point A and at the right end by tension member BC. The applied service loads to B1 are D = 80 psf (including the weight of the framing and concrete deck) and S = 1 psf. Determine the following considering the load combination 1.2D + 1.0S only:

a. Applied factored uniform load to B1.
b. Factored tension in member BC.
c. Determine if the threaded rod and clevis are adequate.
d. Determine the required plate thickness for P1.

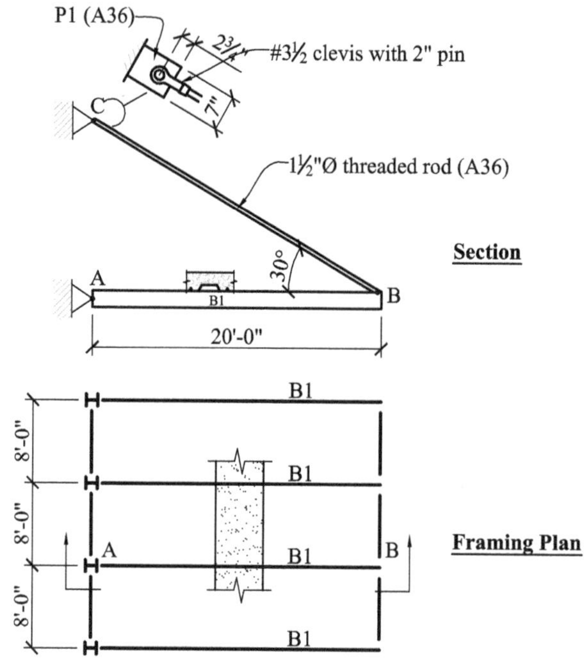

FIGURE 4-41 Details for Exercise 4-17

4-18. Determine the axial tension capacity or strength, ϕP_n, of a **5" wide × 3/8" thick** plate hanger considering only yielding and fracture failure modes (i.e., neglect block shear). Assume the plate is connected to the gusset plate at the ends with a single row of 7/8-inch-diameter bolts. If the hanger supports a service dead load of 12 kips, what is the service floor live load in psf that can be supported by the hanger if the hanger has a tributary floor area of 300 ft^2. Assume the shear lag factor $U = 1.0$ and use ASTM A992 steel.

4-19. For the two hanger configurations shown in **Figure 4-42a (as-designed)** and **Figure 4-42b (as-built)** supporting three floor levels, answer the following questions assuming the *factored* vertical load, P, contributed by the floor beam to the hanger at each floor level is 22 kips.

a. Calculate the load on hanger CD for the **as-designed** condition and select a hanger rod, diameter assuming A36 steel.
b. Calculate the load on hanger CD for the **as-built** condition. Neglect the offset distance between the hangers.
c. Calculate the load on the hanger connection at C (i.e., the nut at C) for the **as-designed** condition.
d. Calculate the load on the hanger connection at C (i.e., the nut at C) for the **as-built** condition. Neglect the offset distance between the hangers.
e. Assuming the floor beams are made up of two channel shapes toe-to-toe, draw an adequate hanger rod-to-box beam connection detail.

f. Assuming ASTM A36 steel, determine the hanger rod diameter that should be specified on the structural drawings to resist the maximum applied loads.

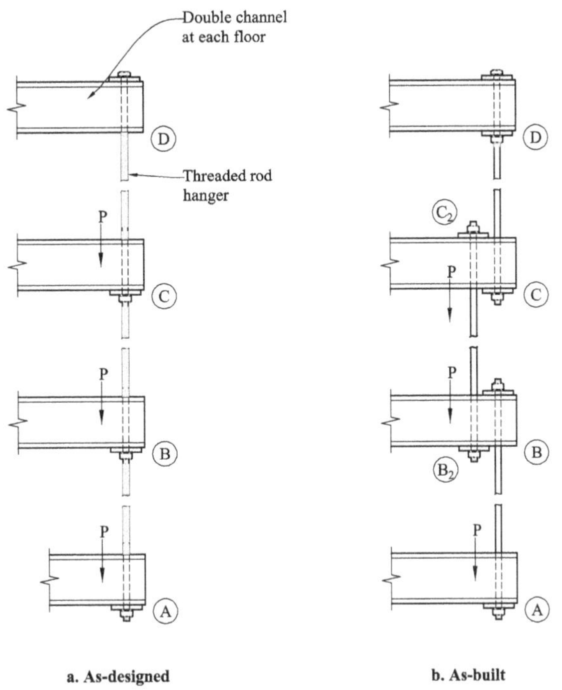

FIGURE 4-42 Details for Exercise 4-19

4-20. Conduct research on the Internet on the 1981 "Hyatt Regency Hotel Collapse" and write a minimum five-page typed report explaining, with the aid of free-body diagrams, the reasons for the collapse, the lessons learned, and ways to prevent similar structural failures in the future. Include in your report at least two different types of hanger rod connection detail that could have prevented the collapse.

4-21. Conduct research on the Internet on "The Fifty-Nine Story Crisis" regarding the Citi-Corp Tower in New York City. Write a minimum five-page typed report explaining—with the aid of drawings, pictures, and free-body diagrams (FBDs)—the reasons for the structural problems and how they were resolved. Discuss the lessons learned from this failure case study.

4-22. Explain why the shear lag factor for axial tension stress for each Case 1 in Figure 4-43 below should be lower than that for each Case 2.

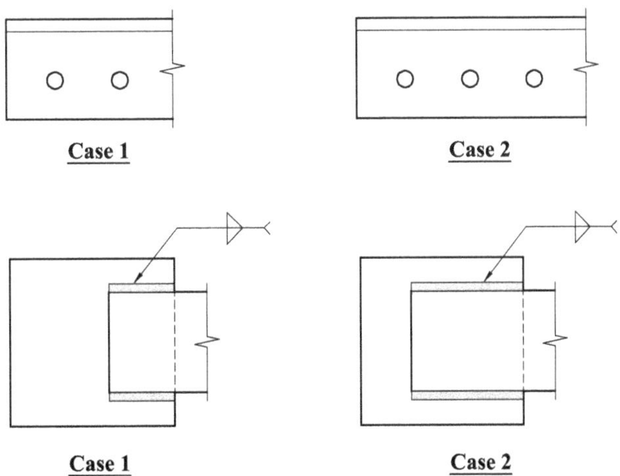

FIGURE 4-43 Details for Exercise 4-22.

4-23

a. Sketch the failure planes on the WT for yielding, rupture, and block shear in Figure 4-44. For block shear, sketch the most logical failure plane for this load.
b. Determine the maximum factored load, T_u, that could be applied based on the strength of the WT in yielding, rupture, and block shear. Steel is ASTM A36

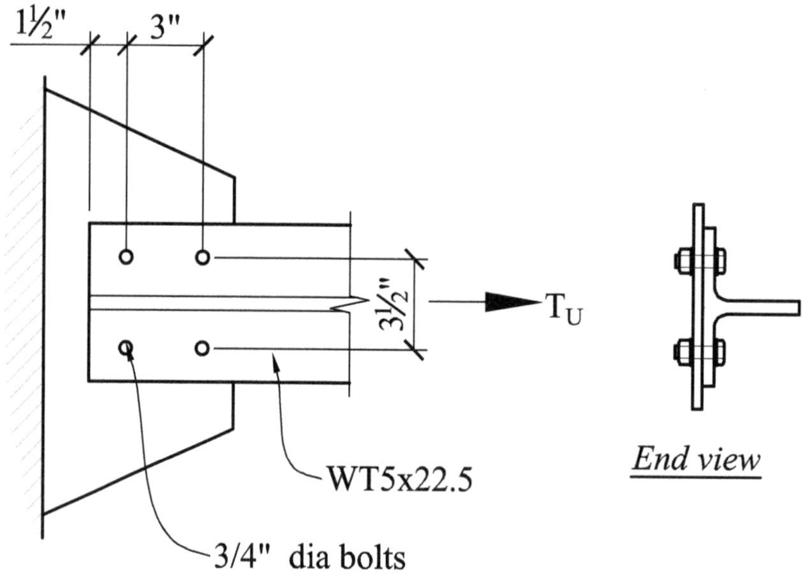

FIGURE 4-44 Details for Exercise 4-23

4-24

a. Sketch the major failure planes for the plate.
b. Determine if the plate is adequate for the applied load based on (a).
c. Determine if the threaded rod, clevis, and turnbuckle are adequate for the applied load.

Plate steel is ASTM A572, grade 50.

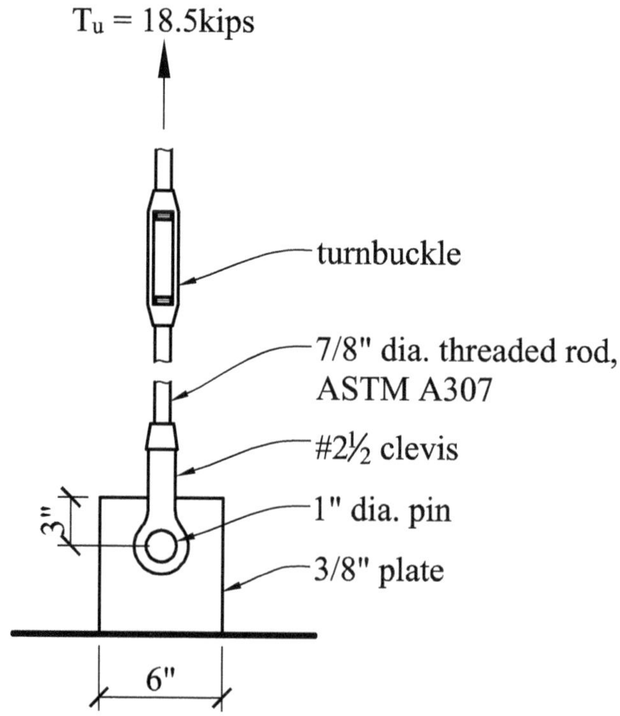

FIGURE 4-45 Details for Exercise 4-24.

4-25. For the steel double plate drag strut in the outrigger-to-concrete core wall connection shown in Figure 4-46, determine the axial tension capacity assuming a 3-in. thick × 24-in. deep plate. Assume ASTM A572 Grade 50. Assume 6 rows of 7/8-in.-diameter bolts in the connection of the plates to the outrigger truss top chord.

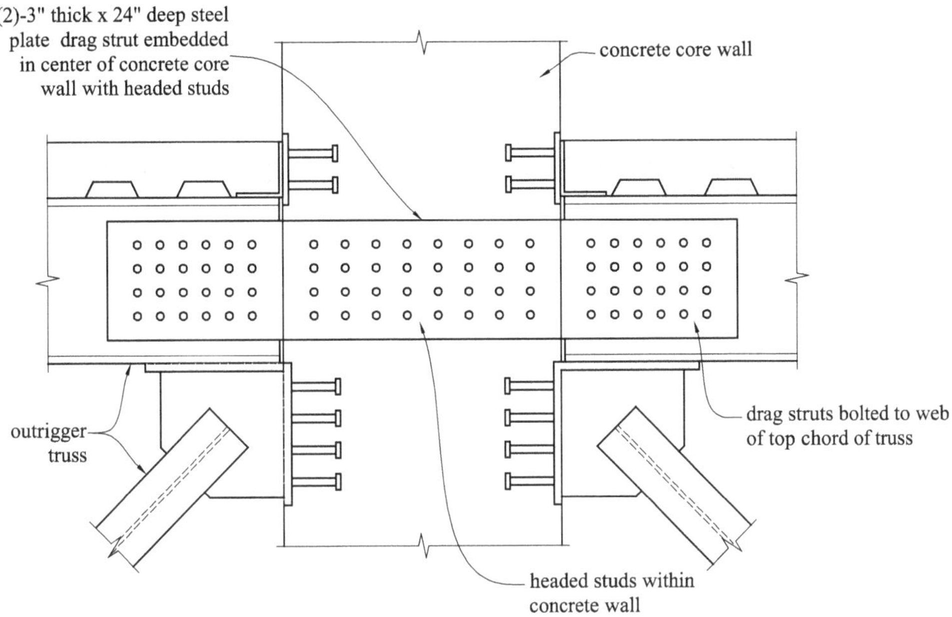

FIGURE 4-46 Steel plate drag strut details for Exercise 4-25.

4-26. Calculate the tension capacity (LRFD) for the L3x3x3/8 hanger in Figure 4-47 connected to the web of the wide flange girder coped at the top flange to make room for the hanger.

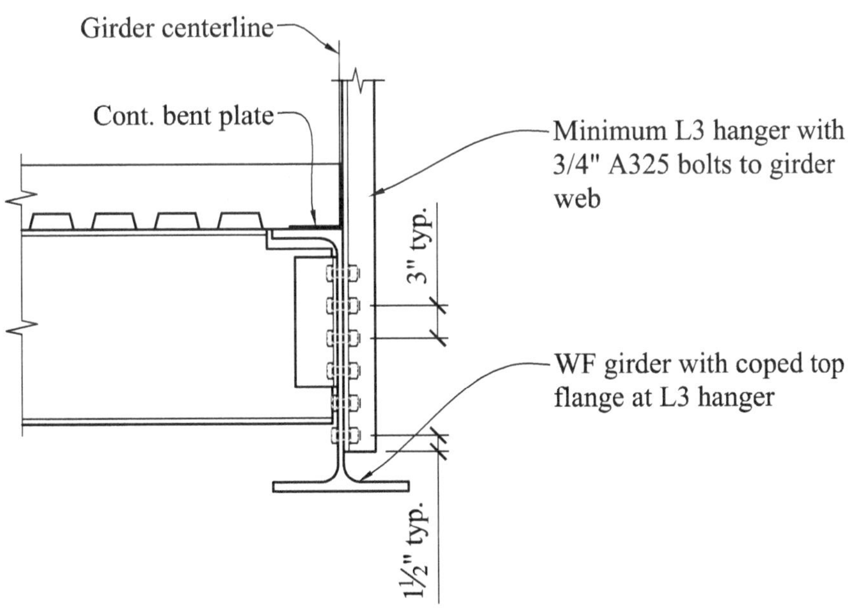

FIGURE 4-47 Steel hanger connection details for Exercise 4-26.

CHAPTER

5

Compression Members Under Concentric Axial Loads

5-1 INTRODUCTION

Structural steel elements are sometimes subjected to concentric compressive axial loads without any accompanying bending moment. Some examples include truss web members, compression chords of some trusses, struts, compression members in concentric and eccentric braced frames such as X-braces and Chevron braces, and some columns in buildings. The web members in trusses resist axial compression and tension forces. Struts—in the form of round HSS or round steel pipes—may be used as horizontal braces to ensure the stability of basement excavations before the basement walls and the basement and ground floor slabs and framing are completed; in addition to the axial compression load, these struts may be subjected to accompanying moments from sagging due to their self-weight. Horizontal struts are also used in buildings to brace inclined or sloped columns. A series of pairs of 4-story-high diagonal struts supporting multi-story columns were used in the award-winning 150 N. Riverside Building in Chicago, designed by Magnusson Klemencic Associates (MKA), to increase the size of the building plan at the upper levels from a much smaller and physically constrained ground floor footprint. The tops of the diagonal struts were braced by tension ties encased within the series of concrete shear walls.

Compression members with light loads—used to support a platform or mezzanine—are called posts. In this chapter, we will cover the analysis and design of structural members subject to concentric axial compression load only, with no accompanying bending moment. In Chapter 8, we will discuss beam columns, or structural steel elements subjected to combined axial compression or tension load plus bending moments, which may occur due to eccentrically applied axial load or transverse loads acting within the length of the member. In structural steel, the common shapes used for columns are wide flange sections (i.e., I-shapes), round and square hollow structural sections (HSS), and built-up sections (e.g., box sections) which are used as columns in high-rise buildings. With the use of high-strength steel (e.g., steel with F_y = 65 or 70 ksi), standard hot-rolled sections as large as W36 × 925 have been used in super-tall buildings. The largest hot-rolled section in the United States is currently W36 × 925 [7]. For truss members, double- or single-angle shapes, as well as round and square HSS and WT-shapes, may be used (see Figure 5-1a(i)). Figures 5-1a(ii) through 5-1a(v) show different structural systems where some structural members may be subjected to concentric axial compression loads. Figure 5-1b shows the struts, tie rods, and columns used to support the roof at an airport entrance canopy. Figures 5-1c(i) to 5-1c(v) also show photos of structural members that may be subjected to axial compression forces.

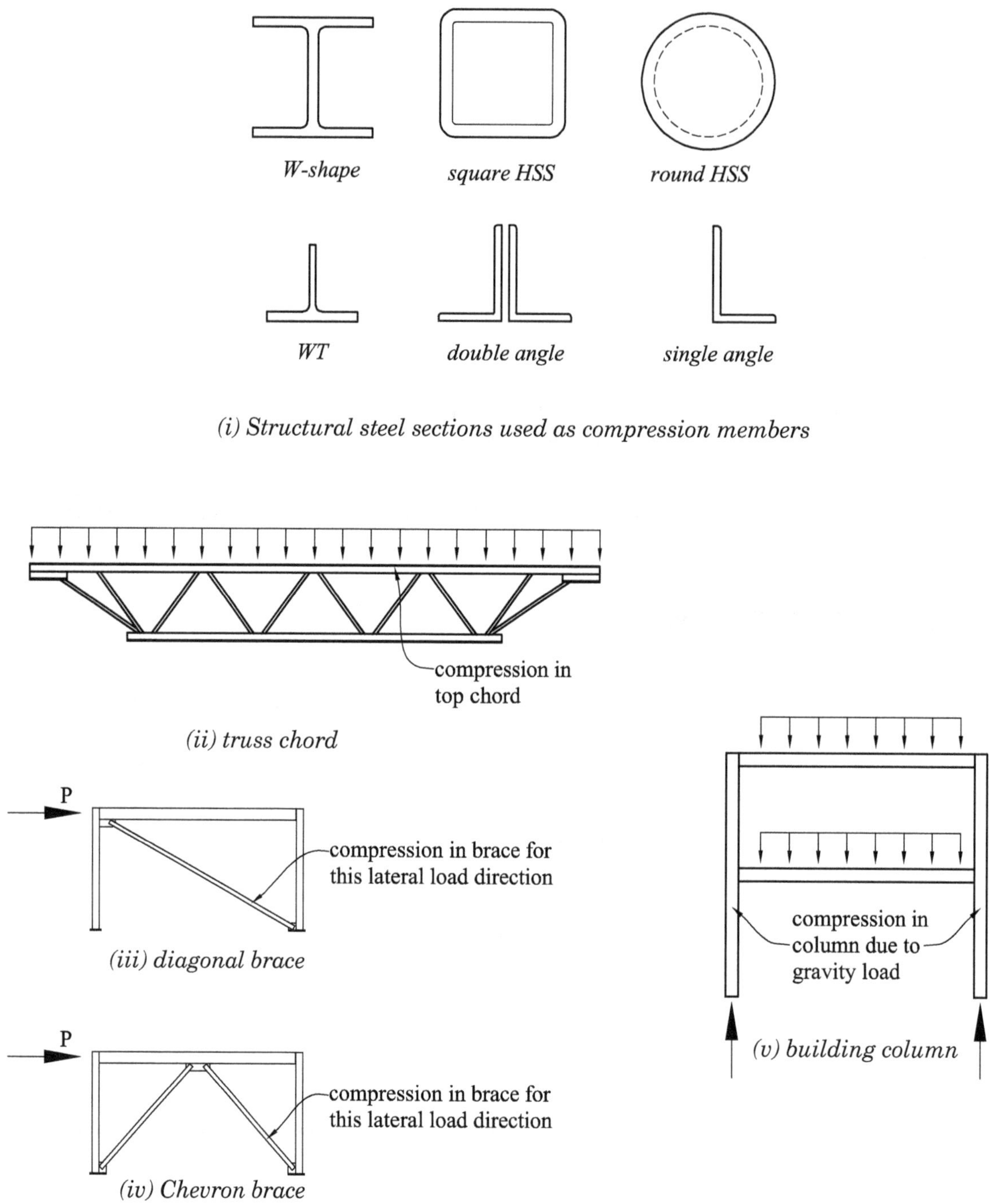

FIGURE 5-1a Basic types of compression members

FIGURE 5-1b Struts, tie rods, and columns at an airport entrance canopy (Photo by Abi Aghayere)

FIGURE 5-1c(i) Wide-flange Chevron braces for an industrial platform in a precast concrete plant

FIGURE 5-1c(ii) Wide-flange Chevron brace connection to a girder with a gusset plate

FIGURE 5-1c(iii) HSS diagonal brace connections to steel girders and columns with gusset plates

FIGURE 5-1c(iv) Achieving direct axial compression load transfer using a sloped stub column

Compression members may fail in one of the following modes (i.e., limit states) [1]:

- **Yielding** of the cross-section in compression, and this occurs at low slenderness ratios.
- **Flexural buckling** is where the member bends globally between lateral supports and buckles about its weaker axis (i.e., the axis with the larger slenderness ratio) because of the presence of the axial compression load. This limit state is usually applicable to compression members comprised of non-slender elements, and the failure mode can be either elastic buckling or inelastic buckling depending on the slenderness ratio of the member. This limit state is covered in AISC Specification Section E3.
  Figure 5-1c(v) shows lateral bracing of the top of the steel column at a deep steel box transfer girder since the girder frames over the top of the steel column rather than framing into the side face of the column.
- **Torsional buckling**, which involves twisting about the longitudinal axis of the member without any lateral displacement. Examples of sections susceptible to torsional buckling include cruciform shapes. See AISC Specification E4.
- **Lateral-torsional or flexural-torsional buckling** is a combination of the above two failure modes and is common in wide flange sections. It is caused by the flexural compression stresses on the compression flange of a beam or column with large unbraced lengths. Torsional and flexural buckling limit states apply to singly symmetric and non-symmetric members and some doubly symmetric members, such as cruciform-shaped columns with non-slender elements. The flexural and torsional buckling limit state for members with non-slender elements is covered in AISC Specification Section E4.
  The first three buckling modes are global buckling modes where the entire length of the compression member is involved.
- **Local buckling** occurs where the component elements of the structural member—such as the web and the flanges—are slender and can buckle locally (i.e., web local buckling and flange local buckling) in contrast to the global buckling that occurs in the first three failure modes discussed previously (see AISC Specification Section E7).

Fig. 5-1c(v) contd.

FIGURE 5-1c(v) Lateral bracing of the top of a steel column supporting a deep steel box transfer girder

5-2 EULER CRITICAL BUCKLING LOAD

The Euler critical buckling load for a pin-ended column is derived in this section. Consider a pin-ended column with a length L subjected to a concentric axial load, P, as shown in Figure 5-2a. The column is assumed to be made of a linearly elastic and homogeneous material and is perfectly straight [2]. As the load, P, on this column is increased, the column remains straight until it fails either by crushing or yielding of the member (material failure) or by buckling (outward bending of the member) or by a combination of these failure mechanisms. For long or slender columns (see Figures 5-2c and 5-2d), failure will occur by buckling.

At the moment of buckling, the lateral deflection of the column is y, which occurs at a distance, x, from the origin at point A (i.e., top of the column). The free body diagram of segment AC of the column is shown in Figure 5-2b.

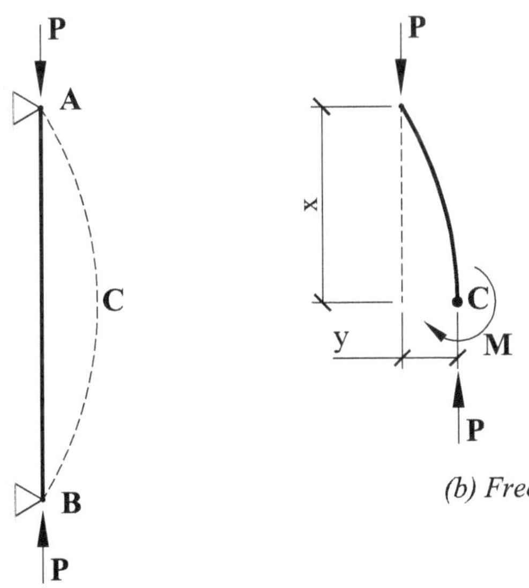

(a) Pin ended column subject to axial compression load

(b) Free body diagram

FIGURE 5-2a-b Pin ended column

FIGURE 5-2c Slender column at an entrance canopy

FIGURE 5-2d Two-story columns in a steel building under construction

Summing moments about point C in Figure 5-2b yields

$$-Py + M = 0$$

Therefore,
$$M = Py \tag{5-1}$$

Assuming small deflections, the relationship between the internal bending moment in the column and its curvature is given by the following differential equation [2, 5, 6]:

$$\frac{d^2y}{dx^2} = -\frac{M}{EI} \tag{5-2}$$

Substituting equation (5-1) into equation (5-2) and rearranging yields the second-order differential equation

$$\frac{d^2y}{dx^2} + \frac{Py}{EI} = 0 \tag{5-3}$$

The solution to equation (5-3) is as follows [3]:

$$y = A_1 \sin\left(\sqrt{\frac{P}{EI}}x\right) + A_2 \cos\left(\sqrt{\frac{P}{EI}}x\right) \tag{5-4}$$

The boundary conditions for the pin-ended column are as follows:

At $x = 0$, the deflection, $y = 0$

At $x =$ L, the deflection, $y = 0$

Substituting the first boundary condition into equation (5-4) means that the constant, $A_2 = 0$, which reduces equation (5-4) to

$$y = A_1 \sin\left(\sqrt{\frac{P}{EI}}x\right) \tag{5-5}$$

Substituting the second boundary condition into equation (5-5) leads to the following equation:

$$0 = A_1 \sin\left(\sqrt{\frac{P}{EI}}L\right)$$

The non-trivial solution to the preceding equation is obtained from

$\sin\left(\sqrt{\frac{P}{EI}}L\right) = 0$, which yields the following solution:

$$\left(\sqrt{\frac{P}{EI}}L\right) = n\pi,$$

Where, n = the number of deflection waves in the column between lateral supports = 1, 2, 3…

Therefore, the solution is $P = \dfrac{n^2\pi^2EI}{L^2}$

The smallest value of n (i.e., n = 1), which indicates a single half-sine wave between the column lateral supports, gives the Euler critical load of the column, P_e,

Where,
$$P_e = \frac{\pi^2EI}{L^2} \tag{5-6}$$

and,

P_e = Euler critical buckling load, lb.,

E = Modulus of elasticity (for steel, E = 29 × 10^6 psi),

I = Moment of inertia, in.^{4}, and

L = Length of the column between lateral supports or brace points, in.

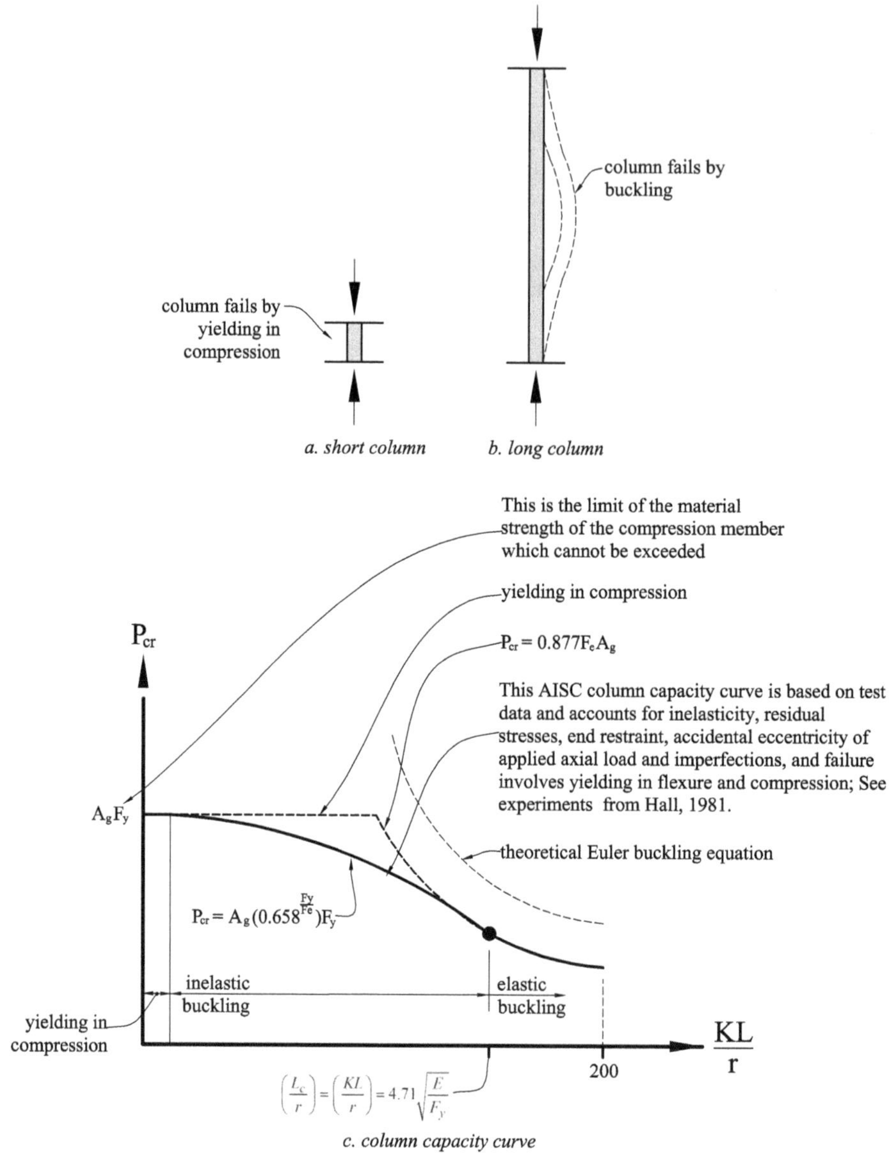

FIGURE 5-3 Column failure modes and capacity curves

Equation (5-6) gives the theoretical Euler critical buckling load for a homogeneous concentrically loaded member, that is, the load at which a perfectly straight compression member will start to bow outwards. This equation was originally derived by the Swiss mathematician Leonhard Euler in 1757. [2, 6]. It assumes a perfectly straight, long, and slender compression member without any initial crookedness or imperfections and no residual stresses. In reality, if the column is sufficiently slender (i.e., a *long column* with a high slenderness ratio), it will buckle elastically (i.e., the column will return to its original shape if the axial compression load is removed); since the column is in the linear elastic range, the modulus of elasticity for this case is the linear elastic modulus, E. Conversely, if the column is very stocky (i.e., a *short column* with a small slenderness ratio), it may not buckle at all, but the cross-section may fail in material yielding at the yield stress, F_y. Somewhere between these two slenderness extremes is the case of the *intermediate column* that has some intermediate slenderness which causes the column to buckle inelastically (i.e., there will be a permanent lateral deformation of the column, and it will not return to its original longitudinal profile when the axial compression load is removed). In this case, the column will be in the non-linear or inelastic range and will have a reduced modulus of elasticity (the tangent modulus) due to the inelastic action. The Euler equation has to be modified to account for the inelasticity. The behavior of slender and stocky axially loaded members is illustrated in Figure 5-3.

In Figure 5-3*a*, the column is short enough that failure occurs by yielding in compression. This is called a *short column*. For the longer column shown in Figure 5-3*b*, failure occurs by a combination of flexure and compression due to the initial imperfections in the member.

It should be noted that the Euler critical buckling load is a “theoretical abstraction” since every real-world structural member will have initial imperfections (i.e., out-of-plumbness or out-of-straightness) and will fail by a combination of flexure and compression due to the initial imperfections—even though the member is subjected to a concentric compression load.

From the mechanics of materials, the moment of inertia is $I = Ar^2$, and the compression stress on any member is $f_c = P/A$ [2]. Therefore, the Euler critical buckling stress can be obtained from equation (5-6) as

$$F_e = \frac{P_e}{A} = \frac{\pi^2 E}{\left(L/r\right)^2}, \tag{5-7}$$

where

F_e = Euler elastic critical buckling stress, psi,

A = Cross-sectional area, in.^{2}, and

r = Radius of gyration, in.

Note that the Euler elastic critical buckling stress is independent of the yield strength and the ultimate tensile strength of the steel member. Equations (5-6) and (5-7) assume that the ends of the column are pinned. For other end conditions, an adjustment or effective length factor, K, is applied to the column length. Thus, the actual column is replaced by an equivalent pin-ended column with an effective length, KL. The effective length of a column, KL, is also denoted as L_c in the AISC Specifications (i.e., $L_c = KL$), and the effective length factor, K, is usually determined by one of two methods as follows:

	(a)	(b)	(c)	(d)	(e)	(f)
Buckling modes	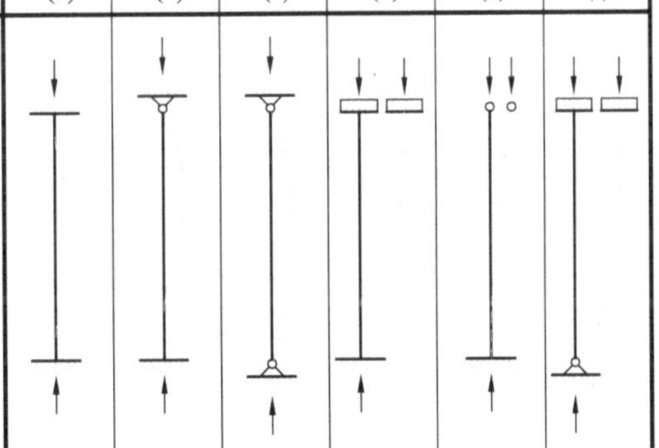					
Theoretical K value	0.50	0.70	1.0	1.0	2.0	2.0
Recommended design K when ideal conditions are approximated	0.65	0.80	1.0	1.2	2.1	2.4
End condition legend	Rotation fixed, translation fixed Rotation free, translation fixed Rotation fixed, translation free Rotation free, translation free					

FIGURE 5-4 Effective length factors, *K*

Adopted from *AISCM* Table C-A-7.1. Copyright ã American Institute of Steel Construction.

1. ***AISCM*, Table C-A-7.1**—The *recommended design values* for the effective length factor from this table are commonly used in design practice to determine the effective lengths of columns because the theoretical values assume idealized end support conditions. This table, reproduced in Figure 5-4, is especially useful for preliminary design when the sizes of the beams, girders, and columns are still unknown. In Figure 5-4, cases *(a)* through *(c)* represent columns in *braced frames* (i.e., buildings or structures laterally braced by X-braces, Chevron braces, or concrete, masonry, steel plate, or steel-concrete composite shear walls), while cases *(d)* through *(f)* represent columns in *unbraced frames* (i.e., buildings or structures laterally braced by moment or "rigid" frames where the lateral load is resisted by the bending resistance of the beam/girders and the columns and their connections). It should be noted that for building columns supported at the top and bottom ends, it is common in design practice to assume that the column is pinned at both ends, resulting in a practical effective length factor, K, of 1.0. For columns fixed at both ends, the recommended design value for the effective length factor, K, is 0.65.
2. **Nomographs or alignment charts (*AISCM*, Figures C-A-7.1 and C-A-7.2)**—The alignment charts use the actual restraints at the girder-to-column connections to determine the effective length factor, K. They provide more accurate values for the effective length factor than *AISCM*, Table C-A-7.1, or Figure 5-4, but the process for obtaining these values is more tedious than the first method, and the alignment charts can only be used if the sizes of the columns and girders are known. The alignment charts (see Figures 5-10 and 5-11) will be discussed later in this chapter.

Equations (5-6) and (5-7) assume that there are no slender elements in the compression member and thus preclude any local buckling. When the column end conditions are other than pinned, equations (5-6) and (5-7) can be modified as follows:

$$P_e = \frac{\pi^2 EI}{(KL)^2} = \frac{\pi^2 EI}{(L_c)^2} \tag{5-8}$$

$$F_e = \frac{\pi^2 E}{(KL/r)^2} = \frac{\pi^2 E}{(L_c/r)^2} \tag{5-9}$$

Equation (5-9), which corresponds to *AISCM* equation E3-4, is the elastic buckling stress that accounts for the end support conditions of the compression member. The term KL/r or L_c/r is called the slenderness ratio, and the AISC Specification recommends limiting the column slenderness ratio such that

$$\frac{KL}{r} = \frac{L_c}{r} \leq 200 \text{ for compression members.} \tag{5-10}$$

Although this slenderness ratio limit is not mandatory (*AISCM* Section E2), it should be adhered to because the load capacity equations in the AISC specifications were obtained empirically and $\frac{KL}{r}$ *or* $\frac{L_c}{r} = 200$ is also the cutoff point for the *AISCM* design tables for compression members. The KL/r limit of 200 was based on the assumption that in the fabrication and erection of steel members, they are handled with care so as not to induce any unintended damage or imperfections in the member [1]. The poor handling of a steel member during fabrication, delivery, and erection could induce unintended imperfections that could further reduce the axial load capacity of the column section, hence the need to adhere to the KL/r limit of 200 for compression members.

Braced Versus Unbraced Frames

In using the alignment charts, or Table C-A-7.1 of the *AISCM*, it is necessary to distinguish between *braced* and *unbraced frames*. *Braced frames* exist in buildings where the lateral loads are resisted by diagonal bracing or shear walls, as shown in Figure 5-5*a*. The beams and girders in braced frames are usually connected to the columns with simple shear connections, and thus there is very little moment restraint at these beam-to-column or girder-to-column connections. The ends of columns in braced frames are assumed to have comparatively small relative lateral sway; therefore, the term *nonsway-* or *side-sway-inhibited* is used to describe braced frames. The effective length factor for columns in braced frames is taken as 1.0. On the other hand, in *unbraced* or *moment frames* (Figure 5-5*b*), the lateral loads are resisted through the bending of the beams, girders, and columns, and thus the girder-to-column and beam-to-column connections are designed as moment connections. In contrast to braced frames, the ends of the columns in unbraced frames undergo relatively appreciable side-sway or lateral deflection; therefore, the term *sway* or *sway-uninhibited* is used to describe unbraced frames. The effective length of columns in moment frames is usually greater than 1.0.

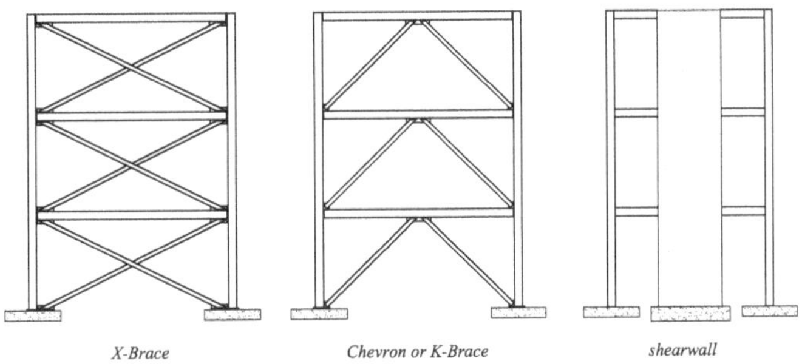

a) braced frames

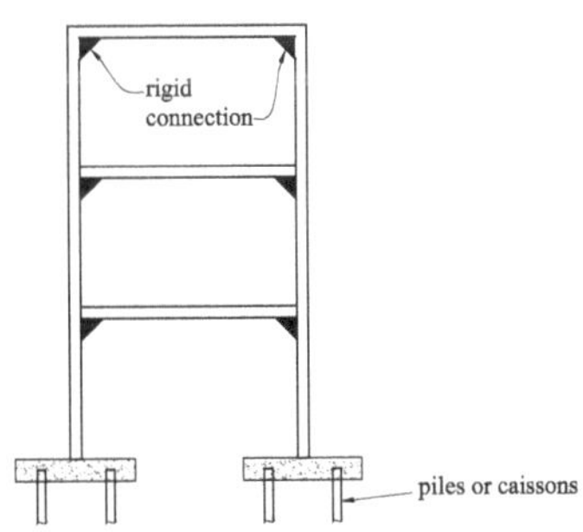

b) unbraced frames

FIGURE 5-5a-b Braced and unbraced frames

EXAMPLE 5-1

Determination of Effective Length Factor, *K*, using Figure 5-4 or the *AISCM* Table C-A-7.1

Determine the effective length factor for the ground floor columns in the following frames (see Figure 5-6).

Solution

a. Braced Frame

Since the building is braced by diagonal braces and shear walls, the K value for all columns in the building is assumed to be 1.0 (i.e., Case "c").

b. Unbraced Frames (Moment Frame with Pinned Column Bases)

Since the bottom ends of the ground floor columns are pinned, the effective length factor, K, for each column at this level in the moment frame is 2.4 (i.e., Case "f").

c. Unbraced Frames (Moment Frame with Fixed Column Bases)

Since the bottom ends of the ground floor columns are fixed, the effective length factor, K, for each column at this level in the moment frame is 1.2 (i.e., Case "d").

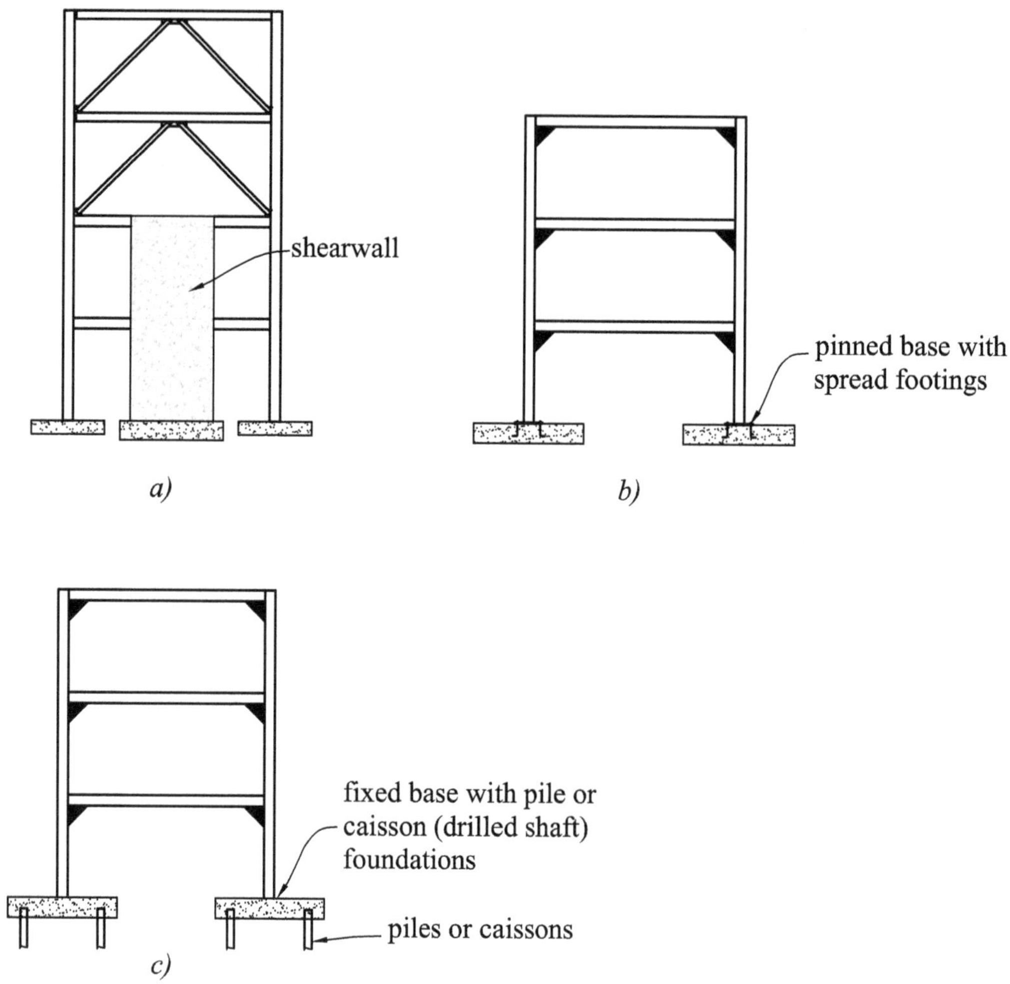

FIGURE 5-6 Details for Example 5-1

5-3 COMPRESSION MEMBER STRENGTH

The assumptions used in the derivation of equations (5-6) and (5-7) consider idealized support conditions that cannot be achieved in real-life structural members. The factors that affect the strength of columns in real-life structures include initial crookedness or geometric imperfections in the column, the presence of residual stresses resulting from welding or hot-rolling of the column section, the shape of the column, and the end support conditions in the compression member. To account for these factors, the AISC Specification Section E3 defines the design compressive strength of a column for the flexural buckling limit state as follows:

The nominal axial compression strength is given as

$$P_n = F_n A_g, \tag{5-11}$$

Where,
P_n = Nominal axial compression strength, kips,
F_n = Nominal stress (see below), ksi
A_g = Gross cross-sectional area of the column, in.^{2}
For the LRFD method, the design axial compressive strength is given as

$$\phi_c P_n = \phi_c F_n A_g, \tag{5-12}$$

For the ASD method, the allowable axial compressive load is given as

$$P_a = \frac{P_n}{\Omega_c} = \frac{F_n A_g}{\Omega_c} \tag{5-13}$$

Where,
ϕ_c= strength reduction factor for compression = 0.90,
Ω_c= factor of safety for compression = 1.67
The slenderness ratio at which the behavior of the compression member changes from inelastic buckling to elastic buckling is

$$\frac{KL}{r} = \frac{L_c}{r} = 4.71\sqrt{\frac{E}{F_y}}$$

- Therefore, the member buckles *elastically* when $\frac{KL}{r} = \frac{L_c}{r} > 4.71\sqrt{\frac{E}{F_y}}$ (i.e., for long columns), and
- The member buckles *inelastically* when $\frac{KL}{r} = \frac{L_c}{r} \leq 4.71\sqrt{\frac{E}{F_y}}$ (i.e., for short and intermediate columns).

The AISC nominal stress, F_n, is determined as follows:
When $\frac{KL}{r} = \frac{L_c}{r} \leq 4.71\sqrt{\frac{E}{F_y}}$ $\left(\text{or when } F_y \leq 2.25F_e \ or \ F_e \geq 0.44F_y\right)$, the compression member buckles *inelastically*, and the nominal stress is

$$F_n = \left[0.658^{\frac{F_y}{F_e}}\right] F_y \tag{5-14}$$

When $\frac{KL}{r} = \frac{L_c}{r} > 4.71\sqrt{\frac{E}{F_y}}$ $\left(\text{or when } F_y > 2.25F_e \ or \ F_e < 0.44F_y\right)$, the compression member buckles *elastically*, and the nominal stress is

$$F_n = 0.877F_e \tag{5-15}$$

Equation (5-14), which corresponds to AISC equation E3-2, accounts for the case where inelastic buckling dominates the column behavior because of the presence of residual stresses in the member and the stockiness of the column, while equation (5-15), which corresponds to AISC equation E3-3, accounts for elastic flexural buckling in long or slender columns. Both critical flexural buckling stress equations have been derived from extensive research on columns and the fitting of the equations to experimental data from column tests [1]. The equations account for residual stresses, initial imperfections, end support conditions, and the shape of the compression member. Equations (5-14) and (5-15) consider global flexural buckling of the column but do not account for local buckling of the component elements of the compression member.

5-4 LOCAL BUCKLING OF COMPRESSION MEMBERS

The strength equations derived in the preceding section were based on the global buckling of the entire compression member and assumes that local buckling of the plate elements of the column section does not occur. In this section, we will look at the local stability of the individual rectangular plate elements that make up a compression member. Local buckling (see Figure 5-7) due to the slenderness of the component elements of a compression member leads to a reduction in the strength of the member and prevents it from reaching its full axial compression load capacity.

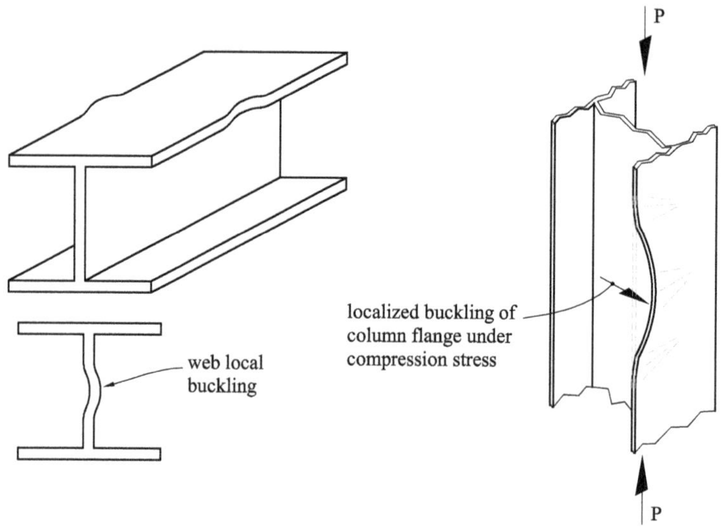

FIGURE 5-7 Local buckling of column under axial compression load

To avoid or prevent local buckling, the AISC specification prescribes limits to the width-to-thickness ratios (λ_p and λ_r) of the plate elements that make up the compression member. These limits are given in Section B4 of the *AISC Specification*.

In Section B4 of the *AISC Specification*, compression members subject to *axial compression* can be classified regarding local stability as follows (see Table 5-1 or *AISCM* Table B4.1a):

- *Nonslender* elements, or
- *Slender* element sections

For compression members subject to *flexure or bending*, there are three possible classifications regarding local stability (see Table 6-2 or *AISCM* Table B4.1b):

- Compact section
- Noncompact section
- Slender-element section.

A *compact* section reaches its cross-sectional material strength before local buckling occurs. In a *noncompact* section, only a portion of the cross-section reaches its yield strength before local buckling occurs. In a *slender-element* section (i.e., sections with slender elements), the cross-section does not yield, and the strength of the member is governed by local buckling. The use of slender sections as compression members is not efficient or economical; therefore, the authors do not recommend their use in design practice, if avoidable.

There are also two types of elements that make up a column section, and these are defined in the *AISCM* as *stiffened* and *unstiffened* elements. Stiffened elements are supported along both edges parallel to the applied axial load. An example of a stiffened element is the web of an I-shaped column because the web is connected to the flanges at both ends of the web. An unstiffened element has only one supported edge parallel to the axial load with the other end unsupported; for example, the outstanding flange of a wide flange or I-shaped column that is connected to the web on one edge and free or unsupported along the outer edge is an unstiffened element (see Figure 5-8). Other examples of sections consisting only of unstiffened elements include angles, tee sections, and cruciform sections. All the components of hollow structural sections (HSS) are stiffened elements.

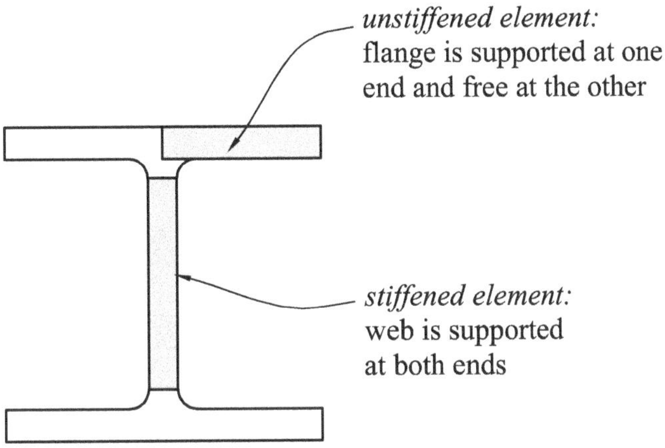

FIGURE 5-8 Stiffened and unstiffened elements

The width-to-thickness ratios, λ, of compression elements affect their axial compression buckling strength. The limiting width-to-thickness ratio, λ_r, of the stiffened and unstiffened elements of different column sections subject to axial compression that determines whether a column element is classified as a *nonslender* or *slender* compression element are given in Table 5-1, which is adopted from *AISCM* Table B4.1a. Compression members subject to axial compression and having width-to-thickness ratios, λ, greater than the limiting values, λ_r, from Table 5-1 are classified as slender-element sections and should be designed taking into account the strength reductions due to local buckling that are specified in AISC Specification Section E7. Note that for the elements of wide flange sections, the width-to-thickness ratio, λ, is h/t_w for the web and $b_f/2t_f$ for the flange. The presence of slender elements leads to a reduced effective area, A_e, of the column cross-section. The determination of the axial load capacity of compression members with slender elements in accordance with AISC Specification Section E7 will be covered later in this Chapter. For compression members with nonslender elements subject to axial compression (i.e., column elements with $\lambda \leq \lambda_r$), there is no reduction in the cross-sectional area, A_g, and equations (5-14) and (5-15) are used directly.

For compression elements in members subject to *flexure*, the limiting width-to-thickness ratios for *compact*, *noncompact*, and *slender* sections are given in Table 6-2 or *AISCM* Table B4.1b (see Chapter 6). Flexural members with compression elements having width-to-thickness ratios less than or equal to λ_p, from Table 6-2 or *AISCM* Table B4.1b, are classified as *compact* sections—these sections can fully develop their plastic moment capacity, M_p. Flexural members with compression elements that have width-to-thickness ratios greater than λ_p but less than λ_r, from Table 6-2 or *AISCM* Table B4.1b, are classified as *noncompact* sections. These sections are only able to develop partial yielding in the compression elements before buckling occurs. Flexural members with any compression elements that have width-to-thickness ratios greater than λ_r are classified as slender-element sections—these elements will buckle elastically before reaching their yield stress. Note that flange local buckling and/or web local buckling reductions have to be considered in the design of noncompact and slender-element sections for bending in accordance with *AISCM* Chapter F.

The effect of local buckling in the compression zone on the flexural or moment capacity of steel sections will be covered in Chapter 6 [1].

TABLE 5-1 Limiting Width-to-Thickness Ratios for Compression Elements Subject to Axial Compression*

	Description	Limiting width–thickness ratio		Details
		λ (actual width-to-thickness ratio of the element)	λ_r ($\lambda \le \lambda_r$ = nonslender; $\lambda > \lambda_r$ = slender)	
Unstiffened	1. Flanges of rolled I-shaped sections (Case 1) 2. Plates projecting from rolled I-shaped sections			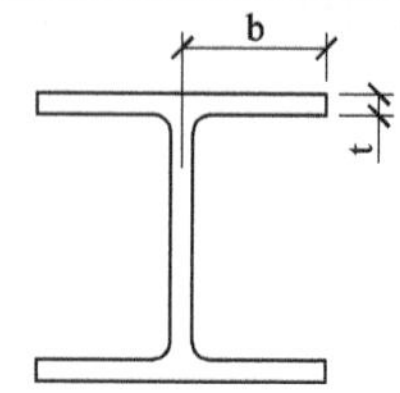
	3. Outstanding legs of pairs of angles in continuous contact (i.e., with no separators) (Case 1)	$\frac{b}{t}$	$0.56\sqrt{\frac{E}{F_y}}$	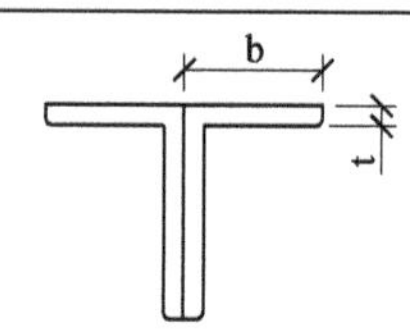
	4. Flanges of C-shapes (Case 1)			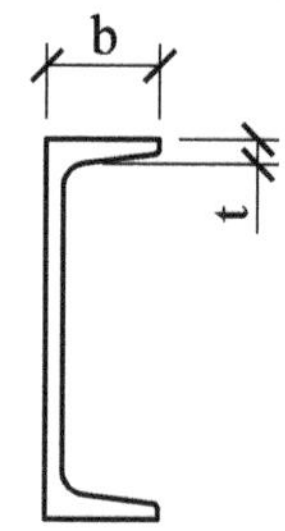
	5. Flanges of WT-shapes (**Case 1**)	$\frac{b}{t}$	$0.56\sqrt{\frac{E}{F_y}}$	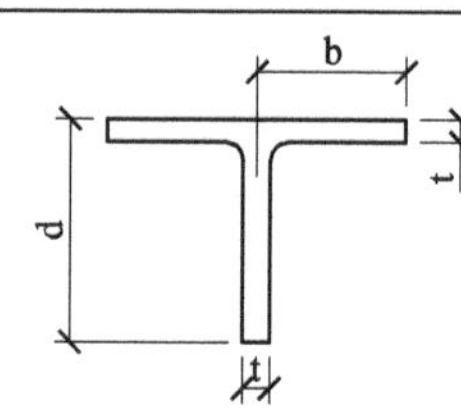
	1. Flanges of built-up I-shaped sections 2. Plates or angle legs projecting from built-up I-shaped sections (**Case 2**)		$0.64\sqrt{\frac{k_c E}{F_y}}$ $k_c = \frac{4}{\sqrt{h/t_w}} \ge 0.35$ ≤ 0.76	
	Stems of WT-shapes (**Case 4**)	$\frac{d}{t}$	$0.75\sqrt{\frac{E}{F_y}}$	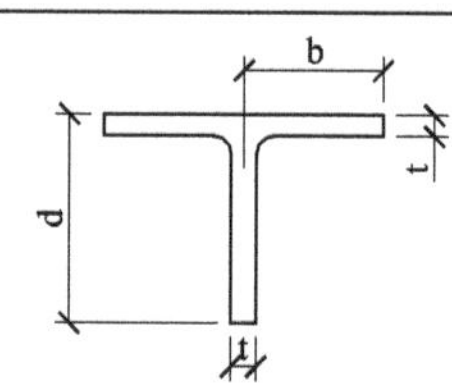
	1. Legs of single angles 2. Legs of double angles with separators (i.e., legs not in continuous contact) 3. All other unstiffened shapes (e.g., cruciform shapes) **Case 3**)	$\frac{b}{t}$	$0.45\sqrt{\frac{E}{F_y}}$	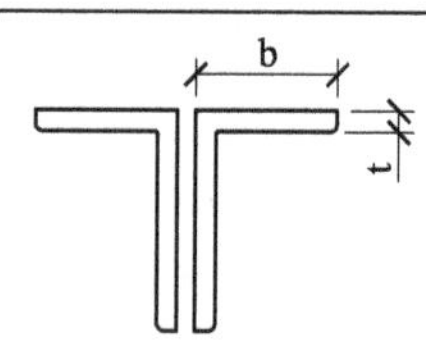 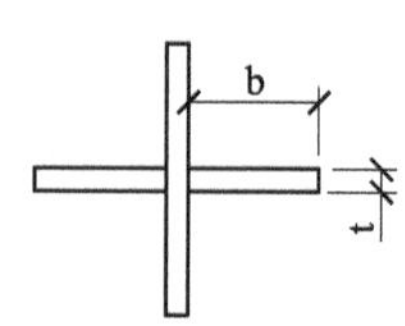

	Description	Limiting width–thickness ratio: λ (actual width-to-thickness ratio of the element)	Limiting width–thickness ratio: λ_r ($\lambda \le \lambda_r$ = nonslender; $\lambda > \lambda_r$ = slender)	Details
Stiffened	Webs of I-shaped sections (**Case 5**)	$\frac{h}{t_w}$	$1.49\sqrt{\frac{E}{F_y}}$	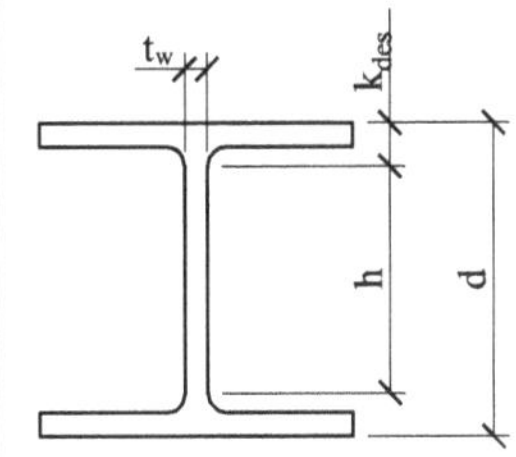
	Webs of C-shapes (**Case 5**)			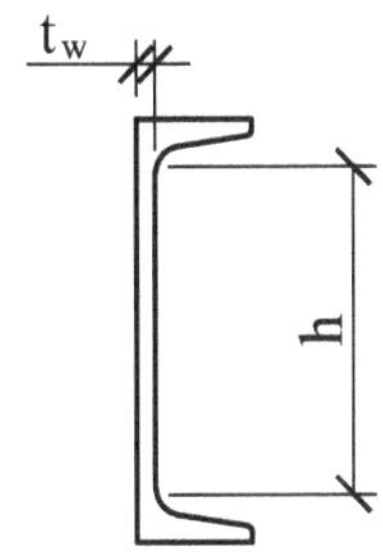
	Flange cover plates between lines of fasteners or welds (**Case 7**)		$1.40\sqrt{\frac{E}{F_y}}$	
	Built-up box sections (**Case 8**) b = clear distance between walls of box section; t = wall thickness	$\frac{b}{t}$	$1.49\sqrt{\frac{E}{F_y}}$	
	Square or rectangular HSS (**Case 6**)	$\frac{b}{t}$	$1.40\sqrt{\frac{E}{F_y}}$	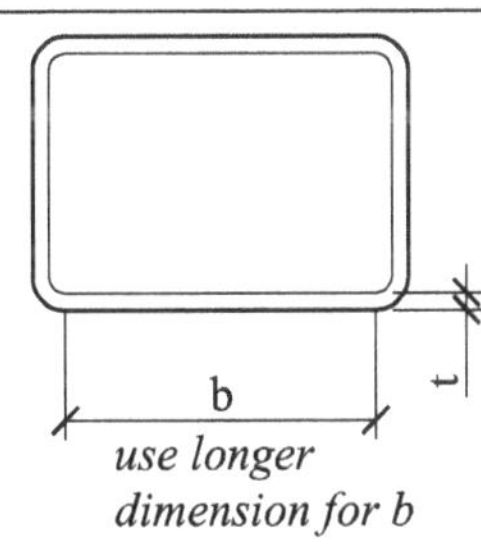
	Round HSS or pipes (**Case 9**)	$\frac{D}{t}$	$0.11\left(\frac{E}{F_y}\right)$	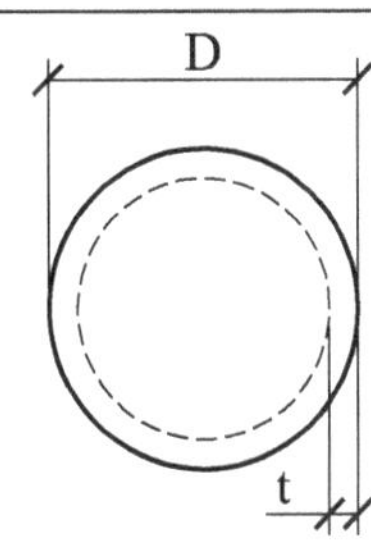

* Adapted from *AISCM* Table B4.1a [1].

In Table 5-1, λ is the actual width-to-thickness ratio of the element. For wide flange sections, λ for the flange and web, respectively, are calculated as follows:

Flange:

$$\lambda_f = \frac{b}{t} = \frac{\left(\frac{1}{2}\right)b_f}{t_f} = \frac{b_f}{2t_f}$$

Web:

$$\lambda_w = \frac{h}{t_w} = \frac{d - 2k_{des}}{t_w}$$

where,

b_f = width of the flange

t_f = thickness of the flange

t_w = Web thickness

d = overall depth of the wide flange section

k_{des} = tabulated decimal value of the distance from the outer face of the flange to the point where the fillet intersects the web.

$h = d - 2k_{des}$

The above geometric dimensions for standard steel sections are tabulated in part 1 of the *AISCM*. Recall that the decimal values of the tabulated dimensions of steel sections from part 1 of the *AISCM* are to be used in design calculations, while the fractional values are to be used for detailing purposes only [1]. For square hollow structural sections (HSS), the width or depth that should be used to calculate the width-to-thickness ratio is the "flat" dimension, which is the clear distance between the webs minus the fillets on both sides. For square HSS with outside dimensions, $b \times b$, and rectangular HSS with outside dimensions, $b \times h$, and wall thickness, t, the width-to-thickness ratio for each element, λ, is calculated according to Section B4.1b(d) of the AISC Specification as follows:

For **square** HSS:
$$\lambda = \frac{b - 3t}{t} \tag{5-16}$$

For **rectangular** HSS:

Short leg:
$$\lambda = \frac{b - 3t}{t} \tag{5-17a}$$

Long leg:
$$\lambda = \frac{h - 3t}{t} \tag{5-17b}$$

For compression members with slender elements (which are therefore susceptible to local buckling), the effect of the slender elements is to reduce the effective cross-sectional area that is available to resist axial compression loads. The nominal axial compression strength, P_n, determined in accordance with AISC Specification Section E7, is the product of the nominal stress, F_n, and the reduced effective area, A_e, where the effective area is the gross cross-sectional area of the column less the loss in cross-sectional area due to local buckling caused by the presence of the slender elements. The axial compressive strength for columns with slender elements according to Section E7 of the AISC Specification is calculated as follows:

$$P_n = F_n A_e \tag{5-18}$$

Where,

F_n is the nominal stress due to flexural buckling, torsional buckling, and flexural-torsional buckling limit states, which ever produces the lowest stress.

The summation of the effective areas of all the component elements of the cross-section is given as

$$A_e = A_g - \sum_{i=1}^{n} (b_i - b_{e,i})t_i \tag{5-18b}$$

For round HSS, refer to AISC Specification Section E7.2 for the expression for the effective area, A_e. The term $\sum_{i=1}^{n} (b_i - b_{e,i})t_i$ in equation (5-18b) represents the loss in cross-sectional area due to local buckling, and the parameters in this equation are defined as follows:

n = number of slender elements

A_g = gross cross-sectional area of the column

t_i = thickness of slender element, i

b_i = width of slender element, i (use d for WT webs and h for webs of wide flanges; for HSS sections, use $b - 3t$ for the short legs and $h - 3t$ for the long legs in accordance with AISC Specification Section B4.1b(d))

$b_{e,i}$ = effective width of the slender element, and is calculated as follows:

$$\text{When } \lambda \leq \lambda_r\sqrt{\frac{F_y}{F_n}}, \qquad b_e = b \tag{5-19a}$$

$$\text{When } \lambda > \lambda_r\sqrt{\frac{F_y}{F_n}}, \qquad b_e = b\left(1 - c_1\sqrt{\frac{F_{el}}{F_n}}\right)\sqrt{\frac{F_{el}}{F_n}} \tag{5-19b}$$

C_1 = effective width imperfection adjustment factor obtained from Table 5-2 (AISC Specification Table E7.1)

$$c_2 = \frac{1-\sqrt{1-4c_1}}{2c_1} \text{ (see Table 5-2)}$$

The elastic local buckling stress is

$$F_{el} = \left(c_2\frac{\lambda_r}{\lambda}\right)^2 F_y \tag{5-19c}$$

λ = width-to-thickness ratio for the element (see Table 5-1 or AISC Specification Tables B4.1a)

λ_r = limiting the width-to-thickness ratio for the element (see Table 5-1 or AISC Specification Table B4.1a)

Table 5-2 Effective Width Imperfection Adjustment Factors, c_1 and c_2*

Case	Slender Element Description	c_1	c_2
(a)	Stiffened elements except walls of square and rectangular HSS	0.18	1.31
(b)	Walls of square and rectangular HSS	0.20	1.38
(c)	All other elements	0.22	1.49

For round HSS, where D is the outside diameter in inches and t is the thickness of the wall, the effective area, A_e, is given as follows (see AISC Specification Section E7.2):

$$\text{When } \frac{D}{t} \le 0.11\frac{E}{F_y}, \qquad A_e = A_g \tag{5-20a}$$

$$\text{When } 0.11\frac{E}{F_y} < \frac{D}{t} < 0.45\frac{E}{F_y}, \qquad A_e = \left(\frac{0.038E}{F_y\left(D/t\right)} + \frac{2}{3}\right)A_g \tag{5-20b}$$

Most wide flange shapes that are listed in the *AISCM* do *not* have slender elements; therefore, there is no reduction in the gross area for most standard wide-flange column sections. There are, in fact, very few sections listed in the *AISCM* that have slender elements, and these are usually indicated by a footnote. On the other hand, some HSS (round, square, and rectangular), double-angle shapes, and WT-shapes consist of slender elements.

5-5 ANALYSIS PROCEDURES FOR COMPRESSION MEMBERS

It is sometimes necessary to determine the axial compression strength of an existing structural steel member for which the size and other design parameters are known; this process is called 'analysis', which contrasts with 'design', where the size of the member is unknown and has to be determined. For the analysis of compression members using the LRFD method, the design axial compression strength, $\phi_c P_n$, is determined and then compared to the factored total axial load or factored axial load demand, P_u.

In the ASD method, the nominal axial compression strength, P_n, is determined and the allowable axial load, P_a, is calculated as $\frac{P_n}{\Omega_c}$, where Ω_c is the factor of safety for the limit states of flexural buckling, torsional buckling, and flexural torsional buckling, and is equal to 1.67 [1]. This allowable load is then compared to the unfactored or service total axial load, P_s. The goal in the LRFD method is to ensure that $P_u \le \phi P_n$, while the goal in the ASD method is to ensure that $P_s \le P_a$.

There are several methods available for the analysis of compression members, and these are discussed here. These methods are used to obtain the nominal axial compressive strength, P_n, from which the design axial strength, ϕP_n, for the LRFD method or the allowable load for the ASD method, $P_a = \frac{P_n}{\Omega_c}$, can be determined. The first step is to determine the effective length, L_c or KL, and the slenderness ratio, KL/r *or* $\frac{L_c}{r}$, for both orthogonal axes of the column (for single angles, there are three axes to consider: x-, y-, and z- axes). For many shapes and column configurations, both the effective length, L_c or KL, and the radius of gyration, r, are different for each orthogonal axis of the column (see Figure 5-9).

Method 1: Using slenderness ratios and the AISC equations

Use equations (5-12) through (5-15), using the larger of $\frac{L_{cx}}{r_x}$ *or* $\frac{K_x L_x}{r_x}$ and $\frac{L_{cy}}{r_y}$ *or* $\frac{K_y L_y}{r_y}$.

Method 2: Using slenderness ratios and AISC Available Critical Stress Tables (*AISCM*, Table 4-14)

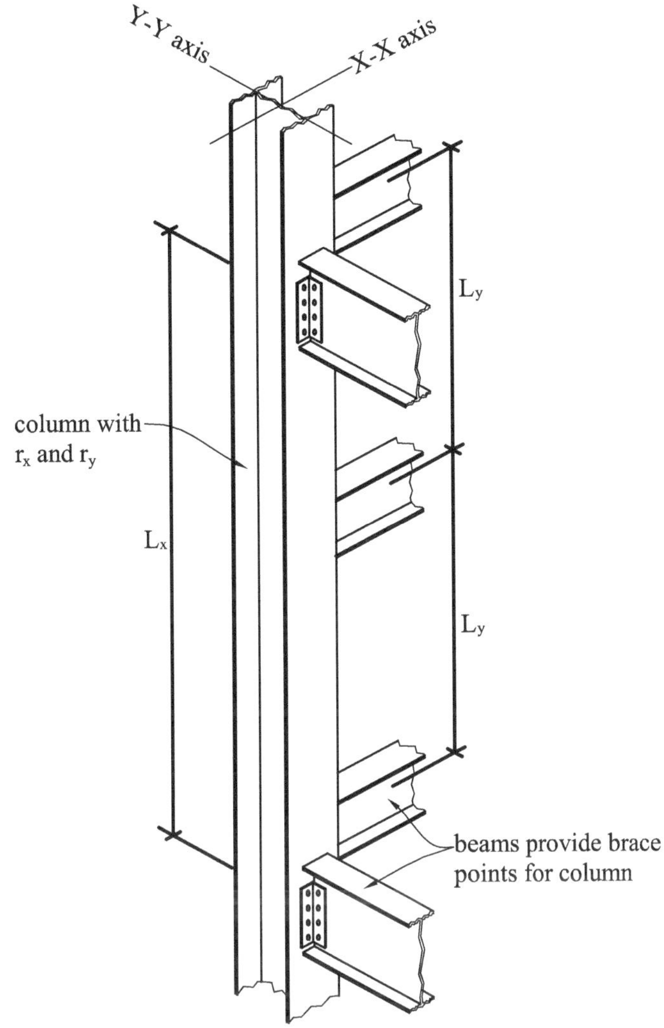

FIGURE 5-9 Effective length and slenderness ratio

This table gives the design compression stress, $\phi_c F_n$ (for LRFD) and allowable compression stress, $\frac{F_n}{\Omega_c}$ (for ASD) as a function of $\frac{L_c}{r}$ *or* $\frac{KL}{r}$ for different values of F_y (35 ksi, 36 ksi, 46 ksi, 50 ksi, 65 ksi, and 70 ksi).

For a given $\frac{L_c}{r}$ *or* $\frac{KL}{r}$, determine $\phi_c F_n$ (for LRFD) and $\frac{F_n}{\Omega_c}$ (for ASD) from the table using the larger of $\frac{L_{cx}}{r_x}$ *or* $\frac{K_x L_x}{r_x}$ and $\frac{L_{cy}}{r_y}$ *or* $\frac{K_y L_y}{r_y}$.

(e.g., when $\frac{L_c}{r} = \frac{KL}{r} = 97$ and F_y = 36 ksi, *AISCM* Table 4-14 gives $\phi_c F_n$ = 19.7 ksi and $\frac{F_n}{\Omega_c}$ = 13.1 ksi).

Knowing the design compression stress, $\phi_c F_n$, the axial design capacity for the LRFD method can be calculated from the equation

$$\phi P_n = \phi F_n A_g$$

where A_g is the gross cross-sectional area of the compression member.

If the ASD method is used, the corresponding allowable load is $P_a = \left(\frac{F_n}{\Omega_c}\right) A_g$

Method 3: Using the *AISCM* Available Strength in Axial Compression Tables (*AISCM* Tables 4-1 through 4-12), also known as the Column Load Capacity tables.

These column load capacity tables give the design strength, $\phi_c P_n$ (for the LRFD method) and the allowable load, $P_a = \dfrac{P_n}{\Omega_c}$ (for the ASD method) for selected shapes at of various effective lengths, L_c or KL, and for selected values of the yield strength, F_y. Enter the appropriate "Available Strength in Axial Compression" table (i.e., the column load capacity table) with the value of L_c or KL, using the larger of

$$\frac{K_x L_x}{\left(\dfrac{r_x}{r_y}\right)} \text{ and } K_y L_y \text{ (i.e., the larger of } \frac{L_{cx}}{\left(\dfrac{r_x}{r_y}\right)} \textit{ and } L_{cy})$$

Notes:

- Ensure that the slenderness ratio for the member is not greater than 200, that is, $\dfrac{L_c}{r} = \dfrac{KL}{r} \leq 200.$
- Check that local buckling will not occur, and if local buckling limits are not satisfied, calculate the reduced axial strength using equations (5-18) through (5-20).
- Use the column load capacity tables (i.e., the "Available Axial Compression Strength" tables) whenever possible. Note that only a few selected sections are listed in these tables; however, the listed sections are typically the most used column sections for conventional building construction.
- Equations (5-9), and (5-12) through (5-15) can be used in all cases for column shapes that have no *slender* elements.

EXAMPLE 5-2

Column Analysis Using the AISC Equations

a. Using the LRFD method, calculate the design compressive strength of a W12 × 65 column, 20 ft. long and pinned at both ends. Use ASTM A992 Grade 50 steel.
b. Using the ASD method, determine the allowable unfactored axial compression load on the column.

Solution

$L_x = L_y = 20$ ft.
$K = 1.0$ (Figure 5-4)
$F_y = 50$ ksi
For W12 × 65, $A_g = 19.1$ in.^{2} (see part 1 of the *AISCM*)

Compute the slenderness ratio about the *x-x* and *y-y* axes and use the larger of the two values (i.e., the larger of $\dfrac{KL_x}{r_x}$ *or* $\dfrac{KL_y}{r_y}$). Alternatively, since $L_x = L_y$ and therefore $KL_x = KL_y$, the smaller radius of gyration will control. For W12 × 65 from *AISCM*, part 1, the radius of gyration about both orthogonal axes is as follows:

$r_x = 5.28$ in., and

$r_y = 3.02$ in., (smaller value controls).

$$\frac{L_{cy}}{r_y} = \frac{KL_y}{r_y} = \frac{(1.0)(20\text{ ft.})(12)}{3.02} = 79.5 \quad (controls)$$

$$< 200 \quad \text{OK.}$$

$$\frac{L_{cx}}{r_x} = \frac{KL_x}{r_x} = \frac{(1.0)(20\text{ ft.})(12)}{5.28in.} = 45.5 \ < 200 \quad \text{OK.}$$

The larger $\frac{L_c}{r}$ value governs.

Check the slenderness criteria for compression elements:

For W12 × 65,

$b_f = 12$ in. $(b = b_f/2 = 12/2 = 6$ in.$)$

$t_f = 0.605$ in.

$t_w = 0.39$ in.

$h = d - 2k_{des} = 12.1 - (2)(1.20) = 9.7$ in.

(*Note:* k_{des} decimal values presented in part 1 of the *AISCM* are used for design, and the larger k_{det} fractional values are used for detailing by the steel detailers. The difference in these values is due to the variation in the fabrication and rolling processes.)

The width-to-thickness ratios for component elements of the column (i.e., the flange and the web) are computed and compared to the limits from Table 5-1:

$$\frac{b}{t} \le 0.56\sqrt{\frac{E}{F_y}}; \quad \frac{b}{t} = \frac{6}{0.605} = 9.92 < 0.56\sqrt{\frac{29{,}000}{50}} = 13.48 \quad \text{OK.}$$

$$\frac{h}{t_w} \le 1.49\sqrt{\frac{E}{F_y}}; \quad \frac{h}{t_w} = \frac{9.7}{0.39} = 24.88 < 1.49\sqrt{\frac{29{,}000}{50}} = 35.88 \quad \text{OK.}$$

Therefore, the component elements are non-slender.

Determine the nominal compression stress, F_n:

The maximum slenderness ratio for inelastic buckling is

$$4.71\sqrt{\frac{E}{F_y}} = 4.71\sqrt{\frac{29{,}000}{50}} = 113.4$$

Since $\frac{L_c}{r} = \frac{KL}{r} = 79.5 < 113.4$, this indicates that the failure mode is by *inelastic* flexural buckling, therefore, use equation (5-14) to determine F_n. From equation (5-9), the elastic buckling stress is

$$F_e = \frac{\pi^2 E}{(KL/r)^2} = \frac{(\pi^2)(29{,}000)}{(79.5)^2} = 45.3 \text{ ksi}$$

From equation (5-14), the nominal stress is

$$F_n = \left[0.658^{\frac{F_y}{F_e}}\right] F_y = \left[0.658^{\frac{50}{45.3}}\right](50) = 31.5 \text{ ksi}$$

LRFD method:

The nominal axial compressive strength of the column is determined from equation (5-11):

$P_n = F_n A_g = (31.5)(19.1) = 601.6$ kips

The design compressive strength of the column is determined:

$\phi_c P_n = (0.90)(601.6) = 541$ kips

From *AISCM* Table 4-14, the design compression stress, ϕF_n, could be obtained directly by entering the table with $\frac{L_c}{r} = \frac{KL}{r} = 79.5$ and F_y = 50 ksi. A value of $\phi_c F_n$ = 28.4 ksi is obtained, which confirms the design axial compression strength as $\phi_c F_n A_g = (28.4\,ksi)(19.1\,in.^2) = 542$ kips.

Alternatively, the design axial compression strength could be obtained directly from Table 4-1 of the *AISCM* (i.e., the column load capacity tables). Enter the table with $L_{cy} = KL_y = 20\ ft.$ and obtain $\phi_c P_n$ = 541 kips.

The design compressive strength is compared to the factored axial load P_u. The column is structurally adequate if the Capacity ≥ the Demand (i.e., if $\phi_c P_n \geq P_u$, or if the Capacity-to-Demand ratio (C/D) is greater than 1.0 (i.e., if $(\phi_c P_n / P_u) \geq 1.0$.

ASD method:

P_n = 601.6 kips (calculated previously)

The allowable axial load is determined from equation (5-13) as $P_a = \frac{P_n}{\Omega_c} = \frac{601.6}{1.67} = 360\ kips$

The allowable axial load, P_a, is compared to the maximum applied total unfactored load, P_s, on the column. The column is structurally adequate if $P_a \geq P_s$

EXAMPLE 5-3

Analysis of Columns Using the *AISCM* Tables

a. Using the LRFD method, determine the design compressive strength for a pin-ended HSS 8 × 8 × $^3/_8$ column of ASTM A500, grade C steel with an unbraced length of 35 ft.

b. Using the ASD method, determine the allowable axial load on the column.

Solution

Unbraced column length, $L = 35$ ft.

Pin-ended column: $K = 1.0$,

$$L_c = KL = (1.0)(35\text{ ft.}) = 35\text{ ft.}$$

ASTM A500 Grade C steel: $F_y = 50$ ksi for HSS

For HSS 8 × 8 × $^3/_8$ from part 1 of the *AISCM*, the cross-sectional properties are as follows:

$A_g = 10.4\text{ in.}^2$

$r_x = r_y = 3.10$ in.

$$\frac{L_c}{r} = \frac{KL}{r} = \frac{(1.0)(35\text{ ft.})(12)}{3.10} = 135.5 < 200 \quad \text{OK.}$$

LRFD method:

From *AISCM* Table 4-14, the design compression stress for LRFD, $\phi_c F_n$, is obtained directly by entering the table with $\frac{L_c}{r} = \frac{KL}{r} = 135.5$ and $F_y = 50$ ksi. A value of $\phi_c F_n = 12.3$ ksi is obtained; therefore, the design axial compressive strength is

$$\phi_c P_n = \phi_c F_n A_g = (12.3)(10.4) = 128 \text{ kips.}$$

Alternatively, the design strength could be obtained directly from the *AISCM* column load capacity tables (Table 4-4), but these are only listed for $F_y = 50$ ksi. Enter the table with $KL = 35$ ft. and interpolate to obtain $\phi_c F_n = 128$ kips for $F_y = 50$ ksi. Note that equations (5-12) through (5-15) could be used to find the axial compression capacity for any F_y.

Alternate Check Using Equations (5-12) through (5-15):

Check the slenderness criteria for compression elements:

$$\frac{b}{t} \leq 1.40\sqrt{\frac{E}{F_y}}; \qquad \text{i.e., } 19.9 \leq 1.40\sqrt{\frac{29{,}000}{50}} = 33.7 \quad \text{OK.}$$

$$\left(\frac{b}{t} = 19.9 \text{ from part 1 of the } AISCM\right)$$

Therefore, there are no slender elements in this section.

Determine the nominal stress, F_n:

The maximum slenderness ratio for *inelastic* buckling is

$$4.71\sqrt{\frac{E}{F_y}} = 4.71\sqrt{\frac{29{,}000}{50}} = 113.4$$

Since $\frac{L_c}{r} = \frac{KL}{r} = 135.5 > 113.4$, this indicates that the failure mode is by *elastic* buckling, therefore, use equation (5-15) to determine the nominal stress, F_n:

The elastic buckling stress is

$$F_e = \frac{\pi^2 E}{(KL/r)^2} = \frac{(\pi^2)(29{,}000)}{(135.5)^2} = 15.6 \text{ ksi}$$

The nominal stress from equation (5-15) is

$$F_n = 0.877F_e = (0.877)(15.6) = 13.7 \text{ ksi}$$

The nominal axial compressive strength is

$$P_n = F_n A_g = (13.7)(10.4) = 142.5 \text{ kips}$$

The design axial compressive strength of the column is then determined for the LRFD method using equation (5-12):

$$\phi_c P_n = (0.90)(142.5) = 128 \text{ kips}$$

ASD method:

Entering *AISCM* Table 4-14 with $\frac{L_c}{r} = \frac{KL}{r} = 135.5$ and $F_y = 50$ ksi, we obtain the allowable compression stress, $\frac{F_n}{\Omega_c} = 8.19 \text{ksi}$.

The allowable axial load on the column is $P_a = \left(\frac{F_n}{\Omega_c}\right)A_g = (8.19 \text{ksi})(10.4 \text{in.}^2) = 85$ kips

Equations (5-12) through (5-15) could also be used to compute the nominal axial compressive strength, P_n, which was obtained previously for the LRFD method as 142.5 kips. Therefore, using equation (5-13), the allowable axial load for ASD is obtained as:

$$P_a = \frac{P_n}{\Omega_c} = \frac{142.5}{1.67} = 85.3 \text{ kips}$$

Where, Ω_c = factor of safety for compression = 1.67

EXAMPLE 5-4

Analysis of HSS Column with Slender Elements

Calculate the design axial compressive strength, $\phi_c P_n$, of an HSS 12 × 8 × 3/16 column, 28 ft. long, and pinned at both ends. Use ASTM A500 Grade C steel. Also, using the ASD method, calculate the allowable axial compression load, $P_{a.}$

Solution

The unbraced lengths are

$L_x = L_y = 28$ ft

The effective length factor is K = 1.0 (Figure 5–4)

$F_y = 50$ ksi

For HSS 12 × 8 × 3/16, the section properties from Part 1 of the AISCM are

$A_g = 6.76 \text{ in.}^2$

$t = 0.174\,in.$ (*use only the decimal values for design*)

$b - 3t = 8 - 3(0.174) = 7.48$ in.

$h - 3t = 12 - 3(0.174) = 11.48$ in.

$$\lambda_{(short\,leg)} = \frac{b-3t}{t} = \frac{7.48}{0.174} = 43$$

$$\lambda_{(long\,leg)} = \frac{h-3t}{t} = \frac{11.48}{0.174} = 66$$

$r_x = 4.56$ in.

$r_y = 3.35$ in.

Compute the slenderness ratio about the *x-x* and *y-y* axes and use the larger of the two values (i.e., use the larger of $\frac{KL_x}{r_x}$ or $\frac{KL_y}{r_y}$).

Since the unbraced length is the same about both axes (i.e., $L_x = L_y$), therefore, the smaller radius of gyration will control.

$$\frac{L_{cy}}{r_y} = \frac{KL}{r} = \frac{(1.0)(28\text{ ft.})(12)}{3.35} = 100.3 < 200 \quad \text{OK}$$

Check the slenderness criteria for compression elements (i.e., check for local buckling):
The limiting width-to-thickness ratios from Table 5-1 are as follows:

$$\lambda_r = 1.40\sqrt{\frac{E}{F_y}} = 1.40\sqrt{\frac{29{,}000}{50}} = 33.7$$

$\lambda\,(short\,leg) = 43 \;>\; \lambda_r = 33.7$

$\lambda\,(long\,leg) = 66 \;>\; \lambda_r = 33.7$

The actual width-to-thickness ratios, λ, for both short and long legs are greater than the limiting width-to-thickness ratio, λ_r, therefore, both the short and long legs of this HSS column are slender and therefore susceptible to local buckling. Hence, the local buckling failure mode must therefore be checked.

Determine the nominal stress, F_n:

The maximum slenderness ratio for inelastic buckling is

$$4.71\sqrt{\frac{E}{F_y}} = 4.71\sqrt{\frac{29{,}000}{50}} = 113.4$$

Since $\frac{KL}{r} = 100.3 < 113.4$, this implies that the failure mode is by inelastic flexural buckling; therefore, use equation (5-14) to determine the nominal stress, F_n.

The elastic buckling stress is

$$F_e = \frac{\pi^2 E}{(KL/r)^2} = \frac{\pi^2(29{,}000)}{(100.3)^2} = 28.45 \text{ ksi}$$

The nominal stress is

$$F_n = \left[0.658^{\frac{F_y}{F_e}}\right]F_y = \left[0.658^{\frac{50}{28.45}}\right](50) = 24 \text{ ksi}$$

Calculate the reduction in effective area due to local buckling:

From Table 5-2 (see also AISC Specification Table E7.1), the effective width imperfection adjustment factors for the walls of the HSS 12×8 are obtained as

$c_1 = 0.2$

$c_2 = 1.38$

Short legs:

The elastic local buckling stress for the short legs is

$$F_{el} = \left(c_2\frac{\lambda_r}{\lambda}\right)^2 F_y = \left(1.38\frac{33.7}{43}\right)^2(50 \text{ ksi}) = 58.5 \text{ ksi}$$

$$\lambda_r\sqrt{\frac{F_y}{F_n}} = 33.7\sqrt{\frac{50}{24}} = 48.6$$

Since $\lambda_{(short\,leg)} = 43 < 48.6$, therefore, use equation (5-19a) to obtain $b_e = b = 7.48$ in.

The reduction in effective area for the short legs of the HSS due to local buckling

$= (2 \text{ short legs})(b - b_e)(t)$

$= (2)(7.48 - 7.48)(0.174) = 0 \text{ in.}^2$

Long legs:

The elastic local buckling stress for the long legs is

$$F_{el} = \left(c_2 \frac{\lambda_r}{\lambda}\right)^2 F_y = \left(1.38 \frac{33.7}{66}\right)^2 (50 \text{ ksi}) = 24.8 \text{ ksi}$$

$$b_{(long\ leg)} = 11.48 \text{ in.}$$

$$\lambda_r \sqrt{\frac{F_y}{F_n}} = 33.7\sqrt{\frac{50}{24}} = 48.6 < \lambda_{(long\ leg)} = 66,$$

Since $\lambda_{(long\ leg)} = 66 > \lambda_r \sqrt{\frac{F_y}{F_n}} = 48.6$, use equation (5-19b) to obtain b_e.

Therefore, $b_e = b\left(1 - c_1\sqrt{\frac{F_{el}}{F_n}}\right)\sqrt{\frac{F_{el}}{F_n}} = 11.48\left(1 - 0.2\sqrt{\frac{24.8}{24}}\right)\sqrt{\frac{24.8}{24}} = 9.3$ in.

The reduction in effective area for the long legs of the HSS due to local buckling

$= (2 \text{ long legs})(b - b_e)(t)$

$= (2)(11.48 - 9.3)(0.174) = 0.76 \text{ in.}^2$

The effective area due to local buckling is,

$A_e = A_g$ – reduction in effective area in the short legs – reduction in effective area in the long legs

Therefore, $A_e = 6.76 \text{ in.}^2 - 0.0 \text{ in.}^2 - 0.76 \text{ in.}^2 = 6.0 \text{ in.}^2$

The nominal axial load capacity of the column, $P_n = F_n A_e = (24 \text{ ksi})(6.0 \text{ in.}^2) = 144$ kips

LRFD method:

The design axial compression load capacity for the LRFD method is

$$\phi_c P_n = (0.9)(144 \text{ kips}) = 130 \text{ kips}$$

Compare the value obtained above to the value of $\phi_c P_n = 129$ kips from the column load tables, *AISCM* Table 4-3, the results are found to be essentially the same.

ASD method:

The allowable axial compression load for the ASD method is $P_a = \frac{P_n}{\Omega_c} = \frac{144}{1.67} = 86$ kips.

Compare to the value of $P_a = 86.1$ kips obtained from *AISCM* Table 4-3, and the results are found to be essentially the same.

5-6 DESIGN PROCEDURES FOR COMPRESSION MEMBERS

The design procedures for compression members are presented in this section, starting with the design procedure for sections that are not listed in the *AISCM* column design tables.

1. For Members Not Listed in the *AISCM* Column Design Tables:

a. (i) For the LRFD method, calculate the factored axial compression load or the required axial strength, P_u, on the column, and assume a value for the design compression stress, $\phi_c F_n$, that is less than the yield stress, F_y.

(ii) For the ASD method, calculate the maximum applied unfactored load, P_s, on the column and assume a value for the allowable compression stress, $\frac{F_n}{\Omega_c}$ that is less than the allowable yield stress, $\frac{F_y}{\Omega_c}$

b. **(i)** For the LRFD method, determine the required gross area, $A_{g\ \text{required}}$, which should be greater than or equal to $\frac{P_u}{\phi_c F_n}$.

(ii) For the ASD method, determine the required gross area, $A_{g\ \text{required}}$, which should be greater than or equal to $\frac{P_s}{\left(F_n/\Omega_c\right)}$.

c. Select a section from part 1 of the *AISCM* that has $A_g > A_{g\ \text{required}}$.

- Check that $L_c/r = KL/r \leq 200$ for each axis of the compression member.

d. **(i)** LRFD Method: For the section selected in step 1c, compute the actual design stress, $\phi_c F_n$, using either *AISCM* Table 4-14 or equations (5-14) or (5-15).

(ii) ASD Method: For the section selected in step 1c, compute the allowable stress, $\frac{F_n}{\Omega_c}$, using either *AISCM* Table 4-14 or equations (5-14) or (5-15).

e. **(i)** For the LRFD method, compute the design compressive strength, $\phi_c P_n = (\phi_c F_n) A_g$, for the selected shape from step 1c:

- If $\phi_c P_n \geq P_u$, the section selected is adequate; go to step 1g.
- If $\phi_c P_n < P_u$, the column is **inadequate**; go to step 1f.

(ii) For the ASD method, the allowable design strength, $P_a = \frac{P_n}{\Omega_c} = \frac{F_n A_g}{\Omega_c}$

- If $P_a \geq P_s$, the section selected is adequate; go to step 1g.
- If $P_a \leq P_s$, the column is **inadequate**; go to step 1f.

f. **(i)** For the LRFD method, use the value of $\phi_c F_n$, obtained in step '*d*,' to repeat steps (a) through (e) until $\phi_c P_n$ is just greater than or equal to the factored load, P_u.

(ii) For the ASD method, use the value of $\frac{F_n}{\Omega_c}$ obtained in step *1d* to repeat steps *1a* through *1e* until $P_a = \frac{P_n}{\Omega_c}$ is just greater than or equal to the total unfactored load, P_s, on the column.

g. Check local buckling (see Table 5-1).

2. **For Members Listed in the *AISCM* Column Design Tables:**

The design of columns using the *AISCM* column load capacity tables is as follows:

a. For the LRFD method, calculate the factored axial load on the column, P_u.
For the ASD method, calculate the total unfactored axial load, P_s, on the column.

b. Obtain the recommended effective length factor, K, from Figure 5-4 and calculate the effective length, $L_c = KL$, for buckling about each axis.

c. Enter the column load capacity tables (i.e., *AISCM* Tables 4-1 through 4-12) with $L_c = KL$, which is the larger of $\frac{K_x L_x}{\left(\frac{r_x}{r_y}\right)}$ and $K_y L_y$, and find the ***lightest*** column section that has a design strength, $\phi_c P_n \geq$ the factored load, P_u for the LRFD method, or

that has an allowable load, $P_a = \frac{P_n}{\Omega_c} \geq P_s$ for the ASD method.

In practice, many designers use the column load capacity tables (i.e., *AISCM* Tables 4-1 through 4-12) whenever possible because they are the easiest design tool to use for the analysis and design of steel columns. Note that these column load capacity tables are derived from equations (5-9) and (5-12) through (5-15).

EXAMPLE 5-5

Design of Axially Loaded Columns Using the *AISCM* Tables

Using both the LRFD and ASD methods, select a W14 column of ASTM A992, Grade 50 steel, 14 ft. long, pinned at both ends, and subjected to the following service loads:

$P_D = 160$ kips

$P_L = 330$ kips

Solution

- A992, Grade 50 steel: $F_y = 50$ ksi
- Pinned at both ends, therefore, $K = 1.0$
- $L = 14$ ft.; $L_c = KL = (1.0)(14) = 14$ ft.

LRFD method:

The factored load, $P_u = 1.2P_D + 1.6P_L = 1.2(160) + 1.6(330) = 720$ kips.

From the column load capacity tables in part 4 of the *AISCM*, find the W14 tables for $F_y = 50$ ksi (i.e., *AISCM* Table 4-1a).

Enter these W14 tables with $L_c = KL = 14$ ft. and find the lightest W14 that has a design compression strength, $\phi P_n \geq P_u$.

We obtain a **W14 × 82** column with an axial compression strength, $\phi_c P_n = 774$ kips $> P_u = 720$ kips.

(*Note:* $\phi_c P_n = 701$ kips for W14 × 74 which is less than P_u, so the W14 × 74 column size is not adequate.)

ASD method:

The total service load, $P_s = P_D + P_L = 160$ kips $+ 330$ kips $= 490$ kips

Enter the column load tables (i.e., *AISCM* Table 4-1a) with $L_c = KL = 14$ ft., and find the lightest W14 that has the allowable compression load, $P_a = \frac{P_n}{\Omega_c} \geq P_s$.

We obtain W14 × 82 at $L_c = 14$ ft, $\frac{P_n}{\Omega_c} = 514$ kips $> P_s = 490$ kips. OK.

EXAMPLE 5-6

Column Design Using Sections Listed in the AISCM Column Design Tables (LRFD method)

Using the *AISCM* column design tables, select the lightest steel column section for a factored compression load, $P_u = 194$ kips, and a column length, $L = 24$ ft. Use ASTM A572, grade 50 steel, and assume that the column is pinned at both ends.

Solution

$K = 1.0$

$L_c = KL = (1.0)(24 \text{ ft.}) = 24 \text{ ft.}$

For $L_c = KL = 24$ ft., we obtain the following axial compression strengths, $\phi_c P_n$, from the column load capacity tables (*AISCM*, Table 4-1):

	Selected Size	$\phi_c P_n$, kips
W8	W8 × 58	205
W10	W10 × 49	254
W12	W12 × 53	261
W14	W14 × 61	293

Always select the lightest column section if other considerations or design constraints (such as architectural constraints on the maximum column size and the maximum interior partition wall thickness) do not restrict the column size.

Therefore, considering the lightest weight section, use a **W10** × **49** column.

EXAMPLE 5-7

Column Design for Sections Not Listed in the *AISCM* Column Design Tables (LRFD method)

Using the LRFD method, select a W18 column of ASTM A36 steel, 26 ft. long, and subjected to a factored axial load of 500 kips. Assume that the column is pinned at both ends.

Solution

Since W18 shapes are *not* listed in the *AISCM* column load capacity tables, we cannot use these tables to design this column. The procedure for members not listed in the *AISCM* column design tables from Section 5-6 will be followed.

$P_u = 500$ kips,

$L_c = KL = 1.0 \times 26 \text{ ft.} = 26 \text{ ft.}$

Cycle 1:

1. Assume the critical compression stress, $\phi_c F_n = 20 \text{ ksi} < F_y = 36 \text{ ksi}$.
2. $A_{g\text{ required}} \geq \dfrac{P_u}{\phi_c F_n} = \dfrac{500}{20} = 25 \text{ in.}^2$
3. Select the W18 section from part 1 of the *AISCM* with $A_g \geq 25 \text{ in.}^2$.
   Try **W18** × **86** with $A_g = 25.3 \text{ in.}^2$,
   $r_x = 7.77$ in., and
   $r_y = 2.63$ in.

$$\frac{KL}{r_{\min}} = \frac{(1.0)(26 \text{ ft.})(12)}{2.63} = 118.6 < 200 \text{ OK}$$

4. Enter the *AISCM* Table 4-14 with the *KL/r* value from step 3 and obtain $\phi_c F_n = 15.5$ ksi.
5. The design axial compressive strength,
   $\phi_c P_n = (\phi_c F_n)(A_g) = (15.5 \text{ ksi})(25.3 \text{ in.}^2) = 392 \text{ kips} < P_u = 500 \text{ kips}$
   The selected column is *not* adequate. Therefore, proceed to cycle 2.

Cycle 2:

1. Assume the design compression stress, $\phi_c F_n$ = 15.5 ksi (from step 4 of the previous cycle).
2. $A_{g\ \text{required}} \geq \dfrac{P_u}{\phi_c F_n} = \dfrac{500}{15.5} = 32.2\ \text{in.}^2$
3. Try **W18 × 119** with $A_g = 35.1\ \text{in.}^2 > A_{g\ \text{required}}$,
   $r_x = 7.90$ in., and
   $r_y = 2.69$ in.

   $$\frac{KL}{r_{\min}} = \frac{(1.0)(26\ \text{ft.})(12)}{2.69} = 116 < 200\ \text{OK}$$
4. Enter the *AISCM* Table 4-14 with KL/r = 116 and obtain $\phi_c F_n$ = 16.0 ksi.
5. The design axial compressive strength is

   $$\phi_c P_n = (\phi_c F_n)(A_g) = (16.0\ \text{ksi})(35.1\ \text{in.}^2) = 561\ \text{kips} > P_u = 500\ \text{kips OK}$$

   Therefore, the **W18 × 119** column is adequate.
6. Check the slenderness criteria for the component elements of the section (i.e., the flanges and the web) to determine if they are slender or not (see Table 5-1):

   $b_f = 11.3\ \text{in.} (b = 11.3/2 = 5.65\ \text{in.})$

   $t_f = 1.06$ in.

   $t_w = 0.655$ in.

   $h = d - 2k_{\text{des}} = 19.0 - (2)(1.46) = 16.08\ \text{in.}$

   $$\frac{b}{t} \leq 0.56\sqrt{\frac{E}{F_y}}; \qquad \frac{b}{t} = \frac{5.65}{1.06} = 5.33 < 0.56\sqrt{\frac{29{,}000}{36}} = 15.9 \qquad \text{OK.}$$

   $$\frac{h}{t_w} \leq 1.49\sqrt{\frac{E}{F_y}}; \qquad \frac{h}{t_w} = \frac{16.08}{0.655} = 24.5 < 1.49\sqrt{\frac{29{,}000}{36}} = 42.2 \qquad \text{OK.}$$

This implies that the elements that make up the W18 x 119 column section are not slender; therefore, the design strength calculated previously does not have to be reduced for local buckling effects.

EXAMPLE 5-8

Analysis of Columns with Unequal Unbraced Lengths Using the *AISCM* Column Design Tables

Using the LRFD and ASD methods, determine the design axial compressive strength and the allowable compression load for the following column with unequal unbraced lengths:

- W14 × 82 column
- ASTM A992, Grade 50 steel
- Unbraced length for strong (x–x) axis bending = 25 ft.
- Unbraced length for weak (y–y) axis bending = 12.5 ft.

Solution

$L_{cx} = K_x L_x = (1.0)(25) = 25$ ft.
$L_{cy} = K_y L_y = (1.0)(12.5) = 12.5$ ft.

From part 1 of the *AISCM*, we find the following properties for W14 × 82:

$A_g = 24.0$ in.^{2}
$r_x = 6.05$ in.
$r_y = 2.48$ in.

LRFD method:

1. Using *AISCM* Table 4-14,

Calculate the slenderness ratios

$$\frac{L_{cx}}{r_x} = \frac{K_x L_x}{r_x} = \frac{(25\text{ ft.})(12)}{6.05} = 49.6 < 200 \quad \text{OK., and}$$

$$\frac{L_{cy}}{r_y}\frac{K_y L_y}{r_y} = \frac{(12.5\text{ ft.})(12)}{2.48} = 60.5 < 200 \quad \text{OK., the larger } \frac{L_c}{r} = \frac{KL}{r} \text{ value governs.}$$

Entering *AISCM* Table 4-14, with $\frac{L_c}{r}$ *or* $\frac{KL}{r}$ = 60.5 and F_y = 50 ksi, we obtain the design compression nominal stress, $\phi_c F_n$ = 34.45 ksi.
Column design compressive strength, $\phi_c P_n = \phi_c F_n A_g$ = (34.45)(24.0 in.^{2}) = **827 kip**

2. Using the column load capacity tables (*AISCM* Table 4-1a), Calculate the effective lengths,

$$\frac{K_x L_x}{\left(\frac{r_x}{r_y}\right)} = \frac{25\text{ ft.}}{\left(\frac{6.05}{2.48}\right)} = 10.25\text{ ft.}$$

(*Note:* r_x/r_y is also listed at the bottom of the column load tables.)

$K_y L_y$ = **12.5**; The larger effective length, KL, value controls.

Entering the column load capacity table for W14 × 82 (F_y = 50 ksi) in part 4 of the *AISCM* with an unbraced length, $L_c = KL$ = **12.5 ft.**, we obtain

$L_c = KL$ (ft.)	$\phi_c P_n$, kips
12.0	844
12.5	**827**
13.0	809

The design axial compression strength, $\phi_c P_n$, is 827 kips. Note that the column load capacity tables also indicate whether the compression member has components that are slender in accordance with Table 5-1. For members with slender components, the column load tables already account for this local buckling effect in the tabulated design strength.

ASD method:

1. Using *AISCM* Table 4-14 for the ASD Method
Entering *AISCM* Table 4-14, with an unbraced length, $L_c = KL/r$ = 60.5 and F_y = 50 ksi, we obtain the allowable compression stress, $\frac{F_n}{\Omega_c} = 22.9$ ksi

Allowable column load, $P_a = \left(\frac{F_n}{\Omega_c}\right) A_g = (22.9)(24.0\,\text{in.}^2) = 550$ kips

2. Column load capacity tables for the ASD method (*AISCM* Table 4-1a): Calculate the effective lengths,

$$\frac{K_x L_x}{\left(\frac{r_x}{r_y}\right)} = \frac{25\text{ ft.}}{\left(\frac{6.05}{2.48}\right)} = 10.25\text{ ft.}$$

(*Note:* r_x/r_y is also listed at the bottom of the column load tables.)

$K_y L_y = \mathbf{12.5}$; The larger effective length, KL, value controls.

Entering the column load capacity table for W14 × 82 (F_y = 50 ksi) in part 4 of the *AISCM* with $L_c = KL$ = 12.5 ft., we obtain the following allowable compression loads for the unbraced lengths of 12 ft and 13 ft, from which the allowable compression load for the unbraced length of 12.5 ft can be obtained by linear interpolation.

$L_c = KL$ (ft.)	$P_a = \frac{P_n}{\Omega_n}$ kips
12.0	562
12.5	**550**
13.0	538

The allowable compression load, $P_a = \frac{P_n}{\Omega_c}$, is 550 kips.

EXAMPLE 5-9

Design of Columns with Unequal Unbraced Lengths Using the *AISCM* Column Design Tables (LRFD method)

Select an ASTM A992, Grade 50 steel column to resist a factored compression load, P_u = 780 kips. The unbraced lengths are L_x = 25 ft. and L_y = 12.5 ft., and the column is pinned at both ends.

Solution

1. P_u = 780 kip
2. $L_{cx} = K_x L_x$ = (1.0)(25 ft.) = 25 ft. (strong axis)
   $L_{cy} = K_y L_y$ = (1.0)(12.5 ft.) = 12.5 ft. (weak axis)
3. Initially, assume that the *weak* axis governs, that is, $L_c = KL = K_y L_y$. Enter the column load capacity table (*AISCM* Table 4-1) with $L_c = KL = K_y L_y$ = 12.5 ft.

   *W12 × 79 $\phi_c P_n$ = 874 kips

   W14 × 82 = 828 kips

   * Try this section because it is the lightest.

4. For W12 × 79, the radius of gyration about the x- and y- axes are

$r_x = 5.34$ in.,

$r_y = 3.05$ in., and

$r_x/r_y = 1.75$ (from *AISCM* Table 4–1);

thus, the effective length for buckling about the x- and y- axes, respectively, are

$$\frac{K_x L_x}{\left(\frac{r_x}{r_y}\right)} = \frac{\mathbf{25\ ft.}}{\mathbf{1.75}} = \mathbf{14.3\ ft.} \quad \text{(the larger value controls)}$$

$K_y L_y = 12.5$ ft.

Therefore, the original assumption in step 3—that the weak axis (i.e., the y-axis) controls—was incorrect; the strong axis (i.e., the x-axis) actually controls.

5. Enter the W12 × 79 column load capacity tables (*AISCM* Table 4-1a) with $\frac{K_x L_x}{\left(\frac{r_x}{r_y}\right)} = \mathbf{14.3}$ ft. and we obtain the following allowable compression loads for the unbraced lengths of 14 ft. and 15 ft., from which the allowable compression load for the unbraced length of 14.3 ft. can be obtained by linear interpolation:

$L_c = KL$ (ft.)	$\phi_c P_n$, (kips)
14.0	836
14.3	**827**
15.0	809

For W12 × 79, $\phi_c P_n = 827$ kips $> P_u = 780$ kips OK

Use a W12 × 79 column.

Note: The same result can be obtained using the AISC equations (i.e., equations (5-9), and (5-12) through (5-15)) by using the larger of the two slenderness ratios: $\frac{\left(\frac{K_x L_x}{r_x / r_y}\right)}{r_y}$ and $\frac{K_y L_y}{r_y}$.

5-7 ALIGNMENT CHARTS OR NOMOGRAPHS

As already discussed in this chapter, the alignment charts, or nomographs (see Figures 5-10 and 5-11 or see *AISCM* Figure C-A-7.2), are an alternate method to Figure 5-4 that can be used to determine the effective length factor, K, for columns. These nomographs account for the momentary or rotational restraints provided at the ends of the column by the beams or girders framing into the columns. They provide more accurate K values but require knowledge of the sizes of the beams, girders, and columns and are more cumbersome to use. Two charts are presented in the *AISCM*: sideways inhibited (i.e., buildings with braced frames or shear walls), reproduced in Figure 5-10, and sideways uninhibited (i.e., buildings with moment frames), reproduced in Figure 5-11. The following assumptions have been used in deriving these alignment charts, or nomographs [1]:

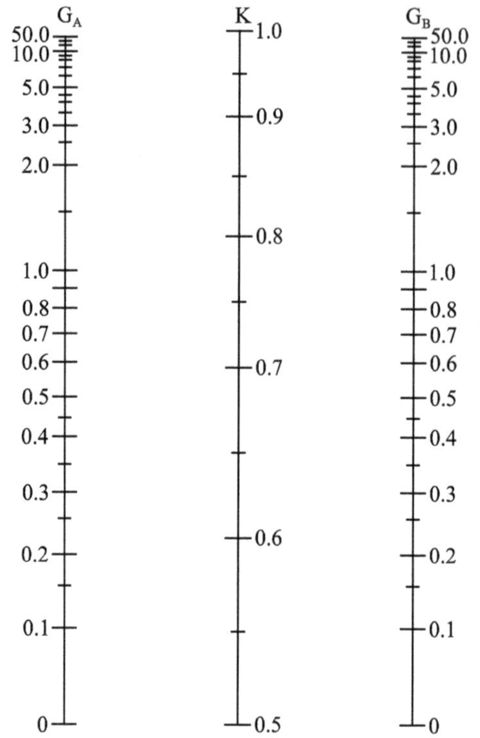

FIGURE 5-10 Alignment chart: sidesway inhibited (braced frames)

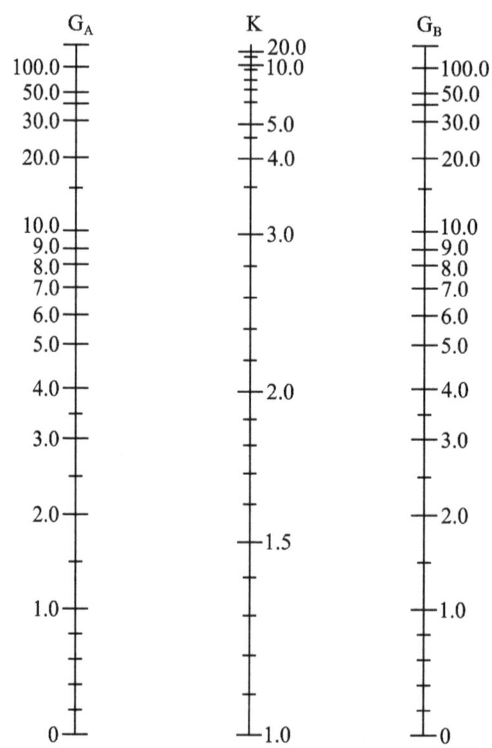

FIGURE 5-11 Alignment chart: sidesway uninhibited (moment frames)

1. The behavior of the frame is purely elastic; that is, elastic buckling of the columns is assumed.
2. All members have a prismatic or constant cross-section.
3. All the beam-to-column or girder-to-column joints are "rigid" or fully restrained (FR) moment connections.
4. For columns in buildings with braced frames or shear walls as the lateral force resisting system (i.e., side-sway *inhibited* frames), but which have "rigid" beam-to-column and girder-to-column connections, the rotations at opposite ends of the restrained beams or girders are equal in magnitude and opposite in direction, resulting in single-curvature bending in the beams and girders.
5. For columns in buildings with moment-resisting frames as the lateral force-resisting system (i.e., sideways *uninhibited* frames), the rotations at opposite ends of the restraining beams or girders are equal in magnitude and direction, resulting in double-curvature bending in the beams and girders.
6. The stiffness parameter, $L\sqrt{P/EI}$, of all columns is equal, where L is the unbraced length of the column.
7. The distribution of the joint restraint is proportional to the stiffness, EI/L, of the column segments above and below the joint.
8. All columns in the structure buckle at the same time.
9. There is no significant axial compression force in the beams or girders.
10. The shear deformations in the frame members are negligible.

It should be noted that for many real-world structures, it would be near impossible to meet all the requirements presented above for the use of the alignment charts. For example, the requirement that all columns in the structure buckle simultaneously will be difficult to satisfy since some columns will have lower axial compression loads than others and therefore will not buckle at the same time as the more heavily loaded columns. Also, most of the columns will buckle inelastically, and there will be axial loads in the beams and girders. As will be discussed in Chapter 8, the use of the Direct Analysis Method for stability analysis and design of columns is strongly encouraged because it obviates the need to calculate the effective length (or K) factors.

To use the alignment charts, the relative stiffness, G, of the columns, compared to the girders (or beams) meeting at the joint at both ends of each column, is calculated using equation (5-21):

$$G = \frac{\text{Total column stiffness at the joint}}{\text{Total girder stiffness at the joint}}$$

$$= \frac{\sum\left[\dfrac{\tau_b E_c I_c}{L_c}\right]}{\sum\left[\dfrac{\tau_g E_g I_g}{L_g}\right]} = \frac{\left[\dfrac{\tau_b E_c I_c}{L_c}\right]_{\text{below joint}} + \left[\dfrac{\tau_b E_c I_c}{L_c}\right]_{\text{above joint}}}{\left[\dfrac{\tau_g E_g I_g}{L_g}\right]_{\text{left}} + \left[\dfrac{\tau_g E_g I_g}{L_g}\right]_{\text{right}}}, \tag{5-21}$$

where

E_c, E_g = Modulus of elasticity of the columns and girders, respectively.

I_c = Moment of inertia of column in the plane of bending of the moment frame.

I_g = Moment of inertia of the girders in the plane of bending of the column.

L_c, L_g = Unsupported or unbraced length of the columns and girders, respectively.

τ_b = stiffness reduction factor to account for inelasticity in the columns.

$$\tau_b = 4\left(\frac{P_u}{P_{ns}}\right)\left(1-\frac{P_u}{P_{ns}}\right) \quad \text{for LRFD}$$

$$= 1.0 \quad \text{if } \frac{P_u}{P_{ns}} \le 0.5$$

$$\tau_b = 4\left(\frac{1.6P_s}{P_{ns}}\right)\left(1-\frac{1.6P_s}{P_{ns}}\right) \quad \text{for ASD}$$

$$= 1.0 \quad \text{if } \frac{1.6P_s}{P_{ns}} \le 0.5$$

The higher the applied load, P_u or P_s, the higher the likelihood for inelasticity in the column. In lieu of using the above equations, the values of τ_b are also tabulated in *AISCM* Table 4-13 for various values of F_y and P_u/A_g for the LRFD method, (P_s/A_g values for the ASD method).

τ_b = 1.0 if all the ten assumptions previously discussed are satisfied

P_{ns} = cross-section compressive strength = F_yA_g for non-slender element sections
= F_yA_e for slender element sections

A_g = gross cross-sectional area of the structural member

A_e = effective area, defined in Section 5-4 of the text (see Section E7 of the AISC Specification).

P_u = total factored compression load on the column using the LRFD load combinations (see Section 2-3).

P_s = total service or unfactored compression load on the column using the ASD load combinations (see Section 2-3).

τ_g = Girder stiffness modification factor to account for girder far end conditions that are *not* fully restrained moment connections (see Table 5-3 or Appendix 7 of the AISC Specification).
= 1.0 if both ends of the girder are connected to the columns with fully restrained moment connections.

TABLE 5-3 Girder Stiffness Modification Factors, τ_g

	Girder far-end condition	**Girder stiffness modification factor, τ_g**
Sidesway uninhibited (i.e., **unbraced or moment frames**)	Far end rotation of girder is fully restrained	2/3
	Far end of girder is pinned	½
Sidesway inhibited (i.e., **braced frames**)	Far end rotation of girder is fully restrained	2.0
	Far end is pinned	1.5

Source: Adapted from Appendix 7 of Ref. [1].

For steel structures, the modulus of elasticity of the column, E_c, is the same value as the modulus of elasticity of the girder or beam, E_g; thus, E_c and E_g will cancel out in equation (5-21). In real-life structures, the assumptions listed previously are usually approximately satisfied. Where assumption No. 1 is not satisfied—implying *inelastic* behavior in the column—the column elastic modulus of elasticity, E_c, in equation (5-21) needs to be replaced by the lower tangent modulus of elasticity, E_t. Thus, the column elastic stiffness in equation (5-21) will be reduced by the ratio E_t/E_c or the column stiffness reduction factor, which is denoted as τ_b.

In the calculation of the relative stiffness, G, fully restrained (FR) moment connections are assumed at the girder-to-column connections at both ends of the girders. For other situations (i.e., where assumptions 4 and 5 discussed earlier are not satisfied), the stiffness of the beams or girders are modified by the adjustment factors given in Table 5-3 for the various far-end support conditions of the girders. Note that when the near end of a girder is pinned (i.e., a simple shear connection), the girder or beam stiffness reduction factor is zero. The use of these modification factors yields conservative values of the effective length factor, K.

Although the theoretical G value for a column with a pinned base (e.g., column supported on spread footing) is infinity, a practical value of 10 is recommended in the *AISCM* because there is no perfect pinned support condition. Similarly, the G value for a column with a fixed base is theoretically zero, but a value of 1.0 is recommended for practical purposes.

The alignment charts assume fully restrained (FR) moment connections at the girder-to-column connections (i.e., rigid joints), and, as such, they are mostly useful for columns in unbraced frames or moment-resisting frames (see Figure 5-11 or *AISCM*, Figure C-A-7.2). The girder-to-column connections in braced frames are usually simple shear connections with no moment restraints; therefore, for a braced frame with pinned column bases, the G factor will be 10 for both the top and the bottom ends of a typical column in the frame. Using these G factors in the sway-inhibited alignment chart (Figure 5-10 or *AISCM*, Figure C-A-7.1) yields an effective length factor, K, of 0.96. This value is not much different from the effective length factor of 1.0 that is obtained for a similar pin-ended column using Figure 5-4. It should be noted that the effective length factor, K, of 1.0 is the recommended practical value for columns in building structures that are laterally supported at both ends. Consequently, the alignment charts in this text will be used only for unbraced frames or moment frames (i.e., , side-sway uninhibited).

The procedure for using the alignment charts to determine the effective length factor, K, for a column is as follows:

1. Calculate the factored axial load, P_u, or the total service load, P_s, on the column. It is assumed at this stage that the girder and beam sizes and the preliminary column sizes have already been determined.
2. Calculate the stiffness of the girders and columns and, where necessary, modify the girder stiffness using the adjustment factors, τ_g, in Table 5-3 based on the support conditions at the far ends of the girders.

   Note: Where a girder is pinned at the joint under consideration (i.e., connected to the column with a simple shear connection at the near joint), that girder stiffness (i.e., EI/L of the girder) will be taken as zero in calculating the relative stiffness factor, G, at that joint.
3. Where necessary, modify the column stiffness by the inelasticity reduction factor, τ_b, using the equations previously presented or using *AISCM* Table 4-13.
4. Calculate the G factors at both ends of the column. Assume that G_A is the G factor calculated at the *bottom* of the column and G_B is the G factor calculated at the *top* of the column.
5. For a pinned column base, use $G_A = 10$; for a fixed column base, use $G_A = 1.0$.
6. Plot the G_A and G_B factors on the corresponding vertical axes of the applicable alignment chart. For unbraced or moment frames, use the alignment chart shown in Figure 5-11.
7. Join the two plotted points (i.e., G_A and G_B) with a straight line, and the point at which the vertical K-axis on the alignment chart is intercepted by the straight line gives the value of the column effective length factor, K.

The alignment charts (or nomographs) yield more accurate values of the effective length factor, K, than Figure 5-4; however, the alignment charts can only be used if the preliminary sizes of the beams, girders, and columns are known, unlike Figure 5-4, where the effective length factors are independent of the member sizes. The recommended design values for the effective length factor, K, from Figure 5-4 can be used for preliminary, as well as final, design.

EXAMPLE 5-10

Effective Length Factor of Columns Using Alignment Charts (LRFD method)

For the two-story moment frame shown in Figure 5-12, the preliminary column and girder sizes have been determined as shown. Assume in-plane bending about the strong axes for the columns and girders, and assume the columns are supported by spread footings. The factored axial loads on columns BF and FJ are 590 kip and 140 kip, respectively, and $F_y = 50$ ksi. Assume the girder stiffness reduction factor, $\tau_g = 1.0$.

1. Determine the effective length factor, K, for columns BF and FJ using the alignment charts, assuming *elastic* behavior in the column (i.e., the column stiffness reduction factor, $\tau_g = 1.0$.).
2. Determine the effective length factor, K, for columns BF and FJ using the alignment charts, assuming *inelastic* behavior in the column.

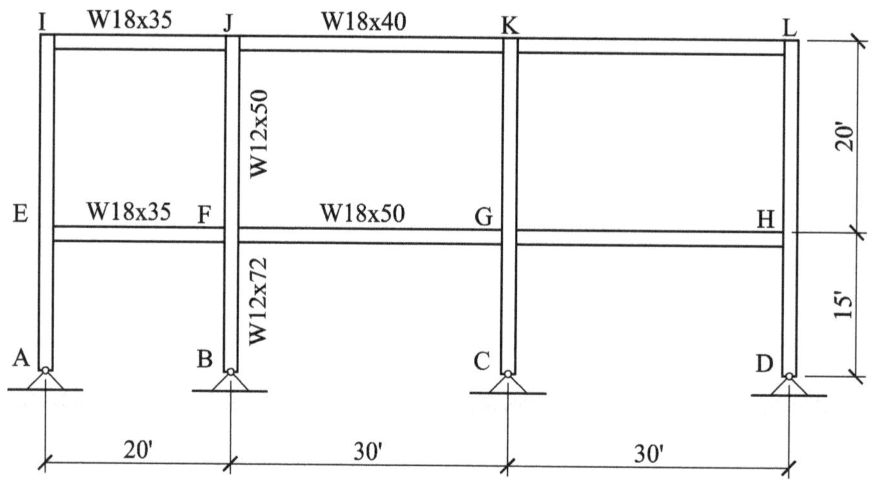

FIGURE 5-12 Moment frame for Example 5-10

Solution

Since this is a moment frame, this implies that the side-sway is *uninhibited*. Therefore, the sideways-uninhibited alignment chart in Figure 5-11 will be used.

The moments of inertia for the given column and girder sections are tabulated as follows:

Member	Section	I_{xx} (in.^{4})	Length (ft.)	I_{xx}/L
FJ	W12 × 50	391	20	19.6
BF	W12 × 72	597	15	39.8
IJ, *EF*	W18 × 35	510	20	25.5
JK	W18 × 40	612	30	20.4
FG	W18 × 50	800	30	26.7

1. Elastic behavior in the column

For elastic behavior in the column, $\tau_b = 1.0$;

$\tau_g = 1.0$ (given)

Column *BF*:

Joint B: The bottom of column *BF* is supported by a spread footing that provides little or no moment restraint to the column; therefore, it is assumed to be pinned. Thus,

$G_A = 10$ (This is the practical value recommended in the *AISCM* as discussed previously.)

***Joint F*:**

The members framing into joint *F*: Girders *FE* and *FG*, and Columns *FB* and *FJ*

$\tau_b = 1.0$ and $\tau_g = 1.0$

$$G_B = \frac{\sum\left(\frac{\tau_b E_c I_c}{L_c}\right)}{\sum\left(\frac{\tau_g E_g I_g}{L_g}\right)} = \frac{E\left[(1.0)(39.8)+(1.0)(19.6)\right]}{E\left[(1.0)(25.5)+(1.0)(26.7)\right]} = 1.14 \text{ (top of column } BF)$$

Entering the alignment chart for unbraced frames (see Figure 5-13a) with a G_A of 10 at the bottom of column *BF* and a G_B of 1.14 at the top of the column and joining these two points with a straight line yields a ***K* value of 1.93** for column BF, as shown in Figure 5-13a.

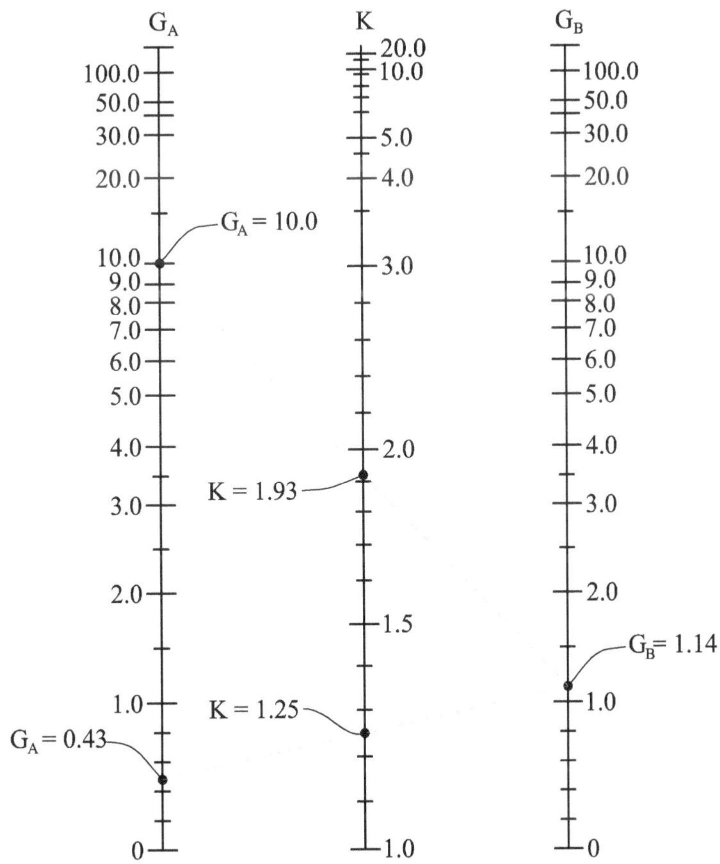

a) elastic behavior

FIGURE 5-13a Alignment chart for Example 5-10

Column *FJ*:

$\tau_b = 1.0$ and $\tau_g = 1.0$

Joint *F*:

The members framing into joint *F*: Girders *FE* and *FG*, and Columns *FB* and *FJ*

$$G_A = \frac{\sum\left(\tau_b \frac{E_c I_c}{L_c}\right)}{\sum\left(\tau_g \frac{E_g I_g}{L_g}\right)} = \frac{E\left[(1.0)(39.8)+(1.0)(19.6)\right]}{E\left[(1.0)(25.5)+(1.0)(26.7)\right]} = 1.14 \quad \text{(bottom of column } FJ\text{)}$$

Joint *J*:

The members framing into joint *J*: Girders *JI* and *JK*, and Column *JF*

$$G_B = \frac{\sum\left(\tau_b \frac{E_c I_c}{L_c}\right)}{\sum\left(\tau_g \frac{E_g I_g}{L_g}\right)} = \frac{E\left[(1.0)(19.6)\right]}{E\left[(1.0)(25.5)+(1.0)(20.4)\right]} = 0.43 \quad \text{(top of column } FJ\text{)}$$

Enter the alignment chart for moment frames (see Figure 5-13a) with a G_A of 1.14 at the bottom of column *FJ* (i.e., at the second-floor level) and a G_B of 0.43 at the top of the column (i.e., at the roof level), and joining these two points on the Alignment Charts with a straight line yields a ***K* value of 1.25** for column FJ, as shown in Figure 5-13a. Note that because of the symmetry of the alignment charts, G_A could be on the left vertical axis while G_B could be on the right vertical axis, or vice versa. Thus, the same value of K of 1.25 is obtained with a G_A of 0.43 and a G_B of 1.14.

2. Inelastic behavior in the column

Column	Size	Area, A_g, in.^{2}	Factored axial load, P_u, kips	P_u/A_g	Column Stiffness reduction factor,* τ_b
BF	W12 × 72	21.1	590	28.0	0.986
FJ	W12 × 50	14.6	140	9.6	1.0

* From *AISCM* Table 4-13 for LRFD and F_y = 50 ksi

Column *BF*:

Joint *B*: The bottom of column *BF* is supported by a spread footing; therefore, it is assumed to be pinned. Thus,

G_A = 10 (This is the practical value recommended in the *AISCM* as discussed earlier.)

Joint *F*:

τ_b = 0.986 for Column BF or FB
τ_b = 1.0 for Column FJ
and τ_b = 1.0

$$G_B = \frac{\sum\left(\frac{\tau_b E_c I_c}{L_c}\right)}{\sum\left(\frac{\tau_g E_g I_g}{L_g}\right)} = \frac{E\left[(0.986)(39.8)+(1.0)(19.6)\right]}{E\left[(1.0)(25.5)+(1.0)(26.7)\right]} = 1.13 \quad \text{(top of column } BF\text{)}$$

Entering the alignment chart (see Figure 5-13b) for unbraced frames with a G_A of 10 at the bottom of column *BF* and a G_B of 1.13 at the top of the column and joining these two points with a straight line yields a ***K*-value of 1.91** for column BF assuming inelastic column behavior, as shown in Figure 5-13b. Note that the difference between the effective length factor calculated assuming column inelasticity and that obtained previously in Part (1) of this example is negligible for this low-rise building example. Thus, the effect of inelasticity is negligible for this example. It is common in design practice to conservatively assume elastic behavior for columns in low-rise buildings.

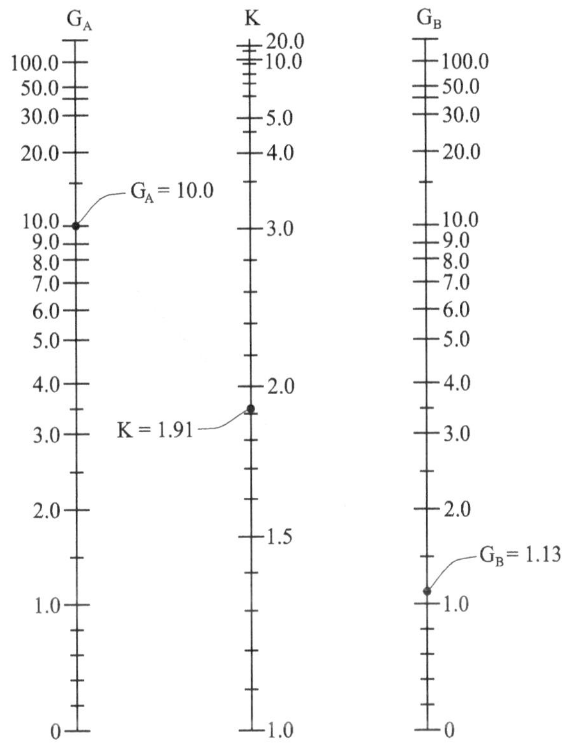

FIGURE 5-13b Alignment chart for Example 5-10

Column *FJ*:

Column *FJ* is unchanged from the solution in part 1 because the stiffness reduction factor for this column is 1.0; therefore, the effective length factor will be as calculated in part 1; that is, ***K* is 1.25** for column *FJ*.

Comparison of Figure 5-4 and the Alignment Charts:

Note that if the effective length factors from Figure 5-4 had been used, the values of the effective length factors for columns *BF* and *FJ* would have been 2.4 (i.e., case f) and 1.2 (i.e., case d), respectively. Compare these effective length factors to the values of 1.93 and 1.25 obtained for columns *BF* and *FJ*, respectively, using the Alignment Charts or Nomograms and assuming elastic column behavior.

5-8 TORSIONAL AND FLEXURAL-TORSIONAL BUCKLING OF COMPRESSION MEMBERS

As already discussed earlier in this chapter, when a structural member (e.g., a doubly symmetric I-shape) is subjected to bending about its stronger axis, the longitudinal compression stresses induced in the compression zone of the structural member result in an out of plane bending of the member about its weaker axis, causing *flexural buckling*. Flexural buckling is dependent on how well the structural member is laterally braced against bending about its weaker axis. *Torsional buckling*, on the other hand, results from the twisting or the torsional rotation of a structural member under load about its longitudinal axis. *Flexural-torsional buckling* involves both torsional buckling (twisting of the member) and flexural buckling (bending of the member about its weaker axis), resulting in an out-of-plane lateral displacement of the member. When a structural member subjected to axial compression load buckles, it fails in one of these three modes (i.e., flexural buckling, torsional buckling, or flexural-torsional buckling). The failure mode that results in the lowest elastic buckling stress, F_e, will control the strength of the member.

In this section, we will cover the calculation of the axial compression load capacity of singly symmetric shapes (e.g., double angles) and unsymmetric shapes (e.g., single angles) that may fail in torsional buckling or flexural-torsional buckling. This section is applicable to the following members:

1. Special doubly symmetric sections such as cruciform-shaped columns or built-up columns without slender elements. Refer to Table 5-1 for the limiting width-to-thickness ratios to qualify as a nonslender element.
2. Single angles with $b/t > 0.71\sqrt{\frac{E}{F_y}}$, where b is the length of the longer leg of the angle and t is the thickness; in accordance with Section E5 of the AISC Specification, the axial load capacity for single angles with $b/t > 0.71\sqrt{\frac{E}{F_y}}$ is determined from Section E4 of the AISC Specification. For ASTM A36 steel, the $0.71\sqrt{\frac{E}{F_y}}$ limit is equal to 20, and is 17 for ASTM Grade 50 steel. Angles listed in the *AISCM* have b/t ratios no greater than 20. Note that for single angles with $b/t \le 0.71\sqrt{\frac{E}{F_y}}$, the flexural-torsional buckling limit state does not need to be considered, and the axial compression load capacity is calculated using the provisions of Section E3 of the AISC Specification; that is, equations (5-9), (5-14), or (5-15) are used to calculate the axial compression load capacity.
3. All doubly symmetric sections where the torsional unbraced length exceeds the flexural unbraced length for buckling about the minor axis of the column.

The axial load capacity for *single angles*, *double angles*, *tee-shapes (WT)*, and *built-up compression members* is calculated as follows:

The nominal axial compressive strength is

$$P_n = A_g F_n \tag{5-22}$$

For LRFD, the design axial compressive strength is

$$\phi_c P_n = \phi_c A_g F_n \tag{5-23}$$

For ASD, the allowable axial compression load is

$$P_a = \frac{P_n}{\Omega_c} \qquad (5\text{-}24)$$

Where,

$\phi_c = 0.9$

$\Omega_c = 1.67$

The elastic buckling stress, F_e, for torsional and torsional-flexural buckling failure modes are obtained using equations (5-25) through (5-29) (see Section E4 of the AISC Specification). These equations have been derived from elastic buckling or structural stability theory [6]. Inelastic buckling behavior is considered by using the F_e from these equations in equation (5-14) or equation (5-15). Note that the elastic flexural buckling stress, F_e, from equation (5-9) for the flexural buckling failure mode should also be checked. The smallest of the elastic buckling stress (F_e) values from equations (5-25) through (5-29) and equation (5-9) will give the elastic buckling stress, F_e, that will be used in either equations (5-14) or (5-15).

A. Doubly symmetric members and Z-shapes twisting about the shear center (*e.g., doubly symmetric I-shape, cruciform shape, Z-shapes*):
The nominal stress, F_n, for the limit states of torsional and lateral-torsional buckling is determined using equations (5-14) or (5-15), and the elastic torsional buckling stress, F_e, is calculated from equation (5-25) or AISC Specification Equation E4-2:

$$F_e = \left[\frac{\pi^2 E C_w}{(L_{cz})^2} + GJ\right]\frac{1}{I_x + I_y} \qquad (5\text{-}25)$$

B. For **singly symmetric members twisting about the shear center and where the *y-axis* is the axis of symmetry,** such as WT and double angle sections, use equations (5-14) or (5-15) with the elastic torsional buckling stress, F_e, calculated from equation (5-26) or AISC Specification Equation E4-3:

$$F_e = \left(\frac{F_{ey} + F_{ez}}{2H}\right)\left[1 - \sqrt{1 - \frac{4F_{ey}F_{ez}H}{(F_{ey} + F_{ez})^2}}\right] \qquad (5\text{-}26)$$

Note: For singly symmetric sections where the *x-axis* is the axis of symmetry, such as channels, replace F_{ey} in equation (5-26) with F_{ex}.

C. For unsymmetric members twisting about its shear center (*e.g., unsymmetric built-up section, and single angles with b/t* > $0.71\sqrt{\frac{E}{F_y}}$ (i.e., b/t > *20 for ASTM A36 steel; note that there are no hot-rolled angle sections of A36 steel in the AISCM for which this condition would apply*)), use equations (5-14) or (5-15), with the elastic buckling stress, F_e, calculated as the smallest root of the following cubic equation from equation (5-27) or AISC Specification Equation E4-4 [1]:

$$(F_e - F_{ex})(F_e - F_{ey})(F_e - F_{ez}) - F_e^2(F_e - F_{ey})\left(\frac{x_o}{\bar{r}_o}\right)^2 - F_e^2(F_e - F_{ex})\left(\frac{y_o}{\bar{r}_o}\right)^2 = 0 \qquad (5\text{-}27)$$

Where,

A_g = Gross cross-sectional area of member, in.^{2}

C_w = Warping torsional constant, in.^{6} (see notes below)

G = Shear modulus of elasticity of steel = 11,200 ksi

I_x, I_y = Moment of inertia about the x–x and y–y axes, respectively, in.^{4}

J = Torsional constant of the member, in.^{4} (**Note**: for double angles, $J = 2 \times J$ for a single angle)

K_x = Effective length factor for flexural buckling about the x–x axis

K_y = Effective length factor for flexural buckling about the y–y axis

K_z = Effective length factor for torsional buckling about the longitudinal axis of the member

L_x = unbraced length for flexural buckling about the x-axis

L_y = unbraced length for flexural buckling about the y-axis

L_z = unbraced length for torsional buckling about the longitudinal axis of the member

$L_{cx} = K_x L_x$ = effective length for flexural buckling about the x-axis

$L_{cy} = K_y L_y$ = effective length for flexural buckling about the y-axis

$L_{cz} = K_z L_z$ = effective length for torsional buckling about the longitudinal axis of the member

$\bar{r}_o$ = Polar radius of gyration about the shear center, in.

r_x = Radius of gyration about the x–axis, in.

r_y = Radius of gyration about the y–axis, in.

x_o, y_o = Coordinates of the *shear center* with respect to the centroid of the section in

$$\bar{r}_o^2 = x_o^2 + y_o^2 + \frac{I_x + I_y}{A_g}$$

$$H = 1 - \frac{x_o^2 + y_o^2}{\bar{r}_o^2} \text{ = a flexural constant}$$

$$F_{ex} = \frac{\pi^2 E}{\left(\frac{L_{cx}}{r_x}\right)^2}$$

$$F_{ey} = \frac{\pi^2 E}{\left(\frac{L_{cy}}{r_y}\right)^2}$$

$$F_{ez} = \left[\frac{\pi^2 E C_w}{(L_{cz})^2} + GJ\right]\frac{1}{A_g \bar{r}_o^2}$$

Notes:

- For **tee-shapes** and **double angles**: Omit the term with C_w in the F_{ez} equation and assume $x_o = 0$.
- For **doubly-symmetric I-shaped** sections: $C_w \approx I_y h_o^2/4$, where h_o is the distance between the centroids of the flanges, i.e., $h_o = d - \frac{t_f}{2} - \frac{t_f}{2} = d - t_f$
- For double angles, $C_w = 2\,C_w$ of a single angle
- The warping torsional constant, C_w for shapes made of thin rectangular plate elements with centerlines that intersect at a common point (e.g., cruciform shapes) is approximately zero and may be neglected for small plate thicknesses [3]. For sections with thick plates, C_w will not be negligible and should be calculated.

- The shear center of a cross-section is the point through which a load can be applied to the member without causing twist in the member. The location of the shear center (x_o, y_o) for a given cross-section can be determined using the procedures described in Ref. [2]. Shear center locations for some unsymmetric and singly symmetric shapes are shown in Figure 5-14 [3, 4].

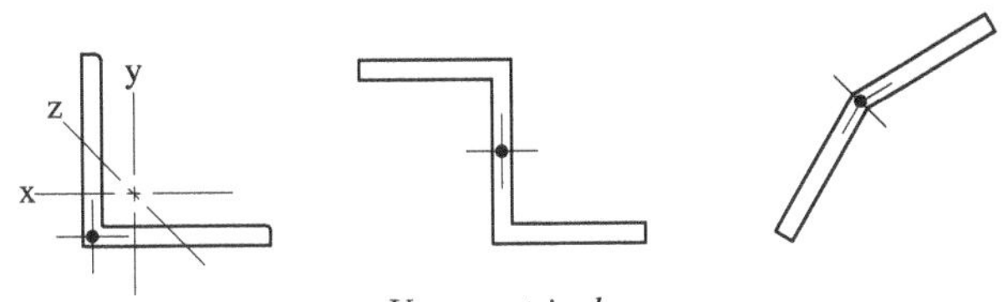

a. Unsymmetric shapes

shear center denoted as "●"

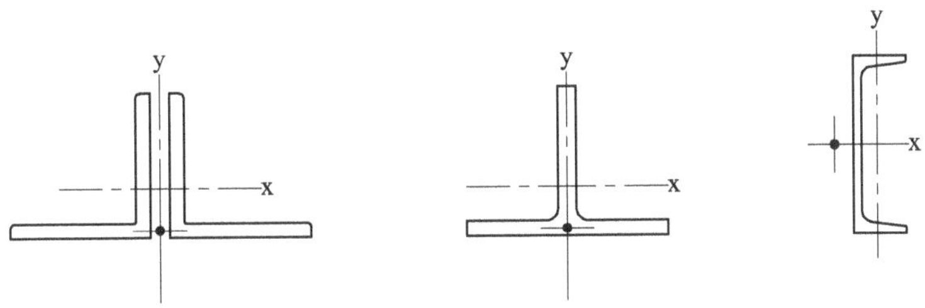

b. Singly symmetric shapes

FIGURE 5-14 Shear centers for unsymmetric and singly symmetric shapes

Torsional Buckling of Members with Lateral Bracing Offset from the Shear Center (AISC Specification Sections E4d and E4e)

There are certain details used in practice where the location of the lateral bracing of a member is offset from their shear center, resulting in twisting about the shear center because the lateral brace only resists lateral movement, and thus the laterally braced member is susceptible to a combination of flexural buckling and constrained-axis torsional buckling. Examples include the following situations:

1. Laterally braced wide flange columns where only *one* flange of the column is laterally braced within the height of the column, thus the lateral brace is offset from the shear center of the column. In this case, if the *lateral bracing were located at the mid-height of the column* and oriented parallel to the x-axis, the unbraced lengths with only one flange laterally braced will be as follows:

   a. Buckling about the x-axis: unbraced length, $L_{cx} = L$, the full height of the column.

   b. Buckling about the y-axis: unbraced length, $L_{cy} = 0.5L$

   c. Torsional buckling about the longitudinal axis, $L_{cz} = L$, the full height of the column.

2. Laterally braced wide flange columns, where a column is laterally braced at the shear center within its height. It is assumed that the lateral brace does *not* possess any in-plane flexural rigidity to restrain the column from torsion or twisting. In this case, if the *lateral bracing were located at the mid-height of the column* at the shear center of the column and parallel to the x-axis, the unbraced lengths of the column will be as follows:

a. Buckling about the x-axis: unbraced length, $L_{cx} = L$, the full height of the column.

b. Buckling about the y-axis: unbraced length, $L_{cy} = 0.5L$

c. Torsional buckling about the longitudinal axis, $L_{cz} = L$, the full height of the column.

3. Laterally braced wide flange columns, where the column is laterally braced at the shear center within its height. It is assumed that the lateral brace has sufficient in-plane flexural rigidity to restrain the column from torsion or twisting about its longitudinal axis. In this case, if the *lateral bracing were located at the mid-height of the column* at the shear center of the column and oriented parallel to the x-axis, the unbraced lengths of the column would be as follows:

   a. Buckling about the x-axis: unbraced length, $L_{cx} = L$, the full height of the column.

   b. Buckling about the y-axis: unbraced length, $L_{cy} = 0.5L$

   c. Torsional buckling about the longitudinal axis, $L_{cz} = 0.5L$

4. Laterally braced wide-flange columns, where the column is laterally braced within its length at *both* flanges. In this case, if the lateral bracing were located at the mid-height of the column at both flanges and oriented parallel to the x-axis, the unbraced lengths of the column would be as follows:

   a. Buckling about the x-axis: unbraced length, $L_{cx} = L$, the full height of the column.

   b. Buckling about the y-axis: unbraced length, $L_{cy} = 0.5L$

   c. Torsional buckling about the longitudinal axis, $L_{cz} = 0.5L$

5. Roof trusses with wide-flange top chord in compression where the lateral bracing of the top chord is connected to the top flange of the top chord , offset from its shear center. The unbraced lengths of the truss top chord would be as follows:

   a. Buckling about the x-axis: unbraced length, $L_{cx} = L$, the full length of the truss top chord between the truss panel points.

   b. Buckling about the y-axis: unbraced length, $L_{cy} = 0.5L$

   c. Torsional buckling about the longitudinal axis, $L_{cz} = L$, the full length of the truss top chord between the truss panel points.

D. For doubly symmetric I-shaped members with ***minor*** axis *lateral bracing offset from the shear center*, the torsional or flexural-torsional elastic buckling stress, F_e, is calculated as follows (*AISC* Specification Equation E4-10):

$$F_e = \left[\frac{\pi^2 EI_y}{(L_{cz})^2}\left(\frac{h_o^2}{4} + y_a^2\right) + GJ\right]\frac{1}{A_g r_o^2} \tag{5-28}$$

Where,

h_o = distance between flange centroids in inches = d – 2k

The polar radius of gyration, $r_o^2 = r_x^2 + r_y^2 + y_a^2 + x_a^2$

x_a = bracing offset distance along x-axis = 0

y_a = bracing offset distance along y-axis in inches

E. For doubly symmetric I-shaped members with ***major*** axis *lateral bracing offset from the shear center*, the torsional or flexural-torsional elastic buckling stress, F_e, is calculated as follows (*AISC* Specification Equation E4-12):

$$F_e = \left[\frac{\pi^2 EI_y}{(L_{cz})^2} \left(\frac{h_o^2}{4} + \frac{I_x}{I_y} x_a^2 \right) + GJ \right] \frac{1}{A_g r_o^2} \qquad (5\text{-}29)$$

Where,

The polar radius of gyration, $r_o^2 = r_x^2 + r_y^2 + y_a^2 + x_a^2$

x_a = bracing offset distance along x-axis in inches

y_a = bracing offset distance along y-axis = 0

According to the *AISC* Specification Commentary on Section E4 [1], "It is important to recognize that as a lateral brace [for a member] is moved away from the shear center [of the member], it permits the flexural buckling stiffness to interact with the torsional buckling stiffness to yield a constrained-axis torsional buckling strength that is less than the pure torsional buckling strength had the brace not existed. Figure C-E4.3 in the *AISCM* [1] shows that, for a column with lateral and torsional restraint at the ends only and lateral bracing at mid-height, the torsional strength is greatest when the lateral brace is located at the shear center and diminishes, approaching the weak-axis flexural buckling strength for a column with an effective length equal to L_{cz}, as the lateral brace moves away from the shear center."

When the bracing is offset from the shear center, the buckling that results is called constrained axis torsional buckling; buckling in this mode will exhibit twisting, though the braced member is restrained laterally.

F. Built-up Members:

Built-up members shall meet the following requirements, and the individual components shall be connected with connectors spaced at a distance, *a*, such that the effective slenderness ratio, $K_i a/r_i$, of each individual component shape between the connectors does not exceed 0.75 times the governing slenderness ratio (i.e., the *maximum* slenderness ratio) of the *built-up* member, where r_i is the smallest or least radius of gyration of each component part.

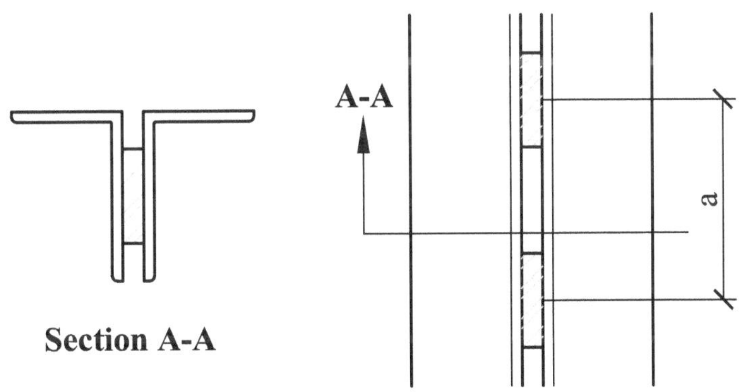

FIGURE 5-14b Double angle with spacers

The capacity of built-up compression members comprised of two shapes (e.g., double angles) that are connected intermittently by filler plates or spacers (see Figure 5-14b) is impacted by the spacing of the filler plates because of the slippage from interface shear that occurs between the angles. Another example are two wide flange columns placed side by side (flange tip to flange tip) and connected with fillet-welded stitch plates to the outer faces of the flanges of both shapes. The slippage from the interface shear that occurs between the two connected shapes is considered in *AISC Specification* Section E6 using the following prescribed effective slenderness ratios, which are dependent on the spacing of the connectors (e.g., the filler plates or stitch plates) and how they are connected to the two shapes that comprise the built-up member.

a. For intermediate connectors that are *bolted* snug-tight in *double-angle* compression members, *Kl*/*r* is replaced by $(KL/r)_m$ as follows:

$$\frac{KL}{r} = \left(\frac{KL}{r}\right)_m = \sqrt{\left(\frac{KL}{r}\right)_o^2 + \left(\frac{a}{r_i}\right)^2} \tag{5-30}$$

b. For intermediate connectors that are *welded* or connected by means of *pre-tensioned bolts (with Class A or B fraying surfaces)* in *double-angle* compression members

i. When $\frac{a}{r_i} \leq 40$

$$\frac{KL}{r} = \left(\frac{KL}{r}\right)_m = \left(\frac{KL}{r}\right)_o \tag{5-31}$$

ii. When $\frac{a}{r_i} > 40$

$$\frac{KL}{r} = \left(\frac{KL}{r}\right)_m = \sqrt{\left(\frac{KL}{r}\right)_o^2 + \left(\frac{K_i a}{r_i}\right)^2}, \tag{5-32}$$

Where,

$\frac{K_i a}{r_i} \leq 0.75\left(\frac{KL}{r}\right)_{max}$ (see *AISCM* Section E6.2a)

$\left(\frac{KL}{r}\right)_m = \left(\frac{L_c}{r}\right)_m$ = Modified slenderness ratio of the built-up member

$\left(\frac{KL}{r}\right)_o = \left(\frac{L_c}{r}\right)_o$ = Slenderness ratio of the built-up member acting as a unit in the buckling direction being addressed.

$L_c = KL$ = effective length of the built-up member
K_i = 0.50 for angles back-to-back
= 0.75 for channels back-to-back
= 0.86 for all other cases
a = longitudinal spacing between connectors, in.
r_i = Minimum radius of gyration of the individual components that make up the built-up member (for double angles, $r_i = r_z$ of one of the angles)
r = smallest radius of gyration of the built-up section = smaller of r_x and r_y of the double angles

EXAMPLE 5-11

Axial Compression Load Capacity of a WT section

Determine the nominal axial load capacity, the LRFD design axial load capacity, and the ASD allowable axial load of a WT18×128 section that is used as the top chord of a steel transfer truss (see the transfer truss in Figure 6-36). Assume the unbraced length between panel points is 6 ft. 3 in. Use ASTM A992 Grade 50 steel.

Solution

F_y = 50 ksi
Unbraced length, L_b = 6.25 ft = 75 inches

The section properties of WT 18×128 from *AISCM* Part 1 are as follows:

$A_g = 37.6 \text{ in.}^2$
$d = 18.7 \text{ in.}$
$S_x = 87.4 \text{ in.}^3$
$Z_x = 156 \text{ in.}^3$
$I_x = 1210 \text{ in.}^4$
$I_y = 264 \text{ in.}^4$
$J = 26.4 \text{ in.}^4$
$C_w = 205 \text{ in.}^6$
$r_x = 5.66 \text{ in.}$
$r_y = 2.65 \text{ in.}$
$\bar{y} = 4.92 \text{ in.}$
$t_f = 1.73 \text{ in.}$

$$\frac{b_f}{2t_f} = 3.53$$

$$\frac{d}{t_w} = 19.5$$

For possible flexural buckling failure mode, calculate the slenderness ratios for x-axis bending and y-axis bending, respectively:

$$\frac{L_{cx}}{r_x} = \frac{K_x L_x}{r_x} = \frac{(1.0)(75 \text{ in.})}{5.66 \text{ in.}} = 13.3$$

$$\frac{L_{cy}}{r_y} = \frac{K_y L_y}{r_y} = \frac{(1.0)(75 \text{ in.})}{2.65 \text{ in.}} = 28.3 < 200 \text{ (}y\text{-}y\text{ axis controls for flexural buckling)}$$

Therefore, if the member is going to fail by flexural buckling failure, the y-axis will control. It remains to be checked to ascertain whether flexural buckling or torsional buckling will control the strength of this WT section.

Check if the flange and stem of the WT are slender or not for axial compression using the limits in Table 5-1 or *AISCM* Table B4.1a:

Limit for Flange of WT (Case 1 of Table 5-1 or *AISCM* Table B4.1a):

$$\lambda_r = 0.56\sqrt{\frac{E}{F_y}} = 0.56\sqrt{\frac{29{,}000 \text{ ksi}}{50 \text{ ksi}}} = 13.5$$

$$\lambda \textit{ for the flange of WT} = \frac{b}{t} = \frac{b_f}{2t_f} = 3.53 < \lambda_r = 13.5$$

Therefore, the flange of the WT is not slender.

Limit of Stem or Web of WT (Case 4 of Table 5-1 or *AISCM* Table B4.1a):

$$\lambda_r = 0.75\sqrt{\frac{E}{F_y}} = 0.75\sqrt{\frac{29{,}000 \; ksi}{50 \; ksi}} = 18.06$$

$$\lambda \text{ for the stem of WT} = \frac{d}{t} = \frac{d}{t_w} = 19.5 > \lambda_r = 18.06$$

Therefore, the stem of the WT is slender.

Use AISC specification Section E7 for the stem to determine the effective area, A_e

Since the WT section is singly symmetric, use the equations from Section 5-8 to check the torsional buckling failure mode. The torsional elastic buckling stress will be compared to the flexural elastic buckling stress (equation (5-9)), and the smaller value will determine the axial compression strength of the WT section:

$$x_o = 0$$

$$y_o = \bar{y} - \frac{t_f}{2} = 4.92 \text{ in.} - \frac{1.73 \text{ in.}}{2} = 4.06 \text{ in.}$$

$$\bar{r}_o^2 = x_o^2 + y_o^2 + \frac{I_x + I_y}{A_g} = (0)^2 + (4.06)^2 + \frac{1210 + 264}{37.6} = 55.7 \text{in.}^2$$

$$\bar{r}_o = 7.46 \text{ in.}$$

$$H = 1 - \frac{x_o^2 + y_o^2}{\bar{r}_o^2} = 1 - \frac{(0)^2 + (4.06)^2}{(7.46)^2} = 0.704$$

The elastic flexural buckling stress about the y-axis and the z-axis, respectively, are

$$F_{ey} = \frac{\pi^2 E}{\left(\frac{L_{cy}}{r_y}\right)^2} = \frac{\pi^2(29{,}000)}{\left(\frac{75}{2.65}\right)^2} = 357.3 \text{ksi}$$

$$F_{ez} = \left[\frac{\pi^2 E C_w}{(L_{cz})^2} + GJ\right]\frac{1}{A_g \bar{r}_o^2} =$$

$$= \left[\frac{\pi^2(29{,}000 \text{ ksi})(205 \text{ in.}^6)}{(75 \text{ in.})^2} + (11200 \text{ ksi})(26.4 \text{ in.}^4)\right]\frac{1}{(37.6 \text{ in.}^2)(7.46 \text{ in.})^2} = 146.3 \text{ ksi}$$

From equation (5-26), the elastic torsional buckling stress is determined as follows:

$$F_e = \left(\frac{F_{ey} + F_{ez}}{2H}\right)\left[1 - \sqrt{1 - \frac{4F_{ey}F_{ez}H}{(F_{ey} + F_{ez})^2}}\right]$$

$$= \left(\frac{357.3\text{ksi} + 146.3\text{ksi}}{(2)(0.704)}\right)\left[1 - \sqrt{1 - \frac{(4)(357.3\text{ksi})(146.3\text{ksi})(0.704)}{(357.3\text{ksi} + 146.3\text{ksi})^2}}\right] = 126 \text{ ksi}$$

Since the elastic torsional buckling stress F_e is smaller than the elastic flexural buckling stresses, F_{ey} and F_{ez}, flexural buckling does not control, and the WT section will fail in torsional buckling. Note that F_{ex} is much larger than F_{ey} since the x-axis had a smaller slenderness ratio than the y-axis. Since the elastic torsional buckling stress, F_e = 126 ksi > $0.44F_y$ = 0.44(50) = 22 ksi, therefore the WT section buckles inelastically, and therefore, use equation (5-14) to determine the nominal stress for the WT section as follows:

$$F_n = \left[0.658^{\frac{F_y}{F_e}}\right]F_y = \left[0.658^{\frac{50}{126}}\right](50) = 42.35 \text{ ksi}$$

Next, use Section 5-4 of the text (i.e., *AISCM* Section E7) to determine the effective area, A_e. Only the WT stem will be checked since the flange is not a slender element.

From Table 5-1 (i.e., AISC Specification Table B4.1a), the limiting width-to-thickness ratio for the WT stem (Case 4) is

$$\lambda_r = 0.75\sqrt{\frac{E}{F_y}} = 0.75\sqrt{\frac{29{,}000 \text{ ksi}}{50 \text{ksi}}} = 18.06$$

Using equation (5-19), we obtain

$$\lambda_r\sqrt{\frac{F_y}{F_n}} = (18.06)\sqrt{\frac{50\,\text{ksi}}{42.35\,\text{ksi}}} = 19.6$$

Since λ for the stem of WT $= \dfrac{d}{t_w} = 19.5 < \lambda_r\sqrt{\dfrac{F_y}{F_n}} = 19.6$,

Therefore, from equation (5-19a) (i.e., AISC Specification Equation E7-2), $d_e = d$.

Since $d_e = d$, the reduction in the area of the slender stem $= (d - d_e)t_w = 0$, therefore there is no local buckling.

The summation of the effective areas of all elements of the WT section is

$A_e = A_g$ – total reduction in area $= A_g - 0 = 37.6\text{ in.}^2 - 0\text{ in.}^2 = 37.6\text{ in.}^2$

The nominal axial compression strength from equation (5-18) (i.e., AISC Specification Equation E7-1) is

$P_n = F_nA_e = (42.35\text{ ksi})(37.6\text{ in.}^2) =$ **1592 kips**

LRFD:

The design axial compression load capacity, $\phi_c P_n = (0.9)(1592) = 1433$ kips.

ASD:

The allowable axial compression load is

$$P_a = \frac{P_n}{\Omega_c} = \frac{1592\text{ kips}}{1.67} = 953\text{kips}.$$

Axial Compression Load Capacity of cruciform Sections

There are two common types of cruciform shapes: the plain cruciform built-up section made up of rectangular steel plates, and the flanged cruciform sections made up of one W- section and two WT sections (see Figure 5-15). Plain cruciform shapes are sometimes used as monumental columns in atrium areas of high-rise buildings and as compression members in transmission towers. Flanged cruciform shapes are used as mega-columns in high-rise buildings when higher axial load capacities are required, which cannot be achieved by a single rolled section. They are also used in orthogonal moment frames where the column resists lateral load moments from moment frames in the two orthogonal directions. The possible failure modes in axial compression for cruciform sections include flexural buckling, torsional buckling, and flexural-torsional buckling. To determine the strength of these sections, the elastic flexural buckling stress is calculated using equation (5-9) and the elastic torsional buckling stress is calculated using equation (5-25). The smaller of the two values will determine the elastic buckling stress that will be used in either equation (5-14) or equation (5-15) to obtain the nominal stress, F_n, which is then used in conjunction with the equations for effective area from Section 5-4 (if the section has slender elements) to calculate the axial compression load capacity. As cruciform shapes gain popularity in their use, some researchers have investigated these sections numerically and experimentally [8, 9, 10], though experimental testing of these sections, especially plain cruciform sections in bending considering lateral torsional buckling, is sparse. In the next two examples, we will examine the axial compression load capacity for a plain plate cruciform shape and a flanged cruciform shape.

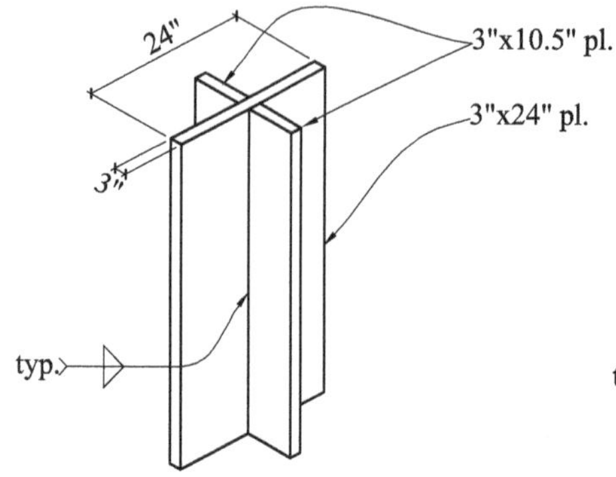

Figure 5-15a: Plain cruciform shape

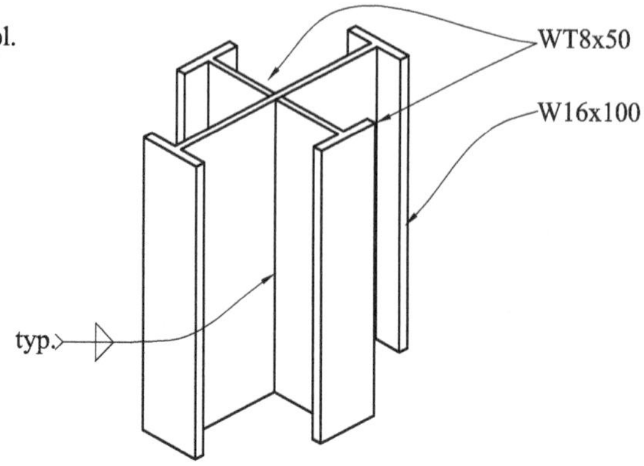

Fig 5-15b: Flanged cruciform shape

FIGURE 5-15 Plain and flanged cruciform sections

EXAMPLE 5-12

Axial Compression Load Capacity of a Plain cruciform Section

A built-up plain cruciform section is made from one 24 in. by 3 in. thick plate welded to two 10.5 in. by 3 in. thick plates (see Figure 5-15a). The built-up section has an unbraced length of 18 ft., and the plates are made of ASTM A992 Grade 50 steel. Determine the nominal axial load capacity, the LRFD design axial load capacity, and the ASD allowable axial load of the section.

Solution

t = 3 in.

h = 24 in.

F_y = 50 ksi

Unbraced length, L_b = 18 ft. = 216 inches

The section properties for the cruciform shape are calculated as follows:

A_g = (24 in.)(3 in.) + 2(10.5 in.)(3 in.) = 135 in.^{2}

C_w for a doubly symmetric section with very thin plates is approximately = 0 in.^{6}, but the 3 in. thick plates are *not* considered thin plates, so we will calculate the actual C_w for this plain cruciform section as follows [8]:

$$C_w = \frac{t^3h^3}{72} = \frac{(3\text{ in.})^3(24\text{ in.})^3}{72} = 5184\text{ in.}^6$$

$$J = \sum_{i=1}^{n}\frac{b_it_i^3}{3} = \frac{1}{3}(24\text{ in.})(3\text{ in.})^3 + \frac{1}{3}(2)(10.5\text{ in.})(3\text{ in.})^3 = 405\text{ in.}.^4$$

$$I_x = \frac{1}{12}(3\text{ in.})(24\text{ in.})^3 + \frac{1}{12}(2)(10.5\text{ in.})(3\text{ in.})^3 = 3503\text{ in.}^4$$

$$I_y = \frac{1}{12}(3\text{ in.})(24\text{ in.})^3 + \frac{1}{12}(2)(10.5\text{ in.})(3\text{ in.})^3 = 3503\text{ in.}^4$$

$$r_x = \sqrt{\frac{I_x}{A_g}} = \sqrt{\frac{3503}{135}} = 5.09\text{ in.}$$

$$r_y = \sqrt{\frac{I_y}{A_g}} = \sqrt{\frac{3503}{135}} = 5.09 \text{ in.}$$

Check if the legs of the cruciform section are slender for axial compression using the limits in Table 5-1 (i.e., *AISCM* Table B4.1a):

Limit for leg of cruciform section (Case 3 of Table 5-1 or *AISCM* Table B4.1a):

$$\lambda_r = 0.45\sqrt{\frac{E}{F_y}} = 0.45\sqrt{\frac{29{,}000\, ksi}{50\, ksi}} = 10.8$$

$$\lambda \text{ for the leg of the cruciform section} = \frac{b}{t} = \frac{(24 \text{ in.} - 3 \text{ in.})/2}{3 \text{ in.}} = 3.5 < \lambda_r = 10.8$$

Therefore, the legs of the cruciform section are not slender.

Since the cruciform sections are not slender, that implies that local buckling does not occur; hence, the gross cross-sectional area of the cruciform section is effective in resisting axial compression loads.

Since the cruciform section is doubly symmetric, use the equations from Section 5-8 to check the torsional buckling failure mode. The elastic torsional buckling stress will be compared to the elastic flexural buckling stress from equation (5-9), and the smaller value will determine the strength of the cruciform section.

For doubly symmetric sections twisting about their shear center, elastic torsional buckling stress is obtained from equation (5-25) as follows:

$$F_e = \left[\frac{\pi^2 EC_w}{(L_{cz})^2} + GJ\right]\frac{1}{I_x + I_y} =$$

$$= \left[\frac{\pi^2(29{,}000 \text{ ksi})(5184 \text{ in.}^6)}{(216 \text{ in.})^2} + (11200 \text{ ksi})(405 \text{ in.}^4)\right]\frac{1}{(3503 \text{in.}^4 + 3503 \text{ in.}^4)}$$

$$= [31{,}802 + 4{,}536{,}000]\frac{1}{(3503 \text{ in.}^4 + 3503 \text{ in.}^4)}$$

$$= 652 \text{ ksi}$$

The reader can observe that the "*GJ*" is much larger than the value for the "EC_w" term in the above equation, indicating that the effect of the torsional warping constant on the elastic torsional buckling stress is negligible for this plain cruciform section. We could have easily neglected the "EC_w" term in the above equation, and the difference in the torsional elastic stress would be less than 1%. The "*GJ*" term is dominant in this example because of the large unbraced length of 216 inches. The effect of the "EC_w" term becomes more pronounced at small, unbraced lengths.

The elastic flexural buckling stress is calculated using equation (5-9) or AISC Equation E3-4 as follows:

$$F_e = \frac{\pi^2 E}{\left(\frac{L_c}{r}\right)^2} = \frac{\pi^2(29{,}000\,\text{ksi})}{\left(\frac{216\text{ in.}}{5.09}\right)^2} = 159\text{ ksi}$$

Since the elastic flexural buckling stress (159 ksi) is smaller than the elastic torsional buckling stress (652 ksi), this means that this cruciform section will fail in flexural buckling only. Therefore, use F_e = 159 ksi in equations (5-14) or (5-15).

Since F_e = 159 ksi > $0.44F_y$ = 0.44(50) = 22 ksi, therefore this WT section will buckle inelastically; hence, use equation (5-14) to determine the nominal stress as follows:

$$F_n = \left[0.658^{\frac{F_y}{F_e}}\right]F_y = \left[0.658^{\frac{50}{159}}\right](50) = 43.8\text{ ksi}$$

The nominal axial compression strength from equation (5-22) or AISC Specification Equation E4-1 is

$P_n = F_n A_g$ = (43.8 ksi)(135 in.^{2}) = **5913 kips**

LRFD:

The design axial load capacity for the plain cruciform section is

$\phi_c P_n$ = (0.9)(5913 kips) = 5322 kips.

ASD:

The allowable axial load for the plain cruciform section is

$$P_a = \frac{P_n}{\Omega_b} = \frac{5913\,\text{kips}}{1.67} = 3541\,\text{kips}$$

EXAMPLE 5-13

Axial Compression Load Capacity of a Flanged Cruciform Section

A built-up flanged cruciform section is made up of one W16 x 100 welded to two WT 8 x 50, as shown in Figure 5-15b. The built-up section has an unbraced length of 20 ft., and the plates are made of ASTM A992 Grade 36 steel. Determine the nominal axial load capacity, the LRFD design axial load capacity, and the ASD allowable axial load of the section.

Solution

F_y = 36 ksi

Unbraced length, L_b = 20 ft. = 240 inches

$d_{(W16\times100)}$ = 17.0 in.

$d_{(WT8\times50)}$ = 8.49 in.

t_w (for both W16 × 100 and WT 8x50) = 0.585 inch

$\frac{b_f}{2t_f}$ = 5.29 (for both W16 × 100 and WT8 × 50)

I_x (W16 × 100) = 1490 in.^{4}

I_y (W16 × 100) = 186 in.^{4}

A_g (W16 × 100) = 29.4 in.^{2}

C_w (W16 × 100) = 11900 in.^{6}

J (W16 × 100) = 7.73 in.^{4}
A_g for the cruciform section ≈ (2 W-sections)(29.4 in.^{2}) = 58.8 in.^{2}
C_w for the cruciform section ≈ (2 W-sections)(11,900 in.^{6}) = 23,800 in.^{6}
J for the cruciform section ≈ (2 W-sections)(7.73 in.^{4}) = 15.46 in.^{4}
I_x of cruciform section = $I_{x\ (W16 \times 100)} + I_{y\ (W16 \times 100)}$ = 1490 in.^{4} + 186 in.^{4} = 1676 in.^{4}
I_y of cruciform section = $I_{y\ (W16 \times 100)} + I_{x\ (W16 \times 100)}$ = 186 in.^{4} + 1490 in.^{4} = 1676 in.^{4}

$$r_x = \sqrt{\frac{I_x}{A_g}} = \sqrt{\frac{1676}{58.8}} = 5.34 \text{ in.}$$

$$r_y = \sqrt{\frac{I_y}{A_g}} = \sqrt{\frac{1676}{58.8}} = 5.34 \text{ in.}$$

Check the flanges of the W16 × 100 and WT 8 × 50 to determine if they are slender, using the limits in Table 5-1 (i.e., Case 1 of *AISCM* Table B4.1a):

$$\lambda_r = 0.56\sqrt{\frac{E}{F_y}} = 0.56\sqrt{\frac{29{,}000 \text{ ksi}}{36 \text{ ksi}}} = 15.9$$

$$\lambda \text{ for the flange of the W16} \times 100 \text{ and WT8} \times 50 \text{ sections} = \frac{b}{t} = \frac{b_f}{2t_f} = 5.29 < \lambda_r = 15.9$$

Therefore, the flange of the W16×100 and WT8×50 sections are not slender.

Check the Web of the W16×100 and the stem of the WT8 × 50 - Case 5 of Table 5-1 or *AISCM* Table B4.1a:

$$\lambda_r = 1.49\sqrt{\frac{E}{F_y}} = 1.49\sqrt{\frac{29{,}000 \text{ ksi}}{36 \text{ ksi}}} = 42.3$$

$$\lambda \text{ for the web of the W16} \times 100 \text{ section} = \frac{(d/2)}{t_w} = \frac{(17.0/2)}{0.585} = 14.5 < \lambda_r = 42.3$$

Therefore, the web of the W16×100 section is not slender. Note that d / 2 is used because the web of the W16×100 is laterally braced at the mid - depth by the orthogonal WT8 sections.

$$\lambda \text{ for the stem of the WT8} \times 50 \text{ section} = \frac{d}{t_w} = \frac{8.49}{0.585} = 14.5 < \lambda_r = 42.3$$

Therefore, the stem of the WT8×50 section is not slender.

Since the flanges of the W16 × 100 and the stem of the WT8 × 50 are not slender, local buckling is not applicable for this flanged cruciform section; hence, the gross area of the cruciform section is effective in resisting axial compression loads.

Since the cruciform section is doubly symmetric, use the equations from Section 5-8 to check the torsional buckling failure mode. The elastic torsional buckling stress will be compared to the elastic flexural buckling stress from equation (5-9) and the smaller value will determine the strength of the cruciform section. For doubly symmetric sections twisting about their shear center, the elastic torsional buckling stress is obtained from equation (5-25) as follows:

$$F_e = \left[\frac{\pi^2 EC_w}{(L_{cz})^2} + GJ\right]\frac{1}{I_x + I_y}$$

$$= \left[\frac{\pi^2(29{,}000 \text{ ksi})(23{,}800 \text{ in.}^6)}{(240 \text{ in.})^2} + (11200 \text{ ksi})(15.46 \text{ in.}^4)\right]\frac{1}{(1676 \text{ in.}^4 + 1676 \text{ in}^4)}$$

$$= [118{,}264 + 173{,}152]\frac{1}{(1676 \text{ in.}^4 + 1676 \text{ in.}^4)}$$

$$= 87 \text{ ksi}$$

The reader can observe that for this flanged cruciform section at the 20 ft. unbraced length, the "GJ" term is of the same order of magnitude as the "EC_w" term in the above equation, indicating that the effect of the torsional warping constant on the elastic torsional buckling stress is significant in this case. As the unbraced length decreases, the effect of the "EC_w" term will become even greater, to the point that it will become much larger than the "GJ" term. For example, at an unbraced length of 10 ft., the "EC_w" term is almost three times as large as the "GJ" term. Conversely, as the unbraced length becomes larger, the effect of the "EC_w" term will decrease significantly. For example, at an unbraced length of 50 ft., the "EC_w" term will become almost negligible with a value of approximately 10% of the "GJ" term.

The elastic flexural buckling stress is calculated using equation (5-9) or AISC Equation E3-4 as follows:

$$F_e = \frac{\pi^2 E}{\left(\frac{L_c}{r}\right)^2} = \frac{\pi^2(29{,}000 \text{ ksi})}{\left(\frac{240 \text{ in.}}{5.34}\right)^2} = 141.7 \text{ ksi}$$

Since the elastic flexural buckling stress (141.7 ksi) is greater than the elastic torsional buckling stress (87 ksi), this means that this cruciform section will fail in torsional buckling. Therefore, use $F_e = 87$ *ksi* in equations (5-14) or (5-15).

Since $F_e = 87$ ksi $> 0.44F_y = 0.44(36) = 15.84$ ksi, therefore the cruciform section will buckle inelastically: hence, use equation (5-14) to determine the nominal stress as follows:

$$F_n = \left[0.658^{\frac{F_y}{F_e}}\right]F_y = \left[0.658^{\frac{36}{87}}\right](36) = 30.3 \text{ ksi}$$

The nominal axial compression strength from equation (5-22) or AISC Specification Equation E4-1 is

$P_n = F_n A_g = (30.3 \text{ ksi})(58.8 \text{ in.}^2) =$ **1782 kips**

LRFD:

The design axial compression load capacity for this flanged cruciform section is, $\phi_c P_n = (0.9)(1782 \text{ kips}) = 1604 \text{ kips}$.

ASD:

The allowable axial compression load of the flanged cruciform section is

$$P_a = \frac{P_n}{\Omega_c} = \frac{1782\,\text{kips}}{1.67} = 1067\,\text{kips}.$$

EXAMPLE 5-14

Axial Load Capacity and Allowable Axial Load of a Built-up Member – Double Angles

Determine the axial compression capacity (LRFD) and the allowable axial load (ASD) of a 12-ft. long double $L5 \times 5 \times \frac{1}{2}$ of A36 steel that is used as a plane truss web member. The double angles are welded to $\frac{3}{8}$" gusset plates at the ends and to $\frac{3}{8}$" filler plates at 4 ft. on centers.

Solution

The cross-sectional properties for double angles with $\frac{3}{8}$ in separation from *AISCM* Table 1-15 are

$A = 9.58$ in.^{2} $\quad r_x = 1.53$ in. $\quad r_y = 2.22$ in. $\quad r_z = r_i = 0.98$ in. $\quad \bar{r}_o = 2.94$ in. $\quad H = 0.842$

$$J(\text{single angle}) = 0.417\ \text{in.}^4$$

$$J(\text{double angles}) = (2)(0.417\ \text{in.}^4) = 0.934\ \text{in.}^4$$

$$C_w = 2C_w \text{ of single angle} = (2)(0.744) = 1.49\ \text{in.}^6$$

Check flexural buckling about the *x*-axis:

$$\frac{K_x L_x}{r_x} = \frac{(1.0)(12\,\text{ft.} \times 12)}{1.53} = 94.1$$

Use equation (5-9) to calculate the elastic buckling stress:

$$F_{ex} = \frac{\pi^2 E}{(K_x L_x / r_x)^2} = \frac{\pi^2 (29{,}000)}{(94.1)^2} = 32.3\ \text{ksi}$$

$$F_{ex} > 0.44F_y = 0.44(36\ \text{ksi}) = 15.8\ \text{ksi}$$

Therefore, use equation (5-14) to calculate the nominal stress:

$$F_n = \left[0.658^{\frac{F_y}{F_e}}\right] F_y = \left[0.658^{\frac{36}{32.3}}\right] (36\ \text{ksi}) = 22.6\ \text{ksi}$$

That is, $F_{nx} = 22.6$ ksi

The nominal axial compression strength for flexural buckling about the x-axis is

$$P_n = A_g F_{nx} = (9.58\ \text{in.}^2)\,(22.6\ \text{ksi}) = 216.5\ \text{kips}$$

LRFD Method:

The design axial compressive strength based on flexural buckling about the ***x*-axis** is

$$\phi_c P_n = (0.9)(216.5\ \text{kips}) = 195\ \text{kips}.$$

ASD Method:

The allowable axial compression load based on flexural buckling about the ***x*-axis** is

$$P_a = \frac{P_n}{\Omega_c} = \frac{216.5\text{kips}}{1.67} = 130\,\text{kips}.$$

Check flexural-torsional buckling about the *y*-axis (i.e., the axis of symmetry):

For flexural-torsional buckling about the *y*-axis, the longitudinal spacing of the connectors, a, has to be considered.

Unbraced length, L = 12 ft. = 144 inches.

The longitudinal spacing of the filler plate connectors, a = 4 ft., = 48 inches.

$$r_i = r_{\min} = r_z = 0.98 \text{ in.}$$

$$\frac{a}{r_i} = \frac{48 \text{ in.}}{0.98 \text{ in.}} = 49 > 40$$

Therefore, use equation (5-32), and the equivalent slenderness ratio is

$$\frac{KL}{r} = \left(\frac{KL}{r}\right)_m = \sqrt{\left(\frac{KL}{r}\right)_o^2 + \left(\frac{K_i a}{r_i}\right)^2},$$

where $\left(\frac{KL}{r}\right)_o = \left(\frac{KL}{r_y}\right) = \frac{1.0(144\,\text{in.})}{2.22\,\text{in.}} = 64.9$, since the buckling being considered here is about the y-axis.

K_i = 0.5 (for angles back-to-back)

$$\left(\frac{KL}{r}\right)_{\max} = 94.1 \quad (\textit{calculated previously})$$

$$\frac{K_i a}{r_i} = \frac{K_i a}{r_z} = \frac{0.5(48 \text{ in.})}{0.98 \text{ in.}} = 24.5$$

$$\leq 0.75\left(\frac{KL}{r}\right)_{\max} = 0.75(94.1) = 70.6 \quad \text{OK.}$$

Therefore, the built-up member will act as one unit, and there will be no "compound buckling." According to Section E6 of the AISC Specification, "compound buckling" is a failure mode that includes both the global buckling failure mode of the member and the localized component buckling mode that occurs between intermediate connectors. [1]. The equivalent slenderness ratio for buckling about the *y-axis* is calculated as

$$\frac{KL}{r} = \left(\frac{KL}{r}\right)_m = \sqrt{\left(\frac{KL}{r}\right)_o^2 + \left(\frac{K_i a}{r_i}\right)^2} = \sqrt{(64.9)^2 + (24.5)^2} = 69.4$$

The elastic buckling stress about the *y*-axis, $F_{ey} = \frac{\pi^2 E}{\left(\frac{KL}{r}\right)_m^2} = \frac{\pi^2(29{,}000)}{(69.4)^2} = 59.4$ ksi

$$L_{cz} = K_z L_z = (1.0)(144\,\text{in.}) = 144\,\text{in.}$$

$$F_{ez} = \left(\frac{\pi^2 E C_w}{L_{cz}^2} + GJ\right)\left(\frac{1}{A_g \bar{r}_o^2}\right)$$

$$= \left(\frac{\pi^2(29000)(1.49)}{(144)^2} + (11200)(0.934)\right)\left(\frac{1}{(9.58)(2.94)^2}\right) = 126.6\text{ ksi}$$

From equation (5-26), the equivalent critical buckling stress based on the limit states of torsional and flexural-torsional buckling is calculated as

$$F_e = \left(\frac{F_{ey} + F_{ez}}{2H}\right)\left[1 - \sqrt{1 - \frac{4F_{ey}F_{ez}H}{\left(F_{ey} + F_{ez}\right)^2}}\right]$$

$$= \left(\frac{59.4 + 126.6}{2 \times 0.842}\right)\left[1 - \sqrt{1 - \frac{4(59.4)(126.6)(0.842)}{(59.4 + 126.6)^2}}\right] = 53.3\,\text{ksi}$$

$$0.44F_y = (0.44)(36\,\text{ksi}) = 15.8\,\text{ksi}$$

$F_e = 53.3$ ksi $> 0.44F_y$, which indicates *inelastic* buckling; therefore, equation (5-14) is used to calculate the nominal stress as follows:

$$F_{ny} = \left[0.658^{\frac{Fy}{Fe}}\right]F_y = \left[0.658^{\frac{36}{53.3}}\right] = 27.1\,\text{ksi}$$

The nominal axial compression strength based on flexural buckling about the ***y*-axis** is

$P_n = A_g F_{ny} = (9.58\text{ in.}^2)\,(27.1\text{ ksi}) = 259.6$ kips

LRFD Method:

The design axial compressive strength based on flexural-torsional buckling about the *y*-axis is

$\phi_c P_n = (0.9)(259.6\text{ kips}) = 234$ kips. $<$ 195 kips obtained for flexural buckling about the x-axis

Therefore, flexural buckling about the x-axis controls and $\phi_c P_n = 195$ kips

ASD Method:

The allowable axial compression load based on flexural buckling about the ***y*-axis** is

$$P_a = \frac{P_n}{\Omega_c} = \frac{259.6\,\text{kips}}{1.67} = 155\,\text{kips.}$$

The design strength for the flexural-torsional buckling mode about the ***y-axis*** is 234 kips for the LRFD method and an allowable load of 155 kips for the ASD method. These values are higher than the corresponding values of 195 kips axial design compressive strength for the LRFD method and 130 kips allowable axial compression load for the ASD method obtained previously for the flexural buckling failure mode about the ***x-axis***. Therefore, the flexural buckling failure mode about the *x-axis* will control the failure strength of this double-angle member, and the design axial compression capacity and the allowable axial compression loads, respectively, are as follows:

- The axial design compressive strength, ϕP_n = **195 kips**. (for LRFD)
- The allowable axial compression load, P_a = **130 kips**. (for ASD)

For the LRFD method, the calculated axial load capacity of 195 kips compares well with the axial load capacity obtained from ***AISCM* Table 4-8,** where we find that ϕP_n = **195 kips** for an effective length, KL = 12 ft. Similarly, for the ASD method, the calculated allowable axial compression load of 130 kips compares well with the allowable load of 130 kips from *AISCM* Table 4-8. Thus, the use of the *AISCM* tables yields the same design capacities as were calculated using the very cumbersome equations presented earlier. The reader is thus encouraged to use the *AISCM* load capacity tables whenever possible because of their efficiency and ease of use.

EXAMPLE 5-15

Axial Compression Load Capacity of a Single Angle with All Legs Connected to the Supports

Determine the axial compression capacity (LRFD) and the allowable axial compression load (ASD) of a 6 ft. long single L 3 × 3 × ¼ of A36 steel, assuming all legs of the angle are connected to the end supports and thus participate in resisting the axial load (i.e., the load is applied concentrically). Assume the effective length factor, K = 1.0.

Solution

The cross-sectional properties for L 3 × 3 × ¼ from *AISCM* Table 1-7 are:

$$A=1.44 \text{ in.}^2 \quad r_x=0.926 \text{ in.} \quad r_y=0.926 \text{ in.} \quad r_z=0.585 \text{ in.} \quad \bar{r}_o=1.65 \text{ in.} \quad J=0.0313 \text{ in.}^4 \quad C_w=0.0206 \text{ in.}^6$$

Check flexural buckling about the weakest axis (i.e., the axis with the largest slenderness ratio):

$$\frac{L_{cx}}{r_x}=\frac{K_xL_x}{r_x}=\frac{(1.0)(6\text{ft.})(12)}{0.926\text{in.}}=77.8$$

$$\frac{L_{cy}}{r_y}=\frac{K_yL_y}{r_y}=\frac{(1.0)(6\text{ft.})(12)}{0.926\text{in.}}=77.8$$

$$\frac{L_{cz}}{r_z}=\frac{K_zL_z}{r_z}=\frac{(1.0)(6\text{ft.})(12)}{0.585\text{in.}}=123.1 \quad \text{(largest value controls)} < 200 \quad \text{OK.}$$

$4.71\sqrt{\frac{29000}{36}}=133.7>123.1$; therefore, the failure mode is inelastic buckling.

Since the slenderness ratio is largest for bending about the z-axis, the single angle will fail by buckling (bending) about the z-axis.

Check Local Buckling (see Table 5-1):

The leg length to thickness ratio, $\frac{b}{t}=\frac{3 \text{ in.}}{0.25 \text{ in.}}=12$

$$< 0.45\sqrt{\frac{E}{F_y}}=0.45\sqrt{\frac{29000}{36}}=12.8 \quad \textit{OK.}$$

Therefore, use equation (5-9) to calculate the elastic flexural buckling stress,

$$F_e=F_{ez}=\frac{\pi^2E}{(K_zL_z/r_z)^2}=\frac{\pi^2(29{,}000)}{(123.1)^2}=18.9 \text{ ksi}$$

$F_{ez} > 0.44F_y = 0.44(36 \text{ ksi}) = 15.8$ ksi, indicating *inelastic* buckling.
Therefore, use equation (5-14) to calculate the nominal stress,

$$F_n = \left[0.658^{\frac{F_y}{F_e}}\right] F_y = \left[0.658^{\frac{36}{18.9}}\right] (36 \text{ ksi}) = 16.2 \text{ ksi}$$

The nominal axial compression strength is
$P_n = A_g F_n = (1.44 \text{ in.}^2)(16.2 \text{ ksi}) = 23.3$ kips

LRFD Method:

The design axial compression strength is
$\phi_c P_n = (0.9)(23.3 \text{kips}) = 21 \text{kips}.$

ASD Method:

The allowable axial compression load is

$$P_a = \frac{P_n}{\Omega_c} = \frac{23.3 \text{kips}}{1.67} = 14 \text{ kips.}$$

These values compare well with the axial load capacities from ***AISCM*** **Table 4-11,** where we obtain the same LRFD axial compression load capacity of 21 kips for an unbraced length, $L_c = KL = 6$ ft. Similarly, the allowable axial compression load for the ASD method from *AISCM* Table 4-11 is 14 kips, which is the same value obtained in this example using the equations presented previously.

5-9 SINGLE ANGLE MEMBERS UNDER ECCENTRIC COMPRESSION LOADS

Where the ends of single angle members are milled and the axial load introduced with base plates at the ends of the angle such that the two legs of the angle participate in transferring the load, as was the case in Example 5-15, the axial compression load will be applied uniformly to the angle. When a single angle is concentrically loaded, the equations presented in Section 5-3 or *AISCM* Section E3 will apply (i.e., use equations (5-9) and (5-14) or (5-15)).

On the other hand, for some situations, the axial compression load on a single angle is applied predominantly through only one leg of the angle (e.g., through a gusset plate connected to only one leg of the angle, as may occur in trusses), which results in significant eccentricity of the axial loading. In these cases, in lieu of designing the angle for biaxial bending plus axial loads using the interaction equations presented in Chapter 8 of this text, the AISC Specifications, Section E5 provides a simplified procedure to account for the load eccentricity by using an *effective slenderness ratio* for the single angle for buckling about the x-axis; the effective slenderness ratios for different angles with different support conditions are presented in equations (5-33) through (5-40). These equations can only be used if certain conditions are met, and they were derived on the presumption that significant restraint to the bending about the y-axis (i.e., the axis perpendicular to the connected leg of the angle) is provided to the angle at the supports [1]. Therefore, only the bending and buckling about the x-axis is considered in the equations. Once the effective slenderness ratio is determined, the axial compression load capacity of the single angle can then be determined as follows:

- Axial load capacity for single angles with **b/t** $\leq 0.71\sqrt{\frac{E}{F_y}}$ (= 20 for ASTM A36 steel and 17 for ASTM Grade 50 steel), where b is the larger leg length:

Flexural torsional buckling need *not* be considered when **b/t** ≤ $0.71\sqrt{\frac{E}{F_y}}$ per Section E5 of the AISC Specification [1].

Use equations (5-9) and (5-14) or equation (5-15) with the effective slenderness ratio, KL/r, obtained from equations (5-33) through (5-40).

- Axial load capacity for single angles with b/t > $0.71\sqrt{\frac{E}{F_y}}$ (i.e., b/t > 20 for ASTM A36 steel and b/t > 17 for ASTM Grade 50 steel), where b is the larger leg length:

  Use equation (5-27) from Section 5-8 of the text with the appropriate slenderness ratios about the x-, y-, and z-axes, and calculate the axial load capacity using equations (5-14) or (5-15).

  To be able to neglect the effect of the eccentricity of axial compression load for single angles and therefore be able to use the *effective slenderness equations* for single angles that are presented below, the following conditions must be satisfied; otherwise, the angle cannot be analyzed or designed using equations (5-33) through (5-40); for such cases, the eccentricity of the axial compression load must be considered, and the angle will have to be designed for combined biaxial bending plus axial compression load using the column interaction equations presented in Chapter 8 (see Section E5 of the AISC Specification [1]):

  1. The single-angle member is loaded at both ends in compression through the same one leg.
  2. The single angle is connected to its supports by welding or bolting with a minimum of two bolts, thus providing some end restraint.
  3. The single angle is subjected to only an axial compression load with no intermediate transverse loads between its supports.
  4. The slenderness ratio of the single angle, L_c/r = KL/r ≤ 200.
  5. For unequal leg angles, the ratio of the lengths of the long leg to the short leg is less than 1.7.

Equivalent Slenderness Ratios of Single Angles:

The equivalent slenderness ratios of single angles that satisfy the five conditions presented above are calculated for different angles and support conditions using the following equations:

1. For **equal leg** angles or **unequal leg** angles that are connected through the *longer* leg and *that are individual members or web members of plane trusses,* with adjacent web members attached to the same side of the gusset plate or chord, use equations (5-33) and (5-34) (i.e., AISC Specification Equations E5-1 and E5-2) to determine the equivalent slenderness ratio:

   a. When $\frac{L}{r_a} \le 80$, $$\frac{KL}{r} = 72 + 0.75\frac{L}{r_a} \quad (5\text{-}33)$$

   b. When $\frac{L}{r_a} > 80$, $$\frac{KL}{r} = 32 + 1.25\frac{L}{r_a} \le 200 \quad (5\text{-}34)$$

2. For **unequal leg** angles that are connected through the *shorter* leg and with the ratio of lengths of the long leg to the short leg less than 1.7 (i.e., $\frac{b_l}{b_s} < 1.7$), and *that are individual members or web membesr of plane trusses,* with adjacent web members attached to the same side of the gusset plate or chord, use equations (5-35) and (5-36) to determine the equivalent slenderness ratio:

   a. When $\frac{L}{r_a} \leq 80$,
   $$\frac{KL}{r} = 4\left[\left(\frac{b_l}{b_s}\right)^2 - 1\right] + \left(72 + 0.75\frac{L}{r_a}\right) \geq \frac{0.95L}{r_z} \qquad (5\text{-}35)$$

   b. When $\frac{L}{r_a} > 80$,
   $$\frac{KL}{r} = 4\left[\left(\frac{b_l}{b_s}\right)^2 - 1\right] + \left(32 + 1.25\frac{L}{r_a}\right) \geq \frac{0.95L}{r_z} \leq 200 \qquad (5\text{-}36)$$

3. For angles that are web members of box or space trusses with adjacent web members attached to the same side of the gusset plate or chord: For **equal leg** angles or **unequal leg** angles that are connected through the *longer* leg, use equations (5-37) and (5-38) (i.e., AISC Specification Equations E5-3 and E5-4) to determine the equivalent slenderness ratio:

   a. When $\frac{L}{r_a} \leq 75$,
   $$\frac{KL}{r} = 60 + 0.8\frac{L}{r_a} \qquad (5\text{-}37)$$

   b. When $\frac{L}{r_a} > 75$,
   $$\frac{KL}{r} = 45 + \frac{L}{r_a} \leq 200 \qquad (5\text{-}38)$$

4. For **unequal leg** angles that are connected through the *shorter* leg and with the ratio of the lengths of the long leg to the short leg less than 1.7 (i.e., $\frac{b_l}{b_s} < 1.7$), and *that are web members of box or space trusses,* with adjacent web members attached to the same side of the gusset plate or chord, use equations (5-39) and (5-40) to determine the equivalent slenderness ratio:

   a. When $\frac{L}{r_a} \leq 75$,
   $$\frac{KL}{r} = 6\left[\left(\frac{b_l}{b_s}\right)^2 - 1\right] + \left(60 + 0.8\frac{L}{r_a}\right) \geq \frac{0.82L}{r_z} \qquad (5\text{-}39)$$

   b. When $\frac{L}{r_a} > 75$,
   $$\frac{KL}{r} = 6\left[\left(\frac{b_l}{b_s}\right)^2 - 1\right] + \left(45 + \frac{L}{r_a}\right) \geq \frac{0.82L}{r_z} \leq 200, \qquad (5\text{-}40)$$

   Where,
   L = Length of angle member between the work points at the truss chord centerlines, in.
   b_l = Length of longer leg of angle, in.
   b_s = Length of shorter leg of angle, in.
   r_a = Radius of gyration of the angle about the *geometric axis* parallel to the connected leg, in.
   r_z = Radius of gyration of the angle about the minor principal axis, in

EXAMPLE 5-16

Axial Load Capacity of a Single Angle with One Leg Connected to the Supports

Recalculate the axial load capacity of the single $L3 \times 3 \times 1/4$ in Example 5-15, assuming the angle is used as a plane truss web member and is connected on the same leg to gusset plates at each end. Assume ASTM A36 steel.

Solution

The cross-sectional properties for $L3 \times 3 \times 1/4$ from *AISCM* Table 1-7 are:

$$A=1.44\text{ in.}^2\quad r_x=0.926\text{ in.}\quad r_y=0.926\text{in.}\quad r_z=0.585\text{ in.}\quad \bar{r}_o=1.65\text{ in.}$$

$$J=0.0313\text{ in.}^4\quad C_w=0.0206\text{ in.}^6$$

$$0.71\sqrt{\frac{E}{F_y}}=0.71\sqrt{\frac{29000}{36}}=20$$

$$\frac{b}{t}=\frac{3\text{in.}}{\left(\frac{1}{4}\text{ in.}\right)}=12<20\qquad \text{O.K.}$$

Therefore, flexural-torsional buckling need not be considered.

Flexural buckling about the *x*-axis:

$$\frac{L_{cx}}{r_x}=\frac{(6\text{ ft.})(12)}{0.926}=77.8<80$$

The equivalent slenderness ratio from equation (5-33) is

$$\frac{KL}{r}=72+0.75\frac{L}{r_x}=72+0.75(77.8)=130.3<200,\qquad \text{OK}$$

Comparing the above effective slenderness ratio to the slenderness ratio corresponding to the least radius of gyration, r_z, we find that

$$\frac{L_{cz}}{r_z}=\frac{KL_z}{r_z}=\frac{(1.0)(6\text{ ft.})(12)}{0.585\text{in.}}=123.1<130.3$$

Therefore, the calculated equivalent slenderness ratio of 130.3 is indeed controlled.

Since the length of the angle leg-to-thickness ratio, $\frac{b}{t}=\frac{3\text{in.}}{0.25\text{in.}}=12<0.71\sqrt{\frac{29000}{36}}=20$,

Local buckling does not occur. Therefore, use equation (5-9) to calculate the elastic buckling stress (see AISC Specification Section E3 [1]),

$$F_{ex}=\frac{\pi^2 E}{(K_x L_x/r_x)^2}=\frac{\pi^2(29{,}000)}{(130.3)^2}=16.9\text{ ksi}$$

Since $F_{ex}>0.44F_y$ = 0.44(36 ksi) = 15.8 ksi, indicating inelastic buckling failure mode.

Therefore, use equation (5-14) to calculate the nominal stress,

$$F_n=\left[0.658^{\frac{F_y}{F_e}}\right]F_y=\left[0.658^{\frac{36}{16.9}}\right](36\text{ ksi})=14.8\text{ ksi}$$

The nominal axial compression strength is

$P_n = A_g F_n = (1.44 \text{ in.}^2)(14.8 \text{ ksi}) = 21.3 \text{ kips}$

LRFD Method:

The design axial compressive strength is

$\phi_c P_n = (0.9)(21.3 \text{kips}) = 19.2 \text{ kips}.$

ASD Method:

The allowable axial compression load is

$$P_a = \frac{P_n}{\Omega_c} = \frac{21.3\text{kips}}{1.67} = 12.8 \text{ kips}.$$

Alternatively, Table 4-14 of the *AISCM* [1] can be used to obtain the design stress,

$\phi_c F_n = 13.24 \text{ ksi}$

The design's compressive strength is

$\phi_c P_n = (13.24 \text{ ksi})(1.44 \text{ in.}^2) = 19.1 \text{ kips}.$

Similarly, for the ASD method,

$$\frac{F_n}{\Omega_c} = 8.82 \text{ ksi}$$

$$P_a = \frac{F_n}{\Omega_c} A_g = (8.82 \text{ ksi})(1.44 \text{ in.}^2) = 12.7 \text{ kips}.$$

The buckling strength about the z-axis has already been calculated as 21 kips in Example 5-15, compared to 19.2 kips obtained in Example 5-16. Similarly, the ASD allowable load for Example 5-15 was calculated as 14 kips, compared to 12.7 kips obtained in Example 5-16.

Recall that in Example 5-15, all the legs of the angle were assumed to be connected to the end supports thus implying concentric loading of the angle, whereas in Example 5-16, only one leg of the angle is connected to the supports, resulting in eccentricity of loading. The axial compression load capacity and the allowable axial compression load calculated in Example 5-16 are smaller than those calculated in Example 5-15 because of the load eccentricity in the single angle in Example 5-16, since only one leg of the angle is connected to the support. In Example 5-15, there is no load eccentricity since all the legs of the angle are connected to the support.

Also, note that the axial compression load capacity of 19.2 kips (LRFD) calculated above differs from the axial load capacity of 13.2 kips obtained from *AISCM* Table 4-12 for the *L*3 × 3 × ¼ with *KL* of 6 ft. Similarly, the allowable axial compression load of 12.8 kips (ASD) calculated above differs from the allowable axial compression load of 8.97 kips obtained from *AISCM* Table 4-12 for the *L*3 × 3 × ¼ with *KL* of 6 ft.

The axial compression load capacities presented in *AISCM* Table 4-12 were derived assuming *biaxial bending* of the single angle without any end restraints (i.e., the angles are pinned at the supports), and the axial compression load is applied at a point located along a line that is perpendicular to, and passes through the mid-length of, the long leg of the angle with an eccentricity of $0.75t$ from the outside face of this long leg, where t is the thickness of the angle leg (see *AISCM* Figure 4-4 [1]). However, end restraints are assumed and required in order to use the equivalent slenderness ratios in equations (5-33) through (5-40) that was used in Example 5-16 (see *AISCM* Section E5 [1]).

If the end restraint conditions assumed and required in Section 5-9 for single angles (i.e., single angles welded to its supports or bolted with a minimum of two bolts) cannot be achieved or assured, then the angle will be eccentrically loaded; thus, without the end restraints identified in Section 5-9 that ensure concentric loading, biaxial bending of the angle becomes the design consideration, and *AISCM* Table 4-12 should be used to determine the axial load capacity of the angle in such situations. The design axial compression capacities and the allowable axial compression loads in *AISCM* Table 4-12 can be obtained by using the method for combined axial compression load and biaxial bending moments presented in Chapter 8 of this text.

References

[1] American Institute of Steel Construction. 2023. *Steel Construction Manual*. 16th ed. Chicago, IL: AISC.

[2] Hibbeler, R. C. 2008. *Mechanics of Materials*. 7th ed. Upper Saddle River, NJ: Prentice Hall.

[3] Seaburg, P. A., and C. J. Carter. 1997. "Torsional Analysis of Structural Steel Members". *Steel Design Guide 9*, Chicago, IL: AISC.

[4] Galambos, T. V., F. J. Lin, and B. G. Johnston. 1996. *Basic Steel Design with LRFD*. Upper Saddle River, NJ: Prentice Hall.

[5] McGuire, W. 1968. "Steel Structures." Englewood Cliff, NJ: Prentice Hall.

[6] Chajes, A. 1974. "Principles of Structural Stability Theory." *Prentice Hall*, Englewood Cliffs, NJ.

[7] Weisenberger, G. 2017. "Above and Beyond". *Modern Steel Construction*, July.

[8] Harris, N., and G. Urgessa. 2018. "Strength of Flanged and Plain Cruciform Members." *Advances in Civil Engineering*, Vol. 2018. https://doi.org/10.1155/2018/8417208

[9] Hawileh, R. A., A. A. Obeidah, F. Abed, and J. A. Abdalla. 2012. "Experimental Investigation of Inelastic Buckling of Built-up Steel Columns." *Steel and Composite Structures*, September Doi: 10.12989/scs.2012.13.3.295.

[10] Dinis, P. B., D. Camotim, and N. Silvestre. 2010. "On the Local and Global Buckling Behavior of Angle, T-section, and Cruciform Thin-walled Members." *Thin-Walled Structures*, 48: 786-797.

Exercises

5-1. Determine the design strength of the columns shown in Figure 5-16 using the following methods:

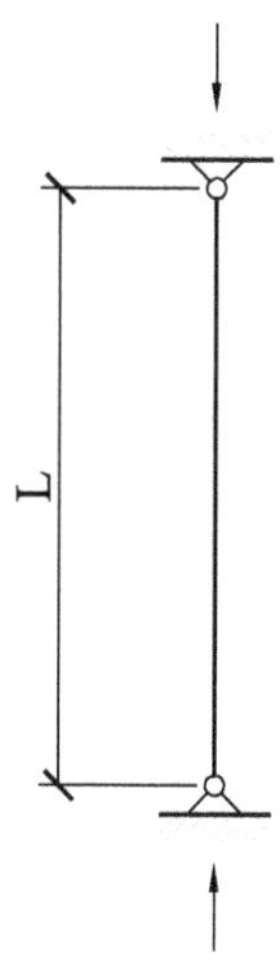

Column Size	Grade	L
a) HSS4x4x1/4	A500, Grade C	16'-0"
b) HSS10x10x3/8	A500, Grade C	24'-0"
c) HSS12x12x5/16	A500, Grade C	17'-6"
d) HSS6x4x3/16	A500, Grade C	12'-0"
e) HSS10x6x1/2	A500, Grade C	15'-0"
e) HSS20x8x3/8	A500, Grade C	21'-0"
f) HSS4.5x0.375	A500, Grade C	10'-0"
g) HSS8.625x0.322	A500, Grade C	14'-6"

FIGURE 5-16 Details for Exercise 5-1

1. Design equations; check local slenderness.
2. Confirm the results from part 1 using *AISCM*, Table 4-14.
3. Confirm the results from part 1 using *AISCM*, Tables 4-3 to Table 4-5, where possible.

5-2. Determine the design axial compression strength (LRFD) and the allowable compression load (ASD) of the columns shown in Figure 5-17 using the following methods:

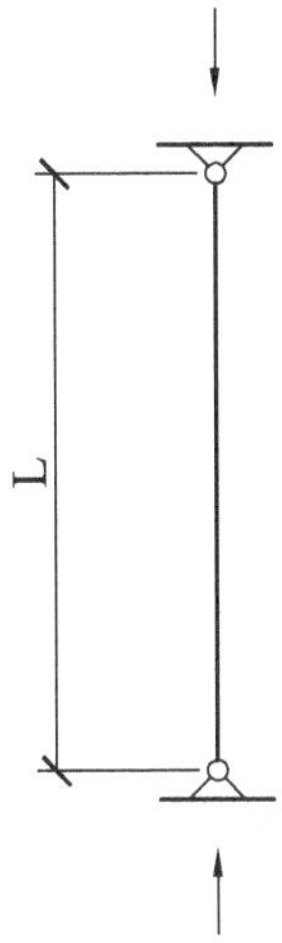

Column Size	Grade	L
a) W6x20	A36, Grade 36	19'-6"
b) W8x48	A992, Grade 50	14'-0"
c) W10x112	A992, Grade 50	18'-6"
d) W12x106	A992, Grade 50	12'-0"
e) W14x132	A36, Grade 36	17'-0"
e) W21x101	A992, Grade 50	21'-0"

FIGURE 5-17 Details for Exercise 5-2

1. Design equations; check local slenderness.
2. Confirm the results from part 1 using *AISCM* Table 4-14.
3. Confirm the results from AISCM Table 4-1, where possible.

5-3. A column with an unbraced length of 18 ft. must resist a factored load of P_u = 200 kips.

a. Select the lightest W-shape to support this load, consider W8, W10, W12, and W14 shapes. The steel is ASTM A992, grade 50.
b. Select the lightest square HSS shape to support this load, consider HSS6, HSS8, HSS10, and HSS12 shapes. The steel is ASTM A500, grade 46.
c. Select the lightest round pipe shape to support this load, consider Pipe 6, Pipe 8, Pipe 10, and Pipe 12 shapes. The steel is ASTM A53, F_y = 35 ksi.

5-4. A pipe column with a factored load of 80 kips has an unbraced length of 10 ft. Select the lightest shape using the following:

1. Standard pipe (std),
2. Extra strong pipe (XS), and
3. Double extra strong pipe (XXS).

Steel is ASTM A53, grade B.

5-5. What is the maximum unbraced length permitted for a 4-in. extra strong (XS) pipe column to support a factored load of 80 kips?

5-6. A W-shape column must support service loads of D = 200 kips and L = 300 kips. The unbraced length in the x-direction is 30 ft., and in the y-direction it is 15 ft. Select the lightest *W*-shape to support this load.

5-7. Determine if a W8 × 28 column is adequate to support a factored axial load of P_u = 175 kips with unbraced lengths, L_x = 24 ft. and L_y = 16 ft., and pinned-end conditions. The steel is ASTM A992, grade 50.

5-8. Refer to the floor framing plan shown in Figure 5-18 and the following assumptions:

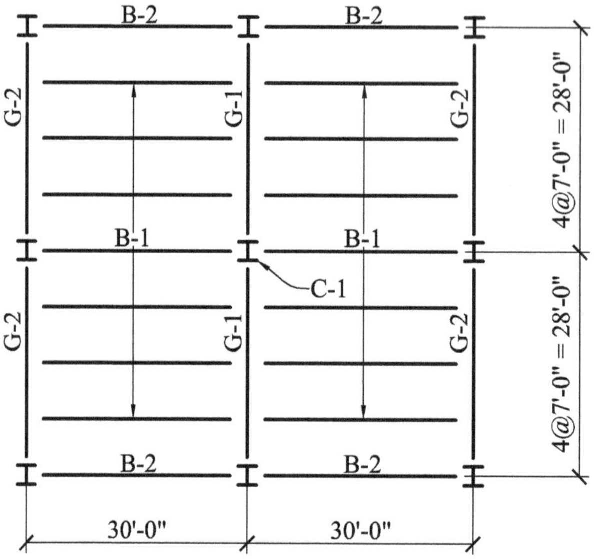

FIGURE 5-18 Floor framing plan for Exercise 5-8

- C-1 is a W8 × 31; Steel is ASTM A992 Grade 50
- Column unbraced length is 15-0"; K=1.0
- Floor dead load is 72 psf (includes self-weight)
- Ignore connection eccentricities

Determine the following:

a. Design axial compressive strength, in kips, for C-1.
b. The design service live load capacity of the floor in psf based on the axial strength of C-1.
c. Using the live load table in ASCE 7, assign a possible occupancy type based on part (b).

5-9. The preliminary column and girder sizes for a two-story moment frame are shown in Figure 5-19. Assuming in-plane bending about the strong axes for the columns and girders, determine the effective length factor, K, for columns CG and GK using the alignment chart and assuming elastic behavior.

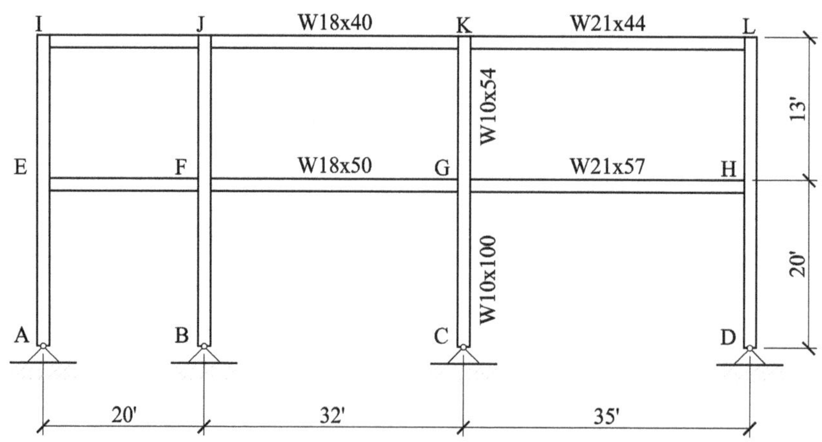

FIGURE 5-19 Detail for Exercise 5-9

5-10 Determine the following for the column shown in Figure 5-20. The steel is ASTM A992, Grade 50, and is supported at the bottom only (encased in concrete).

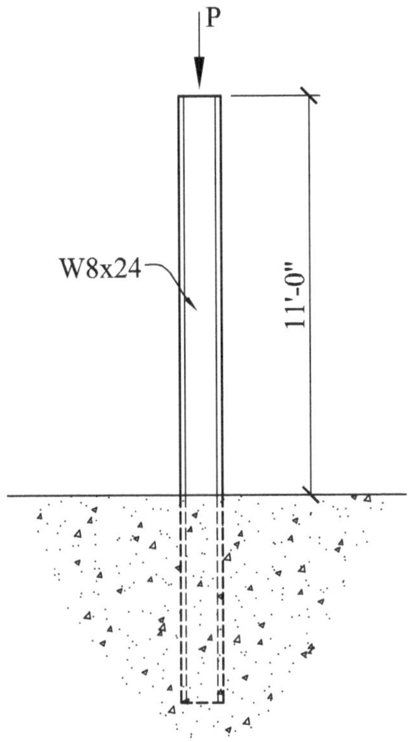

FIGURE 5-20 Detail for Exercise 5-10

a. Determine the effective length factor, K, and provide an explanation for your choice of the K value.
b. Find the design axial compressive strength, ϕP_n, in kips. Find the service live load that the column could carry if the service dead load is 12 kips. Consider the load combination $1.2D + 1.6L$ only.

5-11 Find the design axial compressive strength (ϕP_{nx}, ϕP_{ny}) in kips in the x- and y-axes for a W18 × 65. Use $K = 1.2$ and unbraced lengths of $L_x = 30$ ft. and $L_y = 16$ ft. Use A36 steel.

5-12 Develop a spreadsheet to determine the axial load capacity in compression for a standard wide flange column shape. The following variables are required:

- Column dimensions, b_f, t_f, A, r_x, r_y
- unbraced length of the column, L_x, L_y
- K, E, F_y
- Slenderness effects
- Submit results for a W10 × 33, $L_x = 25$ ft., $L_y = 15$ ft., $K = 1.0$, $F_y = 36$ ksi

5-13 Develop a spreadsheet to determine the axial load capacity in compression for a standard HSS square or round column shape. The following variables are required:

- Column dimensions, B or D, t, A, r
- unbraced length of the column, L_x, L_y
- K, E, F_y
- Slenderness effects
- Submit results for an HSS12 × 12 × $\frac{5}{16}$, L= 17'-6", $F_y = 46$ ksi, and a Pipe 4 STD, L = 15'-0", $F_y = 35$ ksi

5-14 Develop a spreadsheet to determine the axial load capacity in compression for a standard HSS rectangular column shape. The following variables are required:

- Column dimensions, B, H, t, A, r_x, r_y
  unbraced length of the column, L_x, L_y
- K, E, F_y
- Slenderness effects
- Submit results for an HSS10 × 6 × 1/2, L_x = 21'-0", L_y = 15'-0", F_y = 46 ksi

5-15 Develop a spreadsheet to determine the axial load capacity in compression for a flat plate. The following variables are required:

a. plate dimensions t, b
b. unbraced length of the plate, L_x, L_y
c. K, E, F_y
d. Local slenderness effects are ignored
e. Submit results for a ¼" × 3" plate, L_x = 4 ft., L_y = 1.5 ft., K = 1.0, F_y = 50 ksi

5-16 Refer to the ladder drawing in Figure 5-21 and the following loading requirements from OSHA:

Loading from OSHA 1910.27(a)(1)(i): The minimum design live load shall be a single concentrated load of 200 lbs.

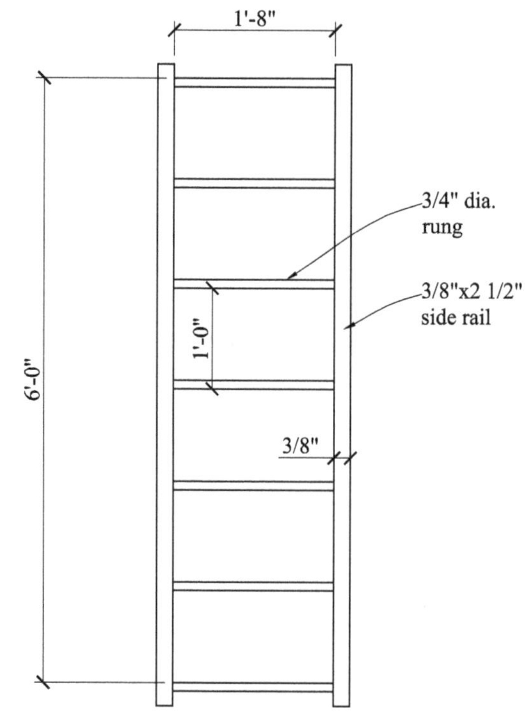

FIGURE 5-21 Detail for Exercise 5-16

F_y = 36 ksi

K = 1.0

Ladder rungs are assumed to be pinned

L_x = 6'-0", L_y = 1'–0"

a. Find the axial capacity of the side rail in the x-axis. Is the side rail adequate in compression?
b. Find the axial capacity of the side rail in the y-axis. Is the side rail adequate in compression?

5-17. Determine the axial compression strength in kips for the x- and y-axes of a (2)-L3 ½ × 3 ½ × 5⁄16 member with an unbraced length of 15 ft. The angles are separated by ⅜ " plates spaced 5'-0" on centers (o.c.). Use A36 steel and verify the results using AISCM Table 4-8.

5-18. Determine the axial compression strength in kips for the x- and y-axes of a (2)-L5 × 3 × 7⁄16 (LLBB, i.e., long legs back-to-back) member with an unbraced length of 16 ft. The angles are separated by ⅜ " plates spaced 4'-0" o.c. Use A36 steel and verify the results using *AISCM* Table 4-9.

5-19. Determine the axial compression strength in kips for a concentrically loaded L4 × 3 × ⅜ member with an unbraced length of 8 ft. Use A36 steel and verify the results using the *AISCM* Load Tables.

5-20. A welded built-up 16 in. x 16 in. box column in the bottom story of a highrise building is made up of 3 in. thick steel plates (A992 Grade 50). Assuming the column has an unbraced length of 14 ft., determine the following:

a. The gross area, A_g, of the column.
b. The moments of inertia about the x-x and y-y axes, I_x and I_y, respectively,
c. The radii of gyration about the x-x and y-y axes, r_x and r_y, respectively.
d. Using the LRFD method, determine the axial compression strength in kips for the column.
e. Using the column load tables, select a wide flange or I- shaped column of A992 Grade 50 steel that has the same axial compression load capacity as the box column and compare the weights of both sections. Which of the column sections (built-up box section or I- shaped) is more economical?

5-21. The detail in Figure 5-22 consists of four-story tall diagonal struts connected to a tension tie that passes through and is connected to the reinforced concrete shear wall – the main lateral force resisting system (LFRS) in a unique 54-story building in the Midwest United States [7]. This detail is spaced at regular intervals at each transverse shear wall location and is used to support perimeter steel columns that terminate at the 4th floor level. The perimeter columns terminate at the 4th floor level because of underground rail lines within the building footprint on the one side of the building and a river on the opposite side of the building within the building footprint; the supported perimeter columns, in turn, support multiple floor levels starting from the 5th floor up to the roof:

a. Using the AISC Equations, calculate the axial compression load capacity (in kip) of the W36 x 925 strut with an unbraced length of 60 ft shown in Figure 5-22. Assume ASTM A913 Grade 65 steel and assume pinned supports at both ends of the strut.
b. Using the axial capacity of the diagonal strut, ϕP_n, obtained from Part (a), and using the geometry in Figure 5-22, calculate the maximum factored column load, P_{max}, that can be supported by the W36 diagonal strut.
c. Calculate the maximum tension force in the horizontal tension tie and determine the gross area required if the tension member fails to yield. Neglect any bending loads.

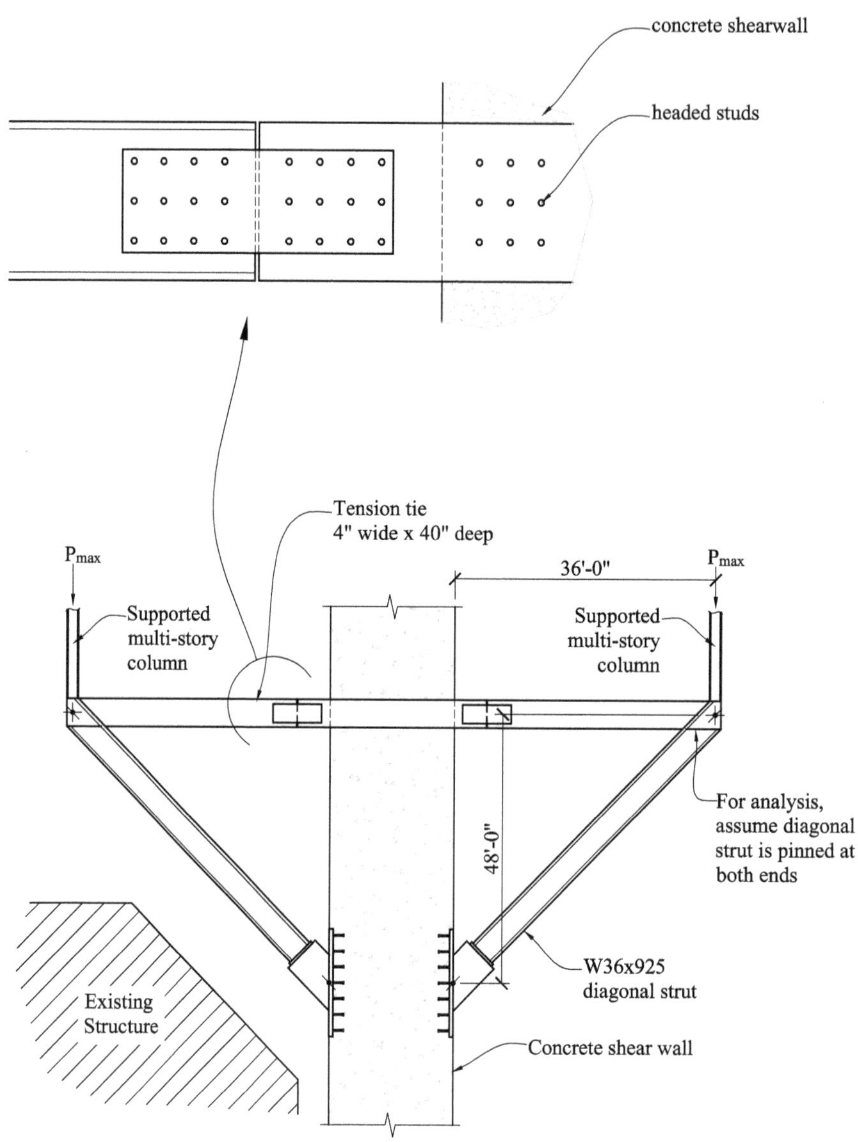

FIGURE 5-22 Detail for Exercise 5-21

5-22. A built-up flanged cruciform section is to be used as a high-rise column section. The cruciform section is made up of one W16 x 100 and two WT 8 × 50, as shown in Figure 5-15b. The plates are made from ASTM Grade 36 steel. Determine the nominal axial load capacity, the LRFD design axial load capacity, and the ASD allowable axial load of the section (in kips) for the following unbraced lengths: 10 ft., 30 ft., and 40 ft.

5-23. A built-up flanged cruciform section is made up of one W24 × 55 and two WT 12 × 27.5, similar to that shown in Figure 5-15b. The plates are made from A992 Grade 50 steel. Determine the nominal axial load capacity, the LRFD design axial load capacity, and the ASD allowable axial load of the section (in kips) if the unbraced length is 40 ft.

5-24. For the 63 ft. super-tall perimeter column in Exercise 13-8, select an I-shape built-up column to support the axial load from the roof and 5th floor levels. Consider axial load only in the column and ignore any connection eccentricities.

5-25. Using the Internet and other sources, conduct research on at least three structural failures that have been caused by buckling (or failure by bending and axial compression) of one or more members in the structure. One such failure is the 1978 Hartford Coliseum

Roof collapse. Write a minimum five-page report discussing a minimum of three collapses that were caused by buckling, explaining, with the aid of diagrams, why the structures failed by buckling, and discuss ways to prevent future failures of the kind.

5-26 Showing calculations, determine the design axial strength for a W12×87 column using grade 50 steel and an unbraced height of 33 ft. Assume simple beam and girder framing at each end of the column.

5-27 Sloped columns between floor levels are sometimes used to accommodate setbacks in a building façade.

a. Design a wide flange sloped exterior column with a 2 ft-6 in, setback between two floors that have a floor-to-floor height of 15 ft., and with the column set back at the lower floor level. The vertical factored load at the top of the inclined column is 1250 kips. Assume ASTM A913 Grade 50 steel.
b. Calculate all the forces in the sloped column and at the column connections and illustrate with free body diagrams.
c. Show the connection details at the end of the column that will be used to resist the applied forces.
Note that the horizontal component of the forces at the ends of the columns must be added to the lateral wind and seismic forces acting on the building.

5-28 Calculate the design axial compression capacity of a built-up mega column in a skyscraper formed by connecting two W36×925 Grade 65 steel tip of flange to tip of flange with 4 in, wide × 60 in. long stitch plates that are fillet-welded to the outer faces of the flanges of the W36. The stitch plates are spaced along the height of the column at 5 ft. 4in. on centers. Assume the column has an unbraced height of 15 ft. Is the column adequate to resist a factored axial compression load of 28,000 kips?

5-29. Calculate the axial compression capacity of pin-ended 2L8×4×7/8 assuming ASTM 992 Grade 50 steel and an unbraced length of 25 ft. The 3/8 in. gusset plate spacers between the double angles are spaced at 75 inches on centers and stitch bolted to the angles.

5-30. A pin-ended W14 x 145 column has unbraced lengths L_x = 22 ft and L_y = 26 ft. Using the AISC Equations, calculate the axil compression load capacity of the column. Assume ASTM A992 Grade 50 steel.

5-31. For the braced frame shown in Figure 5-23, analyze and design the diagonal brace using a W14 section. Assume ASTM A913 Grade 50 steel.

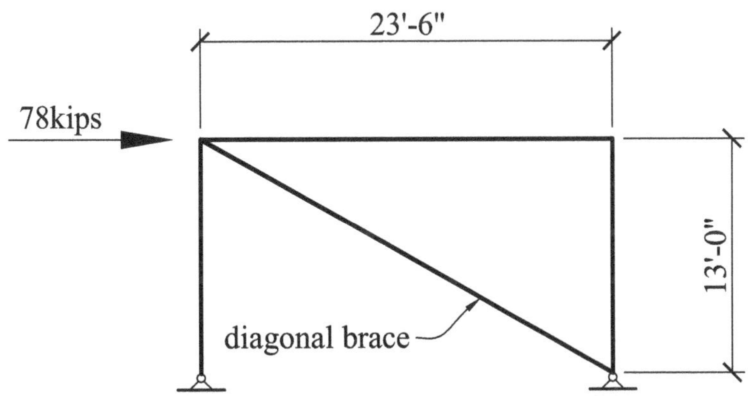

FIGURE 5-23 Diagonal brace for Exercise 5-31

5-32. Using the AISC Equations, select a W14 column to support a factored load of 8100 kips, assuming a floor-to-floor height of 18 ft. Verify your result using the AISC Manual tables. Assume ASTM A913 Grade 65 Steel.

5-33. To avoid the use of a cantilevered transfer girder at the 4th floor of a multi-story building to accommodate a 2 ft. setback in the curtainwall of the building at the 3rd floor level, the exterior column in the building is sloped between the 4th and 3rd floors of the building. The floor-to-floor height is 12 ft. and the factored vertical load on the column immediately above the 4th floor is 1200 kips. Assume ASTM A992 Grade 50 steel.

a. Calculate the factored axial compression load in the sloping column between the 4th and 3rd floors.

b. Calculate the factored tension force to be resisted at the 4th floor level and the factored compression force to be resisted at the 3rd floor level.

c. For the sloping column, select the most economical W12 column to resist the factored axial compression load in the column. Repeat the exercise for a W14 column and determine which column size is more economical.

d. Select an appropriate W-shape tension member at the 4th floor level and draw a detail of the connection of the tension member at the 4th floor level to the W12 and W14 sloping columns.

CHAPTER

6

Flexural Members

6-1 INTRODUCTION

Flexural members occur in steel structures in the form of beams and beam-columns. Beams are primarily horizontal members subjected to only bending or flexure about their primary or strong axis due to gravity loads, whereas beam columns (see Chapter 8) are usually vertical members that support axial compression or tension loads in addition to being subjected to flexure or bending from eccentric axial loads and/or lateral wind loads.

Beams that have no shear connectors (e.g., shear studs) at the interface between the steel beam top flange and the steel deck/concrete slab are called non-composite beams. Composite beams, on the other hand, have at least the AISC Code-prescribed minimum amount of shear connectors at the interface between the steel beam top flange and the concrete slab; the shear connectors enable the concrete slab and the steel beam to engage in composite action acting in concert to support the gravity loads. Roof beams are usually non-composite since concrete slabs are typically not used for roof framing, while floor beams can either be non-composite or composite. Non-composite beams will be the focus of this chapter, while composite beams will be covered in Chapter 7, and beam columns will be covered in Chapter 8. Some common types of beams are illustrated in Figure 6-1. A typical framing layout showing flexural members (beams and girders) is shown in Figure 6-1h, and Figure 6-1i shows metal deck on steel framing. Figures 6-1j and 6-1k show composite metal deck with steel studs on the beam and girder framing. Figure 6-1m shows cantilevered roof and floor girders with moment connections at the columns.

The flexural member components used in steel-framed buildings are as follows:

- **Non-composite metal deck (see Figures 7-1e and 7-1f)**
  - Metal deck without protrusions on the interior of the deck.
  - Used for roof decks.
  - Also occur as form decks that can be used as floor decks.
- **Composite metal deck (see Figures 6-1j, 6-1k, 7-1a and 7-1d)**
  - Metal deck with protrusions on the interior of the deck.
  - The protrusions enable both the deck and the concrete to act in concert or compositely.
  - Used for floor decks.

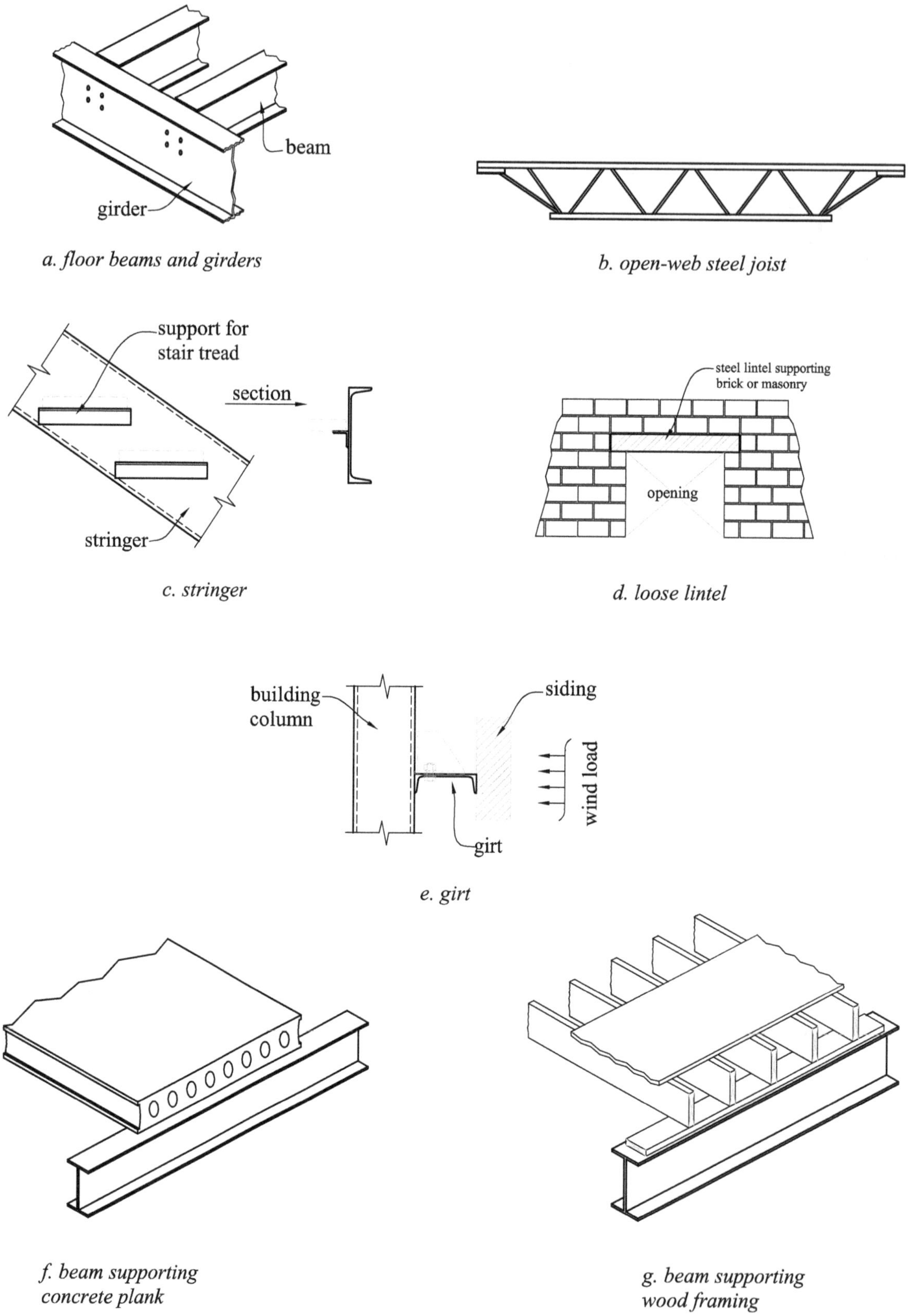

FIGURE 6-1a-g Common beam members

FIGURE 6-1h Typical steel framing layout

FIGURE 6-1l Metal deck on steel beams

FIGURE 6-1j Composite metal deck with headed steel studs

FIGURE 6-1k Composite metal deck with steel studs along beam and girder lines

FIGURE 6-1m Cantilevered roof and floor girders with moment connections

- **Composite metal deck (see Figures 6-1j, 6-1k, 7-1a and 7-1d)**
  - Metal deck with protrusions on the interior of the deck.
  - The protrusions enable both the deck and the concrete to act in concert or compositely.
  - Used for floor decks.
- **Noncomposite beams or girders (covered in Chapter 6)**
  - Non-composite metal deck supported by beams or girders *without* shear studs (typically used for roofs).
  - Composite metal deck supported on beams/girders *without* shear studs (used for floors).
  - The beam and girder act alone (non-compositely) in resisting the loads.
- **Composite beams and girders (see Chapter 7)**
  - Composite metal deck supported by beams/girders *with* shear studs welded to the top flange of the beams or girders.
  - Due to the shear studs, the concrete slab and the steel beams/girders act together (i.e., compositely) to resist the loads in bending.

Joist, beams, and girders which are covered in this chapter can be further delineated as follows:

- **Open-Web Steel Joists (OWSJ):**
  - Open-web steel joists (OWSJ) are truss-like members that are used in lieu of roof beams in steel-framed roofs
  - They are not frequently used in steel-framed floors because of their susceptibility to undesirable floor vibrations (see Chapter 12)
  - Their advantage is that the open webs allow passage of mechanical ducts
- **Beams:**
  - Bending only about the strong axis (x-x), or
  - Bending about both strong (x-x) and weak (y-y) axes
  - Usually subject to uniform loads from the roof or floor
- **Girders:**
  - Bending only about the strong axis (x-x), or
  - Bending about both strong (x-x) and weak (y-y) axes
  - Usually supports the beams and is subjected to concentrated loads from the beam reactions, though girders may also support uniform loads due to cladding loads, roof or floor edge loads, or operable partition wall loads hung from the girder.
- **Beam-columns (see Chapter 8):**
  - Axial tension or compression plus bending about the strong axis (***x-x***), or
  - Axial tension or compression plus bending about **both** strong (x-x) and weak (y-y) axes

Beams can be further classified by the function that they serve. A girder is a member that is typically larger in size than beams and supports other beams or framing members. A joist is typically a lighter section than a beam—such as an open-web steel joist. A stringer is a sloping flexural member that is the main support beam for a stair. A lintel (or loose lintel) is usually a smaller section that frames over a wall opening, supporting the wall above the door or window openings. A girt is a horizontal flexural member that provides horizontal support to exterior cladding or siding in resisting lateral wind loads perpendicular to the face of the cladding or siding.

The basic design for beams includes checking bending, shear, deflection, and vibrations. The loading conditions and beam configuration will dictate which of the preceding design parameters controls the size of the beam.

We will now review some of the basic principles of the mechanics of flexure from the mechanics of materials. When a beam is subjected to loads that cause flexure or bending, the bending stress in the extreme fiber is defined as

$$f_b = \frac{Mc}{I} = \frac{M}{S}, \tag{6-1}$$

and the yield moment is defined as

$$M_y = F_y S, \tag{6-2}$$

where

f_b = Maximum bending stress,
M_y = Yield moment,
F_y = Yield stress,
M = Applied bending moment,
c = Distance from the neutral axis to the extreme fiber,
I = Moment of inertia, and
S = Elastic section modulus.

The above formulation is based on the elastic behavior of the beam. Figure 6-2 illustrates the behavior of a simply supported steel beam as it is loaded to failure. It shows the change in stress distribution at the point of maximum moment as the load on the beam is increased until the applied moment is large enough that a plastic hinge forms at the point of maximum moment. Initially, the beam is in the elastic stage, where the extreme fiber stresses at the point of maximum moment are less than the yield stress. With increasing load, the extreme fiber stresses reach the yield stress, F_y, and the section reaches the yield point. The moment at which the extreme fibers first reach the yield stress is known as the yield moment, M_y. This is stage "a" in Figure 6-2. Since the extreme fiber stresses cannot exceed the yield stress, any further increase in the load on the beam causes the yielding of the beam to spread from the extreme fibers towards the neutral axis of the beam. This is stage "b" in Figure 6-2, where the outer parts of the critical section have yielded, but the inner parts closer to the neutral axis are still in the elastic state. As the load is further increased up to the maximum load, this causes all the fibers in the cross-section at the point of maximum moment to reach the yield stress (see stage "c" in Figure 6-2). At this stage, the yielding has spread through the whole cross-section of the flexural member at the point of maximum moment, and a plastic hinge is formed. Note that the plastic hinge has a finite width. For a simply supported beam, after the plastic hinge forms, there can no longer be any further increase in the load since with the hinge formation at the point of maximum moment, a collapse mechanism has formed. A plastic hinge forms when the entire cross section of the beam, not just the extreme fibers, has reached the yield stress, F_y. The preceding discussion assumes that the beam is fully braced against lateral torsional buckling, and therefore, lateral torsional buckling is not a design consideration. Also, it is assumed that there is no local buckling of the component elements that make up the flexural member. As previously discussed in Chapter 1, the moment at which a plastic hinge is developed in a beam is called the plastic moment, M_p., and is defined as

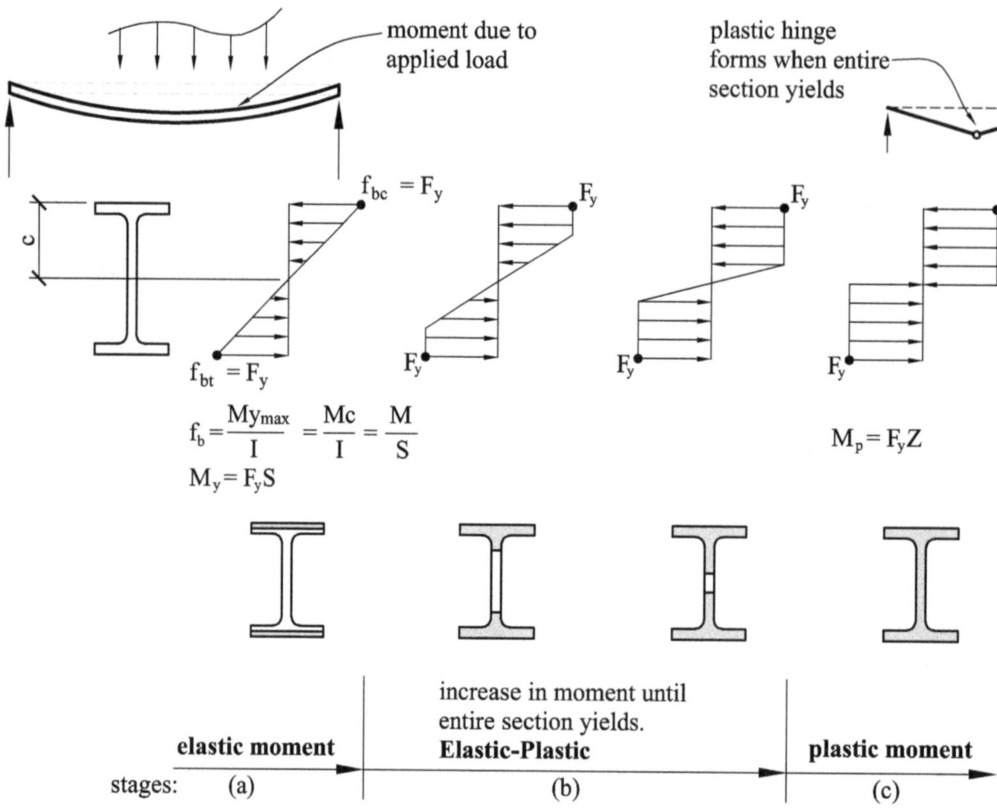

FIGURE 6-2 Internal Stress distribution in flexural members under external load

$$M_p = F_y Z, \tag{6-3}$$

where

M_p = Plastic moment, and

Z = Plastic section modulus.

The plastic moment is the maximum moment, or nominal bending strength, of a beam with full lateral stability. For standard wide flange shapes, the ratio of the plastic moment, M_p, to the yield moment, M_y, usually varies from 1.10 to 1.25 for strong axis bending (Z_x, S_x), and 1.50 to 1.60 for weak axis bending.

The design parameters for shear and deflection will be discussed later in this chapter. A summary of the basic load effects for common beam loading conditions is presented in Table 6-1. *AISCM* Table 3.22, has several other loading conditions beyond what is shown here.

6-2 CLASSIFICATION OF FLEXURAL MEMBERS

All flexural members are classified as follows depending on the *width-to-thickness ratios* of the individual elements that make up the beam section:

- **Compact**,
- **Non-compact**, or
- **Slender**

There are two types of elements defined in the AISC Specification as follows:

- ***Stiffened*** elements, and
- ***Unstiffened*** elements.

Stiffened elements are supported along both edges parallel to the load. An example of this is the web of an I-shaped beam because it is connected to the flanges at either end of the web. An unstiffened element has only one unsupported edge parallel to the load; an example of this is the outstanding flange of an I-shaped beam that is connected to the web on one side and free on the other.

TABLE 6-1 Summary of Shear, Moment, and Deflection Formulas

Loading condition	Loading diagram	Maximum shear	Maximum moment	Maximum deflection
Uniformly loaded simple span	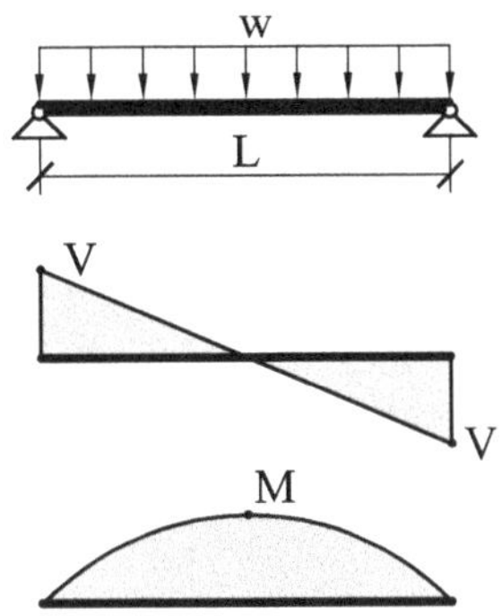	$V = \frac{wL}{2}$	$M = \frac{wL^2}{8}$	$\Delta = \frac{5wL^4}{384EI}$
Concentrated load at midspan	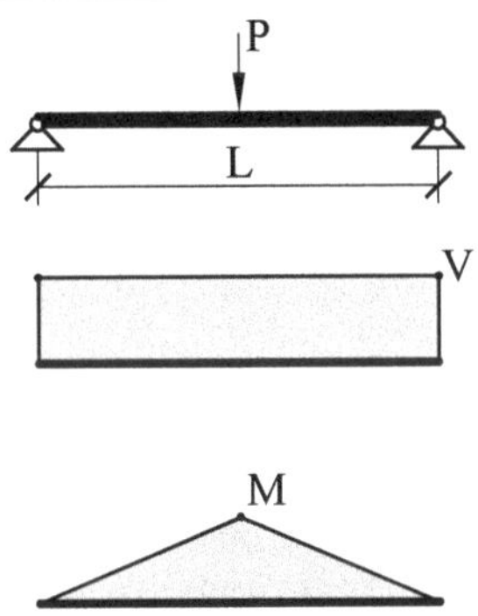	$V = \frac{P}{2}$	$M = \frac{PL}{4}$	$\Delta = \frac{PL^3}{48EI}$
Concentrated loads at ⅓ points	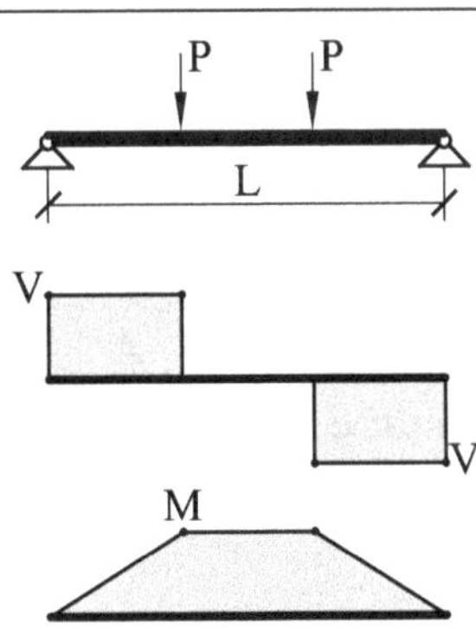	$V = P$	$M = \frac{PL}{3}$	$\Delta = \frac{PL^3}{28EI}$
Concentrated loads at ¼ points	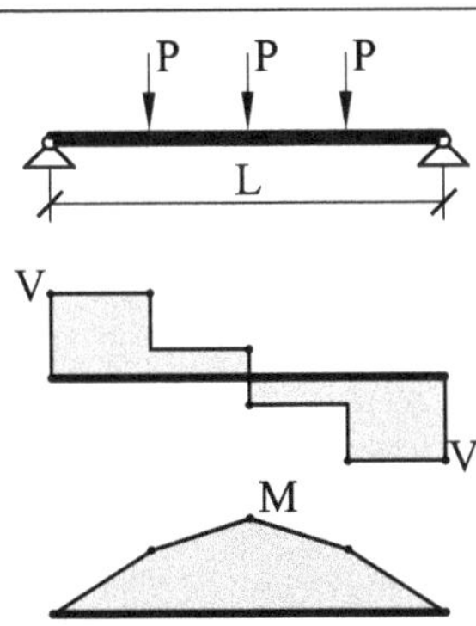	$V = 1.5\,P$	$M = \frac{PL}{2}$	$\Delta = \frac{PL^3}{20EI}$

(Continued)

Loading condition	Loading diagram	Maximum shear	Maximum moment	Maximum deflection
Uniformly loaded, cantilever	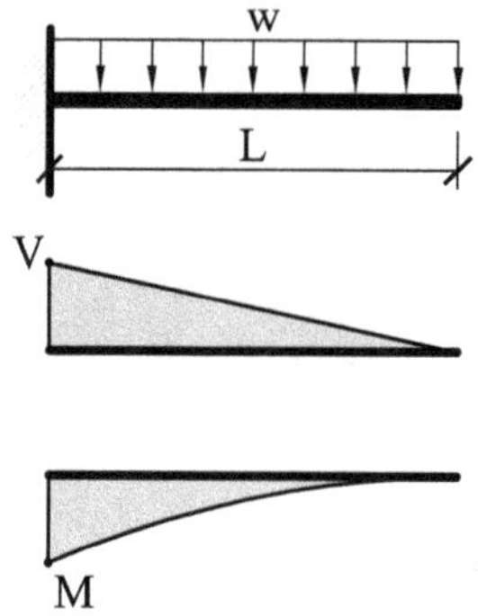	$V = wL$	$M = \frac{wL^2}{2}$	$\Delta = \frac{wL^4}{8EI}$
Concentrated load at the end of the cantilever	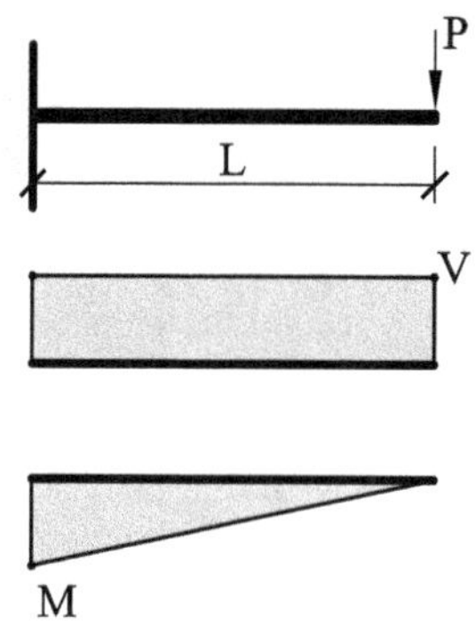	$V = P$	$M = PL$	$\Delta = \frac{PL^3}{3EI}$

Table 6-2 gives the upper limits for the width-to-thickness ratios for the individual elements of flexural sections. These ratios provide the basis for the classification of flexural sections.

- When the width-to-thickness ratio is less than λ_p, the section is classified as *compact*.
- When the ratio is greater than λ_p but less than λ_r, the section is classified as *noncompact*.
- When the ratio is greater than λ_r, the section is classified as *slender*.

The classification of flexural members with respect to the width-to-thickness ratio of its component elements is necessary because the flexural design strength or moment capacity is a function of this classification. The width-to-thickness ratios are tabulated in Part 1 of the *AISCM* for structural shapes, so there is usually no need to calculate this ratio.

EXAMPLE 6-1

Classification of a W-Shape

Determine the classification of a W18 × 35 and a W21 × 48 for F_y = 50 ksi. Check both the flange and the web.

Solution

From part 1 of the *AISCM*,

W18 × 35	W21 × 48
$\frac{b_f}{2t_f} = 7.06$	$\frac{b_f}{2t_f} = 9.47$

$$\frac{h}{t_w} = 53.5 \qquad \frac{h}{t_w} = 53.6$$

Flange:
The maximum flange width-to-thickness ratios to ensure a compact flange are calculated as follows:

$$\lambda_{pf} = 0.38\sqrt{\frac{E}{F_y}} = 0.38\sqrt{\frac{29{,}000}{50}} = 9.15 > 7.06 \text{ for W}18 \times 35$$

$$< 9.47 \text{ for W}21 \times 48$$

$\therefore$ the W18 × 35 flange is compact.

TABLE 6-2 Limiting Width–Thickness Ratios for Flexural Elements

	Description	Limiting width–thickness ratio: λ_p, λ_{pf} (compact flange), λ_{pw} (compact web)	Limiting width–thickness ratio: λ_r, λ_{rf} (non-compact flange), λ_{rw} (non-compact web)	Details
Unstiffened	Flanges of I-shaped rolled sections (Case 10)			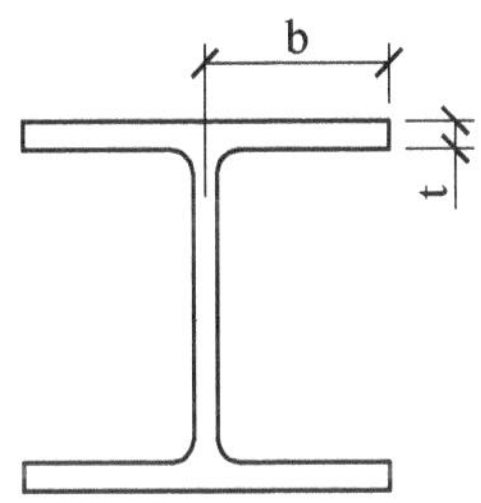
Unstiffened	Flanges of C-shapes (Case 10)	$\frac{b}{t} \le 0.38\sqrt{\frac{E}{F_y}}$	$\frac{b}{t} \le 1.0\sqrt{\frac{E}{F_y}}$	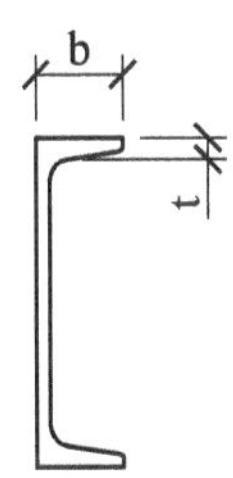
Unstiffened	Flanges of WT-shapes (Case 10)			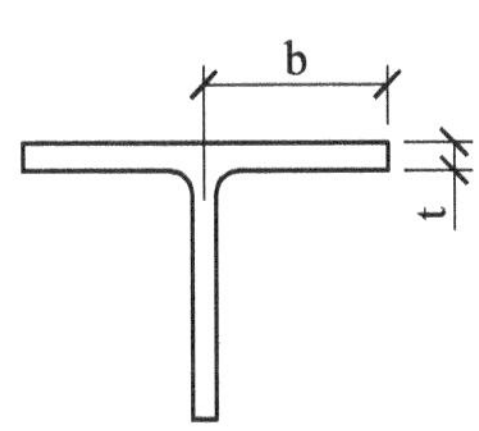
Unstiffened	Outstanding legs of single angles (Case 12)	$\frac{b}{t} \le 0.54\sqrt{\frac{E}{F_y}}$	$\frac{b}{t} \le 0.91\sqrt{\frac{E}{F_y}}$	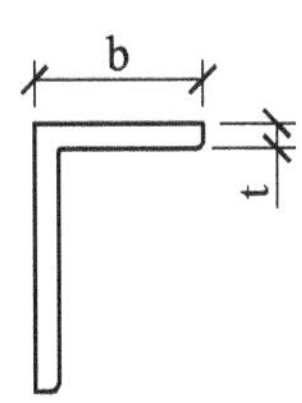

(Continued)

	Description	Limiting width–thickness ratio		Details
		λ_p λ_{pf} (flange) λ_{pw}(web)(compact)	λ_r λ_{rf} (flange) λ_{rw}(web)(noncompact)	
Stiffened	Webs of I-shaped sections (Case 15)	$\frac{h}{t_w} \leq 3.76\sqrt{\frac{E}{F_y}}$	$\frac{h}{t_w} \leq 5.70\sqrt{\frac{E}{F_y}}$	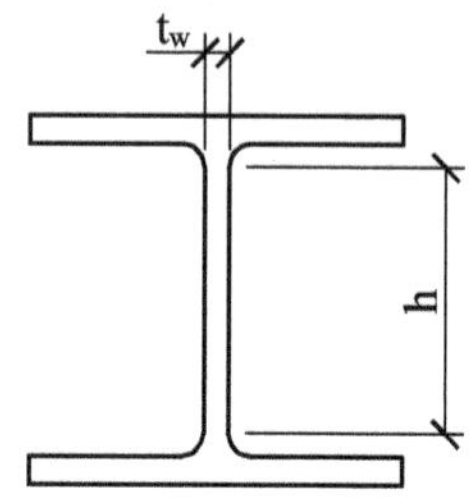
	Webs of C-shapes (Case 15)			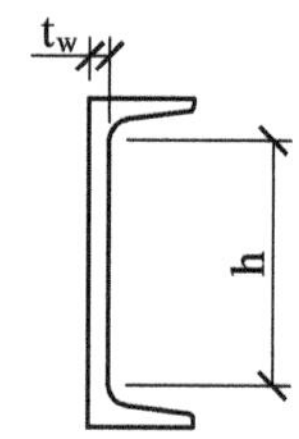
	Flanges of rectangular HSS (Case 17)	$\frac{b}{t} \leq 1.12\sqrt{\frac{E}{F_y}}$	$\frac{b}{t} \leq 1.40\sqrt{\frac{E}{F_y}}$	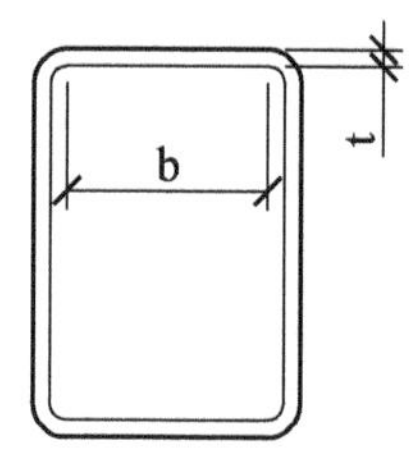
	Webs of rectangular HSS and Box sections (Case 19)	$\frac{h}{t} \leq 2.42\sqrt{\frac{E}{F_y}}$	$\frac{h}{t} \leq 5.70\sqrt{\frac{E}{F_y}}$	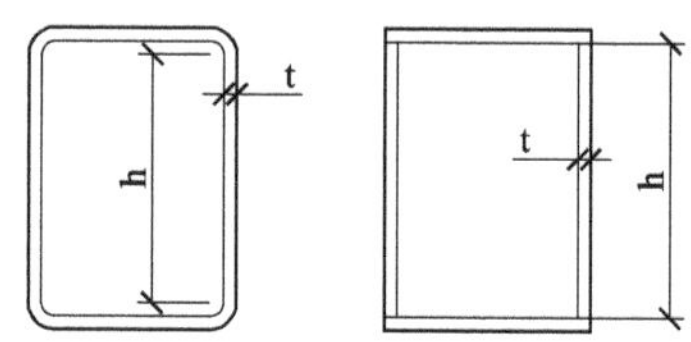
	Flanges of Box sections (Case 21)	$\frac{b}{t} \leq 1.12\sqrt{\frac{E}{F_y}}$	$\frac{b}{t} \leq 1.49\sqrt{\frac{E}{F_y}}$	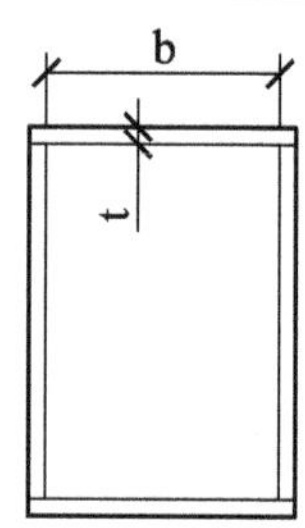
	Round HSS or pipes (Case 20)	$\frac{D}{t} \leq 0.07\frac{E}{F_y}$	$\frac{D}{t} \leq 0.31\frac{E}{F_y}$	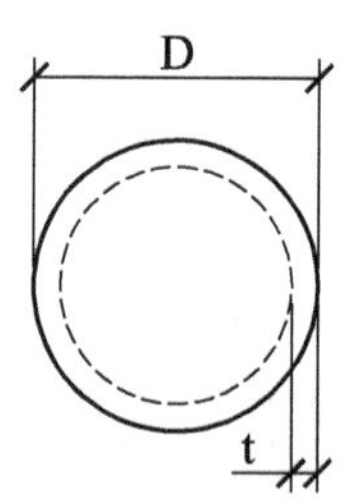

Source: Adapted from Table B4.1b of the *AISCM*.[1].

The maximum flange width-to-thickness ratio beyond which the flange is slender and below which the flange is noncompact is

$$\lambda_{rf} = 1.0\sqrt{\frac{E}{F_y}} = 1.0\sqrt{\frac{29{,}000}{50}} = 24.0 > 9.47 \text{ for W21} \times 48$$

∴ the W21 × 48 flange is noncompact.

Web:

The maximum web depth-to-thickness ratio to ensure a compact web is

$$\lambda_{pw} = 3.76\sqrt{\frac{E}{F_y}} = 3.76\sqrt{\frac{29{,}000}{50}} = 90.5 > 53.5 \text{ for W18} \times 35;$$

$$> 53.6 \text{ for W21} \times 48$$

∴ both the W18 × 35 and W21 × 48 webs are compact.

Note that shapes that are noncompact for bending are denoted by subscript "*f*" in Part 1 of the *AISCM*.

6-3 FAILURE MODES OF FLEXURAL MEMBERS AND DESIGN STRENGTH FOR COMPACT SHAPES

Flexural members can fail in one of the following modes, and the nominal bending strength, M_n, of a flexural member is a function of these failure modes:

1. Yielding of the entire cross-section at the point of maximum moment
2. Lateral–torsional buckling (LTB)
3. Flange local buckling (FLB) in compression
4. Web local buckling (WLB) in flexural compression

Note that flexural members are also subject to elastic buckling (just like compression members) because some components of the flexural member are usually subjected to compression stresses due to the bending moment applied to the member. The thinner the component elements—such as the flanges and webs—are, the greater the tendency for the elements in compression to buckle locally. Flange local buckling and web local buckling are localized failure modes and are only of concern with shapes that have noncompact webs or flanges, which will be discussed in further detail later.

Lateral–torsional buckling — a global buckling failure mode — occurs when the distance between lateral brace points is large enough that the flexural member (e.g., a beam) fails by lateral displacement perpendicular to the longitudinal axis of the member, in combination with twisting of the section plus vertical displacement of the member in the direction of the applied vertical load (see Figure 6-3). The twisting occurs because the compression flange experiences buckling due to the compression stress, while the tension flange does not experience buckling but restrains the web; as the compression flange displaces laterally, the section twists in order to maintain compatibility. Flexural members with wider flanges are less susceptible to lateral–torsional buckling because the wider flanges provide more resistance to lateral displacement. In general, adequate restraint against lateral–torsional buckling is accomplished by the addition of a brace or similar restraint somewhere between the centroid of the member and the compression flange (see Figure 6-3b). For simple-span beams supporting normal gravity loads, the top flange will be in compression, but the bottom flange could be in compression in the following

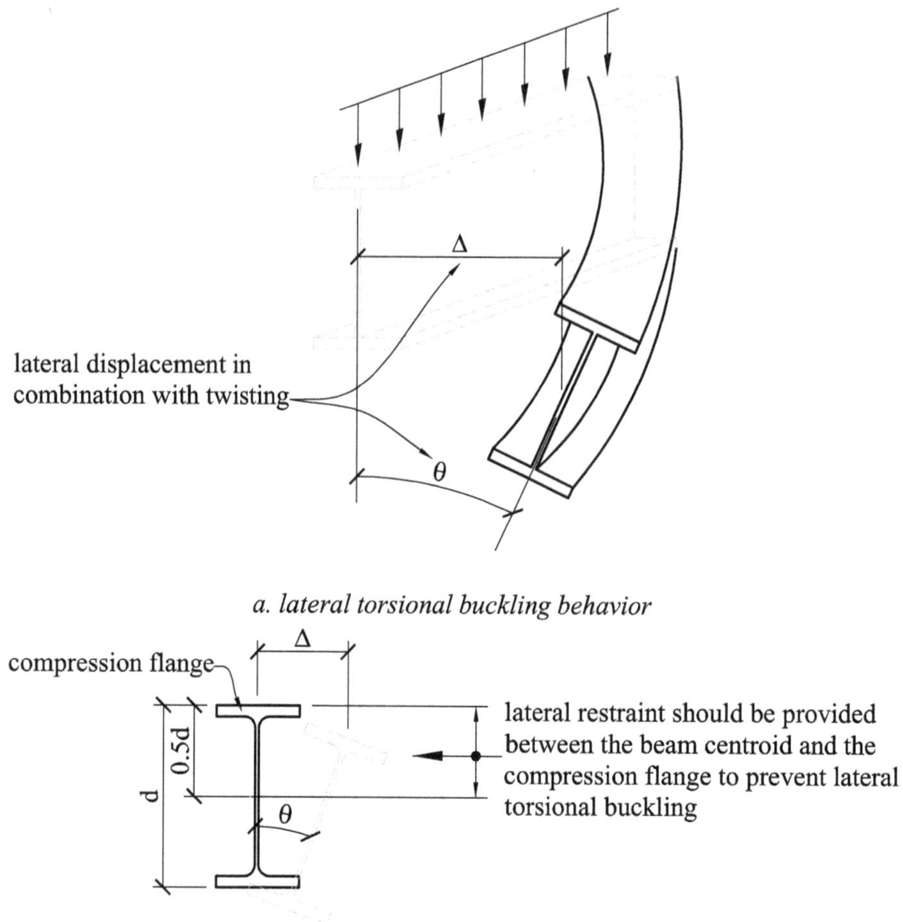

a. lateral torsional buckling behavior

b. lateral torsional buckling restraint

FIGURE 6-3 Lateral torisonal buckling

scenarios: at the support of continuous beams, or beams in moment frames, or roof beams subjected to net uplift loads due to wind suction (see Chapter 3).

Lateral torsional buckling could also be an issue during the erection of beams and girders. A rule of thumb used in practice indicates that beams and girders with a span-to-flange width ratio (i.e., L/b) less than 80 will be laterally stable during erection, where b is one-half the flange width of wide flange sections. Beams with L/b greater than 80 would need some form of lateral support during erection to mitigate lateral instability, and camber in a beam would exacerbate the lateral instability.

Lateral–torsional buckling can be controlled in several ways, but it is usually dependent on the actual construction details used. Beams with metal deck ribs oriented perpendicular to the longitudinal axis of the beam are considered fully braced (Figure 6-4a), whereas the girder in Figure 6-4b is not considered braced by the deck because the ribs are parallel to the girder and have very little stiffness to prevent the lateral displacement of the girder top flange. The girder in Figure 6-4c would be considered braced by the intermediate framing members and would have an unbraced length , L_b , equal to the maximum spacing between the infill beams. The factors affecting the bending strength of flexural members are:

- Unbraced length, L_b, of the compression flange or zone of the flexural member
- The width-to-thickness ratio of the flanges, $\lambda_{(flange)}$

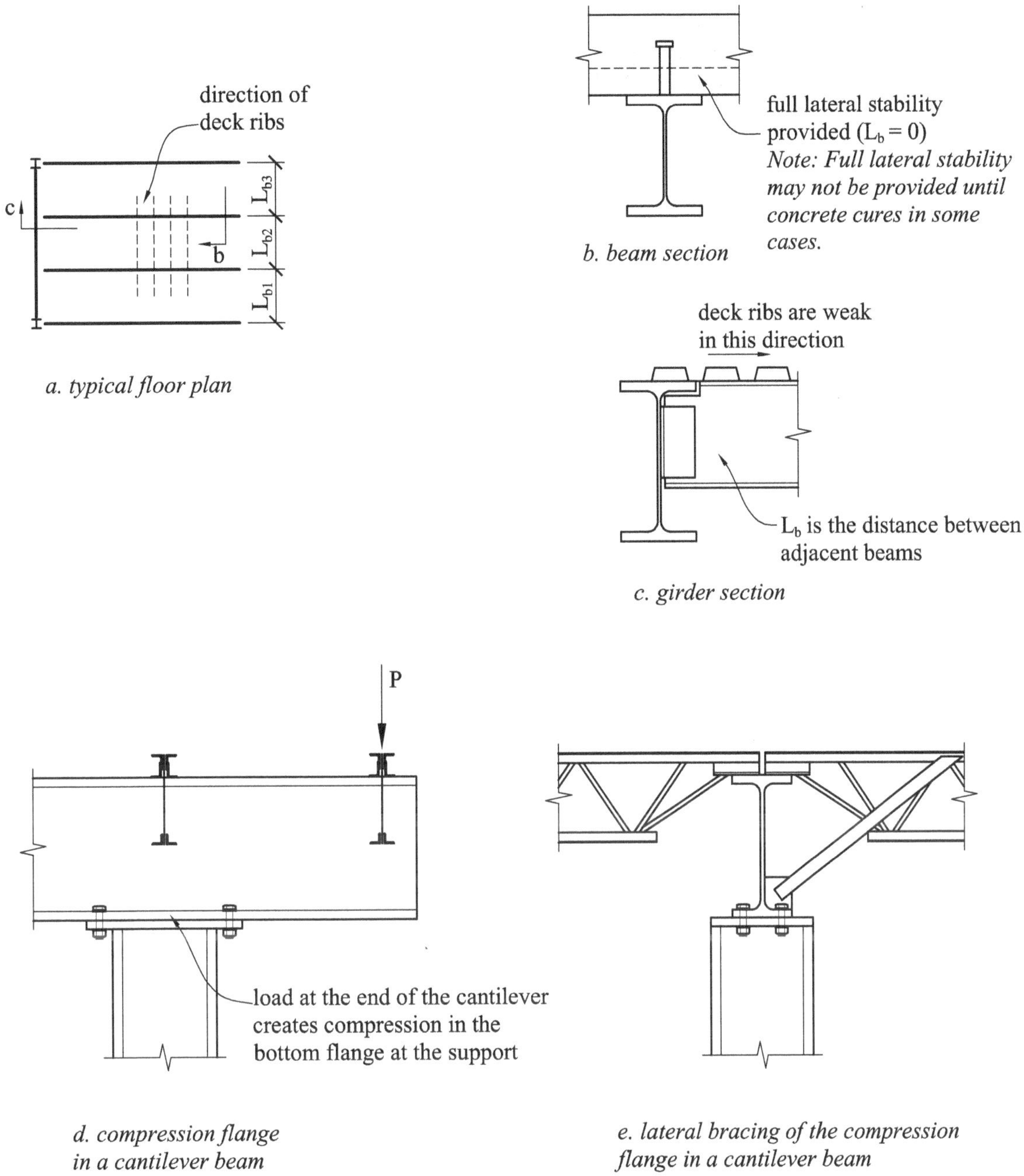

FIGURE 6-4 Lateral bracing details

Unbraced Lengths of Infill Beams and Girders:

- **Infill Beams:**
  The unbraced length of a beam is the length of the compression flange over which the beam is laterally unbraced. Since the roof or floor deck are usually welded to the top flange of the beam (i.e., the compression flange for a simply supported beam), the beams will be continuously braced laterally; therefore, their unbraced length will be zero (i.e., $L_b = 0$). If the floor had an opening on both sides of a beam with a length, $L_{opening}$, the unbraced length of the beam (again, that is the length over which the compression flange is laterally unbraced) will be equal to the length of the opening parallel to the beam, that is, $L_b = L_{opening}$. Some beams are laterally unbraced for their entire lengths and are therefore subject to elastic lateral torsional buckling;

the unbraced length for these beams will be equal to the span of the beam between supports, and examples include:

- Elevator hoist beams - they are usually unbraced for their entire span and support the maximum weight of the elevator.
- Lintel beams that do not have any floor or roof level laterally bracing the beam.
- Laterally unsupported beams (see Figure 6-4f) supporting masonry walls or cladding loads, and with their top flange not laterally braced by a floor level.

FIGURE 6-4f A skewed laterally unsupported beam

- **Girders:**
  The girders usually support the in-fill beams that frame into the girders. The deck's lateral stiffness parallel to the direction of the ribs of the deck is much higher than the lateral stiffness perpendicular to the direction of the deck ribs. Since the deck has very little lateral or in-plane stiffness perpendicular to the direction of the deck ribs, the deck is not considered adequate to laterally brace the girder. What braces the girder compression or top flange laterally are the beams framing into the girders. Therefore, the maximum unbraced length for the girder will be the spacing between the in-fill beams. That is, for the girder: $L_b = s$.
- **Cantilevered Beams and Girders:**
  The unbraced length of cantilever beams and girders is typically taken as equal to the cantilever span. In practice, the tip of the cantilevered girder is usually laterally braced, in addition to the lateral bracing of the girder at the support. For cantilevered beams and girders laterally braced at the support to prevent warping, and with the free end unbraced, the lateral torsional buckling stability factor, or the moment gradient factor, C_b, is taken as 1.0 (AISC specification Section F1.c). This factor will be discussed later in Section 6-3

Bending strength of doubly symmetric compact wide flanges or I-sections and channels

In the following discussion, the bending strength of **doubly symmetric compact wide flange or I-sections, and channels** bending about their major or strong axis is presented. The basic design equations for bending are as follows:

$$M_u \leq \phi_b M_n \quad \text{for LRFD} \tag{6-4a}$$

$$M_s \leq \frac{M_n}{\Omega_b} \quad \text{for ASD} \tag{6-4b}$$

Where,

M_u = Factored moment,

M_s = Service or unfactored Moment,

ϕ_b = 0.9,

Ω_b = 1.67,

M_n = Nominal bending strength,

$\frac{M_n}{\Omega_b}$ = available bending strength,

$\phi_b M_n$ = Design bending strength,

When full lateral stability is provided for a beam, the nominal moment strength is the plastic moment capacity of the beam ($M_p = F_y Z_x$, see Section 6-1). Once the unbraced length reaches a certain upper limit, lateral–torsional buckling will occur, and therefore the nominal bending strength will likewise decrease. The failure mode of lateral–torsional buckling can be either inelastic or elastic. The AISC specification Equation F2-5 defines the unbraced length limit for the limit state of yielding, or the unbraced length above which inelastic lateral–torsional buckling occurs (see Figure 6-5) as

$$L_p = 1.76 r_y \sqrt{\frac{E}{F_y}}. \tag{6-5}$$

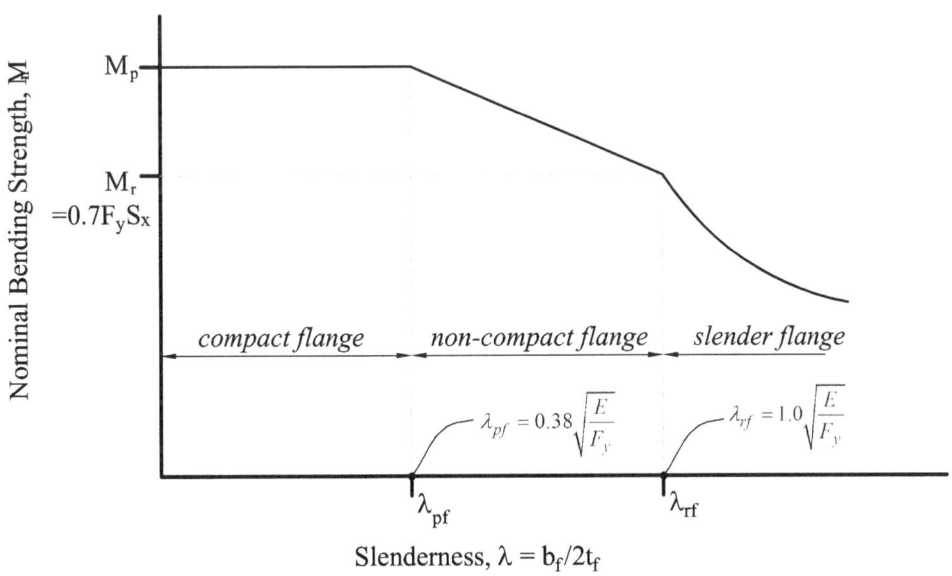

a) *Nominal moment strength variation with flange slenderness for I shape (Adapted from AISCM Figure C-F1.1)*

Figure 6-5 *Continued*

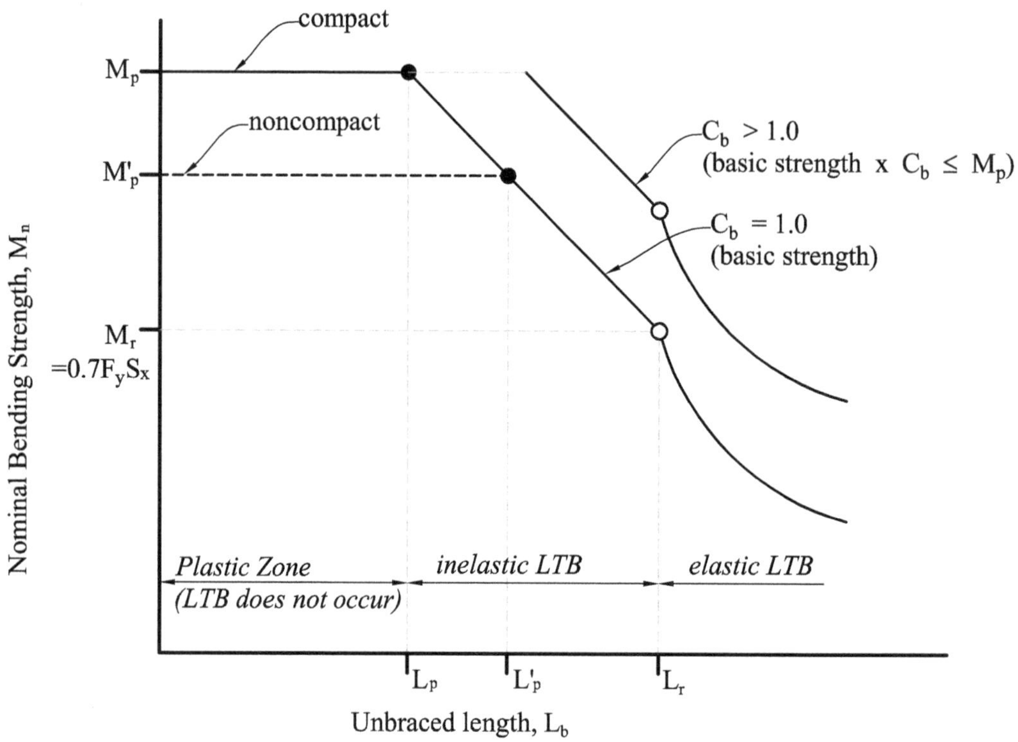

a) Nominal moment strength variation with unbraced length (Adapted from AISCM Figure C-F1.2)

FIGURE 6-5 Moment strength with respect to unbraced length and local slenderness

L_p is also the maximum unbraced length below which the yielding failure mode dominates and at which the nominal bending strength equals the plastic moment capacity. The unbraced length, L_r, at which elastic lateral–torsional buckling begins (i.e., the maximum unbraced length for inelastic lateral torsional buckling), is obtained by equating the stress at the transition point between inelastic and elastic lateral-torsional buckling (i.e., $0.7F_y$) to the elastic lateral-torsional buckling stress, F_{cr}, giving:

$$L_r = 1.95 r_{ts} \frac{E}{0.7F_y} \sqrt{\left(\frac{Jc}{S_x h_o}\right)} \sqrt{1 + \sqrt{1 + 6.76 \left(\frac{0.7F_y}{E} \frac{S_x h_o}{Jc}\right)^2}} \tag{6-6}$$

Where,

$$r_{ts} = \left(\frac{\sqrt{I_y C_w}}{S_x}\right)^{1/2} \tag{6-7}$$

$c = 1.0$ (for doubly symmetric I-shapes i.e., W-shapes, M-shapes, or S-shapes),

$$c = \frac{h_o}{2} \sqrt{\frac{I_y}{C_w}} \quad \text{(for channel, S or C-shapes),} \tag{6-8}$$

F_y = Yield strength,
E = Modulus of elasticity,
J = Torsional constant,
S_x = Elastic section modulus for bending about the major axis (i.e., x-axis),

I_y = Moment of inertia for bending about the minor axis (i.e., y-axis),
C_w = Warping constant, and
h_o = Distance between flange centroids.
r_{ts} = effective radius of gyration.

For doubly symmetric I-shapes with rectangular flanges,

$$C_w = \frac{I_y h_o^2}{4}$$

Therefore, Equation (6-7) becomes $r_{ts} = \sqrt{\dfrac{I_y h_o}{2S_x}}$

r_{ts} can conservatively be approximated as $r_{ts} = \dfrac{b_f}{\sqrt{12\left(1 + \dfrac{ht_w}{6b_f t_f}\right)}}$

Three zones are identified in Figure 6-5 with regard to the effect of the unbraced length of the compression zone of a flexural member on its bending moment capacity. These are

- Plastic zone (zone 1)
- Inelastic lateral torsional buckling zone (zone 2)
- Elastic lateral torsional buckling zone (zone 3)

Plastic (zone 1) - $L_b \le L_p$:
When lateral–torsional buckling is not a concern (i.e., when the unbraced length, $L_b \le L_p$), the failure mode is flexural yielding. The nominal bending strength for flexural yielding failure mode is

$$M_n = M_p = F_y Z_x \tag{6-9a}$$

where
F_y = Yield strength, and
Z_x = Plastic section modulus (tabulated for various sections in Part 1 of the AISCM).

For sections that are not listed in the *AISCM*, the plastic section modulus and the elastic section modulus can be calculated using the method introduced and discussed in Section 1-4.

The design moment capacity is

$$\phi M_n = \phi M_p = \phi F_y Z_x \tag{6-9b}$$

where $\phi = 0.9$.

Inelastic Lateral-Torsional Buckling or Inelastic LTB (zone 2) - $L_p \le L_b \le L_r$:
For compact I-shapes and C-shapes, when $L_p \le L_b \le L_r$, the nominal flexural strength is obtained by interpolation from Figure 6-5

$$M_n = C_b\left[M_p - \left(M_p - 0.7F_y S_x\right)\left(\frac{L_b - L_p}{L_r - L_p}\right)\right] \le M_p. \tag{6-10a}$$

The design moment capacity is

$$\phi M_n = C_b\left[\phi M_p - \left(\phi M_p - 0.7\phi F_y S_x\right)\left(\frac{L_b - L_p}{L_r - L_p}\right)\right] \le \phi M_p \tag{6-10b}$$

where $\phi = 0.9$

In equation 6-10, the term $0.7F_y S_x$, also denoted as M_r, accounts for the effects of residual stresses, which are assumed to be $30\%F_y$. Thus, $M_r = S_x(F_y - 30\%F_y) = 0.7F_yS_x$, which corresponds to the limiting buckling moment when $L_b = L_r$ and is the transition point between inelastic and elastic lateral–torsional buckling (see the location of M_r in Figure 6-5). M_r is the moment strength of a non-compact section, accounting for the $0.3F_y$ residual stress.

Elastic Lateral-Torsional Buckling (Elastic LTB) Zone (zone 3) - $L_b > L_r$:
The elastic lateral torsional buckling strength for a wide flange section, M_{cr}, is derived from solving the differential equation for lateral torsional buckling [13] and is defined as:

$$M_{cr} = \frac{C_b \pi}{L} \sqrt{EI_y GJ + \left(\frac{\pi E}{L}\right)^2 I_y C_w}$$

The preceding equation for the lateral torsional buckling of doubly symmetric sections was used in earlier AISC codes. For compact I-shapes and C-shapes, when $L_b > L_r$, the nominal flexural strength for the lateral torsional buckling limit state is now given as

$$M_n = F_{cr} S_x \leq M_p, \tag{6-11a}$$

The design moment capacity is

$$\phi M_n = \phi F_{cr} S_x \leq \phi M_p, \tag{6-11b}$$

Where,
$\phi = 0.9$

$$F_{cr} = \frac{C_b \pi^2 E}{\left(\frac{L_b}{r_{ts}}\right)^2} \sqrt{1 + 0.078 \frac{Jc}{S_x h_o} \left(\frac{L_b}{r_{ts}}\right)^2} \tag{6-12}$$

$$C_b = \frac{12.5M_{\max}}{2.5M_{\max} + 3M_A + 4M_B + 3M_C}, \tag{6-13}$$

Where,
F_{cr} = lateral torsional buckling stress,
C_b = lateral torsional buckling modification factor,
L_b = Distance between points of lateral bracing of the compression flange,
$M_{\max}$ = Absolute value of the maximum moment in the unbraced segment,
M_A = Absolute value of the moment at the ½ point of the unbraced segment,
M_B = Absolute value of the moment at the centerline of the unbraced segment,
M_C = Absolute value of the moment at the ¾ point of the unbraced segment,
I_y = moment of inertia about the principal y-axis,
I_{yc} = Moment of inertia of the compression flange about the y-axis.

For doubly symmetric shapes, I_{yc} is approximately equal to $I_y/2$. For reverse-curvature bending, I_{yc} is the moment of inertia of the smaller flange.

The lateral torsional buckling modification factor, C_b, is also referred to as the moment gradient factor, and it accounts for the moment gradient or the non-uniform moment distribution within the unbraced length of the beam. The moment strength equations in the AISC Code were derived assuming a uniform moment within the unbraced length of a beam. When a flexural member is subjected to a non-uniform bending moment distribution or a moment gradient within its unbraced length, the moment capacity of the flexural member is increased, and the

C_b factor is used to account for this increase. It is conservative to assume that $C_b = 1.0$ for any loading condition, which implies that the applied moment is constant across the entire beam. In lieu of using equation (6-13), the value of C_b can also be obtained from Table 6-3, which is based on equation (6-13).

The variation of the nominal bending moment strength with respect to the unbraced length is shown in Figure 6-5a, which summarizes the preceding discussion of bending strength for beams. This figure shows the bending strength for both compact and noncompact shapes. The bending strength of noncompact shapes will be discussed in a later section but is indicated here for completeness.

There are three distinct zones shown in the figure, the first being where lateral–torsional buckling does not occur, and the bending strength is a constant value of M_p. In this zone, the section is fully plastic, meaning that the entire cross-section at the point of maximum moment has reached the yield stress. The second and third zones show how the bending strength decreases due to inelastic and elastic lateral–torsional buckling as the unbraced length increases.

The point on the curve at which the bending strength starts to decrease (i.e., when the unbraced length, L_b, equals L_p) is indicated with a dark-shaded circle. The point at which the bending strength undergoes a transition from inelastic to elastic lateral–torsional buckling (i.e., when the unbraced length, L_b, equals L_r and when the nominal bending strength, M_n, equals M_r) is indicated with an unshaded circle. The use of these *AISCM* moment capacity curves will be discussed in a later section.

TABLE 6-3 Values of Moment Gradient Factor, C_b, for Simple-Span Beams

Load description	Lateral bracing	C_b
Concentrated load at midspan	None	1.32
	At load point	1.67
Concentrated load at ⅓ points	None	1.14
	At all load points	1.67 1.0 1.67
Concentrated load at ¼ points	None	1.14
	At all load points	1.67 1.11 1.67

Load description	Lateral bracing	C_b
Uniformly loaded	None	1.14
	At midspan	1.30
	At ⅓ points	1.45 1.01 1.45
	At ¼ points	1.52 1.06 1.52

EXAMPLE 6-2

Moment Gradient Factor

Determine the moment gradient factor, C_b, for the beam shown in Figure 6-6.

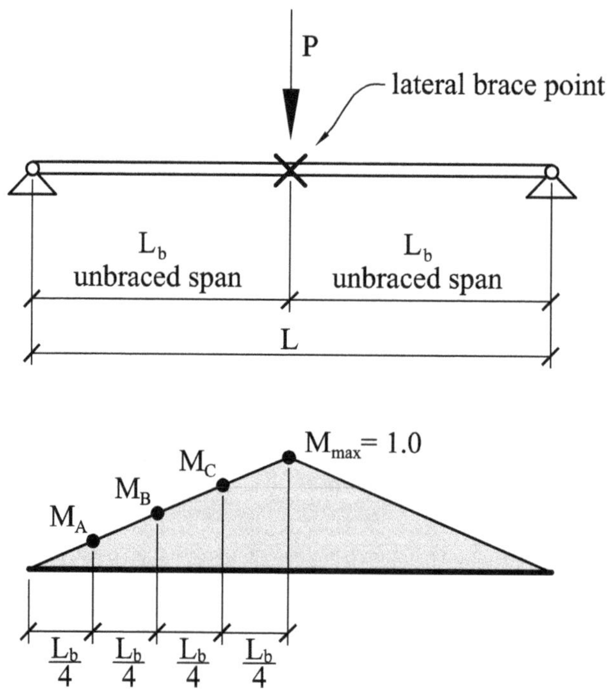

FIGURE 6-6 Detail for Example 6-2

Solution

Assuming a maximum unit moment of $M_{max} = 1.0$ within the unbraced length, the moments at various locations can be determined by linear interpolation for this moment diagram.

From equation (6-13),

$$C_b = \frac{12.5M_{max}}{2.5M_{max} + 3M_A + 4M_B + 3M_C} \leq 3.0,$$

$$C_b = \frac{12.5(1.0)}{2.5(1.0)+3(0.25)+4(0.5)+3(0.75)} \le 3.0,$$

$$= 1.67 \text{ (agrees with Table 6-3).}$$

EXAMPLE 6-3

Bending Strength of a Wide Flange Beam

Determine the available bending strength (M_n/Ω_b) and the design bending strength ($\phi_b M_n$), for a W14 × 74 flexural member of ASTM A992, grade 50 steel, assuming

1. Continuous lateral support.
2. Unbraced length = 15 ft., C_b = 1.0; and
3. Unbraced length = 15 ft., C_b = 1.30.

Solution

From part 1 of the *AISCM*, the section properties for W14 × 74 are

Z_x = 126 in.^{3}

J = 3.87 in.^{4}

r_{ts} = 2.83 in.

S_x = 112 in.^{3}

h_o = 13.4 in.

r_y = 2.48 in.

Check the compact section criteria for the W14 × 74 flexural member:

Flange:

The actual flange width-to-thickness ratio, $\frac{b_f}{2t_f}$, for W14×74 is 6.41.

The maximum flange width-to-thickness ratio to ensure a compact flange is

$$\lambda_{pf} = 0.38\sqrt{\frac{E}{F_y}} = 0.38\sqrt{\frac{29{,}000}{50}} = 9.15 > 6.41,$$

$\therefore$ the W14 × 74 flange is compact.

Web:

The actual web depth-to-thickness ratio, $\frac{h}{t_w}$, for W14×74 is 25.4.

The maximum web depth-to-thickness ratio to ensure a compact web is

$$\lambda_{pw} = 3.76\sqrt{\frac{E}{F_y}} = 3.76\sqrt{\frac{29{,}000}{50}} = 90.5 > 25.4,$$

$\therefore$ the W14 × 74 web is compact.

1. Continuous lateral support, L_b = 0 (upper linear zone in Figure 6-5)

$$M_n = M_p = F_y Z_x$$

$$= (50)(126) = 6300 \text{ in. kips} = 525 \text{ ft.-kips}$$

$$\phi_b M_n = (0.9)(525) = \mathbf{473\ ft.\ kips} \text{ for LRFD}$$

$$\frac{M_n}{\Omega_b} = \frac{525}{1.67} = \mathbf{314\ ft.kips} \text{ for ASD}$$

2. L_b = 15 ft., and C_b = 1.0

$$L_p = 1.76 r_y \sqrt{\frac{E}{F_y}} = 1.76(2.48)\sqrt{\frac{29{,}000}{50}} = 105.1 \text{ in.} = 8.76 \text{ ft.} < L_b = 15\text{ft.}$$

$$L_r = 1.95 r_{ts} \frac{E}{0.7F_y}\sqrt{\left(\frac{Jc}{S_x h_o}\right)}\sqrt{1+\sqrt{1+6.76\left(\frac{0.7F_y}{E}\frac{S_x h_o}{Jc}\right)^2}}$$

$$= 1.95(2.83)\frac{29{,}000}{0.7(50)}\sqrt{\left(\frac{(3.87)(1.0)}{(112)(13.4)}\right)}\sqrt{1+\sqrt{1+6.76\left(\frac{0.7(50)}{29{,}000}\frac{(112)(13.4)}{(3.87)(1.0)}\right)^2}}$$

$$= 372.5 \text{ in.} = 31.0 \text{ ft.}, > L_b = 15\text{ft}$$

Note that c = 1.0 for I-shapes.

Since $L_p < L_b < L_r$, the nominal moment strength is found from equation (6-10):

$$M_n = C_b\left[M_p - \left(M_p - 0.7F_y S_x\right)\left(\frac{L_b - L_p}{L_r - L_p}\right)\right] \le M_p$$

$$= (1.0)\left[6300 - \left(6300 - (0.7)(50)(112)\right)\left(\frac{15 - 8.76}{31.0 - 8.76}\right)\right] \le 6300$$

$$M_n = 5632 \text{ in.-kips} = 469 \text{ ft.-kips}$$

$$\phi_b M_n = (0.9)(469) = \mathbf{422 \text{ ft.-kips}} \text{ for LRFD}$$

$$\frac{M_n}{\Omega_b} = \frac{469}{1.67} = \mathbf{280 \text{ ft.-kips}} \quad \text{for ASD}$$

3. L_b = 15 ft., and C_b = 1.3

$$M_n = C_b\left[M_p - \left(M_p - 0.7F_y S_x\right)\left(\frac{L_b - L_p}{L_r - L_p}\right)\right] \le M_p$$

$$= 1.3\left[6300 - \left(6300 - (0.7)(50)(112)\right)\left(\frac{15 - 8.76}{31.0 - 8.76}\right)\right] \le 6300$$

$M_n = 7321$ in.-kips > 6300 in.-kips, $\therefore$ flexural yielding controls.

$$M_n = M_p = 6300 \text{ in.-kips} = 525 \text{ ft.-kips}$$

$$\phi_b M_n = (0.9)(525) = \mathbf{473 \text{ ft.-kips}} \text{ for LRFD}$$

$$\frac{M_n}{\Omega_b} = \frac{525}{1.67} = \mathbf{314 \text{ ft-kips}} \text{ for ASD}$$

6-4 DESIGN STRENGTH IN BENDING FOR NONCOMPACT AND SLENDER SHAPES

In the previous section, we considered the flexural strength of *doubly symmetric compact wide flange sections and channel shapes*. In this section, we will consider the strength of noncompact shapes, that is, ***doubly symmetric I-shaped members with compact webs but noncompact or slender flanges, bending about their major axis***. There are a few noncompact shapes listed in the *AISCM*, **but there are no slender standard shapes**. Furthermore, all the available sections in the *AISCM* have compact webs, so the web depth-to-thickness ratio does not have to be checked for standard rolled sections. Built-up plate girders can have slender flanges and slender webs, and the design strength of these sections will be covered in Chapter 13. The following list indicates the available sections that have noncompact flanges for F_y = 50 ksi (also denoted with a footnote "*f*" in the *AISCM*): M4 × 6, W6 × 8.5, W6 × 9, W6 × 15, W8 × 10, W8 × 31, W10 × 12, W12 × 65, W14 × 90, W14 × 99, W21 × 48.

For sections with noncompact or slender flanges, the nominal moment strength, M_n, will be the smaller of the moment strengths from lateral torsional buckling (see equations (6-10) through (6-13)) and compression flange local buckling (see Section F3 of the AISC Specifications [1]).

For compression flange local buckling of noncompact shapes, the nominal flexural strength is determined as follows: For sections with *noncompact flanges*, the nominal moment strength is

$$M_n = M'_p = \left[M_p - \left(M_p - 0.7F_y S_x \right) \left(\frac{\lambda - \lambda_{pf}}{\lambda_{rf} - \lambda_{pf}} \right) \right], \tag{6-14a}$$

The design moment capacity is

$$\phi M_n = \phi M'_p = \phi \left[M_p - \left(M_p - 0.7F_y S_x \right) \left(\frac{\lambda - \lambda_{pf}}{\lambda_{rf} - \lambda_{pf}} \right) \right], \tag{6-14b}$$

For sections with *slender flanges* (common in plate girders), the nominal moment strength is

$$M_n = \frac{0.9Ek_c S_x}{\lambda^2} \tag{6-14c}$$

The design moment capacity is

$$\phi M_n = \frac{\phi 0.9Ek_c S_x}{\lambda^2} \tag{6-14d}$$

Where,

$k_c = \dfrac{4}{\sqrt{\dfrac{h}{t_w}}}$ and $0.35 \le k_c \le 0.76$

h = distance as defined in Table 6-2

$\lambda = \dfrac{b_f}{2t_f}$ = Width-to-thickness ratio of flange (see Table 6-2),

b_f = flange width,
t_f = flange thickness,
λ_{pf} = Limiting slenderness ratio for a compact flange (see Table 6-2), and
λ_{rf} = Limiting slenderness ratio for a noncompact flange (see Table 6-2).
$\lambda \le \lambda_{pf}$ implies a *compact flange,*
$\lambda_{pf} < \lambda \le l_{rf}$ implies a *non-compact flange,*
$\lambda > \lambda_{rf}$ implies a *slender flange.*

Referring to Figure 6-5a, the limiting unbraced length for noncompact sections at which lateral–torsional buckling occurs is denoted as $L_p¢$, which is greater than L_p (for compact sections), but less than L_r:

$$L_p' = L_p + \left(L_r - L_p \right) \left(\frac{M_p - M_p'}{M_p - M_r} \right), \tag{6-15}$$

Where,

$$M_r = 0.7F_yS_x \tag{6-16}$$

For noncompact sections, when $L_b > L_p'$, lateral torsional buckling controls, and therefore Equation (6-11) is used to obtain the nominal moment capacity of the section. Non-compact and slender sections cannot attain the plastic moment capacity, M_p, because of elastic local buckling of the component elements of the section, but they can only attain the yield moment of the section, M_y. On the other hand, compact sections can develop the plastic moment capacity since they are not susceptible to local buckling. The nominal moment strength can also be expressed as a function of flange slenderness ratio, as shown in Figure 6-5b. The nominal strength, M_n, is equal to the plastic moment, M_p, for values of λ less than or equal to λ_{pf}, which is the limiting slenderness ratio for a compact flange. The nominal strength reduces linearly until the flange width-to-thickness ratio reaches, λ_{rf}, which is the limiting slenderness ratio for a non-compact flange. Beyond this limit, the flange is slender.

EXAMPLE 6-4

Bending Strength of a Noncompact Shape

Determine the design moment for a W10 × 12 beam with (1) $L_b = 0$, and (2) $L_b = 10$ ft. The yield strength is $F_y = 50$ ksi and $C_b = 1.0$.

Solution

From part 1 of the *AISCM*, the section properties for W10 × 12 are

$Z_x = 12.6 \text{ in.}^3$
$J = 0.0547 \text{ in.}^4$
$r_{ts} = 0.983$ in.
$S_x = 10.9 \text{ in.}^3$
$h_o = 9.66$ in.
$r_y = 0.785$ in.

1. $L_b = 0$ and $C_b = 1.0$:

A W10 × 12 is noncompact. Therefore, equation (6-14a) is used to determine the nominal moment capacity.

From Table 6-2, the maximum flange width-to-thickness ratio for a compact flange is

$$\lambda_{pf} = 0.38\sqrt{\frac{E}{F_y}} = 0.38\sqrt{\frac{29{,}000}{50}} = 9.15$$

The maximum flange width-to-thickness ratio for a noncompact flange is

$$\lambda_{rf} = 1.0\sqrt{\frac{E}{F_y}} = 1.0\sqrt{\frac{29{,}000}{50}} = 24.0$$

From *AISCM* Table 1-1, the actual flange width-to-thickness ratio for W10×12 is

$$\lambda = \frac{b_f}{2t_f} = 9.43,$$

Since $\lambda_{pf} < \lambda < \lambda_{rf}$. ∴ the W10×12 flange is noncompact

$M_p = F_yZ_x = (50)(12.6) = 630$ in.-kips

$$M_n = M_p' = \left[M_p - \left(M_p - 0.7F_yS_x \right)\left(\frac{\lambda - \lambda_{pf}}{\lambda_{rf} - \lambda_{pf}} \right) \right]$$

$$= \left[630 - \left(630 - 0.7(50)(10.9) \right)\left(\frac{9.43 - 9.15}{24.0 - 9.15} \right) \right]$$

$= 625$ in.-kips $= 52.1$ ft.-kips

$\phi_b M_n = (0.9)(52.1) = \mathbf{46.9}$ **ft.-kips** for LRFD

$\dfrac{M_n}{\Omega_b} = \dfrac{52.1}{1.67} = \mathbf{31.1}$ **ft.-kips** for ASD

2. $L_b = 10$ ft., and $C_b = 1.0$:

Determine L_p' and L_r:

$$L_p = 1.76 r_y \sqrt{\frac{E}{F_y}} = 1.76(0.785)\sqrt{\frac{29{,}000}{50}} = 33.3 \text{ in.} = 2.77 \text{ ft.}$$

$$L_r = 1.95 r_{ts} \frac{E}{0.7F_y} \sqrt{\left(\frac{Jc}{S_x h_o} \right)} \sqrt{1 + \sqrt{1 + 6.76\left(\frac{0.7F_y}{E} \frac{S_x h_o}{Jc} \right)^2}}$$

$$= 1.95(0.983)\frac{29{,}000}{0.7(50)} \sqrt{\left(\frac{(0.0547)(1.0)}{(10.9)(9.66)} \right)} \sqrt{1 + \sqrt{1 + 6.76\left(\frac{0.7(50)}{29{,}000} \frac{(10.9)(9.66)}{(0.0547)(1.0)} \right)^2}}$$

$$L_r = 1.95(0.983)\frac{29{,}000}{0.7(50)} \sqrt{\left(\frac{(0.0547)(1.0)}{(10.9)(9.66)} \right)} \sqrt{1 + \sqrt{1 + 6.76\left(\frac{0.7(50)}{29{,}000} \frac{(10.9)(9.66)}{(0.0547)(1.0)} \right)^2}},$$

$= 96.6$ in. $= 8.05$ ft.

Note that c = 1.0 for I-shapes.

$$M_r = 0.7F_yS_x = (0.7)(50)(10.9) = 382 \text{ in.-kips}$$

$$L_p' = L_p + \left(L_r - L_p \right)\left(\frac{M_p - M_p'}{M_p - M_r} \right)$$

$$= 2.77 + (8.05 - 2.77)\left(\frac{630 - 625}{630 - 382} \right) = 2.87 \text{ ft}$$

Since L_b = 10 ft. > L_r, lateral–torsional buckling controls and equation (6-11) will be used to determine the bending strength:

$$F_{cr} = \frac{C_b \pi^2 E}{\left(\frac{L_b}{r_{ts}} \right)^2} \sqrt{1 + 0.078 \frac{Jc}{S_x h_o} \left(\frac{L_b}{r_{ts}} \right)^2}$$

$$= \frac{(1.0)\pi^2 (29{,}000)}{\left(\frac{(10)(12)}{0.983} \right)^2} \sqrt{1 + 0.078 \frac{(0.0547)(1.0)}{(10.9)(9.66)} \left(\frac{(10)(12)}{0.983} \right)^2} = 24.3 \text{ ksi}$$

$$M_n = F_{cr}S_x \le M_p = 630 \text{ in.-kips}$$

$$= (24.3)(10.9) = 265.1 \text{ in.-kips} = \mathbf{22.1\ ft.\text{-}kips}$$

$$\phi_b M_n = (0.9)(22.1) = \mathbf{19.9\ ft.\text{-}kips} \text{ for LRFD}$$

$$\frac{M_n}{\Omega_b} = \frac{22.1}{1.67} = \mathbf{13.2\ ft.\text{-}kips} \text{ for ASD}$$

6-5 DESIGN OF FLEXURAL MEMBERS FOR SHEAR

The bending of a flexural member results in horizontal shear in the interfaces between the horizontal layers of the beam or girder. This horizontal shear between the horizontal layers of the beam must be resisted to ensure that all the layers of the beam work together as a solid unit rather than as individual layers to resist the applied load. From the mechanics of materials, it can be shown that for equilibrium, the horizontal shear is equal to the magnitude of the vertical shear (see Figure 6-7a). In the design process for steel beams, shear rarely controls the design; therefore, most beams need to be designed only for bending and deflection. Special loading conditions, such as heavy concentrated loads close to the beam support or heavy loads on a short span beam, might cause shear to control the design of beams.

From the mechanics of materials, the general formula for the shear stress in a beam is

$$f_v = \frac{VQ}{Ib}, \qquad \text{(6-17)}$$

Where,

f_v = Shear stress at the point under consideration,
V = Vertical shear at a point along the beam under consideration,
Q = First moment about the neutral axis of the area between the location of interest on the cross section and the extreme fiber at the top or bottom of the cross section,
I = Moment of inertia about the neutral axis, and
b = width of the section at the point under consideration.

The variation in shear stress across the depth of a wide flange shape (i.e., an I-shape) is shown in Figure 6-7b. Note here that the shear stress in the flange is much smaller than the stress in the web because the variable, b, in equation 6-17 would be equal to the flange width when calculating the shear stress within the depth of the beam flange and equal to the web thickness for calculating the shear stress within the depth of the beam web. For common wide flange-shapes, the flange width can range from 10 to 20 times the thickness of the beam web. Additionally, the distribution of shear stress in the beam flange does not occur as indicated in equation 6-17 because of the low aspect ratio as measured by the low flange thickness-to-flange width ratio [8]. Equation 6-17 is more directly applicable to steel sections with a high aspect ratio with respect to the direction of the load (e.g., a beam under gravity loads bending about its strong axis). For this reason, the AISC specification allows the design for shear to be based on an approximate or average shear stress distribution as shown in Figure 6-7c, where the shear stress is concentrated only in the vertical section of the beam, for which the aspect ratio between the beam depth, d, and the web thickness, t_w, is generally high, where $f_v = \frac{V}{A_w}$ or $V = f_v A_w$.

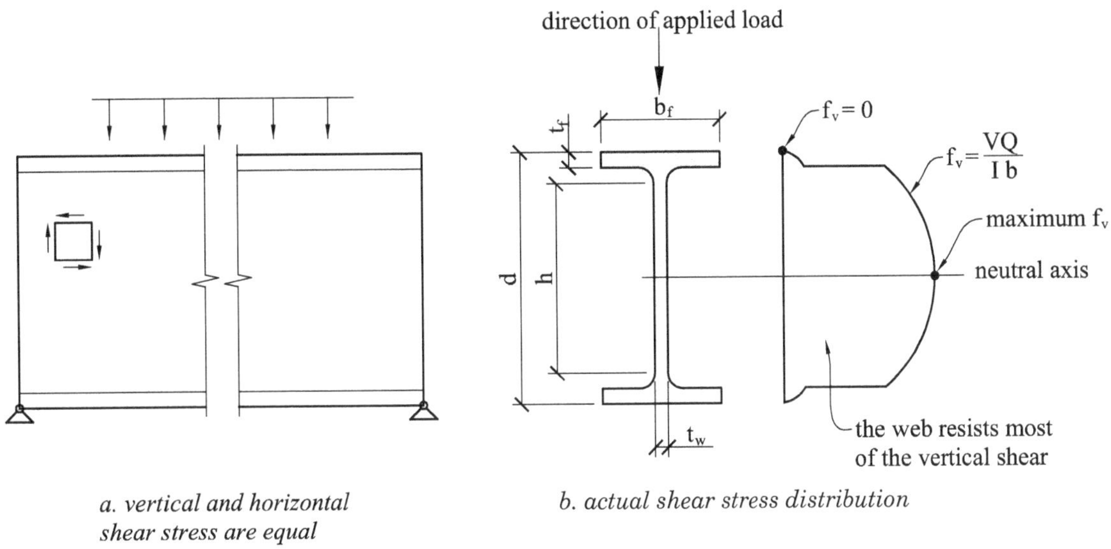

FIGURE 6-7 (a-b) Shear in a beam under gravity loads

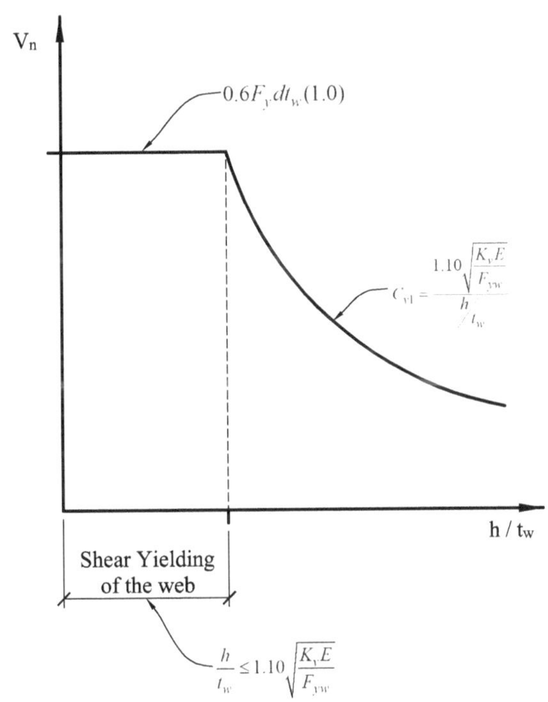

FIGURE 6-7c Variation of Nominal Shear strength with h/t_w

In the AISC specification, the shear yield stress is taken as 60% of the tension yield stress, F_y. The design shear strength of a beam web without tension field action, which accounts for shear yielding of the web and some post-buckling strength of the web (see Chapter 13 or Section G2 of the AISC Specifications), is defined as follows:

$$\phi_v V_n = \phi_v 0.6 F_y A_w C_{v1} \quad \text{for LRFD} \tag{6-18}$$

$$\frac{V_n}{\Omega_v} = \frac{0.6 F_y A_w C_{v1}}{\Omega_v} \quad \text{for ASD} \tag{6-19}$$

Where,

F_y = Yield stress,
A_w = Area of the web = dt_w,
C_{v1} = Web shear coefficient without tension field action (see below), and
ϕ_v = 1.0, except as noted below
Ω_v = 1.5, except as noted below

Since the shear stress is concentrated in the beam web, localized buckling of the web needs to be checked. A web slenderness limit for local web buckling in rolled I-shaped members is defined as

$$\frac{h}{t_w} \leq 2.24\sqrt{\frac{E}{F_y}}. \tag{6-20}$$

When equation (6-20) is satisfied, local web buckling *does not* occur and shear yielding is the controlling limit state, and the strength reduction factor, ϕ_v = 1.0 (LRFD), and Ω_v = 1.5 (ASD). Most I-shaped members that are listed in the *AISCM* meet the criteria in equation (6-20), except for the following shapes for F_y = 50 ksi: W12 × 14, W16 × 26, W24 × 55, W30 × 90, W33 × 118, W36 × 135, W40 × 149, and W44 × 230.

In part 1 of the *AISCM*, shapes that do not meet the web slenderness criteria of equation (6-20) are denoted with a superscript "*v*." For the I-shaped members just listed and for all other doubly and singly symmetric shapes and channels (excluding round HSS), the shear strength reduction and safety factors, respectively, are ϕ_v = 0.9 (LRFD) and Ω_v =1.67 (ASD).

The web shear coefficient *without* tension field action, C_{v1}, is given as follows as a function of the $\frac{h}{t_w}$ ratio (see Figure 6-7c):

$$\text{For } \frac{h}{t_w} \leq 1.10\sqrt{\frac{k_v E}{F_y}}, \qquad C_{v1} = 1.0. \text{ (web yielding)} \tag{6-21a}$$

$$\text{For } \frac{h}{t_w} > 1.10\sqrt{\frac{k_v E}{F_y}}, \quad C_{v1} = \frac{1.10\sqrt{\frac{k_v E}{F_y}}}{\frac{h}{t_w}} \tag{6-21b}$$

Where, the shear buckling coefficient, k_v = 5.34 for unstiffened webs of I-shapes.

For all ASTM A6 W, S, M, and HP shapes listed in the *AISCM*, C_{v1} = 1.0, except for the following shapes when F_y = 50 ksi: M10 × 7.5, M10 × 8, M12 × 10, M12 × 10.8, M12 × 11.8, M12.5 × 11.6, and M12.5 × 12.4.

EXAMPLE 6-5

Shear Strength of a Steel Shape

Determine the design shear strength of the following sections, using F_y = 50 ksi for the W-shapes and F_y = 36 ksi for the C-shape:

1. W16 × 26, and
2. W18 × 50.
3. C12 × 20.7

Solution

1. W16 × 26

From *AISCM*, Table 1.1,

$t_w = 0.25$ in.

$d = 15.7$ in.

$$\frac{h}{t_w} = 56.8$$

$$\frac{h}{t_w} \le 2.24\sqrt{\frac{E}{F_y}}$$

$$2.24\sqrt{\frac{29{,}000}{50}} = 54.0 < h/t_w = 56.8, \therefore \phi_v = 0.9.$$

Since $\frac{h}{t_w} = 56.8 > 2.24\sqrt{\frac{E}{F_y}} = 54$, therefore, $\phi_v = 0.9$

Determine C_{v1}:

$$\text{If } \frac{h}{t_w} \le 1.10\sqrt{\frac{k_v E}{F_y}}, C_{v1} = 1.0$$

$$1.10\sqrt{\frac{(5.34)(29{,}000)}{50}} = 61.2 > h/t_w = 56.8, \therefore C_{v1} = 1.0.$$

For W16 × 26, $\phi_v = 0.9$ and $\Omega_v = 1.67$.

From equation (6-18),

$\phi_v V_n = \phi_v 0.6 F_y A_w C_{v1} = (0.9)(0.6)(50)(0.25)(15.7)(1.0) =$ **106 kips**. for LRFD

From equation (6-19),

$\frac{V_n}{\Omega_v} = \frac{0.6 F_y A_w C_{v1}}{\Omega_v} = \frac{(0.6)(50)(0.25)(15.7)(1.0)}{1.67} =$ **70.5 kips**. for ASD

2. W18 × 50

From *AISCM*, Table 1.1,

$t_w = 0.355$ in.

$d = 18.0$ in.

Since W18 × 50 meets the requirement of equation (6-20a); therefore, $C_{v1} = 1.0$, $\Omega_v = 1.5$ and $\phi_v = 1.0$

From equation (6-18),

$\phi_v V_n = \phi_v 0.6 F_y A_w C_{v1} = (1.0)(0.6)(50)(0.355)(18.0)(1.0) =$ **192 kips**. for LRFD

From equation (6-19),

$\frac{V_n}{\Omega_v} = \frac{0.6 F_y A_w C_{v1}}{\Omega_v} = \frac{(0.6)(50)(0.355)(18.0)(1.0)}{1.5} =$ **128 kips.** for ASD

3. C12 × 20.7

From *AISCM*, Table 1.5,

$t_w = 0.282$ in.

$d = 12$ in.

For a C12 × 20.7 which is a channel, C_{v1} = 1.0, Ω_v =1.67 and ϕ_v = 0.9 as discussed previously.

From equation (6-18),

$$\phi_v V_n = \phi_v 0.6 F_y A_w C_{v1} = (0.9)(0.6)(36)(12)(0.282)(1.0) = \mathbf{65.8\ kips.}\ \text{for LRFD}$$

From equation (6-19),

$$\frac{V_n}{\Omega_v} = \frac{0.6 F_y A_w C_{v1}}{\Omega_v} = \frac{(0.6)(36)(0.282)(12)(1.0)}{1.67} = \mathbf{43.8\ kips.}\ \text{for ASD}$$

6-6 *AISCM* BEAM DESIGN TABLES

The design bending strength of wide flange shapes (i.e., I-shapes or also sometimes called W-shapes) and C-shapes with respect to the unbraced length are given in *AISCM* Tables 3-10 and 3-11, respectively. These tables assume a moment gradient factor of C_b = 1.0, which is conservative for all cases, and yield strengths of F_y = 50 ksi for W-shapes and C-shapes. These curves are similar to the curve shown in Figure 6-5, except that the *AISCM* beam curves or tables have the strength reduction factor, ϕ, and the safety factor, Ω, already incorporated into the design strength.

In the beam design tables or curves, the sections that are shown in bold font are the lightest and, therefore, the most economical sections available for a given group of section shapes; these sections should be used especially for small unbraced lengths, where possible. For beams with large unbraced lengths, the **bold font** sections will no longer be the most economical or the lightest section available.

For beams with C_b greater than 1.0, multiply the moment capacity calculated using the tables and curves by the C_b value to obtain the actual design moment capacity of the beam for design moments that correspond to unbraced lengths greater than L_p. Since the moment capacity of a flexural member cannot exceed its plastic moment capacity, note that $C_b \phi M_n$ must always be less than ϕM_p for LRFD; and $C_b \frac{M_n}{\Omega}$ must always be less than $\frac{M_p}{\Omega}$ for ASD.

AISCM Tables 3-2 through 3-5 can be used to select the most economical beam based on section properties. *AISCM* Table 3-2 lists the plastic section modulus, Z_x, for a given series of shapes, with the most economical in one series at the top of the list in bold font. The most economical shapes for I_x, Z_y, and I_y are provided in *AISCM* Tables 3-3, 3-4, and 3-5, respectively.

AISCM Table 3-6 provides a useful summary of the beam design parameters for W-shapes. For any given shape, the lower part of the table provides the values for $\frac{M_p}{\Omega}$ and ϕM_p; $\frac{M_{pr}}{\Omega}$ and ϕM_r; $\frac{V_n}{\Omega}$ and $\phi_v V_n$, L_p, and L_r. The upper portion of the table provides the maximum possible load that a beam may support based on either shear or bending strength.

AISCM Table 3-6 can also be used to determine the design bending strength for a given beam if the unbraced length is between L_p and L_r. When the unbraced length is within this range, the design bending strength from equation (6-10) can be simplified to

$$\phi_b M_n = \phi_b M_p - \phi_b BF(L_b - L_p),\ \text{for LRFD} \qquad \text{(6-22a)}$$

and

$$\frac{M_n}{\Omega_b} = \frac{M_p}{\Omega_b} - \frac{BF}{\Omega_b}\left(L_b - L_p\right), \text{ for ASD} \tag{6-22b}$$

where $\phi_b BF$ and $\frac{BF}{\Omega_b}$ are constants found from *AISCM* Table 3-2 or Table 3-6. Note that equation (6-22) is a simpler version of equation (6-10), where $BF = \frac{M_p - 0.7F_y S_x}{L_r - L_p}$. Thus, $\phi_b BF = \frac{\phi_b M_p - 0.7\phi_b F_y S_x}{L_r - L_p}$.

The following examples will illustrate the use of the *AISCM* beam design tables.

EXAMPLE 6-6

Design Bending Strength Using the *AISCM* Tables (Compact Shape)

Confirm the results from Example 6-3 using *AISCM*, Tables 3-2 or 3-10.

Solution

The section is W14 × 74 with F_y = 50 ksi

1. Continuous lateral support, i.e., unbraced length, $L_b = 0$
   From *AISCM* Table 3-10, for the given unbraced length,
   $\phi_b M_n$ = 473 ft.-kips for LRFD
   $\frac{M_n}{\Omega_b}$ = 314 ft.-kips for ASD
2. Unbraced length, L_b = 15 ft., C_b = 1.0
   From *AISCM* Table 3-10, for the given unbraced length,
   $\phi_b M_n$ = 422 ft.-kips for LRFD
   $\frac{M_n}{\Omega_b}$ = 281 ft.-kips for ASD
3. L_b = 15 ft., C_b = 1.3
   From *AISCM*, Table 3-10,
   $\phi_b M_n$ = 422 ft.-kips for LRFD
   $\frac{M_n}{\Omega_b}$ = 281 ft.-kips for ASD;

   Multiplying by C_b yields

   $C_b \phi_b M_n = (1.3)(422) = 548$ ft.-kips for LRFD $\leq \phi_b M_p = 473$ ft.-kips (see AISCM Table 3-2). Since $C_b \phi_b M_n$ is greater than $\phi_b M_p$, the plastic moment capacity controls for LRFD. Similarly, for ASD,

   $C_b \frac{M_n}{\Omega_b}$ = (1.3)(281) = 366 ft-kips for ASD $\leq \frac{M_p}{\Omega_b} = 314$ ft.-kips (see *AISCM* Table 3-2).

   Therefore, the design moment capacity and the allowable moment, respectively, are

   $\phi_b M_n$ = 473 ft.-kips for LRFD
   and
   $\frac{M_n}{\Omega_b}$ = 314 ft.-kips for ASD

4. **Alternate Solution for Part (2) using *AISCM* Table 3-2:**
The LRFD solution is:

$$\phi_b M_p = 473 \text{ ft.-kips;} \qquad L_p = 8.76 \text{ ft.}$$

$$\varphi_b M_r = 294 \text{ ft.-kips;} \quad L_r = 31.0 \text{ ft.}$$

$$\phi_b BF = 8.03 \qquad \text{(LRFD)}$$

Using equation (6-22a) gives

$$\phi_b M_n = \phi_b M_p - \phi_b BF\left(L_b - L_p\right) = 473 - (8.03)(15 - 8.76)$$

$$= 422 \text{ ft.-kips (same value as obtained previously for part 2)}$$

The ASD Solution is:

$$\frac{M_p}{\Omega_b} = 314 \text{ ft.-kips} \qquad L_p = 8.76 \text{ ft.;}$$

$$\frac{M_r}{\Omega_b} = 196 \text{ ft.-kips;} \quad L_r = 31.0 \text{ ft.}$$

$$\frac{BF}{\Omega_b} = 5.34$$

Using equation (6-22b) gives

$$\frac{M_n}{\Omega_b} = \frac{M_p}{\Omega_b} - \frac{BF}{\Omega_b}\left(L_b - L_p\right) = 314 - (5.34)(15 - 8.76)$$

$$= 281 \text{ ft.-kips (same value as obtained previously for part 2)}$$

All of the above values agree with the answers obtained previously for Example 6-3.

EXAMPLE 6-7

Design Bending Strength Using the *AISCM* Tables (Noncompact Shape)

Confirm the results from Example 6-4 using *AISCM* Table 3-10:

Solution

The section is $W10 \times 12$, with $F_y = 50$ ksi; and assume $C_b = 1.0$

1. Continuous lateral support, i.e., unbraced length, $L_b = 0$
From *AISCM* Table 3-10, $\phi_b M_n = 46.9$ ft.-kips for LRFD,
and $\dfrac{M_n}{\Omega_b} = 31.2$ ft.-kips for ASD

   From *AISCM* Table 3-10, the values of $L_p = 2.8$ ft., and $L_r = 8$ ft., are obtained, which corresponds to the location of the solid, dark shaded circle, and the unshaded circle, respectively, on the moment capacity curves.

2. Unbraced length, $L_b = 10$ ft.

From *AISCM* Table 3-10, for the given unbraced length,

$\phi_b M_n = 20$ ft.-kips   for LRFD

$\dfrac{M_n}{\Omega_b} = 13.3$ ft.-kips   for ASD

All the above values agree with the answers previously obtained for Example 6-4.

Alternate Solution

Alternatively, *AISCM* Table 3-6 can also be used to determine the available moment strength:

$\phi_b M_p = 46.9$ ft.-kips   for LRFD

$\dfrac{M_p}{\Omega_b} = 31.2$ ft.-kips   for ASD

$L_p = 2.87$ft.

$L_r = 8.05$ft.

AISCM Table 3-6 can be used for part (1) of this example because the unbraced length is less than L_r (i.e., $L_b = 0 < L_r$); however, for part (2) of this example where the unbraced length, $L_b = 10$ ft., is greater than L_r, indicating that the lateral torsional buckling limit state controls, *AISCM* Table 3-6 ***cannot*** be used to determine the design moment capacity $(\phi_b M_n)$ or the allowable moment $\left(\dfrac{M_n}{\Omega_b}\right)$. Therefore, only equation (6-11) or *AISCM* Table 3-10 can be used for part (2) of this example since lateral torsional buckling controls.

EXAMPLE 6-8

Design Shear Strength Using the *AISCM* Tables

Confirm the results from Example 6-5 using the *AISCM* Tables.

Solution

1. W16 × 26

From *AISCM* Table 3-6,

$\phi_v V_n = 106$ kips for LRFD

$\dfrac{V_n}{\Omega_v} = 70.5\text{kips}$ for ASD

2. W18 × 50

From *AISCM* Table 3-6,

$\phi_v V_n = 192$ kips for LRFD

$\dfrac{V_n}{\Omega_v} = 128\text{kips}$ for ASD

3. C12 × 20.7

From *AISCM* Table 3-8,

$\phi_v V_n = 65.8$ kips   for LRFD

$\dfrac{V_n}{\Omega_v} = 43.8\text{kips}$ for ASD

All the above values agree with the answers previously obtained for Example 6-5.

6-7 SERVICEABILITY REQUIREMENTS FOR BEAMS

In addition to designing for bending and shear, beams and girders also need to be checked for serviceability. There are two main serviceability requirements: deflection and floor vibrations. Floor vibrations are covered in Chapter 12. For beams, deflections must be limited for several reasons as discussed in Chapter 2 (see Section 2-15). In addition to assuring the functionality of deflection sensitive finishes, the occupants of the structure must not perceive that the structure is unsafe because of excessive deflections. Excessive deflections will also often lead to vibration problems.

The deflection equations for common loading conditions are given in Table 6-1. The basic deflection limits are found in Section 1604 of the International Building Code (IBC) [2] and are summarized in Table 6-4. Note that only service level loads are used for serviceability considerations, as discussed in Chapter 2.

The deflection limits in Table 6-4 do not consider the effects of ponding due to rain loads, which is discussed further in Chapter 14. In some cases, a beam can be cambered upward to counteract the dead load deflection such that the beam will be in a somewhat flat position prior to the application of live or other superimposed loads. The amount of camber typically varies approximately between 75% and 85% of the actual dead load deflection in order to prevent the possibility of over-cambering of the beam because the end connections provide more end restraint than the simply supported end conditions typically assumed in the deflection calculations.

Where concrete masonry unit (CMU) walls or brick veneer walls run perpendicular to steel floor framing, a beam should be placed directly underneath, and parallel to, the CMU or brick wall, to control the vertical deflection of the CMU wall. For members that support masonry walls, ACI 530 [18] requires a deflection limit of $L/600$ or a maximum of 0.3 in., where L is the beam span. When designing members that support cranes, the vertical deflection limit varies from $L/600$ for light cranes to $L/1000$ for heavy cranes [11], where the applied load is the crane lifting capacity. For lateral loads on cranes, the deflection limit is $L/400$, where the lateral load is taken as 20% of the crane lifting capacity.

For cantilever beams (or beams with overhangs), the length, L, that should be used in the deflection limit equations is *twice* the span of the cantilever (i.e., $2L_c$, where L_c is the length of the cantilever), since the deflection at the end of a cantilever beam is equivalent to the midspan deflection of a simple span beam with a span equal to $2L_c$ [2].

In practice, it is common to develop approximate deflection equations in order to allow for quicker selection of a flexural member based on deflection limitations.

TABLE 6-4 Deflection Limits for Beams and Girders

Member description	Live load	Snow load or wind load	Dead plus live load
Roof members			
Supporting plaster ceiling	$L/360$	$L/360$	$L/240$
Supporting non-plaster ceiling	$L/240$	$L/240$	$L/180$
Not supporting ceiling	$L/180$	$L/180$	$L/120$
Floor Members	$L/360$	N/A	$L/240$

Uniformly Distributed Load - Deflections and Required Moment of Inertia Expressed in Terms of the Maximum Unfactored Moment:

The deflection for a uniformly loaded, simple-span beam is

$$\Delta = \frac{5wL^4}{384EI}$$

and the maximum moment is

$$M = \frac{wL^2}{8}$$

Converting the units such that the moment is in ft.-kips and the beam length is in ft., the preceding two equations can be combined as follows:

$$8M = wL^2$$

$$\Delta = \frac{5wL^4}{384EI} = \frac{5(wL^2)L^2}{384EI} = \frac{5(8M)L^2}{384EI} = \frac{(5)(8)(M)(L^2)(12^3)}{(384)(29{,}000)I} = \frac{ML^2}{(161.1)I}$$

Therefore, $\Delta = \dfrac{ML^2}{(161.1)I}$, (6-23)

Where,
Δ = maximum deflection, inches,
w = uniform load, kips/ft.
M = maximum unfactored moment, ft.-kips
L = span of the simply supported beam, ft.
I = moment of inertia of the beam, in.^{4}.

Note that M in equation (6-23) is the maximum unfactored moment in ft.-kips. For the total load deflection calculation, M is the maximum unfactored moment due to dead plus live loads; for calculating the live load deflection, M in equation (6-23) is the maximum unfactored moment due to live load only. Using the two basic deflection limits of $L/240$ for total dead plus live load deflection and $L/360$ for live load deflection, equation (6-23) can be rearranged to obtain the required minimum moment of inertia for a uniformly loaded, simply supported beam in terms of the maximum moment in ft.-kips and the span of the beam, L, in ft., as follows:

For the $L/240$ deflection limit under dead plus live loads,

$$\Delta_{D+L} = \frac{(L)(12)}{(240)} = \frac{M_{D+L}L^2}{(161.1)I}$$

Therefore, $I_{\text{required}} = \dfrac{M_{D+L}L}{8.056}$ (required moment of inertia for $L/240$) (6-24)

Similarly, for the $L/360$ deflection limit under live load only,

$$\Delta_L = \frac{(L)(12)}{(360)} = \frac{M_L L^2}{(161.1)I}$$

Therefore, $I_{\text{required}} = \dfrac{M_L L}{5.37}$ (required moment of inertia for $L/360$), (6-25)

Note that M_{D+L} in equation (6-24) is the maximum unfactored moment (in ft.-kips) due to dead plus live loads, and M_L in equation (6-25) is the maximum unfactored moment (in ft.-kips) due to live load only. Equations (6-23), (6-24), and (6-25) allow for the quick selection of a steel section based on a known moment and beam span. These equations can also be used for the approximate sizing of a beam with nonuniform loads by substituting the maximum moment due to the nonuniform loads as M in the preceding equations. Note that by inspection, it is evident that when the live load is more than twice the dead load (i.e., L > 2D), the live load deflection limit (i.e., $L/360$) will control. When the dead load is more than half of the live load (i.e., D > ½L), then the total load deflection limit (i.e., $L/240$) will control.

Single Concentrated Load at Midspan - Deflections and Required Moment of Inertia Expressed in Terms of the Concentrated Load:

For the case of a single concentrated load at the midspan of a simply supported beam or girder, similar equations can be developed for the quick selection of a steel section using the elastic deflection equation, $\Delta = \dfrac{PL^3}{48EI} = \dfrac{PL^3(12^3)}{(48)(29{,}000)I} = \dfrac{PL^3}{806I}$, where P is in kips and the span, L, is in ft.:

For the $L/240$ deflection limit under dead plus live loads,

$$\Delta_{D+L} = \frac{(L)(12)}{(240)} = \frac{P_{D+L}L^3}{48EI} = \frac{P_{D+L}L^3(12^3)}{(48)(29{,}000)I}$$

Solving for I yields,

$$I_{\text{required}} = \frac{P_{D+L}L^2}{40.28} \quad \text{(required moment of inertia for } L/240\text{)}. \qquad (6\text{-}26)$$

Similarly, for the $L/360$ deflection limit under live load only,

$$\Delta_L = \frac{(L)(12)}{(360)} = \frac{P_L L^3}{48EI} = \frac{P_L L^3(12^3)}{(48)(29{,}000)I}$$

Therefore, $I_{\text{required}} = \dfrac{P_L L^2}{26.85}$ (required moment of inertia for $L/360$). (6-27)

Where,

P_{D+L} = maximum unfactored concentrated *dead plus live load* on the girder, kips.

P_L = maximum unfactored concentrated *live load* on the girder, kips.

L = beam span, in ft.

Uniformly Distributed Load - Deflections and Required Moment of Inertia Expressed in Terms of the Uniform Load:

For the uniformly loaded simply supported beam, the equations for the required moment of inertia in terms of the uniform load, w, on the beam can be derived using the maximum deflection expression $\Delta = \dfrac{5wL^4}{384EI} = \dfrac{5wL^4(12^3)}{384(29{,}000)I} = \dfrac{wL^4}{1289I}$ as follows:

For the $L/240$ deflection limit under dead plus live loads,

$$\Delta_{D+L} = \frac{(L)(12)}{(240)} = \frac{5w_{D+L}L^4}{384EI} = \frac{5w_{D+L}L^4(12^3)}{384(29{,}000)I}.$$

Solving for I gives

$$I_{\text{required}} = \frac{w_{D+L}L^3}{64.44} \quad \text{(required moment of inertia for } L/240\text{)}. \qquad (6\text{-}28)$$

Similarly, for the $L/360$ deflection limit under live load only,

$$\Delta_L = \frac{(L)(12)}{(360)} = \frac{5w_L L^4}{384EI} = \frac{5w_L L^4(12^3)}{384(29{,}000)I}$$

$$I_{\text{required}} = \frac{w_L L^3}{42.96} \quad \text{(required moment of inertia for } L/360\text{)}, \tag{6-29}$$

Where,

w_{D+L} = unfactored uniformly distributed dead plus live load, in kips/ft.

w_L = unfactored uniformly distributed live load, in kips/ft.

L = girder span, ft.

I = moment of inertia, in.^{4}

The preceding deflection and required minimum moment of inertia equations are summarized in Table 6-5.

TABLE 6-5 Summary of Common Deflection Equations and Required Moment of Inertia

Loading	Deflection	L/240^	L/360^^
Variable moment†	$\Delta = \frac{ML^2}{(161.1)I}$	$I_{\text{required}} = \frac{ML}{8.056}$	$I_{\text{required}} = \frac{ML}{5.37}$
Concentrated load at midspan	$\Delta = \frac{PL^3}{806I}$	$I_{\text{required}} = \frac{PL^2}{40.28}$	$I_{\text{required}} = \frac{PL^2}{26.85}$
Uniformly loaded	$\Delta = \frac{wL^4}{1289I}$	$I_{\text{required}} = \frac{wL^3}{64.44}$	$I_{\text{required}} = \frac{wL^3}{42.96}$

^For total load deflection,

- M is the maximum unfactored moment in ft.-kips due to dead plus live load,
- P is the concentrated dead plus live load in kips, and
- w is the uniformly distributed dead plus live load in kips/ft.;
- The span, L, is in ft., and I is in in.^{4}.

^^For live load deflection,

- M is the maximum unfactored moment in ft.-kips due to live load,
- P is the concentrated live load in kips, and
- w is the uniformly distributed live load in kips/ft.
- The span, L, is in ft., and I is in in.^{4}.

† Loading is based on a uniformly distributed load on a simple-span beam, though the deflections and required moment of inertia are expressed in terms of the maximum unfactored moment. Therefore, the equations can also be used to calculate the approximate deflections for other load cases with a known maximum moment.

6-8 BEAM DESIGN PROCEDURE

The typical design procedure for beams involves selecting a member that has adequate strength in bending and adequate stiffness for serviceability. Shear typically does not control, but it should be checked. The beam or girder design process using the LRFD method is as follows:

1. Determine the service and factored loads on the beam. Service loads are used for deflection calculations, and factored loads are used for strength design. The weight of the beam is unknown at this stage, but the self-weight can be initially estimated and is usually comparatively small enough that a change in its value will not affect the design.
2. Determine the factored shear and moments on the beam.
3. Select a shape that satisfies strength and deflection criteria. Factored loads are used to check the strength criteria and service loads are used to check the deflection criteria. One of the following methods can be used:

   a. For the steel sections listed in the AISC beam design tables, select the most economical beam to support the factored moment. Then check deflection and shear for the selected shape.

b. Determine the required minimum moment of inertia using Table 6-5. Select the most economical shape based on the calculated moment of inertia and check the selected section for bending and shear.

c. For shapes not listed in the AISC beam design tables, an initial size must be assumed. An estimate of the available bending strength can be made for an initial beam selection; then check shear and deflection. A more accurate method might be to follow the procedure in step 3b.

4. Check floor vibrations (see Chapter 12).
The beam or girder design process using the ASD method would follow a similar approach as the preceding, but the ASD load combinations in Section 2-3 are used.

Notes:

- For beams with small unbraced lengths, the **bold font** sections in *AISCM* Table 3-2 will be the most economical (i.e., they are the lightest sections).
- For larger unbraced lengths ($L_b > L_r$), the bold font sections will no longer be the lightest shape for resisting bending; the shallower members will be more economical for this case.

EXAMPLE 6-9

Floor Beam and Girder Design

For the floor plan shown in Figure 6-8, design members B1 and G1 for bending, shear, and deflection. Compare deflections with $L/240$ for total loads and with $L/360$ for live loads. The steel is ASTM A992, grade 50; assume that $C_b = 1.0$ for bending. The dead load (including the beam weight) is assumed to be 85 psf and the live load 150 psf. Assume that the floor deck provides full lateral stability to the top flange of B1. Ignore live load reduction. Use the design tables in the *AISCM* where appropriate.

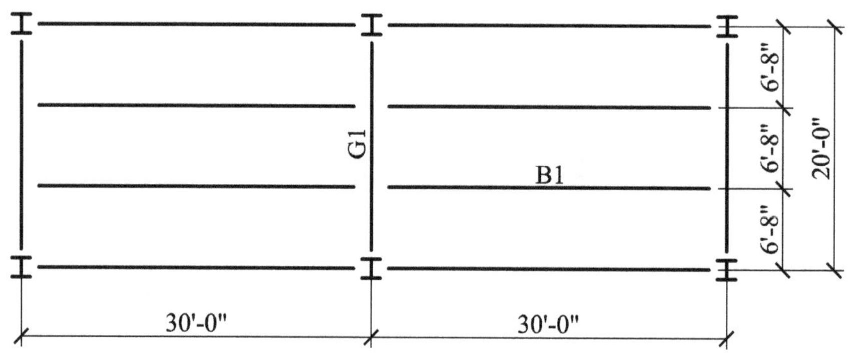

FIGURE 6-8 Floor plan for Example 6-9

Solution

Since the dead load is more than half of the live load, the total load deflection of $L/240$ will control.

Summary of Loads (see Figure 6-9):

L = 150 psf

$$p_s = 85 + 150 = 235 \text{ psf} = 0.235 \text{ ksf (total service load)}$$

$$p_u = (1.2)(85) + (1.6)(150) = 342 \text{ psf} = 0.342 \text{ ksf (total factored load)}$$

Beam B1

Tributary width = 6 – 8″ = 6.67 ft.

Span of the beam, L = 30 ft.

$$\text{Tributary area} = (6.67')(30') = 200 \text{ ft.}^2$$

$$w_L = (6.67)(0.150) = 1.0 \text{kips / ft}$$

$$w_s = w_{D+L} = (6.67)(0.235) = 1.57 \text{ kips / ft.}$$

$$w_u = (6.67)(0.342) = 2.28 \text{ kips / ft.}$$

$$V_u = \frac{w_u L}{2} = \frac{(2.28)(30)}{2} = 34.2 \text{ kips}$$

Factored beam reaction = 34.2 kips

$$M_u = \frac{w_u L^2}{8} = \frac{(2.28)(30)^2}{8} = 257 \text{ ft.-kips}$$

Girder G1

Span of girder = 20 ft.

$$P_L = (200)(0.150) = 30 \text{ kips}$$

$$P_s = P_{D+L} = \left(\frac{200}{2} + \frac{200}{2}\right)(0.235) = 47 \text{ kips}$$

$$P_u = \left(\frac{200}{2} + \frac{200}{2}\right)(0.342) = 68.4 \text{ kips}$$

$$V_u = P_u = 68.4 \text{ kips}$$

$$M_u = \frac{P_u L}{3} = \frac{(68.4)(20)}{3} = 456 \text{ ft.-kips}$$

The loading diagrams for B-1 and G-1 are shown in Figure 6-9.

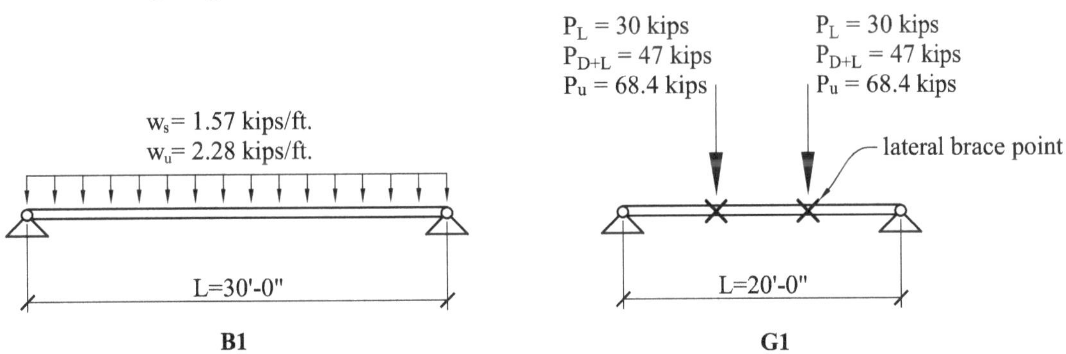

FIGURE 6-9 Loading for B1 and G1

1. **Design of Beam B1**

   We will use the *AISCM* beam design charts and the required minimum moment of inertia method to select a beam size.

   M_u = 257 ft.-kips; and the unbraced length, L_b = 0, since the compression flange of the beam is fully
   laterally braced by the floor deck.

   From *AISCM* Table 3-10, a W16 × 40 is selected as the most economical size for bending, with $\phi_b M_n$ = 274 ft.-kips. On the other hand, note that a W18 × 40 has the same beam weight and, therefore, virtually the same cost, but provides more strength ($\phi_b M_n$ = 294 ft.-kips) and stiffness (I = 612 in.^{4} versus I = 518 in.^{4}). Therefore, a W18 × 40 is initially selected.

   From Table 6-5 (total load controls deflection),

$$I_{\text{required}} = \frac{w_{D+L} L^3}{64.44} = \frac{(1.57)(30)^3}{64.44} = 658 \text{ in.}^4$$

Alternatively, the beam deflection can be calculated from first principles (similar to the girder) using well-known deflection equations from *AISCM* Table 3-21a. For the uniformly loaded beam, the maximum deflection is given as

$$\Delta_{D+L} = \frac{5w_{D+L}L^4}{384EI} = \frac{0.013PL^3}{EI} \leq \frac{L}{240}$$

$$\Delta_L = \frac{5w_L L^4}{384EI} = \frac{0.013PL^3}{EI} \leq \frac{L}{360}$$

Where, $P = wL$ for the uniformly loaded beam.

Therefore,

$P_{D+L} = w_{D+L}L = (0.131\,kip/in)(360\,in.) = 47.16\,kips$

$P_L = w_L L = (0.083\,kip/in)(360\,in.) = 29.9\,kips$

Solving for the required moment of inertia using the deflection equations and the deflection limits above, we obtain the required moment of inertia as 655 in.^{4} for dead plus live load and 628 in.^{4} for only live load. Therefore, the higher value of 655 in.^{4} will control. This value is very close to the required moment of inertia of 658 in.^{4} obtained previously using Table 6-5.

The required moment of inertia is greater than the moment of inertia of the W18 × 40, which is 612 in.^{4}; therefore, a new beam size needs to be selected.

From *AISCM* Table 1-1,

W16 × 50, $I = 659$ in.^{4}

W18 × 46, $I = 712$ in.^{4}

W21 × 44, $\boldsymbol{I = 843}$ in.^{4} ← *Select*

W24 × 55, $I = 1350$ in.^{4}

The W21 × 44 has the lightest weight of all the sections above, so this beam size is selected. Alternatively, from *AISCM* Table 3-3, a W21 × 44 is found to be the lightest section with a moment of inertia greater than 658 in.^{4} Checking the moment capacity with *AISCM* Table 3-10, we find that $\phi_b M_n$ = 358 ft.-kips > M_u = 257 ft.-kips.

Checking shear, note that a W21 × 44 (t_w = 0.35 in.; d = 20.7 in.) does not have a slender web; therefore, the design shear strength is determined from equation (6-18), with C_{v1} = 1.0 and ϕ_v = 1.0:

$$\phi_v V_n = \phi_v 0.6 F_y A_w C_{v1} = (1.0)(0.6)(50)(0.35)(20.7)(1.0) = \mathbf{217\ kips} > \boldsymbol{V_u} = \mathbf{34.2\ kips} \quad \text{OK.}$$

Alternatively, the shear strength can be found in *AISCM* Table 3-6 ($\phi_v V_n$ = 217 kips, same as above).

A W21 × 44 is selected for member B1.

2. **Design of Girder G1**

   We will use the *AISCM* beam design charts and the required moment of inertia method to select a girder size. From Figure 6-9, the maximum factored moment in the girder is calculated as

   $M_u = 456$ ft.-kips and the unbraced length, $L_b = 6.67$ ft.

   That is, the girders are assumed to be laterally braced only at the beam locations.

   From *AISCM* Table 3-10, a W24 × 55 is selected as the most economical size for bending, with ϕM_n = 460 ft.-kips.

Check the total load deflection of the girder using the total unfactored concentrated dead plus live loads, P_{D+L} = 47 kips. The deflection equation for a girder with two equal and symmetrically located concentrated loads [1] is:

$$\Delta_{D+L} = \frac{P_{D+L}L^3}{28EI} = \frac{0.036P_{D+L}L^3}{EI} = \frac{0.036(47)\left[(20)(12)\right]^3}{(29{,}000)(1350)} = 0.6 \text{ in.} < \frac{L}{240} = \frac{(20)(12)}{240} = 1 \text{ in. OK}$$

As an alternate approximate check, use equation (6-23) to determine the deflection: The total maximum unfactored moment due to P_{D+L} = 47 kips is

$$M_{D+L} = \frac{P_{D+L}L}{3} = \frac{(47)(20)}{3} = 313.3 \text{ ft.-kips}$$

With beam span, L = 20 ft, the deflection in inches is computed from equation (6-23) as

$$\Delta_{D+L} = \frac{M_{D+L}L^2}{(161.1)I} = \frac{(313.3)(20)^2}{(161.1)(1350)} = 0.589 \text{ in.}$$

Recall that equation (6-23) was developed for the uniform load case, but the results for the concentrated loads at $^1/_3$ points are close to the actual value of Δ = 0.593 in. (less than a 3% difference). Therefore, equations (6-24) and (6-25), which are also expressed as a function of the moment in a beam, could also have reasonably been used for a quick size selection based on the required stiffness or moment of inertia.

Check Shear:

From Figure 6-9, the maximum factored shear is calculated as 68.4 kips. Note that from Section 6-5, the design shear strength is determined from equation (6-18), with C_{v1} = 1.0 and ϕ_v = 0.9 (recall that a W24×55 (t_w = 0.395 in., d = 23.6 in.) does not meet the web depth-to-thickness ratio from equation (6-20)):

$$\phi_v V_n = \phi_v 0.6 F_y A_w C_{v1} = (0.9)(0.6)(50)(0.395)(23.6)(1.0) = \mathbf{251\ kips} > V_u = \mathbf{68.4\ kips} \text{ OK.}$$

Alternatively, the shear strength can be found from *AISCM* Table 3-6 ($\phi_v V_n$ = 251 kips, same as above).

A W24 × 55 is selected for member G1.

6-9 BIAXIAL BENDING AND TORSION

Biaxial bending of a flexural member occurs when there is flexure about both principal axes (i.e., the x- and y- axes) when the loads along each axis are applied directly through the shear center of the beam; the shear center is the point within a section through which loads can be applied without resulting in twisting of the member. When the applied loads do not pass through the shear center, as is often the case with singly symmetric shapes, torsion will occur. Examples of beams or girders where the loads are not applied through the shear center and are therefore subject to torsion or twisting include:

- Crane girders
- Purlins for roof framing
- Angles or channels supporting gravity loads
- Unbraced beams providing lateral support to exterior cladding
- Beams that are curved in plan.

Each of these examples has an applied load along the x- and y- axes, but since the applied loads in each case do not always pass through the shear center, torsional stresses (and torsional rotation) will occur in addition to bending stresses and deflections due to bending (see Figures 6.10 and 6-11). The two types of torsion that occur in steel members are discussed below:

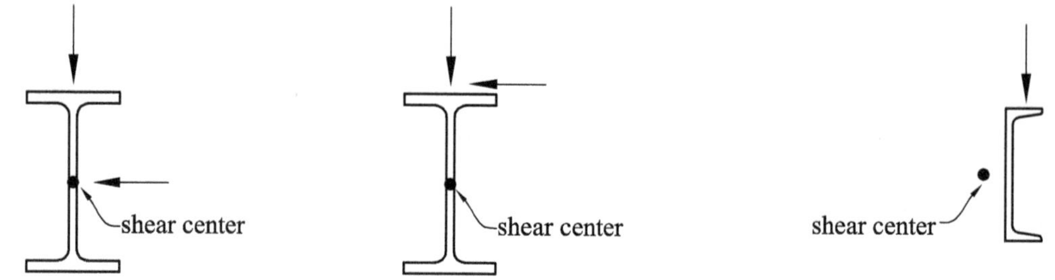

a. pure biaxial bending   *b. biaxial bending with torsion*   *c. torsion in a singly symmetric shape*

FIGURE 6-10 Biaxial bending and torsion loading

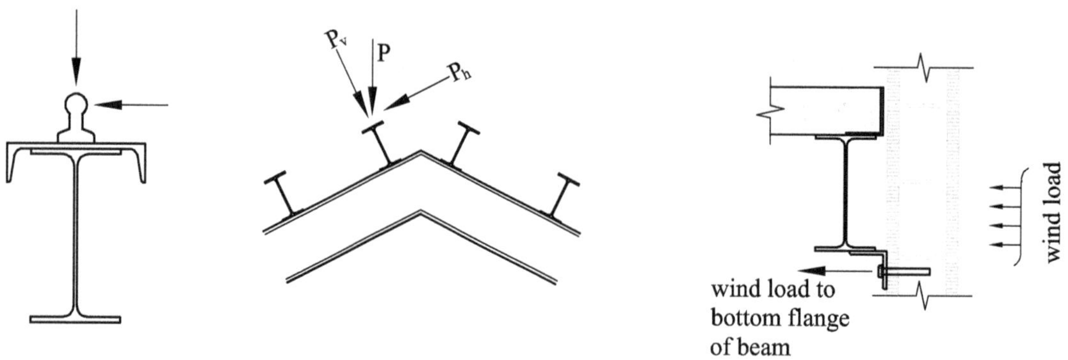

a. crane girder   *b. roof purlins*   *c. beams supporting cladding for lateral loads*

FIGURE 6-11 Biaxial bending and torsion examples

1. **Type 1 - St. Venant or Pure Torsion:**
   - Rotation of elements only and plane sections remain plane after twisting.
2. **Type 2 - Warping Torsion:**
   - Lateral or out-of-plane displacement of the top and bottom flanges of the member and in opposite directions. Warping of the cross-section, if restrained, causes additional non-uniform axial (i.e., normal or longitudinal) stresses in the member in addition to the axial (or normal) stresses caused by bending.
   - Plane sections do *not* remain plane after twisting because of the warping or distortion of the cross-section.
3. In circular shafts, only Type 1 torsion occurs; that is, no warping occurs, and plane sections remain plane after twisting.
4. Types 1 and 2 occur in other closed sections and in open sections such as I-shapes or W-shapes.

5. Closed or box sections (e.g., HSS) have higher torsional stiffness than open sections such as I-shapes, Wide flange, or W-shapes.
   - Therefore, closed sections are more efficient in resisting torsion.

When torsion occurs in a steel section, the effect of warping must be considered. Warping occurs primarily in open sections such as W-, C-, and L-shapes and will result in longitudinal and normal shear stresses. Warping of a W-shape is a condition in which the top and bottom flanges of the cross section deflect laterally in such a way that they are no longer parallel to each other (see Figure 6-12), and plane sections will not remain plane after rotation. The torsional resistance of an open section is the sum of the torsional stiffness of the component elements of the beam section. On the other hand, when a closed section such as a circular tube or HSS is subjected to torsion, each component element of the section rotates without warping; that is, plane sections remain virtually plane after rotation. The torsional resistance of a closed section is much larger than that of an open section since the torsional stresses can be equally distributed in a closed section. For this reason, closed sections are highly recommended when any significant torsion is to be resisted.

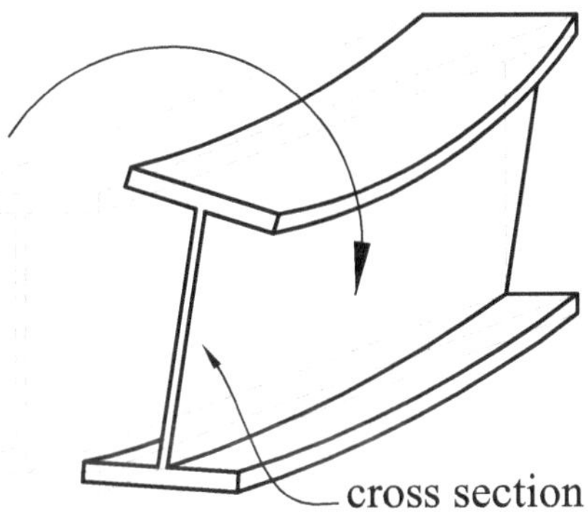

FIGURE 6-12 Warping action

The general relationship between the torsional moment and the angle of twist [12] is

$$\theta = \frac{TL}{GJ}, \qquad (6\text{-}30)$$

Where,
θ = Angle of twist, radians,
T = Torsional moment,
L = Unit length,
G = Shear modulus of elasticity = 11,200 ksi for steel, and
J = Torsional resistance constant.

The torsional resistance constant, J, is equal to the polar moment of inertia for circular sections. For rectangular sections, J is slightly less than the polar moment of inertia; for open sections, J is *much* less than the polar moment of inertia. The value of J for any section can be calculated [14, 15] but tabulated values for standard sections are found in Part 1 of the *AISCM*. Since closed sections are not subject to warping, the general relationship given in equation 6-30 can be applied directly to various torsional loading conditions on closed sections. Table 6-6 provides the formulas for the rotation of closed sections subject to various torsional loads. Note that, for closed sections, the end conditions are usually assumed to be torsionally fixed since warping in closed sections is negligible.

TABLE 6-6 Torsion Displacement Equations for Closed Sections

Loading	θ (radians)	Detail
Cantilever—concentrated torsion at the end	$\theta = \dfrac{PeL}{GJ}$	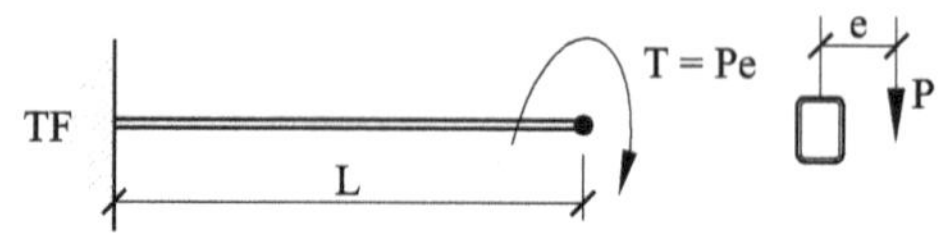
Cantilever—uniform torsion	$\theta = \dfrac{weL^2}{2GJ}$	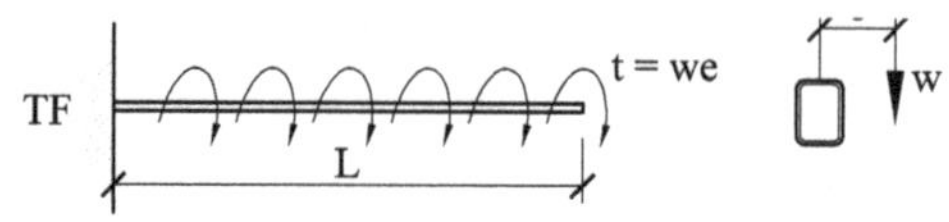
Simple span—concentrated torsion	$\theta = \dfrac{Peab}{LGJ}$ at a = b: $\theta = \dfrac{PeL}{4GJ}$	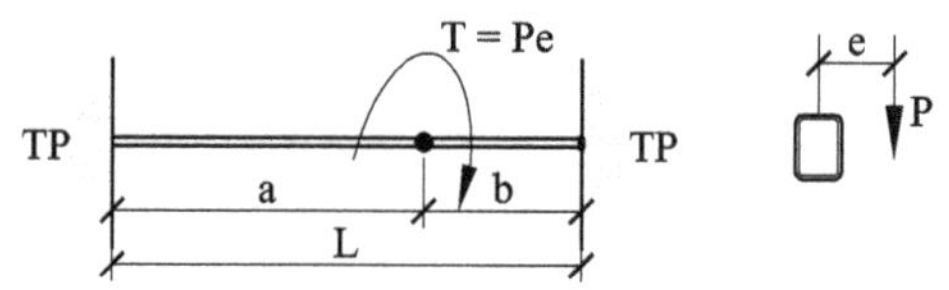
Simple span—uniform torsion	$\theta = \dfrac{weL^2}{8GJ}$	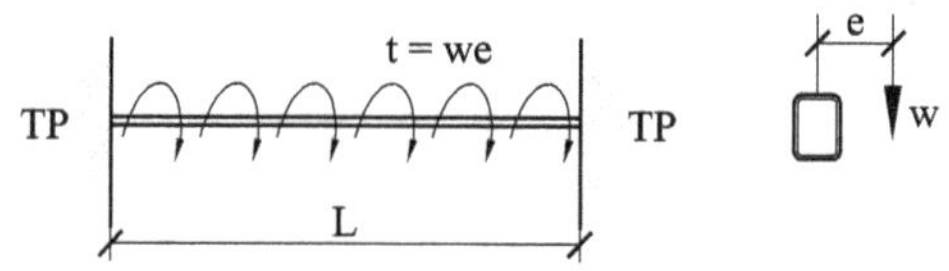

Note: T = Concentrated torsion = Pe
t = Uniformly distributed torsion = we

To account for the effects of warping in open sections, tabulated values of a warping constant, C_w, are given in Part 1 of the *AISCM*. When a W-shaped beam is subjected to torsional loading, the supported ends of the beam are generally restrained against rotation with standard shear connections, and the torsional stresses are concentrated in the beam flanges. In the middle portion of the beam span, rotation is generally unrestrained, and torsional stresses are distributed to the beam web as well. A torsional bending constant, a, is suggested in Ref. [14] to determine the approximate location along the length of the beam where the effects of torsional restraint are negligible, and this constant is defined as

$$a = \sqrt{\frac{EC_w}{GJ}} \tag{6-31}$$

Table 6-7 [14] provides the approximate values of the flange moment and rotation in W-shapes. (The concept of flange moment will be discussed in the next section.) Note that the end conditions for the cases shown in Table 6-7 are assumed to be torsionally fixed for cantilever beams, a requirement for equilibrium, and torsionally pinned for simple span conditions.

A torsionally pinned (TP) end support does not restrain the section from warping (i.e., the section is allowed to warp, and therefore warping stresses develop within the beam span but they are zero at the ends of the beam), but torsional rotation is restrained at both ends of the beam (see Figure 6-13a), whereas a torsionally fixed (TF) end support restrains the section from warping (therefore, warping stresses develop within the beam span and at the ends of the beam) and also restrains the torsional rotation at the ends of the beam. For a W-shape to be torsionally fixed, the top and bottom flanges must have full restraint at the supported ends against rotation about its longitudinal axis; this can be achieved by welding vertical side plates—on both sides of the web and parallel to the web—to the edges of the top and bottom flanges to form a tube or closed section at the ends of the beam (see Figure 6-13b). To be effective, these side plates should extend a length equal to the beam depth from the end of the beam [16]. Figure 6-13 shows examples of torsionally pinned (TP) and torsionally fixed (TF) end conditions for W-shapes. Note that

TABLE 6-7 Approximate Torsional Rotation and Flange Moment Equations for W-shapes

Loading		Flange moment, M_f	θ (radians)	Detail
Cantilever—concentrated torsion at end	$L/a < 0.5$	$M_f = \dfrac{PeL}{h_o}$	$\theta = 0.32\left(\dfrac{Pea}{GJ}\right)\left(\dfrac{L}{a}\right)^3$	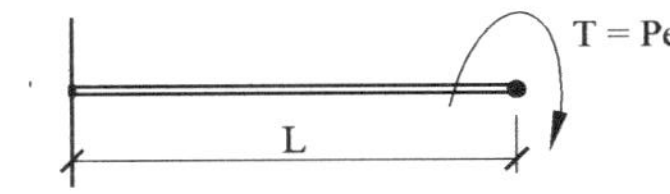
	$0.5 < L/a < 2.0$	$M_f = \dfrac{Pea}{h_o}\left[0.05 + 0.94\left(\dfrac{L}{a}\right) - 0.24\left(\dfrac{L}{a}\right)^2\right]$	$\theta = \dfrac{Pea}{GJ}\left[-0.029 + 0.266\left(\dfrac{L}{a}\right)^2\right]$	
	$2.0 < L/a$	$M_f = \dfrac{Pea}{h_o}$	$\theta = \left(\dfrac{Pe}{GJ}\right)(L - a)$	
Cantilever—uniform torsion	$L/a < 0.5$	$M_f = \dfrac{weL^2}{2h_o}$	$\theta = 0.114\left(\dfrac{weLa}{GJ}\right)\left(\dfrac{L}{a}\right)^3$	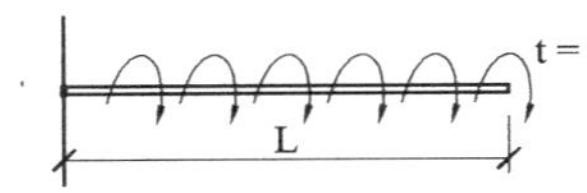
	$0.5 < L/a < 3.0$	$M_f = \dfrac{weLa}{h_o}\left[0.041 + 0.423\left(\dfrac{L}{a}\right) - 0.068\left(\dfrac{L}{a}\right)^2\right]$	$\theta = \dfrac{weLa}{GJ}\left[-0.023 + 0.029\left(\dfrac{L}{a}\right) + 0.86\left(\dfrac{L}{a}\right)^2\right]$	
	$3.0 < L/a$	$M_f = \dfrac{weLa}{h_o}\left(1 - \dfrac{a}{L}\right)$	$\theta = \left(\dfrac{weLa}{GJ}\right)\left(\dfrac{L}{2a} - 1 + \dfrac{a}{L}\right)$	
Simple span—concentrated torsion	$L/a < 1.0$	$M_f = \dfrac{PeL}{4h_o}$	$\theta = 0.32\left(\dfrac{Pea}{GJ}\right)\left(\dfrac{L}{2a}\right)^3$	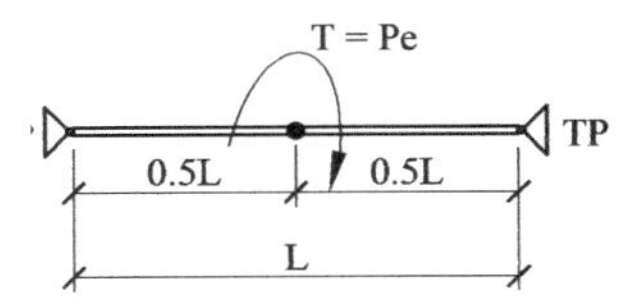
	$1.0 < L/a < 4.0$	$M_f = \dfrac{Pea}{2h_o}\left[0.05 + 0.94\left(\dfrac{L}{2a}\right) - 0.24\left(\dfrac{L}{2a}\right)^2\right]$	$\theta = \dfrac{Pea}{GJ}\left[-0.029 + 0.266\left(\dfrac{L}{2a}\right)^2\right]$	
	$4.0 < L/a$	$M_f = \dfrac{Pea}{4h_o}$	$\theta = \left(\dfrac{Pe}{GJ}\right)\left(\dfrac{L}{2} - a\right)$	
Simple span—uniform torsion	$L/a < 1$	$M_f = \dfrac{weL^2}{8h_o}$	$\theta = 0.094\left(\dfrac{weLa}{GJ}\right)\left(\dfrac{L}{2a}\right)^3$	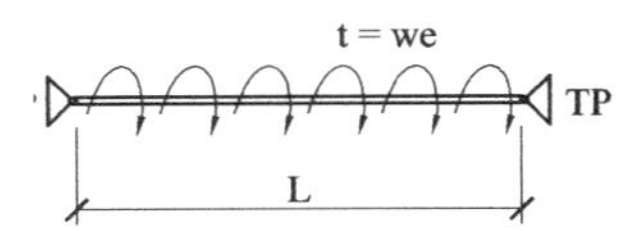
	$1 < L/a < 6.0$	$M_f = \dfrac{weLa}{h_o}\left[0.097 + 0.094\left(\dfrac{L}{2a}\right) - 0.0255\left(\dfrac{L}{2a}\right)^2\right]$	$\theta = \dfrac{weLa}{GJ}\left[-0.032 + 0.062\left(\dfrac{L}{2a}\right) + 0.052\left(\dfrac{L}{2a}\right)^2\right]$	
	$6.0 < L/a$	$M_f = \dfrac{wea^2}{h_o}$	$\theta = \left(\dfrac{weLa}{GJ}\right)\left(\dfrac{L}{8a} - \dfrac{a}{L}\right)$	

Adapted from GALAMBOS, THEODORE V., BASIC STEEL DESIGN WITH LRFD, 1st Edition, © 1996 Reprinted by permission of Pearson Education, Inc. [14]

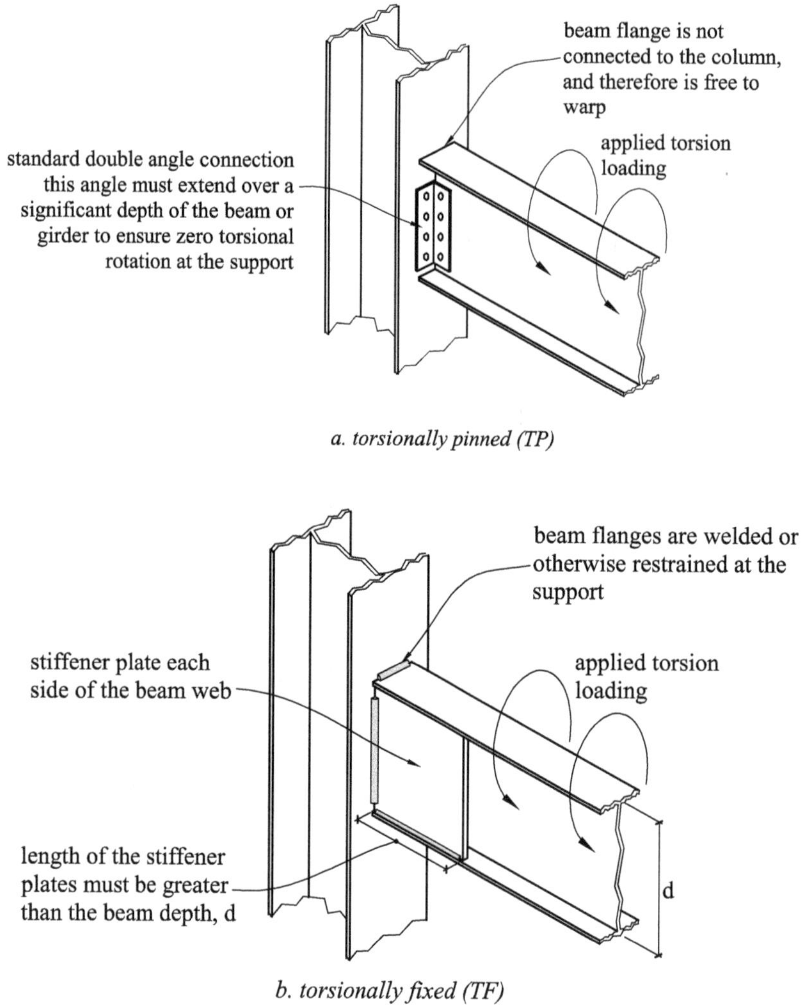

FIGURE 6-13 Torsionally pinned and torsionally fixed ends for W-shapes

beams intended to resist torsion should preferably not be coped because coping of the beam will reduce the torsional restraint provided at the end of the beam. A commonly used rule of thumb assumes that there is sufficient torsional restraint for a beam if the depth of the beam connection element (i.e., the double angles, shear plate, or shear tab) is at least *one-half the beam depth* (i.e., $d/2$) and the beam is *un-coped*. Extended single-plate shear connections should not be used if torsional restraint is required [19].

The AISC Design Guide 9 [8] provides more detailed coverage of how to determine the torsional displacement for various loading and boundary conditions. A summary of the support conditions for torsion are given below:

- **Torsionally PINNED end support (TP):**
  - Warping of the cross-section at the end supports is allowed/unrestrained.
  - Torsional rotation of the cross-section at the end supports is restrained (i.e., θ or $\phi = 0$).
  - Warping stresses at the ends of the beam = 0, and warping stresses within the beam span are not zero.
  - See Figure 6-13
- **Torsionally FIXED end support (TF):**
  - Warping of the cross-section at the end supports is restrained.

- Torsional rotation, θ or ϕ, of the cross-section at the end supports is restrained (i.e., θ or $\phi = 0$).
- Warping stresses occur at the fixed ends of the beam and within the beam span.
- See Figure 6-13.

- **Torsionally FREE end (TFR):**
  - Warping of the cross-section at the free end is allowed/unrestrained.
  - Torsional rotation of the cross-section at the free end is allowed/unrestrained.
  - Warping stresses at the free end of the beam = 0.
  - Example: free end of a cantilever

THE FLEXURE ANALOGY METHOD OF TORSIONAL ANALYSIS

In this section, we will discuss the Flexure Analogy method, a conservative analysis procedure for beams subjected to torsion [22]. There are more exact methods of torsional analysis that could be used [8], but they involve the solution of complex differential equations, which are time-consuming, and it is often desirable in practice to use a quicker, more conservative method to expedite the design process. The Flexure Analogy method resolves the torsional moment from any torsional loading on a wide flange section into an equivalent lateral force-couple acting on the top and bottom flanges in a direction perpendicular to the longitudinal axis of the flange; thus, the Flexure Analogy method of torsional analysis results in analyzing the member for bi-axial bending moments: that is, bending about the strong axis (i.e., x-axis) from the gravity loading and bending about the weak axis (i.e., y-axis) of the beam caused by the lateral force-couple arising from the torsional moment. For wide flange sections, the equivalent equal and opposite lateral forces acting on the top and bottom flanges of the section result in bending of the flanges about their strong axis (i.e., bending about the *y-y* axis of the section), with both flanges bending laterally in opposite directions.

Three cases of bi-axial bending will be considered:

- **Case 1**: Beams where the load passes *through* the shear center (see Figure 6-14).
- **Case 2**: Beams in which the resultant load *does not* pass through the shear center, but the vertical component of the load does pass through the shear center. The *shear center* (SC) is the point through which the load must act if there is to be no twisting of the beam section. The location of the shear center is given in AISC Design Guide 9 [8]. (See Figure 6-15 for the Case 2 loading diagram.)
- **Case 3**: Vertical load that *does not* pass through the shear center (see Figure 6-16).

For bi-axial bending **Case 1**, where the inclined load passes through the shear center, the load can be resolved into a horizontal and a vertical component, each of which passes through the shear center. This will result in no twisting of the beam and will cause simple bending about both the *x*- and the *y*-axes.

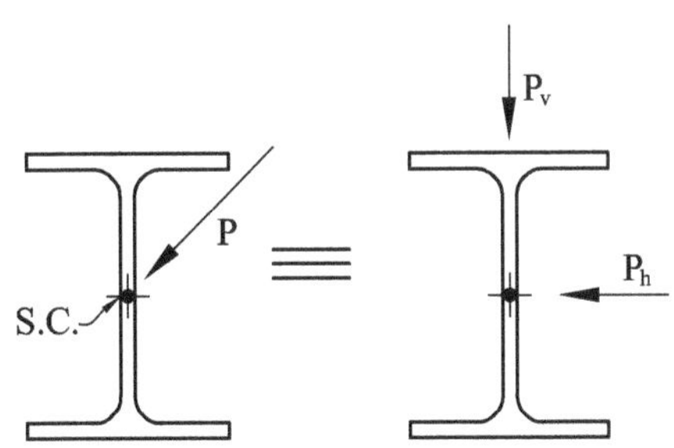

FIGURE 6-14 Inclined load passing through the shear center

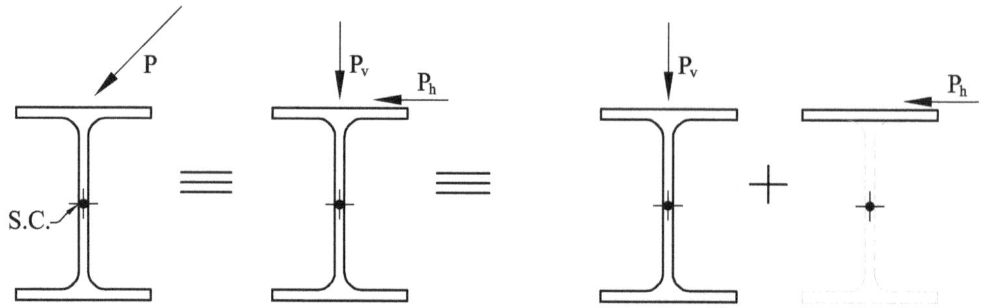

FIGURE 6-15 Inclined load *not* passing through the shear center

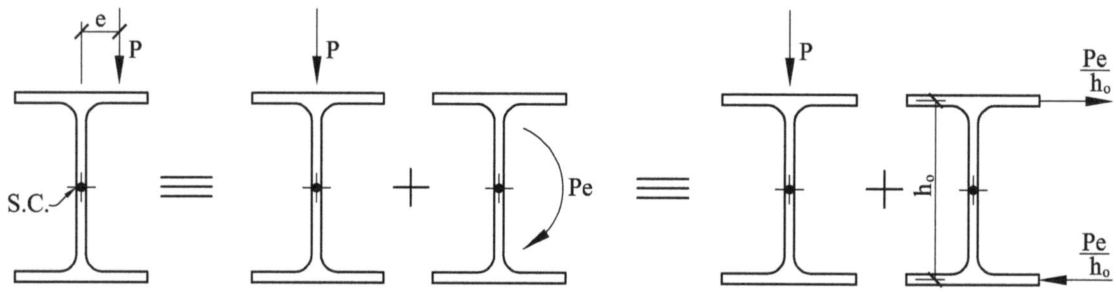

FIGURE 6-16 Vertical load eccentric to the shear center

An interaction equation is used to determine whether the member is adequate for combined bending. The biaxial bending interaction equation for the LRFD method for the Case 1 loading is given in the AISC specification as

$$\frac{M_{ux}}{\phi_b M_{nx}} + \frac{M_{uy}}{\phi_b M_{ny}} \leq 1.0, \tag{6-31}$$

Where,
M_{ux} = Factored moment about the x-axis,
M_{uy} = Factored moment about the y-axis,
M_{nx} = Nominal bending strength for the x-axis,
M_{ny} = Nominal bending strength for the y-axis, and
ϕ_b = 0.9

Bending Strength About the Weak or Y-Y Axis

The nominal bending strength for the x-axis has been discussed previously in this chapter. For the bending about the minor or y-axis, lateral-torsional buckling is not a limit state since the member does not buckle about the strong axis when the bending is about the weaker axis.

For shapes with ***compact* flanges** (i.e., $\lambda \leq \lambda_{pf}$), the nominal bending strength about the y-y axis is

$$M_{ny} = M_{py} = F_y Z_y \leq 1.6 F_y S_y, \tag{6-33a}$$

The design moment capacity for bending about the y-y axis using the LRFD method is

$$\phi M_{ny} = \phi M_{py} = \phi F_y Z_y \leq 1.6 \phi F_y S_y \tag{6-33b}$$

Where,
M_{ny} = Nominal bending strength about the y-axis,
M_{py} = Plastic bending strength about the y-axis,
F_y = Yield stress,
Z_y = Plastic section modulus about the y-axis, and
S_y = Elastic section modulus about the y-axis.

For shapes with ***non-compact* flanges** (i.e., shapes that are susceptible to local buckling and $\lambda_{pf} \le \lambda \le \lambda_{rf}$), the nominal bending strength about the y-y axis is

$$M_{ny} = M_{py} - \left(M_{py} - 0.7F_yS_y\right)\left(\frac{\lambda - \lambda_{pf}}{\lambda_{rf} - \lambda_{pf}}\right) \tag{6-34a}$$

The design moment capacity for bending about the y-y axis is

$$\phi M_{ny} = \phi\left[M_{py} - \left(M_{py} - 0.7F_yS_y\right)\left(\frac{\lambda - \lambda_{pf}}{\lambda_{rf} - \lambda_{pf}}\right)\right] \tag{6-34b}$$

Note that equation (6-34), which applies to weak axis (y-y axis) bending, is similar in format to equation (6-14a), which applies to strong axis (x-x axis) bending for sections with noncompact flanges.

For shapes with ***slender* flanges** (i.e., shapes that are susceptible to local buckling), the nominal bending strength about the y-y axis is

$$M_{ny} = F_{cr}S_y \tag{6-34c}$$

The design moment capacity for bending about the y-y axis is

$$\phi M_{ny} = \phi F_{cr}S_y \tag{6-34d}$$

Where,

$$F_{cr} = \frac{0.70E}{\left(\frac{b}{t_f}\right)^2}$$

For I-shapes, $b = \frac{1}{2}b_f$

For Channels, $b = b_f$ of the channel

$$\lambda = \frac{b}{t_f}$$

For **biaxial bending Case 2**, where the inclined load does not pass through the shear center but the vertical component does, the load is resolved into a vertical component and a horizontal component located at the top flange (see Figure 6-15). The horizontal component of the inclined load is concentrated at the top flange and is assumed to be resisted only by the top flange bending about the strong axis of the flange. Thus, the moment capacity of the beam about its weak axis (i.e., y-axis) considering only the top flange is approximately equal to one-half the bending moment capacity about the y-axis of the entire beam. The interaction equation for this case is

$$\frac{M_{ux}}{\phi_b M_{nx}} + \frac{M_{uy}}{0.5\left(\phi_b M_{ny}\right)} \le 1.0 \tag{6-35}$$

For **biaxial bending Case 3**, where there is a vertical load eccentric to the shear center, the load is resolved into a vertical load coincident with the shear center and horizontal forces located at the top and bottom flanges acting in opposite directions perpendicular to the longitudinal axis of the beam (see Figure 6-16).

The moment about the y-axis (i.e., the flange bending moment) can be calculated by determining the equivalent flange force-couple, which can be taken as $P_{uf} = \frac{Pe}{h}$. The moment about the y-axis can then be calculated once this flange force is determined, and the interaction equation (6-35) is also used for Case 3 since the top and bottom flanges are bending and deflecting laterally in opposite directions. This is the warping behavior that was previously discussed, so

Case 3 is equivalent to beams loaded in pure torsion. For W-shaped beams, the formulas for flange moment and rotation in Table 6-7 are recommended. Practical examples of Case 3 loading are illustrated in Figure 6-17. References [8, 9, 14, 15] provide more detailed coverage of torsion in steel design, but a simplified approach (i.e., Case 3 loading) is often used in design practice.

In practice, it is common to provide adequate details that directly resist the forces arising from the torsional moment rather than allowing any significant torsion and twist in the flexural member, since the analysis of flexural members for torsion can be quite cumbersome. The two most common methods for controlling torsion in practice (see Figures 6-18a through 6-18d) are:

- Providing a closed steel section such as a hollow structural section or a built-up box section, and
- Providing adequate lateral diagonal bracing (i.e., kickers).

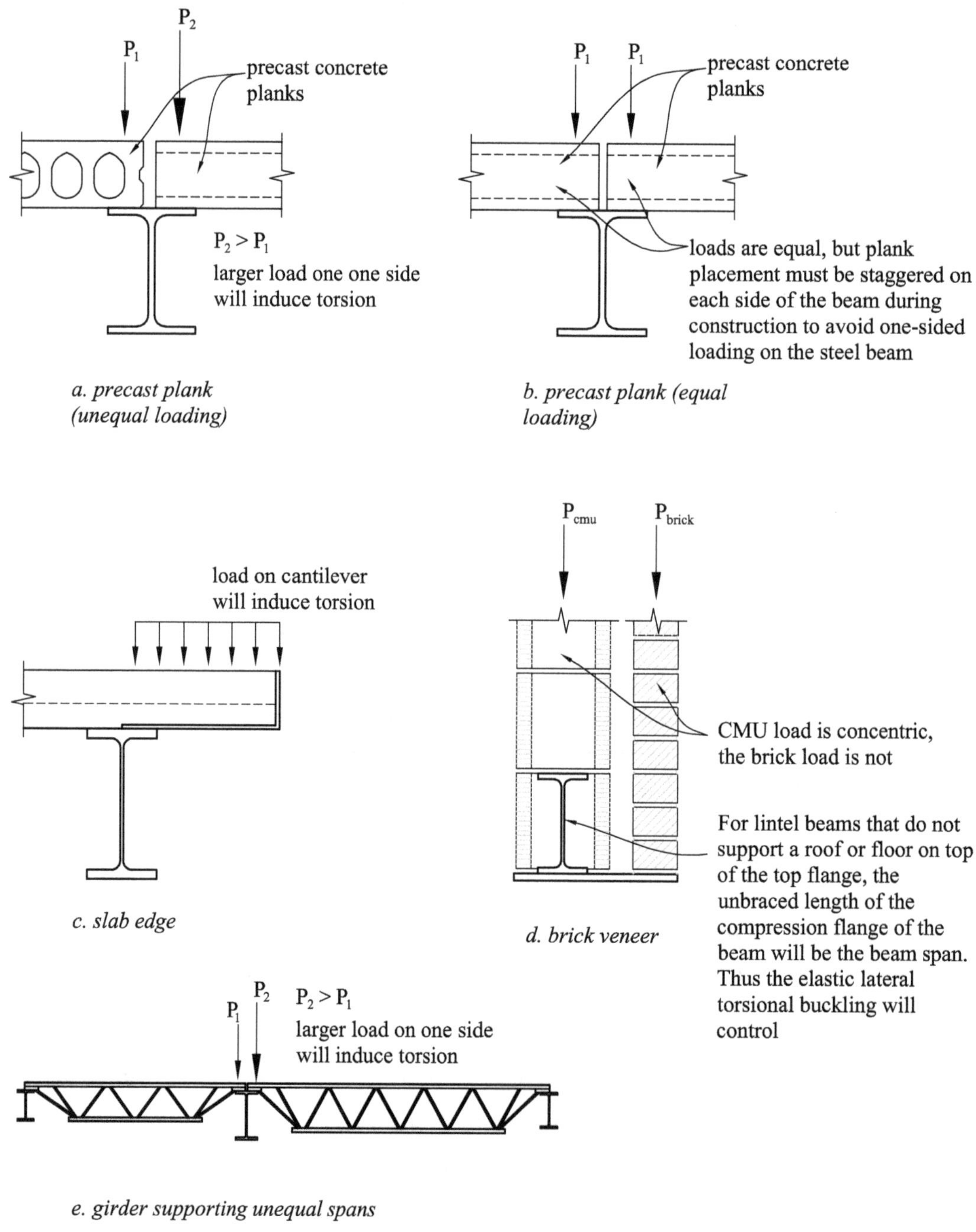

FIGURE 6-17 Common torsion examples

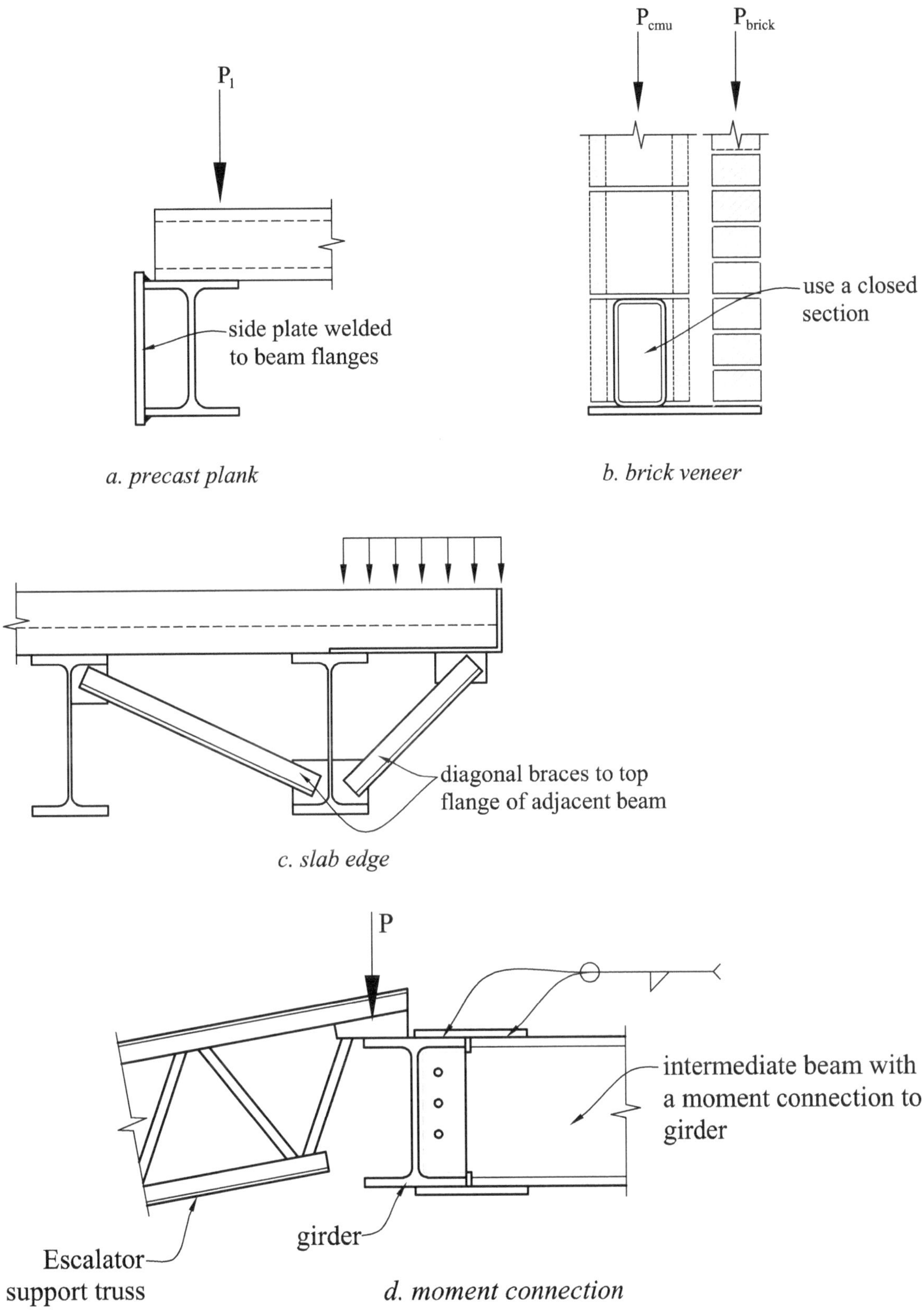

FIGURE 6-18 (a-d) Details used to control torsion

Closed sections are very effective in distributing torsional stresses around the perimeter of the section, whereas open sections, such as W-shapes, rely on the stiffness of the individual components (i.e., the web and flanges) that make up the section to resist torsion. The primary measure of the torsional stiffness of a member is the torsional constant, J, which is listed for each shape in Part 1 of the *AISCM*. To provide a comparative example, the torsional stiffness of a W8 × 31 is listed as $J = 0.536$ in.^{4} An HSS member of equivalent size and weight would be an HSS 8 × 8 × 5/16, which has torsional stiffness of $J = 136$ in.^{4}, nearly 250 times the value obtained for the W8 × 31. The torsional stiffness constant, J, is directly proportional to the rotation, so for the same loading, the W8 × 31 would undergo a torsional rotation that is much greater than the torsional rotation of an equivalent HSS or closed section.

Providing adequate lateral bracing can also help to control torsion because the addition of lateral braces will reduce the length of the member, L, used in the torsional analysis and therefore result in a smaller torsional moment in the member (see Tables 6-6 and 6-7).

EXAMPLE 6-10

Torsion in a Spandrel Beam

Determine whether the beam shown in Figure 6-19 is adequate for combined bending loads. The floor-to-floor height is 12 ft.; the beam span is 15 ft., and is unbraced for this length on both the x- and y-axes. The loads shown are service loads. Use the LRFD method.

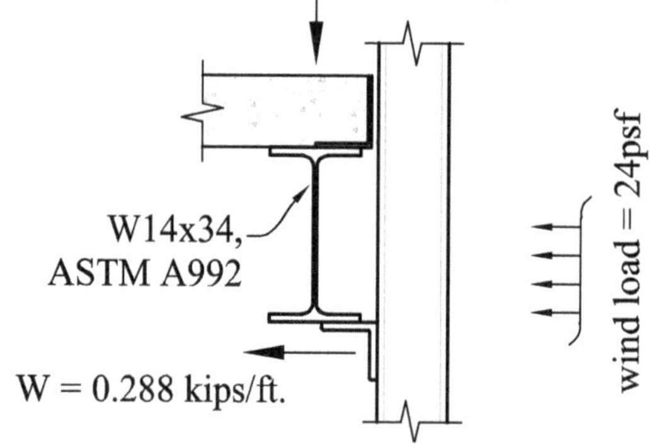

FIGURE 6-19 Detail for Example 6-10

Solution

Since only the bottom flange of the beam is subjected to y-y axis bending, this corresponds to Case 2 biaxial bending. The beam is subjected to biaxial bending from the vertical gravity loads and the horizontal wind loads. Thus,

Lateral wind load, $W = 24 \text{ psf} \times 12 \text{ ft. tributary height} = 0.288 \text{ kips / ft.}$

The critical LRFD load combination for these loads is $1.2D + 1.0W + 0.5L$ (see Section 2-3).

Vertical Load and Moment:

$$w_{ux} = 1.2(1.25) + 0.5(1.75) = 2.38 \text{ kips / ft.;}$$

$$M_{ux} = \frac{w_{ux}L^2}{8} = \frac{(2.38)(15)^2}{8} = \textbf{67 ft.-kips}$$

Horizontal Load and Moment (caused by 1.0W):

$$w_{uy} = 1.0(0.288) = 0.288 \text{ kips / ft.} \quad \text{(wind load)}$$

$$M_{uy} = \frac{w_{uy}L^2}{8} = \frac{(0.288)(15)^2}{8} = \textbf{8.1 ft.-kips}$$

From *AISCM* Table 3-10 for a W14x34 with an unbraced length, L_b = 0 ft. (since the compression flange is laterally braced by the floor for bending about the strong or x-x axis), the design moment capacity about the x-axis is $\phi_b M_{nx} = \phi_b M_{px} = 205$ ft.-kips.

A W14 × 34 has compact flanges, so the design bending strength about the *y*-axis is found from equation (6-33a) as:

$$M_{ny} = M_{py} = F_y Z_y \leq 1.6 F_y S_y$$

$$= (50)(10.6) \leq (1.6)(50)(6.91) = 530 \text{ in-kips} \leq 553 \text{ in.-kips}$$

Therefore, M_{ny} = 530 in-kips

$$\phi_b M_{ny} = \frac{(0.9)(530)}{12} = 39.7 \text{ ft.-kips.}$$

Checking the interaction equation for Case 2 using equation (6-35),

$$\frac{M_{ux}}{\phi_b M_{nx}} + \frac{M_{uy}}{0.5(\phi_b M_{ny})} \leq 1.0;$$

$$\frac{67}{205} + \frac{8.1}{0.5(39.7)} = 0.73 \leq 1.0 \quad \text{OK.}$$

Therefore, the W14 × 34 beam is adequate for torsion and the equivalent bi-axial bending.

EXAMPLE 6-11

Torsion in a Spandrel Beam Supporting Brick Veneer

Determine whether the beam shown in Figure 6-20 is adequate for combined bending loads and torsion. The floor-to-floor height is 12 ft., and the beam span is 17 ft. The loads shown are service loads. Use a unit weight of 40 psf for the brick veneer and assume that the top flange has continuous lateral support. Use the LRFD method.

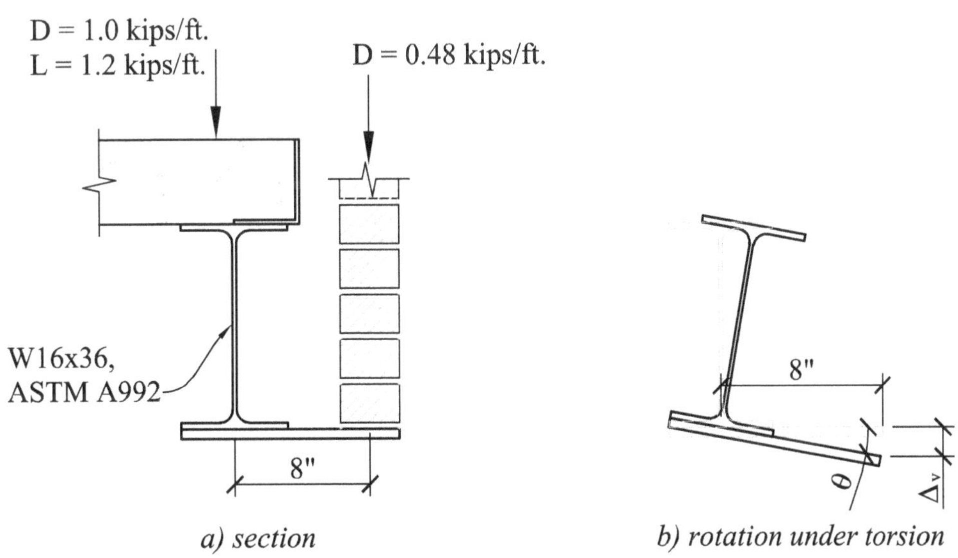

FIGURE 6-20 Detail for Example 6-11

Solution

The loading shown is equivalent to Case 3. We will use both methods presented for Case 3 loading to compare the results.

$$\text{Veneer weight, } w_v = (0.040 \text{ ksf})(12 \text{ ft.}) = 0.48 \text{ kips / ft.}$$

The critical load combination for these loads is $1.2D + 1.6L$ (see Section 2-3).

Vertical Load and Moment:

$$w_{ux} = 1.2(1.0 + 0.48) + 1.6(1.2) = 3.70 \text{ kips / ft.;}$$

$$M_{ux} = \frac{w_{ux}L^2}{8} = \frac{(3.70)(17)^2}{8} = \mathbf{134\ ft.\text{-}kips}$$

Horizontal Load and Moment (caused by the eccentric brick veneer load):
The torsional moment is $Pe = (1.2)(0.48)(8 \text{ in.}/12) = 0.384$ ft-kips/ft.
The distance between the centroid of the top and bottom flanges is,

$$h_o = d - t_f = 15.9 \text{ in.} - 0.43 \text{ in.} = 15.47 \text{ in.} = 1.29 \text{ ft.}$$

Resolving the torsional applied moment as a couple in the top and bottom flanges of the beam, the lateral force in the flanges in kips/ft., is calculated as

$w_{uy} = p_{uf} = \frac{Pe}{h} = \frac{0.384}{1.29} = 0.299 \text{kips / ft.}$ (flange force is applied as a uniformly distributed horizontal load that causes y-y axis bending in the beam, but only one flange is active; hence, the torsional effects can be approximated as biaxial bending Case 3)

$$M_{uy} = \frac{w_{uy}L^2}{8} = \frac{(0.299)(17)^2}{8} = 10.8 \text{ft.-kips}$$

From *AISCM* Table 3-6 for a W16 × 36 and $L_b = 0$ ft., since for bending about the x-x axis, the is braced against lateral torsional buckling by the floor, therefore, $\phi_b M_{nx} = 240$ ft.-kips

$$M_{ny} = M_{py} = F_y Z_y \leq 1.6 F_y S_y$$

$$= (50)(10.8) \leq (1.6)(50)(7)$$

$$= 540 \text{ in-kips} \leq 560 \text{ in.-kips}$$

Therefore, $M_{ny} = 540$ in-kips

$$\phi_b M_{ny} = \frac{(0.9)(540)}{12} = \mathbf{40.5\ ft.\text{-}kips}$$

Check the interaction equation for **biaxial bending Case 3** using equation (6-35):

$$\frac{M_{ux}}{\phi_b M_{nx}} + \frac{M_{uy}}{0.5(\phi_b M_{ny})} \leq 1.0$$

$$\frac{134}{240} + \frac{10.8}{0.5(40.5)} = 1.09 > 1.0$$

Therefore, the W16x36 beam is not adequate for torsion and the equivalent biaxial bending using the flexure analogy method.

As a comparison, we will use the approximate equation from Table 6-7 [14] to determine if the preceding approximate flexure analogy method of analysis for torsion was too conservative.

From part 1 of the *AISCM*,

$C_w = 1460 \text{ in.}^6$

$J = 0.545 \text{ in.}^4$

$$a = \sqrt{\frac{EC_w}{GJ}} = \sqrt{\frac{(29{,}000)(1460)}{(11{,}200)(0.545)}} = 83.3 \text{ in.}$$

$$\frac{L}{a} = \frac{(17)(12)}{83.3} = 2.45, \text{ therefore } 1 < \frac{L}{a} < 6$$

From Table 6-7, the weight of the eccentric brick veneer that is causing the torsion is,

$$w = (1.2)(0.48 \text{ kips / ft.}) = 0.576 \text{ kips / ft.} = 0.048 \text{ kips / in.}$$

$e = 8$ in.

$L = (17)(12) = 204$ in.

$$M_f = \frac{weLa}{h_o}\left[0.097 + 0.094\left(\frac{L}{2a}\right) - 0.0255\left(\frac{L}{2a}\right)^2\right]$$

$$M_f = \frac{(0.048)(8)(204)(83.3)}{15.47}\left[0.097 + 0.094\left(\frac{204}{(2)(83.3)}\right) - 0.0255\left(\frac{204}{(2)(83.3)}\right)^2\right]$$

$M_f = 73.3$ in.-kips $= 6.11$ ft. kips (compare to 10.8 ft-kips from the flexure analogy method)

This value of M_f is less than the value of 10.8 ft.-kips previously calculated. Checking the interaction equation using the new value of M_f gives

$$\frac{M_{ux}}{\phi_b M_{nx}} + \frac{M_{uy}}{0.5(\phi_b M_{ny})} \le 1.0; \quad \frac{134}{240} + \frac{6.11}{0.5(40.5)} = 0.86 < 1.0$$

Therefore, the W16 × 36 beam is adequate to resist the torsion and the equivalent biaxial bending.

Check Deflection:

For deflections, the combined weight of the *veneer* and the *live load* will be used (it is assumed that the deflection due to floor dead load would have already occurred before the veneer is built); the deflection will be compared to a deflection limit of the smaller of *L*/600 and 0.3 in. (the 0.3 in. deflection limit controls only for $L > 15$ ft.). Furthermore, the Brick Industry Association [5] recommends a maximum deflection, caused by the torsional rotation, of 1/16 in., which will also be checked.

The combined load on the beam due to the floor live load and the weight of the brick veneer is,

$$w = 1.2 + 0.48 = 1.68 \text{ kips / ft.}$$

The vertical deflection due to the floor live load and cladding weight is computed as

$$\Delta = \frac{5wL^4}{384El} = \frac{5(1.68/12)(17 \times 12)^4}{(384)(29 \times 10^6)(448)}$$

$$= 0.24 \text{ in.} \quad < \frac{L}{600} = \frac{(17)(12)}{600} = 0.34 \text{ in.} \quad \text{OK.}$$

$$< 0.3 \text{ in.} \quad \text{OK.}$$

Check torsional rotation:

The brick veneer load which causes the torsional rotation is,

$w = 0.480 \text{ kips / ft.} = 0.04 \text{ kips / in.}$

From Table 6-7, the torsional rotation is computed as

$$\theta = \frac{weLa}{GJ} = \left[-0.032 + 0.062\left(\frac{L}{2a}\right) + 0.052\left(\frac{L}{2a}\right)^2\right]$$

$$\theta = \frac{(0.04)(8)(204)(83.3)}{(11{,}200)(0.545)}\left[-0.032 + 0.062\left(\frac{204}{(2)(83.3)}\right) + 0.052\left(\frac{204}{(2)(83.3)}\right)^2\right]$$

$\theta = 0.109$ radians

Therefore, $\theta = (0.109)\left(\frac{180}{\pi}\right) = 6.22°$, this is quite a large rotation.

With reference to Figures 6-20a and 6-20b, the vertical displacement at the exterior tip of the bottom plate caused by this torsional rotation is

$\Delta_v = (\tan 6.22°)(8 \text{ in.}) = 0.872 \text{ in.}$

This amount of twist causes a vertical deflection that far exceeds the Brick Institute of America (BIA) deflection limit of 1/16 in., and therefore the W16×36 lintel beam is not adequate. The reader should confirm that adding lateral support at $^1/_3$ points would decrease the torsional rotation of the lintel beam to about 0.11°, resulting in Δ_v = 0.015 in., which is less than the 1/16 in. limit. Therefore, adding lateral bracing would be recommended here.

An alternate solution would be to use the more torsionally efficient section for the lintel beam (e.g., HSS). As a comparison, we will now consider the torsional rotation of an equivalent closed section such as HSS16 × 8 × $^1/_4$.

From part 1 of the *AISCM*, J = 300 in.^{4} From Table 6-6, the torsional rotation is

$$\theta = \frac{weL^2}{8GJ} = \frac{(0.04)(8)(17 \times 12)^2}{(8)(11{,}200)(300)} = 0.0005 \text{ radians}$$

Therefore, $\theta = (0.0005)\left(\frac{180}{\pi}\right) = 0.028°$

The vertical displacement at the exterior tip of the bottom plate caused by this rotation is

$\Delta_v = (\tan 0.028°)(8 \text{ in.}) = 0.004 \text{ in.} < 1/16 \text{ in. OK.}$

This displacement is more than 200 times smaller than the calculated torsion-induced displacement of the W16 × 36 lintel beam for the same span. The maximum permissible deflection is $^1/_{16}$ in., or 0.0625 in. [5], so the HSS16 × 8 × $^1/_4$ would be adequate for this loading. This is another indication that closed sections are far more efficient in torsion than open sections; therefore, open sections such as W-shapes and channels should preferably not be used to resist torsion.

6-10 BEAM BEARING FAILURE MODES AND DESIGN STRENGTH

In typical steel structures, steel beams and girders are connected to other steel members by some combination of gusset plates, clip angles, welds, and bolts to transfer the end reactions (see Chapters 9 and 10). In some cases, the end reaction of a beam is transferred in direct bearing onto a masonry or concrete wall or onto another steel member such as when a beam or girder is continuous over the top of a steel column. Another bearing situation is where a steel beam or girder supports a discontinuous steel column on its top flange (see Figures 6-21a through 6-21f for these common beam bearing conditions). In the above cases, the beam or girder must be laterally braced at the concentrated load or reaction locations. When steel beams are supported in bearing, a steel bearing plate is used to spread the load over a larger horizontal surface area. In the case of

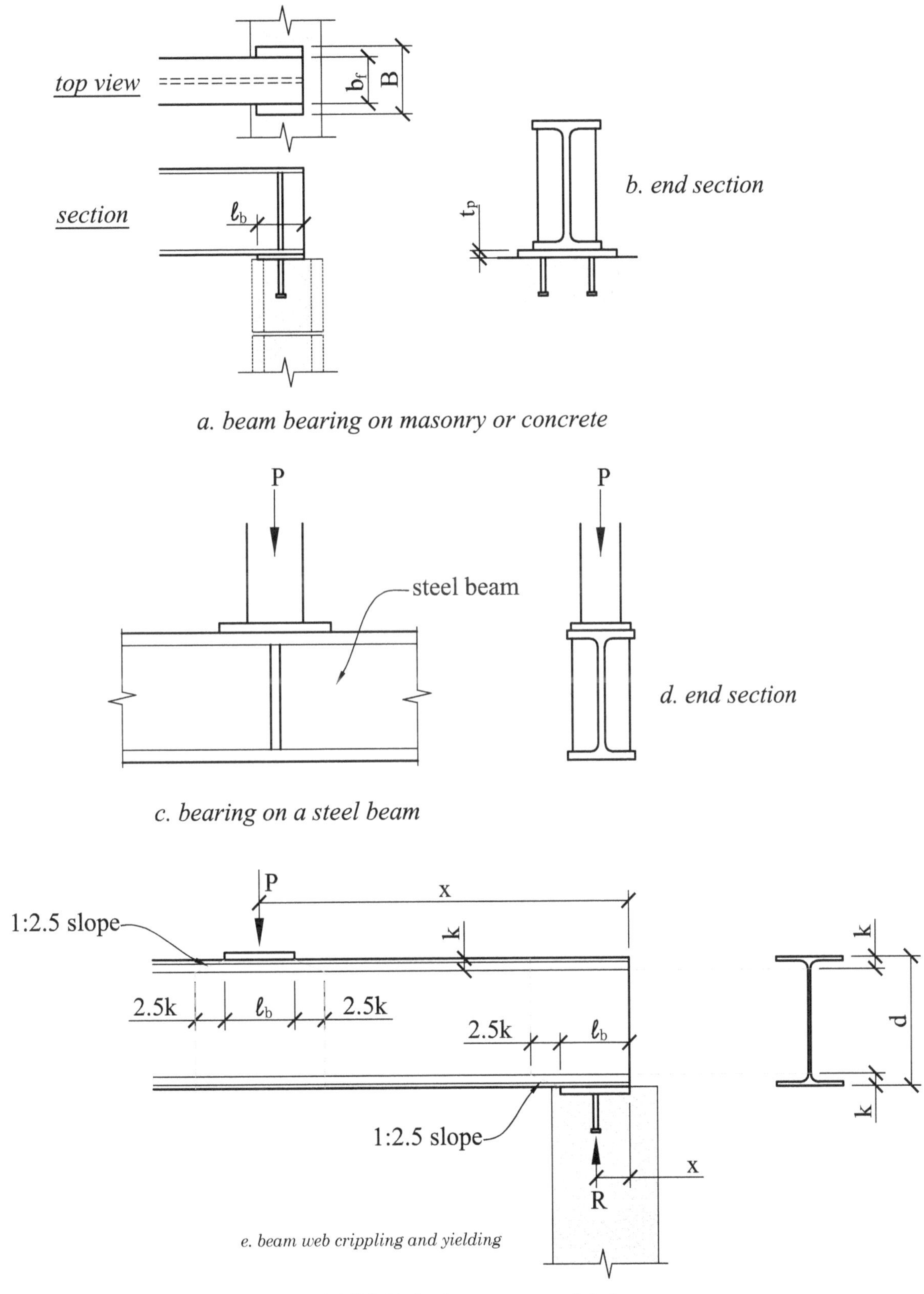

FIGURE 6-21 Bearing on masonry and steel

FIGURE 6-21f Steel beams bearing within pockets in a masonry wall

bearing on concrete or masonry, the bearing plate needs to have a sufficient plan area to distribute the bearing stress so that the bearing strength of the concrete or masonry wall is not exceeded. In the case of a steel beam or column bearing on another steel section, the bearing plate is designed to have sufficient plan area so that local buckling does not occur in the supporting steel section.

The dimensions of the beam bearing plate are ℓ_b which is parallel to the longitudinal axis of the beam, and B, which is perpendicular to the beam. For practical purposes, ℓ_b is usually a minimum of 6 in., and B is usually greater than or equal to the beam flange width, b_f (see Figures 6-21a and 6-21e). This allows for reasonable construction tolerances in placing the bearing plate and the beam. Both the plate dimension B and ℓ_b should be selected in increments of 1 in., and the plate thickness, t_p, is usually selected in increments of $^1/_4$ in.

The possible failure modes for beam bearing include (see AISC Specification Sections J10):

- Beam web local yielding
- Beam web local crippling
- Plate bending due to the bearing stress, and
- Bearing stress in the supporting concrete or masonry wall.

Web Local Yielding

Web yielding is the crushing of a beam web subjected to compression stress due to a concentrated load. When a concentrated load occurs at or near the end of the beam, the compression stress distribution is less than if the load was placed on the interior portion of the beam (see Figure 6-21e).

The compression stress is assumed to be dispersed or distributed longitudinally in the beam in a vertical-to-horizontal dimension ratio of 1:2.5 through the beam flange and inner radius (see Figure 6-21e). This dispersion angle is dependent on the degree of inelasticity in the member at failure; the more inelastic the member is, the wider the spread or dispersion parallel to the longitudinal axis of the beam. This gives the effective width parallel to the longitudinal direction of the beam, and multiplying this effective width by the web thickness and the yield stress gives the following design capacity for **web local yielding** (see Figure 6-21e and AISC Specifications Section J10.2):

For $x > d$:

(i.e., the concentrated load is applied at a distance from the end of the member *greater* than the member depth, d)

$$\phi_{wy} R_n = \phi_{wy} (5\,k + \ell_b) F_{yw} t_w, \quad \text{for LRFD} \tag{6-36a}$$

$$\frac{R_n}{\Omega_{wy}} = \frac{(5\,k + \ell_b) F_{yw} t_w}{\Omega_{wy}}, \quad \text{for ASD} \tag{6-36b}$$

Thus, the compression load is assumed to act over an effective width of the beam web (parallel to the longitudinal axis of the beam) equal to $5\,k + \ell_b$.

For $x \le d$:

(i.e., the concentrated load is applied at a distance from the end of the member *less than or equal to* the member depth, d)

$$\phi_{wy} R_n = \phi_{wy} (2.5\,k + \ell_b) F_{yw} t_w, \quad \text{for LRFD} \tag{6-37a}$$

$$\frac{R_n}{\Omega_{wy}} = \frac{(2.5\,k + \ell_b) F_{yw} t_w}{\Omega_{wy}}, \quad \text{for ASD} \tag{6-37b}$$

Thus, for this case that occurs at the end of a beam or girder, the compression load acts over an effective length of web equal to $2.5\,k + \ell_b$.

Where,

ϕ_{wy} = 1.0 (i.e., resistance factor for web yielding)
Ω_{wy} = 1.5 (i.e., safety factor for web yielding),
R_n = Nominal design strength, kips
t_w = Beam web thickness,
ℓ_b = Bearing length,
d = Beam depth,
x = Distance from the end of the beam to the concentrated load (see Figure 6-21e),
k = vertical distance between the outer face of the top or bottom flange and the toe of the fillet in the web (see *AISCM*, part 1).
F_y = yield strength, ksi
E = 29,000 ksi

Web Local Crippling

This occurs when the concentrated load causes a local buckling of the beam web. The design strengths for web crippling for wide flange sections and for HSS (with connecting surface in tension) are obtained from *AISC* Specifications, Section J10.3 as follows:

For $x \ge \dfrac{d}{2}$:

$$\phi_{wc} R_n = \phi_{wc} 0.8 t_w^2 \left[1 + 3\left(\frac{\ell_b}{d}\right)\left(\frac{t_w}{t_f}\right)^{1.5}\right] \sqrt{\frac{E F_{yw} t_f}{t_w}}\, Q_f \quad \text{for LRFD} \tag{6-38a}$$

$$\frac{R_n}{\Omega_{wc}} = \frac{0.8 t_w^2 \left[1 + 3\left(\frac{\ell_b}{d}\right)\left(\frac{t_w}{t_f}\right)^{1.5}\right] \sqrt{\frac{E F_{yw} t_f}{t_w}}\, Q_f}{\Omega_{wc}} \quad \text{for ASD} \tag{6-38b}$$

For $x < \dfrac{d}{2}$ and $\dfrac{\ell_b}{d} \le 0.2$:

$$\phi_{wc} R_n = \phi_{wc} 0.4 t_w^{\,2} \left[1 + 3\left(\frac{\ell_b}{d}\right)\left(\frac{t_w}{t_f}\right)^{1.5} \right] \sqrt{\frac{E F_{yw} t_f}{t_w}}\, Q_f \qquad \text{for LRFD} \qquad (6\text{-}39a)$$

$$\frac{R_n}{\Omega_{wc}} = \frac{0.4 t_w^{\,2} \left[1 + 3\left(\frac{\ell_b}{d}\right)\left(\frac{t_w}{t_f}\right)^{1.5} \right] \sqrt{\frac{E F_{yw} t_f}{t_w}}}{\Omega_{wc}}\, Q_f \qquad \text{for ASD} \qquad (6\text{-}39b)$$

For $x < \dfrac{d}{2}$ and $\dfrac{\ell_b}{d} > 0.2$,

$$\phi_{wc} R_n = \phi_{wc} 0.4 t_w^{\,2} \left[1 + \left(\frac{4\ell_b}{d} - 0.2\right)\left(\frac{t_w}{t_f}\right)^{1.5} \right] \sqrt{\frac{E F_{yw} t_f}{t_w}}\, Q_f \qquad \text{for LRFD} \qquad (6\text{-}40a)$$

$$\frac{R_n}{\Omega_{wc}} = \frac{0.4 t_w^{\,2} \left[1 + \left(\frac{4\ell_b}{d} - 0.2\right)\left(\frac{t_w}{t_f}\right)^{1.5} \right] \sqrt{\frac{E F_{yw} t_f}{t_w}}}{\Omega_{wc}}\, Q_f \qquad \text{for ASD} \qquad (6\text{-}40b)$$

Where,

ϕ_{wc} = 0.75 (i.e., resistance factor for web crippling),

Ω_{wc} = 2.0 (i.e., safety factor for web crippling),

Q_f = chord-stress interaction parameter,

= 1.0 for wide flange sections, channels, box sections, and the connecting surface of HSS in tension,

R_n = Nominal design strength, kips,

t_w = Beam web thickness,

t_f = Beam flange thickness,

ℓ_b = Bearing length,

d = Beam depth,

x = Distance from the end of the beam to the concentrated load,

k = vertical distance between the outer face of the top or bottom flange and the toe of the fillet in the web (see *AISCM*, part 1),

E = 29000 ksi,

F_{yw} = Yield strength of the web, ksi.

For the limit states of web yielding and web crippling, a pair of transverse stiffeners or a web doubler plate can be added to reinforce the beam section when the design strength is less than the applied loads. This topic, including the design of the stiffener and doubler plates, is also covered in Chapter 11. Note that it is advisable to avoid the use of stiffeners or doubler plates by using a beam with thicker web or flanges, whenever possible, because installation of stiffeners is labor-intensive and therefore results in increased fabrication cost.

Note that the web compression buckling provisions of AISC Specification Section J10.5 do not apply to beam bearing since the concentrated load or reaction is only applied to one flange, whereas Section J10.5 is only applicable when a pair of compressive single concentrated loads are applied at both the top and bottom flanges of the beam or girder and the beam or girder is laterally braced against sideways. Similarly, web side-sway buckling does not apply in this case since the beam is assumed to be laterally braced at the concentrated load or reaction point.

One situation where beam or girder web compression buckling is applicable is when a beam or girder is continuous over a column, but the column also continues above the girder top flange (see Figure 6-21g); in this case, both the top and bottom flanges of the girder are subjected to a pair of compressive forces (a downward load from the column above the girder and an upward reaction from the column below the girder). In Figure 6-21g, at the column support, the girder web is susceptible to web compression buckling; and at the tip of the cantilever, the girder web is susceptible to web side-sway buckling. For situations where web side-sway buckling or web compression buckling (see Figure 6-21g) are applicable, the reader should refer to Section 6-11 and Chapter 11 of this text, or to *AISC* Specification Sections J10.4 and J10.5.

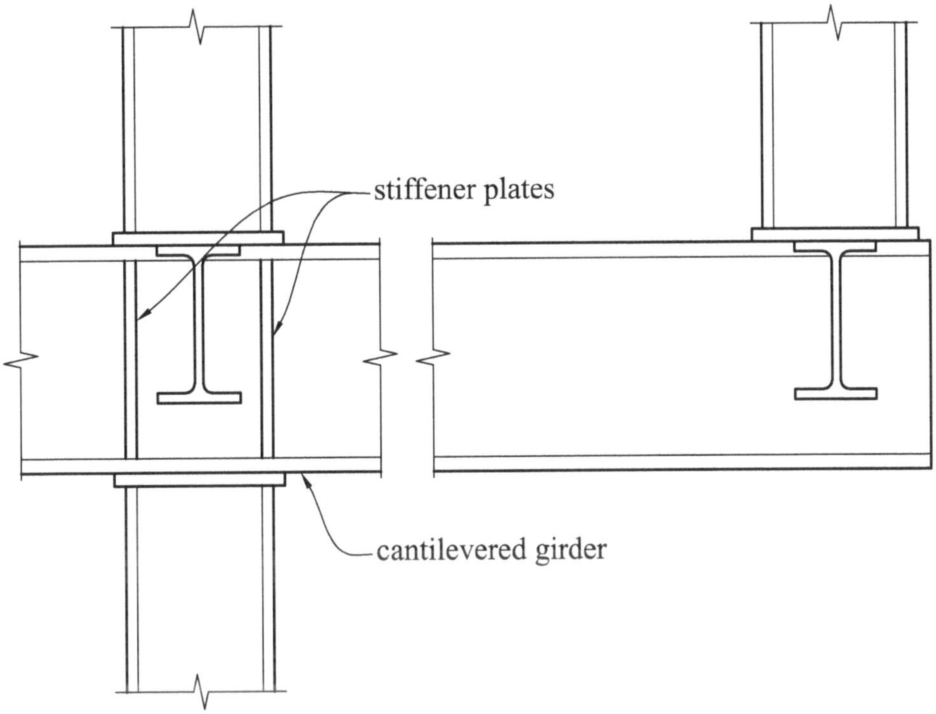

FIGURE 6-21g Web side-sway buckling and web compression buckling

Beam Bearing on Concrete or Masonry

The bearing strength of the supporting concrete or masonry in crushing on the full support area is obtained from Section J8 of the AISC Specifications,

$$\phi_{cb} P_p = \phi_{cb} 0.85 f'_c A_1 \qquad \text{for LRFD} \qquad (6\text{-}41a)$$

$$\frac{P_p}{\Omega_{cb}} = \frac{0.85 f'_c A_1}{\Omega_{cb}} \qquad \text{for ASD} \qquad (6\text{-}41b)$$

When the bearing is on less than the full area of concrete support, the bearing strength is

$$\phi_{cb} P_p = \phi_{cb} 0.85 f'_c A_1 \sqrt{\frac{A_2}{A_1}} \leq \phi_{cb} 1.7 f'_c A_1, \qquad \text{for LRFD} \qquad (6\text{-}42a)$$

$$\frac{P_p}{\Omega_{cb}} = \frac{0.85 f'_c A_1 \sqrt{\dfrac{A_2}{A_1}}}{\Omega_{cb}} \leq \frac{1.7 f'_c A_1}{\Omega_{cb}}, \qquad \text{for ASD} \qquad (6\text{-}42b)$$

Where,

$\phi_{cb} = 0.65$

$\Omega_{cb} = 2.31$

P_p = Nominal design strength,

f'_c = 28-day compressive strength of the concrete or masonry,

A_1 = Area of steel bearing = $B\ell_b$, and

A_2 = Maximum area of the support geometrically similar and concentric with the loaded area

= $(B + 2e)(\ell_b + 2e)$. Note that the dimension e is the minimum distance from the edge of the plate to the edge of the concrete support.

The strength reduction factor given in the AISC specification for bearing on concrete is 0.65, similar to that given in the ACI Code [4].

The dimensional parameters for A_1 and A_2 are shown in Figure 6-22.

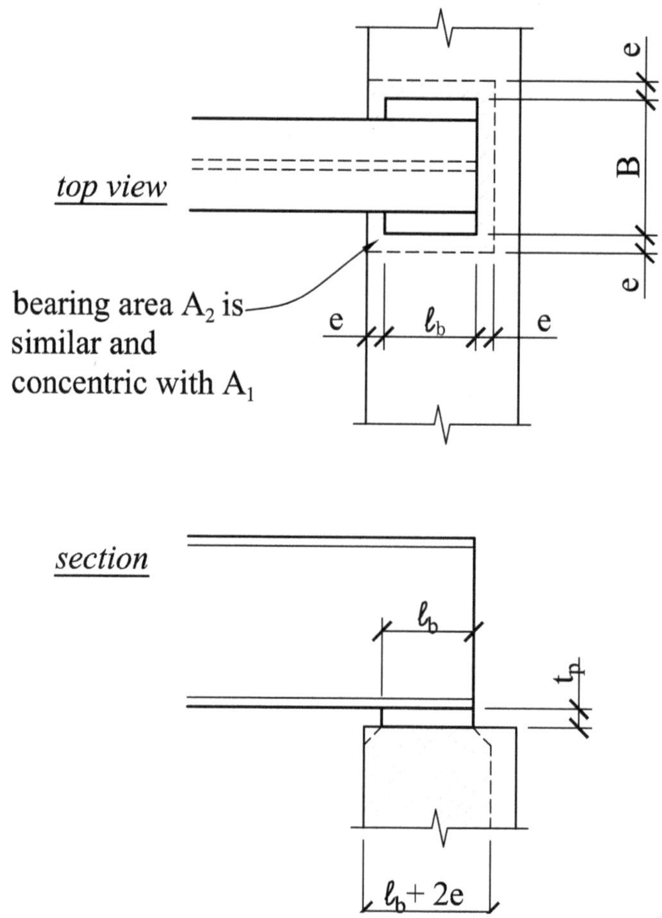

FIGURE 6-22 A_1 and A_2 parameters

Strength of the Bearing Plate in Bending

The strength of the bearing plate in bending must also be checked. Using equation (6-33b), the design moment strength for a plate bending about its weak axis (i.e., y-axis) is

$$\phi_b M_n = \phi_b M_p = \phi_b F_y Z_y \leq 1.6\phi_b F_y S_x \quad \text{for LRFD} \tag{6-43a}$$

$$\frac{M_n}{\Omega_b} = \frac{M_p}{\Omega_b} = \frac{F_y Z_y}{\Omega_b} \leq \frac{1.6 F_y S_x}{\Omega_b} \quad \text{for ASD} \tag{6-43b}$$

Where,

$\phi_b = 0.9$

$\Omega_b = 1.67$

F_y = Yield stress, and

$$Z_y = (\ell_b)\left(\frac{t_p}{2}\right)\left(\frac{t_p}{4}\right)(2) = \frac{\ell_b t_p^{\,2}}{4} \text{ (plastic section modulus for a bearing plate).} \tag{6-44}$$

From Figure 6-23, the maximum factored moment is

$$M_u = \left(\frac{R_u}{B}\right)(\ell)\left(\frac{\ell}{2}\right) = \frac{R_u \ell^2}{2B} \tag{6-45}$$

Combining equations (6-43), (6-44), and (6-45) yields

$$\phi_b M_n \geq M_u$$

$$(0.9)(F_y)\left(\frac{\ell_b t_p^{\,2}}{4}\right) = \frac{R_u \ell^2}{2B} \qquad \text{for LRFD}$$

$$\frac{M_n}{\Omega_b} \geq M_s$$

$$\frac{(F_y)\left(\frac{\ell_b t_p^{\,2}}{4}\right)}{1.67} = \frac{R_s \ell^2}{2B} \qquad \text{for ASD}$$

Solving for the plate thickness, t_p, yields

$$t_p \geq \sqrt{\frac{2R_u \ell^2}{0.9B\ell_b F_y}}, \qquad \text{for LRFD} \tag{6-46a}$$

$$t_p \geq \sqrt{\frac{2(1.67)R_s \ell^2}{B\ell_b F_y}}, \qquad \text{for ASD} \tag{6-46b}$$

Where,

t_p = bearing plate thickness,
$B\ell_b$ = Bearing plate plan dimensions,
R_s = Service load reaction,
R_u = Factored reaction,

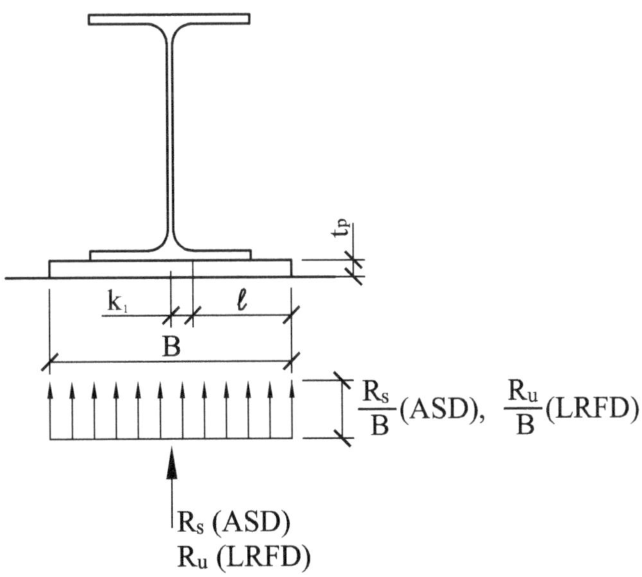

FIGURE 6-23 Bearing plate bending

ℓ = Moment arm for plate bending $= \dfrac{B - 2k_1}{2}$ (6-47)

k_1 = horizontal distance between the centroid of the beam web and the toe of the flange fillet (from part 1 of the *AISCM*),

F_y = Yield stress.

The design procedure for bearing plates can be summarized as follows:

1. Determine the location of the load relative to the nearest end of the beam (i.e., dimension x in Figure 6-21e).
2. Assume a value for the bearing plate length, ℓ_b.
3. Check the beam for web yielding and web crippling for the assumed value of ℓ_b; adjust the value of ℓ_b as required.
4. Determine the bearing plate width, B, such that the bearing plate area, $A_1 = B\ell_b$, is sufficient to prevent crushing of the concrete or masonry support.
5. Determine the thickness, t_p, of the beam bearing plate so that the plate has adequate strength in bending.

EXAMPLE 6-12

Web Yielding and Web Crippling in a Beam

Check web yielding and web crippling for the beam shown in Figure 6-24. The steel is ASTM A992, Grade 50.

Solution

For W18 × 50, we obtain the following properties from *AISCM* Table 1-1:

$k = 0.972$ in.

$t_w = 0.355$ in.

$d = 18$ in.

$t_f = 0.57$ in.

We will now follow the steps in the bearing plate design procedure from the preceding discussion:

Step 1. x = 3'-0 > d and > $d/2$

Step 2. ℓ_b = 6 in. (given)

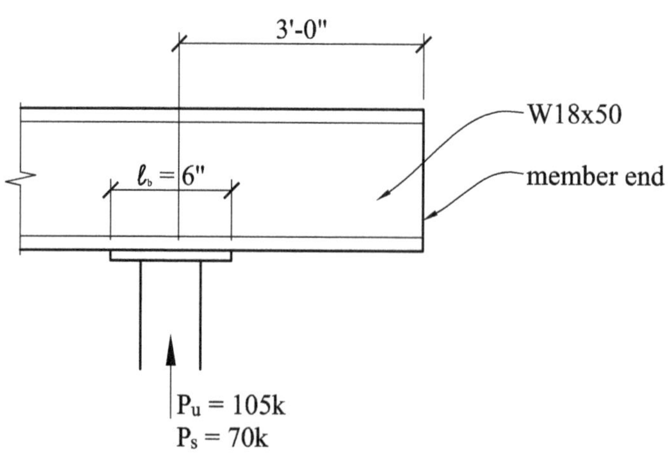

FIGURE 6-24 Detail for Example 6-12

LRFD Solution:

Step 3a. *Web Yielding*

$$\phi_{wy}R_n = \phi_{wy}(5k + \ell_b)F_y t_w = 1.0[(5)(0.972) + 6](50)(0.355)$$

$$= \mathbf{192\ kips} > P_u = 105\,\text{kips} \quad \text{OK.}$$

Step 3b. *Web Crippling*

$$\phi_{wc}R_n = \phi_{wc}0.8t_w^{\,2}\left[1 + 3\left(\frac{\ell_b}{d}\right)\left(\frac{t_w}{t_f}\right)^{1.5}\right]\sqrt{\frac{EF_y t_f}{t_w}}Q_f$$

$$= (0.75)(0.8)(0.355)^2\left[1 + 3\left(\frac{6}{18.0}\right)\left(\frac{0.355}{0.57}\right)^{1.5}\right]\sqrt{\frac{(29{,}000)(50)(0.57)}{0.355}}(1.0)$$

$$\phi_{wc}R_n = \mathbf{172\ kips} > P_u = 105\ \text{kips} \quad \text{OK.}$$

The W18 × 50 beam is adequate for web yielding and web crippling.

Note that **web compression buckling** consideration in AISC Specification Section J10.5 *does not* apply in this case since we do not have a pair of compressive forces acting on the beam web. Similarly, **web side-sway buckling** considerations in AISC Specification Section J10.5 *do not* also apply in this example, since there are no pair of concentrated forces applied at the bottom and top flanges of the W18x50 beam; there is only one column reaction applied at the bottom flange of the beam. For web side-sway buckling, see Section 6-11 and Chapter 11.

ASD Solution:

Step 3a. *Web Yielding*

$$\frac{R_n}{\Omega_{wy}} = \frac{(5k + \ell_b)F_y t_w}{\Omega_{wy}} = \frac{[(5)\ (0.972) + (6)](50)(0.355)}{1.5},$$

$$\frac{R_n}{\Omega_{wy}} = \mathbf{128\ kips} > P_s = 70\,\text{kips} \quad \text{OK.}$$

Step 3b. *Web Crippling*

$$\frac{R_n}{\Omega_{wc}} = \frac{0.8t_w^{\,2}\left[1 + 3\left(\frac{\ell_b}{d}\right)\left(\frac{t_w}{t_f}\right)^{1.5}\right]\sqrt{\frac{EF_y t_f}{t_w}}}{\Omega_{wc}}Q_f$$

$$= \frac{0.8(0.355)^2\left[1 + 3\left(\frac{6}{18.0}\right)\left(\frac{0.355}{0.57}\right)^{1.5}\right]\sqrt{\frac{(29000)(50)(0.57)}{(0.355)}}}{(2.0)}(1.0)$$

$$\frac{R_n}{\Omega_{wc}} = \mathbf{114\ kips} > P_s = 70\ \text{kips} \quad \text{OK.}$$

The W18 × 50 beam is adequate for web yielding and web crippling.

EXAMPLE 6-13

Beam Bearing on a Concrete Wall

A W18 × 50 beam is simply supported on 10-in.-thick concrete walls at both ends, as shown in Figure 6-25. Using the LRFD method, design a beam-bearing plate at the concrete wall supports, assuming the following:

Beam span, $L = 20$ ft. center-to-center of support

Use ASTM A36 steel for the beam and bearing plate

Dead load, $D = 1.5$ kips / ft.

Live load, $L = 2$ kips / ft.

$f'_c = 4000$ psi

Solution

For a W18 × 50, we obtain the following properties from *AISCM* Table 1-1:

$k = 0.972$ in.

$t_w = 0.355$ in.

$d = 18$ in.

$t_f = 0.57$ in.

$k_1 = 13/16$ in.

The factored loads and reactions at each end of the beam are

$$w_u = 1.2D + 1.6L = (1.2)(1.5) + (1.6)(2) = 5.0 \text{ kips / ft.}$$

$$R_u = \frac{w_u L}{2} = \frac{(5.0)(20)}{2} = 50 \text{ kips}$$

Design Steps:

1. $x < 5$ in., (half of the wall thickness)
Therefore, $x < d$ and $x < d/2$.

2. Assume $\ell_b = 6$ in. (recommended practical value)

3a. *Web Yielding*

$$\phi_{wy} R_n = \phi_{wy}(2.5\,k + \ell_b) F_y t_w = 1.0[(2.5)(0.972) + 6](36)(0.355)$$

$$= \mathbf{107\ kips} > \mathrm{R_u} = 50 \text{ kips} \qquad \text{OK.}$$

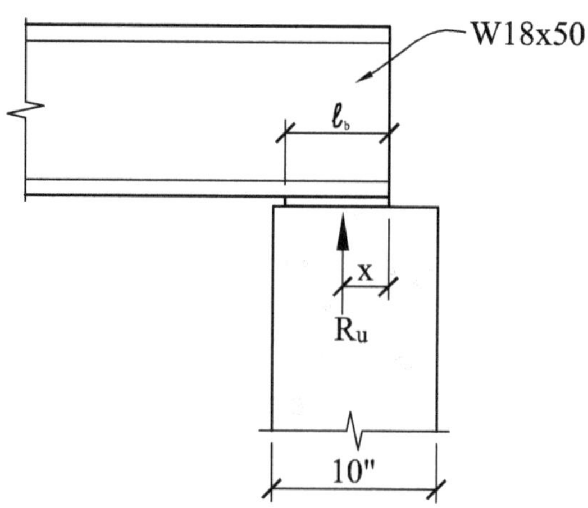

FIGURE 6-25 Detail for Example 6-13

3b. *Web Crippling*

$$\frac{\ell_b}{d} = \frac{6}{18.0} = 0.33 > 0.2; \text{ equation } (6\text{-}40a) \text{ is used :}$$

$$\phi_{wc}R_n = \phi_{wc}0.4t_w^{\,2}\left[1+\left(\frac{4\ell_b}{d}-0.2\right)\left(\frac{t_w}{t_f}\right)^{1.5}\right]\sqrt{\frac{EF_yt_f}{t_w}}\,Q_f$$

$$= (0.75)(0.4)(0.355)^2\left[1+\left(\frac{(4)(6)}{18}-0.2\right)\left(\frac{0.355}{0.57}\right)^{1.5}\right]\left[\sqrt{\frac{(29{,}000)(36)(0.57)}{0.355}}\right](1.0)$$

$\phi_{wc}R_n = \mathbf{72.0\ kips} > R_u = 50\text{ kips}$ OK.

4. *Plate Bearing*

$f'_c = 4\text{ ksi}$

From equation (6-41a), $\phi_{cb}P_p$ can be conservatively taken as

$\phi_{cb}P_p = \phi_{cb}0.85f'_cA_1 = \phi_{cb}0.85f'_c\ell_bB \geq R_u$

That is, $(0.65)(0.85)(4)(6\text{ in.})B \geq 50$ kips

Solving for B gives $B_{\min} = 3.78$ in.

It is practical to use a value of B at least equal to or greater than the beam flange width.

$b_f = 7.495$ in.; therefore, Try **B = 8 in.**

Therefore, the trial bearing plate size ($B \times \ell_b$) is 8 in. × 6 in.

5. Determine the plate thickness using equation (6-46a):

$$\ell = \frac{B-2k_1}{2} = \frac{8-(2)\left(\frac{13}{16}\right)}{2} = 3.19\text{ in.}$$

$$t_p \geq \sqrt{\frac{2R_u\ell^2}{0.9B\ell_bF_y}} = \sqrt{\frac{(2)(50)(3.19)^2}{(0.9)(8)(6)(36)}} = 0.81\text{ in.}$$

Select plate thickness in increments of $^1/_4$ in.; therefore,
Use an 8-in. wide × 6-in. long × 1-in. thick bearing plate.

6-11 BEARING STIFFENERS

Bearing stiffeners are the vertical plates provided at the bearing locations of a beam when the beam does not have sufficient strength in the web to support the end reaction or concentrated load. The limit states for this condition are web local yielding, web crippling, and web sidesway buckling. The design provisions for web local yielding and web crippling were covered in the previous section, and the design of the stiffener plates for these two limit states is covered in Chapter 11.

Web Side-sway Buckling

Web sidesway buckling can occur when a concentrated *compressive* force is applied to the compression flange of a beam within its span, where the compression flange is laterally braced, but the tension flange is not (see AISC specification Figure C-J10.2). The concentrated compressive force could be applied at a point within the span of the beam (see Figure 6-21c), or the compression force could be the beam end reaction (see Figures 6-21a or 6-21e).

Web side-sway buckling can also occur at a cantilever beam or girder supporting a discontinued column (see Figure 6-21g) if the tension flange (i.e., the top flange of the cantilever beam) is laterally braced, but the compression flange (i.e., the bottom flange of the cantilever beam) is not. In this case, the compression flange will be susceptible to web side-sway buckling if it is not laterally braced.

Web side-sway buckling occurs because the *relative lateral movement* between the compression flange and the tension flange *is not restrained* at the location of the concentrated force. The relative lateral movement between the loaded compression flange and the tension flange of a beam can be prevented either by providing lateral bracing to both flanges at the load point, or by providing a pair of vertical transverse stiffeners or doubler plates at the compressive force location. Note that stiffeners and doubler plates are only effective if the rotation of the loaded flange is restrained from rotation about the longitudinal axis of the member (see *AISCM* Section C-J10.4). When the loaded flange is laterally restrained by horizontal braces, the braces should be designed for an axial tension and compression force equal to 1% of the concentrated compressive force on the flange [1], or use the point bracing design provisions discussed in Chapter 14.

When the **compression flange *is restrained* against rotation about the longitudinal axis of the beam** (e.g., when restrained from rotation because of the connection of the loaded flange to a floor slab), the limit state of web sidesway buckling (see Section J10.4 of AISC Specification) is as follows:

When $\dfrac{h/t_w}{L_b/b_f} \leq 2.3$,

$$\phi R_n = \phi \frac{C_r t_w^3 t_f}{h^2}\left[1 + 0.4\left(\frac{h/t_w}{L_b/b_f}\right)^3\right] \quad \text{for LRFD} \qquad (6\text{-}48a)$$

$$\frac{R_n}{\Omega} = \frac{\dfrac{C_r t_w^3 t_f}{h^2}\left[1 + 0.4\left(\dfrac{h/t_w}{L_b/b_f}\right)^3\right]}{\Omega} \quad \text{for ASD} \qquad (6\text{-}48b)$$

For $\dfrac{h/t_w}{L_b/b_f} > 2.3$, web sidesway buckling does *not* occur and therefore does not have to be checked.

When the **compression flange is *not* restrained against rotation**, the limit state of web sidesway buckling is as follows:

For $\dfrac{h/t_w}{L_b/b_f} \leq 1.7$,

$$\phi R_n = \phi \frac{C_r t_w^3 t_f}{h^2}\left[0.4\left(\frac{h/t_w}{L_b/b_f}\right)^3\right] \quad \text{for LRFD} \qquad (6\text{-}49a)$$

$$\frac{R_n}{\Omega} = \frac{\dfrac{C_r t_w^3 t_f}{h^2}\left[0.4\left(\dfrac{h/t_w}{L_b/b_f}\right)^3\right]}{\Omega} \quad \text{for ASD} \qquad (6\text{-}49b)$$

For $\dfrac{h/t_w}{L_b/b_f} > 1.7$, the limit state of web sidesway buckling does *not* occur and therefore does not have to be checked.

The parameters in the preceding equations (6-48) and (6-49) are defined as follows:

$\phi =$ 0.85

$\Omega =$ 1.76

$h =$ Clear distance between the flanges for built-up shapes,

$=$ Clear distance between the flanges less the fillets for rolled shapes

$=$ T for wide flange sections (see Part 1 of the *AISCM*)

$t_w =$ Web thickness,

h/t_w (from part 1 of the *AISCM*, for standard sections),

$L_b =$ Largest unbraced length along either the top or the bottom flange at the load point,

$b_f =$ Flange width,

$t_f =$ Flange thickness,

$C_r =$ 960,000 ksi when $M_u < M_y$ for LRFD or $1.5M_s < M_y$ for ASD

$=$ 480,000 ksi when $M_u > M_y$ for LRFD or $1.5M_s \geq M_y$ for ASD,

$M_y = F_y S_x$,

$F_y =$ Yield stress, and

$S_x =$ Elastic section modulus,

$M_u =$ Factored moment in kips-inch from the LRFD load combination in Section 2-3 of Chapter 2,

$M_s =$ Service or unfactored total moment in kips-inch from the ASD load combination in Section 2-3 of Chapter 2.

When the demand concentrated compressive force is greater than the design strength of the web, either **bearing stiffeners** or **lateral bracing** are required at the location of the concentrated force, provided the loaded flange is restrained against rotation. Recall that lateral bracing of both flanges are always effective against web side-sway buckling, but stiffeners are *not* effective to mitigate the effects of web side-sway buckling if the loaded flange is *not* restrained against rotation. When bearing stiffeners are provided to resist the full compressive force, the limit states of *web local buckling*, *web crippling*, and **web side-sway buckling** do not have to be checked.

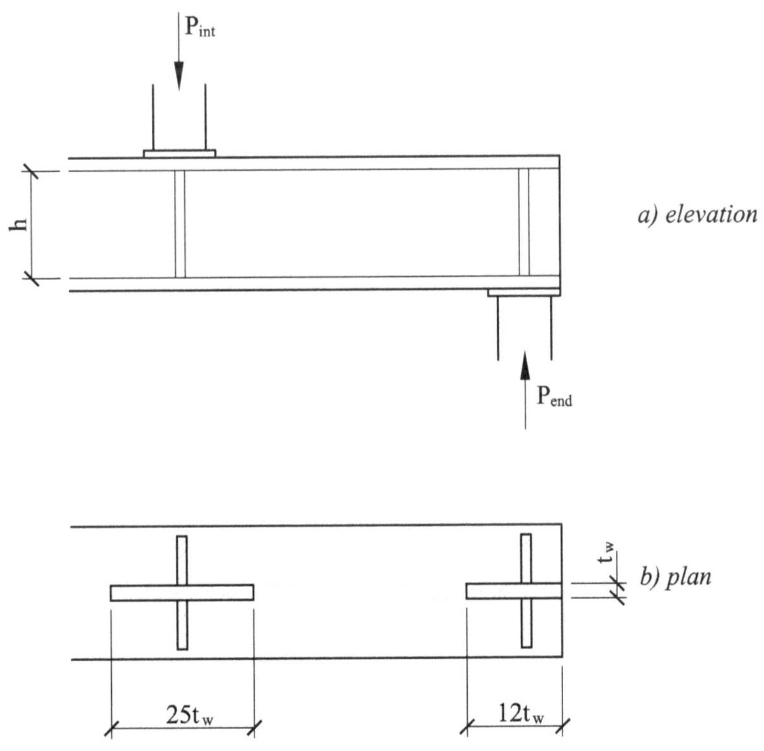

FIGURE 6-26 Section properties for bearing stiffeners

Bearing stiffeners are designed as short columns when they are provided to reinforce the web of a beam subjected to concentrated loads or the web of a beam at an end reaction. The section properties of the stiffened beam section are as shown in Figure 6-26. The AISC specification allows a portion of the web to be included in the calculation of the design compressive strength of the localized section. For bearing stiffeners at the end of a member, the section of web included has a maximum length of $12t_w$; for interior stiffeners, the maximum length is $25t_w$. The effective length factor for stiffeners is $K = 0.75$.

The calculated section properties are then used to determine the design strength of the stiffener in compression. For connection elements such as bearing stiffeners, AISC Specification Section J4.4 permits the design compressive strength to be determined as follows:

For $\dfrac{KL}{r} \leq 25$,

$$\phi_c P_n = \phi F_y A_g, \quad \text{for LRFD} \quad (6\text{-}50a)$$

$$\frac{P_n}{\Omega_c} = \frac{F_y A_g}{\Omega_c}, \quad \text{for ASD} \quad (6\text{-}50b)$$

Where,

$\phi_c = 0.9$,

$\Omega_c = 1.67$,

F_y = Yield stress,

A_g = Gross area of the bearing stiffener section,

$K = 0.75$ for bearing stiffeners,

$L = h$, and

r = Least radius of gyration for the bearing stiffener section.

For $KL / r > 25$, the provisions from Chapter 5 apply, or *AISCM* Table 4-14 can be used to determine $\phi_c F_{cr}$ for any value of KL / r.

The limit state of bearing strength also needs to be checked, but this rarely controls the design of stiffeners. The design bearing strength for stiffeners is

$$\phi_{pb} R_n = \phi_{pb} 1.8 F_y A_{pb}, \quad \text{for LRFD} \quad (6\text{-}51a)$$

$$\frac{R_n}{\Omega_{pb}} = \frac{1.8 F_y A_{pb}}{\Omega_{pb}}, \quad \text{for ASD} \quad (6\text{-}51b)$$

Where,

$\phi_{pb} = 0.75$

$\Omega_c = 2.0$

F_y = Yield stress

A_{pb} = Cross-sectional area of the bearing stiffeners.

The bearing stiffeners are usually welded to the flanges and the web of the beam. On the other hand, the stiffener is not required to be welded to the compression flange. For the weld connecting the stiffener to the web, the difference between the total concentrated force and the smallest design strength for web local yielding, web crippling, and web sidesway buckling can be used to determine the weld size.

Web Compression Buckling

Web compression buckling, covered in AISC specification Section J10.5, apply only when a pair of concentrated forces are applied at both flanges of a member at the same cross-section (e.g., see the beam at the column location in Figure 6-21g). The design strength and allowable strength in web compression buckling are given in equations (6-51c) and (6-51d), respectively.

$$\phi R_n = \phi\left(\frac{24(t_w)^3\sqrt{EF_y}}{h}\right)Q_f \qquad \text{(6-51c)}$$

$$\frac{R_n}{\Omega} = \frac{1}{\Omega}\left(\frac{24(t_w)^3\sqrt{EF_y}}{h}\right)Q_f \qquad \text{(6-51d)}$$

Where,

ϕ = 0.90

Ω = 1.67

Q_f = 1.0 for wide flange sections, channels, box sections, and the connecting surface of HSS in tension, For all other HSS conditions, see AISC specification Section K1.3.

R_n = Nominal strength, kips,

t_w = Beam web thickness,

Note: If the pair of concentrated forces are applied at a distance from the end of the beam of less than one-half the depth of the beam (i.e., d/2), the design strength and allowable strength will be 50% of the values from equations (6-51c) and (6-51d).

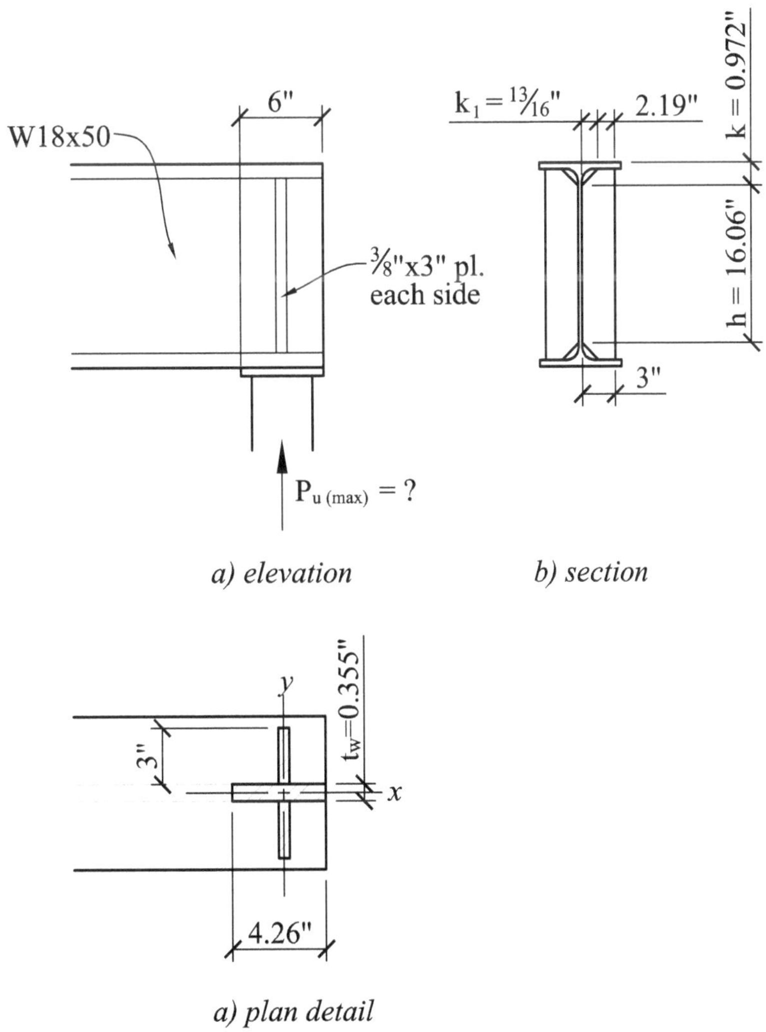

FIGURE 6-27 Detail for Example 6-14

EXAMPLE 6-14

Bearing Stiffeners

Determine the design bearing strength at the end of the W18 × 50 beam shown in Figure 6-27. The steel is ASTM Grade 50.

Recall that when bearing stiffeners are provided, web local yielding, web crippling, and web sidesway buckling do not have to be checked. (Web local yielding and web crippling were checked in the previous examples.)

Solution

From *AISCM* Table 1-1, we find the following properties for W18×50:

$k = 0.972$ in., $k_1 = 13/16$ in., $t_w = 0.355$ in.

$d = 18$ in., $t_f = 0.57$ in., $b_f = 7.495$ in.

Section properties of the stiffened area about the web are

$L_w = 12t_w = (12)(0.355) = 4.26$ in.

$h = d - 2k = 18'' - (2)(0.972\text{ in.}) = 16.06$ in.

Area of stiffener plates = (2 plates)(3 in.)(3/8 in.) = 2.25 in.^{2}
Area of effective web = (4.26 in.)(0.355 in.) = 1.51 in.^{2}

Shape	A, in.^{2}	I, in.^{4}	d, in.	Ad^2, in.^{4}	$I + Ad^2$, in.^{4}
Stiff. plates	2.25	1.69	1.68	6.33	8.02
Web	1.51	0.016	0	0	0.016
$\sum =$	**3.76**				**8.03 = I_{xx}**

$$I_{yy} = \frac{0.355(4.26)^3}{12} + \frac{3(0.375)^3}{12} + \frac{3(0.375)^3}{12} = 2.31 \text{ in.}^4$$

$$r = \sqrt{\frac{I_{min}}{A}} = \sqrt{\frac{2.31}{3.76}} = 0.78 \text{ in.}$$

Slenderness Ratio:

$$\frac{KL}{r} = \frac{(0.75)(16.06)}{0.78} = 15.4 < 25$$

Therefore, $F_n = F_y$

LRFD solution:

Since the slenderness ratio is less than 25, we can use equation (6-50a) to compute the axial compression capacity of the stiffened area about the web (with a gross area, A_g = 3.76 in.^{2}):

$$\phi_c P_n = \phi_c F_n A_g = \phi_c F_y A_g = (0.9)(50)(3.76) = 169 \text{ kips}$$

Bearing:

The bearing area will be the area of the two stiffener plates, excluding the fillets:

$$A_{pb} = (2)(3/8)(3 - k_1) = (2)(3/8)(3 - 13/16) = 1.64 \text{ in.}^2$$

$$\phi_{pb} R_n = \phi_{pb} 1.8 F_y A_{pb} = (0.75)(1.8)(50)(1.64) = \mathbf{110\,kips}.$$

The capacity in the bearing is less than $\phi_c P_n$ of 169 kips; therefore, the bearing strength controls, so the maximum factored reaction is 110 kips.

The design of welds is covered in Chapter 10, but the design of the weld connecting the stiffener plate to the beam web is covered here for completeness.

Weld to the web:

$t_{plate} = 3/8 \text{ in.} = 0.375 \text{ in.} (\text{thicker part})$

$t_w = 0.355 \text{ in.} (\text{thinner part})$

From Table 10-2,

Minimum weld size = 3/16 in.

Maximum weld size = 3/8 in. – 1/16 in. = 5/16 in.

The ultimate limit state design (LRFD) equation is

$$\phi R_n = 1.392 DL \geq R_u$$

Where,

ϕR_n = Design weld strength (kips),

D = Weld size in sixteenths of an inch (e.g., D = 5 for a $^5/_{16}$-in. weld),

L = Weld length, in.

= $2h$ = (2)(16.06 in.) = 32.12 in.,

R_u = 110 kips

Therefore, $\phi R_n = (1.392)(D)(32.12) \geq 110 \text{ kips}$

Solving for D gives D = 2.47; therefore, use a $^3/_{16}$ in. weld to connect the stiffener plate to the beam web.

ASD solution:

Since the slenderness ratio is less than 25, we can use equation (6-50b):

$$\frac{P_n}{\Omega_c} = \frac{(50)(3.76)}{1.67} = 112 \text{ kips}$$

Bearing:

The bearing area will be the area of the two stiffener plates, excluding the fillets:

$$A_{pb} = (2)(3/8)(3 - k_1) = (2)(3/8)(3 - 13/16) = 1.64 \text{ in.}^2$$

$$\frac{R_n}{\Omega_{pb}} = \frac{1.8(50)(1.64)}{2.0} = \mathbf{73.8\ kips}$$

Bearing strength controls, so the maximum service load reaction is 73.8 kips.

Weld strength is covered in Chapter 10 of this text, but the weld connecting the stiffener plate to the web will be covered here for completeness (see also Section J2.2 of the AISC specification).

$t_{plate} = 3/8 \text{ in.} = 0.375 \text{ in.} (\text{thicker part})$

$t_w = 0.355 \text{ in.} (\text{thinner part})$

From Table 10-2,

Minimum weld size = 3/16 in.
Maximum weld size = 3/8 in. – 1/16 in. = 5/16 in.
The ASD design equation is

$$\frac{R_n}{\Omega} = 0.928DL \geq R_s$$

Where,

$\frac{R_n}{\Omega}$ = Allowable weld strength (kips),
R_S = unfactored maximum total load = 73.8 kips
D = Weld size in sixteenths of an inch (e.g., D = 5 for a $^5/_{16}$-in. weld),
L = Weld length, in.,
= $2h$ = (2)(16.06 in.) = 32.12 in.

Therefore, $(0.928)(D)(32.12) = 73.8$ kips

Solving for D gives D = 2.47; therefore, use a 3/16 in. weld to connect the stiffener plate to the beam web.

6-12 OPEN-WEB STEEL JOISTS

Open-web steel joists (OWSJ) are pre-engineered manufactured structural steel members that can be used in lieu of steel beams in either floor or roof framing, but they are more commonly used for roof framing. These joists are sometimes called bar joists because, at one time, many were fabricated with round bars as web members. Most joists today are manufactured with double-angle members used for the top and bottom chords and either single or double-angle web members. Some joists are manufactured with proprietary, nonstandard steel sections, and the load-carrying capacity of these sections would have to be determined from the joist manufacturers' load tables.

The following are the main advantages of open-web steel joists when compared to rolled wide flange sections:

1. They are lighter in weight than rolled steel shapes for a given span and thus can span longer distances.
2. The open webs allow for easy passage of mechanical duct work and electrical conduits, compared to beams where the ducts and conduits have to be located below the beams. This increases the overall plenum depth (i.e., the depth from the top of the floor or roof to the underside of the ceiling), which increases the overall building height and thus the construction cost.
3. They may be more economical than rolled shapes, depending on the span length.

The main disadvantages of open-web steel joists are as follows:

1. They cannot easily support concentrated loads at locations away from the panel points (see Figure 6-28).
2. The light weight of the OWSJ makes them more susceptible to vibration problems when they are used for floor framing (see Chapter 12 for further discussion).
3. They may not be economical for floor framing because of the closer joist spacing that is required due to the heavy floor loads.
4. Future structural modifications are not as easy to accomplish with joists compared to wide flange sections.

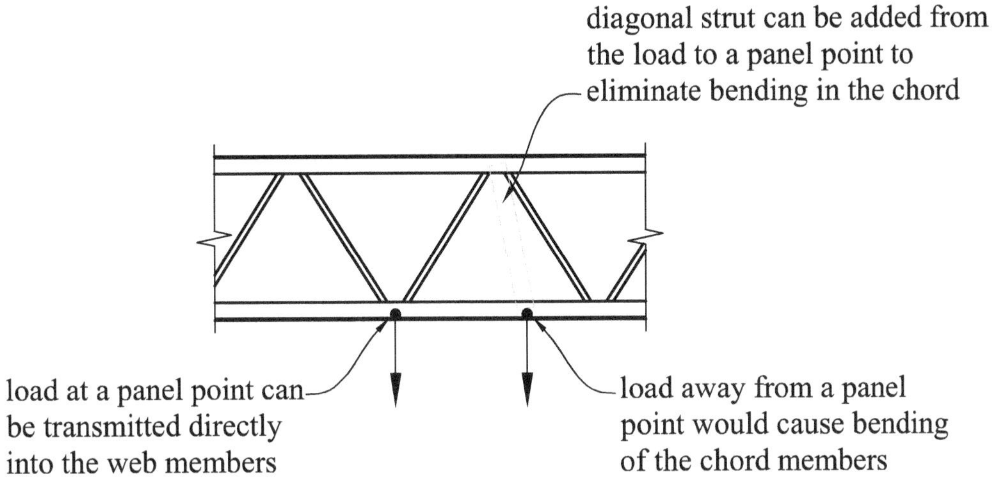

FIGURE 6-28 Loading at and away from joist panel points

FIGURE 6-28b Open-web steel joists for roof framing in lieu of roof beams

Note that open-web steel joists are not typically used along the perimeter of a building. Instead, wide flange beams and girders are used along the exterior column lines as well as along column lines that lie on the same vertical plane as the lateral force resisting systems (LFRS), such as X-braced frames, Chevron braced frames, shear walls, or moment frames. Along the interior column lines where there are no LFRS, wide flange beams or girders or joists with bottom chords extended and connected to the columns (these are called tie joists) are typically used. See Figure 6-28b for roofs framed with open-web-steel-joists, in lieu roof of beams.

There are several manufacturers of open-web steel joists, but the most commonly used publication for the selection of these members is the *Standard Specifications and Load Tables for Steel Joists and Joist Girders*, published by the Steel Joist Institute (SJI) [7].

The load tables in the SJI catalog are identified by a designation that corresponds to a certain strength and stiffness. Many individual manufacturers will provide joists based on these load tables and the corresponding SJI specifications.

K-Series Joists

K-series joists are the most commonly used joists, and they serve as the primary members for roof or floor framing. They are selected from the SJI catalog by a number designation (e.g., 14K1). The first number—14—represents the overall depth of the joist, and the last number—1—is the series designation. 14K3 would have more strength than the 14K1. They are listed in the catalog with a certain load-carrying capacity as a function of the span. The load-carrying capacity has two numbers listed in pounds per linear foot. For the LRFD tables, the first, or upper, number is the total factored load-carrying capacity of the joist. The second, or lower, number (often in red font) is the service live load that will produce a deflection of L/360 for floors and $L/240$ for roof members. The standard joist seat depth for K-series joists is $2\frac{1}{2}$ in. (see Figure 6-29a). Larger depths or inclined seats can be used for special conditions.

K-series joists are generally economical for spans of up to 50 ft., and vary in overall depth from 8 in. to 30 in.

KCS-Series Joists

KCS joists are K-series joists with the ability to support a constant shear across the span (hence the CS designation). KCS joists are designed to support a constant moment across all interior panel points and a constant shear across the entire length. KCS joists are used for nonuniform loading conditions such as equipment loads or trapezoidal snowdrift loads. The designer simply needs to calculate the applied maximum shear and moment for a given special joist loading and select a KCS joist with corresponding larger or equal shear and moment capacities. The joist seat depth for KCS joists is also 2½ in.

Type S and R Extensions

Top chord extensions (often designated as TCX) are used at perimeter conditions where a cantilever is desired. These extensions are often designed for the same load as the joist in the back span; however, two standard cantilevered sections are provided in the SJI specification—the S-type and the R-type. The S-type implies that only the upper seat angles are extended, whereas the R-type is a stronger section where the entire depth of the joist seat is extended (see Figures 6-29c and 6-29d).

LH-Series and DLH-Series

These are long-span joists that are used for spans of up to about 130 ft. The letter D in the DLH designation indicates a deeper section than the LH-series. The designation is similar to that of

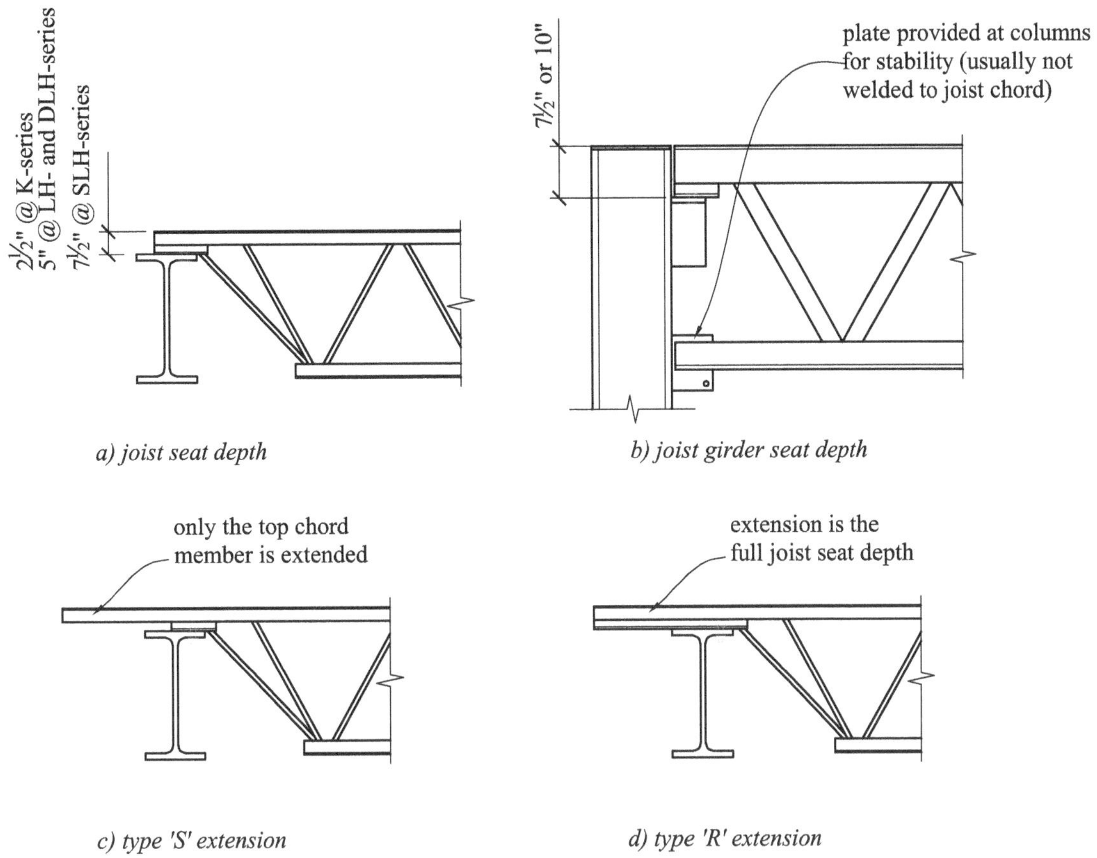

FIGURE 6-29 Joist seat types

K-series joists (e.g., 32LH06). The number "32" is the overall depth in inches, and "06" is a series designation. LH- and DLH-series joists vary in depth from 18 in. to 72 in. The joist seat depth is 5 in. for LH and most DLH joists. Some of the larger DLH joists require a $7\frac{1}{2}$-in. joist seat depth.

Joist Girders

Joist girders are of open-web construction and are used to support steel joists (see Figure 6-30a and 6-30b). Joist girders are designed as steel trusses that support concentrated loads from the joists that frame into them. They are designated by their overall depth, the number of panel

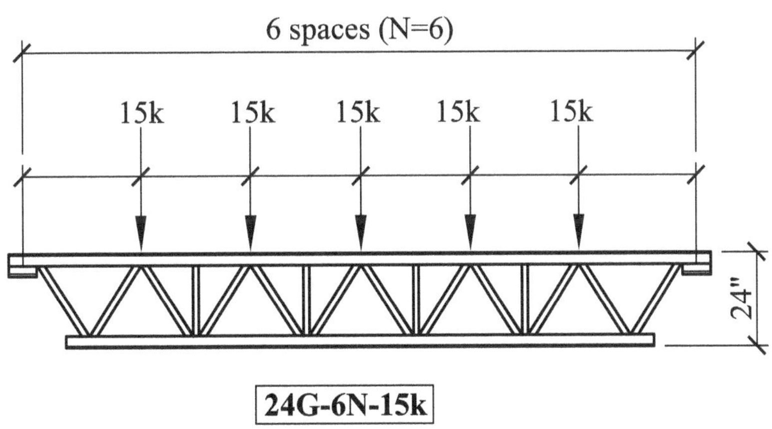

FIGURE 6-30a Joist girder designation

points, and the loads at each panel point (e.g., 24G-6N-15k) (see Figure 6-30a). The number "24" is the overall depth, "6N" indicates the number of panel spaces, and "15k" indicates a total load of 15 kips at each panel point (factored load for the LRFD tables and service loads for the ASD tables). The seat depth for joist girders is either $7\frac{1}{2}$ in. or 10 in., depending on the loading and configuration of the joist girder (see Figure 6-29b).

FIGURE 6-30b Joist girders supporting open web steel joists

When a required joist design does not fit one of these categories, a special joist is often designed where the designer provides a special joist loading diagram for the joist manufacturer to use in designing the joist. Examples of this would include joists supporting concentrated loads, joists and joist girders supporting nonuniform loads, and joists and joist girders that have end moments, such as would be the case for joists in moment frames. See Ref. [11] for more detailed coverage of the design of joists and joist girders.

EXAMPLE 6-15

Selection of a K-series Joist

Select a K-series using the SJI specifications to support the following loads for the framing shown in Figure 6-31.

Roof dead load = 30 psf, snow load = 40 psf

Solution

Joist tributary width = 5.5 ft.

Joist span = 25 ft.

LRFD Solution:

$$\text{Total load (factored)} = [(1.2)(30) + (1.0)(40)](5.5 \text{ ft.}) = 418 \text{ lb./ft.}$$

$$\text{Live load (service)} = (40)(5.5 \text{ ft.}) = 220 \text{ lb./ft.}$$

From the SJI load tables for K-series joists, the joist selections are as shown in Table 6-8.

LRFD Solution:

TABLE 6-8 Joist Selection (LRFD Method)

Joist selection	Total load capacity, lb./ft.	Live load capacity, lb./ft.	Joist weight lb./ft.	
16K6	576	238	8.1	
18K5	600	281	7.7	
20K4	**594**	**312**	**7.2**	← *Select*
22K4	657	381	7.3	

Note: 20K4 is the most economical joist for the given loads since it is the least weight joist.

ASD Solution:

$$\text{Total load (unfactored)} = [30 + 40](5.5 \text{ ft.}) = 385 \text{ lb./ft.}$$

$$\text{Live load (service)} = (40)(5.5 \text{ ft.}) = 220 \text{ lb./ft.}$$

From the SJI load tables for K-series joists, the joist selections are as shown in Table 6-9.

ASD Solution:

TABLE 6-9 Joist Selection (ASD Method)

Joist selection	Total load capacity, lb./ft.	Live load capacity, lb./ft.	Joist weight lb./ft.	
16K6	428	263	8.1	
18K5	400	281	7.7	
20K4	**396**	**312**	**7.2**	← *Select*
22K4	438	381	7.3	

Note: 20K4 is the most economical joist for the given loads since it is the least weight joist.

EXAMPLE 6-16

Selection of a Joist Girder (LRFD Method)

Select a joist girder using the SJI specifications for member JG1 in Figure 6-31 to support a total roof dead load of 20 psf and a snow load of 64 psf.

Solution

$$\text{Total factored load} = [(1.2)(20) + (1.0)(64)] = 88 \text{ psf}$$

Concentrated factored load at each panel point of the joist girder:

$$P_u = (88\text{ psf})(5.5\text{ ft.})\left(\frac{25\text{ ft.}}{2} + \frac{25\text{ ft.}}{2}\right) = 12.1\text{ kips}$$

The factored load used in Table 6-9 was P_u = 12.0 kips, which is close to the actual load of P_u = 12.1 kips. In general, larger joist girder depths will lead to a lighter overall section. The actual designation for this joist girder is **28G-4N-12.1k**. The largest depth joist girder was selected here, but architectural or other design constraints might dictate a shallower section.

TABLE 6-9 Joist Girder Selection (LRFD Method)

Joist girder span	Joist spaces, N	Depth	Joist weight, lb./ft.	
22 ft.	4N @ 5.5	20	19	
		24	17	
		28	**16**	← *Select*

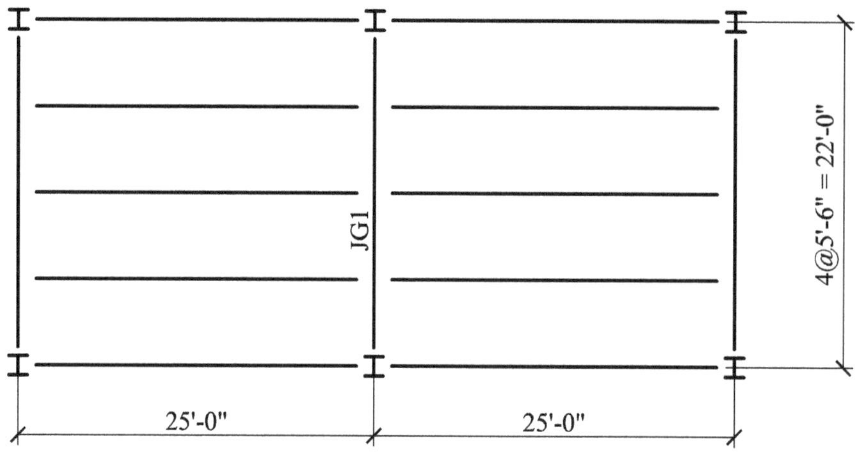

FIGURE 6-31 Roof plan for Example 6-16

6-13 FLOOR PLATES, RECTANGULAR PLATES, AND GRATING

Floor plates that are used in a flatwise orientation with the plate bending about its weak axis can be used as decking material for industrial applications such as a floor deck for mezzanines or similar types of structures. There are several other types of decking materials, which are generally proprietary, but only the floor plate is addressed in the *AISCM* (see Figure 6-32a). The selection of proprietary decking is done by using data provided by the specific manufacturer. The most common floor plate has a raised pattern and is often called a *diamond plate* because of the shape of the raised pattern.

A floor plate should conform to ASTM A786, which has a minimum yield stress of F_y = 27 ksi. Higher grades could be specified, such as ASTM A36, but availability should be considered. For deflection, a relatively low limit of $L/100$ is recommended by AISC.

AISCM Table 3-17a (for deflection-controlled applications) and *AISCM* Table 3-17b (for flexural-strength-controlled applications) are selection tables for floor plates of various thicknesses and superimposed surface load capacities for spans ranging from 18 in. to 7 ft. These tables are based conservatively on a simple-span condition for the floor plate. The plate is selected in 1/8-in. increments for thicknesses less than 1 in. and in ¼ in. increments for thicknesses greater than 1 in.

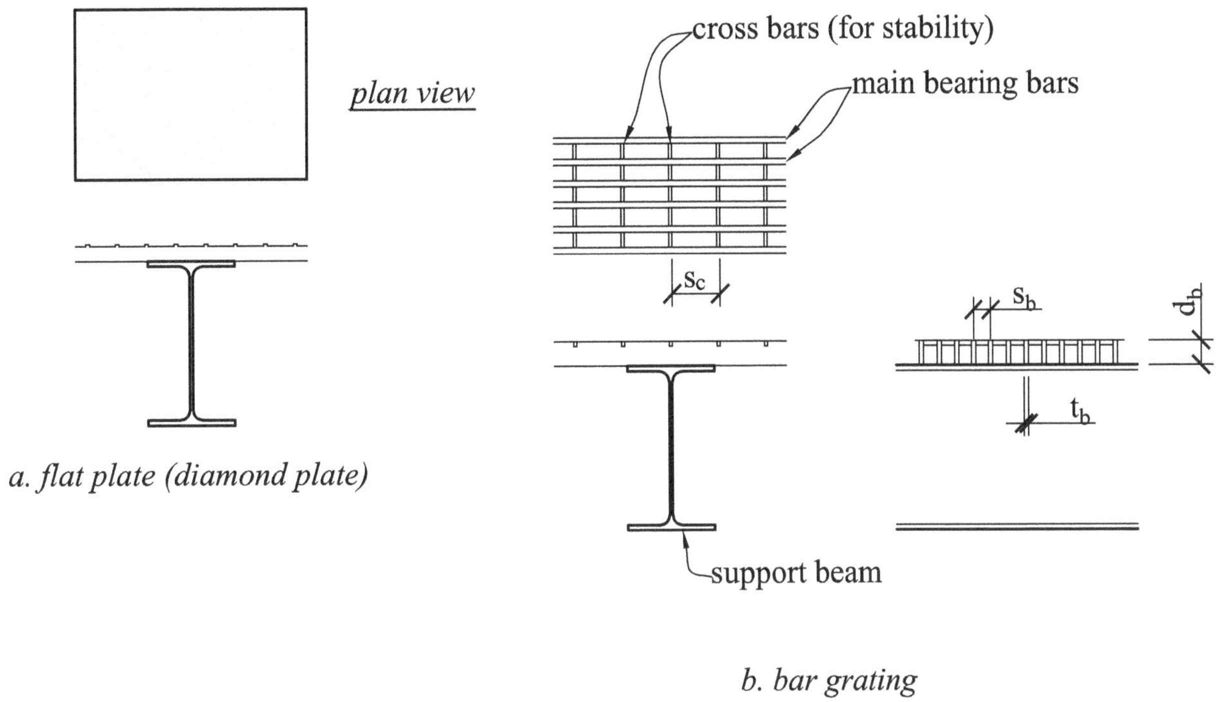

FIGURE 6-32 Various floor deck types for industrial applications

Another common steel product used as a decking material is bar grating (Figure 6-32b). Bar grating is made up of main bearing bars that are closely spaced, usually under 2 in. (s_b). The depth of the main bearing bars, d_b, varies from ¾ in. to 2 in. for usual applications with pedestrian and light equipment loading. For applications with heavier point loads, the depth can range from above 2 inches to as high as 6 inches, and these are more commonly used in bridges with truck loading. Cross bars are added for lateral stability, and their spacing commonly varies between 2 in. to 4 in. The thickness of the cross bars usually varies between 1/8 in. to 3/8 in.

For plates that are loaded in the ***flat-wise* direction** resulting in bending about its weaker (i.e., y-y) axis, lateral torsional buckling is *not* a design consideration, and thus the design moment strength is based solely on the plastic moment capacity about the weaker (i.e., y-y) axis (see equations (6-43a) and (6-43b)).

For rectangular bars or plates with a thickness, t, and a depth, d, that are loaded in the ***edge-wise* direction** resulting in bending about its strong (i.e., x-x) axis, the plate bending moment strength accounting for lateral torsional buckling is given as follows (see *AISCM* Section F11):

1. For rectangular bars and plates bent about their major (i.e., x-x) axis and with $\frac{L_b d}{t^2} \leq \frac{0.08E}{F_y}$, the limit state of lateral torsional buckling is not applicable; the design moment capacity and the allowable moment capacity, respectively, are calculated as follows:

$$\phi_b M_n = \phi_b M_p = \phi_b F_y Z_x \leq \phi_b 1.5 F_y S_x \quad \text{for LRFD} \tag{6-52a}$$

$$\frac{M_n}{\Omega_b} = \frac{M_p}{\Omega_b} = \frac{F_y Z_x}{\Omega_b} \leq \frac{1.5 F_y S_x}{\Omega_b} \quad \text{for ASD} \tag{6-52b}$$

2. For rectangular bars and plates bent about their major (i.e., x-x) axis and with $\frac{0.08E}{F_y} < \frac{L_b d}{t^2} \leq \frac{1.9E}{F_y}$, the design moment capacity and the allowable moment capacity, respectively, are calculated as follows:

$$\phi_b M_n = \phi_b C_b \left[1.52 - 0.274\left(\frac{L_b d}{t^2}\right)\left(\frac{F_y}{E}\right)\right] M_y \le \phi_b M_p \quad \text{for LRFD} \tag{6-53a}$$

$$\frac{M_n}{\Omega_b} = \frac{C_b \left[1.52 - 0.274\left(\frac{L_b d}{t^2}\right)\left(\frac{F_y}{E}\right)\right] M_y}{\Omega_b} \le \frac{M_p}{\Omega_b} \quad \text{for ASD} \tag{6-53b}$$

3. For rectangular bars and plates bent about their major (i.e. *x*-*x*) axis and with $\frac{L_b d}{t^2} > \frac{1.9E}{F_y}$, the allowable moment and design moment capacity, respectively, are calculated as follows:

$$\phi_b M_n = \phi_b F_{cr} S_x \le \phi_b M_p \quad \text{for LRFD} \tag{6-54a}$$

$$\frac{M_n}{\Omega_b} = \frac{F_{cr} S_x}{\Omega_b} \le \frac{M_p}{\Omega_b} \quad \text{for ASD} \tag{6-54b}$$

The parameters in the preceding equations (6-52) through (6-54) are as follows:

ϕ_b = 0.9,
Ω_b = 1.67,
L_b = unbraced length,
C_b = moment gradient factor,
d = bar or plate depth (i.e., the plate or bar dimension *perpendicular* to the axis of bending),
t = bar or plate thickness (i.e., the plate or bar dimension *parallel* to the axis of bending),

$$S_x = \text{bar elastic section modulus} = \frac{td^2}{6}$$

$$Z_x = \text{bar plastic section modulus} = \frac{td^2}{4}$$

M_y = yield moment for bending about the strong (i.e., *x*-*x*) axis = $F_y S_x$

$$F_{cr} = \frac{1.9EC_b}{\frac{L_b d}{t^2}}$$

EXAMPLE 6-17

Steel Floor Plate Selection (LRFD)

Determine the required thickness of a steel plate floor deck to support a live load of 125 psf. The plate conforms to ASTM A786 and the span between the supports is 3 ft. 6 inches.

Solution

The dead load of the floor plate will have to be assumed. Assuming a ⅜-in.-thick plate, the dead load is

$$\left(\frac{3/8}{12}\right)(490 \text{ lb.}/\text{ft.}^3) = 16 \text{ psf.}$$

The factored uniform load on the plate floor deck is

$$(1.2)(16) + (1.6)(125) = 220 \text{ psf} = 0.220 \text{ ksf}$$

The service total load = (16 + 125) = 141 psf = 0.141 ksf = 0.141 kips/ft./ft. width
Considering a 1 ft wide strip of steel plate, the factored moment is calculated as

$$M_u = \frac{(1\,\text{ft.})(0.220\,\text{ksf})(3.5\,\text{ft.})^2}{8} = 0.34\,\text{ft.-kips/ft. width} = 4.04\,\text{in.-kips/ft. width}$$

Check the strength using equations (6-43a) and (6-43b) since the plate is used in a flatwise orientation, and therefore, *lateral torsional buckling is not a design consideration.* The design moment capacity is the plastic moment capacity about the weak (or y-y) axis, and the LRFD relation is written as follows (note that the yield strength of the steel plate, F_y = 27 ksi for ASTM A786 steel):

$$\phi_b M_p = \phi_b F_y Z_y \geq M_u$$

That is, $(0.9)(27\,\text{ksi})Z_y \geq 4.04\,\text{in.-kips/ft. width}$

Solving for Z_y yields,

$$Z_y = \frac{4.04}{(0.9)(27)} = 0.167\,\text{in.}^3\text{/ ft. width of plate}$$

Since the plate is bent about its weak or minor axis, the plastic section modulus is given as

$$Z_y = \frac{bt^2}{4} = \frac{(12)\,t^2}{4} = 0.167, \text{ where } b = 1 \text{ ft width} = 12 \text{ inches}$$

Therefore, t_{min} = 0.236 in. = ¼-in.-thick plate

Using *AISCM* Table 3-17a (for flexural-strength-controlled applications), we find that a ¼-in.-thick plate has a factored load capacity of **245 psf**, which is greater than the total applied factored load of 220 psf. It also has less weight than the assumed value of 16 psf.

Check the deflection of the plate:

$$\Delta = \frac{5wL^4}{384EI} \leq \frac{L}{100} \text{ (see the preceding discussions for the AISC recommended deflection limit)}$$

Minimum required moment of inertia, *I* (use service load for deflection),

The total service load = 16 psf + 125 psf = 141 psf = 0.141 ksf

For b = 1 ft., the total service load = 0.141 kips/ft., per foot width of plate

Span of plate = 3.5 ft.

Solving for the minimum required moment of inertia, I, using the deflection equation above gives

$$I = \frac{(100)(5)\left(\frac{0.141}{12}\right)\left[(3.5)(12)\right]^3}{(384)(29{,}000)} = 0.0391\,\text{in.}^4\text{/ ft. width}$$

$$I = \frac{bt^3}{12} = \frac{(12)\,t^3}{12} = 0.0391$$

t_{min} = 0.34 in., therefore, select $\frac{3}{8}$-in. plate.

Using *AISCM* Table 3-17a (for deflection-controlled applications), we find that a 3/8-in.-thick plate has a service load capacity of **190 psf**, which is greater than the demand or applied unfactored load of 141 psf.

Note that a ¼-in.-thick plate is adequate for bending but inadequate for deflections; it has a service load capacity of only 56.4 psf for deflection considerations, which is lower than the applied service load of 141 psf.

EXAMPLE 6-18

Determine the adequacy of a bar grating to support service live loads of either 100 psf or a concentrated load of 300 Ib (these live loads are *not* applied concurrently). The bar grating has 1-1/2" × 3/16" bearing bars (i.e., d = 1.5 in. and t = 0.1875 in.) and is a style 19-W-4 spanning 5 ft. Assume the dead load, including the weight of the grating, is 20 psf. Use F_y = 30 ksi and limit deflections to the smaller of ¼-in. or L/360. The concentrated load is assumed to spread out over a width of 12 inches. Use the ASD method.

Solution

The designation 19-W-4 is commonly used in the industry to identify grating. The '19' indicates the spacing, s_b, of the bearing bars in 16ths of an inch, which in this case would be 19/16 in. or 1.19 in. The '4' represents the spacing of the cross bars in inches. This 4 in. spacing is the unbraced length, $L_{b,}$ of the main bearing bars.

Section properties of one bar (depth, d = 1.5 in. and thickness, t = 0.1875 in.):

$$S_x = \frac{td^2}{6} = \frac{(0.1875)(1.5)^2}{6} = 0.0703 \text{in.}^3$$

$$Z_x = \frac{td^2}{4} = \frac{(0.1875)(1.5)^2}{4} = 0.105 \text{in.}^3$$

$$I_x = \frac{td^3}{12} = \frac{(0.1875)(1.5)^3}{12} = 0.0527 \text{in.}^4$$

Assuming a 12 in. width of load spread for the concentrated load, the section properties **per lineal foot**, with the spacing of the bearing bars, s_b = 19/16 in. = 1.19 in., are as follows:

$$S_{x12} = S_x\left(\frac{12 \text{ in.}}{s_b}\right) = \left(0.0703 \text{ in.}^3\right)\left(\frac{12 \text{ in.}}{1.19 \text{ in.}}\right) = 0.709 \text{ in.}^3\text{/ft., width}$$

$$Z_{x12} = Z_x\left(\frac{12 \text{ in.}}{s_b}\right) = \left(0.105 \text{in.}^3\right)\left(\frac{12 \text{ in.}}{1.19 \text{ in.}}\right) = 1.06 \text{ in.}^3\text{/ft., width}$$

$$I_{x12} = I_x\left(\frac{12 \text{ in.}}{s_b}\right) = \left(0.0527 \text{ in.}^3\right)\left(\frac{12 \text{ in.}}{1.19 \text{ in.}}\right) = 0.532 \text{ in.}^3\text{/ft., width}$$

Check if lateral-torsional buckling is a factor (see equations (6-52) through (6-54)):

L_b = 4 in. (unbraced length of the main bearing bars)

d = 4 in.

t = 0.1875 in.

$$\frac{L_b d}{t^2} = \frac{(4)(1.5)}{(0.1875)^2} = 170$$

$$\frac{0.08E}{F_y} = \frac{0.08(29000)}{(30)} = 77.3$$

$$\frac{1.9E}{F_y} = \frac{1.9(29000)}{(30)} = 1837$$

Since $\frac{0.08E}{F_y} < \frac{L_b d}{t^2} \leq \frac{1.9E}{F_y}$, lateral torsional buckling must be considered.

Therefore, use equation (6 - 53b) to compute the allowable moment as follows :

The plastic moment, $M_p = Z_{x12} F_y = (1.06 \text{ in.}^3)(30 \text{ ksi}) = 31.8 \text{ in.-kip/ft.width}$

The yield moment, $M_y = S_{x12} F_y = (0.709 \text{ in.}^3)(30 \text{ ksi}) = 21.27 \text{ in.-kip/ft.width}$

Using equation (6-53b), the allowable moment is calculated as

$$\frac{M_n}{\Omega_b} = \frac{(1.0)\left[1.52 - 0.274(170)\left(\frac{30}{29{,}000}\right)\right](21.27)}{1.67} \leq \frac{31.8}{1.67}$$

That is, $\frac{M_n}{\Omega_b} = \mathbf{18.75 in.\text{-}kip/ft.width} \leq 19.04 \text{ in.-kip/ft.width}$

Therefore, $\frac{M_n}{\Omega_b} = 18.75 \text{ in.-kip/ft.width} = \left(\frac{18.75}{12}\right)(1000) = \mathbf{1563 ft.\text{-}Ibs/ft.width}$

Determine the applied moments per unit width:

Uniform loads:

$w_D = (20 \text{ psf})(1 \text{ ft}) = 20 \text{ plf}$

$w_L = (100 \text{ psf})(1 \text{ ft}) = 100 \text{ plf}$

Span of plate, L = 5 ft.

$$M_{s1} = \frac{w_{D+L}L^2}{8} = \frac{(20+100)(5)^2}{8} = 375 \text{ ft.-lb/ft. width}$$

Concentrated load plus uniform dead load:

P = 300 Ib (this concentrated load is spread out over a 1 ft., width of grating)

$w_D = 20 \text{ plf}$

Span of plate, $L = 5$ ft

$$M_{s2} = \frac{PL}{4} + \frac{w_D L^2}{8} = \frac{(300)(5)}{4} + \frac{(20)(5)^2}{8} = 438 \text{ ft.-lb/ft. width (controls)}$$

$$\frac{M_n}{\Omega_b} = 1563 \text{ ft.-lb/ft. width} > 438 \text{ ft.-lb/ft. width}$$

Therefore, the grating is adequate for bending.

Check deflection:

$w_L = 100 \text{ plf}$

The deflection due to the uniform live load is

$$\Delta_{L1} = \frac{5wL^4}{384EI_x} = \frac{5(100/12)[(5)(12)]^4}{384(29x10^6)(0.532)} = 0.091 \text{in.}$$

The deflection due to the concentrated live load is

$$\Delta_{L2} = \frac{PL^3}{48EI_x} = \frac{(300)[(5)(12)]^3}{48(29x10^6)(0.532)} = 0.087 \text{ in.}$$

The uniform live load deflection controls, so

$\Delta_L = 0.091 \text{in.}$

The maximum allowable deflection is the larger of ¼-in. or L/360 = (5 ft.)(12)/360 = 0.167 in. (controls). Since 0.091 in. < 0.167 in., therefore the grating is adequate for deflection. **Note that the 12 in. width of spread assumed for the concentrated load may be unconservative.** In design practice, an effective spread width of twice the spacing of the main bearing bars is recommended for concentrated loads; this assumes the concentrated load will be spread over three main bearing bars.

6-14 CANTILEVERED GIRDER SYSTEMS

The typical framing system for gravity loads can be made more economical if the girders along the column lines could be framed in such a way that they cantilever over the columns and thus allow some of the bending moment that would occur at the midspan of the girder to be distributed to the girder over their supports, thus potentially allowing for smaller girder sizes.

Consider the beams shown in Figure 6-33, both with uniform loading and with different support conditions. In the simple span case (Figure 6-33a), a series of simply supported beams support a uniform gravity load, and thus each beam has a maximum bending moment at midspan varying to a zero moment at the supports. While this type of framing is common and reasonable, the second case shows an alternate framing system that reduces the maximum positive moment in the beams. In this second case (Figure 6-33b), the beams are cantilevered over the supports some distance to allow balancing of the maximum positive and negative bending moments. In the simple span case, the maximum moment in the beam is 18 kips-ft., whereas a cantilevered system reduces this maximum positive moment down to 13.8 kips-ft., which could lead to a smaller beam size. This framing system, also called the "Gerber roof framing system," is often used in the roofs of one-story buildings where the top of the column cap plate is underneath the bottom flange of the roof beam to allow the beam to cantilever over the top of the column (see Figure 6-34) and thus obviate the need for an expensive moment connection for the beam. Note that the columns in cantilevered framed roof construction have to be properly braced to ensure lateral stability in the direction perpendicular to the beams or girders framing over the columns. The free end of the cantilever or overhang (preferably the tension flange) should also be laterally braced, in addition to laterally bracing the compression flange (i.e., bottom flange) of the beam or girder at the column supports. The girder must be laterally supported at the column supports to prevent torsional rotation or twisting of the girder about its longitudinal axis. A perpendicular beam framing into the girder at the column line can be used to achieve this lateral bracing, and preferably, stiffener plates should be provided in the girder on both sides of the web at the column support.

Similarly, during demolition or deconstruction of structures with cantilevered construction, the engineer of record should ensure that the lateral stability of the structure is not compromised at any stage of the demolition process. Several tragic collapses have occurred in these types of structures during demolition because of faulty sequencing of the demolition process that left the structure unstable and consequently led to collapse.

Table 3-21b in the AISC Manual is a design aid that provides dimensions and design moments for idealized framing systems that use cantilevered construction. This table assumes equal spans and equal spacing between load points or uniformly loaded conditions. The table provides solutions for moments and reactions for different beam configurations and cantilevered span dimensions, x, as shown in Figure 6-35. Note that the cantilevered span dimensions in *AISCM* Table 3-21b are denoted with the symbols "a" through "f." Example 6-19 illustrates the use of this design aid.

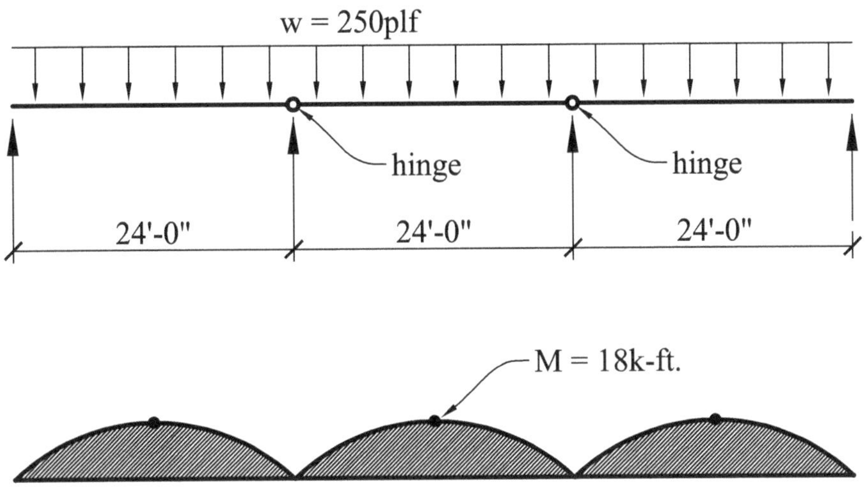

FIGURE 6-33a Loading and moment diagram for simple span beams or girders

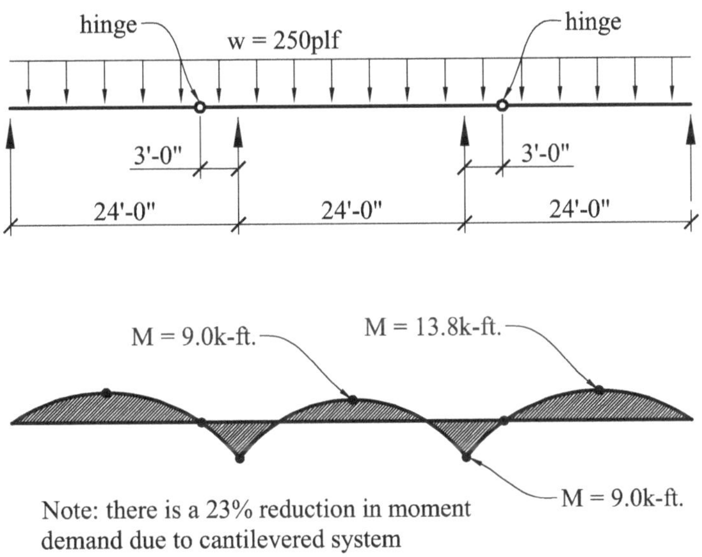

FIGURE 6-33b Loading and moment diagram for cantilevered beams or girders

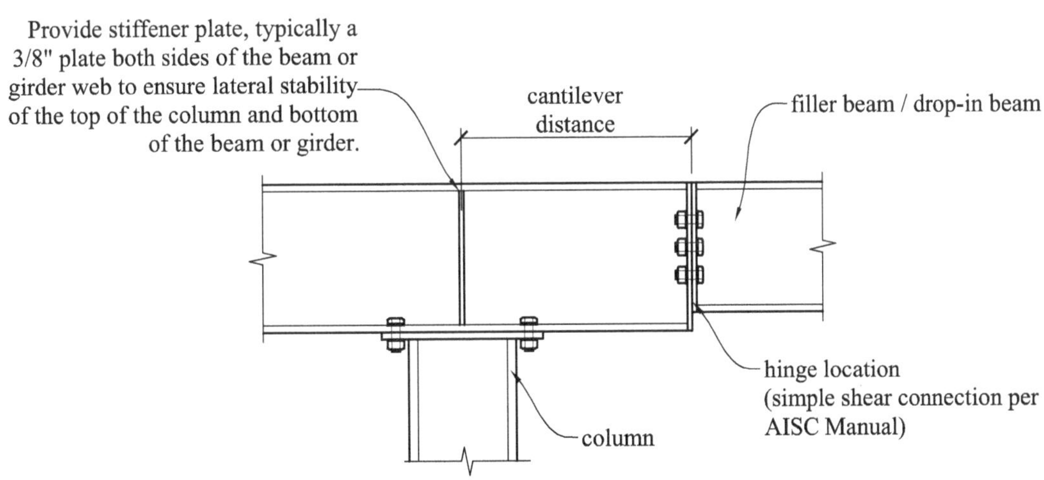

FIGURE 6-34 Loading and moment diagram for cantilevered beams or girders (Gerber System)

EXAMPLE 6-19

Cantilevered beam system

Determine the ideal cantilever distance "*x*" for the 3-span beam shown in Figure 6-35 and draw the shear and moment diagrams. Use *AISCM* Table 3-21b.

Solution

The following are obtained from *AISCM* Table 3-21b for the uniformly loaded case:

Total span of beam between adjacent supports, L = 24 ft.
Cantilever dimension, b = x = 0.125L = (0.125)(24 ft.) = 3 ft.
Total load with the span of the beam between adjacent supports,
P = (250 plf)(24 ft.) = 6000 lb.

The moments are calculated from *AISCM* Table 3-21b as follows:
$M_2 = 0.096PL$ = (0.096)(6000)(24) = 13,781 ft.-lb.
$M_3 = 0.063PL$ = (0.063)(6000)(24) = 9,000 ft.-lb.

The beam support reactions are calculated from *AISCM* Table 3-21b as follows:
R_C = 0.438P = (0.438)(6000) = 2625 lb.
R_D = 1.063P = (1.063)(6000) = 6375 lb.
R_{Dm} = (250)(24)/2 = 3000 lb.

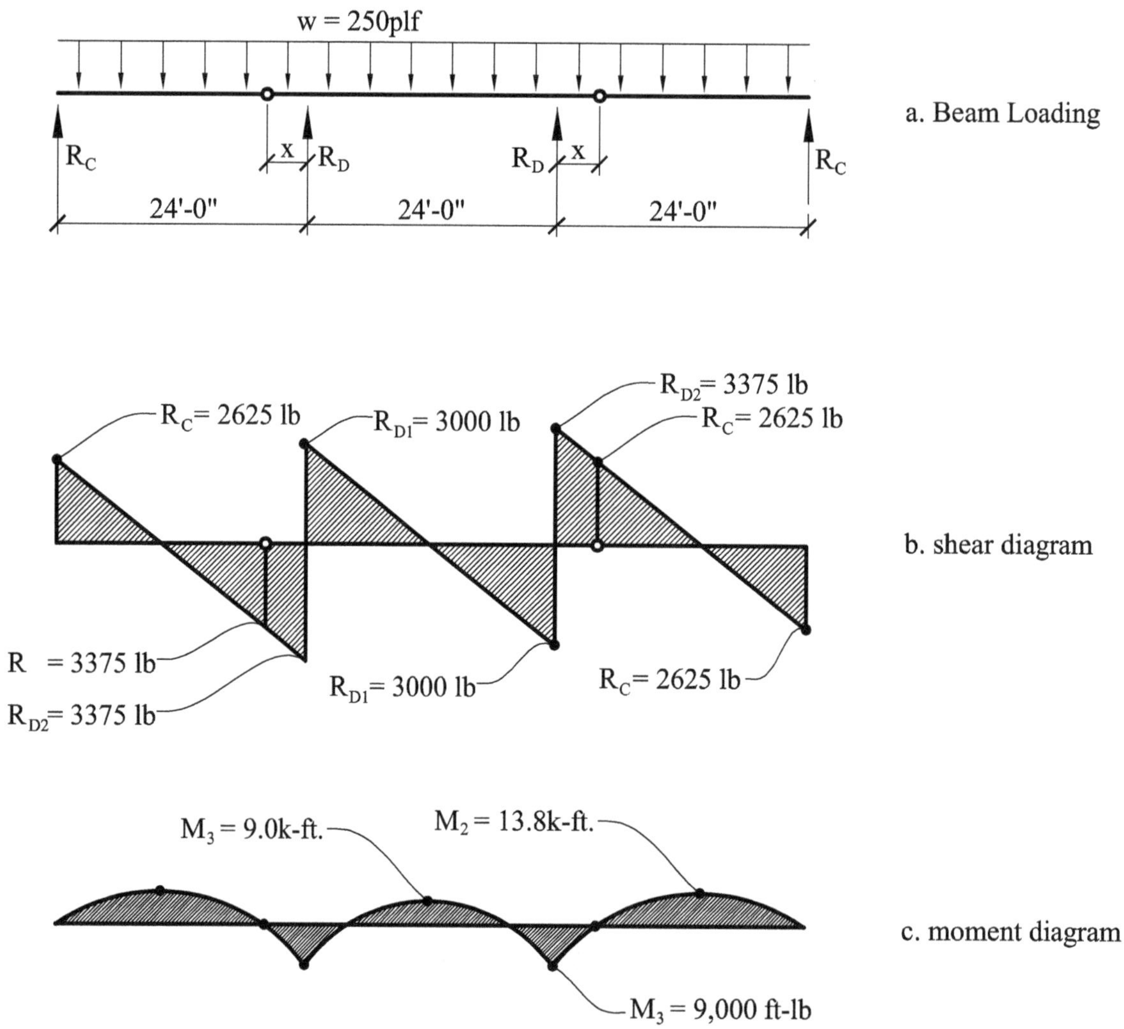

FIGURE 6-35 Loading and moment diagram for cantilevered beams

Note that the cantilever dimensions obtained from *AISCM* Table 3-21b is an ideal value, but other values could be used depending on the actual conditions. For example, if the beam spans were not equal, then an iterative process would likely be needed to arrive at the optimal cantilever distance. On the other hand, the values obtained from the *AISCM* tables would prove to be close to what is ideal, so the use of this table can be helpful in the initial layout of cantilevered framing systems.

6-15 FLEXURAL STRENGTH OF WT-SECTIONS AND DOUBLE ANGLES

Tee-sections (i.e., WT) and double-angle sections are often used as the top and bottom chords of trusses (see Figure 6-36). WT sections are typically used in long-span roof trusses spanning over large column-free areas or in transfer floor trusses supporting discontinued columns over large column-free areas, such as auditoriums, ballrooms, exhibition halls, and atrium areas. Trusses with double-angle chord members are used to support lighter loads over shorter spans. When

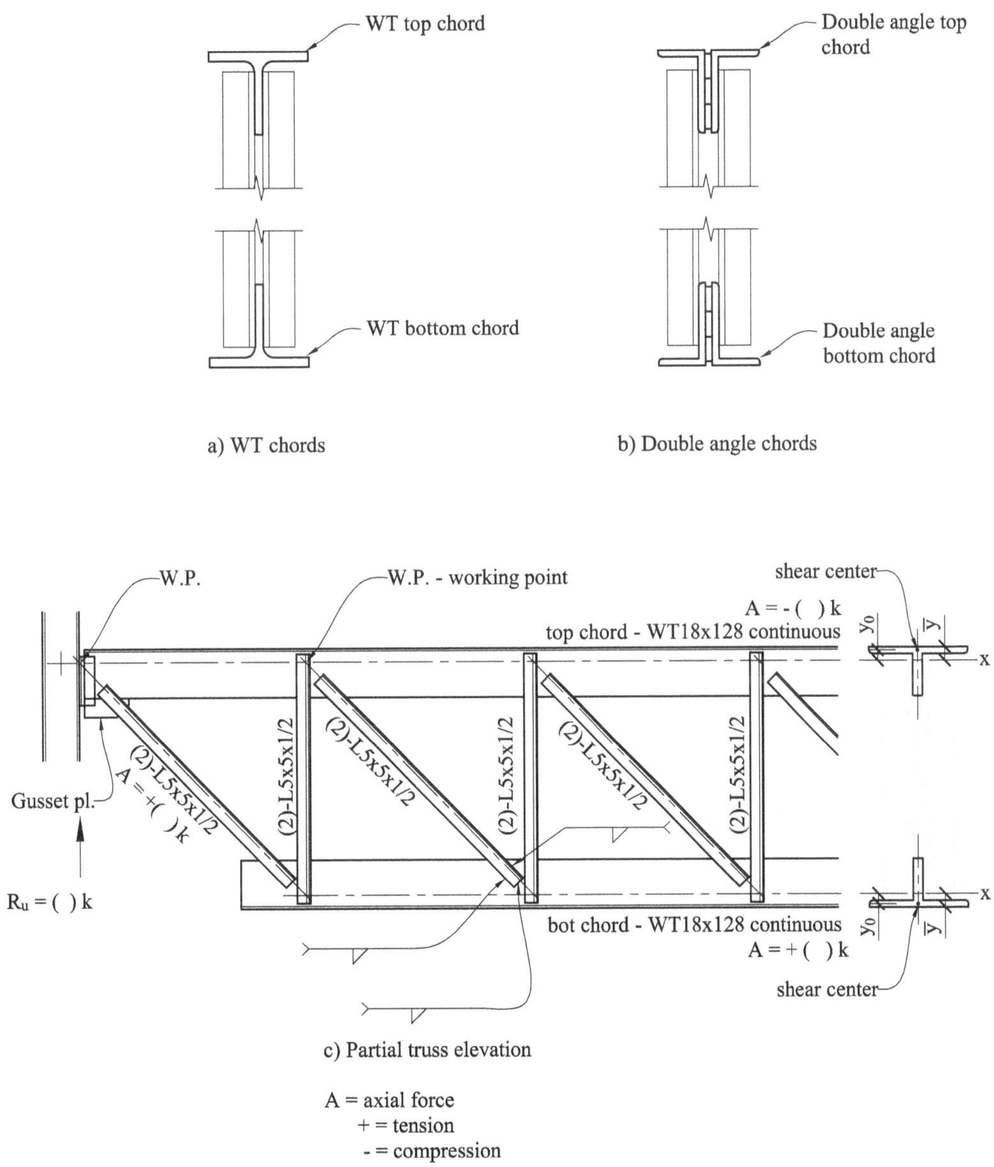

FIGURE 6-36 WT sections and double angles used as top and bottom chords of trusses

used as top chords, the WT and double angle sections are subjected to both axial compression forces and flexure due to the roof or floor uniform loads. When used as bottom chords, they are subjected to combined axial tension plus flexure, where the bending moment is due to dead loads from the ceiling, and the mechanical and electrical equipment. The bottom chords of trusses may also support the dead and live loads from suspended catwalks and from theater stage rigging loads (the theater stage rigging loads to be used in the design of rigging trusses should be obtained from the theater consultant). The axial tension and compression capacity of WT sections and double angles can be determined using the methods presented in Chapters 4 and 5, but the flexural strength of these sections will be discussed here.

For WT sections and double angles bending about the X-X axis (which is perpendicular to the Y-Y axis or the axis of symmetry), the potential limit states to be considered are as follows:

1. The limit states of yielding (where the section attains the plastic moment, M_p),
2. The limit state of lateral torsional buckling,
3. The limit state of flange local buckling,
4. The limit state of local buckling of the stem of the WT section and the legs of the double angles.

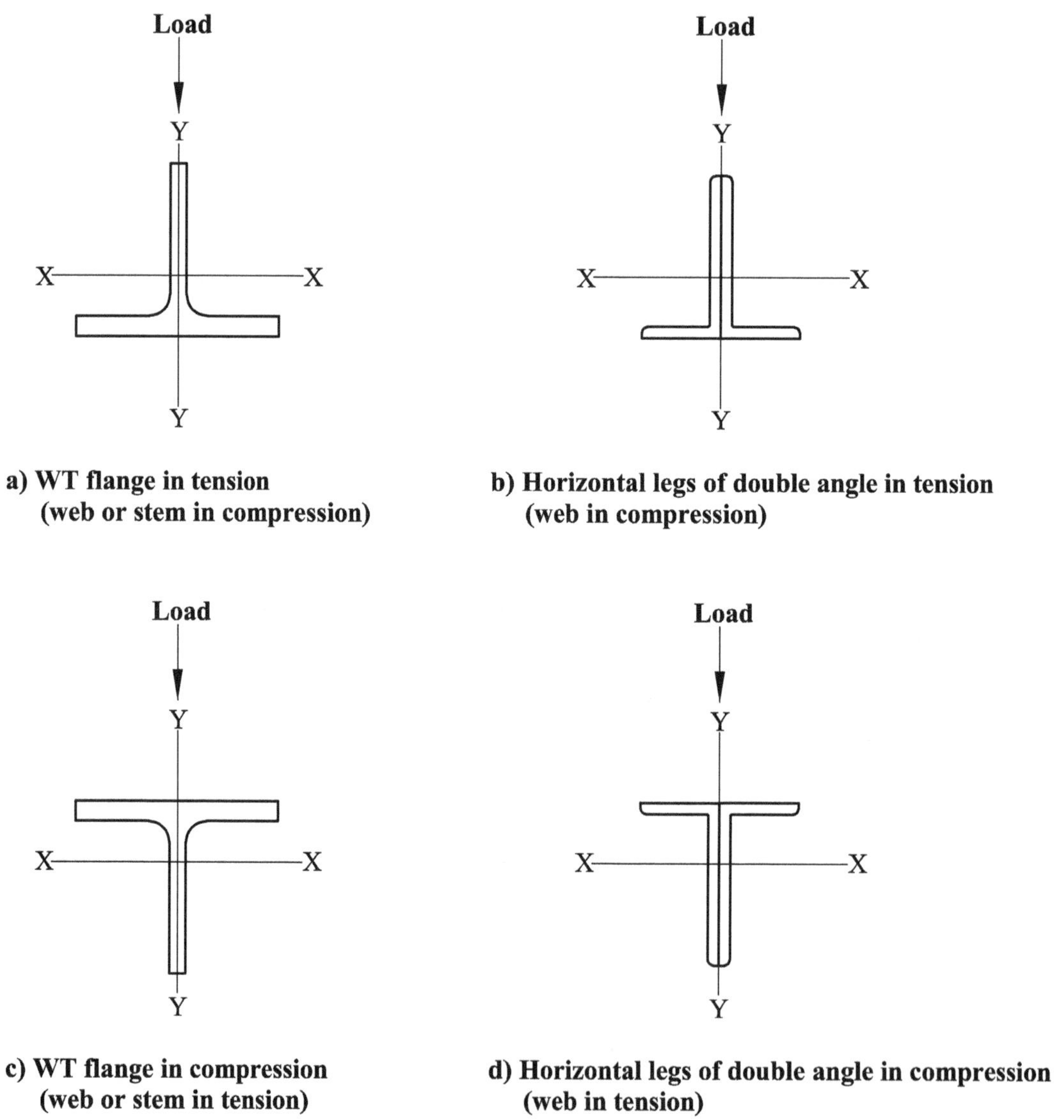

FIGURE 6-37 Loading Configurations for WT and Double Angles

The moment capacity will depend on whether the flanges of the WT and the legs of the double angles are in tension or compression (see Figure 6-37 for the loading configuration). The strength reduction factor, ϕ_b , and safety factor, Ω_b , are 0.90 (for LRFD) and 1.67 (for ASD), respectively. The nominal moment strength for each of the limit states are as follows (see Section F9 of the AISC Specifications):

1. **Yielding Limit State**

A. For stems of the WT section in **tension,** that is, when the flange of the WT section is in compression **(e.g., Figure 6-37c)**, the nominal moment strength for bending about the X-X axis from *AISC* Specifications Section F9.1a is

$$M_n = F_y Z_x \leq 1.6 F_y S_x \tag{6-55}$$

B. For web legs of double angles in **tension,** that is, when the two flange legs of the double angles are in compression **(e.g., Figure 6-37d)**, use equation (6-55) to determine the nominal moment strength.

C. For stems of the WT section in **compression,** that is, when the flange of the WT is in tension **(e.g., Figure 6-37a)**, the nominal moment strength (see *AISC* Specifications Section F9.1b) is:

$$M_n = F_y S_x \tag{6-56}$$

D. For web legs of double angles in **compression,** that is, when the two flange legs of the double angles are in tension **(e.g., Figure 6-37b)**, the nominal moment strength (see *AISC* Specifications Section F9.1c) is:

$$M_n = 1.5 F_y S_x \tag{6-57}$$

Note that the latter two cases above - where the stem of the **WT** or the web legs of the **double angles** are in **compression (e.g., Figures 6-37a and 6-37b)** - are more critical than the former two cases.

2. **Lateral-torsional Buckling Limit State**

A. For stem or web of WT section and web legs of double angles in *tension* (**e.g., Figures 6-37c and 6-37d**):

(i) When the unbraced length of the compression flange, $L_b \leq L_p$, the WT section and the double angles will not fail by lateral torsional buckling, so the limit state of lateral torsional buckling is not applicable for this case (see *AISC* Specification Section F9.2(a)(1)).

(ii) For $L_p \leq L_b \leq L_r$, the nominal moment strength (see *AISC* Specifications Section F9.2(a)(2)) is:

$$M_n = \left[M_p - (M_p - M_y) \left(\frac{L_b - L_p}{L_r - L_p} \right) \right] \tag{6-58}$$

(iii) For $L_b > L_r$, the nominal moment strength (see *AISC* Specifications Section F9.2(a)(3)) is:

$$M_n = M_{cr} \tag{6-59}$$

Where,

$$L_p = 1.76 r_y \sqrt{\frac{E}{F_y}} \qquad (6-60)$$

$$L_r = 1.95\left(\frac{E}{F_y}\right)\frac{\sqrt{I_y J}}{S_x}\sqrt{2.36\left(\frac{F_y}{E}\right)\frac{d S_x}{J}+1} \qquad (6-61)$$

$$M_{cr} = \frac{1.95E}{L_b}\sqrt{I_y J}\left(B+\sqrt{1+B^2}\right) \qquad (6-62)$$

$$B = 2.3\left(\frac{d}{L_b}\right)\sqrt{\frac{I_y}{J}} \qquad (6-63)$$

d = depth of WT section or width of web leg of the double angles in *tension*, inches.

b. WT stems and web legs of double angles in **compression** within the unbraced length, L_b; that is, when the flange of the WT and the two flange legs of the double angles are in tension (**e.g., Figure 6-37a and 6-37b**).
(i) For WT stems in **compression**:
The nominal moment strength (see *AISC* Specifications Section F9.2(b)(1)) is calculated as follows:

$$M_n = M_{cr} \leq F_y S_x \qquad (6\text{-}64)$$

Where, M_{cr} is obtained from equation (6-62) and B is obtained from equation (6-67).
(ii) For web legs of double angles in flexural **compression**, the nominal moment strengths in accordance with *AISC* Specifications Section F9.2(b)(2) are to be calculated using *AISC* Specifications Equations F10-2 and F10-3 as follows:

$$When\ \frac{M_y}{M_{cr}} \leq 1.0,\ M_n = \left(1.92 - 1.17\sqrt{\frac{M_y}{M_{cr}}}\right)M_y \leq 1.5M_y \qquad (6-65)$$

$$When\ \frac{M_y}{M_{cr}} > 1.0,\ M_n = \left(0.92 - \frac{0.17M_{cr}}{M_y}\right)M_{cr} \leq 1.5M_y \qquad (6-66)$$

Where, M_{cr} is obtained from equation (6-62),

$$B = -2.3\left(\frac{d}{L_b}\right)\sqrt{\frac{I_y}{J}} \qquad (6-67)$$

$$M_y = F_y S_x \qquad (6-68)$$

d = depth of WT section or width of the web legs of the double angles in *compression*, in.

3a. Flange Local Buckling - WT flange

A. For WT sections with **compact flange** in compression due to bending (see Table 6-2), flange local buckling does *not* occur, so flange local buckling is not applicable.

B. For WT sections with **non-compact** flange in compression due to bending (see Table 6-2), the nominal moment strength (see *AISC* Specifications Section F9.3(a)(2)) is:

$$M_n = \left[M_p - (M_p - 0.7F_y S_{xc}) \left(\frac{\lambda - \lambda_{pf}}{\lambda_{rf} - \lambda_{pf}} \right) \right] \leq 1.6M_y \qquad (6-69)$$

C. For WT sections with a **slender** flange in compression due to bending, the nominal moment strength (see *AISC* Specifications Section F9.3(a)(3)) is:

$$M_n = \frac{0.7ES_{xc}}{\left(\frac{b_f}{2t_f} \right)^2} \qquad (6-70)$$

Where,

S_{xc} = elastic section modulus of the WT section with reference to the *compression* flange, in.^{3}

$\lambda = \frac{b_f}{2t_f}$

λ_{pf} = the slenderness limit for a compact flange from *AISCM* Table B4.1b

λ_{rf} = the slenderness limit for a non-compact flange from *AISCM* Table B4.1b

3b. Flange Local Buckling - flange legs of the double angles

A. For **compact** flange legs of double angles that are in compression due to bending (see Table 6-2), local buckling of the legs does *not* occur, so leg local buckling is not applicable.

For other conditions below (see *AISC* Specifications Section F9.3(b)), the nominal moment strengths are obtained from Equations F10-6 through F10-8 from *AISC* Specifications Section F10.3 as follows in Parts B and C below:

B. For double angles with **non-compact** flange legs in compression due to bending (see Table 6-2), the nominal moment strength (see *AISC* Specifications Section F10.3(b)) is:

$$M_n = F_y S_c \left[2.43 - 1.72 \left(\frac{b}{t} \right) \sqrt{\frac{F_y}{E}} \right] \qquad (6-71)$$

C. For double angles with **slender** flange legs in compression due to bending (see Table 6-2), the nominal moment strength (see *AISC* Specifications Section F10.3(c)) is:

$$M_n = F_{cr} S_c \qquad (6-72)$$

Where,

S_c = elastic section modulus of the double angles about the axis of bending with respect to the toe in compression, in.^{3}

$$F_{cr} = \frac{0.71E}{\left(\frac{b}{t} \right)^2} \qquad (6-73)$$

b = full width of the double angle leg in compression, in.

4a. Local Buckling of WT stems

The nominal moment strength (see *AISC* Specifications Section F9.4(a)) is

$$M_n = F_{cr} S_x \qquad (6-74)$$

Where,

S_x = elastic section modulus about the x-axis, in.^{3}

F_{cr} = the critical stress which is calculated as follows:

$$When\ \frac{d}{t_w} \le 0.84\sqrt{\frac{E}{F_y}}, \qquad F_{cr} = F_y \tag{6-75}$$

$$When\ 0.84\sqrt{\frac{E}{F_y}} < \frac{d}{t_w} \le 1.52\sqrt{\frac{E}{F_y}}, \qquad F_{cr} = \left(1.43 - 0.515\frac{d}{t_w}\sqrt{\frac{F_y}{E}}\right)F_y \tag{6-76}$$

$$When\ \frac{d}{t_w} > 1.52\sqrt{\frac{E}{F_y}}, \qquad F_{cr} = \frac{1.52E}{\left(\frac{d}{t_w}\right)^2} \tag{6-77}$$

4b. Local Buckling of the web legs of double angles

For this failure mode, the strength of the double angles is assumed to be equal to the strength of two single angles.

A. For **compact** web legs of double angles in *compression* due to bending, local buckling of the legs does *not* occur; therefore, local buckling of the web legs is *not* applicable.

B. For double angles with **non-compact** web legs in *compression* due to bending (see Table 6-2), the nominal moment strength (see *AISC* Specifications Section F10.3(b)) is:

$$M_n = F_y S_c \left[2.43 - 1.72\left(\frac{b}{t}\right)\sqrt{\frac{F_y}{E}}\right] \tag{6-71}$$

C. For double angles with **slender** web legs in compression due to bending (see Table 6-2), the nominal moment strength (see *AISC* Specifications Section F10.3(c)) is:

$$M_n = F_{cr} S_c \tag{6-72}$$

Where,

S_c = elastic section modulus of the double angles at the toe in compression with respect to the axis of bending, in.^{3}

$$F_{cr} = \frac{0.71E}{\left(\frac{b}{t}\right)^2} \tag{6-73}$$

b = full width of the double angle leg in compression, in.

EXAMPLE 6-20

WT Section in Bending

Calculate the nominal moment capacity, M_n, of a WT 18 × 128 section that is used as the top and bottom chords of a steel transfer truss (see Figure 6-36c). The WT section is made from ASTM A992 Grade 50 steel and has an unbraced length of 6 ft. 3 in (the distance between the truss panel points). Calculate the LRFD design moment capacity, ϕM_n, and the ASD allowable moment, M_a.

Solution

$F_y = 50$ ksi

Unbraced length, $L_b = 6$ ft. 3 in. = 75 inches

The section properties for WT 18x128 from *AISCM* Part 1 are as follows:

$d = 18.7$ in.

$S_x = 87.4$ in.^{3}

$Z_x = 156$ in.^{3}

$I_y = 264$ in.^{4}

$J = 26.4$ in.^{4}

$r_x = 5.66$ in.

$r_y = 2.65$ in.

$$\frac{b_f}{2t_f} = 3.53$$

$$\frac{d}{t_w} = 19.5$$

At the top chord of a transfer truss, the flange of the WT18x128 will be in **compression between the panel points** (see Figure 6-36c) due to flexure from the gravity loads on the top chord, but the flange of the WT will be in tension at the panel points due to the continuity of the top chord.

At the bottom chord of the transfer truss, the flange of the WT will be in tension between the panel points due to flexure from the gravity loads on the bottom chord (see Figure 6-37a) and in compression at the panel points due to the continuity of the bottom chord.

1. **Yielding Limit State**

   For stems of WT section in **compression**, $M_n = F_y S_x$ = (50ksi)(87.4 in^3) = 4370 in.-kips

   For stems of WT section in **tension,** that is when the flange of the WT section is in compression (e.g., see Figure 6-37c),

$$M_n = F_y Z_x = (50\,\text{ksi})(156\,\text{in.}^3) = 7800\ \text{in.-kips}$$

$$\leq 1.6 F_y S_x = (1.6)(50\,\text{ksi})(87.4\,\text{in.}^3) = 6992\ \text{in.-kips}$$

$$\therefore\ \mathbf{M_n = 6992\ in.\text{-}kips}$$

2. **Lateral-torsional Buckling Limit State**

   For stems or webs of the WT section in **tension**:

From equation (6-60), $L_p = 1.76 r_y \sqrt{\dfrac{E}{F_y}} = 1.76(2.65\,\text{in.})\sqrt{\dfrac{29{,}000}{50}} = 112.3$ in.

From equation (6-61),

$$L_r = 1.95\left(\frac{E}{F_y}\right)\frac{\sqrt{I_y J}}{S_x}\sqrt{2.36\left(\frac{F_y}{E}\right)\frac{d S_x}{J} + 1}$$

$$= 1.95\left(\frac{29{,}000}{50}\right)\frac{\sqrt{(264)(26.4)}}{87.4}\sqrt{2.36\left(\frac{50}{29{,}000}\right)\frac{(18.7)(87.4)}{26.4} + 1} = 318\,\text{in.}$$

From equation (6-63), $B = 2.3\left(\frac{d}{L_b}\right)\sqrt{\frac{I_y}{J}} = 2.3\left(\frac{18.7}{75}\right)\sqrt{\frac{264}{26.4}} = 1.813$

From equation (6-62),

$$M_{cr} = \frac{1.95E}{L_b}\sqrt{I_y J}\left(B + \sqrt{1 + B^2}\right)$$

$$= \frac{1.95(29{,}000)}{75}\sqrt{(264)(26.4)}\left(1.813 + \sqrt{1 + (1.813)^2}\right) = 244{,}455 \text{ in.-kips}$$

Since the maximum unbraced length of the compression flange of the WT section, $L_b = 75$ in., is less than $L_p = 112.3$ in., the WT section will *not* fail by lateral torsional buckling; therefore, the limit state of lateral torsional buckling is **not applicable** for the case when the webs of the WT section are in **tension**.

B(i). WT stems in **compression** within the unbraced length, L_b:

From equation (6-67), $B = -2.3\left(\frac{d}{L_b}\right)\sqrt{\frac{I_y}{J}} = 2.3\left(\frac{18.7}{75}\right)\sqrt{\frac{264}{26.4}} = -1.813$

Substituting B = -1.813 into equation (6-62) yields,

$$M_{cr} = \frac{1.95E}{L_b}\sqrt{I_y J}\left(B + \sqrt{1 + B^2}\right)$$

$$= \frac{1.95(29{,}000)}{75}\sqrt{(264)(26.4)}\left(-1.813 + \sqrt{1 + (-1.813)^2}\right) = 16{,}209 \text{ in.-kips}$$

From equation (6-64), $M_n = M_{cr} = 16{,}209$ in.-kips

$$\leq F_y S_x = (50\,\text{ksi})(87.4\,in.^3) = 4370 \text{ in.-kips},$$

$\therefore$ **M_n = 4370 in.-kips**

3a. Local Buckling of the WT flange

A. For WT sections with **compact flange** in compression due to bending, flange local buckling does *not* occur, so flange local buckling is not applicable.

The slenderness limit for a compact flange from *AISCM* Table B4.1b (Case 10) or Table 6-2 is

$$\lambda_p = 0.38\sqrt{\frac{E}{F_y}} = 0.38\sqrt{\frac{29{,}000}{50}} = 9.15$$

$\frac{b_f}{2t_f} = 3.53$ (from *AISCM* Part 1) < 9.15, therefore the WT flange is compact.

Since the WT flange is compact, flange local buckling is **not applicable**.

4a. Local Buckling of WT stems

Calculate the design parameters from equations (6-75) through (6-77).

$$0.84\sqrt{\frac{E}{F_y}} = 0.84\sqrt{\frac{29{,}000}{50}} = 20.23$$

$$1.52\sqrt{\frac{E}{F_y}} = 1.52\sqrt{\frac{29{,}000}{50}} = 36.61$$

From *AISCM* Part 1 for WT18x128, $\frac{d}{t_w} = 19.5 < 20.23$,

Therefore $F_{cr} = F_y$ (see equation (6-75))

From equation (6–74), $M_n = F_{cr}S_x = F_yS_x = (50\,\text{ksi})(87.4\,\text{in.}^3)$

$= \mathbf{4370\ in.\text{-}kip}$

The smallest value of all the nominal moment capacities, M_n, calculated (i.e., 6992 in.-kips; 4370 in.-kips; and 4370 in.-kips) is used.

Therefore, M_n **= 4370 in.-kip**

LFRD Solution:

The design moment capacity, $\phi_b M_n = (0.9)(4370) = 3933$ in.-kips = 328 ft.-kips

ASD Solution:

The allowable moment is

$$M_a = \frac{M_n}{\Omega_b} = \frac{4370\ \text{in.-kips}}{1.67} = 2617\ \text{in.-kips} = 218\ \text{ft.-kips}$$

6-16 FLEXURAL STRENGTH OF SQUARE AND RECTANGULAR HSS, AND BOX-SHAPED SECTIONS

Hollow structural sections (HSS) and box-shaped sections (see Figure 6-37e) are commonly used as columns and beam-columns in various structural systems due to their higher torsional rigidity and their greater lateral torsional stability compared to open sections such as I-shapes and channels. HSS members are also frequently used as curved beams or as lintels over large spans where the lintel beam is laterally unbraced between supports. When used as curved beams or as lintels, HSS members are subjected to torsional moments and torsional rotation or twist from

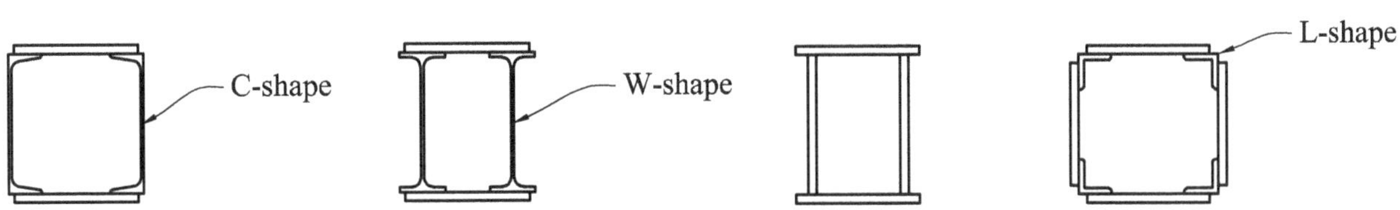

FIGURE 6-37e Built-up Box Sections

the eccentricity of the supported loads (e.g., the block or cold-formed steel backup wall is usually centered on the lintel while the brick veneer or cladding is usually eccentric with respect to the longitudinal axis of the lintel beam). Built-up box sections can also be used as flexural members in a transfer arch to support loads from existing columns in a structural retrofit where the existing column may need to be cut off below a particular level to provide additional headroom [21]. As discussed earlier in this chapter, closed sections such as HSS and box-shaped members are far more effective at resisting twist and torsion than open sections such as W-shapes and channels. The use of wide flange shapes and channels to resist torsion should be avoided since open steel sections are very inefficient in resisting torsion.

Whenever possible, the thickness of the flanges and the webs of built-up box sections should be chosen so that a compact section is assured (see Table 6-2 or *AISCM* Table B4.1b (Cases 17, 19 and 21)). For built-up box sections, a compact web is assured when $\frac{h}{t_w} \le 2.42\sqrt{\frac{E}{F_y}}$ and a compact flange is assured when $\frac{b}{t_f} \le 1.12\sqrt{\frac{E}{F_y}}$, where h is the clear height of the web, and b is the clear width of the flange, and t_w and t_f are the web and flange thicknesses, respectively.

The design strength in flexure of HSS and box-shaped members is covered in Section F7 of the AISC Specifications [1] as follows:

1. **Yielding Limit State**

   The nominal moment strength is

   $$M_n = M_p = F_y Z \tag{6-78}$$

   Where Z = plastic section modulus about the axis of bending.

2. **Flange Local Buckling Limit State**

   a. For *compact* HSS or box-shaped sections (see Table 6-2), the limit state of flange local buckling does *not* occur and is therefore not applicable.

   b. For sections with *non-compact flanges*, the nominal moment strength is

   $$M_n = M_p - (M_p - F_y S)\left(\frac{\lambda - \lambda_{pf}}{\lambda_{rf} - \lambda_{pf}}\right) \le M_p \tag{6-79}$$

   Where,
   S = elastic section modulus about the axis of bending,
   b = width of the compression flange of the HSS or box section as defined in Table 6-2.

   c. For sections with *slender flanges*, the nominal moment strength is

   $$M_n = F_y S_e \tag{6-80}$$

   Where,
   S_e = effective elastic section modulus about the axis of bending determined using the effective width, b_e, of the compression flange, where,

   $$\text{For } \textbf{HSS}: \quad b_e = 1.92 t_f \left(\sqrt{\frac{E}{F_y}}\right)\left(1 - \frac{0.38}{b/t_f}\sqrt{\frac{E}{F_y}}\right) \le b \tag{6-81}$$

   $$\text{For } \textbf{Box sections}: \quad b_e = 1.92 t_f \left(\sqrt{\frac{E}{F_y}}\right)\left(1 - \frac{0.34}{b/t_f}\sqrt{\frac{E}{F_y}}\right) \le b \tag{6-82}$$

3. Web Local Buckling Limit State

a. For *compact* HSS or box-shaped sections (see Table 6-2), the limit state of web local buckling does *not* occur and is therefore not applicable.

b. For sections with *non-compact webs*, web local buckling may occur; the nominal moment strength is

$$M_n = M_p - (M_p - F_y S)\left(\frac{\lambda - \lambda_{pw}}{\lambda_{rw} - \lambda_{pw}}\right) \leq M_p \tag{6-83}$$

Where,
h = depth of the web of the HSS or box-section as defined in Table 6-2

c. For sections with *slender webs* and compact or non-compact flanges, web local buckling will occur, and the nominal moment strengths are obtained as follows:

$$M_n = R_{pg} F_y S \tag{6-84}$$

R_{pg} is obtained from equation (13-7) in Chapter 13 of this text, where,

$$a_w = \frac{2ht_w}{bt_f}$$

Note: There are no listed hollow structural sections (HSS) in the *AISCM* with slender webs, so part 3(c) would only apply to built-up box sections, and not HSS.

4. Lateral Torsional Buckling Limit State

a. When $L_b \leq L_p$, the HSS or box-shaped section in flexure reaches its plastic moment capacity because lateral torsional buckling does *not* occur and is therefore not applicable, except for HSS or box sections with high depth-to-width ratios.

b. When $L_p \leq L_b \leq L_r$, the nominal moment strength is

$$M_n = C_b\left[M_p - \left(M_p - 0.7F_y S_x\right)\left(\frac{L_b - L_p}{L_r - L_p}\right)\right] \leq M_p \tag{6-85}$$

c. When $L_b > L_r$, the nominal moment strength is

$$M_n = 2EC_b \frac{\sqrt{JA_g}}{L_b / r_y} \leq M_p \tag{6-86}$$

Where,

$$L_p = 0.13Er_y \frac{\sqrt{JA_g}}{M_p} \tag{6-87}$$

$$L_r = 2Er_y \frac{\sqrt{JA_g}}{0.7F_y S_x} \tag{6-88}$$

A_g = gross cross-sectional area of the HSS or box-shaped section.

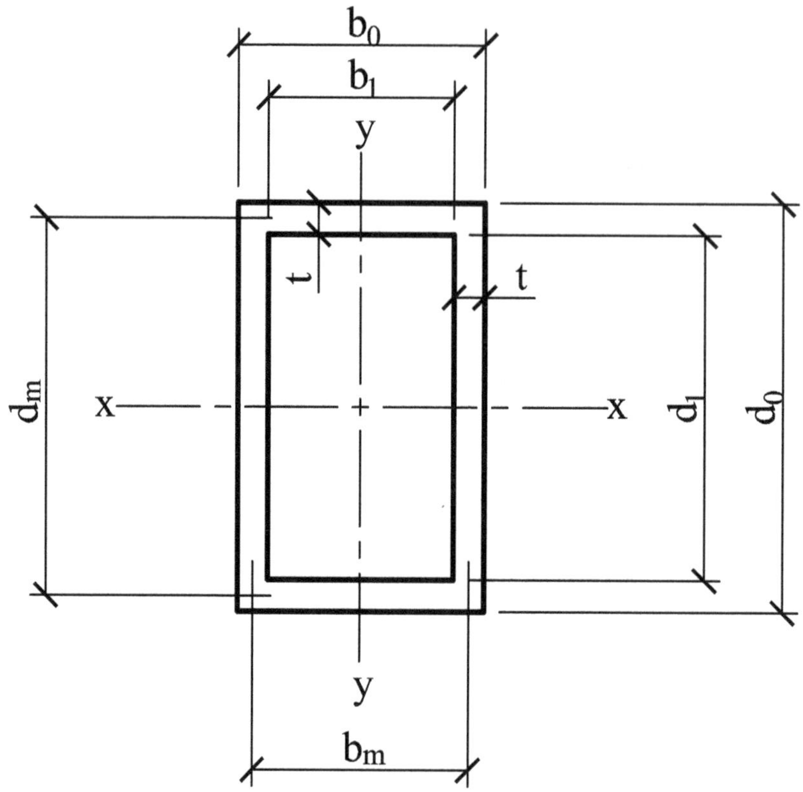

FIGURE 6-37f Box section properties

For bending of round HSS members, refer to AISC Specifications Section F8.

As discussed previously, L_p is the unbraced length limit below which the flexural failure mode of the HSS and box-shaped sections will be yielding, and L_r is the unbraced length limit above which the failure mode is elastic lateral torsional buckling; between L_p and L_r, the failure mode is inelastic lateral torsional buckling. Square HSS and square box-shaped sections are not susceptible to lateral torsional buckling. Only rectangular HSS or rectangular box-shaped sections with high web depth-to-thickness ratios bending about their strong axis are susceptible to lateral torsional buckling [1]. Note that for built-up box sections where the flange and web thicknesses are both equal (i.e., $t_f = t_w = t$), the torsional constant, J, is given as [23]:

$$J = \frac{2tb_m^2 d_m^2}{b_m + d_m}$$

Where,

b_m = mean width = $b_o - t = b_1 + t$

d_m = mean depth = $d_o - t = d_1 + t$

b_o, d_o = overall width and depth of the box section, respectively (see Figure 6-37f)

b_1, d_1 = clear dimensions of the flange and web of the box section, respectively (see Figure 6-37f)

t = flange and web wall thickness

Comparing the built-up box section in Table 6-2 to Figure 6-37, $h = d_1$ and $b = b_1$.

The other section properties for the built-up box section in Figure 6-37f are as follows:

Cross-sectional area, $A = b_o d_o - b_1 d_1$

$$I_x = \frac{b_o d_o^{\,3} - b_1 d_1^3}{12}$$

$$S_x = \frac{b_o d_o^{\,3} - b_1 d_1^3}{6 d_o}$$

$$Z_x = \frac{b_o d_o^{\,2} - b_1 d_1^2}{4}$$

$$r_x = \sqrt{\frac{I_x}{A}}$$

$$c = \frac{d_o}{2}$$

Moment of inertia for bending about the weak axis, $I_y = \dfrac{d_o b_o^{\,3} - d_1 b_1^3}{12}$

Elastic section modulus for bending about the weak axis, $S_y = \dfrac{d_o b_o^{\,3} - d_1 b_1^3}{6 b_o}$

Nominal Shear Strength

For HSS and built-up box sections, the nominal shear strength from Section G4 of the *AISC Specifications* is

$$V_n = 0.6 F_y A_w C_{v2}$$

Where,

C_{v2} = Web shear buckling coefficient (See Table 13-9)

$A_w = 2ht$

$\dfrac{h}{t_w} = \dfrac{h}{t}$

$k_v = 5$

t = design wall thickness
h = clear distance between the flanges of the HSS or box section
= $d_o - 3t$ for HSS sections, because of the fillets at the corners
= $d_o - 2t$ for Built-up box section = d_1 in Figure 6-37f

References

[1] American Institute of Steel Construction. 2023. *Steel Construction Mannual.* 16th ed., Chicago, IL: AISC.

[2] International Codes Council. 2024. *International Building Code—2024.* Falls Church, VA: ICC.

[3] American Society of Civil Engineers (ASCE). 2022. ASCE 7-22: *Minimum Design Loads for Buildings and Other Structures*, Reston, VA.

[4] American Concrete Institute. 2019. *Building Code Requirements for Structural Concrete and Commentary, ACI 318*, Farmington Hills, MI: ACI.

[5] Brick Industry Association (BIA). 1987. *Technical Notes on Brick Construction: Structural Steel lintels #31B (Revised),* Reston, VA: BIA.

[6] Vulcraft Steel Deck Product Catalog. 2020/2021. NUCOR. *https://vulcraft.com/products/deck* (accessed September 16, 2024).

[7] Steel Joist Institute. 2017. *Standard Specifications—Load Tables and Weight Tables for Steel Joists and Joist Girders*. 44th ed. Myrtle Beach, SC: SJI.

[8] American Institute of Steel Construction. 2003. *Steel Design Guide Series 9: Torsional Analysis of Structural Steel Members*, Chicago, IL: AISC.

[9] Lin, P. H. 1977. "Simplified Design of Torsional Loading of Rolled Steel Members." *Engineering Journal*, 3rd Quarter.

[10] American Institute of Steel Construction. 2005. *Steel Design Guide Series 7: Industrial Buildings—Roofs to Anchor Rods*, Chicago, IL: AISC.

[11] Fisher, J., M. West, and J. Van de Pas. 2002. *Designing with Vulcraft Steel Joists, Joist Girders, and Steel Deck*. 2nd ed. Milwaukee, WI: Nucor.

[12] Blodgett, O. 1966. *Design of Welded Structures*. Cleveland, OH: The James F. Lincoln Arc Welding Foundation.

[13] Timoshenko, S. P., and J. M. Gere. 1961. *Theory of Elastic Stability*. 2nd ed. New York: McGraw-Hill.

[14] Galambos, T. V., F. J. Lin, and B. G. Johnston. 1996. *Basic Steel Design with LRFD*. Upper Saddle River, NJ: Prentice Hall.

[15] Johnston, B. G. 1982. "Design of W-Shapes for Combined Bending and Torsion." *Engineering Journal,* AISC, 2nd Quarter: 65–85.

[16] Hotchkiss, J. G. 1966. "Torsion of Rolled Sections in Building Structures." *Engineering Journal,* AISC: 19–45.

[18] Concrete Institute. 2013. *Building Code Requirements for Masonry Structures, ACI 530*. Farmington Hills, MI.

[19] Muir, L. S. 2017. "A Shear Connection Extends its Reach." *Modern Steel Construction*, December: 16-21.

[20] American Institute of Steel Construction. 2008. *Steel Design Guide Series 23: Constructability of Structural Steel Buildings*, Chicago, IL: AISC.

[21] Savery, A. B. 2006. "Over and Above." *Modern Steel Construction*, August.

[22] Salmon, C. G. and J. E. Johnson. 1996. *Steel Structures – Design and Behavior*, 4th ed. New York, NY: HarperCollins.

[23] Blodgett, O. W. 1966. *Design of Welded Structures*. Cleveland, OH: The James F. Lincoln Arc Welding Foundation.

Exercises

6-1. Draw a design moment, $\phi_b M_n$, versus unbraced length, L_b, curve for a W21 × 50 beam for ASTM A992 Grade 50 steel. Include the following points and calculations:

a. Web and flange slenderness ratios
b. L_p and L_r
c. Design moments for $L_b < L_p$; $L_p < L_b < L_r$; and $L_b = 15$ ft.

6-2. Determine the design moment for a W14 × 22 beam with (a) $L_b = 0$ and (b) $L_b = 14$ ft. The yield strength is $F_y = 36$ ksi and $C_b = 1.0$.

6-3. Determine the following for the cantilever beam shown below in Figure 6-38. The steel is ASTM A36. Assume $C_b = 1.0$ and the unbraced length, L_b = 4'-0".

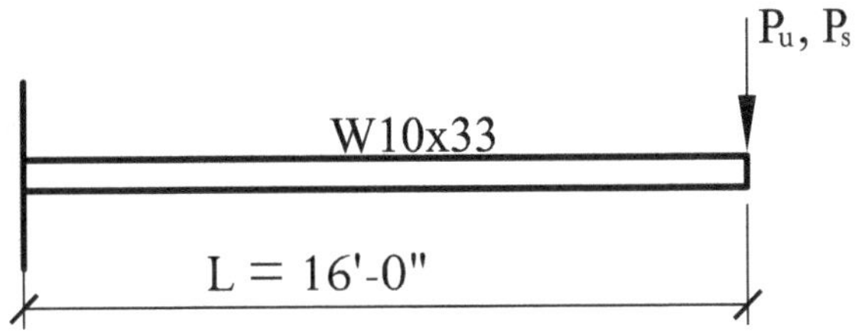

FIGURE 6-38 Details for Exercise 6-3

a. Determine the maximum load, P_u, that could be applied based on shear.
b. Determine the maximum load, P_u, that could be applied based on bending strength.
c. Determine the maximum load, P_s, that could be applied based on deflection (compare with the L/240 limit).
d. Draw the $\phi_b M_n$ vs. L_b curve for this beam, showing the following points:
   i. $\phi_b M_p$
   ii. $\phi_b M_r$
   iii. L_p
   iv. L_r
   v. $\phi_b M_n$ at L_b = 4'-0"

6-4. Determine the maximum value of "P" (service load) that can be applied to the cantilever beam below in Figure 6-39 considering bending strength and deflection only. Use the following parameters:

ASTM A36 steel

Beam is W12 × 35

$L_b = L = 12$ ft.

$C_b = 1.0$

$d_{\text{allowable}} = L/720$

All loads are live; use a load factor of 1.6.

Ignore self-weight

Loads shown are service levels.

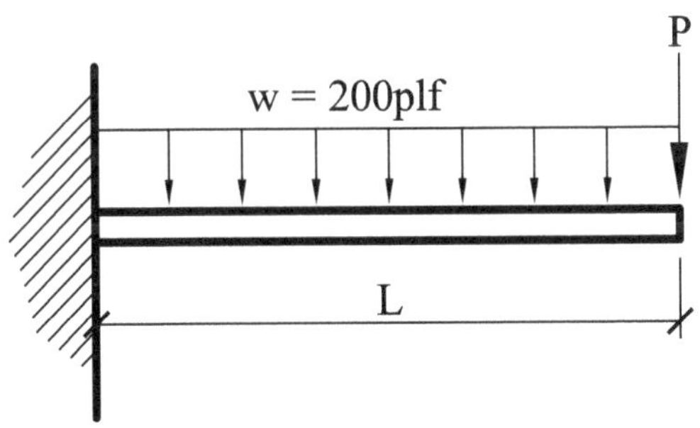

FIGURE 6-39 Details for Exercise 6-4

6-5. For the beam diagram shown in Figure 6-40, find the maximum value of "P" in kips based on:

a. Shear strength, $\phi_v V_n$
b. Bending strength, $\phi_b M_n$
c. Deflection using a limit of $L/400$

The value of "x" is variable, and the beam has full lateral stability. Use A36 steel, and the beam is W21 × 62.

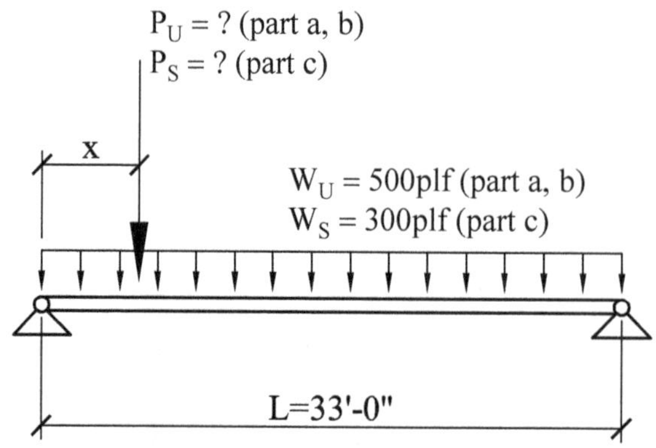

FIGURE 6-40 Details for Exercise 6-5

6-6. For the floor framing shown in Figure 6-41, select the most economical W-shape for beam B1 and girder G1. The floor dead load is 75 psf (including the weight of the framing), and the live load is 80 psf. Use ASTM A992, grade 50 steel. Check bending, shear, and deflection. Assume that beam B1 has full lateral stability and girder G1 is braced at the beam joints.

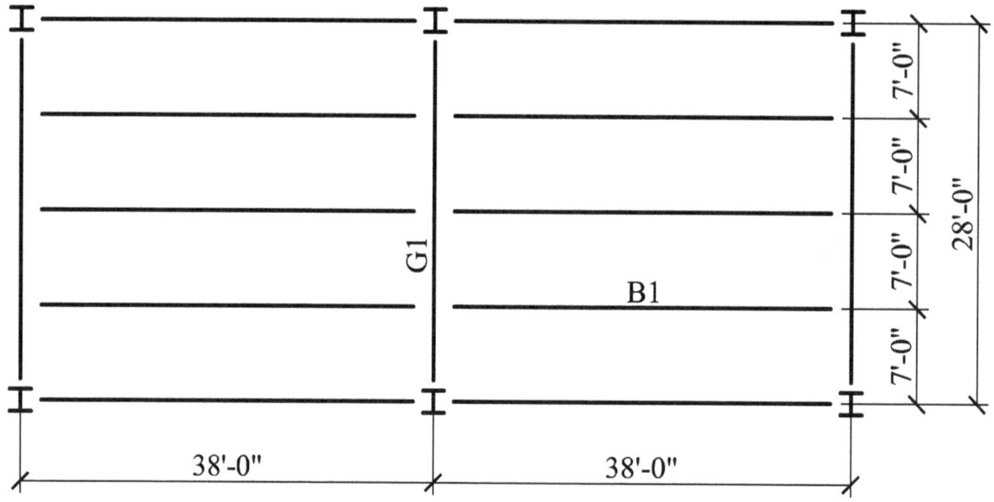

FIGURE 6-41 Floor plan for Exercise 6-6

6-7. Determine the most economical size for the wide flange beams supporting the floor loads shown in Figure 6-42 based on bending and deflection. Provide calculations for the minimum required section modulus and moment of inertia. The loads shown are service loads, and the steel is ASTM A992, grade 50. Assume C_b = 1.0 and L_b = 0. Limit the live load deflection to $L/360$ and the total load deflection to $L/240$.

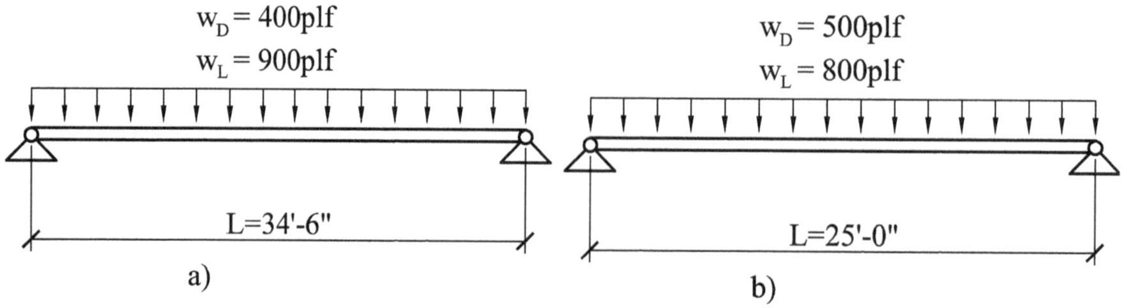

FIGURE 6-42 Details for Exercise 6-7

6-8. Determine whether the beams shown in Figure 6-43 are adequate for the given loads considering bending, shear, and deflection. The loads shown are service loads, and the steel is ASTM A36. Limit the live load deflection to *L*/360 and the total load deflection to *L*/240.

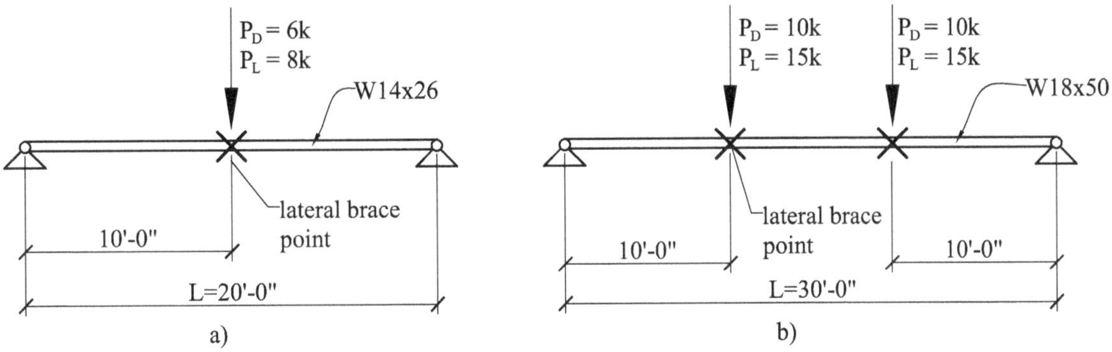

FIGURE 6-43 Details for Exercise 6-8

6-9. Determine the maximum factored loads that can be applied to the beam shown in Figure 6-44 based on web crippling and web yielding. The steel is ASTM A992, grade 50.

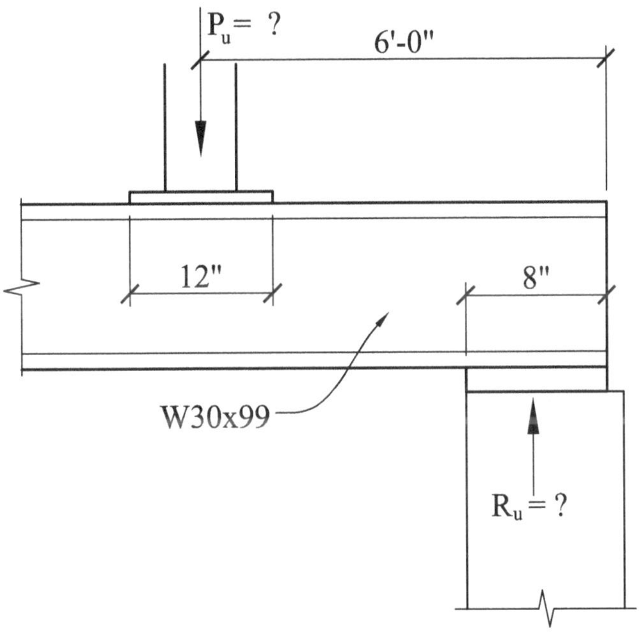

FIGURE 6-44 Detail for Exercise 6-9

6-10. Determine whether the following is adequate for the joint shown in Figure 6-45. The beam is ASTM A992, grade 50, and the bearing plate is ASTM A36. The concrete strength is f'_c = 3.5 ksi.

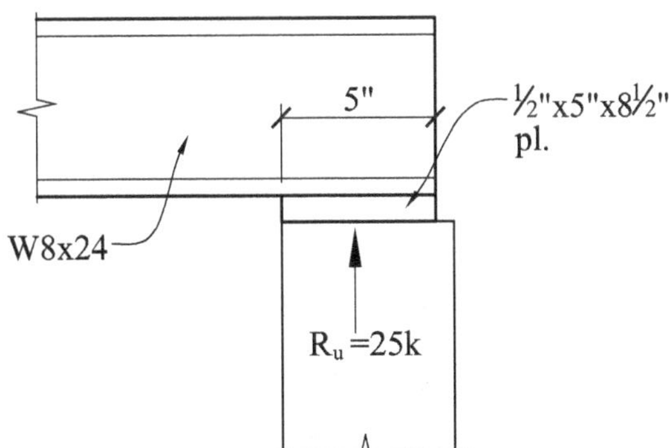

FIGURE 6-45 Detail for Exercise 6-10

6-11. Design a bearing plate using ASTM A572, grade 50 steel, for a factored reaction of $R_u = 65$ kips in the beam shown in Figure 6-46. Check web crippling and web yielding in the beam. Use $f'_c = 3$ ksi.

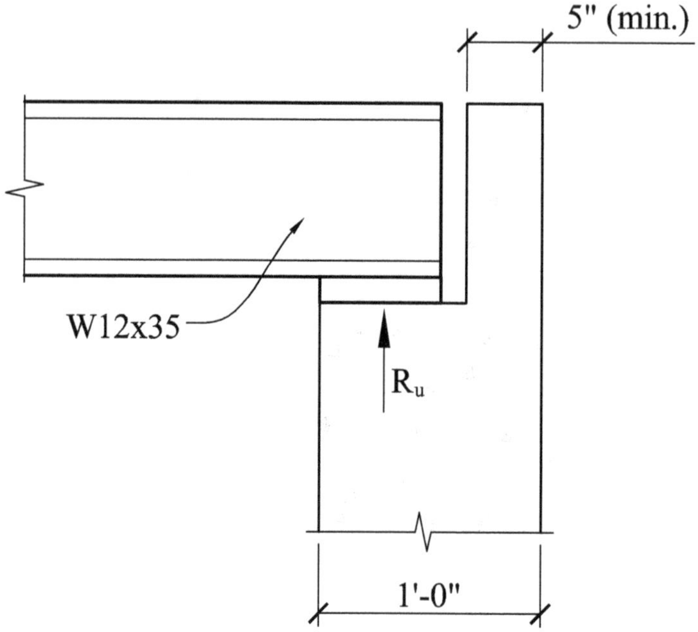

FIGURE 6-46 Detail for Exercise 6-11

6-12. The beam shown in Figure 6-47 is a W10 × 22, made from ASTM Grade 36 steel with a simple span between A and B and cantilevered to C. Determine the following:

a. The design strength in web crippling and web yielding in kips at point B.
b. Based on the answer from part (a), what is the maximum possible value of P_u in kips?
c. For a factored applied load of $P_u = 40$ kips, design the bearing plate thickness for point B. Use Grade 36 steel. Ignore the strength of the concrete and ignore the results from parts (a) and (b).

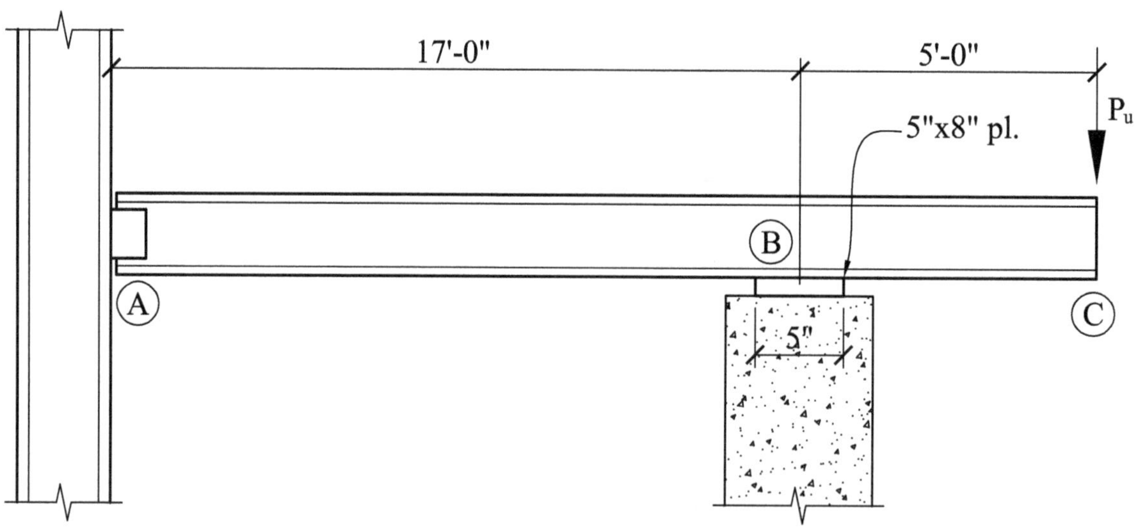

FIGURE 6-47 Detail for Exercise 6-12

6-13. For the details shown in Figure 6-48, determine if the bearing plate shown is adequate in bearing and bending. The steel is ASTM A36, and the concrete strength, $f'_c = 4{,}000$ psi.

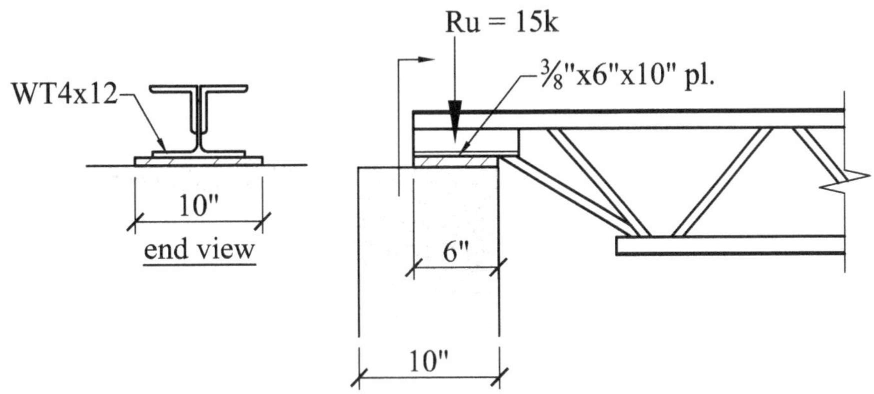

FIGURE 6-48 Detail for Exercise 6-13

6-14. Select the most economical open-web steel joist J1 for the floor framing plan shown in Figure 6-49 and select the most economical W-shape for member G1. The dead load is 65 psf and the floor live load is 80 psf. Consider bending, deflection, and shear. The steel is ASTM A992, grade 50. Assume that the unbraced length, L_b, is 3 ft., for member G1.

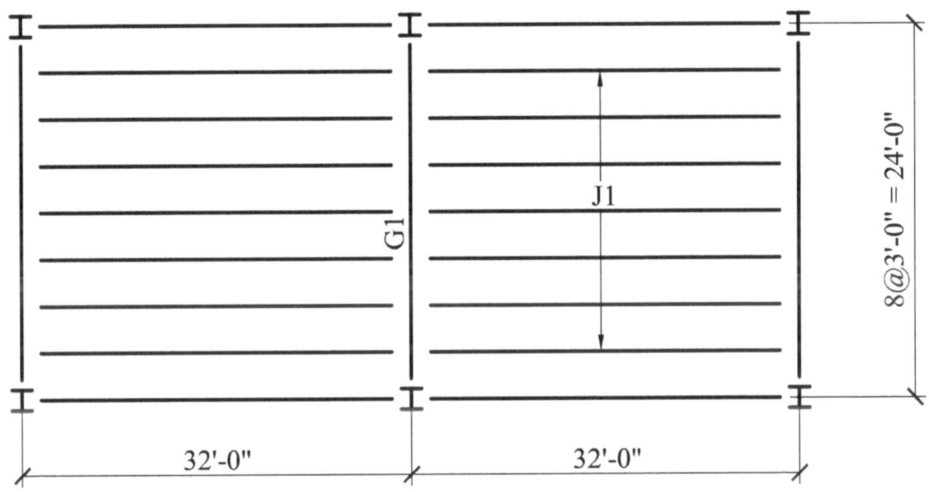

FIGURE 6-49 Floor plan for Exercise 6-14

6-15. Select the most economical open-web steel joist J1 for the roof framing plan shown in Figure 6-50 and specify a joist girder for member JG1. The dead load is 25 psf and the flat-roof snow load is 60 psf.

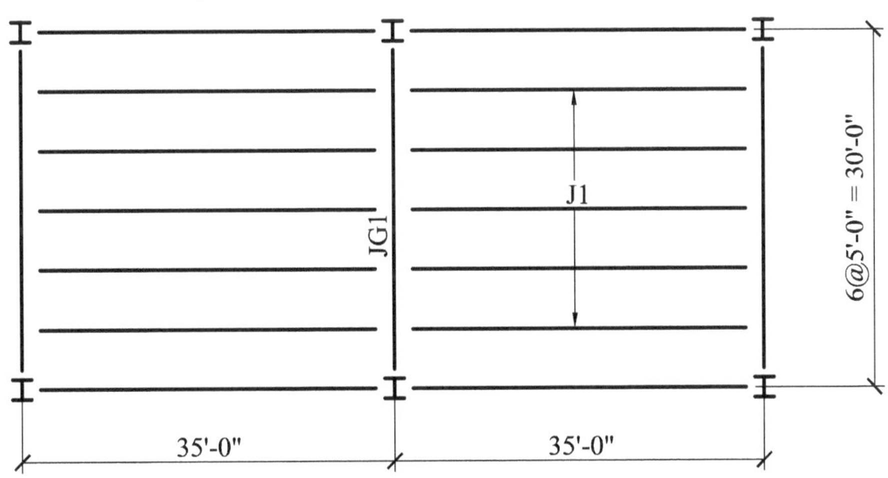

FIGURE 6-50 Roof plan for Exercise 6-15

6-16. For the roof framing plan shown in Figure 6-51, the service dead load is 27 psf. Consider the load combination as 1.2D + 1.0S only. Load designations shown are factored.

a. Determine the service snow load capacity in psf of the roof based on the strength of J1.
b. Determine the service snow load capacity in psf of the roof based on the strength of JG1.
c. What is the service snow load capacity of the roof based on (a) and (b)?

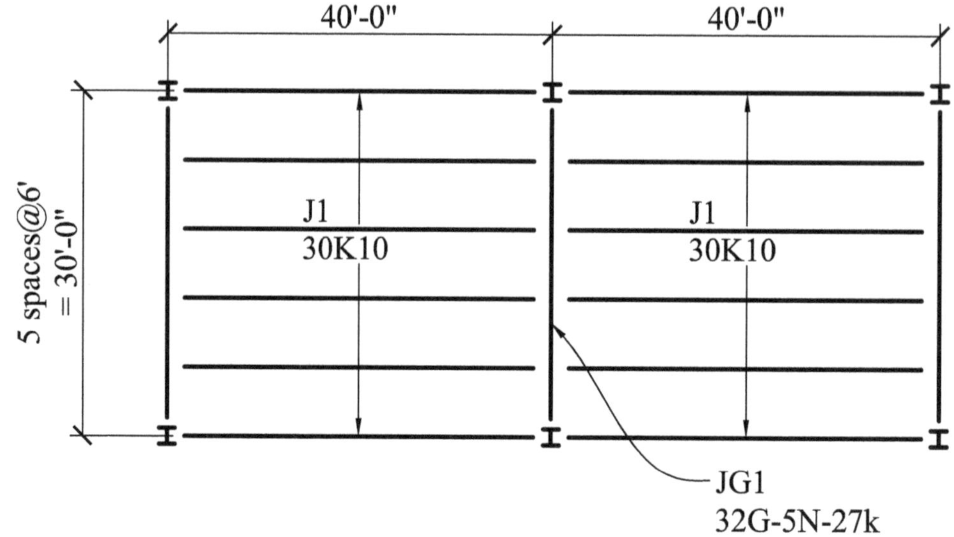

FIGURE 6-51 Roof plan for Exercise 6-16

6-17. For the floor framing plan shown in Figure 6-52, the service dead load is 80 psf. Consider the load combination to be 1.2D + 1.6L only. Load designations shown are factored.

a. Determine the service live load capacity in psf of the floor based on the strength of J1.
b. Determine the service live load capacity in psf of the floor based on the strength of JG1.
c. What is the service live load capacity of the floor based on (a) and (b)?

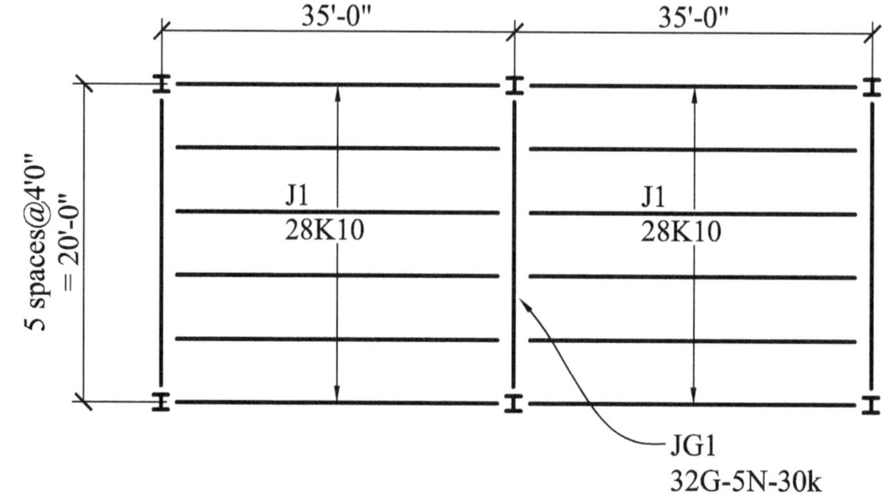

FIGURE 6-52 Floor framing plan for Exercise 6-17

6-18. Determine the service live load capacity of the floor shown in Figure 6-53 based on the capacity of B1 and G1. Consider the strength of B1 and G1 in bending and deflection only using a deflection limit of $L/240$ for total loads. Use the following:

Steel is grade 50.

B1 is a W16 × 36 and is braced continuously.

G1 is a W16 × 40 and is braced at one-third points.

Floor dead load is 85 psf and includes the self-weight.

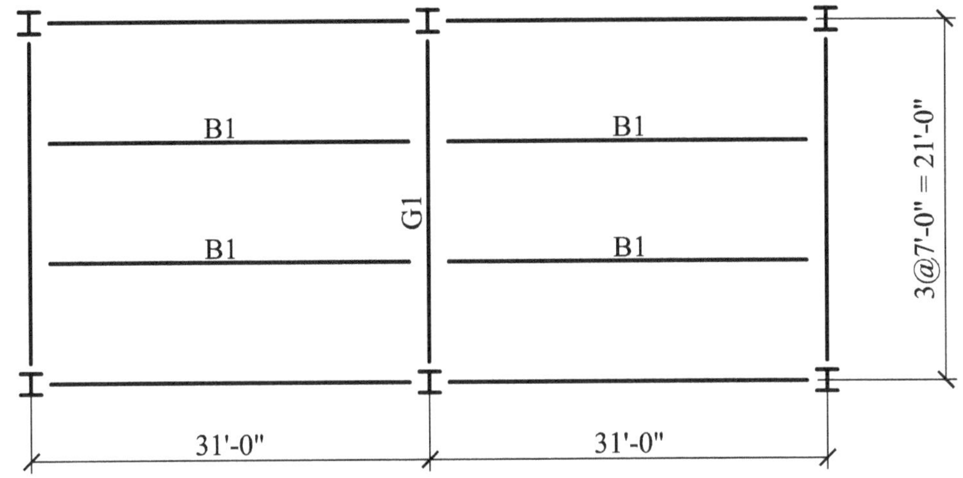

FIGURE 6-53 Floor plan for Exercise 6-18

6-19. For the beam diagram shown in Figure 6-54:

w = 500 plf

L = 25 ft.

a. Draw the shear and moment diagrams for beams ABCDEF, assuming x = 3.5 ft.
b. Determine the value of "x" using *AISCM* Table 3-21b.
c. Draw the shear and moment diagrams for beams ABCDEF using the value of "x" from part (b). Use the coefficients from *AISCM* Table 3-21b, directly for this part.
d. What value of "x" should be used in the final design? Justify your recommendation based on the above analysis.

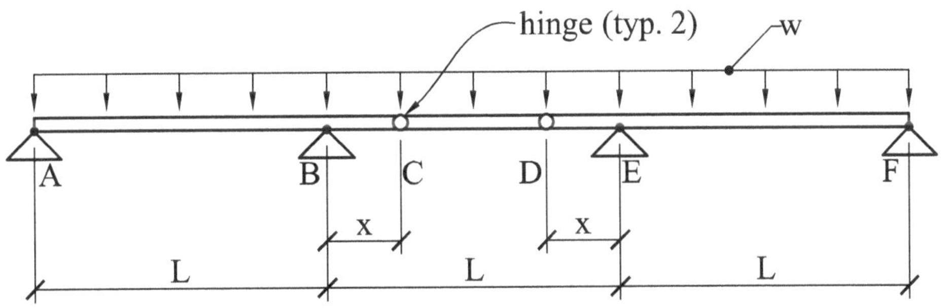

FIGURE 6-54 Details for Exercise 6-19

6-20. Determine the maximum span allowed for a ¼-in.-thick floor plate with a superimposed live load of 100 psf for strength and deflection. Compare the results with *AISCM* Tables 3-17a and 3-17b. The plate conforms to ASTM A786.

6-21. Develop a spreadsheet to determine the design bending strength of a wide flange (WF) shape for any unbraced length. Submit the output for a W12 × 26 using F_y = 50 ksi and L_b = 18 ft.

The following calculations are to be included:

- ϕM_p, ϕM_r for any L_b
- L_p, L_r

6-22. Develop a spreadsheet to determine the design bending strength of a channel shape for any unbraced length. Submit the output for a C12 × 30 using F_y = 36 ksi and L_b = 12 ft.

The following calculations are to be included:

ϕM_p, ϕM_r for any L_b, L_p, L_r

6-23. A series of non-composite floor beams spaced at 7 ft., on centers are simply supported over a span of 56 ft., above an auditorium and support a floor dead load of 90 psf (including the self-weight of framing) and a live load of 50 psf. The compression flange of the beam is fully laterally braced by the floor deck. Due to ceiling height limitations, the beam depth is limited to a maximum of 20 inches.

a. Select a beam size that will satisfy the bending moment, shear, and deflection criteria (live load deflection limit = L/600) if the beam will be cambered upwards an amount equal to the dead load deflection. Assume ASTM A992 Grade 50 steel.
b. Determine the required upward camber in inches that would be specified on the structural drawings.

6-24. Calculate the nominal moment capacities, M_n, of the following beams at the given unbraced lengths. Assume the moment gradient factor, C_b, is 1.0:

a. W18 × 35, $L_b = 0$ ft.
b. W24 × 55, $L_b = 0$ ft
c. W24 × 55, $L_b = 5$ ft
d. W24 × 55, $L_b = 13.9$ ft.
e. W21 × 50, $L_b = 13.6$ ft.

6-25. Two lintel beams above the bay windows in an existing building are used to support 8 ft., height of masonry walls consisting of nominal 8 in. thick block backup wall that weighs 55 psf and nominal 4 in. thick brick veneer that weighs 40 psf. The lintel beams are W18 × 46 that spans 60 ft., and W16 × 26 that spans 30 ft. Investigate the structural adequacy of these lintel beams to support the masonry loads. What is the limit state that controls the behavior of these beams? For this exercise, ignore the eccentricity of the masonry loads on the lintel beams. Assume ASTM A992 Grade 50 steel.

6-26. A 20 in. wide × 34 in. deep built-up box section has flange and web thicknesses of 1-in. Determine the moment capacity about the strong axis, assuming an unbraced length of 30 ft. Assume ASTM A992 Grade 50 steel and use the LRFD method.

6-27. Two roof girders are spaced at 32 ft., apart and span 150 ft., between column supports (assume simple supports). The roof supports HSS 12×6×1/4 (LSV) trellis framing without a deck or glass roofing and bending about their strong axis. The trellis beams are spaced at 6 ft., on centers and supported on top of the roof girders and cantilevers 5 ft., beyond the girders at both ends. (LSV = long side vertical)

a. Assuming a snow load of 30 psf on the trellis framing and assuming a roof dead load of 10 psf, check if the HSS trellis framing is adequate. Assume ASTM A500 Grade B (i.e., $F_y = 46$ ksi and $F_u = 58$ ksi) for the HSS.
b. What is the lightest W44 section that can be used for the girder to satisfy both the strength and deflection criteria? Assume a total load deflection limit of span/240 with a maximum upward camber of 3 inches. Assume ASTM A992 Grade 50 steel.
c. Check the HSS trellis framing.
d. Due to the length of the girder in this exercise, the girder will be shipped to the site in four equal segments that will be joined together in the field with bolted splices. Using the methods from Chapter 11, design the bolted splice.

6-28. A series of W44 × 290 floor girders spaced at 25 ft., on centers spans 20 ft., between column supports and cantilevers 20 ft., beyond the column. Assume the load on the girders can be modeled as a uniformly distributed load. Check the typical interior girder for the following loads:

D = 110 psf (does not include the girder self-weight)

L = 150 psf

- Assume the top of the girders between the columns (i.e., compression flange) is fully laterally braced, and assume the bottom flange of the girders in the cantilever (i.e., compression flange) is also fully laterally braced.
- Use ASTM A992 Grade 50 steel.
- Use deflection criteria for both total load and live load of span/360.

Hint: Recall that the deflection criteria for a cantilever span uses a different span length than the deflection criteria in the backspan between the simple supports.

6-29. Conduct Internet research on the 1988 "Supermarket Roof Collapse — Burnaby, B. C., Canada" and write a minimum five-page report explaining, with the aid of free-body diagrams, the reasons for the collapse and the lessons learned. How has this collapse influenced structural steel design practice?

6-30. Calculate the following section properties for a built-up box section with 20-in. × 34-in. outside dimensions and a 1-in. wall thickness: A, S_x, I_x, Z_x, and J. By how much percentage do these differ from the exact values?

6-31 The beam below is a W12 × 30, grade 50 steel. It is unbraced for the full length with concentrated loads at midspan. The loads shown are service level.

Showing calculations:

a. Find L_p and L_r and the design bending strength, ϕM_n. Is the beam adequate for bending?
b. Determine if the beam is adequate for deflection considering the typical limits of L/360 for live loads and L/240 for total loads.
c. Explain and sketch a practical condition where a concentrated load could occur at midspan but not provide lateral stability to act as a brace point.

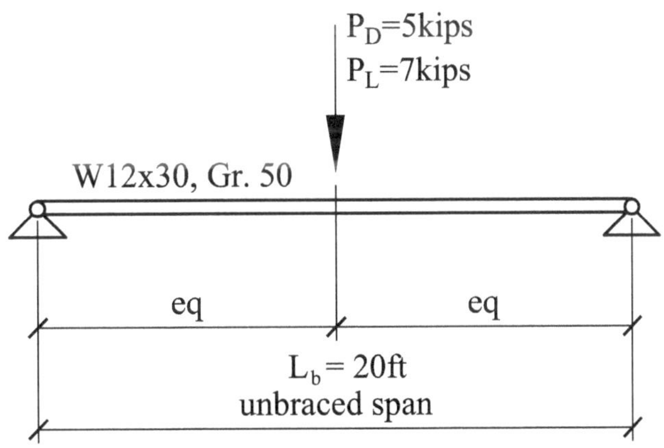

FIGURE 6-55 Details for Exercise 6-31

6-32. An elevator hoist beam spans 9.5 ft., between the elevator core walls and supports a concentrated load of 5000 Ib, the maximum weight of the elevator. Design a W10 hoist beam assuming a deflection limit of span/1666 and note that the hoist beam is usually unbraced for its entire span.

6-33. Rooftop equipment on the roof of tall buildings is often hidden from view with a screen wall (cladding) supported laterally by a series of HSS posts cantilevered off the roof beams and girders.

a. Determine the maximum uniform wind pressure in psf that can be supported by a 20-ft., tall screen wall if the screen wall is supported by HSS 8 × 8 × ½ posts spaced at 8 ft., on centers. Assume ASTM A500 Grade C (F_y = 50 ksi).
b. Provide the details for the moment connection at the base of the posts.

CHAPTER

7

Composite Beams and Columns

7-1 INTRODUCTION

In steel-framed building construction, the gravity load-resisting system can consist of steel or metal roof deck on open web steel joists or roof beams and girders, and concrete floor slab on composite steel or metal deck, or reinforced concrete floor slab, or wood panels (e.g., cross-laminated timber or CLT panels) supported on floor beams and girders. Floor decks for industrial structures and catwalks may also consist of bar grating or a flat steel plate (see Chapter 6), with the supporting steel beams spaced at closer intervals. A concrete slab on metal or steel deck supported on steel beams and girders is the most commonly used floor system in steel-framed buildings (see Figures 7-1a, 7-1d, and 7-1e), while a corrugated metal or steel deck (see Figure 7-1f) supported on beams or open web steel joists (OWSJ) and girders is the most commonly used system for the roof of steel-framed buildings.

The corrugated metal deck used in floor construction serves several purposes: it acts as a formwork for the wet concrete, and the protrusions in the inside face of composite metal deck engage the hardened concrete, creating composite action between the concrete and the metal deck in resisting the floor loads as it spans between the beams. A *form* deck (see Figure 7-1d) acts as a form for the wet concrete but reinforcement must be provided in the concrete to provide adequate strength to enable the deck span between supporting beams, since a form deck usually does not have enough strength to support more than the weight of the concrete. The reinforcement in the slab is usually a welded wire fabric (WWF). A *composite* deck (see Figure 7-1e) is usually strong enough to support more than just the weight of the concrete and is often used in composite construction (i.e., steel beams and girders with shear studs). The concrete slab supported by a composite deck often has a layer of WWF reinforcement to control shrinkage cracking in the slab. Additional top mild steel reinforcement is needed over and perpendicular to the girders to mitigate the potential cracking at the top of the slab over and parallel to the girders. The available thicknesses of metal decks vary from a thin 28 gage to a much thicker 16 gage deck as noted in Figure 7-1f, with the 20 gage and 22 gage decks being the most commonly specified.

A reinforced concrete slab without the corrugated metal deck (see Figure 7-1b) can also be used as the floor deck; this system, which requires formwork, is more commonly used in bridge construction. In the past, steel floor beams and columns were commonly encased in concrete, as shown in Figure 7-1c. The steel framing was encased in concrete, with the concrete providing adequate fire protection for the steel beams. In current practice, it is generally more economical to provide passive fire protection to the steel framing with spray-on cementitious fireproofing material or by providing a ceiling system below the steel framing to achieve the desired fire protection; sometimes, the corrugated metal deck may

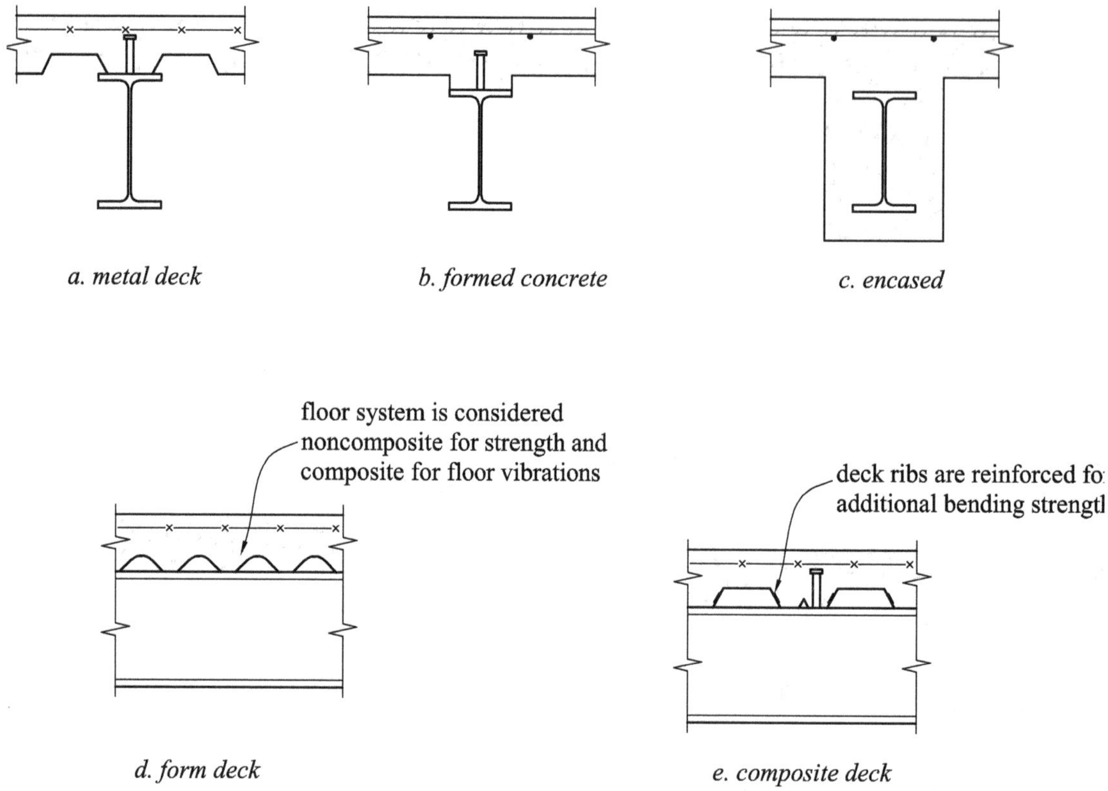

FIGURE 7-1a-e Types of composite beams

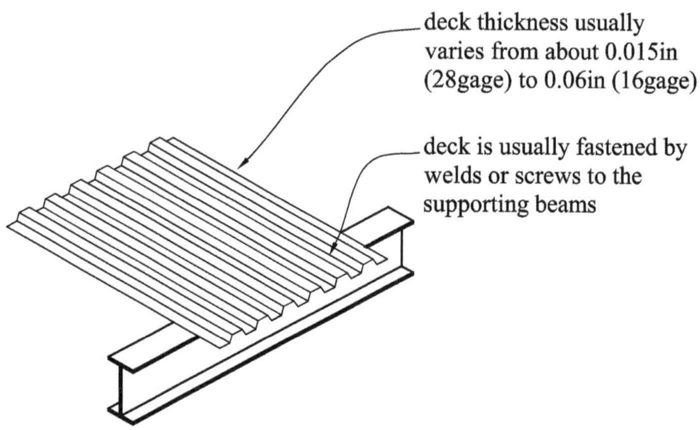

FIGURE 7-1f Corrugated metal deck in a steel beam

also need to be sprayed with fireproofing material. Intumescent paint, applied to the structural framing and the deck, and which expands as it is heated is also sometimes used as a fireproofing material when the structure is exposed for aesthetics purposes. Consequently, steel framing encased in concrete is not commonly used today for fire protection.

However, steel framing encased in concrete is sometimes used in some long span concrete structures where the long span and a constraint on the maximum depth of the beam may necessitate using a steel-concrete composite beam to increase the stiffness and strength of the beam. In this case, the steel beam is encased in the reinforced concrete beam, with steel studs welded to the flanges and web of the steel beam to achieve composite action between the concrete beam and the steel beam.

The foci of this chapter will be on composite columns (which will be discussed later in this chapter), and concrete floor slabs on a composite metal deck supported by composite beams and

FIGURE 7-2a Steel beams and girders with composite metal deck and steel studs (Courtesy of Magnusson Klemencic Associates

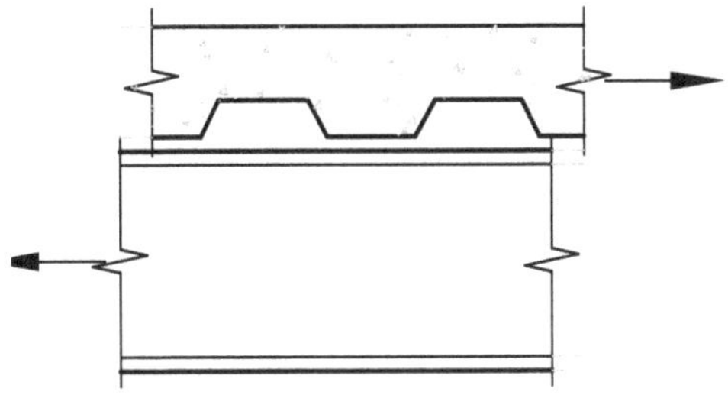

FIGURE 7-2b Slippage at the interface between a steel beam and a composite floor deck

girders – that is, beams with steel shear studs welded to the top flange through the metal deck and welded to the top flange of the girder. The shear studs facilitate composite action between the concrete slab and the steel beams and girders (see Figure 7-1d). Figures 7-2a and 7-2d show composite floors under construction. The composite action in a floor system, facilitated by the shear studs, increases the moment capacity of the combined steel beam/girder and concrete section, with the concrete in compression acting as the compression flange of the composite section resisting the internal compression force, and most of the steel section resisting the internal tension force, thus increasing the strength and stiffness of the composite section. This combination of dissimilar materials (concrete slab and structural steel) to form an equivalent singular structural element, acting together in concert to resist gravity loads, is called *composite construction*.

A slab system with a *form* deck (see Figure 7-1e) does not usually have headed studs to engage the concrete with the steel beam (see Section 7-2 for a further discussion on headed studs); therefore, beams supporting form decks are designed as non-composite, which means that the beams support the full weight of the concrete slab and the superimposed dead and live loads without any contribution from the concrete slab or steel deck. In contrast, the composite slab system (i.e., concrete slab on corrugated composite metal deck) with the headed studs welded to the top flange of the steel beams and girders is designed as a composite system. The headed studs resist the horizontal interface shear forces between the concrete slab and the steel beam or girder, thus, the hardened concrete and the steel beam or girder act together in unison to resist the gravity load moments while the shear is resisted solely by the steel beam or girder.

In steel construction, the floor systems can either be *unshored* or *shored*, with unshored construction more commonly specified. Note that for the composite system with unshored construction, the steel beams and girders have to support the weight of the wet concrete and any construction live loads since the composite action will only exist after the concrete has hardened. Therefore, the steel beams and girders in the composite system during the construction phase - when the concrete is still wet - act as non-composite beams in resisting the gravity dead loads and construction live loads (see Chapter 6) .

For the evaluation of floors for human-induced vibrations, the floor system with a metal deck and concrete slab is always analyzed as if composite action occurs even if headed studs or other shear connectors are not used (see Chapter 12 for more discussions).

7-2 SHEAR STUDS

For the steel beams and girders to work in concert with the concrete slab on composite metal deck in resisting gravity loads, there needs to be an adequate horizontal force transfer mechanism at the interface where the steel beam/girder and the concrete slab meet, to prevent slippage between the surfaces (see Figure 7-2b). This horizontal force transfer between the concrete slab and the steel beam or girder is accomplished by using shear connectors that are welded through the deck to the top of the beam flange or directly to the top of the girder top flange; these shear connectors are typically headed studs (see Figure 7-2a and 7-2c), but channels or some other types of deformed connectors could also be used. Headed shear studs are almost exclusively used in bridge and building construction due to their ease of installation, and the focus in this chapter will be on these types of shear connectors.

Headed studs should conform to ASTM A29 and the mechanical requirements for Type B studs in the American Welding Society (AWS) Code, Section D1.1. These studs are field welded to the top flange of a steel beam or girder and spaced at regular intervals to adequately transfer the horizontal shear at the interface between the steel beam and the concrete slab on metal deck (see Figure 7-3). Shop welding of studs to the top flanges of beams, girders, or beam attachments is not permitted by OSHA safety regulations Section 1926.754(c)(1) because of the trip hazard they create. For steel beams, the studs are field welded through the metal deck to the beam top flange, while for the girders, the studs are field welded directly to the top flange of the girder. The welding process of headed studs to the flange of the beam or girder typically involves a stud welding gun and an electric arc that melts the stud and the surface of the beam. Ceramic rings or ferrules centered on the headed studs at the bottom of the stud surround the weld area to protect the molten weld from oxidation, and prevent it from spreading outward on the top of the steel beam or girder flange; the ferrules confine the molten metal and the heat generated by the arc. The ceramic rings or ferrules should be chipped away after the welding process - to expose the finished stud weld for proper inspection.

The importance of removing the ceramic rings or ferrules after welding the studs was further highlighted by the deadly collapse of the Mexico City Metro Bridge in May 2021 that killed 26 people. One of the root causes of this failure was attributed to the non-removal of ceramic rings around the headed studs welded to the top flange of the bridge beams, and the sloppy welding of the headed studs. A number of the steel studs were found to have sheared off the beam [12].

FIGURE 7-2c Composite deck with headed steel studs, and showing the welded wire fabric (WWF) reinforcement

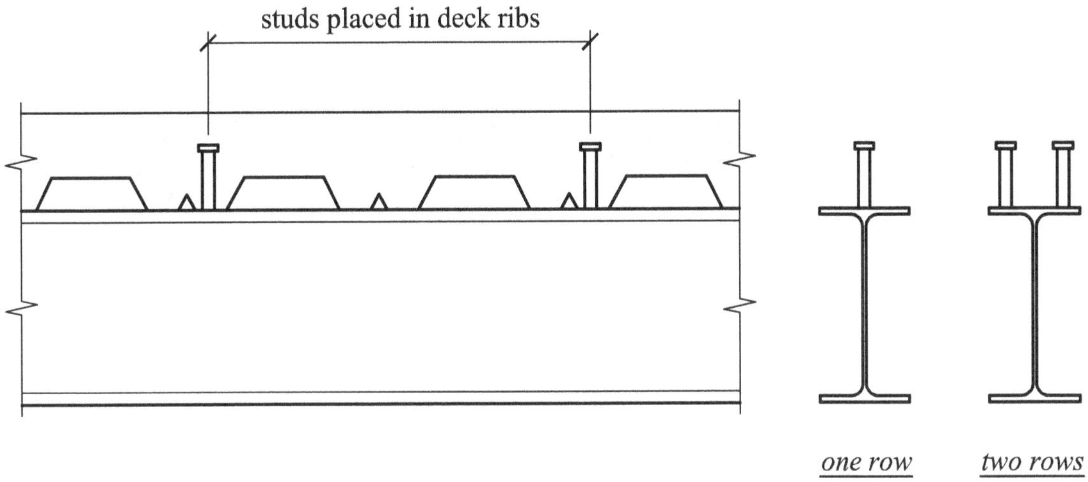

FIGURE 7-3 Headed studs

For beams supporting corrugated metal deck and concrete slab, the spacing between the studs coincides with multiples of the spacing between the ribs of the decking. More than one row of shear connectors can be provided, but there are some dimensional limitations such as the width of the beam or girder flange that often limit the use of multiple rows of studs. These limitations will be discussed in detail later in this chapter.

The number of shear connectors provided will determine how much of the concrete slab is engaged or acting in compression in combination with the steel beam (which will be in tension for the most part) to resist the flexural moments on the beam. The shear is assumed to be resisted solely by the steel beam or girder. When 100% of the concrete slab in compression in a floor system is not engaged in composite action with the steel beam or girder and thus requires a smaller number of shear connectors, this system is called *partially composite* construction. A fully composite system is one in which enough shear connectors are provided to completely engage the entire cross-sectional area of the concrete slab in compression. In this case, there is an upper limit to the number of studs that can be provided because adding more studs beyond this limit will not contribute to the strength of the composite floor system.

In building construction, the floor framing members that are oriented perpendicular to the direction of the span of the slab-deck system are usually referred to as beams and the floor framing members that support the beams and are oriented parallel to the span of the slab-deck system are called girders (see Figure 7-4). That is, the deck ribs are oriented perpendicular to the longitudinal axis of the steel beams and parallel to the longitudinal axis of the girders.

The AISC Specification requirements [1] for headed stud placement are summarized below and illustrated in Figure 7-5.

1. Except for the corrugated metal deck, the minimum lateral cover around headed studs is 1 in.
2. The maximum stud diameter, D_s, is $2.5t_f$ (t_f is the beam flange thickness); if the studs are located directly over the beam web, then this provision does not apply. For studs placed in a formed steel deck, the maximum stud diameter is limited to $^3/_4$ in.
3. The minimum stud spacing along the longitudinal axis of the beam is $6D_s$ (D_s is the stud diameter).
4. The maximum stud spacing along the longitudinal axis of the beam must be less than $8Y_{con}$, or 36 in. (Y_{con} is the total slab thickness.)

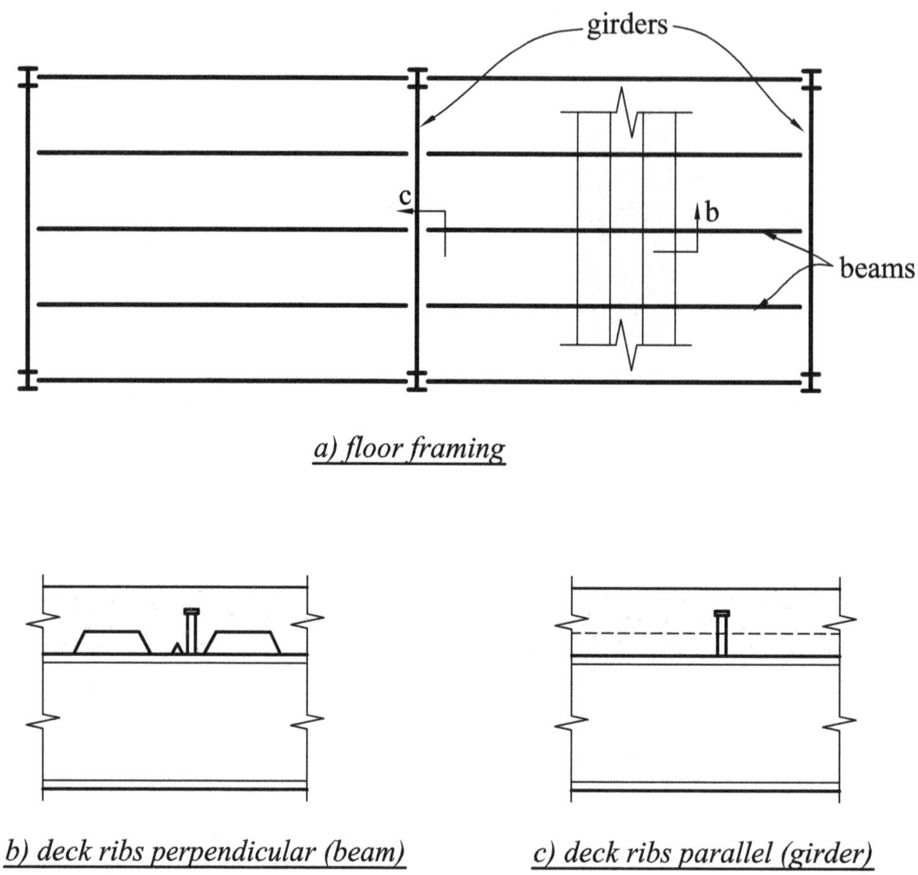

FIGURE 7-4 Floor beams and girders

5. The minimum stud spacing across the flange width is $4D_s$.
6. The minimum stud length is $4D_s$.
7. For a formed steel deck, the rib height, h_r, must be less than or equal to 3 in.
8. For a formed steel deck, the minimum rib width, w_r, must be greater than 2 in. but shall not be taken as less than the clear width across the top of the deck.
9. For a formed steel deck, the studs should extend at least 1½ in. above the top of the deck, with at least ½ in. of concrete cover over the top of the stud.
10. For a formed steel deck, the deck must be anchored to the supporting steel beams at intervals not exceeding 18 in. The anchorage can be provided by some combination of spot welds (also called puddle welds), mechanical connectors, or welding the headed stud through the deck.

The number of shear studs that are provided is a function of the required strength of the composite section, which will be discussed later. When studs are placed in a corrugated metal deck, it is ideal to place the stud in the middle of the deck rib. However, the deck ribs are usually reinforced or stiffened in the center, thus forcing the stud to be offset from the center of the deck rib (see Figure 7-6). Studs should be placed on the side of the deck rib *closest to the nearest end* of the span of the beam because more interface shear can be transmitted to the stud through the concrete due to the additional concrete cover within the deck rib pushing against the stud in the direction of the interface shear force, V_h, in the beam (see Figure 7-6 and Ref. [4]). This is called the "strong" or "favorable" position of the stud. When the studs are located within the deck rib

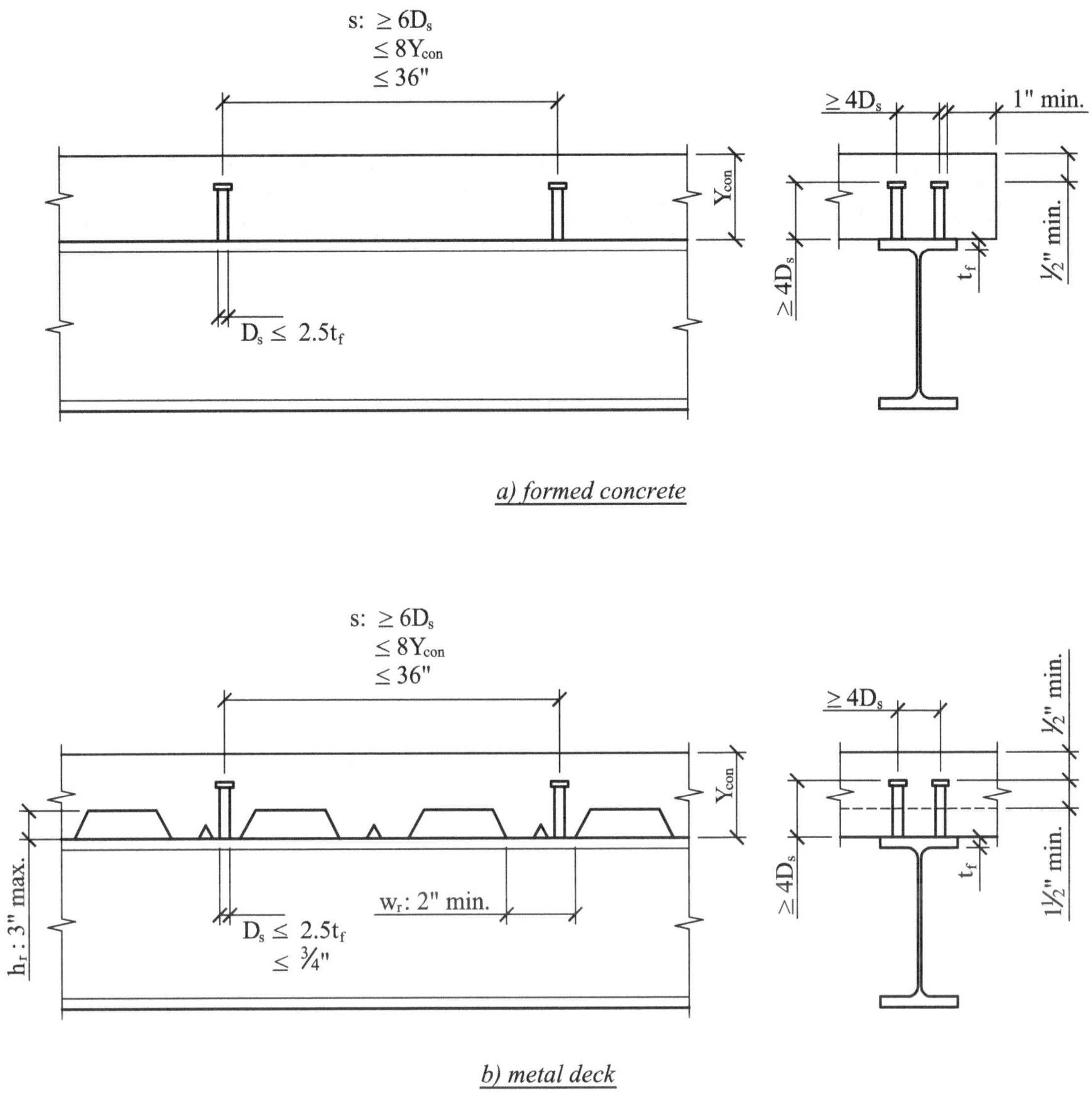

FIGURE 7-5 Dimensional requirements for headed studs

in the "weak" or "unfavorable" position, there is a lesser amount of concrete within the deck rib pushing against the studs in the direction of the interface shear force in the beam, V_h, and the shear strength is decreased by about 25% to 33%. Note that the direction of V_h in Figure 7-6 is the direction of the interface shear flow under gravity loads.

When shear studs are required by design, the number of studs required between the point of maximum moment and the nearest point of zero moment (i.e. within the shear span) is denoted as N_s. For the case of a uniformly loaded beam (see Figure 7-7a), the maximum moment occurs at midspan, and therefore a total of N_s studs are provided on each side of the beam centerline.

For a girder with two symmetrically placed concentrated loads (Figure 7-7b), a total of N_s studs are provided between each end of the girder and the location of the nearest concentrated load. The middle third of the beam has a constant moment (i.e., zero shear) and therefore does not require shear studs; however, a nominal number of shear studs are commonly provided in practice at the discretion of the designer to account for slight variations in loading that may create a small moment gradient (and therefore some shear) in this middle third region. A common layout of studs within this middle third region is to provide studs at 24 in. spacing. When three or more symmetrically placed concentrated loads are present, the moment diagram approaches

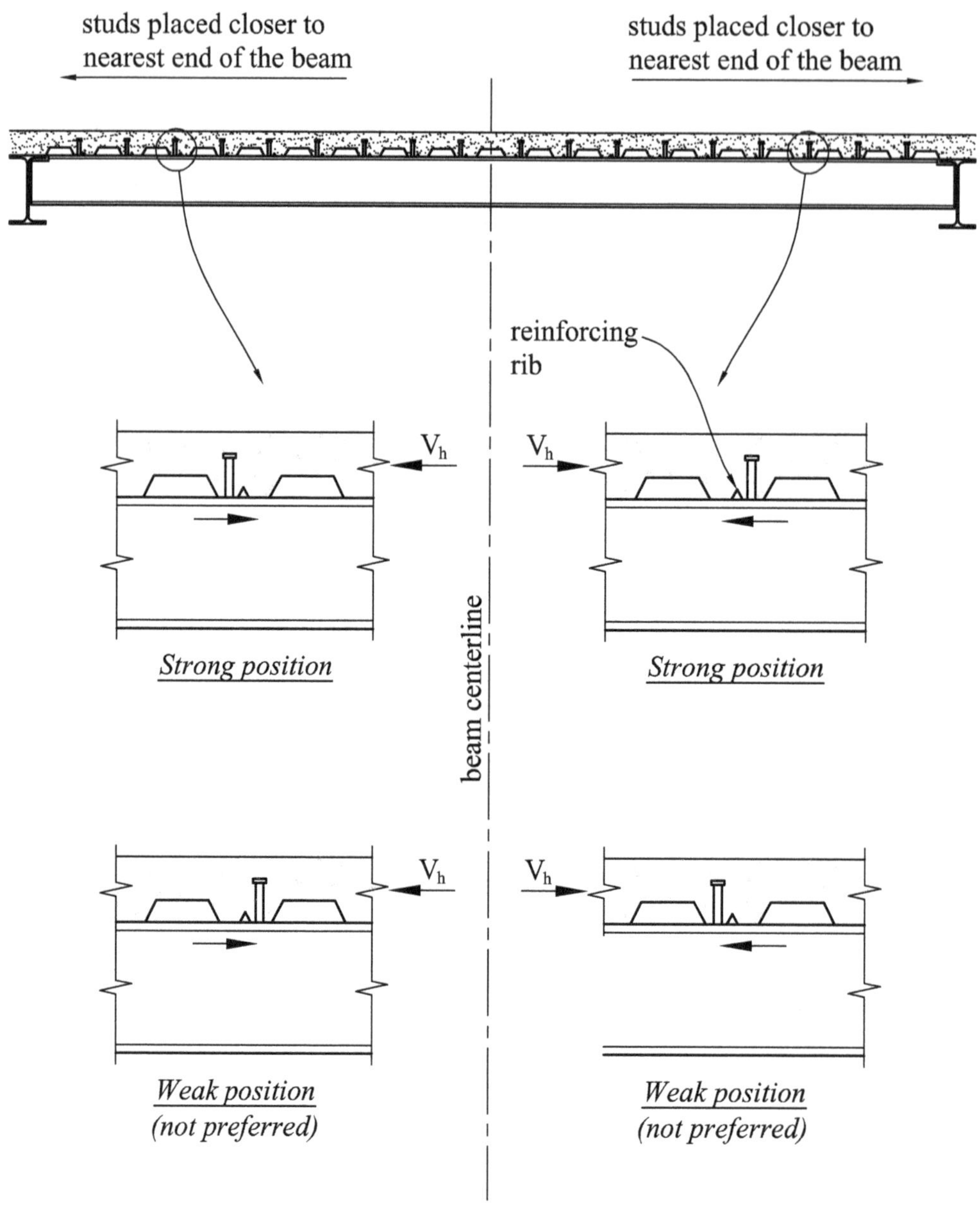

FIGURE 7-6 Weak and strong stud positions

the uniformly loaded case shown in Figure 7-7a, and the stud spacing is uniform along the girder length (see Figure 7-7a).

For beams with nonsymmetrical loads, a total of N_s studs are still provided between the point of maximum moment and the nearest point of zero moment (see Figure 7-7c). This creates a situation in which the required stud layout is not symmetrical about the beam midspan; thus, in some cases, the designer will add more studs so that the spacing is uniform to avoid errors in the field placement of the studs. There is an additional provision for concentrated loads, which requires that the number of studs placed between a concentrated load and the nearest point of zero moment shall be sufficient to develop the moment at the concentrated load. This is illustrated in Figure 7-7c, where a total of N_1 studs are provided between the point of maximum moment, M_1, and the nearest point of zero moment. A total of N_2 studs are required between the location of the concentrated load, where the moment is M_2, and the nearest point of the zero moments. In this case, M_1 is the maximum moment in the beam, and M_2 is the moment at the concentrated load.

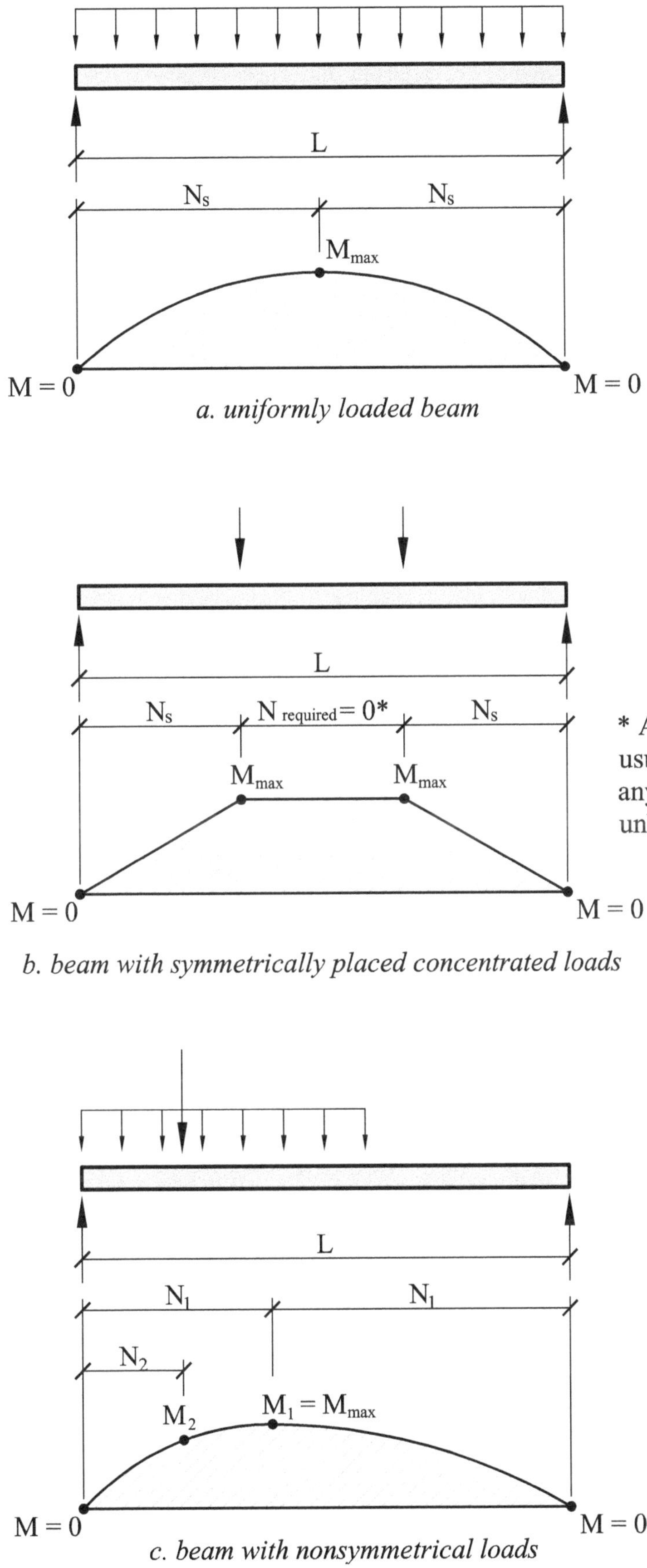

* A nominal number of studs are usually provided to account for any induced shear due to unbalanced loads

FIGURE 7-7 Stud placement

7-3 COMPOSITE BEAM STRENGTH

In order to analyze a composite beam section made of different materials (i.e., steel and concrete), we need to develop an equivalent model to determine the behavior of the composite section under loads. Figure 7-8a shows the typical composite beam section with a steel beam and a concrete deck. In order to analyze this composite section, we need to transform the concrete section into an equivalent steel section using principles from the mechanics of materials because the modulus of elasticity, and therefore the behavior of the two materials, are different under loading.

Once the concrete section is transformed into an equivalent steel section, the section properties of the composite beam can be determined. The area of concrete is transformed by dividing the area of the concrete by the modular ratio, n, defined as follows:

$$n = \frac{E_s}{E_c}, \tag{7-1}$$

Where,

n = Modular ratio,

E_s = Modulus of elasticity for steel (29×10^6 psi),

E_c = Modulus of elasticity for concrete

$$= w_c^{1.5}\sqrt{f'_c}, \tag{7-2}$$

w_c = Unit weight of the concrete, lb./ft.^{3}, and

f'_c = 28-day compressive strength of the concrete, ksi, and it is permitted to be between 3000 psi and 10,000 psi for normal weight concrete, and 3000 psi and 6,000 psi for lightweight concrete (see AISC Specifications Section I1.3).

Note: ACI 318 defines the modulus of elasticity for concrete, E_c, as $w_c^{1.5}33\sqrt{f'_c}$, where f'_c is in pounds per square inch; however, *AISC* has adopted an approximate value (see equation (7-2)) where the f'_c is in units of kips per square inch, or ksi. The *AISC* value will be used in this chapter for consistency. The area of the concrete section (Figure 7-8a) is then transformed into an equivalent steel section (Figure 7-8b) by dividing the area of the concrete by the modular ratio:

$$A_{ct} = \frac{A_c}{n}, \tag{7-3}$$

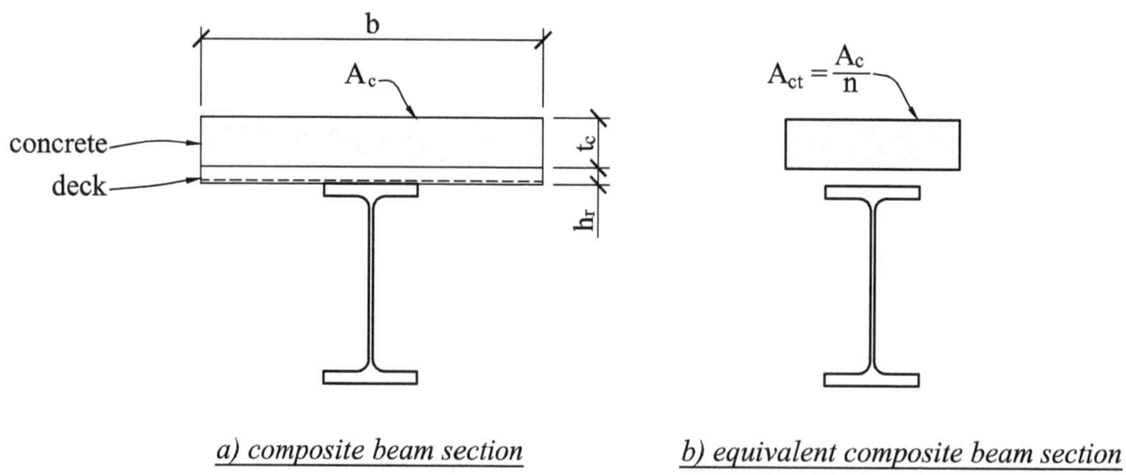

FIGURE 7-8 Transformed composite beam section

where

A_{ct} = Transformed concrete area, and

A_c = Concrete area

= bt_c.

b = effective width of the concrete slab

t_c = thickness of the concrete slab above the top of the metal deck

Once the concrete slab is transformed into an equivalent steel section, the section properties of the composite section can be determined, but this would depend on the location of the plastic neutral axis (PNA). Since we now have a homogeneous section, the PNA is the axis of equal area; that is, the cross-sectional area above the PNA equals the cross-sectional area below the PNA. There are three possible cases that must be considered, each corresponding to the location of the plastic neutral axis (PNA) of the composite section. For positive bending, the composite section area below the PNA is in tension and the area above the PNA is in compression. We will first look at the horizontal shear strength of the shear connectors. The horizontal shear due to the compression force above the PNA is assumed to be resisted by the shear connectors. The horizontal shear in the shear connectors is taken as the lowest of the strength for the following three failure modes:

Crushing of the concrete:

$$V' = 0.85f'_cA_c \tag{7-4}$$

Tensile yielding of the steel beam:

$$V' = F_yA_s \tag{7-5}$$

Strength of the shear connectors:

$$V' = \Sigma Q_n, \tag{7-6}$$

where

V' = Force in the shear connectors,

f'_c = 28-day compressive strength of the concrete, ksi

A_c = Concrete area

= bt_c,

A_s = Area of the steel beam,

F_y = Minimum yield stress in the steel beam (shall not exceed 75 ksi per *AISC* Specification Section I1.3),

Q_n = Nominal strength of the shear connectors between the points of maximum positive and zero bending moment.

The nominal strength of a single shear stud (see AISC Specification Section I8) is

$$Q_n = 0.5A_{sa}\sqrt{f'_cE_c} \le R_gR_pA_{sa}F_u, \tag{7-7}$$

where

A_{sa} = Cross-sectional area of the shear stud,

R_g = Reduction coefficient for a corrugated deck (see Table 7-1)

= 1.0 for formed concrete slabs (no deck),

R_p = Reduction coefficient for a corrugated deck (see Table 7-1)

= 0.75 for formed concrete slabs (no deck), and

F_u = Minimum tensile strength of the shear connector

= 65 ksi for ASTM A29 (see *AWS D1.1 for the mechanical requirements of Type B headed steel studs [1]*).

TABLE 7-1 Reduction Coefficients R_g and R_p

Framing condition		R_g	R_p	Notes
No deck interference (or no deck for formed concrete slabs)		1.0	0.75	Shear connectors are welded directly to the beam top flange; miscellaneous deck fillers can only be placed over less than 50% of the beam flange
Deck ribs oriented parallel to beam (i.e., girders)	$\frac{w_r}{h_r} \geq 1.5$	1.0	0.75	
	$\frac{w_r}{h_r} < 1.5$	0.85	0.75	The value for $R_g = 0.85$ only applies to a single stud
Deck ribs oriented perpendicular to beam	1 stud per rib	1.0	0.6	Values for R_p may be increased to 0.75 when $e_{mid\text{-}ht} \geq 2$ in. (i.e., when studs are placed in the "strong" position, see Figure 7-6)
	2 studs per rib	0.85	0.6	
	3 or more shear studs per rib	0.7	0.6	

Source: Adapted from Section I8.2a in Ref [1].

The shear stud strength, Q_n, can also be determined from *AISCM* Table 3-20.

The number of shear connectors provided is a function of how much of the concrete slab needs to be engaged in composite action with the steel beam to provide the required moment design strength. In many cases, it is economical to provide only enough shear studs to resist the applied moments such that the concrete slab is only partially engaged in composite action, resulting in what is termed *partially composite action.*, as previously mentioned in Section 7-2.

For any composite condition, the section properties can be determined using the parallel axis theorem from the mechanics of materials [9]. Recall from the mechanics of materials that the location of the neutral axis of a composite shape is given by equation (7-8).

Notes:

w_r = Average width of the deck rib

h_r = Nominal deck rib height

$e_{mid\text{-}ht}$ = Horizontal distance from the face of the shank of the shear stud to the mid-height of the adjacent deck rib in the direction of the load (i.e., in the direction of maximum moment for a simply supported composite beam),

$$\bar{y} = \frac{\Sigma A_i y_i}{\Sigma A_i} \tag{7-8}$$

Where,

A_i = area of element i

y_i = location of element i from an assumed origin

The composite moment of inertia, which is called the transformed moment of inertia here since the concrete slab will be transformed into an equivalent steel section, is defined as

$$I_{tr} = \Sigma\left(I_i + A_i d_i^2\right) \tag{7-9}$$

Where,

I_i = moment of inertia of element i above its centroid

$d_i = y_i - \bar{y}$

The transformed section properties calculated using the full slab depth assume that there is full or 100% composite action between the concrete slab and the steel beam and that enough shear connectors are provided across the concrete slab-to-steel interface to achieve this condition. For this to be the case, the shear strength of the shear connectors, ΣQ_n, must be equal to or greater than the compression force in the concrete slab. When the strength of the shear connectors is less than the maximum compression force in the concrete slab, a partially composite section results, and the section properties are reduced. The Commentary to Section I of the *AISCM* gives the reduced effective moment of inertia of a composite section for the partially composite condition as

$$I_{eff} = I_s + \sqrt{\frac{\Sigma Q_n}{C_f}}(I_{tr} - I_s), \tag{7-10}$$

where

I_s = Moment of inertia of the steel section,

I_{tr} = Moment of inertia of the fully composite section,

C_f = the compression force in the concrete slab

= the smaller of $0.85 f'_c A_c$ and $F_y A_s$ (i.e., V' from equations (7-4) and (7-5)), and $\frac{\Sigma Q_n}{C_f} \geq 0.25$.

The reduction term $\frac{\Sigma Q_n}{C_f}$ in equation (7-10) represents the degree of composite action of a section. The degree of composite action should not be less than 50% to ensure ductile failure of the shear connection and to obviate the need to investigate the ductility requirements of the shear connection (see Section I8 of the Commentary to the *AISC* Specifications). However, this requirement is not mandatory, and the *AISC* Specifications allow a minimum degree of composite action as low as 25% without having to investigate the ductility requirements of the shear connectors, provided at least one of the following requirements are satisfied (*AISC* Specification Commentary Section I3.2.2d):

- The beam span does not exceed 30 ft. or
- The shear stud capacity is at least 16 kips per ft. on average within the shear span (i.e., equivalent to ¾-in. diameter studs at 12-in. maximum spacing).

The ductility or slip capacity of the shear connectors is their ability to deform under the interface horizontal shear loads without fracturing, and this allows the interface shear load to be uniformly distributed among the shear connectors within the shear span (i.e., between the point of maximum moment and the nearest point of zero moment) [1]. Composite action should be ignored and the beam treated as a non-composite member when the degree of composite action is less than 25% because of excessive slippage at the concrete-steel interface. Equation (7-10) cannot be used for $\frac{\Sigma Q_n}{C_f}$ values less than 25%.

Effective Width of Composite Beams and Girders

The width of the concrete slab that is effective in the composite section is a function of the beam spacing and the length of the beam. The effective slab width on each side of the beam centerline (see Figure 7-9) is the smallest of

- $^1/_8$ of the beam span,
- ½ of the distance to the adjacent beam, or
- The edge of the slab distance (for edge beams).

For a formed-steel deck, the effective thickness of the concrete slab is reduced to account for the deck ribs. When the deck ribs are oriented perpendicular to the beam, the concrete below the top of the deck (i.e., the concrete within the deck flutes) is neglected. When the deck ribs are oriented parallel to the beam (i.e., the girder), the concrete below the deck ribs may be included in the composite girder analysis. The effective concrete slab thickness is usually taken as the average thickness for this case (see Figure 7-10).

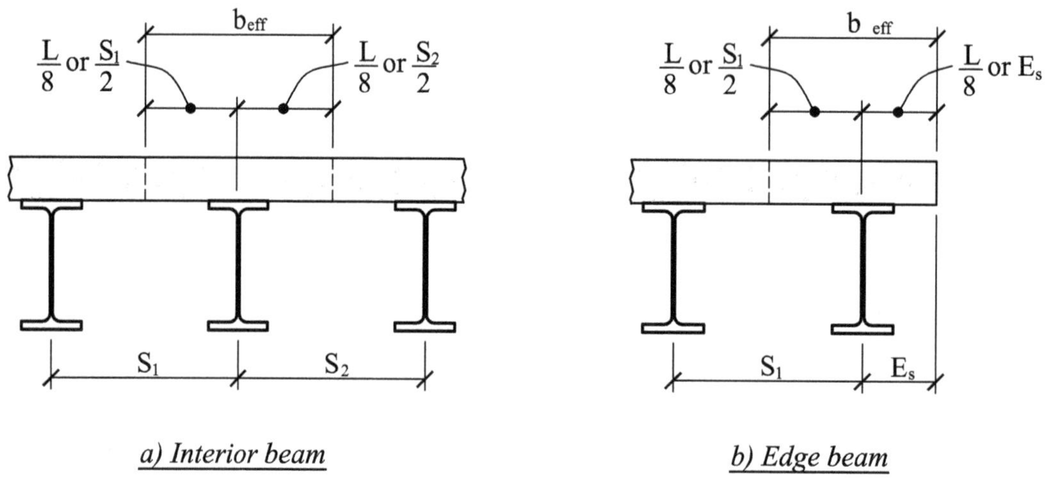

FIGURE 7-9 Effective slab width

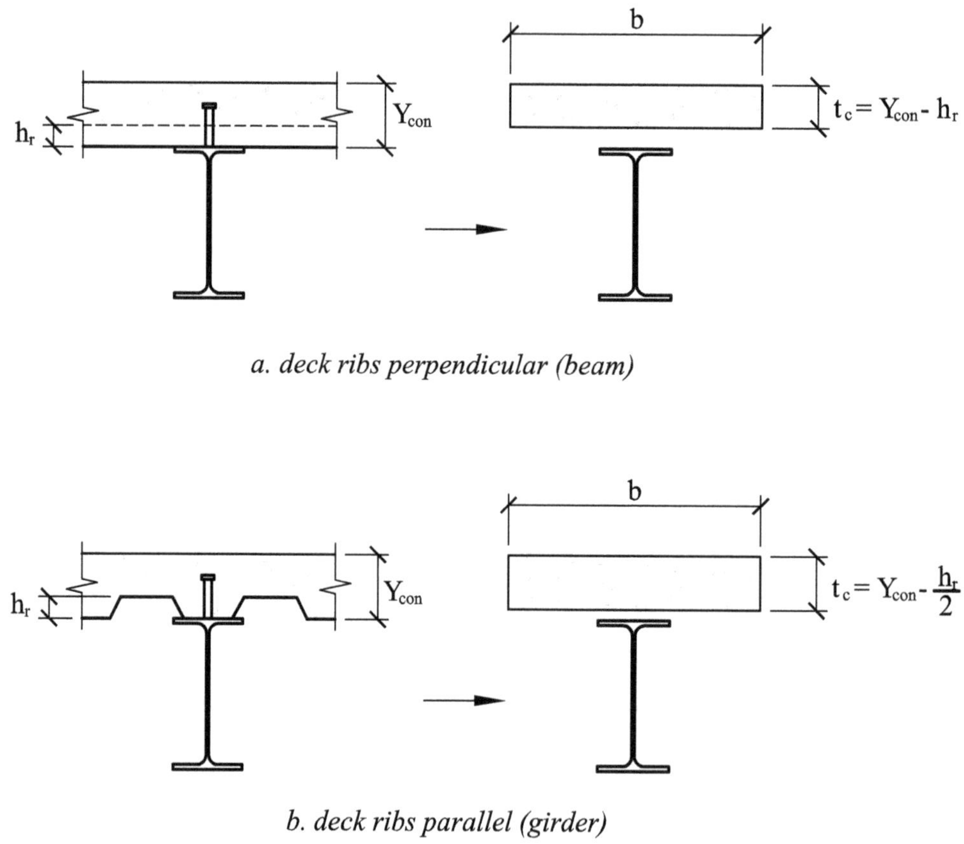

FIGURE 7-10 Effective slab depth

Bending Strength of Composite Sections

We will now consider the bending strength of a composite section. There are three possible locations of the plastic neutral axis (PNA):

- PNA within the concrete slab depth,
- PNA within the thickness of the beam top flange, and
- PNA within the beam web

With partial-composite action, the PNA is usually located within the steel section (in the beam flange or web). This is because a partially composite section engages a smaller amount of the concrete slab in compression; therefore, any additional cross-sectional area needed for compression is provided by the upper portion of the steel beam. A fully composite or 100% composite section, on the other hand, utilizes the steel member fully in tension.

PNA located within the concrete slab depth:

When the PNA is within the concrete slab depth (Figure 7-11), the nominal moment can be determined by taking a summation of moments as follows:

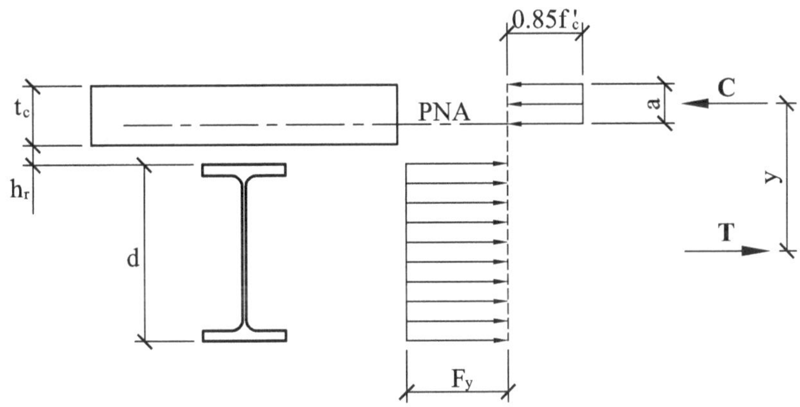

FIGURE 7-11 PNA is within the concrete slab

$$M_n = Ty \ (\text{or } C_c y), \tag{7-11}$$

where

T = Tension force in the steel,

C_c = Compression force in the concrete, and

y = Distance between T and C (see Figure 7-11).

The depth of the compression stress block is

$$a = \frac{A_s F_y}{0.85 f_c' b}, \tag{7-12}$$

where

A_s = Area of the steel section,

F_y = Yield stress,

f_c' = 28-day compressive strength of the concrete, and

b = Effective slab width. (Note that this symbol is used interchangeably with b_{eff} in this text to define the effective slab width.) The effective width accounts for the effect of shear lag in the compression zone of the composite beam. Shear lag indicates that the distribution of the normal compression stress due to bending over the full width of the slab tributary to the beam or girder is not uniform. The effective width is the equivalent or reduced width of slab across which the normal compression stress is assumed to be uniform.

The equations for the design moment capacity and the allowable moment, respectively, are given as

$$\phi_b M_n = \phi_b A_s F_y \left(\frac{d}{2} + h_r + t_c - \frac{a}{2} \right), \text{ for LRFD} \quad (7\text{-}13)$$

$$\frac{M_n}{\Omega_b} = \frac{A_s F_y \left(\frac{d}{2} + h_r + t_c - \frac{a}{2} \right)}{\Omega_b}, \text{ for ASD} \quad (7\text{-}14)$$

Where,

h_r = Deck thickness (i.e., height of the deck ribs),
t_c = Concrete thickness above the deck, and
d = Beam depth.
ϕ_b = 0.9
Ω_b = 1.67
Y_{con} = total slab thickness = $h_r + t_c$.

PNA located within the thickness of the steel beam top flange:

When the PNA is within the top flange (Figure 7-12), the compression force in the concrete slab is at its maximum value. Therefore,

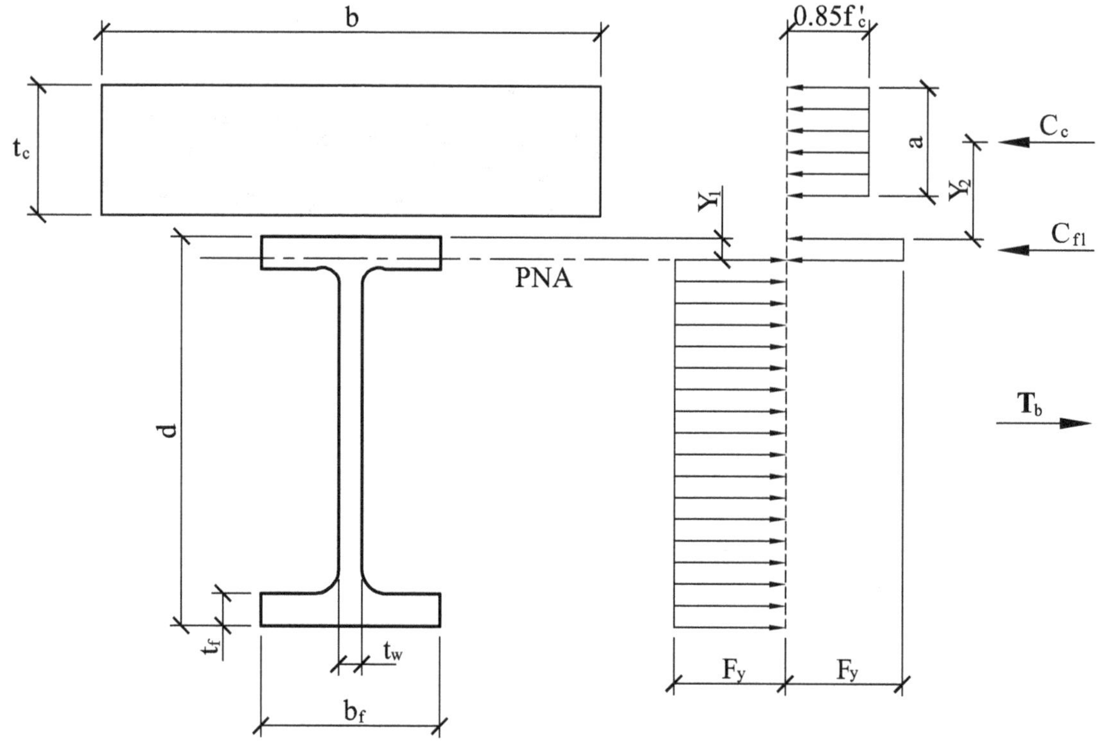

FIGURE 7-12 PNA is within the flange of the steel beam

$$C_c = 0.85 f_c' A_c, \quad (7\text{-}15)$$

where C_c is the compression force in the concrete slab and A_c is the area of concrete. The compression force in the top flange of the steel beam is

$$C_{fl} = b_f Y_1 F_y, \quad (7\text{-}16)$$

Where,

b_f = Beam flange width, and

Y_1 = Distance from the PNA to the top of the top flange of the steel member.

To determine the tension in the bottom portion of the beam, T_b, we assume that the entire beam has yielded in tension, and we subtract the portion that is in compression:

$$T_b = A_s F_y - b_f Y_1 F_y \quad (7\text{-}17)$$

For equilibrium, the summation of horizontal forces yields

$$T_b = C_c + C_{fl}. \quad (7\text{-}18)$$

Combining equations (7-15) through (7-18) yields an expression for Y_1 as follows:

$$A_s F_y - b_f Y_1 F_y = 0.85 f'_c A_c + b_f Y_1 F_y$$

Solving the above equation gives

$$Y_1 = \frac{A_s F_y - 0.85 f'_c A_c}{2 F_y b_f}, \quad (7\text{-}19)$$

Where,

Y_1 is the distance from the top of the top flange to the PNA.

The design moment capacity and the available moment, respectively, can be determined by summing moments about the PNA, which yields

$$\phi_b M_n = \phi_b \left[0.85 f'_c A_c (Y_1 + Y_2) + 2 F_y b_f Y_1 \left(\frac{Y_1}{2} \right) + A_s F_y \left(\frac{d}{2} - Y_1 \right) \right], \text{ for LRFD} \quad (7\text{-}20)$$

$$\frac{M_n}{\Omega_b} = \frac{\left[0.85 f'_c A_c (Y_1 + Y_2) + 2 F_y b_f Y_1 \left(\frac{Y_1}{2} \right) + A_s F_y \left(\frac{d}{2} - Y_1 \right) \right]}{\Omega_b}, \text{ for ASD} \quad (7\text{-}21)$$

where,

Y_2 = the distance from the top of the beam top flange to the centroid of the concrete compressive force , $C_c = Y_{con} - \frac{a}{2} = h_r + t_c - \frac{a}{2}$

$\phi_b = 0.9$

$\Omega_b = 1.67$

PNA located within steel beam web:

When the PNA is located within the web of the steel beam, a similar analysis can be carried out to determine Y_1 (see Figure 7-13). The compression force in the slab, C_c, is found from equation (7-15). The compression force in the beam flange is

$$C_{fl} = b_f t_f F_y, \quad (7\text{-}22)$$

where t_f is the beam flange thickness.

The compression force in the upper part of the beam web is

$$C_w = t_w F_y (Y_1 - t_f), \quad (7\text{-}23)$$

where t_w is the beam web thickness.

The tension force in the bottom portion of the beam is

$$T_b = A_s F_y - b_f t_f F_y - t_w F_y (Y_1 - t_f) \quad (7\text{-}24)$$

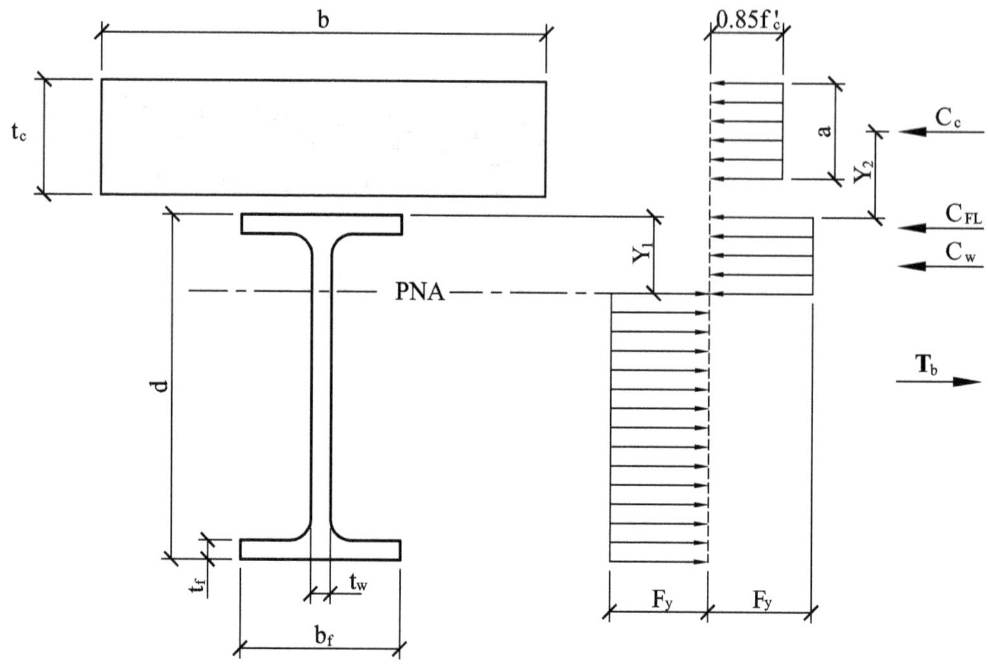

FIGURE 7-13 PNA is within the web of the steel beam

$$T_b = A_s F_y - b_f t_f F_y - t_w F_y (Y_1 - t_f) \tag{7-24}$$

For equilibrium, the summation of horizontal forces yields

$$T_b = C_c + C_{fl} + C_w \tag{7-25}$$

Combine equations (7-24) and (7-25) and solve for Y_1 as follows:

$$A_s F_y - b_f t_f F_y - t_w F_y (Y_1 - t_f) = 0.85 f'_c A_c + b_f t_f F_y + t_w F_y (Y_1 - t_f)$$

$$Y_1 = \frac{A_s F_y - 0.85 f'_c A_c - 2 b_f t_f F_y}{2 t_w F_y} + t_f \tag{7-26}$$

The nominal moment strength can be determined by summing moments about the PNA, which yields

$$M_n = \left[0.85 f'_c A_c (Y_1 + Y_2) + 2 b_f t_f F_y \left(Y_1 - \frac{t_f}{2} \right) \right.$$

$$\left. + 2 t_w F_y (Y_1 - t_f) \left(\frac{Y_1 - t_f}{2} \right) + A_s F_y \left(\frac{d}{2} - Y_1 \right) \right] \tag{7-27}$$

For LRFD, the design moment capacity is:

$$\phi_b M_n = \phi_b \left[0.85 f'_c A_c (Y_1 + Y_2) + 2 b_f t_f F_y \left(Y_1 - \frac{t_f}{2} \right) + 2 t_w F_y (Y_1 - t_f) \left(\frac{Y_1 - t_f}{2} \right) + A_s F_y \left(\frac{d}{2} - Y_1 \right) \right] \tag{7-28}$$

For ASD, the allowable moment capacity is:

$$\frac{M_n}{\Omega_b} = \frac{\left[0.85f'_c A_c (Y_1 + Y_2) + 2b_f t_f F_y \left(Y_1 - \frac{t_f}{2}\right) + 2t_w F_y (Y_1 - t_f)\left(\frac{Y_1 - t_f}{2}\right) + A_s F_y \left(\frac{d}{2} - Y_1\right)\right]}{\Omega_b} \tag{7-29}$$

Where,
$\phi_b = 0.9$
$\Omega_b = 1.67$

Several examples will follow to illustrate how to calculate the section properties, the nominal bending strength, the design moment capacity (for LRFD), and the allowable moment (for ASD) for each possible PNA location.

EXAMPLE 7-1

Transformed Section Properties: Full-Composite Action

For the composite section shown in Figure 7-14, determine the transformed moment of inertia. Assume that the section is fully composite and that the concrete has a density of 145 pcf and a 28-day concrete strength of 3.5 ksi.

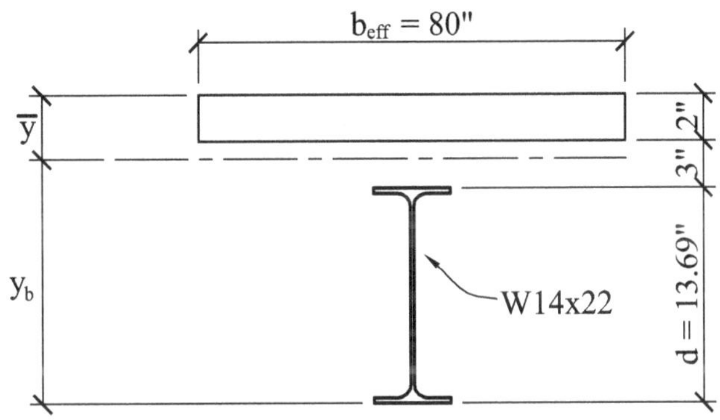

FIGURE 7-14 Detail for Example 7-1

Solution

From *AISCM* Table 1-1, the section properties of $W14 \times 22$ are

$A = 6.49 \text{ in.}^2$
$d = 13.7 \text{ in.}$
$I = 199 \text{ in.}^4$

Transforming the concrete section into an equivalent steel section,

$$E_c = w_c^{1.5}\sqrt{f'_c}$$
$$= (145)^{1.5}\sqrt{3.5} = 3266 \text{ ksi}$$
$$n = \frac{E_s}{E_c}$$
$$n = \frac{29{,}000}{3266} = 8.87$$
$$A_{ct} = \frac{A_c}{n}$$
$$= \frac{(80)(2)}{8.87} = 18.02 \text{ in.}^2$$

Using the top of the concrete slab as the datum, we develop Table 7-2.

TABLE 7-2 Transformed Section Properties

Element	A	y	Ay	I	$d = y - \bar{y}$	$I + Ad^2$
Slab	18.02	1	18.02	6.01	−2.87	154.4
W14 × 22	6.49	11.85	76.9	199	7.98	612.2
Σ =	**24.51**		**94.92**			I_{tr} **= 766.6**

$$I_{ct} = \frac{bt_c^3}{12n}$$

$$= \frac{(80)(2)^3}{(12)(8.87)} = 6.01 \text{ in.}^4$$

$$\bar{y} = \frac{\Sigma Ay}{\Sigma A}$$

$$= \frac{94.92}{24.51} = 3.87 \text{ in.}$$

$$I_{tr} = \sum\left(I + Ad^2\right)$$

$$= 766.6 \text{ in.}^4 \ \text{(see Table 7-2)}$$

EXAMPLE 7-2

Design Bending Strength of a Composite Section

Determine the design bending strength of the composite section given in Example 7-1, assuming full-composite action and ASTM A992 steel.

Solution

Since there is full-composite action, the compressive force in the concrete is the smaller of $0.85 f_c' A_c$ (crushing of the concrete) and $F_y A_s$ (tensile yielding of the steel beam).

$$C_c = 0.85 f_c' A_c$$

$$= (0.85)(3.5)(2)(80) = 476 \text{ kips}$$

$$C_c = F_y A_s$$

$$= (50)(6.49) = \mathbf{324.5\,kips}$$

Tensile yielding in the steel controls, which means that only a portion of the concrete slab is needed to develop the compressive force. Figure 7-15 shows the stress distribution in the slab.

From equation (7-4), the effective slab depth can be determined as follows:

$$C_c = 0.85 f_c' A_c = 0.85 f_c' ab$$

$$324.5 = (0.85)(3.5)(a)(80)$$

Therefore, $a = 1.36$ in.

The moment arm between the tensile and compressive forces is

$$y = \frac{d}{2} + 3 + 2 - \frac{a}{2}$$

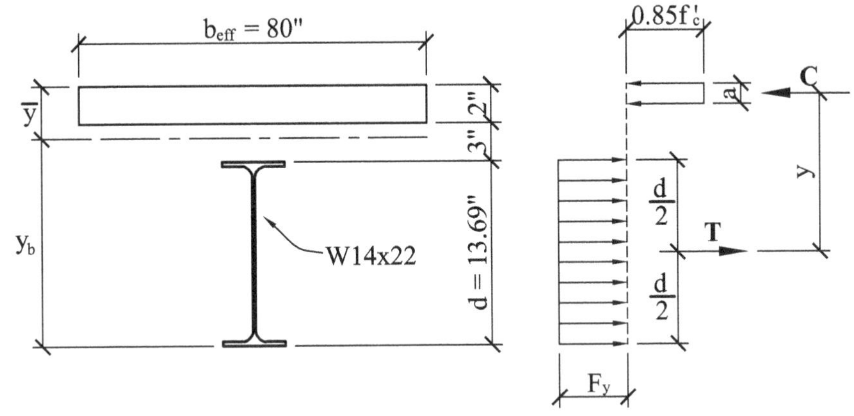

FIGURE 7-15 Stress distribution for Example 7-2

$$y = \frac{13.7}{2} + 3 + 2 - \frac{1.36}{2} = 11.16 \text{ in.}$$

The design bending strength (for LRFD) and the allowable bending moment (for ASD) are found using equation (7-11) as follows:

For LRFD:

$$\phi_b M_n = \phi_b Ty \ (\text{or } \phi_b C_c y) = (0.9)(324.5)(11.16) = 3260 \text{ in.-kips} = 271 \text{ ft.-kips.}$$

For ASD:

$$\frac{M_n}{\Omega_b} = \frac{Ty}{\Omega_b}\left(\text{or } \frac{C_c y}{\Omega_b}\right) = \frac{(324.5)(11.16)}{1.67} = 2169 \text{ in.-kips} = 180 \text{ ft.-kips.}$$

EXAMPLE 7-3

Transformed Section Properties: Partial-Composite Action

Determine the section properties and design moment capacity of the composite section given in Example 7-1, assuming that (8)- $^3/_4$" ASTM A29-headed studs (Type B with F_u = 65 ksi) are provided between points of maximum and zero moments in the deck profile shown in Figure 7-16. Assume that the studs are placed in the "strong" or "favorable" position within the deck flutes or ribs.

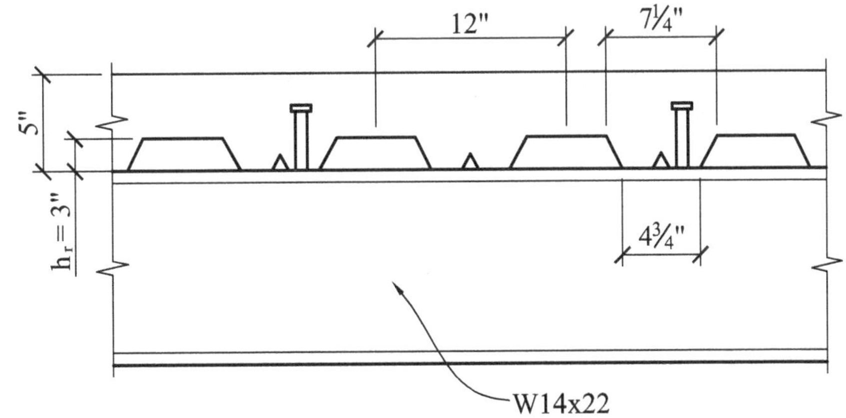

FIGURE 7-16 Metal deck profile for Example 7-3

Solution

From *AISCM* Table 1-1, the section properties for W14x22 are as follows:

$A = 6.49 \text{ in.}^2$

$d = 13.7 \text{ in.}$

$I = 199 \text{ in.}^4$

$b_f = 5.00 \text{ in.}$

$t_f = 0.335 \text{ in.}$

$t_w = 0.23 \text{ in.}$

The horizontal force in the shear connectors (V') is the smaller of the following three values:

$$V' = 0.85 f'_c A_c$$

$$= (0.85)(3.5)(2)(80) = 476 \text{ kips, or}$$

$$V' = A_s F_y$$

$$= (50)(6.49) = 324.5 \text{ kips.}$$

$V' = \sum Q_n$, and this will now be calculated.

From Table 7-1, for one row of shear studs, and with the deck ribs perpendicular to the beam,

$R_g = 1.0$

$R_p = 0.75$("strong" position of the stud in the deck rib)

$$A_{sa} = \frac{\pi d^2}{4} = \frac{\pi(0.75)^2}{4} = 0.441 \text{ in.}^2$$

$$Q_n = 0.5 A_{sa} \sqrt{f'_c E_c} \leq R_g R_p A_{sa} F_u$$

$$= (0.5)(0.441)\sqrt{(3.5)(3266)} \leq (1.0)(0.75)(0.441)(65)$$

$$= 23.5 \text{ kips} \leq 21.5 \text{ kips}$$

Therefore, $Q_n = 21.5$ kips (this value agrees with *AISCM* Table 3-20 for one stud per deck rib in a strong position).

Note: If the weak stud position was assumed (and thus obviating the need to have to verify the strong stud position on site), R_p will be 0.6 instead of 0.75, and the stud shear capacity, Q_n, will be 17.2 kips instead of 21.5 kips. Some designers assume the weak stud position in practice.

Since there are 8 studs between the point of maximum moment and the nearest point of zero moment, therefore, N_s = 8, and

$$V' = \Sigma Q_n = N_s Q_n = 8 Q_n$$

$$= (8)(21.5) = 172 \text{ kips} < 476 \text{ kips or } 324.5 \text{kips}$$

Therefore, C = 172 kips

The degree of composite action is found from the reduction term in equation (7-10), with C_f being 324.5 kips (the smaller of equations (7-4) and (7-5)):

$$\frac{\Sigma Q_n}{C_f} = \frac{172}{324.5} = 0.53, \text{ or about 53\% composite action.}$$

This section is at least 25% composite, so the ductility of the shear stud connection does not need to be investigated, so we can proceed to calculate the reduced section properties:

The effective moment of inertia (see equation (7-10)) is

$$I_{eff} = I_s + \sqrt{\frac{\Sigma Q_n}{C_f}}\left(I_{tr} - I_s\right)$$

$$= 199 + \sqrt{\frac{172}{324.5}}\left(766.6 - 199\right) = 612 \text{ in.}^4$$

Recall that the transformed moment of inertia, I_{tr}, from Example 7-1 was calculated as 766.6 in.^{4} Note that ΣQ_n = 172 kips represents the maximum horizontal force that can be transferred through the concrete slab-steel beam interface.

Since $\Sigma Q_n < A_s F_y$, this indicates that the steel beam is not fully yielded in tension; therefore, some portion of the steel beam will be in compression in addition to the compression in the concrete slab. Therefore, there is a compressive force within the beam section, and hence the PNA lies somewhere within the steel beam depth. Using equation (7-4), the effective slab depth can be determined by equating the compression force in the concrete to the shear capacity of the shear studs (i.e., $C_c = \Sigma Q_n$):

$$C_c = 0.85 f'_c A_c = 0.85 f'_c ab$$

$$172 = (0.85)(3.5)(a)(80)$$

Therefore, a = 0.722 in.

Assuming that the PNA is within the beam flange, Y_1 is determined from equation (7-19):

$$Y_1 = \frac{A_s F_y - 0.85 f'_c A_c}{2 F_y b_f}$$

$$Y_1 = \frac{(6.49)(50) - (0.85)(3.5)(0.722)(80)}{(2)(50)(5.00)} = 0.305 \text{ in.}$$

$Y_1 = 0.305\ in. < t_f = 0.335\ in.$; therefore, the PNA is within the beam flange depth as assumed.

The design bending strength (for LRFD) and the allowable bending moment (for ASD) are then determined from equations (7-20) and (7-21), respectively, as follows:

The distance from the top of the steel beam top flange to the centroid of the concrete compression force is

$$Y_2 = Y_{con} - \frac{a}{2} = h_r + t_c - \frac{a}{2}$$

$$= 3 + 2 - \frac{0.722}{2} = 4.64 \text{ in.}$$

For LRFD:

$$\phi_b M_n = \phi_b \left[0.85 f'_c A_c (Y_1 + Y_2) + 2F_y b_f Y_1 \left(\frac{Y_1}{2}\right) + A_s F_y \left(\frac{d}{2} - Y_1\right)\right]$$

$$\phi_b M_n = 0.9 \left[\left[(0.85)(3.5)(0.722)(80)(0.305 + 4.64)\right] + \left[(2)(50)(5.00)(0.305)\left(\frac{0.305}{2}\right)\right] + \left[(6.49)(50)\left(\frac{13.7}{2} - 0.305\right)\right]\right]$$

$$\phi_b M_n = 2697 \text{ in.-kips} = 225 \text{ ft.-kips}$$

For ASD:

$$\frac{M_n}{\Omega_b} = \frac{\left[0.85 f'_c A_c (Y_1 + Y_2) + 2F_y b_f Y_1 \left(\frac{Y_1}{2}\right) + A_s F_y \left(\frac{d}{2} - Y_1\right)\right]}{\Omega_b}$$

$$= \left(\frac{1}{1.67}\right)\left[\left[(0.85)(3.5)(0.722)(80)(0.305+4.64)\right]\right.$$

$$\left. + \left[(2)(50)(5.00)(0.305)\left(\frac{0.305}{2}\right)\right] + \left[(6.49)(50)\left(\frac{13.7}{2} - 0.305\right)\right]\right]$$

$$\frac{M_n}{\Omega_b} = 1798 \text{ in.-kips} = 150 \text{ ft.-kips}$$

EXAMPLE 7-4

PNA in the Concrete Slab

Determine the design moment strength for the beam shown in Figure 7-17. The concrete has a 28-day compressive strength of 4 ksi. Assume full-composite action, and use F_y = 50 ksi for the steel beam.

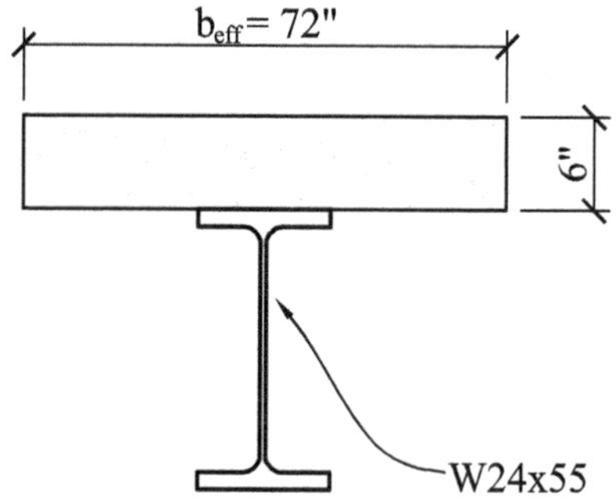

FIGURE 7-17 Detail for Example 7-4

Solution

From *AISCM* Table 1-1, the section properties for W24x55 are

$A = 16.2$ in.^{2}

$d = 23.6$ in.

$I = 1350$ in.^{4}

Determine the location of the PNA starting with equation (7-12):

$$a = \frac{A_s F_y}{0.85 f'_c b_e} = \frac{(16.2)(50)}{(0.85)(4)(72)} = 3.31 \text{ in.} < Y_{con} = 6 \text{ in.};$$

Therefore, the PNA is within the slab depth.

The design bending strength (for LRFD) and the allowable bending moment (for ASD) are found from equations (7-13) and (7-14) as follows:

For LRFD:

$$\phi_b M_n = \phi A_s F_y \left(\frac{d}{2} + h_r + t_c - \frac{a}{2}\right) = (0.9)(16.2)(50)\left(\frac{23.6}{2} + 0 + 6 \text{ in.} - \frac{3.31}{2}\right)$$

$$= 11{,}769 \text{ in.-kips} = 980 \text{ ft.-kips}$$

For ASD:

$$\frac{M_n}{\Omega_b} = \frac{A_s F_y \left(\frac{d}{2} + h_r + t_c - \frac{a}{2}\right)}{\Omega_b} = \frac{(16.2)(50)\left(\frac{23.6}{2} + 0 + 6 \text{ in.} - \frac{3.31}{2}\right)}{1.67} = 7846 \text{ in.-kips} = 653 \text{ ft.-kips}$$

EXAMPLE 7-5

PNA in the Beam Web

Determine the design moment strength for the beam shown in Figure 7-18. The concrete has a 28-day strength of 4 ksi. Assume 25% composite action, and use F_y = 50 ksi for the steel beam.

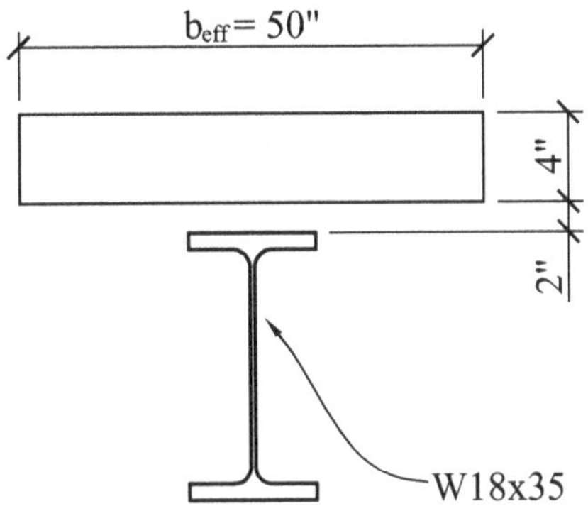

FIGURE 7-18 Detail for Example 7-5

Solution

From *AISCM* Table 1-1, the section properties for W18 × 35 are

$A = 10.3 \text{ in.}^2$, $t_f = 0.425$ in.

$d = 17.7$ in., $b_f = 6.00$ in.

$I = 510 \text{ in.}^4$, $t_w = 0.300$ in.

$V' =$ Smaller of $0.85f'_c A_c$ or $F_y A_s$ (i.e., see equations (7-4) and (7-5))

$$0.85f'_c A_c = (0.85)(4)(50)(4) = 680 \text{ kips}$$

$$F_y A_s = (50)(10.3) = 515 \text{ kips (controls)}$$

Assuming 25% composite action, C_c = (0.25)(515) = 129 kips < 680 kips or 515 kips. Therefore, C_c = 129 kips.

Since $C_c < A_s F_y$, therefore, the PNA is in the steel beam web.

$$C_c = 0.85f'_c ab,$$

i.e., $129 \text{ kips} = (0.85)(4)(a)(50 \text{ in.})$

Therefore, a = 0.757 in.

$$Y_2 = h_r + t_c - \frac{a}{2} = 2 + 4 - \frac{0.757}{2} = 5.62 \text{ in.}$$

Recall that Y_2 is the vertical distance between the top of the top flange of the steel beam or girder and the centroid of the concrete compression stress block, while Y_1 is the distance from the PNA to the top of the steel beam top flange. Use equations (7-26) and (7-27) to determine Y_1 as follows (if Y_1 is greater than t_f, then PNA is, in fact, in the beam web as assumed):

$$Y_1 = \frac{A_s F_y - 0.85f'_c A_c - 2b_f t_f F_y}{2t_w F_y} + t_f$$

$$= \frac{(10.3)(50) - (0.85)(4)(50)(0.757) - (2)(6.0)(0.425)(50)}{(2)(0.300)(50)} + 0.425 = 4.80 \text{ in.}$$

Since Y_1 is greater than the beam flange thickness ($Y_1 = 4.80$ in. > $t_f = 0.425$ in.), the PNA cannot be located within the steel beam flange, but rather, it is located within the depth of the beam web. The design bending strength (for LRFD) and the allowable bending moment (for ASD) are then found from equations (7-28) and (7-29) as follows:

For LRFD,

$$\phi_b M_n = \phi_b\left[0.85f'_c A_c(Y_1 + Y_2) + 2b_f t_f F_y\left(Y_1 - \frac{t_f}{2}\right) + 2t_w F_y(Y_1 - t_f)\left(\frac{Y_1 - t_f}{2}\right) + A_s F_y\left(\frac{d}{2} - Y_1\right)\right]$$

$$= (0.9)\left[(0.85)(4)(50)(0.757)(4.80 + 5.62) + (2)(6.0)(0.425)(50)\left(4.80 - \frac{0.425}{2}\right)\right] +$$

$$(0.9)\left[(2)(0.300)(50)(4.80 - 0.425)\left(\frac{4.80 - 0.425}{2}\right) + (10.3)(50)\left(\frac{17.7}{2} - 4.80\right)\right]$$

$= 4396$ in.-kips

$\phi_b M_n =$ **367 ft.-kips.**

For ASD,

$$\frac{M_n}{\Omega_b} = \frac{\left[0.85f'_c A_c(Y_1 + Y_2) + 2b_f t_f F_y\left(Y_1 - \frac{t_f}{2}\right) + 2t_w F_y(Y_1 - t_f)\left(\frac{Y_1 - t_f}{2}\right) + A_s F_y\left(\frac{d}{2} - Y_1\right)\right]}{\Omega_b}$$

$$= \left(\frac{1}{1.67}\right)\left[(0.85)(4)(50)(0.757)(4.80 + 5.62) + (2)(6.0)(0.425)(50)\left(4.80 - \frac{0.425}{2}\right)\right] +$$

$$\left(\frac{1}{1.67}\right)\left[(2)(0.300)(50)(4.80 - 0.425)\left(\frac{4.80 - 0.425}{2}\right) + (10.3)(50)\left(\frac{17.7}{2} - 4.80\right)\right]$$

$= 2930$ in.-kips

$\frac{M_n}{\Omega_b} = 244$ **ft.-kips.**

7-4 SHORING OF COMPOSITE BEAMS

Before discussing the deflection of composite beams, we need to look at the concept of shored versus unshored construction. Shoring can be provided under the beams and girders to allow the concrete floor to cure and harden, and reach its design strength prior to imposing any load on the steel beams and girders (see Figure 7-19a). Once the shores are removed, the steel framing (i.e., the beams and girders) will undergo an instantaneous deflection due to the weight of the concrete floor slab, but at this time, the composite action of the steel framing with the hardened concrete has already developed, and thus, the load will be jointly supported by the steel framing and the concrete slab, acting compositely. On the other hand, with unshored construction, the weight of the wet concrete is superimposed on the steel floor framing — while it is still non-composite — prior to the concrete curing, thus causing the non-composite steel beams and girders to deflect under the weight of the concrete and any construction live loads. In unshored construction, the steel beams and girders alone — without help from the concrete and thus acting as a non-composite beams and girders — support the construction phase dead and live loads.

The construction phase dead load will include the weight of the wet concrete, including any concrete ponding loads (i.e., the additional concrete weight) arising from the deflection profile of the beam or girder (see Figure 7-19b), while the construction phase live loads include the weight of the construction crew and equipment on the slab during the concrete pour (see Table 2-13). Recommended construction live loads (see ASCE 37-14 [10]) are as follows:

- 20 psf for very light duty (frequently used for slab construction);
- 25 psf for light duty; 50 psf for medium duty; and
- 75 psf for heavy duty construction

Most designers use a construction live load of 20 psf for the construction phase design check [5]. (Recall that construction live loads were previously discussed in Chapter 2.)

The instantaneous deflection of the beams when shored construction is used will not be as much as it is in the unshored construction scheme since the beam in the former case will be in a composite state and thus have greater flexural stiffness from the get-go before it begins to resist any loads.

From a design standpoint, there is no conclusive data that favors either form of construction. From a construction standpoint, the unshored scheme is usually preferred because it avoids the labor required to install and remove the shores. However, the disadvantage of unshored construction is that because the beams deflect under the weight of the freshly poured concrete, additional concrete will have to be poured to achieve a flat floor surface, and this could result in concrete ponding. As discussed in Chapter 2, concrete ponding occurs when the deflected shape of a beam loaded with wet concrete allows additional concrete to accumulate (see Figure 7-19b). This creates a situation where the builder or contractor must account in their bid price for the additional volume of concrete that is needed to be placed to achieve a flat floor due to the beam/girder deflections, and the designer must also account for the extra concrete weight in their design loads, as previously discussed in Chapter 2. For common floor framing systems, the additional concrete volume required to account for ponding can range from 10% to 15% (see Figure 7-19c). One way to mitigate concrete ponding is to camber the beams; this is an upward curvature fabricated into the beam or girder to negate the downward deflection of the member, and the minimum camber should be at least $^3/_4$ in. It would be ideal to camber the beams by an amount equal to the beam deflection due to the wet concrete, but this is not recommended because if there is too much upward camber in the beam, then the slab might end up being too thin at the middle of the beam, and there would not be enough concrete cover for the headed studs (see Figure 7-19d) that is needed for fire and corrosion protection and for bonding of the stud with the surrounding concrete. For this reason, most designers provide a camber that is equivalent to 75% to 85% of the calculated deflection due to the dead load of the concrete. This reduction accounts for the possibility of overestimating the dead load, as well as the fact that the calculated deflection assumes pinned supports and does not account for any partial restraint at the beam supports. Beams and girders that are part of the lateral force resisting system or beams/girders with end moment connections, or cantilevered beams, or spandrel beams, and beams or girders that are less than 25 ft. long should not be cambered. These limits ensure economy in fabrication and cambering processes. The tolerances for camber of beams presented in Section 11.2.4 of the AISC Code of Standard Practice [1] allow beams or girders with spans up to 50 ft to be over-cambered by up to ½ inch but do not allow for a smaller amount of camber than specified; that is, the tolerance for camber of beams/girders is -0/+½ in.

With unshored construction, the beams must be designed to support the concrete slab, as well as the temporary construction dead and live loads present while the slab is being placed. In floor systems with a formed-steel deck, the floor deck is usually adequate to fully brace the top flange of the beams against lateral–torsional buckling because the deck ribs are oriented in their strong orthogonal direction, with the ribs perpendicular to the beam. However, the floor deck is usually not adequate to provide lateral stability to the girders because the deck is oriented

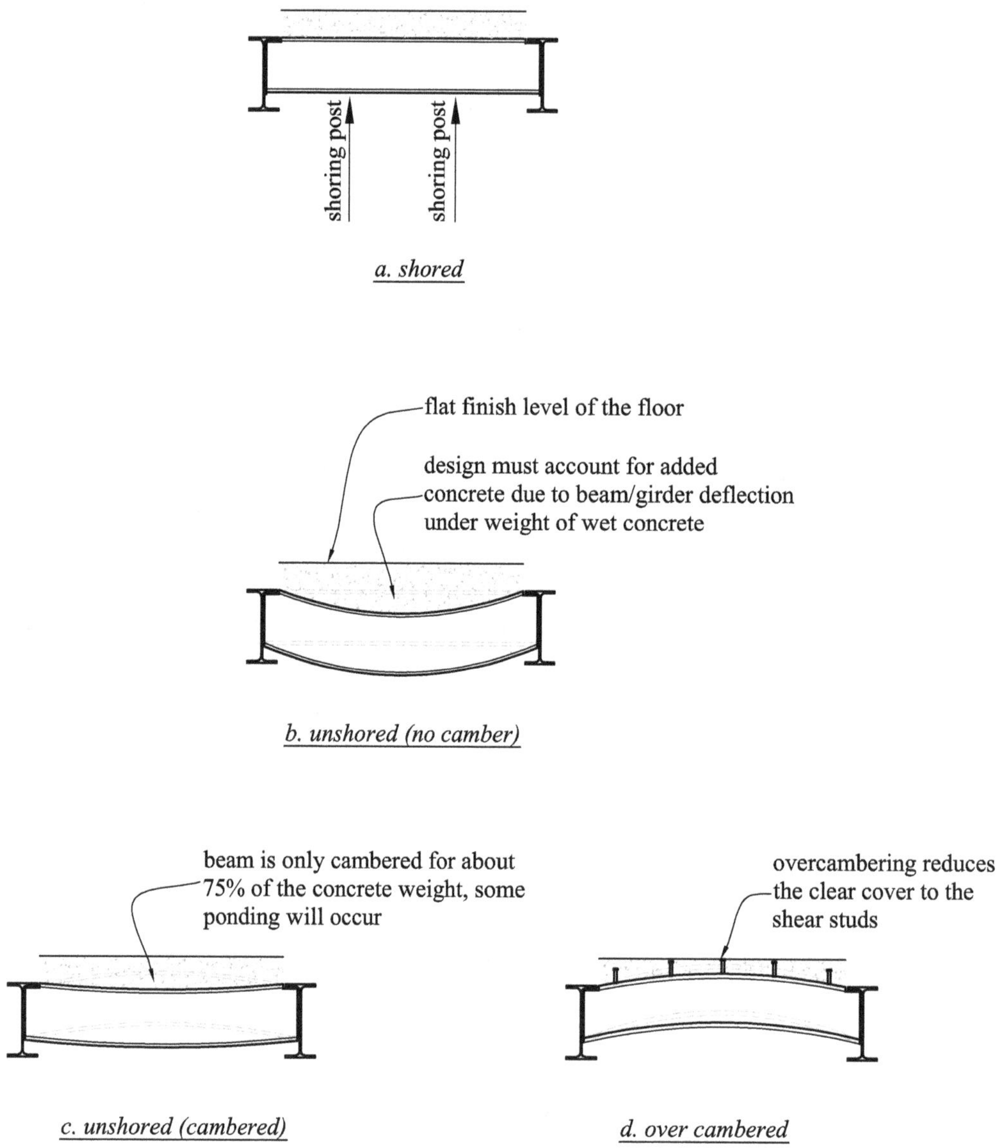

FIGURE 7-19 Shored and unshored beams

in its weak orthogonal direction, with the deck ribs parallel to the girder. In this orientation, the lateral load that causes lateral torsional buckling in the girder will be perpendicular to the deck ribs, and the inplane load capacity of the deck is smallest in this direction since the deck will simply crumble or fold up under this inplane load. The designer must therefore consider the unbraced length of the beams and girders during slab construction phase loading. The compression flange of the beams is assumed to be fully braced by the deck; the girders are laterally braced only at the locations where the infill beams frame into the girder.

For shored construction, the advantages are that all of the strength and deflection checks are based on the composite condition, and the strength of the steel beam alone is not a factor when the concrete is still wet. Aside from the fact that added labor and material will be required for shored construction, one key disadvantage is that cracks are likely to occur over the supporting girders and sometimes over the beams as well. One way to mitigate this occurrence is to add reinforcement over the beams and girders to control the cracking (see Figure 7-20). Even with unshored construction, cracking is somewhat common over the supporting girders, so steel reinforcement (rebar) is often added over (and perpendicular to) the girders in either construction scheme.

Another way to mitigate cracking over the beams and girders in a shored scheme is to place the shoring in such a way that some amount of deflection in the beams will occur while minimizing the amount of ponding. Figure 7-21 shows one recommended shoring placement that allows some deflection and minimizes ponding. In this scheme, shores are placed under the girders at the points where a beam intersects and at a distance of $L/5$ from the ends of the beam.

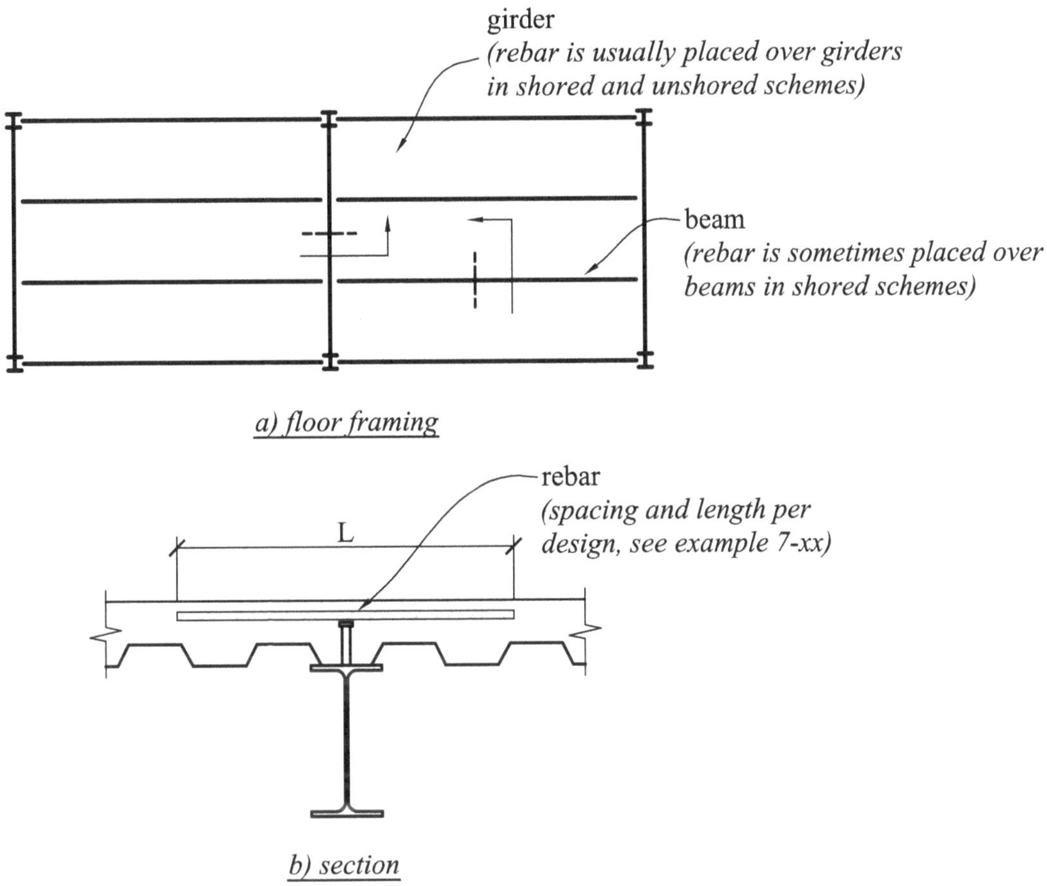

FIGURE 7-20 Crack control over beams and girders

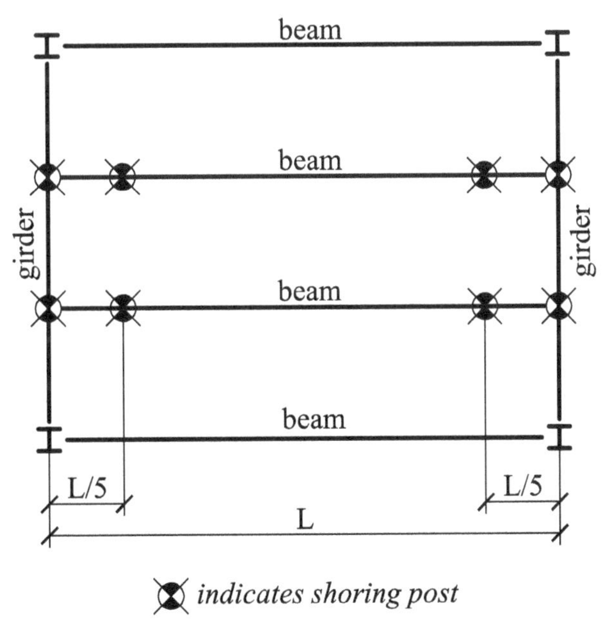

FIGURE 7-21 Shoring to allow some deflection and minimize ponding

EXAMPLE 7-6

Construction Phase Loading

For the floor framing shown in Figure 7-22, determine the following, assuming ASTM A992 Grade 50 steel and a slab weight of 75 psf including concrete ponding load (neglect the self-weight of the framing):

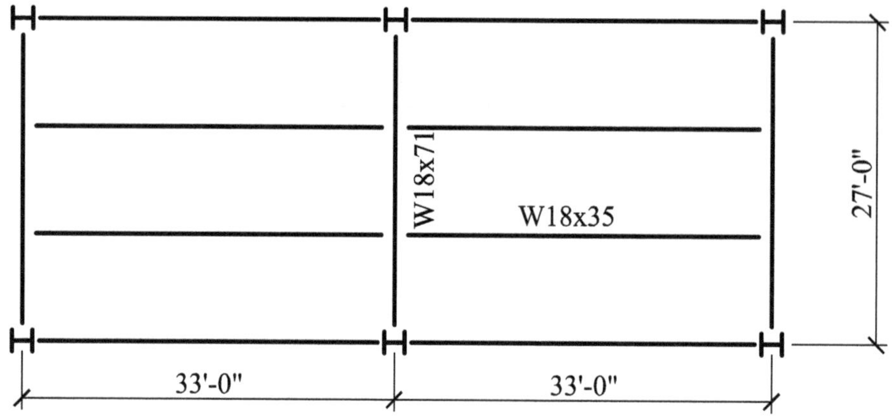

FIGURE 7-22 Floor framing for Example 7-6

1. Required camber in the W18 × 35 beam for an unshored scheme,
2. Required camber in the W18 × 71 girder for an unshored scheme,
3. Adequacy of the W18 × 35 beam for construction phase loading
4. Adequacy of the W18 × 71 girder for construction phase loading.

Assume very light construction without any motorized vehicles on the floor.

Solution

Recall from Chapter 2 that the recommended construction live load for very light construction without motorized vehicles on the floor during construction is 20 psf.

The following load combinations are applicable for LRFD for the construction phase:

$$1.4D = (1.4)(75 \text{ psf}) = 105 \text{ psf}$$

$$1.2D + 1.6L = (1.2)(75 \text{ psf}) + (1.6)(20 \text{ psf}) = 122 \text{ psf} \leftarrow \text{ Controls}$$

Therefore, the factored design floor load (construction phase), $w_u = 122$ psf
Uniform load on beam in kips/ft. and kips/in. (construction phase):
Tributary width of the beam = 9 ft.

$$w_D = (75 \text{ psf})(9 \text{ ft.}) = 675 \text{ lb./ft.} \ (= 56.25 \text{ lb./in.}) \qquad \text{Unfactored dead load}$$

$$w_u = (122 \text{ psf})(9 \text{ ft.}) = 1098 \text{ lb./ft.} \ (= 1.098 \text{ kips/ft.}) \qquad \text{Factored construction phase load}$$

Concentrated loads on the girder (construction phase):
Span of the beam = 33 ft.

The beam reactions on the typical interior girder for the construction phase are calculated as follows:

$$P_D = (75 \text{ psf})(33 \text{ ft.})(9 \text{ ft.}) = 22{,}275 \text{ lb. or } 22.3 \text{ kips}$$

$$P_L = (20 \text{ psf})(33 \text{ ft.})(9 \text{ ft.}) = 5940 \text{ lb. or } 5.94 \text{ kips}$$

$$P_u = (1.2)(22.3) + (1.6)(5.94) = 36.3 \text{ kips}$$

Camber

For the required camber, 75% of the slab weight will be assumed. The moments of inertia of the W18 × 35 beam and the W18 × 71 girder are 510 in.^{4} and 1170 in.^{4}, respectively (see Part 1 of the *AISCM*). For deflection calculations, the reader can use the deflection equations presented in *AISCM* Tables 3-21a or 3-22.

The camber for the beam under a uniform unfactored dead load of 56.25 Ib/in. is

$$\Delta_{cB} = (0.75)\frac{5w_D L^4}{384EI}$$

$$= (0.75)\frac{(5)(56.25)[(33)(12)]^4}{(384)(29\times 10^6)(510)} = 0.913 \text{ in. Use a } 7/8\text{-in. camber.}$$

The camber for the girder under two equal and symmetrically located unfactored concentrated loads of 22.3 kips is

$$\Delta_{cG} = (0.75)\frac{P_D L^3}{28EI}$$

$$= (0.75)\frac{(22.3)(27\times 12)^3}{(28)(29\times 10^6)(1170)} = 0.60 \text{ in.} < 3/4 \text{ in. Therefore, no camber required.}$$

With no camber required, the construction phase deflection will have to be added to the deflections caused by the superimposed dead and live loads to obtain the total load defection. Recall that the minimum recommended camber (if provided) is $^3/_4$ in. and the minimum recommended beam span is 25 ft. for cambered beams.

Construction Phase Strength

The factored bending moment due to construction phase loading for the beam is

$$M_{cB} = \frac{w_u L^2}{8} = \frac{(1.098)(33)^2}{8} = 150 \text{ ft.-kips.}$$

The factored bending moment due to construction phase loading for the girder is

$$M_{cG} = \frac{P_u L}{3} = \frac{(36.3)(27)}{3} = 327 \text{ ft.-kips.}$$

The unbraced length of the beam = 0 ft. (since the beam is fully braced by the deck)
The unbraced length of the girder = the spacing between the beams = 9 ft
From *AISCM* Table 3-10, the design bending strength of the beam and girder, respectively, are

$$\phi_b M_n = 250 \text{ ft.-kips} > 150 \text{ ft.-kips (at } L_b = 0 \text{ ft. for W18} \times 35) \quad \text{OK.}$$

$$\phi_b M_n = 500 \text{ ft.-kips} > 327 \text{ ft.-kips (at } L_b = 9 \text{ ft. for W18} \times 71) \quad \text{OK.}$$

Therefore, the infill beams and the girders are adequate for the construction phase loading.

7-5 DEFLECTION OF COMPOSITE BEAMS

Composite beams have a larger moment of inertia than non-composite beams of the same size once the concrete slab cures and reaches its required strength because of the interaction between the concrete slab and the steel beams and girders made possible by the shear studs. The designer must first consider whether or not the slab is shored (see Section 7-4). For shored construction, all of the deflection will occur "post-composite" and will be a function of the transformed moment of inertia. For unshored construction, the weight of the concrete will cause an initial deflection based on the moment of inertia of the non-composite steel beam or girder; then any live load and superimposed and sustained dead load deflections will be based on the transformed moment of inertia of the concrete-steel composite section (i.e., the moment of inertia of the composite section).

When a composite beam is subjected to sustained loads, the concrete slab will be in a constant state of compression and is susceptible to *creep*. Creep is the deformation that occurs slowly over a period of time due to the constant presence of compression loads or stresses. It is difficult to quantify the amount of creep in a composite section, but designers estimate this deflection by using a modular ratio of $2n$ to $3n$ instead of the n value calculated in equation (7-1). The *AISCM* recommends a modular ratio of $2n$ for calculating long-term deflections due to creep [3]. The creep deflection due to sustained dead loads is not significant in typical steel buildings. Creep should be accounted for when there is a large amount of sustained live load. A more detailed coverage of this topic is found in Ref. [6].

In Section 7-3, we discussed the calculation of the transformed moment of inertia, I_{tr}, for a composite section, as well as the effective moment of inertia, I_{eff}, for a partially composite section. In many cases, the terms I_{tr} and I_{eff} are used interchangeably to describe the stiffness of a composite section. In this text, we will use the term I_{eff} to describe the moment of inertia for either a fully composite or partially composite section (in equation (7-10), note that $I_{eff} = I_{tr}$ for a fully composite section). Previous editions of the AISC specification prior to 2016 had recommended a reduction factor of 0.75 for the effective moment of inertia, but this reduction factor has been removed from the AISC Specifications starting with the 2016 edition.

Alternatively, in lieu of using the transformed moment of inertia, the AISC Specifications allow the use of the lower-bound moment of inertia tabulated in the lower-bound elastic moment of inertia table (Table 3-19 in the *AISCM*). The lower-bound moment of inertia (I_{LB}) is the moment of inertia at the ultimate limit state, which is less than the moment of inertia at the serviceability limit state, where deflections are usually calculated. Thus, the lower bound moment of inertia yields a conservative estimate of the actual deflections. To obtain I_{LB} from *AISCM* Table 3-19, look for a Y_2 value from the ultimate design stage and for a Y_1 value corresponding to ΣQ_n, obtained in the ultimate design stage. The I_{LB} obtained is then used to calculate the deflection.

The deflection limits for beams are found in the International Building Code (IBC), Section 1604.3. Composite beams are normally used as floor members, and from the IBC Table 1604.3, live load deflection is limited to $L/360$, and the total load deflection is limited to $L/240$. For *shored construction*, the deflection limits are as follows:

$$\Delta_{LL} \leq \frac{L}{360}, \text{ and} \tag{7-30}$$

$$\Delta_{TL} \leq \frac{L}{240}, \tag{7-31}$$

where

Δ_{LL} = Live load deflection,

D_{TL} = Total load deflection (dead plus live load), and

L = Beam span.

For *unshored construction*, the live load deflection limit is as indicated in equation (7-30), but the total load deflection must also account for the construction phase deflection. Therefore, for unshored construction,

$$\Delta_{TL} = \Delta_{SDL} + \Delta_{LL} + \Delta_{CDL}, \tag{7-32}$$

where

Δ_{SDL} = Superimposed dead load deflection (post-composite condition),

Δ_{LL} = Live load deflection, and

Δ_{CDL} = Construction phase dead load deflection less any camber (pre-composite condition).

Note that Δ_{SDL} and Δ_{LL} will be a function of the composite moment of inertia (from equation (7-24)) and Δ_{CDL} will be a function of the moment of inertia of the steel beam only (i.e., non-composite moment of inertia). Note also that the construction dead load, Δ_{CDL}, can be minimized by specifying a camber. More stringent deflection limits than the ones used in this section may be specified if the floor supports deflection-sensitive elements such as operable partition walls or sensitive equipment (see Section 2-15)

Floor vibrations are an additional serviceability consideration for composite as well as non-composite floor members, and this topic is covered in Chapter 12.

7-6 COMPOSITE BEAM ANALYSIS AND DESIGN USING THE *AISCM* TABLES

The analysis process presented in the previous sections can be tedious when performing calculations manually, which is why most composite beam designs are carried out with the aid of a computer program. As an alternative, there are several design aids in the *AISCM* that can be used. *AISCM* Table 3-18 is used to determine the design bending strength of a composite section, and *AISCM* Table 3-19 is used to determine the lower bound moment of inertia, I_{LB}. Both tables are a function of Y_1 (location of the PNA), Y_2 (location of the compression force in the slab), and ΣQ_n (magnitude of the compression force, which is a function of the number of shear studs between the point of maximum moment and the nearest point of zero moment).

Referring to Figure 7-23, there are seven possible locations of the PNA, starting from the top flange of the steel beam to some point within the web of the beam. If it is found that the PNA is above the top of the beam flange, then the values obtained from PNA location 1 are conservatively used. Full-composite action occurs at PNA location 1 (i.e., $\Sigma Q_n = A_sF_y$), whereas PNA location 7 represents 25% composite action. Note that a minimum degree of composite action of 25% (i.e., $\Sigma Q_n = 0.25A_sF_y$) was recommended in previous editions of the AISC Specifications for all

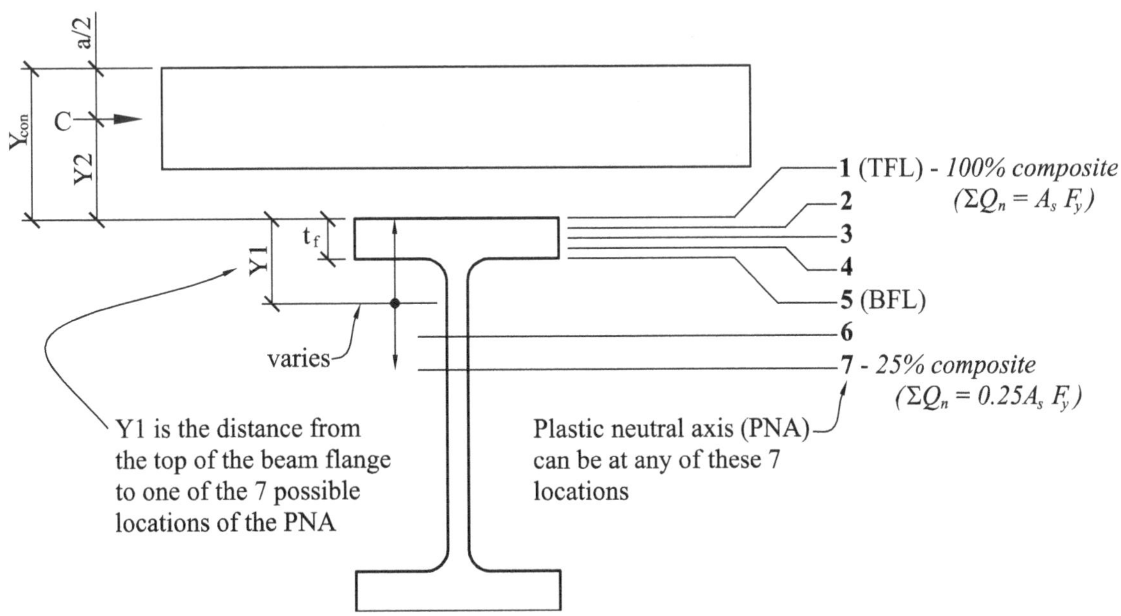

FIGURE 7-23 PNA location for using AISC tables

composite beams, but due to the non-ductile failures of the shear studs that have been observed in long span composite beams with low percentages of composite action, a 50% minimum composite action is now recommended, but not mandatory, in the current AISC Specifications [1]. For degrees of composite action between 25% and 49%, the ductility of the shear stud connections must be investigated except for composite beams that satisfy the requirements previously presented in Section 7-3. If the degree of composite action is lower than 25%, the beam or girder should be treated and designed as a non-composite flexural member (see Chapter 6). In the author's experience, degrees of composite action between 50% and 75% typically result in the most economical composite floor construction.

The examples that follow illustrate the use of composite beam design tables in the *AISCM*.

EXAMPLE 7-7

Composite Beam Deflections

Determine whether the composite beam in Example 7-3 is adequate for immediate and long-term deflections, assuming the following conditions:

Slab dead load = 50 psf (ignore self-weight of the beam)

Superimposed dead load = 25 psf (partitions, mechanical and electrical)

Live load = 50 psf (assume that 20% of this live load is sustained)

Beam span = 28′ – 6″

Beam spacing = 8 ft.

Solution

Loads to Beam:

$$w_{CDL} = (50\text{ psf})(8\text{ ft.}) = 400\text{ plf} \quad (= 33.3\text{ lb./in.})$$

$$w_{SDL} = (25\text{ psf})(8\text{ ft.}) = 200\text{ plf} \quad (= 16.7\text{ lb./in.})$$

$$w_{LL} = (50\text{ psf})(8\text{ ft.}) = 400\text{ plf} \quad (= 33.3\text{ lb./in.})$$

From Example 7-3,

$I = 199\text{ in.}^4$ (noncomposite moment of inertia)

$I_{eff} = 612\text{ in.}^4$ (the calculated effective moment of inertia)

The moment of inertia used for long-term deflections is

$I_{eff} = 612\text{ in.}^4$

As an alternate check, recall from Example 7-3 that $Y_1 = 0.305$ in., $Y_2 = 4.64$ in., and $\Sigma Q_n = 172$ kips. From *AISCM* Table 3-19, we find that $I_{LB} = 496\text{ in.}^4$ (by interpolation). Recall that the lower bound moment of inertia is the value of the moment of inertia at the ultimate limit state, which is conservative.

Construction Phase Deflection:

$$\Delta_{CDL} = \frac{5w_{CDL}L^4}{384EI}$$

$$= \frac{(5)(33.3)\left[(28.5)(12)\right]^4}{(384)(29\times10^6)(199)} = 1.03\text{ in.}$$

If this beam were to be cambered, we would take 75% of the construction phase dead load deflection as the camber, as follows:

$\Delta_c = (0.75)(1.03 \text{ in.}) = 0.771 \text{ in.}$ Use $^3/_4$-in. camber.

The construction phase deflection would then be reduced by $^3/_4$ in. and recalculated as follows:

$\Delta_{CDL} = 1.03 \text{ in.} - 0.75 \text{ in.} = 0.28 \text{ in.}$

A construction phase deflection of 1.03 in. would require the addition of more concrete in the floor slab due to ponding to achieve a flat floor surface, so a camber of $^3/_4$ in. will be specified. The post-composite deflection using I_{eff} is

$$\Delta_{SDL} = \frac{5w_{SDL}L^4}{384EI}$$

$$= \frac{(5)(16.7)\left[(28.5)(12)\right]^4}{(384)(29\times10^6)(612)} = 0.168 \text{ in.}$$

$$\Delta_{LL} = \frac{5w_{LL}L^4}{384EI}$$

$$= \frac{(5)(33.3)\left[(28.5)(12)\right]^4}{(384)(29\times10^6)(612)} = 0.334 \text{ in.}$$

$$\Delta_{TL} = \Delta_{SDL} + \Delta_{LL} + \Delta_{CDL}$$

$$= 0.168 \text{ in.} + 0.334 \text{ in.} + 0.28 \text{ in.} = 0.782 \text{ in.}$$

The deflection limits are found from equations (7-30) and (7-31):

$$\Delta_{LL} \le \frac{L}{360}$$

$$0.334 \text{ in.} \le \frac{(28.5)(12)}{360} = 0.95 \text{ in., OK for live load deflection}$$

$$\Delta_{TL} \le \frac{L}{240}$$

$$0.782 \text{ in.} \le \frac{(28.5)(12)}{240} = 1.43 \text{ in., OK for total load deflection.}$$

It can be seen by inspection that the $^3/_4$-in. camber is needed to meet the total load deflection limit.

We must also consider the long-term deflection to account for creep effects. To do this, the effective moment of inertia must be recalculated using a modular ratio of $2n$. From Example 7-1, n = 8.87; therefore,

$n' = 2n = 17.74$ (to account for creep)

$$A_{ct} = \frac{A_c}{n'} = \frac{(80)(2)}{17.74} = 9.01 \text{ in.}^2$$

$$I_{ct} = \frac{bt_c^3}{12n'} = \frac{(80)(2)^3}{(12)(17.74)} = 3.0 \text{ in.}^4$$

Using the top of the concrete slab as the datum, we develop Table 7-3.

TABLE 7-3 Composite Beam Section Properties

Element	A	y	Ay	I	$d = y - \bar{y}$	$I + Ad^2$
Slab	9.01	1	9.01	3.0	−4.54	188.7
W14 × 22	6.49	11.85	76.9	199	6.31	457.4
Σ =	**15.5**		**85.91**			$\boldsymbol{I_{tr} = 646.1}$

$$\bar{y} = \frac{\Sigma Ay}{\Sigma A} = \frac{85.91}{15.5} = 5.54 \text{ in.}$$

$$I_{tr} = \Sigma\left(I + Ad^2\right)$$

$$I_{tr} = 646.1 \text{ in.}^4 \ (\text{see Table 7-3})$$

Accounting for the partial-composite behavior (see Example 7-3), the effective moment of inertia is

$$I_{eff} = I_s + \sqrt{\frac{\Sigma Q_n}{C_f}}\left(I_{tr} - I_s\right) = 199 + \sqrt{\frac{172}{324.5}}\left(646.1 - 199\right) = 524.5 \text{ in.}^4$$

The ratio of the effective moment of inertia used for total loads versus that used for sustained loads is

$$\frac{612 \text{ in.}^4}{524.53 \text{ in.}^4} = 1.168.$$

This ratio can be used to calculate the actual sustained load deflection. The amount of sustained live load specified for this example is 20% of the total live load, and the remaining 80% is assumed to be transient. The total sustained load deflection then is

$$\Delta_{TL(sust)} = 1.168\left(\Delta_{SDL} + 0.2\Delta_{LL}\right) + \Delta_{CDL} + 0.8\Delta_{LL}$$

$$= 1.168\left[0.168 + (0.2)(0.334)\right] + 0.28 + (0.8)(0.334) = 0.82 \text{ in.}$$

This is less than the $L/240$ = 1.43 in. calculated previously, so the section is adequate for long-term deflections due to sustained loads.

EXAMPLE 7-8

Composite Design Strength Using the *AISCM* Tables

Determine the design strength of the composite beam in Example 7-2 using *AISCM* Table 3-18.

Solution

From example 7-2, the steel beam is W14 × 22;

$C_c = 324.5$ kips, and

$a = 1.36$ in.;

therefore,

$$Y_2 = h_r + t_c - \frac{a}{2} = 3 \text{ in.} + 2 \text{ in.} - \frac{1.36}{2} = 4.32 \text{ in.}$$

From *AISCM* Table 3-18 (by linear interpolation), the design moment strength and the allowable bending moment, respectively, can be obtained as follows:

For LRFD:

$\phi_b M_n$ = **274 ft.-kips**

For ASD:

$$\frac{M_n}{\Omega_b} = \mathbf{182\ ft.-kips.}$$

EXAMPLE 7-9

Composite Design Strength Using the *AISCM* Tables

Determine the design strength of the W14 × 22 beam in Example 7-3 using *AISCM* Table 3-18.

Solution

From Example 7-3, for the W14 × 22 beam

$$C_c = 172 \text{ kips}$$

$$a = 0.722 \text{ in.}$$

$$Y_1 = 0.305 \text{ in.}$$

$$Y_2 = 4.64 \text{ in.}$$

From *AISCM* Table 3-18 (by linear interpolation), the design moment strength and the allowable bending moment, respectively, can be obtained as follows:

For LRFD:

$\phi_b M_n$ = **225 ft.-kips**

For ASD:

$$\frac{M_n}{\Omega_b} = 150\ \mathbf{ft.-kips.}$$

EXAMPLE 7-10

Composite Design Strength Using the *AISCM* Tables

Determine the design strength of the W18 × 35 beam in Example 7-5 using *AISCM* Table 3-18.

Solution

From Example 7-5, for the W18 × 35 beam

$$C_c = 129 \text{ kips}$$

$$Y_2 = 5.62 \text{ in.}$$

From *AISCM* Table 3-18 (by linear interpolation), the design moment strength and the allowable bending moment, respectively, can be obtained as follows:

For LRFD:

$\phi_b \boldsymbol{M_n}$ = **368 ft.-kips**

For ASD:

$$\frac{M_n}{\Omega_b} = 245\ \mathbf{ft.-kips.}$$

7-7 COMPOSITE BEAM DESIGN PROCEDURE

There are several factors that must be considered in the design of composite floors. One key constraint is that the required floor structure depth is usually specified early in the design stage. In some cases, the depth of the steel beams will be limited if the floor-to-floor height of a building is limited. The beam spacing, another factor that should be considered early in the design stage, is generally a function of the slab strength, but tighter beam spacing could be required to minimize the steel beam depth. The slab design is mainly a function of the beam spacing (i.e., thinner slabs can be used with smaller beam spacing, and thicker slabs are required for larger beam spacing). The fire rating of the floor structure must also be considered. Many occupancy categories will require that a floor structure have a certain fire rating, usually measured in hours. The Underwriters Laboratory (UL) has tested several floor assembly types and assigned ratings for each assembly type. The factors affecting the fire rating of a concrete slab-on-metal deck supported on steel beams and girders include the following:

- The concrete slab thickness above the metal deck.
- The type of concrete – normal weight or lightweight concrete.
- Whether the deck is fireproofed or not.
- The fireproofing of the supporting steel beams and girders.

Fireproofing was first considered in Section 1-10 in our discussions of roof and floor framing layout, and will be covered in more detail in Chapter 14.

In order to determine the most economical floor framing system for any steel building, the designer may have to perform several iterations of framing schemes, considering all of the variables already noted, to determine the most suitable system. We will assume at this point that the slab spacing and the beam spacing have already been determined so that we can proceed with the composite beam or girder design.

The following steps are given as a guide for the design of composite beams:

1. Tabulate the design loads (dead load, superimposed dead load, live load, and sustained live load). Tabulate the service loads to be used for the deflection calculations, and tabulate the factored loads to be used for the strength design checks.
2. Compute the maximum shear demand, V_s or V_u, and moment demand, M_s or M_u, for design, and the maximum moment demand during the construction phase, M_{sc} or M_{uc}, for unshored construction.
3. Estimate the beam weight. Assuming that the PNA is in the concrete slab, we can use equation (7-11) to determine the allowable bending moment and the design moment capacity, respectively:

   For LRFD:

   $\phi_b M_n = \phi_b T y$

   Using $M_u = \phi_b M_n$,

   $T = A_s F_y$, and $y = \left(\frac{d}{2} + Y_{con} - \frac{a}{2}\right)$,

   we can solve for the required area of steel as follows:

$$A_s = \frac{M_u}{\phi_b F_y\left(\frac{d}{2} + Y_{con} - \frac{a}{2}\right)} \tag{7-33}$$

For ASD:

$$\frac{M_n}{\Omega_b} = \frac{Ty}{\Omega_b}$$

Using $M_s = \dfrac{M_n}{\Omega_b}$,

$T = A_s F_y$, and $y = \left(\dfrac{d}{2} + Y_{con} - \dfrac{a}{2}\right)$,

we can solve for the required area of steel as follows:

$$A_s = \frac{\Omega_b M_s}{F_y\left(\dfrac{d}{2} + Y_{con} - \dfrac{a}{2}\right)} \tag{7-34}$$

An assumed value of the depth of the concrete compression stress block or the effective concrete flange depth, $a = 0.4t_c$ is recommended here (see Figure 7-10 for t_c).

In equations (7-33 and 7-34), Y_{con} is the distance from the top of the steel beam to the top of the slab. Knowing the density of steel is 490 lb./ft.^{3} and using equations (7-33) and (7-34), we can solve for the beam weight for the LRFD and ASD methods, respectively, as follows:

For LRFD:

$$w = \left(\frac{A_s}{144}\right)(490) = \frac{3.4M_u}{\phi_b F_y\left(\dfrac{d}{2} + Y_{con} - \dfrac{a}{2}\right)}, \tag{7-35}$$

For ASD:

$$w = \left(\frac{A_s}{144}\right)(490) = \frac{3.4\Omega_b M_s}{F_y\left(\dfrac{d}{2} + Y_{con} - \dfrac{a}{2}\right)} \tag{7-36}$$

where w is in pounds per linear foot (i.e., Ib/ft.). For various assumed values of the nominal beam depth, d, in inches, the required minimum beam weight, in pounds per linear foot, is calculated, and the lightest weight beam is selected.

4. Conduct the construction phase strength and deflection check (unshored beams only):

 a. Check the selected steel beam as a non-composite section to support the following loads:

 - Weight of wet concrete including the concrete ponding loads,
 - Weight of metal deck,
 - Steel beam self-weight, and
 - Construction live load = 20 psf (weight of workers and equipment).

 b. Calculate the deflection of the non-composite beam under construction phase dead load (CDL). A percentage of this deflection can be the specified camber. It is not recommended to camber steel beams or girders when the span is less than 25 ft. or if the construction phase dead load deflection is less than $^3/_4$ in.

5. Calculate $A_s F_y$ for the selected steel section, and then calculate the degree of composite action. For 100% composite action (i.e., fully composite), $\Sigma Q_n = A_s F_y$. The minimum degree of composite action of 25% (i.e., $\Sigma Q_n = 0.25A_s F_y$) was recommended in previous editions of the AISC Specifications for all composite beams, but non-ductile failures of the shear studs have been observed in some research to occur with low percentages of composite action; therefore, a 50% minimum composite action is now recommended, but not mandatory, in the current AISC Specifications [1]. This ensures ductile behavior of the horizontal interface connection between the steel beam/girder and the concrete slab made possible by the headed shear studs. If the degree of composite action is between 25% and 49%, the shear stud connections must be investigated for ductility. See the exceptions presented previously in Section 7-3. Note that when the degree of composite action is lower than 25%, the beam should be designed as a non-composite beam (see Chapter 6).

6. Calculate the effective flange width, b, of the concrete slab which is the sum of the effective width on each side of the beam centerline. The effective width on each side of the beam centerline should not be greater than the smallest of the following:
   - $^1/_8$ of the beam span,
   - $^1/_2$ of the distance to the adjacent beam, or
   - The edge of slab distance (for edge beams).
7. Calculate the actual depth of the effective concrete flange, a (this might be different from the value assumed in step 3):

$$a = \frac{\Sigma Q_n}{0.85 f'_c b} \tag{7-37}$$

8. Compute the distance from the beam top flange to the centroid of the effective concrete flange:

$$Y_2 = Y_{con} - \frac{a}{2} = h_r + t_c - \frac{a}{2} \tag{7-38}$$

9. Use the value of Y_2 from step 8 and the assumed value of ΣQ_n from step 5. Go to the composite beam selection table (Table 3-18 in the *AISCM*) corresponding to the beam size chosen in step 3. Determine the design bending strength of the composite section, linearly interpolating if necessary:
   a. If $\phi_b M_n > M_u$, the beam section is adequate for LRFD

   If $\frac{M_n}{\Omega_b} > M_s$, the beam section is adequate for ASD

   b. If $\phi_b M_n < M_u$ (for LRFD) or if $\frac{M_n}{\Omega_b} < M_s$ (for ASD), the beam section is inadequate, and the following options should be considered:
   - Increase the degree of composite action (i.e., increase ΣQ_n up to $\leq A_s F_y$).
   - Use a larger beam size.
10. Check the shear strength of the steel beam:

    $\phi_v V_n > V_u$, where $\phi_v V_n = \phi_v 0.6 F_y A_w C_{v1}$ for LRFD

    $\frac{V_n}{\Omega_v} > V_s$, where $\frac{V_n}{\Omega_v} = \frac{0.6 F_y A_w C_{v1}}{\Omega_v}$ for ASD

    Recall that for webs of I-shaped members with $\frac{h}{t_w} \leq 2.24\sqrt{\frac{E}{F_y}}$; $\phi_v = 1.0$ and $C_{v1} = 1.0$

    (see Chapter 6 for other values of C_{v1}). Alternatively, $\phi_v V_n$ or $\frac{V_n}{\Omega_v}$ can be obtained for W-shapes from the maximum total uniform load table (Table 3-6 in the *AISCM*).
11. Select the shear stud spacing (see Section 7-2 and Figure 7-7). Recall that $N_s = \frac{\Sigma Q_n}{Q_n}$, where N_s is the number of studs between the point of maximum moment and the

nearest point of zero moment. Next, check the stud spacing and other requirements to ensure compliance with the AISC Specifications. Refer to the bulleted list of requirements in Section 7-2.

12. Check deflections. There are two methods for computing the actual moment of inertia of the composite section required in the deflection calculations:

    a. Use the lower-bound moment of inertia tabulated in the lower-bound elastic moment of inertia table (see *AISCM* Table 3-19).

    b. Determine the transformed moment of inertia and calculate I_{eff} from equation (7-10), which is then used to calculate the deflections.

    Recall that the modular ratio (i.e., the ratio of the modulus of elasticity of steel to that of the concrete) used for short-term deflections is n; for long-term (sustained load) deflection, the modular ratio is $2n$. Deflections are generally limited to $L/360$ for live loads and $L/240$ for total dead plus live loads, but more stringent deflection limits may be required if the floor framing supports deflection sensitive equipment or operable partition walls.

13. Check floor vibrations (see Chapter 12).

14. Compute the reinforcement required in the concrete slab and over the girders (see Section 7-8).

EXAMPLE 7-11 (LRFD)

Composite Beam and Girder Design

Given the floor plan shown in Figure 7-24, design a typical infill beam B1 and girder G1 using the LRFD method. The floor consists of a 3.5-in. normal-weight concrete slab on a 1.5-in. × 20-ga. galvanized composite metal deck with 6 × 6 – W2.9 × 2.9 WWF to reinforce the slab. Assume ASTM A992, Grade 50 steel, and a concrete strength, $f_c' = 3.5$ ksi. Use a floor live load of 150 psf. The concrete ponding load is assumed to be approximately 10% of the self-weight of the concrete slab (see Chapter 2).

Solution

For ASTM A992 steel, $F_y = 50$ ksi and $F_u = 65$ ksi.

Floor Loads:

Weight of the concrete slab	= 56 psf (51 psf selfwt + 5 psf concrete ponding load)
Weight of the steel deck	= 2 psf
Beam self-weight (35 lb./ft./6.67 ft.)	= 5 psf (assumed)
Girder self-weight (50 lb./ft./30 ft.)	= 2 psf (assumed)
Partitions	= 15 psf
Ceiling + Mechanical/Electrical	= 5 psf
Construction phase dead load (CDL)	= 56 + 2 + 5 + 2 = 65 psf
Superimposed dead load (SDL)*	= 15 + 5 = 20 psf
Total floor dead load (D)	= 85 psf
Floor live load (L)	**= 150 psf**

*(i.e. partitions, ceiling+mechanical/electrical)

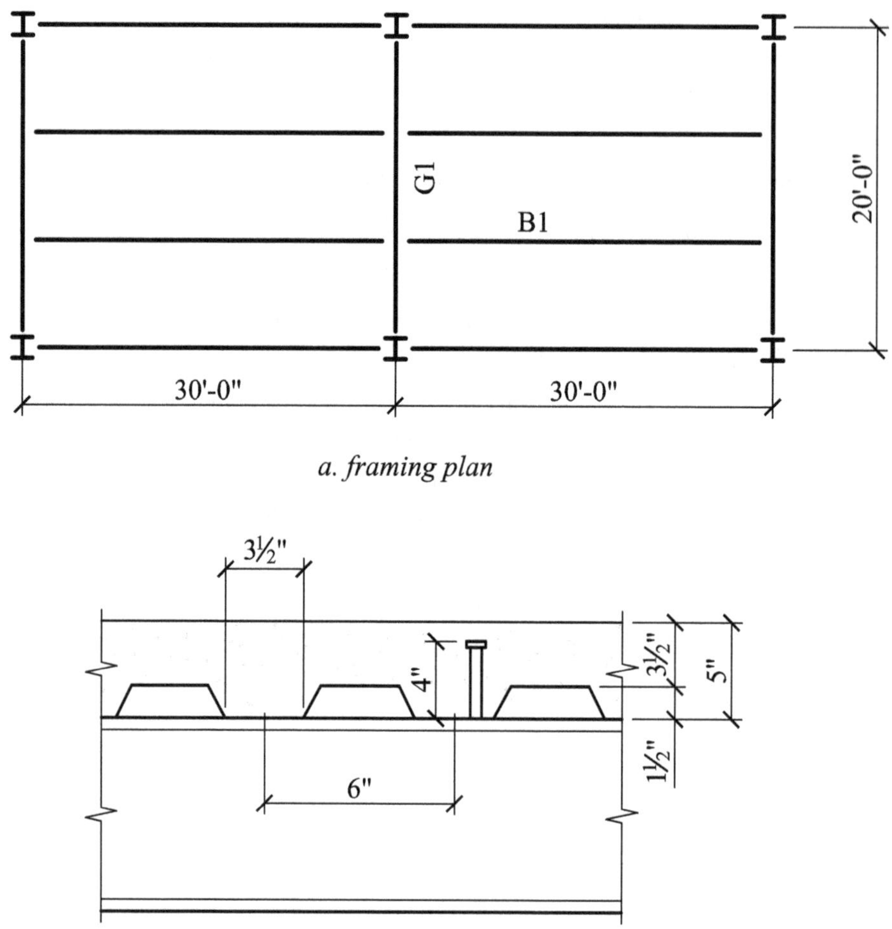

a. framing plan

b. floor deck details

FIGURE 7-24 Details for Example 7-11

For 1.5-in. metal deck, deck rib width, w_r = 3.5 in. > 2 in. OK, and
Deck rib depth or height, h_r = 1.5 in.

Note: The above two deck parameters must be obtained from the deck manufacturer's catalog.

For 3.5-in. concrete slab on top of deck, t_c = 3.5 in. > 2-in. minimum. OK
Stud diameter, $D_s = {}^3/_4$ in. $\leq 2.5\, t_f$. Therefore, minimum flange thickness = 0.75/2.5 = 0.3 in.

Total slab depth, $Y_{con} = t_c + h_r$ = 3.5 in. + 1.5 in. = 5 in.

Assuming a 1-in. clear concrete cover over the head of the stud, we have

Stud length, $H_s = Y_{con}$ − 1 in. = 4 in.

Maximum longitudinal stud spacing = the smaller of $8Y_{con}$ = 40 in. *or* 36 in.
Minimum longitudinal stud spacing = $6D_s$ = 6(¾ in.) = 4.5 in.
Minimum transverse spacing of stud = $4D_s$ = 4(¾ in.) = 3 in.

Design of composite beam B1:

1. Composite design (applied loads and moments):

   Floor dead load = 85 psf
   Floor live load = 150 psf

   Tributary width, TW = 6.67 ft.

Service live load, $w_L = (150 \text{ psf})(6.67 \text{ ft.}) = 1.0 \text{ kips/ft.} \quad (= 0.084 \text{ kips/in.})$

Superimposed dead load, $w_{SDL} = (20 \text{ psf})(6.67 \text{ ft.}) = 0.13 \text{ kips/ft.} \quad (= 0.011 \text{ kips/in.})$

Ultimate factored total load, $w_u = [(1.2)(85 \text{ psf}) + (1.6)(150 \text{ psf})](6.67 \text{ ft.})$

$= 2.28 \text{ kips/ft.}$

2. Factored moment, $M_u = (2.28 \text{ kips/ft.})(30 \text{ ft.})^2/8 = 257 \text{ ft.-kips}$

Factored shear, $V_u = \dfrac{(2.28 \text{ kips/ft.})(30 \text{ ft.})}{2} = 34.2 \text{ kips}$

Service live load reaction, $R_{LL} = \dfrac{(1 \text{ kip/ft.})(30 \text{ ft.})}{2} = 15 \text{ kips}$

Superimposed dead load reaction, $R_{SDL} = \dfrac{(0.13 \text{ kips/ft.})(30 \text{ ft.})}{2} = 1.95 \text{ kips}$

3. Assume the depth of the concrete compression stress block or the effective concrete flange depth, $a = 0.4t_c = (0.4)(3.5 \text{ in.}) = 1.4 \text{ in.}$

$Y_{con} = 5 \text{ in.}$

$F_y = 50 \text{ ksi}$

$\phi_b = 0.90$

For $d = 10$ in.,

$$w = \frac{3.4M_u}{\phi_b F_y\left(\dfrac{d}{2} + Y_{con} - \dfrac{a}{2}\right)} = \frac{(3.4)(257 \times 12)}{(0.9)(50)\left(\dfrac{10}{2} + 5 - \dfrac{1.4}{2}\right)} = 26 \text{ lb/ft.}$$

For $d = 12$ in., beam weight = 23 lb/ft.

For $d = 14$ in., beam weight = **21 lb/ft.**

Try a W14 × 22 beam for which $A_s = 6.49 \text{ in.}^2$ and the moment of inertia, $I = 199 \text{ in.}^4$.

4. Construction phase strength and deflection check:

Construction phase dead load (CDL) = 65 psf
Construction phase live load = 20 psf
Tributary width of the beam, TW = 6.67 ft.

The construction phase factored total load on the infill beam is

$w_u = 1.4\text{D} \times \text{TW} = (1.4)(65 \text{ psf})(6.67 \text{ ft.}) = 0.61 \text{ kips/ft.}$ or

$w_u = (1.2\text{D} + 1.6\text{L}) \times \text{TW} = (1.2)(65 \text{ psf})(6.67 \text{ ft.}) + (1.6)(20 \text{ psf})(6.67 \text{ ft.}) = 0.73 \text{ kips/ft.}$

Therefore, $w_u = 0.73 \text{ kip/ft.}$

The construction phase dead load on the infill beam is

$w_{CDL} = (65 \text{ psf})(6.67 \text{ ft.}) = 0.43 \text{ kip/ft.} \quad (= 0.036 \text{ kip/in.}).$

Construction phase dead load reaction, $R_{CDL} = \dfrac{(0.43 \text{ kips/ft.})(30 \text{ ft.})}{2} = 6.5 \text{ kips.}$

The construction phase factored shear, $V_u = \frac{(0.73 \text{ kips/ft.})(30 \text{ ft.})}{2} = 11$ kips.

The construction phase-factored moment in the infill beam is

$M_u = \frac{(0.73)(30)^2}{8} = 82$ ft.-kips (top flange of beam is assumed to be fully braced by the deck; therefore, $L_b = 0$).

From the beam design selection table (*AISCM* Table 3-6 or 3-10), we obtain the design moment capacity of the non-composite beam:

$\phi_b M_n$ for W14 × 22 = 125 ft.-kips > M_u = 82 ft.-kips. OK

The construction phase dead load deflection is

$$\Delta = \frac{5w_{CDL}L^4}{384EI} = \frac{(5)(0.036)[(30)(12)]^4}{(384)(29{,}000)(199)} = 1.36 \text{in.} > \tfrac{3}{4} \text{ in.}$$

∴ Camber is required.

The required camber is, $\Delta_c = (0.75)(1.36 \text{ in.}) = 1.0$ in.

Therefore, use 1-in. camber.

Hence, $\Delta_{CDL} = 1.36$ in. − 1 in. (Camber) = 0.36 in.

(This is the remaining dead load deflection after cambering the beam).

5. A_sF_y for W14 × 22 = (6.49 in.^{2})(50 ksi) = 325 kips

ΣQ_n must be $\leq A_sF_y$ (Choose a value between 25% A_sF_y and 100% A_sF_y)

Assume $\Sigma Q_n = 325$ kips ⇒ 100% composite action

6. Effective concrete flange width, b, is the smaller of
($^1/_8$)(30 ft.) + ($^1/_8$)(30 ft.) = 7.5 ft., or
($^1/_2$)(6.67 ft.) + ($^1/_2$)(6.67 ft.) = **6.67 ft.** (**80 in.**), (controls).

7. The depth of the effective concrete flange (i.e., the depth of the compression stress block) is

$$a = \frac{\sum Q_n}{0.85f'_c b} = \frac{325}{(0.85)(3.5)(80)} = 1.37 \text{ in.}$$

8. Distance from the top of the steel beam top flange to the centroid of the effective concrete flange is

$$Y_2 = Y_{con} - 0.5a = 5 \text{ in.} - 0.5 \times 1.37 \text{ in.} = 4.32 \text{ in.}$$

9. Using the composite beam selection table (*AISCM* Table 3-18) with Y_2 = 4.32 in. and the ΣQ_n = 325 kips assumed in step 5, we obtain, by linear interpolation,

$\phi_b M_n = 273$ ft.-kips > $M_u = 257$ ft.-kips OK, and

Y_1 = 0.0 in. = Distance from top of steel beam to the PNA. (Note the value of ΣQ_n = 325 kips corresponds to Y_1 = 0.0 in *AISCM* Table 3-19)

If $\phi_b M_n$ had been much less than M_u, we would have had to increase the beam size to W14 × 26 or W16 × 26 since we are already at 100% composite action and cannot get any more composite action from the W14 x 22 beam.

10. Factored shear, $V_u = 34.2$ kips, and recall that $A_w = dt_w$.
The design shear strength is from Chapter 6, or *AISCM* Table 3-6:
$\phi_v V_n = \phi_v 0.6 F_y A_w C_{v1} = (1.0)(0.6)(50)(0.23)(13.7)(1.0)$
$= 94.5 \text{ k} > V_u = 34.2$ kips. OK

11. $N_s = \dfrac{\Sigma Q_n}{Q_n}$ = Number of studs between the point of maximum moment and the nearest point of zero moment.

- Deck rib depth, $h_r = 1.5$ in.
- Stud length, $H_s = 4$ in. $< h_r + 3$ in. $= 4.5$ in.
- $N_r = 1$ (assuming a single row of studs)
- Deck rib width, $w_r = 3.5$ in.

For metal deck ribs perpendicular to the beam and placed in the strong position, the shear stud capacity reduction factors from Table 7-1 are:
$R_g = 1.0$

$R_p = 0.75$

The shear stud capacity is
$Q_n = 0.5 A_{sa} \sqrt{f'_c E_c} \leq R_g R_p A_{sa} F_u$

$E_c = w_c^{1.5} \sqrt{f'_c} = (145)^{1.5} \sqrt{3.5} = 3266$ ksi

$$Q_n = 0.5\left(\frac{\pi(0.75)^2}{4}\right)\sqrt{(3.5)(3266)} \leq (1.0)(0.75)\left(\frac{\pi(0.75)^2}{4}\right)(65)$$

$Q_n = 23.6$ kips ≤ 21.5 kips

Therefore, $\mathbf{Q_n = 21.5}$ **kip**
Or, from *AISCM* Table 3-20, with the deck ribs perpendicular to the beam and strong stud per rib and one row of studs per deck rib, the shear stud capacity, $Q_n = 21.3$ kip.
The number of studs between the point of maximum moment and the nearest point of zero moment is

$$N_s = \frac{\Sigma Q_n}{Q_n} = \frac{325}{21.5} = 15.1; \qquad \therefore \text{ use 16 studs,}$$

Since the loading on the beam is symmetrical, the total number of studs on the beam is $N = 2N_s = (2)(16) = 32$ studs.

$$\text{Stud spacing, } s = \frac{(30\text{ ft.})(12)}{32} = 11.25 \text{ in.} < 36 \text{ in. OK}$$

> 4.5 in. OK

Note: For composite beams, the actual stud spacing will depend on the spacing of the deck flutes. This implies that the final spacing of the studs will need to be less than or equal to the 11.25 in. obtained above, depending on the spacing of the deck flutes (i.e., the final stud spacing in a composite beam must be a multiple of the deck flute spacing). For example, if the deck flute spacing on this beam were 12 in., it would not be physically possible to place 32 studs in one row on the beam that is 30 ft. long (i.e., the beam would have no more than 30 deck flutes or 30×12in./12in. available). We would then have to consider two rows of studs or an increase in the beam size since we are already at 100% composite action. For $1^1/_2$-in. composite metal deck, the usual rib spacing is approximately 6 in., so 32 studs could be placed in one row on this 30 ft. long beam.

12. We will now check deflections. We need to calculate the lower-bound moment of inertia, I_{LB}, of the composite beam using the values of $Y_1 = 0.0$ in. and $Y_2 = 4.32$ in. obtained from step 9. Using the lower bound moment of inertia table for a W14 × 22 (*AISCM* Table 3-19), we obtain $I_{LB} = 606$ in.^{4} by linear interpolation.

The live load deflection is

$$\Delta = \frac{5w_L L^4}{384EI} = \frac{(5)(0.084)[(30)(12)]^4}{(384)(29{,}000)(606)} = 1.04 \text{ in.}$$

$$\frac{L}{360} = \frac{(30)(12)}{360} = 1 \text{ in. (close to 1.04 in. therefore, OK)}$$

Since the beam is cambered, the total deflection, Δ_{TL}, which is the sum of the deflections due to superimposed dead load ($w_{SDL} = 0.011$ kip/in.), live load ($w_L = 0.084$ kip/in.), and the deflection due to the dead load not accounted for in the camber, is

$$\Delta_{TL} = \Delta_{SDL} + \Delta_L + \Delta_{CDL}$$

$$\Delta_{TL} = \frac{5(w_{SDL} + w_L)L^4}{384EI} + \Delta_{CDL} = \frac{(5)(0.011 + 0.084)[(30)(12)]^4}{(384)(29{,}000)(606)} + 0.36 \text{ in.} = 1.54 \text{ in.}$$

$$\frac{L}{240} = \frac{(30)(12)}{240} = 1.5 \text{ in.} < 1.54 \text{ in.,}$$ but it is close enough since it exceeds the limit by only 2.7%.

Therefore, use a W14 × 22 beam ($N = 32$ and Camber, c = 1 in.)

Design of Composite Girder G1:

1. Loads on the girder (see Figure 7-25):

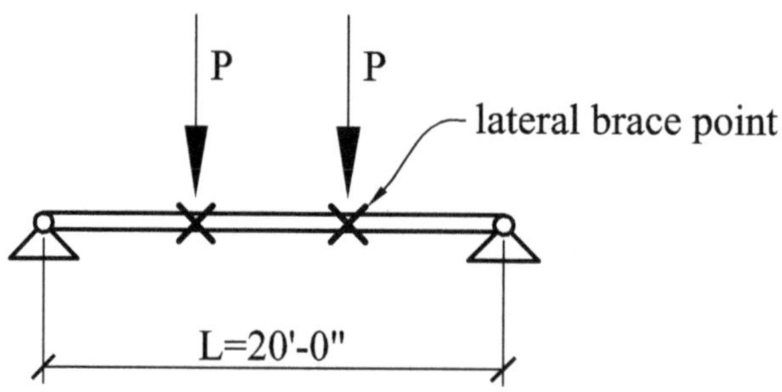

FIGURE 7-25 Loading for Girder G1

Girder tributary width = 30 ft.

Factored load, $P_u = 2 \text{ beams} \times 34.2 \text{ kips} = 68.4 \text{ kips}$

Service live load, $P_L = 2 \text{ beams} \times 15 \text{ kips} = 30 \text{ kips}$

Service superimposed dead load $P_{SDL} = 2 \text{ beams} \times 1.95 \text{ kips} = 3.9 \text{ kips}$

2. Factored moment, M_u = (68.4 kips) (6.67 ft.) = 456 ft.-kips
Factored shear, V_u = 68.4 kips

3. Assume the depth of the concrete compression stress block or the effective flange depth, $a = 0.4\, t_c$

$$t_c = Y_{con} - \frac{h_r}{2} \text{ (see Figure 7-10)}$$

$$= 5 - \frac{1.5}{2} = 4.25 \text{ in.}$$

$$a = (0.4)(4.25) = 1.7 \text{ in.}$$

$Y_{con} = 5$ in., $F_y = 50$ ksi, $\phi_b = 0.90$

For $d = 16$ in.,

$$w = \frac{3.4M_u}{\phi_b F_y\left(\frac{d}{2} + Y_{con} - \frac{a}{2}\right)} = \frac{(3.4)(456\times 12)}{(0.9)(50)\left(\frac{16}{2} + 5 - \frac{1.7}{2}\right)} = 35 \text{ lb./ft.}$$

For $d = 18$ in., girder self-weight = 32 lb. / ft.

For $d = 21$ in., girder self-weight = 29 lb. / ft.

Try a W18 × 35 girder, for which cross-sectional area, A_s = 10.3 in.^{2} and moment of inertia, I = 510 in.^{4}

4. Construction phase strength and deflection check for the girder:
Construction phase factored load, $P_u = 2$ beams $\times 11$ kips $= 22$ kips

Construction phase dead load, $P_{CDL} = 2$ beams $\times 6.5$ kips $= 13$ kips

Construction phase factored moment, $M_u = (22 \text{ kips})(6.67 \text{ ft.})$

$= 147$ ft.-kips (unbraced length, $L_b = 6.67$ ft.)

From the beam design selection table (*AISCM* Table 3-10), we obtain the design moment capacity of the non-composite girder:

ϕM_n for $W18\times 35 = 220$ ft.-kips $> M_u = 147$ ft.-kips O.K.

The construction phase dead load deflection of the non-composite girder is

$$\Delta_{CDL} = \frac{PL^3}{28EI} = \frac{(13)\left[(20)(12)\right]^3}{(28)(29{,}000)(510)} = 0.43 \text{ in.} < {}^3/_4 \text{ in.}$$

∴ No camber is required.
Since no camber is required, the calculated construction phase dead load deflection will be added in step 12 to the superimposed dead load and live load deflections to obtain the total load deflection, Δ_{TL}.

5. A_sF_y for W18 × 35 = (10.3 in.^{2}) (50 ksi) = 515 kips

ΣQ_n must be $\le A_sF_y$ (Choose a value between 25% A_sF_y and 100% A_sF_y)

Assume $\Sigma Q_n = 515$ kips (i.e., 100% composite action)

6. Effective concrete flange width, b, for the girder is the smaller of
$({}^1/_8)$(20 ft.) + $({}^1/_8)$(20 ft.) = 5 ft. (60 in.), ←Controls
Or $({}^1/_2)$(30 ft.) + $({}^1/_2)$(30 ft.) = 30 ft. (360 in.).

7. The depth of the concrete flange (i.e., the depth of the compression stress block) is

$$a = \frac{\Sigma Q_n}{0.85 f'_c b} = \frac{515}{(0.85)(3.5)(60)} = 2.89 \text{ in.}$$

8. Distance from the top of the steel girder to the centroid of the effective concrete flange is

$$Y_2 = Y_{con} - 0.5a = 5 - (0.5)(2.89) = 3.55 \text{ in.}$$

9. Using the composite beam selection table for the W18 × 35 girder (i.e., *AISCM* Table 3-18) with Y_2 = 3.55 in. and ΣQ_n = 515 kips, we obtain the girder design moment capacity as

$\phi_b M_n = 479$ ft.-kips $> M_u = 456$ ft.-kips OK, and

$Y_1 = 0.0$ in. = Distance from top of steel beam to PNA.

10. V_u = 68.4 kips. The design shear strength can be obtained using the design shear strength equation below that was introduced in Chapter 6 or directly from *AISCM* Table 3-6:

$$\phi_v V_n = \phi_v 0.6 F_y A_w C_{v1} = (1.0)(0.6)(50)(0.30)(17.7)(1.0)$$

$$= 159 \text{ kips} > V_u = 68.4 \text{ kips. OK}$$

11. $N_s = \dfrac{\Sigma Q_n}{Q_n}$ = Number of studs between the point of maximum moment and the nearest point of zero moment

$w_r / h_r = 3.5/1.5 = 2.33$

From Table 7-1, for deck ribs oriented parallel to the girder, the shear stud capacity reduction factors are

$R_g = 1.0$

$R_p = 0.75$

The shear stud capacity is

$$Q_n = 0.5 A_{sa} \sqrt{f'_c E_c} \leq R_g R_p A_{sa} F_u$$

$$E_c = w_c^{1.5} \sqrt{f'_c} = (145)^{1.5} \sqrt{3.5} = 3266 \text{ ksi}$$

$$Q_n = 0.5 \left(\frac{\pi (0.75)^2}{4} \right) \sqrt{(3.5)(3266)} \leq (1.0)(0.75) \left(\frac{\pi (0.75)^2}{4} \right) (65)$$

$$= 23.6 \text{ kips} \leq 21.5 \text{ kips.}$$

Therefore, $Q_n = 21.5\,kips$

Or, from *AISCM* Table 3-20, with deck ribs parallel to the girder, Q_n = 21.3 kips. The number of studs between the point of maximum moment and the nearest point of zero moment (i.e., within 6'-8" from each end of the girder) is

$$N_s = \frac{\Sigma Q_n}{Q_n} = \frac{515}{21.5} = 23.9; \quad \therefore \text{ use 24 studs.}$$

The stud spacing, $s = \dfrac{(6.67 \text{ ft.})(12)}{24} = 3.33 \text{ in.} \leq 36 \text{ in. O.K.}$

≤ 4.5 in.; Not good.

Therefore, the stud spacing is not adequate.

The required shear stud spacing is too small, so either the girder size needs to be increased since we are already at 100% composite action with the W18 x 35 girder previously chosen, or two rows of studs could be used since that would effectively double the required stud spacing from 3.33 in. to 6.67 in., which is greater than the minimum AISC Code-specified spacing of 4.5 in. This particular design already has a relatively high number of studs, so a larger girder size will have to be selected. Note that a rule of thumb is that *one shear stud equates to 10 pounds of steel* [7]. We will examine this relationship later in the design of the girder.

Revise the girder size to W18 × 46 (I_x = 712 in.^{4}).

A_sF_y for W18 × 46 = (13.5 in.^{2})(50 ksi) = 675 kips.

Therefore, for 100% composite action, $\sum Q_n = 675$ kips.

We will now do a quick check before proceeding to a more detailed design check of the W18x46. From *AISCM* Table 3-18, it can be seen that the smallest design moment capacity for a W18 × 46 at approximately 25% composite action (i.e., the lowest degree of composite action allowed in the AISC specifications if the beam is to be designed as a composite beam) is $\phi_b M_n$ = 450 ft.-kips ($\approx M_u$ = 456 ft.-kips). Since a 25% degree of composite action (i.e., PNA case 7 where $\sum Q_n = 169\,\text{kips}$) resulted in a slightly undersized girder, we will assume a 40% degree of composite action for the revised W18 x 46 girder.

At 40% composite action for the W18 x 46 girder, $\sum Q_n$ = (40%)(675 kips) = 270 kips.

The depth of the concrete compression stress block, $a = \dfrac{\Sigma Q_n}{0.85 f_c' b} = \dfrac{(270)}{(0.85)(3.5)(60)} = 1.52$ in.

$Y_2 = Y_{con} - 0.5a = 5\text{ in.} - 0.5 \times 1.52\text{ in.} = 4.24\text{ in.}$

The actual moment capacity of the girder with Y_2 = 4.24 in. and $\sum Q_n = 270$ kips are found by linear interpolation from *AISCM* Table 3-18 to be **532 ft. kips** > M_u = 456 ft.-kips.

$Y_1 = 1.61$ in. (by linear interpolation from *AISCM* Tables 3-18 or 3-19)

$$N_s = \frac{\Sigma Q_n}{Q_n} = \frac{(270)}{21.5} = 12.6 \quad \therefore \text{ use 13 studs}$$

$$\text{Stud spacing, } s = \frac{(6.67\text{ ft.})(12)}{13} = 6.15\text{ in.} < 36\text{ in. OK.}$$

$$> 4.5\text{ in. OK.}$$

Since we have concentrated loads acting on the girder and the maximum moments occur at these concentrated loads, N_s is the number of studs on the girder between the concentrated load location (or the point of maximum moment) to the nearest point of zero moment (i.e., the number of studs within the shear span of the girder). Only a nominal number of studs (e.g., 4 studs or studs at 24 in. on centers) are typically provided between the two concentrated loads on the girder, since the moment gradient or horizontal shear between these points is negligible. Therefore, the total number of studs provided on the girder is specified as N = 13, 4, 13.

Note that for girders, the spacing of the shear studs does not have to be a multiple of the deck flute spacing because the deck flutes or ribs are parallel to the longitudinal axis of the girder and therefore independent of the deck flute spacing. In contrast, the spacing of studs for beams is dependent on, and has to be a multiple of, the deck flute spacing.

12. We will now check the deflections. Using values of $Y_2 = 4.24$ in. and $Y_1 = 1.61$ in. in the lower-bound moment of inertia table for W18 × 46 (*AISCM* Table 3-19), we obtain the lower-bound moment of inertia for the composite section, $I_{LB} = 1389$ in.^{4} Recall that the service loads on the girder are as follows:

$P_{LL} = 30$ kips, $\quad P_{SDL} = 3.9$ kips, $\quad P_{CDL} = 13$ kips

Construction phase dead load deflection for the W18 × 46 (use the moment of inertia of the non-composite girder) is

$$\Delta_{CDL} = \frac{PL^3}{28EI} = \frac{(13)\left[(20)(12)\right]^3}{(28)(29{,}000)(712)} = 0.31 \text{ in.}$$

Live load deflection (use the moment inertia of the composite girder, I_{LB}):

$$\Delta_{LL} = \frac{PL^3}{28EI} = \frac{(30)\left[(20)(12)\right]^3}{(28)(29{,}000)(1389)} = 0.37 \text{ in.}$$

$$\Delta_{SDL} = \frac{PL^3}{28EI} = \frac{(3.9)\left[(20)(12)\right]^3}{(28)(29{,}000)(1389)} = 0.05 \text{ in.}$$

Note that the superimposed dead loads and the live loads are applied after the concrete has cured and the concrete has hardened, in which case the composite moment of inertia is applicable.

$$\frac{L}{360} = \frac{(20)(12)}{360} = 0.67 \text{ in.} > 0.37 \text{ in.} \quad \text{O.K.}$$

Total load deflection (use the moment inertia of the composite girder, I_{LB}):

$$\Delta_{TL} = \Delta_{SDL} + \Delta_{LL} + \Delta_{CDL}$$
$$= 0.05 + 0.37 + 0.31 = 0.73 \text{ in.}$$

The total initial dead load deflection, Δ_{CDL}, is included here since no camber was specified in step 3.

$$\frac{L}{240} = \frac{(20)(12)}{240} = 1.5 \text{ in.} > \Delta_{TL} = 0.73 in., \text{ OK.}$$

Use a W18 × 46 girder (N = 13, 4, 13).

See Chapter 12 for floor vibrations.

The size of Girder G1 was increased to W18x46 (Option 2) in lieu of using more shear studs that would have had to be placed in two rows on a smaller W18x35 girder size (Option 1), so the difference in steel weight, if any, between the two options will now be compared. The increase in girder selfweight for Option 2 (i.e., using W18x46 section with fewer studs) is (46 plf – 35 plf) (20 ft.) = 220 lb. The decrease in the number of required shear studs is 48 – 26 = 22 studs. Given that one shear stud is equivalent to 10 pounds of steel (see Section 7-8), there is no net increase in the weight of steel used between the two options. Therefore, for this floor framing, there is no difference in the total steel weight between the two options, though the ease of placement of the fewer shear studs in Option 2 may result in some economy.

13. Check floor vibrations (see Chapter 12).

14. Determine additional rebar over the girders (see Figure 7-26a).

Concrete strength, $f'_c = 3.5$ ksi

Yield strength of rebar, $f_y = 60$ ksi

Effective concrete flange width of a composite girder, $b = 60$ in. (see step 6 of the girder design); therefore, the span of the cantilevered slab on both sides of the girder is 60 in./2 = 30 in. = 2.5 ft.
Factored load on slab, w_u = (1.2)(85 psf) + (1.6)(150 psf) = 342 psf = 0.342 kip/ft.^{2}

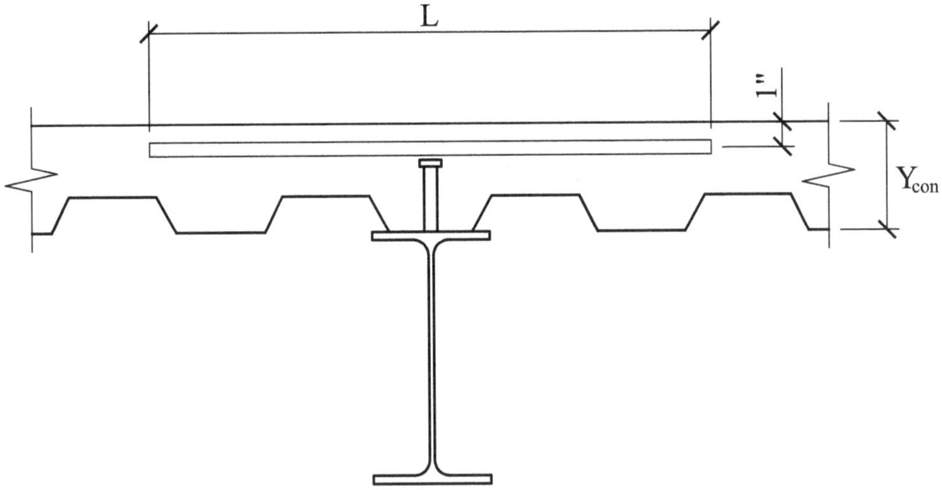

FIGURE 7-26a Rebar over Girder G1

The maximum negative moment in the concrete slab over the girder is,

$$M_u\ (\text{neg.}) = \frac{(0.342\ \text{kips / ft.})(2.5\ \text{ft.})^2}{2} = 1.07\ \text{ft.-kips/ft. width of the slab}$$

The approximate required area of rebar (in square inches per foot width of slab) can be derived using reinforced concrete design principles [2, 13] as follows:
Area of reinforcing steel required, $A_s \approx M_u/4d > A_s$ min.
Where,

d = Effective depth of concrete slab in inches, and
M_u = Factored moment in ft.-kips/ft. width of slab.

Note: The inconsistency of the units in the equation for A_s has already been accounted for by the constant "4" in the denominator.
$d = Y_{con}$ − 1 in. cover = 5 in. − 1 in. = 4 in.,
$A_{s\ min}$ = (0.0018)(12 in.)(Y_{con}) = (0.0018)(12 in.)(5 in.) = 0.11 in.^{2}/ft. width of slab, and

$$A_{s\ \text{required}} = \frac{1.07\ \text{ft.-kips/ft.}}{(4)(4\ \text{in.})} = 0.07\ \text{in.}^2\ /\ \text{ft. width of slab} < A_{s\ \min}$$

Therefore, use $A_s = A_{s\ min}$ = 0.11 in.^{2}/ft. width of slab.
Use #3 top bars @ 12 in. on centers (o.c.) × 7 ft. long over girder G1.
The length of the rebar is a function of the required development length of the reinforcement (see Ref. [2, 13]). The required development length of a #3 bar is less than 24 in. past the point of maximum moment but is also 12 in. past any point of stress, so the total length of the #3 bar is the effective slab width, 60 in. plus 12 in. for each side, or 7 ft. long. Figure 7-26b summarizes the design for Example 7-11.

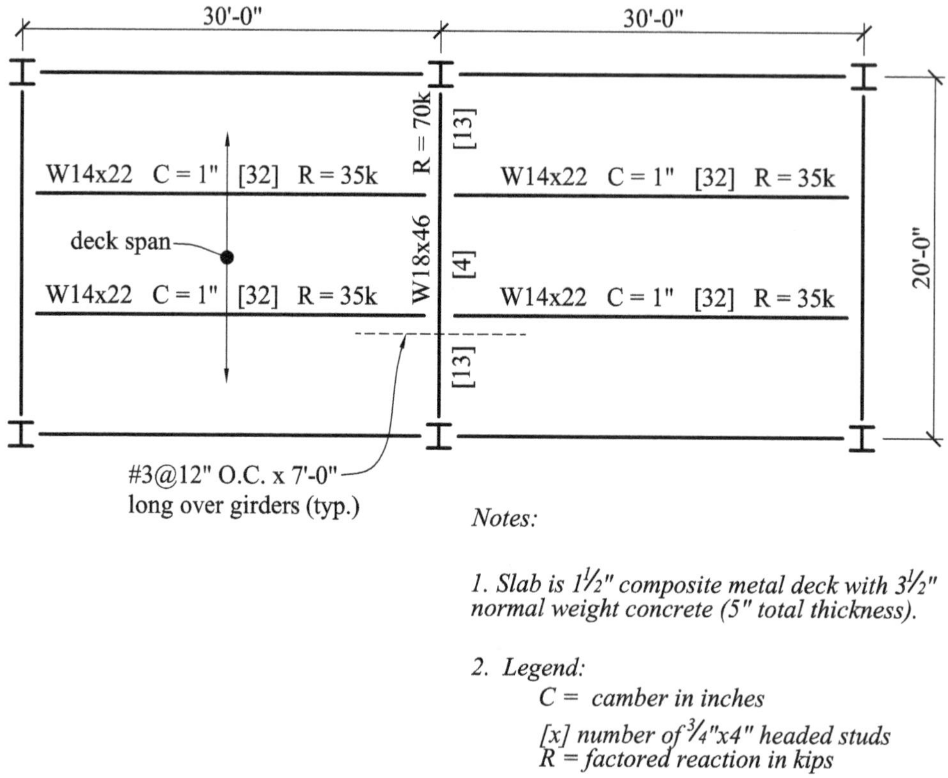

FIGURE 7-26b Floor framing design for example 7-11

7-8 PRACTICAL CONSIDERATIONS FOR COMPOSITE BEAM AND GIRDER CONSTRUCTION

1. In the United States, composite beams and girders are usually *unshored*, with the deck and the bare steel beam supporting the construction phase dead and live loads. It is more expensive to shore the beams and girders, but shored construction results in smaller size beams and girders. When using computer design software, ensure that the proper design option (e.g., unshored construction) is selected.
2. Camber refers to the upward bowing of the beam or girder. Steel beams can be cambered in one of two ways:
   - **Cold cambering:** The unheated beam is forced into a cambered shape by passing the beam through guides that have been set at the predetermined radius to achieve the specified camber. This is the most commonly used method.
   - **Heat cambering:** Similar to cold cambering, except that the beam is heated before cambering; heat cambering is the more expensive of the two cambering methods. Cambering of beams using local application of heat should be done at temperatures not exceeding approximately 1100–1200 degrees Fahrenheit to avoid damage to the steel member according to AISC Specification, Section M2.1.
3. For uncambered beams and girders, limit the construction phase dead load deflection, Δ_{CDL}, to $L/360$ to minimize the effect of concrete ponding.
   Because of concrete ponding, an additional volume of concrete is needed to achieve a level floor due to the deflection of the floor beams and girders. As concrete is first poured, the beams and girders deflect, resulting in more concrete being required to achieve a level

or flat floor, which in turn results in increased loading, which in turn leads to more deflections and thus leads to even more concrete being required to achieve a level floor. This process continues until the beams and girders (if they are capable of resisting the additional loads) reach equilibrium and the floor becomes level. The more flexible the beams and girders are, the greater the effect of concrete ponding on the floor system.

4. For uncambered beams and girders, add 10% to 15% of the concrete slab selfweight to account for the additional weight that will result from concrete ponding due to the deflection of the steel beam or girder during the construction phase (see Chapter 2).
5. A camber of 80%–85% of the dead load deflection on the floor is common, but it could be as high as 100% with more stringent concrete placement requirements. The cambering of composite beams results in less slab cracking over the girders than shoring the beams because the dead load deflection occurs before the slab is set. Shoring of the composite beams, which is not common in design practice, means that the dead load deflection will occur only after the concrete slab is cured and acting compositely with the steel beam or girder. In any case, the cracks in the slab can be controlled with top or negative steel reinforcement over the girders, parallel to the beams.
6. Calculation of the construction phase deflection, or the required camber, usually assumes pinned supports for the beams and girders, but some restraint against deflection is inherent in the simple shear connections typically used at the ends of the beams and girders, and thus, the actual construction phase deflection may be less than that calculated. This could lead to an overestimation of the required camber, which is undesirable.
7. Overestimating the required camber could lead to problems, resulting in a floor slab with less than adequate depth at the critical sections. This could lead to inadequate cover for the shear studs at these sections. In order to avoid this situation and to account for the restraint at the beam or girder shear connections, it is advisable to specify 75% to 85% of the construction phase dead load deflection, Δ_{CDL}, as the required camber. Underestimating the camber or not cambering beams when they should have been cambered or shored can lead to costly reinforcing details, such as coverplating the beams or even the addition of columns. Due to the resulting excessive beam deflections, the concrete slab might require a self-leveling compound on top of the slab, which adds weight to the structure and thus the potential need for steel beam reinforcing.
8. Metal decks are usually available in widths of 2 ft. to 3 ft. and lengths of up to 42 ft. Specify 3-span decks whenever possible (i.e., decks that are long enough to span over four or more beams in order to achieve the maximum strength of the deck due to the deck continuity over the beam supports). Avoid single-span decks whenever possible because it is more susceptible to ponding and it is not as strong as the 2-span or 3-span decks.
9. Shear stud diameters could be $^1/_2$ in., $^5/_8$ in., or $^3/_4$ in., but $^3/_4$ in.-diameter shear studs are the most commonly used in design practice.
10. Where shear studs cannot be placed at the center of the deck flute, offset the stud toward the nearest end support of the beam or girder (this is the strong stud position). Note that in practice, some engineers assume a weak stud position, thus the stud will have a lower shear strength, but the stud position will not need to be verified on site. This is convenient and conservative.
11. If the number of shear studs required in the beam or girder exceeds the number that can be placed in each flute in a single row, lay out the balance of studs in double rows starting from both ends of the beam or girder. Note the reduction in the nominal horizontal shear strength of shear studs when more than one stud is placed in a deck flute (i.e., 2 or 3 rows of studs) for deck ribs perpendicular to the beams (see *AISCM* Table 3-20).

12. If the number of studs required exceeds the number that can be placed in every second flute, place the studs in every second flute and add the remaining studs to the deck flutes in between, starting from both ends of the beam or girder. This is an alternative to uniformly spacing the studs in every flute throughout the beam or girder.
13. A rule of thumb to maintain economy in the balance between adding shear studs and increasing the beam size is that one shear stud is approximately equivalent to 10 pounds of steel.
14. The vertical support reactions of composite beams or girders are usually higher than those of comparable non-composite beams or girders. Steel fabricators usually design the end connections for one-half of the maximum total factored uniform load (see *AISCM* Table 3-6), but this is often not adequate for composite beams or girders. To account for the higher-end reactions in composite beams or girders, either specify on the plan the actual reactions at the ends of the composite beams or girders or specify that the composite beam or girder joints be designed for three-quarters of the maximum total factored uniform load.
15. The effect of floor openings on the composite action of beams and girders is a function of the size and location of the openings. Ref. [8] provides an analytical approach for calculating the effective width of composite beams with floor openings. A conservative but quick approach for considering the effect of floor openings on composite beams is as follows:
    - If the floor opening is located on *only one side* of the beam, the beam is considered an L-shaped beam for calculating the effective width at the location of the opening, but the beam is still assumed to support the full tributary width of the floor on both sides of the beam (less the reduction in load due to the floor opening).
    - In the case where there are floor openings at the same location on *both sides* of the beam, the beam is considered to be non-composite at that location, and the load on the beam will include the load from the tributary width on both sides of the beam (less the reduction in load due to the floor openings).

    It is recommended that additional reinforcement be provided in the slab at the edges of the floor openings to control cracking.
16. Consider the layout of the beams with respect to the camber and the layout of the metal deck. It is ideal for the metal deck to generally lay flat across a series of beams, but this uniform position of the bottom deck surface could be interrupted and be problematic if adjacent cambered beams have different camber profiles, which might occur at a column line or at a perimeter beam since perimeter beams are not usually cambered (see Figure 7-27a). It is also good practice not to camber floor girders since their deflections are usually smaller than the beam deflections, and the presence of a girder camber might have a negative impact on the overall deck layout across several cambered beams with different camber profiles, as well as complicating the connections of the cambered beams to the girders.

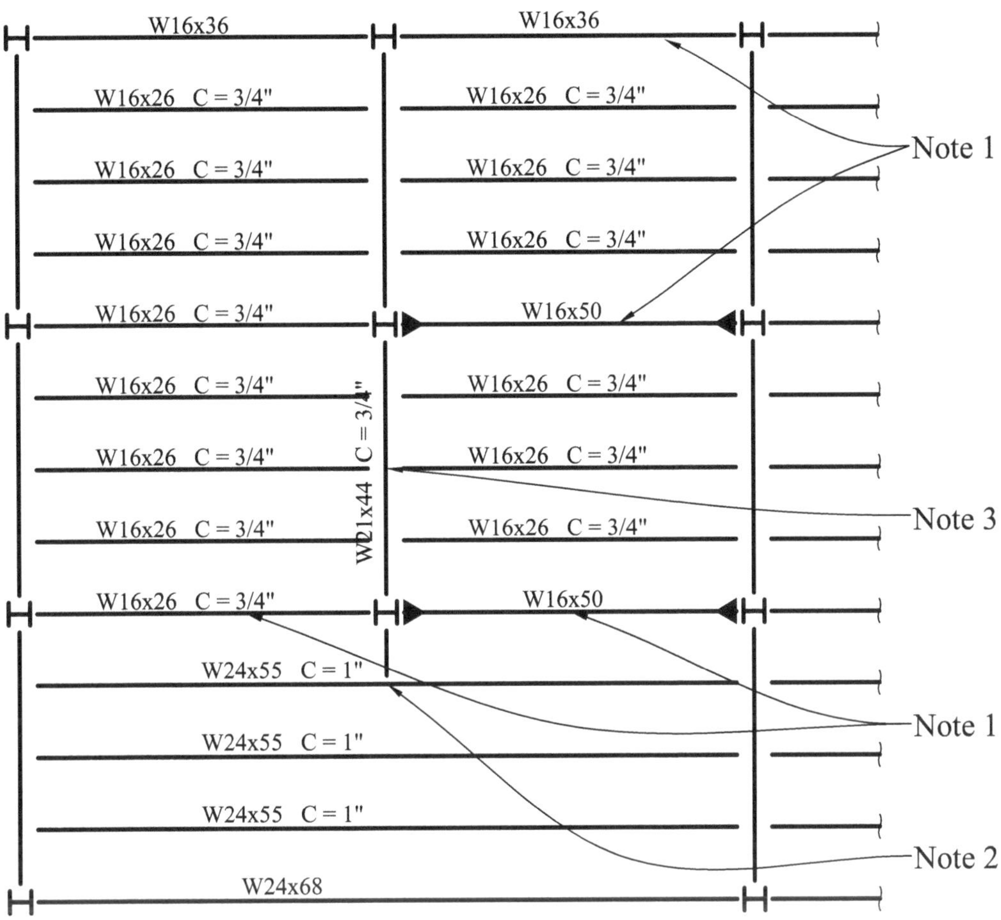

Note 1: Beams that are adjacent to other beams with a differential in camber will create issues for the placement of the metal deck; the deck will not be allowed to lay flat at these locations, so additional design consideration is needed to avoid the above conditions.

Note 2: This is the same issue with a camber adjacent to a position with zero camber, which in this case is a column.

Note 3: Camber in a girder could create alignment issues with the bolt holes from beams that frame into the girder

FIGURE 7-27a Effect of cambered beams on the floor deck layout

17. During the fabrication process, bolt holes are typically placed prior to cambering. This may be a factor for beams that frame into girders nearer to the mid-span of the girder, where the camber in the girder would be the greatest. It is advisable not to camber the girders to avoid alignment issues with the bolt holes in the cambered beams.
18. Consider the effect of cambered beams on the exterior façade and attachments to the spandrel beams. It is therefore good practice *not* to camber the spandrel or edge beams to avoid issues with the cladding connections. Beams that support cranes and cantilevered beams should also *not* be cambered.

19. For large steel deck/concrete composite floor areas where the concrete has to be placed in more than one concrete pour, construction joints will be required as follows:

- For construction joints that are *parallel to the infill beams* (i.e., perpendicular to the girders), the construction joint should be located near the middle-third of the slab between infill beams, near one end of the girder. This will "allow [the girder] to deflect completely before concrete at [the girder] hardens." [11]
- For construction joints that are *parallel to the girders*, locate the construction joint near the girder at a distance from the girder equal to one-third the span of the infill beam, but not less than 4 ft. from the girder (see the guidelines in Ref. [11]). The initial concrete pour will include the infill beams up to the 4 ft distance from the girder, but the girder and the infill beams on the opposite side of the girder will be included in the second concrete pour. Locating the construction joints as noted above ensures the following [11]:

  a. "..allows sufficient distance for the development of the flange width of the girder."

  b. "...allows nearly the full dead load from concrete placement to be applied to [the infill beams] that are included in the initial concrete placement. The [girder] should generally be included in the second concrete placement at the construction joint. This will allow [the girder] to deflect completely before concrete at [the girder] hardens."

Figure 7-27b shows the detail of the reinforcement dowels across the construction joints at the top of the slab (e.g., # 4 dowels at 12 in. on centers × 4 ft. long), with adequate development length provided on both sides of the construction joint.

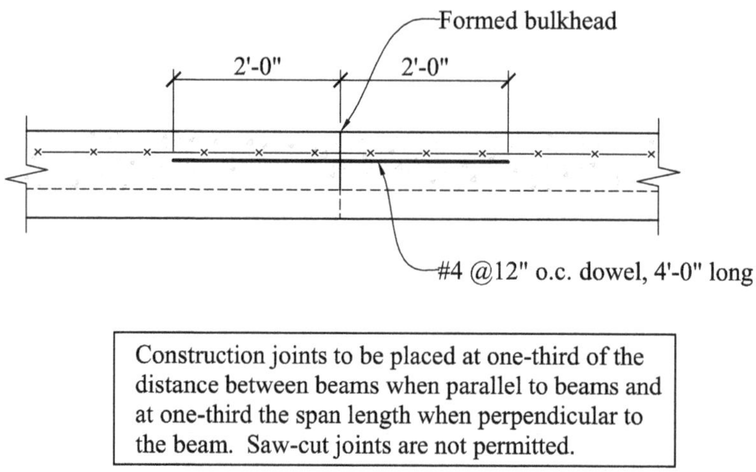

FIGURE 7-27b Dowels at construction joints in a concrete slab on composite metal deck

20. For industrial buildings, non-composite floor construction may be preferred (i.e., floor beams and girders without steel shear studs) because it provides the building owner some flexibility in cutting openings in the floor in the future for installation of new equipment, while for office, residential or other commercial buildings, it is usually more economical to use composite floor construction (i.e., floor beams and girders with steel shear studs).

7-9 OTHER TYPES OF COMPOSITE CONSTRUCTION

There are other types of composite construction, such as steel beams or girders encased in concrete, where the shear studs are welded to the top and bottom flanges and to the web of the girder to engage the concrete in composite action. This type of construction may be used in reinforced concrete construction where there are beam or girder depth restrictions due to architectural constraints and where post-tensioned concrete beams cannot be used. The concrete encasing the steel girder also provides fire protection to the steel girder. The design of such encased composite girders is covered in Sections I1 and I3 of the AISC Specifications.

The axial load capacity of steel columns can also be increased by encasing them in concrete. Shear studs are also welded to the flanges and web of the steel column to engage the concrete in composite action with the steel column. The design of concrete-encased composite steel columns and concrete-filled HSS and steel box sections are covered in Sections I1 and I2 of the AISC Specifications, and will covered in this text in the remaining sections of this chapter.

7-10 AXIALLY LOADED COMPOSITE COLUMNS

To increase the strength of steel or concrete columns, composite columns such as steel columns encased in concrete may be used. Hollow structural sections (HSS) and steel box sections filled with concrete are also sometimes used as columns; in some cases, the steel is exposed and coated with intumescent paint to achieve the required fire-resistance rating. The design of these composite columns is covered in AISC Specifications Sections I1.4, I1.5, and I2; the load transfer requirements, the force transfer mechanisms between the concrete and the steel member, and the detailing requirements are covered in AISC Specifications Section I6. *AISC* Specification Section I1.3 limits the strengths of the normal-weight concrete and the structural steel as follows:

Concrete: 3.0 ksi $\le f'_c \le 10$ ksi and Steel: $F_y \le 75 ksi$

If the methods presented in *AISC* Specification Appendix 2 are used to determine the design compressive strength, the design flexural strength, or the combined flexural and axial design strength for *rectangular*-filled composite members, the limits on the concrete and steel strengths are as follows:

Concrete: $f'_c \le 15$ksi and Steel: $F_y \le 100 ksi$

Note that the *AISC* Specification Appendix 2 requirement applies only to rectangular-filled composite members.

The methods provided in Section I of the *AISC* Specifications will be used for the design of composite members in this text. Note that the strength reduction factor, ϕ_c, for bare steel from Chapter E of the AISC Specifications (i.e., from Chapter 5 of the text) is 0.9, whereas the strength reduction factor for composite columns from Chapter I of the AISC Specifications is 0.75. Consequently, the AISC Specification sets a lower limit for the strength calculated for composite columns to be the strength for the bare steel section. That is:

$$\phi_c P_n \, (composite\ column) \ge \phi_c P_n \, (non-composite\ or bare steel\ column) \tag{7-39}$$

Axially Loaded Concrete-Encased Composite Members

Encased composite steel columns are sometimes used in mid- and high-rise buildings. The following requirements from AISC Specification Section I2 must be met:

1. The cross-sectional area of the steel core, A_s, must be at least 1% of the total cross-sectional area of the composite section, A_g.
2. The minimum amount of continuous longitudinal reinforcement, A_{sr}, must be at least 0.4% of the gross area of the composite section, A_g. That is, $A_{sr} \ge 0.004A_g$.
3. Transverse reinforcement in the form of ties or hoops must be provided around the longitudinal reinforcement and the encased steel core. This reinforcement shall be a minimum of either No. 3 bars at 12-inch maximum center to center spacing or a minimum of No. 4 bars at 16-inch maximum center to center spacing. The spacing of the transverse reinforcement shall not exceed one-half the smaller dimension of the column.

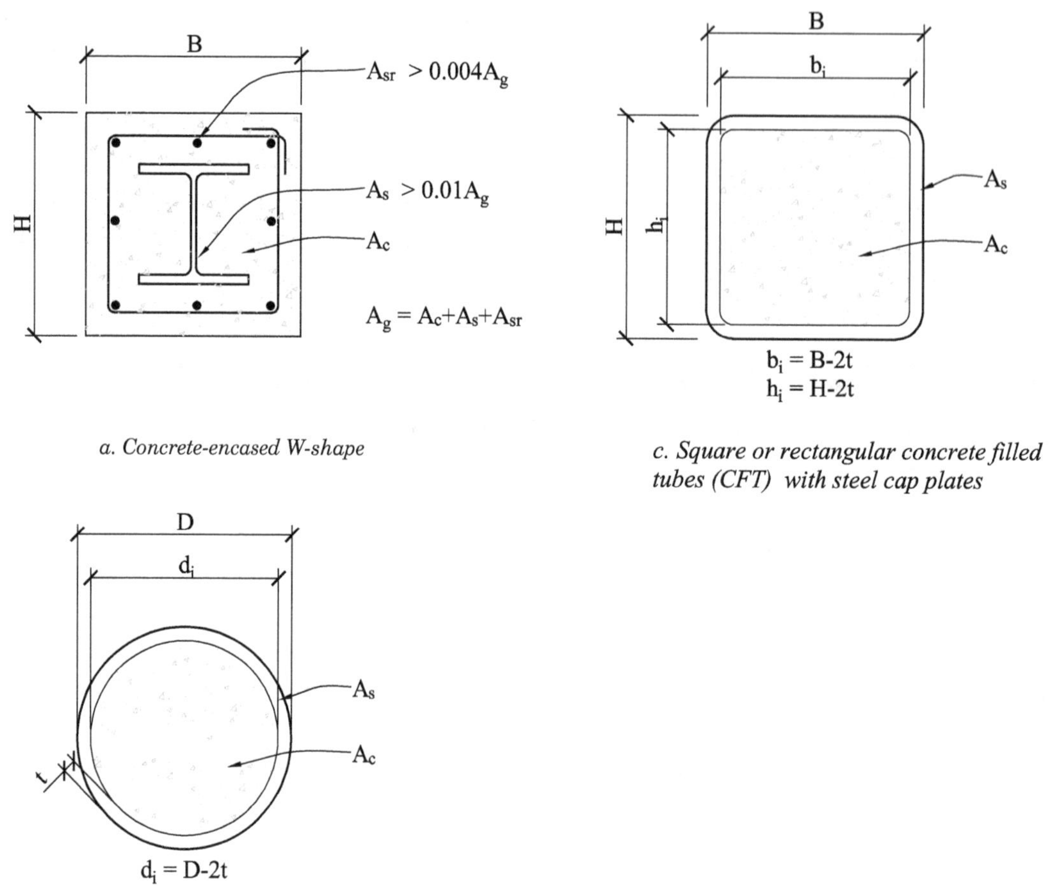

a. Concrete-encased W-shape

b. Round concrete filled tubes (CFT) with steel cap plates

c. Square or rectangular concrete filled tubes (CFT) with steel cap plates

FIGURE 7-28 Composite columns

4. The maximum continuous longitudinal reinforcement shall conform to the ACI 318 requirements — that is, $A_{sr} \leq 0.08A_g$, including at the rebar splice locations.
5. The minimum clear spacing between the longitudinal rebar and the structural steel shape is 1.5 x longitudinal reinforcing bar diameters ≥ 1.5 in.
6. If the encased steel is made up of two or more encased sections, the sections shall be interconnected with lacing or tie plates to prevent buckling of the individual steel shapes.
7. It is assumed that the axial compression load is carried by both the steel and the concrete concurrently. This can be achieved either through the use of a rigid cap plate welded to the top of the steel section and extending over the entire plan area of the column or through the use of headed studs welded to the flanges of the W-shape.

Note that local buckling does *not* need to be considered for concrete-encased composite members. The procedure for determining the axial compressive strength of the concrete-encased composite column from AISC Specification Section I2.1b is as follows:

1. Determine the axial compression capacity of the gross section, neglecting slenderness or length effects,

$$P_{no} = F_y A_s + F_{ysr} A_{sr} + 0.85 f'_c A_c \tag{7-40}$$

2. Calculate the effective stiffness of the composite section, $(EI)_{eff}$, and the elastic critical buckling load, P_e .

$$(EI)_{eff} = E_s I_s + E_{sr} I_{sr} + C_1 E_c I_c \tag{7-41}$$

$$P_e = \frac{\pi^2 (EI)_{eff}}{L_c^2} \tag{7-42}$$

3. The coefficient, C_1, for calculating the effective rigidity of the encased section is given as

$$C_1 = 0.25 + 3\left(\frac{A_s + A_{sr}}{A_g}\right) \leq 0.7 \tag{7-43}$$

4. Calculate the nominal axial compressive strength from AISC Specification Equations I2-2 and I2-4 as follows:

$$When\ \frac{P_{no}}{P_e} \leq 2.25, \quad P_n = P_{no}\left(0.658^{\frac{P_{no}}{P_e}}\right) \tag{7-44}$$

$$When\ \frac{P_{no}}{P_e} > 2.25, \quad P_n = 0.877 P_e \tag{7-45}$$

For LRFD, the design axial compressive strength (LRFD) is ϕP_n

For ASD, the allowable axial compressive strength is $\frac{P_n}{\Omega}$

Where,

ϕ_c = 0.75 (LRFD)
Ω = 2.0 (ASD)

$E_c = w_c^{1.5}\sqrt{f'_c}$, ksi

f'_c = specified compressive concrete strength, ksi
w_c = density of concrete, pcf $(90 \leq w_c \leq 155\ Ib / ft^3)$
E_s = modulus elasticity of steel, ksi = 29,000 ksi
I_c = moment of inertia of the concrete section about the elastic neutral axis of the composite section
I_s = moment of inertia of the steel shape about the elastic neutral axis of the composite section
I_{sr} = moment of inertia of the longitudinal reinforcing bars about the elastic neutral axis of the composite section
$L_c = KL$ = effective length of the member, in.
A_c = area of concrete, in.^{2} = $A_g - A_{sr} - A_s$
A_g = gross area of the composite member, in.^{2} = BH
B = overall width of the composite column
H = overall depth of the composite column
A_s = cross-sectional area of structural steel, in.^{2}
F_y = specified minimum yield stress of structural steel section, ksi
F_{ysr} = specified minimum yield stress of reinforcing steel, ksi

As previously noted, the axial compressive strength of the concrete-encased composite column, $\phi_c P_{no}$, must be greater than or equal to the axial compressive strength, $\phi_c P_n$, for the bare steel member calculated using the principles discussed earlier in Chapter 5; see equation (7-39).

Axially Loaded Concrete-Filled Composite Members or Concrete-filled Tubes (CFT)

Filled composite steel columns are sometimes used in mid- and high-rise buildings. The following requirements from AISC Specification Section I2.2 must be satisfied:

Table 7-4 Limiting Width-to-Thickness Ratios for Compression Steel Elements in Composite Members Subjected to Axial Compression*

Description of Element	Width-to-Thickness Ratio^	λ_p Compact Composite/ Non-Compact Composite	λ_r Non-Compact Composite/Slender-Element Composite	Maximum Permitted
Walls of rectangular HSS and box sections of uniform thickness	b/t	$2.26\sqrt{\frac{E}{F_y}}$	$3.0\sqrt{\frac{E}{F_y}}$	$5.0\sqrt{\frac{E}{F_y}}$
Round HSS	D/t	$\frac{0.15E}{F_y}$	$\frac{0.19E}{F_y}$	$\frac{0.31E}{F_y}$

*Adapted from AISC Specification Table I1.1a
^For rectangular HSS: b = B - 3t; h = H - 3t
^For rectangular built-up box section: b = B - 2t; h = H - 2t
Refer to Figures 7-28a, b, and c for the definitions of width, B, and depth, H, and thickness, t for rectangular HSS and box sections.

1. The cross-sectional area of the structural steel section, A_s, must be at least 1% of the total cross-sectional area of the composite section, A_g.
2. Local buckling must be considered for concrete-filled composite members. Classification of filled composite members for local buckling for members subjected to axial compression load is according to AISC Specification Table I1.1a, which is reproduced in Table 7-4, and spelt out as follows:
   a. *Round* hollow structural section (HSS) filled with concrete is **compact composite** if the width-to-thickness ratio, $\frac{D}{t} \le \frac{0.15E}{F_y}$;
   b. *Round* HSS filled with concrete is **non-compact composite** if the $\frac{0.15E}{F_y} < \frac{D}{t} \le \frac{0.19E}{F_y}$;
   c. *Round* HSS filled with concrete is **slender-element composite** if $\frac{0.19E}{F_y} < \frac{D}{t} \le \frac{0.31E}{F_y}$,
   and $\frac{D}{t}$ must not be greater than $\frac{0.31E}{F_y}$;
   where D = outside diameter of the round HSS.
   d. *Rectangular* HSS and box sections of uniform thickness; the section is **compact composite** if the width-to-thickness ratio, $\frac{b}{t} \le 2.26\sqrt{\frac{E}{F_y}}$ and $\frac{h}{t} \le 2.26\sqrt{\frac{E}{F_y}}$.
   e. *Rectangular* HSS and box sections of uniform thickness, the section is **non-compact composite** if $2.26\sqrt{\frac{E}{F_y}} < \frac{b}{t} \le 3.0\sqrt{\frac{E}{F_y}}$ and $2.26\sqrt{\frac{E}{F_y}} < \frac{h}{t} \le 3.0\sqrt{\frac{E}{F_y}}$.

f. *Rectangular* HSS and box sections of uniform thickness, the section is **slender-element composite** if $3.0\sqrt{\frac{E}{F_y}} < \frac{b}{t} \leq 5.0\sqrt{\frac{E}{F_y}}$ and $3.0\sqrt{\frac{E}{F_y}} < \frac{h}{t} \leq 5.0\sqrt{\frac{E}{F_y}}$.

where,

t = design wall thickness

For HSS: $t = 0.93t_{nominal}$ for ASTM A500

$= 1.0t_{nominal}$ for ASTM A1065 and A1085

D = outside diameter of round HSS

d_i = inside diameter of round HSS = $D - 2t$

B = outside width of the rectangular HSS or built-up box section

H = outside depth of the rectangular HSS or built-up box section

b_i = clear distance between the *webs* of the HSS or box sections = $B - 2t$

h_i = clear distance between the *flanges* of the HSS or box sections = $H - 2t$

b = width of the stiffened *flange* element in compression for the **HSS** = $B - 3t$

h = width of the stiffened *web* element in compression for the **HSS** = $H - 3t$

For built-up box sections:

b = width of the stiffened *flange* element in compression of the **box section** $= B - 2t$

h = width of the stiffened *web* element in compression of the **box section** $= H - 2t$

Note that b and h are the dimensions used for calculating the wall slenderness ratios and for checking local buckling.

3. Continuous longitudinal reinforcement is *not* required.
4. Transverse reinforcement is *not* usually provided in concrete-filled tubes. If used, the transverse reinforcement shall be a minimum of either No. 3 bars at 12 inches maximum on centers or a minimum of No. 4 bars at 16 inches maximum on centers.
5. The maximum continuous longitudinal reinforcement, if provided, shall conform to the ACI 318 requirements based on the gross area of the concrete—that is, $A_{sr} \leq 0.08A_g$, including at the rebar splice locations.
6. Where provided, the minimum clear spacing between the inside face of the HSS or box section and the continuous longitudinal reinforcement must be the larger of 1.5 × longitudinal bar diameter and 1.5 in.
7. It is assumed that the axial compression load is carried by both the steel and the concrete concurrently. This can be achieved through the use of a **rigid cap plate welded to the top of the steel section.**

The procedure for determining the nominal axial compressive strength of the concrete-filled tubes (CFT) from AISC Specification Section I2.2b is as follows:

a. **Compact composite** sections:

The nominal axial compressive strength without considering slenderness effects, $P_{no} = P_p$.

Where, the nominal plastic axial compressive strength, $P_p = F_y A_s + C_2 f'_c \left(A_c + A_{sr}\frac{E_s}{E_c} \right)$ (7-46)

C_2 = 0.85 for rectangular sections

= 0.9 for round sections

b. **Non-compact composite** sections:

$$P_{no} = P_p - \frac{P_p - P_y}{(\lambda_r - \lambda_p)^2}(\lambda - \lambda_p)^2 \quad (7\text{-}47)$$

Where,

$$\lambda_p = 2.26\sqrt{\frac{E}{F_y}} \text{ for rectangular HSS and box sections of uniform wall thickness}$$

$$= \frac{0.15E}{F_y} \text{ for round HSS}$$

$$\lambda_r = 3.0\sqrt{\frac{E}{F_y}} \text{ for rectangular HSS and box sections of uniform wall thickness}$$

$$= \frac{0.19E}{F_y} \text{ for round HSS}$$

$$P_y = F_y A_s + 0.7 f_c' \left(A_c + A_{sr} \frac{E_s}{E_c} \right) \tag{7-48}$$

c. Slender-element composite sections**:**

$$P_{no} = F_n A_s + 0.7 f_c' \left(A_c + A_{sr} \frac{E_s}{E_c} \right) \tag{7-49}$$

The critical buckling stresses for filled composite members are as follows:
For **rectangular-concrete-filled composite sections**, the critical buckling stress (see AISC Specification Equation I2-10) is:

$$F_n = \frac{9E_s}{\lambda^2}, \tag{7-50}$$

$$\lambda = \text{larger of } \frac{b}{t} \text{ or } \frac{h}{t}$$

For **round-concrete-filled composite sections**, the critical buckling stress (see AISC Specification Equation I2-11) is: $F_n = \dfrac{0.72F_y}{\left[\left(\dfrac{D}{t}\right)\dfrac{F_y}{E_s}\right]^{0.2}}$ (7-51)

Where,

B = outside width of the rectangular HSS or built-up box section
H = outside depth of the rectangular HSS or built-up box section
D = outside diameter of round HSS
b = $B - 3t$ for rectangular concrete-filled **HSS**
= $B - 2t$ for rectangular concrete-filled **built-up box sections**
h = $H - 3t$ for rectangular concrete-filled **HSS**
= $H - 2t$ for rectangular concrete-filled **built-up box sections**
d_i = inside diameter of **round HSS** = $D - 2t$
t = design wall thickness

For HSS: $t = 0.93t_{nominal}$ for ASTM A500
= $1.0t_{nominal}$ for ASTM A1065 and A1085

Design Axial Compressive Strength of CFT:

The effective stiffness or rigidity of the concrete-filled composite member (AISC Specification Equation I2-12) is

$$(EI)_{eff} = E_s I_s + E_s I_{sr} + C_3 E_c I_c \tag{7-52}$$

For **rectangular sections filled with concrete**, the effective stiffness should be calculated for bending about both the x-x and y-y axes as follows:

$$(EI)_{eff,xx} = E_s I_{s,xx} + E_s I_{sr,xx} + C_3 E_c I_{c,xx} \tag{7-53}$$

$$(EI)_{eff,yy} = E_s I_{s,yy} + E_s I_{sr,yy} + C_3 E_c I_{c,yy} \tag{7-54}$$

The coefficient for calculating the effective stiffness of a concrete-filled composite section (AISC Specification Equation I2-13) is

$$C_3 = 0.45 + 3\left(\frac{A_s + A_{sr}}{A_g}\right) \leq 0.9 \tag{7-55}$$

The elastic critical buckling load, P_e, is calculated from AISC Specification Equation I2-4 as

$$P_e = \frac{\pi^2 (EI)_{eff}}{L_c^2} \tag{7-56}$$

For **rectangular sections filled with concrete**, the elastic critical buckling load should be calculated for bending about both the x-x and y-y axes, and the smaller value will be selected. Thus,

$$P_{e,xx} = \frac{\pi^2 (EI)_{eff,xx}}{L_{c,xx}^2} \tag{7-57}$$

$$P_{e,yy} = \frac{\pi^2 (EI)_{eff,yy}}{L_{c,yy}^2} \tag{7-58}$$

The axial compression strength of the concrete-filled composite column without considering slenderness effects must be greater than or equal to the axial compressive strength for the bare steel member, calculated using the principles discussed earlier in Chapter 5; that is, recalling equation (7-39),

$$\phi_c P_{no}\,(composite\ column) \geq \phi_c P_n\,(non-composite\ or\,bare\,steel\ column) \tag{7-39}$$

Calculate the nominal axial compressive strength considering slenderness effects from AISC Specification Equations I2-2 and I2-4 as follows:

$$When\ \frac{P_{no}}{P_e} \leq 2.25, \quad P_n = P_{no}\left(0.658^{\frac{P_{no}}{P_e}}\right) \tag{7-59}$$

$$When\ \frac{P_{no}}{P_e} > 2.25, \quad P_n = 0.877 P_e \tag{7-60}$$

For LRFD, the design axial compressive strength (LRFD) is ϕP_n

For ASD, the allowable axial compressive strength is $\frac{P_n}{\Omega}$

$$\phi_c P_n\,(composite\ column) \geq \phi_c P_n\,(non-composite\ or\,bare\,steel\ column)$$

Where,

ϕ_c = 0.75 (LRFD)

ϕ_c (bare steel) for LRFD = 0.9 (see Chapter 5)

Ω = 2.0 (ASD)

$E_c = w_c^{1.5}\sqrt{f_c'}\ ,ksi$

f_c' = specified compressive concrete strength, ksi

w_c = density of concrete, pcf $\left(90 \leq w_c \leq 155\ Ib/ft^3\right)$

E_s = modulus elasticity of steel, ksi = 29,000 ksi

I_c = moment of inertia of the concrete section about the elastic neutral axis of the composite section

I_s = moment of inertia of the steel shape about the elastic neutral axis of the composite section

I_{sr} = moment of inertia of the longitudinal reinforcing bars about the elastic neutral axis of the composite section

$L_c = KL$ = effective length of the member, in.

A_c = area of concrete, in.^{2}

$= \dfrac{\pi d_i^2}{4}$ for concrete-filled round HSS or tube

$= b_i h_i$ for concrete-filled rectangular built-up box section

$= b_i h_i - \left(4r^2 - \pi r^2\right) = b_i h_i - \left(4t^2 - \pi t^2\right) = b_i h_i - t^2\left(4 - \pi\right)$ for concrete-filled rectangular HSS;

The equation above for the A_c for concrete-filled rectangular HSS accounts for the fillet radius. Note that there is not much accuracy lost by neglecting the effect of the fillet radius. That is, the concrete area can be calculated using $A_c = b_i h_i$. In calculating the moments of inertia, the effect of the fillet radius can also be neglected without any loss in accuracy.

A_s = cross-sectional area of structural steel, in.^{2}

F_y = specified minimum yield stress of structural steel section, ksi

F_{ysr} = specified minimum yield stress of reinforcing steel, ksi

EXAMPLE 7-12

Axially Load Capacity of Concrete-Encased Composite Member

Determine the axial compression load capacity of a W14 x 342 encased in a 27 in. x 27 in. concrete column, assuming the compression load is applied concurrently to the steel and concrete. The column is 12.5 ft. long and reinforced with 12 No. 10 longitudinal bars and No. 4 ties at 12 in. on centers. The concrete compressive strength is 10 ksi, and the column is pinned at both ends. Assume ASTM A913 Grade 65 structural steel and Grade 75 rebar.

Solution

Since the W14 × 342 is encased in concrete, local buckling does not need to be checked.

For W14 × 342:

$A_s = 101 \text{ in}^2$

$I_{s,xx} = 4900 \text{ in.}^4$

$I_{s,yy} = 1810 \text{ in.}^4$

$F_{ys} = 65$ ksi

$F_{ysr} = 75$ ksi

$f'_c = 10$ ksi

$E_c = w_c^{1.5}\sqrt{f'_c}$, ksi $= \left(145 pcf\right)^{1.5}\sqrt{10 \text{ ksi}} = 5521$ ksi

$kL_x = KL_y = (1.0)(12.5 \text{ ft})(12) = 150$ in.

$B = H = 27$ in.

$A_g = BH = (27 \text{ in.})(27 \text{ in.}) = 729 \text{ in.}^2$

$A_s = 101 \text{ in.}^2 = 0.14A_g >> 1\%A_g$ OK

The longitudinal reinforcement = 12 No. 10 bars (i.e., 4 bars on each face of the column).

A_{sr} (12 No. 10) = 15.2 in.^{2} = 0.021A_g > 0.004A_g OK.

< 0.08A_g OK.

Area of concrete, $A_c = A_g - A_s - A_{sr}$ = 729 – 101 – 15.2 = 612.8 in.^{2}

Spacing of No. 4 ties = 12 in. ≤ 16 in. OK.

< (½)(27 in.) = 13.5 in. OK.

Ensure that the minimum clear spacing between the longitudinal rebar and the structural steel shape is 1.5 x longitudinal reinforcing bar diameters ≥ 1.5 in.

Determine the axial compression capacity of the gross section, neglecting slenderness or length effects:

$$P_{no} = F_y A_s + F_{ysr} A_{sr} + 0.85 f'_c A_c$$

$$= (65 ksi)(101 in.^2) + (75 ksi)(15.2 in.^2) + 0.85(10 ksi)(612.8 in.^2) = 12{,}914 \, kips$$

Calculate the effective stiffness of the composite section, $(EI)_{eff}$, and the elastic critical buckling load, P_e:

E_s = 29000 ksi

I_{sx} = 4900 in.^{4}

I_{sy} = 1810 in.^{4}

Distance from the face of concrete to the center of closest longitudinal bar = 1.5 in. cover + ½ in. tie + (½)(1.27 in.) = 2.64 in.

Distance between the centroid of corner rebars = 27 in. – (2.64 in.) – (2.64 in.) = 21.72 in.

Since there are 4 bars on each face of the column, the distance between the centroid of adjacent bars = $\dfrac{21.72 \text{ in.}}{(4 \text{ bars} - 1)} = \dfrac{21.72 \text{ in.}}{3} = 7.24$ in.

With 4 No. 10 longitudinal bars on each face of the column, the area of rebar on each face = (4)(1.27 in.^{2}) = 5.08 in.^{2}

The moment of inertia of the vertical rebar about the centroid of the composite section is

$$I_{sr,xx} = I_{sr,yy} = (2)\left[(4 \text{ bars})(1.27 \text{ in.}^2)\left(\frac{21.32 \text{ in.}}{2}\right)^2\right] + (2)\left[(2 \text{ bars})(1.27 \text{ in.}^2)\left(\frac{7.24 \text{ in.}}{2}\right)^2\right]$$

$$= 1221 \text{ in.}^4$$

The moment of inertia of the concrete about the centroid of the composite section can be calculated as

$$I_{c,xx} = \left[\frac{(27 \text{ in.})(27 \text{ in.})^3}{12}\right] - I_{s,xx} - I_{sr,xx}$$

$$= \left[\frac{(27 \text{ in.})(27 \text{ in.})^3}{12}\right] - 4900 - 1221 = 38{,}166 \text{ in.}^4$$

$$I_{c,yy} = \left[\frac{(27 \text{ in.})(27 \text{ in.})^3}{12}\right] - I_{s,yy} - I_{sr,yy}$$

$$= \left[\frac{(27 \text{ in.})(27 \text{ in.})^3}{12}\right] - 1810 - 1221 = 41{,}256 \text{ in.}^4$$

The coefficient, C_1, for calculating the effective rigidity of the concrete-encased section is given as

$$C_1 = 0.25+3\left(\frac{A_s + A_{sr}}{A_g}\right) \le 0.7$$

$$= 0.25+3\left(\frac{101 \text{ in.}^2 + 15.2 \text{ in.}^2}{729 \text{ in.}^2}\right) = 0.74 > 0.7$$

Therefore, use $C_1 = 0.7$

Calculate the effective moment of inertia about both x- and y-axes as follows:

$$(EI)_{eff,xx} = E_s I_{s,xx} + E_{sr} I_{sr,xx} + C_1 E_c I_{c,xx}$$

$$= (29{,}000 \text{ ksi})(4900 \text{ in.}^4) + (29{,}000 \text{ ksi})(1221 \text{ in.}^4) + 0.7(5521 \text{ ksi})(38{,}166 \text{ in.}^4)$$

$$= 3.23x10^8 \text{ kip.in.}^2$$

$$(EI)_{eff,yy} = E_s I_{s,yy} + E_{sr} I_{sr,yy} + C_1 E_c I_{c,yy}$$

$$= (29{,}000 \text{ ksi})(1810 \text{ in.}^4) + (29{,}000 \text{ ksi})(1221 \text{ in.}^4) + 0.7(5521 \text{ ksi})(41{,}256 \text{ in.}^4)$$

$$= 2.47x10^8 \text{ kip.in.}^2$$

Use the smaller of the effective stiffnesses since that will produce the smaller critical buckling load, which is calculated as

$$P_e = \frac{\pi^2 (EI)_{eff}}{L_c^2} = \frac{\pi^2 (2.47x10^8)}{(150 \text{ in.})^2} = 108{,}346 \text{ kips}$$

$P_{no} = 12{,}914$ kips (previously calculated)

$$\frac{P_{no}}{P_e} = \frac{12{,}914 \text{ kips}}{108{,}346 \text{ kips}} = 0.119 < 2.25$$

Calculate the nominal axial compressive strength from AISC Specification Equations I2-2 as follows:

$$When\ \frac{P_{no}}{P_e} \le 2.25,\quad P_n = P_{no}\left(0.658^{\frac{P_{no}}{P_e}}\right) = (12{,}914 \text{kips})(0.658^{0.119})$$

$$= 12{,}287 \text{ kips}$$

For LRFD, the design axial compressive strength (LRFD) is

$\phi P_n = 0.75(12{,}287 \text{kips}) = 9{,}215$ kips

For ASD, the allowable axial compressive strength is $\frac{P_n}{\Omega} = \frac{12{,}287 \text{kips}}{2.0} = 6{,}144 \text{kips}$

Note that for the bare steel alone (i.e., W14 × 342), the design compressive strength with slenderness considered is 5245 kips (see Chapter 5, and note that the strength reduction factor for the bare steel column is 0.90). That is,

$$\phi_c P_{no}\ (composite\ column) \ge \phi_c P_n\ (non-composite\ or bare steel\ column)$$

Encasing the W14 × 342 in concrete with a compressive strength of 10 ksi and reinforcing the column with 12 No. 10 longitudinal rebar increases the design's axial compressive strength by 76% compared to the axial compression capacity of the bare steel column.

EXAMPLE 7-13

Axially Loaded Concrete-Filled Composite Members or Concrete-filled Tubes

Determine the axial compression load capacity of a concrete-filled HSS 6 x 6 x ¼ column that is 25 ft. long, assuming the compression load is applied concurrently to the steel and concrete. The concrete compressive strength is 5 ksi, and the column is pinned at both ends. Assume ASTM A500 Grade C with a yield strength F_y = 50 ksi.

Solution

$L_c = KL = (1.0)(25\text{ft})(12) = 300$ in.

A_s for HSS 6 x 6 x ¼ = 5.24 in.^{2}

Design wall thickness for ASTM A500 Grade C, t = 0.93(¼ in.) = 0.233 in.

B = H = 6 in.

$b_i = B - 2t = 6\text{ in.} - (2)(0.233\text{ in.}) = 5.534\text{ in.}$

$h_i = H - 2t = 6\text{ in.} - (2)(0.233\text{ in.}) = 5.534\text{ in.}$

$b = B - 3t = 6\text{ in.} - (3)(0.233\text{ in.}) = 5.30\text{ in.}$

$h = H - 3t = 6\text{ in.} - (3)(0.233\text{ in.}) = 5.30\text{ in.}$

A_c = area of concrete, in.^{2}

$$= b_i h_i - t^2(4-\pi) = (5.534\text{ in.})(5.534\text{ in.}) - (0.233)^2(4-\pi) = 30.58\text{ in.}^2$$

This is the concrete area for concrete-filled rectangular HSS accounting for the fillet radius. If the fillet radius is neglected, the concrete area is $A_c = b_i h_i = (5.534\text{ in.})(5.534\text{ in.}) = 30.63\text{ in.}^2$, a difference of less than 1%.

A_{sr} = 0 in.^{2} (no longitudinal rebar provided)

$A_g = A_s + A_c + A_{sr}$ = 5.24 in.^{2} + 30.58 in.^{2} + 0 = 35.82 in.^{2}

A_s = 5.24 in.^{2} = $0.146A_g \geq 1\%A_g$ OK.

$E_s = 29{,}000\text{ ksi}$

$E_c = w_c^{1.5}\sqrt{f'_c}$, ksi $= (145\text{pcf})^{1.5}\sqrt{5\text{ksi}} = 3904\text{ ksi}$

Note that the moments of inertia, I_s, I_c, and I_{sr} about both the x-x and y-y axes are the same since this is a square HSS column. If this were a rectangular column, the moments of inertia as well as the effective stiffness, $(EI)_{eff}$, and hence the elastic critical buckling load, P_e, for buckling about each axis (x-x and y-y), would have had to be calculated for each axis.

$I_{s,xx} = I_{s,yy} = 28.6\text{in.}^4$

$I_{sr,xx} = I_{sr,yy} = 0$ in.^{4} (no longitudinal reinforcement)

$$I_{c,xx} = I_{c,yy} = \frac{b_i h_i^3}{12} = \frac{(5.534)(5.534)^3}{12} = 78.2\text{ in.}^4$$

For rectangular steel tubes filled with concrete, the HSS column is a compact section if

$$\frac{b}{t} \leq 2.26\sqrt{\frac{E}{F_y}} = 2.26\sqrt{\frac{29{,}000}{50}} = 54.4$$

$$\frac{b}{t} = \frac{h}{t} = \frac{5.30\text{ in.}}{0.233\text{ in.}} = 22.8 < 54.4 \quad OK.$$

Therefore, HSS 6 x 6 x ¼ is a **compact composite** section.

Note that the $\frac{b}{t}$ and $\frac{h}{t}$ ratios for HSS are tabulated in Part 1 of the AISC Manual. Recall that for checking local stability in HSS members, b = B - 3t and h = H - 3t.

The procedure for determining the nominal axial compressive strength of the concrete-filled tubes (CFT) from AISC Specification Section I2.2b is as follows:

a. Compact Composite Sections:

The nominal axial compressive strength without considering slenderness effects, $P_{no} = P_p$. Where, the nominal plastic axial compressive strength,

$$P_p = F_y A_s + C_2 f_c'\left(A_c + A_{sr}\frac{E_s}{E_c}\right)$$

$$=(50\,\text{ksi})(5.24\text{ in.}^2)+0.85(5\text{ ksi})\left(30.58\text{ in.}^2+(0)\frac{E_s}{E_c}\right)$$

$$= 392\text{ kips}$$

$C_2 = 0.85$ for rectangular sections

Therefore, $P_{no} = P_p = 392$ kips

Design Axial Compressive Strength of CFT:

The coefficient for calculating the effective stiffness of a concrete-filled composite section (AISC Specification Equation I2-13) is

$$C_3 = 0.45+3\left(\frac{A_s + A_{sr}}{A_g}\right) \le 0.9$$

$$= 0.45+3\left(\frac{5.24\text{in.}^2+0}{35.82\text{in.}^2}\right)=0.89<0.9$$

Therefore, use $C_3 = 0.89$

The effective stiffness or rigidity of the concrete-filled composite member (AISC Specification Equation I2-12) is

$$(EI)_{eff} = E_s I_s + E_s I_{sr} + C_3 E_c I_c$$

$$=(29{,}000)(28.6)+(29{,}000)(0)+0.89(3904)(78.2)$$

$$=1.1x10^6\text{ Kip}-\text{in.}^2$$

The elastic critical buckling load, P_e , is calculated from AISC Specification Equation I2-4 as

$$P_e = \frac{\pi^2 (EI)_{eff}}{L_c^2} = \frac{\pi^2\left(1.1x10^6\text{ Kip}-\text{in.}^2\right)}{(300\text{in.})^2} = 120.6\text{kips}$$

For the bare steel (HSS 6 x 6 x ¼) at 25 ft height, $\phi P_n = 72$ kips

Therefore, $P_n = \frac{72\text{ kips}}{0.9} = 80$ kips

Note that the axial compressive strength of the concrete-filled composite column without considering slenderness effects must be greater than or equal to the axial compressive strength, for the bare steel member, calculated using the principles discussed earlier in Chapter 5; that is,

$$\phi_c P_{no}\,(composite\ column) \ge \phi_c P_n\,(non-composite\ or bare steel\ column)$$

$$\phi_c P_{no} = (0.75)(392 \text{ kips}) = 294 \text{kips} \geq \phi_c P_n \text{ (bare steel)} = (0.9)(80 \text{ kips}) = 72 \text{ kips, } OK.$$

Calculate the nominal axial compressive strength considering slenderness effects from AISC Specification Equations I2-2 and I2-4 as follows:

$$\frac{P_{no}}{P_e} = \frac{392 \text{ kips}}{120.6 \text{ kips}} = 3.25 > 2.25$$

Therefore, $P_n = 0.877 P_e = 0.877(120.6 \text{kips}) = 106$ kips

For LRFD, the design axial compressive strength (LRFD) is

$\phi P_n = 0.75(106 \text{ kips}) = 79.5 \text{kips} > \phi P_n$ of the bare HSS 6 x 6 x ¼ = 72 kips OK.

Therefore, the axial compression capacity of a 25-ft long concrete-filled HSS 6 x 6 x ¼ column with $f_c' = 5 \text{ksi}$ is $\phi P_n = 79.5 \text{kips}$

This compares favorably with the 79.3 kips value obtained from AISC Volume 16.0 Design Table 4.D [14].

Note that the strength reduction factor, ϕ_c, for bare steel from Chapter E of the AISC Specifications (i.e., from Chapter 5 of the text) is 0.9, whereas the strength reduction factor for composite columns from Chapter I of the AISC Specifications is 0.75. Consequently, the AISC Specification sets a lower limit for the strength calculated for composite sections as the strength for the bare steel section as indicated by equation (7-39).

$$\phi_c P_n \text{ (composite column)} \geq \phi_c P_n \text{ (non-composite or bare steel column)}$$

7-11 FLEXURAL STRENGTH OF CONCRETE-FILLED COMPOSITE COLUMNS

The flexural strength of composite columns can be determined using the provisions of Section I3 of the AISC Specifications. Figure 7-28d shows the different stress blocks used to calculate the nominal moment strength for concrete-filled composite rectangular box sections. The equations for the neutral axis locations for the three cases in Figure 7-28d are given below. The provisions for shear are given in AISC Specification Section I4.1 for encased composite members and AISC Specification Section I4.2 for concrete-filled composite columns. For combined axial compression and bending moment, the beam-column interaction equations (8-1) and (8-2) in Chapter 8 can be used.

Neutral Axis Location for Figure 7-28d(a):

$$a_p = \frac{2F_y H t_w + 0.85 f_c' b_i t_f}{4 t_w F_y + 0.85 f_c' b_i} \tag{7-61}$$

Neutral Axis Location for Figure 7-28d(b):

$$a_y = \frac{2F_y H t_w + 0.35 f_c' b_i t_f}{4 t_w F_y + 0.35 f_c' b_i} \tag{7-62}$$

Neutral Axis Location for Figure 7-28d(c):

$$a_{cr} = \frac{F_y H t_w + (0.35 f_c' + F_y - F_n) b_i t_f}{t_w (F_n + F_y) + 0.35 f_c' b_i} \tag{7-63}$$

Concrete-Filled Composite Columns

The stress blocks for determining the flexural strength of concrete-filled composite columns with compact composite section, non-compact composite section, and slender-element composite section, as adapted from AISC Specification Section I3.4 and Figure C-I3.8, are shown in Figure 7-28d. Given the neutral axis location and the tension and compression forces acting on the sections, the nominal moment capacity, M_n, can easily be obtained by summing moments of forces about the neutral axis, and thus the design moment, $\phi_b M_n$, can be obtained, where the resistance factor, ϕ_b, is equal to 0.9 (AISC Specification Section I3.4.4b).

Compact Composite Sections:

The nominal flexural moment capacity is

$$M_n = M_p \tag{7-64}$$

Where,

M_p = moment corresponding to the plastic stress distribution over the composite cross-section, in in.-kip. This is the moment obtained by summing the steel and concrete forces about the neutral axis in Figure 7-28d(a).

Non-Compact Composite Sections:

The nominal flexural moment capacity is

$$M_n = M_p - \left(M_p - M_y\right)\left(\frac{\lambda - \lambda_p}{\lambda_r - \lambda_p}\right) \tag{7-65}$$

Where,

λ = actual width-to-thickness ratio of the section.

λ_p, λ_r = limiting width-to-thickness ratios determined from Table 7-5 (adapted from AISC Specification Table I1.1b).

M_y = yield moment corresponding to the yielding of the tension flange and the first yield of the compression flange, in in.-kip. This is the moment obtained by summing the steel and concrete forces about the neutral axis in Figure 7-28d(b).

F_n = critical buckling stress in Figure 7-28d(c) is determined using the following equations

- For rectangular-filled composite sections, the critical buckling stress (see AISC Specification Equation I2-10) is:

$$F_n = \frac{9E_s}{\lambda^2}, \tag{7-66}$$

$$\lambda = \text{larger of } \frac{b}{t} \text{ or } \frac{h}{t}$$

- For round-filled composite sections, the critical buckling stress (see AISC Specification Equation I2-11) is:

$$F_n = \frac{0.72F_y}{\left[\left(\frac{D}{t}\right)\frac{F_y}{E_s}\right]^{0.2}} \tag{7-67}$$

Where,

B = outside width of the rectangular HSS or built-up box section

H = outside depth of the rectangular HSS or built-up box section

D = outside diameter of round HSS

$b = B - 3t$ for rectangular concrete-filled HSS

$= B - 2t$ for rectangular concrete-filled built-up box sections

$h = H - 3t$ for rectangular concrete-filled HSS

$= H - 2t$ for rectangular concrete-filled built-up box sections

d_i = inside diameter of round HSS = $D - 2t$

t = design wall thickness

For HSS: $t = 0.93t_{nominal}$ for ASTM A500

$= t_{nominal}$ for ASTM A1065 and A1085

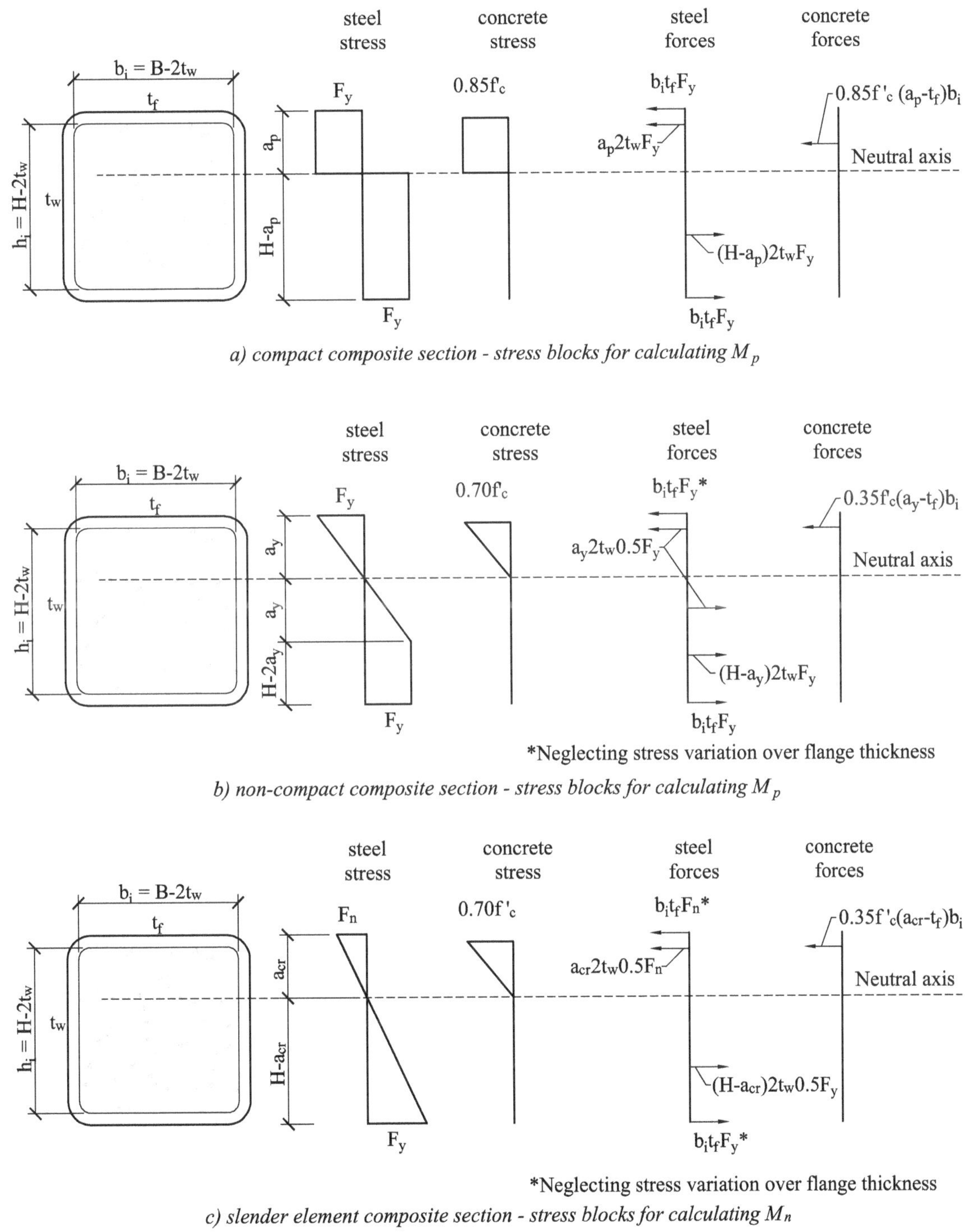

FIGURE 7-28d Stress blocks for calculating nominal flexural strengths of concrete-filled rectangular box sections (Adapted from AISC Specification Fig. C - I3.8)

Slender-Element Composite Sections:

The nominal flexural moment capacity, M_n, is obtained by summing the steel and concrete forces about the neutral axis in Figure 7-28d(c).

Note that the strength reduction factor, ϕ_b, for all three cases is 0.9.

Table 7-5 Limiting Width-to-Thickness Ratios for Compression Steel Elements in Composite Members Subjected to Flexure*

Description of Element	Width-to-Thickness Ratio^	λ_p Compact Composite/ Non-Compact Composite	λ_r Non-Compact Composite/Slender-Element Composite	Maximum Permitted
Flanges of rectangular HSS and box sections of uniform thickness	b/t	$2.26\sqrt{\frac{E}{F_y}}$	$3.0\sqrt{\frac{E}{F_y}}$	$5.0\sqrt{\frac{E}{F_y}}$
webs of rectangular HSS and box sections of uniform thickness	h/t	$3.0\sqrt{\frac{E}{F_y}}$	$5.7\sqrt{\frac{E}{F_y}}$	$5.7\sqrt{\frac{E}{F_y}}$
Round HSS	D/t	$\frac{0.09E}{F_y}$	$\frac{0.31E}{F_y}$	$\frac{0.31E}{F_y}$

*Adapted from AISC Specification Table I1.1b

^For rectangular HSS: b = B - 3t; h = H - 3t

^For rectangular built-up box section: b = B - 2t; h = H - 2t

Refer to Figures 7-28a, b, and c for the definitions of width, B, and depth, H, and thickness, t, for rectangular HSS and box sections.

7-12 FLEXURAL STRENGTH OF CONCRETE-ENCASED COMPOSITE COLUMNS

The flexural strength of composite columns can be determined using the provisions of Section I3.3 of the AISC Specifications. In this case, it is required that shear anchors in the form of shear studs be provided along the column section so that the concrete and the steel section will be engaged in composite action. The longitudinal reinforcement and the ties in the concrete encasement must meet the minimum detailing requirements of the AISC Specification Section I2.1a. The AISC Specification permits three design methods for calculating the flexural strengths of *concrete-encased* steel members (see AISC Specification Section I3.3.3a).

a. **Method 1** assumes an elastic stress distribution with first yield in the tension flange of the composite section.
b. **Method 2** assumes a flexural strength based on the plastic moment capacity of the encased steel section alone. Note that the steel section and its components are not limited for local or lateral buckling because they are assumed to be adequately braced against local and lateral buckling by the concrete encasement.
c. **Method 3** assumes a plastic stress distribution in the composite section or the strain compatibility method. It requires steel stud anchors welded to the concrete-encased steel section.

The authors prefer Method 2 for determining the flexural moment capacity because it is more straightforward and easier to use.

For shear requirements for encased composite members, see AISC Specification Section I4.1.

References

[1] American Institute of Steel Construction. 2023. *Steel Construction Mannual*. 16th ed. Chicago, IL: AISC.

[2] American Concrete Institute. 2019. *ACI-318: Building Code Requirements for Structural Concrete and Commentary*. Farmington Hills, MI: ACI.

[3] American of Steel Construction. 2005. *Steel Design Guide Series 5: Low- and medium-rise steel buildings*, AISC, Chicago, IL: AISC.

[4] Easterling, S., D. Gibbings, and T. Murray. 1993. "Strength of Shear Studs in Steel Deck on Composite Beams and Joists." *AISC Engineering Journal*, 2nd Quarter.

[5] Hansell, W. C., T. V. Galambos, M. K. Ravindra, and I. M. Viest. 1978. "Composite Beam Criteria in LRFD." *Journal of the Structural Division*, ASCE 104 (ST9).

[6] Viest, I. M., J. P. Colaco, R. W. Furlong, L. G. Griffis, R. T. Leon, and L. A. Wyllie. 1997. *Composite Construction Design for Buildings*. New York: McGraw Hill.

[7] Carter, C. J., T. M. Murray, and W. A. Thornton. 2000. "Economy in Steel". *Modern Steel Construction*, April.

[8] Weisner, K. B. 1996. "Composite Beams with Slab Openings." *Modern Steel Construction*, March: 26–30.

[9] Hibbeler, R. C. 2008. *Mechanics of Materials*.7th ed. New Jersey: Pearson.

[10] ASCE. 2015. *ASCE 37-14: Design Loads on Structures During Construction*, Reston, VA: ASCE.

[11] American Concrete Institute. 2015. *ACI-302.1R-15: Guide to Concrete Floor and Slab Construction*, Farmington Hills, MI: ACI.

[12] YahooNews. 2021. Report blames poor welds for Mexico City subway collapse. *https://www.yahoo.com/news/report-blames-poor-welds-mexico-182851550.html* Last accessed December 24, 2024

[13] Aghayere, A. and Vigil, J. 2024. *Reinforced Concrete Design*.10th ed. Dulles, VA: Mercury Learning and Information.

[14] American Institute of Steel Construction. 2023. *Manual Companion to the Steel Construction Mannual (Design Tables)*. V16.0 Chicago, IL: AISC.

Exercises

7-1. Determine the transformed moment of inertia for the sections shown in Figure 7-29. The concrete has a density of 115 pcf and a 28-day strength of 3 ksi. Assume full-composite action.

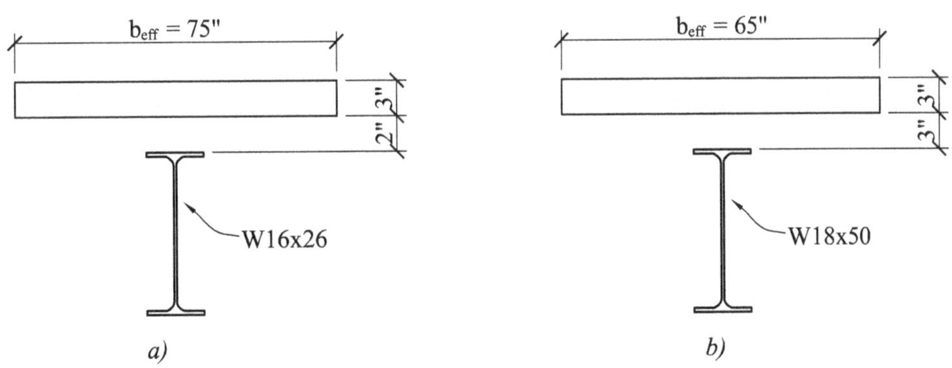

FIGURE 7-29 Detail for Exercise 7-1

7-2. Determine the design strength of the composite section given in Exercise 7-1, assuming full-composite action and ASTM A992 Grade 50 steel. Confirm the results with *AISCM* Table 3-18.

7-3. Determine the following for the composite beam section shown in Figure 7-30. The steel is ASTM A992, Grade 50, and the concrete is normal weight (145 pcf).

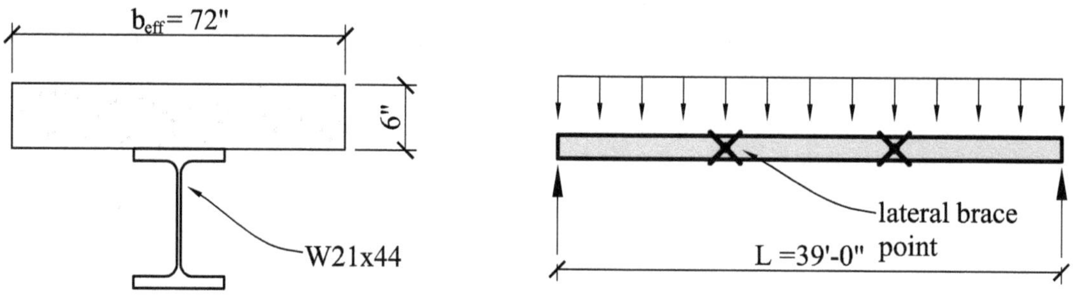

FIGURE 7-30 Details for Exercise 7-3

a. Determine the required camber, if applicable.

b. Determine if the section is adequate for construction phase loading assuming $C_b = 1.0$, a uniform construction live load of 120 plf (service), and lateral support at one-third points.

7-4. Determine the following for girder G-1 shown in Figure 7-31. The steel is ASTM A992, Grade 50, and the service loads to consider are as follows:

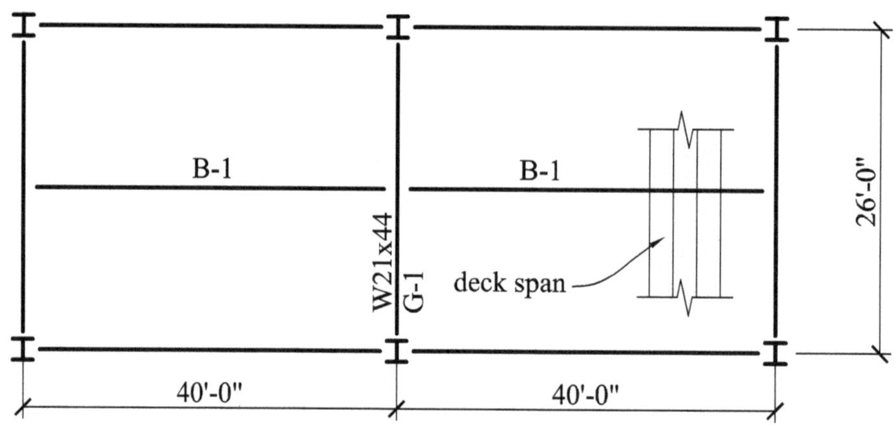

FIGURE 7-31 Details for Exercise 7-4

Dead load = 90 psf
Construction live load = 25 psf

a. Determine the required camber, if applicable.

b. Determine if the section is adequate for construction phase loading. The girder is only laterally braced at midspan. Use a value for $C_b = 1.0$ justified by analysis.

7-5. Determine the following for the W16 × 45 composite beam section shown in Figure 7-32. The steel is ASTM A992, Grade 50, and the concrete is normal weight (145 pcf), with a 28-day strength of 3 ksi. Assume full-composite action. Consider only the loads shown in the figure.

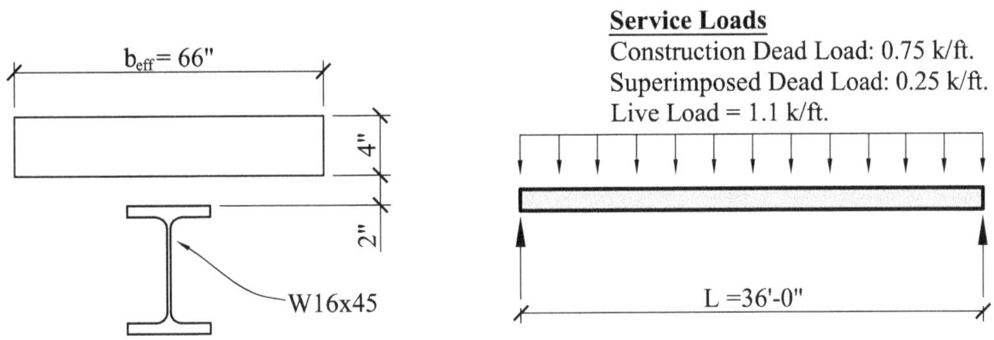

FIGURE 7-32 Details for Exercise 7-5

a. Determine the required camber, if applicable.

b. Determine the design bending strength in ft.-kips k-ft. of the composite section. Compare results with the AISC manual (*AISCM*) tables. Is the section adequate for bending?

c. Determine I_{LB} from *AISCM* Table 3-19. Is the section adequate for post-composite deflections (compared with L/480)?

d. Determine the number of ¾" studs required for the entire beam. Assume one stud is in deck ribs in the strong direction.

7-6. Assume the following for a W16 × 36 composite beam:

- Beam length is 20 ft. -8 in.
- There are 62 total studs on the beam, equally spaced in a single row.
- Deck is 3.5" of normal weight concrete slab on 1.5-in. composite metal deck.
- Deck runs perpendicular to the beam.
- Studs are in the weak location.
- The tributary width of the beam is 9 ft.
- Concrete is normal weight with $f_c' = 3$ ksi.
- Total dead load is 120 psf (includes framing and concrete).
- $F_y = 50$ ksi.

Find:

a. The design bending strength for the composite section (use *AISCM* Table 3-18).

b. The design live load (service) in psf that the beam could support based on (a).

c. The live load deflection is based on (b) (use *AISCM* Table 3-19).

7-7. Determine the following for the section shown in Figure 7-33. The concrete has a density of 145 pcf and a 28-day strength of 3.5 ksi. The steel is ASTM A992 Grade 50. Assume $Q_n = 21.5$ kips for one stud.

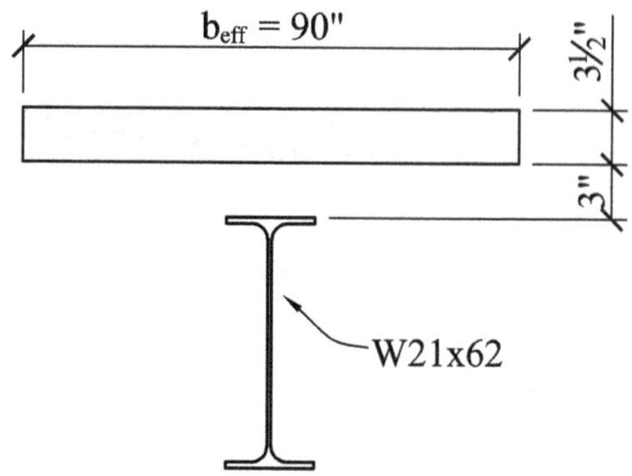

FIGURE 7-33 Detail for Exercise 7-7

a. Transformed moment of inertia.

b. Design strength, $f_b M_n$, assuming full-composite action.

c. Design moment, $f_b M_n$, assuming 40% action.

d. The number of $^3/_4$-in. ASTM A29 studs (Type B with F_u = 65 ksi) required between the points of maximum moment required for 100% and 40% composite action.

Confirm the results from Parts b, c, and d using the *AISCM* tables.

7-8. For the beam shown in Figure 7-34, determine the effective moment of inertia and the design moment capacity for 40% composite action and compare the results with *AISCM* Tables 3-18 and 3-19. Use f'_c = 3.5 ksi and a concrete density of 145 pcf.

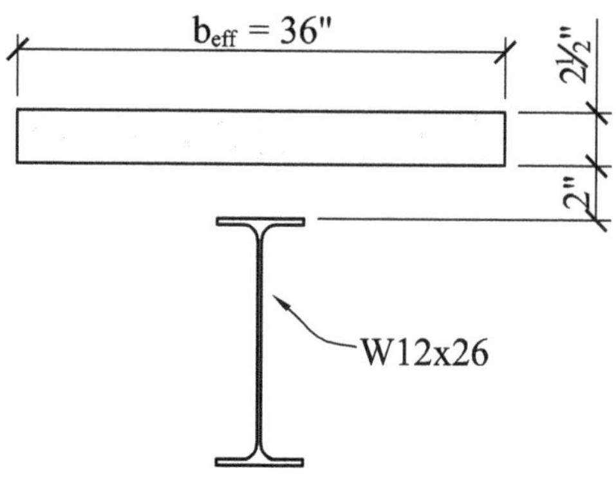

FIGURE 7-34 Detail for Exercise 7-8

7-9. For the floor framing shown in Figure 7-35, design the composite members B1 and G1. The floor construction is $3^1/_2$-in. normal weight concrete slab on 3-in. composite metal deck ($6^1/_2$ in. total thickness). The steel is ASTM A992, Grade 50, and the concrete has a 28-day strength of 3.5 ksi. Design for flexure, shear, and deflection, considering dead and live loads.

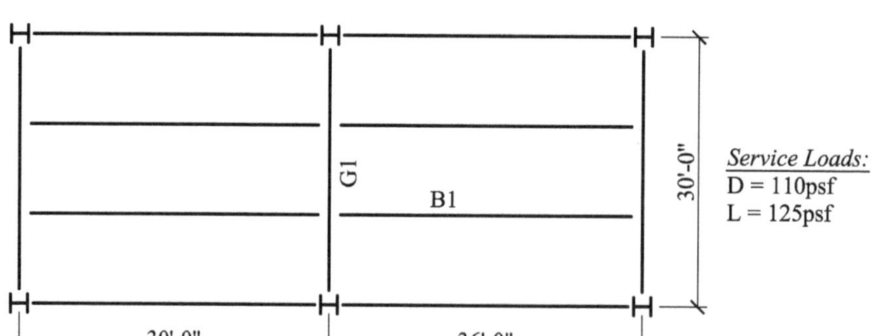

FIGURE 7-35 Detail for Exercise 7-9

7-10. For the beam loading shown in Figure 7-36, design the lightest W21 composite beam for flexural loading and deflection. The floor construction is $4^1/_2$-in. normal weight concrete slab on 2-in. composite deck ($6^1/_2$ in. total thickness). The steel is ASTM A992, Grade 50, and the concrete has a 28-day strength of 3.0 ksi. Use an effective flange width of 42 inches. The service or unfactored loads are as follows:

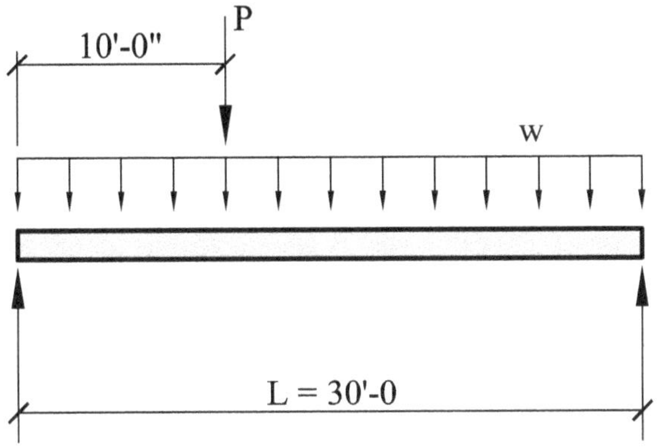

FIGURE 7-36 Detail for Exercise 7-10

Concentrated load, P: Dead = 1.2 kips (pre-composite)
Dead = 2.4 kips (post composite)
Live = 7.5 kips

Uniform load, w: Dead = 350 plf (pre-composite)
Dead = 100 plf (post-composite)
Live = 600 plf

7-11. For the beam loading shown in Figure 7-37, design the lightest W16 composite beam for flexural loading and deflection. The floor construction is $4^1/_2$-in. normal weight concrete slab on 2-in. composite deck ($6^1/_2$ in. total thickness). The steel is ASTM A992, Grade 50, and the concrete has a 28-day strength of 3.0 ksi. Use an effective flange width of 36 inches. The service or unfactored loads are as follows:

Concentrated, P: Dead = 1.0 kips (pre-composite)
Dead = 2.0 kips (post composite)
Live = 6.0 kips

Uniform, w: Dead = 300 plf (pre-composite)
Dead = 300 plf (post-composite)
Live = 750 plf

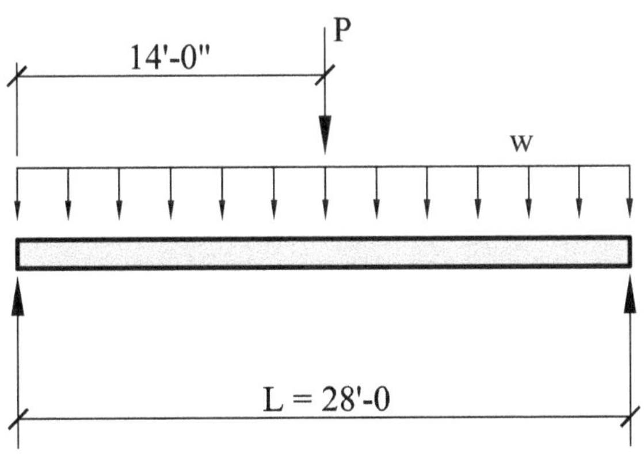

FIGURE 7-37 Detail for Exercise 7-11

7-12. The 5th floor typical interior beam in Exercise 13-8 is a W16 x 40 composite beam with a total of 60- ¾-inch diameter studs x 4 inches long (in two rows) and with an upward camber of 1-inch; is the beam adequate for strength and deflections? Use *AISCM* Tables 3-18 and 3-19.

7-13. Develop a spreadsheet to calculate the transformed section properties of a beam and slab assembly with unknown beam and slab dimensions. Calculate the transformed moment of inertia, I_{tr}, and submit your output for Example 7-1.

7-14 The beam below is a composite W12x14, grade 50 steel. It is fully braced, and the loads shown are service level. Any superimposed dead load is included in the live load. The beams are equally spaced.

Showing calculations:

a. Determine an appropriate camber, if applicable.

b. Assuming 100% composite action, determine the design bending strength, ϕM_n. Is the composite beam section adequate for bending?

c. Determine the required number of ¾" studs for the entire beam length. Use ¾" studs and a 28-day concrete strength of 4,000 psi. Assume studs are in the strong position of the 2" composite metal deck. The *AISCM* Table 3-20 may be used for the stud strength.

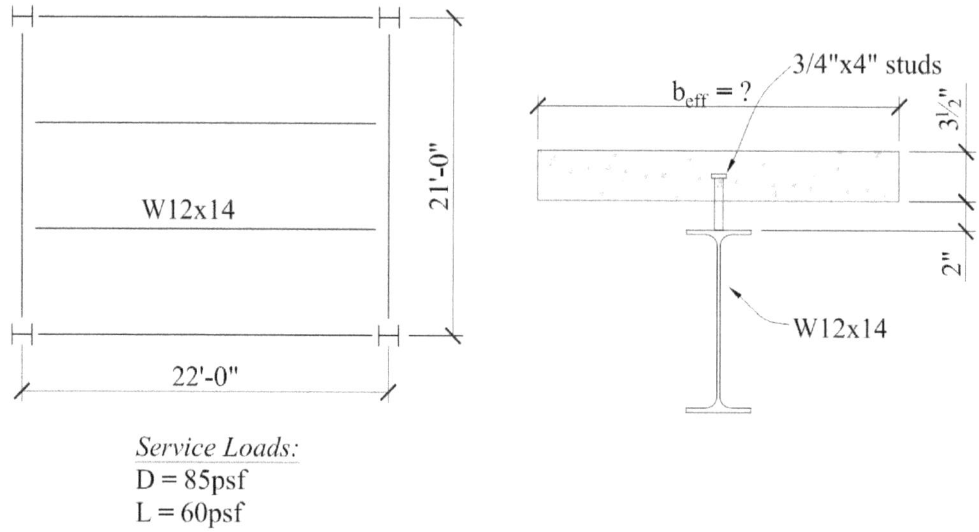

7-15 Determine the axial compression load capacity of a 20 ft. long concrete-encased composite cruciform column, formed from one W21 x 201 and two WT 10.5 x 100.5 welded to the web of the W21 × 201 at its mid-depth. The steel cruciform column is encased in a 38-in. diameter concrete column reinforced with 12 No. 10 vertical bars with #5 ties at 12 inches on centers. The outside flanges of the two WT 10 and the W21 are two studs at 12 inches on centers along the length of the column to ensure that the compression load is applied concurrently to the steel and concrete. The concrete compressive strength is 10 ksi, and the column is pinned at both ends. Assume ASTM A913 Grade 65 structural steel and Grade 75 rebar.

CHAPTER

8

Members Under Combined Axial Load and Bending Moment

8-1 INTRODUCTION TO BEAM-COLUMNS

Structural members that are subjected to combined axial load and bending moment are called beam-columns (or simply columns in practice). Beam-columns could be part of braced frames or unbraced frames (i.e., moment frames), and the design of these columns will differ depending on whether the building frame is braced or unbraced.

In this chapter, we will discuss the analysis and design of beam-columns in general, with particular applications to typical steel-framed buildings. In Chapter 5, we covered the design of compression members under concentric compression loads (i.e., without bending), which rarely exists in real-life structures due to the presence of inherent geometric imperfections. Generally speaking, most building columns are actually beam-columns because of the way they are loaded, so the majority of this chapter will focus on members subjected to combined axial load and flexure.

Braced Frames

In buildings with braced frames, the lateral loads are resisted by diagonal bracing or shear walls. Braced frames are also referred to as nonsway frames or sidesway-inhibited frames. In braced frames, the beams and girders are connected to the columns with simple shear connections that resist shear only, with practically little or no moment restraint. The moments in the columns, M_{nt} (i.e., **n**o **t**ranslation moments), are nonsway moments that result from the eccentricity of the beam and girder reactions relative to the centroid of the columns. For these frames, the sway moment, M_{lt} (i.e., **l**ateral **t**ranslation moment), is zero.

Unbraced Frames or Moment Frames

Unbraced frames or moment frames (also referred to as sway frames or sidesway-uninhibited frames) resist lateral loads through bending in the columns and the beams or girders and the rigidity of the girder-to-column moment connections. The moments in these frames are a combination of no-translation moments, M_{nt}, and lateral-translation moments, M_{lt}.

Types of Beam-Columns

The conditions under which beam-columns might occur in building structures are discussed in this section.

1a. Typical Columns in Buildings with Braced Frames: In this case, the moments result from the eccentricity of the girder and beam reactions relative to the centroid of the columns. Therefore, the moment due to the reaction eccentricity is

$$M = Pe,$$

where e is the eccentricity of the girder or beam reactions as shown in Figure 8-1.

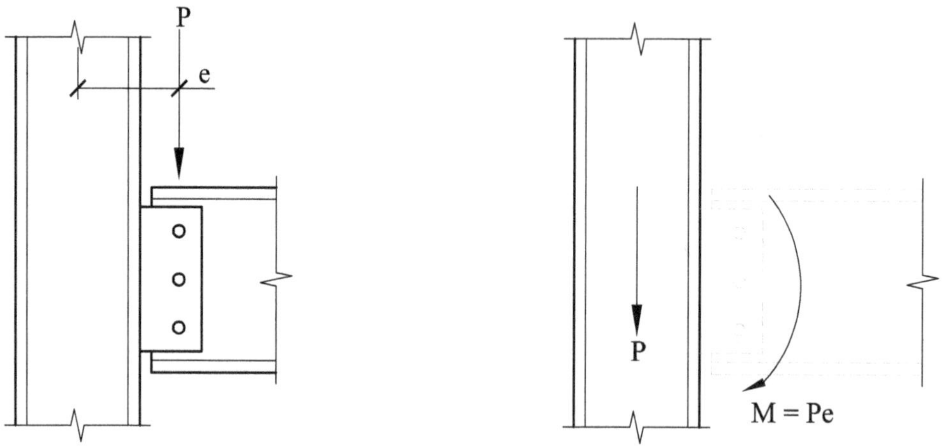

FIGURE 8-1a Case 1 columns in braced frames

1b. Columns Supporting Crane Girders in Industrial Buildings: In this case, the moment in the column results from the following components (see Figures 8-1b and 8-1c):

a. Eccentricity of the vertical reaction, P_V, from the crane-support girder
b. The eccentricity of the horizontal force, P_H, from the crane.
c. Bending of the column about its strong axis due to the horizontal force, P_H, from the crane.

(Figure 8-1b continued)

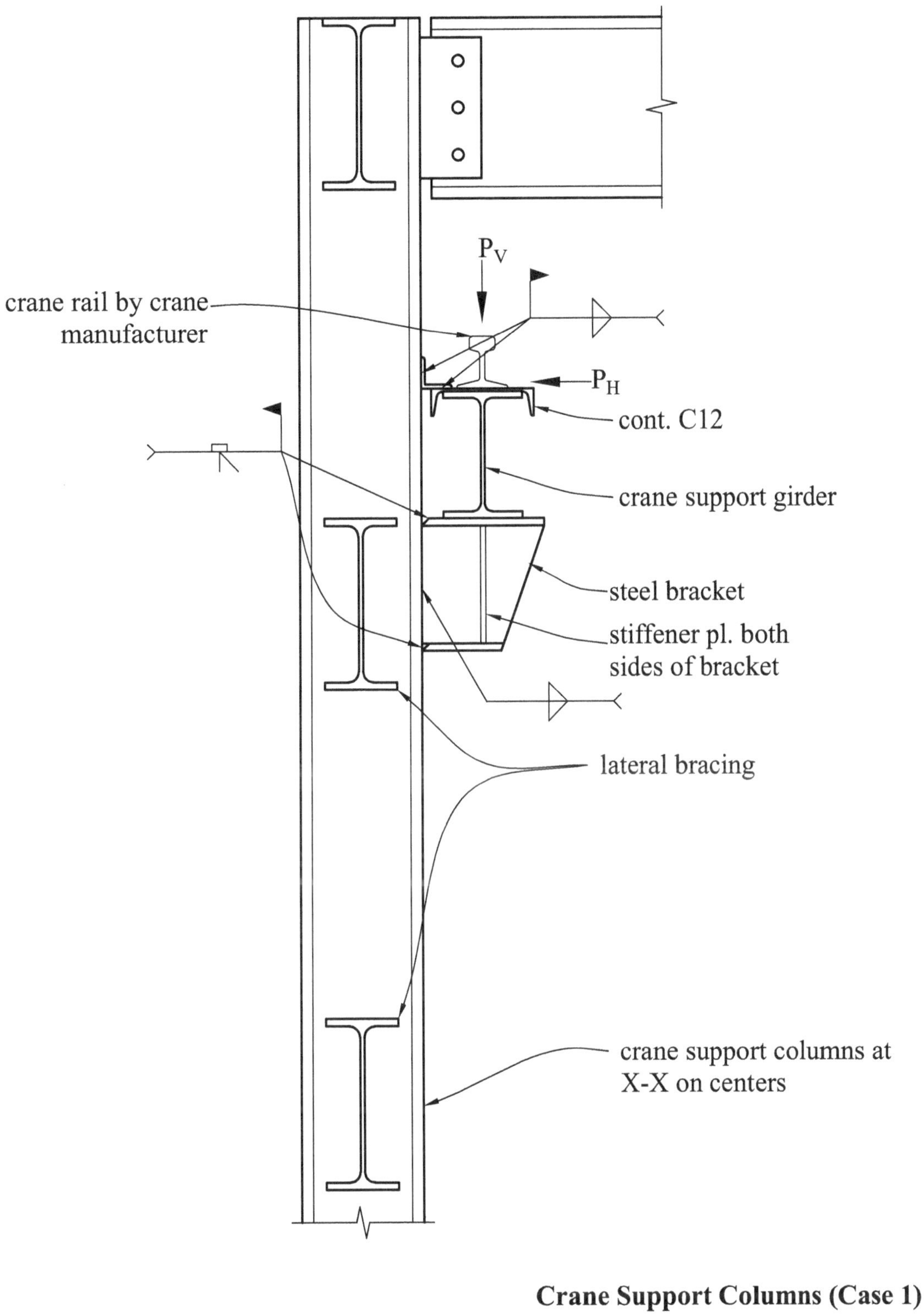

FIGURE 8-1b Crane-support columns

FIGURE 8-1c Crane-support girders on columns with brackets

2. **Exterior Columns and Girts:** For buildings with large story heights, the cladding system may not have the flexural capacity to span vertically between the floor levels to resist the wind load perpendicular to the face of the cladding; therefore, it may be necessary to provide beams in the vertical plane of the cladding to reduce the vertical span of the exterior cladding system. These beams, which are known as girts (see Figure 8-2), are subject to bending in the horizontal plane due to wind loads acting perpendicular to the face of the cladding. These girts usually consist of HSS or channels with their webs parallel to the horizontal plane and their toes pointing downwards. The channel girts are often oriented this way so that debris or moisture does not accumulate on the girt. The channel girts, because of their orientation, are also subjected to weak axis bending due to their weight. To minimize vertical deflection due to self-weight, sag rods are provided at regular intervals, as shown in Figure 14-17a. It should also be noted that the exterior columns in the plane of the cladding will be subjected to flexure from the wind pressure perpendicular to the face of the cladding due to the horizontal reactions from the girts, in addition to the axial loads on the column. Where the cladding system spans horizontally between the exterior columns, additional columns that do not support axial gravity loads may be located between the column grid lines to reduce the horizontal span of the cladding. These additional columns are called wind columns. These are discussed further in Chapter 14 (see Section 14-6).

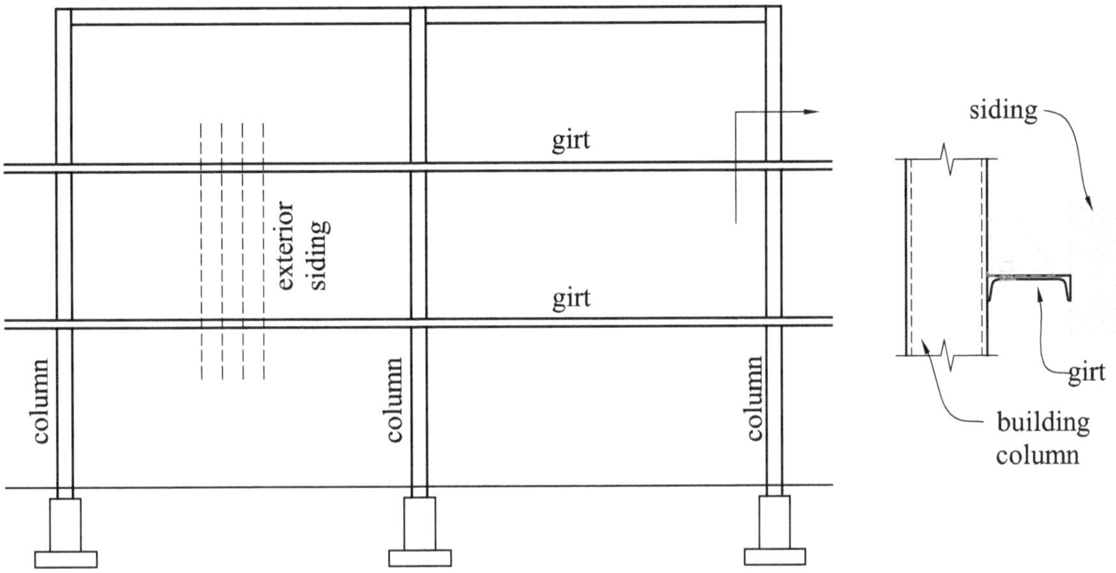

FIGURE 8-2 Wall elevation showing exterior columns and girts

3. **Truss Chords:** The top and bottom chords of trusses (see Figure 8-3a) behave as beam columns. These members are subjected to combined axial load and bending moment; the bending moment is caused by floor or roof loads applied directly to the top chord between the panel points of the truss and by the ceiling and mechanical and electrical equipment loads applied directly to the bottom chord, as well as the moments induced in the top and bottom chords due to their continuity, since most trusses are built with continuous top and bottom chords (see Figure 8-3b). The bottom chords of long-span trusses spanning over auditoriums and theater halls may also support dead and live loads from hanging catwalks.

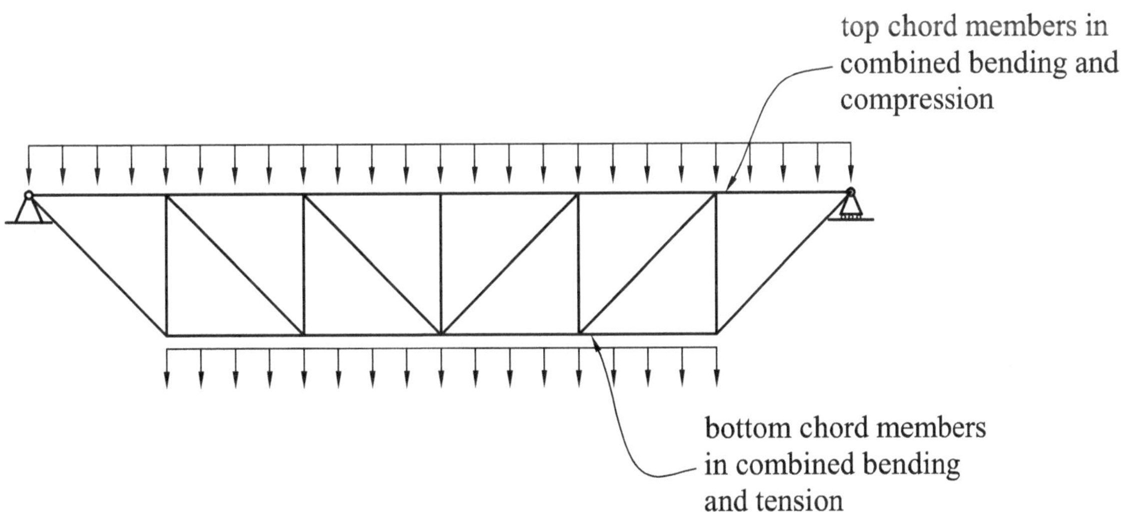

FIGURE 8-3a Loads on top and bottom chords of trusses

FIGURE 8-3b Bolted truss with continuous top and bottom chords (photo courtesy of Joe Messner, JGM, Inc.)

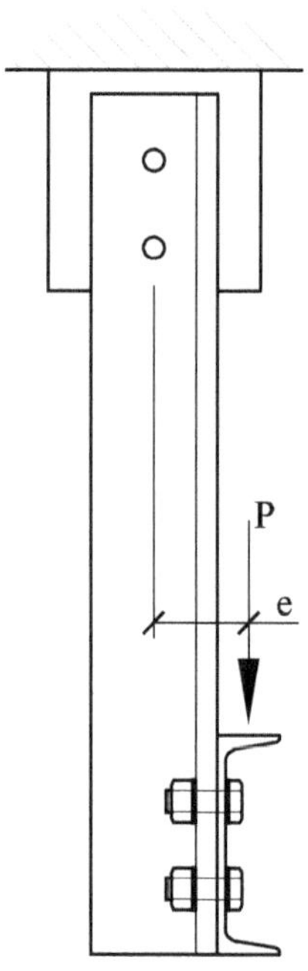

FIGURE 8-4 Hanger with eccentric axial load

4. **Hangers with Eccentric Axial Loads:** Hangers with eccentric axial loads where the structural member is subjected to combined axial tension and bending moment also behave as beam-columns, and they occur in lighter structures, such as catwalks or mezzanines (see Figure 8-4).
5. **Moment frames or unbraced frames:** Moment frames or moment-resisting frames consist of columns and beams or girders connected at the columns with moment connections (see Figure 8-5), and they are used to resist lateral wind and seismic loads through the flexure of the columns and the connecting girders/beams. The columns in moment frames are subject to axial load plus the moment due to the lateral loads (i.e., the sway moments M_{lt},) and the moment caused by the eccentricity of the gravity loads (i.e., the non-sway moments, M_{nt}). For high-rise buildings where column axial loads are high and moment frames may be used in both orthogonal directions, some columns or beam-columns may be part of two orthogonal moment frames (i.e., the columns that occur at the intersection of the moment frames in the two orthogonal directions). These columns will be subjected to bi-axial bending moments due to lateral loads from both orthogonal directions, in addition to axial loads. In order to increase the axial compression load capacity and the flexural capacity of these hot-rolled beam-column sections about both orthogonal axes, the beam-columns that are part of orthogonal moment frames are sometimes made of flanged cruciform-shaped sections (see Figures 1-14h and 5-15b).

Typical Column Sizes in Steel Buildings

Column sizes in building structures typically consist of HSS members, wide flange sections, or built-up non-composite or composite sections. For HSS columns, HSS 6x6 and larger are typically specified. The wide flange sections typically specified for columns in buildings are as follows:

- W8 (i.e., **W8 × 31 and heavier**), or
- W10 (i.e., **W10 × 33 and heavier**), or
- W12 (i.e., **W12 × 40 and heavier**), or
- W14 (i.e., **W14 × 43 and heavier**).

W12 and W14 columns are preferred in practice if there are no architectural restrictions on the column size because W12 and W14 columns are easier to connect to, especially when double-angle connections are used, due to the adequate space or clearance available between the inside faces of their flanges, which allows for easier connection of the beams to the W12 or W14 column web (unlike W8 and W10, where the clearance is limited). W14 columns have the additional advantage that their splices are always end-bearing because the distance between the inside faces of all W14 column flanges is similar, which thus assures that the flanges of a W14 column above a column splice will always have full contact bearing on the flanges of the larger W14 column below the splice.

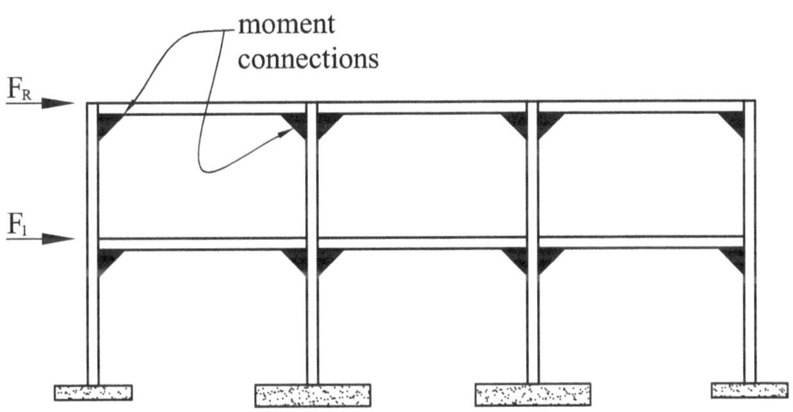

FIGURE 8-5 Moment or unbraced frames

8-2 EXAMPLES OF BEAM–COLUMNS IN BRACED FRAMES

In this section, we present several examples of beam-columns in building structures that are laterally braced with braced frames or shearwalls; these include buildings braced laterally with anyone, or a combination, of the following lateral force resisting systems: concrete, masonry, steel plate, or concrete filled steel composite shear walls, concentric and eccentrically braced frames such as X-braced frames, chevron braced frames (or inverted V-braced frames), and single diagonal braced frames. The connection eccentricities result in unbalanced moments in the columns due to unequal gravity load reactions from the girders or the beams framing into a column on opposite sides of the column.

Unbalanced moments occur primarily in columns in a building with braced frames due to the eccentricity of the reactions that are transferred to the column at the girder-to-column and beam-to-column connections. Thus, moments about two orthogonal axes of the column will exist in a typical building column, and these moments will be more critical for corner columns and slender columns. The different types of beam-to-column and girder-to-column connections and the resulting unbalanced moments from the beam or girder reaction eccentricities are shown in Figures 8-6 through 8-9.

It should be noted that some engineers ignore these connection eccentricities and design all steel columns for only axial loads because of their belief that the unbalanced moments due to the beam and girder reactions acting at these connection eccentricities will not be resisted by the column alone, but instead the moment will be distributed to all the members meeting at the girder-to-column or beam-to-column joints because of some moment restraint provided by the beam-to-column or girder-to-column shear connections since they are not truly hinged connections. Thus, they opine that the effect of the load eccentricity will be negated by the connection restraint [1]. The AISC takes no official position on this issue but indicates that the decision as to whether or not column connection eccentricity is considered or neglected should be based on the judgment of the engineer of record (EOR) [2]. The authors recommend that column connection eccentricity be included in the design of columns since the moment restraint alluded to in the preceding discussion cannot be assured by the simple shear connections used at typical beam-to-column and girder-to-column connections; this philosophy is followed in this text. The load eccentricity for several common structural connections are as follows:

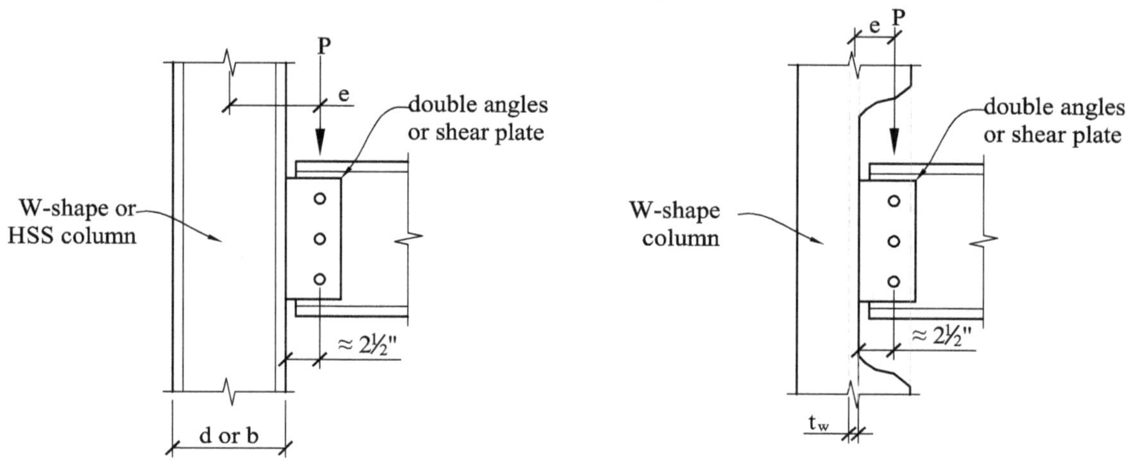

a. connection to column flange

b. connection to column web

$e = \frac{d}{2} + 2.5"$ for beams framing into column flange

$e = \frac{t_w}{2} + 2.5"$ for beams framing into column web

$e = \frac{b}{2} + 2.5"$ for beams framing into face of HSS

FIGURE 8-6 Simple shear connection eccentricity

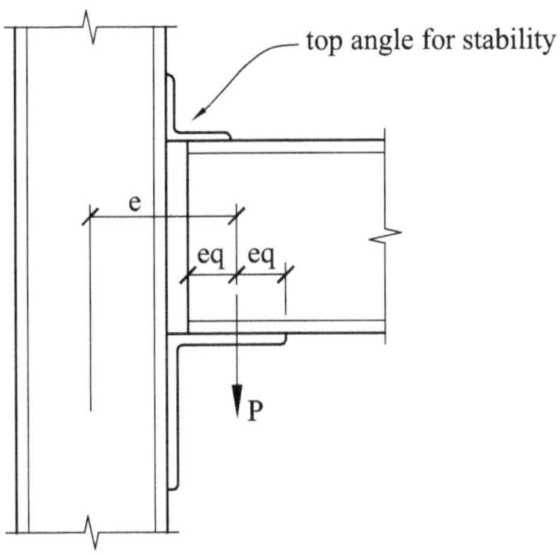

FIGURE 8-7 Seated connection eccentricity

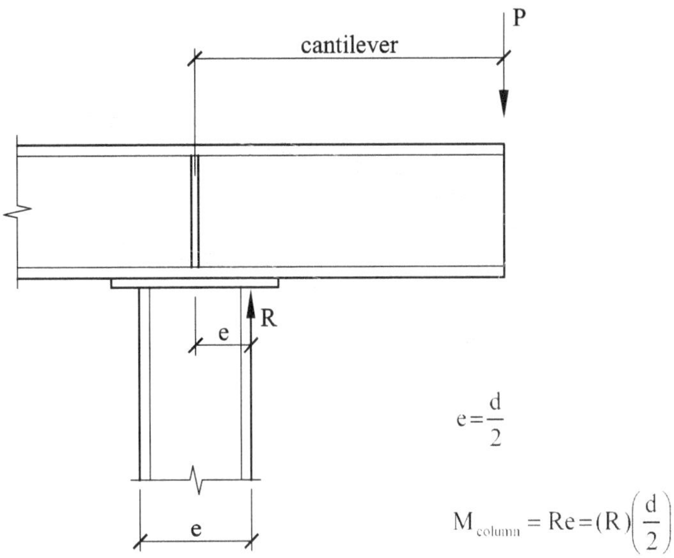

FIGURE 8-8a Top-connected connection eccentricity

FIGURE 8-8b Steel girder supported on column cap plate

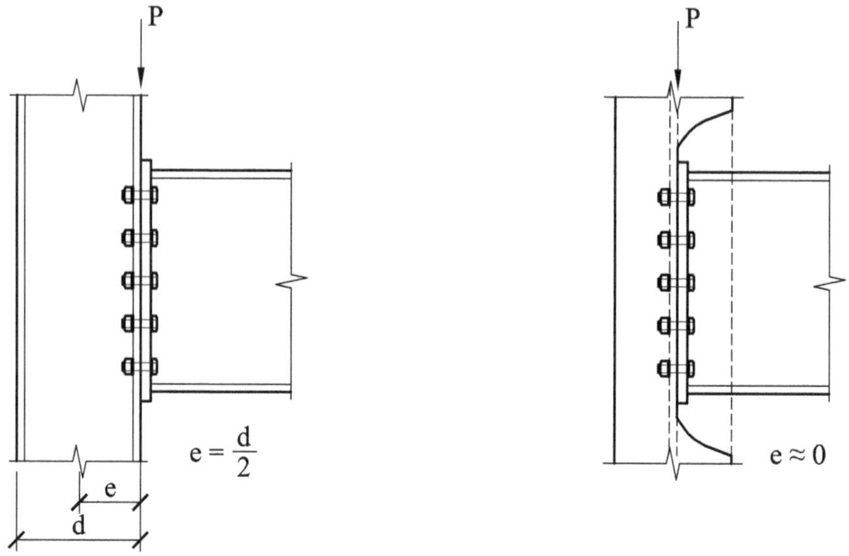

a. connection to the column flange *b. connection to the column web*

FIGURE 8-9 End-plate connection eccentricity

1. **Simple shear connection eccentricity** (see Figure 8-6).
   The majority of the connections in steel buildings are standard simple shear connections that consist of double angles or a shear plate welded to the column and bolted to the beam or girder web. The eccentricity, *e*, is the distance between the centroid of the column and the centroid of the bolt group in the supported beam or girder.
2. **Seated connection eccentricity** (see Figure 8-7).
3. **Top-connected connection eccentricity** (see Figures 8-8a and 8-8b).
4. **End-plate connection eccentricity** (see Figure 8-9).

For end-plate connections, the connection eccentricity for strong axis bending is taken as one-half the distance from the face of the column flange to the centroid of the column (see Figure 8-9). The eccentricity for weak axis bending for this end plate connection will be equal to one-half the web thickness for wide flange columns; this eccentricity is practically negligible for wide flange columns and therefore can be ignored in the design of columns with end plate connections. However, where a shear plate or shear tab connected to the column web is used, the eccentricity will be similar to that shown in Figure 8-6(b).

8-3 BEAM—COLUMN DESIGN

The beam-column design interaction equations from Chapter H of the AISC specification [3] for *doubly and singly symmetric members subject to biaxial bending and axial compression load* are introduced in this section. These are normalized and versatile interaction curves that account for the following factors:

- unbraced length of the column.
- the limit states of elastic or inelastic buckling of the column under axial load.
- the limit states of yielding, lateral-torsional, and local buckling.
- moment amplification in the column due to both P-δ and P-Δ effects (see Chapter 5 and Section 8-4).

The column interaction equation for *large* axial compression loads plus bending in beam-columns using the LRFD method is given in equation 8-1 (i.e., AISC equation H1-1a in AISC Specification Section H1.1):

$$\text{If } \frac{P_u}{\phi_c P_n} \geq 0.20, \qquad \frac{P_u}{\phi_c P_n} + \frac{8}{9}\left(\frac{M_{ux}}{\phi_b M_{nx}} + \frac{M_{uy}}{\phi_b M_{ny}}\right) \leq 1.0; \tag{8-1}$$

The column interaction equation for *small* axial loads on beam-columns using the LRFD method is given in equation 8-2 (i.e., AISC equation H1-1b in AISC Specification Section H1.1):

$$\text{If } \frac{P_u}{\phi_c P_n} < 0.20, \qquad \frac{P_u}{2\phi_c P_n} + \left(\frac{M_{ux}}{\phi_b M_{nx}} + \frac{M_{uy}}{\phi_b M_{ny}}\right) \leq 1.0, \tag{8-2}$$

where

P_u = Factored axial compression load or the required axial compressive strength calculated using the LRFD load combinations (see Section 2-3).

$\phi_c P_n$ = Compression design strength accounting for elastic or inelastic buckling

- For *compression* members in *braced frames*, $\phi_c P_n$ is calculated using an effective length factor, K, that is typically less than or equal to 1.0; K = 1.0 is commonly used in practice.
- For *compression* members in *moment frames*, $\phi_c P_n$ is calculated using K = 1.0 if a Direct Analysis Method (DAM) is used for the stability design of columns in moment frames. If the Effective Length Method (ELM) is used for the stability design, the effective length factor, K, is typically greater than 1.0;

K = 2.1 is an approximate value that is commonly used for preliminary design in this case.

K values can also be determined using the Nomograghs or alignment charts introduced in Chapter 5.

Note that the axial compression strength reduction factor, ϕ_c =0.9.

M_{ux} = Factored second-order bending moment about the x-axis (i.e., the strong axis) of the member calculated using the LRFD load combinations and accounting for both P-δ and P-Δ effects.

M_{uy} = Factored second-order bending moment about the y-axis (i.e., the weak axis) of the member calculated using the LRFD load combinations and accounting for both P-δ and P-Δ effects.

$\phi_b M_{nx}$ = Design moment capacity for bending about the x-axis or strong axis of the member, taking into account the unbraced length of the member, lateral-torsional buckling, and local buckling (see flexural member design in Chapter 6).

$\phi_b M_{ny}$ = Design moment capacity for bending about the y-axis or weak axis of the member ($\phi_b Z_y F_y \leq 1.6\phi_b S_y F_y$, where ϕ_b = 0.9).

Note that for beam-columns with axial compression loads plus bending moments, the factored moments about the x–x and y–y axes (i.e., M_{ux} and M_{uy}, respectively) must include the effect of the slenderness of the compression member (i.e., the so-called P-delta effects). This will be discussed in the next section. The interaction equations 8-1 and 8-2 are represented graphically in Figure 8-10. The combinations of axial compression load and moment that lie on the interaction line or between the interaction line and the vertical and horizontal axes are considered safe. Combinations of axial compression load and moments that fall outside of the interaction diagram are unsafe.

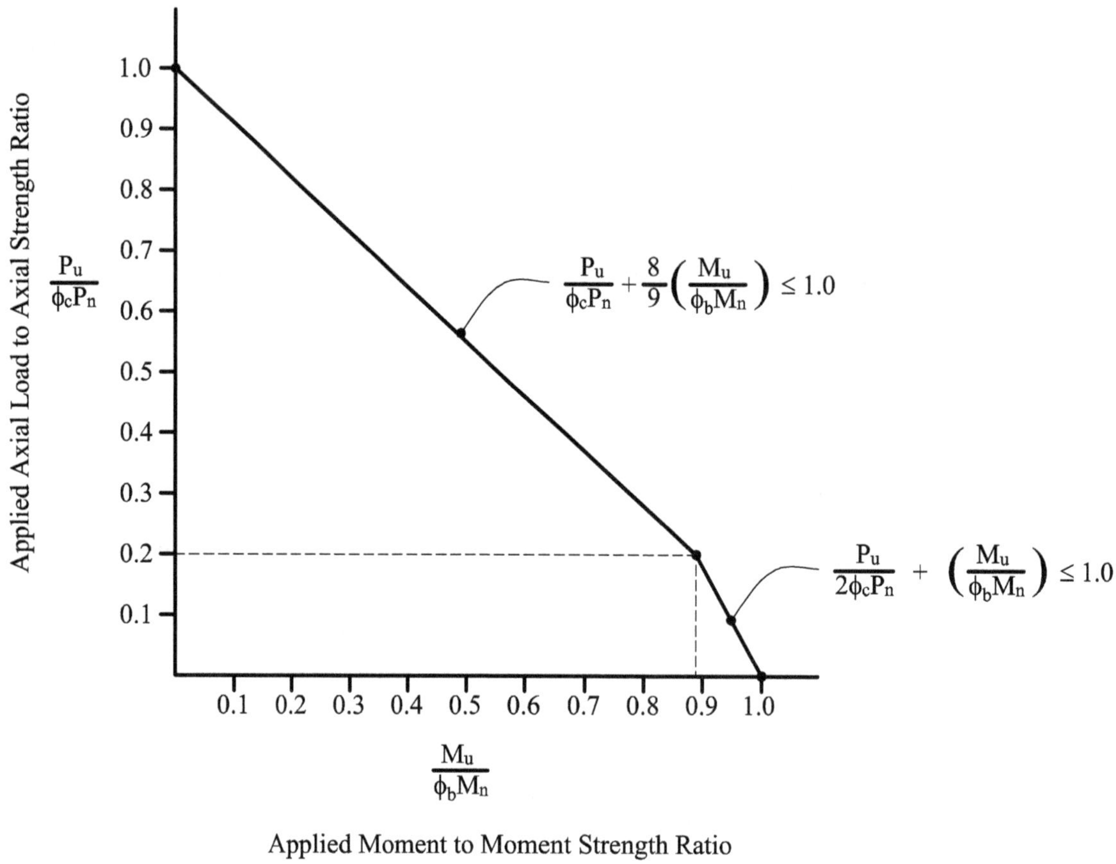

FIGURE 8-10 Interaction diagram for steel beam-columns

For the **ASD method**, the corresponding interaction equations for beam-columns subject to combined axial compression plus bending are as follows [3]:

$$\text{If } \frac{P_s}{P_n/\Omega_c} \geq 0.20, \quad \frac{P_s}{P_n/\Omega_c} + \frac{8}{9}\left(\frac{M_{sx}}{M_{nx}/\Omega_b} + \frac{M_{sy}}{M_{ny}/\Omega_b}\right) \leq 1.0 \tag{8-1a}$$

$$\text{If } \frac{P_s}{P_n/\Omega_c} < 0.20, \quad \frac{P_s}{2P_n/\Omega_c} + \left(\frac{M_{sx}}{M_{nx}/\Omega_b} + \frac{M_{sy}}{M_{ny}/\Omega_b}\right) \leq 1.0 \tag{8-2a}$$

Where,

P_s = unfactored total axial compression load on the beam-column calculated using the ASD load combinations (see Section 2-3)

P_n = is the nominal axial compression capacity

Ω_c = factor of safety for compression = 1.67

M_{sx} = total service load moment about the x- axis (i.e., the strong axis) of the beam-column calculated using the ASD load combinations and accounting for both P-δ and P-Δ effects.

M_{sy} = total service load moment about the y- axis (i.e., the weak axis) of the beam-column calculated using the ASD load combinations and accounting for both P-δ and P-Δ effects.

M_{nx} = nominal moment capacity of the beam-column about the x- axis, taking into account the unbraced length, lateral-torsional buckling, and local buckling (see flexural member design in Chapter 6).

M_{ny} = nominal moment capacity of the beam-column about the y- axis as previously discussed in Chapter 6

Ω_b = factor of safety for bending or flexure = 1.67

The AISC Manual or *AISCM* [3] provides Table 6-1 (Available Strength for Members Subject to Axial, Shear, Flexural, or Combined Forces) that can be used directly to obtain the axial compression and tension strengths and the moment capacities needed in the analysis and design of beam-columns under combined axial compression or axial tension load and bending moments. The moment magnification factors, B_{1x} and B_{1y}, would still need to be calculated for beam-columns with axial compression loads; note also that *AISCM* Table 6-1 assumes a moment gradient factor, C_b, of 1.0 and an effective area for tension rupture of 75% of the gross area of the tension member (i.e., $A_e = 0.75A_g$). These caveats need to be considered when using *AISCM* Table 6-1.

8-4 MOMENT AMPLIFICATION (P-DELTA) EFFECTS

When an axial *compression* load is applied to a beam-column that has some initial crookedness or deflection and that is laterally supported at both ends (i.e., nonsway, meaning the ends of the column do not displace laterally), additional moments are produced in the compression member due to the presence of the axial compression load. This is the first type of P-delta (***P*-δ**) effect), which accounts for the effect of the axial compression load acting on the deflected shape of the compression member between its end supports (see Figure 8-11a).

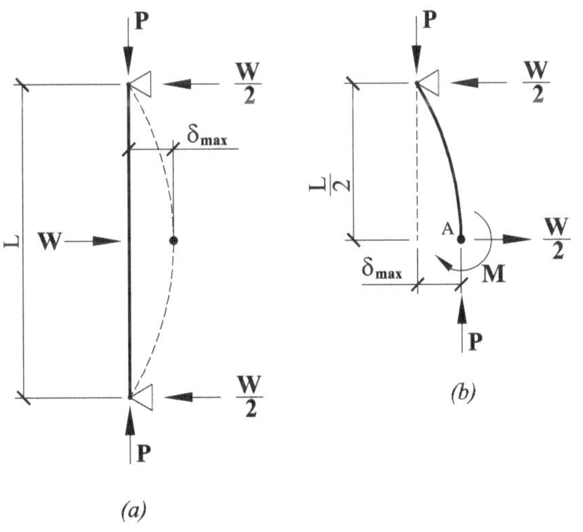

FIGURE 8-11a Moment amplification from P - δ effects

Summing the moments of the free-body diagram in Figure 8-11a about point A results in the following equilibrium equation:

$$\Sigma M_A = M - P\delta - (W/2)(L/2) = 0$$

Therefore, $M = \dfrac{WL}{4} + P\delta.$ (8-3)

The "$WL/4$" term in equation (8-3) is known as the first-order moment, while the second term is known as the P-δ moment. The total second-order moment, M, in equation (8-3) can be rewritten as

$$M = B_1\left[M_{unt}\right] \tag{8-4}$$

where

B_1 = Moment amplification factor due to the column deflection between the laterally supported ends of the column (the so-called ***P*-δ** effect). This applies to individual beam-columns in nonsway frames or braced frames (see AISC Specification, Section 8.1.2 of Appendix 8).

M_{unt} = factored first-order nonsway moments.

The first-order moment, M_{unt}, may be caused by lateral loads applied between the supported ends of the column or due to the eccentricity of the beam and girder gravity load reactions discussed previously.

When an axial *compression* load is applied to a beam-column that is subjected to relative lateral sway at the ends of the member, additional moments are also produced due to the destabilizing effect of the axial load as it undergoes the lateral translation, Δ. This is the second type of P-delta (or P-Δ) effect and is applicable to beam-columns in moment frames.

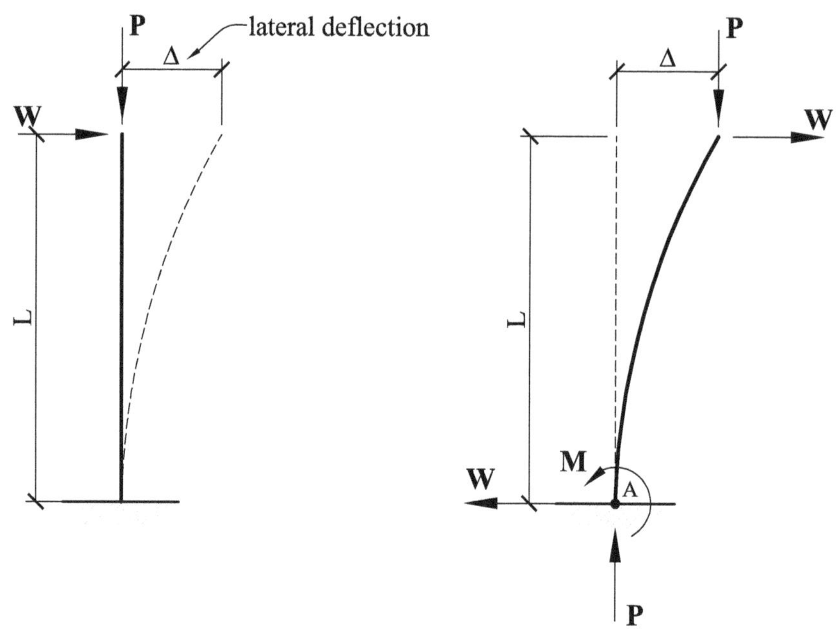

FIGURE 8-11b Moment amplification from P-Δ effects

Summing the moments of the free-body diagram in Figure 8-11b about point A results in the following equilibrium equation:

$$\Sigma M_A = -M + P\Delta + WL = 0$$

$$\text{Therefore, } M = WL + P\Delta. \tag{8-5}$$

The "WL" term in equation (8-5) is the first-order sway moment, while the second term is the sway or P-Delta (P-Δ) moment. The second-order moment, M, from equation (8-5) can either be obtained directly from a second-order frame analysis that accounts for both $P-\delta$ and $P-\Delta$ effects (i.e., using the direct analysis method in Chapter C of the AISC Specification), or alternatively, the first-order moments from a first-order frame analysis can be magnified (see

AISC Specification, Section 8.1.3 of Appendix 8). In this latter case, which uses the B_2 factor to account for the $P-\Delta$ effect, the approximate second-order moment is

$$M = B_2\left[M_{ult(\text{1st order})}\right] \tag{8-6}$$

where

B_2 = Moment amplification factor due to lateral deflection of the top end of the column relative to the bottom end (the so-called P-Δ effect). This magnification factor is applicable to all columns in moment frames for the story under consideration and for each global direction of displacement of the structure; and

$M_{ult\,(\text{1st order})}$ = factored first-order lateral sway moments in the column. These moments are typically caused by wind or seismic lateral loads or by unbalanced gravity loads.

In summary, second-order effects lead to moment amplification in beam-columns, and the two types of second-order or P-delta effects previously discussed are:

- P-δ effects for columns in **Braced Frames:**
  - Accounts for the effect of the axial compression load on the lateral deflection of the column between its top and bottom supports.
- P-Δ effects for columns in **Unbraced or Sway frames (i.e., Moment Frames):**
  - Accounts for the effect of the axial compression load on the relative lateral deflection of the ends of the column.

8-5 STABILITY ANALYSIS AND CALCULATION OF THE REQUIRED STRENGTHS OF BEAM–COLUMNS

The effects of the following factors on the stability of a structure need to be considered [3]:

1. Flexural deformations and any other deformations that contribute to the displacement of the structure.
2. The nonsway second-order effect (i.e., ***P*-δ** effect) is due to the local displacement of the member between its end supports, and the sway second-order effects are due to the inter-story lateral sway displacement of the structure and the corresponding destabilizing effect of the laterally displaced gravity loads on the global stability of the structure (i.e., ***P*-Δ** effect). The effect of the gravity loads in the columns that are not part of the lateral load resisting system (i.e., the leaning columns) on the global stability of the structure should also be considered.
3. The geometric imperfections are due to the initial out-of-plumbness or out-of-straightness of the structural element or system. Every structure has some initial imperfections, and these effects can either be directly modeled or represented through the use of notional loads.
4. Stiffness reduction due to inelasticity caused by residual stresses in the structural member, which is taken into account through the use of stiffness reduction factors.
5. Uncertainty in stiffness and strength which are taken into account through the use of stiffness reduction factors.
6. The AISC Specification presents four methods that can be used for the stability analysis and design of steel building frames, and for the calculation of the required strengths or

factored loads and moments in beam-columns [3, 4]. In the AISC Specifications Section C1, the direct analysis method (DAM) is presented. Three alternate methods are also presented in the AISC Specifications [3, 4]:

a. Effective length method (ELM) presented in Appendix 7, Section 7.2.
b. First-order analysis method (FOAM) presented in Appendix 7, Section 7.3.
c. Amplified first-order analysis method (AFOAM) presented in Section 8.1 of Appendix 8 in the AISC Specifications.

These methods will be discussed in the following sections, after the introduction of leaning columns.

Leaning Columns

Leaning columns are beam-columns that are laterally braced by the moment frames or other lateral force-resisting systems in the plane of bending; these leaning columns do not participate in any way in resisting the lateral loads on the structure, but they "just go for the ride" when the structure deflects laterally. They possess no lateral stiffness but depend on the moment frames for their lateral support. In designing leaning columns, an effective length factor, K, of 1.0 is usually assumed; therefore, the axial loads on the leaning columns must be considered in analyzing the moment frames that provide lateral bracing to these columns. The effect of leaning columns is automatically considered in the direct analysis method. If the effective length method is used for the stability design, the effect of leaning columns on the moment frames could be included by designing the moment frame columns for additional axial loads *in the plane of bending* for which the moment frame provides lateral support to the leaning columns [3]. The additional axial load on the restraining columns is the total axial load on all the leaning columns distributed to the moment frame columns in proportion to the plan tributary area of the moment frame columns. In this chapter, the direct analysis method will be used, so the effect of the leaning columns will automatically be included in the analysis since the gravity loads on the leaning columns are included either in a rigorous second-order analysis or in the calculation of the sway amplification factor, B_2. In summary,

- **Leaning Columns** are columns that are not part of the LFRS; that is, they are not part of the braced frames or the moment-resisting frames.
- Leaning columns depend on the LFRS (i.e., the moment frame, braced frames, or shear walls) for lateral support; therefore, an effective length factor of K = 1.0 is used in the design of the leaning columns.
- The total axial loads acting on the leaning columns must be considered in analyzing the moment frames for P-Δ effects:
  - In the direct analysis method (DAM), the effect of the leaning columns is automatically included in the rigorous 2nd order analysis.
  - In the approximate second-order analysis (i.e., AFOAM), the effect of leaning columns is included in the calculation of the moment magnification factor, $\boldsymbol{B_2}$.

8-6 DIRECT ANALYSIS METHOD (DAM)

The direct analysis method (DAM) is the default method prescribed in the AISC specifications (*AISCM* Chapter C) for the analysis and design of moment frames, and it uses a direct second-order analysis method. The DAM is applicable to all structures without any limitations. It is the only rational method for designing steel structures for stability. The direct second-order structural analysis is used to obtain directly the second-order moments and axial load in a moment frame; the analysis accounts for axial, bending, and shear deformations of the members of the frame as well as the rotational deformation at the member connections and all geometric and material nonlinearities, including imperfections and inelasticity [3,4]. Both the P-δ effect (due to the lateral displacement of a member within its length) and the P-Δ effect (due to the lateral displacement between two consecutive levels of a moment frame) are taken into account in the direct second-order analysis. In the analysis, the members are modeled with nodes within the member length in order to capture the $P - \delta$ effects that occur due to the lateral deflection within the length of the compression member.

Since the design moments obtained from this second-order analysis method are actually the second-order moments, M_u, the design of the columns is carried out without any further moment magnification. This direct second-order analysis must include **all of the gravity loads** tributary to the moment frame that is being analyzed, including the axial loads on all ***leaning columns***. In the direct analysis method, geometric imperfections are accounted for by direct modeling of the imperfections or by the application of a notional lateral load, which is usually a defined percentage of the factored gravity loads on the frame. Inelasticity can be taken into account by using reduced flexural and axial stiffness for the columns and girders. A reduced *flexural stiffness* of $0.8EI\,\tau_b$ (see Chapter 5) and a reduced *axial stiffness* of $0.8EA$ can be used for all members of the moment frame, including those that do not contribute to the stability of the structure (*AISCM* Section C2.3). See Chapter 5 for the calculation of τ_b. The AISC Code allows τ_b to be taken as 1.0 if an additional notional load of $0.001\alpha Y_i$ is applied to the structure at each individual floor level for all load combinations (α and Y_i are defined below in the following section). In the direct analysis method (DAM), an effective length factor, $K = 1.0$, is used for both nonsway and sway buckling in the design of all members. The use of $K = 1.0$ in the DAM is a significant and welcome simplification in the stability design of steel columns, especially given the inaccuracies and unrealistic assumptions associated with the calculation of the K values from the alignment charts.

Although a number of structural analysis software capable of performing a direct second-order analysis are available in practice, many engineers are more familiar with *first-order* structural analysis software, which is also the analysis software that students are introduced to in many undergraduate civil engineering programs. As an alternative to using the direct second-order analysis method that is required for the DAM, a simplified second-order structural analysis can be used in conjunction with the B_1 and B_2 moment amplification factors in a method called the amplified first-order analysis method (AFOAM), and this will be discussed later in this chapter (see Section 8.1 of Appendix 8 of the AISC Specifications). Note that for the DAM that uses direct second-order structural analysis, the moment amplification factors, B_1 and B_2, are set equal to 1.0.

Notional Loads

The initial geometric imperfections due to the out-of-plumpness and out-of-straightness of the structural members must be modeled when using a direct second-order analysis. The effect of the initial imperfections or out-of-plumpness in the columns can be modeled in structural analysis software by displacing laterally (i.e., horizontally) the top end of each column a distance of $L/500$ relative to its bottom end, where L is the height of the column. The direct modeling of these initial imperfections for every member in the structure in the structural analysis can be cumbersome, but fortunately, the AISC code allows and specifies notional lateral loads that can be used to model and account for the effect of these initial geometric imperfections. The notional lateral load is 0.2% of the total factored gravity load at each level, based on an initial story imperfection or out-of-plumbness of $L/500$ (or $0.002L$), which is the maximum tolerance on column plumbness that is specified in the AISC Code [3]. Note that the notional load will vary depending on the load combination and will be additive to the other lateral loads for the load combinations with gravity plus lateral loads. For the DAM, the notional lateral load applied at a level i in addition to other lateral loads is given in *AISCM* Equation C2-1 as

$$N_i = 0.002\alpha Y_i,$$

Where,

Y_i = Total factored gravity load, in kips, applied at level *i only (for LRFD)*
Or total unfactored gravity load, in kips, applied at level *i only (for ASD).*

$\alpha = 1.0$ *for LRFD, and* $\alpha = 1.6$ *for ASD*

The notional lateral loads at each diaphragm level act through the center of gravity of the loading area and the notional loads should be applied in a positive or negative sense in the two global orthogonal directions so as to produce the most destabilizing effect. A simple way of accounting for the notional lateral loads in a direct second-order analysis is to apply the notional loads as lateral point loads at each level of the columns as follows [5]:

1. Calculate the total factored gravity load applied at each level of each column in the structure for each load combination. Note that this axial load, Y_i, is *not* cumulative; it is the load applied at *each* level of the structure.
2. Apply a notional lateral load of 0.2% of the factored column load at each level. The notional load coefficient of 0.2% is based on a maximum story out-of-plumbness ratio of 1/500 specified in the AISC code. If stricter tolerances are used (not recommended), the notional lateral load coefficient will decrease proportionally.
3. The lateral point load in each column at each level is then added to the other lateral loads acting on the structure in the 3-D direct second-order analysis so as to produce the most destabilizing effect. All load combinations (gravity and lateral) and all the columns in the structure should be included. Including all the gravity load of all the columns in the structure allows the effect of the leaning columns to be included directly in the 3-D second-order structural analysis.

When the ratio of second-order lateral drift to the first-order lateral drift is less than or equal to 1.7 (i.e., the sway amplifier, $B_2 \leq 1.7$), the notional loads may be applied only in the gravity load combinations and not in the load combinations that have lateral loads. In this case, there would be no need to analyze the structure for the notional loads if the lateral wind or seismic loads are larger than the notional loads. However, when $B_2 > 1.7$, the notional loads will be additive to the lateral loads [3].

If the stiffness reduction factor, τ_b, is assumed to be 1.0 for all members, an additional notional lateral load of $0.001\alpha Y_i$ must be used (*AISCM* Section C2.3(c)); this additional notional load is additive to the other lateral loads in the load combinations and the other notional loads, if any, that are used to account for geometric imperfections and must be included in all load combinations and applied at all levels. In summary,

- Notional loads are **additional lateral loads** caused by the geometric imperfections or out-of-plumbness of the top of the columns in each story relative to the bottom of the columns:
  - It applies only to sway or moment frames.
  - Notional loads can be modeled directly by displacing the top of a column relative to its bottom by an amount equal to the column height/500.
  - It is cumbersome to model notional loads directly.

- AISC 360-22 allows an approximate method to model notional loads by assuming a notional lateral load of 0.2% of the total factored gravity load at each level (based on an initial story imperfection of L/500 or 0.002L), where L is the story height.

8-7 AMPLIFIED FIRST-ORDER ANALYSIS METHOD (AFOAM) AND AMPLIFICATION FACTORS

In the direct analysis method (DAM), the P-delta effects can be accounted for through the use of a direct second-order analysis. Four methods for stability analysis are presented in Section C1 and Appendices 7 and 8 of the AISC Specifications as follows:

- The **direct analysis method** (DAM) that uses a general second-order elastic structural analysis that directly accounts for geometric and material non-linearities is presented in Chapter C of the AISC Specifications.
- The **effective length method** (ELM) is presented in Appendix 7.
- The **first-order analysis method** (FOAM) is presented in Appendix 7.
- The approximate second-order elastic analysis method, known as the ***Amplified First-order Analysis*** *Method* (AFOAM). that accounts for the P-delta effects in lieu of using a direct second-order analysis is presented in Section 8.1 of Appendix 8 of the AISC Specifications.

The ***Amplified First-order Analysis*** *Method* (AFOAM), which will be used in this text, accounts for the second-order stability effects by amplifying the required moment and axial strengths or the factored load effects obtained from a first-order structural analysis. Note that the AFOAM is a direct analysis method (DAM) except that it uses an approximate second-order analysis that accounts for P-delta effects using the moment amplification factors, B_1 and B_2. The B_1 factor accounts for the $P-\delta$ or moment magnification effects on the column non-sway moments, while the B_2 factor accounts for the $P-\Delta$ or moment magnification effects on the sway moments in the column. The equations for calculating the amplification or magnification factors, B_1 and B_2, according to the AISC specification, will now be presented.

Nonsway Moment Amplification Factor, B_1

The nonsway moment amplification factor, B_1, is calculated as follows from the AISC specification using the gross (unreduced) stiffness of the member (see AISC Specification, Section 8.1.2 of Appendix 8):

$$B_1 = \frac{C_m}{1 - \dfrac{\alpha P_u}{P_{e1}}} \geq 1.0 \quad (for\ LRFD),$$

or (8-7)

$$B_1 = \frac{C_m}{1 - \dfrac{\alpha P_s}{P_{e1}}} \geq 1.0 \quad (for\ ASD)$$

Where,

C_m = factor that accounts for the moment variation or distribution within the column. Note that for a uniform moment, C_m = 1.0.

P_u = The cumulative factored axial load in the beam-column using the LRFD load combinations (see Section 2-3).

P_s = The cumulative unfactored axial load in the beam-column using the ASD load combinations (see Section 2-3).

The theoretical elastic buckling load of the column, P_{e1}, is calculated as

$$P_{e1} = \frac{\pi^2 EI^*}{(KL)^2} = \frac{\pi^2 EA^*}{\left(\frac{KL}{r}\right)^2} = \frac{\pi^2 EA^*}{\left(\frac{L_c}{r}\right)^2}, \tag{8-8}$$

Where,

$L_c/r = KL/r$ = Slenderness ratio about the **axis of bending** of the member.

$L_c = KL$

K = 1.0 for all columns when using the DAM

≈ 1.0 for columns in braced frames when using the ELM and FOAM.

(K may be ≤ 1.0, but it is practical to use a value of 1.0 for columns in braced frames).

EI^* = Flexural rigidity used in the analysis

= $0.80\, \tau_b\, EI$ for the DAM

= Gross EI for the ELM and the FOAM

EA^* = Axial rigidity used in the analysis

= $0.80EA$ for the DAM

= EA for the ELM and the FOAM

A = Gross cross-sectional area of the beam-column, in^2.

$\alpha = 1.0$ (for LRFD) *and* $\alpha = 1.6$ (*for ASD*)

τ_b is a flexural stiffness reduction factor that accounts for inelasticity in the beam-column due to residual stresses, and this factor was defined in Chapter 5. The AISC specification allows τ_b to be taken as 1.0, provided an additional notional load of $0.001\alpha Y_i$ is applied at all levels of the structure and in all the load combinations.

Alternatively, P_{e1x} and P_{e1y} can also be obtained for wide flange columns or I-shapes from the bottom rows of the column load capacity tables in Section 4 of the *AISCM* (see *AISCM*, Table 4-1) as follows:

$$P_{e1}, \text{ in kips} = \frac{10^4}{(L_c)^2} \times \text{Corresponding value from the column load tables}, \tag{8-9}$$

where

$L_c = KL$ is in inches in equation (8-9),

The moment reduction coefficient, C_m, used in equation (8-7) accounts for the effect of moment gradient in the column and is obtained as follows:

For braced beam-columns with **no** transverse loads between the supports, the moment distribution coefficient, C_m is given as

$$C_m = 0.6 - 0.4\frac{M_1}{M_2}, \tag{8-10}$$

where

$\frac{M_1}{M_2}$ = Absolute ratio of first-order bending moments at the ends of the member (M_1 is the **smaller** end moment, M_2 is the **larger** end moment)

Note:

- The sign of $\frac{M_1}{M_2}$ in equation (8-10) is taken as *negative* (i.e., -ve) for *single curvature* bending of the member (see Figure 8-12a), and
- The sign of $\frac{M_1}{M_2}$ in equation (8-10) is taken as *positive* (i.e., +ve) for *double curvature* bending of the member (see Figure 8-12a).

Examples of columns that are subject to double-curvature bending include exterior columns and columns in moment frames; for these columns, the maximum possible C_m value from equation (8-10) is 0.6. The exterior and interior columns at the lowest level of a building will be subjected to single curvature bending and will have a maximum possible C_m value of 0.6 if the bases of the columns are assumed to be pinned. On the other hand, interior columns above the second-floor level may be subjected to single-curvature or double-curvature bending, depending on the live load pattern on the beams and girders framing into the top and bottom ends of the column at the floor levels. The maximum possible C_m value for these interior columns is 1.0. For beam-columns with transverse loads between the supports, C_m = 1.0, as shown in Figure 8-12b. Single and double curvature bending are illustrated in Figure 8-12a.

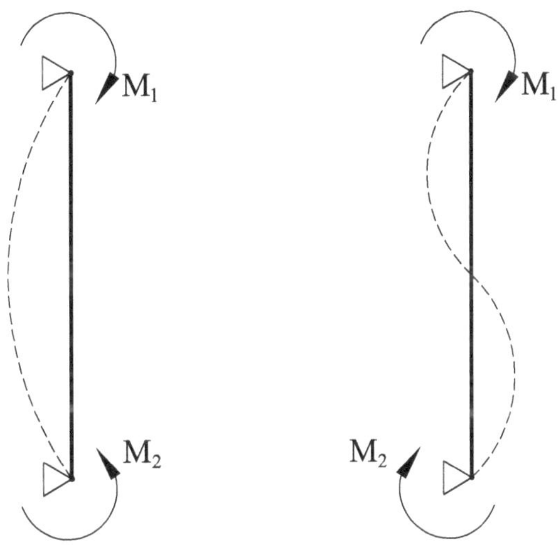

FIGURE 8-12a Single and double curvature bending

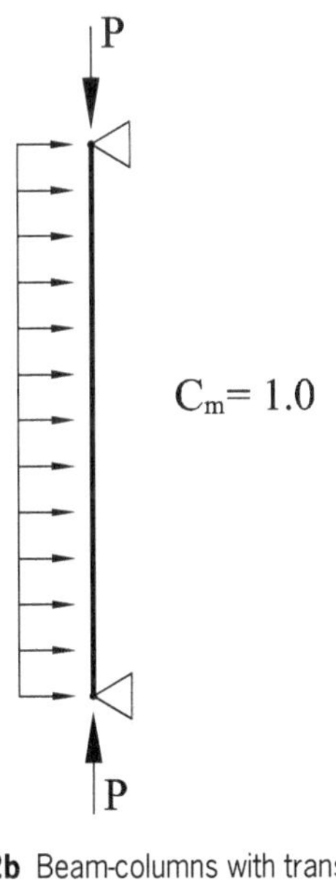

FIGURE 8-12b Beam-columns with transverse loads

Since several combinations of column end moments, M_1 and M_2, are possible, C_m can take on many different values for the same column, depending on the load combinations and live load patterns considered at the floor levels at the top and bottom of the column. To simplify the determination of the C_m factor, and because the nonsway moment amplification factor, B_1, is almost always equal to 1.0 for many practical cases, the suggested approximate values for C_m, presented in Figure 8-13, can be used, and these values will yield conservative results [6, 7].

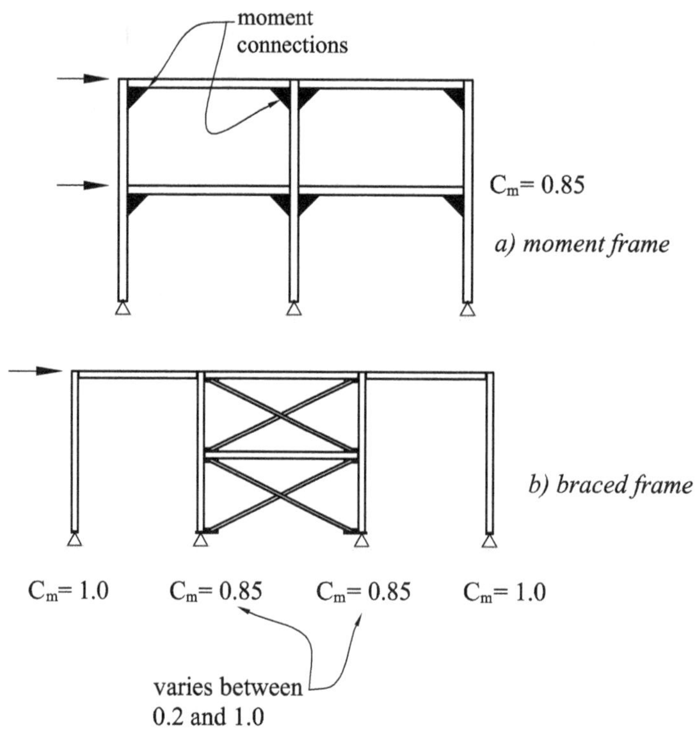

FIGURE 8-13 Approximate values of C_m for beam-columns

Sway Moment and Axial Load Amplification Factor, B_2, for Unbraced Frames or Moment Frames

The sway amplification factor, B_2, is calculated (see AISC Specification, Section 8.1.3 of Appendix 8), for all the columns in **each story** of a moment frame, using the approximate AISC specification method presented below in equations (8-11a) and (8-11b). Thus, all columns in a given story will have the same sway magnification factor as follows:

$$B_2 = \frac{1}{1 - \alpha \dfrac{P_{story}}{P_{e,story}}} \geq 1.0, \tag{8-11a}$$

or

$$B_2 = \frac{1}{1 - \dfrac{\Delta_H}{L} \alpha \dfrac{P_{story}}{H R_M}} \geq 1.0, \tag{8-11b}$$

Where,

For LRFD:

P_{story} = Sum of the *cumulative factored* gravity loads supported by the **story** under consideration, including axial loads in the columns that are not part of the lateral force-resisting system (i.e., the leaning columns). This gravity load is cumulative; that is, it includes loads at and above the level under consideration, and uses the LRFD load combinations (see Section 2-3).

For ASD:

P_{story} = Sum of the *cumulative unfactored* gravity loads supported by the **story** under consideration, including axial loads in the columns that are not part of the lateral force-resisting system (i.e., the leaning columns). This gravity load is cumulative; that is, it includes loads at and above the level under consideration, and uses the ASD load combinations (see Section 2-3).

$P_{e,story}$ = Elastic buckling strength of the **story** under consideration in the direction of translation.

$$P_{e,story} = R_M \frac{HL}{\Delta_H} \tag{8-12}$$

$\alpha = 1.0$ (for LRFD) *and* $\alpha = 1.6$ (*for ASD*)

H = Factored horizontal or lateral shear in the story under consideration for LRFD, or unfactored horizontal or lateral shear in the story under consideration for ASD.

L = Story height or floor-to-floor height of the moment frame,

Δ_H = First-order inter-story drift caused by the lateral shear, H, and

$\frac{\Delta_H}{L}$ = Allowable drift index due to the lateral loads (Typical allowable range of values are as follows:

- 1/300 to 1/240 under *factored* wind loads
- 1/500 to 1/400 under *unfactored* wind loads;
- See ASCE 7, Table 12.12.1 for drift limits for factored seismic loads.

R_M = stiffness reduction coefficient that accounts for the effect of member P-δ on the entire structure P-Δ effects and is calculated as

$$= 1 - 0.15\left(\frac{P_{mf}}{P_{story}}\right)$$

= 0.85 when all the columns in the story participate as part of the moment frames in resisting lateral loads

= 1.0 if there are no moment frames in the story (i.e., braced frames only, which means $P_{mf} = 0$)

P_{mf} = Total cumulative factored gravity load (for LRFD) or total cumulative unfactored gravity load (for ASD) in the moment frame columns in the story.

See Section 2-15 for discussions of drift limits under wind and seismic loads. For practical situations, the allowable drift index used in the calculation of the sway moment amplification factor can be assumed to be 1/500 for *service* lateral wind loads using the ASD load combinations from Section 2-3 (i.e., wind load = 0.6W); this inter-story drift index limit is commonly used in design practice to satisfy serviceability requirements under wind loads. Since factored gravity and lateral loads are used in equation (8-11b), the allowable drift index should be modified to the factored wind load level (1.0W). Therefore, an allowable drift index of 1/(0.6 × 500), or 1/300 of the story height, may be used at the *factored* wind load level for moment frames subjected to wind loads. For seismic loads, the drift index limits given in Table 12.12.1 of the ASCE 7 load specification should be used [8].

Total Factored Moment, M_u, and Factored Axial Load, P_u, in a Beam-Column (LRFD)

The total factored second-order moment and the total factored second-order axial load, or the required moment and axial strengths (for LRFD), are given as

$$M_u = B_1 M_{u,nt} + B_2 M_{u,lt} \tag{8-13a}$$

$$P_u = P_{u,nt} + B_2 P_{u,lt} \tag{8-13b}$$

Where,

P_u = Total cumulative second-order factored axial load in the beam-column,

$P_{u,nt}$ = Total cumulative factored first-order axial load in the beam-column with lateral translation of the structure *restrained* (i.e., braced condition),

$P_{u,lt}$ = Total cumulative factored first-order axial load in the beam-column with lateral translation of the structure *unrestrained* (i.e., unbraced condition),

$M_{u,nt}$ = Factored moments in the beam-column when no appreciable sidesway occurs (nt = no translation) or when the structure is restrained against lateral translation. *The M_{unt} moments are caused by **gravity loads** acting at the simple-shear beam-to-column connection eccentricities for columns in braced frames or the moments in the columns of unbraced frames when the frame is restrained against lateral translation.*

$M_{u,lt}$ = Factored lateral translation moments in the beam-column caused by any of the following:

- Wind or earthquake lateral loads on the moment frame,
- The restraining force necessary to prevent sidesways in a symmetrical frame loaded with asymmetrically placed gravity loads, and
- The restraining force necessary to prevent sidesways in an asymmetrical frame loaded with symmetrically placed gravity loads.

("***lt***" = Lateral translation and "***nt***" = no translation)

For most reasonably symmetric moment frames, the $M_{u,lt}$ moments are caused only by the lateral wind or seismic loads.

We can rewrite the equations (8-13a) and (8-13b) for both the x- and y-axes of bending as follows:

$$M_{ux} = B_{1x} M_{u,ntx} + B_{2x} M_{u,ltx} \tag{8-14a}$$

$$M_{uy} = B_{1y} M_{u,nty} + B_{2y} M_{u,lty} \tag{8-14b}$$

$$P_u = P_{u,nt} + B_2 P_{u,lt} \tag{8-14c}$$

Note that for *braced frames*, there are no lateral translation moments and no lateral translation axial loads, which implies that

$M_{u,ltx} = 0$, $M_{u,lty} = 0$, and $P_{u,lt} = 0$ for braced frames

Therefore, for *braced frames*,

$M_{ux} = B_{1x} M_{u,ntx}$, $M_{uy} = B_{1y} M_{u,nty}$, and $P_u = P_{u,nt}$.

Note that B_{1x} and B_{1y} relate to bending about the individual beam-column member axes, but B_{2x} and B_{2y} are related to the global axes of translation of the structure and not necessarily to the local axes of bending of the individual members of the structure.

- B_{2x} is applicable to all load effects—shears, moments, and axial forces—produced by the story translation in the global x-direction (i.e., in the X-direction).

- Similarly, B_{2y} is applicable to all load effects—shears, moments, and axial forces—produced by the story translation in the global y-direction (i.e., in the Y-direction).

The sway amplification factors (B_{2x} and B_{2y}) should be applied to the appropriate local moments and axial loads acting on the individual moment frame columns depending on the orientation of the columns relative to the global direction of the lateral loads under consideration [3].

8-8 FIRST-ORDER ANALYSIS METHOD (FOAM)

A first-order analysis is a structural analysis of a building frame where the effects of geometric nonlinearities (or P-delta effects) are **not** included. This method uses ***unreduced* stiffness and cross-sectional areas** for the columns and girders. The AISC specification allows the first-order analysis loads and moments to be used for the design of beam-columns if the following conditions are satisfied [3]:

1. The gravity loads in the structure are supported by vertical columns, walls, or frames.
2. The factored axial compression loads (or required compression strength) are not greater than 50% of the cross-section compressive strength (i.e., $\boldsymbol{\alpha P_u \leq 0.5\, P_{ns}}$) for all members whose flexural stiffnesses are considered to contribute to the lateral stiffness of the frame (α = 1.0 for LRFD). For ASD, α = 1.6, and the unfactored axial compression load multiplied by α should not be greater than 50% of the cross-section compressive strength (i.e., $\boldsymbol{\alpha P_s \leq 0.5\, P_{ns}}$).
   - **a.** P_{ns}, the cross-section compressive strength, is equal to F_yA for sections without slender elements and
   - **b.** P_{ns}, is equal to F_yA_e for sections with slender elements, where A_e is the effective area of the slender element sections as presented previously in Section 5-4 of this text (i.e., see Section E7 of the *AISC* Specifications). That is, $P_{ns} \leq F_yA \leq F_yA_e$.
3. The ratio of the maximum second-order lateral drift to the first-order lateral drift in all stories (determined for the LRFD load combinations or for 1.6 multiplied by the ASD load combinations) is less than or equal to 1.5.
   That is, for all stories, the sway amplifier, $B_2 \leq 1.5$.
4. For the *horizontal members* in moments frames, the required axial compressive strengths should satisfy the following requirement:
   - **a.** For LRFD: $P_u \leq 0.08P_e$
   - **b.** For ASD: $1.6P_s \leq 0.08P_e$

     Where,
     P_u and P_s are obtained using the LRFD and ASD load combinations, respectively
     $P_e = \pi^2 EI/L^2$
     I = moment of inertia in the plane of bending, in.^{4}
     L = Story height in inches.

If these conditions are satisfied, an effective length factor, $\boldsymbol{K} = \mathbf{1.0}$, is used for the design of the beam-columns, and the total member moments should conservatively be amplified by multiplying them by the nonsway moment magnification factor, B_1 (see AISC Specification Appendix 7.3).

For all load combinations, an additional *notional lateral load* must also be applied in both orthogonal directions, in addition to any applied lateral loads. This notional load is given as

$$N_i = 2.1\,\alpha(\Delta/L)Y_i \geq 0.0042Y_i,$$

where

Y_i = Total factored gravity load, in kips, applied at level *i only (for LRFD)* or total unfactored gravity load, in kips, applied at level *i only (for ASD)*

$\alpha = 1.0$ (for LRFD) *and* $\alpha = 1.6$ (*for ASD*)

Δ/L = Maximum ratio of the inter-story drift (Δ) to the story height (L) for all stories in the structure,

Δ = First-order inter-story drift, in inches, due to factored (LRFD) load combinations or unfactored (ASD) load combinations, and

L = Story height in inches.

The first-order analysis method (FOAM) from *AISCM* Appendix 7.3 will *not* be used in this text.

8-9 EFFECTIVE LENGTH METHOD (ELM)

The effective length method is an indirect second-order analysis procedure (see Appendix 7.2 of the AISC Specification) where the first-order moments are amplified by the B_1 and B_2 moment magnification factors as demonstrated in the previous section. The method uses ***unreduced* stiffnesses and cross-sectional areas** for the columns and girders, and the analysis is carried out (for the LRFD method) at the factored load level or for the ASD method using 1.6 times the ASD load combinations. The effective length method must satisfy the following requirements:

1. The gravity loads in the structure are supported by vertical columns, walls, or frames.
2. This method is limited to building frames where the sway moment magnification factor, B_2, does not exceed 1.50. Where B_2 exceeds 1.50, the AISC specification requires that the direct analysis method (DAM) be used (see Section 8-6).
3. For braced frames, the effective length factor, K, is taken as 1.0 for all columns, and for sway or moment frames, the effective length factor, K, which will be greater than or equal to 1.0, is determined from Figure 5-4 or the alignment charts presented in Figures 5-10 and 5-11.
4. Also, when $B_2 \leq 1.1$, all the columns can be designed with an effective length factor, $K = 1.0$.
5. In addition, the effective length method (ELM) requires that all the load combinations with ***gravity loads-only*** must include a minimum *notional lateral load* of $\mathbf{0.002\alpha Y_i}$ applied at each level in both orthogonal directions, where Y_i is the total factored gravity load applied at the story under consideration; $\alpha = 1.0$ for LRFD and $\alpha = 1.6$ for ASD. For the load combinations with only gravity loads, the lateral translation will be small and therefore negligible in many practical situations, even with the notional lateral load included, except for highly asymmetrical frames; so, it is reasonable to assume the nonsway case for the load combinations with only gravity loads. Thus, for the gravity-load-only combinations, the sway moment amplification factor, B_2, (see equation (8-11)) can be assumed to be 1.0, and the sway moment, or lateral translation moment, M_{lt}, is assumed to be negligible and taken as zero.

Table 8-1 summarizes the requirements from the preceding discussions of the four methods of stability analysis prescribed in the AISC specifications (see *AISCM* Table 2-1 [3]).

TABLE 8-1 Summary of AISC Specification Requirements for the Stability Analysis and Design of Moment Frames*

	Direct analysis method (DAM)	First-order analysis method (FOAM)	Effective length method (ELM)
Limitations of Use	None	Yes $B_2 \le 1.5$ $(1.0)P_u/P_{ns} \le 0.5$ (**LRFD**): $(1.0)P_u/P_{ns} \le 0.5$ $(1.0)P_u \le 0.08P_e$ (**ASD**): $(1.6)P_s/P_{ns} \le 0.5$ $(1.6)P_s \le 0.08P_e$	Yes $B_2 \le 1.5$
Type of structural analysis	General second-order elastic analysis or the **Amplified First Order Analysis Method (AFOAM)**—i.e., approximate second-order analysis using B_1 and B_2 amplification† factors.	First-order analysis (member moments amplified using the B_1 factor)	General second-order elastic analysis or approximate second-order analysis using the B_1 and B_2 moment amplification factors
Member stiffness used in the structural analysis	Reduced EI and EA to account for inelasticity from residual stresses ($0.80\tau_b EI$ and 0.80EA)**	Gross EI and EA to	Gross EI and EA
Is a notional load required?	Yes; Minimum is $0.002\alpha Y_i$††	Yes; Additional lateral load is $2.1\alpha(\Delta/L)Y_i$ $\ge 0.0042Y_i$	Yes; Minimum is $0.002\alpha Y_i$
Effective length factor, K, used in column design	$K = 1.0$ (all frames)	$K = 1.0$ (all frames)	Braced frames, $K=1.0$ Moment frames, $K > 1.0$ Determine K using the alignment charts. For $B_2 < 1.1$, use K = 1.0
Reference Section in the AISC Specifications	Chapter C for DAM, and Appendix 8, Section 8.1 for AFOAM	Section 7.3 of Appendix 7	Section 7.2 of Appendix 7

Note:

E = Modulus of elasticity
A = Cross-sectional area
I = Moment of inertia
Δ = first-order inter-story drift
α = 1.0 for LRFD and 1.6 for ASD
EA = Axial stiffness
EI = Bending stiffness
L = story height
Δ/L = maximum drift ratio for all stories
Y_i = Gravity load applied at level i

P_{ns} = Cross-section compressive strength

$= F_y A_g$ for members with non-slender element sections

$= F_y A_e$ for slender element sections

A_e = Effective area defined in Section 5-4 of the text (AISC Specification Section E7)

† The LRFD load combinations must be used in the structural analyses for DAM and AFOAM.

** The reduced stiffness is applied to all members. The stiffness reduction factor, τ_b, can be taken as 1.0 if an additional notional load is applied at each level equal to 0.001 times the total gravity load applied at that level.

†† The notional load in the DAM is *additive* to other lateral loads when $B_2 > 1.5$;
When $B_2/B_1 < 1.7$, apply the notional load *only in the gravity-only load combinations*.

EXAMPLE 8-1 (LRFD)

Columns in Braced Frames

Using the LRFD method, determine the adequacy of a 20 ft. long W14 × 109 column in a building that is braced by shear walls or diagonal braces (i.e., a Braced Frame). The column is pinned at both ends and subjected to the following factored load case obtained from a first-order structural analysis:

Top of the column:
P_u = 520 kips,
M_{ux} = 210 ft.-kips, and
M_{uy} = 100 ft.-kips.

Bottom of the column:
M_{ux} = 105 ft.-kips
M_{uy} = 100 ft.-kips

The moments act in the clockwise direction at the top of the column and in the counterclockwise direction at the bottom of the column, resulting in single curvature bending in the column for the given load case. Assume ASTM A992 Grade 50 steel. Use the AISC equations to determine the axial compression capacity of the column and use the actual C_m factor for the column that corresponds to the given load case.

Solution (LRFD)

The section properties for W14 × 109 from *AISCM* Table 1-1 are as follows:

$A = 32 \text{ in.}^2,\ r_x = 6.22 \text{ in.},\ r_y = 3.73 \text{ in.},\ Z_x = 192 \text{ in.}^3,\ Z_y = 92.7 \text{ in.}^3,\ S_y = 61.2 \text{ in.}^3$

From *AISCM* Table 3-2, the flexural properties of W14 × 109 for strong axis bending are $\phi_b M_{px} = 720 \text{ ft.-kips},\ \phi_b BF = 7.54,\ L_p = 13.2 \text{ ft.},\ L_r = 48.4 \text{ ft.}$

Axial Compression Capacity, $\phi_c P_n$:

$$\frac{L_{cx}}{r_x} = \frac{K_x L_x}{r_x} = \frac{(1.0)(20\text{ ft.})(12)}{6.22\text{ in.}} = 38.6 < 200 \quad OK.$$

$$\frac{L_{cy}}{r_y} = \frac{K_y L_y}{r_y} = \frac{(1.0)(20\text{ ft.})(12)}{3.73\text{ in.}} = 64.3 < 200 \quad OK$$

The larger slenderness ratio controls, therefore, use $\frac{L_c}{r} = \frac{KL}{r} = 64.3$

To determine if the column will buckle in elastic or inelastic buckling, calculate the slenderness ratio at the transition point between inelastic and elastic buckling as follows:

$$4.71\sqrt{\frac{E}{F_y}} = 4.71\sqrt{\frac{29{,}000\text{ ksi}}{50\text{ ksi}}} = 113.4 > \frac{KL}{r} = 64.3$$

Therefore, this column fails in *inelastic* buckling, and equation (5-14) applies. The Euler critical buckling stress is

$$F_e = \frac{\pi^2 E}{\left(KL/r\right)^2} = \frac{\pi^2(29{,}000\text{ ksi})}{(64.3)^2} = 69.2 \text{ ksi}$$

The nominal stress for *inelastic* buckling from equation (5-14) is calculated as

$$F_n = \left[(0.658)^{F_y/F_e}\right]F_y = \left[(0.658)^{50/69.2}\right](50 \text{ ksi}) = 36.95 \text{ ksi}$$

The nominal axial compression load capacity is

$P_n = F_n A = (36.95 \text{ ksi})(32 \text{ in.}^2) = 1182 \text{ kips}$

The design axial compression load capacity is

$\phi_c P_n = \phi_c F_n A = (0.9)(36.95 \text{ ksi})(32 \text{ in.}^2) = 1064 \text{ kips}$

This value compares well with the design strength of 1060 kips at KL = 20 ft. obtained from *AISCM* Table 4-1a (i.e., the column load capacity tables), or from *AISCM* Table 6-1.

Moment Capacity about the *x-x* or strong axis:

The unbraced length of the column for bending about the *x-x* axis, L_b = 20 ft.
Since $L_p < L_b < L_r$, inelastic lateral-torsional buckling must be considered (see Chapter 6).
The design moment capacity for bending about the *x-x* or strong axis is

$$\phi_b M_{nx} = C_b\left[\phi_b M_{px} - \phi_b BF\left(L_b - L_p\right)\right] \le \phi_b M_{px}$$

$$= (1.0)\left[720 - (7.54)(20 - 13.2)\right] = \mathbf{669\ ft.\text{-}kips} < \phi_b M_{px} = 720 \text{ ft.-kips}$$

Therefore, $\phi_b M_{nx}$ = 669 ft.-kips.
The nominal moment capacity, $M_{nx} = (669/0.9) = 743$ ft.-kips
It is conservative to use a moment gradient factor, C_b = 1.0 as assumed above. The C_b factor accounts for the non-uniform moment distribution within the unbraced length of the flexural member. The actual value of C_b can be calculated using the moment gradient factor equation from Chapter 6 as follows:

$$C_b = \frac{12.5M_{max}}{2.5M_{max} + 3M_A + 4M_B + 3M_C}$$

M_{max} = absolute maximum moment in the member within the unbraced length
M_A = absolute value of the moment in the member at the quarter point of the unbraced length
M_B = absolute value of the moment in the member at the mid-point of the unbraced length
M_C = absolute value of the moment in the member at the three-quarter point of the unbraced length

With the given applied moments about the *x-x* or strong axis at the top and bottom of the column and noting the linear distribution of moments between the top and bottom of the column, the moments at quarter-points within the unbraced length of the member can be calculated as follows:

M_{max} = 210 ft.-kips
M_A = ½(105 + 157.5) = 131.3 ft.-kips
M_B = ½(105 + 210) = 157.5 ft.-kips
M_C = ½(157.5 + 210) = 183.8 ft.-kips

Therefore, the moment gradient factor or the lateral-torsional buckling factor is calculated as

$$C_b = \frac{12.5(210)}{2.5(210) + 3(131.3) + 4(157.5) + 3(183.8)} = 1.25$$

Therefore, recalculating the bending moment capacity about the *x-x* or strong axis using the actual moment gradient factor, we obtain

$$\phi_b M_{nx} = C_b\left[\phi_b M_{px} - \phi_b BF\left(L_b - L_p\right)\right] \le \phi_b M_{px}$$

$$= (1.25)\left[720-(7.54)(20-13.2)\right] = 836\ ft.-kips \le \phi_b M_{px} = \mathbf{720\ ft.-kips}$$

Therefore, $\phi_b M_{nx} = \mathbf{720\ ft.-kips}$

The nominal moment capacity about the *x-x* axis, $M_{nx} = (720/0.9) = 800\text{ ft.-kips}$

Moment Capacity about the *y-y* or weak axis:

The design moment capacity for bending about the *y-y* or weak axis is

$$\phi_b M_{ny} = \phi_b Z_y F_y \le 1.6\phi_b S_y F_y$$

$$= (0.9)(92.7\text{ in.}^3)(50\text{ ksi}) \le (1.6)(0.9)(61.2\text{ in.}^3)(50\text{ ksi})$$

$$= 4172\text{ in.-kips} \le 4406\text{ in.-kips}$$

$$= \mathbf{348\ ft.-kips} \le 367\text{ ft.-kips}$$

Therefore, $\phi_b M_{ny} = 348$ ft.-kips.

The nominal moment capacity about the *y-y* axis, $M_{ny} = (348/0.9) = 387\text{ ft.-kips}$

Moment Amplification or *P-δ* effect:

Determine the nonsway moment amplification factor, B_{1x}, for *x-x* or strong axis buckling: The unbraced length for buckling about the *x-x* axis, $L_x = 20$ ft.

$$K_x = 1.0$$

$$\frac{K_x L_x}{r_x} = \frac{(1.0)(20\text{ft.})(12)}{6.22\text{in.}} = 38.6 < 200 \quad OK.$$

$$P_{e1x} = \frac{\pi^2 EA}{\left(K_x L_x / r_x\right)^2} = \frac{\pi^2(29{,}000\text{ksi})(32\text{in.}^2)}{(38.6)^2} = 6147\text{kips}$$

An approximate C_{mx} value is typically used in this text to account for the effect of the non-uniform moments in columns, but since the top and bottom moments are known for this problem, the actual C_{mx} value for the given load case will be computed as follows:

$M_{1x} = 105$ ft.-kips (smaller end moment about *x-x* or strong axis)

$M_{2x} = 210$ ft.-kips (larger end moment about *x-x* or strong axis)

$$C_{mx} = 0.6 - 0.4\frac{M_1}{M_2}$$

$\dfrac{M_1}{M_2}$ = negative for single curvature bending

$\dfrac{M_1}{M_2}$ = positive for double curvature bending

Since the moments at the top and bottom of the column result in single curvature bending, therefore, M_1/M_2 = negative and,

$$C_{mx} = 0.6 - 0.4\left[-\frac{(105\,\text{ft.-kips})}{(210\,\text{ft.-kips})}\right] = 0.6 + (0.4)(0.5) = 0.80$$

The nonsway moment amplification factor for buckling about the x - x axis is

$$B_{1x} = \frac{C_{mx}}{1 - \dfrac{P_u}{P_{e1x}}} \geq 1.0$$

$$= \frac{0.8}{1 - \dfrac{520\,\text{kips}}{6147\,\text{kips}}} = 0.88 \geq 1.0$$

Therefore, use $B_{1x} = 1.0$

The magnified factored moment about the x-x or strong axis is calculated using the larger end moment. Therefore,

$$M_{ux} = B_{1x} M_{ux,nlt} = (1.0)(210\,\text{ft.-kips}) = 210\ \text{ft.-kips}$$

The implication of $B_{1x} = 1.0$ indicates that there is no P-delta effect for this column for bending about the x-x axis, and therefore, there is no amplification of the factored bending moment of the x-x axis.

Determine the nonsway moment amplification factor, B_{1y}, for y-y or weak axis buckling:

The unbraced length for buckling about the y-y axis, $L_y = 20\,\text{ft.}$

$$K_y = 1.0$$

$$\frac{K_y L_y}{r_y} = \frac{(1.0)(20\ \text{ft.})(12)}{3.73\ \text{in.}} = 64.3 < 200 \quad \text{O.K.}$$

$$P_{e1y} = \frac{\pi^2 EA}{\left(\dfrac{K_y L_y}{r_y}\right)^2} = \frac{\pi^2(29{,}000\,\text{ksi})(32\ \text{in.}^2)}{(64.3)^2} = 2215\,\text{kips}$$

An approximate C_m value is typically used in this text to account for the effect of the nonuniform moments in columns in this text, but since the top and bottom moments are known for this problem, the actual C_{my} value for the given load case will be computed as follows:

$M_{1y} = 100\,\text{ft.-kips}$ (smaller end moment about y - y or weak axis)

$M_{2y} = 100\,\text{ft.-kips}$ (larger end moment about y - y or weak axis)

$$C_{my} = 0.6 - 0.4\frac{M_{1y}}{M_{2y}}$$

$\dfrac{M_{1y}}{M_{2y}}$ = negative for single curvature bending

$\dfrac{M_{1y}}{M_{2y}}$ = positive for double curvature bending

Since the moments at the top and bottom of the column result in single curvature bending, therefore, M_1/M_2 = negative, and,

$$C_{my} = 0.6 - 0.4\left[-\frac{(100\,\text{ft.-kips})}{(100\,\text{ft.-kips})}\right] = 0.6 + (0.4)(1.0) = 1.0$$

The nonsway moment amplification factor for buckling about the y - y axis is

$$B_{1y} = \frac{C_{my}}{1 - \dfrac{P_u}{P_{e1y}}} \geq 1.0$$

$$= \frac{1.0}{1 - \dfrac{520\,\text{kips}}{2215\,\text{kips}}} = 1.31 \geq 1.0$$

Therefore, use $B_{1y} = 1.31$

The magnified factored moment of the *y-y* or weak axis is calculated using the larger end moment. Therefore,

$$M_{uy} = B_{1y} M_{uy,nlt} = (1.31)(100\,\text{ft.-kips}) = 131\ \text{ft.-kips}$$

Column Interaction Equation:

$$\frac{P_u}{\phi_c P_n} = \frac{520\ \text{kips}}{1064\ \text{kips}} = 0.49 > 0.2$$

Therefore, use the Column Interaction equation (8-1):

$$\frac{P_u}{\phi_c P_n} + \frac{8}{9}\left(\frac{M_{ux}}{\phi_b M_{nx}} + \frac{M_{uy}}{\phi_b M_{ny}}\right) \leq 1.0$$

$$i.e.\ 0.49 + \frac{8}{9}\left(\frac{210}{720} + \frac{131}{348}\right) = \mathbf{1.08} > 1.0\ \ (Not\ Good)$$

Therefore, the W14 × 109 column is overstressed and inadequate. A slightly larger column size, such as a W14 × 120 column, should easily suffice for this column, and the reader should recalculate all the variables for this column and verify that the value of the interaction equation for a W14 × 120 column will be less than or equal to 1.0.

Note that *AISCM* Table 6-1 (Available Strength for Members Subject to Axial, Shear, Flexural, or Combined Forces) [3] could also be used to directly obtain the design strengths ($\phi_c P_n$, $\phi_t P_n$, $\phi_b M_{nx}$, *and* $\phi_b M_{ny}$) for wide flange sections for any unbraced length that can be used directly in the beam-column interaction equations. The moment magnification factors, B_{1x} and B_{1y}, would still need to be calculated and applied, and note also that *AISCM* Table 6-1 assumes a moment gradient factor, C_b of 1.0. The values from the table would have to be modified to account for the actual values of C_b , B_{1x} and B_{1y}.

EXAMPLE 8-2 (ASD)

Columns in Braced Frames

Using the ASD method, check the column from Example 8-1, assuming the following unfactored load case from a first-order structural analysis:

Top of the column:
P_s = 385 kips,
M_{sx} = 156 ft.-kips, and
M_{sy} = 74 ft.-kips at the top of the column.

Bottom of the column:
M_{sx} = 78 ft.-kips and
M_{sy} = 74 ft.-kips.

The moments act in the clockwise direction at the top of the column and in the counterclockwise direction at the bottom of the column, resulting in single curvature bending in the column for the given load case. Assume ASTM A992 Grade 50 steel. Use the AISC equations to determine the axial compression capacity of the column and use the actual C_m factor for the column corresponding to the given load case.

Solution (ASD)

The section properties of W14 × 109 from *AISCM* Table 1-1 are as follows:

$$A = 32 \text{ in.}^2,\ r_x = 6.22 \text{ in.},\ r_y = 3.73 \text{ in.},\ Z_x = 192 \text{ in.}^3,\ Z_y = 92.7 \text{ in.}^3,\ S_y = 61.2 \text{ in.}^3$$

From *AISCM* Table 3-2, the flexural properties of W14 × 109 for strong axis bending are

$$\frac{M_{px}}{\Omega_b} = 479 \text{ ft.-kips},\quad \frac{BF}{\Omega_b} = 5.01,\ L_p = 13.2 \text{ ft.},\ L_r = 48.4 \text{ ft.}$$

Nominal Axial Compression Load Capacity, P_n:

$$\frac{K_x L_x}{r_x} = \frac{(1.0)(20 \text{ ft.})(12)}{6.22 \text{ in.}} = 38.6 < 200 \quad \text{O.K.}$$

$$\frac{K_y L_y}{r_y} = \frac{(1.0)(20 \text{ ft.})(12)}{3.73 \text{ in.}} = 64.3 < 200 \quad \text{O.K.}$$

The larger slenderness ratio controls, therefore, use $\frac{KL}{r} = 64.3$

To determine if the column will buckle in elastic or inelastic buckling, calculate the slenderness ratio limit at the transition point between inelastic and elastic buckling as follows:

$$4.71\sqrt{\frac{E}{F_y}} = 4.71\sqrt{\frac{29{,}000 \text{ ksi}}{50 \text{ ksi}}} = 113.4 > \frac{KL}{r} = 64.3$$

Therefore, this column fails in inelastic buckling, and equation (5-14) applies. The Euler buckling stress is

$$F_e = \frac{\pi^2 E}{\left(KL/r\right)^2} = \frac{\pi^2 (29{,}000 \text{ ksi})}{(64.3)^2} = 69.2 \text{ ksi}$$

The nominal stress for inelastic buckling from equation (5-14) is calculated as

$$F_n = \left[(0.658)^{F_y/F_e}\right]F_y = \left[(0.658)^{50/69.2}\right](50\,\text{ksi}) = 36.95\text{ ksi}$$

The nominal axial compression load capacity is

$$P_n = F_n A = (36.95\text{ ksi})(32\text{ in.}^2) = 1182\text{ kips}$$

$$\frac{P_n}{\Omega_c} = \frac{1182}{1.67} = 708\,\text{kips}$$

The value of the allowable axial compression load, $\frac{P_n}{\Omega_c} = 708\,\text{kips}$ compares with the allowable axial compression load capacity of 708 kips at KL = 20 ft. from *AISCM* Table 4-1a or from *AISCM* Table 6-1.

Allowable Moment Capacity about the *x-x* or strong axis:

The unbraced length for bending about the x-x axis, L_b = 20 ft.
Since $L_p < L_b < L_r$, inelastic lateral-torsional buckling must be considered (see Chapter 6).

$$\frac{M_{nx}}{\Omega_b} = C_b\left[\frac{M_{px}}{\Omega_b} - \frac{BF}{\Omega_b}(L_b - L_p)\right] \le \frac{M_{px}}{\Omega_b}$$

$$= (1.0)\left[479 - (5.01)(20 - 13.2)\right] \le 479\,\text{ft.-kips}$$

$$= 445\,\text{ft.-kips} \le 479\text{ ft.-kips}$$

Therefore, the allowable moment, $\frac{M_{nx}}{\Omega_b} = \mathbf{445\,ft.\text{-}kips}$

It is conservative to use a moment gradient factor, C_b = 1.0 as assumed above. The C_b factor accounts for the non-uniform moment along the length of the member. The actual value of C_b can be calculated using the moment gradient factor equation from Chapter 6 as follows:

$$C_b = \frac{12.5M_{max}}{2.5M_{max} + 3M_A + 4M_B + 3M_C}$$

M_{max} = absolute maximum moment in the member within the unbraced length
M_A = absolute value of the moment in the member at the quarter point of the unbraced length
M_B = absolute value of the moment in the member at the mid-point of the unbraced length
M_C = absolute value of the moment in the member at the three-quarter point of the unbraced length

With the given moments of the x-x or strong axis, we find that

M_{max} = 156 ft.-kips
M_A = ½(78 + 117) = 97.5 ft.-kips
M_B = ½(78 + 156) = 117 ft.-kips
M_C = ½(117 + 156) = 136.5 ft.-kips

Therefore, the moment gradient factor or lateral-torsional buckling factor is

$$C_b = \frac{12.5(156)}{2.5(156)+3(97.5)+4(117)+3(136.5)} = 1.25$$

Therefore, recalculating the bending moment capacity about the x-x or strong axis, we obtain

$$\frac{M_{nx}}{\Omega_b} = C_b \left[\frac{M_{px}}{\Omega_b} - \frac{BF}{\Omega_b}(L_b - L_p) \right] \le \frac{M_{px}}{\Omega_b}$$

$$= (1.25)\left[479 - (5.01)(20-13.2)\right] \le 479 \text{ ft.-kips}$$

$$= 556 \text{ ft.-kips} \le \mathbf{479 \text{ ft.-kips}}$$

Therefore, the allowable moment about the x - x axis, $\frac{M_{nx}}{\Omega_b} = \mathbf{479 \text{ ft.-kips}}$

Allowable Moment Capacity about the *y-y* or weak axis:

$$M_{ny} = Z_y F_y \le 1.6 S_y F_y$$

$$= (92.7 \text{ in.}^3)(50 \text{ ksi}) \le (1.6)(61.2 \text{ in.}^3)(50 \text{ ksi})$$

$$= \mathbf{4635 \text{ in.-kips}} \le 4896 \text{ in.-kips}$$

Therefore, the nominal moment capacity about the y - y axis, $M_{ny} = 387$ ft.-kips

The allowable moment about the y - y axis, $\frac{M_{ny}}{\Omega_b} = \frac{387}{1.67} = \mathbf{232 \text{ ft.-kips}}$

Moment Amplification or *P-δ* effect:

Determine the nonsway moment amplification factor, B_{1x}, for strong (x-x) axis buckling: The unbraced length for buckling about the x-x axis, $L_x = 20$ ft.

$$K_x = 1.0$$

$$\frac{K_x L_x}{r_x} = \frac{(1.0)(20 \text{ ft.})(12)}{6.22 \text{ in.}} = 38.6 < 200 \quad \text{O.K.}$$

$$P_{e1x} = \frac{\pi^2 EA}{\left(K_x L_x / r_x\right)^2} = \frac{\pi^2 (29{,}000 \text{ ksi})(32 \text{ in.}^2)}{(38.6)^2} = 6147 \text{ kips}$$

An approximate C_m value is typically used in this text to account for the effect of the non-uniform moments in columns in this text, but since the top and bottom moments are known for this problem, the actual C_m value for the given load case will be computed as follows:

$M_{1x} = 78$ ft.-kips (smaller end moment about x - x or strong axis)

$M_{2x} = 156$ ft.-kips (larger end moment about x - x or strong axis)

$$C_{mx} = 0.6 - 0.4\frac{M_1}{M_2}$$

$$\frac{M_1}{M_2} = \textit{negative for single curvature bending}$$

$$\frac{M_1}{M_2} = \textit{positive for double curvature bending}$$

Since the moments at the top and bottom of the column result in single curvature bending, therefore, M_1/M_2 = negative, and,

$$C_{mx} = 0.6 - 0.4\left[-\frac{(78\,\text{ft.-kips})}{(156\,\text{ft.-kips})}\right] = 0.80$$

The nonsway moment amplification factor for buckling about the x - x axis is

$$B_{1x} = \frac{C_{mx}}{1 - \frac{\alpha P_u}{P_{e1x}}} \geq 1.0\,,$$

$$= \frac{0.80}{1 - \frac{(1.6)(385\,\text{kips})}{6147\,\text{kips}}} = 0.89 \;\; \geq 1.0$$

and $\alpha = 1.6$ (*for ASD*)

Therefore, use $B_{1x} = 1.0$

The magnified unfactored moment about the strong (x-x) axis is calculated using the larger end moment. Therefore,

$$M_{sx} = B_{1x} M_{sx,nlt} = (1.0)(156\,\text{ft.-kips}) = 156\ \text{ft.-kips}$$

Determine the nonsway moment amplification factor, B_{1y}, for *y-y* or weak axis buckling: The unbraced length for buckling about the *y-y* axis, $L_y = 20$ ft

$K_y = 1.0$

$$\frac{K_y L_y}{r_y} = \frac{(1.0)(20\,\text{ft.})(12)}{3.73\,\text{in.}} = 64.3 < 200 \quad \text{O.K.}$$

$$P_{e1y} = \frac{\pi^2 EA}{\left(K_y L_y / r_y\right)^2} = \frac{\pi^2 (29{,}000\,\text{ksi})(32\ \text{in.}^2)}{(64.3)^2} = 2215\,\text{kips}$$

An approximate C_m value is typically used in this text to account for the effect of the non-uniform moments in columns in this text, but since the top and bottom moments are known for this problem, the actual C_{my} value for the given load case will be computed as follows:

$M_{1y} = 74$ ft.-kips (unfactored smaller end moment about y - y or weak axis)

$M_{2y} = 74$ ft.-kips (unfactored larger end moment about y - y or weak axis)

$$C_{my} = 0.6 - 0.4\frac{M_{1y}}{M_{2y}}$$

$$\frac{M_{1y}}{M_{2y}} = \textit{negative for single curvature bending}$$

$$\frac{M_{1y}}{M_{2y}} = \textit{positive for double curvature bending}$$

Since the moments at the top and bottom of the column result in single curvature bending, therefore, M_1/M_2 = negative, and,

$$C_{my} = 0.6 - 0.4\left[-\frac{(74\,\text{ft.-kips})}{(74\,\text{ft.-kips})}\right] = 0.6 + (0.4)(1.0) = 1.0$$

The nonsway moment amplification factor for buckling about the y - y axis is

$$B_{1y} = \frac{C_{my}}{1 - \dfrac{\alpha P_s}{P_{e1y}}} \geq 1.0$$

$$= \frac{1.0}{1 - \dfrac{(1.6)(385\,\text{kips})}{2215\,\text{kips}}} = 1.39 \geq 1.0$$

$\alpha = 1.6$ (*for ASD*)

Therefore, use $B_{1y} = 1.39$

The magnified unfactored moment about the *y-y* or weak axis is calculated using the larger end moment. Therefore,

$$M_{sy} = B_{1y} M_{sy,nlt} = (1.39)(74\,\text{ft.-kips}) = 103\ \text{ft.-kips}$$

Column Interaction Equation:

The factor of safety for compression, $\Omega_c = 1.67$
The factor of safety for bending or flexure, $\Omega_b = 1.67$

$$\frac{P_s}{P_n/\Omega_c} = \frac{385\,\text{kips}}{(1182\,\text{kips}/1.67)} = 0.54 > 0.2$$

Therefore, use the Column Interaction equation (8-1a):

$$\frac{P_s}{P_n/\Omega_c} + \frac{8}{9}\left(\frac{M_{sx}}{M_{nx}/\Omega_b} + \frac{M_{sy}}{M_{ny}/\Omega_b}\right) \leq 1.0$$

$$\text{i.e. } 0.54 + \frac{8}{9}\left(\frac{156}{(479)} + \frac{103}{(232)}\right) = \mathbf{1.22} > 1.0 \ \ (\textit{Not Good})$$

Therefore, the W14 × 109 column is overstressed and inadequate. A larger column size needs to be selected, and as an exercise, the reader should verify that a W14 × 132 column would be adequate.

Note that *AISCM* Table 6-1 (Available Strength for Members Subject to Axial, Shear, Flexural, or Combined Forces) [3] could also be used to directly obtain the nominal strengths - P_n/Ω_c, P_n/Ω_t, M_{ny}/Ω_b, and M_{nx}/Ω_b - for wide flange sections for any unbraced length that can be used directly in the beam-column interaction equations. The moment amplification factors, B_{1x}

and B_{1y}, would still need to be calculated and applied, and note that *AISCM* Table 6-1 assumes a C_b value of 1.0. The values from the table would have to be modified to account for the actual values of C_b, $B_{1x,}$ and B_{1y}.

8-10 UNBALANCED MOMENTS, M_{nt}, FOR COLUMNS IN BRACED FRAMES DUE TO THE ECCENTRICITY OF THE GIRDER AND BEAM REACTIONS

Unbalanced moments from gravity loads occur in columns due to the unbalanced reactions from the beams and girders framing into the column from opposite sides (see Figure 8-14). In Examples 8-1 and 8-2, the unbalanced moments were given. The procedure for calculating these unbalanced moments in columns is discussed in this section. The differences in the girder and beam reactions framing onto opposite sides of a column may occur due to the following reasons:

- The differences in the span of the beams and girders framing into the column from opposite sides
- The differences in the dead and live loads on the beams and girders framing into opposite sides of a column
- The likelihood of pattern live loading on the beams and girders framing into opposite sides of a column. That is, for each orthogonal axis of a column section, the live loads on the girders or the beams framing onto one side of the column are skipped, absent, or removed, while the full live load is applied on the girders and beams framing onto the opposite side of the column.

Ordinarily, live load skipping should be considered in the girders and beams framing into each column at the top and bottom of the column; however, this would result in too many potential load cases or too many possible combinations of moments and axial load on a column that would needlessly complicate the column design.

To minimize the number of load cases, the authors have chosen in this text to consider two load cases that will yield conservative results:

- **Load Case 1** - which *maximizes the factored* cumulative *axial load* on the column, and
- **Load Case 2** - which *maximizes the unbalanced moments* at each level about both orthogonal axes of the column.

This is accomplished by applying and removing the entire live load from the beams and girders framing into one or both sides of the column to achieve the maximum axial load or the maximum unbalanced moment. It should be remembered that the dead load always remains on the beams and girders at all times and cannot be skipped; the pattern loading or load skipping only applies to live loads. It is also assumed in this simplified column analysis that the unbalanced moment at one end of a column is not affected by the unbalanced moment at the other end (i.e., there is no moment distribution or carryover of a moment from one end of a column to its other end).

In calculating the axial loads and unbalanced moments for columns in braced frames as shown in Figure 8-14, only the LRFD option is illustrated here for brevity, but the principle also applies to the ASD method. With reference to Figure 8-14, the following points should be noted:

- All loads are **factored**, and the LRFD load combinations from Chapter 2 should be used.

- The loading conditions are illustrated for a three-story building, but the principle is applicable to any building with any number of stories or levels.
- In summing up the column axial loads, the maximum factored dead plus live loads from the floors/roof above must be included in the cumulative axial loads using the appropriate load combinations (see Section 2-3) as indicated in Table 8-2.

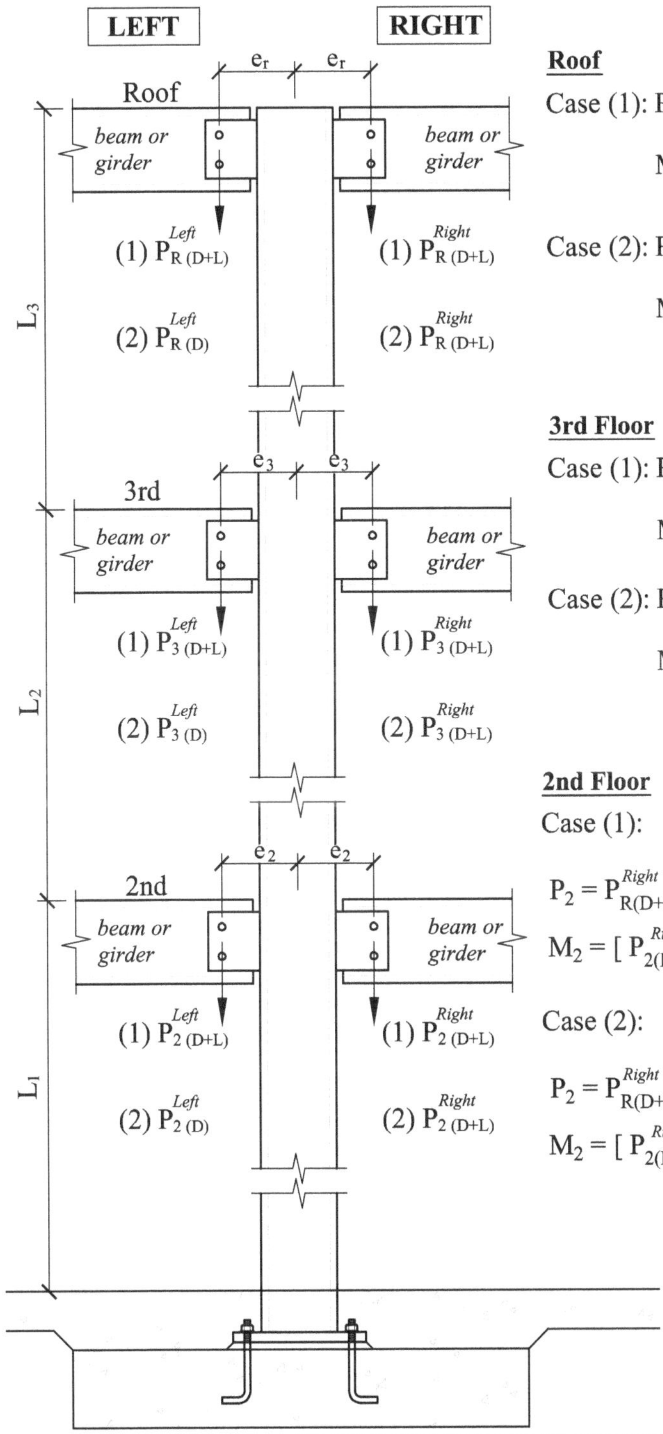

Roof

Case (1): $P_R = P^{Right}_{R(D+L)} + P^{Left}_{R(D+L)}$

$$M_R = [\,P^{Right}_{R(D+L)} - P^{Left}_{R(D+L)}\,]e_r$$

Case (2): $P_R = P^{Right}_{R(D+L)} + P^{Left}_{R(D)}$

$$M_R = [\,P^{Right}_{R(D+L)} - P^{Left}_{R(D)}\,]e_r$$

3rd Floor

Case (1): $P_3 = P^{Right}_{R(D+L)} + P^{Left}_{R(D+L)} + P^{Right}_{3(D+L)} + P^{Left}_{3(D+L)}$

$$M_3 = [\,P^{Right}_{3(D+L)} - P^{Left}_{3(D+L)}\,]e_3$$

Case (2): $P_3 = P^{Right}_{R(D+L)} + P^{Left}_{R(D+L)} + P^{Right}_{3(D+L)} + P^{Left}_{3(D)}$

$$M_3 = [\,P^{Right}_{3(D+L)} - P^{Left}_{3(D)}\,]e_3$$

2nd Floor

Case (1):

$$P_2 = P^{Right}_{R(D+L)} + P^{Left}_{R(D+L)} + P^{Right}_{3(D+L)} + P^{Left}_{3(D+L)} + P^{Right}_{2(D+L)} + P^{Left}_{2(D+L)}$$

$$M_2 = [\,P^{Right}_{2(D+L)} - P^{Left}_{2(D+L)}\,]e_2$$

Case (2):

$$P_2 = P^{Right}_{R(D+L)} + P^{Left}_{R(D+L)} + P^{Right}_{3(D+L)} + P^{Left}_{3(D+L)} + P^{Right}_{2(D+L)} + P^{Left}_{2(D)}$$

$$M_2 = [\,P^{Right}_{2(D+L)} - P^{Left}_{2(D)}\,]e_2$$

Load Case (1) maximizes axial load on the column
Load Case (2) maximizes unbalanced moment in the column

FIGURE 8-14 Axial loads and un-balanced moments in columns

- The moments in Figure 8-14 are the $M_{u,nt}$ moments at **each floor level**; it should be reiterated that unlike the axial loads, which are cumulative (i.e., they accumulate on the column) from the roof to the lowest floor level, the moments are ***not*** cumulative but apply only at the level at which they occur, and it is assumed that there is no moment transfer or carry over moment in the columns between levels (i.e., the unbalanced moment at each level of the building is independent of the unbalanced moment at any other levels). The $M_{u,nt}$ moments at any floor level are caused by the **unbalanced** reactions from the beams and girders framing into the column at that level; these moments are the *first-order moments*, and they will have to be amplified to account for the P-δ effects. The reader should recall that there are no sway *P*-Δ effects for columns in braced frames (i.e., buildings that are laterally braced with concentric or eccentric braced frames or shear walls).
- The factored axial loads at **each** floor level are defined as follows, and note that these are ***not*** the cumulative axial loads on the column, but they are the loads contributed to the column by each individual floor/roof level:

Roof Level:

$$P_{roof\ D+L} = 1.2P_{roof\ D} + \left(1.6P_{L_r}\ or\ 1.0P_S\ or\ 1.6P_R\right)$$
$$P_{roof\ D} = 1.2P_{roof\ D}$$

Third-Floor Level:

$$P_{3D+L} = 1.2P_{3D} + 1.6P_{3L}$$
$$P_{3D} = 1.2P_{3DL}$$

Second-Floor Level:

$$P_{2D+L} = 1.2P_{2D} + 1.6P_{2L}$$
$$P_{2D} = 1.2P_{2D},$$

Where,

D = Service or unfactored dead load,
L = Service or unfactored live load,
L_r = Unfactored roof live load,
S = Unfactored snow load,
R = Unfactored rain load,
$P_{roof(D+L)}$ = Factored axial load on the column contributed by the roof level due to dead load plus the controlling roof live load (S, L_r, or R)
$P_{roof(D)}$ = Factored axial load on the column contributed by the roof level due to dead load only
$P_{3(D+L)}$ = Factored axial load on the column contributed by the 3rd floor level due to dead load plus floor live load
$P_{3(D)}$ = Factored axial load on the column contributed by the 3rd floor level due to dead load only
$P_{2(D+L)}$ = Factored axial load on the column contributed by the 2nd floor level due to dead load plus floor live load
$P_{2(D)}$ = Factored axial load on the column contributed by the 2nd floor level due to dead load only

TABLE 8-2 Applicable Load Combinations for Cumulative Axial Load in Columns

Level	Cumulative axial load in the column	Applicable load combination for the cumulative axial load in the column
Roof level	P_{roof}	$1.2D + (1.6L_r$ or $1.0S$ or $1.6R)$
nth floor	P_n	$1.2D + 1.6L + (0.5L_r$ or $0.3S$ or $0.5R)$
Third floor	P_3	$1.2D + 1.6L + (0.5L_r$ or $0.3S$ or $0.5R)$
Second or *Lowest floor	P_2 or $*P_{lowest\ floor}$	$1.2D + 1.6L + (0.5L_r$ or $0.3S$ or $0.5R)$

*For buildings with basement levels.

Moment Distribution Between Columns Segments Above and Below Each Level

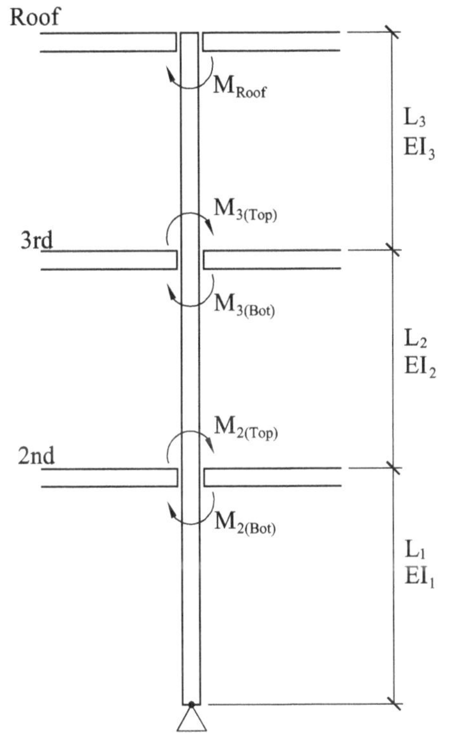

The moments at each level which are distributed between the column segments above and below each level are caused by the eccentricities of the beam and girder reactions relative to the centroidal axis of the column

FIGURE 8-15 Moment distribution in columns

The distribution of the no-translation moments (i.e., moment split) between the column segments above and below a given floor level (see Figure 8-15) is a function of the following factors:

- The continuity of the column and the type of column splice (if any),
- The floor-to-floor heights, and
- The moments of inertia of the columns above and below the floor level under consideration.

The column just below the roof level has to resist 100% of the unbalanced moment at the roof level since there is no column segment above that level with which to split or share the unbalanced moment, and we have assumed that the typical girder-to-column or beam-to-column shear connections are incapable of resisting moment. The moment distribution or moment split to the column segments above and below each level are obtained as follows:

Roof:

$M = M_{roof}$ (i.e., the column below the roof level resists all the unbalanced moments at the girder/beam-to-column connections at the roof level.)

Third Floor:

The total unbalanced moment at the third-floor girder/beam-to-column connections is resisted by the column segments above and below the third-floor level, and the moment is split between these column segments in proportion to the ratio of the column stiffnesses as follows:

$$\frac{M_{top}}{M_{bot}} = \frac{\frac{EI_{top}}{L_{top}}}{\frac{EI_{bot}}{L_{bot}}} = \frac{\frac{EI_3}{L_3}}{\frac{EI_2}{L_2}}$$

To simplify the analysis, it can be assumed that the column segments above and below the 3rd floor level have the same moment of inertia; that is, I_{bot} (or I_2) = I_{top} (or I_3). Also, note that the column splices provide moment restraints, so the continuity of the columns at the splices is assumed. For many practical situations, there is not much loss in accuracy with the preceding assumptions. In fact, these assumptions are widely used in practice in the design of steel buildings. Therefore, the moment split between the upper and lower column segments at any floor level is inversely proportional to the length of the column segments above and below that floor level.

$$i.e. \quad \frac{M_{top}}{M_{bot}} = \frac{1/L_3}{1/L_2}$$

$$M_3 = M_{3top} + M_{3bot}$$

$$M_{3bot} = \left(\frac{1/L_2}{1/L_2 + 1/L_3}\right) M_3$$

$$M_{3top} = M_3 - M_{3bot}$$

Second Floor:

$$M_{2bot} = \left(\frac{1/L_1}{1/L_1 + 1/L_2}\right) M_2$$

$$M_{2top} = M_2 - M_{2bot},$$

Where,

M_{3bot} = Column moment just *below* the third-floor level,
M_{3top} = Column moment just *above* the third-floor level,
M_{2bot} = Column moment just *below* the second-floor level, and
M_{2top} = Column moment just *above* the second-floor level.

EXAMPLE 8-3

Design of Beam–Columns in a *Braced* Frame (LRFD)

The typical roof and floor plans for a three-story braced-frame building are shown in Figures 8-16 and 8-17, respectively. The beams and girders are connected to the columns with simple shear connections (e.g., double angles or shear plates). Using the LRFD method and assuming a floor-to-floor height of 10 ft., design columns C1 and C2 for the gravity loads shown.

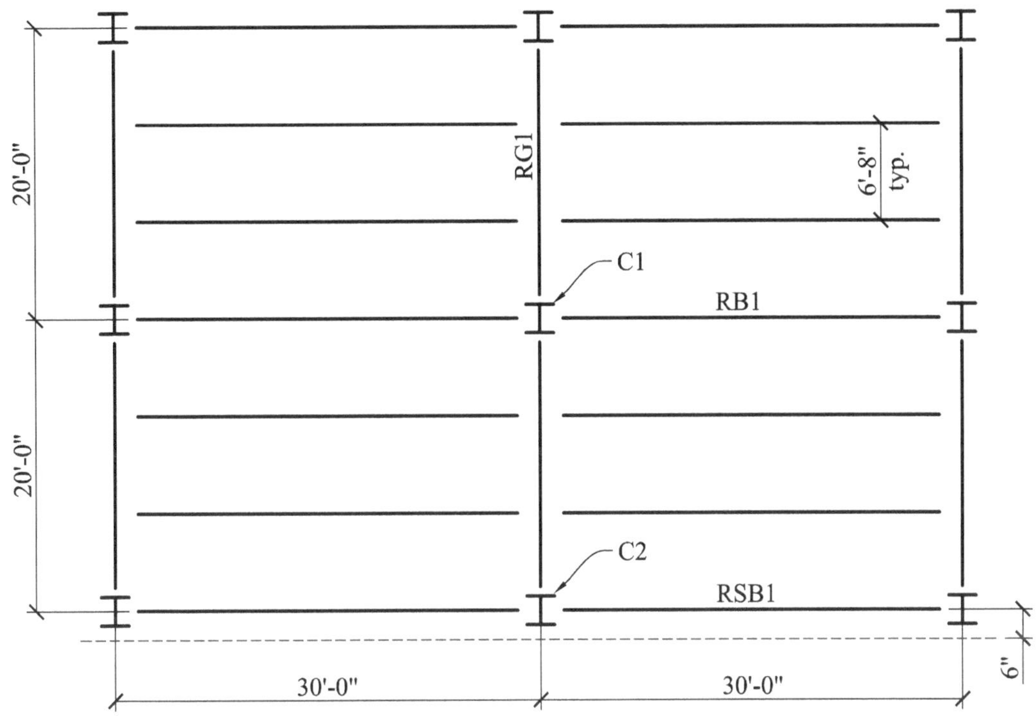

Roof Loads

Dead Load = 30 psf
Live Load = 56 psf (snow)

Factored Roof Load:

$$w_u = (1.2)(30) + (1.0)(56) = 0.092 \text{ ksf}$$

Service Roof Load:

$$w_{s(D+S)} = 30 + 56 = 0.086 \text{ ksf (Dead + Snow)}$$
$$w_{s(S)} = 56 = 0.056 \text{ ksf (Snow)}$$

Factored Roof Dead Load:

$$w_{u(D)} = (1.2)(30) = 0.036 \text{ ksf}$$

FIGURE 8-16 Roof plan

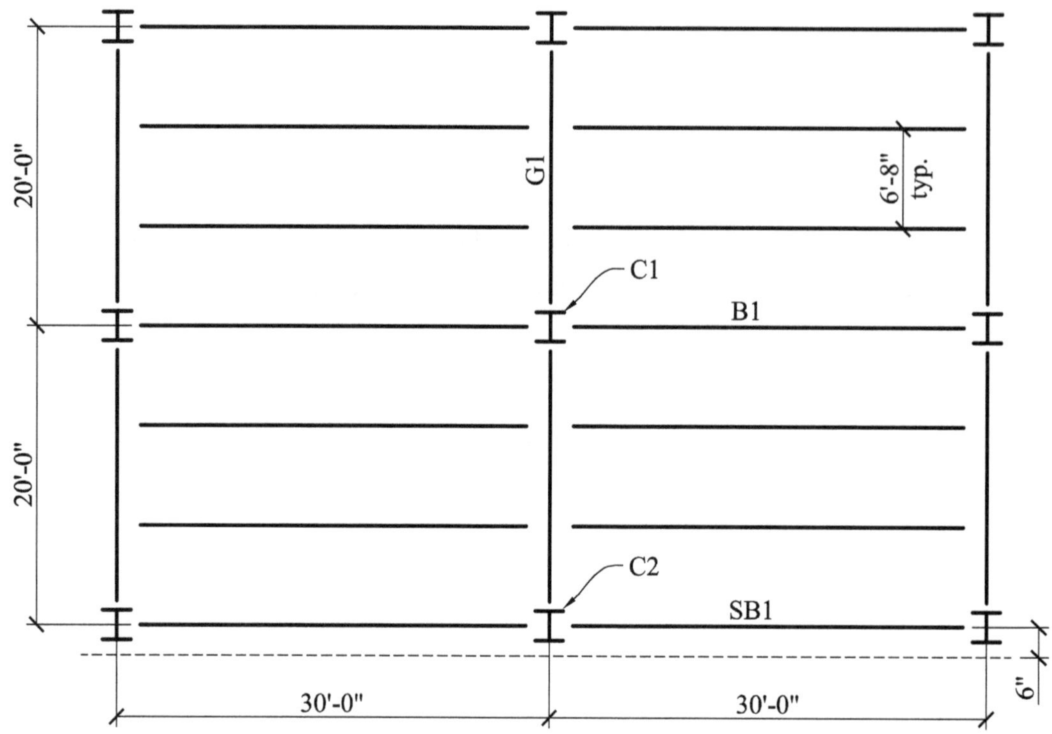

Floor Loads

Dead Load = 85psf
Live Load = 150psf

Factored Floor Load:

$$w_u = (1.2)(85) + (1.6)(150) = 0.342\text{ksf}$$

Service Floor Load:

$$w_{s(D-L)} = 85 + 150 = 0.235\text{ksf (Dead + Live)}$$
$$w_{s(L)} = 150 = 0.150\text{ ksf (Live)}$$

Factored Floor Dead Load:

$$w_{u(D)} = (1.2)(85) = 0.10\text{ ksf}$$

FIGURE 8-17 Typical floor plan

Solution

The first step in the solution process is to determine the governing moments and the factored cumulative axial loads at each level of the column for the two load cases considered (i.e., Load cases 1 and 2) in Figure 8-14.

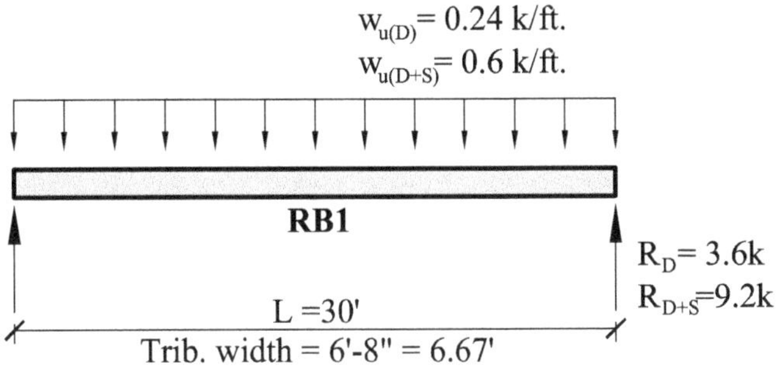

$$w_{u(D)} = (0.036\text{ksf})(6.67\text{ft}) = 0.24\text{k/ft.}$$
$$w_{u(D+S)} = (0.092\text{ksf})(6.67\text{ft}) = 0.6\text{k/ft.}$$

$$R_D = \frac{w_{u(D)}L}{2} = \frac{(0.24)(30)}{2} = 3.6\text{k}$$

$$R_{D+S} = \frac{w_{u(D+S)}L}{2} = \frac{(0.6)(30)}{2} = 9.2\text{k}$$

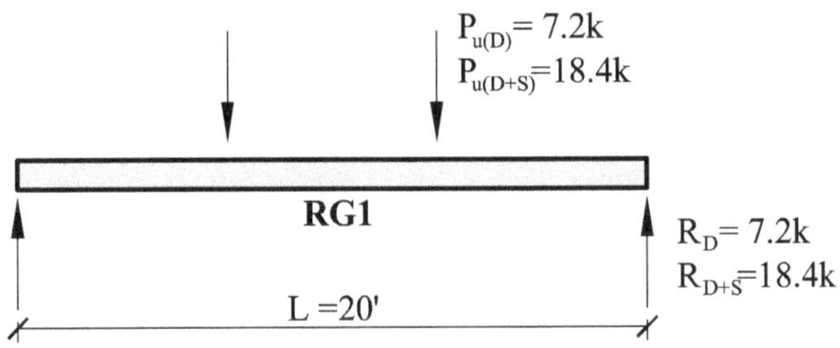

$$P_{u(D)} = (3.6\text{k})(2 = 7.2\text{k}$$
$$P_{u(D+S)} = (9.2\text{k})(2) = 18.4\text{k}$$

FIGURE 8-18 Calculation of roof beam and girder reactions

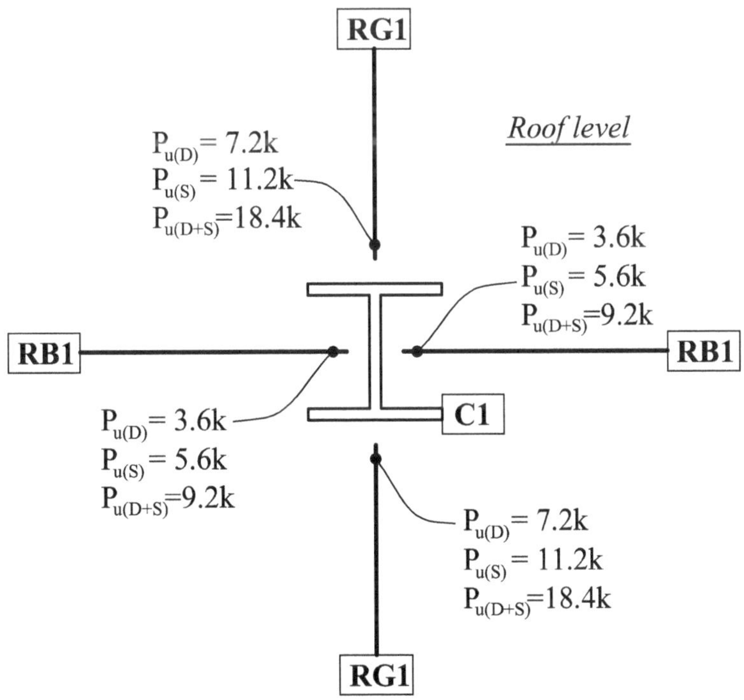

FIGURE 8-19 Reactions of roof beams and girders framing into column C1 at the Roof level

Load Calculations for Column C1

All Loads are Factored

ROOF: *Loads to Column C1 at the **Roof** level* (see Figures 8-16, 8-18, and 8-19.)

Girder and beam connection eccentricities relative to the centroid of the column:

e_r **girders** $= 2^1/_2$ in. + $^1/_2$ Column depth
$= 2^1/_2$ in. + $^1/_2$ (8 in.) Assuming at least a W8 column with 8 in. nominal depth
$= 6^1/_2$ in. = **0.54 ft.**

e_r **beams** $= 2^1/_2$ in. + $^1/_2$ Column web thickness
$= 2^1/_2$ in. + $^1/_2$ (0.5 in.) assumed
= 2.75 in. ≈ **0.25 ft.**

A W8 section is the minimum wide-flange column size typically used in design practice. Smaller column sizes are not frequently used because the flange width of these columns does not provide enough room to accommodate the double-angle connections used to connect the girders to the columns. Where the girders are connected to the columns using shear tabs or shear plates, a column size smaller than W8 × 31 may be used, provided it is adequate for resisting the applied loads and moments.

Factored Loads at the Roof Level (see Figure 8-18):

The loads on a typical interior roof beam are calculated as follows:

$$w_u = 0.092 \text{ ksf}(6.67 \text{ ft.}) = 0.613 \text{ kips/ft.}$$

$$w_{u(S)} = (1.0)(0.056 \text{ ksf})(6.67 \text{ ft.}) = 0.374 \text{ kips/ft.}$$

$$w_{u(D)} = 0.036 \text{ ksf}(6.67 \text{ ft.}) = 0.24 \text{ kips/ft.}$$

The typical interior beam reactions are:

$$P_{D,beam} = (0.24 \text{ kips / ft.})\left(\frac{30 \text{ ft.}}{2}\right) = 3.6 \text{ kips}$$

$$P_{S,beam} = (0.374 \text{ kips / ft.})\left(\frac{30 \text{ ft.}}{2}\right) = 5.6 \text{ kips}$$

The concentrated loads on the typical interior roof girder from the roof beam reactions at each load point are calculated as follows:

$$P_D = 3.6 \text{ kips} + 3.6 \text{ kips} = 7.2 \text{ kips}$$

$$P_S = 5.6 \text{ kips} + 5.6 \text{ kips} = 11.2 \text{ kips}$$

$$P_{D+S} = 7.2 \text{ kips} + 11.2 \text{ kips} = 18.4 \text{ kips}$$

Load Case 1: At the roof level, the axial load is maximized by applying all the dead and live loads on the roof beams and roof girders framing into column C1 in both orthogonal directions (see Figures 8-18 and 8-19).

$$P_{roof} = \underset{(P^R_{D+S}beams)}{9.2} + \underset{(P^L_{D+S}beams)}{9.2} + \underset{(P^R_{D+S}girders)}{18.4} + \underset{(P^L_{D+S}girders)}{18.4} = 55 \text{kips}$$

$$M_{rx-x} = (18.4 - 18.4)\underset{\uparrow (e_r \text{ girder})}{(0.54 \text{ ft.})} = 0 \text{ ft.-kips}$$

$$M_{ry-y} = (9.2 - 9.2)\underset{\uparrow (e_r \text{ beams})}{(0.25 \text{ ft.})} = 0 \text{ ft.-kips}$$

$$P = 55 \text{ kips}$$

$$M_{rx-x} = 0 \text{ ft.-kips}$$

$$M_{ry\text{-}y} = 0 \text{ ft.-kips}$$

For this case, the column is designed for axial load only since the unbalanced moments due to beam and girder reaction eccentricities are zero.

Load Case 2: At the roof level, the unbalanced moments are maximized by skipping the live loads on one roof beam and one roof girder framing into column C1 and keeping the full dead and live load on the other roof beam and roof girder framing into the opposite sides of column C1 (see Figures 8-18 and 8-19).

$$P_{roof} = \underset{(P^R_{D+S}beams)}{9.2} + \underset{(P^L_{D}beams)}{3.6} + \underset{(P^R_{D+S}girders)}{18.4} + \underset{(P^L_{D}girders)}{7.2} = 38 \text{ kips}$$

$$M_{rx-x} = (18.4 - 7.2)(0.54 \text{ ft.}) = 6 \text{ ft.-kips}$$
$$M_{ry-y} = (9.2 - 3.6)(0.25 \text{ ft.}) = 1.4 \text{ ft.-kips}$$

Note: The superscripts "*R*" and "*L*" refer to the right-hand side and left-hand side, respectively.

$$P = 38 \text{ kips}$$
$$M_{rx-x} = (100\%)(6 \text{ ft.-kips}) = 6 \text{ ft.-kips}$$
$$M_{ry-y} = (100\%)(1.4 \text{ ft.-kips}) = 1.4 \text{ ft.-kips}$$

For this load case, the column is designed for combined axial load plus bending.

The reader can observe that the column moments at the roof level are resisted solely by the column below the roof level since there is no column segment above the roof level. For the roof loads that will be cumulatively added to the floor loads below, the appropriate load factor must be applied to the live loads. Recall that the load combination $1.2D + (1.6L_r$ or $1.0S$ or $1.6R)$ applies to the column load only at the roof level. For cumulative loads at the lower levels, the applicable load combination that maximizes the live loads at the floor levels is $1.2D + 1.6L + (0.5L_r$ or $0.3S$ or $0.5R)$ or $1.2D + 1.6L + 0.3S$, since snow load is more critical than rain or roof live load for this example problem. Therefore, the contribution from the roof level to the cumulative column axial loads at the lower levels (i.e., third- and second-floor levels) is calculated as follows:

Using the $1.2D + 1.0S$ *roof beam and girder reaction values from Figure 8-19, we calculate*

$$\begin{aligned} P_{roof} &= 1.2D + 0.3S \\ &= \left[3.6\,\text{kips} + \frac{0.3}{1.0}(5.6\,\text{kips})\right] + \left[3.6\,\text{kips} + \frac{0.3}{1.0}(5.6\,\text{kips})\right] \\ &\quad + \left[7.2\,\text{kips} + \frac{0.3}{1.0}(11.2\,\text{kips})\right] + \left[7.2\,\text{kips} + \frac{0.3}{1.0}(11.2\,\text{kips})\right] \\ &= 32 \text{ kips.} \end{aligned}$$

Note that the "0.3" and "1.0" coefficients in the preceding equation are the load factors for snow load in load combinations 2 and 3 from Section 2-3 of the text.

The above value could also be obtained using the tributary area method as follows:

$$\begin{aligned} P_{roof} &= (1.2D + 0.3S)(\textit{Tributary area of Column C1}) \\ &= \frac{[(1.2)(30\,psf) + (0.3)(56\,psf)](20\,ft)(30\,ft)}{1000} \\ &= 32 \text{ kips.} \end{aligned}$$

THIRD FLOOR: *Loads to Column C1 at the **Third** Floor:* (See Figures 8-17, 8-20, and 8-21)

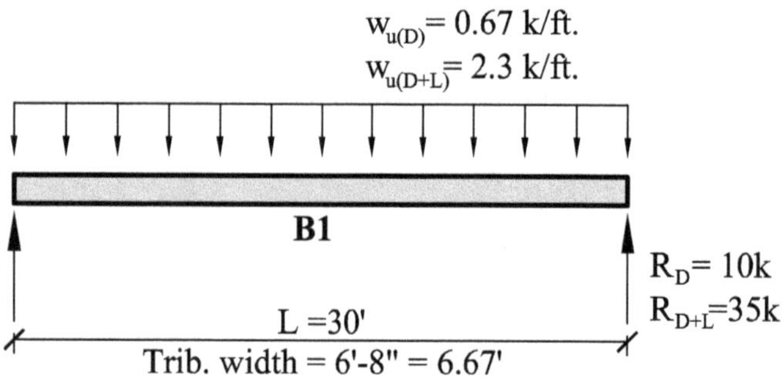

$$w_{u(D)} = (0.10\text{ksf})(6.67') = 0.67 \text{ k/ft.}$$

$$w_{u(D+L)} = (0.342\text{ksf})(6.67') = 2.3 \text{ k/ft.}$$

$$R_D = \frac{w_{u(D)}L}{2} = \frac{(0.67)(30)}{2} = 10\text{k}$$

$$R_{D+L} = \frac{w_{u(D+L)}L}{2} = \frac{(2.3)(30)}{2} = 35\text{k}$$

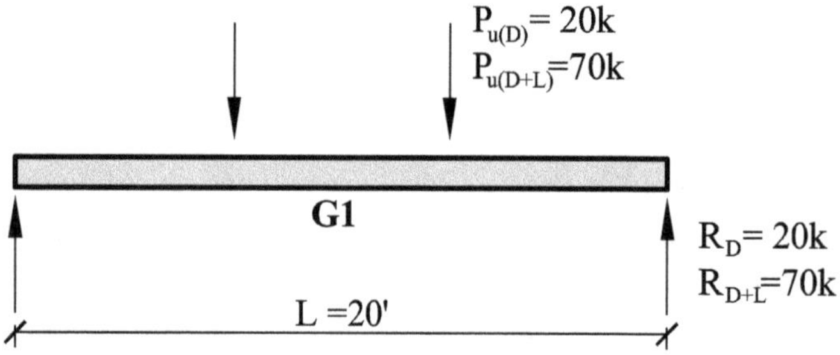

$$P_{u(D)} = (10\text{k})(2) = 20\text{k}$$

$$P_{u(D+L)} = (35\text{k})(2) = 70 \text{ k}$$

FIGURE 8-20 Calculation of floor beam and girder reactions at the 3rd floor level

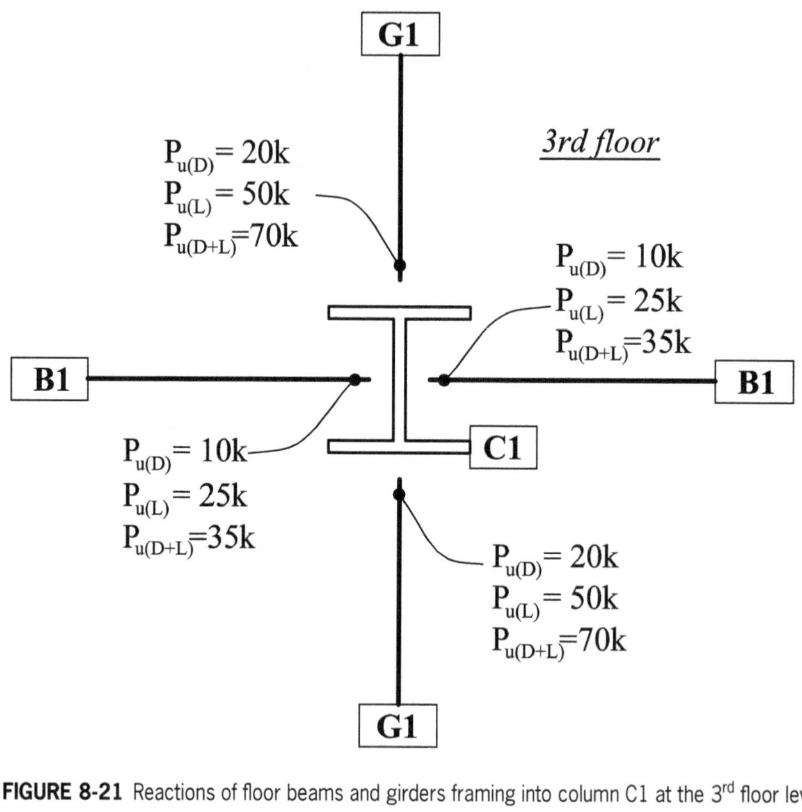

FIGURE 8-21 Reactions of floor beams and girders framing into column C1 at the 3rd floor level

Girder and beam connection eccentricities relative to the centroid of the column:

$\mathbf{e_3}$ **girders** $= 2\frac{1}{2}$ in. + $\frac{1}{2}$ Column depth
$= 2\frac{1}{2}$ in. + $\frac{1}{2}$ (8 in.) assuming at least a W8 column
$= 6\frac{1}{2}$ in. = 0.54 ft.

$\mathbf{e_3}$ **beams** $= 2\frac{1}{2}$ in. + $\frac{1}{2}$-in. Column web thickness
$= 2\frac{1}{2}$ in. + $\frac{1}{2}$ (0.5 in.) assumed
= 2.75 in. ≈ **0.25 ft.**

Factored Loads at the 3rd Floor Level (see Figure 8-20):

$$w_u = 0.342 \text{ ksf } (6.67 \text{ ft.}) = 2.3 \text{ kips/ft.}$$

$$w_{u(D)} = 0.10 \text{ ksf } (6.67 \text{ ft.}) = 0.67 \text{ kips/ft.}$$

$$P_D = 10 \text{ kips } (2 \text{ beams}) = 20 \text{ kips}$$

$$P_{D+L} = 35 \text{ kips } (2 \text{ beams}) = 70 \text{ kips}$$

Moment Split in the Column segments above and below the 3rd floor level:

$$M_{3bot} = \left(\frac{1/L_2}{1/L_2 + 1/L_3}\right) M_3; \qquad L_2 = 10 \text{ ft. and } L_3 = 10 \text{ ft.}$$

$$= \left(\frac{1/10}{1/10 \text{ ft.} + 1/10 \text{ ft.}}\right) M_3$$

$$= 0.5\ M_3$$

Load Case 1: The axial load is maximized by applying all dead and live loads on the beams and girders framing into column C1 in both orthogonal directions (see Figures 8-20 and 8-21).

$$P_3 = \underset{(\max P_{roof})}{32^*} + \left(\underset{(P^R_{D+L}beam)}{35} + \underset{(P^L_{D+L}beam)}{35}\right) + \left(\underset{(P^R_{D+L}girder)}{70} + \underset{(P^L_{D+L}girder)}{70}\right) = 242 \text{ kips}$$

$$M_{3x-x} = (70\text{kips} - 70\text{kips})\,(e_3\ \textbf{girder}) = (0 \text{ kips})(0.54 \text{ ft.}) = 0 \text{ ft.-kips}$$

$$M_{3y-y} = (35\text{kips} - 35\text{kips})\,(e_3\ \textbf{beam}) = (0 \text{ kips})(0.25 \text{ ft.}) = 0 \text{ ft.-kips}$$

$$P_3 = 242 \text{ kips}$$

$$M_{3botx-x} = 0.5(0 \text{ ft.-kips}) = 0 \text{ ft.-kips}$$

$$M_{3boty-y} = 0.5(0 \text{ ft.-kips}) = 0 \text{ ft.-kips}$$

$*1.2D + 0.3S = (1.2x30psf + 0.3x56psf) = 52.8psf$

$(52.8psf\ x\ 30ft\ x\ 20ft)/1000 = 32kips$

For this load case, design the column for only axial load since the unbalanced moments due to the beam and girder eccentricities are zero. Note that there would have been unbalanced moments in column C1 if the spans of the beams framing into opposite sides of the column web and/or the spans of the girders framing into the two flanges of the column were unequal.

Load Case 2: The moments at the 3rd floor level are maximized by skipping the live loads on one 3rd floor beam and one 3rd floor girder framing into column C1 and keeping the full dead and live loads on the other 3rd floor beam and girder framing into the opposite sides of column C1 (see Figures 8-20 and 8-21).

$$P_3 = \underset{(\max P_{roof})}{32^*} + \left(\underset{(P^R_{D+L}beam)}{35} + \underset{(P^L_{D+L}beam)}{10}\right) + \left(\underset{(P^R_{D+L}girder)}{70} + \underset{(P^L_{D}girder)}{20}\right) = 167 \text{ kips}$$

$$M_{3x-x} = (70\text{kips} - 20\text{kips})\,(e_3\ \textbf{girder}) = (50 \text{ kips})(0.54 \text{ ft.}) = 27 \text{ ft.-kips}$$

$$M_{3y-y} = (35\text{kips} - 10\text{kips})\,(e_3\ \textbf{beam}) = (25 \text{ kips})(0.25 \text{ ft.}) = 6.3 \text{ ft.-kips}$$

$*1.2D + 0.3S = (1.2 \times 30 \text{ psf} + 0.3 \times 56 \text{psf}) = 52.8 \text{psf}$

$(52.8 \text{ psf} \times 30 \text{ ft} \times 20 \text{ ft}) / 1000 = 32 \text{ kips}$

The unbalanced moment at the 3rd floor level (M_3) will be distributed between the column segments above and below the 3rd floor level in a ratio inversely proportional to the length of the column segments above and below the 3rd floor level. Therefore, the load and moments on the column just below the 3rd floor level for **load case 2** are calculated as follows:

$$P_3 = 167 \text{ kips}$$

$$M_{3botx-x} \cong 0.5(27 \text{ ft.-kips}) = 14 \text{ ft.-kips}$$

$$M_{3boty-y} \cong 0.5(6.3 \text{ ft.-kips}) = 3.2 \text{ ft.-kips}$$

The column will be designed for this combined axial load and bending moment.

The "0.5" factor in the preceding moment equations is the moment distribution or moment split factor between the column segments below and above the 3rd floor level.

SECOND FLOOR: *Loads to Column C1 at the **Second** Floor:* (See Figures 8-17, 8-22, and 8-23)

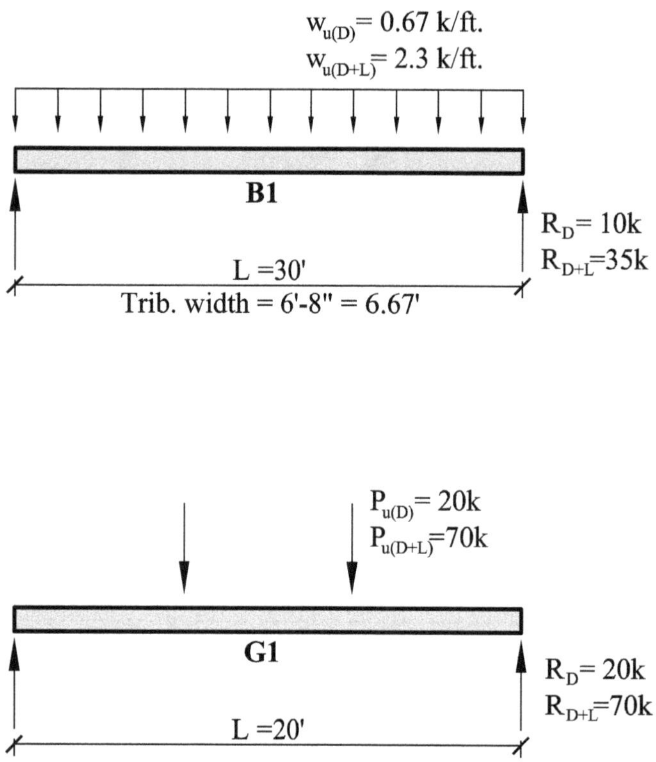

$$w_{u(D)} = (0.10\text{ksf})(6.67') = 0.67 \text{ k/ft.}$$

$$w_{u(D+L)} = (0.342\text{ksf})(6.67') = 2.3 \text{ k/ft.}$$

$$R_D = \frac{w_{u(D)}L}{2} = \frac{(0.67)(30)}{2} = 10\text{k}$$

$$R_{D+L} = \frac{w_{u(D+L)}L}{2} = \frac{(2.3)(30)}{2} = 35\text{k}$$

$$P_{u(D)} = (10\text{k})(2) = 20\text{k}$$

$$P_{u(D+L)} = (35\text{k})(2) = 70 \text{ k}$$

FIGURE 8-22 Calculation of floor beam and girder reactions at the 2nd floor level

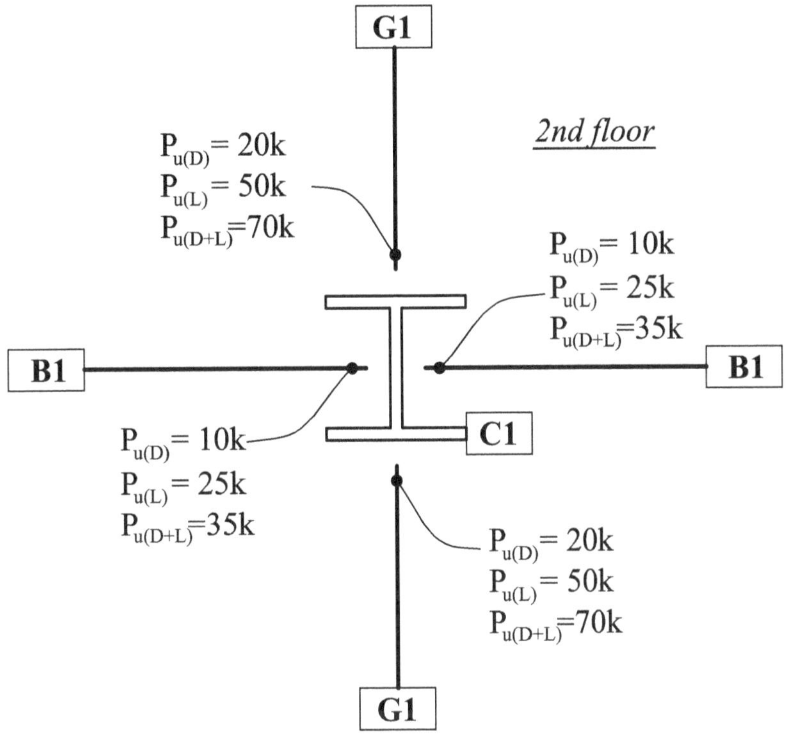

FIGURE 8-23 Reactions of floor beams and girders framing into column C1 at the 2nd floor level

Girder and beam connection eccentricities relative to the centroid of the column:

e_2 **girders** $= 2^1/_2$ in. + $^1/_2$ (Column depth)
$= 2^1/_2$ in. + $^1/_2$ (10 in.) assuming W10 column, with 10 in. nominal depth
$=$ 7.5 in. = **0.63 ft.**

e_2 **beams** $= 2^1/_2$ in. + $^1/_2$ (Column web)
$= 2^1/_2$ in. + $^1/_2$ (0.5 in.) assumed
$=$ 2.75 in. ≈ **0.25 ft.**

Factored Loads at the 2nd Floor Level (see Figure 8-22):

$$w_u = 0.342 \text{ ksf}(6.67 \text{ ft.}) = 2.3 \text{ kips/ft.}$$

$$w_{u(D)} = 0.10 \text{ ksf}(6.67 \text{ ft.}) = 0.67 \text{ kips/ft.}$$

$$P_D = 10 \text{ kips } (2 \text{ beams}) = 20 \text{ kips}$$

$$P_{D+L} = 35 \text{ kips } (2 \text{ beams}) = 70 \text{ kips}$$

Moment Split in the Column segments above and below the 2nd floor level:

$$M_{2bot} = \left(\frac{1/L_1}{1/L_1 + 1/L_2}\right) M_2, \text{ where } L_1 = 10 \text{ ft. and } L_2 = 10 \text{ ft.}$$

$$= \left(\frac{1/10 \text{ ft.}}{1/10 \text{ ft.} + 1/10 \text{ ft.}}\right) M_2$$

$$= 0.5 M_2$$

This implies that 50% of the unbalanced moment at the 2nd floor girder/beam-to-column shear connections (M_2) will be resisted by the column segment below the 2nd floor level, and 50% of the unbalanced moment is resisted by the column segment above the 2nd floor level.

Load Case 1: The axial load at the 2nd floor level is maximized by applying all dead and live loads on the 2nd floor beams and girders framing into column C1 in both orthogonal directions (see Figures 8-22 and 8-23).

$$P_2 = \underset{(\max P_3)}{242} + \underset{(P^R_{D+L}beam)}{(35} + \underset{(P^L_{D+L}beam)}{35)} + \underset{(P^R_{D+L}girder)}{(70} + \underset{(P^L_{D+L}girder)}{70)} = 452 \text{ kips}$$

$$M_{2x-x} = (70kips - 70kips)(e_2 \textbf{ girder}) = (0 \text{ kips})\ (0.63 \text{ ft.}) = 0 \text{ ft.-kips}$$
$$M_{2y-y} = (35kips - 35kips)(e_2 \textbf{ beam}) = (0 \text{ kips})\ (0.25 \text{ ft.}) = 0 \text{ ft.-kips}$$

$$P_2 = 452 \text{ kips}$$
$$M_{2botx-x} = 0.5(0 \text{ ft.-kips}) = 0 \text{ ft.-kips}$$
$$M_{2boty-y} = 0.5(0 \text{ ft.-kips}) = 0 \text{ ft.-kips}$$

For this load case, design the column for only axial load.

Load Case 2: The moments at the 2nd floor level are maximized by skipping the live loads on one 2nd floor beam and one girder framing into column C1, and keeping the full dead and live loads on the other 2nd floor beam and girder framing into the opposite sides of column C1 (see Figures 8-22 and 8-23).

$$P_2 = \underset{(\max P_3)}{242} + \underset{(P^R_{D+L}beam)}{(35} + \underset{(P^L_{D}beam)}{10)} + \underset{(P^R_{D+L}girder)}{(70} + \underset{(P^L_{D}girder)}{20)} = 377 \text{ kips}$$

$$M_{2x-x} = (70 \text{ kips} - 20 \text{ kips})(e_2 \textbf{ girder}) = (50 \text{ kips})(0.63 \text{ ft.}) = 32 \text{ ft.-kips}$$
$$M_{2y-y} = (35 \text{ kips} - 10 \text{ kips})(e_2 \textbf{ beam}) = (25 \text{ kips})(0.25 \text{ ft.}) = 6.3 \text{ ft.-kips}$$

The unbalanced moment at the 2nd floor level will be distributed between the column segments above and below the 2nd floor level in a ratio inversely proportional to the length of the column segments above and below the 2nd floor level. Therefore, the load and moments on the column segment just below the second floor for **load case 2** are calculated as

$$P_2 = 377 \text{ kips}$$
$$M_{2botx-x} = 0.5(32 \text{ ft.-kips}) = 16 \text{ ft.-kips}$$
$$M_{2boty-y} = 0.5(6.3 \text{ ft.-kips}) = 3.2 \text{ ft.-kips}$$

The column will be designed for this combined axial load and bending moment. The “0.5” factor in the preceding moment equations is the moment distribution or moment split factor between the column segments below and above the 2nd floor level.

The summary of the factored loads and the factored first-order moments for column C1 are shown in Table 8-3.

TABLE 8-3 Summary of Factored Loads and First-Order Moments for Column C1 (LRFD)

Level	Load case 1	Load case 2
Roof	P = 55 kips	P = 38 kips
	$M_{r\text{-}xx}$ = 0 ft.-kips	$M_{r\text{-}xx}$ = 6 ft.-kips
	$M_{r\text{-}yy}$ = 0 ft.-kips	$M_{r\text{-}yy}$ = 1.4 ft.-kips
Third floor	P = 242 kips	P = 167 kips
	M_{3botxx} = 0 ft.-kips	M_{3botxx} = 14 ft.-kips
	M_{3botyy} = 0 ft.-kips	M_{3botyy} = 3.2 ft.-kips
Second floor	P = 452 kips	P = 377 kips
	M_{2botxx} = 0 ft.-kips	M_{2botxx} = 16 ft.-kips
	M_{2botyy} = 0 ft.-kips	M_{2botyy} = 3.2 ft.-kips

Using similar principles as in Example 8-1, the design of Column C1 for the above-tabulated loads and moments at each level can be carried out.

Load Calculations for Column C2

The cumulative factored axial loads and the unbalanced factored moments at each floor/roof level for **Column C2** will now be calculated:

All loads are factored (see the LRFD load combinations in Section 2-3).

ROOF: *Loads to Column C2 at the **Roof** Level:* (See Figures 8-16, 8-24, and 8-25)

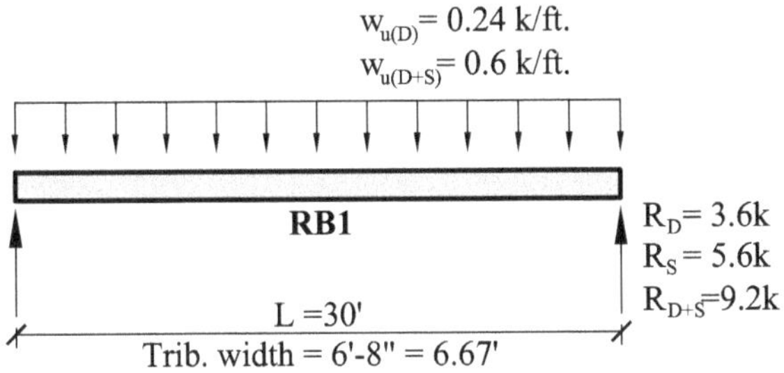

$$w_{u(D)} = (0.036\text{ksf})(6.67\text{ft}) = 0.24\text{k/ft.}$$
$$w_{u(D+S)} = (0.092\text{ksf})(6.67\text{ft}) = 0.6\text{k/ft.}$$

$$R_D = \frac{w_{u(D)}L}{2} = \frac{(0.24)(30)}{2} = 3.6\text{k}$$

$$R_{D+S} = \frac{w_{u(D+S)}L}{2} = \frac{(0.6)(30)}{2} = 9.2\text{k}$$

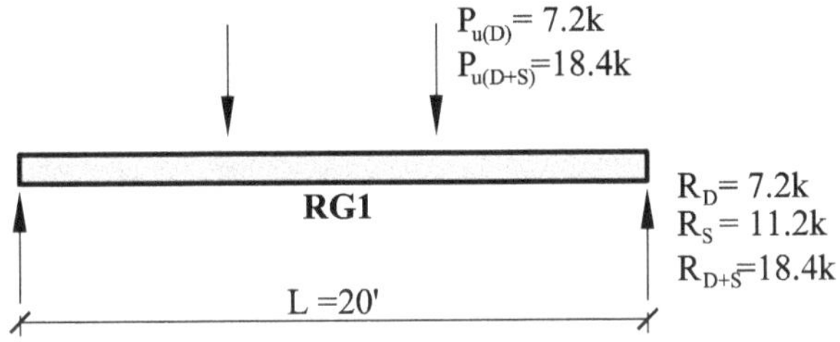

$$P_{u(D)} = (3.6\text{k})(2 = 7.2\text{k}$$
$$P_{u(D+S)} = (9.2\text{k})(2) = 18.4\text{k}$$

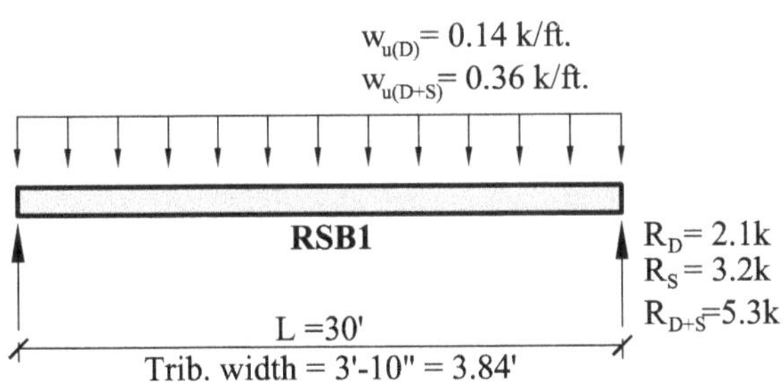

$$w_{u(D)} = (0.036\text{ksf})(3.84\text{ft}) = 0.14\text{k/ft.}$$
$$w_{u(D+S)} = (0.092\text{ksf})(3.84\text{ft}) = 0.36\text{k/ft.}$$

$$R_D = \frac{w_{u(D)}L}{2} = \frac{(0.14)(30)}{2} = 2.1\text{k}$$

$$R_{D+S} = \frac{w_{u(D+S)}L}{2} = \frac{(0.36)(30)}{2} = 5.3\text{k}$$

FIGURE 8-24 Calculation of roof beam and girder reactions framing into column C2

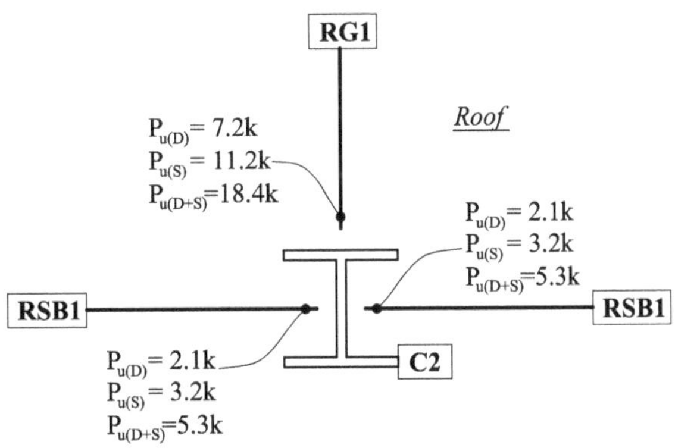

FIGURE 8-25 Reactions of roof beams and girders framing into column C2

Girder and beam connection eccentricities relative to the centroid of the column:

e_r **girders** $= 2\frac{1}{2}$ in. + $\frac{1}{2}$ (Column depth)
$= 2\frac{1}{2}$ in. + $\frac{1}{2}$ (8 in.) assuming minimum W8 column
$= 6\frac{1}{2}$ in. = **0.54 ft.**

e_r **beams** $= 2\frac{1}{2}$ in. + $\frac{1}{2}$ (Column web thickness)
$= 2\frac{1}{2}$ in. + $\frac{1}{2}$ (0.5 in.) assumed
= 2.75 in. ≈ **0.25 ft.**

Factored Loads at the Roof Level (see Figure 8-24):

The factored loads on a **typical interior roof beam** are calculated as follows:

$$w_u = 0.092 \text{ ksf}(6.67 \text{ ft.}) = 0.613 \text{ kips/ft.}$$
$$w_{u(S)} = 0.056\text{ksf}(6.67 \text{ ft.}) = 0.374 \text{ kips/ft}$$
$$w_{u(D)} = 0.036 \text{ ksf}(6.67 \text{ ft.}) = 0.24 \text{ kips/ft.}$$

The typical interior beam reactions are:

$$P_{D,beam} = (0.24 \text{kips / ft.})\left(\frac{30 \text{ft.}}{2}\right) = 3.6 \text{ kips}$$

$$P_{S,beam} = (0.374 \text{kips / ft.})\left(\frac{30 \text{ft.}}{2}\right) = 5.6 \text{ kips}$$

The factored concentrated loads on the typical interior roof girder from the roof beam reactions at each load point are calculated as follows:

$$P_D = 3.6 \text{ kips} + 3.6 \text{ kips} = 7.2 \text{ kips}$$
$$P_S = 5.6 \text{ kips} + 5.6 \text{ kips} = 11.2 \text{ kips}$$
$$P_{D+S} = 7.2 \text{ kips} + 11.2 \text{ kips} = 18.4 \text{ kips}$$

For **perimeter or spandrel beams**, the tributary width, TW = (6.67 ft./2) + 0.5-ft. edge distance = 3.84 ft.

Therefore, the factored loads on a **typical spandrel roof beam** are calculated as follows:

$$w_u = 0.092 \text{ ksf}(3.84 \text{ ft.}) = 0.36 \text{ kips/ft.}$$
$$w_{u(S)} = 0.056\text{ksf}(3.84 \text{ ft.}) = 0.21 \text{ kips/ft}$$
$$w_{u(D)} = 0.036 \text{ ksf}(3.84 \text{ ft.}) = 0.14 \text{ kips/ft.}$$

The typical spandrel beam reactions that are supported directly by column C2 are:

$$P_{D,beam} = (0.14 \text{kips / ft.})\left(\frac{30 \text{ft.}}{2}\right) = 2.1 \text{ kips}$$

$$P_{S,beam} = (0.21 \text{kips / ft.})\left(\frac{30 \text{ft.}}{2}\right) = 3.2 \text{ kips}$$

Load Case 1: The axial load at the **roof level** is maximized by applying all dead and live loads on the beams and girders framing into column C2 in both orthogonal directions (see Figures 8-24 and 8-25).

$$P_{roof} = \underset{(P^R{}_{D+S}beams)}{5.3} + \underset{(P^L{}_{D+S}beams)}{5.3} + \underset{(P^R{}_{D+S}girders)}{18.4} + \underset{(P^L{}_{D+S}girders)}{0} = 29 \text{ kips}$$

$$M_{rx-x} = (18.4 \text{ kips} - 0 \text{ kips})(0.54 \text{ ft.}) = 10 \text{ ft.-kips}$$
$$\uparrow (e_r, \text{girder})$$

$$M_{ry-y} = (5.3 \text{ kips} - 5.3 \text{ kips})(0.25 \text{ ft.}) = 0 \text{ ft.-kips}$$
$$\uparrow (e_r, \text{beams})$$

$$P = 29 \text{ kips}$$
$$M_{rx-x} = 10 \text{ ft.-kips}$$
$$M_{ry-y} = 0 \text{ ft.-kips}$$

For this case, the column will be designed for combined axial load plus bending moment.

Load Case 2: The moments at the roof level are maximized by skipping the live loads on one perimeter beam framing into column C2, and keeping the full dead and live loads on the other perimeter beam and the interior girder framing into the other sides of column C2 (see Figures 8-24 and 8-25).

$$P_{roof} = \underset{(P^R{}_{D+S}beams)}{5.3} + \underset{(P^L{}_{D}beams)}{2.1} + \underset{(P^R{}_{D+S}girders)}{18.4} + \underset{(P^L{}_{D}girders)}{0} = 26 \text{ kips}$$

$$M_{rx-x} = (18.4 \text{ kips} - 0 \text{ kips})(0.54 \text{ ft.}) = 10 \text{ ft.-kips}$$
$$M_{ry-y} = (5.3 \text{ kips} - 2.1 \text{ kips})(0.25 \text{ ft.}) = 1.0 \text{ ft.-kips}$$

$$P = 29 \text{ kips}$$
$$M_{rx-x} = 10 \text{ ft.-kips}$$
$$M_{ry-y} = 1.0 \text{ ft.-kips}$$

For this load case, the column will be designed for combined axial load plus bending moment. The unbalanced column moments at the roof level are resisted solely by the column segment below the roof level (since there is no column segment above the roof level). For the roof loads that will be cumulatively added to the floor loads below, the appropriate load factor must be applied to the live loads. Recall that the load combination $1.2D$ + (1.6 L_r, or $1.0S$, or $1.6R$) applies to the column load only at the roof level. For cumulative loads at the lower levels, the applicable load combination that maximizes the live loads at the floor levels is $1.2D + 1.6L$ + (0.5 L_r, or $0.3S$, or $0.5R$) or $1.2D + 1.6L + 0.3S$, since the snow load is more critical than the rain or roof live load for this example. Therefore, the contribution from the roof level to the cumulative column axial loads at the lower levels (i.e., third- and second-floor levels) is calculated as follows:

Using the $1.2D + 1.0S$ *roof beam and girder reaction values from Figure 8-25, we calculate*

$$P_{roof} = 1.2D + 0.3S$$

$$= \left[2.1 \text{ kips} + \frac{0.3}{1.0}(3.2 \text{ kips})\right] + \left[2.1 \text{ kips} + \frac{0.3}{1.0}(3.2 \text{ kips})\right] + \left[7.2 \text{ kips} + \frac{0.3}{1.0}(11.2 \text{ kips})\right]$$

$$= 17 \text{ kips.}$$

Note that the "0.3" and "1.0" coefficients in the preceding equation are the load factors (see Section 2-3 of the text).

The above value could also be obtained using the tributary area method as follows:

$$P_{roof} = (1.2D + 0.3S)(Tributary\,area\,of\,Column\,C2)$$

$$= \frac{\left[(1.2)(30\,psf)+(0.3)(56\,psf)\right]\left(0.5\,ft + \frac{20\,ft}{2}\right)(30\,ft)}{1000}$$

$$= 17 \text{ kips.}$$

THIRD FLOOR: *Loads to Column C2 at the* ***Third*** *Floor:* (See Figures 8-17, 8-26, and 8-27)

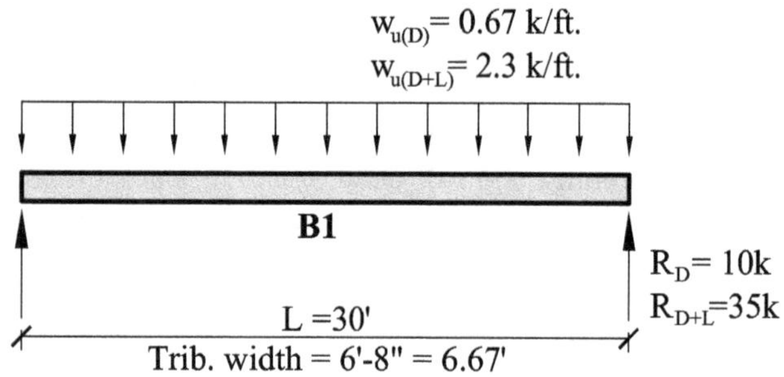

$w_{u(D)} = (0.10\text{ksf})(6.67') = 0.67$ k/ft.

$w_{u(D+L)} = (0.342\text{ksf})(6.67') = 2.3$ k/ft.

$$R_D = \frac{w_{u(D)}L}{2} = \frac{(0.67)(30)}{2} = 10\text{k}$$

$$R_{D+L} = \frac{w_{u(D+L)}L}{2} = \frac{(2.3)(30)}{2} = 35\text{k}$$

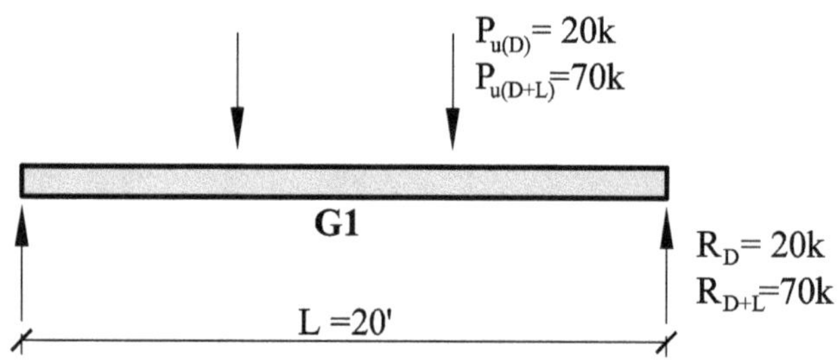

$P_{u(D)} = (10\text{k})(2) = 20$k

$P_{u(D+L)} = (35\text{k})(2) = 70$ k

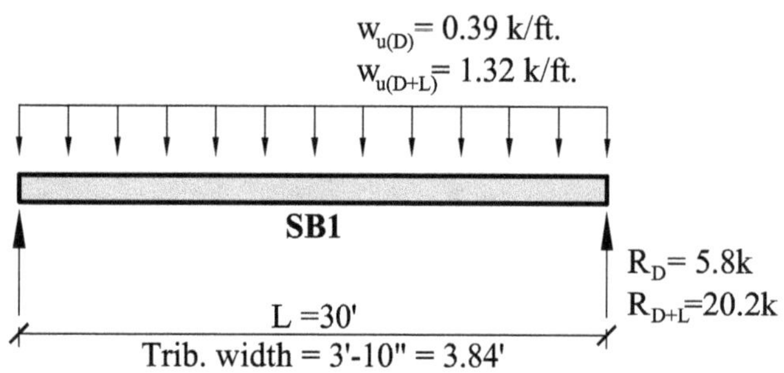

$w_{u(D)} = (0.10\text{ksf})(3.84') = 0.39$ k/ft.

$w_{u(D+L)} = (0.342\text{ksf})(3.84') = 1.32$ k/ft.

$$R_D = \frac{w_{u(D)}L}{2} = \frac{(0.39)(30)}{2} = 5.8\text{k}$$

$$R_{D+L} = \frac{w_{u(D+L)}L}{2} = \frac{(1.32)(30)}{2} = 20.2\text{k}$$

FIGURE 8-26 Calculation of floor beam and girder reactions at the 3rd floor level

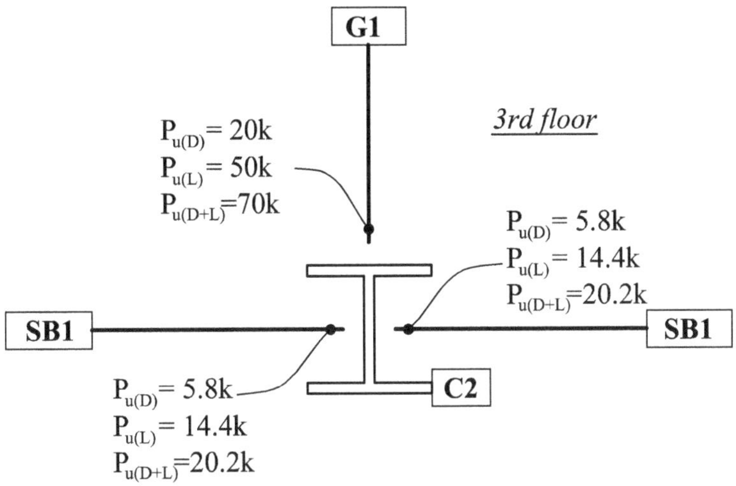

FIGURE 8-27 Reactions of floor beams and girders framing into column C2 at the 3rd floor level

Girder and beam connection eccentricities relative to the centroid of the column:

e_3 **girders** $= 2\tfrac{1}{2}$ in. + $\tfrac{1}{2}$ (Column depth)
$= 2\tfrac{1}{2}$ in. + $\tfrac{1}{2}$ (8 in.) assuming minimum W8 column
$= 6\tfrac{1}{2}$ in. = **0.54 ft.**

e_3 **beams** $= 2\tfrac{1}{2}$ in. + $\tfrac{1}{2}$ (Column web thickness)
$= 2\tfrac{1}{2}$ in. + $\tfrac{1}{2}$ (0.5 in.) assumed
= 2.75 in. ≈ **0.25 ft.**

Factored Loads at the 3rd Floor Level (see Figure 8-26):

$$w_u = 0.342 \text{ ksf } (6.67 \text{ ft.}) = 2.3 \text{ kips/ft.}$$
$$w_{u(D)} = 0.10 \text{ ksf } (6.67 \text{ ft.}) = 0.67 \text{ kips/ft.}$$

For the reactions for girder G1, refer to the load calculations for column C1.
For **perimeter or spandrel beams**, the tributary width,

$$\text{TW} = (6.67 \text{ ft.}/2) + 0.5\text{-ft. edge distance} = 3.84 \text{ ft.}$$

Therefore, the loads on the spandrel beams are

$$w_u = 0.342 \text{ ksf } (3.84 \text{ ft.}) = 1.32 \text{ kips/ft.}$$
$$w_{u(D)} = 0.10 \text{ ksf } (3.84 \text{ ft.}) = 0.39 \text{ kips/ft.}$$

Moment Split in the Column segments above and below the 3rd floor level:

$$M_{3bot} = \left(\frac{1/L_2}{1/L_2 + 1/L_3}\right) M_3; \qquad L_2 = 10 \text{ ft.; and } L_3 = 10 \text{ ft.}$$
$$= \left(\frac{1/10}{1/10 \text{ ft.} + 1/10 \text{ ft.}}\right) M_3$$
$$= 0.5\, M_3$$

Load Case 1: The axial load is maximized by applying all dead and live loads on the beams and girders framing into column C2 in both orthogonal directions (see Figures 8-26 and 8-27).

$$P_3 = \underset{(\max P_{roof})}{17^*} + \underset{(P^R{}_{D+L}beam)}{(19.8} + \underset{(P^L{}_{D+L}beam)}{19.8)} + \underset{(P^R{}_{D+L}girder)}{(70} + \underset{(P^L{}_{D+L}girder)}{0)} = 127 \text{ kips}$$

$$M_{3x-x} = (70\,\text{kips} - 0\,\text{kips})(0.54 \text{ ft.}) = 38 \text{ ft.-kips}$$
$$M_{3y-y} = (19.8\,\text{kips} - 19.8\,\text{kips})(0.25 \text{ ft.}) = 0 \text{ ft.-kips}$$

$$P_3 = 127 \text{ kips}$$
$$M_{3botx-x} = 0.5(38 \text{ ft.-kips}) = 19 \text{ ft.kips}$$
$$M_{3boty-y} = 0.5(0 \text{ ft.-kips}) = 0 \text{ ft.-kips}$$

For this load case, the column will be designed for combined axial load plus bending moment.

Recall that the "17" in the equation above (and recalculated below) is the cumulative load in kips from the roof considering the appropriate load combinations that maximizes the live loads at the floor levels.

$$*1.2D + 0.3S = (1.2x30psf + 0.3x56psf) = 52.8psf$$

$$(52.8psf)\left[(30ft)\left(0.5ft + \frac{20ft}{2}\right)\right]/1000 = 17\,kips$$

Load Case 2: The moments at this floor level are maximized by skipping the live loads on one perimeter beam framing into column C2, and keeping the full dead and live loads on the other perimeter beam and the interior girder framing into the other sides of column C2 (see Figures 8-26 and 8-27).

$$P_3 = \underset{(\max P_{roof})}{17^*} + \underset{(P^R{}_{D+L}beam)}{(5.8} + \underset{(P^L{}_{D}beam)}{19.8)} + \underset{(P^R{}_{D+L}girder)}{(70} + \underset{(P^L{}_{D+L}girder)}{0)} = 113 \text{ kips}$$

$$M_{3x-x} = (70\,\text{kips} - 0\,\text{kips})(e_3 \textbf{ girder}) = (70 \text{ kips})(0.54 \text{ ft.}) = 38 \text{ ft.-kips}$$
$$M_{3y-y} = (19.8\,\text{kips} - 5.8\,\text{kips})(e_3 \textbf{ beam}) = (14 \text{ kips})(0.25 \text{ ft.}) = 4 \text{ ft.-kips}$$

Recall that the "17" in the equation above (and recalculated below) is the cumulative load in kips from the roof considering the appropriate load combinations that maximizes the live loads at the floor levels.

$$*1.2D + 0.3S = (1.2x30psf + 0.3x56psf) = 52.8psf$$

$$(52.8psf)\left[(30ft)\left(0.5ft + \frac{20ft}{2}\right)\right]/1000 = 17\,kips$$

The unbalanced moment at the 3rd floor level (M_3) will be distributed between the column segments above and below the 3rd floor level in a ratio inversely proportional to the length of the column segments above and below the 3rd floor level. Therefore, the load and moments on the column just below the 3rd floor level for **load case 2** are calculated as

$$P_3 = 113 \text{ kips}$$
$$M_{3botx-x} = 0.5(38 \text{ ft.-kips}) = 19 \text{ ft.-kips}$$
$$M_{3boty-y} = 0.5(4 \text{ ft.-kips}) = 2 \text{ ft.-kips}$$

The column will be designed for combined axial load plus bending moment.

The "0.5" factor in the preceding moment equations is the moment distribution or moment split factor.

SECOND FLOOR: *Loads to Column C2 at the* ***Second*** *Floor:* (See Figures 8-17, 8-28, and 8-29)

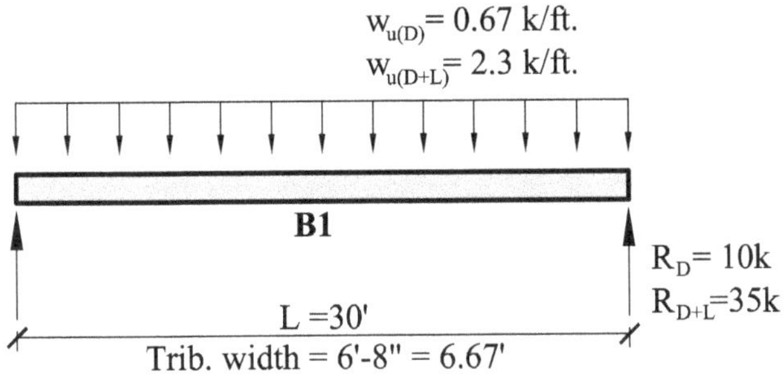

$$w_{u(D)} = (0.10\text{ksf})(6.67') = 0.67 \text{ k/ft.}$$
$$w_{u(D+L)} = (0.342\text{ksf})(6.67') = 2.3 \text{ k/ft.}$$

$$R_D = \frac{w_{u(D)}L}{2} = \frac{(0.67)(30)}{2} = 10\text{k}$$
$$R_{D+L} = \frac{w_{u(D+L)}L}{2} = \frac{(2.3)(30)}{2} = 35\text{k}$$

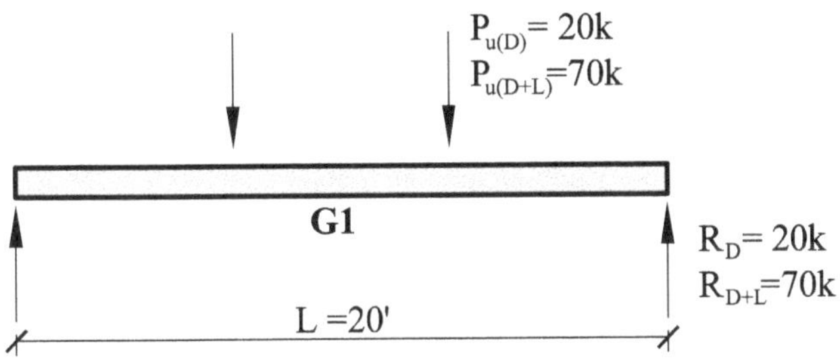

$$P_{u(D)} = (10\text{k})(2) = 20\text{k}$$
$$P_{u(D+L)} = (35\text{k})(2) = 70 \text{ k}$$

$$w_{u(D)} = (0.10\text{ksf})(3.84') = 0.39 \text{ k/ft.}$$
$$w_{u(D+L)} = (0.342\text{ksf})(3.84') = 1.32 \text{ k/ft.}$$

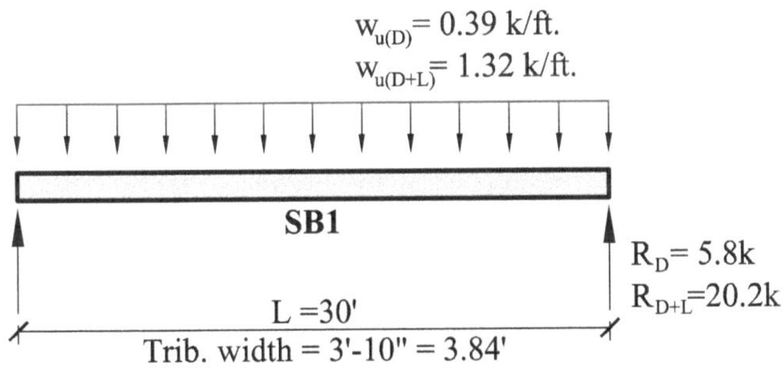

$$R_D = \frac{w_{u(D)}L}{2} = \frac{(0.39)(30)}{2} = 5.8\text{k}$$
$$R_{D+L} = \frac{w_{u(D+L)}L}{2} = \frac{(1.32)(30)}{2} = 20.2\text{k}$$

FIGURE 8-28 Calculation of floor beam and girder reactions at the 2nd floor level

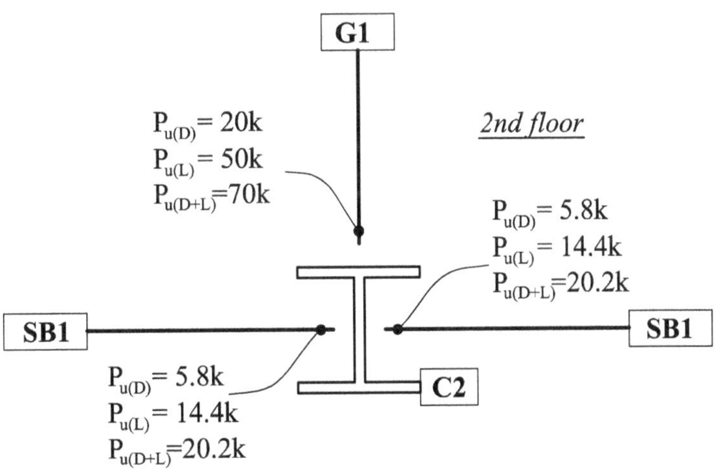

FIGURE 8-29 Reactions of floor beams and girders framing into column C2 at the 2nd floor level

Girder and beam eccentricities relative to the centroid of the column:

e_2 girders $= 2\frac{1}{2}$ in. + $\frac{1}{2}$ (Column depth)

$= 2\frac{1}{2}$ in. + $\frac{1}{2}$ (10 in.) assuming W10 column, with 10 in. nominal depth

$= 7.5$ in.

$=$ **0.63 ft.**

e_2 beams $= 2\frac{1}{2}$ in. + $\frac{1}{2}$ (Column web)

$= 2\frac{1}{2}$ in. + $\frac{1}{2}$ (0.5 in.) assumed

$= 2.75$ in. $\approx$ **0.25 ft.**

Factored Loads at the 2nd Floor Level (see Figure 8-28):

$$w_u = 0.342 \text{ ksf } (6.67 \text{ ft.}) = 2.3 \text{ kips/ft.}$$

$$w_{u(D)} = 0.10 \text{ ksf } (6.67 \text{ ft.}) = 0.67 \text{ kips/ft.}$$

For the reactions in girder G1, refer to the load calculations for column C1.
For **perimeter or spandrel beams**, the tributary width,

$$\text{TW} = 6.67 \text{ ft.}/2 + 0.5\text{-ft. edge distance}$$
$$= 3.84 \text{ ft.}$$

Therefore, the loads on the spandrel beams are

$$w_u = 0.342 \text{ ksf } (3.84 \text{ ft.}) = 1.32 \text{ kips/ft.}$$

$$w_{u(D)} = 0.10 \text{ ksf } (3.84 \text{ ft.}) = 0.39 \text{ kips/ft.}$$

Moment Split in the Column segments above and below the 2nd floor level:

$$M_{2bot} = \left(\frac{1/L_1}{1/L_1 + 1/L_2}\right) M_2$$

$$= \left(\frac{1/10 \text{ ft.}}{1/10 \text{ ft.} + 1/10 \text{ ft.}}\right) M_2,$$

where $L_1 = 10$ ft. and $L_2 = 10$ ft.

$$= 0.5 M_2$$

Load Case 1: The axial load is maximized by applying all dead and live loads on the beams and girders framing into column C2 in both orthogonal directions (see Figures 8-28 and 8-29).

$$P_2 = \underset{(P_3 \max)}{127} + (19.8 + 19.8) + (70 + 0) = 237 \text{ kips}$$

$$M_{2x-x} = (70 \text{ kips} - 0 \text{ kips})(0.63 \text{ ft.}) = 44 \text{ ft.-kips}$$

$$M_{2y-y} = (19.8 \text{ kips} - 19.8 \text{ kips})(0.25 \text{ ft.}) = 0 \text{ ft.-kips}$$

$$P_2 = 237 \text{ kips}$$

$$M_{2botx-x} = 0.5(44 \text{ ft.-kips}) = 22 \text{ ft.-kips}$$

$$M_{2boty-y} = 0.5(0 \text{ ft.-kips}) = 0 \text{ ft.-kips}$$

For this load case, the column will be designed for combined axial load plus bending moment.

Load Case 2: The moments at this floor level are maximized by skipping the live loads on one perimeter beam framing into column C2, and keeping the full dead and live loads on the other perimeter beam and the interior girder framing into the other sides of column C2 (see Figures 8-28 and 8-29).

$$P_2 = 127 + (5.8 + 19.8) + (70 + 0) = 223 \text{ kips}$$
$$M_{2x-x} = (70 \text{kips} - 0 \text{kips})(0.63 \text{ ft.}) = 44 \text{ ft.-kips}$$
$$M_{2y-y} = (19.8 \text{kips} - 5.8 \text{kips})(0.25 \text{ ft.}) = 4 \text{ ft.-kips}$$

The moment at the second-floor level is assumed to be distributed between the column segments above and below the 2nd floor level in a ratio inversely proportional to the length of the column segments above and below the 2nd floor level. Therefore, the load and moments on the column just below the second-floor level for **load case 2** are calculated as

$$P_2 = 223 \text{ kips}$$
$$M_{2botx-x} = 0.5(44 \text{ ft.-kips}) = 22 \text{ ft.-kips}$$
$$M_{2boty-y} = 0.5(4 \text{ ft.-kips}) = 2 \text{ ft.-kips}$$

The column will be designed for combined axial load plus bending moment.

The "0.5" factor in the preceding moment equations is the moment distribution or moment split factor. A summary of the factored loads and first-order moments for column C2 is shown in Table 8-4.

TABLE 8-4 Summary of Factored Loads and First-Order Moments for Column C2 (LRFD)

Level	Load case 1	Load case 2
Roof	$P = 29$ kips $M_{roofx-x} = 10$ ft.-kips $M_{roofy-y} = 0$ ft.-kips	$P = 26$ kips $M_{roofx-x} = 10$ ft.-kips $M_{roofy-y} = 1.0$ ft.-kips
Third floor	$P = 127$ kips $M_{3botx-x} = 19$ ft.-kips $M_{3boty-y} = 0$ ft.-kips	$P = 113$ kips $M_{3botx-x} = 19$ ft.-kips $M_{3boty-y} = 2$ ft.-kips
Second floor	$P = 237$ kips $M_{2botx-x} = 22$ ft.-kips $M_{3boty-y} = 0$ ft.-kips	$P = 223$ kips $M_{2botx-x} = 22$ ft.-kips $M_{2boty-y} = 2$ ft.-kips

We will now proceed to design Column C1 using a column design template:

Design of Column C1 (Roof to Third Floor) - LRFD

Load case 1 (with $M = 0$)	Load case 2 or Load case 1 with moments
$P_u = 55$ kips: $KL \approx 1 \times 10$ ft. $= 10$ ft. <u>From **column load tables** (*AISCM* Table 4-1)</u> **Try W8 × 31 (A = 9.12 in.²)** $\phi_c P_n = 317$ kips $> P_u$ OK	**Check W8 × 31 column for** $P_u = 38$ kips $M_{rx-x} = M_{ntx} = 6$ ft.-kips (first-order moment) $M_{ry-y} = M_{nty} = 1.4$ ft.-kips (first-order moment) Column unbraced length, $L_b = 10$ ft. $\boldsymbol{r_x} = 3.47$ in., $\boldsymbol{r_y} = 2.02$ in.
	LRFD Beam Design Tables (*AISCM*, Table 3-2; F_y = 50 ksi) $\phi_b M_p = 114$ ft.-kips; $L_p = 7.18$ ft.; $\phi_b BF = 2.37$ kips; $L_r = 24.8$ ft. $> L_b$ If $L_b \le L_p$, $\phi_b M_{nx} = \phi_b M_{px}$ $\phi_b M_{nx} = \phi_b M_p - \phi_b BF\,(L_b - L_p) = [114 - (2.37)(10 - 7.18)] =$ **107 ft.-kips** $\le \phi_b M_p$ From *AISCM*: $S_x = 27.5$ in.³; $Z_y = 14.1$ in.³; $S_y = 9.27$ in.³ $\phi_b M_{ny} = \phi_b Z_y F_y \le 1.6\phi_b S_y F_y$ $\phi_b M_{ny} = (0.9)(14.1)(50)/12 \le 1.6(0.9)(9.27)(50)/12$ $\phi_b M_{ny} =$ **54 ft.-kips** ≤ 56 ft.-kips $\phi_c P_n = 317$ kips; $\phi_b M_{nx} = 107$ ft.-kips; $\phi_b M_{ny} = 54$ ft.-kips

(continued)

Load case 1 (with $M = 0$)	Load case 2 or Load case 1 with moments
	$M_{ux} = B_{1x}M_{ntx} + B_{2x}M_{ltx}$; $M_{ltx} = 0$ (for braced frames) $M_{uy} = B_{1y}M_{nty} + B_{2y}M_{lty}$; $M_{lty} = 0$ (for braced frames) <u>P-δ; Effect for x–x axis</u> $C_{mx} = 0.6 - 0.4(M_{1x}/M_{2x})$; but conservatively use $C_{mx} = 0.85$ $P_{elx} = \dfrac{\pi^2 EA}{(KL/r_x)^2} = 2183$ kips $B_{1x} = \dfrac{C_{mx}}{(1 - P_u/P_{elx})} = \dfrac{0.85}{[1-(38/2183)]} = 0.87 \geq 1.0 \;\therefore\; \mathbf{B_{1x} = 1.0}$ $\therefore M_{ux} = B_{1x}M_{ntx} = 1.0(6) = 6$ ft.-kips (2nd order moment) <u>P-δ Effect for y–y axis</u> $C_{my} = 0.6 - 0.4(M_{1y}/M_{2y})$; but conservatively use $C_{my} = 0.85$ $P_{ely} = \dfrac{\pi^2 EA}{(KL/r_y)^2} = 740$ kips $B_{1y} = \dfrac{C_{my}}{(1 - P_u/P_{ely})} = \dfrac{0.85}{[1-(38/740)]} = 0.90 \geq 1.0 \;\therefore\; \mathbf{B_{1y} = 1.0}$ $\therefore M_{uy} = B_{1y}M_{nty} = 1.0(1.4) = 1.4$ ft.-kips (2nd order moment)

Column Interaction Equation

$$P_u/\phi_c P_n = 38/317 = 0.12 < 0.20, \textit{therefore use equation } (8-2)$$

$$\frac{P_u}{2\phi_c P_n} + \left(\frac{M_{ux}}{\phi_b M_{nx}} + \frac{M_{uy}}{\phi_b M_{ny}}\right) = \frac{0.12}{2} + \left(\frac{6}{107} + \frac{1.4}{54}\right) = 0.14 < 1.0 \quad \text{OK} \Rightarrow$$

W8 × 31 column is adequate for Roof to 3rd Floor.

***Note*:** For building structures, the minimum practical column size recommended is a W8 column to ensure sufficient flange width to accommodate the girder-to-column double-angle connections.

Design of Column C1 (Third Floor to Second Floor) - LRFD

Load case 1 (with $M = 0$)	Load case 2 or Load case 1 with moments
$P_u = 242$ kips: $KL \approx 1 \times 10$ ft. $= 10$ ft. <u>From **column load tables (*AISCM* Table 4-1)**</u> **Try W8 × 31 ($A = 9.12$ in.^{2})** $\phi_c P_n = 317$ kips $> P_u$ OK	**Check W8 × 31 column for** $P_u = 167$ kips $M_{3x-x} = M_{ntx} = 14$ ft.-kips (first-order moment) $M_{ry-y} = M_{nty} = 3.2$ ft.-kips (first-order moment) Column unbraced length, $L_b = 10$ ft.; $r_x = 3.47$ in., $r_y = 2.02$ in.
	LRFD Beam Design Tables (*AISCM* Table 3-2; $F_y = 50$ ksi) $\phi_b M_p = 114$ ft.-kips; $L_p = 7.18$ ft.; $\phi_b BF = 2.37$ kips; $L_r = 24.8$ ft. $> L_b$ If $L_b \leq L_p$, $\phi_b M_{nx} = \phi_b M_{px}$ $\phi_b M_{nx} = \phi_b M_p - \phi_b BF\,(L_b - L_p) = [114 - (2.37)(10 - 7.18)] =$ **107 ft.-kips** $\leq \phi_b M_p$ From *AISCM*: $S_x = 27.5$ in.^{3}; $Z_y = 14.1$ in.^{3}; $S_y = 9.27$ in.^{3}

(continued)

Load case 1 (with $M = 0$)	Load case 2 or Load case 1 with moments
	$\phi_b M_{ny} = \phi_b Z_y F_y \le 1.6\phi_b S_y F_y$ $\phi_b M_{ny} = (0.9)(14.1)(50)/12 \le 1.6(0.9)(9.27)(50)/12$ $\phi_b M_{ny} = \mathbf{54\ ft.\text{-}kips} \le 56$ ft.-kips $\phi_c P_n = 317$ kips; $\phi_b M_{nx} = 107$ ft.-kips; $\phi_b M_{ny} = 54$ ft.-kips $M_{ux} = B_{1x} M_{ntx} + B_{2x} M_{ltx}$; $M_{ltx} = 0$ (for braced frames) $M_{uy} = B_{1y} M_{nty} + B_{2y} M_{lty}$; $M_{lty} = 0$ (for braced frames) <u>P-δ Effect for x–x axis</u> $C_{mx} = 0.6 - 0.4(M_{1x}/M_{2x})$; but conservatively use $C_{mx} = 0.85$ $P_{elx} = \dfrac{\pi^2 EA}{(KL/r_x)^2} = 2183$ kips $B_{1x} = \dfrac{C_{mx}}{(1 - P_u/P_{elx})} = \dfrac{0.85}{[1-(167/2183)]} = 0.92 \ge 1.0 \therefore \mathbf{B_{1x} = 1.0}$ $\therefore M_{ux} = B_{1x} M_{ntx} = 1.0(14) = 14$ ft.-kips (2nd order moment) <u>P-δ Effect for y–y axis</u> $C_{my} = 0.6 - 0.4(M_{1y}/M_{2y})$; but conservatively use $C_{my} = 0.85$ $P_{ely} = \dfrac{\pi^2 EA}{(KL/r_y)^2} = 740$ kips $B_{1y} = \dfrac{C_{my}}{(1 - P_u/P_{ely})} = \dfrac{0.85}{[1-(167/740)]} = 1.1 \ge 1.0 \therefore \mathbf{B_{1y} = 1.1}$ $\therefore M_{uy} = B_{1y} M_{nty} = 1.1\,(3.2) = 4.0$ ft.-kips (2nd order moment)

Column Interaction Equation

$$P_u/\phi_c P_n = 167/317 = 0.53 \ge 0.20, \textit{ therefore use equation } (8-1)$$

$$\frac{P_u}{\phi_c P_n} + \frac{8}{9}\left(\frac{M_{ux}}{\phi_b M_{nx}} + \frac{M_{uy}}{\phi_b M_{ny}}\right) = 0.53 + \frac{8}{9}\left(\frac{14}{107} + \frac{4.0}{54}\right) = 0.71 < 1.0 \qquad \textbf{OK}$$

W8 × 31 column is adequate for 3rd Floor to 2nd Floor.

Design of Column C1 (Second Floor to Ground Floor) - LRFD

Load case 1 (with $M = 0$)	Load case 2 or Load case 1 with moments
$P_u = 452$ kips: $KL \approx 1 \times 10$ ft. $= 10$ ft. <u>From **column load tables (*AISCM* Table 4-1)**</u> **Try W8 × 48 (A = 14.1 in.²)** $\phi_c P_n = 497$ kips $> P_u$ OK	**Check W8 × 48 column for** $P_u = 377$ kips $M_{3x\text{-}x} = M_{ntx} = 16$ ft.-kips (first-order moment) $M_{3y\text{-}y} = M_{nty} = 3.2$ ft.-kips (first-order moment) Column unbraced length, $L_b = 10$ ft.; $\boldsymbol{r_x} = 3.61$ in., $\boldsymbol{r_y} = 2.08$ in.

(continued)

Load case 1 (with $M = 0$)	Load case 2 or Load case 1 with moments
	LRFD Beam Design Tables (*AISCM* Table 3-2; F_y = 50 ksi) $\phi_b M_p$ = 184 ft.-kips; L_p = 7.35 ft.; $\phi_b BF$ = 2.53 kips; L_r = 35.2 ft. $> L_b$ If $L_b \le L_p$, $\phi_b M_{nx} = \phi_b M_{px}$ $\phi_b M_{nx} = \phi_b M_p - \phi_b BF\,(L_b - L_p)$ = [184 – (2.53)(10 – 7.35)] = **177 ft.-kips** $\le \phi_b M_p$ From *AISCM*: S_x = 43.2 in.^{3}; Z_y = 22.9 in.^{3}; S_y = 15.0 in.^{3} $\phi_b M_{ny} = \phi_b Z_y F_y \le 1.6\phi_b S_y F_y$ $\phi_b M_{ny}$ = (0.9)(22.9)(50)/12 ≤ 1.6(0.9)(15.0)(50)/12 $\phi_b M_{ny}$ = **86 ft.-kips** ≤90 ft.-kips Governs $\phi_c P_n$ = 497 kips; $\phi_b M_{nx}$ = 177 ft.-kips; $\phi_b M_{ny}$ = 86 ft.-kips $M_{ux} = B_{1x} M_{ntx} + B_{2x} M_{ltx}$; $M_{ltx} = 0$ (for braced frames) $M_{uy} = B_{1y} M_{nty} + B_{2y} M_{lty}$; $M_{lty} = 0$ (for braced frames) P-δ Effect for x–x axis $C_{mx} = 0.6 - 0.4(M_{1x}/M_{2x})$; but conservatively use C_{mx} = 0.85 $P_{e1x} = \dfrac{\pi^2 EA}{(KL/r_x)^2} = 3652$ kips $B_{1x} = \dfrac{C_{mx}}{(1 - P_u/P_{e1x})} = \dfrac{0.85}{[1-(377/3652)]} = 0.95 \ge 1.0 \therefore \boldsymbol{B_{1x} = 1.0}$ $\therefore M_{ux} = B_{1x} M_{ntx}$ = 1.0(16) = 16 ft.-kips (2nd order moment) P-δ Effect for y–y axis $C_{my} = 0.6 - 0.4(M_{1y}/M_{2y})$; but conservatively use C_{my} = 0.85 $P_{e1y} = \dfrac{\pi^2 EA}{(KL/r_y)^2} = 1212$ kips $B_{1y} = \dfrac{C_{my}}{(1 - P_u/P_{e1y})} = \dfrac{0.85}{[1-(377/1212)]} = 1.23 \ge 1.0 \therefore \boldsymbol{B_{1y} = 1.23}$ $\therefore M_{uy} = B_{1y} M_{nty}$ = 1.23(3.2) = 4.0 ft.-kips (2nd order moment)

Column Interaction Equation

$$P_u/\phi_c P_n = 377/497 = 0.76 \ge 0.20, \textit{ therefore use equation } (8-1)$$

$$\frac{P_u}{\phi_c P_n} + \frac{8}{9}\left(\frac{M_{ux}}{\phi_b M_{nx}} + \frac{M_{uy}}{\phi_b M_{ny}}\right) = 0.76 + \frac{8}{9}\left(\frac{16}{177} + \frac{4.0}{86}\right) = 0.88 < 1.0 \qquad \textbf{OK}$$

Therefore, **W8 × 48 column is adequate for 2nd Floor to Ground Floor.**

8-11 STUDENT PRACTICE EXERCISE AND COLUMN DESIGN TEMPLATES

As an exercise, the reader should now design column C2, following the same procedure that was used for the design of column C1. To aid the reader, column design templates for wide-flanged or I-shaped sections and HSS columns are presented on the following pages. The moment capacity equations in the LRFD column design templates presume that the selected column will have an unbraced length, L_b, in bending that is less than or equal to the unbraced length, L_r, at the transition point between inelastic and elastic lateral-torsional buckling. Where this is not the case and elastic lateral-torsional buckling controls the bending moment capacity of the column, the reader should calculate the moment capacity using the method presented in Chapter 6 for the case where $L_b > L_r$.

Note that *AISCM* Table 6-1 provides values of axial compression strength, axial tension strength, and moment capacities about both the strong and weak axes of a column for different wide flange column sections and for varying unbraced lengths; the capacities from *AISCM* Table 6-1 can be used directly in the column design templates presented in this section. As a reminder, note that *AISCM* Table 6-1 assumes a moment gradient factor, C_b of 1,0, and also, the moment magnification factor, B_1, is not yet included in the tabulated values. The actual values of these two variables must be used to modify the strength values from *AISCM* Table 6-1.

W-Shape Column Design Template - LRFD

Design of Column (Floor to Floor)

<table>
<tr><th>Load case 1 (with $M = 0$)</th><th>Load case 2 or Load case 1 with moments</th></tr>
<tr><td>$P_u =$ kips; $KL =$
From Column Load Tables (AISCM, Table 4-1)
Try W × (A = in.^{2})
Is $\phi_c P_n =$ kips $> P_u$ OK?</td><td>Check W × column for
$P_u =$ kips
$M_{x\text{-}x} = M_{ntx} =$ ft.-kips (first order moment)
$M_{y\text{-}y} = M_{nty} =$ ft.-kips (first order moment)
Column unbraced length, $L_b =$; $\boldsymbol{r_x} =$, $\boldsymbol{r_y} =$</td></tr>
<tr><td></td><td>LRFD Beam Design Tables (AISCM Table 3-2; F_y = 50 ksi)
$\phi_b M_p =$ ft.-kips; $L_p =$; $\phi_b BF =$ kips
If $L_b \leq L_p$, $\phi_b M_{nx} = \phi_b M_{px}$
$\phi_b M_{nx} = \phi_b M_p - \phi_b BF\,(L_b - L_p) = [\quad - (\quad)(\quad - \quad)] =$ ft.-kips $\leq \phi_b M_p$
From AISCM: $S_x =$ in.^{3}; $Z_y =$ in.^{3}; $S_y =$ in.^{3}

$\phi_b M_{ny} = \phi_b Z_y F_y/12 \leq 1.6\phi_b S_y F_y/12$
$\phi_b M_{ny} = (0.9)(\quad)(\quad)/12 \leq 1.6(0.9)(\quad)(\quad)/12$
$\phi_b M_{ny} =$ ft.-kips $\leq$ ft.-kips
$\phi_c P_n =$ kips; $\phi_b M_{nx} =$ ft.-kips; $\phi_b M_{ny} =$ ft.-kips
$M_{ux} = B_{1x} M_{ntx} + B_{2x} M_{ltx}$; $M_{ltx} = 0$ (for braced frames)
$M_{uy} = B_{1y} M_{nty} + B_{2y} M_{lty}$; $M_{lty} = 0$ (for braced frames)

<u>P-δ Effect for x–x axis</u>
$C_{mx} = 0.6 - 0.4(M_{1x}/M_{2x})$; but conservatively use $C_{mx} = 0.85$
$P_{e1x} = \dfrac{\pi^2 EA}{(KL/r_x)^2} =$ kips</td></tr>
</table>

(continued)

Load case 1 (with $M = 0$)	Load case 2 or Load case 1 with moments
	$B_{1x} = \dfrac{C_{mx}}{(1 - P_u/P_{e1x})} = \dfrac{0.85}{[1-(\ \ /\ \ )]} = \quad \geq 1.0 \therefore \boldsymbol{B_{1x}} =$ $\therefore M_{ux} = B_{1x}M_{ntx} = \ (\ \ ) = $ ft.-kips (2nd order moment) <u>P-δ Effect for y–y axis</u> $C_{my} = 0.6 - 0.4(M_{1y}/M_{2y})$; but conservatively use $C_{my} = 0.85$ $P_{e1y} = \dfrac{\pi^2 EA}{(KL/r_y)^2} = $ kips $B_{1y} = \dfrac{C_{my}}{(1 - P_u/P_{e1y})} = \dfrac{0.85}{[1-(\ \ /\ \ )]} = \quad \geq 1.0 \therefore \boldsymbol{B_{1y}} =$ $\therefore M_{uy} = B_{1y}M_{nty} = \ (\ \ ) = $ ft.-kips (2nd order moment

Column Interaction Equation (check that the interaction equation value ≤ 1.0; if not, the beam-column is inadequate.)

$$\frac{P_u}{\phi_c P_n} = - = \quad ; \quad \text{If } \frac{P_u}{\phi_c P_n} \geq 0.20, \textit{therefore use equation } (8-1), \textit{otherwise use equation } (8-2)$$

$$(8\text{-}1): \quad \frac{P_u}{\phi_c P_n} + \frac{8}{9}\left(\frac{M_{ux}}{\phi_b M_{nx}} + \frac{M_{uy}}{\phi_b M_{ny}}\right) = - + \frac{8}{9}\left(- + -\right) = \quad \leq 1.0$$

$$(8\text{-}2): \quad \frac{P_u}{2\phi_c P_n} + \left(\frac{M_{ux}}{\phi_b M_{nx}} + \frac{M_{uy}}{\phi_b M_{ny}}\right) = - + \left(- + -\right) = \quad \leq 1.0$$

HSS Column Design Template - LRFD

Design of Column (Floor to Floor)

Load case 1 (with $M = 0$)	Load case 2 or Load case 1 with moments
$P_u =$ kips: $KL =$ From **Column Load Tables** (*AISCM* Table 4-3) **Try HSS × (A = in.²)** Is $\phi_c P_n =$ kips > P_u kips OK?	**Check HSS × column for** $P_u =$ kips $M_{x\text{-}x} = M_{ntx} =$ ft.-kips (first-order moment) $M_{y\text{-}y} = M_{nty} =$ ft.-kips (first-order moment) Column unbraced length, $L_b =$; $\boldsymbol{r_x} =$; $\boldsymbol{r_y} =$
	From Table 1-2, F_y for HSS = From *AISCM*: $S_x =$ in.³; $Z_x =$ in.³ ; $Z_y =$ in.³; $S_y =$ in.³ $\phi_b M_{nx} = \phi_b Z_x F_y/12 \leq 1.6\phi_b S_x F_y/12 =$ ft.-kips $\phi_b M_{nx} = (0.9)(\ \ )(\ \ )/12 \leq 1.6(0.9)(\ \ )(\ \ )/12$ = ft.-kips ≤ ft.-kips $\phi_b M_{ny} = \phi_b Z_y F_y/12 \leq 1.6\phi_b S_y F_y/12$ $\phi_b M_{ny} = (0.9)(\ \ )(\ \ )/12 \leq 1.6(0.9)(\ \ )(\ \ )/12$ $\phi_b M_{ny} =$ ft.-kips ≤ ft.-kips

(continued)

Load case 1 (with $M = 0$)	Load case 2 or Load case 1 with moments
	$\phi_c P_n =$ kips; $\phi_b M_{nx} =$ ft.-kips; $\phi_b M_{ny} =$ ft.-kips $M_{ux} = B_{1x}M_{ntx} + B_{2x}M_{ltx}$; $M_{ltx} = 0$ (for braced frames) $M_{uy} = B_{1y}M_{nty} + B_{2y}M_{lty}$; $M_{lty} = 0$ (for braced frames) <u>P-δ Effect for x–x axis</u> $C_{mx} = 0.6 - 0.4(M_{1x}/M_{2x})$; but conservatively use $C_{mx} = 0.85$ $P_{e1x} = \dfrac{\pi^2 EA}{(KL/r_x)^2} =$ kips $B_{1x} = \dfrac{C_{mx}}{(1 - P_u/P_{e1x})} = \dfrac{0.85}{[1-(\ \ /\ \ )]} = \quad \geq 1.0 \therefore B_{1x} =$ $\therefore M_{ux} = B_{1x}M_{ntx} = \quad (\quad) =$ ft.-kips (2nd order moment) <u>P-δ Effect for y–y axis</u> $C_{my} = 0.6 - 0.4(M_{1y}/M_{2y})$; but conservatively use $C_{my} = 0.85$ $P_{e1y} = \dfrac{\pi^2 EA}{(KL/r_y)^2} =$ kips $B_{1y} = \dfrac{C_{my}}{(1 - P_u/P_{e1y})} = \dfrac{0.85}{[1-(\ \ /\ \ )]} = \quad \geq 1.0 \therefore B_{1y} =$ $\therefore M_{uy} = B_{1y}M_{nty} = \quad (\quad) =$ ft.-kips (2nd order moment)

Column Interaction Equation (check that the interaction equation value ≤ 1.0; if not, the beam-column is inadequate)

$$P_u/\phi_c P_n = / = ;\ \text{If}\ \frac{P_u}{\phi_c P_n} \geq 0.20,\ \textit{therefore use equation}\ (8-1), \textit{otherwise use equation}\ (8-2)$$

$$(8\text{-}1):\quad \frac{P_u}{\phi_c P_n} + \frac{8}{9}\left(\frac{M_{ux}}{\phi_b M_{nx}} + \frac{M_{uy}}{\phi_b M_{ny}}\right) = - + \frac{8}{9}\left(- + -\right) = \quad \leq 1.0$$

$$(8\text{-}2):\quad \frac{P_u}{2\phi_c P_n} + \left(\frac{M_{ux}}{\phi_b M_{nx}} + \frac{M_{uy}}{\phi_b M_{ny}}\right) = - + \quad \left(- + -\right) = \quad \leq 1.0$$

8-12 ANALYSIS OF MOMENT FRAMES USING THE AMPLIFIED FIRST-ORDER METHOD (AFOAM)

In lieu of using a general-purpose finite element analysis (FEA) software program that accounts for both *P*-Δ (frame slenderness) and *P*-δ (member slenderness) effects directly and gives the total factored second-order moments, M_u, and factored axial loads, P_u, acting on the columns of a moment frame or unbraced frame, approximate analysis methods are allowed in the AISC Specification, as previously discussed. The amplified first-order analysis method, or AFOAM, presented in Section 8.1 of Appendix 8 of the AISC Specifications is an approximate direct analysis method that can be used to account for both the *P*-Δ and *P*-δ effects. As discussed in Section 8-7, this approximate method uses a first-order frame analysis, which many engineers and civil engineering students are exposed to and familiar with. The procedure that is used to determine the no-translation moments, M_{nt}; the lateral translation moments, M_{lt}; and the axial loads, P_{lt}, is based on the principle of superposition and is described as follows [9]:

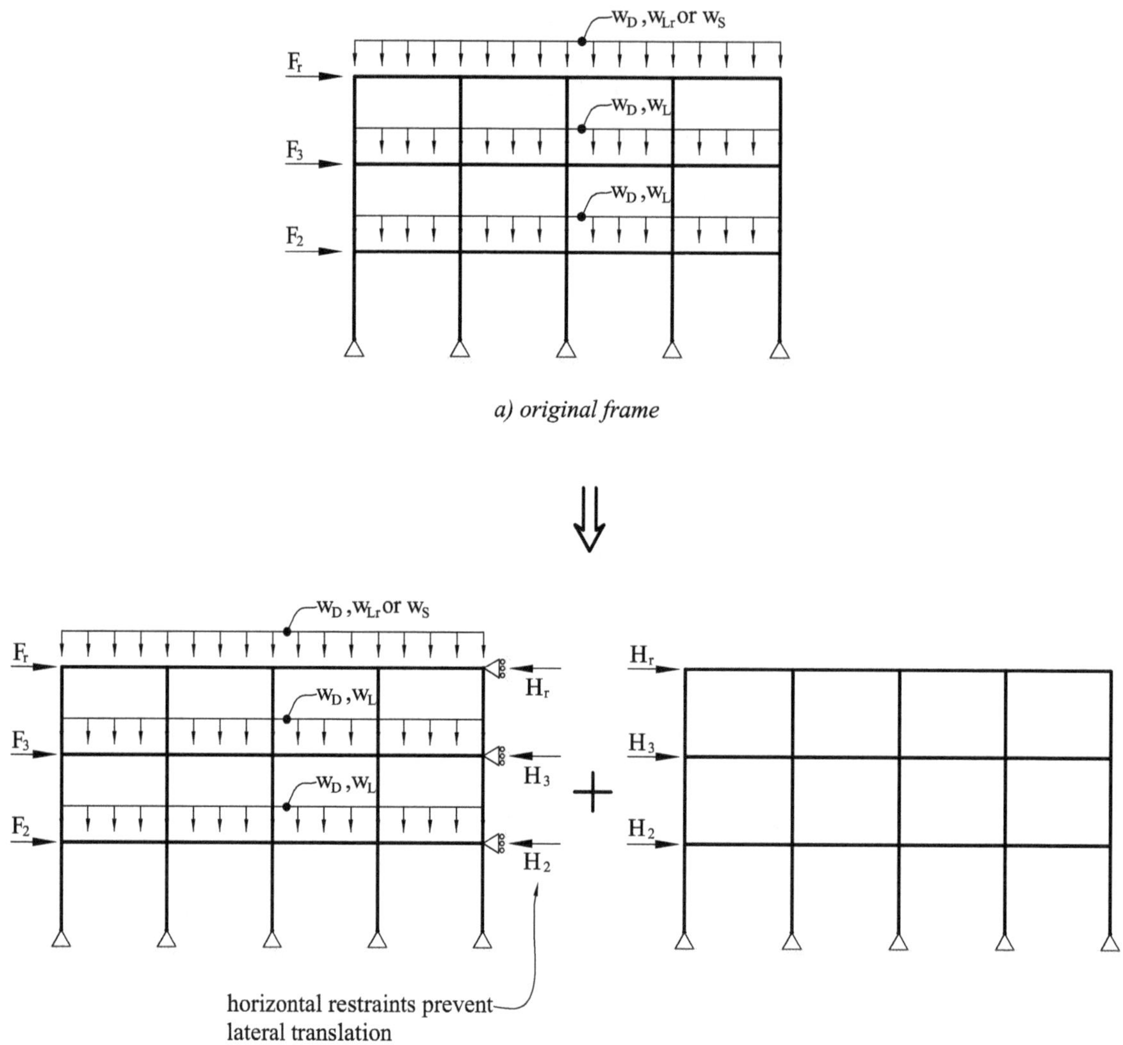

b) No translation Frame: M_{nt} moments (to be amplified by B_1)

c) Lateral translation Frame: M_{lt} moments (to be amplified by B_2)

FIGURE 8-30 a, b, c Determination of M_{nt} and M_{lt} moments in Moment Frames

1. Given the original frame shown in Figure 8-30a, perform a first-order analysis of the frame for all the applicable load combinations (see Section 2-3) that include gravity and lateral loads, and assume that horizontal restraints (i.e., vertical rollers) are added to the structure at each floor/roof level, as shown in Figure 8-30b, to prevent any sidesway of the frame. Therefore, the moments and axial loads obtained are the no-translation moments, M_{nt}, and the no-translation axial loads, P_{nt}. The resulting horizontal reactions, H_2, H_3, ... H_n, and H_{roof}, at the vertical rollers obtained at each floor and roof level are noted (see Figure 8-30b).
2. The fictitious horizontal restraints introduced in step 1 are removed, and for each load combination from step 1, lateral loads are applied at each level to the frames that are equal and opposite to the horizontal reactions obtained in step 1, as shown in Figure 8-30c. Note that according to the AISC specification, notional loads must be applied at

each floor level if the horizontal reactions obtained from step 1 are smaller than the notional loads (see Section 8-6). The frame is reanalyzed for the lateral forces that are equal and opposite to the horizontal reactions obtained in step 1, and the moments and axial loads obtained in this step are the *lateral translation moments*, $M_{u,lt}$, and the *lateral translation axial load*, $P_{u,lt}$.

The analyses in steps 1 and 2 above are carried out using reduced stiffnesses ($0.80\tau_b EI$ and $0.80EA$) for all members of the frame. The conditions under which the stiffness reduction factor, τ_b, may be taken as 1.0 have already been discussed in Section 8-6.

The rationale behind the preceding procedure will now be discussed. The amplified first-order analysis uses the principle of superposition to decompose the total moment that would have been obtained if the original frame (Figure 8-30a) with all of the gravity and lateral loads was analyzed using a second-order frame analysis software that accounts for both geometric and material nonlinearities. Because we have different magnifiers for the *no-translation moment*, $M_{u,nt}$, and the *lateral translation moment*, $M_{u,lt}$, the issue is how to separate the total moment into the two types of moments so that M_{nt} can be amplified by B_1 and M_{lt} can be amplified by B_2. The analysis in step 1, with the horizontal restraints added, represents the braced-frame portion of the original frame so that the moments obtained from step 1— the M_{nt} moments — can then be amplified by the non-sway moment amplification factor, B_1.

For the analysis in step 2, lateral loads that are equal and opposite to the fictitious restraint reactions obtained from step 1 are applied to the frame. Using the principle of superposition, it can be seen that the original frame (Figure 8-30a) has not really been altered, but instead, we have only decomposed the original frame into two component frames (Figure 8-30b and Figure 8-30c), which when added together are equivalent to the original frame. Of course, linear elastic behavior and small displacements are implicitly assumed, which allows the principle of superposition to be valid and applicable to this moment frame. The moments from step 2 are the M_{lt} moments that can then be amplified by the sway moment amplification factor, B_2. Therefore, the final second-order moments, $\boldsymbol{M_u = B_1 M_{nt}}$ (from the step 1 analysis) + $\boldsymbol{B_2 M_{lt}}$ (from the step 2 analysis). The total second-order factored moment, M_u, from the amplified first-order analysis method (AFOAM) usually compares very favorably with results from a second-order finite element analysis that directly calculates M_u and accounts directly for all types of geometric and material nonlinearities [9].

Design of Columns in Moment Frames (i.e., Unbraced Frames)

The procedure for designing columns in moment frames is as follows:

1. Determine the factored axial loads, P_u, on each column in the frame from the gravity loads only.
2. Obtain the preliminary column size based on the axial load from step 1 using the column design template presented earlier. It is assumed that the preliminary sizes of the beams and girders have already been obtained for gravity loads only, assuming simple support conditions.
3. Using the preliminary column and girder sizes, perform the amplified first-order analysis as described in the previous section for all applicable load combinations. This will yield the M_{nt} and M_{lt} moments for each column, as well as the factored axial loads, P_{nt} and P_{lt}. This analysis is carried out in both orthogonal directions of the building if moment frames are used in both directions. Where braced frames are used in one

direction, the analysis has to be carried out only in a direction parallel to the moment frames.

4. Compute the moment magnification factors, B_1 and B_2, and calculate the factored second-order moments and axial load as follows:

$$M_{ux-x} = B_{1x-x} M_{ntx-x} + B_{2x-x} M_{ltx-x}$$
$$M_{uy-y} = B_{1y-y} M_{nty-y} + B_{2y-y} M_{lty-y}, \text{and}$$
$$P_u = P_{u,nt} + B_2 P_{u,lt}$$

5. Using the column design template, design the column for the factored axial loads and the factored moments from step 4.
6. In designing columns in moment frames, the effect of the leaning columns (i.e., columns that do not participate in resisting the lateral loads) is implicitly taken into account since the cumulative gravity loads on all columns in the story (including the leaning columns) are included in P_{story} in the calculation of the B_2 amplification factor in equations 8-11a and 8-11b.

EXAMPLE 8-4 (LRFD)

Calculation of the Sway Amplification Factor, B_2

Calculate the sway amplification factor for the columns in the first story (i.e., the ground floor columns) of the typical moment frame in Figure 8-31 with the service gravity loads and the design lateral wind loads shown. Assume that the direct analysis method (DAM) will be used, and $R_M = 0.85$.

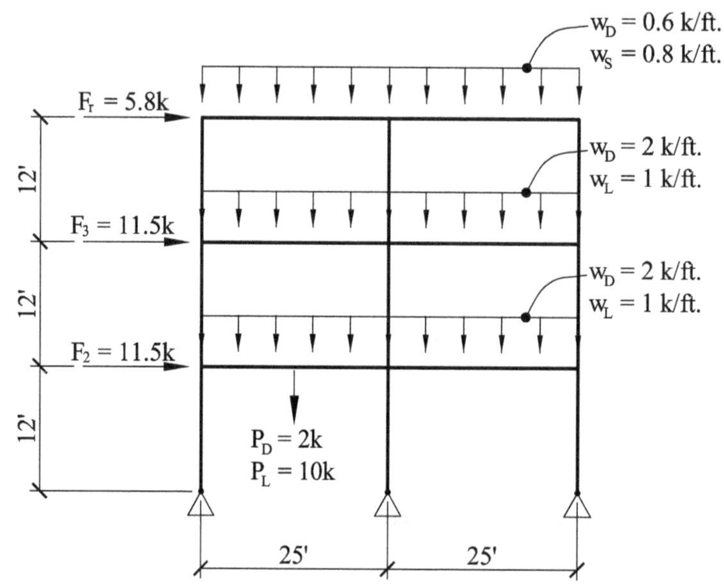

FIGURE 8-31 Unbraced frame for Example 8-4

Solution

The applicable load combinations from Chapter 2 are:

$$1.2D + 1.0W + L + (0.5L_r \text{ or } 0.3S \text{ or } 0.5R), \text{ and}$$
$$0.9D + 1.0W.$$

It should be obvious that the $0.9D + 1.0W$ load combination will not be critical for calculating the sway amplification factor since the higher the cumulative axial load in the story, the higher the amplification; therefore, only the first load combination needs to be checked.

Calculate the cumulative maximum factored total axial load, P_{story}, in the ground-floor columns by summing up the components of the critical load combination as follows:

$$1.2D = 1.2\left[(0.6\text{ kip/ft.})(50\text{ ft.}) + (2\text{ kip/ft.})(50\text{ ft.}) + (2\text{ kip/ft.})(50\text{ ft.}) + (2\text{ kips})\right]$$
$$= 279\text{ kips}$$

$$1.0L = 1.0\left[(1\text{ kip/ft.})(50\text{ ft.}) + (1\text{ kip/ft.})(50\text{ ft.}) + (10\text{ kips})\right] = 110\text{ kips}$$

$$0.3S = 0.3(0.8\text{ kip/ft.})(50\text{ ft.}) = 12\text{ kips}$$

$$P_{story} = 279 + 110 + 12 = \mathbf{401\text{ kips}}.$$

The factored lateral wind shear at the ground-floor columns is

$$H = 1.0W = 1.0(5.8\text{ kips} + 11.5\text{ kips} + 11.5\text{ kips}) = 28.8\text{ kips}.$$

A drift index for *factored* wind loads of $\frac{1}{300}$ is assumed (since a drift index of between $\frac{1}{500}$ to $\frac{1}{400}$ is commonly used in practice for building frames under *service* or unfactored wind loads). Therefore,

$$\frac{\Delta_H}{L} = \frac{1}{300}.$$

The sway moment magnification factor is calculated as

$$B_2 = \frac{1}{1 - \frac{\Delta_H}{L}\alpha\left(\frac{P_{story}}{H\,R_M}\right)} \geq 1.0$$

$$= \frac{1}{1 - \frac{1}{300}(1.0)\left(\frac{401}{(28.8)(0.85)}\right)} = 1.06 < 1.5.\text{ OK}$$

Where, $\alpha = 1.0$ for LRFD

Therefore, the sway amplification factor, B_2 = **1.06**

EXAMPLE 8-5 (LRFD)

Design of Columns in Moment Frames

The typical floor and roof plans for a two-story office building laterally braced with moment frames in both orthogonal directions are shown in Figure 8-32a. The floor-to-floor height is 12 ft., and the uniform roof and floor loads and the lateral wind load have been determined as follows:

Roof dead load = 30 psf   Snow load = 56 psf

Floor dead load = 100 psf   Floor live load = 50 psf

Assume a design uniform lateral wind load, W, of 32 psf. The building is located in a region where seismic loads do not control.

1. Using commercially available structural analysis software, perform an amplified first-order analysis for the East-West moment frame along line A to determine the M_{nt} and M_{lt} moments in the ground-floor columns. For this example, consider only the load

combination $1.2D + 1.0W + L + (0.5L_r$ or $0.3S$ or $0.5R)$. Nevertheless, in practice, the load combination (0.9D + 1.0W) that could cause uplift would also need to be checked. Note that if the notional lateral load is greater than the wind load W, then the notional load should be used as the minimum lateral load.

2. Design column A-3 at the ground-floor level for the combined effects of axial load and moments. Use the amplified first-order analysis method (AFOAM).

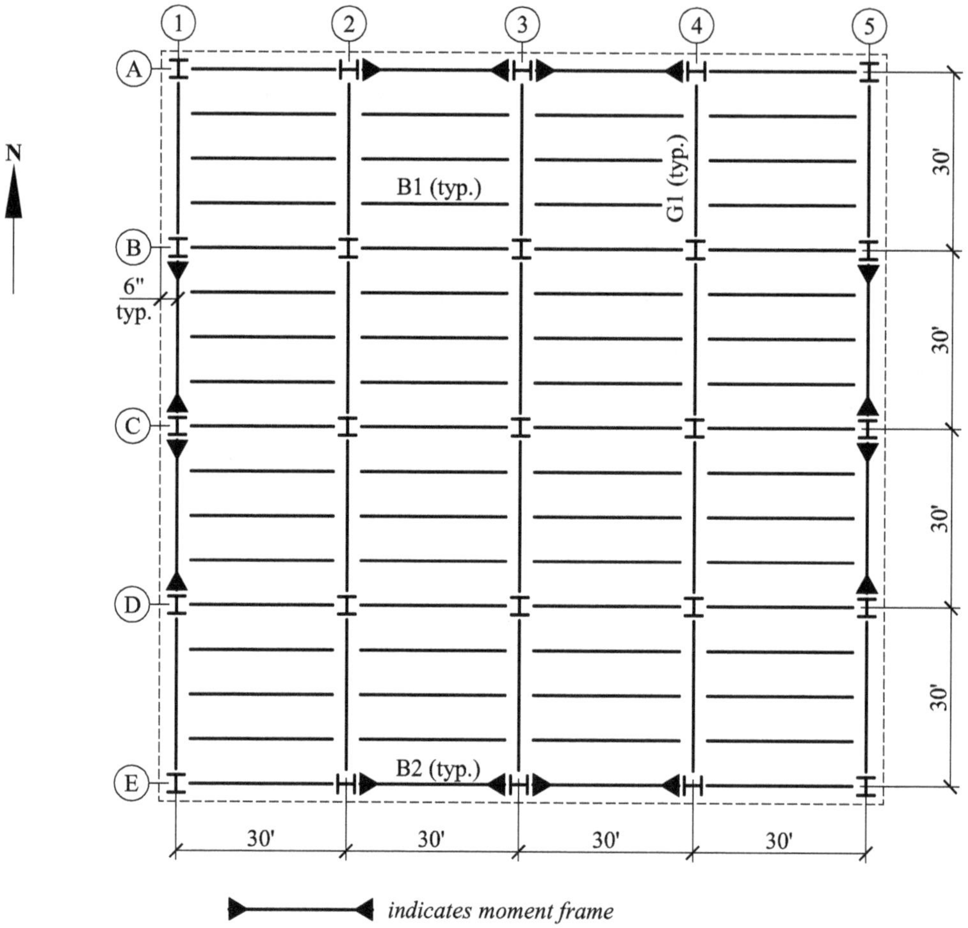

FIGURE 8-32a Framing plan for Example 8-5

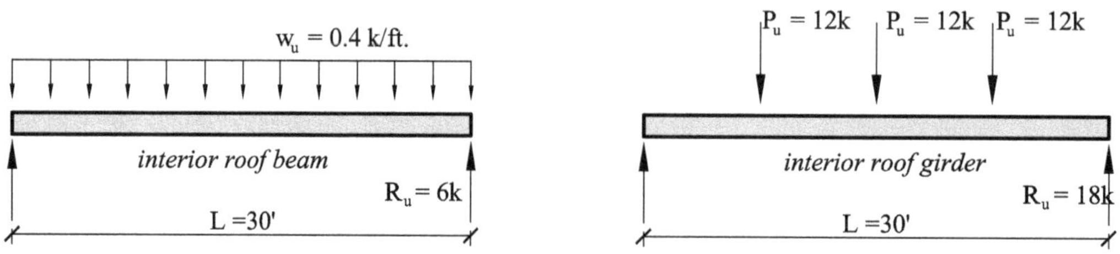

FIGURE 8-32b Roof beam and girder reactions

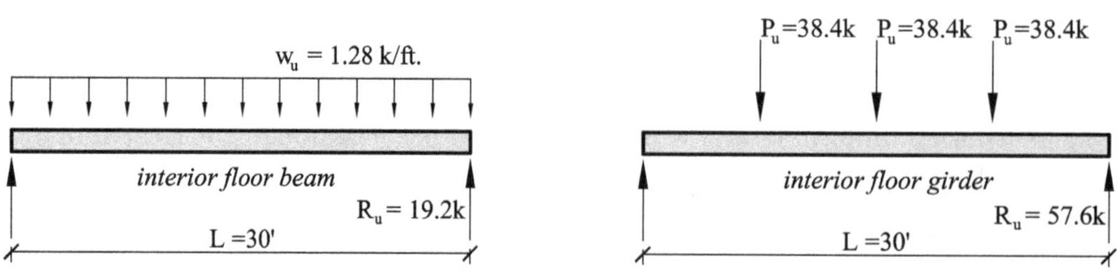

FIGURE 8-32c Floor beam and girder reactions

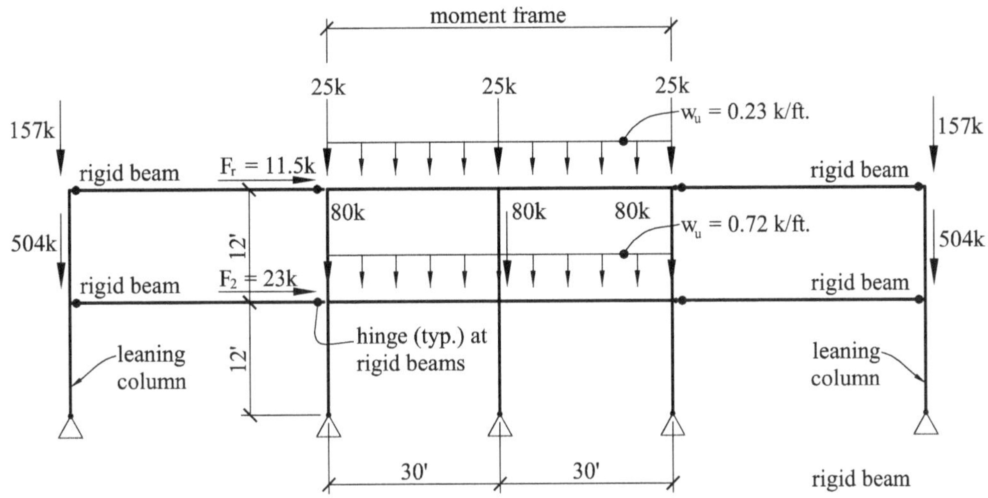

FIGURE 8-32d Factored gravity and lateral loads on East-West moment frame along gridline A

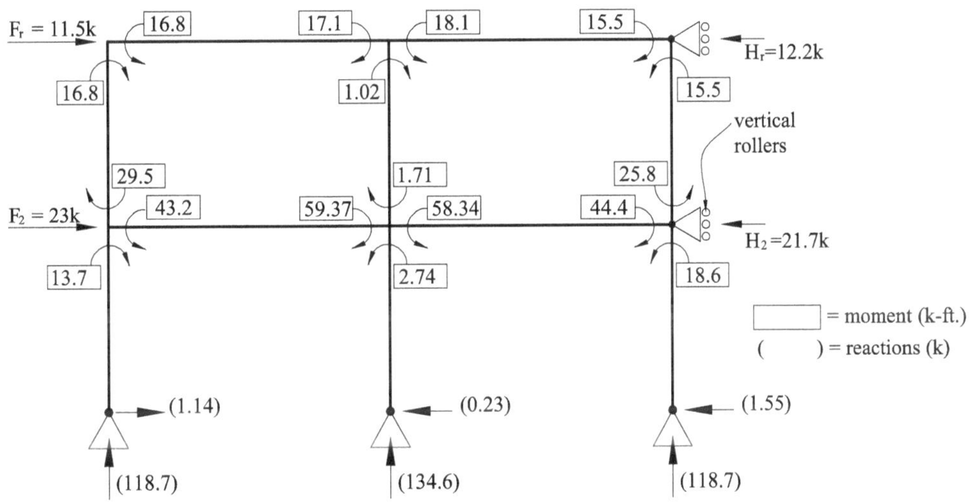

FIGURE 8-32e No translation frame: M_{nt} moments

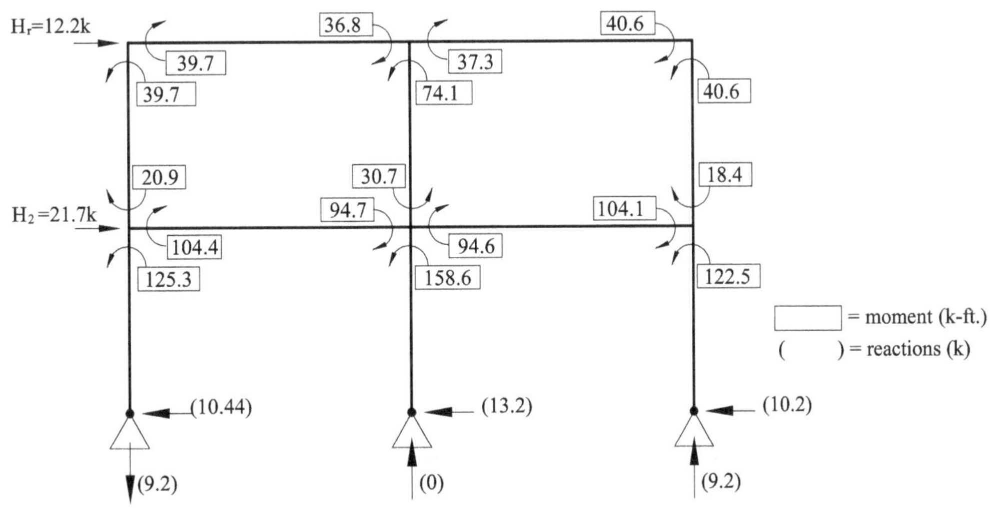

FIGURE 8-32f Lateral translation frame: M_{lt} moments

Solution

Calculate Beam and Girder Reactions: (See Figures 8-32b and 8-32c)

Roof Level -

Spandrel roof beam, **RB2**:

$$\text{Tributary width, TW} = \left(\frac{7.5\text{ ft.}}{2} + 0.5\text{-ft. edge distance}\right) = 4.25\text{ ft.}$$

$$w_D = (30\text{ psf})(4.25\text{ ft.}) = 128\text{ lb/ft.}$$

$$w_S = (56\text{ psf})(4.25\text{ ft.}) = 238\text{ lb/ft.}$$

$$w_u = 1.2D + 0.3S = 1.2(128) + 0.3(238) = 225\text{ lb/ft.} = 0.23\text{ kips/ft.}$$

This uniform load acts on the East-West moment frame spandrel roof beams.

$$\text{Spandrel Roof beam factored reaction, } R_u = (0.23\text{ kips/ft.})\left(\frac{30\text{ ft.}}{2}\right) = 3.4\text{ kips}$$

Interior roof beam, **RB1**:

$$\text{Tributary width, TW} = 7.5\text{ ft.}$$

$$w_D = (30\text{ psf})(7.5\text{ ft.}) = 225\text{ lb./ft.}$$

$$w_S = (56\text{ psf})(7.5\text{ ft.}) = 420\text{ lb./ft.}$$

$$w_u = 1.2D + 0.3S = 1.2(225) + 0.3(420) = 396\text{ lb/ft.} \approx 0.4\text{ kips/ft.}$$

$$\text{Roof beam factored reaction, } R_u = (0.4\text{ kips/ft.})\left(\frac{30\text{ ft.}}{2}\right) = 6\text{ kips}$$

Interior roof girder, **RG1**:

Since two interior infill roof beams frame into the interior roof girder at each of the three load points,

$$\text{Interior Roof Girder factored reaction, } R_u = \frac{1}{2}(12kip + 12kip + 12kip) = 18\text{ kips}$$

Second-Floor Level -

Spandrel floor beam, **B2**:

$$\text{Tributary width, TW} = \left(\frac{7.5\text{ ft.}}{2} + 0.5\text{ft. edge distance}\right) = 4.25\text{ ft.}$$

$$w_D = (100\text{ psf})(4.25\text{ ft.}) = 425\text{ lb./ft.}$$

$$w_S = (50\text{ psf})(4.25\text{ ft.}) = 213\text{ lb./ft.}$$

$$w_u = 1.2D + 1.0L = 1.2(425) + 1.0(213) = 723\text{ lb./ft.} = 0.72\text{ kips/ft.}$$

This uniform load acts on the East-West moment frame spandrel floor beams.

Interior floor beam, **B1**:

$$\text{Tributary width, TW} = 7.5\text{ ft.}$$

$$w_D = (100\text{ psf})(7.5\text{ ft.}) = 750\text{ lb./ft.}$$

$$w_L = (50\text{ psf})(7.5\text{ ft.}) = 375\text{ lb./ft.}$$

$$w_u = 1.2D + 1.0L = 1.2(750) + 1.0(375) = 1275\text{ lb./ft.} = 1.28\text{ kips/ft.}$$

$$\text{Floor beam factored reaction, } R_u = (1.28\text{ kips/ft.})\left(\frac{30\text{ ft.}}{2}\right) = 19.2\text{ kips}$$

Interior floor girder, **G1**:

Since two interior infill floor beams frame into the interior floor girder at each of the three load points,

$$\text{Interior Floor Girder factored reaction, } R_u = \frac{1}{2}(38.4\,\text{kip} + 38.4\,\text{kip} + 38.4\,\text{kip}) = 57.6 \text{ kips}$$

Factored Lateral Wind Loads on the East-West Moment Frame along Grid Line A or E:

The lateral wind loads on each moment frame are:

$$F_{roof} = (1.0)(32 \text{ psf})\left(\frac{12 \text{ ft.}}{2}\right)\left(\frac{120 \text{ ft.} + 0.5 \text{ ft.} + 0.5 \text{ ft.}}{2}\right) = 11.5 \text{ kips}$$

$$F_2 = (1.0)(32 \text{ psf})\left(\frac{12 \text{ ft.}}{2} + \frac{12 \text{ ft.}}{2}\right)\left(\frac{120 \text{ ft.} + 0.5 \text{ ft.} + 0.5 \text{ ft.}}{2}\right) = 23 \text{ kips}$$

Factored Vertical Load on the Columns of East-West Moment Frames:

Roof Load at A-2, A-3, and A-4: 3.4 kips + 18 kips + 3.4 kips = 25 kips

Floor Load at A-2, A-3, and A-4: 10.9 kips + 57.6 kips + 10.9 kips = 80 kips

Leaning Columns:

Total factored Roof loads on leaning columns, LC:

For the East-West moment frames being considered, all the columns except A-2, A-3, A-4, E-2, E-3, and E-4 are lean columns. Since there are two East-West moment frames, the total tributary area of the roof per each East-West moment frame =

(0.5 ft + 120 ft + 0.5 ft) (0.5 ft + 120 ft + 0.5 ft)/2 frames = 7,321ft.^{2}

The tributary area of the columns of each East-West moment frame =

(0.5 x 30 ft + 30 ft + 30 ft + 0.5 x 30 ft) (0.5 ft + 0.5 x 30 ft) = 1395 ft.^{2}

Therefore, the total roof area supported by the leaning columns for each East-West moment frame = 7,321 ft.^{2} - 1395 ft.^{2} = 5,926 ft.^{2}

Roof level:

For the load combination of 1.2D + 1.0W + 1.0L + 0.3S, the total factored gravity load on the leaning columns at the **roof level** per each East-West moment frame = [(1.2)(30 psf) + (0.3)(56 psf)](5,926 ft.^{2})/1000 = 313 kips

Due to symmetry, the vertical factored load at the **roof level** = ½(313 kips) = 157 kips on the leaning columns to the left of the frame and 157 kips to the leaning columns on the right of the frame.

2nd floor level:

Therefore, the **total floor area supported by the leaning columns** for each East-West moment frame = 7,321 ft.^{2} - 1395 ft.^{2} = 5,926 ft.^{2}

For the load combination of 1.2D + 1.0W + 1.0L + 0.3S, the total factored gravity load on the leaning columns at the **floor level** per each East-West moment frame = [(1.2)(100 psf) + (1.0)(50 psf)](5,926 ft.^{2})/1000 = 1008 kips

Due to symmetry, the vertical factored load on the LC at the 2nd floor level = ½(1008 kips) = 504 kips on the leaning columns to the left of the frame and 504 kips to the leaning columns on the right of the frame.

To determine the nonsway moments and axial loads (M_{nt} and P_{nt}) and the sway moments and axial loads (M_{lt} and P_{lt}) using the amplified first-order analysis (i.e., the approximate second-order analysis) method of beam-column design, we replace the original frame in Figure 8-32d with two constituent frames (Figure 8-32e and Figure 8-32f) such that when the load effects of these constituent frames are summed up, we obtain the actual load effects on the original frame. In Figure 8-32e, we restrain the lateral movement at each level of the building structure by adding vertical rollers at each floor and roof levels in the structural analysis, resulting in fictitious horizontal reactions at the roof and floor levels, H_{roof} and H_2. The first-order analysis of this frame (Figure 8-32e), using a typical structural analysis software, also yields the *no-translation* moments, $M_{nt\,x\text{-}x}$ (frame columns are bending about their strong axis), and P_{nt} as shown in Figure 8-32e. Note that the structural analysis in this example is performed with reduced member stiffnesses ($0.8\tau_b EI$ and $0.8EA$). To determine the sway moments and axial loads, the horizontal reactions H_{roof} and H_2 from the first analysis are applied as lateral loads to the frame in a direction opposite the direction of the reactions from the first analysis, and without any other gravity or lateral loads and no horizontal restraints. This second frame (Figure 8-32f) is then analyzed, and this second analysis yields the *lateral translation* moments, $M_{lt\,x\text{-}x}$, and the *lateral translation* axial loads, P_{lt}. (Note that the East-West moment frame columns in Figure 8-32a along grid lines A and E are bending about their strong axis for lateral loads in the East-West direction.) Figures 8-32e and 8-32f show the results of the computer-aided structural analyses.

The cumulative maximum total factored axial load, P_u or P_{story} (for the LRFD load combination $1.2D + 1.0W + L + (0.5L_r$ or $0.3S$ or $0.5R)$, in the ground-floor columns tributary to **each East-West moment frame** (including the axial loads on the leaning columns) is

$$1.2D = 1.2(30\text{ psf} + 100\text{ psf})\frac{(120\text{ ft.} + 0.5\text{ft.} + 0.5\text{ft.})(120\text{ ft.} + 0.5\text{ft.} + 0.5\text{ft.})}{2\ East - West frames} = 1142\text{ kips}$$

$$1.0L = 1.0(50\text{ psf})\frac{(120\text{ ft.} + 0.5\text{ft.} + 0.5\text{ft.})(120\text{ ft.} + 0.5\text{ft.} + 0.5\text{ft.})}{2\ East - West frames} = 366\text{ kips}$$

$$0.3S = 0.3(56\text{ psf})\frac{(120\text{ ft.} + 0.5\text{ft.} + 0.5\text{ft.})(120\text{ ft.} + 0.5\text{ft.} + 0.5\text{ft.})}{2\ East - West frames} = 123\text{ kips}$$

$$P_u = P_{story}\ per\ East - West\ moment\ frame = 1142 + 366 + 123 = 1631\text{ kips.}$$

Note: This P_{story} or P_u value for each East-West moment frame could also have been obtained by summing the loads on the moment frame columns and the leaning columns tributary to each East-West moment frame as follows (see Figure 8-32d):

$$P_u = P_{story}\ per\ East - West\ moment\ frame$$
$$= (3)(25 kips) + (3)(80 kips) + 157 kips + 157 kips + 504 kips + 504 kips = 1637\text{ kips.}$$

The value of 1637 kips is approximately equal to the 1631 kips obtained above using the tributary area method.

The factored lateral wind shear at the ground-floor columns is

$$H = 1.0W = (1.0)(11.5 + 23) = 34.5\text{ kips.}$$

An allowable drift index of $\frac{1}{300}\left(\text{i.e., } \frac{1}{(500)(0.6)}\right)$ for *factored* wind loads (1.0W) is assumed (because an allowable drift index of between $\frac{1}{500}$ to $\frac{1}{400}$ is commonly used in practice for frames under *service* or unfactored wind loads of 0.6W). Therefore, at the factored wind load level (i.e., at 1.0W), the allowable drift index is

$$\frac{\Delta_H}{L} = \frac{1}{300}.$$

$$R_M = 1 - 0.15\left(\frac{P_{mf}}{P_{story}}\right)$$

$$R_M = 1 - 0.15\left(\frac{P_{mf}}{P_{story}}\right)$$

$$= 1 - 0.15\left(\frac{[P_u \text{ for moment frame columns only}]}{P_{story} \text{ for all columns tributary to the moment frame}}\right)$$

$$= 1 - 0.15\left(\frac{[(3.4\,kips + 18\,kips + 3.4\,kips) + (10.9\,kips + 57.6\,kips + 10.9\,kips)](3\,columns)}{1631\,kips}\right)$$

$$= 1 - 0.15\left(\frac{{}^{*}313\,kips}{1631\,kips}\right)$$

$$= 0.97$$

*This load could also have been calculated by summing the moment frame column loads from Figure 8-32d: (3 columns)(25 kips) + (3 columns)(80 kips) = 315 kips which is approximately equal to the 313 kips obtained previously.

There is also an alternate way to calculate R_M. Since the total factored axial load in the East-West moment frame columns, P_{mf}, and the total factored axial load in each story, P_{story}, for all the columns that are tributary to the moment frame are proportional to the total cumulative tributary areas supported by the moment frame columns and the tributary columns in the story, R_M can be rewritten as

$$R_M = 1 - 0.15\left(\frac{\text{Trbutary area of moment frame columns}}{\text{Trbutary area of columns in the story that are tributary to the frame}}\right)$$

$$= 1 - 0.15\left(\frac{[(0.5)(30\,\text{ft}) + 30\,\text{ft} + 30\,\text{ft} + (0.5)(30\,\text{ft})]\left(\frac{30\,\text{ft}}{2} + 0.5\,\text{ft}\right)}{\frac{(0.5\,\text{ft} + 120\,\text{ft} + 0.5\,\text{ft})(0.5\,\text{ft} + 120\,\text{ft} + 0.5\,\text{ft})}{2\ \text{East} - \text{West frames}}}\right)$$

$$= 1 - 0.15\left(\frac{1395\,\text{ft}^2}{7321\,\text{ft}^2}\right)$$

$$= 0.97$$

The sway amplification factor for each East-West moment frame is calculated as

$$B_2 = \frac{1}{1 - \frac{\Delta_H}{L}\alpha\frac{P_{story}}{H R_M}} \geq 1.0$$

$$= \frac{1}{1 - \frac{1}{300}(1.0)\left(\frac{1631\,kips}{(34.5\,kips)(0.97)}\right)} = 1.20 \quad \geq 1.0$$

Therefore, $B_{2x-x} = 1.20$

Where, α = 1.0 for LRFD.

The value of B_{2x-x} could also be obtained by using $\Delta_H = \dfrac{L}{300}$ and the value of

$\alpha \dfrac{P_{story}}{H R_M} = (1.0)\left(\dfrac{1631\,\text{kips}}{(34.5\,\text{kips})(0.97)}\right) = 48.74$ in *AISCM* Table 2-2, yielding $B_{2x-x} \approx 1.20$

Note that the sway amplification is caused by the lateral loading in the East-West direction, which results in the bending of the East-West moment frame columns A-2, A-3, A-4, E-2, E-3, and E-4 about their strong (*x-x*) axis; hence, the amplification factor has been designated as B_{2x-x}. This amplification factor will be used later for the design of Column A-3.

The notional lateral loads at each level of the structure (**which represents the effects of initial geometric imperfections**) can be calculated as follows for the load combination, 1.2*D* + 1.0*W* + *L* + (0.5L_r or 0.3*S* or 0.5*R*):

$$N_{roof} = (0.002)\left[1.2(30\text{ psf}) + (0.3)(56\text{ psf})\right]\frac{\left[(120\text{ ft.} + 0.5\text{ft.} + 0.5\text{ft.})(120\text{ ft.} + 0.5\text{ft.} + 0.5\text{ft.})\right]}{2\textit{ East} - \textit{West frames}} = 0.8\text{ kips}$$

$$N_2 = (0.002)\left[1.2(100\text{ psf}) + (1.0)(50\text{ psf})\right]\frac{\left[(120\text{ ft.} + 0.5\text{ft.} + 0.5\text{ft.})(120\text{ ft.} + 0.5\text{ft.} + 0.5\text{ft.})\right]}{2\textit{ East} - \textit{West frames}} = 2.5\text{ kips}$$

Since $\boldsymbol{B_2} < \mathbf{1.7}$, the notional loads should only be applied to the gravity load combination and not in any of the lateral load combinations. However, since the notional load at each level is less than the corresponding wind load (1.0*W*), the notional loads are not critical and therefore can be neglected in this case (see *AISCM* Section C2.2b(d)).

Weak Axis (*y–y*) Bending Moment in East-West Moment Frame Columns A-2, A-3, and A-4

In the North-South direction, the East-West moment frame columns A-2, A-3, and A-4 will behave as leaning columns. They will be laterally braced at the floor levels for bending in the North-South direction (i.e., about their weak axis) by the North-South moment frames. Since these leaning columns do not participate in resisting the lateral load in the North-South direction, the sway amplification factor for columns A-2, A-3, and A-4 for bending about their weak axis, $\boldsymbol{B_{2y\text{–}y}} = \mathbf{1.0}$, and the sway moments for bending about their weak axis, $M_{lt\,y\text{–}y} = 0$.

The only moments in columns A-2, A-3, and A-4 acting about their weak (*y–y*) axis will occur due to the eccentricity of the girder or beam reactions at the floor level under consideration caused by the gravity loads.

For **column A-3**, the reaction of second-floor girder G1 = 57.6 kips (see Figure 8-32c and previous load calculations).

(**Note:** Only the load combination 1.2D + 1.0W + L + (0.5L_r or 0.3S or 0.5R) was required to be checked in this example exercise)

The eccentricity of girder G1 reaction, $e_{y\text{–}y}$ = $2\tfrac{1}{2}$ in. + $\tfrac{1}{2}$ (column web thickness)
= $2\tfrac{1}{2}$ in. + $\tfrac{1}{2}$ (0.5 in.) assumed
= 2.75 in. ≈ **0.25 ft.**

The unbalanced moment on column A3 about the *y-y* axis at the 2nd floor is $M_{2u,nt\,y\text{-}y}$, and is due to connection eccentricity, $e_{y\text{-}y}$, at the second (2nd) floor level

$M_{2u,nt\,y\text{-}y} = 57.6(0.25\text{ ft.}) = \mathbf{14.4\ ft.\text{-}kips}$.

With 50% moment split between the column segments above and below the 2nd floor level since the floor-to-floor heights are constant at 12 ft., therefore, the factored moment in the column segment just below the 2nd floor level is

$M_{2u,nt\,y-y} = 50\%(14.4 \text{ ft.-kips}) = 7.2 \text{ ft.-kips}$

Effects of Leaning Columns on the East-West Moment Frames

In the East-West direction, the only moment frames providing lateral stability for the building are the frames along grid lines A and E (between grid lines 2 and 4). Since all other East-West girders along the column lines are assumed to be connected with simple shear connections to the columns along grid lines B, C, and D, therefore, the columns along grid lines B, C, and D, as well as the corner columns at A-1, A-5, E-1, and E-5, will behave as leaning columns. These leaning columns are restrained or braced laterally in the East-West direction by the six East-West moment frame columns (A-2, A-3, A-4, E-2, E-3, and E-4) along grid lines A and E.

It should be noted that the leaning columns have to be designed for the factored axial load that they directly support, plus the $M_{nt\,x\text{–}x}$ and $M_{nt\,y\text{–}y}$ moments (i.e., the non-sway moments) resulting from the eccentricities of the beam and girder reactions framing into the leaning columns in the two orthogonal directions. For the East-West direction, the leaning columns are those columns that are not part of the East-West moment frames along grid lines A and E. Therefore, only the East-West moment frame columns (i.e., A-2, A-3, and A-4; and E-2, E-3, and E-4) are not leaning columns. All other columns in Figure 8-32a are leaning columns and will be restrained by the East-West moment frame columns.

The effective length factor, K, for the leaning columns is 1.0, and the sway moment magnifier, B_2, and the lateral translation moment, M_{lt}, for these leaning columns for bending in the East–West direction will be 1.0 and zero, respectively. Thus, for bending in the East–West direction, the only moments for which the leaning columns have to be designed are the moments due to the eccentricities of the beam and girder reactions from the gravity load. In effect, the leaning columns are designed similar to columns in braced frames, as was done in Example 8-3.

Table 8-5 summarizes the different types of columns in this example and the source of the bending moments in the columns, and Table 8-6 provides a summary of the factored axial loads and moments just below the second-floor level for the East-West moment frame columns along grid line A (i.e., columns A-2, A-3, and A-4). These were obtained from a first-order structural analysis with a reduced flexural stiffness of 0.80EI (with τ_b assumed to be 1.0) and an axial stiffness of 0.80EA for all members of the frame.

The design of column A-3 at the ground-floor level is carried out next using a modified form of the column design template introduced earlier in this chapter. Since the direct analysis method is used, the stiffness factor for all columns is $K = 1.0$, and the design yields a **W10 × 68** column as shown in the column design template for column A-3.

Note that τ_b was assumed to be 1.0 in the design of the column, and according to *AISCM* Section C.2.3(c), this necessitates adding an additional notional load of $0.001\alpha Y_i$ at each level of the building ($\alpha = 1.0$ for LRFD and $\alpha = 1.6$ for ASD). This load is additive to the lateral loads and should be applied to all load combinations. Note that this notional load, which represents the effect of assuming τ_b to be 1.0, is different from the notional load that represents the effects of initial geometric imperfections in the structure, which was calculated previously.

The additional notional lateral loads (representing the effect of assuming τ_b to be 1.0) at each level of the structure is calculated as follows for the load combination: $1.2D + 1.0W + L + (0.5L_r$ or $0.3S$ or $0.5R)$:

$$N_{roof(add)} = (0.001)\left[1.2(30\text{ psf}) + (0.3)(56\text{ psf})\right]\frac{\left[(120\text{ ft.} + 0.5\text{ft.} + 0.5\text{ft.})(120\text{ ft.} + 0.5\text{ft.} + 0.5\text{ft.})\right]}{2\ East - West\ frames} = 0.4\text{ kip}$$

$$N_{2(add)} = (0.001)\left[1.2(100\text{ psf}) + (1.0)(50\text{ psf})\right]\frac{\left[(120\text{ ft.} + 0.5\text{ft.} + 0.5\text{ft.})(120\text{ ft.} + 0.5\text{ft.} + 0.5\text{ft.})\right]}{2\ East - West\ frames} = 1.25\text{ kips}$$

Comparing the total shear due to these additional notional loads to the total wind shear indicates that the effect of the notional loads would increase the moments and shears by less than 5%, as calculated below:

$$\frac{0.4\text{ kips} + 1.25\text{ kips}}{11.5\text{ kips} + 23\text{ kips}} = 0.048 \Rightarrow 4.8\%\ increase\ in\ moments\ and\ forces.$$

Multiplying the interaction value of 0.77 for the W10 × 68 column obtained in the design of column A-3 (see Column Design Template for A-3) by 1.048 (i.e., 1 + 4.8%) yields a revised interaction value of 0.81, which is still less than 1.0, thus implying that the W10 × 68 column is still adequate even with the additional notional loads considered.

It should also be noted that even though only one load combination, $1.2D + 1.0W + L + (0.5L_r$ or $0.3S$ or $0.5R)$, was considered in this example, a number of other load combinations (see Section 2-3) would have to be investigated to determine which combination would produce the largest moments and forces in the members. The load combination $0.9D + 1.0W$ should always be investigated to check for possible uplift or overturning at the base of the columns of the lateral force resisting system, and the reader should repeat the stability analysis and design above for this and other applicable load combinations.

TABLE 8-5 Types of Columns (Leaning or Moment Frame) and the Source of Bending Moments in Figure 8-32a

	North–South bending		East–West bending	
Moment frame column	**Type of column**	**Source of bending moments**	**Type of column**	**Source of bending moments**
A-2, A-3, A-4	Leaning	Interior girder eccentricity	Moment frame	Moment frame
E-2, E-3, E-4	Leaning	Interior girder eccentricity	Moment frame	Moment frame
B-2, B-3, B-4,C-2, C-3, C-4,D-2, D-3, D-4	Leaning	Interior girder eccentricity	Leaning	Beam eccentricity
A-1, B-1, C-1, D-1, E-1	Moment frame	Moment frame	Leaning	Beam eccentricity
A-5, B-5, C-5, D-5, E-5	Moment frame	Moment frame	Leaning	Beam eccentricity

TABLE 8-6 Summary of Factored Moments and Axial Loads in Ground-Floor Columns in East–West Moment Frame

Moment frame column	$M_{nt,x\text{-}x}$ Moments, ft.-k	$M_{lt,x\text{-}x}$ Moments, ft.-k	$P_{u\ nt}$, kips	$P_{u\ lt}$, kips
A-2	–13.7	125.3	118.7	+ or –9.2
A-3	2.74	158.6	134.6	0
A-4	18.6	122.5	118.7	+ or –9.2

Design of Column A-3 (Second Floor to Ground Floor) - LRFD

Load Case 1 (with $M = 0$)	Load Case 2 or Load Case 1 with Moments
P_u = 134.6 kips: $KL \approx 1 \times 12$ ft. = 12 ft. From **column load tables** (*AISCM* Table 4-1a) **Try W10 × 68 (A = 20 in.^{2})** $\phi_c P_n$ = 714 kips > P_u kips OK	**Check W10 × 68 column for the following forces from Table 8-6.** P_{unt} = 134.6 kips; P_{ult} = 0 kips $M_{nt,x\text{-}x}$ = 2.74 ft.-kips; $M_{lt,x\text{-}x}$ = 158.6 ft.-kips (computer results) $M_{nt,y\text{-}y}$ = 7.2 ft.-kips; $M_{lt,y\text{-}y}$ = 0 ft.-kips (A-3 is a leaning column in North-South) Column unbraced length, L_b = 12 ft.; r_x = 4.44 in.; r_y = 2.59 in.
Unbraced frame column $K_x = 1.0$ $K_y = 1.0$	**LRFD Beam Design Tables (*AISCM* Table 3-2; F_y = 50 ksi)** $\phi_b M_p$ = 320 ft.-kips; L_p = 9.15 ft.; $\phi_b BF$ = 3.85 kips; L_r = 40.6 ft. > L_b If $L_b \le L_p$, $\phi_b M_{nx} = \phi_b M_{px}$ $\phi_b M_{nx} = \phi_b M_p - \phi_b BF\,(L_b - L_p)$ = [320 – (3.85)(12 – 9.15)] = **309 ft.-kips** $\le \phi_b M_p$ From *AISCM*: S_x = 75.7 in.^{3}; Z_y = 40.1 in.^{3}; S_y = 26.4 in.^{3} $\phi_b M_{ny} = \phi_b Z_y F_y \le 1.6\phi_b S_y F_y$ $\phi_b M_{ny} = (0.9)(40.1)(50)/12 \le 1.6(0.9)(26.4)(50)/12$ $\phi_b M_{ny}$ = **150 ft.-kips** ≤ 158 ft.-kips $\boxed{\phi_c P_n = 714 \text{ kips}; \phi_b M_{nx} = 309 \text{ ft.-kips}; \phi_b M_{ny} = 150 \text{ ft.-kips}}$ **Sway Amplification Factors:** $B_{2x\text{-}x}$ = 1.20; $B_{2y\text{-}y}$ = 1.0 (see previous calculations/discussions) $P_{u\,total} = P_{unt} + B_{2x\text{-}x} P_{ult} = 134.6^* + 1.20(0)^\dagger$ = 134.6 kips (see Table 8-6) *P-δ Effect for x–x axis* $C_{mx} = 0.6 - 0.4(M_{1x}/M_{2x})$; but conservatively use C_{mx} = 0.85 $P_{e1x} = \dfrac{\pi^2 EA^*}{(K_x L/r_x)^2} = \dfrac{\pi^2(0.80)(EA)}{(K_x L/r_x)^2} = 2264 \text{ kips}$ ($K_x = 1.0$ for column in East □ West moment frame for DAM) τ_b *is assumed* = 1.0, ∴ *add an additional notional load of* $0.001\alpha Y_i$. $B_{1x} = \dfrac{C_{mx}}{(1 - P_u/P_{e1x})} = \dfrac{0.85}{[1-(134.6/2264)]} = 0.90 < 1.0 \Rightarrow$ $\boldsymbol{B_{1x} = 1.0}$ $\boxed{\therefore B_{1x} M_{ntx} = 1.0(2.74) = 2.8 \text{ ft.-kips}}$ (2nd order moment) *P-δ Effect for y–y axis* $C_{my} = 0.6 - 0.4(M_{1y}/M_{2y})$; but conservatively use C_{my} = 0.85 $P_{e1y} = \dfrac{\pi^2 EA^*}{(K_y L/r_y)^2} = \dfrac{\pi^2(0.80)(EA)}{(K_y L/r_y)^2} = 1482 \text{ kips}$ ($K_y = 1.0$ for bending about weak axis in DAM) $B_{1y} = \dfrac{C_{my}}{(1 - P_u/P_{e1y})} = \dfrac{0.85}{[1-(134.6/1482)]} = 0.94 < 1.0 \Rightarrow$ $\boldsymbol{B_{1y} = 1.0}$

(continued)

Load Case 1 (with $M = 0$)	Load Case 2 or Load Case 1 with Moments
	$\boxed{\therefore B_{1y}M_{nty} = 1.0(7.2) = 7.2 \text{ ft.-kips}}$ (2nd order moment) $M_{ux} = B_{1x}M_{ntx} + B_{2x}M_{ltx} = 2.8 + (1.20)(158.6) = 193$ ft.-kips $M_{uy} = B_{1y}M_{nty} + B_{2y}M_{lty} = 7.2 + (1.0)(0) = 7.2$ ft.-kips See Table 8-6 for the lateral translation moments, M_{lt}.

* See axial load from the *no-translation* analysis from Figure 8-32e and Table 8-6.
† Axial load in column A-3 from lateral translation of frame (see Table 8-6 for $P_{u,lt}$).

Column Interaction Equation

$$P_u/\phi_c P_n = 134.6/714 = 0.19 < 0.20, \text{ therefore use equation (8-2)}$$

$$\frac{P_u}{2\phi_c P_n} + \left(\frac{M_{ux}}{\phi_b M_{nx}} + \frac{M_{uy}}{\phi_b M_{ny}}\right) = \frac{0.19}{2} + \left(\frac{193}{309} + \frac{7.2}{150}\right) = 0.77 < 1.0 \qquad \textbf{OK}$$

$\therefore$ **W10 × 68 is adequate for column A**-3 *from 2nd Floor to Ground Floor.*

8-13 ANALYSIS AND DESIGN OF BEAM-COLUMNS FOR COMBINED AXIAL TENSION AND BENDING

Members subjected to combined axial tension load and lateral bending moment are common in building structures, and some examples include:

- The bottom chord of trusses (see Figure 8-3),
- Hangers with eccentric axial load (see Figure 8-4),
- Multi-story glass curtainwall mullions hung from roof framing and connected at the base to the floor framing or slab with vertically-slotted connections (VSC) [22]. The mullion is subjected to axial tension load from the weight of the glass curtainwall plus the self-weight of the mullions, in addition to lateral bending moment from lateral wind and seismic loads.

Members subjected to axial tension loads are obviously not susceptible to column buckling, and therefore, the moments in these members are not amplified because there are no P-delta effects in tension members. Also, depending on the amount of axial tension in a member, lateral-torsional buckling due to lateral bending may not occur if the member is subjected to a net tension stress under combined loads. To account for the beneficial effect of axial tension in reducing the tendency for lateral torsional buckling to occur, the AISC Specification Section H1.2 allows the moment gradient factor, C_b, for doubly symmetric members to be multiplied by an amplification factor, which is calculated as follows:

This moment gradient amplification factor is

$$\sqrt{1 + \frac{(1.0)P_u}{P_{ey}}} \qquad \text{for LRFD}$$

$$\sqrt{1 + \frac{(1.6)P_s}{P_{ey}}} \qquad \text{for ASD}$$

Where,

$$P_{ey} = \frac{\pi^2 EI_{y-y}}{L_b^2} = \text{Euler buckling load of the tension member about the } y\text{-}y \text{ axis}$$

L_b = unbraced length of the compression flange of the tension member

I_{y-y} = moment of inertia of the tension member about the y - y or weak axis

Recall from Chapter 6 that the moment gradient factor magnifies the design moment capacity about the strong axis of the wide flange member; thus, the final design moment capacity about the x-x axis is $C_b\phi_b M_{nx}$. The column interaction equations used for combined axial compression plus bending moment can be adapted for the analysis and design of beam-columns under combined axial tension plus bending moment by neglecting the moment amplification factors (see AISC Specification Section H1-2).

The interaction equations for the LRFD method for combined axial tension and bending are written as follows:

If $\frac{P_u}{\phi_t P_n} \geq 0.20$, therefore use equation (8 - 1b), otherwise use equation (8 - 2b)

$$\frac{P_u}{\phi_t P_n} + \frac{8}{9}\left(\frac{M_{ux}}{\phi_b M_{nx}} + \frac{M_{uy}}{\phi_b M_{ny}}\right) \leq 1.0 \qquad (8\text{-}1b)$$

$$\frac{P_u}{2\phi_t P_n} + \left(\frac{M_{ux}}{\phi_b M_{nx}} + \frac{M_{uy}}{\phi_b M_{ny}}\right) \leq 1.0 \qquad (8\text{-}2b)$$

Where,

P = nominal axial tension capacity as previously obtained in Chapter 4.

$\phi_t P_n$ = design axial tension capacity as previously discussed in Chapter 4. For *tension* members, $\phi_t P_n$ is the smaller of the tensile yielding capacity ($0.9A_gF_y$), the tensile rupture capacity ($0.75A_eF_u$), or the block shear capacity as presented in Chapter 4.

P_u = factored axial tension load on the beam-column that occurs in combination with the bending moment, calculated using the LRFD load combinations (see Section 2-3).

ϕ_t = strength reduction factor
= 0.9 for tension yielding
= 0.75 for rupture

M_{ux} = factored moment about the x- axis on the beam-column calculated using the LRFD load combinations. Note that there are no P-delta effects for tension members, so there is no amplification of moments.

M_{uy} = total factored moment about the y- axis on the beam-column calculated using the LRFD load combinations. Note that there are no P- delta effects for tension members, so there is no amplification of moments.

$\phi_b M_{nx}$ = design moment capacity of the beam-column about the x- axis as previously obtained in Chapter 6. This moment capacity is amplified by the C_b factor.

$\phi_b M_{ny}$ = design moment capacity of the beam-column about the y- axis as previously obtained in Chapter 6 $\left(\phi_b M_{ny} = \phi_b Z_y F_y \leq 1.6\phi_b S_y F_y\right)$.

For the **ASD method**, the corresponding interaction equations for beam-columns subject to combined axial tension plus bending are

If $\frac{P_s}{P_n/\Omega_t} \geq 0.20$, therefore use equation (8-1c), otherwise use equation (8-2c)

$$\frac{P_s}{P_n/\Omega_t} + \frac{8}{9}\left(\frac{M_{sx}}{M_{nx}/\Omega_b} + \frac{M_{sy}}{M_{ny}/\Omega_b}\right) \leq 1.0 \qquad (8\text{-}1c)$$

$$\frac{P_s}{2P_n/\Omega_t} + \left(\frac{M_{sx}}{M_{nx}/\Omega_b} + \frac{M_{sy}}{M_{ny}/\Omega_b}\right) \leq 1.0 \qquad (8\text{-}2c)$$

Where,

P_{ps} = unfactored total axial tension load on the beam-column that occurs in combination with the bending moment, calculated using the ASD load combinations (see Section 2-3)

P_n = is the nominal axial tension capacity presented in Chapter 4

Ω_t = factor of safety for tension
= 1.67 for tension yielding
= 2.0 for tension rupture

M_{sx} = total service load moment about the x- axis on the beam-column calculated using the ASD load combinations.

M_{sy} = total service load moment about the y- axis on the beam-column calculated using the ASD load combinations.

M_{nx} = nominal moment capacity of the beam-column about the x- axis as previously discussed in Chapter 6.

M_{ny} = nominal moment capacity of the beam-column about the y- axis as previously discussed in Chapter 6.

Ω_b = factor of safety for bending or flexure = 1.67.

EXAMPLE 8-6

Analysis of Beam-Columns for Axial Tension and Bending (LRFD)

a. A W10 × 33 welded tension member, 15 ft. long, is subjected to a factored axial tension load of 105 kips and factored moments about the strong and weak axes of 30 ft.-kips and 18 ft.-kips, respectively. Assuming ASTM A572, Grade 50 steel, and the member is braced only at the supports, check if the beam-column is adequate. Initially assume Cb = 1.0 and neglect the moment gradient amplification factor.

b. Include the moment gradient amplification factor and recalculate the value of the interaction equation.

Solution

Part (a):

Factored axial tension load, $P_u = 105$ kips

Factored moments, $M_{ux} = 30$ ft.-k and $M_{uy} = 18$ ft.-kips

Unbraced length of tension member, $L_b = 15$ ft.

Calculate the Axial Tension Capacity of the Member:

Since there are *no bolt holes*, only the tensile *yielding* limit state is possible for this tension member:

$$\text{Axial tension design strength, } \phi_t P_n = \phi_t F_y A_g$$
$$= (0.9)(50 \text{ ksi})(9.71 \text{ in.}^2) = \mathbf{437 \text{ kips}}$$

Calculate the bending moment capacities about both the *x–x* and *y–y* axes:

For a W10 × 33, from the LRFD beam design tables (Table 3-2 of the *AISCM*), we obtain the moment capacity about the strong (*x-x*) axis as follows:

$$\phi_b M_p = 146 \text{ ft.-kips; } \phi_b BF = 3.59 \text{ kips; } L_p = 6.85 \text{ ft.} < L_b = 15 \text{ ft.;}$$
$$L_r = 21.8 \text{ ft.} > L_b = 15 \text{ ft.}$$

Therefore, the member is susceptible to inelastic lateral-torsional buckling, and with C_b = 1.0, the design moment capacity for bending about the x-x or strong axis is calculated as

$$\phi_b M_{nx} = \phi_b M_p - \phi_b BF(L_b - L_p) = 146 - 3.59(15 \text{ ft.} - 6.85 \text{ ft.}) \le \phi_b M_p$$
$$= \mathbf{116 \text{ ft.-kips}} \le 146 \text{ ft.-kips}$$

From *AISCM* part 1, $Z_y = 14 \text{ in.}^3$ and $S_y = 9.2 \text{ in.}^3$ for W10 x 33

The design moment capacity for bending about the *y-y* or weak axis is obtained as follows:

$$\phi_b M_{ny} = \phi_b Z_y F_y/12 \le 1.6\phi_b S_y F_y/12$$

$$\phi_b M_{ny} = (0.9)(14)(50)/12 \le 1.6(0.9)(9.2)(50)/12$$

$$\phi_b M_{ny} = \mathbf{52.5 \text{ ft.-kips}} \le 55 \text{ ft.-kips}$$

Therefore, $\phi_b M_{ny} = 52.5$ ft.-kips

$P_u/\phi_t P_n = 105/437 = 0.24 > 0.2 \therefore$ Use interaction equation (8-1*b*).

Using the column interaction equation (8-1b), we obtain

$$\frac{P_u}{\phi_t P_n} + \frac{8}{9}\left(\frac{M_{ux}}{\phi_b M_{nx}} + \frac{M_{uy}}{\phi_b M_{ny}}\right) = \frac{105}{437} + \frac{8}{9}\left(\frac{30}{116} + \frac{18}{52.5}\right) = 0.78 < 1.0 \qquad O.K.$$

Therefore, **W10 × 33** is adequate in combined axial tension plus bending.

Part (b) - Including the Moment Gradient Factor Amplification Factor:

The presence of axial tension reduces the effect of lateral torsional buckling in a flexural member. The amplification factor for the moment gradient factor is calculated as follows:

The Euler buckling load of the tension member about the y-y axis is

$$P_{ey} = \frac{\pi^2 EI_{y-y}}{L_b^2} = \frac{\pi^2(29000 \text{ ksi})(36.6 \text{ in.}^3)}{(15 ft. x 12)^2} = 323 \text{ kips}$$

The amplificator factor is

$$\sqrt{1 + \frac{(1.0)P_u}{P_{ey}}} = \sqrt{1 + \frac{(1.0)(105 \text{ kips})}{323 \text{ kips}}} = 1.15$$

The revised moment gradient factor, C_b, is (1.15)(1.0) = 1.15.

If this revised moment gradient factor is used, the recalculated design moment capacity for bending about the x-x axis is (1.15)(116 ft.-kips) = **133 ft.-kips.**

Therefore, the interaction equation (8-1b) is recalculated as follows:

$$\frac{P_u}{\phi_t P_n}+\frac{8}{9}\left(\frac{M_{ux}}{\phi_b M_{nx}}+\frac{M_{uy}}{\phi_b M_{ny}}\right)=\frac{105}{437}+\frac{8}{9}\left(\frac{30}{133}+\frac{18}{52.5}\right)=0.75<1.0 \qquad O.K.$$

Note the reduction in the value of the interaction equation, however small, due to the positive effect of the axial tension load on lateral torsional buckling. The higher the axial tension load, the greater the amplification factor, and the higher the value of the moment gradient factor because the increased tension load reduces the likelihood of lateral torsional buckling from the lateral bending loads; this increases the design moment capacity of the beam-column about the x-x axis.

8-14 DESIGN OF BEAM-COLUMNS FOR COMBINED AXIAL TENSION AND BENDING

The procedure for designing steel members for combined axial tension plus bending is a trial-and-error process that involves an initial guess of the size of the member, after which the member is analyzed as described in Section 8-13. The design procedure using the LRFD method is outlined below:

1. Determine the factored axial tension load, P_u, and the factored moments about both axes, M_{ux} and M_{uy}.
   If the connection eccentricities, e_x and e_y, about the x- and y-axes are known, the applied factored moments about the orthogonal axes of the column can be calculated as $M_{ux} = P_u e_x$ and $M_{uy} = P_u e_y$.
2. Make an **initial guess** of the member size and calculate the design tension strength, $\phi_t P_n$.
3. Calculate the design bending strength about both axes, $\phi_b M_{nx}$ and $\phi_b M_{ny}$.
   The unbraced length of the tension member will affect the bending moment capacity about the x-x axis, $\phi_b M_{nx}$. In addition, the presence of axial tension will reduce the impact of lateral-torsional buckling on the bending capacity of the member. Calculate the moment gradient amplification factor, and the revised moment capacity about the x-x axis.
4. If $P_u/\phi_t P_n \geq 0.2 \Rightarrow$ Use the interaction equation (8-1b); otherwise, use equation (8-2b).
   Note that for members subjected to combined axial tension and bending, there are no buckling or P-delta effects; therefore, the moment amplification factors are not applicable. $\phi_t P_n$ is the axial tension capacity previously presented in Chapter 4.
5. If the interaction equation yields a value ≤ 1.0, the tension member is deemed adequate. If the interaction equation yields a value > 1.0, the tension member is **not** adequate; therefore, increase the member size and repeat the analysis process.

For design with the ASD method, use unfactored axial tension load and unfactored moments, and use the ASD beam-column interaction equations (i.e., equations (8-1c) and (8-2c)) previously presented in Section 8-13.

8-15 COLUMN BASE PLATES

Base plates are flat steel sections shop welded to the bottom of steel columns to provide bearing for the columns and to help in transferring and spreading the column axial loads to the concrete pier or footing that supports the column. The base plate, which is bent into a bowl shape (see Figure 8-33c), helps to distribute the bearing stresses due to the column loads and prevents crushing of the concrete underneath the column; the base plate also provides temporary support to the column during steel erection by allowing the column (in combination with the anchor rods) to act temporarily as a vertical cantilever. The base plate is usually connected to the column with fillet welds (up to $^3/_4$ in. in size) on both sides of the web and flanges, and they are usually shop welded. The base plate welds are usually sized to develop the full tension capacity of the anchor rods. For wide flange columns, fillet welds that wrap around the tips of the flanges and the curved fillet at the intersection of the web and the flanges are not recommended because they add very little strength to the capacity of the column-to-base-plate connection and yet result in high residual stresses in the welds that could cause cracking [10]. Full-penetration groove welds are also not typically used for column-to-base plate connections because of the high cost of such welds, except for column-to-base plate connections subjected to very large moments. The thickness of base plates vary from 3/8 in. to 6 in., and the preferred material specification for base plates is now ASTM Grade 50 steel, but ASTM A36 steel has for many years been the preferred material specification for base plates. Steel plate availability is shown in Table 8-7 [11]. The base plate is typically larger than the column size (depending on the shape of the column) by as much as 3 in. to 4 in. all around to provide adequate room for the placement of the anchor rod holes outside of the column footprint. For I-shaped or wide flange columns where the anchor rods can be located within the column footprint on either side of the web, the plan size of the base plate may only need to be just a little larger than the column size to allow for the fillet welding of the column to the base plate, but the actual plate size is still dependent on the applied load and the concrete bearing stress (see Figure 8-33).

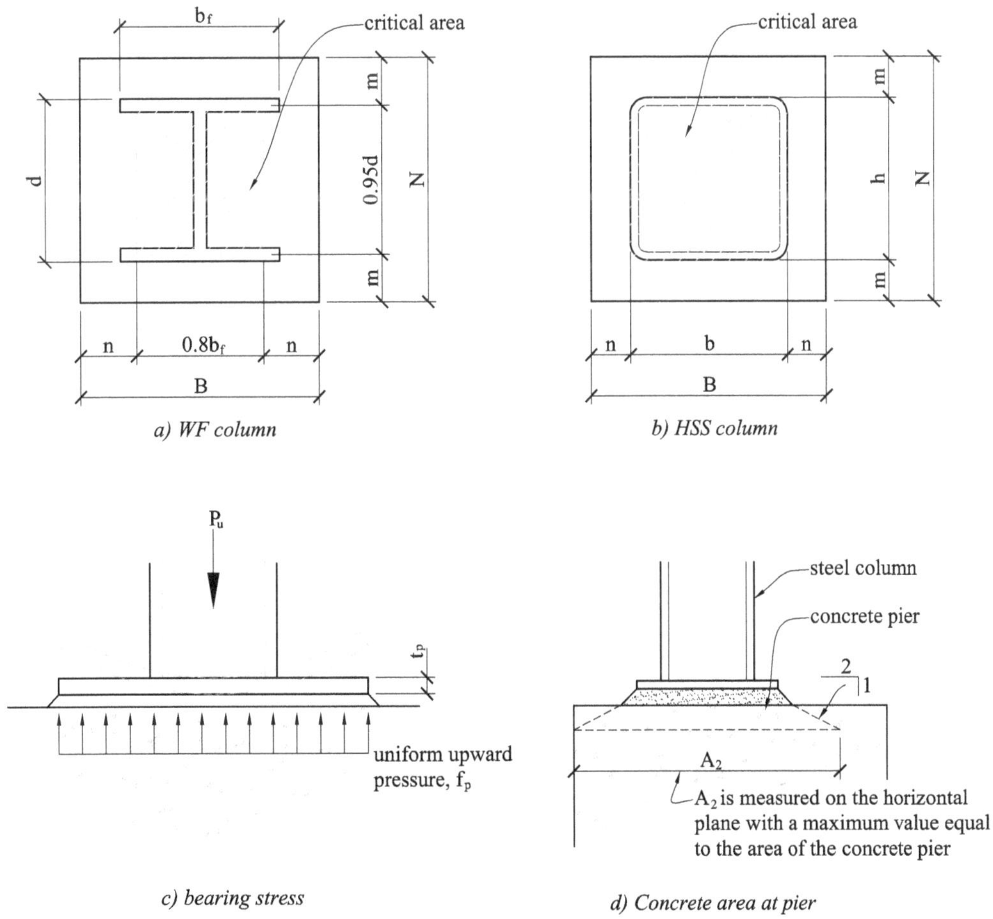

FIGURE 8-33 Column base plate, bearing stresses, and critical areas

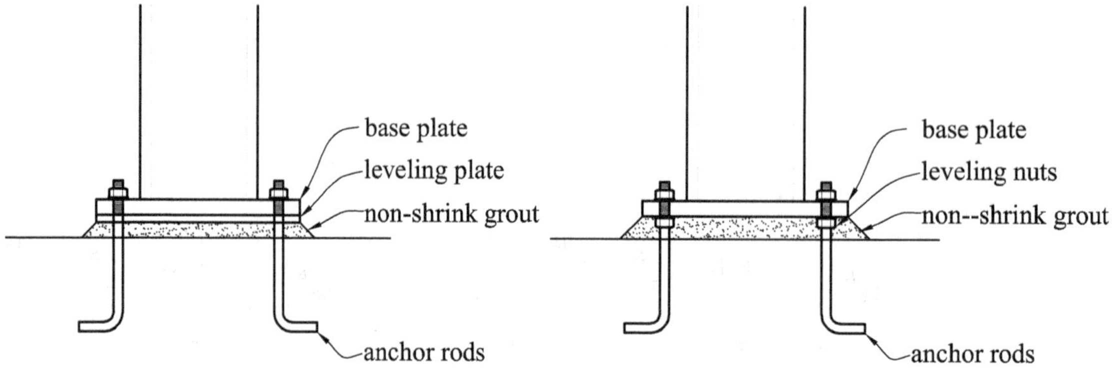

a) leveling plate

b) leveling nuts

FIGURE 8-34 Leveling plate and leveling nut

FIGURE 8-34c Column base plate at diagonal brace locations

The column base plate often bears on a layer of $^3/_4$-in. to $1^1/_2$-in. non-shrink grout that provides a uniform bearing surface and, in turn, the grout is supported on a concrete pier or directly on a concrete footing or pile cap. The compressive strength of the grout should be at least equal to the compressive strength of the concrete used in the pier or footing; however, a grout compressive strength of twice the concrete pier or footing compressive strength is recommended when concrete confinement is assumed in calculating the design bearing stress in the concrete pier or foundation element supporting the column base plate [12]. During steel erection, $^1/_4$-in.-thick leveling plates, which are slightly larger in plan than the base plates or leveling nuts, are used at the underside of the column base plates to align or plumb the column (see Figure 8-34). For base plates larger than 24 in. in width or breadth, or for base plates supporting heavy loads, leveling nuts (see Figure 8-34b) should be used instead of leveling plates for setting or plumbing the columns [12]. Some contractors use shim packs to level steel columns during erection. "*Steel shim packs approximately 4 in. wide are set at the four edges of the base plate. The areas of the steel shim stacks are typically large enough to carry substantial dead load prior to grouting of the base plate.*" [12]. Note that wood shims should never be used to align or plumb steel columns during steel erection, as this could lead to instability and structural collapse as the wood shims rot.

The concrete piers should be larger than the base plate by at least twice the grout thickness to prevent interference between the anchor rods and the pier reinforcing, and the exterior column piers should be poured or cast integrally with the perimeter foundation walls. In the design of steel column base plates, the bearing stresses below the plate are assumed to be uniform, and the base plate is assumed to bend in two orthogonal directions into a bowl-shaped surface (i.e., the base plate is assumed to cantilever beyond the critical area of the column and bend about the two orthogonal axes).

TABLE 8-7 Availability of Base Plate Materials

Base plate thickness, in.	Plate availability
$t_p \le 4$ in.	ASTM A36* ASTM A572, grade 42 or 50 ASTM A588, grade 42 or 50
4 in. $< t_p \le 6$ in.	ASTM A36* ASTM A572, grade 42 or 50 ASTM A588
$t_p > 6$ in.	ASTM A36

*ASTM A36 is the preferred material specification for steel plates.

The design strength of concrete in bearing from ACI 318 [13] is given as

$$\phi_{cb} P_p = \phi_{cb} (0.85 f'_c) A_1 \sqrt{\frac{A_2}{A_1}} \ge P_u \tag{8-15}$$

Where,

P_p = nominal bearing strength = $(0.85 f'_c) A_1 \sqrt{\frac{A_2}{A_1}}$

B = Width of base plate,

N = Length of base plate,

$\sqrt{\frac{A_2}{A_1}}$ = concrete confinement factor, and is such that,

$$1 \le \sqrt{\frac{A_2}{A_1}} \le 2 \tag{8-16}$$

A_1 = Base plate area = $B \times N$,

A_2 = Area of the concrete pier concentric with the base plate area, A_1, projected at a slope of 1-vertical to 2-horizontal from the top of the concrete pier (or the top of the concrete footing when the column base plate is supported directly on the footing) without extending beyond the edges of the concrete pier or footing (see Figure 8-33d).

f'_c = 28-day compressive strength of the concrete in the pier or footing, ksi, and

ϕ_{cb} = Strength reduction factor for concrete in bearing = 0.65 (ACI 318 Table 21.2.1) [13].

P_u = factored axial load in the column at the base plate location

The $\sqrt{\dfrac{A_2}{A_1}}$ term in equations (8-15) and (8-16) accounts for the beneficial effect of confinement of the concrete below the base plate when the concrete pier area (or the area of the footing when the footing directly supports the column) is greater than the base plate area. The ratio $\sqrt{\dfrac{A_2}{A_1}}$ varies between 1 and 2.

In determining the base plate thickness, the base plate is assumed to be rigid enough to ensure a uniform bearing stress distribution at the bottom of the base plate, and the portion of the plate beyond the critical area of the column bends upwards as a cantilever. The uniform bearing stress at the bottom of the base plate due to the column axial load is calculated as

$$f_{pu} = P_u / A_1 \quad \text{(for LRFD)} \tag{8-17a}$$

or

$$f_p = P_s / A_1 \quad \text{(for ASD)} \tag{8-17b}$$

Where,

P_u = the factored total column axial load calculated using the LRFD load combinations

P_s = unfactored total column axial load calculated using the ASD load combinations

The base plate extends or cantilevers beyond the outline or edges of the critical area by the critical distance, ℓ (see Figures 8-33a and 8-33b), and the maximum applied moment in the base plate at the edge of the critical area due to this uniform bearing stress (see Figure 8-33c) is calculated as follows:

$$\textit{For LRFD}: \; M_u = (f_{pu}\ell)\left(\frac{\ell}{2}\right) = \frac{f_{pu}\ell^2}{2} = \frac{P_u\ell^2}{2A_1}, \quad \text{kip-in./in. width} \tag{8-18a}$$

$$\textit{For ASD}: \; M_s = (f_p\ell)\left(\frac{\ell}{2}\right) = \frac{f_p\ell^2}{2} = \frac{P_s\ell^2}{2A_1}, \quad \text{kip-in./in. width} \tag{8-18b}$$

The nominal bending strength of the plate bending about its weak axis is

$$M_n = Z_y F_y = \left(\frac{b_p t_p^2}{4}\right) F_y \tag{8-19}$$

The design bending strength for LRFD and the allowable bending strength for ASD are given as follows:

$$\text{For LRFD}: \; \phi_b M_n = \phi_b Z_y F_y = \phi_b \left(\frac{b_p t_p^2}{4}\right) F_y \tag{8-20a}$$

$$\text{For ASD}: \; M_{allowable} = \frac{M_n}{\Omega_b} = \left(\frac{b_p t_p^2}{4}\right)\frac{F_y}{\Omega_b} \tag{8-20b}$$

where the unit width of the plate, $b_p = 1$ in.

Equating the design bending strength to the applied factored moment (i.e., $\phi_b M_n = M_u$) for LRFD or equating the allowable moment to the applied unfactored moment ($M_{allowable} = M_s$) for ASD gives the required base plate thickness for the LRFD and ASD methods, respectively, as follows:

$$t_p = \sqrt{\frac{2P_u \ell^2}{\phi_b A_1 F_y}} = \ell\sqrt{\frac{2P_u}{0.90F_y BN}} \quad \text{(for LRFD)} \qquad (8\text{-}21a)$$

$$t_p = \sqrt{\frac{2\Omega_{cb} P_s \ell^2}{A_1 F_y}} = \ell\sqrt{\frac{2(1.67)P_s}{F_y BN}} \quad \text{(for ASD)} \qquad (8\text{-}21b)$$

The practical minimum base plate thickness used for columns in steel buildings is $\frac{3}{8}$ in., (i.e., $t_p \geq 3/8$ in.) with thickness increments of $\frac{1}{8}$ in. for base plate thickness less than or equal to $1\frac{1}{4}$ in., and $\frac{1}{4}$-in. increments for base plate thickness greater than $1\frac{1}{4}$ in. The thickness of a base plate should not be less than the column flange thickness to ensure adequate rigidity of the base plate (i.e., $t_p \geq t_{f\,column}$). However, for lightly loaded columns such as posts and wind columns, $\frac{1}{4}$-in.-thick base plates may be used.

The length, N, and the width, B, of the base plate should be specified in 1-inch increments. Square base plates are normally preferred and are more commonly used in practice because the directions of the length and width of the plate do not have to be specified on the structural drawings and verified on site. The use of square base plates also helps minimize the likelihood of construction errors. The critical base plate cantilever length, ℓ, is the largest of the following cantilever lengths beyond the edges of the critical section of the column: m, n, and $\lambda n'$

For **wide-flange columns**, the critical cantilever length of the base plate is calculated as

$$\ell \geq \begin{cases} m = \dfrac{N - 0.95d}{2}, \\ n = \dfrac{B - 0.80b_f}{2}, \text{ and} \\ \lambda n' = \lambda \dfrac{1}{4}\sqrt{db_f}. \end{cases} \qquad (8\text{-}22)$$

For a **square HSS column,** the critical cantilever length of the base plate is calculated as

$$\ell \geq \begin{cases} m = \dfrac{N - b}{2}, \\ n = \dfrac{B - b}{2}, \text{ and} \\ \lambda n' = \lambda \dfrac{b}{4}, \end{cases} \qquad (8\text{-}23)$$

Where,

$\phi_b = 0.9$ (strength reduction factor for bending of the base plate),

$\Omega_b = 1.67$ (factor of safety for bending of base plate),

F_y = Yield strength of the base plate,

d = Depth of wide flange column,

b_f = Flange width of wide flange column,

b = width of square HSS (for square HSS, $d = b_f = b$)

ϕ_{cb} = strength reduction factor for bearing = 0.65
Ω_{cb} = factor of safety for bearing = 1.67
$\phi_{cb}P_p$ = design bearing strength from equation (8-15)

$$\lambda = \frac{2\sqrt{x}}{1+\sqrt{1-x}} \le 1.0 \tag{8-24}$$

where,

$$x = \left[\frac{4db_f}{(d+b_f)^2}\right]\frac{P_u}{\phi_{cb}P_p} \quad \text{(for LRFD)} \tag{8-25a}$$

$$x = \left[\frac{4db_f}{(d+b_f)^2}\right]\frac{P_s}{\left(P_p/\Omega_{cb}\right)} \quad \text{(for ASD)} \tag{8-25b}$$

Since $\frac{P_u}{\phi_{cb}P_p}$ *or* $\frac{P_s}{\left(P_p/\Omega_{cb}\right)}$ are usually less than or equal to 1.0, the value of x from equations (8-25a) and 8-25b) will take on a maximum value of 1.0 or less for both LRFD and ASD. Therefore, it is conservative to assume $\boldsymbol{\lambda}$ **= 1.0** in equations (8-22) and (8-23) [12]. This value is adopted in this text.

The procedure for the design of column base plates can be summarized as follows:

1. Check the bearing capacity of the concrete below the base plate.
2. Check the thickness of the base plate for bending due to the bearing stresses; the thickness varies from ½ in. to 6 inches.
   a. Base plate thickness increments:
      i. ⅛ in. increment for base plates up to 1¼ in. thick
      ii. ¼ in. increment for base plates greater than 1¼ in. thick
   b. Base plate thickness should not be less than the flange thickness for W-shape columns.
3. Plan Dimensions of Base Plate:
   a. For **HSS columns**, select the base plate plan dimension to be larger than the column plan dimensions by 3 in. to 4 in. all around to allow room for placement of the anchor rods outside of the column footprint.
   b. For **W-shape columns**, the anchor rods can be located within the footprint of the column or outside the footprint of the column (typically preferred).
4. Design the anchor rods.

8-16 ANCHOR RODS

Anchor rods are used to safely attach the column base plates to concrete piers, pile caps, or footings and to prevent the overturning of columns during erection. They are also used to resist the base moments and the uplift forces that a column base plate may be subjected to from lateral wind or earthquake forces.

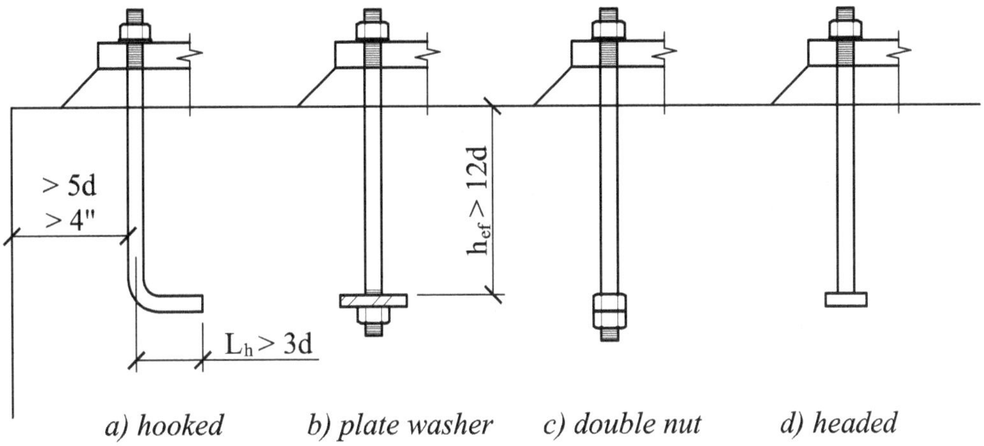

FIGURE 8-35 Types of anchor rods

The most commonly used specification for anchor rods is ASTM F1554 [14]; it covers hooked and headed anchor rods and anchor rods with nuts (see Figure 8-35). This specification provides for anchor rods in grades 36, 55, and 105, with grade 36 being the most commonly used in design practice. Anchor rods in grades 36 and 55 are available in diameters ranging from ¼ in. to 4 in. ASTM F1554, grade 36 anchor rod has a tensile strength, F_u, of 58 ksi and is weldable; the weldability of grade 55 (F_u = 75 ksi) can be enhanced by limiting the carbon content using a supplementary requirement to enhance weldability (i.e., with Supplement S1). Grade 55 anchor rods are used for resisting large tension forces due to uplift from overturning moments or moment connections at the column base plate.

The weldability of anchor rods becomes a desirable property if and when field repairs are required for existing anchor rods with insufficient projection that involve welding a rod extension to the existing anchor rods. The use of grade 105 anchor rods is not recommended because of difficulty with weldability, so it is advisable to use a larger diameter rod size in lieu of grade 105. Thus, F1554 Grade 36 and Grade 55 anchor rods with Supplement S1 are weldable, but Grade 55 without Supplement S1 and Grade 105 are not.

Prior to 1999, ASTM A36 and A307 were the commonly used material specifications for anchor rods, but these have now been largely replaced by ASTM F1554. Note that ASTM F3125 A325 and A490 should not be specified for anchor rods because these specifications are only valid for bolts used in steel-to-steel connections and are only available in lengths of not more than 8 inches. For anchor rods with nuts at the bottom end embedded in the concrete, the heavy-hex nut is most commonly used and is usually tack welded at the bottom of the nut to the threaded rod to prevent the rod from turning loose from the nut when tightening the top nut above the base plate. The preferred material specification for the heavy-hex nuts is ASTM A563, where the required finish and grade of heavy-hex nuts corresponding to the various grades of anchor rods are given.

It is preferable to specify base plates with symmetrical anchor rod layout to prevent any fabrication or erection errors. The minimum and most commonly used anchor rod size in design practice is ¾-in. diameter with a minimum recommended embedment length in the concrete of 12*d* and a minimum embedded edge distance of 5*d* or 4 in., whichever is greater, where *d* is the anchor rod diameter. Therefore, a typical hooked ¾ in. diameter ASTM F1554 anchor rod requires a minimum embedment of 9 inches plus a 3 in. long hook and a minimum edge distance of 4 inches. The projection of the top of the anchor rod above the underside of the non-shrink grout is typically about 4½ inches [23]. Hooked anchor rods (Figure 8-35a) are used at the base of columns that resist gravity loads only, while anchor rods with a washer plate tack welded to a nut (Figure 8-35b) or anchor rods with tack welded double nuts (Figure 8-35c) or headed anchor rods (Figure 8-35d) are used to resist tension forces due to net uplift from overturning moments and/or column base moments; they are typically used at braced-frame column locations or at the base of moment frame columns. The required embedment depths for these nutted and headed anchor rods to resist the applied uplift forces can be determined using Chapter 17 of ACI 318 [13], and the minimum embedment depths for anchor rods are as follows:

Anchor rod diameter, in.	Minimum embedment, in.
¾	9
7/8	10.5
1	12
1½	18

Although a hooked anchor rod does have some nominal tension capacity, they are typically only used for axially loaded columns with compression loads only. In typical building columns that are not part of the lateral load resisting system, four anchor rods with 9-in. minimum embedment and 3-in. long hook (i.e., a total length of 12 inches) are usually specified. To ensure sufficient moment restraint at the column base—thus ensuring safety during steel erection—the Occupational Safety and Health Administration (OSHA) requires that a minimum of four anchor rods be specified at each steel column base plate. The minimum moment to be resisted during the erection of a column, according to OSHA's requirements, is an eccentric 300-lb. axial load (equivalent to the weight of an iron worker with tools) located 18 in. from the extreme outer face of the column in each orthogonal direction [12]. For a typical W12 column (i.e., $d \approx$ 12 inches), this means that the anchor rods have to be capable of resisting a minimum moment about each orthogonal axis of (0.3 kip)(18 in. + (½)(12 in.)) = 7.2 in.-kips. For a typical axially loaded W12 column with nominal $^3/_4$-in.-diameter anchor bolts, the tension pullout capacity with 9-in. minimum embedment and 3-in. hook is approximately 5.7 kips (see Section 8-15). If the anchor rods were laid out on a minimum 3-in. × 3-in. grid, the moment capacity of the four nominal anchor rods is (2 rods)(5.7 kips)(3 in.) = 34.2 in.-kips, which is much greater than the applied moment of 7.2 in.-kips per OSHA requirements.

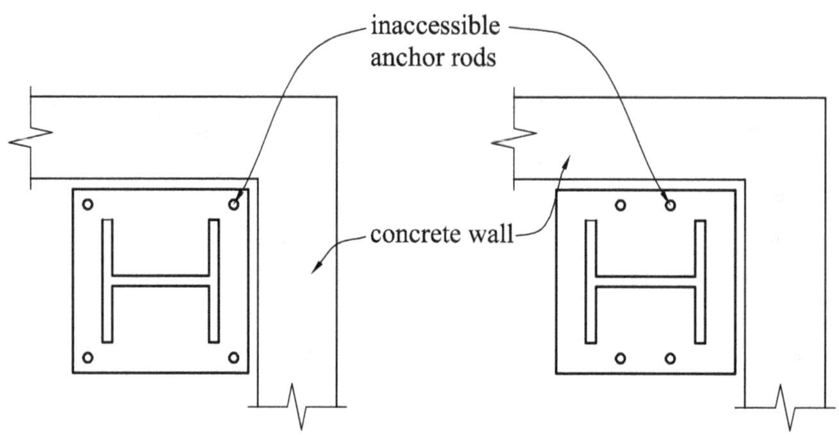

FIGURE 8-36 Column base plate detail with inaccessible anchor rods

For columns that are adjacent to intersecting foundation or basement walls, a special anchor rod layout may be required to provide accessibility to all of the anchor rods because of the presence of a wall, or a construction sequence must be specified that ensures that the column and the anchor rods are in place before the concrete wall is poured [15]. Figure 8-36 shows some details with inaccessible anchor rods that should be avoided in practice [11].

The accurate layout of the anchor rods before the concrete piers or footings are poured is of paramount importance. The plan location of the anchor rods should match the location of the anchor rod holes in the column base plates and must project a sufficient distance above the top of the pier (or footing) to accommodate the nut and the washer. The tolerances for the placement of anchor rods are specified in Section 7.5.1 of the AISC Code of Standard Practice for Steel Buildings and Bridges [3], and the horizontal and tolerances or variation in plan from the specified anchor rod location and the maximum vertical tolerance are as follows:

- ¼ in. *horizontal* tolerance for ¾ in. and 7/8 in. diameter anchor rods;
- 3/8 in. *horizontal* tolerance for 1 in., 1¼ in., and 1½ in. diameter anchor rods; and
- ½ in. *horizontal* tolerance for 1¾, 2 in., and 2½ in. diameter anchor rods.
- The maximum *vertical* tolerance variation between the as-built location of the top of the anchor rod and the specified location is ±½ inch.

When the projection of an anchor rod above the top of a concrete pier is insufficient to accommodate the grout, base plate, nut. and washer because of misplacement of the anchor rod or because of faulty design (see Figure 8-37a), one possible remedial measure is to extend the anchor rod by groove-welding a threaded anchor rod (of similar material) to the existing anchor rod and providing filler or shim plates at the weld location as shown in Figure 8-37b. The filler plates should be tack welded to each other and to the base plate, and the washer plate should also be tack welded to the filler plate. In lieu of groove welding the anchor rods, a threaded coupler that matches the anchor rod size could also be used, and the filler or shim plates would still be required. Note that "*plug welding a partially engaged nut to an anchor rod [with insufficient projection] is not considered an effective means of attachment*" [24].

FIGURE 8-37a Insufficient anchor rod projection at a column base plate

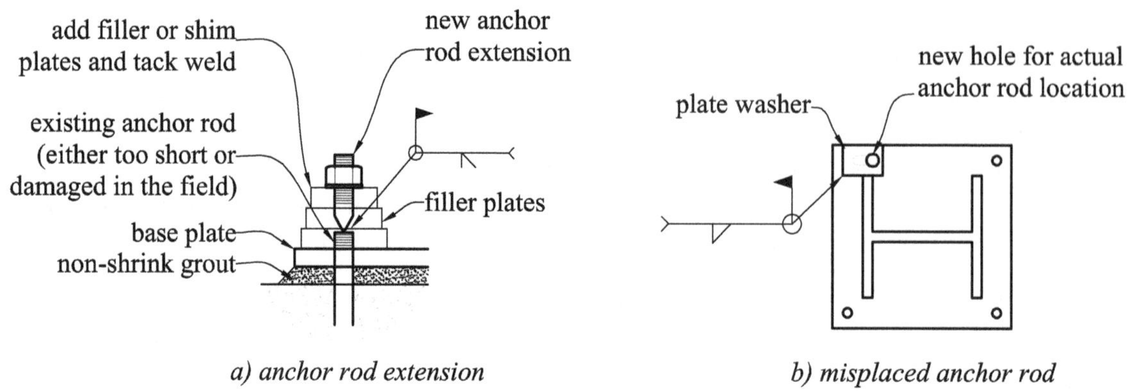

FIGURE 8-37b Repair of misplaced anchor rods

In some situations where the plan locations of anchor rods have been laid out incorrectly, the use of drilled-in epoxy anchors may be the only effective remedy, but care must be taken to follow the epoxy anchor manufacturer's recommendations and design criteria with regard to edge distances and minimum spacing, especially for column base plates that have to resist moments or uplift forces. Smaller anchor spacings and edge distances would lead to a reduction in the pullout capacity of the epoxy anchors. Due to the larger tolerances required for the placement of anchor rods in base plates, the holes in the base plate are typically drilled larger than oversized holes (see Table 9-2); thus, steel washers made of ASTM A36 steel are usually provided for the nuts above the base plate; the washers are located between the underside of the nuts and the top of the base plates. Similarly, steel washers are also provided between the top of the leveling nuts and the underside of the base plates [3].

Some contractors might want to wet-install the anchor rods immediately after the concrete is poured instead of having the anchor rods tied in place together with the pier or footing reinforcement before the pier or footings are poured, but wet-placing of anchor rods should not be allowed because the necessary bond between the concrete and the anchor rods cannot be assured (see Section 7.5 of Ref. [13]). The anchor rods must be set and tied in place within the formwork for the concrete pier or footing before the concrete is poured.

Anchor rods that are inadvertently bent from their vertical alignment due to construction operations or other causes should not be cold-bent back into shape, but rather they should be heat-straightened by heating up the rod to up to 1200°F. The exception to heat straightening repair are ASTM A36 anchor rods of 1 in. or smaller in diameter that are inadvertently bent to an angle not exceeding 45 degrees from the vertical; these rods can be cold-bent back into a vertical alignment [12]. In summary, the purpose of anchor rods is as follows:

- To safely attach the steel column base plates to concrete piers, pile caps, or footings.
- To resist overturning on the column base or steel frame during erection.
- To resist tension forces from uplift and overturning at the column base due to lateral wind and seismic forces.

Some additional general information about anchor rods include the following:

- Anchor rods should **not** be counted on to transfer axial compression loads.

- The most commonly used anchor rod specification is ASTM F1554, which is available in Grades 36, 55, and 105:

	Diameter	Fu (ksi)	Weldability
Grade 36	½ in. to 4 inches	58	**Yes**
Grade 55	½ in. to 4 inches	75	Can improved weldability by limiting carbon content
Grade 105			Grade 105 is NOT recommended because of difficulty with weldability

- It is preferable to specify base plates with **symmetrical layout** of anchor rods.
- The most commonly used anchor rod diameter is ¾ **in.**
- A minimum of **4 anchor rods** per column base are required per OSHA requirements.
- Typical anchor rod specification for columns *without uplift*:
  - (4) ¾ in. diameter anchor rod with 9 in. minimum embedment plus 3 in. long hook (see Figures 8-35a).
- Anchor rods for columns *with uplift and/or moment*:
  - Use anchor rods with anchor plates (see Figures 8-35b), or
  - Use headed anchors (see Figures 8-35c and 8-35d), adequately embedded into the concrete.

8-17 UPLIFT FORCE AT COLUMN BASE PLATES

In this section, we will consider the effects of uplift forces on anchor rods. For the three-story braced frame shown in Figure 8-38, the *unfactored* overturning moment at the base of the building is

$$OM = F_r h_r + F_3 h_3 + F_2 h_2 \tag{8-26}$$

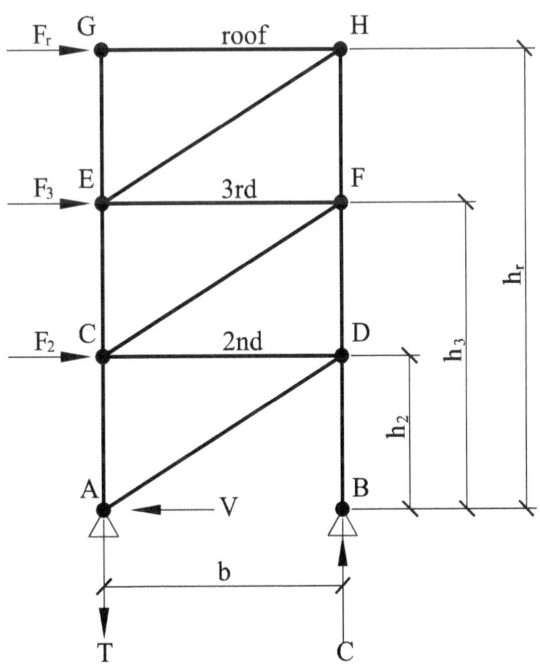

FIGURE 8-38 Typical braced frame

Where,

F_2, F_3, and F_r = the unfactored wind or seismic lateral loads on the braced frame.
h_r, h_3, and h_2 = heights of the roof, third-, and second-floor levels above the ground floor.

Considering only the lateral loads and neglecting the gravity loads for now, and summing up the moments about point B on the braced frame gives

$\Sigma M_B = 0 \Rightarrow \text{OM} - T(b) = 0$

∴ Unfactored uplift force, T = OM/b.

Vertical equilibrium yields the unfactored compression force, C = OM/b,

where

T = Tension force at the base of the column due to lateral loads (negative),
C = Compression force at the base of the column due to lateral loads (positive), and
b = Distance between the braced frame columns.

Assuming that the unfactored cumulative dead load on column AC is P_D and using the load combinations from Section 2-3,

the factored net uplift in column AC = $0.9\,P_D + 1.0\,T_{wind}$ (8-27a)

$0.9\,P_D + 1.0\,T_{earthquake}$ (8-27b)

Note that in equations (8-27a) and (8-27b), the tension force, T, is assumed to take on a negative sign since the dead load, which counteracts the upward tension force, acts downward and is taken as positive.

- If equations (8-27a) or (8-27b) yield a ***net positive value***, then *no uplift* actually exists in the column and, therefore, only nominal anchor rods are required (i.e., four $^3/_4$-in.-diameter anchor rods with 9-in. minimum embedment plus 3-in. hook).
- If equations (8-27a) or (8-27b) yield a ***net negative value***, then a net uplift force exists in the column, base plate, and anchor rods. Therefore, the column, column base plate, anchor rods, concrete pier, and concrete footing or piles have to be designed for this **net** uplift force. Headed anchor rods with adequate embedment into the concrete pier or footing are normally used to resist uplift forces.

The base plate will be subjected to upward bending due to the uplift force that results in a shape of the base plate resembling an upside-down bowl; the plate thickness must be checked for this upward bending moment as shown in Figure 8-39. The bearing elevation of the column footing should also be embedded deep enough into the ground so as to provide adequate dead weight of soil to resist the net uplift force; or in the case of foundations bearing on rock, the footing or drilled pile or caisson should be anchored to the rock bed with high-strength corrosion-protected anchor rods to resist the uplift. The base plate is assumed to be rigid, and thus, prying forces in the anchor rods are assumed to be negligible.

Two cases of uplift will be considered:

Case 1: Net uplift force on base plates with anchor rods outside the column footprint

For anchor rods located outside of the column footprint (see Figure 8-39a), the plate thickness is determined assuming that the base plate cantilevers from the face of the column are due to the concentrated tension forces at the anchor rod locations. The applied moment per unit width in the base plate due to the uplift force is

$$M_u = \frac{\left(\frac{T_u}{2}\right)x}{B} \tag{8-28}$$

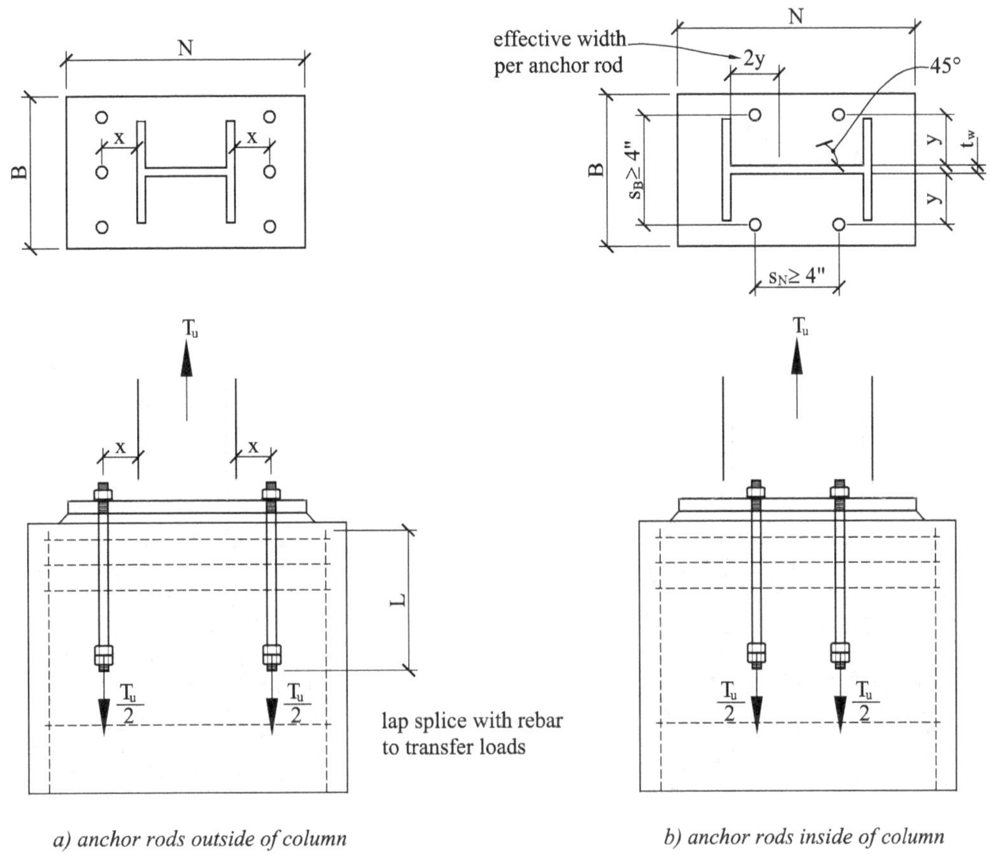

FIGURE 8-39 Bending in base plates due to uplift force

Where,

x = Distance from the centroid of the anchor rod to the nearest face of the steel column,
T_u = Total factored *net uplift force* on the column from the LRFD load combinations, and
B = Width of base plate (usually parallel to the column flange)

The nominal bending strength of the plate bending about its weak axis is

$$M_n = Z_y F_y = \left(\frac{b_p t_p^2}{4}\right) F_y \qquad (8-29)$$

The design bending strength for LRFD and the allowable bending strength for ASD are given, respectively, as

$$\phi_b M_n = \phi_b Z_y F_y = \phi_b \left(\frac{b_p t_p^2}{4}\right) F_y \quad \text{(for LRFD)} \qquad (8\text{-}30a)$$

$$M_{allowable} = \frac{M_n}{\Omega_b} = \left(\frac{b_p t_p^2}{4}\right) \frac{F_y}{\Omega_b} \quad \text{(for ASD)} \qquad (8\text{-}30b)$$

Where,

$b_p = 1$ in.

Equating the design bending strength to the applied factored moment (i.e., $\phi_b M_n = M_u$) for LRFD and equating the allowable moment to the applied unfactored moment $(M_{allowable} = M_s)$ for ASD gives the required base plate thickness for LRFD and ASD, respectively, as

$$t_p = \sqrt{\frac{2T_u x}{\phi_b B F_y}} \quad \text{(for LRFD)} \tag{8-31a}$$

$$t_p = \sqrt{\frac{2\Omega_b T_s x}{B F_y}} \quad \text{(for ASD)} \tag{8-31b}$$

T_s = Total unfactored net uplift force on the column from the ASD load combinations
$\phi_b = 0.9$
$\Omega_b = 1.67$

Case 2: Net uplift force on base plates with anchor rods within the column footprint

For anchor rods located within the column footprint as shown in Figure 8-39b, the plate thickness is determined assuming that the base plate cantilevers from the face of the column web due to the concentrated tension forces at the anchor rod locations. If the distance from the centroid of the anchor rod to the face of the column web is designated as **y**, the effective width of the base plate resisting the bending caused by each anchor rod will be the width of the plate at the face of the column web, assuming a 45-degree dispersion angle as shown in Figure 8-39b. This assumption yields an effective base plate width of **2y** for each anchor rod at the face of the column web. Therefore, the applied moment per unit width in the base plate due to the uplift force is

$$M_u = \frac{\left(\frac{T_u}{n}\right)y}{2y} = \frac{T_u}{2n}, \tag{8-32}$$

where

y = Distance from the centroid of the anchor rod to the web of the wide flange column,
T_u = Total factored net uplift force on the column, and
n = Total number of anchor rods in tension.

Equation 8-32 assumes that the anchor bolts are spaced far enough apart that the influence lines that define the effective plate width for each anchor bolt do not overlap those of the adjacent anchor rod. The design bending strength (LRFD) of the base plate bending about its weak axis is

$$\phi_b M_n = \phi_b Z_y F_y = \phi_b \left(\frac{b_p t_p^2}{4}\right) F_y, \tag{8-33}$$

Where,
$b_p = 1$ in.

Equating the bending strength to the applied moment (i.e., $\phi_b M_n = M_u$) gives the required base plate thickness (for LRFD) as

$$t_p = \sqrt{\frac{2T_u}{\phi_b n F_y}} \quad \text{(for LRFD)} \tag{8-34a}$$

Similarly, for ASD, equating the allowable moment, $M_a = \frac{M_n}{\Omega_b}$, to the applied unfactored moment caused by the net uplift, T_s, gives the required base plate thickness (for ASD) as

$$t_p = \sqrt{\frac{2\Omega_b T_s}{n F_y}} \quad \text{(for ASD)} \tag{8-34b}$$

8-18 TENSION CAPACITY OF ANCHOR RODS EMBEDDED IN CONCRETE

The failure of anchor rods embedded in plain concrete can occur either by the tensile rupture failure of the anchor rod, or by the pullout of the anchor rod from the concrete, or by the breakout of the concrete. The uplift capacity of the anchor rod is the smaller of the following:

- Tension capacity of the anchor rod in tensile rupture
- Pullout Strength in Tension of a Hooked Anchor rod in Concrete

Tensile Capacity of the Anchor Rod:

The calculation of the tension capacity of steel threaded rods in tension was discussed previously in Chapter 4. The tension capacity of the anchor rod in tensile rupture is given as

$$\phi R_n = (0.75)(\phi) F_u A_b, \quad (8\text{-}35)$$

Where,
A_b = Gross area of the anchor rod = $\pi d_b^2/4$,
F_u = Ultimate tensile strength of the anchor rod (58 ksi for grade 36 steel), and
$\phi = 0.75$

For a nominal ¾-in.-diameter anchor bolt with 9-in. minimum embedment and 3-in. hook into a concrete footing or pier (ASTM F1554, grade 36 steel) in 4000-psi concrete, the uplift capacity is the smaller of

$$\phi R_n = (0.75)(\phi) F_u A_b = (0.75)(0.75)(58 \text{ ksi})\frac{(\pi)\left(\frac{3}{4}\right)^2}{4} = 14.4 \text{ kips}$$

Pullout Strength in Tension of a Hooked Anchor Rod in Concrete:

From equation (17.6.3.2.2b) in Chapter 17 of ACI 318 [13], the pullout strength in tension of a hooked anchor rod i.e., J- or L-bolts) embedded in plain concrete is:

$$\phi R_n = \phi N_p = \phi(0.9 f'_c e_h d_b), \quad (8\text{-}36)$$

where
$\phi = 0.7$ (ACI 318 Table 17.5.3(c)),
e_h = Hook extension, where $3d_a \le e_h \le 4.5d_a$
d_a = Diameter of anchor rod
f'_c = Concrete compressive strength at 28 days in psi.

For a nominal ¾-in.-diameter anchor bolt with 9-in. minimum embedment and 3-in. hook into a concrete footing or pier (ASTM F1554, grade 36 steel) in 4000-psi concrete, the pullout strength in tension of a hooked anchor rod (see ACI 318 Section 17.6.3.2.2) is:

$\phi R_n = \phi(0.9 f'_c e_h d_a) = (0.7)(0.9)(4000)(3 \text{ in.})\left(\frac{3}{4}\right)$ = **5.7 kips** (this smaller value controls)

Where,

$d_a = d_b$ = anchor rod diameter = ¾ in.

$3d_a = 3\left(\frac{3}{4}\right) = 2.25$ in. *and* $4.5d_a = 4.5\left(\frac{3}{4}\right) = 3.38$ in.
and $3d_a \le e_h \le 4.5d_a$ *or* $2.25 \text{ in.} \le e_h \le 3.38$ in.

Since the hook extension, $e_h = 3$ in., which falls within the ACI Code prescribed limits, therefore OK.

In Section 8-14, this anchor rod pullout capacity was used to determine the overturning moment capacity of a column base plate and was found to meet the OSHA steel column erection requirements.

To calculate the pullout and breakout strengths of a single or a group of headed or nutted anchor rods in plain concrete, the reader should refer to Chapter 17 of ACI 318 [13]. It should be emphasized that hooked anchor rods are not recommended for resisting uplift loads or for column base plates subjected to moments or lateral loads. The hooked anchor rods should be used only for leaning columns and to provide temporary stability for steel columns during erection. Only headed anchor rods or anchor rods with nuts should be used to resist uplift forces or moments at column bases.

Alternatively, the uplift tension forces can be transferred through bond from the anchor rods to the vertical reinforcement in the concrete pier or pile cap, thus obviating the need to use Chapter 17 of ACI 318 [13], where concrete breakout is the failure mode. For this tension force transfer mechanism through bond to be realized, the anchor rods have to be tension lap spliced with the vertical reinforcement in the concrete pier using a tension lap splice length of 1.3 times the tension development length of the vertical reinforcement in the pier; this required lap-splice length will determine the minimum height of the concrete pier (or minimum depth of the pile cap) and hence the bearing elevation of the concrete footing. For example, if a 4000 psi concrete pier is reinforced with No. 8 rebars (f_y = 60,000 psi), the required tension splice length to transfer the tension force from the anchor rods into the steel reinforcement in the concrete will be approximately 37 inches [16]; allowing for concrete cover, the minimum requirement height of the concrete pier would be approximately 42 inches, or 3 ft. 6 in [16]. For anchor rods that are lap spliced with pier reinforcement, the tension-yielding capacity of the anchor rod [12] is

$$\phi_t R_n = \phi_t (0.75 A_b) F_y, \tag{8-37}$$

Where,

$\phi_t = 0.9$

$0.75A_b$ = Tensile stress area of the threaded anchor rod (see Chapter 4),

A_b = Nominal or unthreaded area of the anchor rod, and

F_y = Yield strength of the anchor rod.

For the base plate and anchor rod examples in this chapter, it is assumed that the full tension capacity of the anchor rods will be developed either by one of the following mechanisms:

- Tension lap splice of the anchor rods with the pier reinforcement or
- Provide adequate concrete pullout capacity ensured through adequate embedment of the nutted or headed anchor rod into the concrete footing or pier, with adequate edge distance and spacing between the anchor rods.

EXAMPLE 8-7

Uplift at the Base of Columns in Braced Frames (LRFD)

The three-story braced frame shown in Figure 8-40 is subjected to the service gravity loads and the design wind loads shown. Determine the maximum factored uplift tension force at the base of the ground-floor column.

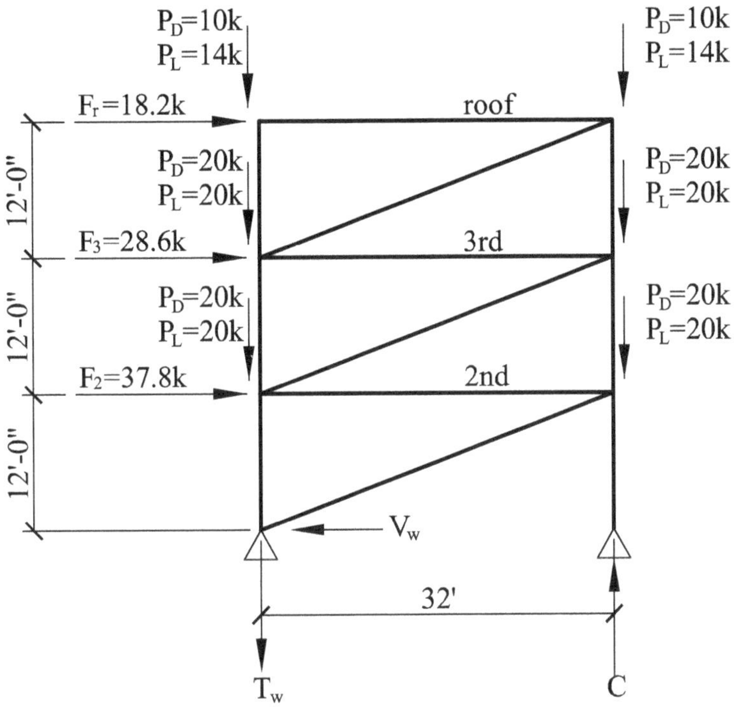

FIGURE 8-40 Braced frame for Example 8-7

Design Wind Load:

Roof level : 18.2 kips

Third-floor level : 28.6 kips

Second-floor level : 37.8 kips

Dead Load on Column:

Roof level : 10 kips

Third-floor level : 20 kips

Second-floor level : 20 kips

Live Load on Column:

Roof level (snow): 14 kips

Third-floor level : 20 kips

Second-floor level : 20 kips

Solution

Using equation (8-26), the maximum factored overturning moment at the base of the column due to wind loads is,

OM = (18.2)(36 ft.) + (28.6)(24 ft.) + (37.8)(12 ft.) = 1795 ft.-kips

Recall from Section 2-3 that the maximum load factor for wind (and seismic) loads is 1.0. Therefore, the factored uplift force due to the overturning moments from wind loads is

$$T_W = \pm \text{OM}/b = \pm 1795 \text{ ft.-kips}/32 \text{ ft.}$$
$$= \pm 56 \text{ kips } (-\text{ve for tension and } +\text{ve for compression}).$$

The cumulative unfactored dead load at the base of the column, P_D = 10 + 20 + 20 = 50 kips

The factored net uplift caused by the tension force due to wind is calculated using equation (8-27a) as

$$0.9D + 1.0W = 0.9(50) + 1.0(-56) = -11 \text{ kips}.$$

The sign of the tension force in this load combination is negative because the tension force acts in an opposite direction to the downward-acting dead load, D.

Therefore, the column, base plate, anchor rods, pier, and footing will be subjected to this net uplift tension force of 11 kips, and they must be designed to resist this force. The designer should also ensure that the concrete footing is embedded deep enough into the ground to engage sufficient dead load from its self-weight, plus the weight of the soil above it, to resist the net uplift force if the weight of the soil is needed to counteract the uplift. Often, the spread footings for braced frame columns may need to be placed at lower elevations compared to other columns in order to engage enough weight of soil to provide adequate resistance to the net uplift.

If drilled shafts (i.e., caissons) or pile foundations are used, they must also be designed to resist the net factored uplift force in addition to the lateral shear force and the accompanying lateral moment in the drilled shafts. The piles or drilled shafts may need to be embedded or socketed into the bedrock to resist the net uplift force. For resisting lateral loads, the point below grade at which fixity may be assumed for piles or drilled shafts (and where the lateral moment is calculated) is usually specified by the Geotechnical Engineer in the Geotechnical Report for the project.

Increased Compression Load on Braced-Frame Columns due to Overturning Moment

The maximum compression force on the braced-frame column due to overturning moments (see equation (8-26)) should also be calculated using the load combinations from Section 2-3. The column, base plate, pier, and footing (or piles and drilled shafts) also need to be designed for this increased compression load.

8-19 RESISTING LATERAL SHEAR AT COLUMN BASE PLATES

The base of columns in braced frames, moment frames, and sloping columns, are usually subjected to lateral shear in addition to moments, axial compression, and uplift tension forces. This lateral shear can be resisted by one, or a combination, of the following lateral force transfer mechanisms (note that in practice, the lateral shear resistance from surface friction due to the bearing of the column base plate on the grout is typically neglected since this resistance will be limited in column base plates with large moments or net tensile uplift forces):

- **Bearing of the steel column against the concrete floor slab (see Figure 8-41a)**
  In this force transfer mechanism, the lateral shear is transferred into the slab by the flanges or the web of the column bearing against the floor slab. The load that can be transferred by this mechanism is limited by the bending in the flange and web of the steel column due to the bearing stress. This appears to be the most commonly assumed lateral shear transfer mechanism in design practice, at least for low-rise buildings. Sometimes, horizontal steel channel struts with headed studs can be welded to the underside of the column base plate and extended for a sufficient length into the concrete slab to transfer the shear into the concrete slab, which, in turn, transfers the lateral shear into the footing through the vertical dowels (see Figure 8-41a).

- **Bending of the anchor rods (see Figure 8-41a)**
  For this force transfer mechanism, the anchor bolts are assumed to bend in double curvature in resisting the lateral shear, with a point of inflection assumed to occur in the anchor rod at the mid-depth of the grout. In addition, the anchor rod will also resist any tension force on the column base. This lateral shear transfer mechanism, which involves double curvature bending of the anchor rods, is only possible if the anchor rod can bear directly against the column base plates; however, if oversized anchor rod holes are used in the base plate (see Table 9.2), slippage of the base plate will occur until the anchor rod bears against the sides of the holes in the base plate. To avoid this slippage, the anchor rod washer and nut must be welded to the base plate to help transmit the shear from the base plate into the anchor rod through the nut and washer. This lateral shear transfer mechanism is not recommended by the authors because of the limited bending capacity of the anchor rods, and the distribution of the lateral shear among the anchor rods may not be uniform or equal.
- **Shear lugs, shear stubs, or shear plates (see Figure 8-41b)**
  For this force transfer mechanism, a shear lug consisting of a steel plate or a wide flange stub welded to the bottom of the column base plate provides the vertical bearing surface required to transfer the lateral shear from the base plate to the concrete. The shear lug plate is subject to bending about its weak axis (see Figure 8-41b). When the shear lug is a wide-flange stub section, it is oriented so that it bends about its strong axis as it transfers the lateral shear from the column base plate into the surrounding concrete. For the design of shear lugs to resist lateral shear, the reader should refer to the AISC *Design Guide 1* [12].
- **Embedded plates (see Figure 8-41c)**
  For this force transfer mechanism, a horizontal plate is embedded into the concrete slab in front of the column parallel to the lateral force-resisting system. A vertical gusset plate is welded to the top of the column base plate and to the top of the embedded horizontal plate; the gusset plate is used to transfer the lateral shear from the column base plate through the embedded plate into the concrete slab. An alternate detail might involve plates with headed studs embedded into the concrete on both sides of the column base plate, and angles (with length parallel to the lateral shear) are field welded to the embedded plates and to the base plate (see Figure 8-41c).
- **Hairpin bars or tie rods in slabs-on-grade for industrial buildings (see Figure 8-41d)**
  Hairpin bars are U-shaped steel reinforcements that are wrapped around the steel columns in light gauge metal buildings to resist the lateral shear at the base of the column (see Figure 8-41d). The reinforcing is designed as a tension member to transfer the lateral shear in the base of the column into the concrete slab-on-grade. The reinforcing has to be developed for a sufficient length into the concrete slab-on-grade to transfer this shear force, and often the reinforcing will cross a control joint in the slab-on-grade. The reinforcing must be developed past this control joint.

Figure 8-41e shows the details at an offset or sloped column for resisting the induced lateral tension and compression forces.

Of all the base details for resisting lateral base shear presented in the preceding discussion, Figures 8-41c (embedded plates) and 8-41d (shear lugs) are the more commonly used.

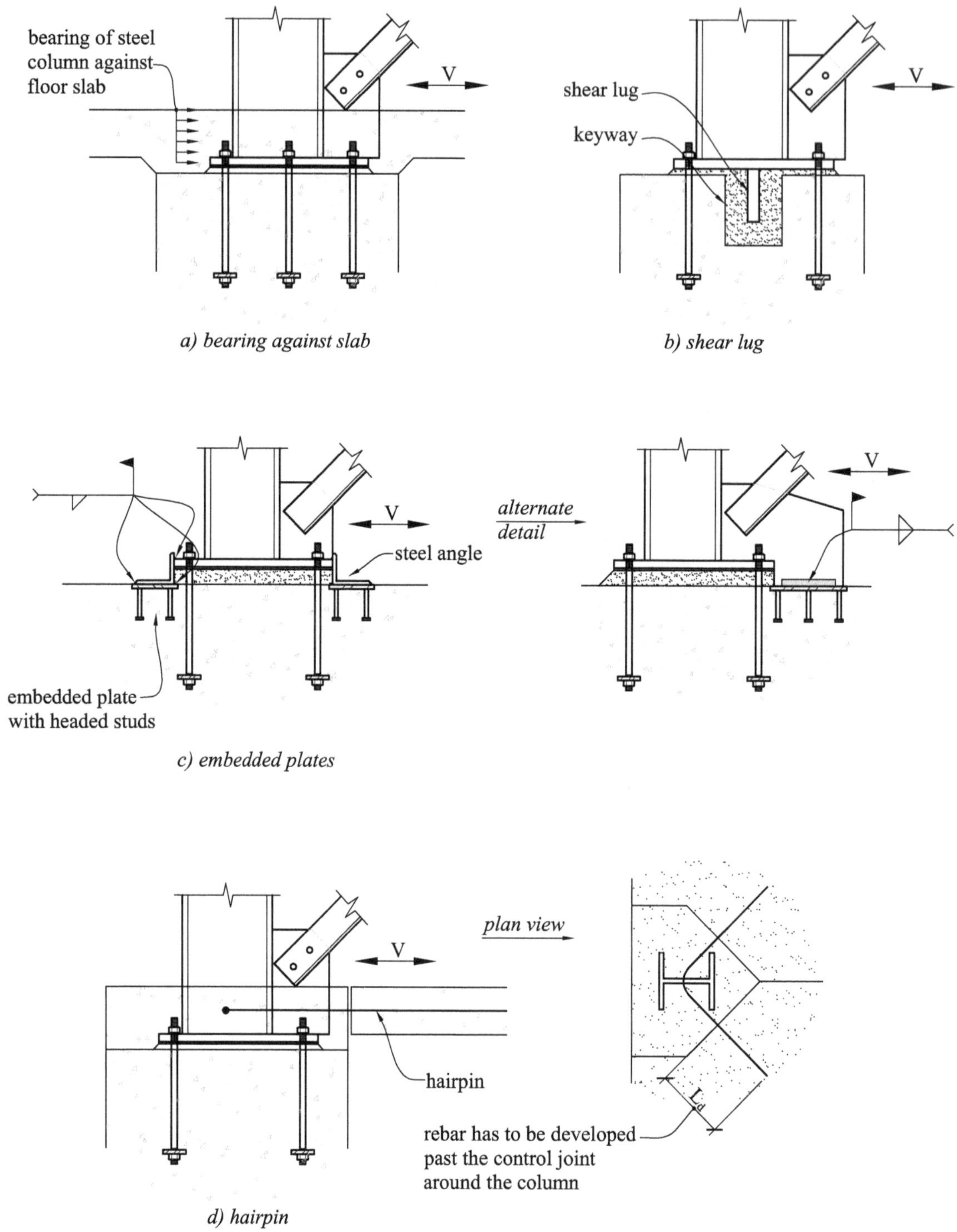

FIGURE 8-41a to d Details for resisting shear at column base plates

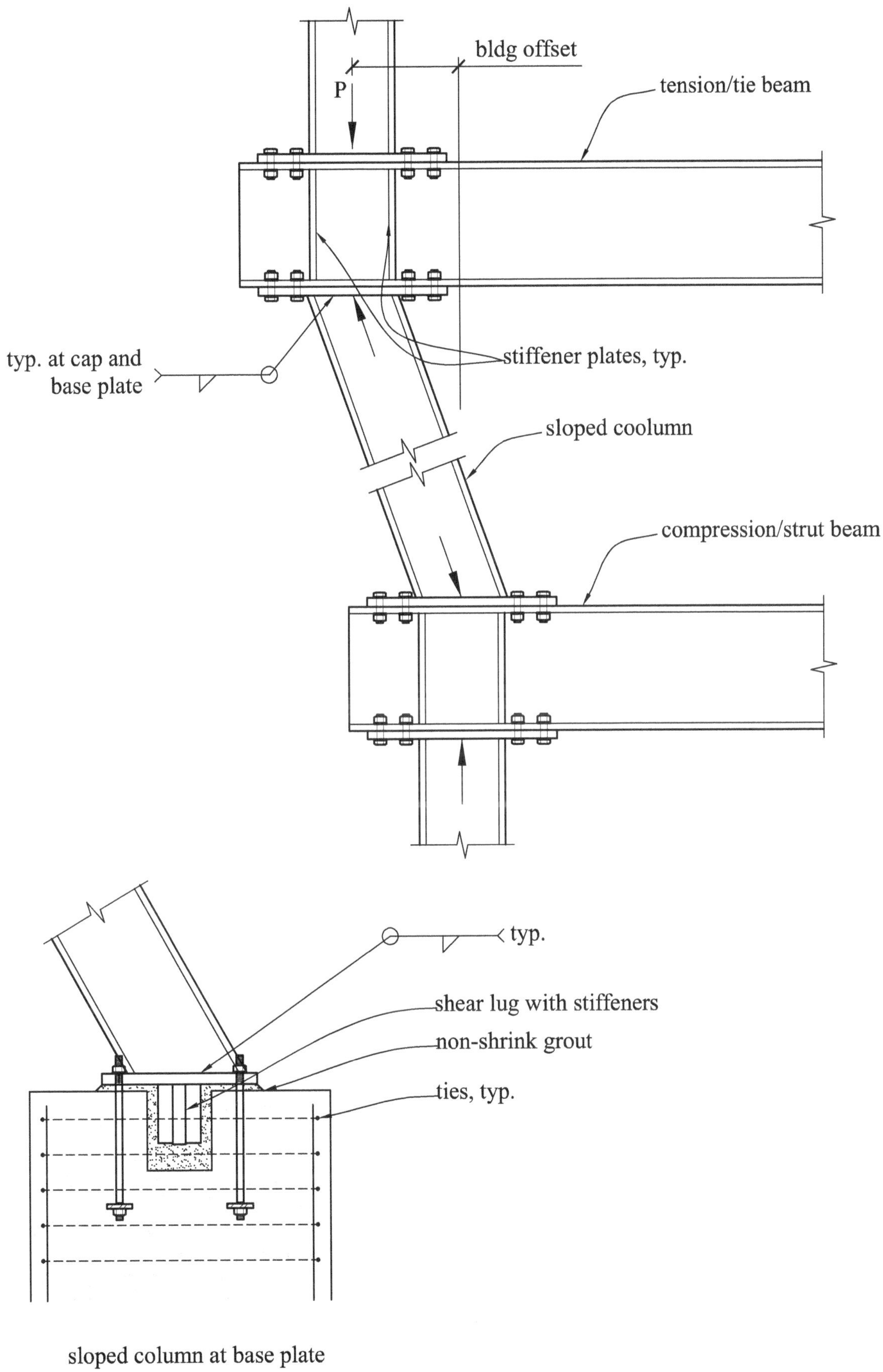

FIGURE 8-41e Sloped column details

8-20 COLUMN BASE PLATES UNDER AXIAL LOAD AND MOMENT

The bases of columns in moment frames are sometimes modeled as fixed supports, requiring that the base plates be designed to resist moments. Other situations with fixed supports include the bases of flagpoles, light poles, handrails, and sign structures (see Figure 8-42). Two different cases will be considered: base plates with axial load plus a small moment and base plates with axial load plus large moments. The authors recommend that the base of columns in moment frames be modeled and designed as pinned bases (i.e., without moment restraints) because of the difficulty of achieving a fully fixed support condition in practice; however, where moment restraint is absolutely needed, for example, because of the need to reduce the lateral drift of the building frame, the column base plates and the anchor bolts, as well as the concrete footing, must be designed to resist the base moments while also limiting the rotation of the footing. In this case, we recommend that the spread footing be sized to ensure that the resultant load lies within the middle third of the footing dimension to reduce undesirable rotations of the footing. Alternatively, the eccentric moment in the column can be resisted by strapping the column to an adjacent column footing using strap beams or grade beams. For a detailed design of eccentrically loaded strap footings, the reader should refer to the authors' reinforced concrete design text [16]. It is usually much easier to achieve a fixed column base with piles or drilled shafts (caissons) or mat foundations than with spread footings.

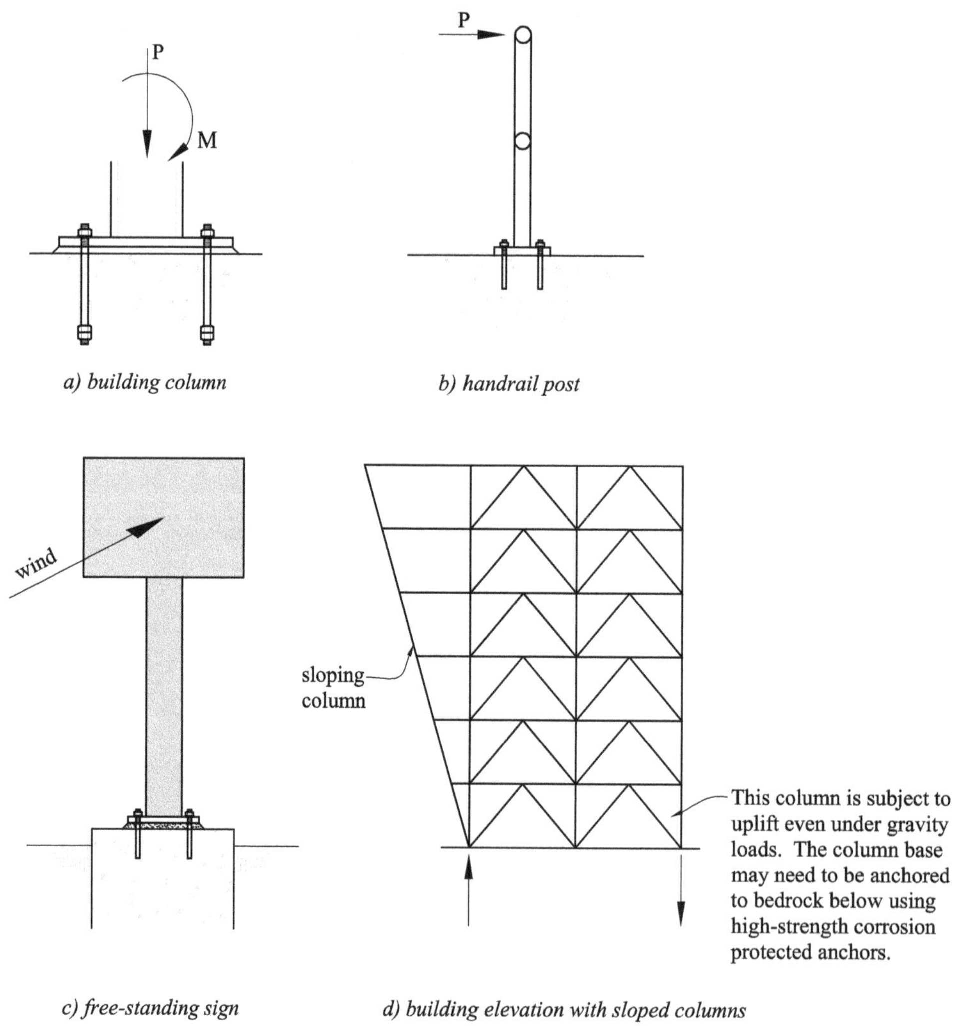

FIGURE 8-42 Examples of column base subjected to axial load plus bending moment

The two load cases that will be considered are as follows, and it is assumed that the LRFD method will be used here:

- axial load with a small moment, and
- axial load with a large moment.

Case 1: Axial Load Plus Small Moment with the Eccentricity of Loading, $e = M_u/P_u \leq N/6$

In this case, the moment is small enough that no tension stresses develop below the base plate. The base plate is subjected to a trapezoidally varying bearing stress that ranges from a minimum compression stress at one edge of the base plate to a maximum value at the opposite edge, as shown in Figure 8-43.

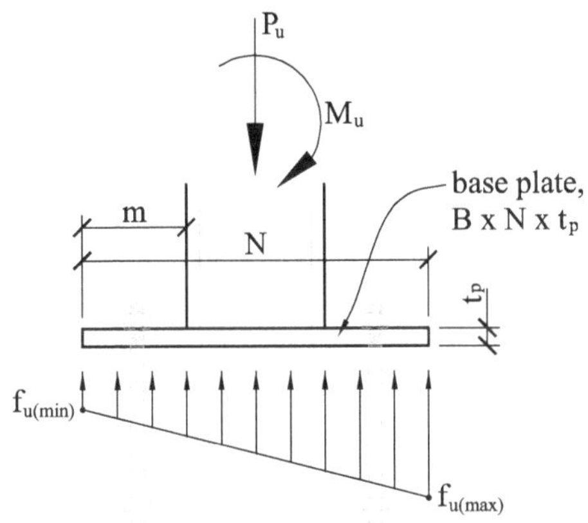

FIGURE 8-43 Column base plate with axial load and moment: Small moment

The minimum and maximum **compression** stresses, respectively, are presented as follows. Note that the maximum stress must not exceed the maximum allowed bearing stress, and the minimum applied bearing stress should be no less than zero (i.e., no tension stresses developed); that is,

$$f_{u,\,min} = \frac{P_u}{BN} - \frac{M_u}{\left(\frac{BN^2}{6}\right)} \geq 0, \text{ and} \tag{8-38}$$

$$f_{u,\,max} = \frac{P_u}{BN} + \frac{M_u}{\left(\frac{BN^2}{6}\right)} \leq \phi_{cb} f_p \tag{8-39}$$

Where,

$$\phi_{cb} f_p = \frac{\phi_{cb} P_p}{A_1} = 0.65\left(0.85 f_c'\right) \tag{8-40}$$

P_u = Applied factored axial compression load at the base of the column, calculated using the LRFD load combinations (see Section 2-3)
B = Width of base plate,
N = Length of base plate,
M_u = Factored applied moment at the base of the column, calculated using the LRFD load combinations (see Section 2-3),
f_c' = Concrete compressive strength at 28 days,
P_p = nominal bearing strength,
ϕ_{cb} = strength reduction factor for concrete bearing = 0.65

The procedure for the design of column base plates with small moments is as follows:

1. Determine the factored axial load, P_u, and the factored moment, M_u, for the column base.
2. Select a trial base plate width, B, and a base plate length, N, such that

   $B \geq b_f + 4$ in., and

   $N \geq d + 4$ in.
3. Determine the load eccentricity, $e = M_u/P_u$.
   If $e \leq N/6$, that indicates "small moments": This implies case 1; therefore, go to step 4.
   If $e > N/6$, that indicates "large moments": Use case 2 (i.e., large moments).
4. Determine the plate cantilever lengths, m and n:

   $$m = (N - 0.95d)/2$$
   $$n = (B - 0.8b_f)/2$$
5. Determine the minimum and maximum bearing pressures using equations (8-31) and (8-32).
   If equations (8-38) and (8-39) are not satisfied, increase the base plate size until these equations are both satisfied.
6. Determine the maximum moment in the plate at the face of the column flange (i.e., at a distance $\boldsymbol{m}$ from the edge of the plate).
   The maximum bearing pressure at a distance $\boldsymbol{m}$ from the edge of the plate (see Figure 8-43) is

   $$f_{u,m} = f_{u,min} + (f_{u,max} - f_{u,min})\frac{(N - m)}{N} \tag{8-41}$$

   The applied maximum bending moment per unit width in the plate at a distance $\boldsymbol{m}$ from the edge of the plate is

   $$M_{u,m} = (f_{u,m})\frac{m^2}{2} + \frac{1}{2}(f_{u,max} - f_{u,m})(m)\left(\frac{2}{3}m\right) \tag{8-42}$$

   The bending strength per unit width of the base plate is

   $$\phi_b M_n = \phi_b Z_y F_y = \phi_b\left(\frac{b_p t_p^2}{4}\right)F_y \quad \text{(for LRFD)} \tag{8-43}$$

   Equating the design bending strength to the factored applied moment (i.e., $\phi_b M_n = M_u$) yields the required minimum base plate thickness for LRFD as

   $$t_p = \sqrt{\frac{4M_{u,m}}{\phi_b b_p F_y}} \quad \text{(for LRFD)} \tag{8-44}$$

   Where,

   $M_{u,m}$ = Maximum factored bending moment per unit width in the base plate, in.-kips/in. width

   b_p = Base plate unit width = 1 in.

   F_y = Yield strength of the base plate.

Case 2: Axial Load Plus Large Moment with the Eccentricity of Loading, $e = M_u/P_u > N/6$

In this case, the moment at the base of the column is large enough that the base plate will lift off the grout bed on the tension side of the base plate, thus resulting in about half of the anchor rods in tension. Similar to the limit states design principles used in reinforced concrete design, the design of column base plates with large moments assumes the following:

1. A uniform concrete stress distribution (i.e., a rectangular concrete stress block is assumed)
2. The moment is large enough for the anchor rods in the tension zone to develop their full tension capacity. The ultimate strength of column base plates was found to be in good agreement with the analytical results when the anchor rod failure controlled the strength [17].

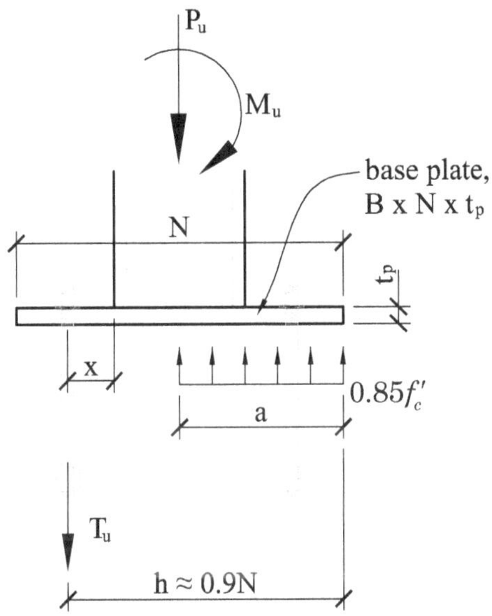

FIGURE 8-44 Column base plate with axial load and moment: Large moment

The maximum stress in the compression zone is assumed to be $0.85f_c'$, acting over the full width, B, of the base plate and over the depth, a., of the compression stress block. The assumed stress distribution is shown in Figure 8-44. Assuming that the total area of the anchor rods in the **tension zone** is A_b, the vertical equilibrium of forces in Figure 8-44 requires that

$$0.85\phi_{cb}f_c'Ba = P_u + T_u \qquad (8\text{-}45)$$

Assuming that all anchor rods in the tension zone reach their tensile strength, then the ultimate tensile strength of the anchor rods is

$T_u = \phi_t R_n = 0.75\phi_t A_b F_u$.

If the tensile strength of the anchor rod, T_u, is limited by its pullout or breakout capacity in the concrete, then T_u will be equal to the smaller of the pullout or breakout capacity of the anchor group, and this value will have to be substituted into equation (8-45). For the case where all the anchor rods in the tension zone yield and the ultimate tensile strength is not limited by the concrete pullout or breakout capacity, the depth of the rectangular concrete stress block, a, can be obtained from equation (8-46).

$$0.85\phi_{cb}f_c'Ba = P_u + 0.75\phi_t A_b F_u, \qquad (8\text{-}46)$$

Where,
$\phi_{cb} = 0.65$ (ACI 318)
$\phi_t = 0.75$
B = Width of base plate,
a = Depth of the rectangular stress block,
P_u = Applied factored axial load on the column,
F_u = Ultimate tensile stress of the anchor rod, and
A_b = Total area of anchor rods in the **tension zone**,
h = distance from the anchor rods in tension to the opposite plate edge.

Summing the moments of force about the centroid of the anchor rods in the tension zone gives

$$P_u(h - 0.5N) + M_u = 0.85\phi_{cb}f'_cBa\left(h - \frac{a}{2}\right) \tag{8-47}$$

Note that the above equations assume that the full anchor rod tensile strength can be developed within the concrete embedment provided. If the anchor rods do not fail in tension, the tensile strength of the rods will be limited by the pullout strength of the anchor rod (see Ref. [11]). Equations (8-45) through (8-47) contain four unknowns that will be determined using an iterative approach as follows:

1. Determine the factored axial load, P_u, and the factored moment, M_u, for the column base, and calculate the load eccentricity, $e = M_u/P_u$.
2. Select a trial base plate width, B, and a base plate length, N, such that
   $B \geq b_f + 4$ in., and
   $N \geq d + 4$ in.
   If $e > N/6 \Rightarrow$ Case 2 (i.e., large moments). Go to step 3.
   If $e \leq N/6 \Rightarrow$ Case 1 (i.e., small moments). Stop and use the method for small moments.
3. Assume an approximate value for the effective depth, h. Assume $h = 0.9N$.
4. Assume a trial value for the area of the anchor rod in the **tension zone**, A_b.
5. Solve equation (8-45) or equation (8-46) for the depth of the rectangular concrete stress block, a, and then solve equation (8-47) for the base plate length, N.
6. Use the larger of the N values obtained from steps 2 and 5.
7. The cantilever length of the base plate in the **compression zone** is the larger of
   $m = (N - 0.95d)/2$, and
   $n = (B - 0.8b_f)/2$.

   The maximum factored moment per unit width in the base plate is

$$M_u = (0.85\phi_{cb}f'_c)\frac{\ell^2}{2} \tag{8-48a}$$

   Where,
   ℓ = larger of (m *and* n).
   The bending strength of the plate per unit width (for LRFD) is

$$\phi_b M_n = \phi_b Z_y F_y = \phi_b\left(\frac{b_p t_p^2}{4}\right)F_y \tag{8-48b}$$

   Equating the design bending strength to the factored applied moment (i.e., $\phi_b M_n = M_u$) yields the required minimum base plate thickness for LRFD due to **compression stresses** in the compression zone as

$$t_p = \sqrt{\frac{4M_u}{\phi_b b_p F_y}} \quad \text{(for LRFD)} \tag{8-48c}$$

Where,
b_p = Base plate unit width = 1 in., and
d = Depth of the column.

8. Check the bending of the base plate in the tension zone caused by the **tension force in the anchor bolts**. The plate is assumed to cantilever from the face of the column due to the tension in the anchor rods. The applied moment per unit width in the base plate due to the tension force in the anchor bolts is

$$M_u = \frac{T_u x}{B} \tag{8-49}$$

where
x = Distance from the centroid of the anchor rod in tension to the nearest face of the column,
T_u = Total factored tension force in the anchor rods = $0.75\phi_t A_b F_u$, and
B = Width of the base plate.

The bending strength of the plate bending about its weak axis (for LRFD) is

$$\phi_b M_n = \phi_b Z_y F_y = \phi_b \left(\frac{b_p t_p^2}{4}\right) F_y \tag{8-50}$$

Where,
$b_p = 1$ in.

Equating the bending strength to the applied moment (i.e., $\phi_b M_n = M_u$) yields the required minimum base plate thickness from the LRFD method as

$$t_p = \sqrt{\frac{4T_u x}{\phi_b B F_y}} = \sqrt{\frac{4M_u}{\phi_b F_y}} \tag{8-51}$$

9. The larger of the base plate thicknesses from steps 7 and 8 governs.

10. If the designer wishes to limit the base plate thickness and also prevent excessive deformation of the base plate, vertical stiffeners may be provided for the base plate in the two orthogonal directions that will prevent the plate from cantilevering the distance, $\boldsymbol{m}$, indicated in step 7 and the distance, $\boldsymbol{x}$, in step 8 (see Figures 8-45a and 8-45b).

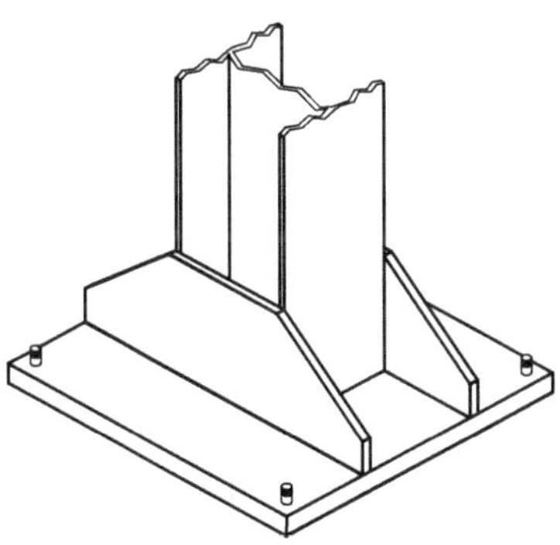

a) reinforced single direction

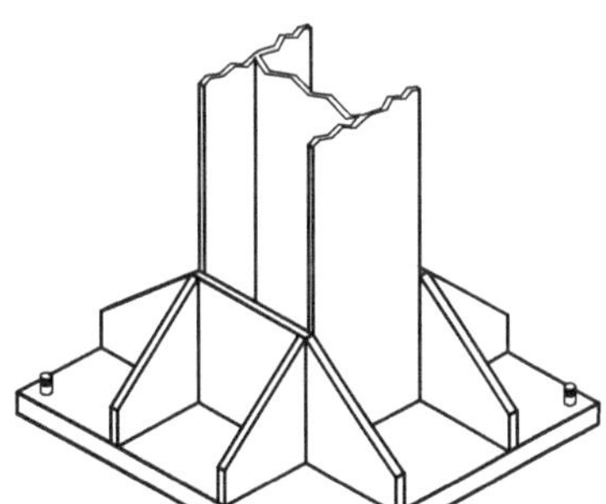

b) reinforced both directions

FIGURE 8-45a, b Column base plate with stiffeners

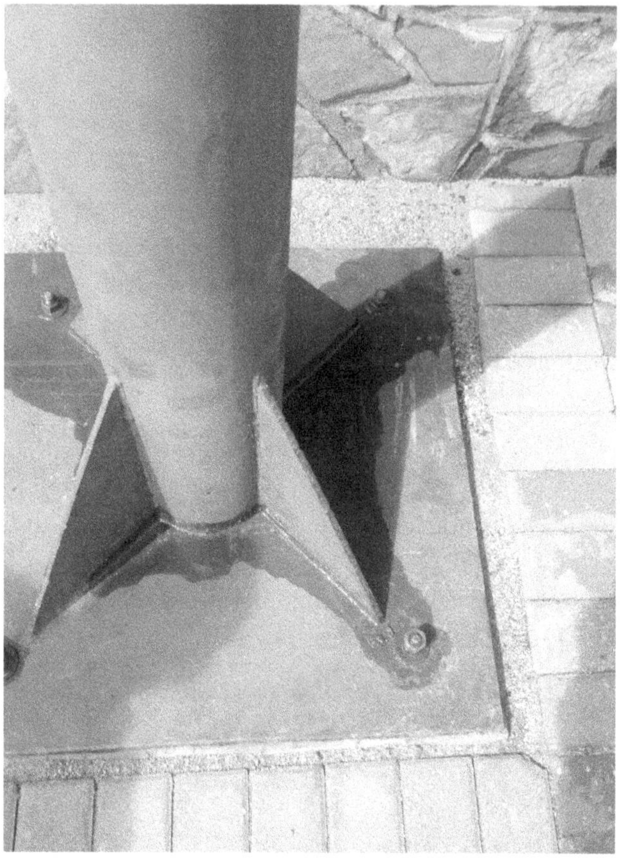

FIGURE 8-45c Base plate with stiffeners (Photo by Abi Aghayere)

In this case, the stiffeners bending about their strong axis will resist the maximum factored plate moments calculated in steps 7 and 8. Nevertheless, in design practice, the use of stiffeners at base plates should be avoided because of the increased cost associated with the increased labor involved in welding stiffener plates to the base plate and the likely interference of the stiffener plates with the anchor rods [10]. Base plates with stiffeners should only be used for columns resisting very large base moments (e.g., columns supporting a jib crane), where the base plate thickness required without stiffeners will be excessive. The stiffeners limit the length of the base plate cantilever beyond the critical column area. The presence of stiffener plates may also complicate or render impossible any field corrections that have to be made to the anchor rods in case of misplacement of the anchor rods.

For typical building columns, the authors recommend that sufficient base plate thickness be provided to obviate the need for stiffeners in base plates. However, stiffener plates on base plates may, on occasion, be required where remedial action is needed to repair or strengthen under-designed column base plates [18].

It should be noted that the capacity of a single isolated column foundation to resist moments due to lateral or eccentric loading is not only dependent on the column or the base plate, or the anchor rod capacities, but it is also dependent on the interaction of the concrete foundation with the soil. The moment on an isolated column foundation will cause rotation of the column base and the footing, and this is undesirable. One approach that is used to limit the rotation of an isolated column foundation due to lateral load moments is to connect the foundation to an adjacent column foundation using a concrete grade beam or strap beam; the grade beam helps to resist and transfer the lateral moments to the soil without causing rotation at the two connected footings (see Ref. [16].

EXAMPLE 8-8

Design of Base Plate for an I-Shaped Column Subject to Axial Load Only (LRFD)

Design the base plate and select the minimum concrete pier size for a W12 × 58 column with a factored axial compression load of 300 kips. Assume a 1-in. grout thickness, concrete compressive strength of 4 ksi, and ASTM A36 steel for the steel plate.

Solution

From part 1 of the *AISCM*,

For W12 × 58, $d = 12.2$ in.; $b_f = 10$ in.

- Try a base plate 4 in. larger than the column in both directions (i.e., 2 in. larger all around than the column) to allow room for the placement of the anchor bolts outside the column footprint.

  Try $B \approx b_f + 4 \text{ in.} = 10 + 4 = 14 \text{ in.}$

  $N = d + 4 \text{ in.} = 12.2 \text{ in.} + 4 \text{ in.} \approx 17 \text{ in., say } 18 \text{ in.}$

  $A_1 = B \times N = (14 \text{ in.})(18 \text{ in.}) = 252 \text{ in.}^2$

  Try a 14-in. by 18-in. base plate (see Figure 8-46).

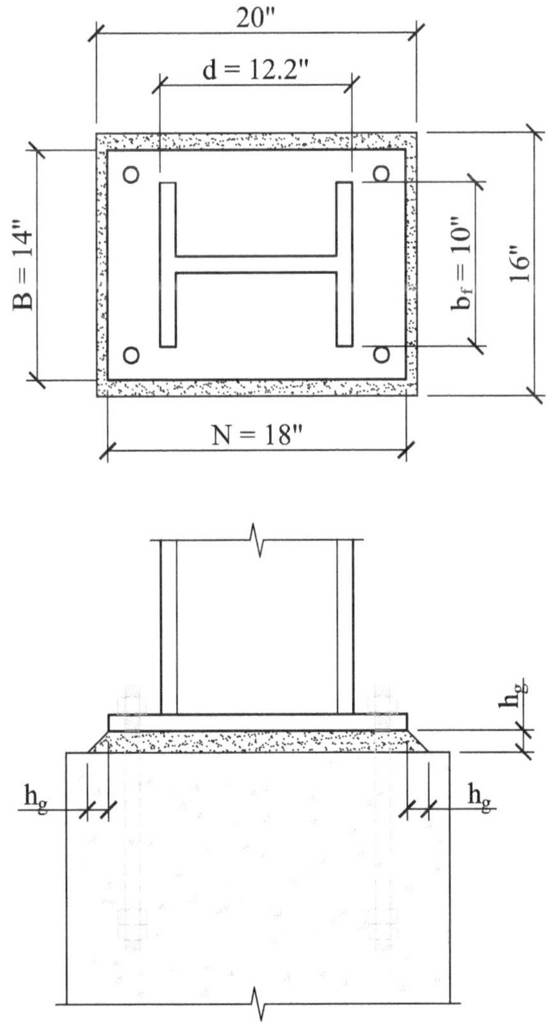

FIGURE 8-46 Column base plate for Example 8-8

- Select the **minimum** pier size:

$$= (B + 2h_g) \times (N + 2h_g)$$

$$= [14 \text{ in.} + (2 \times 1 \text{ in.})][(18 \text{ in.} + (2 \times 1 \text{ in.})]$$

$$= 16 \text{ in.} \times 20 \text{ -in. pier} \quad \therefore A2 = (16 \text{ in.})(20 \text{ in.}) = 320 \text{ in.2}$$

$$1.0 < \sqrt{A_2/A_1} = \sqrt{320/252} = 1.13 < 2 \quad \text{O.K.}$$

The base plate extensions or overhang beyond the critical area will now be calculated (see Figure 8-33a)

$$\ell \geq \begin{cases} m = \dfrac{N - 0.95d}{2} = \dfrac{18 - 0.95(12.2)}{2} = 3.21 \text{ in. (Largest value governs)} \\ n = \dfrac{B - 0.80b_f}{2} = \dfrac{14 - 0.80(10)}{2} = 3 \text{ in.} \\ n' = \dfrac{1}{4}\sqrt{db_f} = \dfrac{1}{4}\sqrt{(12.2)(10)} = 2.8 \text{ in.} \end{cases}$$

Therefore, $\ell = (La\text{rgest value of m,n,and n'}) = 3.21$ in.

From equation (8-21a), the required minimum base plate thickness (for LRFD) is given as

$$t_p = \ell\sqrt{\frac{2P_u}{\phi_b BNF_y}} = 3.21 \text{ in.}\sqrt{\frac{(2)(300 \text{ kips})}{0.9(252 \text{ in.}^2)(36 \text{ ksi})}}$$

$$= 0.87 \text{ in.} \approx 7/8 \text{ in.}$$

Use $^1/_8$-in. increments for base plate thicknesses of less than $1^1/_4$ in. and $^1/_4$-in. increments for others.

Check the Bearing Capacity of the Concrete Pier:

$$\phi_{cb}P_p = \phi_{cb}(0.85f_c')A_1\sqrt{A_2/A_1} = (0.65)(0.85)(4 \text{ ksi})(252)(1.13)$$

$$= 628 \text{ kips} > P_u = 300 \text{ kips} \quad \text{OK}$$

Use a 14-in. × 7/8-in. × 18-in. base plate

For this rectangular base plate, the orientation of the 14-in. and 18-in. dimensions would have to be indicated on the column schedule or on the foundation plan, and verification on site may be necessary to ensure accurate placement of the base plate. Where the bearing capacity of a pier, $\phi_{cb}P_b$, is less than the factored axial load, P_u, on a column, an increase in the bearing capacity can be achieved far more efficiently by increasing the pier size (i.e., area A_2) than by increasing the base plate size (i.e., area A_1). The reader should verify that if a square base plate is used, the required base plate size will be $16 \text{ in} \times 1\frac{1}{8} \text{in} \times 16 \text{ in}$.

EXAMPLE 8-9

HSS Column Base Plate Subject to Axial Load Only (LRFD)

Design the base plate for an HSS 12 × 12 × $^5/_{16}$ column with a factored axial compression load of 450 kips. Assume a 2-in. grout thickness, a concrete compressive strength of 4 ksi for the pier, and ASTM A36 steel for the base plate. Determine the minimum pier size for this column.

Solution

From part 1 of the *AISCM*,

For HSS $12 \times 12 \times {}^5/_{16}$, $b = b_f = d = 12$ in.

- Try a base plate 4 in. larger than the column in both directions (i.e., 2 in. larger all around than the column) to allow room for the placement of the anchor bolts outside the column footprint.

  Try $B = b_f + 4 \text{ in.} = 12 + 4 = 16$ in.

  $N = d + 4 \text{ in.} = 12 + 4 = 16$ in.

  $A_1 = B \times N = (16 \text{ in.})(16 \text{ in.}) = 256 \text{ in.}^2$

Try a 16-in. × 16-in. base plate (see Figure 8-47).

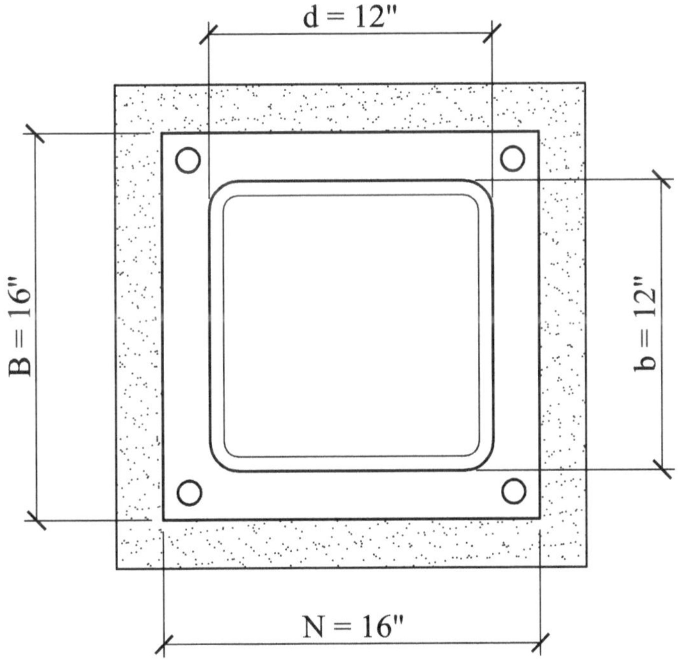

FIGURE 8-47 Column base plate for Example 8-9

- Select the **minimum** size of concrete pier

  $= (B + 2h_g) \times (N + 2h_g)$

  $= [(16 \text{ in.}) + (2)(2 \text{ in.})][(16 \text{ in.}) + (2)(2 \text{ in.})]$

  $= 20\text{-in.} \times 20\text{-in. pier} \therefore A_2 = (20 \text{ in.})(20 \text{ in.}) = 400 \text{ in.}^2$

  $1.0 < \sqrt{A_2/A_1} = \sqrt{400/256} = 1.25 < 2$ O.K.

The base plate extensions or overhang beyond the critical area will now be calculated as follows (see Figure 8-33b):

$$\ell \geq \begin{cases} m = \dfrac{N-b}{2} = \dfrac{16-12}{2} = 2 \text{ in.} \\[2ex] n = \dfrac{B-b}{2} = \dfrac{16-12}{2} = 2 \text{ in.} \\[2ex] n' = \dfrac{b}{4} = \dfrac{12}{4} = 3 \text{ in.} \left(largest\, value\ controls \right) \end{cases}$$

Therefore, $\ell = (\text{Largest value of m,n,and n}') = 3$ in.

From equation (8-21a), the required minimum base plate thickness (for LRFD) is given as

$$t_p = \ell\sqrt{\frac{2P_u}{\phi_b BNF_y}} = (3\text{ in.})\sqrt{\frac{(2)(450\text{ kips})}{(0.9)(256\text{ in.}^2)(36\text{ ksi})}}$$

$$= 0.99\text{ in.} \qquad \text{Use 1-in. thick base plate.}$$

(Use 1⁄8-in. increments for base plate thicknesses of less than 11⁄4 in. and 1⁄4-in. increments for others.)

Check the Bearing Capacity of the Concrete Pier:

$$\phi_{cb}P_b = \phi_{cb}(0.85f_c')A_1\sqrt{A_2/A_1} = 0.65(0.85)(4\text{ ksi})(256)(1.25)$$
$$= 707\text{ kips} > P_u = 450\text{ kips OK}$$

Use a 16-in. × 1-in. × 16-in. base plate

EXAMPLE 8-10

Design of Column Base Plate with Axial Compression Load and Moment (LRFD)

The base plate for a W12 × 96 column is subjected to the following service or unfactored axial compression loads and moments. If the compressive strength of the concrete pier is 4000 psi, design the base plate and anchor rods for this column.

$P_D = 110$ kips

$P_L = 150$ kips

$M_D = 950$ in.-kips

$M_L = 1600$ in.-kips

Solution

The section properties for a W12 × 96 column are d = 12.7 in. and b_f = 12.2 in. (*AISCM*, part 1).

For ASTM A36 anchor rods, the ultimate strength F_u = 58 ksi and the tensile strength F_y = 36 ksi.

Assume, at this stage, that the moment is large and that load case 2 will govern. This will have to be verified later. The LRFD method is used in this example; therefore, the factored load and moment are calculated as follows (see Section 2-3):

1. $P_u = 1.2(110) + 1.6(150) = 372$ kips

$M_u = 1.2(950) + 1.6(1600) = 3700$ in.-kips

$e = M_u / P_u = 3700/372 = 9.95$ in.

2. Select a trial base plate width, B, and base plate length, N, such that

   $B \geq b_f + 4$ in. $= 12.2 + 4 = 16.2$ in. Try $B = 21$ in., and

   $N \geq d + 4$ in. $= 12.7 + 4 = 16.7$ in. Try $N = 21$ in.

   Check if $e > N/6$:

   $e = 9.95$ in. $> N/6 = 21/6 = 3.5$ in. Therefore, use load case 2 (i.e., large moments), in which case a rectangular concrete stress block is assumed and the anchor rods in the tension zone are assumed to develop their full tension capacity.

3. Assume an approximate value for the effective depth, h. Assume that $h = 0.9N$.

   Assume that $h = 0.9N = (0.9)(21 \text{ in.}) = 18.9$ in.

4. Assume a trial value for the area of the anchor rod, A_b, in the **tension zone**.

   Assume three $1\frac{1}{2}$-in.-diameter anchor rods with hex nuts in the tension zone, therefore, $A_b = 5.3 \text{ in.}^2$

5. Solve equation (8-46) for the depth of the rectangular concrete stress block, a, and then solve equation (8-47) for the base plate length, N.

   Solving equation (8-46) for the depth, a, of the rectangular concrete stress block gives

   $$(0.85)(0.65)(4 \text{ ksi})(21 \text{ in.})(a) = (372 \text{ kips}) + (0.75)(0.75)(5.3 \text{ in.}^2)(58 \text{ ksi})$$

   Therefore, $a = 11.74$ in. (see Figure 8-48)

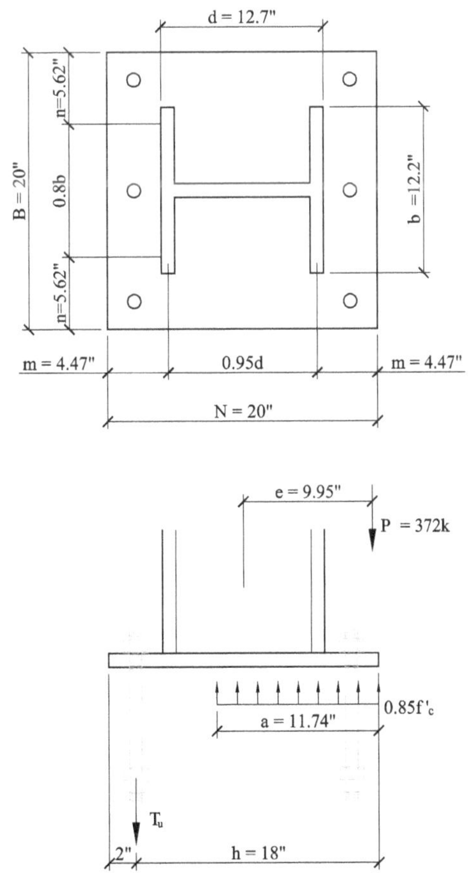

FIGURE 8-48 Column base plate for Example 8-10

Solving equation (8-47) for the length, N, of the base plate gives

$$(372 \text{ kips})(0.9N - 0.5N) + 3700 \text{ in.-kips}$$

$$= 0.85(0.65)(4 \text{ ksi})(21 \text{ in.})(11.74 \text{ in.})\left(0.9N - \frac{11.74 \text{ in.}}{2}\right).$$

Therefore, N = 20.2 in., say 21 in.

6. Use the larger of the N values obtained from steps 2 and 5.
Using the larger of the values from steps 2 and 5 gives $N \approx 21$ in.
The base plate trial size will be 21 in. × 21 in. (as depicted in Figure 8-48).

7. The cantilever length of the base plate in the **compression zone** is the larger of

$$m = (N - 0.95d)/2 = [(21 \text{ in.}) - (0.95)(12.7 \text{ in.})]/2 = 4.47 \text{ in., and}$$

$$n = (B - 0.8b_f)/2 = [(21 \text{ in.}) - (0.80)(12.2 \text{ in.})]/2 = 5.62 \text{ in. (controls)}$$

$$n' = \frac{1}{4}\sqrt{db_f} = \frac{1}{4}\sqrt{(12.7 \text{ in.})(12.2 \text{ in.})} = 3.11 \text{ in.}$$

From equation (8-48a), the maximum factored moment per unit width in the base plate is

$$M_u = (0.85)(0.65)(4 \text{ ksi})\frac{(5.62 \text{ in.})^2}{2} = 35 \text{ in.-kips/in. width of the base plate,}$$

where
l = Larger of (m, n, *and* n').
The required minimum base plate thickness due to **compression stresses** in the compression zone from equation (8-48c) is

$$t_p = \sqrt{\frac{4(35 \text{ in.-kips/in.})}{(0.9)(1 \text{ in.})(36 \text{ ksi})}} = 2.08 \text{ in.,}$$

Where,
b_p = Plate unit width = 1 in.
Check the bending of the base plate in the tension zone caused by the tension force, T_u, in the anchor bolts. The plate is assumed to cantilever from the face of the column due to the tension in the anchor bolts.

$$T_u = C_u - P_u = (0.85)(0.65)(4 \text{ ksi})(21 \text{ in.})(11.74 \text{ in.}) - (372 \text{ kips}) = 173 \text{ kips}$$

Assume that the edge distance of the tension zone bolts is 1.5 in. Therefore,
x = Distance from the centroid of the anchor bolts in tension to the nearest face of the column
= m – 1.5 in. = 4.47 in. – 1.5 in. = 2.97 in.
The applied moment per unit width in the base plate due to the tension force in the anchor bolts (see equation (8-49)) is

$$M_u = \frac{T_u x}{B} = \frac{(173 \text{ kips})(2.97 \text{ in.})}{21 \text{ in.}} = 24.5 \text{ in.-kips/in. width of base plate.}$$

The required thickness of the base plate from equation (8-51) is

$$t_p = \sqrt{\frac{4(24.5 \text{ in.-kips/in.})}{(0.9)(1 \text{ in.})(36 \text{ ksi})}} = 1.74 \text{ in.}$$

8. The larger of the base plate thicknesses from step 7 controls, and in this example, the failure of the base plate in the compression zone controls the design. Recall that the minimum practical column base plate thickness is $\frac{1}{2}$ in., with increments of $\frac{1}{4}$ in. up to a 1-in. thickness and increments of $\frac{1}{8}$ in. for base plate thicknesses greater than 1 in. Comparing steps 6 and 7, and using the larger plate thickness and rounding to the nearest $\frac{1}{8}$ in. implies that

t_p **= 2-1/8 in.**

Therefore, use a 21-in. × 2-$\frac{1}{8}$ in. × 21-in. base plate with (six) 1$\frac{1}{2}$-in. diameter threaded anchor rods with hex nuts.

Note: It has been assumed in this example that the anchor rods have sufficient embedment depth into the foundation or pier plus adequate edge distance and spacing between the anchors to ensure that the full tensile strength of the anchor rods can be developed in the concrete pier or footing. If that is not the case, the tensile force, T_u, in the tension zone anchor bolts will be limited by the concrete pullout and breaking strengths, and this reduced value would need to be used in step 5. Alternatively, the anchor rods could be tension lap-spliced with the vertical reinforcement in the pier to achieve the full tension capacity of the anchor rods, but this will increase the required minimum height of the concrete pier to ensure an adequate tension splice length for the anchor rod and the vertical steel reinforcement in the concrete pier.

8-21 COLUMN SCHEDULE

In the design of building structures, there are usually many different column sizes and configurations. Some columns may extend to a different elevation or level compared to other columns, and they also support varying axial loads. It is usually difficult to show on the structural plans all the different column configurations and sizes—even for small building projects. The goal is to group as many similar columns together as possible, and an efficient way of conveying all the pertinent column design information to the contractor is to present the information in a tabular format known as a *column schedule.* The column schedule (see Table 8-8) is an organized way of presenting the design information for all the columns in a building structure. Different columns in the building with identical loadings and heights are usually grouped together as shown in Table 8-8, and the column design information presented typically includes the following: column elevations, column sizes, cumulative factored axial loads at each level, the location of column splices, the elevation of the underside of the column base plates, anchor rod, and base plate sizes. The lower section of the column schedule shows the base plate and anchor bolt sizes and the distance from the floor datum to the underside of the column base plate, which may vary from column to column.

In typical low- to mid-rise buildings, the steel columns are usually spliced every two or four floors since the maximum steel column length that can be transported safely is approximately 60 ft., and the practical column heights that can be erected safely on the construction site with guy wire bracing before the beams and girders are erected is also limited. The two-story tier splice is the most commonly used splice arrangement. For high-rise buildings, it is not uncommon to splice columns every four floors to achieve greater economy (see Figure 8-49), though having such long column lengths between splices results in difficulty in stabilizing the columns and achieving plumpness, which makes it more difficult to make the connections at each floor level. The typical height of the column splice from the floor level is 2 ft. 6 in. for interior columns and 4 ft. 0 in. for exterior columns. The minimum 4 ft. height of splice at the exterior or perimeter columns is an OSHA requirement (OSHA safety regulations Section 1926.756(e)) that allows perimeter safety cables to be strung through holes in the column just below the splice locations at a height of 3 ft.-6 in. to 3 ft.-9 in. above the floor level, thus serving as a fall protection measure. Additionally, the 4 ft. splice height is a good working height for the steel erector erecting the next column tier, and therefore, some engineers prefer to specify the 4 ft. splice height for both interior and exterior columns. Some engineers recommend locating the exterior column splice at a height of 5 ft. above the floor level to provide better tolerance for locating the perimeter safety cables [20]. It has been noted that the weight of one bearing splice connection in a column is equivalent to approximately 500 lbs. of steel, while the weight of a moment-resisting column splice is equivalent to approximately 2000 lbs. of steel; therefore, the fewer the column splices used, the lower the material and labor cost [20]. To save on steel tonnage, some designers and contractors sometimes opt for splice-free columns when the height of the building does not exceed approximately 60 ft. See Chapter 11 for further discussion of column splices.

TABLE 8-8 Column schedule

Level / elevation	Grid Line A-1 A-2	B-1 B-2	C-1 C-2	D-1 D-2
(distance from roof datum to the top of the column)	5"	11"	5"	7"
Roof +52'-0" T/ steel (deck bearing)				
(column load supporting roof)	W8x24 40 kips	W8x24 40 kips	W10x39 65 kips	W10x39 65 kips
4th Flr. +39'-0" T/ conc. slab				
(column load supporting roof and 4th floor)	W8x24 80 kips	W8x24 80 kips	W10x39 130 kips	W10x39 130 kips
splice location (column splice)	4'-0"			
3rd Flr. +26'-0" T/ conc. slab				
(column load supporting roof, 4th, and 3rd floors)	W8x40 120 kips	W8x40 120 kips	W10x49 195 kips	W10x49 195 kips
2nd Flr. +13'-0" T/ conc. slab				
(column load supporting roof, 4th, 3rd, and 2nd floors)	W8x40 160 kips	W8x40 160 kips	W10x49 260 kips	W10x49 260 kips
Gnd. Flr. +0'-0" T/ slab-on-grade				
underside of base plate elevation (distance from ground level to bottom of base plate)	7"	7"	11"	11"
Base Plate B x N x t	14" x 14" x ¾"	14" x 14" x ¾"	16" x 18" x ¾"	16" x 18" x ¾"
Anchor rods	(4)-¾" diam. x 12" embedment with 4" hook	(4)-¾" diam. x 12" embedment with 4" hook	(4)-1" diam. x 24" embedment with heavy hex nut	(4)-1" diam. x 24" embedment with heavy hex nut

grid intersection

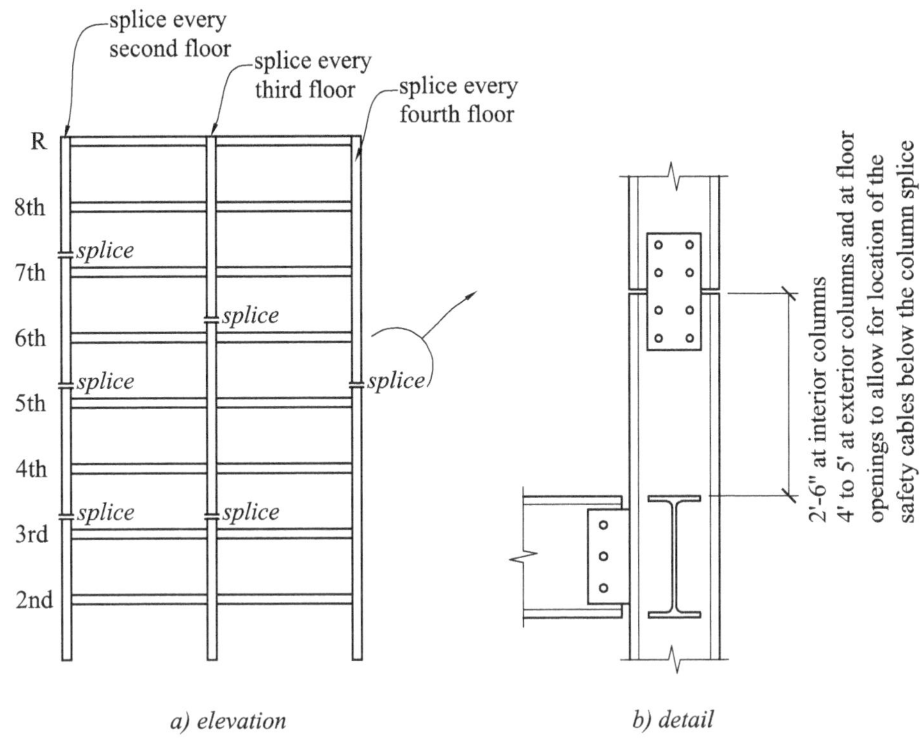

FIGURE 8-49a Column splice locations

FIGURE 8-49b Column splice locations

References

[1] Ioannides, S. A. 1995. "Minimum Eccentricity for Simple Columns". *ASCE Structures Congress Proceedings,* Volume I.

[2] Lini, C. 2016. "Eccentricity on Columns". *The Steel Interchange, Modern Steel Construction*, February.

[3] American Institute of Steel Construction, 2023. *Steel Construction Manual*, 16th ed. Chicago, IL: AISC.

[4] Nair, R. S. 2009. "Simple and Direct." *Modern Steel Construction,* January: 43–44.

[5] Ericksen, J. 2011. "A How-To Approach to Notional Loads." *Modern Steel Construction*, January.

[6] Abu-Saba, E.G. 1995. *Design of Steel Structures*. New York: Chapman and Hall.

[7] Hoffman, E. S., S. G. Albert, P. G. David, and F. R. Paul. 1996. *Structural Design Guide to the AISC (LRFD) Specification for Buildings.* 2nd ed. New York: Chapman and Hall.

[8] American Society of Civil Engineers. 2022. *ASCE 7: Minimum Design Loads for Buildings and Other Structures*. Reston, VA: ASCE.

[9] Kulak, G. L., and G.Y. Grondin. 2006. *Limit States Design in Structural Steel*. Willowdale, ON, Canada: Canadian Institute of Steel Construction.

[10] Shneur, V. 2006. "24 Tips for Simplifying Braced Frame Connections." *Modern Steel Construction*, May: 33–35.

[11] Honeck, W. C., and W. Derek. 1999. Practical Design and Detailing of Steel Column Base Plates. Steel Tips. Moraga, CA: Structural Steel Educational Council.

[12] Amit Kanvinde, A., M. Maamouri. and J. Buckholt. 2024. *AISC Design Guide 1: Base Connection Design for Steel Structures.* 3rd ed. Chicago, IL: AISC.

[13] American Concrete Institute. 2019. *ACI 318-19: Building Code Requirements for Structural Concrete and Commentary*. Farmington Hills, MI: ACI.

[14] Carter, C. J. 2004. "Are You Properly Specifying Materials?" *Modern Steel Construction*, January.

[15] Swiatek, D., and W. Emily. 2004. Anchor Rods–Can't Live With 'Em, Can't Live Without 'Em." *Modern Steel Construction*, December: 31–33.

[16] Aghayere, A. 2024. *Reinforced Concrete Design*. 10th ed. Boston, MA: Mercury Learning.

[17] Kavinde, A. M., and G. G. Deierlein. 2011. "Recent Research on Column Base Connections." *Modern Steel Construction*, April.

[18] Post, N. M. 1997. "Structural Fix: VA Expects Disputes Over Delay." *Engineering News Record*, February 10: 10–11.

[19] Boulanger, S., C. Drucker, L. F. Kruth, D. K. Miller, and T. M. Boake. 2016. "The Splice is Right." *Modern Steel Construction*, March: 48-51.

[20] Fisher, J. M., and M. A. West. 2010. "98 tips for designing structural steel." *Modern Steel Construction*, September.

[21] Savery, A. B. 2006. Over and Above, Modern Steel Construction, August.

[22] Blanchard, A., and J. Abramson. 2019. "When Science Becomes Transparent." *STRUCTURE Magazine*, November: 30-32.

[23] Klinger, J. E., E. F. Connolly, C. D. Charlson, O. R. Antommattei, and B. A. Suprenant. 2024. "Constructability of Anchor Bolts (Rods) in Foundations." *Concrete International*, July: 38-41.

[24] Marstellar, B. 2003. "Specifying & Designing with Steel." *Structural Engineer*, July: 18-20.

Exercises

8-1. Determine the adequacy of a 15-ft.-long W12 × 72 column in a braced frame to resist a factored axial load of 200 kips and factored moments of 100 ft.-kips and 65 ft.-kips about the x- and y- axes, respectively. The column is assumed to be pinned at both ends. Use ASTM A992 Grade 50 steel.

8-2. A 15-ft.-long W12 × 96 column is part of a moment frame with column bases that are pinned. The factored axial load on the column is 250 kips, and the factored moment is 120 ft.-kips about the x-axis. The building is assumed to be braced in the orthogonal direction. Determine whether this column is adequate to resist the applied loads. Use ASTM A992 steel.

8-3. For the mezzanine floor plan shown in Figure 8-50, assume HSS 7 × 7 columns, and a floor-to-floor height of 15 ft. The floor dead load is 100 psf and the live load is 250 psf.

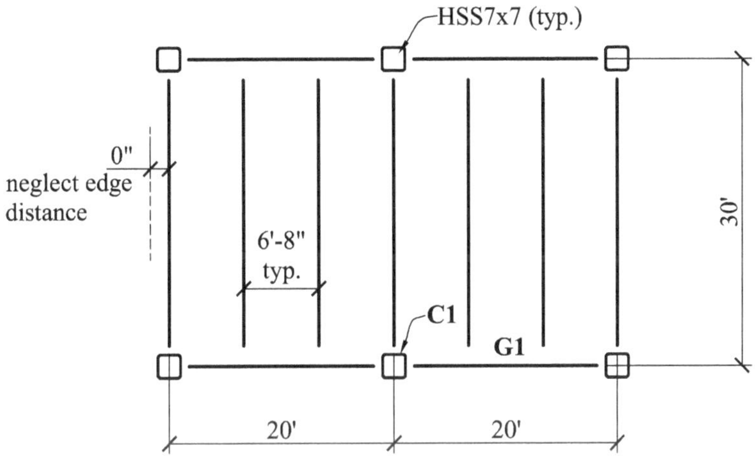

FIGURE 8-50 Detail for Exercise 8-3

a. Calculate the factored axial loads and moments (for load case 1 and load case 2) for the exterior column C1.

b. Using the HSS beam column design templates, design the most economical or lightest HSS 7 × 7 column size for column C1 that is adequate to resist the loads and moments. Assume ASTM A500 Grade C (F_y = 50 ksi) steel.

8-4. For the floor and roof framing plan shown in Example 8-5 (Figure 8-32a), analyze the North-South moment frames using the amplified first-order analysis method and design column B1, including the effect of the leaning columns.

8-5. A W8 × 40 welded tension member, 16 ft. long, is subjected to a factored axial tension load of 150 kips and factored moments about the strong and weak axes of 30 ft.-kips and 20 ft.-kips, respectively. Assuming ASTM A992, Grade 50 steel, and that the member is braced only at the supports, determine whether the beam-column is adequate.

8-6. Design the base plate for an HSS 10 × 10 × 5⁄16 column with a factored axial compression load of 400 kips. Assume a 1-in. grout thickness, a concrete compressive strength of 4 ksi for the pier, and ASTM A36 steel for the base plate.

8-7. Design the base plate and select the concrete pier size for a W12 × 72 column with a factored axial compression load of 550 kips. Assume a 1-in. grout thickness, a concrete compressive strength of 4 ksi, and ASTM A36 steel for the base plate.

8-8. a. Design the base plate for an HSS 8 × 8 column with a factored axial load, P_u, of 500 kips. Assume ASTM A36 steel for the base plate.

b. Select the most economical concrete pier size required. Assume a 1-in. non-shrink grout and concrete 28-day strength, f_c', of 3000 psi.

8-9. A 15-ft.-long HSS 6 × 6 × ½ hanger supports a factored axial tension load of 70 kips and factored moments, M_{ux} = 40 ft.-kips and M_{uy} = 20 ft.-kips. Assuming that the hanger is fully welded at the beam support above, check if the hanger is adequate. Use ASTM A500, Grade C (F_y = 50 ksi) steel.

8-10. Select the lightest 8-ft.-long W10 hanger to support a factored tension load of 90 kips applied with an eccentricity of 6 in. with respect to the strong (*x–x*) axis of the section and an eccentricity of 3 in. with respect to the weak (*y–y*) axis of the section. The member is fabricated from ASTM A36 steel, is fully welded at the connections, and is braced laterally at the supports.

8-11. For the three-story braced frame building shown in Figure 8-51, design column C2 for the axial loads and no-translation bending moments resulting from the beam-to-column and beam-to-girder reaction eccentricities. Use the W-shape column design template and present your results in a column schedule. Use the same roof and floor loads shown in Exercise 8-14. Assume that the column will be spliced at 4 ft. above the second-floor level. Use ASTM A992 Grade 50 steel.

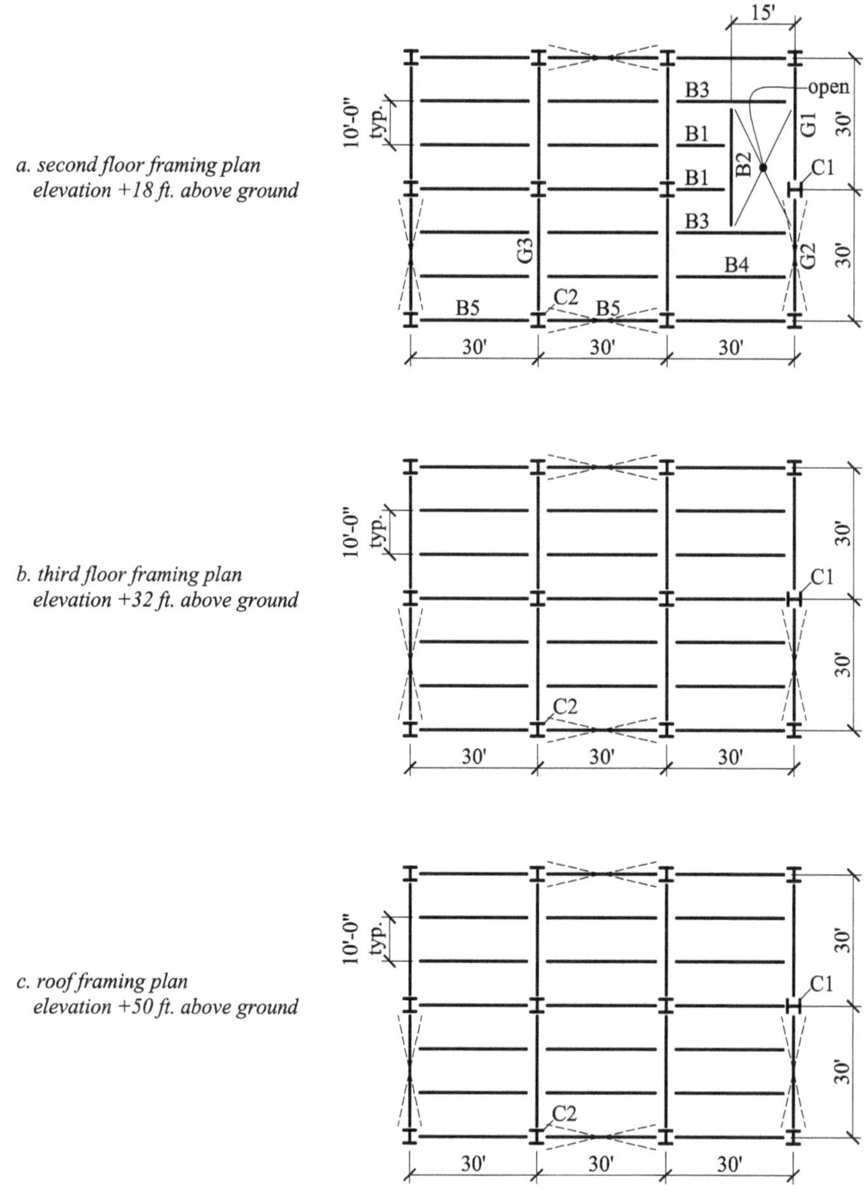

FIGURE 8-51 Framing plans for Exercise 8-11

8-12. Determine if column C1 in the one-story building shown in Figure 8-52 is adequate for combined bending and compression loads. The steel is ASTM A992, Grade 50. Assume $K = 1.0$, $C_m = 0.85$, and $B_2 = 0$. Consider only the loads shown and use the given eccentricities.

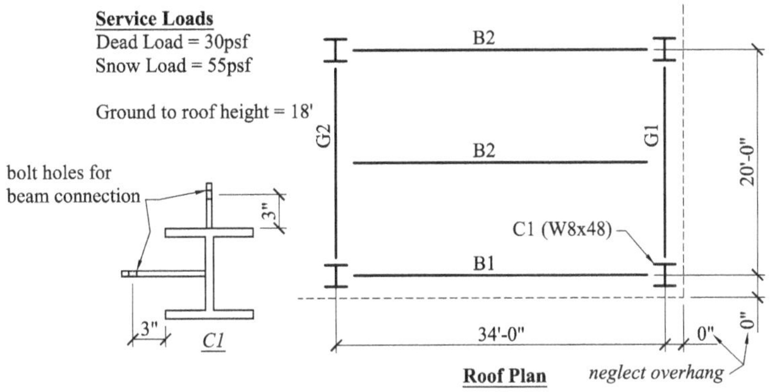

FIGURE 8-52 Details for Exercise 8-12

8-13. Refer to the roof framing and details in Figure 8-53 for this exercise, and assume ASTM A992 Grade 50 steel:

a. Determine the applied axial load and unbalanced moments on the column.
b. Determine the moment and axial capacities, ϕM_{nx}, ϕM_{ny}, and ϕP_n for the column.
c. Determine if the column is adequate for the maximum axial load.
d. Determine if the column is adequate for the worst-case combined axial load and bending.

Use the following parameters:
$K = 1.0$ $F_y = 50$ ksi
Service loads: Dead = 30 psf, Snow = 75 psf

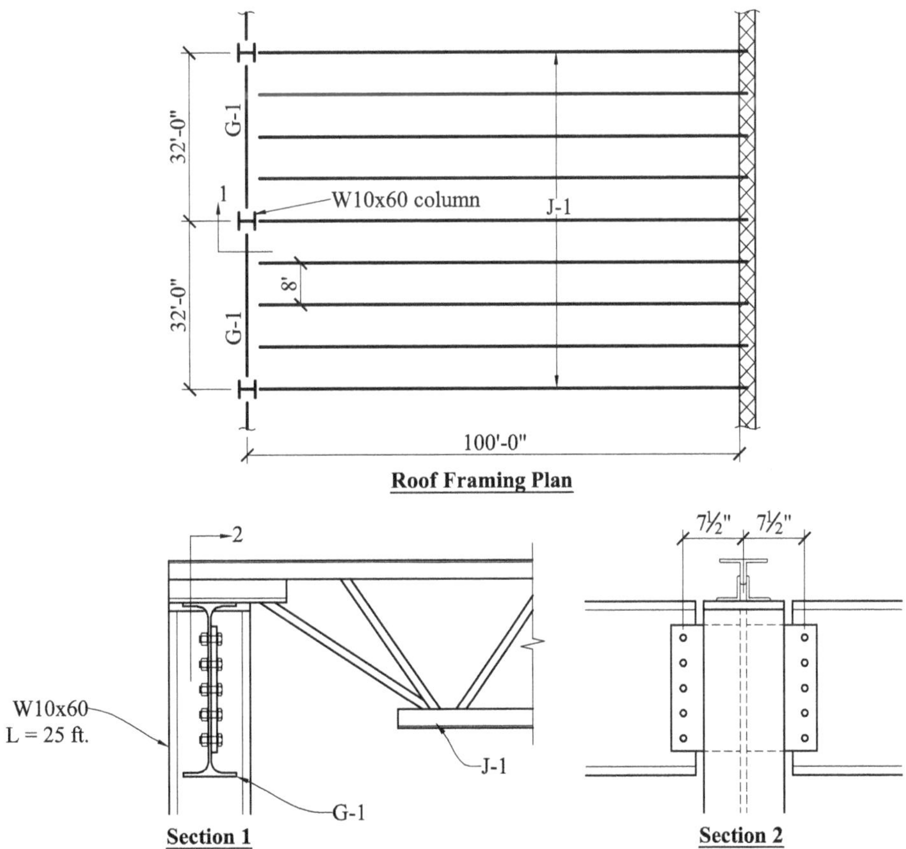

FIGURE 8-53 Details for Exercise 8-13

8-14. Refer to the framing plans in Figure 8-54 for a two-story building and assume ASTM Grade 50 steel. Analyze the ground floor columns (i.e., the lower level column segment between the ground and second floors) as follows:

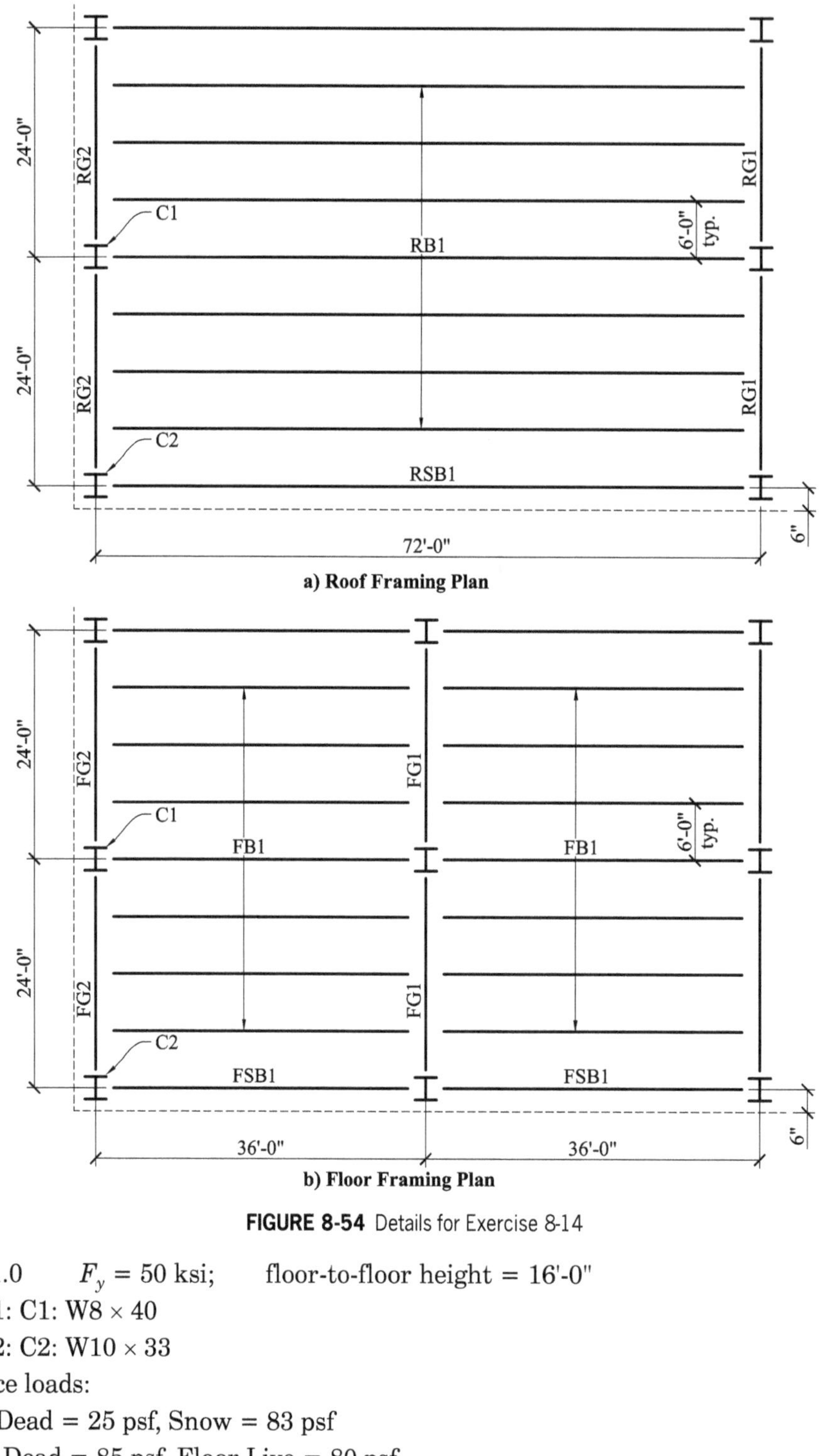

FIGURE 8-54 Details for Exercise 8-14

$K = 1.0$ $F_y = 50$ ksi;   floor-to-floor height = 16'-0"

Part 1: C1: W8 × 40

Part 2: C2: W10 × 33

Service loads:

Roof Dead = 25 psf, Snow = 83 psf

Floor Dead = 85 psf, Floor Live = 80 psf

a. Determine the applied axial load and unbalanced moments on the column.

b. Determine the moment and axial capacities, ϕM_{nx}, ϕM_{ny}, and ϕP_n for the column.

c. Determine if the column is adequate for the maximum axial load.

d. Determine if the column is adequate for the worst-case combined axial load and bending.

8-15. Determine if the exterior W12 × 26 column (ASTM A992 Grade 50 steel) supporting gravity and wind loads is adequate for combined loading (see Figure 8-55). Use only the load combination 1.2*D* + 1.6W + 0.3S. The loads shown in Figure 8-54 are unfactored. Assume the gravity loads are concentric (no moments) and the bending moment is only due to wind load, which is transferred to the columns through the girts.

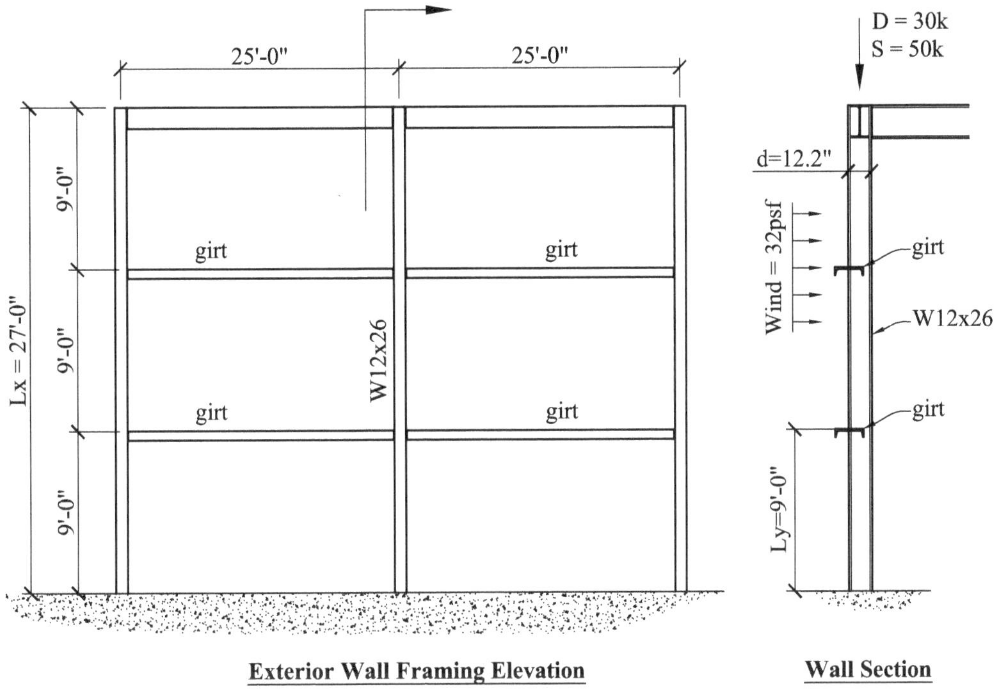

FIGURE 8-55 Details for Exercise 8-15

8-16. Determine the following for the truss top chord loaded in compression and bending (see Figure 8-56):

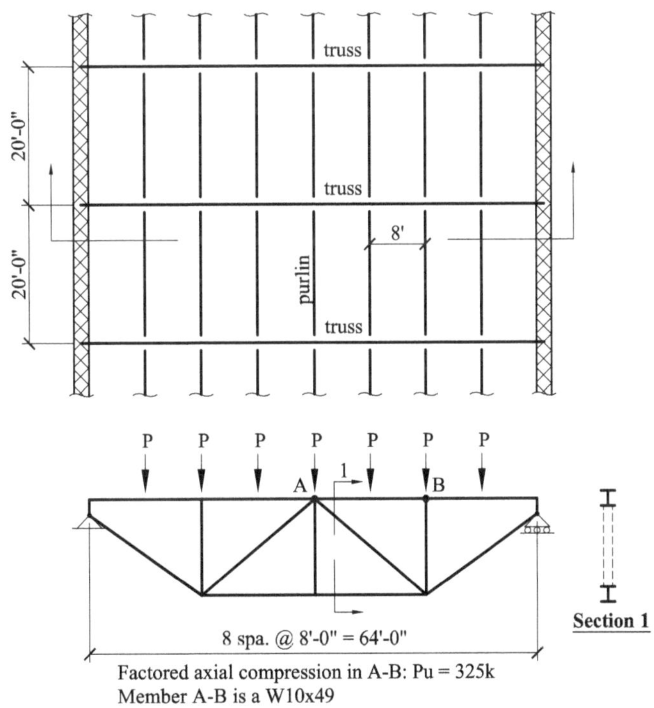

FIGURE 8-56 Details for Exercise 8-16

a. The maximum factored moment in member A-B.

b. The moment and axial capacities, ϕM_{nx} and ϕP_n for member A-B.

c. Determine if member A-B is adequate for the worst-case combined axial load and bending.

Use the following parameters:

$K = 1.0 \quad F_y = 50$ ksi, $\quad C_m = 0.85$

Service loads are: Dead = 25 psf, Snow = 96 psf

Consider the load combination $1.2D+1.0S$ only

Assume beam A-B (span, L=16'-0") is simply supported and is laterally braced only at midspan.

8-17. The top chord of a truss is subjected to a uniform factored load of 3.0 kips/ft. and an axial compression load of 200 kips. If the unbraced length between panel points is 15 ft., determine adequacy for combined loading. The top chord is a W10 × 45 (loaded in the strong axis in bending) and is ASTM A992 steel. Use $K = 1.0$ and $B_2 = 0$.

8-18. The bottom chord of a truss is subjected to a uniform factored load of 2.5 kips/ft. and an axial tension load of 150 kips. If the unbraced length between panel points is 20 ft., determine adequacy for combined loading. The bottom chord is a W12 × 35 (loaded in the strong axis in bending). Assume ASTM A992 Grade 50 steel.

8-19. Determine the maximum axial load that can be applied to the column based on the strength of the base plates shown in Figure 8-57.

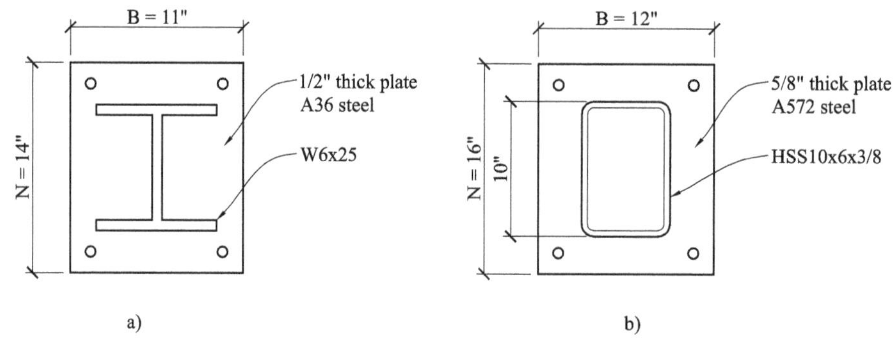

FIGURE 8-57 Details for Exercise 8-19

8-20. Design a base plate to support an HSS 8 × 4 × ½" column to support a factored load of 195 kips. Use $f'_c = 2.5$ ksi. Neglect the size of the foundation. Use ASTM A572 Grade 50 steel for the base plate.

8-21. Determine if the base plate shown in Figure 8-58 is adequate for the given factored loads. Check the bearing strength of the concrete pier, assuming $f'_c = 4$ ksi. The steel used for the base plate is ASTM A572, Grade 50 steel.

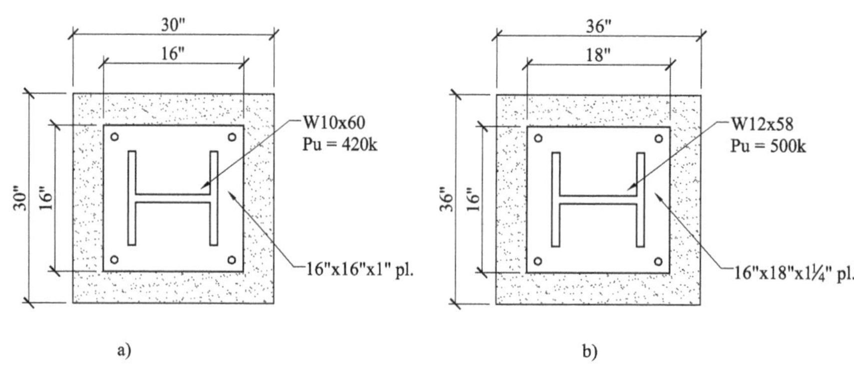

FIGURE 8-58 Details for Exercise 8-21

8-22. Develop a spreadsheet to determine the adequacy of a column loaded with combined axial and bending loads. Submit results for Exercise 8-14, Part 1.

8-23. Develop a spreadsheet to determine the base plate thickness for a given load, P_u, and the plate dimensions, *B and N*. Include variables for the column dimensions, d and b_f, and also the yield strength, F_y.

Submit the design for a W10 × 88 with factored axial load, P_u = 350 kips, base plate dimensions, B = 12" and N = 16", and assuming the yield strength of steel, F_y = 50 ksi.

Provide a generic detail with dimensions.

8-24. Design a typical roof truss to span 80 ft. over a concert hall in Philadelphia, PA. The trusses are spaced at 20 ft. on centers, and support 12 inch thick precast concrete ceiling panels with plaster ceiling with a total weight of 150 psf, in addition to the roof snow load, and the weight of the truss. Due to the long span of the truss, the deflection criteria for the truss are limited as follows:

Maximum deflection due to live load = 3/8 inch.

Maximum total dead plus live load deflection = L/600

The elevation of the typical roof truss showing the nominal sizes of the truss members is shown in Figure 8-59. Assume ASTM A992 Grade 50 steel.

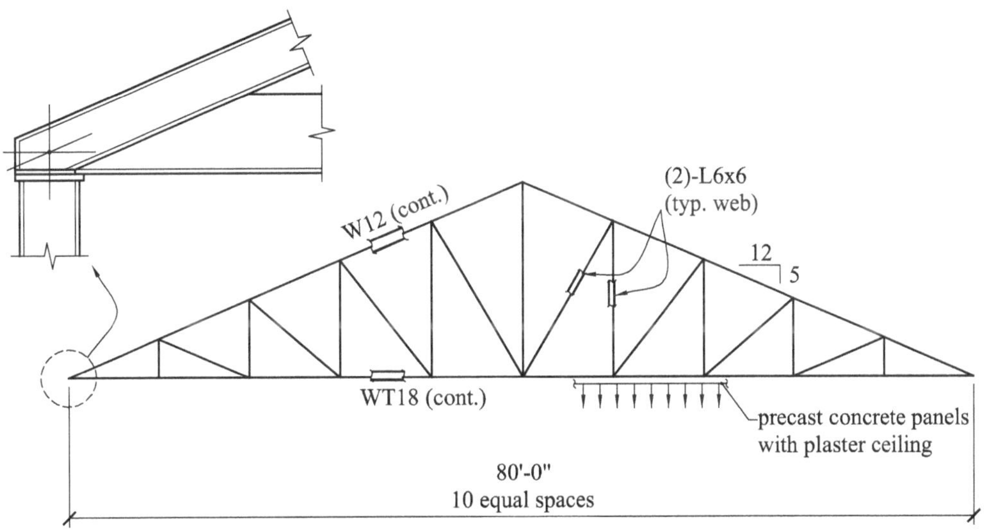

FIGURE 8-59 Roof truss elevation for Exercise 8-24

8-25. Figure 8-60 shows an existing W21×132 column with the axial load (as shown) that is to be transferred to a new tied 2-pinned arch consisting of a built-up box section similar to that used in Ref. [21]. The portion of the column below the arch is to be cut off after the tied arch is installed to create adequate plan space and headroom for new equipment. The plate tension tie will be installed and concealed in a slot to be cut in the existing slab-on-grade. Assume the arch is pinned at the supports A and B and laterally supported by diagonal kickers at the crown, C, and the kickers are connected to the existing roof framing. Determine the following assuming ASTM A992 Grade 50 steel and the total load deflection is limited to 0.30 inches:

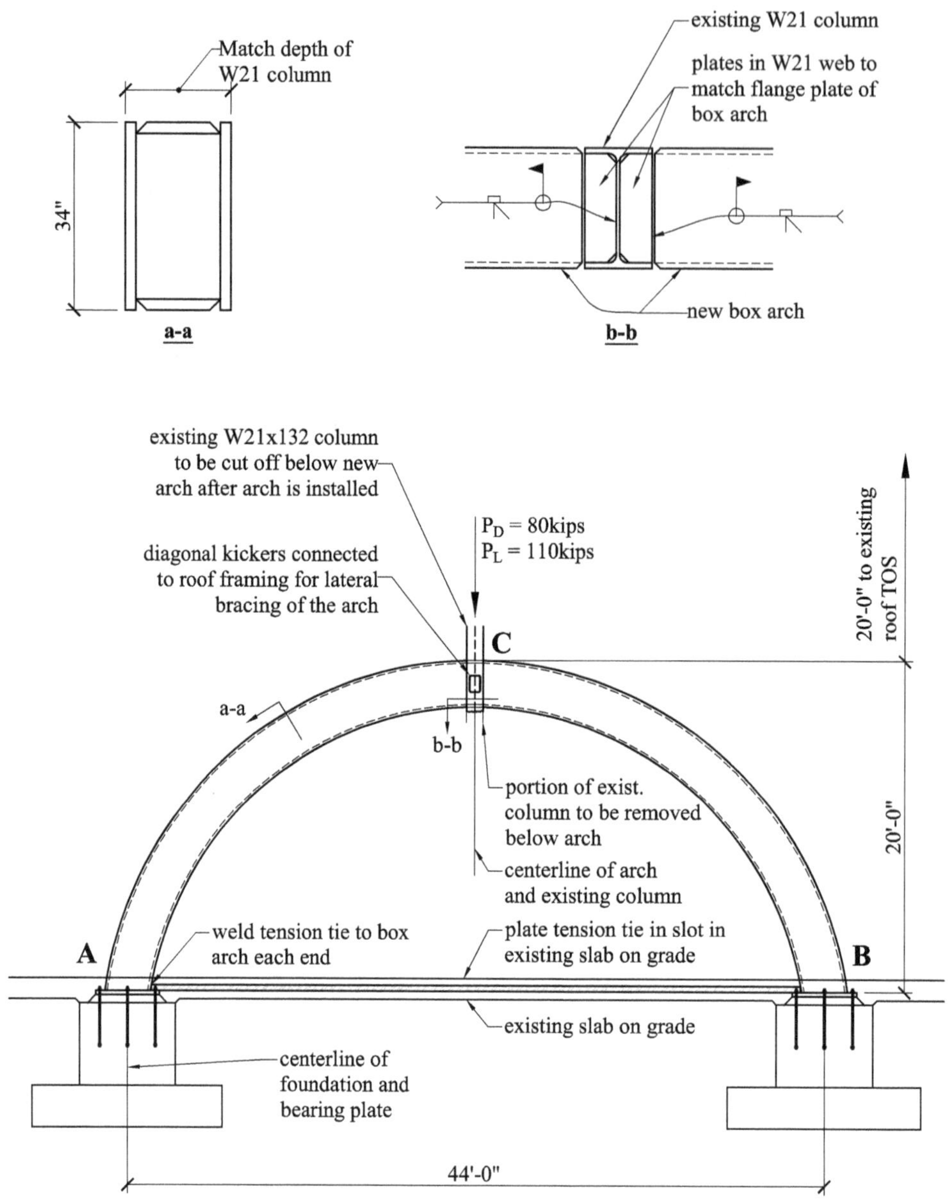

FIGURE 8-60 Transfer box arch for Exercise 8-25

a. For the loads shown and the self-weight of the existing column, carry out a structural analysis of the 2-pinned arch using structural analysis software and determine the forces in the arch and the steel plate tension tie.
b. Determine the minimum thickness of the flanges and webs of the box-shaped arch to satisfy the flexural (see Chapter 6) and axial strength (see Chapter 8) requirements, and the deflection requirements.
c. Determine the minimum width and thickness of the steel plate tension tie.
d. Determine the size of the HSS diagonal kickers or lateral bracing assuming a 45 degree angle for the kickers.

   Hint: See Section 14-10 for the discussion of strength and stiffness requirements for column *point* bracing.
e. Using bullet format, develop a specification or a set of instructions for the contractor for the installation of the arch, the transfer of the existing column load to the arch, and the cutting of the existing column.

8-26. With the aid of free body diagrams, briefly explain the two types of P-delta effects and their impact on the moments in beam-columns subjected to axial compression loads. Briefly explain how the AISC Specification takes the P-delta effects into account in the design of beam-columns.

8-27 A plan view of a 15ft. tall, W10 x 88 column is shown below. Factored dead and snow loads are given at the eccentricities noted from beam connections. Steel is grade 50. You may use any table in the *AISCM* but provide the page number in the manual for any of this data in your solution.

a. Determine the design axial strength and design bending strength in both axes.
b. Determine the worst-case axial and bending loads in both axes.
c. Determine if the column is adequate for combined axial and bending loads.

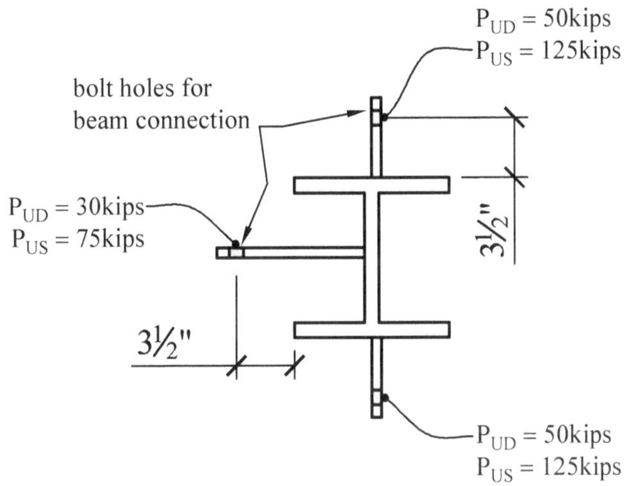

FIGURE 8-61 Column Bending for Exercise 8-27

8-28 a. A column base plate is shown below with a factored axial load of 325 kips. Determine the maximum length of the plate, N, to support this load. Ignore concrete bearing strength.

b. Explain why a rectangular base plate might be used instead of a square base plate and whether or not a square base plate with a similar contact area is more or less efficient than a rectangular one.

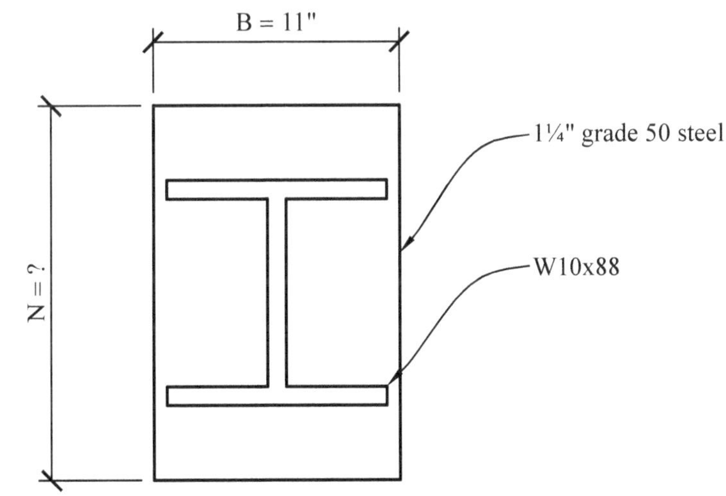

FIGURE 8-62 Column Base Plate for Exercise 8-28

8-29. The transfer truss shown below supports the factored load and unfactored (service dead plus live) load shown and spans over an auditorium. The truss consists of WT18 top and bottom chords and 2L 5 x 5 (back-to-back) web members. The self-weight of the truss is not yet included in the loads shown but must be included in the analysis of the truss. Assume ASTM A913 Grade 50 steel and an allowable truss deflection limit of span/600 under the service dead plus live load plus truss self-weight.

a. Analyze the truss
b. Design the WT 18 top and bottom chords
c. Design the first diagonal web member
d. Design the first vertical web member.

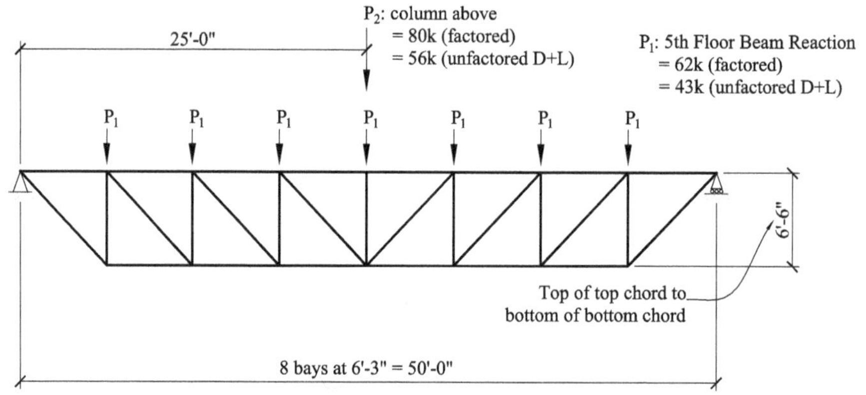

FIGURE 8-63 Transfer Truss for Exercise 8-29

8-30. A series of braced steel arch frames in Figure 8-64 used in a sports arena are spaced at 20 ft. on centers and support a dead load of 50 psf and a snow load of 40 psf.

a. Analyze and design the W36 two-pinned arch.
b. Design the W36 columns.
c. Design the HSS X-bracing
   Assume A913 Grade 50 steel.

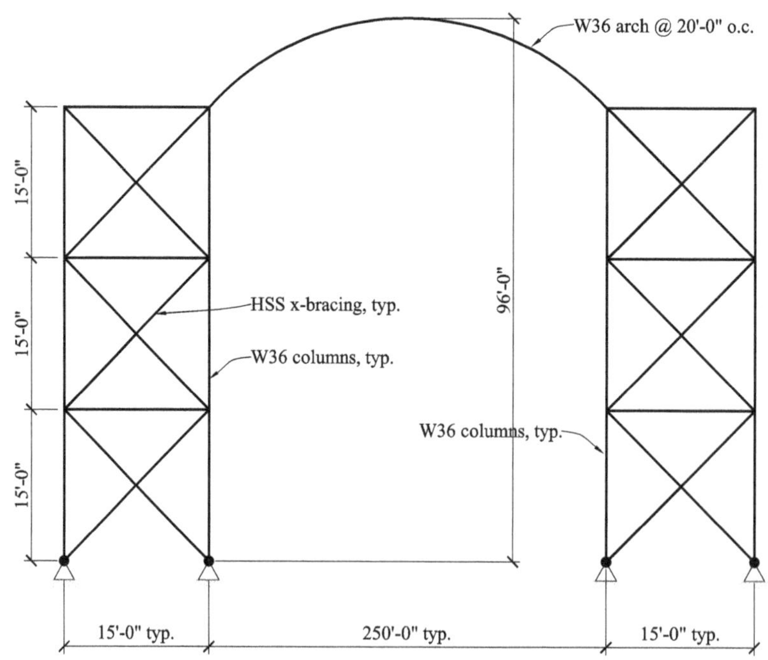

FIGURE 8-64 Braced Arch for Exercise 8-30

8-31. Design a 15 ft. wide pedestrian bridge over a roadway spanning between two buildings. The bridge cantilevers 20 ft. at both ends and spans 100 ft. between columns. Assume a 10 ft deep Warren truss with WT 12 top and bottom chords and HSS 6 x 6 web members and assume 10 ft. wide truss panels and ASTM A913 Grade 65 steel.

Roof: D* = 30 psf, S = 40 psf

Floor: D *= 100 psf, L = 100 psf

Assuming a glass curtain wall load = 20 psf

* Note: The weight of the truss, which must be included in the analysis and design, has not been included in these dead loads.

8-32. 35 ft. deep transfer trusses spaced 40 ft. apart are continuous over two spans with a span of 200 ft. between adjacent column supports and cantilevered beyond the end columns 60 ft. (i.e., the total length of truss = 520 ft.) The truss panel width is 50 ft. within the two spans and 30 ft. within the 60 ft. cantilevers. Assuming W14 ASTM A913 Grade 65 steel, design the top and bottom chords of the truss and the diagonal and vertical web members. The loads on the truss are as follows:

The roof at the top chord of truss: D = 100 psf, L = 180 psf (includes rooftop mechanical equipment and roof live load)

Intermediate floor at mid-depth of truss: D = 100 psf, L = 50 psf

Bottom floor at bottom chord of truss: D = 100 psf, L = 50 psf

CHAPTER

9

Bolted Connections

9-1 INTRODUCTION

In the preceding chapters, we covered the analysis and design process for basic steel members, such as beams, columns, and tension members. In this chapter and in Chapters 10 and 11, we will cover the design of the connections of these members to each other and to other supporting members. In any structure, the individual components are only as strong as their connections. Consequently, designers will often specify that some connections (such as shear connections in beams) be designed for the full capacity of the connected member to avoid creating a "weak link" at the connection. A few bolted connections are shown in Figure 9-1. Bolts are usually lubricated by the manufacturer to enable field installation. See Research Council on Steel Connections (RCSC) Section 2 in the *AISCM*. The commonly used bolt sizes are ¾ in., 7/8 in., and 1 in. diameter bolts, and the two most common grades are A325 and A490 bolts. Due to the tension capacity limitations of available installation equipment, the maximum bolt diameter is typically limited to 1 in. It is recommended to use only one grade of bolt on a project or for each level to avoid the installation of the wrong grade of bolt. Note that A325N or A490N bolts indicate bolts that have threads in the shear plane, whereas A325X and A490X bolts indicate bolts where the connected member thicknesses are such that there are no bolt threads in the shear planes [1].

In practice, there are three basic options for designing and detailing bolted or welded connections in accordance with Sections 3.2.3 and 3.2.4 of the AISC Code of Standard Practice [2].

1. In the first option, the engineer of record (EOR) fully designs and details the connections, and provides connection details on the structural drawings that are complete with connection type, dimensions, and component sizes sufficient for the steel fabricator to build the connection. The main advantage of this method is that the design is completely communicated to the steel fabricator. The disadvantage of this method is that the fabricator, whose input might result in a more economical connection, is left out of the connection design process. For example, some fabricators may prefer to use ⅞ " diameter bolts as a shop standard, whereas ¾" bolts might be specified by the EOR on a given project.
2. In the second option for designing connections, the EOR provides schematic or conceptual details to the steel fabricator, and the fabricator selects the connection details, usually from the standardized tables in Chapter 10 of the AISC manual. In this method, the effort required by the EOR is minimized while allowing the steel fabricator greater flexibility in selecting the most economical connections. One disadvantage of this method is that unique details arise that are not covered in the standardized tables, which can create confusion on a project. It is also important to note that with this option, the EOR has to provide specific loads or reactions and

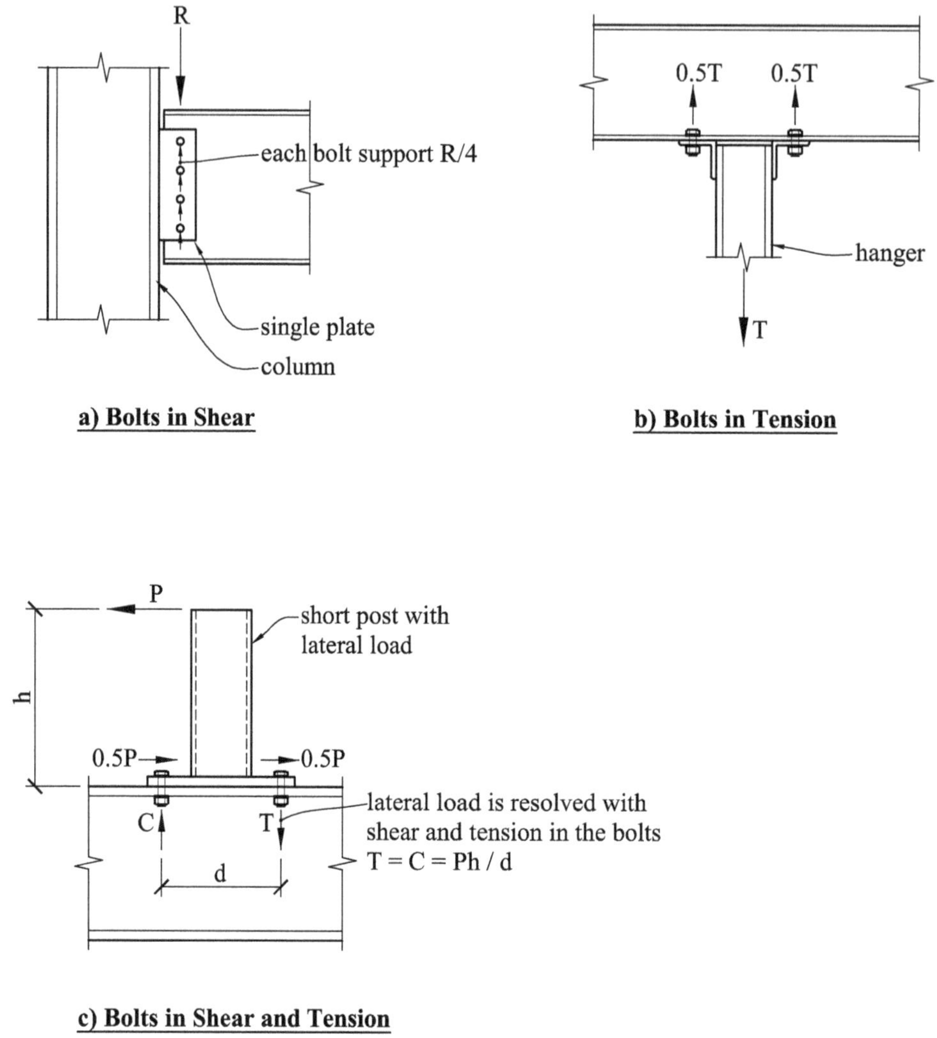

FIGURE 9-1 Connections types using bolts

any other restrictions or parameters sufficient for the fabricator to properly select the connections, but the fabricator is not responsible for providing the engineering services for the connections.

3. In the third option for designing connections, the EOR delegates the connection design to the fabricator's licensed professional engineer. The EOR provides the beam and girder end reactions and any dimensional parameters or other pertinent data and information (such as whether LRFD or the ASD method is used in the design and whether the connection forces shown in the drawings are at the service or factored load level) to allow the fabricator's engineer to design and detail the connection. In this scenario, which is called 'delegated design,' the fabricator is allowed the greatest flexibility in the design and detailing of the connection, but they have to provide calculations that are certified by a licensed professional engineer showing that the connection is adequate to support the specified loads. In some instances, an EOR might require that in addition to providing certified (signed and sealed) calculations, the shop drawings also be signed and sealed. The practice of certifying the shop drawings is discouraged since it creates confusion as to who is ultimately responsible for the design of the structure. In all three cases, it is important for the EOR to be clear as to what is required at the connections. If the connections are designed by the EOR, then sufficient details and all pertinent information must be provided on the structural drawings to minimize the number of

questions or requests for information (RFIs) queries during construction. If the connection designs are delegated to the fabricator, then the loading and any other critical information needs to be properly conveyed to the fabricator, and the EOR is required in accordance with the AISC Code of Standard Practice to review the connection design calculations, and "either approve them or approve them as noted or corrected." In practice, options 1 and 3 are more common in the United States and Canada, but the delegation of connection design (i.e., option 3) is more prevalent in the Eastern United States. In any case, it is always the ultimate responsibility of the EOR to review the design and detailing of the connections in their project to ensure that they conform to the original design intent for the structure.

Bolted connections are the most common and economical connections used in steel structures. They require less labor than welded connections and can be erected in the field in cold weather more easily and efficiently than welded connections, which require special preparations and equipment. Riveted connections were used prior to the advent of bolted and welded connections in the 1950s. The use of riveted connections in structural steel has essentially become obsolete. Rivets required more skilled laborers for installation, as well as more inspection. They were also somewhat more dangerous to install because the rivets had to be heated and installed at a very high temperature (about 1000°F). While high-strength bolts have a greater material cost, they are installed with a greater degree of safety and with less labor.

There are two basic types of bolts—unfinished bolts (also called machine, common, or ordinary bolts) and high-strength bolts. Unfinished bolts conform to ASTM A307 and are generally used in secondary structures, such as handrails, light stairs, service platforms, and other similar structures that are not subject to cyclical loads. Unfinished bolts have a lower load-carrying capacity than high-strength bolts; therefore, their use should be limited to secondary structures that typically have lighter loads. High-strength bolts are the most common type of bolts used in steel structures, and they possess more than twice as much tensile strength as unfinished bolts. High-strength bolts conform to ASTM F3125, F3043, F3111, and ASTM F3148. This specification covers the following grades of bolts:

- Group 120 ($F_u = 120$ ksi): Grades A325, F1852, A354 Grade BC
- Group 144 ($F_u = 144$ ksi): Grade F3148
- Group 150 ($F_u = 150$ ksi): Grades F3125, A490, F2280, A354 BD
- Group 200 ($F_u = 200$ ksi): Grade F3043 and F3111

These are the most commonly specified bolt materials, and they can be used in bearing connections as well as in slip-critical connections (connections where slip does not occur; see Section 9-4). Some bolts are referred to as twist-off, tension-controlled bolts, and these bolts should meet the requirements of ASTM F3125 Grades F1852 or F2280. These bolts are fabricated with a splined end and installed with a special wrench such that the splined end breaks off once the required torque is obtained. The inspection of these bolts is simplified in that only visual inspection is required. Bolts that conform to ASTM F3125 Grade F1852 are equivalent to ASTM F3125 Grade A325, and bolts that conform to ASTM F3125 Grade F2280 are equivalent to ASTM F3125 Grade A490 for strength and design purposes. ASTM F3148 bolts can be mechanically galvanized.

9-2 BOLT INSTALLATION

There are three basic types of bolted joints or connections addressed in the AISC Specifications and by the Research Council on Structural Connections (RCSC), and these are snug tight, pretensioned, and slip-critical bolted connections. The majority of joints are snug tight connections.

The difference between these bolted joints is essentially in the amount of clamping force that is present when tightening the bolts and the degree to which the connected parts can move relative to each other while in service. The contact area between the connected parts is called the *faying surface*. In any project, the engineer must specify the joint type and the faying surface that are to be used for any given connection.

A ***snug-tight*** condition occurs when the bolts are in direct bearing and the plies of a connection (i.e., the members being connected) are in firm contact, and the bolts are in direct bearing against the connected members in resisting the applied load, and they have been tightened enough so that they cannot be removed without the use of a wrench. This can be accomplished by a few impacts of an impact wrench, or by the application of an electric torque wrench until the wrench begins to slow, or by the full effort of a worker using a spud wrench, which is an open-ended wrench approximately 16 in. long. The opposite end of the wrench is tapered to a point, which an ironworker uses to align the bolt holes of the connected parts. A snug-tight joint can be specified for most simple shear connections, as well as tension-only connections. Snug-tight joints have faying surfaces that can either be uncoated, painted, or galvanized but must be free of dirt. They are not permitted to be used for connections supporting dynamic loads, nor are they permitted to be used with A490 bolts loaded in tension. Snug-tight connections are used for simple shear connections (i.e., bearing-type connections), tension-only connections, connections with combined tension and shear using Group 120 bolts only, and connections where loosening of the nut or where fatigue due to vibration or cyclic loading does not occur (see AISC Specification, Section J3.2). Snug-tight bolts are typically used in most building structures unless otherwise noted on structural drawings or the contract documents.

A ***pretensioned*** joint has a greater amount of clamping force than the snug-tight condition and therefore provides a greater degree of slip resistance in the joint. Pretensioned joints are used for joints that are subject to cyclic loads or fatigue loads. They are also required for joints with A490 bolts in tension. Some specific examples where pretension bolted connections should be specified are as follows (see RCSC Section 4.2):

- Connections subject to load reversals (i.e., impact or cyclic loading) such as crane-support structures and bridges.
- Connections that are subject to fatigue loading without reversal of the loading direction.
- Connections with Group 120 bolts that are subject to tensile fatigue
- Connections with Group 144 or 150 bolts that are subject to tension or combined tension and shear.

Other conditions in which pretensioned bolted connections are recommended include the following:

- Connections for lateral bracings in tall buildings.
- Column splices in buildings with high aspect ratios (i.e., buildings with high height-to-width ratios).

It is important to note that the design strength of a pretensioned joint is similar to that of a snug-tight joint. Pretensioned joints do not require prepared faying surfaces able to resist slippage. In a pretensioned joint, slip is prevented until the friction force provided by the clamping force is exceeded. Once the friction force is exceeded because the pretension or clamping force has essentially become zero (i.e., equivalent to a snug-tight condition), the bolts slip into direct bearing. For both snug-tight and pretensioned bolts, the faying surface is permitted to be uncoated, painted, or galvanized but must be free of dirt and other foreign material.

TABLE 9-1 Minimum Bolt Pretension (pretensioned and slip-critical joints)

	Minimum bolt pretension, $= 0.70R_n$* (kips)			
Bolt size, in.	**Group 120 bolts F_u = 120 ksi (ASTM F3125 Grades A325, F1852, and ASTM A354 Grade BC bolts)**	**Group 144 bolts F_u = 144 ksi (ASTM F3148 Grade 144 bolts)**	**Group 150 bolts F_u = 150 ksi (ASTM F3125 Grades A490, F2280, and ASTM A354 Grade BD bolts**	**Group 200 bolts F_u = 200 ksi (ASTM F3043 and ASTM F3111 bolts)**
½	12	15	15	
⅝	19	24	24	
¾	28	35	35	
⅞	39	49	49	
1	51	64	64	90
1⅛	64	80	80	113
1¼	81	102	102	143
1⅜	97	121	121	
1½	118	148	148	

Source: Adapted from Table J3.1 of the *AISCM* [2]. *

From equation J3-1, $R_n = F_{nt}A_b$.
F_{nt} = 90 ksi (Group 120 bolts) = **$0.75F_u$
F_{nt} = 108 ksi (Group 144 bolts)
F_{nt} = 113 ksi (Group 150 bolts)
F_{nt} = 150 ksi (Group 200 bolts)
A_b = Nominal unthreaded body area of bolt
$F_{nt} = 0.75F_u$

When pretensioned bolts are installed, they must be tightened such that a minimum clamping force is achieved between the connected parts. The AISC specification stipulates that the minimum required clamping force should be at least 70% of the nominal tensile strength, R_n, of the fastener (i.e., $0.70F_uA_b$). Table 9-1 indicates the minimum required tension clamping force for various bolt types.

Note that the "0.75" in F_{nt} equation in Table 9-1 (i.e., $F_{nt} = 0.75F_u$) "accounts for the approximate ratio of the effective tension area of the threaded portion of the bolt to the area of the shank of the bolt for common sizes between ½ in. and 2 in., with ratios ranging from 0.72 and 0.80." (see AISC Specification Section C-J3.7). For larger bolt sizes and threaded rods, this equation may be too conservative, and thus, $F_{nt} = F_u$ may be used together with the actual tensile stress area of the bolt, A_t. The net stress area of the bolt in tension from equation (A-3-7) in Appendix 3 of the AISC Specification is

$$A_t = \frac{\pi}{4}\left(d_b - \frac{0.9743}{n}\right)^2$$

Where,

n = number of threads per inch (see *AISCM* Table 7-17)

d_b = nominal diameter of the unthreaded portion of the bolt

As an example, for Group 120 1½ in. diameter bolts in tension ($n = 6$ from *AISCM* Table 7-17), the actual tensile stress area is

$$A_t = \frac{\pi}{4}\left(1.5 - \frac{0.9743}{6}\right)^2 = 1.405 \text{ in.}^2,$$

and the minimum pretension is calculated as

$0.7F_{nt}A_t = 0.7F_uA_t = 0.7(120)(1.405) = 118$ kips (which is the value indicated in Table 9-1).

If the more conservative value of $0.75F_u$ had been used, the minimum pretension would have been $0.7(F_{nt})A_b = 0.7(0.75F_u)A_b = 0.7[(0.75)(120)][(\pi)(1.5)^2/4] = 111.3$ kips, a 6% difference from the actual value of 118 kips in Table 9-1.

In order to achieve the minimum tensile force in pretensioned and slip-critical bolted connections indicated in Table 9-1, the bolts must be pre-tensioned by one of the following methods, starting from a snug-tight condition (see Section 8.2 of the Research Council on Structural Connections (RCSC) 2020 Specifications in Chapter 16-2 of the *AISCM*):

1. **Turn-of-Nut:** Each turn-of-nut is calibrated to represent a certain amount of tensile clamping force. When a nut is advanced along the length of a bolt, each turn corresponds to a certain amount of tensile force in the bolt. Therefore, there is a known relationship between the number of turns and the amount of tension in the bolt. The starting point (i.e., a point where the tensile force in the bolt is just above zero) is defined as the snug-tight condition. In Table 8-2 of the Research Council on Structural Connections (RCSC) Specifications [2], the required number of turns as a fraction of a full rotation is given as a function of the bolt length-to-diameter ratio.
2. **Calibrated Wrench Tightening:** In this method, calibrated wrenches are used so that a minimum torque is obtained, which corresponds to a specific tensile force in the bolt. On any given project, the calibration has to be done daily for each size and grade of bolt.
3. **Twist-off Tension control Bolts:** These bolts have a splined end that breaks off when the bolt is tightened with a special wrench to a certain tensile force (see Figure 9-2a). As discussed in Section 9-1, these bolts conform to ASTM F3125 Grade F1852 (equivalent to ASTM F3125 Grade A325 for strength and design) and ASTM F3125 Grade F2280 (equivalent to ASTM F3125 Grade A490).
4. **Direct-Tension Indicator (DTI):** These bolts have compressible washers that conform to ASTM F959, and they have ribbed protrusions on the bearing surface that compress in a controlled manner when the tension force in the bolt or the clamping force reaches a certain value, such that the clamping force is proportional to the tension in the bolt (see Figure 9-2b). The deformation in the ribs of the washer is measured to determine whether the proper tension has been achieved.
5. **Combined method pretensioning:** After applying an initial torque to the bolts to bring the connected plies or members in contact firmly with each other, additional nut rotation in accordance with Table 8-2 of the RCSC is applied.

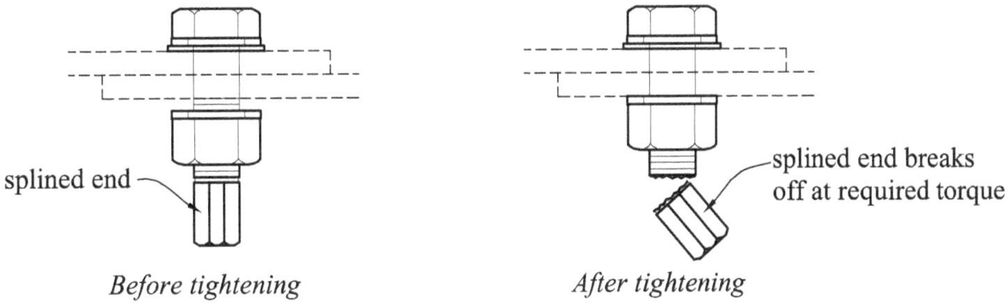

a) Twist-off-type tension-control bolts

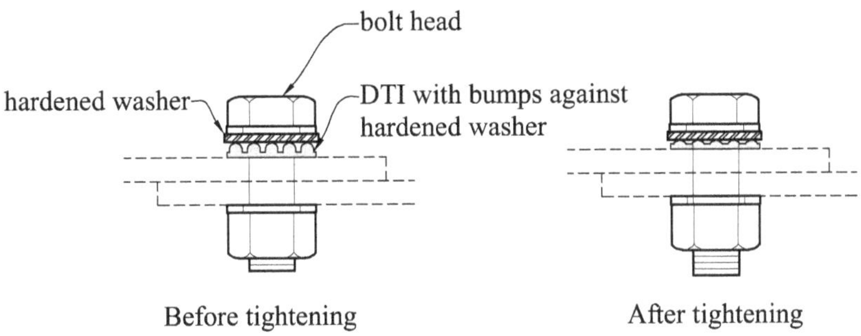

b) Direct-Tension-Indicator (DTI) washer

FIGURE 9-2 Bolts types based on tightening

FIGURE 9-2c Twist-off tension-controlled bolts

One unique condition pertaining to the installation of bolted connections occurs when the connection slips into a bearing condition when the building is in service. When this happens, occupants might hear what sounds like a gunshot, which may be alarming, as the bolt suddenly bangs against the sides of the bolt hole. However, this event, which is known as "banging bolts", is not indicative of a structural failure, but to prevent it from happening, the bolts should be snug-tight or, if possible, the steel erector should not tighten the bolts until the drift pins have been released and the bolts are allowed to slip into bearing before tightening (see Section 9-4 for further discussion on bolt bearing).

The third type of bolted joint is the ***slip-critical*** connection. This type of connection is similar to a pretensioned joint except for the preparation required for the faying surface that ensures a calculable or known amount of slip resistance, and failure occurs when the connected members slip relative to each other when the friction force or slip resistance in the faying surface(s) is overcome by the applied load. As with pretensioned joints, slip-critical joints are used for joints subjected to cyclic loads or fatigue loads. They should also be used in connections that have slotted bolt holes parallel to the direction of the load or in shear connections that use a combination of welds and bolts along the same or common faying surface (see RCSC, Section 4.3). Slip-critical connections are used where slippage of the faying surface would negatively impact the structural performance, such as the connection of a cantilevered sloped roof beam or rafter to the top of a column cap plate, or where slippage of the faying surfaces at bolted connections would amplify the vertical deflection of long-span cantilevered stadium roof canopy plate girders or the lateral deflections of braced frames or moment frames in tall buildings.

The amount of pretension or clamping force for a slip-critical bolt is the same as that for pretensioned joints (see Table 9-1). The design strength of a slip-critical joint is generally smaller than that of a similar bearing-type connection since the friction resistance is usually lower than the strength for any other bolted connection failure mode, such as direct shear or bearing.

The main difference between pretensioned and slip-critical bolted joints is the type of faying surface between the connected parts. The faying or slip surfaces for pretensioned bolted joints or snug-tight connections do not require special preparation, but they have to be free of dirt. They can be uncoated, painted, or galvanized. However, for slip-critical bolted connections, two types of faying surfaces (with different surface preparations) are identified in Section J3.9 of the AISC Specifications: Class A and Class B. Each type has a specific surface preparation and coating requirement that corresponds to a minimum coefficient of friction on the faying or slip surface.

- Class "**A**" faying surface has either unpainted clean mill-scale surfaces or surfaces with Class "A" coatings on blast-cleaned steel or hot-dipped galvanized and roughened surfaces. The Class "A" coating has a minimum mean slip coefficient, $\mu = 0.30$.
- Class **B** faying surface has either an unpainted blast-cleaned steel surface or a surface with Class B coating on blast-cleaned steel. The minimum mean slip coefficient for a Class B surface is $\mu = 0.50$. The mean slip coefficient for any faying surface can also be established by testing for special coatings and steel surface conditions.

9-3 BOLT HOLE TYPES AND SPACING REQUIREMENTS

Some common methods for forming bolt holes include punching, drilling, and thermal cutting. The four basic types of bolt holes recognized in the AISC specification are

- **Standard holes:** Standard bolt holes are the most common and are generally used for bolts in direct bearing.
- **Oversized holes:** Oversized bolt holes are permitted only for slip-critical connections, and they are used where a larger tolerance for erection is needed.
- **Short-slotted holes:** Short-slotted and long-slotted bolt holes independent of the load direction relative to the direction of the slot are permitted for slip-critical connections.
- **Long-slotted bolt holes:** long-slotted and short-slotted bolt holes with the slot direction perpendicular to the load direction are permitted for bearing connections (see *AISCM* Section J3.3).

Note that long slotted or oversized holes are typically specified to ensure adequate tolerances and proper fit-ups in the field, and slip-critical connections are preferred with these types of holes. Table J3.3 in the *AISCM* [2] lists the actual bolt hole sizes for each bolt diameter and bolt hole type. Each bolt hole type offers varying degrees of flexibility and tolerance in the construction of the connections. Figure 9-3 shows the nominal dimensions of *standard, oversized, short-slotted, and long-slotted* holes for bolts, adapted from the *AISCM* Table J3.3. The *AISCM* Section J3.3 or RCSC Section 3.3.1 [2] indicates that when calculating the net area for shear and tension, an additional $\frac{1}{16}$ in. ($\frac{1}{8}$ in. for $d_b \geq 1$ in.) should be added to the bolt hole size to account for the roughened or damaged edges around the hole that result from the punching or drilling process:

- For standard holes, the AISC Specifications specifies bolt hole diameters of $d_b + \frac{1}{16}$ for bolt diameters less than 1 in.
- A standard bolt hole diameter of $d_b + \frac{1}{8}$ is specified for bolt diameters 1-in. and larger (see AISC Specification Section J3.4 or Figure 9-3).
- To be consistent with the hole diameter used in Chapter 4 for computing the net area for tension and shear, it is slightly conservative to use the same hole diameter. Thus, the bolt hole size or diameter used for strength calculations is taken as $d_b + \frac{1}{8}$ in. for standard bolt holes with bolt diameter less than 1-in, and $d_b + \frac{3}{16}$ in. for 1-in diameter bolts in standard holes, where d_b is the bolt diameter
- In this text, the symbols d_b *or* d are used interchangeably to represent bolt diameter.

For column base plates, it has been recognized that the embedment of anchor rods into a concrete foundation generally does not occur within desirable tolerances and thus has led to numerous errors in the alignment of columns, base plates, and anchor rods. One way to mitigate this problem is to provide larger bolt hole sizes in the column base plates to allow for misaligned anchor rods. Table 9-2 shows the recommended maximum bolt hole sizes in column base plates.

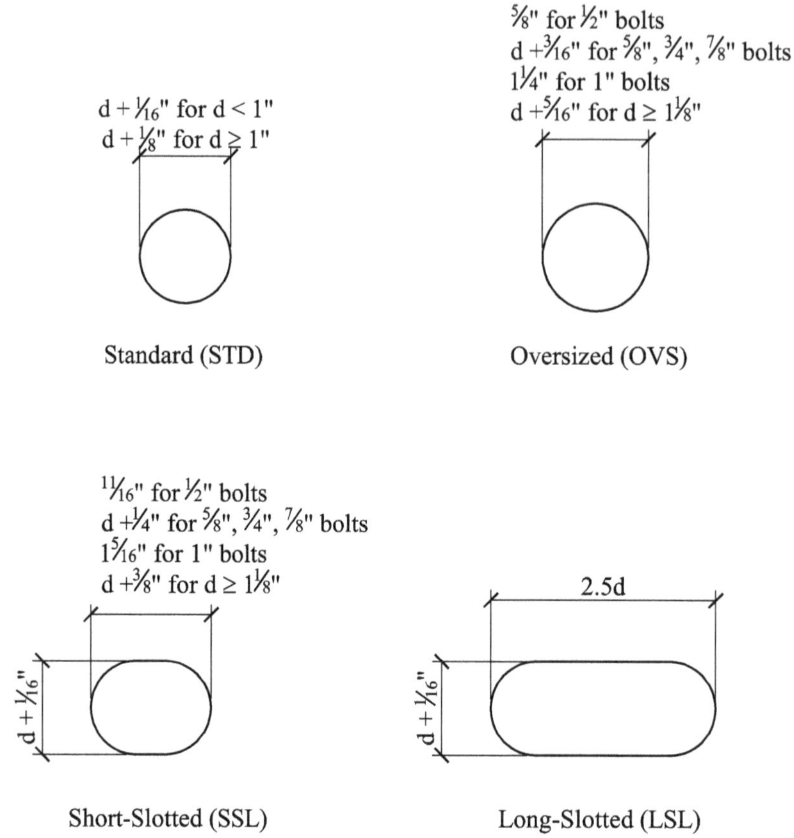

FIGURE 9-3 Bolt Hole Types and Sizes

TABLE 9-2 Recommended Sizes for Anchor Rod Holes in Column Base Plates

Anchor rod diameter, in.	Maximum hole diameter, in.
$^{3}/_{4}$	$1^{5}/_{16}$
$^{7}/_{8}$	$1^{9}/_{16}$
1	$1^{7}/_{8}$
$1^{1}/_{4}$	$2^{1}/_{8}$
$1^{1}/_{2}$	$2^{3}/_{8}$
$1^{3}/_{4}$	$2^{7}/_{8}$
2	$3^{1}/_{4}$
$2^{1}/_{2}$	$3^{3}/_{4}$

Source: Adapted from Table 14-1, *AISC Manual of Steel Construction,* 16th ed.

To allow for standard fabrication procedures as well as workmanship tolerances, the *AISCM* Section J3.4, recommends that the minimum spacing between adjacent bolts be at least $3d$, with an absolute minimum spacing of $2^{2}/_{3}d$, where d is the bolt nominal diameter. Note that a spacing between the centerline of adjacent bolt holes of $3d$ is preferred. The minimum clear distance between bolt holes or slots is d. The minimum distance in any direction from the center of a standard bolt hole to an edge is given in the *AISCM* Table J3.4. In general, the minimum edge distance is approximately $1.25d$ to $1.5d$ for bolts near a sheared edge (see Figure 9-4). One exception is that the workable gages in angle legs (*AISCM* Table 1-7) may be used for single and double angles.

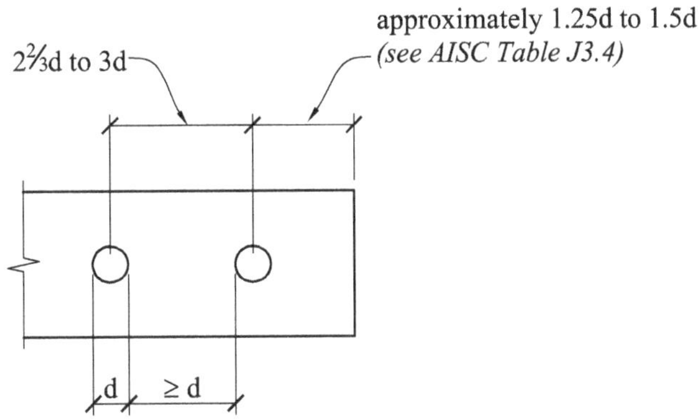

FIGURE 9-4 Edge and spacing requirement for bolts

9-4 FAILURE MODES AND STRENGTH OF BOLTS

The failure modes of bolted connections (see Figure 9-5) include:

1. Shear failure of the bolts
2. Tension failure of the bolts
3. Bearing failure of the connected elements or materials (i.e., the main material and the connection material) that the bolts bear against. In this failure mode, the bolt deforms the edge of the material that it bears against. If the deformation at the bolt hole is a design consideration (i.e., the deformation is limited), the strength of the bolt in bearing is smaller compared to when the deformation at a bolt hole is not limited and therefore is not a design consideration.

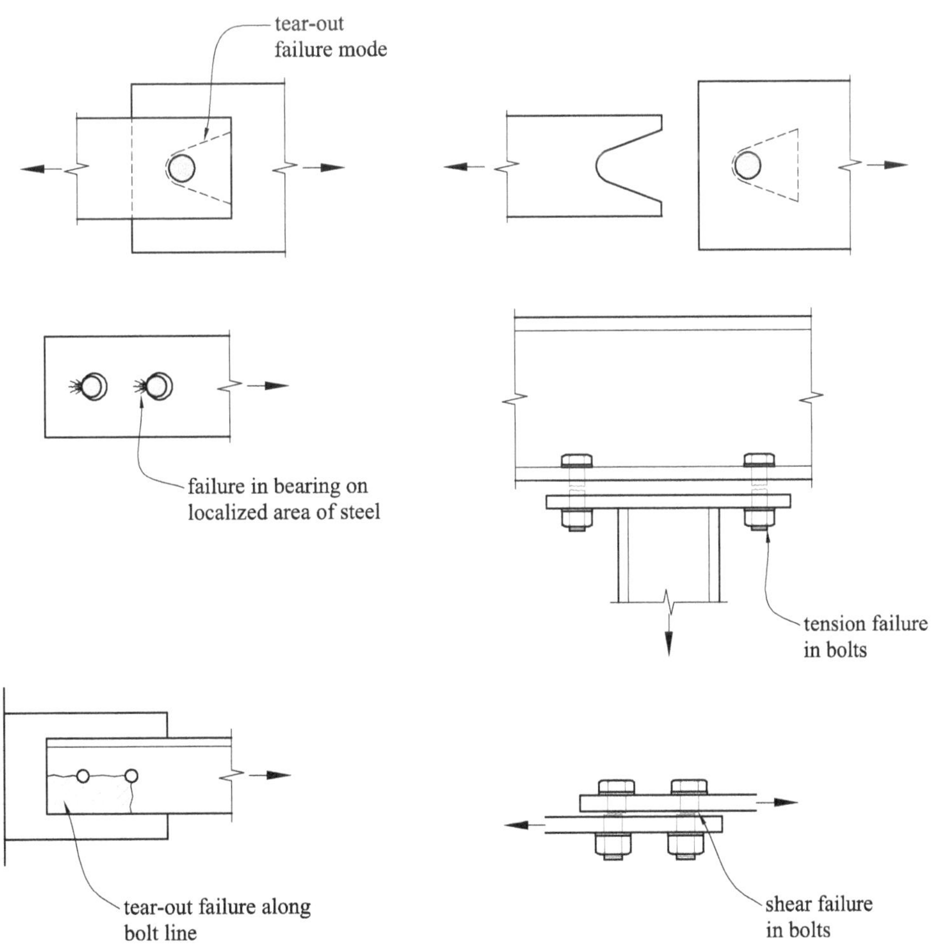

FIGURE 9-5 Bolt failure modes

4. Tearout failure of the connected elements (i.e., the main material and the connection material) due to block shear rupture of the connected parts that the bolts bear against. This is block shear failure of the material in front of the bolt against which the bolt bears in the direction of the applied load [3]. In this case, due to excessive bearing, the bolt tears through the material, leaving an elongated slot in the connected material. Tearout can occur at an edge of the connected part or between adjacent bolt holes, but this later failure mode rarely occurs. Tearout at an edge will generally not occur if the thickness of the connected part is greater than or equal to the bolt diameter, since direct shear of the bolts is more likely to occur [3]. Note that in the 16th Edition *AISCM* all connection material specifications are now ASTM A572 Grade 50 as opposed to the ASTM A36 in previous editions of the *AISCM*. Thus, the strength of all connection materials in the *AISCM* is now 50 ksi.

In this section, we will consider the strength of the fasteners as well as the strength of the connected sections. Most basic shear connections are fastened in such a way that the bolts bear directly on the connected parts. For this to happen, there must be some nominal amount of displacement or slip to allow the bolts to bear directly on the connected parts. A slip-critical connection is a bolted connection where any amount of displacement is not desirable, and therefore the bolts must be pretensioned to the loads indicated in Table 9-1. For bearing connections, the bolts can be loaded either in single shear or double shear (see Figure 9-6). A lapped connection with two members has one shear plane, and therefore the bolt is loaded in a single shear. A three-member connection consists of two shear planes; therefore, the bolts are loaded in double shear. The additional shear plane reduces the amount of load on the bolts in each shear plane, as shown in Figure 9-6.

The design shear and tension strengths of a snug-tight or pretensioned bolt are calculated as follows:

$$\phi R_n = \phi F_n A_b, \quad \text{for LRFD} \tag{9-1a}$$

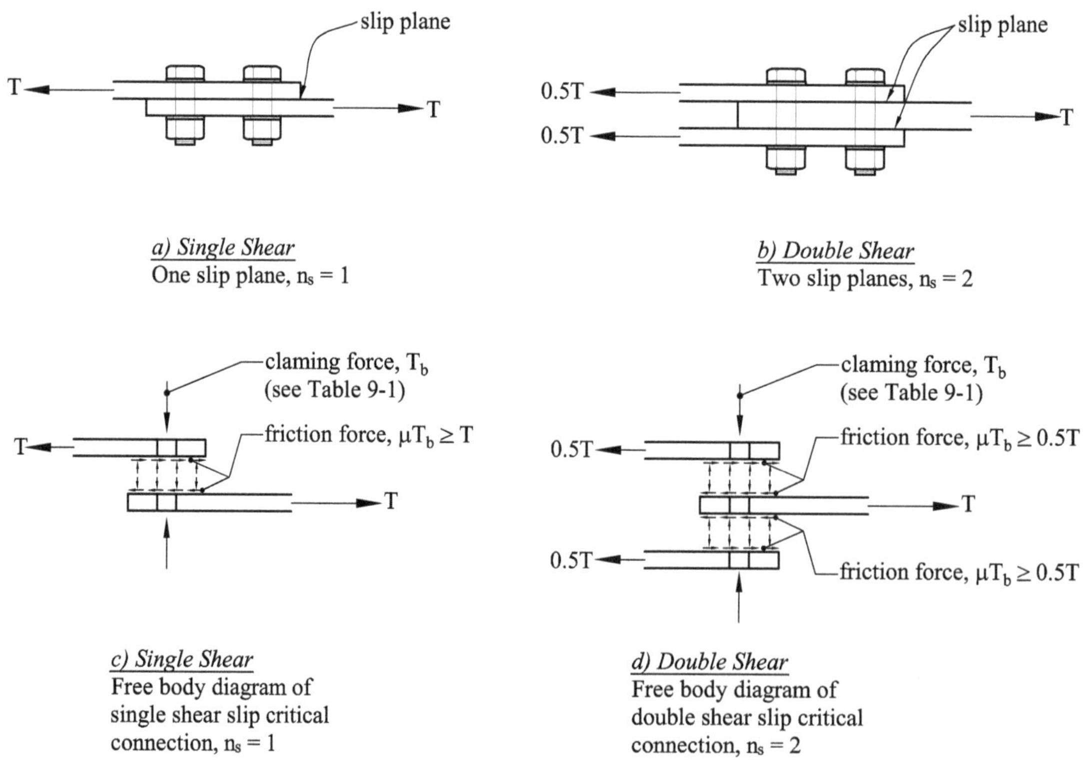

FIGURE 9-6 Bolts loaded in single and double shear

$$\frac{R_n}{\Omega} = \frac{F_n A_b}{\Omega}, \quad \text{for ASD} \tag{9-1b}$$

Where,

ϕ = 0.75

Ω = 2.0

R_n = Nominal shear strength or nominal tension strength of the bolt, kips

F_n = Nominal shear stress (F_{nv}) or nominal tension stress (F_{nt}), ksi (see Table 9-3)

A_b = Nominal unthreaded body area of bolt, in.^{2}

AISCM Table 2-6 lists the applicable ASTM specifications *and* the ultimate tensile strengths and diameter range for different types of structural fasteners. From the *AISCM* Table 2-6, the ultimate tensile strength, F_u, is as follows:

- 60 ksi for A307 grade A bolts
- 120 ksi for Group 120 bolts—A325 bolts (0.5 in. to 1.5 in. diameter) and F1852 bolts (0.5 in. to 1.25 in. in diameter).
- 144 ksi for Group 144 bolts.
- 150 ksi for Group 150 bolts—A490 bolts (0.5 in. to 1.5 in. in diameter) and F2280 bolts (0.5 in. to 1.25 in. in diameter).
- 200 ksi for Group 200 bolts, ASTM F3043 and ASTM F3111, F_u is 200 ksi for bolt sizes between 1 in. and 1.25 in. in diameter.

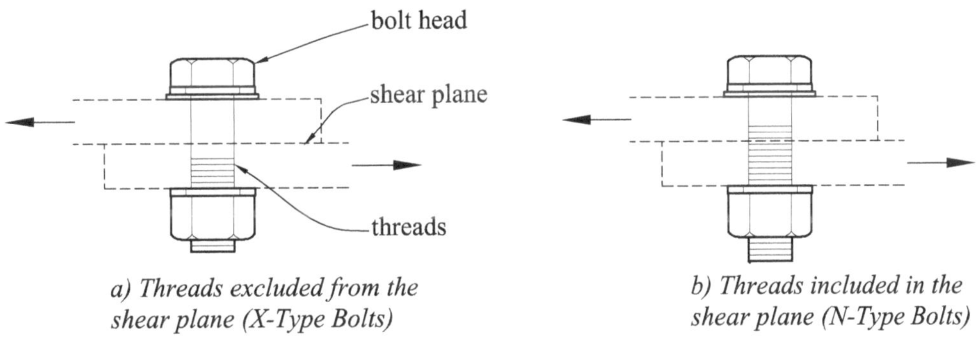

FIGURE 9-6e Bolt threads included and excluded from the shear plane

The nominal tensile strength, F_{nt}, is taken as $0.75F_u$ for all cases. A bolt in tension fails at the threaded section, and since the nominal bolt diameter is used in the calculations, the "0.75" factor accounts for the reduction in cross-sectional area at the threaded section. Thus, the factor "0.75" is the ratio of the cross-sectional area of the threaded section of a ¾ in. diameter bolt to the gross unthreaded cross-sectional area of the bolt. Though the factor will be larger for larger bolt diameters, the Code conservatively uses the lowest reduction factor of "0.75" since ¾ in. diameter bolts are the smallest bolt diameter used in structural steel construction [1].

The nominal shear strength, F_{nv}, is a function of whether or not the bolt threads are in the shear plane (see Figure 9-6e). When the threads are *excluded* from the shear plane (i.e., X-type bolt), the nominal shear strength is taken as $0.563F_u$, which is 90% of the experimentally observed value of $0.625F_u$, where the 90% is the length reduction factor for bolted connections not exceeding 38 inches in length. Both the 90% reduction factor and the strength reduction factor, ϕ, account for second-order effects and differential strain in connections not exceeding 38 inches in length. When the threads are *included* in the shear plane (i.e., N-type bolt), the nominal shear strength is reduced to $0.45F_u$, which is 80% of $0.563F_u$ (see Commentary to AISC Specification Section J3.7). The 80% reduction factor accounts for the reduced area of the threaded portion of the bolt. When bolt threads are intended to be excluded from the shear plane, they are designated as A325X or

A490X. When the bolt threads are intended to be included in the shear plane, the proper designation is A325N or A490N, or just simply A325 or A490. Note that whether the bolt threads will be included or excluded from the shear plane is basically dependent on the thickness of the connected parts. If A325X or A490X bolts are required or desired, which means the threads need to be excluded from the shear planes, then the thicknesses of the connected parts or plies must be selected so that the threads of the bolts will indeed be excluded from the shear planes. Table 9-3 summarizes the nominal shear and tensile strengths for various types of bolts.

TABLE 9-3 Nominal Tensile and Shear Stress Stress of Bolts

Bolt type	**^^Nominal tensile strength, F_{nt}, ksi = $0.75F_u$**	**+^Nominal shear strength, F_{nv}, ksi = $0.563F_u$ for threads <u>*excluded*</u> from shear plane (X)**	**+^Nominal shear strength, F_{nv}, ksi = = $0.450F_u$ when threads <u>*included*</u> in shear plane (N)**
A307	45	27	27
Group **120** bolts (e.g., A325)	90	54	68
Group **144** bolts (e.g., F3148)	108	65	81
Group **150** bolts (e.g., A490)	113	68	84
Group **200** bolts (e.g., F3043)	150	90	113

Source: Adapted from Table J3.2, *AISC Manual of Steel Construction,* 16th ed. [2] Copyright © American Institute of Steel Construction. Reprinted with permission. All rights reserved.

^$F_{nv} = 0.563F_u$ for threads *excluded from* the shear plane
$F_{nv} = 0.45F_u$ for threads *included in* the shear plane
$F_{nt} = 0.75F_u$ for bolts in tension
+F_{nv} shall be multiplied by a reduction factor of 83.3% for end-loaded connections where the length of the connection (i.e., the distance between the outermost bolts in the direction of the axial load) is greater than 38 inches.

For end-loaded connections where the length of the connection (i.e., the distance between the outermost bolts in the direction of the axial load) is greater than 38 inches, the nominal shear strength, F_{nv}, in Table 9-3 shall be multiplied by a reduction factor of 83.3% to account for the uneven or non-uniform distribution of the shear force to the bolt group in end-loaded connections, where the bolts nearest to the end of the connection resist a higher portion of the total load. (see the footnote in Table J3.2 of the AISC Specifications [2]). Some examples of *end-loaded bolted connections* include the following:

- An angle or WT hanger connected with a line of bolts to the web of a floor beam where the top flange is coped;
- A vertical hanger connected with a line of bolts to a gusset plate that is welded to the top flange of a wide flange beam; and
- A wide flange diagonal brace member connected to a gusset plate with a WT or plate section with two lines of bolts parallel to the axial load.

For more examples of end-loaded connections, see Figure C-J3.1 in the AISC Specifications [2]. The length of most bolted connections used in steel buildings will be less than 38 inches; therefore, this reduction factor will not apply in most cases.

- **Bearing failure mode:**
  This applies to connections with:
  - Standard bolt holes
  - Oversized bolt holes,

- Short slotted bolt holes independent of load direction
- Long-slotted bolt holes with the slots parallel to the direction of the applied force.

For the bolt to adequately transfer loads from one connected part to another, the connected members must have adequate strength in bearing. For a bolt in a connection with the above bolt hole types - where the deformation at the bolt hole at service load *is* a design consideration (i.e., the elongation of the bolt hole under a high tensile stress at the net cross-section does not exceed ¼ in.), the design strength in bearing (see *AISCM* Section J3.11a.(1)(a)) is a function of the tensile strength, the thickness of the connected part, and the bolt diameter, and is given as:

$$\phi R_n = \phi 2.4dtF_u, \quad \text{for LRFD} \tag{9-2a}$$

$$\frac{R_n}{\Omega} = \frac{2.4dtF_u}{\Omega}, \quad \text{for ASD} \tag{9-2b}$$

When the deformation at the bolt hole at service load *is* ***not* a design consideration** (i.e., the elongation of the bolt hole under a high tensile stress at the net cross-section can exceed ¼ in. and is not limited), the design strength in bearing is

$$\phi R_n = \phi 3.0dtF_u. \quad \text{for LRFD} \tag{9-3a}$$

$$\frac{R_n}{\Omega} = \frac{3.0dtF_u}{\Omega}, \quad \text{for ASD} \tag{9-3b}$$

Note that the F_u in the bearing and tearout failure mode equations is the nominal tensile strength of the connection member.

- **Tearout failure mode:**
  This applies to connections with:
  - Standard bolt holes
  - Oversized bolt holes,
  - Short slotted bolt holes independent of load direction
  - Long-slotted bolt holes with the slots parallel to the direction of the applied force.

For the tearout failure mode (see *AISCM* Section J3.11a.(1)(b)) in connections with the above bolt hole types - when deformation at the bolt hole at service load ***is* a design consideration** (i.e., the elongation of the bolt hole under a high tensile stress at the net cross-section does not exceed ¼ in.), the design tearout strength is given by equations (9-4a) and (9-4b):

$$\phi R_n = \phi 1.2L_c tF_u, \quad \text{for LRFD} \tag{9-4a}$$

$$\frac{R_n}{\Omega} = \frac{1.2L_c tF_u}{\Omega}, \quad \text{for ASD} \tag{9-4b}$$

When the deformation at the bolt hole at service load is ***not* a design consideration** (i.e., the elongation of the bolt hole under a high tensile stress at the net cross-section can exceed ¼ in. and is not limited), the design tearout strength is given by equations (9-5a) and (9-5b):

$$\phi R_n = \phi 1.5L_c tF_u, \quad \text{for LRFD} \tag{9-5a}$$

$$\frac{R_n}{\Omega} = \frac{1.5L_c tF_u}{\Omega}, \quad \text{for ASD} \tag{9-5b}$$

The following are the AISC Specification design strength equations for bolts with long slotted holes with slots *perpendicular* to the direction of the applied load.

- **Bearing failure mode in long slotted bolt holes with the slots *perpendicular* to the direction of the applied force:**
  The design strength in bearing is (see AISC Specification Section J3.11a(2)):

$$\phi R_n = \phi 2.0 d t F_u, \quad \text{for LRFD} \tag{9-6a}$$

$$\frac{R_n}{\Omega} = \frac{2.0 d t F_u}{\Omega}, \quad \text{for ASD} \tag{9-6b}$$

- **Tearout failure mode in long slotted bolt holes with the slots *perpendicular* to the direction of the applied force:**
  The design tearout strength is (see AISC Specification Section J3.11a(2)):

$$\phi R_n = \phi 1.0 L_c t F_u, \quad \text{for LRFD} \tag{9-7a}$$

$$\frac{R_n}{\Omega} = \frac{1.0 L_c t F_u}{\Omega}, \quad \text{for ASD} \tag{9-7b}$$

Where,

$\phi = 0.75$,

$\Omega = 2.0$,

R_n = Nominal bearing strength or tearout strength, kips,

L_c = Clear distance between the edge of the bolt hole and the edge of an adjacent bolt hole, or the edge of the connected member, in the direction of the applied load (see Figure 9-7a).

t = Thickness of the connected material, in.

d = Bolt diameter, in.

F_u = Minimum tensile strength of the connected member.

Techniques for bolting HSS members to other structural members - where access to the inside of the HSS is not required, include:

- Through-bolts,
- Single-sided bolts such as blind bolts, and
- Threaded studs welded to the outside face of the HSS.

For more information on hollow structural section connections, refer to AISC Design Guide 24: Hollow Structural Section Connections [10]. An example of a bolted HSS brace connection is shown in Figure 9-7b.

The following are the AISC Specification design strength equations for bolts or rods that pass completely through an unstiffened box member or HSS (i.e., through-bolts or through-rods):

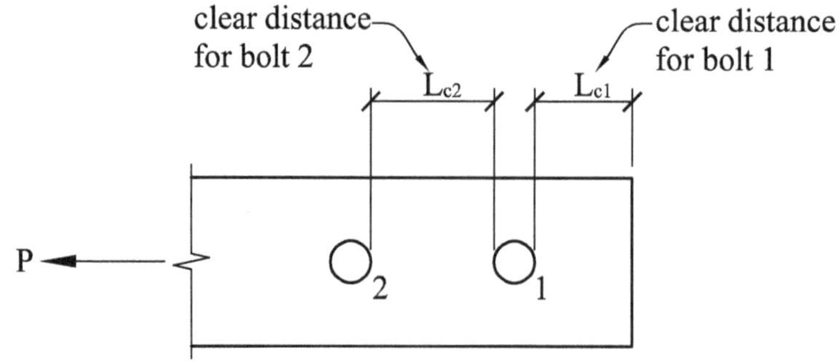

FIGURE 9-7a Clear distance for bolt bearing and tearout

FIGURE 9-7b Bolted HSS Brace Connection

- **Bearing failure mode for through-bolts or through-rods in Box member or HSS:** (see Equation J7-1):

$$\phi R_n = \phi 1.8 F_y A_{pb}, \quad \text{for LRFD} \tag{9-7c}$$

$$\frac{R_n}{\Omega} = \frac{1.8 F_y A_{pb}}{\Omega}, \quad \text{for ASD} \tag{9-7d}$$

A_{pb} = projected area in bearing, in.^{2}

F_y = specified yield stress of the HSS or box member

- **Tearout failure mode for through-bolts or through-rods in Box member or HSS:**
  For a through-bolt or through-rod in a box member of HSS with a standard hole or a short-slotted hole with the slot perpendicular to the direction of the force (see AISC Specification Section J3.11b and Equations J3-6g and J3-6h):

  When the deformation at the bolt hole at service load *is* a design consideration,

$$\phi R_n = \phi 1.2 L_c t F_u, \quad \text{for LRFD} \tag{9-7e}$$

$$\frac{R_n}{\Omega} = \frac{1.2 L_c t F_u}{\Omega}, \quad \text{for ASD} \tag{9-7f}$$

When the deformation at the bolt hole at service load is *not* a design consideration,

$$\phi R_n = \phi 1.5 L_c t F_u, \quad \text{for LRFD} \tag{9-7g}$$

$$\frac{R_n}{\Omega} = \frac{1.5 L_c t F_u}{\Omega}, \quad \text{for ASD} \tag{9-7h}$$

For a through-bolt or through-rod in a box member of HSS with long slotted holes where the slots are *perpendicular* to the direction of the force (see Section J3.11b and Equation J3-6i):

When the deformation at the bolt hole at service load *is* a design consideration,

$$\phi R_n = \phi 1.0 L_c t F_u, \quad \text{for LRFD} \tag{9-7i}$$

$$\frac{R_n}{\Omega} = \frac{1.0 L_c t F_u}{\Omega}, \quad \text{for ASD} \tag{9-7j}$$

Where,

ϕ = 0.75,

Ω = 2.0,

R_n = Nominal bearing strength or tearout strength, kips,

L_c = Clear distance between the edge of the bolt hole and the edge of an adjacent bolt hole, or the edge of the connected member, in the direction of the applied load (see Figure 9-7a).

t = Thickness of the connected material, in.

d = Bolt diameter, in.

F_u = Minimum tensile strength of the connected member.

Note that the condition where the applied force is parallel to the direction of short or long slotted holes or the use of oversized holes is not permitted for bearing-type connections; they are only permitted when slip-critical connections are used (see AISC Specification Sections J3.11b and J3.3).

Slip-critical Connections:

For slip-critical connections, the load and the movement between the connected elements are resisted by friction between the connected parts. The factors affecting the strength of slip-critical connections include:

- Type of faying surface - Class A or B,
- The pretension in the bolts,
- Type of bolt holes, and
- Number of slip planes.

Since bolt bearing against the sides of the bolt holes in the connected parts is assumed not to occur, the strength of the fastener derives entirely from the friction between the faying surfaces of the connected parts, as shown in Figures 9-5c and 9-5d. However, the *AISCM* Section J3.9, still requires that the bolts meet the strength requirements for bearing on the connected parts and shear in the bolts, in case the faying surface friction is overcome and the bolts transition to a bearing-type connection. For these types of connections, it is important to ensure that the connected parts are in firm contact and that the faying surface is properly identified. The design slip resistance of a slip-critical bolted connection is given as follows:

$$\phi R_n = \phi \mu D_u h_f T_b n_s, \quad \text{for LRFD} \tag{9-8a}$$

$$\frac{R_n}{\Omega} = \frac{\mu D_u h_f T_b n_s}{\Omega}, \quad \text{for ASD} \tag{9-8b}$$

Where,

ϕ = 1.0 for standard size and short-slotted bolt holes *perpendicular* to the direction of the load

= 0.85 for oversized and short-slotted bolt holes *parallel* to the direction of the load

= 0.7 for long-slotted bolt holes

Ω = 1.5 for slip critical connections with bolts in standard and short-slotted holes *perpendicular* to the direction of the load
= 1.76 for slip critical connections with bolts in oversized and short-slotted bolt holes *parallel* to the direction of the load
= 2.14 for slip critical connections with bolts in long-slotted bolt holes

R_n = Nominal shear strength, kips,

μ = Mean slip coefficient (see discussion below)
= 0.30 for Class A surfaces,
= 0.50 for Class B surfaces,

D_u = 1.13 (the constant value that represents the ratio between the mean installed bolt pretension force to the minimum specified bolt pretension from Table 9-1; alternate values can be used if it is verified and approved by the engineer of record),

h_f = factor for fillers (see Figure 9-8)
= 1.0 for one or no filler plates between connected parts
= 0.85 for two or more filler plates between connected parts

n_s = Number of slip planes or surfaces.

T_b = Minimum bolt pretension (see Table 9-1).

Filler plates are needed in some connections as depicted in Figure 9-8, and they are also used in column splices when smaller depth columns are spliced with larger depth columns below.

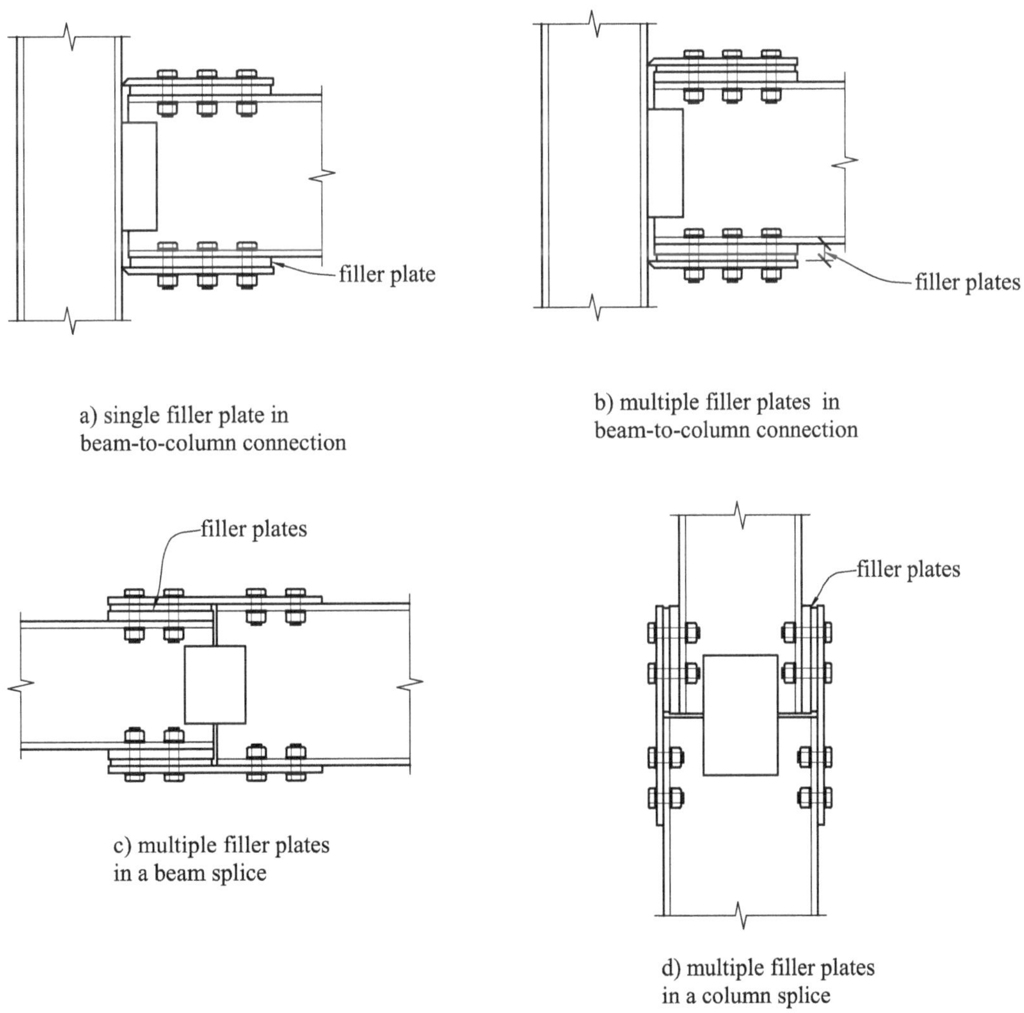

FIGURE 9-8 Filler plates

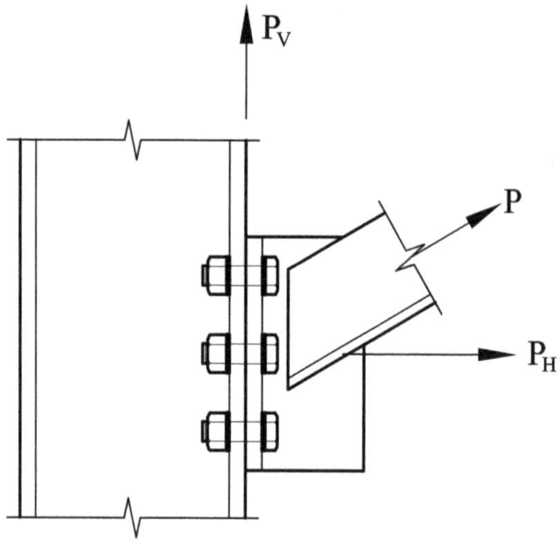

FIGURE 9-9 Connection subjected to combined loading

Note that if the filler plates are welded to the connected members, the reduction factor h_f = 1.0. A slip-critical connection with a single filler plate of any thickness with the proper surface preparation will have h_f = 1.0. Similarly, slip-critical connections with multiple filler plates with either Class B faying surfaces on all surfaces or Class A faying surfaces with turn-of-nut tensioning will have h_f = 1.0.

As previously discussed, there are two types of faying surfaces for slip-critical connections: Class A and Class B faying surfaces, with Class A being the most typically assumed and specified faying surface. Refer to Section 9-2 for the definition of these two faying surfaces. The mean slip coefficients for Class A and Class B faying surfaces are given above. The mean slip coefficient for faying surfaces with other types of coatings different than Classes A or B is permitted to be determined experimentally if approved by the engineer of record (see RCSC Section 3.2.2.2) [2]. Class B faying surfaces are typically specified where slippage needs to be very controlled in order to limit the deflections of the structural member or system (e.g., in long-span stadium roof canopy structures where vertical deflection is critical) or in tall building moment frames and braced frames where limiting the lateral deflection is critical. It is also generally good practice to use standard bolt holes, since oversized and slotted bolt holes are typically not necessary. Slip-critical (SC) bolts are designated as A325SC or A490SC on the structural drawings.

Combined Loading of Bolted Connections:

When fasteners are loaded such that they resist combined shear and tension forces (see Figure 9-9), an interaction equation or curve is used for the design of the fasteners.

For LRFD, research indicates that the interaction curve (see *AISCM* equation C-J3-4a) is described by the following elliptical interaction relationship:

$$\left(\frac{f_t}{\phi F_{nt}}\right)^2 + \left(\frac{f_v}{\phi F_{nv}}\right)^2 = 1.0, \tag{9-9a}$$

For ASD, the corresponding elliptical interaction relationship (see *AISCM* equation C-J3-4b) is

$$\left(\frac{\Omega f_t}{F_{nt}}\right)^2 + \left(\frac{\Omega f_v}{F_{nv}}\right)^2 = 1.0, \tag{9-9b}$$

Where,

f_t = Applied tensile stress $= \dfrac{T_u}{A_b n}$ for LRFD, and $\dfrac{T_s}{A_b n}$ for ASD

F_{nt} = nominal tensile stress

F_{nv} = nominal shear stress
f_v = Applied or required shear stress = $\frac{V_u}{A_b n}$ for LRFD, and $\frac{V_s}{A_b n}$ for ASD
T_u = factored tension force (LRFD)
V_u = factored shear force (LRFD)
T_s = total service or unfactored tension force (ASD)
V_s = total service or unfactored shear force (ASD)
n = number of bolts in the connection
A_b = nominal unthreaded cross-sectional area of bolt

These interaction equations are represented graphically in Figure 9-10; any design that falls under (or on) the interaction curve is acceptable. The *AISCM* Section J3.8 approximates this elliptical interaction curve using the following straight-line relationship that accounts for the presence of the applied shear stress to calculate the modified design tension stress in each bolt (for LRFD), and the allowable tension stress in each bolt (for ASD), as follows:

$$F'_{nt} = 1.3F_{nt} - \frac{F_{nt}}{\phi F_{nv}} f_v \leq F_{nt}, \quad \text{for LRFD} \tag{9-10a}$$

$$F'_{nt} = 1.3F_{nt} - \frac{\Omega F_{nt}}{F_{nv}} f_v \leq F_{nt}, \quad \text{for ASD} \tag{9-10b}$$

Where,
F'_{nt} = Nominal tension stress modified to include the effects of shear force, ksi,
F_{nt} = Nominal tensile stress, ksi (see Table 9-3),
F_{nv} = Nominal shear stress, ksi (see Table 9-3),
f_v = Applied or required shear stress using the LRFD or ASD load combinations, ksi
$\phi = 0.75$
$\Omega = 2.0$

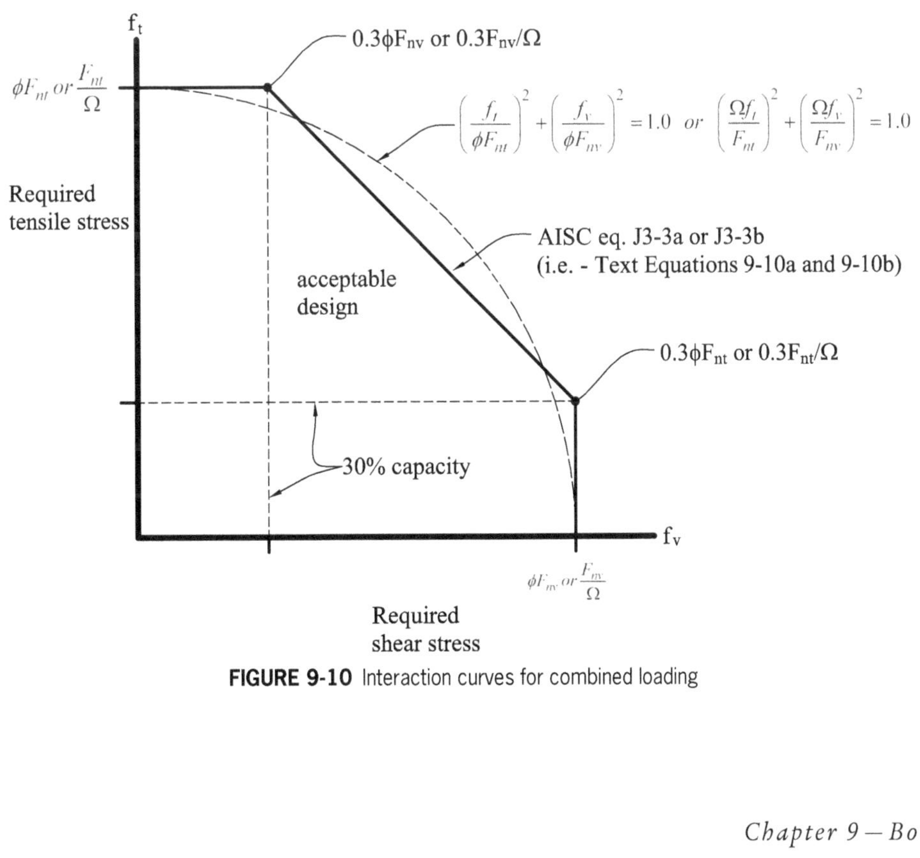

FIGURE 9-10 Interaction curves for combined loading

The modified design tensile strength and the available tensile strength of each bolt under combined tension and shear then become

$$\phi R_n = \phi F'_{nt} A_b \qquad \text{for LRFD} \tag{9-11a}$$

$$\frac{R_n}{\Omega} = \frac{F'_{nt} A_b}{\Omega} \qquad \text{for ASD} \tag{9-11b}$$

Equations (9-10a) and (9-10b) which are expressed in terms of the modified tension strength can also be rewritten to determine the modified shear strength per unit area and the available shear strength per unit area as a function of the applied tensile stress, f_t, as follows (see AISC specification Section C-J3.8) [2]:

$$F'_{nv} = 1.3F_{nv} - \frac{F_{nv}}{\phi F_{nt}} f_t \leq F_{nv}, \qquad \text{for LRFD} \tag{9-12a}$$

$$F'_{nv} = 1.3F_{nv} - \frac{\Omega F_{nv}}{F_{nt}} f_t \leq F_{nv}, \qquad \text{for ASD} \tag{9-12b}$$

Where,

F'_{nv} = Nominal shear stress modified to include the effects of tension force, ksi,

F_{nv} = Nominal shear stress, ksi (see Table 9-3),

F_{nt} = Nominal tension stress, ksi (see Table 9-3),

f_t = Applied or required tension stress using the LRFD or ASD load combinations, ksi

$\phi = 0.75$

$\Omega = 2.0$

The modified design shear strength can be obtained by substituting F'_{nv} in place of F'_{nt} in equations (9-11a) and (9-11b).

It should be noted that when the applied shear stress is less than 30% of the available shear strength, or when the applied tension stress is less than 30% of the available tension strength, the effects of the combined stress do not have to be investigated. That is, the effect of combined stresses can be neglected if the following equations are satisfied (see *AISC* Specification Section J3.8):

$$\frac{f_t}{\phi F_{nt}} \leq 0.30 \quad \text{or} \quad \frac{f_v}{\phi F_{nv}} \leq 0.30 \qquad \text{for LRFD}$$

$$\frac{f_t}{F_{nt}/\Omega_t} \leq 0.30 \quad \text{or} \quad \frac{f_t}{F_{nv}/\Omega_v} \leq 0.30 \qquad \text{for ASD}$$

For LRFD, f_t and f_v are calculated using the LRFD load combinations from Section 2-3, and for ASD, f_t and f_v are calculated using the ASD load combinations from Section 2-3.

Combined Loading of Slip-critical Connections:

When slip-critical connections are subjected to combined shear and tension loads, the tension load reduces the amount of clamping force and therefore reduces the available slip resistance in each bolt. Therefore, the design shear strength per bolt in a slip-critical (SC) connection previously calculated in equations (9-8a) and (9-8b) shall be reduced by the following factor (see *AISC* Specification Section J3.10) [2]:

$$k_{sc} = 1 - \frac{T_u}{D_u T_b n_b} \geq 0 \qquad \text{for LRFD} \tag{9-13a}$$

$$k_{sc} = 1 - \frac{1.5T_a}{D_u T_b n_b} \geq 0 \qquad \text{for ASD} \tag{9-13b}$$

Where,
k_{sc} = Reduction factor,
T_u = Factored tension force (LRFD), kips,
T_a = Unfactored tension force (ASD), kips,
D_u = 1.13 (see Section 9-4 for discussion),
T_b = Minimum bolt pretension (see Table 9-1),
n_b = Number of bolts resisting the applied tension.

The example problems that follow will cover the design strength of bolts in common types of connections.

EXAMPLE 9-1

High-Strength Bolts in Shear and Bearing

Using both the LRFD and ASD methods, determine whether the connection shown in Figure 9-11 is adequate to support the applied loads; consider the strength of the bolts in shear and bearing on the plate. The plates are ASTM A36, and the fasteners are $^3/_4$-in.-diameter ASTM F3125 A325N bolts in standard bolt holes. The applied factored load is P_u = 60 kips (LRFD), and the applied service load is P_s = 40 kips (ASD). Assume deformation at the bolt hole at service loads is a design consideration.

The reader should repeat the example assuming the connection plates are ASTM Grade 50 steel.

Solution (LRFD)

The possible failure modes to be investigated include:

- Bearing failure of the plate due to the bolts bearing against the plate,
- Tearout failure of the plate,
- Direct shear failure of the bolt,
- Tension failure of the tension members (i.e., the 5/8-in. thick plates).

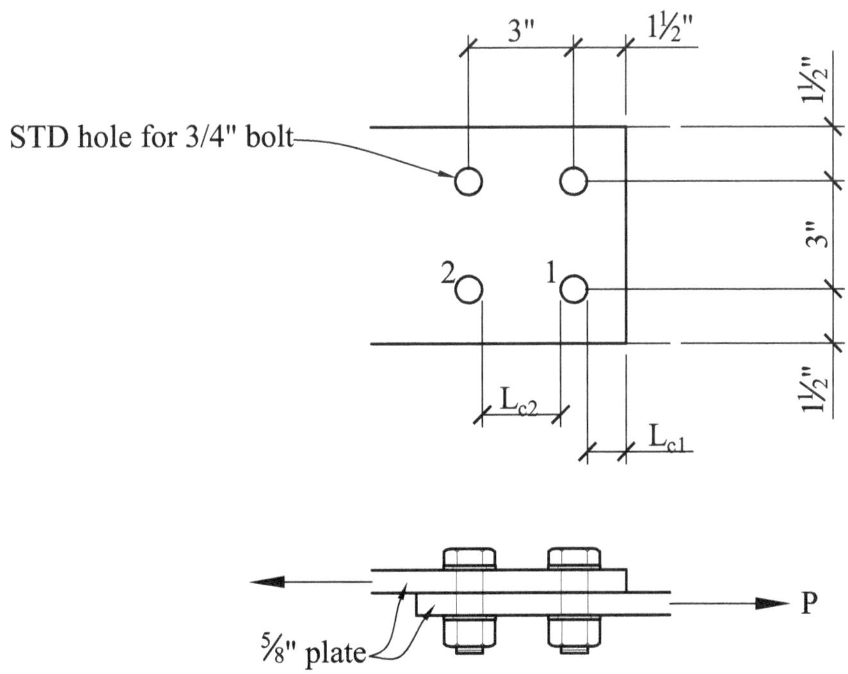

FIGURE 9-11 Connection detail for Example 9-1

Note that in this text, the symbols d or d_b are used interchangeably to indicate the bolt diameter.

$^3/_4$-in.-diameter ASTM F3125 A325N bolts = Group 120 bolts with threads included

F_{nv} = 54 ksi (see Table 9-3)

$$A_b = \frac{\pi (3/4)^2}{4} = 0.442 \text{ in.}^2$$

Plate: F_y = 36 ksi; F_u = 58 ksi

Check bolt bearing and tearout failure modes:

The clear distance between adjacent bolt holes is

$$L_{c1} = 1.5 \text{ in.} - (0.5)\left(\frac{3}{4} + \frac{1}{8}\right) = 1.06 \text{ in.}$$

$$L_{c2} = 3 \text{ in.} - \left(\frac{3}{4} + \frac{1}{8}\right) = 2.13 \text{ in.}$$

The strength of the bolts will be smaller due to the tearout failure mode and the strength due to the bearing failure mode.

For bolt 1,

$$\phi R_n = \phi 1.2 L_c t F_u \le \phi 2.4 d t F_u$$

$$= (0.75)(1.2)(1.06 \text{ in.})(0.625 \text{ in.})(58 \text{ ksi})$$

$$\le (0.75)(2.4)(0.75)(0.625 \text{ in.})(58 \text{ ksi})$$

$$= 34.6 \text{ kips} \le 48.9 \text{ kips}$$

$$= 34.6 \text{ kips.}$$

For bolt 2,

$$\phi R_n = (0.75)(1.2)(2.13 \text{ in.})(0.625 \text{ in.})(58 \text{ ksi})$$

$$\le (0.75)(2.4)(0.75)(0.625 \text{ in.})(58 \text{ ksi})$$

$$= 69.5 \text{ kips} \le 48.9 \text{ kips}$$

$$= 48.9 \text{ kips.}$$

The total strength of the connection in bearing and tearout, considering all the bolts, is

$\phi R_n = (2)(34.6 \text{ kips}) + (2)(48.9 \text{ kips}) =$ **167 kips**

$> P_u =$ **60 kips.** O.K. for bearing and tearout.

Check direct shear on the bolts:

For a single bolt, the design shear strength is

$$\phi R_n = \phi F_{nv} A_b$$

= (0.75)(54 ksi)(0.442 in.^{2})

= 17.9 kips/bolt *(agrees with AISCM Table 7-1).*

The total shear strength of the bolt group is

ϕR_n = (4)(17.9 kips) = **71.6 kips > 60 kips *OK for shear.***

Therefore, the design strength of the connection in shear is smaller than the design bearing strength of the bolt group (167 kips) and the design strength in direct shear (71.6 kips); therefore, the design strength of the connection is **71.6 kips**.

Solution (ASD)

Check bolt bearing and tearout failure modes:

The clear distance between adjacent bolt holes is

$$L_{c1} = 1.5\text{ in.} - (0.5)\left(\frac{3}{4} + \frac{1}{8}\right) = 1.06\text{ in.}$$

$$L_{c2} = 3\text{ in.} - \left(\frac{3}{4} + \frac{1}{8}\right) = 2.13\text{ in.}$$

For bolt 1,

$$\frac{R_n}{\Omega} = \frac{1.2L_c tF_u}{\Omega} \le \frac{2.4dtF_u}{\Omega}$$

$$= \frac{(1.2)(1.06\text{ in.})(0.625\text{ in.})(58\text{ ksi})}{2.0}$$

$$\le \frac{(2.4)(0.75)(0.625\text{ in.})(58\text{ ksi})}{2.0}$$

$$= 23.0\text{ kips} \le 32.6\text{ kips}$$

Therefore, $\frac{R_n}{\Omega} = 23.0$ kips.

For bolt 2,

$$\frac{R_n}{\Omega} = \frac{(1.2)(2.13\text{ in.})(0.625\text{ in.})(58\text{ ksi})}{2.0}$$

$$\le \frac{(2.4)(0.75)(0.625\text{ in.})(58\text{ ksi})}{2.0}$$

$$= 46.3\text{ kips} \le 32.6\text{ kips}$$

Therefore, $\frac{R_n}{\Omega} = 32.6$ kips.

The total strength of the connection in bearing and tearout, considering all the bolts, is

$$\frac{R_n}{\Omega} = (2)(23.0\text{ kips}) + (2)(32.6\text{ kips}) = \mathbf{111\ kips}$$

$> P_S = 40$ kips. O.K. for bearing for tearout failure mode.

Check direct shear on the bolts:

$$\frac{R_n}{\Omega} = \frac{F_{nv}A_b}{\Omega} = \frac{(54\text{ ksi})(0.442\text{ in.}^2)}{2.0}$$

$$= \frac{(54\text{ ksi})(0.442\text{ in.}^2)}{2.0}$$

= 11.9 kips/bolt *(agrees with AISCM Table 7-1).*

The total allowable shear strength of the bolt group is

$$\frac{R_n}{\Omega} = (4)(11.9) = 47.6\text{ kips} > 40\text{ kips} \quad \text{OK. for shear.}$$

Therefore, the allowable strength of the connection in shear is smaller than the allowable bearing strength of the bolt group (111 kips) and the allowable strength in direct shear (47.6 kips); therefore, the allowable strength of the connection is **47.6 kips**.

EXAMPLE 9-2
High-Strength Bolts in a Slip-Critical Connection

Repeat Example 9-1 with ASTM F3125 Grade A325SC bolts with a Class A faying surface.

Solution (LRFD)

Check bolt bearing:

The clear distance between adjacent bolt holes is

$$L_{c1} = 1.5 \text{ in.} - (0.5)\left(\frac{3}{4} + \frac{1}{8}\right) = 1.06 \text{ in.}$$

$$L_{c2} = 3 \text{ in.} - \left(\frac{3}{4} + \frac{1}{8}\right) = 2.13 \text{ in.}$$

For bolt 1,

$$\begin{aligned}\phi R_n &= \phi 1.2 L_c t F_u \le \phi 2.4 d t F_u \\ &= (0.75)(1.2)(1.06")(0.625")(58 \text{ ksi}) \le (0.75)(2.4)(0.75)(0.625")(58 \text{ ksi}) \\ &= 34.6 \text{ kips} \le 48.9 \text{ kips}\end{aligned}$$

Therefore, ϕR_n (bolt 1) = 34.6 kips

For bolt 2,

$$\begin{aligned}\phi R_n &= (0.75)(1.2)(2.13")(0.625")(58 \text{ ksi}) \\ &\le (0.75)(2.4)(0.75)(0.625")(58 \text{ ksi}) \\ &= 69.5 \text{ kips} \le 48.9 \text{ kips}\end{aligned}$$

Therefore, ϕR_n (bolt 2) = 48.9 kips

Since there are *two* bolt 1 and *two* bolt 2 (see Figure 9-11), the total strength of the connection in bearing and tearout, considering all the bolts, is

$$\begin{aligned}\phi R_n &= (2)(34.6 \text{ kips}) + (2)(48.9 \text{ kips}) = 167 \text{ kips} \\ &> P_u = 60 \text{ kips.} \quad \text{OK for bearing and tearout}\end{aligned}$$

Note that the bearing and tearout strength obtained are similar to those obtained in Example 9-1.

Check direct shear on the bolts in the slip critical connection:

$n_s = 1$ (one slip plane)

T_b = 28 kips for ¾ in. diameter A325 bolt (Table 9-1)

h_f = 1 (no filler plate)

D_u = 1.13

μ = 0.30 (Class A faying surface)

ϕ = 1.0 (bolt in standard bolt hole in a slip-critical connection)

The slip-critical design shear strength for one bolt is

$$\begin{aligned}\phi R_n &= \phi \mu D_u h_f T_b n_s \\ &= (1.0)(0.30)(1.13)(1.0)(28)(1) \\ &= 9.49 \text{ kips/bolt (agrees with } \textit{AISCM} \text{ Table 7-3)}\end{aligned}$$

The total slip-critical shear strength of the bolt group (with a total of 4 bolts) is

$\phi R_n = (4)(9.49 \text{ kips}) =$ **38.0 kips** < 60 kips. the connection is not good for shear

The design shear strength of this slip-critical connection is 38 kips. Therefore, the connection is not adequate. Compare the design tension strength of this slip-critical connection (38 kips) to the strength (71.6 kips) for the snug-tight bolted connection in Example 9-1.

Solution (ASD)

Check bolt bearing:

$\Omega = 2.0$ (bolt bearing)

The clear distance between adjacent bolt holes is

$$L_{c1} = 1.5 \text{ in.} - (0.5)\left(\frac{3}{4} + \frac{1}{8}\right) = 1.06 \text{ in.}$$

$$L_{c2} = 3 \text{ in.} - \left(\frac{3}{4} + \frac{1}{8}\right) = 2.13 \text{ in.}$$

For bolt 1,

$$\frac{R_n}{\Omega} = \frac{1.2L_c tF_u}{\Omega} \le \frac{2.4dtF_u}{\Omega}$$

$$= \frac{(1.2)(1.06 \text{ in.})(0.625 \text{ in.})(58 \text{ ksi})}{2.0}$$

$$\le \frac{(2.4)(0.75)(0.625 \text{ in.})(58 \text{ ksi})}{2.0}$$

$$= 23.0 \text{ kips} \le 32.6 \text{ kips}$$

Therefore, $\frac{R_n}{\Omega}(\text{bolt 1}) = 23.0$ kips.

For bolt 2,

$$\frac{R_n}{\Omega} = \frac{(1.2)(2.13 \text{ in.})(0.625 \text{ in.})(58 \text{ ksi})}{2.0}$$

$$\le \frac{(2.4)(0.75)(0.625 \text{ in.})(58 \text{ ksi})}{2.0}$$

$$= 46.3 \text{ kips} \le 32.6 \text{ kips}$$

Therefore, $\frac{R_n}{\Omega}(\text{bolt 2}) = 32.6$ kips.

Since there are *two* bolt 1 and *two* bolt 2 (see Figure 9-11), the total allowable strength of the connection in bearing and tearout, considering all the bolts, is

$$\frac{R_n}{\Omega} = (2)(23.0 \text{ kips}) + (2)(32.6 \text{ kips}) = 111 \text{ kips}$$

$> P_S = 40$ kips. O.K. for bearing and tearout.

Check direct shear on the bolts in the slip-critical connection:

Ω (for standard bolt hole for bolt in slip-critical connection) = 1.5

For a single bolt, the available shear strength is

$$\frac{R_n}{\Omega} = \frac{\mu D_u h_{sc} T_b n_s}{\Omega}$$

$$= \frac{(0.30)(1.13)(1.0)(28)(1)}{1.5}$$

$= 6.33$ kips/bolt (agrees with *AISCM* Table 7-3)

The total shear strength of the bolt group (with a total of 4 bolts) is

$\frac{R_n}{\Omega} = (4)(6.33 \text{ kips}) =$ **25.3 kips** < 40 kips. , the connection not good for shear

The strength of the connection is smaller than the allowable bearing strength (111 kips) and the allowable slip-critical shear strength (25.3 kips), yielding a connection allowable strength of 25.3 kips. Therefore, the connection is not adequate to support the service load $P_S = 40$ kips.

EXAMPLE 9-3

Bolted Splice Connection

For the splice connection shown in Figure 9-12, determine the maximum factored load, P_u, and the maximum allowable load, P_s, that can be applied; consider the strength of the bolts only. The plates are ASTM A36, and the bolts are 1-in.-diameter Group 120 A325X in standard bolt holes.

Solution (LRFD)

1-in. diameter ASTM F3125 A325X bolts means Group 120 bolts with threads excluded.

Check bolt bearing on **the outermost top and bottom plates** (i.e., the two ½-in. thick plates):

The clear distance between adjacent bolt holes is

$$L_{c1} = 2 \text{ in.} - (0.5)(1 + 3/16) = 1.41 \text{ in.}$$

$$L_{c2} = L_{c3} = 3 \text{ in.} - (1 + 3/16) = 1.81 \text{ in.}$$

Recall that for 1-in. diameter bolts and larger, $d_{hole} = d_b + 3/16$

For bolt 1,

The strength of the bolt will be smaller due to the tearout and bearing failure modes.

$\phi = 0.75$

$$\begin{aligned}\phi R_n &= \phi 1.2 L_c t F_u \leq \phi 2.4 d t F_u \\ &= (0.75)(1.2)(1.41 \text{ in.})(0.5 \text{ in.})(58 \text{ ksi}) \\ &\leq (0.75)(2.4)(1 \text{ in.})(0.5 \text{ in.})(58 \text{ ksi}) \\ &= 36.8 \text{ kips} \leq 52.2 \text{ kips}\end{aligned}$$

Therefore ϕR_n (bolt 1) = 36.8 kips.

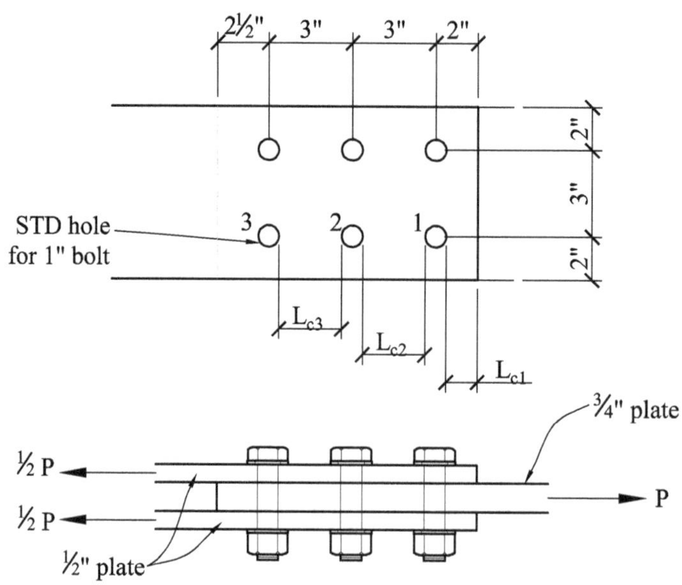

FIGURE 9-12 Connection detail for Example 9-3

For bolts 2 and 3,

$\phi = 0.75$

$$\phi R_n = (0.75)(1.2)(1.81 \text{ in.})(0.5 \text{ in.})(58 \text{ ksi})$$
$$\leq (0.75)(2.4)(1 \text{ in.})(0.5 \text{ in.})(58 \text{ ksi})$$
$$= 47.2 \text{ kips} \leq 52.2 \text{ kips}$$

Therefore ϕR_n (bolts 2 and 3) = 47.2 kips.

The total strength of all the bolts in bearing and tearout on the two outside plates is

$\phi R_n = (2\,plates)(2\,bolts)(36.8 \text{ kips}) + (2\,plates)(4\,bolts)(47.2 \text{ kips}) =$ **525 kips.**

Check bolt bearing and tearout on the **inside ¾-in. thick plate**:
The clear distance between adjacent bolt holes is

$$L_{c3} = 2.5 \text{ in.} - (0.5)(1 + 3/16) = 1.91 \text{ in.}$$

$$L_{c2} = L_{c1} = 3 \text{ in.} - (1 + 3/16) = 1.81 \text{ in.}$$

For bolt 3, the shear strength is the smaller of the strengths in bearing and tearout failure modes:

$\phi R_n = \phi 1.2 L_c t F_u \leq \phi 2.4 d t F_u$ (see equation (9-2a) and (9-4a))

$$= (0.75)(1.2)(1.91 \text{ in.})(0.75 \text{ in.})(58 \text{ ksi})$$
$$\leq (0.75)(2.4)(1 \text{ in.})(0.75 \text{ in.})(58 \text{ ksi})$$
$$= 74.8 \text{ kips} \leq 78.3 \text{ kips}$$

Therefore, ϕR_n (bolt 3) = 74.8 kips

For bolts 1 and 2, the shear strength is the smaller of the strengths in bearing and tearout failure modes:

$$\phi R_n = (0.75)(1.2)(1.81 \text{ in.})(0.75 \text{ in.})(58 \text{ ksi})$$
$$\leq (0.75)(2.4)(1 \text{ in.})(0.75 \text{ in.})(58 \text{ ksi})$$
$$= 70.9 \text{ kips} \leq 78.3 \text{ kips}$$

Therefore, ϕR_n (bolts 1 and 2) = 70.9 kips.

The total strength of all the bolts in bearing and tearout on the one inside plate is

$\phi R_n = (2\,bolts)(74.8 \text{ kips}) + (4\,bolts)(70.9 \text{ kips}) =$ **433 kips** $\geq P_u$

Therefore, $P_u = 433$ kips (considering bearing on the inside plate only)

Check direct shear on the bolts:

$$A_b \text{ (1-in. diameter bolt)} = \frac{\pi (1.0)^2}{4} = 0.785 \text{ in.}^2$$

$\phi = 0.75$

$F_{nv} = 68\ ksi$ (see Table 9-3 or *AISCM* Table J3.2 for F_{nv} for A325X bolt)

For a single bolt, the direct shear strength is

$$\phi R_n = \phi F_{nv} A_b$$
$$= (0.75)(68 \text{ ksi})(0.785 \text{ in.}^2)$$

$= (0.75)(68 \text{ ksi})(0.785 \text{ in.}^2)$

$= 40.0$ kips/bolt *(agrees with AISCM Table 7-1)*

The design shear capacity of the bolt group considering double shear is

P_u = (2 shear planes)(6 bolts)(40.0 kips) = **480 kips** *(considering direct shear on the bolts only).*

The design strength of the connection is the smallest of 525 kips, 433 kips, and 480 kips, which yields a capacity of 433 kips. Therefore, the strength of the bolt's bearing against the inside plate controls the design of this connection, so the maximum factored load that can be applied to the connection is $\boldsymbol{P_u}$ **= 433 kips.**

Solution (ASD)

Check bolt bearings on the outermost top and bottom plates:

The clear distance between adjacent bolt holes is

$$L_{c1} = 2 \text{ in.} - (0.5)(1 + 3/16) = 1.41 \text{ in.}$$

$$L_{c2} = L_{c3} = 3 \text{ in.} - (1 + 3/16) = 1.81 \text{ in.}$$

For bolt 1,

Bearing failure mode: $\dfrac{R_n}{\Omega} = \dfrac{1.2L_c tF_u}{\Omega}$

Tearout failure mode: $\dfrac{R_n}{\Omega} = \dfrac{2.4dtF_u}{\Omega}$

Therefore, $\dfrac{R_n}{\Omega} = \dfrac{1.2L_c tF_u}{\Omega} \leq \dfrac{2.4dtF_u}{\Omega}$

$$= \frac{(1.2)(1.41 \text{ in.})(0.5 \text{ in.})(58 \text{ ksi})}{2.0}$$

$$\leq \frac{(2.4)(1 \text{ in.})(0.5 \text{ in.})(58 \text{ ksi})}{2.0}$$

$$= 24.5 \text{ kips} \leq 34.8 \text{ kips}$$

Therefore, $\dfrac{R_n}{\Omega}(\text{bolt } 1) = 24.5\,kips.$

For bolts 2 and 3,

$$\frac{R_n}{\Omega} = \frac{(1.2)(1.81 \text{ in.})(0.5 \text{ in.})(58 \text{ ksi})}{2.0}$$

$$\leq \frac{(2.4)(1 \text{ in.})(0.5 \text{ in.})(58 \text{ ksi})}{2.0}$$

$$= 31.5 \text{ kips} \leq 34.8 \text{ kips}$$

Therefore, $\dfrac{R_n}{\Omega}(\text{bolts 2 and 3}) = 31.5 \text{ kips}.$

The total strength, considering all the bolts in bearing and tearout on the two outside plates, is

$$\frac{R_n}{\Omega} = (2\,plates)(2\,bolts)(24.5 \text{ kips}) + (2\,plates)(4\,bolts)(31.5 \text{ kips})$$

$$= \mathbf{350\ kips} \geq P_S$$

Therefore,

$P_S = 350$ kips (considering bearing and tearout on the two outside ½-in. thick plates only)

Now, calculate the strength of the bolt bearing on the inner ¾ in. thick plate.
Check bolt bearing and tearout on the one inside plate:

$$L_{c3} = 2.5 \text{ in.} - (0.5)(1 + \tfrac{3}{16}) = 1.91 \text{ in.}$$

$$L_{c2} = L_{c1} = 3 \text{ in.} - (1 + \tfrac{3}{16}) = 1.81 \text{ in.}$$

For bolt 3,

$$\frac{R_n}{\Omega} = \frac{1.2 L_c t F_u}{\Omega} \le \frac{2.4 d t F_u}{\Omega}$$

$$= \frac{(1.2)(1.91 \text{ in.})(0.75 \text{ in.})(58 \text{ ksi})}{2.0}$$

$$\le \frac{(2.4)(1 \text{ in.})(0.75 \text{ in.})(58 \text{ ksi})}{2.0}$$

$$= 49.9 \text{ kips} \le 52.2 \text{ kips}$$

Therefore, $\frac{R_n}{\Omega}\text{(bolt 3)} = 49.9$ kips.

For bolts 1 and 2,

$$\frac{R_n}{\Omega} = \frac{(1.2)(1.81 \text{ in.})(0.75 \text{ in.})(58 \text{ ksi})}{2.0}$$

$$\le \frac{(2.4)(1 \text{ in.})(0.75 \text{ in.})(58 \text{ ksi})}{2.0}$$

$$= 47.2 \text{ kips} \le 52.2 \text{ kips}$$

Therefore, $\frac{R_n}{\Omega}\text{(bolt 1 and 2)} = 47.2$ kips.

The total strength of all the bolts in bearing and tearout on the one inside ¾-in. thick plate is

$$\frac{R_n}{\Omega} = (2 bolts)(49.9 \text{ kips}) + (4 bolts)(47.2 \text{ kips}) = \mathbf{289 \text{ kips}} \ge P_s$$

Therefore, $P_s = 289$ kips (considering bearing and tearout on the one inside plate)

Check direct shear on the bolts:

$$A_b \text{ (1-in. diameter bolt)} = \frac{\pi(1.0)^2}{4} = 0.785 \text{ in.}^2$$

$$\Omega = 2.0$$

$F_{nv} = 68$ *ksi* (see Table 9-3 or *AISCM* Table J3.2 for F_{nv} for A325X bolt)

For a single bolt, the available direct shear strength is

$$\frac{R_n}{\Omega} = \frac{F_{nv} A_b}{\Omega}$$

$$= \frac{(68 \text{ ksi})(0.785 \text{ in.}^2)}{2.0}$$

$= 26.7$ kips/bolt *(agrees with AISCM Table 7-1)*

The allowable shear capacity of the bolt group considering *double shear* is

P_s = (2 shear planes)(6 bolts)(26.7 kips) = **320 kips** *(considering direct shear on the bolts only)*. The allowable shear strength of the connection is the smallest of the allowable strength in bearing/tearout (350 kips and 289 kips) and the allowable direct shear strength (320 kips), yielding an allowable connection strength of 289 kips. The bolt bearing strength controls the design, so the maximum service load that can be applied to the connection is P_s = **289 kips.**

EXAMPLE 9-4

Bolted Connection Loaded in Combined Shear and Tension

For the connection shown in Figure 9-13, determine whether the bolts are adequate to support the applied load under combined loading. The bolts are $^7/_8$-in.-diameter A490X in standard bolt holes. Assume that each bolt resists an equal amount of the applied shear and tension loads.

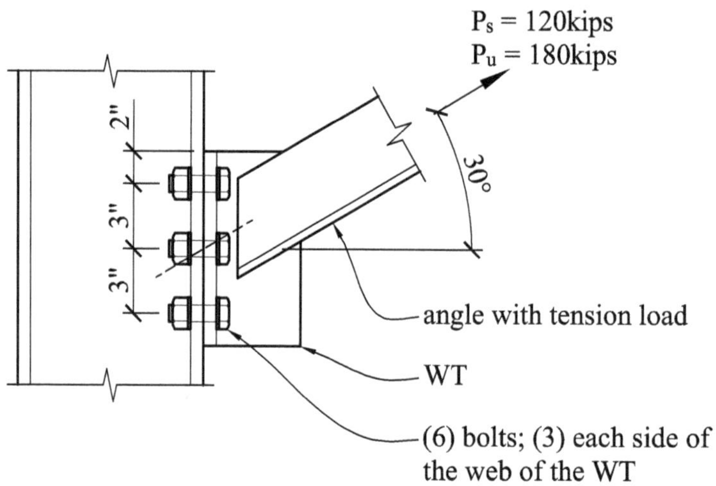

FIGURE 9-13 Connection detail for Example 9-4

Solution (LRFD)

The applied factored load will be resolved into horizontal and vertical components, respectively, as follows:

$$P_H = 180(\cos 30) = 159 \text{ kips, and}$$

$$P_V = 180(\sin 30) = 90 \text{ kips.}$$

$$n = \text{number of bolts} = 6$$

$$A_b \text{ (7/8 -in. diameter bolt)} = \frac{\pi (7/8)^2}{4} = 0.601 \text{ in.}^2$$

$$\phi = 0.75$$

The applied factored shear and tension stresses in each bolt are calculated as follows, assuming each bolt resists an equal amount of load:

$$f_v = \frac{P_V}{A_b n} = \frac{90}{(0.601)(6)} = 25.0 \text{ ksi, and}$$

$$f_t = \frac{P_H}{A_b n} = \frac{159}{(0.601)(6)} = 44.1 \text{ ksi, respectively.}$$

The design strength in shear and tension for each bolt is

$\phi R_n = \phi F_n A_b$, (see Table 9-3 or *AISCM* Table J3.2 for F_{nv} and F_{nt} for A490X bolt)

ASTM F3125 A490X bolts = Group B bolts with threads excluded from the shear plane, an F_{nv} = 84 ksi, and F_{nt} = 113 ksi.

For shear, $\phi R_{nv} = \phi F_{nv} A_b$ = (0.75)(84)(0.601) = 37.9 kips *(agrees with AISCM Table 7-1)*

For tension, $\phi R_{nt} = \phi F_{nt} A_b = (0.75)(113)(0.601) = 51.0$ kips (agrees with *AISCM* Table 7-2)

$$\frac{f_v}{\phi F_{nv}} = \frac{25.0}{(0.75)(84)} = 0.40 > 0.30$$

$$\frac{f_t}{\phi F_{nt}} = \frac{44.1}{(0.75)(113)} = 0.52 > 0.30.$$

Since the ratio of the actual factored applied stress to the available strength is greater than 30% (0.30) for both shear and tension, the combined stress effects must be considered. The interaction equation (9-10a) will be used to determine the modified allowable tension in each bolt:

$$F'_{nt} = 1.3F_{nt} - \frac{F_{nt}}{\phi F_{nv}} f_v \le F_{nt}$$

$$F'_{nt} = (1.3)(113) - \frac{(113)}{(0.75)(84)}(25.0) \le 113$$

F'_{nt} = 102 ksi ≤ 113 ksi

Therefore, use F'_{nt} = 102 ksi

The modified design tension stress is

$\phi F'_{nt}$ = (0.75)(102 ksi) = 76.5 ksi

$\phi F'_{nt}$ = **76.5 ksi** > f_t = 44.1 ksi, OK.

The bolts are adequate for combined tension and shear loading.

Solution (ASD)

The applied service or unfactored load will be resolved into horizontal and vertical components, respectively, as follows:

$P_H = 120(\cos 30) = 104$ kips, and

$P_V = 120(\sin 30) = 60$ kips.

$n = number\,of\,bolts = 6$

$$A_b = \frac{\pi \left(\frac{7}{8}\right)^2}{4} = 0.601\,\text{in.}^2$$

F_{nv} (A490 bolt) = 84 ksi (see Table 9-3)

F_{nt} (A490 bolt) = 113 ksi (see Table 9-3)

$\Omega = 2.0$

The unfactored shear and tension stresses in each bolt are

$$f_v = \frac{P_V}{A_b n} = \frac{60}{(0.601)(6)} = 16.6 \text{ ksi, and}$$

$$f_t = \frac{P_H}{A_b n} = \frac{104}{(0.601)(6)} = 28.8 \text{ ksi, respectively.}$$

The allowable strength in shear and tension for each bolt is

$$\frac{R_n}{\Omega} = \frac{F_n A_b}{\Omega}, \text{ (see Table 9-3 or } \textit{AISCM} \text{ Table J3.2 for } F_{nv} \text{ and } F_{nt} \text{ for A490X bolt)}$$

$$\frac{R_{nv}}{\Omega} = \frac{F_{nv} A_b}{\Omega} = \frac{(84)(0.601)}{2.0} = 25.2 \text{ kips } \textit{(agrees with AISCM Table 7-1)}$$

$$\frac{R_{nt}}{\Omega} = \frac{F_{nt} A_b}{\Omega} = \frac{(113)(0.601)}{2.0} = 34.0 \text{ kips (agrees with } \textit{AISCM} \text{ Table 7-2)},$$

$$\frac{f_v}{F_{nv}/\Omega} = \frac{16.6}{84/2.0} = 0.40 > 0.30$$

$$\frac{f_t}{F_{nt}/\Omega} = \frac{28.8}{113/2.0} = 0.51 > 0.30$$

Since the ratio of the unfactored applied stress to the available strength is greater than 30% (0.30) for both shear and tension, the combined stress effects must be considered. The interaction equation (9-10b) will be used to determine the modified allowable tension in each bolt:

$$F'_{nt} = 1.3F_{nt} - \frac{\Omega F_{nt}}{F_{nv}} f_v \le F_v,$$

$$F'_{nt} = (1.3)(113) - \frac{(2.0)(113)}{84} 16.6 \le 113$$

$$F'_{nt} = 102 \text{ ksi} \le 113 \text{ ksi}$$

Therefore, use $F'_{nt} = 102$ ksi

The modified allowable tension stress is

$$\frac{F'_{nt}}{\Omega} = \frac{102}{2.0} = 51 \, ksi$$

$$\frac{F'_{nt}}{\Omega} = \mathbf{51.0 \ ksi} > f_t = 28.8 \text{ ksi, OK.}$$

The bolts are adequate for combined shear and tension loading.

EXAMPLE 9-5

Bolted Slip-Critical Connection Loaded in Combined Shear and Tension

For the connection shown in Figure 9-13, determine the maximum factored load, P_u, and the maximum service load, P_s, that can be applied considering the strength of the bolts only. The bolts are $^7/_8$-in.-diameter A490SC in standard bolt holes with a Class A faying surface, and there are no fillers. Assume that each bolt takes an equal amount of shear and tension.

Solution (LRFD)

ϕ = 1.0 for slip-critical bolts in standard bolt holes
μ = 0.30 (coefficient of friction for Class A faying surface)
D_u = 1.13
h_f = 1.0 (no filler plates)
T_b = 49 kips (see Table 9-1 for T_b for 7/8 in. diameter A490 bolts)

n_s = 1 (number of slip planes)
n_b = total number of bolts in the slip-critical connection = 6 bolts

The design shear strength of each bolt in the slip-critical connection (see equation (9-8a)) is

$$\phi R_{nv} = \phi \mu D_u h_f T_b n_s$$
$$= (1.0)(0.30)(1.13)(1.0)(49)(1) = 16.6 \text{ kips}$$

The design shear strength per bolt must be reduced to account for the applied tension as follows:

$$k_{sc} = 1 - \frac{T_u}{D_u T_b n_b} \geq 0$$

Therefore, the reduced design shear strength per bolt in the slip critical connection due to the presence of the factored total tension force, $T_u = P_u \cos(30°)$, is equal to $k_{sc}(\phi R_{nv}) = 16.6 k_{sc}$, where P_u is the factored total inclined load on the connection.

An expression can be developed as follows to solve for the maximum load using the LRFD requirement that for each bolt, the design shear strength must be greater than or equal to the factored shear load per bolt:

Therefore, for each bolt: $(16.6 \text{ kips})(k_{sc}) \geq \left(\frac{P_u}{n_b}\right)(\sin 30)$

$$(16.6)\left(1 - \frac{(P_u)(\cos 30)}{D_u T_b n_b}\right) \geq \left(\frac{P_u}{n_b}\right)(\sin 30)$$

That is, $(16.6)\left(1 - \frac{(P_u)(\cos 30)}{(1.13)(49)(6)}\right) \geq \left(\frac{P_u}{6}\right)(\sin 30)$

Solving the above equation for the maximum factored load, P_u, yields in the limit, P_u = 131 kips

Therefore, the maximum total factored load that can be supported by this slip-critical connection is **131 kips**.

Solution (ASD)

Ω = 1.5 for slip-critical bolts in standard bolt holes
μ = 0.30 (coefficient of friction for Class A faying surface)
D_u = 1.13
h_f = 1.0 (no filler plates)
T_b = 49 kips (see Table 9-1 for T_b for 7/8 in. diameter A490 bolts)
n_s = 1 (number of slip planes)
n_b = 6 bolts

The allowable shear strength of each bolt in the slip-critical connection (see equation (9-8b)) is

$$\frac{R_{nv}}{\Omega} = \frac{\mu D_u h_f T_b n_s}{\Omega} = \frac{(0.30)(1.13)(1.0)(49)(1)}{1.5} = 11.1 \text{ kips}$$

The allowable shear strength *per bolt* must be reduced to account for the applied tension as follows:

$$k_{sc} = 1 - \frac{1.5 T_a}{D_u T_b n_b} \geq 0$$

Therefore, the reduced allowable shear strength *per bolt* in the slip critical connection due to the presence of the unfactored total tension force on the connection, $T_a = P_s \cos(30°)$, is

$$k_{sc}\frac{R_{nv}}{\Omega} = k_{sc}(11.1) = 11.1k_{sc} \text{ kips / bolt},$$

That is, $k_{sc}\frac{R_{nv}}{\Omega} = 11.1k_{sc}$ kips / bolt ,

Where P_s is the applied total inclined force and T_a is the applied total tension force on the connection.

An expression can be developed as follows to solve for the maximum total load, P_s, that the connection can support using the ASD requirement that for each bolt, the allowable shear strength must be greater than or equal to the unfactored applied shear load per bolt:

Therefore, $(11.1 \text{ kips})(k_{sc}) \geq \left(\frac{P_s}{n_b}\right)(\sin 30) = \left(\frac{P_s}{6}\right)(\sin 30)$

Substituting $k_{sc} = 1 - \frac{1.5T_a}{D_u T_b n_b}$ in the above equation yields

$$(11.1)\left(1 - \frac{1.5(P_s)(\cos 30)}{D_u T_b n_b}\right) \geq \left(\frac{P_s}{6}\right)(\sin 30)$$

$$(11.1)\left(1 - \frac{1.5(P_s)(\cos 30)}{(1.13)(49)(6)}\right) \geq \left(\frac{P_s}{6}\right)(\sin 30)$$

Solving the above equation for the maximum unfactored load, P_s, yields in the limit,
P_s = 87.6 kips

Therefore, the maximum total unfactored load that can be supported by this slip-critical connection is **87.6 kips**.

9-5 ECCENTRICALLY LOADED BOLTS: SHEAR

When bolts are loaded such that the load is eccentric to the bolt group in the plane of the faying surface (see Figure (9.14)), the shear in the bolts will be a combination of the direct shear caused by the applied load and the additional shear caused by the inplane eccentricity of the load. There are two analytical approaches that can be used to determine the forces in the bolts:

1. The instantaneous center (IC) of the rotation method, and
2. The elastic method.

In both cases, the connection is designed to resist the applied shear, *P*, and the additional shear generated from the moment, *Pe*, due to the applied shear acting at an eccentricity, *e*. Both analytical methods are relatively complex and are generally not used in practice without the aid of computers. The IC method is more accurate but requires an iterative solution. The elastic method is less accurate and more conservative because both the ductility of the bolt group and the redundancy (i.e., load distribution) are ignored. It assumes a linear load-deformation behavior in the bolts, contrary to the non-linear relationship observed in tests [2]. *AISCM* Tables 7-7 through 7-14, which were derived using the IC method, are design aids for these types of connections, and they are more commonly used in practice and will be discussed later.

Instantaneous Center (IC) Method:

In an eccentrically loaded bolt group, both translation and rotation are induced in one connected element relative to another. This combined deformation effect is equivalent to a rotation about a

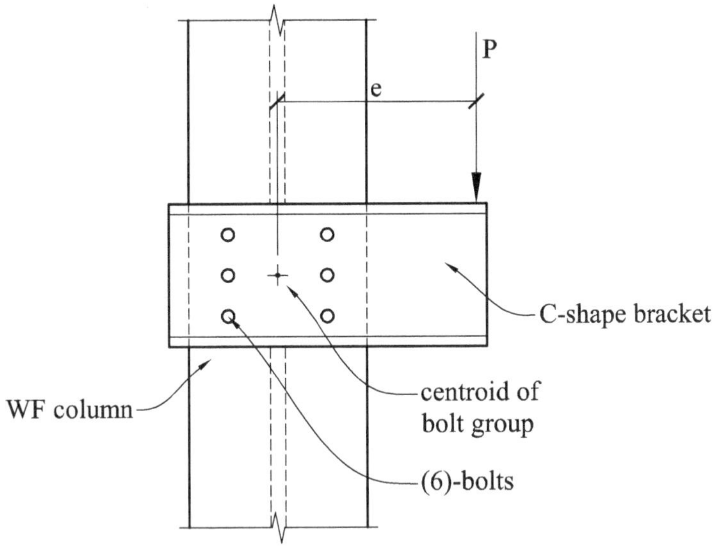

FIGURE 9-14 Eccentrically loaded bolt group

point called the instantaneous center, or IC. The location of the IC(see Figure 9-15) is a function of the geometry and the direction and orientation of the applied load, and requires an iterative solution. Note that, in the connection shown in Figure 9-15, the resultant force in each bolt is perpendicular to the radial line joining the instantaneous center (IC) to the centroid of the bolt, and it will have a vertical component due to the applied vertical load and a horizontal component due to the eccentricity of the applied load.

Based on test data for ¾ in. diameter ASTM F3125 Grade A325 bolts in double shear, the load-deformation relationship for one bolt is given by the following equation [2]:

$$R = R_{ult}\left(1 - e^{-10\Delta}\right)^{0.55} \tag{9-14}$$

where

R = Nominal shear strength of one bolt in single shear at a deformation Δ, kips,

R_{ult} = Ultimate shear strength of one bolt
= 74 kips for $^3/_4$-in.-diameter A325 bolts, and

Δ = Total deformation, including shear, bearing, and bending in the bolt, as well as bearing deformation in the connected elements
= 0.34 in. for $^3/_4$-in.-diameter A325 bolts.

By inspection, the bolt that is most remote or farthest from the IC will have the greatest load and deformation and therefore is assumed to have the maximum load and deformation (i.e.,

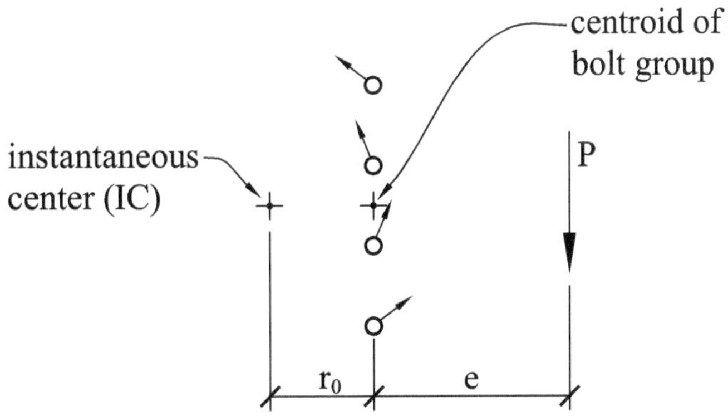

FIGURE 9-15 Eccentrically loaded bolt group: IC Method

R_{ult} = 74 kips and Δ = 0.34 in.), while the nominal shear in the other bolts in the group varies linearly with respect to their distances from the IC. The constant values of R_{ult} = 74 kips and Δ = 0.34 in. are based on test data for $^3/_4$-in.-diameter A325 bolts, but this value can conservatively be used for bolts of other sizes and grades [2]. Tables 7-7 through 7-14 in the *AISCM*, which use the IC method, are based on these constant values.

Using the free-body diagram shown in Figure 9-16, the forces in each bolt can be obtained from the following in-plane static equilibrium equations:

$$\Sigma F_x = \sum_{i=1}^{n} (R_x)_i - P_x = 0, \tag{9-15}$$

$$\Sigma F_y = \sum_{i=1}^{n} (R_y)_i - P_y = 0, \tag{9-16}$$

$$\sum M_{IC} = P(r_0 + e) - \sum_{i=1}^{n} r_i R_i = 0, \tag{9-17}$$

$$(R_x)_i = \frac{r_y}{r_i} R_i, \quad \text{and} \tag{9-18}$$

$$(R_y)_i = \frac{r_x}{r_i} R_i, \tag{9-19}$$

Where,

P = Applied load, with components P_x, and P_y

R_i = Resisting shear force in fastener, i, with components $(R_x)_i$, $(R_y)_i$, along the X- and Y-axes, respectively,

r_i = Distance from the fastener to the IC, with components r_x, r_y, along the X- and Y-axes, respectively,

r_0 = Distance from the IC to the centroid of the bolt group,

e = Load eccentricity; distance from the load to the centroid of the bolt group with components e_x, e_y,

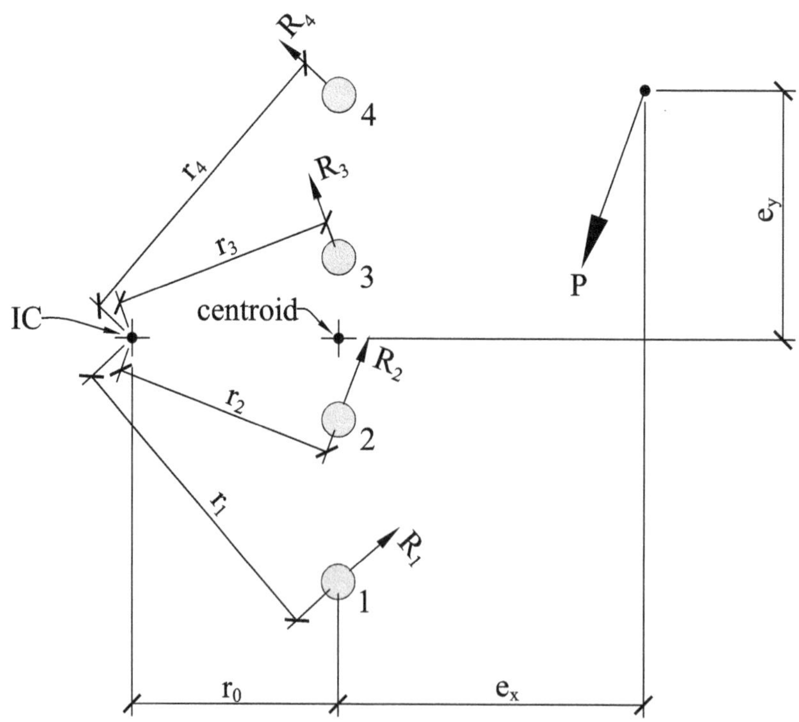

FIGURE 9-16 Free body diagram of eccentrically loaded bolt group

i = Subscript indicating the individual fastener number, and

n = Total number of fasteners.

The deformation at any bolt, i, is given as $\Delta_i = \dfrac{r_i}{r_{max}}\Delta_{max} = \dfrac{r_i}{r_{max}}(0.34 \text{ in.})$

In order to solve the equilibrium equations (9-15) through (9-17), the location of the IC must first be assumed. If the equilibrium equations are not satisfied, then a new value must be assumed for the location of the IC until the equilibrium equations are satisfied. For most cases, the applied load, P, is vertical, and therefore equation (9-15) is eliminated by inspection.

For bolts other than $^3/_4$-in.-diameter A325N, the load capacity is also determined from a linear relationship in equation (9-14). The total capacity of the connection is then the sum of the capacity of each individual fastener. Example 9-6 illustrates this method in further detail, as well as the use of the aforementioned *AISCM* design aids (see *AISCM* Tables 7-7 through 7-14). Note that these design aids can only be used for bolt arrangements with the standard spacing indicated in the designs. For other bolt arrangements that differ from those shown in the design aids, the Instantaneous Center (IC) method should be used, and the step-by-step iterative procedure is as follows:

Step-by-Step Iterative Procedure for the Instantaneous Center Method:

1. Initially assume a location for the Instantaneous Center (IC) of the bolt group. For a bolt group with a symmetrical layout and subjected to a vertical load, P, the IC will be located along a line that is perpendicular to the load, P, and passing through the centroid of the bolt group.
2. Determine the coordinates (r_x, r_y, r_i) of all the bolts in the group relative to the assumed IC location.
3. Determine the maximum value of r_i (i.e., r_{max}), where the maximum load and deformation in the farthest bolt are assumed for all bolt sizes and grades as 74 kips and 0.34 in., respectively.
4. Determine the deformation at each bolt relative to the maximum deformation Δ_{max} of 0.34 in. at the farthest bolt. That is, $\Delta_i = \dfrac{r_i}{r_{max}}\Delta_{max} = \dfrac{r_i}{r_{max}}(0.34 \text{ in.})$
5. Determine the resisting shear force, R_i, in each bolt using Equation (9-14), and then determine the x- and y- components of the resisting shear force in each bolt, R_x and R_y. The shear in the bolts in a group varies linearly with their respective distance from the instantaneous center (IC) (see Figure 9-15).
6. Solve the equilibrium equations (9-15) through (9-17) to obtain the force, P.
7. If all the equilibrium equations are *not* satisfied, then assume a new location for the IC and repeat the process until all the equilibrium equations are satisfied.
8. Recall that for the IC method, the maximum shear strength and deformation of R_{ult} = 74 kips per bolt, and Δ = 0.34 in., respectively, were obtained from tests [2] and are used as the baseline values. To obtain the nominal shear strength of the connection for the actual fastener size and grade used in the connection, linear interpolation is conservatively used. The nominal total shear strength of the connection is calculated as

   $$R_{nv,total} = P\left(\frac{R_{nv} \text{ per bolt}}{R_{ult}}\right),$$

The total design shear strength of the connection is then calculated as $\phi R_{nv,total}$,
Where $\phi = 0.75$

P = the force obtained in step 6 using the equilibrium equations
R_{ult} = 74 kips/bolt

$R_{nv} = F_{nv} A_b$ (see equation (9-1a))
A_b = nominal unthreaded cross-sectional area of the bolt, in.^{2}
F_{nv} = nominal shear strength of the bolt obtained from Table 9-3

Elastic Method:

A simplified and conservative approach for eccentrically loaded bolts in shear is called the *elastic method*. In this method (see Figure 9-17), each bolt is assumed to resist an equal proportion of the applied load, P, and a portion of the shear induced by the moment, Pe, proportional to its distance from the centroid of the bolt group. The shear in each bolt due to the applied load is

$$r_{px} = \frac{P_x}{n}, \text{ and} \tag{9-20}$$

$$r_{py} = \frac{P_y}{n}, \text{ respectively,} \tag{9-21}$$

where
r_p = Force in each bolt due to applied load with horizontal and vertical components r_{px}, r_{py}, respectively
P = Applied load on the connection, with horizontal and vertical components P_x and P_y, respectively
n = Total number of bolts.

The shear in the bolt that is most remote or farthest from the centroid of the bolt group due to the applied moment is

$$r_{mx} = \frac{Mc_y}{I_p}, \text{ and} \tag{9-22}$$

$$r_{my} = \frac{Mc_x}{I_p}, \tag{9-23}$$

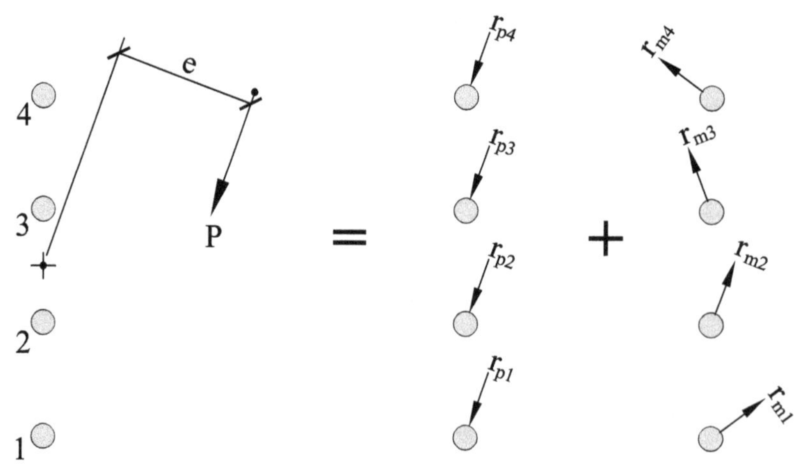

FIGURE 9-17 Eccentrically loaded bolt group: Elastic method

Where,

r_m = Shear force in each bolt due to applied moment with horizontal and vertical components r_{mx}, r_{my}, respectively

M = Resulting moment on the connection due to the eccentrically applied load
= $P_x e_y + P_y e_x$,

c = Radial distance from the centroid of the bolt group with components c_x, c_y, in the X- and Y-axes, respectively,

e = Load eccentricity (i.e., distance from the load to the centroid of the bolt group with components e_x, e_y),

I_p = Polar moment of inertia of the bolt group (in^4/in^2)
= $\Sigma(I_x + I_y)$, where $I = Ad^2$
= $\Sigma(c_x^2 + c_y^2)$ for bolts with the same cross-sectional area within a bolt group.

In the elastic method, the critical fastener force is determined, and this is the basis for the connection design. The maximum applied shear force in the critical fastener is usually found in the fastener located most remote from the centroid of the bolt group. This critical fastener shear force or the maximum applied resultant shear force in the critical fastener is calculated as

$$r = \sqrt{\left(r_{px} + r_{mx}\right)^2 + \left(r_{py} + r_{my}\right)^2} \leq \phi F_{nv} A_b \tag{9-24}$$

The value of r above is used to determine the required shear strength of each fastener in the bolt group. See Example 9-6 for an illustration of the elastic method. In summary,

- The Elastic Method is a simplified, conservative alternative to the more accurate but cumbersome IC method.
- In the Elastic Method, each bolt is assumed to resist an equal proportion of the applied load and a portion of the shear induced by the moment, Pe, proportional to its distance from the centroid of the bolt group.
- The goal is to determine the maximum applied resultant force in the critical fastener (i.e., the fastener that is farthest from the centroid or the fastener with the largest load).
- The shear force in the critical fastener is then used for the design of the bolts.
- The maximum applied shear force in the critical fastener usually occurs in the fastener that is located most remote or farthest from the centroid of the bolt group.
- The shear in each bolt due to the applied load is obtained using Equations (9-20) and (9-21), and the shears due to moments are obtained using Equations (9-22) and (9-23).
- The maximum resultant shear force in the most critical fastener is obtained using Equations (9-24).

EXAMPLE 9-6

Eccentrically Loaded Bolts in Shear

For the bracket connection shown in Figure 9-18, determine the design strength of the connection. Compare the results from the ultimate strength (or IC) method, the elastic method, and the appropriate design aid from *AISCM* Tables 7-7 through 7-14. The bolts are $^3/_4$-in.-diameter A490N.

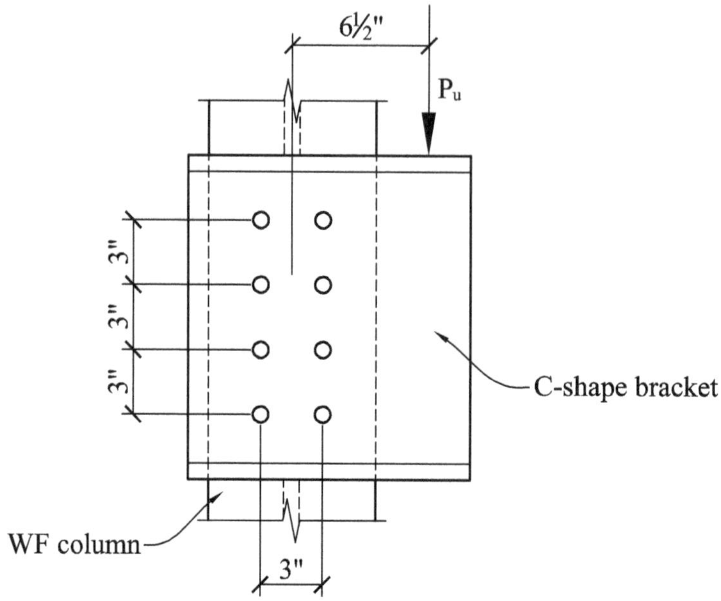

FIGURE 9-18 Connection detail for Example 9-6

Solution

Ultimate Strength, or IC, Method:

A location for the IC must first be assumed after which the coordinates of each bolt relative to the assumed IC (i.e., r_x and r_y) and $r_i = \sqrt{r_x^2 + r_y^2}$ are determined, from which the maximum value of r_i (i.e., r_{max}) is determined. Next, the deformation of each bolt relative to the maximum deformation of 0.34 in. at the most remote bolt (i.e., the bolt with $r_i = r_{max}$) is calculated as $\Delta_i = \frac{r_i}{r_{\max}}(0.34 \text{ in})$.

Then, R_i for each bolt is calculated using equation (9-14), and the components R_x and R_y for each bolt can be obtained. Finally, the equilibrium equations (9-15) through (9-17) are solved, and if they are all satisfied, that would imply that the assumed location of the IC (i.e., the assumed r_0) is correct; if not, a different location for the IC (i.e., a different value of r_0) is chosen, and the above process is repeated until all the equilibrium equations (9-15) through (9-17) are satisfied.

For the current example, using the iterative process previously described, the location of the IC has been finally determined such that the distance from the IC to the centroid of the bolt group, r_0 = 1.90 in. (see point A in Figure 9-19). A free-body diagram of the bolt group is also shown in Figure 9-19.

From the free-body diagram, the distance of each bolt from the IC can be determined:

Bolt	**Distance from the IC**				
i	r_x, **in.**	r_y **in.**	$r_i = \sqrt{r_x^2 + r_y^2}$, **in.**	c_x **in.**	c_y **in.**
1	0.40	−4.5	4.52	−1.5	−4.5
2	3.40	−4.5	**5.64**	1.5	−4.5
3	0.40	−1.5	1.55	−1.5	−1.5
4	3.40	−1.5	3.72	1.5	−1.5
5	0.40	1.5	1.55	−1.5	1.5
6	3.40	1.5	3.72	1.5	1.5
7	0.40	4.5	4.52	−1.5	4.5
8	3.40	4.5	**5.64**	1.5	4.5

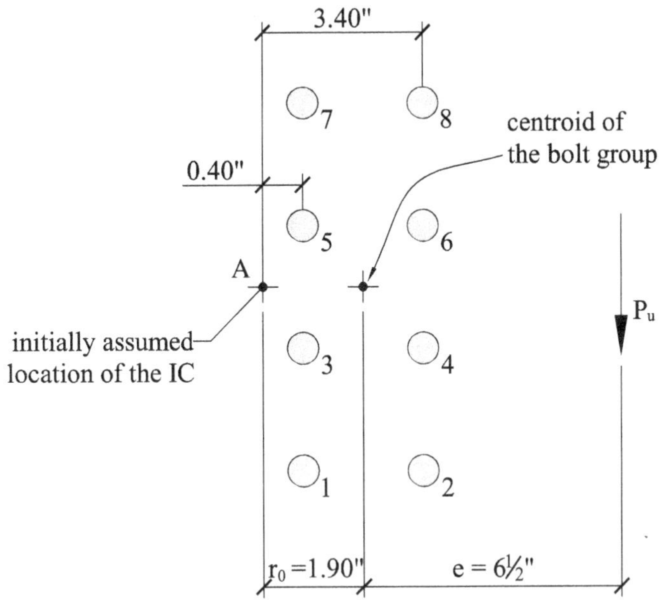

FIGURE 9-19 Free-body diagram for Example 9-16 (IC method)

For the IC method, the bolt or bolts that are most remote from the IC are assumed to be stressed and deformed to failure; the deformation in the remaining bolts is assumed to vary linearly. The bolts at locations 2 and 8 are the most remote from the IC location or point A in Figure 9-19 (i.e., $r_2 = r_8 = 5.64$ in.); therefore, the deformation of each of the most remote or farthest bolts from the IC is assumed to be $\Delta = 0.34$ in. The deformation of the remaining bolts is determined using the relationship $\Delta_i = \frac{r_i}{r_{max}}(0.34 \text{ in})$ as follows:

$$\Delta_1 = \Delta_7 = \frac{4.52 \text{ in.}}{5.64 \text{ in.}}(0.34 \text{ in.}) = 0.272 \text{ in.}$$

$$\Delta_3 = \Delta_5 = \frac{1.55 \text{ in.}}{5.64 \text{ in.}}(0.34 \text{ in.}) = 0.094 \text{ in.}$$

$$\Delta_4 = \Delta_6 = \frac{3.72 \text{ in.}}{5.64 \text{ in.}}(0.34 \text{ in.}) = 0.224 \text{ in.}$$

$$\Delta_2 = \Delta_8 = 0.34 \text{ in.}$$

The nominal shear strength in each bolt can be determined using equation (9-14) as follows:

$$R = R_{ult}\left(1 - e^{-10\Delta}\right)^{0.55}$$

$$R_1 = R_7 = (74)\left(1 - e^{-(10)(0.272)}\right)^{0.55} = 71.3 \text{ kips}$$

$$R_3 = R_5 = (74)\left(1 - e^{-(10)(0.094)}\right)^{0.55} = 56.3 \text{ kips}$$

$$R_4 = R_6 = (74)\left(1 - e^{-(10)(0.224)}\right)^{0.55} = 69.6 \text{ kips}$$

$$R_2 = R_8 = (74)\left(1 - e^{-(10)(0.34)}\right)^{0.55} = 72.6 \text{ kips}$$

The vertical component of the shear force, R, in each bolt, is determined from equation (9-19):

$$\left(R_y\right)_i = \frac{r_x}{r_i}\left(R_i\right)$$

$$R_{y1} = R_{y7} = \frac{0.40 \text{ in.}}{4.52 \text{ in.}}(71.3 \text{ k}) = 6.33 \text{ kips}$$

$$R_{y3} = R_{y5} = \frac{0.40 \text{ in.}}{1.55 \text{ in.}}(56.3 \text{ k}) = 14.5 \text{ kips}$$

$$R_{y4} = R_{y6} = \frac{3.40 \text{ in.}}{3.72 \text{ in.}}(69.6 \text{ k}) = 63.6 \text{ kips}$$

$$R_{y2} = R_{y8} = \frac{3.40 \text{ in.}}{5.64 \text{ in.}}(72.6 \text{ k}) = 43.8 \text{ kips}$$

From the equilibrium equation (9-16),

$$\Sigma F_y = \sum_{i=1}^{n}(R_y)_i - P_y = 0$$

$$(2)(6.33 \text{ kips} + 14.5 \text{ kips} + 63.6 \text{ kips} + 43.8 \text{ kips}) = P_y$$

$$\boldsymbol{P_y = P = 256 \text{ kips.}}$$

Since there is no horizontal component to the applied load, equation (9-15) is satisfied, and $P = P_y$. To verify this value, the in-plane moment equilibrium equation (9-17) will be used:

$$\sum M_{IC} = P(r_0 + e) - \sum_{i=1}^{n} r_i R_i = 0$$

$$P(1.9 \text{ in.} + 6.5 \text{ in.}) =$$
$$2[(4.52 \text{ in.})(71.3 \text{ kips}) + (1.55 \text{ in.})(56.3 \text{ kips}) + (3.72 \text{ in.})(69.6 \text{ kips}) + (5.64 \text{ in.})(72.6 \text{ kips})]$$

Therefore, $\boldsymbol{P = 256}$ **kips** (same as the value obtained from the equilibrium of the vertical forces).

Since all the equilibrium equations have been satisfied, the assumed location for the IC (i.e., r_o = 1.90 in.) is correct. Note that this total load of P = 256 kips corresponds to Rult of 74 kips per bolt.

Determine the design strength of the connection

For the IC method, the maximum shear strength and deformation of R_{ult} = 74 kips per bolt, and Δ = 0.34 in., respectively, were obtained from tests [2] and are used as the baseline values. For fasteners of other sizes and grades, linear interpolation is conservatively used. The nominal shear strength of the connection is therefore calculated as

$$P\left(\frac{R_{nv} \text{ per bolt}}{R_{ult}}\right),$$

Where,

R_{nv} = the nominal shear strength of one bolt in the group, and
P = total load corresponding to R_{ult} of 74 kips per bolt = 256 kips
F_{nv} = 68 ksi for one ¾ in. diameter A490N bolt (see Table 9-3)

The cross-sectional area of the ¾ in. diameter bolt used in this connection is

$$A_b = \pi \frac{(3/4)^2}{4} = 0.442 \text{ in.}^2$$

Using equation (9-1a), the nominal shear strength for one ¾ in. diameter A490N bolt is

$$R_{nv} \text{ per bolt} = F_{nv} A_b$$
$$= (68 \text{ ksi})(0.442 \text{ in.}^2) = 30 \text{ kips.}$$

Therefore, the total shear nominal strength of this connection corresponding to R_n is calculated using linear interpolation as follows:

$$R_{nv,total} = (P)\left(\frac{R_{nv}}{R_{ult}}\right) = (256\,\text{kips})\left(\frac{(30\,\text{kips})}{74\text{ kips}}\right) = 104.0\text{ kips.}$$

The total design shear strength of the connection, ϕR_n = (0.75)(104.0) = **78.0 kips** $\geq P_{u\,(\max)}$
Therefore, $P_{u\,(\max)}$ = **78 kips**

Elastic Method:

For this method, the critical fastener force is determined from the bolt most remote or farthest from the centroid of the bolt group. For this connection, the bolts at locations 1, 2, 7, and 8 are the most remote or farthest from the centroid. The capacity of one of these bolts will be used to determine the strength of the connection. The shear is distributed equally to all the bolts in the group. Since there is no horizontal component for the applied load, equation (9-20) becomes

$$r_{px} = \frac{P_x}{n} = \frac{0}{8} = 0$$

From equation (9-21),

$$r_{py} = \frac{P_y}{n} = \frac{P}{8} \quad \text{(since number of bolts in the connection = 8)}$$

The shear in the bolts due to the eccentricity is determined from equations (9-22) and (9-23):

$e = 6.5$ in.

For the farthest bolt from the centroid (e.g., bolt 8), c_y = 4.5 in. and c_x = 1.5 in.

F_{nv} = 68 ksi for ¾ in. diameter A490N bolts (see Table 9-3)

$$A_b = \pi\frac{\left(3/4\right)^2}{4} = 0.442\text{ in.}^2$$

$\phi = 0.75$

The polar moment of inertia of the bolt group is,

$$I_p = \Sigma\left(c_x^2 + c_y^2\right)$$
$$= \begin{bmatrix}(-1.5)^2 + (-4.5)^2 + (1.5)^2 + (-4.5)^2 + (-1.5)^2 + (-1.5)^2 + (1.5)^2 + (-1.5)^2 + \\ (-1.5)^2 + (1.5)^2 + (1.5)^2 + (1.5)^2 + (-1.5)^2 + (4.5)^2 + (1.5)^2 + (4.5)^2\end{bmatrix}$$
$$= 108\text{ in.}^4$$

For the farthest or most remote bolt from the centroid of the bolt group (i.e., with coordinates, c_x = 1.5 in. and c_y = 4.5 in.), the x- and y- components of the shear in the bolt due to the applied moment is

$$r_{mx} = \frac{Mc_y}{I_p} = \frac{Pec_y}{I_p} = \frac{P(6.5)(4.5)}{108} = 0.271P$$

$$r_{my} = \frac{Mc_x}{I_p} = \frac{Pec_x}{I_p} = \frac{P(6.5)(1.5)}{108} = 0.0902P$$

Equation (9-24) will be used to determine the capacity of the bolt that is farthest from the centroid. The LRFD method requires that the applied resultant force, r, be less than or equal to the design strength, ϕr_n, resulting in the following equation:

$$r = \sqrt{\left(r_{px} + r_{mx}\right)^2 + \left(r_{py} + r_{my}\right)^2} \leq \phi F_{nv} A_b$$

$$\sqrt{(0 + 0.271P)^2 + \left(\frac{P}{8} + 0.0902P\right)^2} \leq (0.75)(68)(0.442)$$

$$\text{That is, } (0.271P)^2 + \left(\frac{P}{8} + 0.0902P\right)^2 \leq \left[(0.75)(68)(0.442)\right]^2$$

Solving for P in the preceding LRFD equation yields, in the limit,
$P = P_u =$ **65.1 kips**
The elastic method is more conservative than the IC method, so as expected, the elastic method connection capacity of 65 kips is lower than the value of 78 kips calculated using the IC method.

AISC Design Aids

Based on the connection geometry, *AISCM* Table 7-8, with angle = 0° and s = 3 in., will be used:
e = 6.5 in., and

n = *number of bolts in one vertical row* = 4

(Note the different or unique definition of n in *AISCM* Table 7-8)
By linear interpolation from *AISCM* Table 7-8, C = 3.48.
The design strength of the connection is

$\phi R_n = \phi C r_n$,

where $\phi r_n = \phi F_{nv} A_b$

$\phi r_n = \phi F_{nv} A_b = (0.75)(68)(0.442) = 22.5$ kips *(agrees with AISCM Table 7-1)*
The total design shear strength of the connection, $\phi R_n = \phi C r_n = (3.48)(22.5)$
= **78.4 kips** $\geq P_u$
Therefore, the maximum $\boldsymbol{P_u}$ **= 78.4 kips.**
This value agrees with the value of P_u obtained from the Instantaneous Center or Ultimate Strength method as expected because the *AISCM* design tables were derived using the IC method.

9-6 ECCENTRICALLY LOADED BOLTS: BOLTS IN SHEAR AND TENSION

When a bolt group is loaded such that the eccentricity of the applied load is normal or perpendicular to the plane of the faying surface, as in Figure 9-20, the bolts will be subjected to combined direct shear and tension [4, 5]. The direct shear is induced in the bolts due to the applied load, P, and is assumed to be equally distributed, and the tension in the bolts is caused by the applied moment, Pe, on the bolt group due to the eccentricity, e, of the applied load. The out-of-plane moment caused by the eccentric load causes tension forces in the bolts above the neutral axis of the bolt group and compression forces in the bolts below the neutral axis of the bolt group.

There are two possible design approaches that can be used for eccentrically loaded bolts: Case I and Case II.

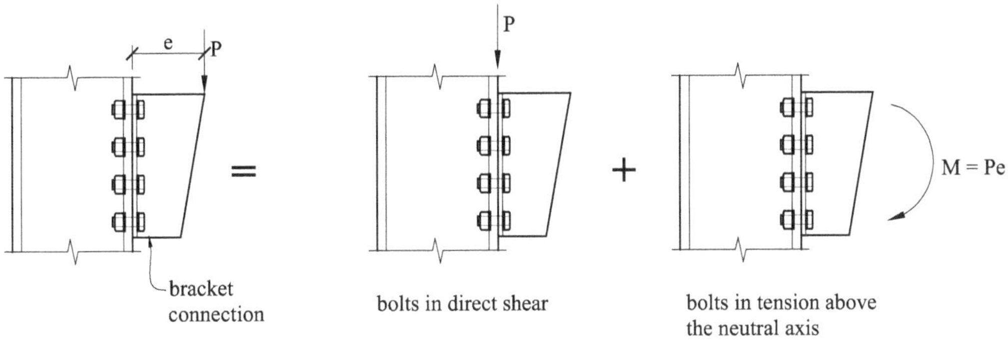

FIGURE 9-20 Eccentrically loaded bolts in shear in tension

- **Case I:** For this case, the neutral axis is **not** necessarily at the centroid of the bolt group and is generally below the centroid of the bolt group.
- **Case II:** This is a more simplified and conservative method in that the neutral axis is assumed to be at the centroid of the bolt group, and only the bolts above the neutral axis resist the tension due to the eccentric load.

In both cases, the shear is distributed equally to each bolt as follows:

$$r_v = \frac{P}{n}, \tag{9-25}$$

where

r_v = Shear force in each bolt due to the applied load,

P = Total applied load on the bolt group,

n = Number of bolts.

For **Case I**, a trial position for the neutral axis has to be assumed. A value of one-sixth of the depth of the connecting element is recommended as a baseline value (see Figure 9-21). The area below the neutral axis is in compression, but only for a certain effective width, b_{eff}. The effective width of the compression zone is defined as

$$b_{eff} = 8t_f \le b_f, \tag{9-26}$$

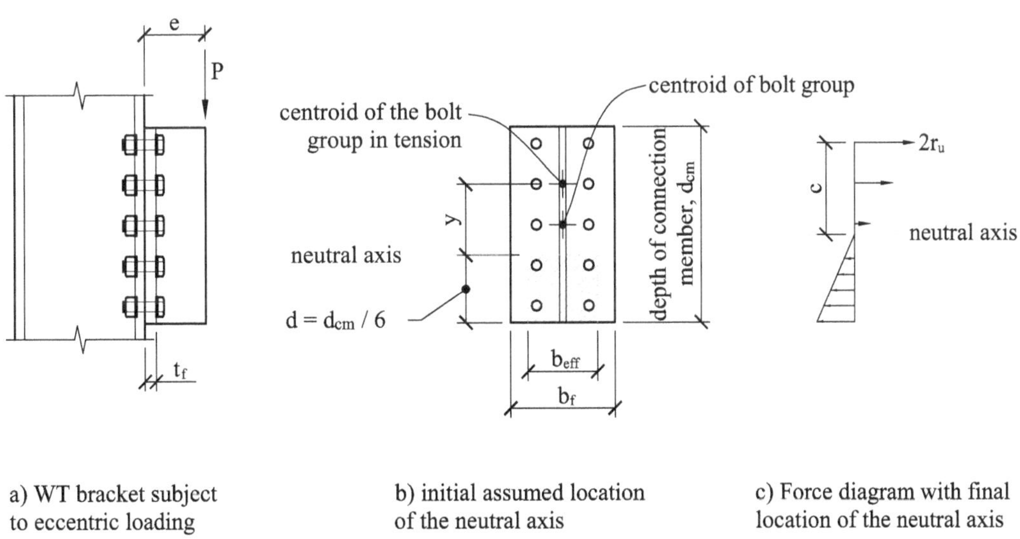

FIGURE 9-21 Eccentrically loaded bolts in shear in tension: Case I

where

b_{eff} = Effective width of the compression stress block below the neutral axis,

t_f = Connecting element thickness (use the average flange thickness of the connecting element where the flange thickness is not constant), and

b_f = Width of the connecting element.

Equation (9-26) is valid for connecting elements of W-shapes, S-shapes, plates, and angles. The location assumed for the neutral axis is verified by summing up moments about the neutral axis. The moment of the bolt areas above the neutral axis taken about the neutral axis is equated to the moment about the neutral axis of the compression stress block below the neutral axis. The summation of moments yields the following equation:

$$\Sigma A_b y = \frac{b_{eff}\, d^2}{2}, \tag{9-27}$$

where

ΣA_b = Sum of the cross-sectional areas of the bolts above the neutral axis (i.e., the bolts in tension)

y = Distance between the centroid of the bolt group above the neutral axis and the neutral axis,

b_{eff} = Effective width of the compression stress block below the neutral axis (see equation (9-26))

d = Depth of the compression stress block (initially, we can assume d = one-sixth of the depth of connection member = $\frac{depth}{6}$).

The location of the neutral axis is then adjusted as required until equation (9-27) is satisfied. Once the exact location of the neutral axis is determined, the tensile force in each bolt is determined as follows:

$$r_t = \frac{Mc}{I_x} A_b, \tag{9-28}$$

where

r_t = Tension force in each bolt due to the applied moment,

M = Moment due to the eccentrically applied load = Pe,

A_b = cross-sectional area of the bolt,

c = Distance from the neutral axis to the most remote (or farthest) bolt in the tension group,

e = Load eccentricity,

I_x = Combined moment of inertia of the areas of the bolt group in tension and the compression stress block about the neutral axis.

Since the bolts are loaded in both shear and tension, these combined forces must be checked for conformance to the interaction equations (9-10a) or (9-10b) (see also equation (J3-3a) in the *AISC specifications*).

For **Case II**, which is the simplified method, the neutral axis is assumed to be at the centroid of the bolt group (see Figure 9-22). Therefore, the bolts above the neutral axis are in tension, and the bolts below the neutral axis are assumed to be in compression. The tension force in each bolt is then defined as

$$r_t = \frac{M}{n' d_m}, \tag{9-29}$$

where

r_t = Force in each bolt due to the applied moment,

n' = Number of bolts above the neutral axis,

d_m = Moment arm between the centroid of the tension group and the centroid of the compression group of bolts,

e = load eccentricity.

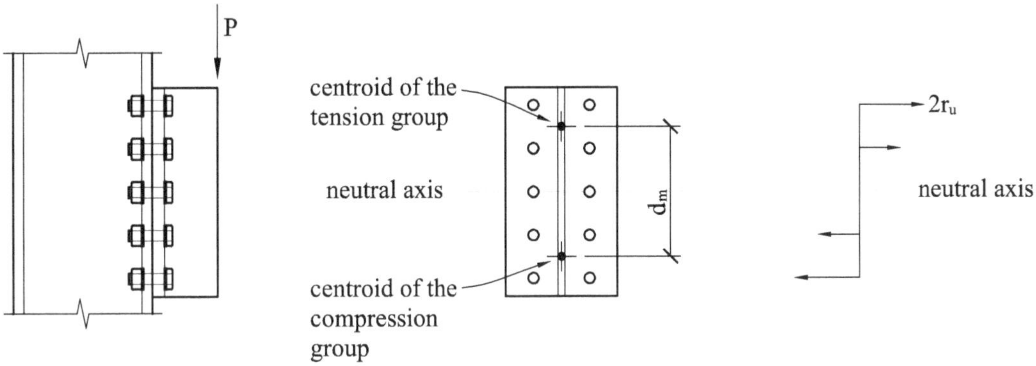

FIGURE 9-22 Eccentrically loaded bolts in shear in tension: Case II

Since the bolts are loaded in both shear and tension, these combined forces must be checked for conformance to the interaction equations (9-10a) or (9-10b) (see also equation (J3-3a) in the *AISC specifications*). Note that the shear stress in each bolt, $f_v = r_v / A_b$, and the tension stress in each bolt, $f_t = r_t / (n' A_b)$.

A summary of the procedures for Case I and Case II are as follows:

Case I – Procedure:

1. Calculate the shear force in each bolt assuming the shear is distributed equally to all the bolts in the connection.
2. Assume a trial position for the neutral axis – a value of one-sixth of the depth of the connecting element is recommended as a base line (see Figure 9-21).
3. Determine the effective width of the area below the neutral axis which is in compression (i.e. the compression zone); see equation (9-26). Note that Equation (9-26) is valid for connecting elements of W-shapes, S-shapes, plates and angles
4. Verify the assumed location of the neutral axis by summing moments of bolt areas about the neutral axis. The moment of the bolt areas above the neutral axis taken about the neutral axis is equated to the moment about the neutral axis of the compression stress block below the neutral axis per equation (9-27).
5. Adjust the location of the neutral axis until equation (9-27) is satisfied.
6. After the location of the neutral axis is determined, the tension force in each bolt is calculated from equation (9-28). Note In Case I, the neutral axis is not necessarily at the centroid of the bolt group and is generally below the centroid of the bolt group.
7. Since the bolts are loaded in both combined shear and tension, check these combined forces using the interaction equations (9-10a) or (9-10b).

Case II – Procedure:

1. Calculate the shear force in each bolt assuming the shear is distributed equally to all the bolts in the connection.
2. The neutral axis is assumed to be at the centroid of the bolt group (see Figure 9-22).
3. The bolts above the neutral axis are in tension and the bolts below the neutral axis are in compression. The tension force in each bolt due to the applied moment is calculated from Equation (9-29).
4. Since the bolts are loaded in both combined shear and tension, check these combined forces using the Interaction equations (9-10a) or (9-10b).

EXAMPLE 9-7

Eccentrically Loaded Bolts in Shear and Tension with the Load Eccentricity Normal to the Faying Surface

Determine whether the connection shown in Figure 9-23 is adequate to support the load, considering the eccentric loading on the bolts only. Compare the results from Cases I and II. The bolts are $^3/_4$-in.-diameter A325N.

Solution

Each bolt is assumed to resist an equal amount of shear for both cases:

n = total number of bolts = 8

$$r_v = \frac{P}{n} = \frac{125 \text{ kips}}{8} = 15.6 \text{ kips}$$

$$A_b = \pi \frac{(3/4)^2}{4} = 0.442 \text{ in.}^2$$

$$f_v = \frac{r_v}{A_b} = \frac{15.6}{0.442} = 35.4 \text{ ksi}$$

Case I:

From Figure 9-24, the neutral axis is determined as follows by assuming that it is located just above the first (or bottom) row of bolts:

Since 2L4 × 4 × 3/8 are used as the connecting element,

t_f = 3/8 in. = 0.375 in.

b_f = 4 in. + ½ in. + 4 in. = 8.5 in.

$$b_{eff} = 8t_f \le b_f$$

$$= (8)(0.375 \text{ in.}) = 3 \text{ in.} < 8.5 \text{ in.}$$

Therefore, the effective width of the compression stress block, $b_{eff} = 3$ in.

Number of bolts in the bolt group above the neutral axis = 6

There are 6 bolts above the assumed neutral axis location (see Figure 9-24). Given that the vertical spacing between each row of bolts is 3 in., the centroid of these 6 bolts measured from the top of the double angles = 1.25 in. + 3 in = 4.25 in.

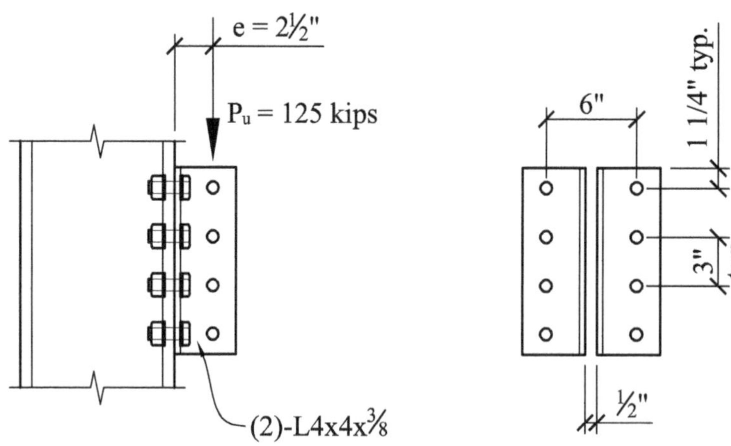

FIGURE 9-23 Connection Detail for Example 9-7

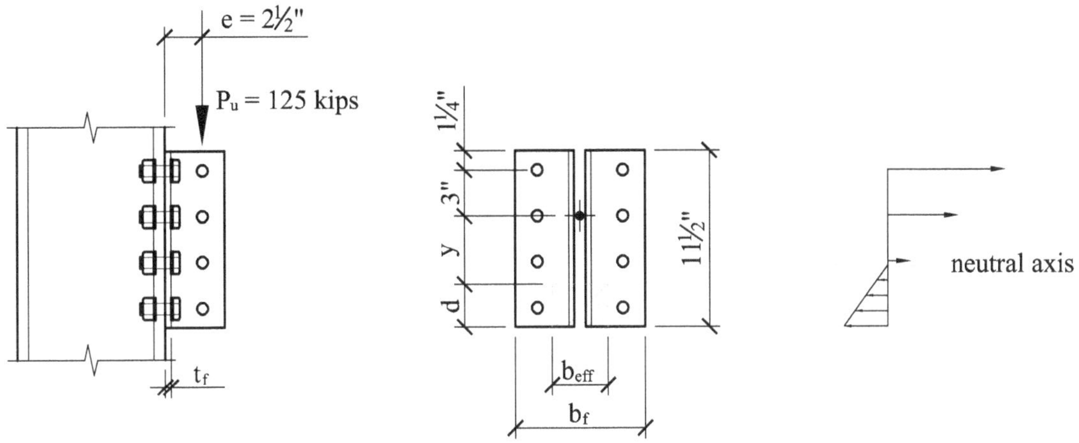

FIGURE 9-24 Free body diagram - Case I

The depth of the neutral axis, measured from the bottom of the double angle = d

Therefore, from Figure 9-24, the distance between the centroid of the bolt group above the neutral axis and the neutral axis is y = 11.5 in. – 4.25 in. – d = 7.25 in. – d

Since there are 6 bolts above the assumed neutral axis, using equation (9-27) yields

$$\Sigma A_b y = \frac{b_{eff} d^2}{2}$$

$$\left[(6)(0.442)\right](7.25 - d) = \frac{3d^2}{2}$$

Solving for d yields,

$d = 2.80$ in.

Since d > 1¼ in. (the edge distance of the outermost bolts as shown in Figure 9-24), the neutral axis is truly located above the first row of bolts as assumed. If this were not the case, the process would have been repeated until the assumed location of the neutral axis is confirmed.

Taking the summation of moments of the compression block and the areas of the bolt group in tension about the neutral axis yields,

$$I_x = \Sigma A y^2 = (b_{eff} d)\left(\frac{d}{2}\right)^2 + \sum A_{b,i}(y_i - d)^2$$

$$(2)(0.442)(4.25-2.8)^2 + (2)(0.442)(7.25-2.8)^2 \cdot \quad 2 \quad 0.442$$

$$+(2)(0.442)(4.25-2.8)^2 + (2)(0.442)(7.25-2.8)^2 + \quad 2 \quad 0.442$$

$I_x = 84.9$ in.^{4}

c = distance from the neutral axis to the most remote or farthest bolt in the tension group = 11.5 – 1.25 – d = 10.25 – 2.80 = 7.45 in.

$M = Pe = (125 \text{ kips})(2.5 \text{ in.}) = 312.5$ in.-kips

The applied tension per bolt in the farthest bolt from the neutral axis is

$$r_t = \frac{M\,c}{I_x} A_b = \frac{(312.5)(7.45)}{84.9}(0.442) = 12.1 \text{ kips}$$

Check the combined shear and tension stresses:

$F_{nt} = 90$ ksi for ¾ in. diameter A325N bolts (see Table 9-3)

$F_{nv} = 54$ ksi for ¾ in. diameter A325N bolts (see Table 9-3)

$\phi = 0.75$

From the LRFD combined stress interaction equation (9-10a),

$$F'_{nt} = 1.3F_{nt} - \frac{F_{nt}}{\phi F_{nv}} f_v \leq F_{nt}$$

$$F_{nt}' = (1.3)(90) - \frac{(90)}{(0.75)(54)}(35.4) \leq 90\,\text{ksi}$$

$= 38.3$ ksi.

The design tensile strength of each bolt is

$$\phi R_n = \phi F'_{nt} A_b = (0.75)(38.3)(0.442)$$

$\phi R_n =$ **12.7 kips** $> r_t = 12.1$ kips. OK.

The connection is found to be adequate using the Case I design approach.

Case II:

As shown in Figure 9-25, only the bolts above the neutral axis ($n' = 4$ bolts) resist the tension. The total depth of connection element (i.e., 2L 4 × 4) = 1.25 + 3 + 3 + 3 + 1.25 = 11.5 in. In this approach, the neutral axis is assumed to be located at the centroid of the bolt group. The location of the centroid of the bolt group from the bottom of the 2L 4 × 4 = (11.5 in.)/2 = 5.75 in.

Distance between the CG of the bolt group *above* the neutral axis (i.e., the bolts in tension) and the CG of the entire bolt group = ½(11.5 in.) – 1.25 in. – ½(3 in.) = 3 in.

Distance between the CG of the bolt group *below* the neutral axis (i.e., the bolts in compression) and the CG of the entire bolt group = ½(11.5 in.) – 1.25 in. – ½(3 in.) = 3 in.

The distance between the centroids of the tension and compression bolt groups is d_m = 3 in. + 3 in. = 6 in.

n' = number of bolts above the neutral axis (i.e., bolts in tension) = 4 bolts

From equation (9-29), the tension force in each bolt due to the applied moment is

$$r_t = \frac{M}{n'd_m} = \frac{312.5}{(4)(6)} = 13 \text{ kips.}$$

Check the combined shear and tension stresses:

The design shear strength considering combined shear and tension stress was already determined for Case I, and the same value applies to Case II as well. The design tensile strength of each bolt is

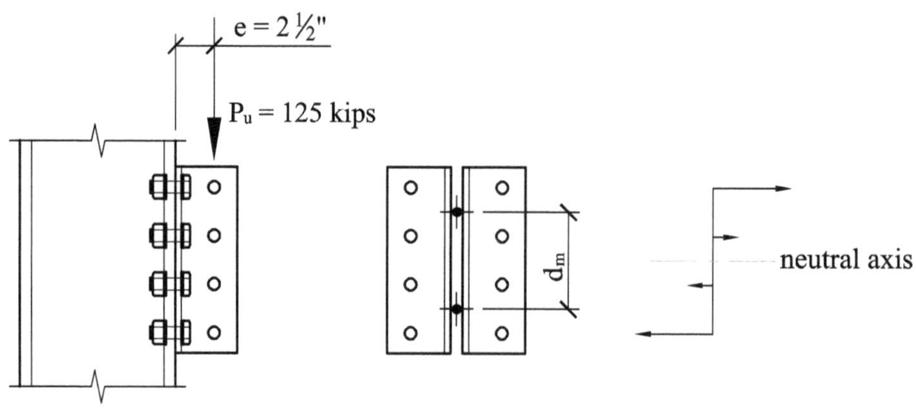

FIGURE 9-25 Free body diagram - Case II

$\phi R_n = \phi F'_{nt} A_b = (0.75)(38.3)(0.442)$

$\phi R_n =$ **12.7 kips** $< r_t = 13$ kips.

The connection is found to be slightly *inadequate* using the Case II design approach (i.e., the simplified method), which is not surprising since the Case II method is a simplified method and is more conservative than the Case I method.

9-7 PRYING ACTION: BOLTS IN TENSION

Prying action is the increase in tension in the bolts supporting a tension member due to the flexibility of the connection elements. There are several types of connections where prying action on the connecting element results in increased tension in the bolts (see Figure 9-26). The connecting element is usually an angle or T-shape, and the critical design factor is the thickness of the outstanding leg of the connecting element that resists the bending or prying action due to the tension force in the connection.

Consider the hanger connection shown in Figure 9-27. If the legs of the connecting element had sufficient stiffness, then the sum of the tension force in the bolts would be equal to the

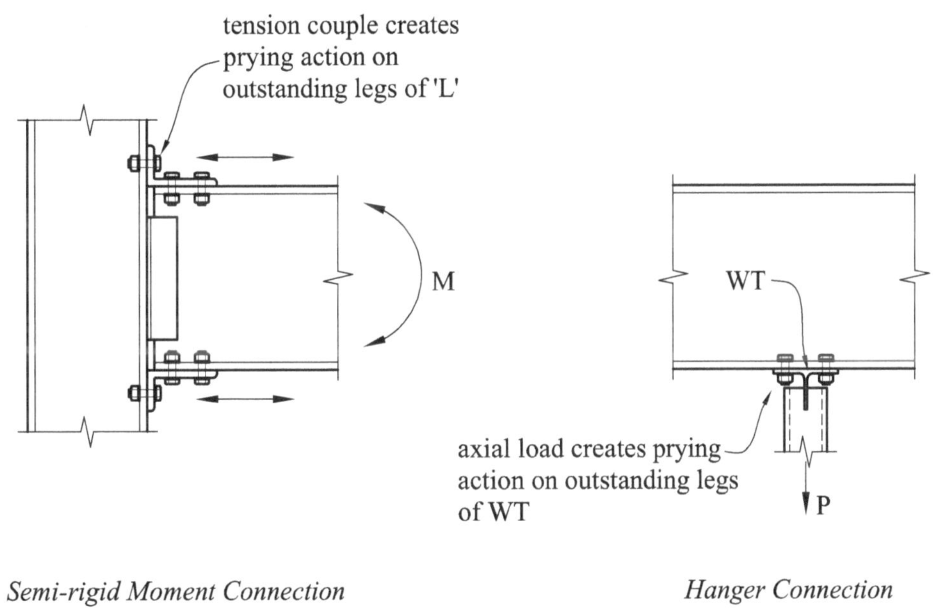

FIGURE 9-26 Connections with prying action

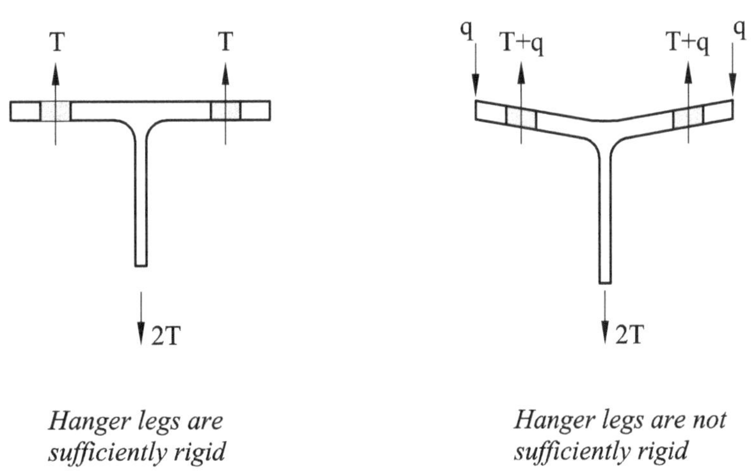

FIGURE 9-27 Prying action on a hanger

applied load, T; however, due to the lack of rigidity of the connecting elements and the resulting prying action, the bolts will experience a total force higher than T. The effect of this prying action results in a compression force near the tip of the outstanding leg of the connecting element, which increases the tension force in the bolt from T to $T + q$.

No Prying Forces

The derivation of the following prying action equations is discussed in Ref. [6] and [7]. Considering no prying forces in the connection will result in the smallest bolt diameter and the largest connecting element thickness. Using a more simplified approach, the minimum required thickness of the connecting element that ensures ***no prying action*** can be obtained as follows [2]:

$$t_{min} = \sqrt{\frac{4T_u b'}{\phi_b p F_u}} = \sqrt{\frac{4.44 T_u b'}{p F_u}} \quad \text{for LRFD} \qquad (9\text{-}30a)$$

$$t_{min} = \sqrt{\frac{4\Omega_b T_a b'}{p F_u}} = \sqrt{\frac{6.68 T_a b'}{p F_u}} \quad \text{for ASD} \qquad (9\text{-}30b)$$

Where,

T_u = factored applied tension force per bolt, *kips*

T_a = Unfactored applied tension force per bolt, *kips*

ϕ_b = 0.9

Ω_b = 1.67

$$b' = b - \frac{d_b}{2}, \qquad (9\text{-}31)$$

b = Distance from the bolt centerline to the face of the tee stem for a T-shape, in. (see Figure 9-28a),

= Distance from the bolt centerline to the angle-leg centerline for an L-shape, in. (see Figure 9-28b),

d_b = Bolt diameter, in.,

p = Tributary length per bolt from yield line theory should be less than g (see Figure 9-28c); and

- For typical *interior* bolts, we can conservatively assume $p = 3.5b \le s$, where s = bolt spacing in inches.
- For *edge* bolts, $p = 1.75b$ + edge distance $\le s/2$ + edge distance,

F_u = Minimum tensile strength of the connecting element.

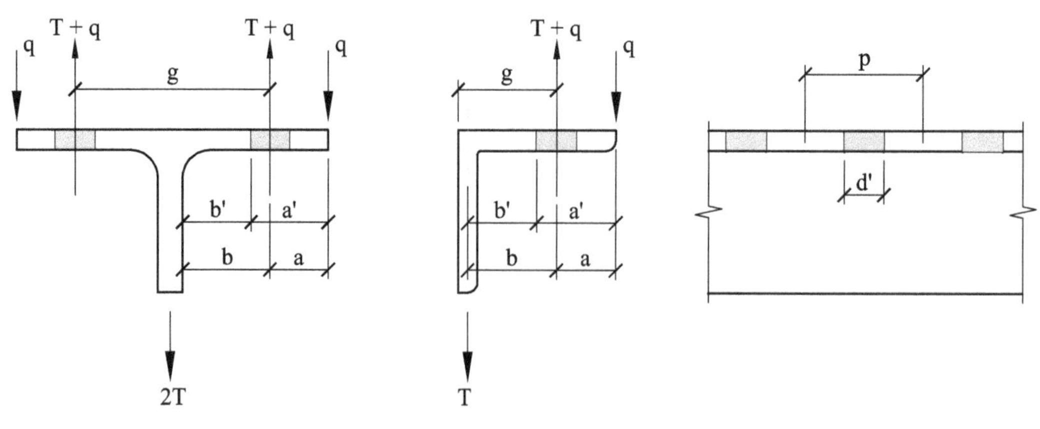

a. WT *b. single angle* *c. side view*

FIGURE 9-28 Prying action variables

In equation 9-30a or 9-30b, the additional force in the bolt due to prying action is nearly zero ($q \approx 0$). If the thickness of the connecting element determined from equation (9-30a or 9-30b) is reasonable and the tension capacity of the bolt is not exceeded, then no further design checks for prying action need to be made.

Prying Forces Considered

A more complex but less conservative approach that allows a thickness of the connecting elements *smaller* than those given in equations (9-30a) or (9-30b) to be used; assumes a value of q that is greater than zero, thus magnifying the tensile force in the bolt. This results in the smallest thickness for the connecting element.

A preliminary connecting element thickness, t_{pr}, for members subject to prying action can be selected from *AISCM* Table 15-1 for F_u = 65 ksi for the connecting elements. The available tensile strength per inch from these tables is multiplied by the bolt tributary length, p, to obtain the allowable tensile strength.

In-depth Analysis Considering Prying Action

We will now proceed with the in-depth analysis considering prying action using equations (9-32a) through (9-37). After the preliminary thickness, t_{pr} , has been selected from *AISCM* Table 15-1, the required thickness, t_{min} , **considering prying action**, is determined using the following equations (see also *AISCM* equations (9-37a) and (9-37b)):

$$t_{min} = \sqrt{\frac{4T_u b'}{\phi_b pF_u(1+\delta\alpha')}} = \sqrt{\frac{4.44T_u b'}{pF_u(1+\delta\alpha')}} \quad \text{for LRFD} \tag{9-32a}$$

$$t_{min} = \sqrt{\frac{4\Omega_b T_a b'}{pF_u(1+\delta\alpha')}} = \sqrt{\frac{6.68T_a b'}{pF_u(1+\delta\alpha')}} \quad \text{for ASD} \tag{9-32b}$$

Where,

ϕ_b = 0.9

Ω_b = 1.67

$$\delta = 1 - \frac{d'}{p} \tag{9-33}$$

d' = Hole width along the length of the fitting (usually the bolt hole diameter) as shown in Figure 9-28c.

α' = 1.0 for $\beta \geq 1$,

$$= \frac{1}{\delta}\left(\frac{\beta}{1-\beta}\right) \leq 1.0 \quad \text{for } \beta < 1, \tag{9-34}$$

$$\beta = \frac{1}{\rho}\left(\frac{T_c}{T_u} - 1\right) \quad \text{for LRFD} \tag{9-35a}$$

$$= \frac{1}{\rho}\left(\frac{T_{c(ASD)}}{T_a} - 1\right) \quad \text{for ASD} \tag{9-35b}$$

$$a' = \left(a + \frac{d_b}{2}\right) \leq \left(1.25b + \frac{d_b}{2}\right) \tag{9-36}$$

$$\rho = \frac{b'}{a'}, \tag{9-37}$$

a = Distance from the bolt centerline to the edge of the connecting element, and

T_c = Available tension strength per bolt for LRFD = $\phi_t r_n$

$T_{c(ASD)}$ = Available tension strength per bolt for ASD = $\frac{r_n}{\Omega_t}$

T_u = Factored applied tension load per bolt for LRFD

T_a = Unfactored applied tension load per bolt for ASD

ϕ_t = 0.75 (see Section 9-4)

Ω_t = 2.0 (see Section 9-4)

- If the final t_{min} obtained from equations (9-32a) or (9-32b) is *less than* the preliminary thickness, t_{pr}, selected for the connecting element, the connection member selected is adequate, and therefore, use $t = t_{pr}$.
- If the final t_{min} obtained from equations (9-32a) or (9-32b) is *greater than* the preliminary thickness, t_{pr}, selected for the connecting element, the leg thickness of the connecting element will have to be increased up to the thickness value, t_{min}, obtained from equations (9-32a) or (9-32b), or the geometry of the connection will have to be changed.

EXAMPLE 9-8

Prying Action Considered (LRFD)

Determine the required angle size for the moment connection shown in Figures 9-29 and 9-30. The bolts are $^3/_4$-in.-diameter A490 (i.e., Group 150) in standard bolt holes, and the steel is ASTM Grade 50.

Solution

A trial size of $L7 \times 4 \times \frac{3}{4}$ LLH (i.e., long leg horizontal) will be selected, therefore,

t_{pr} = ¾ in.

d for W18 × 40 = 17.9 in.

The total factored tension force due to the applied factored moment is calculated as follows:

$$T_u = \frac{M_u}{d} = \frac{(60)(12)}{17.9 \text{ in.}} = 40.2 \text{ kips.}$$

Since there are two bolts to resist the tension in the angle at the column, the factored tension force in each bolt is $\frac{40.2 \text{ kips}}{2} = 20.1$ kips

Therefore, T_u per bolt = 20.1 kips

Using the geometry from Figure 9-30 in conjunction with Figure 9-29, the minimum angle thickness to ensure that there is no prying action can be calculated using the simplified approach (see equation (9-30a)):

$a = 1.5$ in.

$a + g = 4$ in. (angle short leg)

Angle thickness = ¾ in.

$$b = 4 \text{ in.} - 1.5 \text{ in.} - \left(\frac{1}{2}\right)\left(\frac{3}{4} \text{ in.}\right) = 2.12 \text{ in.}$$

$$b' = b - \frac{d_b}{2} = 2.12 \text{ in.} - \frac{0.75 \text{ in.}}{2} = 1.75 \text{ in.}$$

Since each of the two bolts is an edge bolt, the edge distance = (7 in. – 4 in.)/2 = 1.5 in.

For edge bolts,

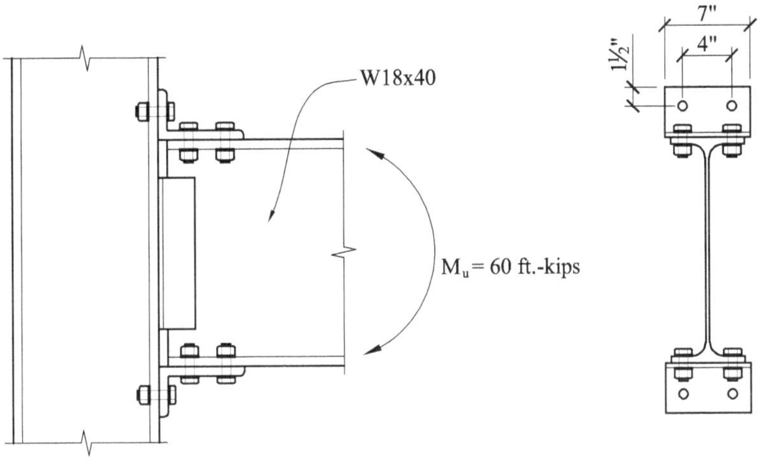

FIGURE 9-29 Connection detail for Example 9-8

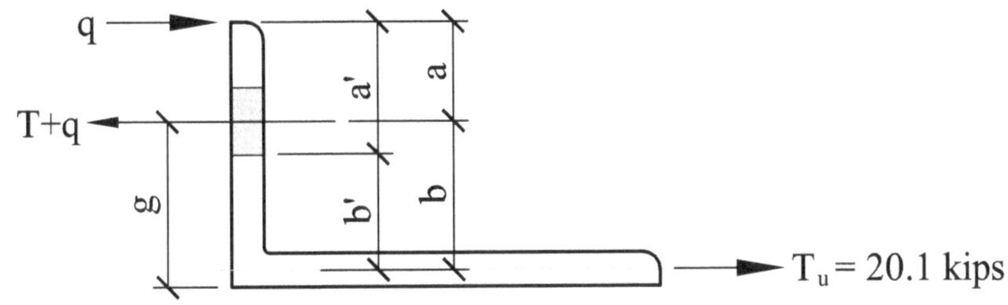

FIGURE 9-30 Prying action variables for Example 9-8

$$p = 1.75b + \text{edge distance} \leq s/2 + \text{edge distance}$$
$$= 1.75(2.12 \text{ in.}) + 1.5 \text{ in.} \leq 4 \text{ in.}/2 + 1.5 \text{in.}$$
$$= 5.21 \text{in.} \leq 3.5 \text{in.}$$

Therefore, use p = 3.5 in.

Using equation (9-30a), the required minimum connection element thickness to ensure no prying action is

$$t_{min} = \sqrt{\frac{4.44T_u b'}{pF_u}} = \sqrt{\frac{(4.44)(20.1)(1.75)}{(3.5)(58)}} = 0.88 \text{ in.}$$

The minimum angle thickness is found to be greater than the thickness of the selected angle, but recall that this approach is conservative. We will also consider a less conservative approach for comparison purposes.

Using the preliminary hanger selection tables (*AISCM* Table 15-1, with F_u = 65 ksi for Grade 50 steel), the required strength is obtained as follows:

The applied factored tension force per unit length per bolt is calculated as

$$\frac{T_u}{p} = \frac{20.1}{3.5} = 5.74 \text{ kips/in.}$$

With b = 2.12 in., we find that the required angle thickness from *AISCM* Table 15-1 is approximately ½ in. for b = 2¼ in., yielding a tension capacity per bolt of **6.50 kips per linear inch**, which is greater than the applied factored tension force per linear inch of 5.74 kips/inch. Based on this required thickness, the L7 × 4 × ¾ angle selected is adequate.

In-depth analysis considering prying action

We will now proceed with the in-depth analysis considering prying action using equations (9-32a) through (9-37).

The minimum angle thickness, from equation (9-32a), is determined as follows:

$$a' = \left(a + \frac{d_b}{2}\right) \leq \left(1.25b + \frac{d_b}{2}\right)$$

$$= \left(1.5 + \frac{0.75}{2}\right) \leq \left((1.25)(2.12) + \frac{0.75}{2}\right)$$

$$= 1.88 \text{ in.} < 3.03 \text{ in.}$$

Therefore, $a' = 1.88$ in.

$$\rho = \frac{b'}{a'} = \frac{1.75}{1.88} = 0.933$$

F_{nt} for A490N bolts or Group 150 bolts = 113 ksi (see Table 9-3)

A_b (¾ in. diameter bolt) = 0.442 in.^{2}

$T_c = \phi_t r_{nt} = \phi_t F_{nt} A_b = (0.75)(113)(0.442) = 37.4$ kips per bolt

(agrees with $\phi_t r_{nt}$ from *AISCM* Table 7-2 for Group150, 3/4in.diameter bolts)

Check that $T_u \leq T_c$, otherwise either increase the number of bolts or use stronger bolts.

$T_u = 20.1$ *kips* $\leq T_c = 37.4$ *kips*, OK.

$$\beta = \frac{1}{\rho}\left(\frac{T_c}{T_u} - 1\right) = \frac{1}{0.933}\left(\frac{37.4}{20.1} - 1\right) = 0.922 < 1.0$$

d' = standard bolt hole width for ¾ in. diameter bolt (see Figure 9-3)

$= \frac{3}{4}$ in $+ \frac{1}{16}$ in $= \frac{13}{16}$ in .

p = 3.5 in. (obtained previously)

$$\delta = 1 - \frac{d'}{p} = 1 - \frac{13/16 \text{ in.}}{3.5 \text{ in.}} = 0.768$$

$$\alpha' = \frac{1}{\delta}\left(\frac{\beta}{1-\beta}\right) \leq 1.0 \text{ for } \beta < 1$$

$$\alpha' = \frac{1}{0.768}\left(\frac{0.922}{1-0.922}\right) \leq 1.0$$

$$\alpha' = 15.39 \leq 1.0$$

Therefore, $\alpha' = 1.0$

From equation (9-32a), the required minimum thickness of the angle leg is

$$t_{\min} = \sqrt{\frac{4.44T_u b'}{pF_u(1+\delta\alpha')}} = \sqrt{\frac{(4.44)(20.1)(1.75)}{(3.5)(58)[1+(0.768)(1.0)]}}$$

$= \mathbf{0.66}$ **in.** $< t_{pr} = \frac{3}{4}$ in. for L7x4x$\frac{3}{4}$ **OK.**

Therefore, L7 × 4 × $\frac{3}{4}$ angle size selected is adequate.

9-8 FRAMED BEAM CONNECTIONS

Framed beam connections are the most common connections used in steel buildings. The three most common variations are beam-to-beam (or girder), beam-to-column, and beam bearing. Figure 9-31 illustrates some of these connections. Any of the connecting elements can be bolted

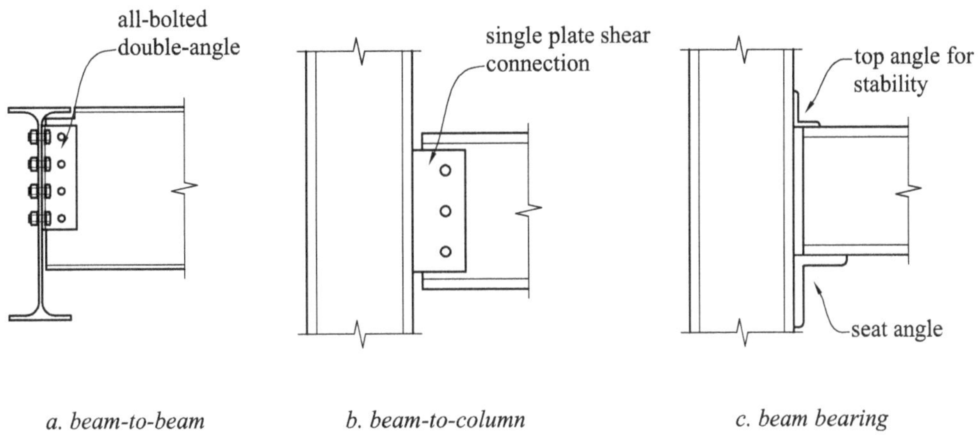

FIGURE 9-31 Common types of framed beam connections

or welded; however, bolted connections are generally preferred because they are the most economical for shear connections; the labor cost and time for welding are usually greater than for bolted connections, and bolted connections are also easier to inspect. In this section, we will present a general overview of the analysis and design process for bolted framed-beam connections; welded connections will be covered in Chapter 10.

With the exception of welds, most of the failure modes that occur in framed beam connections have been discussed previously in this chapter. In lieu of performing each of these design checks for every type of connection, it is more practical to use the design aids in the *AISCM*, and this is what is commonly done in design practice. The tabulated capacities for eccentrically loaded bolts and welds in framed beam connections from the design aids in *AISCM* Tables 10-1 through 10-12 were obtained using the instantaneous center method. Table 9-4 summarizes the various framed-beam connection design and detailing aids in the *AISCM*. These connections are intended to provide vertical reaction to the supported beams or girders, and they must also have sufficient rotational ductility to accommodate the rotation expected from the structural analysis of the structure (see AISC Specifications Sections B3.4a and J1.2). [1]. For the seated beam connections (*AISCM* Tables 10-5 through 10-8), a clip angle connecting the supported beam to the supporting column should be provided at the top flange of the supported beam to ensure lateral support for the beam top flange. In summary, the two general types of framed beam or girder connections are shear connections and moment connections:

1. **Shear Connections:**
    - The most common connections used in steel buildings to resist shear are framed beam connections (i.e., shear connections), similar to those shown in Figure 9-31.
    - Examples of Shear Connections (see Table 9-4):
        - Beam-to-girder connections (i.e., beam to girder web)
        - Beam-to-column connections (i.e., beam to column web)
        - Girder-to-column connections (i.e., girder to column flange)
        - Beam bearing connections
    - Any of the connecting elements can be bolted or welded.
    - These connections are intended to provide a vertical reaction to the supported beams or girders.
    - Since these are shear connections, they must provide a rotational degree of freedom, therefore, the connection must have sufficient rotational ductility to accommodate

the rotation expected from the structural analysis. (see AISC Specification Sections B3.4a and J1.2)

- Bolted connections are easier to inspect. The labor cost and construction time for welded connections are usually greater than for bolted connections.
- Most of the failure modes (e.g., tearout failure, bearing failure, or direct shear failure, etc.) that occur in framed beam connections have already been discussed in Chapter 9 and will also be discussed in Chapter 10 for welded connections.
- For seated beam connections, provide lateral support to the supported beam by connecting the top flange of the supported beam with a clip angle to the supporting column.
- Framed beam connections are easily selected or designed using the *AISCM* Design Aids (see Table 9-4).

TABLE 9-4 Shear Connection Design Aids in the *AISCM*

AISCM table	Connection types
10-1a	All-bolted, double-angle connections: Available Angle Strength based on shear yielding, block shear, and shear rupture (F_y = 50 ksi for angles)
10-1b	All-bolted, double-angle connections—Available Shear Strength at Bolt Holes—bolt shear, bolt bearing, and tearout
10-1c	All-bolted, double-angle connections—Coped Beam web Available Shear Strength in block shear and shear rupture
10-2	Available Weld Strength of Bolted/Welded, Double-Angle connections
10-3	Available Weld Strength of All-welded, Double-Angle connections
10-4a	Shear end-plate connections—Bolted/welded—Available End-Plate Strength
10-4b	Shear end-plate connections—Bolted/welded—Available Shear Transfer Strength at Bolt Holes
10-4c	Shear end-plate connections—Bolted/welded—Weld and Beam web Available Strength
10-5	All-bolted unstiffened seated connections
10-6	All-welded unstiffened seated connections
10-7	Stiffened seated connections – Bolted stiffening elements
10-8	Stiffened seated connections – Welded stiffening elements
10-10a	Single-plate (or shear tab) connections—Plate and Weld Available Strength
10-10b	Single-plate (or shear tab) connections—Available Shear Transfer Strength at Bolt Holes
10-10c	Single-plate (or shear tab) connections—Coped Beam Web Available Shear Strength per inch thickness, kip/in.
10-11	All-bolted single-angle connections
10-12a	Bolted/welded single-angle connections—Angle and Weld Available Strength
10-12b	Bolted/welded single-angle connections—Available Shear Transfer Strength at Bolt Holes
10-12c	Bolted/welded single-angle connections—Coped Beam Web Available Shear Strength per inch thickness, kip/in.

- To provide **adequate lateral stability** in framed beam connections with single plates or angles, the length of the plate or angle should be **at least** equal to **one-half** of the depth, T, of the wide-flange beam (note that $T = d - 2k$). The depth and length of cope in the connected beam will have to be checked, but it generally does not control the design unless the amount of cope is excessive; see Chapter 11 for further discussion of coped beams.
- Shear connections are used in braced frames.

2. **Moment Connections:**
   - Used in Moment Frames or Moment-resisting Frames (see Chapter 11)
   - There are two main types of moment connections:
     - Partially restrained (**PR**) moment connections
     - Fully restrained (**FR**) moment connections
   - For PR or Flexible Moment Connections (FMC), the seismic response factor,R, which is a measure of the ductility of the lateral force resisting system (previously discussed in Chapter 3) must be ≤ 3.0.
   - Gravity load effects do *not* have to be included in the structural analysis of partially restrained or flexible moment frames.
   - For FR or Fully Restrained Moment Connections, the seismic response modification factor, $R \geq 3.0$.
   - Gravity load effects *must* be included in the structural analysis of fully restrained moment frames.

Some examples of beam-to-column connections are shown in Figure 9-31d through 9-31g.

FIGURE 9-31d Beam-to-column shear connections

FIGURE 9-31e Beam and Girder-to-column moment connections (fully restrained)

FIGURE 9-31f Girder-to-column moment connections (fully restrained)

All-Bolted Double-Angle Shear Connections

Each *AISCM* framed beam connection design table noted in Table 9-4 is a compilation of several tables for nearly every common. All-bolted double-angle connections are shown in Figures 9.31g and 9.32. For this type of connection, the failure modes that must be considered in the design of All-bolted double angle connections include:

- Shear failure in the bolts
- Bearing failure in the bolts
- Shear yielding of the connecting angle.
- Tearout failure of the connecting angle.
- Block shear failure of the connecting angle.
- Shear yielding of the supported beam web
- Block shear failure of the supported beam web

AISCM Tables 10-1a, 10-1b, and 10-1c should be used for the design of these types of connections. Due to clearance issues, it is not practical to use double-angle connections to a web of W8 or smaller columns. In practice, when double-angle connections are specified, W10, W12, and W14 columns are more suitable—to avoid the bolting and welding clearance issues with W8 columns.

FIGURE 9-31g All-bolted double angle beam-to-girder shear connection

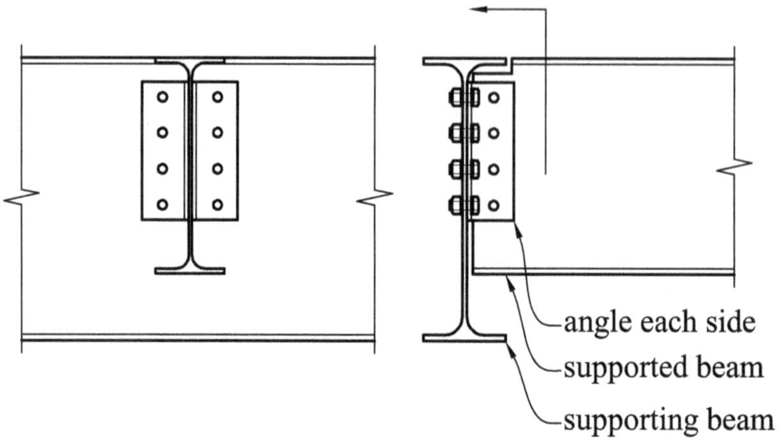

FIGURE 9-32 All-bolted double angle connection

EXAMPLE 9-9

All-Bolted Double-Angle Connection

Determine the shear capacity of the connection shown in Figure 9-33 using *AISCM* Tables 10-1a, b, and c. The W18 × 35 beam frames into a W24 × 55 girder. The bolts are ¾ in. diameter ASTM F3125 Grade F1852N bolts in standard bolt holes. The angles are ASTM A572 Grade 50, and the beam is ASTM A992, Grade 50 steel.

Solution

From *AISCM* Table 10-1a, with the number of horizontal rows of bolt, $N = 4$ for $^3/_4$-in.-diameter Group 120 bolts with thread condition "*N*" and an angle thickness of $^3/_8$ in., the design strength is found to be 170 kips. The capacity of the bolt bearings on the beam and the block shear on the beam also must be checked.

Bolt bearing and tearout strength:

With reference to Figure 9-34, the L_{eh} value of 1¾ in. will be reduced by $^1/_4$ in. to allow for fabrication tolerances. The capacity of the bolts bearing on the beam web is

$$L_{c1} = 2 \text{ in.} - 0.5\left(\frac{3}{4} + \frac{1}{8}\right) = 1.56 \text{ in.}$$

The clear distance between the bolts is:

$$L_{c2} = L_{c3} = L_{c4} = 3 \text{ in.} - \left(\frac{3}{4} + \frac{1}{8}\right) = 2.12 \text{ in.}$$

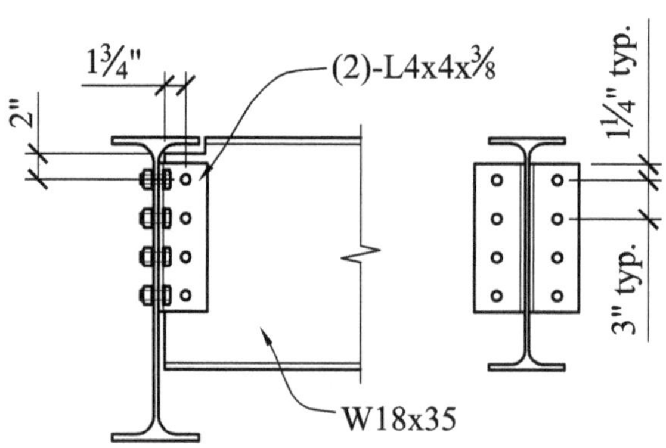

FIGURE 9-33 Connection detail for Example 9-9

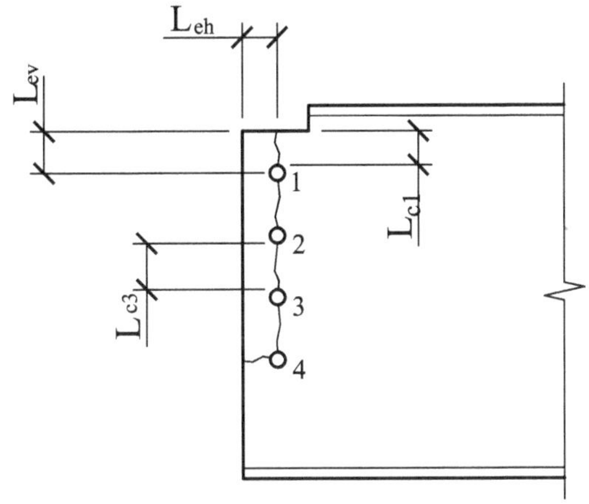

FIGURE 9-34 Bolt bearing and block shear in beam web

For W18 × 35 ASTM A992 Grade 50, t_w = 0.3 in.; F_u = 65 ksi
Diameter of bolt = ¾ in.
For bolt 1 (i.e., the bolt closest to the cope),

$$\phi R_n = \phi 1.2 L_c t F_u \leq \phi 2.4 d t F_u$$
$$= (0.75)(1.2)(1.56 \text{ in.})(0.3 \text{ in.})(65 \text{ ksi})$$
$$\leq (0.75)(2.4)(0.75 \text{ in.})(0.3 \text{ in.})(65 \text{ ksi})$$
$$= 27.4 \text{ kips} \leq 26.3 \text{ kips}$$

Therefore, $\phi R_{n1} = 26.3$ kips.

For bolts 2, 3, and 4,

$$\phi R_n = (0.75)(1.2)(2.12 \text{ in.})(0.3 \text{ in.})(65 \text{ ksi})$$
$$\leq (0.75)(2.4)(0.75 \text{ in.})(0.3 \text{ in.})(65 \text{ ksi})$$
$$= 37.2 \text{ kips} \leq 26.3 \text{ kips}$$

Therefore, $\phi R_{n2} = \phi R_{n3} = \phi R_{n4} = 26.3$ kips.
The total strength of all the bolts in bearing is
$\phi R_n = 26.3 \text{ kips} + (3)(26.3 \text{ kips}) =$ **105 kips.**

Note that since the web thickness for the W24 × 55 is 0.395 in., the bolt bearing on the supported W18 × 35 (with t_w = 0.3 in.) will control.

Block shear in the W18 × 35:
Number of bolt holes on the shear plane = 3.5 holes
Number of bolt holes on the tension plane = 0.5 hole

$$A_{gv} = [2 \text{ in.} + (3)(3 \text{ in.})](0.3 \text{ in.}) = 3.3 \text{ in.}^2$$

$$A_{nv} = \left[2 \text{ in.} + (3)(3 \text{ in.}) - 3.5\left(\frac{3}{4} + \frac{1}{8}\right)\right] 0.3 \text{ in.} = 2.38 \text{ in.}^2$$

$$A_{gt} = (L_{eh} - tolerance) = (1.75 \text{ in.} - 0.25 \text{ in.})(0.3 \text{ in.}) = 0.45 \text{ in.}^2$$

$$A_{nt} = \left[(L_{eh} - tolerance) - 0.5\left(\frac{3}{4} + \frac{1}{8}\right)\right](0.3 \text{ in.})$$
$$= \left[(1.75 \text{ in.} - 0.25 \text{ in.}) - 0.5\left(\frac{3}{4} + \frac{1}{8}\right)\right](0.3 \text{ in.}) = 0.318 \text{ in.}^2$$

Using equation (4-11a), to determine the block shear strength,

$$\phi R_n = \phi\left(0.6F_u A_{nv} + U_{bs} F_u A_{nt}\right) \leq \phi\left(0.6F_y A_{gv} + U_{bs} F_u A_{nt}\right)$$

$$= 0.75\left[(0.6)(65)(2.38) + (1.0)(65)(0.318)\right]$$

$$\leq 0.75\left[(0.6)(50)(3.3) + (1.0)(65)(0.318)\right]$$

$$= 85.1 \text{ kips} \leq 89.8 \text{ kips}$$

Therefore, $\phi R_n =$ **85.1 kips** $<$ 105 kips;

That is, the block shear capacity of the supported W18x35 beam (85.1 kips) controls over bolt bearing or tearout (105 kips) and the shear capacity of the double angle connection elements from AISCM Table 10 - 1a (170 kips).

Therefore, the design shear strength of this connection, ϕR_n = **85.1 kips.**

Alternative Method: Using *AISCM* Tables 10-1a, 10-1b, and 10-1c:

The geometric and design parameters of this connection are

L_{eh} = 1.75 in.

$L_{ev,t}$ = 2 in.

$L_{ev,b}$ >> 2 in.

n = 4

Bolt diameter = ¾-in.

Angle thickness = 3/8-in.

t_w for W18 × 35 = 0.3-in.

***AISCM* Table 10-1a:**

Available angle strength = 170 kips

***AISCM* Table 10-1b:**

Available bearing and tearout strength at *non-edge* bolt = 87.8 kip/in. × 0.3 in. × (n-1) = = 87.7 kip/in. × 0.3 in. × (4-1) = 79 kips

Available bearing and tearout strength at *edge* bolt = 87.8 kip/in. × 0.3 in. × 1 bolt = 26.3 kips

Therefore, total bearing and tearout strength = 79 + 26.3 = **105.3 kips.**

Available bolt shear strength (Group 120 bolts) = 17.9 kip (2)(4) = **143 kips.**

***AISCM* Tables 10-1c:**

With L_{eh} = 1.75 in., $L_{ev,t}$ = 2 in., and $L_{ev,b}$ >> 2 in., obtain the individual strength contributions from the top edge hole, the center hole, and the bottom edge hole from *AISCM* Table 10-1c to the coped beam web available shear strength for the connection with ¾-in. diameter bolts spaced at 3 in. on centers:

Top edge hole ($L_{ev,t}$ = 2 in.) : 31.1* kip/in. × 0.3 in. = 9.3 kips

Center holes (bolt vertical spacing = 3 in.) : 62.2 kip/in. × (n – 1) × 0.3 in. = 62.2 kip/in. × 3 × 0.3 in. = 56 kips

Bottom edge hole ($L_{ev,b}$ >> 2 in.): 51.8 kip/in. × 0.3 in. = 15.6 kips

Total coped beam web available shear strength for the double angle connection = 9.3* + 56 + 15.6 = **80.9 kips.**

*Note that *AISCM* Table 10-1c does not provide values for $L_{ev,t}$ > 1½-in., but calculation of the actual strength with $L_{ev,t}$ = 2-in. gives 45.7 kip/in., not the 31.1 kip/in. from *AISCM* Table 10-1c that corresponds to $L_{ev,t}$ = 1½-in. With a strength for the top hole of 45.7 kip/in, the total strength for the top edge hole becomes 45.7 kip/in. × 0.3 in. or 13.7 kips, giving a total shear strength for the connection of **85.3 kips** (i.e., 13.7 kips + 56 kips + 15.6 kips), which is less than the strength of the angle (170 kips), the direct shear strength of the bolts (143 kips), and the bearing and tearout strength of the W18 × 35 web (105.3 kips), and the capacity compares well with the previously calculated value of ϕR_n = **85.1 kips** for the connection.

Bolted/Welded Shear End-Plate Connections

A bolted/welded shear end-plate connection is shown in Figure 9-35. For this type of connection, the failure modes that must be considered are

- Shear in the bolts,
- Bearing of the bolts,
- Direct shear and block shear in the end plate, and
- Shear in the weld connecting the supported beam to the end plate (welds are covered in Chapter 10).

AISCM Tables 10-4a, 10-4b, and 10-4c should be used for the design of these types of connections.

EXAMPLE 9-10

Bolted/Welded Shear End-Plate Connection

Determine the capacity of the connection shown in Figure 9-36 using *AISCM* Tables 10-4a, 10-4b, and 10-4c. The bolts are $^7/_8$-in.-diameter Group 120 (A325N) in standard bolt holes. The end plate is ASTM A572 Grade 50, and the beam and column are ASTM A992, Grade 50 steel.

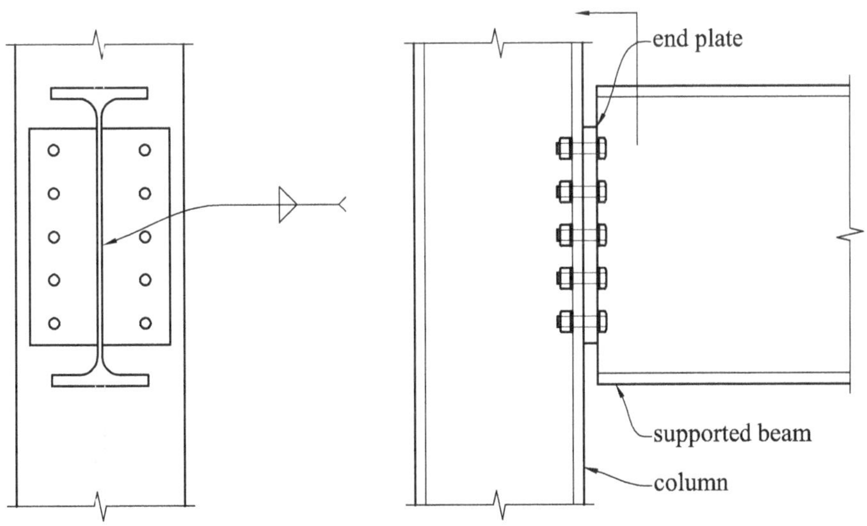

FIGURE 9-35 Bolted/welded shear end-plate connections

Solution

AISCM Tables 10-4a, 10-4b, and 10-4c are the design aids that will be used to calculate the connection strength.

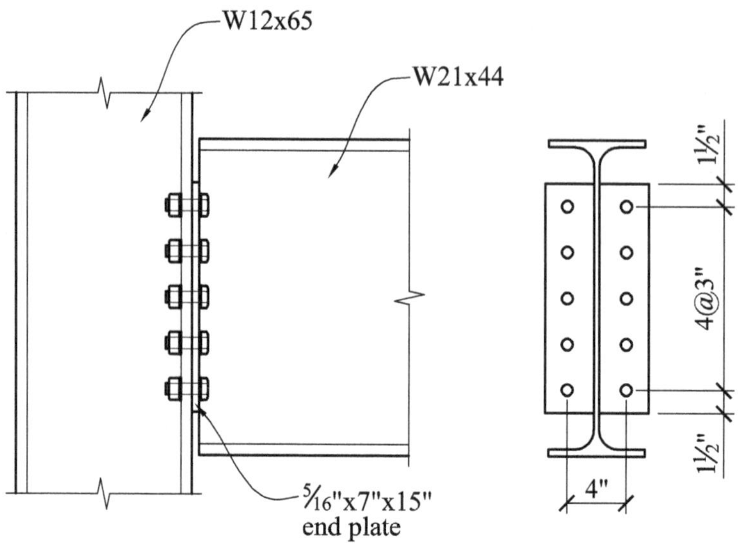

FIGURE 9-36 Connections detail for Example 9-10

***AISCM* Table 10-4a:**

The design strength of the connection, using $^7/_8$-in.-diameter ASTM F3125 Grade A325N Group 120 bolts in STD holes, with number of horizontal rows of bolts, n = 5, with end plate thickness, t_p = 5/16 in.,

ϕR_n = **174 kips**

***AISCM* Table 10-4b:**

Bolt direct shear strength: 10, 7/8-in. diameter bolts in STD holes loaded in single shear, ϕR_n = 10 bolts × 24.3 kip/bolt = **243 kips**

Bearing and Tearout Strength (the 5/16-in thick end plate will control over the 0.605-in thick flange of the W12 × 65 column).

Edge bolt (L_{ev} = 1½-in.; 2, 7/8-in. diameter bolt in STD holes; 5/16-in. end plate): ϕR_n = 14.3 kips × 2 bolts = 28.6 kips

Non-edge bolt (8, 7/8-in. diameter bolt in STD holes, 5/16-in. end plate): ϕR_n = 32 kips × 8 bolts = 256 kips

Total bearing and tearout strength = 28.6 + 256 = **285 kips**

Note: A *non-edge* **bolt** is one that is adjacent to another bolt in the direction of the applied force, while an **edge bolt** is one that is adjacent to the edge of the plate or connected part in the direction of the applied force. Hence, there are two *edge bolts* and 8 *non-edge bolts* in Figure 9-36.

***AISCM* Table 10-4c (Weld and Beam web Available Strength):**

The 5/16-in (i.e., 0.3125 in.) thick end plate is welded with a fillet weld to the web of the W21 × 44 beam, which has a web thickness of t_w = 0.35 in., so the smaller plate thickness controls. The minimum and maximum fillet weld thickness for a 5/16-in plate (see Table 10-2) are 3/16-in and ¼ in., respectively. Using a 3/16-in. fillet weld size and number of bolt rows, n = 5 in *AISCM* Table 10-4c, ϕR_n = **118 kips**

The shear capacity of the end-plate connection is the smallest of the following strengths: 174 kips, 243 kips, 285 kips, and **118 kips**.

Therefore, the shear capacity of the end-plate connection, ϕR_n = **118 kips**.`

The strength of the weld will be discussed in Chapter 10.

As an exercise, we will now verify the bearing and tearout capacity obtained above from *AISCM* Table 10-4b by calculating the bearing and tearout strength using equations (9-2) and (9-4):

The bearing strength is the same for each non-edge bolt.

$L_c = 3 \text{ in.} - (\tfrac{7}{8} + \tfrac{1}{8}) = 2.0 \text{ in.}$

$$\phi R_n = \phi 1.2 L_c t F_u \leq \phi 2.4 d t F_u$$
$$= (0.75)(1.2)(2.0 \text{ in.})(0.3125 \text{ in.})(65 \text{ ksi})$$
$$\leq (0.75)(2.4)(0.875 \text{ in.})(0.3125 \text{ in.})(65 \text{ ksi})$$
$$= 36.5 \text{ kips} \leq 32 \text{ kips}$$

Therefore, $\phi R_n = 32$ kips per bolt (for all the *non-edge* bolts)

The L_c for the edge bolts in the direction of the applied force is 1.5 in. - $\tfrac{1}{2}(\tfrac{7}{8} + \tfrac{1}{8})$ = 1.0 in. Therefore the design strength per *edge bolt* using equations (9-2) and (9-4) = 18.3 kips

With 8 *non-edge bolts* and 2 edge bolts, the total strength of all the bolts in bearing and tearout is

ϕR_n = (8)(32 kips) + (2)(18.3 kips) = 293 kips.

This value is very close to the 286 kips obtained above from *AISCM* Table 10-4b. The small discrepancy is because the actual edge distance in Figure 9-36 is 1.5 in., whereas the strength values in *AISCM* Table 10-4b were derived assuming 1.25 in. edge distance.

Single-Plate or Shear Tab Shear Connections

A single-plate shear connection (also sometimes called a shear tab connection) is shown in Figure 9-37. The plate is fillet-welded to the supporting column and bolted to the supported beam or girder. The failure modes for this connection are

- Shear in the bolts
- Bearing and tearout of the bolts
- Shear yielding and block shear in the plate,
- Failure of the weld to the connecting elements, and
- Shear yielding and block shear in the beam.

AISCM Tables 10-10a, 10-10b, and 10-10c are design aids that can be used for the design of single plate or shear tab connections. The dimensional limitations for this connection type

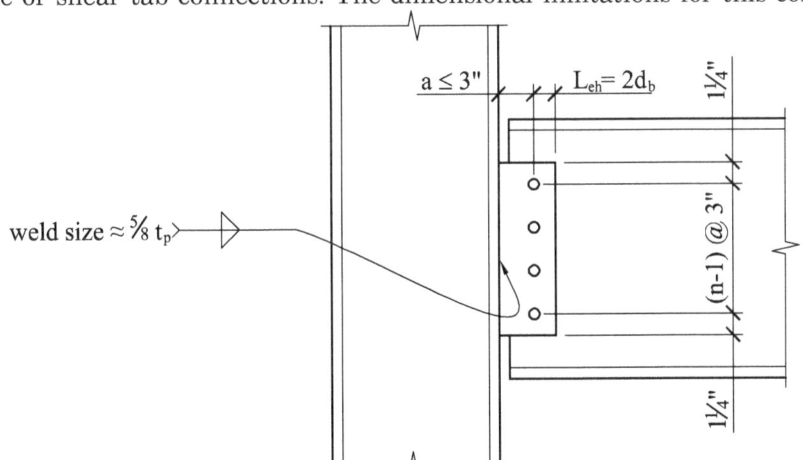

FIGURE 9-37 Single-plate connections

when using *AISCM* Table 10-10a or 10-10b, are noted in Figure 9-37. To ensure a simple pinned connection, thus allowing rotation at the beam supports, the *AISCM* [2] limits the shear plate thickness, t_p, as follows:

- For connections with 2 to 5 bolts in one vertical row in standard bolt holes, the maximum thickness shear plate in inches, $t_p = \frac{d_b}{2} + \frac{1}{16}$, where d_b is the bolt diameter (see *AISCM* Table 10-9).
- For connections with 6 to 12 bolts in one vertical row, the maximum shear plate thickness, $t_p = \frac{d_b}{2} - \frac{1}{16}$.

For beam or girder-to-HSS column connections, shear tabs can also be used where the plate is fillet-welded to the wall of the HSS column. Some engineers also use a through-plate shear connection for beam or girder-to-HSS column connections, though this detail is expensive and not widely recommended.

EXAMPLE 9-11

Single-Plate Shear Connection

Determine the capacity of the connection shown in Figure 9-38 using the *AISCM* Tables. The bolts are $^3/_4$-in.-diameter A490N Group 150 in short-slotted bolt holes transverse to the direction of the load (SSLT). The plate is ASTM A36, and the beam is ASTM A992, grade 50 steel.

Solution

AISCM Table 10-10a gives the plate and weld available strength for shear plates with grade 50 steel. The following design parameters are known:

- With the number of horizontal rows of bolts, n = 3 in Figure 9-38
- Depth of the shear plate, ℓ = 1¼ + 3 +3 + 1¼ = $8^1/_2$ in.
- ¾ in. diameter ASTM F3125 Grade A490N bolts (Group 150) in short slotted bolt holes (SSLT)
- Shear plate thickness, $t_p = {}^3/_8$ in.

The Plate and Weld Available Strength for the connection from *AISCM* Table 10-10a is ϕR_n = **64.4 kips.**

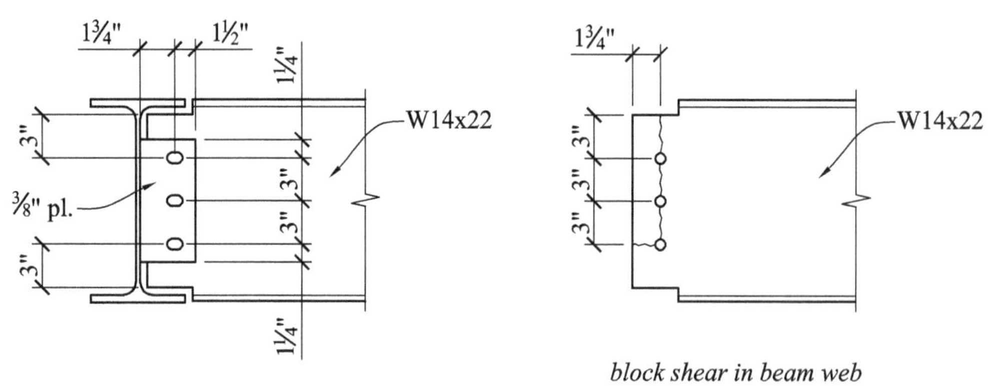

FIGURE 9-38 Connections detail for Example 9-11

Block shear:

The strength of the W14 × 22 beam web (t_w = 0.23 in.) needs to be checked for block shear. Assume a block shear failure mode in the W14 × 22 beam web with a shear failure plane that extends vertically from the top of the beam cope to the centroid of the bottom bolt hole and a tension failure plane that extends horizontally from the centroid of the bottom bolt hole to the edge of the beam web:

Number of bolt holes on the shear plane = 2.5 holes

Number of bolt holes on the tension plane = 0.5 hole

$$A_{gv} = [3 \text{ in.} + (2)(3 \text{ in.})](0.23 \text{ in.}) = 2.07 \text{ in.}^2$$

$$A_{nv} = \left[3 \text{ in.} + (2)(3 \text{ in.}) - (2.5)\left(\frac{3}{4} + \frac{1}{8}\right)\right](0.23 \text{ in}). = 1.57 \text{ in.}^2$$

$$A_{gt} = (L_{eh} - tolerance)(t_w) = (1.75 \text{ in.} - 0.25 \text{ in.})(0.23 \text{ in.}) = 0.345 \text{ in.}^2$$

$$A_{nt} = \left[(1.75 \text{ in.} - 0.25 \text{ in.}) - (0.5)\left(\frac{3}{4} + \frac{1}{8}\right)\right](0.23 \text{ in}). = 0.244 \text{ in.}^2$$

Using equation (4-11a), to determine the block shear strength,

$$\phi R_n = \phi(0.6F_u A_{nv} + U_{bs} F_u A_{nt}) \le \phi(0.6F_y A_{gv} + U_{bs} F_u A_{nt})$$

$$= 0.75[(0.6)(65)(1.57) + (1.0)(65)(0.244)]$$

$$\le 0.75[(0.6)(50)(2.07) + (1.0)(65)(0.244)]$$

$$= 57.8 \text{ kips} \le 58.5 \text{ kips}$$

Therefore, ϕR_n = **57.8 kips** for block shear strength

Bearing and tearout:

The strength of the bolts bearing on the W14 × 22 beam web (t_w = 0.23 in.) also must be checked.

The clear distance from the edge of the top bolt hole to the edge of the beam cope is

$$L_{c1} = 3 - \frac{1}{2}\left(\frac{3}{4} + \frac{1}{8}\right) = 2.56 \text{ in}$$

The clear distance between the bolt holes is

$$L_{c2} = L_{c3} = 3 - \frac{1}{2}\left(\frac{3}{4} + \frac{1}{8}\right) - \frac{1}{2}\left(\frac{3}{4} + \frac{1}{8}\right) = 2.12 \text{ in.}$$

For bolt 1, the capacity in tearout and bearing failure modes are

$$\phi R_n = \phi 1.2 L_c t F_u \le \phi 2.4 d t F_u$$

$$= (0.75)(1.2)(2.56 \text{ in.})(0.23 \text{ in.})(65 \text{ ksi})$$

$$\le (0.75)(2.4)(0.75 \text{ in.})(0.23 \text{ in.})(65 \text{ ksi})$$

$$= 34.4 \text{ kips} \le 20.1 \text{ kips}$$

Therefore, $\phi R_{n1} = 20.1$ kips,

For bolts 2 and 3, the capacity in tearout and bearing failure modes are

$$\phi R_n = \phi 1.2 L_c t F_u \le \phi 2.4 d t F_u$$

$$= (0.75)(1.2)(2.12 \text{ in.})(0.23 \text{ in.})(65 \text{ ksi})$$

$$\leq (0.75)(2.4)(0.75 \text{ in.})(0.23 \text{ in.})(65 \text{ ksi})$$

$$= 28.5 \text{ kips} \leq 20.1 \text{ kips}$$

Therefore, $\phi R_{n2} = \phi R_{n3} = 20.1$ kips

The total bearing strength of all the 3 bolts is

$\phi R_n = (3)(20.1 \text{ kips}) = \mathbf{60.3\ kips.}$

Direct shear in the bolts:

The strength of the bolts in direct shear (see *AISCM* Table 10-10a) = (22.5 kips/bolt)(3 bolts) = **67.5 kips.**

In summary, the strengths for the various possible failure modes are as follows:

- Plate and Weld Available strength = 64.4 kips
- Block shear strength of the web of the W14 × 22 beam = **57.8 kips**
- Bearing and tearout failure of the web of the W14 × 22 beam = 60.3 kips
- Direct shear strength of the bolts = 67.5 kips

The strength of the single plate connection is the smallest of the above strengths. Therefore, the block shear strength of the web of the W14 × 22 beam controls, so the design shear capacity of the connection is ϕR_n = **57.8 kips**.

References

[1] Muir, L. S. 2015. "The Nuts and Bolts of Nuts and Bolts." *Modern Steel Construction,* May: 17-25.

[2] American Institute of Steel Construction. 2023. *Steel Construction Mannual*. 16th ed. Chicago, IL: AISC.

[3] Muir, L. S. 2017. "A Tales of Tearouts." *Modern Steel Construction,* May: 17-21.

[4] Crawford, S. F., and G. L. Kulak. 1971. "Eccentrically Loaded Bolted Connections." *Journal of the Structural Division (ASCE)* 97, no. ST3: 765–783.

[5] Chesson, E., F. Norberto, and M. William. 1965. "High-Strength Bolts Subjected to Tension and Shear." *Journal of the Structural Division (ASCE)* 91, no. ST5: 155–180.

[6] Swanson, J. A. 2002. "Ultimate-Strength Prying Models for Bolted T-stub Connections." *Engineering Journal (AISC)* 39, no. 3: 136–147.

[7] Thornton, W. A. 1992. "Strength and Serviceability of Hanger Connections." *Engineering Journal (AISC)* 29, no. 4: 145–149.

[8] American Institute of Steel Construction. 2002. *Steel Design Guide Series 17: High-Strength Bolts—A Primer for Structural Engineers*. Chicago, IL: AISC.

[9] Dekker, B. and M. Shripka. 2011. "Connection Design Options in the Real World." *Modern Steel Construction,* January.

[10] Packer, J., Sherman, D., and Lecce, M. 2024. *Steel Design Guide Series 24: Hollow Structural Section Connections*. Chicago, IL: AISC

Exercises

9-1. Determine whether the connection shown in Figure 9-39 is adequate in bearing. Check the edge and spacing requirements. The steel is ASTM A36, and the bolts are in standard bolt holes.

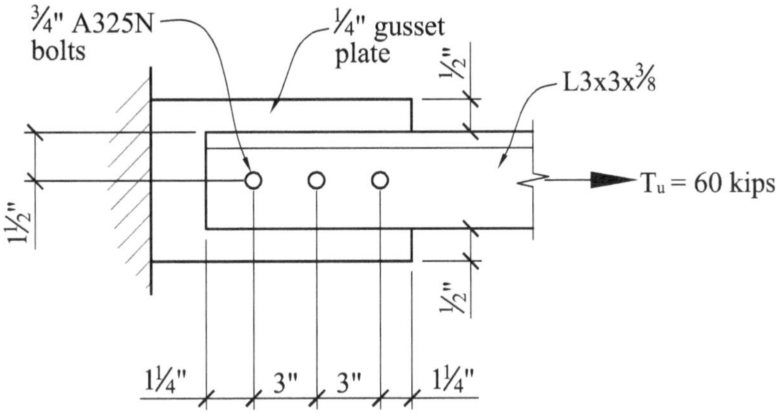

FIGURE 9-39 Detail for Exercise 9-1

9-2. Determine the maximum tensile force, P_u, that can be applied to the connection shown in Figure 9-40 based on the bolt strength; assume ASTM A36 steel for the following bolt types in standard bolt holes:

1. $^7/_8$-in.-diameter A325N,
2. $^7/_8$-in.-diameter A325SC (Class A faying surface, no fillers), and
3. $^3/_4$-in.-diameter A490X.

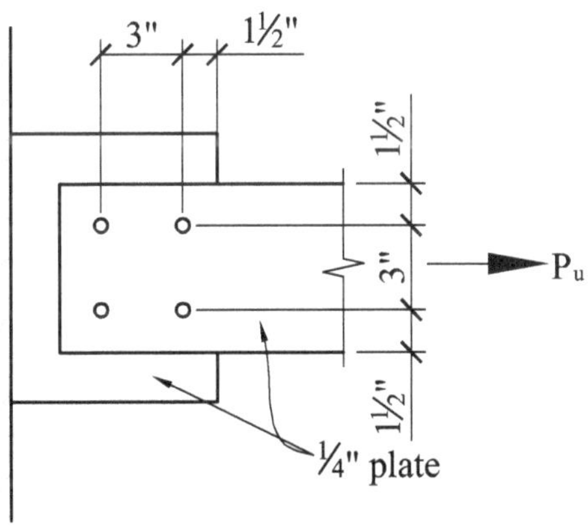

FIGURE 9-40 Detail for Exercise 9-2

9-3. Determine the maximum tensile force, P_u, that can be applied to the connection shown in Figure 9-41 based on the bolt strength and check the bolt bearing; assume ASTM A36 steel for the following bolt types in standard bolt holes:

1. 1-in.-diameter A325N,
2. 1-in.-diameter A325SC (Class B faying surface, no fillers), and
3. $^7/_8$-in.-diameter A490X.

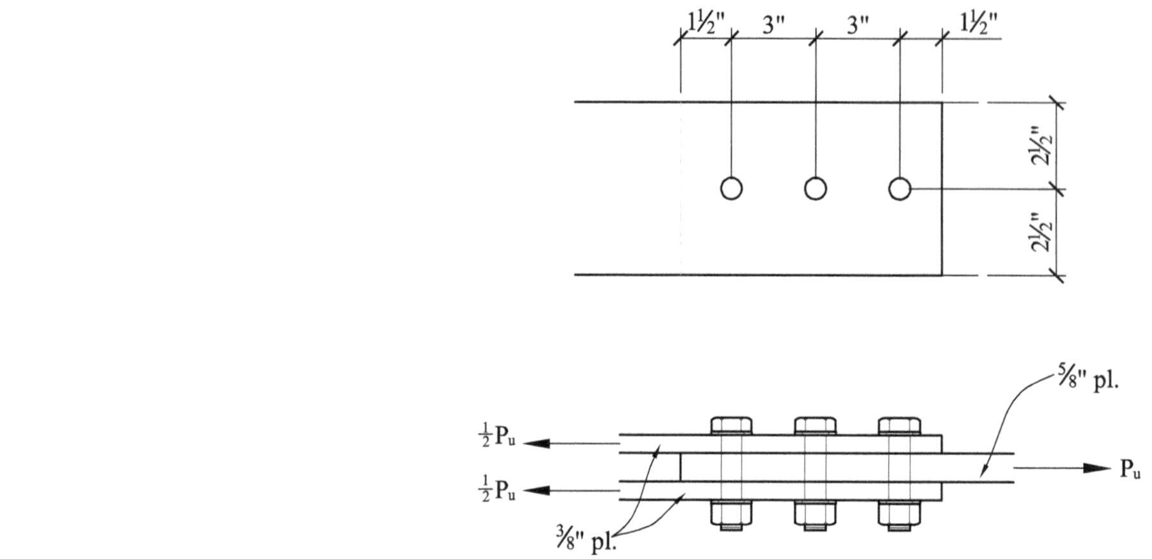

FIGURE 9-41 Detail for Exercise 9-3

9-4. Determine the required bolt diameter for the connection shown in Figure 9-42. The steel is ASTM A36, and the bolts are in standard bolt holes. Check the edge and spacing requirements. Use the following bolt properties:

1. A490N, and
2. A490SC.

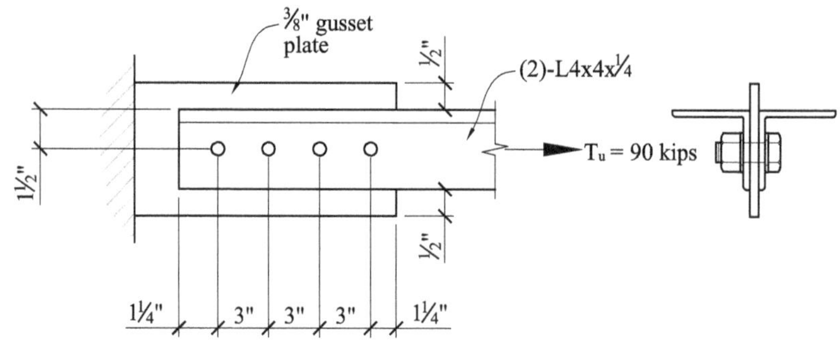

FIGURE 9-42 Detail for Exercise 9-4

9-5. Determine whether the connection shown in Figure 9-43 is adequate to support the applied moment considering the strength of the bolts in shear and bearing. The steel is ASTM A36, and the bolt holes are standard (STD).

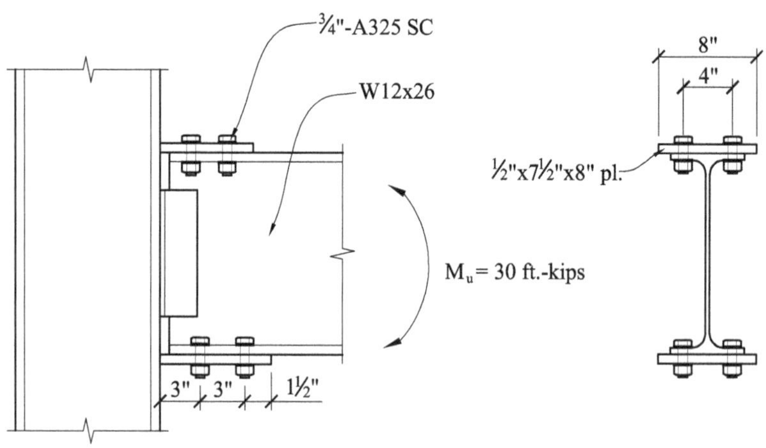

FIGURE 9-43 Detail for Exercise 9-5

9-6. Determine the capacity of the connection shown in Figure 9-44, considering the strength of the bolts only. Assume that each bolt takes an equal amount of shear and tension. There are (8) $^3/_4$-in.-diameter A325N bolts in standard bolt holes.

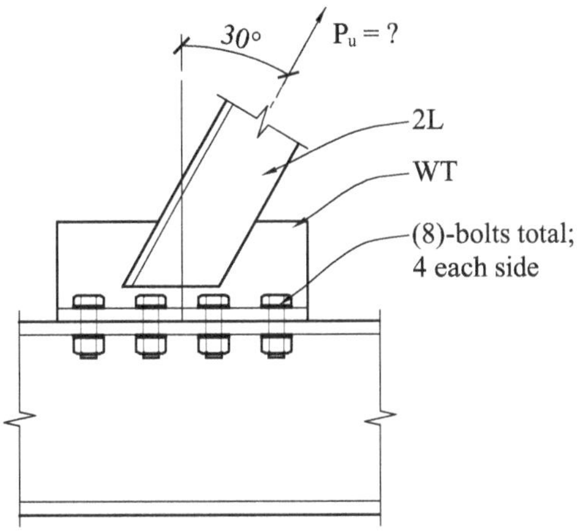

FIGURE 9-44 Connection detail for Exercise 9-6

9-7. Determine the capacity of the connection shown in Figure 9-45, considering the strength of the bolts only. Assume $^3/_4$-in.-diameter A325X bolts in standard bolt holes. Use the following methods and compare the results:

1. Instantaneous center method,
2. Elastic method, and
3. AISC design aids.

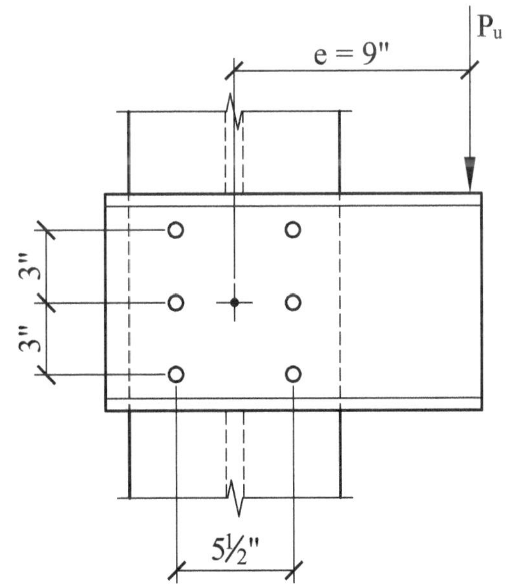

FIGURE 9-45 Connection detail for Exercise 9-7

9-8. Determine the capacity of the connection shown in Figure 9-46, considering the strength of the bolts only. Assume $^3/_4$-in.-diameter A490N bolts in standard bolt holes. Use the following methods and compare the results:

1. Case I,
2. Case II, and
3. AISC design aids.

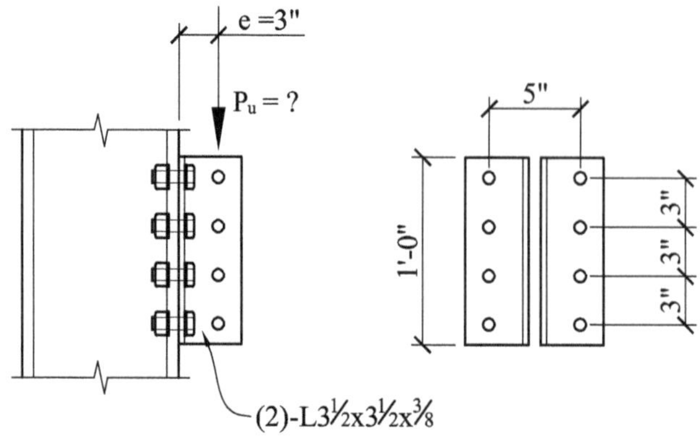

FIGURE 9-46 Connection detail for Exercise 9-8

9-9. Determine whether the moment connection shown in Figure 9-47 is adequate considering the prying action on the angle, shear in the bolts, and bearing in the bolts. The steel is ASTM A36, and the bole holes are oversized (OVS).

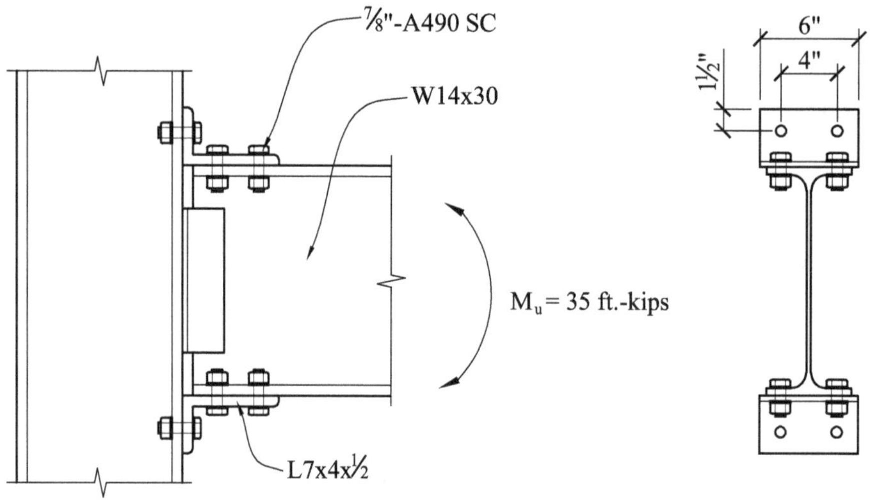

FIGURE 9-47 Connection detail for Exercise 9-9

9-10. Use the following parameters for the floor framing shown in Figure 9-48:

- Connection material is ASTM A36
- Beams are ASTM A992, grade 50
- Dead load is 75 psf, live load is 100 psf; ignore live load reduction.
- Use ¾", A325 bolts in STD bolt holes
  B-1: W18 × 35, B-2: W16 × 26
  G-1: W24 × 55, B-2: W21 × 44

Determine the following:

a. Find the factored reaction at B-1, B-2, G-1, and G-2.
b. Select an all-bolted, double-angle connection from *AISCM* Table 10-1 and sketch the detail for each end connection, and include copes where needed.
c. Check bolt bearings at each member.
d. For beams with copes, check the capacity at the cope.

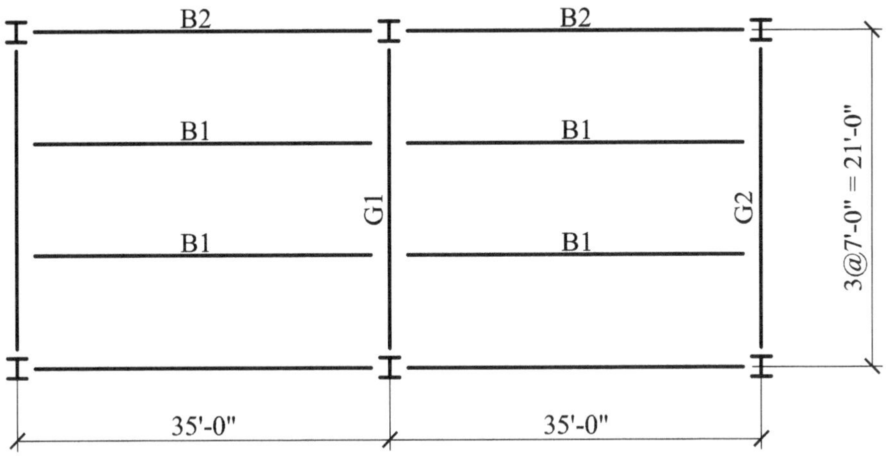

FIGURE 9-48 Framing plan for Exercise 9-10

9-11. Repeat exercise 9-10 using single plate shear connections per *AISCM* Table 10-10a.

9-12. Develop a spreadsheet to check exercise 9-9 to include the following parameters:

- Variable beam and angle sizes,
- Variable bolt size and grade,
- Check prying action, bolt bearing, and block shear in the angle.

9-13. Develop a spreadsheet to check exercise 9-1, but include the following additional variables and design components:

- Check block shear in the angle,
- Allow for variable angle size, bolt diameter, and bolt quantity.

9-14 A W12 × 19 beam frames into a W16 × 26 girder with a single plate shear connection with STD holes for a factored reaction of 42 kips. All steel is grade 50.

a) Select a connection using *AISCM* Table 10-10a. Check the bolt shear strength per Table 10-10b.

b) Sketch the complete connection, showing both beams and all dimensions and weld size callout.

c) For the weld size selected, determine the strength of the weld alone using *AISCM* Table 8-4.

CHAPTER

10

Welded Connections

10-1 INTRODUCTION

Welding is a process in which two steel members are locally fused or connected together by heating the base metals to very high temperatures, with or without the use of a filler metal. Welding in structural steel construction is usually carried out in accordance with the specifications of the Structural Welding Code published by the American Welding Society (AWS). The prequalified joint welding procedures for field and shop welds are covered in Section D1.1 of the AWS Code [1]. In structural steel buildings, connections often incorporate both welds and bolts. The use of bolts or welds in any connection is a function of many factors, such as cost, construction sequence, constructability, and the contractor's preference.

Types of Welding Processes

The heat energy needed to fuse two steel members together in the welding process could be electrical, mechanical, or chemical, but an electrical power source is typically used for welding structural steel. In the welding process, temperatures can reach up to 6,500°F, which results in the melting and fusing of the base metal and the filler metal or electrode [8]. The different types of welding processes [3, 8] are summarized below, with the two most common welding processes being *shielded metal arc welding* (SMAW) and *submerged arc welding* (SAW):

- ***Shielded Metal Arc Welding*** **(SMAW)**
  - A manual process that is also referred to as "stick" welding or rod welding because it uses an electrode that looks like a stick.
  - This is the most commonly used and specified type of weld for steel building construction.
  - The SMAW is used for field welding.
  - The stick electrode contains a steel rod at the center that is surrounded by an outside coating material called flux. The electrode acts as a conductor for electric current.
  - The electrode is connected through an electrical clamp to an electrode holder, and the electricity flows from the welding machine up through the steel rod (the electrode). The electric current forms an arc between the electrode and the steel members that are being connected. This arc heats up the base metal and melds off the electrode, adding it to the base metal as it moves along the welding path, creating a bead of crown-shaped weld. Thus, the electrode is consumed in the welding process, and would eventually need to be replaced.
  - The flux or outer coating material of the electrode melts off, forming a gaseous shield that protects the molten weld from impurities and contaminants in the atmosphere, such as

oxygen and nitrogen that could cause embrittlement of the weld, leaving a coating on the weld called slag. The slag coating is chipped off with a chipping hammer and then brushed clean to expose the weld. If additional passes of the electrode are required, the slag layer is chipped and brushed clean before the next weld pass.

- The basic welding process for the SMAW is shown in Figure 10-1.

- ***Submerged Arc Welding*** **(SAW)**
    - The submerged arc welding (SAW) process is a semi-automatic or automatic welding process typically used for shop welding.
    - The process is similar to SMAW, but an uncoated electrode is used.
    - A granular flux, which is fusable and forms a protective gaseous shield, is deposited over the joint to be welded, while submerging the electrode, the arc, and the weld zone. Thus, the arc and the molten weld are covered and protected by the granular flux.
    - This process is faster than the SMAW process and results in a weld with a deeper penetration and thus a higher weld strength.
- ***Flux Core Arc Welding*** **(FCAW)**
    - The process is similar to stick welding or the SMAW process.
    - The electrode is deposited through an arc, similar to the SMAW process, but unlike the SMAW process, the "tubular wire" electrode is fed continuously or automatically into a welding gun.
    - Similar to the stick welding process, a slag coating layer over the weld is also present and will need to be chipped off and brushed clean.
    - The gaseous shield over the weld zone is provided by "gas fed through the tubular wire [electrode], and additional shielding may be provided by externally supplied CO_2 gas" [8].
    - It is a faster process compared to the SMAW process and thus reduces the labor cost. The automatic feeding of the electrode facilitates continuous welding without

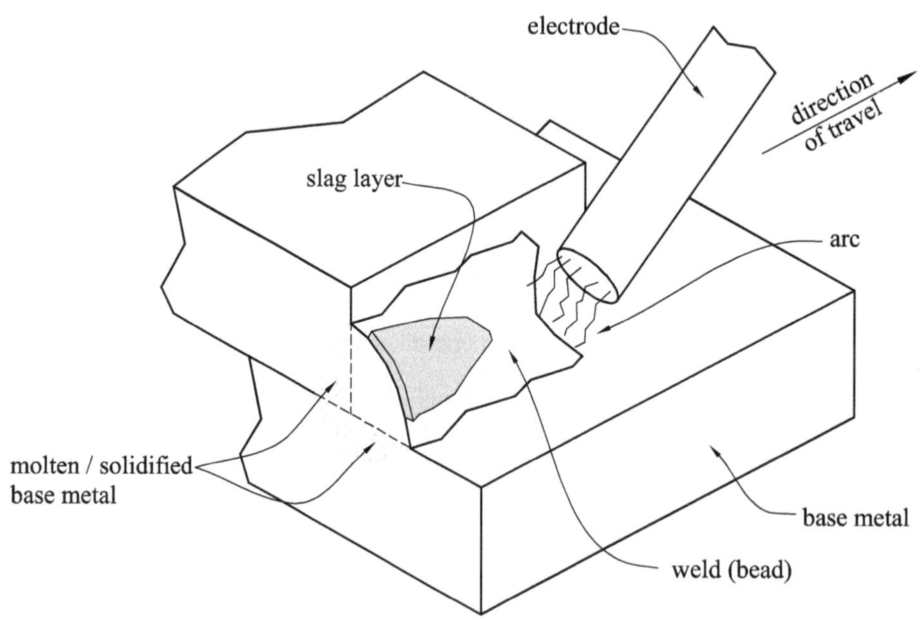

FIGURE 10-1 Basic welding process (SMAW)

having to stop to change to new electrodes when the old one runs out, as is the case in the stick welding or SMAW process.
    - The FCAW is the process of choice for field welding on large projects.
- ***Gas Metal Arc Welding*** **(GMAW)**
    - Also called metal inert gas (MIG) welding, and the process is typically used for shop welding.
    - There is no flux coating on the electrode.
    - The electrode consists of steel wire coated with copper to protect the steel and is fed continuously and automatically through a specialized welding gun that has a nozzle around the outside through which the shielding CO_2 or Argon gas from a gas cylinder is externally supplied, which protects the molten weld from impurities in the atmosphere.
    - There is no slag coating over the weld in the MIG welding process.
    - The process uses more expensive equipment compared to FCAW or SMAW processes.
    - Due to the use of the gas cylinder, the GMAW or MIG welding process is less portable compared to the FCAW or the SMAW processes.
- ***Gas Tungsten Arc Welding*** **(GTAW)**
    - Also called TIG welding, it is typically used for shop welding.
    - Consists of an electrode made of tungsten that does not melt,
    - The filler metal is added separately as the weld is formed.
    - A CO_2 or Argonne gas cylinder—similar to that used for MIG welding—is required.
    - There is no slag coating over the weld in the TIG welding process.
    - Similar to the MIG welding process, the TIG welding process also uses more expensive equipment compared to FCAW or SMAW processes, so the cost is comparable to the MIG welding process.
    - Due to the use of the gas cylinder, the GMAW process is less portable compared to the FCAW or the SMAW processes.

Other types of welding processes not discussed in this text include electroslag welding (ESW) and electrogas welding (EGW).

Common Weld Defects

Some common defects in welds [2, 7] include:

- Surface cracks
- Internal cracks that could be caused by the restraint to weld shrinkage due to the difference in the rates of cooling of the weld and the base metal.
- Porosity of the weld which could be caused by gas bubbles that are trapped in the weld as it cools.
- Impurities in the weld from oxygen and nitrogen in the atmosphere.
- Incomplete fusion of the weld or the lack of appropriate fusion of the connected steel members
- Lack of adequate penetration of the weld into the base metal

Welded connections offer some advantages over bolted connections, but they also have some disadvantages as well. The following summarizes the advantages and disadvantages of welded connections:

Advantages of welded connections

1. Welded connections can be adapted to almost any connection configuration in which bolts are used. This is especially advantageous when construction problems are encountered with a bolted connection (e.g., misaligned bolt holes) and a field-welded connection is the only reasonable solution. Connections to existing steel structures are sometimes easier with welds because greater dimension tolerances are often needed that might not be possible with bolts.
2. The full design strength of a member can be more easily developed with a welded connection using an all-around weld. For example, a beam splice would need to develop the shear and bending capacity of the member in question, and although a bolted connection would be possible for this condition (see Chapter 11), a welded connection requires less material and space to develop the design strength of the beam and would create a beam splice that is completely continuous across the joint. A beam that is fabricated too short and therefore needs additional length to accommodate field conditions would often use a welded splice instead of a bolted splice for just that reason.
3. Welded connections take up less material and space. Consider a bolted moment connection from a beam to a column; the beam would usually have a top plate, as well as bolts protruding into the space that would normally be occupied by a metal deck and concrete slab. The floor deck and slab would have to be modified and reinforced in this area to allow for the placement of the bolted connection. A welded moment connection in this case would allow the floor deck to align flush with the top of the beam.
4. Welded connections are typically more rigid or stiffer and subject to less deformation than bolted connections.
5. Welded connections are often preferred in exposed conditions where aesthetics is a requirement because the welds can be ground smooth to have a cleaner appearance.

Disadvantages of Welded Connections

1. Welded connections require greater skill. A welder is often certified not just for welding in general but also for certain welding techniques (e.g., overhead welding).
2. Welded connections often require more time to construct than equivalent bolted connections. A greater amount of time equates to higher cost because of the labor required, especially for welds that have to be performed in the field. Special preparations are also required for field welds in cold weather conditions. The welding process generates a lot of heat and can lead to undesirable residual stresses if the cooling of the connected members is not controlled and thus leads to uneven cooling of the members; this can cause brittle fracture in the members. In many cases, contractors prefer to avoid field-welded connections as much as possible because of the associated labor cost, even if it leads to using rather large and seemingly oversized bolted connections.
3. The required inspection of welded connections is more extensive than that of bolted connections.
4. In existing structures, welded connections may be difficult or even impossible to use due to the occupancy of the structure. For example, welding in a warehouse with paper

or other flammable material has an obvious element of danger associated with it. Special protective measures would then be required that might interrupt the regular business operation of the facility; in which case an alternate bolted connection might have to be used. With other structures, such as hospitals, a special permit might be required to ensure the safety of the occupants during the welding operation.

Weld Inspection Methods

Weld inspections can be carried out visually or by using more sophisticated weld inspection and non-destructive testing methods [2]. Discontinuities and other internal defects in welds cannot easily be found by visual inspection. The various weld inspection methods are as follows [2]:

- **Visual testing (VT)**
  - This is the most widely specified nondestructive inspection method to ascertain the quality of welds.
  - It is effective for detecting surface defects.
  - For visually inspected fillet welds, AWS D1.1 allows a maximum reduction in weld throat size of 33.3% for a 3/16 in. fillet weld to approximately 40% reduction for a 5/16 in. fillet weld over a length of weld not exceeding 10% of the weld length [1]. Any fillet weld not meeting this requirement on site is deemed inadequate.
  - Example: Assuming a 10-in. weld length, the above AWS requirement implies the following:
    - Minimum allowable reduction in the throat size for a 3/16 in. weld = 0.0624 in. (i.e., 0.333 × 3/16 in.) over a weld length of 1 in. (i.e., 10% of 10 in. = 1 in).
    - Minimum allowable reduction in the throat size for a 5/16 in. weld = 0.125 in. (i.e., 0.4 × 5/16 in.) over a weld length of 1 in. (i.e., 10% of 10 in. = 1 in).
- **Penetrant testing (PT)**
  - This method uses a red dye penetrant that is pressure-sprayed on the weld which penetrates any defect on the surface of the weld. A white developer that is then sprayed on the surface leads to the dye seeping out of the defects – i.e., the cracks or crevices, and appearing as a red image.
  - This method is only useful for detecting surface defects
- **Magnetic particle testing (MT)**
  - Magnetizing current is introduced into the weld with a yoke, resulting in a magnetic field.
  - Any cracks or defects will be indicated by a distortion in the magnetic field that is made obvious when a dry magnetic powder is sprayed on the weld.
- **Ultrasonic testing (UT)**
  - This method, which uses a calibrated pulse echo instrument, involves a short pulse of high-frequency sound waves introduced into the weld, and any discontinuity in the reflection of the sound waves gives an indication of the presence of weld defects.
  - A smooth surface is required.
- **Radiographic testing (RT)**
  - Involves X-ray of the weld and processing of the X-ray film.
  - This is the most expensive weld inspection method.
  - There is a hazard of exposure to radiation.

Each of the above methods (PT, MT, UT, and RT), with the exception of the visual inspection method, requires more labor and equipment, which adds to the cost of the connection. The MT, UT, and RT methods are effective for detecting internal cracks or voids or defects in welds and are also used to validate the integrity of welds where visual inspection indicates the presence of defects. Some welds, such as full-penetration welds, require continuous inspection, which would mean that the welding inspector has to be on site whenever this type of welding is being done, also adding to the cost of the connection. The general contractor or the owner of the structure must arrange for the weld inspection to be carried out. The joints to be welded are typically inspected before and after welding, and any defective weld must be repaired.

10-2 TYPES OF WELDS AND WELDED JOINTS

The types of welds include fillet welds, groove welds, plug welds, and slot welds. There are numerous possible welded joint configurations, but the most common are as follows (see Figure 10-2):

- Lap joint
- Butt joint
- Corner joint
- Tee joint

In Part 8 of the *AISCM*, these joints are designated as B (butt joint), C (corner joint), and T (T-joint). The following combinations are also recognized: BC (butt or corner joint), TC (T- or corner joint), and BTC (butt, T-, or corner joint). These designations are a shorthand form for specifying certain weld types. The common weld types are shown in Figures 10-3 and 10-4, and these are as follows:

Fillet welds

There are a variety of factors that determine what type of weld to use, but the most common weld is the fillet weld (approximately 80% of all welds) [8]. Fillet welds are generally triangular in shape, and they are used to join members together that are usually at right angles to, or parallel to, each other. They are usually deposited on the outside of the connected members. Fillet welds are the most economical type of weld because they require very little surface preparation

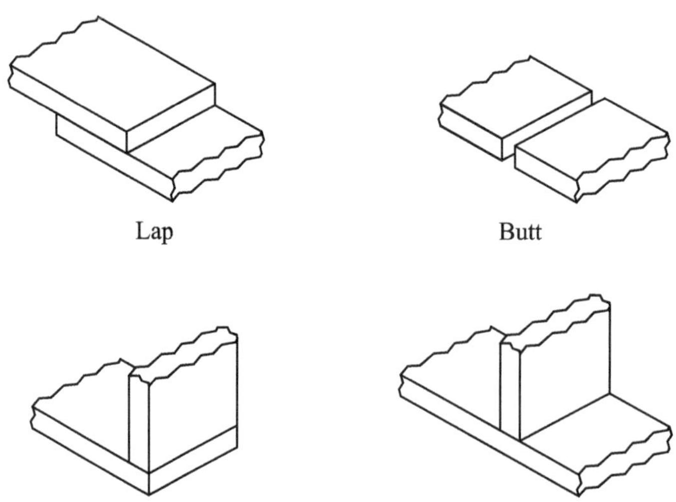

FIGURE 10-2 Joint types

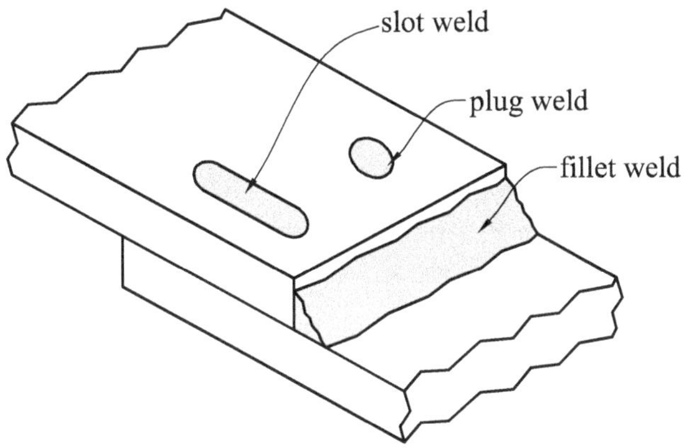

FIGURE 10-3 Weld types

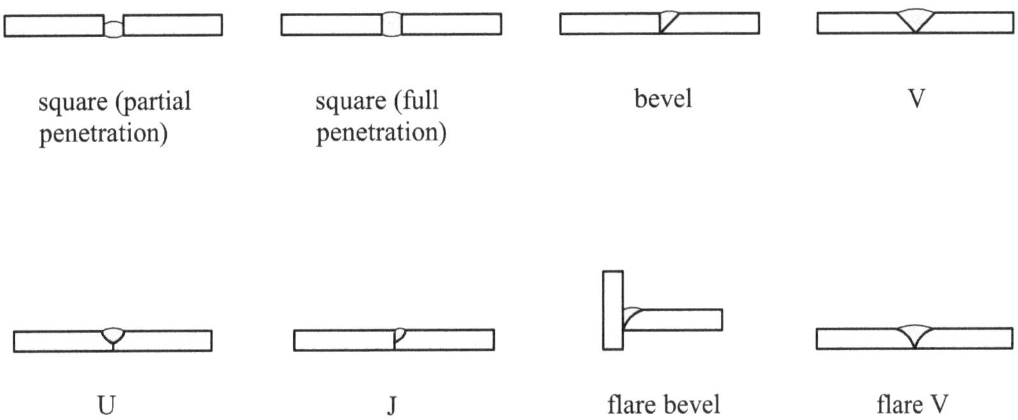

FIGURE 10-4 Groove welds

and can be used in virtually any connection configuration [8]. The size of fillet welds joining members with a fit-up gap or root opening between them should be increased to account for the gap, with the gap no larger than 3/16 inch (see Section 7.21.1 of Ref. [1]).

To minimize the effect of weld shrinkage resulting from the heat generated during the welding process and the differential cooling after welding, fillet welds could be specified as short weld segments of finite length at regular spacings along the length of the connected members rather than being continuous; these types of fillet welds are called intermittent or stitch welds. The weld symbol in Figure 10-7 with ¼ in. fillet weld, 3 inches long at 8 inches on centers is an example of an intermittent weld.

When used on both sides of a T-joint connection (see Figure 10-2), the intermittent welds are usually staggered. To minimize the effect of weld shrinkage and the distortion of the welded members, continuous fillet welds can also be deposited in two stages as two independent sets of staggered intermittent welds, with the second set of intermittent welds deposited later to fill the gaps between the weld segments in the first set of intermittent welds.

Built-up column sections consisting of two wide flange members located flange tip to flange tip can be connected with stitch plates that are welded to the outside faces of the flanges of the built-up section with continuous fillet welds; the design of such a built-up compression member with intermediate connectors is carried out in accordance with *AISCM* Section E6 (see Chapter 5 of this text).

Plug and slot welds

Plug and slot welds are used to transmit shear in lap joints or to connect components of built-up members (such as web doubler plates) to prevent buckling. They are also used to conceal the connections for steel members that are exposed for architectural reasons. In addition, they can be used to add strength to connections with fillet welds (see Example 10-5). Neither plug welds nor slot welds should be used where a tension force is normal to the plane of the faying surface. Nor should they be used to support cyclical loads. Plug welds are placed in round holes, and slot welds are placed in elongated holes. In each case, the weld metal is placed in the hole up to a certain depth (partial penetration or full penetration). The penetration depth of a plug or slot weld is difficult to inspect visually, so often such welds are not preferred.

Groove welds

Groove welds (see Figure 10-4) are used to fill the gap between the ends of two abutting members at a joint and are deposited within the thickness of the connected members. Groove welds (Figure 10-4), which could either be full penetration groove welds or partial penetration groove welds, can be classified as

- square,
- bevel,
- V (or double-bevel),
- U,
- J, or
- flare V (or flare bevel).

In order to contain the weld metal, a backing bar is placed at the bottom of groove welds. In some cases, the backing bar should be removed so that the welding can be inspected. In other cases, the backing bar is removed so that additional weld metal can be added on the weld root side of the joint. In this case, any part of the weld at the root side that is incomplete is gouged or removed prior to depositing additional backing weld. This process, called *backgouging* [1], and which facilitates "complete fusion and complete joint penetration," is generally limited to connections for joints specially detailed for seismic resistance. The backing bar for groove welds can also be left in place where special seismic detailing is not required, but this decision is usually left to the designer.

Groove welds can penetrate the connected member for a portion of the member thickness, or they can penetrate the full thickness of the connected member. These are called partial joint penetration (PJP) and complete-joint penetration (CJP) welds, respectively (see further discussion on groove weld strength in Section 10-7). Complete joint penetration (CJP) welds (also called full-penetration or "full-pen" welds) fuse the entire depth of the ends of the connected members, and they are the most expensive weld type. Partial joint penetration (PJP) welds are more cost-effective than CJP welds, and they are used when the applied loads are such that a full-penetration weld is not required. They can also be used where access to the groove is limited to one side of the connection.

Welding installation positions

Whenever possible, welding should be performed in the shop where the quality of the weld is usually better controlled. Shop welds are not affected by weather conditions, and access to the joint is not limited. Welds can be classified according to their installation position as follows (see Figure 10-5):

- Flat: Easiest to install and most preferred
- Overhead: They are usually done in the field and should be avoided where possible because they are difficult and more time-consuming, and therefore more costly to install.
- Horizontal,
- Vertical.

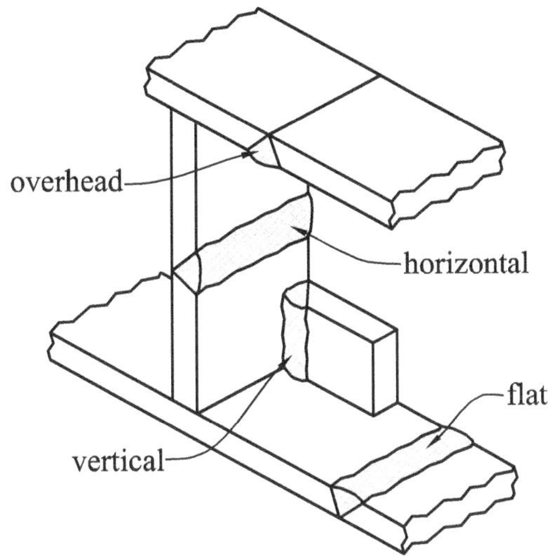

FIGURE 10-5 Weld positions

10-3 WELD SYMBOLS

Weld symbols are commonly used to identify the required weld properties used in connection design. These symbols have been standardized by the American Welding Society [1], and they are summarized in Figure 10-6 or Table 8-2 of the *AISCM* [2]. All the welds in Figure 10-6, except for the first two welds, are groove welds.

Fillet welds are the most specified, and will be used as the basis for the following discussion (see Figures 10-6 and 10-7). The standard symbol is an arrow pointing to the weld or joint, with a horizontal line forming the tail of the arrow. The triangular shape indicates a fillet weld, and the vertical line of the fillet weld symbol is always shown on the left side of the weld symbol. For fillet and other weld types, the information about the weld is given above and below the horizontal line. If the information is below the horizontal line, then the welded joint is on the near side of the arrow. If the information is above the horizontal line, then the welded joint is on the opposite or far side of the arrow. The size of the weld is stated first on the left side of the weld-type symbol, then the length and spacing of the weld are placed to the right of the weld-type

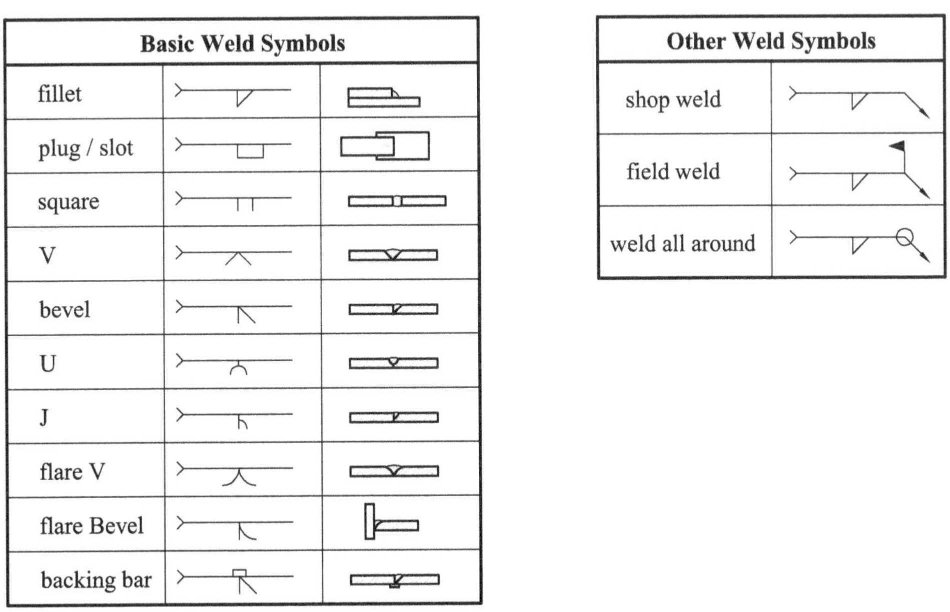

FIGURE 10-6 Basic weld symbols

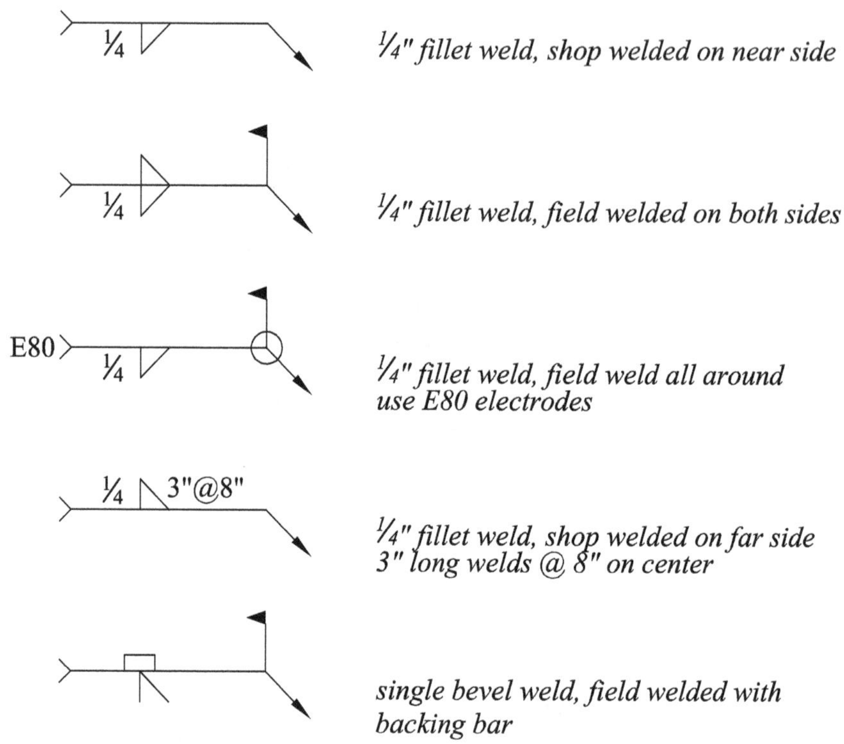

FIGURE 10-7 Weld symbols examples

symbol. A circle at the intersection of the horizontal line and the tail of the arrow indicates that the weld is placed all around the joint; a shaded triangular flag on a vertical line at this location indicates that the weld is to be installed in the field (i.e., a field weld). The absence of the flag indicates that the weld should be done in the shop (see Figures 10-6 and 10-7 for weld symbols). At the end of the horizontal line, any special notes can be added. As noted in Chapter 8, the weld-all-around type of weld is usually discouraged for wide flange column-to-base plate connections but is recommended for HSS column-to-base plate connections.

Table 8-2 of the *AISCM* also lists certain prequalified weld symbols that can be used to identify PJP or CJP weld types. Table 10-1 summarizes the basic notation used.

As an example, in the designation BU-4a used in *AISCM* Table 8-2 [2], the letter B indicates a butt joint, and the letter U indicates that the thickness of the connected parts is not limited and that the weld is a complete joint penetration (CJP) weld. The number 4 indicates that the joint has a single-bevel groove, which means that one connected part is flat or unprepared, whereas the other connected part has a beveled edge. The letter "a" indicates something unique about this joint, which, in this case, means that a backing bar is used. The letter "a" also differentiates this joint type from a BU-4b weld designation, where the letter "b," in this case, indicates that the underside of the joint in question must be backgouged and reinforced with additional weld metal.

TABLE 10-1 Prequalified weld notation

Joint type symbols	B	butt joint
	C	corner joint
	T	T-joint
	BC	butt or corner joint
	TC	T- or corner joint
	BTC	butt, T-, or corner joint

(Continued)

Base metal thickness and penetration symbols	L	limited thickness, complete-joint penetration
	U	unlimited thickness, complete-joint penetration
	P	partial-joint penetration
Weld-type symbols	1	square groove
	2	single-V groove
	3	double-V groove
	4	single-bevel groove
	5	double-bevel groove
	6	single-U groove
	7	double-U groove
	8	single-J groove
	9	double-J groove
	10	flare-bevel groove

10-4 DIMENSIONAL REQUIREMENTS FOR WELDS

There are minimum and maximum dimensional limitations for welds given in the *AISCM* Section J2.2b. Table 10-2 gives the maximum and minimum fillet weld sizes, which are a function of the thickness of the connected parts. The minimum weld size is intended to prevent rapid cooling of the weld, which could lead to loss of ductility and can cause cracking of the welds if it is restrained by a thick material. The maximum weld size ensures easy inspection of the welds.

Some additional dimensional requirements of fillet, plug, and slot welds are presented below:

- The minimum total length of a fillet weld must be at least four times the nominal size of the weld $\left(i.e.\ L \geq 4w\right)$, or else the size of the weld used to determine its design strength shall be assumed to be one-fourth of the weld length (i.e., $\frac{1}{4}L$).
- When longitudinal welds are used to connect the ends of tension members as shown in Figure 10-8, the length of each longitudinal weld shall not be less than the distance between adjacent longitudinal welds (i.e., L ≥ b). This is to minimize shear lag, which occurs when all parts of the connected tension member are not fully engaged in resisting the tension load, leading to a non-uniform or uneven axial tension stress distribution (see Chapter 4).
- Figure 10-8 shows a balanced weld where the center of gravity of the weld group coincides with the center of gravity of the tension member. Note that balanced welds are required for members subject to cyclic loading.

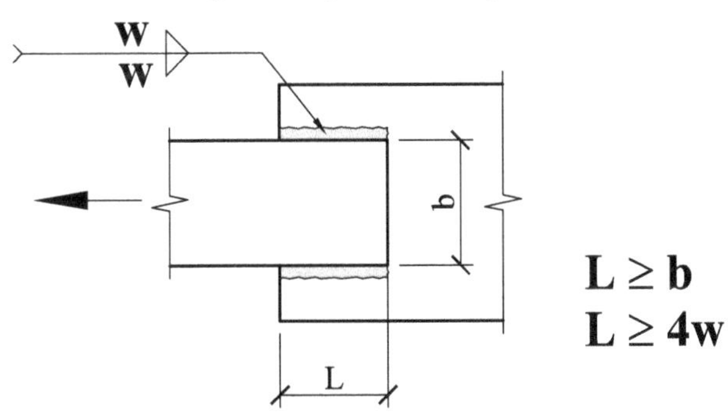

FIGURE 10-8 Longitudinal weld

- The maximum length of a fillet weld is unlimited except for end-loaded welds. End-loaded welds are longitudinal welds in axially-loaded tension members and they are limited as follows:
    - When the length-to-weld size ratio, $\left(\frac{L}{w}\right)$, of an end-loaded fillet weld does not exceed 100, the effective weld length is equal to the actual length of the weld. Thus, the maximum lengths of 3/16 in. and ¼in. fillet welds are 18.75 in. and 25 in., respectively.
    - When the length-to-weld size ratio, $\left(\frac{L}{w}\right)$, of an end-loaded fillet weld exceeds 100, the design strength is adjusted by the following factor (see *AISC* Specification Section J2.2b(d)(2)):

$$\beta = 1.2 - 0.002\left(\frac{L}{w}\right) \leq 1.0, \quad \geq 0.6 \tag{10-1}$$

    - The second part of equation (10-1) indicates that when the length-to-weld size ratio, $\left(\frac{L}{w}\right)$, of an end-loaded fillet weld exceeds 300, the effective length of the weld = 180w (i.e., $\beta = 0.6$)

Where,
β = Weld strength adjustment factor (varies from 0.60 to 1.0),
L = Weld length
w = Weld size

- The minimum diameter of a plug weld or width of a slot weld is the thickness of the part containing it plus $^5/_{16}$ in. rounded up to the next larger odd $^1/_{16}$ in. (see Figure 10-9).
- The maximum diameter of a plug weld or width of a slot weld is $2^1/_4$ times the thickness of the weld.
- Plug welds should be spaced a minimum of four times the hole diameter (4*d*).
- The length of a slot weld shall be less than or equal to 10 times the weld thickness.
- Slot welds should be spaced a minimum of four times the width of the slot in the direction transverse to their length. In the longitudinal direction, slot welds should be spaced a minimum of twice the length of the slot [2].
- For a connected part whose thickness less than or equal to $^5/_8$ in., the minimum thickness of plug or slot weld is the thickness of the material.
- For connected parts whose thickness is greater than $^5/_8$ in., the thickness of plug or slot weld should be at least half the thickness of the material, but not less than $^5/_8$ in.

Figure 10-9 shows the basic dimensional requirements for plug and slot welds presented in the preceding discussion.

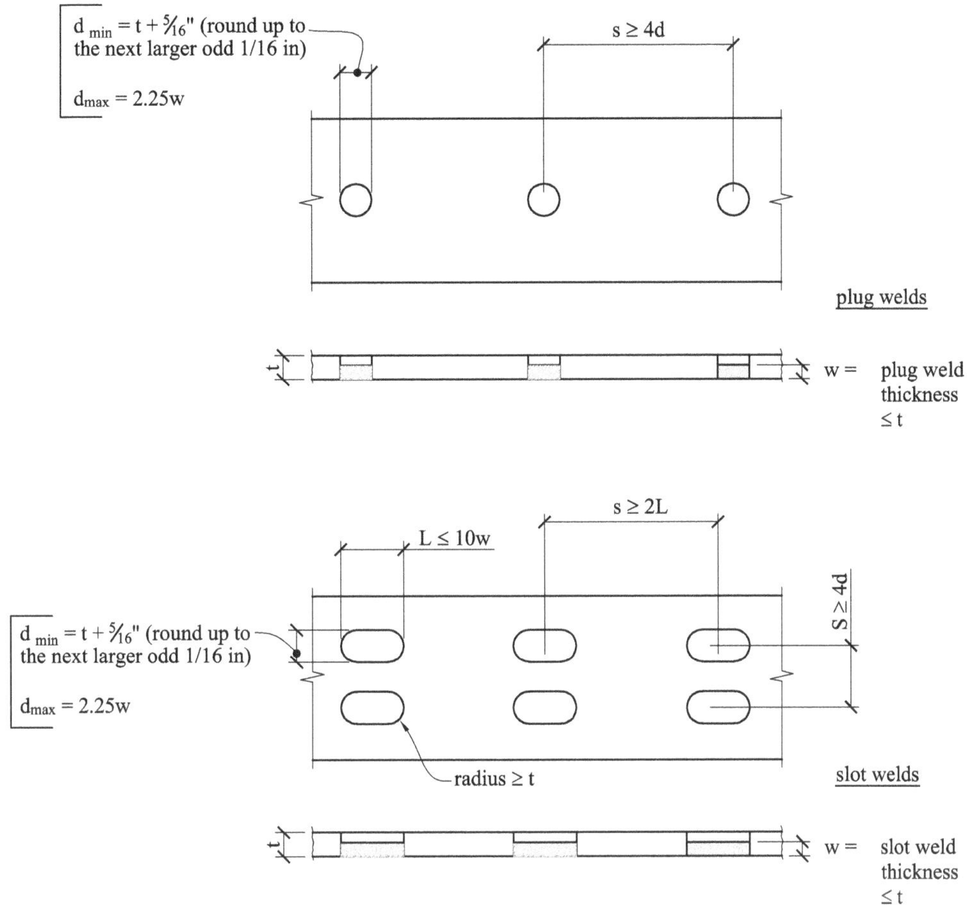

for plug and slot welds,

$w = t$ for $t \le 5/8$ in.

$w \ge t/2 \ge 5/8$in. for $t > 5/8$in.

FIGURE 10-9 Plug and slot welds dimensional limitations (AISC Specification J3.3b)

TABLE 10-2 Maximum and minimum fillet weld sizes

Maximum fillet weld size	
Thickness of thicker connected part, t*	***Maximum weld size, w***
$t < {}^1/_4$"	$w = t$
$t \ge {}^1/_4$"	$w = t - {}^1/_{16}$"
Minimum fillet weld size†	
Thickness of thinner connected part, t	***Minimum weld size, w***
$t \le {}^1/_4$"	${}^1/_8$
${}^1/_4" < t \le {}^1/_2$"	${}^3/_{16}$
${}^1/_2" < t \le {}^3/_4$"	${}^1/_4$
$t > {}^3/_4$"	${}^5/_{16}$

Adapted from *AISCM* Table J2.4. Copyright © American Institute of Steel Construction. Reprinted with permissions. All rights reserved.

* The term *t* is the thickness of the thicker connected part.

† Single-pass welds should be used where possible. **The maximum weld size that can be made in a single pass is ${}^5/_{16}$ in.**

10-5 FILLET WELD STRENGTH

The strength of a fillet weld is based on the assumption that the weld forms a right triangle with a one-to-one slope between the connected parts (see Figure 10-10). Welds can be loaded in any direction, but they are weakest in *shear* and are therefore assumed to fail in shear. The shortest distance across the weld is known as the throat, where the shear failure plane (i.e., a plane parallel to the longitudinal axis of the fillet weld) is assumed to occur. Any additional weld beyond the throat (represented by the dashed line in Figure 10-10) is not considered when calculating the strength of the weld.

Based on the geometry in Figure 10-10, the strength of the weld can be calculated as

$$\begin{aligned} R_n &= k_{ds} A_{we} F_{nw} \\ &= k_{ds}(\text{effective area of weld}) \times (\text{nominal stress of the weld metal}) \\ &= k_{ds}(\text{weld throat thickness}) \times (\text{weld length}) \times (\text{nominal stress of the weld metal}) \\ &= k_{ds}(\text{w sin } 45)(L)(F_{nw}) \end{aligned} \tag{10-2}$$

Where,

R_n = Nominal weld strength, kips,

w = Weld size, in.,

$\text{w}(\sin 45)$ = size of throat of weld (see Figure 10-10),

L = total weld length, in.,

k_{ds} = directional strength increase factor (AISC Specification Equation J2-5)

= $(1 + 0.50\sin^{1.5}\theta)$ for fillet welds where strain compatibility is considered [2]

= 1.0 for fillet welds at the ends of *rectangular* HSS members loaded in tension

= 1.0 for all other conditions

θ = angle between the direction of the applied load and the longitudinal axis of the weld.

- When $\theta = 90°$ (i.e., the applied load is *perpendicular* to the longitudinal axis of the fillet weld, which means the weld is in tension), $k_{ds} = 1.5$,
- When $\theta = 0°$ (i.e., the applied load is *parallel* to the longitudinal axis of the fillet weld, which means the weld is subjected to shear), $k_{ds} = 1.0$.

F_{nw} = Nominal stress of the weld metal, ksi.

The nominal weld strength, F_w, is a function of the weld metal or electrode used. The ultimate electrode tensile strength can vary from 60 ksi to 120 ksi (i.e., this is represented as E60,

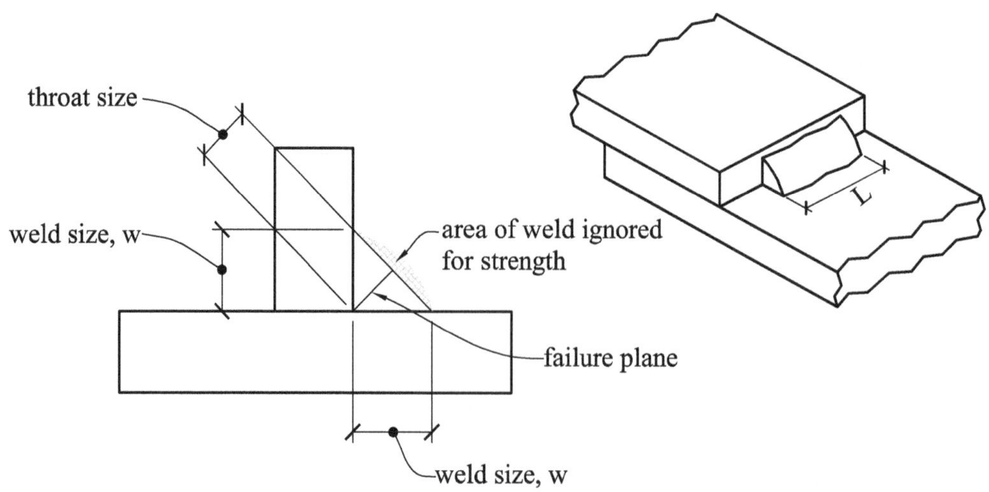

FIGURE 10-10 Fillet weld geometry

E70, E80, E90, E100, E110, or E120), but the most commonly used electrode tensile strength is 70 ksi. The type of electrode is denoted by the letter *E,* followed by two digits that represent the electrode tensile strength and two additional numbers that indicate the welding process. The designation E70XX is commonly used, where the first two digits after "E" indicate the nominal tensile strength of 70 ksi for the electrodes. The third digit "X" indicates the weld position (such as overhead or flat welding), and the fourth digit "X" indicates the welding current and other weld properties not directly pertinent to the weld strength. An example of a complete weld designation is E7028, where the third digit "2" indicates a horizontal and flat weld, and the fourth digit "8" indicates an electrode with a low hydrogen coating. The use of low hydrogen electrodes such as E7018 or E7028, etc., prevents hydrogen embrittlement and cracking of welds, and results in "quality welds with excellent toughness and high ductility."

From *AISCM* Table J2.5 [2], the nominal strength of a fillet weld in shear is

$$F_w = 0.6F_{\text{EXX}}, \tag{10-3}$$

where F_{EXX} is the electrode strength.

Single Fillet Weld:

The fillet weld is weakest when the applied load is parallel to the longitudinal axis of the weld, which indicates a *longitudinal weld in shear* with $\theta = 0°$ and $k_{ds} = 1$. Therefore, the nominal strength of a single fillet weld is found by com- bining equations (10-2) and (10-3) as follows:

$$R_n = k_{ds}(\text{w} \sin 45)(L)(F_{nw})$$

$$= (1)(0.707\text{w})(L)(0.6F_{EXX}) = (0.6)(F_{\text{EXX}})0.707wL \tag{10-4a}$$

The design strength and the allowable strength, respectively, are then given as follows:

$$\phi R_n = \phi(0.6)(F_{\text{EXX}})0.707wL \quad \text{for LRFD} \tag{10-4b}$$

$$\frac{R_n}{\Omega} = \frac{(0.6)(F_{\text{EXX}})0.707wL}{\Omega} \quad \text{for ASD} \tag{10-4c}$$

Where, $\phi = 0.75$ and $\Omega = 2.0$.

The weld size is commonly expressed in terms of the number of sixteenths of an inch. For example, a fillet weld size of $w = {}^1/_4$ in. could be expressed as four-sixteenths of an inch, indicating that the number, $D = 4$. Similarly, a weld size of $w = 3/16$ in. could be expressed as three-sixteenths of an inch, giving the number, $D = 3$. From *AISCM* Table J2.5, the strength reduction factor for shear is $\phi = 0.75$. Combining these with a commonly used electrode strength of $F_{\text{EXX}} = 70$ ksi, the design strength becomes

$$\phi R_n = (0.75)(0.6)(70)(0.707)\left(\frac{D}{16}\right)L = 1.392DL, \quad \text{for LRFD} \tag{10-5a}$$

$$\frac{R_n}{\Omega} = \frac{(0.6)(70)(0.707)\left(\frac{D}{16}\right)L}{2.0} = 0.928DL, \quad \text{for ASD} \tag{10-5b}$$

Where,

R_n = Nominal weld strength, kips,

ϕR_n = Design weld strength, kips,

D = Weld size in *sixteenths* (i.e., 1/16) of an inch, and

L = total weld length, in.

For LRFD and E70XX electrodes, the fillet weld strengths are as follows:

- $^3/_{16}$ in. fillet weld: D = 3 and ϕRn = 1.392(3)(1 in.) = **4.18 kips/in**. length of weld.
- $^1/_4$ in. fillet weld: D = 4 and ϕRn = 1.392(4)(1 in.) = **5.57 kips/in**. length of weld.
- $^5/_{16}$ in. fillet weld: D = 5 and ϕRn = 1.392(5)(1 in.) = **6.96 kips/in**. length of weld.

It should be noted that the strength of the connected part must also be checked for strength in shear, tension, or shear rupture, whichever is applicable to the connection (see Chapter 4).

The above formulation for weld strength is conservative because the shear failure mode, which has the lowest strength of all the possible failure modes, was assumed. Note that the use of single-sided fillet welds is not recommended for connections or joints that are subject to rotation about the toe of the weld [2].

Fillet Weld Group:

For a linear weld group loaded in the plane of the weld, the nominal weld strength, R_n, as a function of the angle (in degrees) between the load direction and the plane or longitudinal axis of the weld is as given in Equation 10-2

$$R_n = k_{ds}(\text{w} \sin 45)(L)(F_{nw})$$

$$R_n = k_{ds}(\text{w} \sin 45)(L)(F_{nw}) = (1 + 0.50\sin^{1.5}\theta)(0.707\text{w})(L)(0.6F_{\text{EXX}}), \tag{10-6}$$

Where the load angle, θ, is the angle in degrees between the direction of the applied load and the longitudinal axis of the fillet weld, as shown in Figure 10-11. Note that a linear weld group is one that has all the welds in line or in parallel to the applied load. Welds loaded perpendicular to the longitudinal axis of the weld (i.e., the load angle, θ = 90 degrees) which are also known as transverse welds are stronger and stiffer, but less ductile, than welds loaded parallel to the longitudinal axis of the weld (also known as longitudinal welds or welds with a load angle, θ = 0 degrees).

For weld groups where the welds are oriented both *longitudinally* and *transversely* to the applied load, the nominal weld strength (see *AISC* Specification Equation J2-6) is:

For LRFD:
$$\phi R_n = 0.85\phi F_{nw}A_{wel} + 1.5\phi F_{nw}A_{wet}, \tag{10-7}$$

For ASD
$$\frac{R_n}{\Omega} = \frac{0.85F_{nw}A_{wel} + 1.5F_{nw}A_{wet}}{\Omega}, \tag{10-8}$$

Where,

$F_{nw} = 0.6F_{\text{EXX}}$
A_{wel} = Effective area of *longitudinal* fillet welds = $0.707wL_l$
A_{wet} = Effective area of *transverse* fillet welds = $0.707wL_t$
w = fillet weld size
L_l = length of longitudinal welds, in.
L_t = length of transverse welds, in.
$\phi = 0.75$
$\Omega = 2.0$

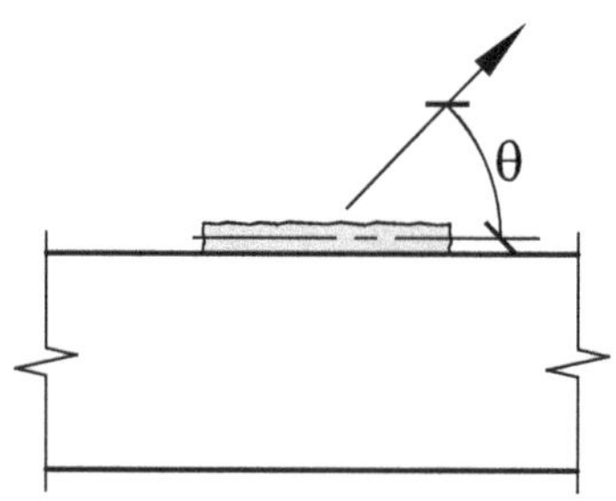

FIGURE 10-11 Weld loaded at an angle

EXAMPLE 10-1

Fillet Weld Strength

Determine the capacity of the connection shown in Figure 10-12 based on weld strength alone. Electrodes are E70XX.

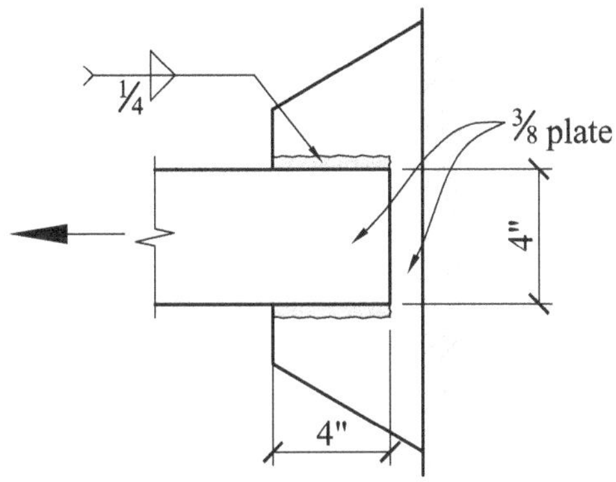

FIGURE 10-12 Detail for Example 10-1

Solution

The AISC requirements for weld size must be checked first.

From Table 10-2,

Minimum weld size = $^{3}/_{16}$ in. < $^{1}/_{4}$ in. provided, O.K.

Maximum weld size = $t_{max} - {}^{1}/_{16}$ in. = $^{3}/_{8}$ in. – $^{1}/_{16}$ in. = $^{5}/_{16}$ in. > $^{1}/_{4}$ in. provided, O.K.

Dimensions of longitudinal fillet welds:

Length of each fillet weld = 4 in.

Distance between longitudinal welds, b = 4 in.

Since the longitudinal fillet weld length is not less than the transverse distance between the welds (i.e., L is not less than b), the spacing between longitudinal fillet welds is adequate.

Weld capacity:

Since E70XX electrodes are specified, equations (10-5a) and (10-5b) can be used:

For ¼ in. (i.e., 4/16 in.) fillet weld, D = 4

Total length of fillet weld, L = 4 in. + 4 in. = 8 in.

$$\phi R_n = 1.392DL$$

$$= (1.392)(4)(8) = \mathbf{44.5\ kips}.\ \text{(LRFD)}$$

$$\frac{R_n}{\Omega} = 0.928DL$$

$$= (0.928)(4)(8) = \mathbf{29.7\ kips}.\ \text{(ASD)}$$

EXAMPLE 10-2

Fillet Weld Strength

Determine the capacity of the connection shown in Figure 10-13 based on weld strength alone. Electrodes are E70XX.

Solution

From Example 10-1, the AISC requirements for weld size are satisfied.

For ¼ in. (i.e., 4/16 in.) fillet weld, D = 4

Total length of longitudinal fillet weld, L_ℓ = 4 in. + 4 in. = 8 in.

Total length of transverse fillet weld, L_t = 6 in.

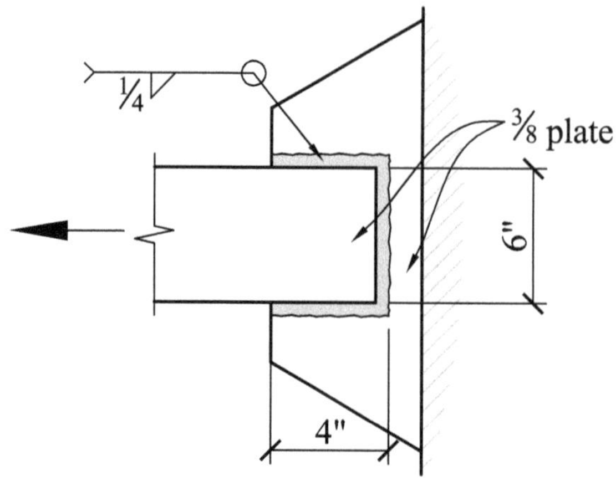

FIGURE 10-13 Detail for Example 10-2

LRFD Solution:

From equation (10-7),

$$\phi R_n = 0.85\phi F_{nw}A_{wel} + 1.5\phi F_{nw}A_{wet},$$

$$\phi F_{nw}A_{wel} = (0.75)[(0.6)(70)]\left[(0.707)\left(\frac{D}{16}\right)L_\ell\right] = 1.392DL_\ell$$

$$= (1.392)(4)(8) = 44.5 \text{ kips}$$

$$\phi F_{nw}A_{wet} = (0.75)[(0.6)(70)]\left[(0.707)\left(\frac{D}{16}\right)L_t\right] = 1.392DL_t$$

$$= (1.392)(4)(6) = 33.4 \text{ kips}$$

$$\phi R_n = (0.85)44.5 + (1.5)33.4 = 87.9 \text{ kips.}$$

The design strength of the weld is ϕR_n = **87.9 kips**.

ASD Solution:

Since both longitudinal and transverse welds are present, equations (10-8a) and (10-8b) are used.

From equation (10-8),

$$\frac{R_n}{\Omega} = \frac{0.85F_{nw}A_{wel} + 1.5F_{nw}A_{wet}}{\Omega},$$

$$\frac{F_{nw}A_{wel}}{\Omega} = \frac{[(0.6)(70)]\left[(0.707)\left(\frac{D}{16}\right)L_\ell\right]}{2.0}$$

$$= 0.928DL_\ell = (0.928)(4)(8) = 29.7 \text{ kips}$$

$$\frac{F_{nw}A_{wet}}{\Omega} = \frac{[(0.6)(70)]\left[(0.707)\left(\frac{D}{16}\right)L_t\right]}{2.0}$$

$$= 0.928DL_t = (0.928)(4)(6) = 22.3 \text{ kips}$$

$$\frac{R_n}{\Omega} = (0.85)29.7 + (1.5)22.3 = 58.7 \text{ kips.}$$

The available strength of the weld is R_n/Ω = **58.7 kips**.

Thus far, only the weld strength has been considered in the analysis of connections. We will now consider the strength of the connected elements or parts. In some cases, the connection is designed for the capacity of the connected elements in order to ensure that the connection does not become the weak link in the load path.

From Chapter 4, the strength of an element in *tensile yielding* is

$$\phi R_n = \phi F_y A_g, \qquad \text{for LRFD} \tag{10-9a}$$

$$\frac{R_n}{\Omega} = \frac{F_y A_g}{\Omega}, \qquad \text{for ASD} \tag{10-9b}$$

Where,

ϕ = 0.90

Ω = 1.67

F_y = Yield stress, and

A_g = Gross area of the connected element.

The strength of an element in *tensile rupture* (see Chapter 4) is

$$\phi R_n = \phi F_u A_e, \qquad \text{for LRFD} \tag{10-10a}$$

$$\frac{R_n}{\Omega} = \frac{F_u A_e}{\Omega}, \qquad \text{for ASD} \tag{10-10b}$$

Where,

ϕ = 0.75

Ω = 2.0

F_u = Ultimate tensile stress,

A_e = Effective area of the connected element

= $A_n U$,

A_n = Net area of the connected element, and

U = Shear lag factor (see Chapter 4 and Table 4-5).

The *block shear* strength from Chapter 4 is given as

$$\phi R_n = \phi\left(0.60F_u A_{nv} + U_{bs} F_u A_{nt}\right) \le \phi\left(0.60F_y A_{gv} + U_{bs} F_u A_{nt}\right), \qquad \text{for LRFD} \tag{10-11a}$$

$$\frac{R_n}{\Omega} = \frac{\left(0.60F_u A_{nv} + U_{bs} F_u A_{nt}\right)}{\Omega} \le \frac{\left(0.60F_y A_{gv} + U_{bs} F_u A_{nt}\right)}{\Omega}, \qquad \text{for ASD} \tag{10-11b}$$

Where,

ϕ = 0.75,

Ω = 2.0

A_{nv} = Net area subject to shear rupture,

A_{nt} = Net area subject to tensile rupture,

A_{gv} = Gross area subject to shear rupture, and

A_{gt} = Gross area subject to tensile rupture

U_{bs} = Block shear coefficient

= 1.0 for uniform tension stress

= 0.5 for nonuniform tension stress.

For welded tension members, the net shear area is equal to the gross shear area since there are no bolt holes, and the net tension area is also equal to the gross tension area. That is, for welded connections, $A_{nv} = A_{gv}$ and $A_{nt} = A_{gt}$, and since F_y is less than F_u, therefore, the block shear capacity from equation (10-11) for welded connections can then be reduced to:

$$\phi R_n = \phi\left(0.60F_y A_{gv} + U_{bs} F_u A_{gt}\right), \qquad \text{for LRFD} \tag{10-12a}$$

$$\frac{R_n}{\Omega} = \frac{\left(0.60F_yA_{gv} + U_{bs}F_uA_{gt}\right)}{\Omega}, \quad \text{for ASD} \tag{10-12b}$$

EXAMPLE 10-3

Design Strength of a Welded Connection

Determine the strength of the connection from Example 10-2, considering the strength of the connected elements. Assume that ASTM A572 Grade 50 steel is used for the steel plate.

Solution

From Example 10-2, the design strength of the weld was found to be ϕR_n = 87.9 kips. The strength of the connected elements in tension, shear, and block shear will now be investigated.

From the geometry of the connection,

$A_{gt} = (3/8 \text{ in.})(6 \text{ in.}) = 2.25 \text{ in.}^2$

$A_{gv} = (0.375)(4 \text{ in.} + 4 \text{ in.}) = 3.0 \text{ in.}^2$

$U = 1.0$ (See Table 4-5)

$A_n = A_g = (3/8 \text{ in.})(6 \text{ in.}) = 2.25 \text{ in.}^2$

$A_e = A_nU = (3/8 \text{ in.})(6 \text{ in.})(1.0) = 2.25 \text{ in.}^2$

$U_{bs} = 1.0$ (the tension stress is assumed to be uniform; see Chapter 4 or *AISCM* Section CJ4.3)

$F_y = 50$ ksi

$F_u = 65$ ksi

LRFD Solution:

Weld failure:

$\phi R_n = \mathbf{87.9}$ *kips* (**controls**). See Example 10-2.

Tensile yielding:

$$\phi R_n = \phi F_yA_g = (0.9)(50)(2.25) = 101.3 \text{ kips}$$

Tensile rupture:

$$\phi_R R_n = \phi_R F_uA_e = (0.75)(65)(2.25) = 109.7 \text{ kips}$$

Block shear:

$$\phi R_n = \phi\left(0.60F_yA_{gv} + U_{bs}F_uA_{gt}\right) = 0.75\left[(0.60)(50)(3.0) + (1.0)(65)(2.25)\right] = 177.2 \text{ kips.}$$

The smallest of the design strengths from the above four failure modes will be the critical failure mode; for this example, it is weld failure (see Example 10-2), where the design strength, ϕR_n = 87.9 kips.

ASD Solution:

From Example 10-2, the allowable strength of the weld was found to be $\frac{R_n}{\Omega} = 58.7$ kips. The strength of the connected elements in tension, shear, and block shear will now be investigated.

Tensile yielding:

$$\frac{R_n}{\Omega} = \frac{F_y A_g}{\Omega}$$

$$= \frac{(50)(2.25)}{1.67} = 67.4 \text{ kips}$$

Tensile rupture:

$$\frac{R_n}{\Omega} = \frac{F_u A_e}{\Omega}$$

$$= \frac{(65)(2.25)}{2.0} = 73.1 \text{ kips}$$

Block shear:

$$\frac{R_n}{\Omega} = \left(\frac{0.60F_y A_{gv} + U_{bs} F_u A_{gt}}{\Omega}\right)$$

$$= \frac{[(0.60)(50)(3.0) + (1.0)(65)(2.25)]}{2.0} = 118.1 \text{ kips.}$$

The smallest of the allowable strengths from the above four failure modes will be the critical failure mode; for this example, it is the weld failure (see Example 10-2), where the allowable strength, $\frac{R_n}{\Omega} = 58.7$ kips.

10-6 PLUG AND SLOT WELD STRENGTH

The strength of a plug or slot weld is a function of the size of the hole or slot that contains the weld as well as the depth or thickness of the weld. The cross-sectional area of the hole or slot in the plane of the connected parts is used to determine the strength of the weld. The strength of plug and slot welds is calculated as follows:

$$\phi R_n = \phi F_w A_w, \quad \text{for LRFD} \qquad (10\text{-}13a)$$

$$\frac{R_n}{\Omega} = \frac{F_w A_w}{\Omega}, \quad \text{for ASD} \qquad (10\text{-}13b)$$

Where,

$\phi = 0.75$

$\Omega = 2.0$

$F_w = 0.6F_{EXX}$, and

A_w = Cross-sectional area of the *plug* weld (= $\frac{\pi d^2}{4}$ for a plug weld of diameter, d.)

A_w = Cross-sectional area of the *slot* weld (= dL, for a slot weld with a width, d, and length, L).

The dimensional requirements for plug and slot welds are given in Section 10-4 (see Figure 10-9).

EXAMPLE 10-4

Plug Weld Strength

Determine the strength of the plug weld in the built-up connection shown in Figure 10-14, considering the strength of the weld only. Electrodes are E70XX.

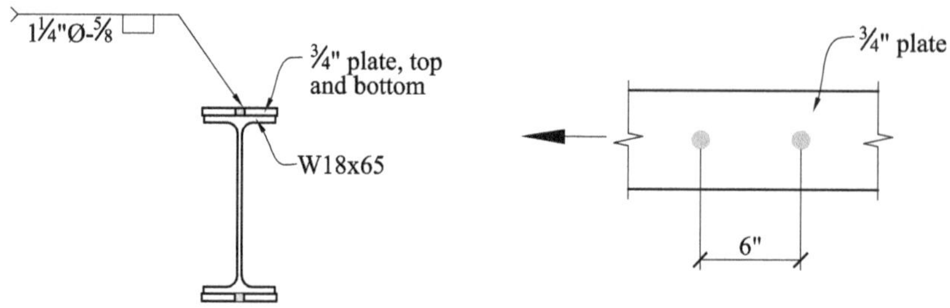

a. section

b. top view showing plug weld spacing

FIGURE 10-14 Detail for Example 10-4

Solution

Weld Dimensions (see Figure 10-9 for plug weld dimensional limitations.)

The dimensions to the left of the plug weld symbol, 1¼"ϕ - 5/8, indicate the following:
Diameter of plug weld = 1¼ in. = 1.25 in.
Thickness or depth of plug weld, $w = {}^5/_8$ in.

Minimum plug weld diameter:

$t_{\text{plate}} = {}^3/_4$ in. (Note: the plug weld is within the plate thickness.)

flange thickness (W18 × 65), $t_f = {}^3/_4$ in.

$d_{\min} = t + {}^5/_{16}\text{ in.} = {}^3/_4\text{ in.} + {}^5/_{16}\text{ in.} = 1^1/_{16}\text{ in.}$

This should be rounded up to the *next odd* ${}^1/_{16}$ in; therefore, use $d_{\min} = 1^3/_{16}\text{ in.} < 1^1/_4\text{ in.}$ OK.

Maximum plug weld diameter:

$$d_{\max} = 2.25w = (2.25)\ ({}^5/_8\text{ in.}) = 1.41\text{ in.} > 1^1/_4\text{ in. OK.}$$

Minimum thickness of the plug weld:

Since $t_{\text{plate}} = {}^3/_4\text{ in.} > w = {}^5/_8$ in., therefore, w_{minimum} is equal to the

Larger of ${}^5/_8$ in. or $\dfrac{t_{plate}}{2}$

$\dfrac{t_{plate}}{2} = \dfrac{{}^3/_4\text{ in.}}{2} = {}^3/_8$ in.; therefore, the required minimum thickness of the plug weld is ${}^5/_8$ in.

Since the plug weld thickness provided $= {}^5/_8$ in., therefore, OK.)

Minimum spacing:

$$s_{min} \geq 4 \times \text{weld diameter} = 4\text{d}$$
$$\geq (4)(1.25) = 5\text{ in.} < 6\text{ in. plug weld spacing provided is OK.}$$

Therefore, all the plug weld dimensional requirements in Figure 10-9 are satisfied.

LRFD:

From equation (10-13a), the design weld strength is

$$\phi R_n = \phi F_w A_w = \phi 0.6 F_{EXX}\left(\frac{\pi d^2}{4}\right)$$

$$= (0.75)(0.6)(70\text{ ksi})\left[\frac{\pi(1.25)^2}{4}\right] = 38.7\text{ kips.}$$

ASD:

From equation (10-13b), the allowable weld strength is

$$\frac{R_n}{\Omega} = \frac{F_w A_w}{\Omega} = \frac{0.6F_{EXX}\left(\frac{\pi d^2}{4}\right)}{\Omega}$$

$$= \frac{(0.6)(70 \text{ ksi})\left[\frac{\pi (1.25)^2}{4}\right]}{2.0} = 25.8 \text{ kips.}$$

EXAMPLE 10-5

Slot Weld Strength

Determine the strength of the connection shown in Figure 10-15, considering the strength of the weld and the steel angle. Electrodes are E70XX, and the steel is ASTM A36.

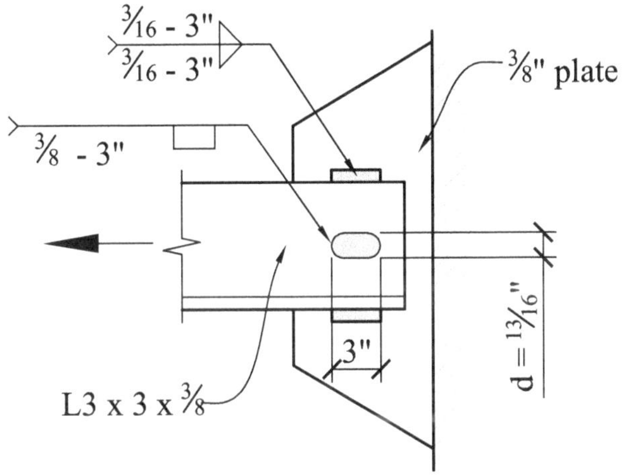

FIGURE 10-15 Detail for Example 10-5

Solution

Weld dimensions (see Figure 10-9 for slot weld dimensional limitations.)

The connection consists of fillet and slot welds.

Slot weld:

The dimensions to the left of the slot weld symbol, $^3/_8$" – 3", indicate the following:
Thickness of slot weld, w = $^3/_8$ in., and
Length of slot weld = 3 in.

Minimum slot weld width (see Figure 10-9):

t for L3 × 3 × $^3/_8$ = $^3/_8$ in.
$d_{min} = t + {}^5/_{16}$ in.= $^3/_8$ in. + $^5/_{16}$ in. = $^{11}/_{16}$ in.

This should be rounded up to the *next odd* $^1/_{16}$ in.; therefore, d_{min} = $^{13}/_{16}$ in. ($^{13}/_{16}$ in. provided). O.K.

Maximum slot weld width (see Figure 10-9):

$d_{max} = 2.25w = (2.25)(^3/_8 \text{ in.}) = 0.85 \text{ in.} > {}^{13}/_{16} = 0.82$ in. OK.

Minimum slot weld thickness (see Figure 10-9):

t for L3 × 3 × $^3/_8$ = $^3/_8$ in. < $^5/_8$ in.
Therefore, from Figure 10-9, the minimum slot weld thickness, w_{min} = t = $^3/_8$ in. OK.
Maximum length of slot weld = $10w$ = (10)($^3/_8$ in.) = 3.75 in. > 3 in. O.K.
Therefore, the dimensional requirements for the slot welds from Figure 10-9 are satisfied.

Fillet weld:

Size of fillet weld = $^3/_{16}$ in.
Length of each fillet weld = 3 in.

Weld Strength: LRFD

From equation (10-13a), the design shear strength for the slot weld, ϕR_n, is calculated as follows:
Slot width width, d = $^{13}/_{16}$ in.
Slot weld length, L = 3 in.
The surface area of the slot weld = (dL)

$$\phi R_n = \phi F_w A_w = \phi 0.6 F_{EXX}(dL)$$
$$= (0.75)(0.6)(70 \text{ ksi})\left(\tfrac{13}{16} \text{ in.}\right)(3 \text{ in.})$$
$$= 76.8 \text{ kips} \qquad (\text{design strength of the slot weld}).$$

From Equation 10-5a, the design shear strength for the fillet weld is calculated as follows:
For $^3/_{16}$ in. weld, D = 3.
Total length of fillet weld, L = 3 in. + 3 in. = 6 in.

$$\phi R_n = 1.392DL$$
$$= (1.392)(3)(6 \text{ in.}) = 25.0 \text{ kips}$$

Total weld strength consists of the slot weld strength plus the fillet weld strength, thus the total weld strength = 76.8 + 25.0 = 101.8 kips.

Tensile Strength of L3 × 3 × 3/8 (see Chapter 4):

Length of longitudinal welds are L_1 = 3 in., and L_2 = 3 in.

Average length of longitudinal fillet welds, $L = \dfrac{L_1 + L_2}{2} = \dfrac{3\text{ in.} + 3\text{ in.}}{2} = 3\text{ in.}$

For L3 × 3 × $^3/_8$, obtain the following properties from part 1 of the *AISCM:*
Gross cross-sectional area, A_g = 2.11 in.^{2}
$\bar{x} = 0.884$ in.

Tensile yielding failure mode:

$$\phi R_n = \phi F_y A_g$$
$$= (0.9)(36)(2.11) = 68.4 \text{ kips}$$

Tensile rupture failure mode:

The shear lag factor is calculated as

$$U = 1 - \frac{\bar{x}}{L} \qquad (\text{See equation (4-9) and Figure 4-5})$$
$$= 1 - \frac{0.884}{3 \text{ in.}} = 0.705$$

The net cross - sectional area of the L3 × 3 × $\frac{3}{8}$ tension member allowing for the hole occupied by the slot weld is

$$A_n = A_g - (d)(t) = 2.11 - \left(\frac{13}{16}\text{in.}\right)\left(\frac{3}{8}\text{in.}\right) = 1.805\,\text{in.}^2$$

The effective area is,

$$A_e = A_n U$$
$$= (1.805)(0.705) = 1.27\ \text{in.}^2$$

$$\phi R_n = \phi F_u A_e$$
$$= (0.75)(58)(1.27) = \mathbf{55.3\ kips}\ \text{Controls.}$$

Allowable Weld Strength: ASD

From equation (10-13b), the allowable shear strength for the slot weld is calculated as follows:
Slot width, d = $^{13}/_{16}$ in.
Slot length = 3 in.

$$\frac{R_n}{\Omega} = \frac{F_w A_w}{\Omega} = \frac{0.6F_{EXX}(d)(L)}{\Omega}$$
$$= \frac{(0.6)(70\ \text{ksi})\left(\frac{13}{16}\text{in.}\right)(3\ \text{in.})}{2.0} = 51.2\,\text{kips}\ \ (\text{strength of the slot weld})$$

From Equation 10-5b, the allowable shear strength for the fillet weld is calculated as follows:
For $^{3}/_{16}$ in. weld, D = 3.
Total length of fillet weld, L = 3 in. + 3 in. = 6 in.

$$\frac{R_n}{\Omega} = 0.928DL$$
$$= (0.928)(3)(6\,\text{in.}) = 16.7\ \text{kips}$$

Total weld strength: 51.2 + 16.7 = 67.9 kips.

Allowable Tensile Strength (ASD) of L3 × 3 × 3/8:

Length of longitudinal welds are $L_1 = 3$ in., and $L_2 = 3$ in.

Average length of longitudinal fillet welds, $L = \frac{L_1 + L_2}{2} = \frac{3\,\text{in.} + 3\,\text{in.}}{2} = 3\,\text{in.}$

For L3 × 3 × $^{3}/_{8}$, obtain the following properties from part 1 of the *AISCM:*
Gross cross-sectional area, A_g = 2.11 in.^{2}
$\bar{x} = 0.884$ in.

Tensile yielding failure mode:

$$\frac{R_n}{\Omega} = \frac{F_y A_g}{\Omega}$$
$$= \frac{(36)(2.11)}{1.67} = 45.5\ \text{kips}$$

Tensile rupture failure mode:

The shear lag factor is calculated as

$$U = 1 - \frac{\bar{x}}{L} \qquad \text{(See equation (4-9) and Figure 4-5)}$$
$$= 1 - \frac{0.884}{3\ \text{in.}} = 0.705$$

The net cross - section area of the L3 × 3 × ⅜ tension member allowing for the hole occupied by the slot weld is

$$A_n = 2.11 - \left(\frac{13}{16}\text{in.}\right)\left(\frac{3}{8}\text{in.}\right) = 1.805\,\text{in.}^2$$

The effective area, $A_e = A_n U$

$$= (1.805)(0.705) = 1.27\text{ in.}^2$$

$$\frac{R_n}{\Omega} = \frac{F_u A_e}{\Omega}$$

$$= \frac{(58)(1.27)}{2.0} = \mathbf{36.8\ kips}\ \text{Controls.}$$

For both the LRFD and ASD methods, tensile rupture controls the design of this welded connection. Note that the strength of the fillet welds alone would not be adequate to match the tensile rupture strength of the angle. More fillet weld could also be added, but the use of a slot weld is shown here for illustrative purposes.

10-7 GROOVE WELD STRENGTH

There are two basic types of groove welds as follows (see Figure 10-16):

- Complete joint penetration (CJP) groove welds and
- Partial joint penetration (PJP) groove welds

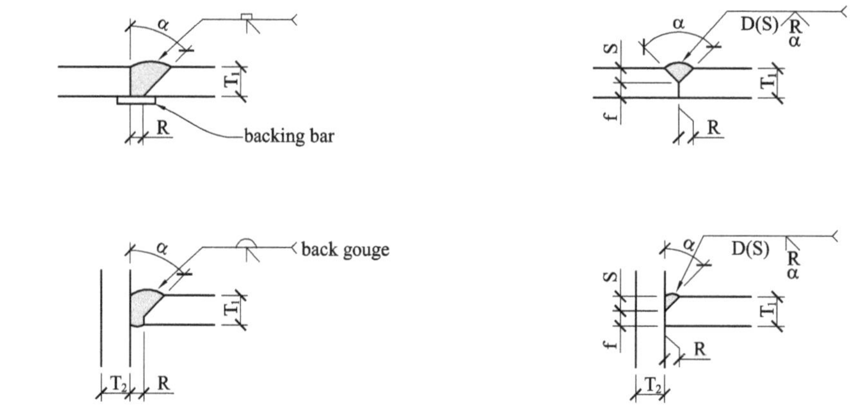

a. Complete Joint Penentration (CJP) *b. Partial Joint Penetration (PJP)*

FIGURE 10-16 Groove welds

Complete Joint Penetration (CJP) Groove Welds

The strength of a joint made with a CJP weld is a function of the strength of the base metal because the CJP weld completely fuses the two connecting parts together as if they were one member. For a CJP weld in *tension*, the nominal strength would be the product of the cross-sectional area of the smaller of the steel elements being joined and the yield strength of the steel. The process of making a CJP weld requires providing weld metal throughout the entire depth of the joint. The best quality CJP weld is one in which the weld metal is applied to one side through the entire base metal thickness and then the opposite side is backgouged (i.e., a cutting or hollowing out of the weld and part of the base metal) and then cleaned with a grinder and prepared for the application of more weld metal (see Figure 10-16a). This ensures proper fusion of the connected base metals. An alternate way to place the weld metal is to provide a backing bar on one side of the joint to contain the weld metal as the weld is placed through the material thickness. The backing bar does not provide any strength to the joint in the final condition, so it can be removed after the weld is placed.

Partial Joint Penetration (PJP) Groove Welds

A PJP weld does not extend through the entire thickness of the connecting parts, so the strength of a PJP weld is a function of the depth of the weld at the connected portion. The nominal strength of a PJP weld is then the cross-sectional area of the weld multiplied by the smaller of either the yield strength, F_y, of the connected parts or the strength of the weld metal, $0.6F_{EXX}$. The "0.6" factor on F_{EXX} is used to "compensate for factors such as the notch effect of the unfused area of the joint and uncertain quality in the root of the weld because PJP groove welds are typically not subjected to nondestructive evaluation" (see Section C-J2.4 of Ref [2]).

There are also some dimensional limitations for PJP welds that impact its design strength. The depth or effective throat of the weld is used to calculate the available cross-sectional area for all weld positions and processes as shown in Table 10-3, and the minimum effective throat of a PJP weld is given in Table 10-4. The use of single-sided partial joint penetration groove welds is not recommended for connections or joints that are subject to rotation about the toe of the weld [2]. In general, single-sided welds should be avoided, and fillet welds should be used whenever possible because they are less expensive than groove welds, which require edge preparation.

TABLE 10-3 Effective throat of partial joint penetration (PJP) groove welds

Welding process	Welding position	Groove type	Effective Throat (S)
SMAW, GMAW, FCAW	All	J or U groove 60° V	Depth of groove, D
SAW	Flat	J or U groove 60° bevel or V	Depth of groove, D
GMAW, FCAW	Flat, Horizontal	45° bevel	Depth of groove, D
SMAW	All	45° bevel	Depth of groove minus ⅛ in., i.e., D – 1/8 in.
GMAW FCAW	Vertical, overhead	45° bevel	Depth of groove minus ⅛ in. i.e., D – 1/8 in.

Source: Adapted from Table J2.1 [2].

TABLE 10-4 Minimum effective throat of partial joint penetration (PJP) groove welds

Thickness of thinner connected part	Minimum effective throat (depth), S (i.e., S_{min})
$t \leq$ ¼ in.	⅛ in.
¼ in. < $t \leq$ ½ in.	3/16 in.
½ in. < $t \leq$ ¾ in.	¼ in.
¾ in. < $t \leq$ 1 ½ in.	5/16 in.
1½ in. < $t \leq$ 2¼ in.	⅜ in.
2¼ in. < $t \leq$ 6 in.	½" in.
Over 6 in.	⅝ in.

Source: Adapted from Table J2.3, [2].

EXAMPLE 10-6

Strength of a Groove Weld

For the given detail in Figure 10-17, the steel is ASTM A572, grade 50, and the weld process is SMAW in the flat position with E70XX electrodes.

a. Determine the strength of the splice connection for a groove weld depth, $D = 1$ in.
b. Determine the minimum groove weld depth required to support an applied factored load of 75 kips.

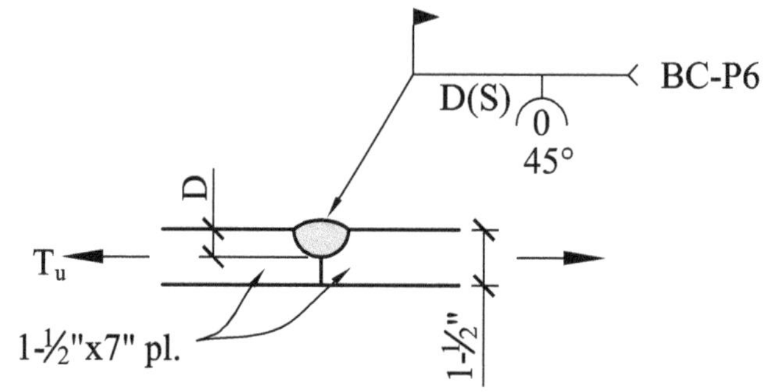

FIGURE 10-17 Weld detail for Example 10-6

Solution

Check weld dimensions:

The depth of groove weld provided, D = 1 in,

From Table 10-3, the weld is in the flat position and is made with a SMAW process; therefore, the effective groove weld depth, $S = \text{D} = 1$ in.

Thickness of thinner part connected, t = 1½ in.

Therefore, from Table 10-4, the minimum effective throat, $S = \frac{5}{16}$ in. < 1 in. O.K.

a. Strength of the splice connection for a groove weld depth, *D* = 1 in.:

(1) Base metal strength:

Groove weld depth, D = 1 in.

Width of plate = 7 in.

The base metal strength is

$\phi_t P_n = \phi_t F_y A_g = \phi_t F_y$ (D)(width of plate) = (0.9)(50 ksi)(1 in.)(7 in.) = 315 kips.

(2) Weld Strength:

$\phi_w R_n = \phi_w 0.6\, F_{EXX} A_w$

$= \phi_w 0.6 F_{EXX}$(D)(width of plate) = (0.75)(0.6)(70 ksi)(1 in.)(7 in.) = 220 kips. (controls)

The smaller of the above two strength values controls, therefore, the weld strength controls; so, the maximum value of T_u is 220 kips.

b. Minimum groove weld depth required to support an applied factored load of 75 kips:

From Part (a), it was determined that the weld strength ($\phi_w R_n$ = 220 kips) controls the design at a specified groove weld depth of 1 in. Therefore, the required minimum groove weld depth to resist the factored load, P_u = 75 kips, is

$$D_{\text{min}} = \left(\frac{75 \text{ kips}}{220 \text{ kips}}\right)(1 \text{ in.}) = 0.34 \text{ in.}, \text{ say } D = \tfrac{3}{8} \text{ in.}$$

Alternatively,

$\phi_w R_n = \phi_w 0.6\, F_{EXX}\, A_w = \phi_w 0.6 F_{EXX}(D_{min})(\text{plate width}) \geq T_u$

$(0.75)(0.6)(70 \text{ ksi})(D_{min})(7 \text{ in.}) \geq 75 \text{ kips}$

$D_{min} = 0.34 \text{ in.} = 3/8 \text{ in.}$

This is greater than the required minimum value of $\frac{5}{16}$ in. for the groove weld depth from Table 10-4, so a partial penetration groove weld with a depth of $\frac{3}{8}$ " is required.

10-8 ECCENTRICALLY LOADED WELDS: SHEAR ONLY

The analysis of eccentrically loaded welds with the eccentricity in the plane of the welds follows the same principles that were used in Chapter 9 for the analysis of eccentrically loaded bolts subjected to in-plane shear. As shown in Figure 10-18, eccentrically loaded welds with the applied load in the same plane as the weld group are subjected to direct shear from the applied load, *P*, plus the additional shear caused by the moment (*P*e) resulting from the eccentric load, *P*. In this case, no tension stresses are induced in the welds, but only shear stresses are present. The three methods that can be used for the analysis of eccentrically loaded weld groups subject to in-plane shear include the following, but only the first two methods will be covered in detail in this text:

- Instantaneous center of rotation (IC) method or the ultimate strength method
- Elastic method
- Plastic methods: conventional and optimum plastic methods.

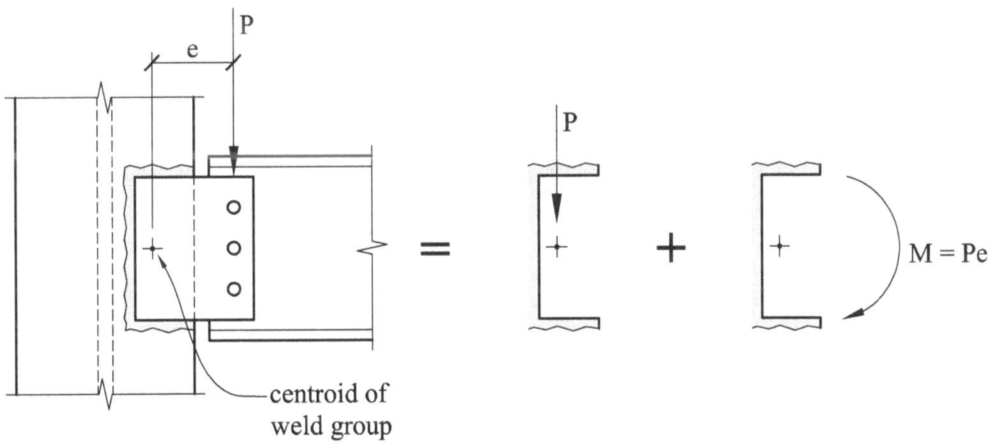

FIGURE 10-18 Eccentrically loaded weld

Instantaneous Center (IC) of Rotation Method or Ultimate Strength Analysis:

The instantaneous center of rotation (IC) method or the ultimate strength analysis method accounts for the strain compatibility of the weld elements and is more accurate than the elastic method, but it requires an iterative solution. As was the case with eccentrically loaded bolted connections, an eccentrically loaded weld will rotate about a point called the instantaneous center (IC) (see Figure 10-19). For analysis, initially the location of the IC has to be assumed, and the IC is assumed as the origin with x- and y- coordinates of (0, 0); knowing the IC location would yield r_o in Figure 10-19. The weld must also be broken down into discrete elements of equal lengths, ℓ_w, for the analysis, and the x- and y- coordinates of the centroid of each discrete weld element relative to the IC are determined, together with the lengths of each element; from this, the radial distance, $r_i = \sqrt{x_i^2 + y_i^2}$, of each weld element from the IC can then be determined. Equations (10-16) through (10-22) are then used to determine the nominal resultant shear strength, R_i , of each weld element (see Figure 10-19). It is recommended that at least 20 elements be selected for the longest weld in the group in order to maintain reasonable accuracy in the solution [2]. The resultant shear strength of each weld element, R_i , is assumed to be perpendicular to the radial line joining the centroid, or CG, of the weld element to the IC.

Using a load-deformation relationship, the nominal strength of the weld group is obtained as follows [2]:

$$R_{ix} = \Sigma F_{wix} A_{wi}, \text{ and} \tag{10-14}$$

$$R_{iy} = \Sigma F_{wiy} A_{wi}, \tag{10-15}$$

Where,

R_{ix} = x component of the nominal weld strength,

R_{iy} = y component of the nominal weld strength,

A_{wi} = Effective weld throat area of element i

= $0.707 w \ell_{wi}$

F_{wi} = nominal weld stress for element i.

$$= 0.60 F_{EXX} \left(1 + 0.5 \sin^{1.5}\theta_i\right)\left[p_i\left(1.9 - 0.9 p_i\right)\right]^{0.3}, \tag{10-16}$$

F_{wix} = x component of the nominal weld stress, F_{wi},

F_{wiy} = y component of the nominal weld stress, F_{wi},

Where,

$$p_i = \frac{\Delta_i}{\Delta_{mi}}, \tag{10-17}$$

$$\Delta_i = \frac{r_i \Delta_{u,cr}}{r_{crit}}, \tag{10-18}$$

r_i = Distance from the IC to the centroid of the discrete weld element i, in. $= \sqrt{x_i^2 + y_i^2}$

x_i, y_i = x- and y-coordinates of the centroid of each discrete weld element measured from the IC,

$\Delta_{u,cr}$ = Deformation, Δ_{ui}, of the weld element that has the smallest $\frac{\Delta_{ui}}{r_i}$ ratio. This usually occurs in the weld element that is farthest from the IC (i.e., the element with the largest r_i)

$\Delta_{u,cr}$ = Deformation, Δ_{ui} , of the weld element that has the smallest $\frac{\Delta_{ui}}{r_i}$ ratio. This usually occurs in the weld element that is farthest from the IC (i.e., the element with the largest r_i)

r_{crit} = Distance from the IC to the weld element with minimum or smallest $\frac{\Delta_{ui}}{r_i}$ ratio, in.

Δ_{ui} = Deformation of weld element i at ultimate stress, in. (This is usually the weld element located farthest from the IC)

$$= 1.087 w \left(\theta_i + 6\right)^{-0.65} \le 0.17 w \tag{10-19}$$

Δ_{mi} = Deformation of weld element i at maximum stress, in.

$$= 0.209 w \left(\theta_i + 2\right)^{-0.32} \tag{10-20}$$

Δ_i = Deformation of weld element, i, at an intermediate stress level

θ_i = Angle in degrees between the longitudinal axis of the weld element, i, and the direction of the resultant force, R_i, on the weld element (see Figure 10-19),

w = Weld leg size, in.

ℓ_w = length of discrete welds

F_{EXX} = Filler metal classification strength (ksi), varies from 60 ksi to 120 ksi, with 70 ksi being the most typical in practice.

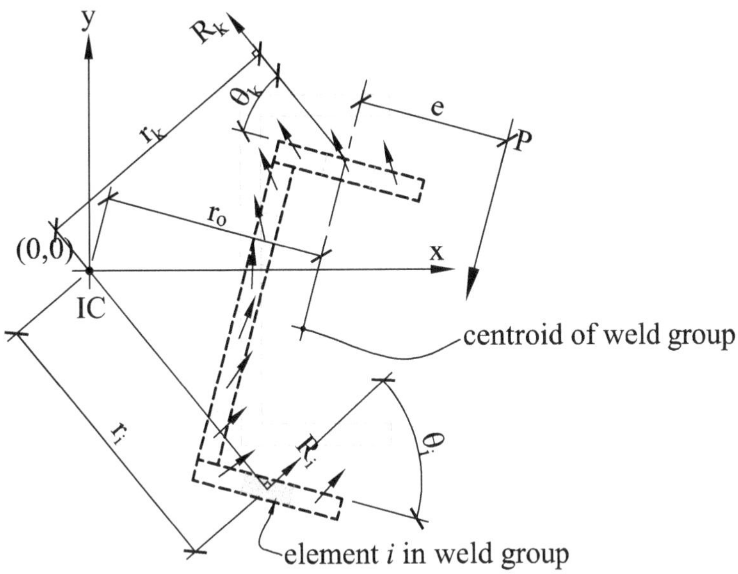

FIGURE 10-19 Eccentrically loaded weld: IC method

The resistance or strength, R_i, of each weld element calculated above is assumed to act perpendicular to a radial line joining the IC to the centroid or CG of the weld element i (see Figure 10-19). When the chosen location of the IC is correct, then all the equilibrium equations will be satisfied (i.e., $\Sigma F_x = 0$, $\Sigma F_y = 0$, and $\Sigma M = 0$). A summary of the step-by-step procedure for using the IC or the ultimate strength analysis method for eccentrically loaded welds subjected to inplane shear is as follows:

1. Divide the welds into equal discrete lengths, ℓ_w (e.g., 0.5 in.). Use at least 20 weld segments along the longest weld length [2]. Note and select the fillet weld size, w.
2. Assume the location of the coordinates of the instantaneous center (IC), and therefore determine r_o. (r_o = radial distance from the IC to the centroid of the weld group).
3. Locate the x- and y- coordinates of all the discrete weld elements relative to the location of the IC
4. Calculate the radial distance from the centroid of each weld element to the IC That is, for weld i, $r_i = \sqrt{x_i^2 + y_i^2}$.
5. The resultant strength or resistance, R_i, of each weld segment or element is perpendicular to the radial line from the centroid of the discrete weld segment to the assumed IC location. Therefore, calculate the angle, θ_i , in degrees between the longitudinal axis of each discrete weld element and the direction of the resultant strength, R_i , in the weld element. Note that unlike bolts under eccentric loads, this angle, θ_i, is required for welds under eccentric loads [2].
6. For each discrete weld element, determine the deformation, Δ_{ui}, from equation (10-19) and calculate $\dfrac{\Delta_{ui}}{r_i}$.
7. For each discrete weld element, determine the deformation, Δ_{mi}, from equation (10-20)
8. Determine the weld element with the smallest $\dfrac{\Delta_{ui}}{r_i}$ ratio, which usually occurs in the weld element that is farthest from the IC Knowing the Δ_{ui} of this weld element, determine $\Delta_{u,cr} = \Delta_{ui}$ for this weld element, and $r_{cr} = r_i$ for this weld element.

9. Using $\Delta_{u,cr}$ and r_{cr} obtained in step 8, determine for each weld Δ_i using equation (10-18)

10. For each weld element, calculate p_i from equation (10-17) using the results from steps 7 and 9.

11. For each weld element, compute F_{wi} from equation (10-16) and calculate R_{nx} from equation (10-14) and R_{ny} from equation (10-15), where, $A_{wi} = 0.707w\ell_{wi}$. Note that the resultant resistance or strength for each weld element is $R_i = \sqrt{R_x^2 + R_y^2}$.

12. Check the following rotational and force equilibrium equations (see Figure 10-19):

$$\sum M_{IC} = \sum_{i=1}^{n} R_i r_i - P(e + r_o) = 0 \qquad (10\text{-}21a)$$

$$\sum F_y = \sum_{i=1}^{n} R_{i,y} - P_y = 0 \qquad (10\text{-}21b)$$

$$\sum F_x = \sum_{i=1}^{n} R_{i,x} - P_x = 0 \qquad (10\text{-}21c)$$

If all the three equilibrium equations (10-21a), (10-21b) and (10-21c) are satisfied, that is, they all produce the same value of *P*, then the IC location assumed in Step 2 is correct. If not, select a new IC location and repeat the process. The above design steps and design parameters can be presented in a tabular format as shown in Table 10-4b.

It is evident that the analytical process involved in the instantaneous center of rotation method is quite tedious and is not practical for common use without the aid of computer software or a spreadsheet. *AISCM* Tables 8-4 through 8-11 are design aids that have been derived using the IC method, and the use of these design aids will be illustrated in Examples 10-8 and 10-9.

TABLE 10-4b Instantaneous center method

Weld element	r_i (step 4)	Δ_{ui} (step 6)	Δ_{ui}/r_i (step 6)	Δ_{mi} (step 7)	$\Delta_{u,cr}/r_{cr}$ = Largest Δ_{ui}/r (step 8)	$\Delta_i = r_i\Delta_{u,cr}/r_{cr}$ (step 9)	$P_i = \Delta_i/\Delta_{mi}$ (step 10)	F_{wi} (step 11)	$R_{i,x}$ and $R_{i,y}$ (step 11)
1									
2									
3									
...									
....									
n									

Elastic Method (In-plane Shear):

In lieu of the more tedious instantaneous center method, the elastic method can be used for the analysis of eccentrically loaded welds subjected to inplane shear.

- The elastic method is more conservative but less tedious than the IC method.
- It ignores the ductility and the redundancy (i.e., the load redistribution capability) of the weld group.

When a weld is subjected to an in-plane eccentric load as shown in Figure 10-20, the forces on the weld can be resolved into a *direct shear* component and an *in-plane torsional shear* component. Considering the direct shear component, the shear in the weld per linear inch is

$$r_p = \frac{P}{L}, \tag{10-22}$$

Where,

r_p = applied shear per linear inch of weld with components r_{px} and r_{py},

$$r_{px} = \frac{P_x}{L}, \tag{10-23}$$

$$r_{py} = \frac{P_y}{L}, \tag{10-24}$$

P = Applied load with components P_x and P_y

L = Total length of the welds in the weld group.

Considering the torsional component, the applied shear in the weld per linear inch due to the inplane moment (M = Pe) is obtained by using the flexure formula (Mc/I) from the mechanics of materials:

$$r_m = \frac{Pec}{I_p}, \tag{10-25}$$

Where,

r_m = applied shear per linear inch of weld with components r_{mx} and r_{my}

$$r_{mx} = \frac{Pec_y}{I_p}, \tag{10-26}$$

$$r_{my} = \frac{Pec_x}{I_p}, \tag{10-27}$$

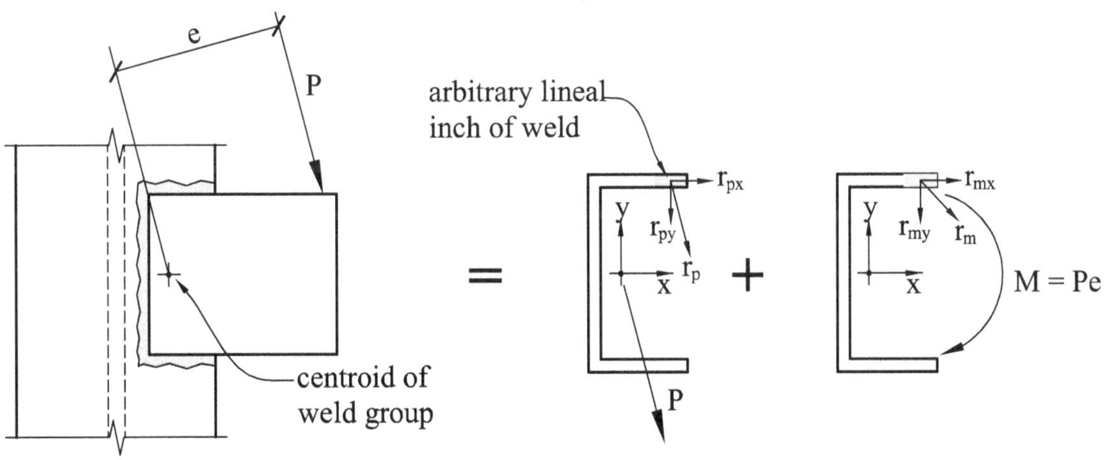

FIGURE 10-20 Eccentrically loaded weld: Elastic method

Where,

e = Distance from the applied load, P, to the centroid of the weld group

c = Distance from the centroid or center of gravity of the weld group to a point most remote from the centroid of the weld group with components c_x and c_y, along the X- and Y-axes, respectively. Note that c_x and c_y are measured from the centroid (CG) of the weld group.

I_p = Polar moment of inertia of the weld group (also commonly noted as J) = $I_x + I_y$; see Table 10-5

TABLE 10-5 Section properties of weld groups

	Weld group	Center of gravity,	Section modulus, in.^{3}		Polar moment of inertia I_p (or J), in.^{4}
		in. $\bar{x}, \bar{y}$	S_{top}	S_{bot}	
a.		$0, \frac{d}{2}$	$\frac{d^2}{6}$		$\frac{d^3}{12}$
b.	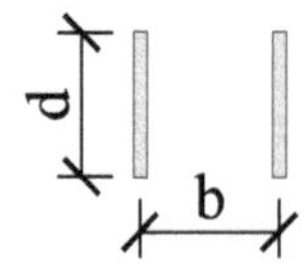	$\frac{b}{2}, \frac{d}{2}$	$\frac{d^2}{3}$		$\frac{d(3b^2+d^2)}{6}$
c.	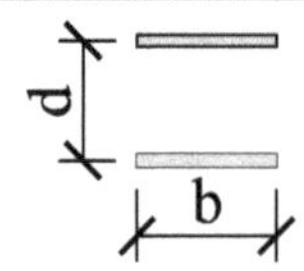	$\frac{b}{2}, \frac{d}{2}$	bd		$\frac{b^3+3bd^2}{6}$
d.	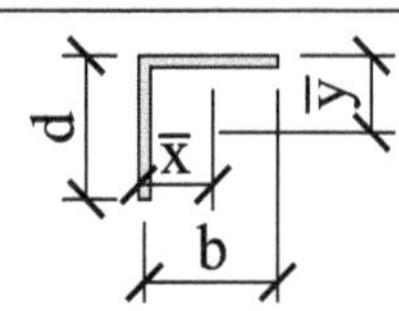	$\frac{b^2}{2(b+d)}, \frac{d^2}{2(b+d)}$	$\frac{4bd+d^2}{6}$	$\frac{d^2(4b+d)}{6(2b+d)}$	$\frac{(b+d)^4-6b^2d^2)}{12(b+d)}$
e.	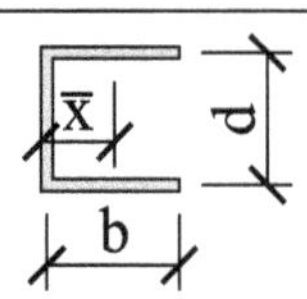	$\frac{b^2}{2b+d}, \frac{d}{2}$	$bd+\frac{d^2}{6}$		$\frac{(2b+d)^3}{12}-\frac{b^2(b+d)^2}{(2b+d)}$
f.	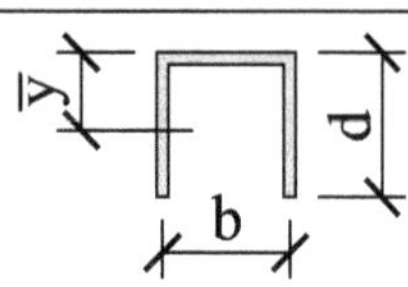	$\frac{b}{2}, \frac{d^2}{b+2d}$	$\frac{2bd+d^2}{3}$	$\frac{d^2(2b+d)}{3(b+d)}$	$\frac{(b+2d)^3}{12}-\frac{d^2(b+d)^2}{(b+2d)}$
g.	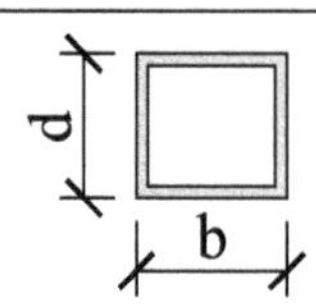	$\frac{b}{2}, \frac{d}{2}$	$bd+\frac{d^2}{3}$		$\frac{(b+d)^3}{6}$
h.	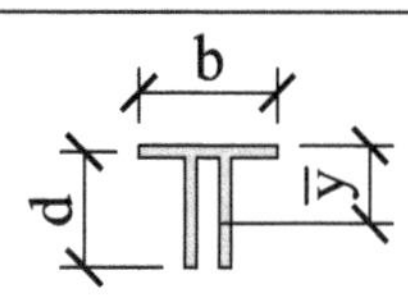	$\frac{b}{2}, \frac{d^2}{b+2d}$	$\frac{2bd+d^2}{3}$	$\frac{d^2(2b+d)}{3(b+d)}$	$\frac{(b+2d)^3}{12}-\frac{d^2(b+d)^2}{(b+2d)}$
i.	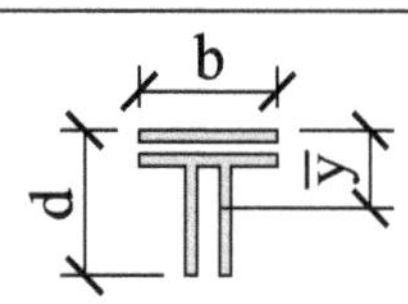	$\frac{b}{2}, \frac{d^2}{2(b+d)}$	$\frac{4bd+d^2}{3}$	$\frac{4bd^2+d^3}{6b+3d}$	$\frac{d^3(4b+d)}{6(b+d)}+\frac{b^3}{6}$

TABLE 10-5 (Continued)

j.	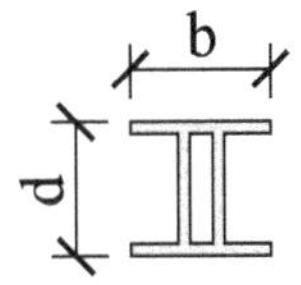	$\frac{b}{2}, \frac{d}{2}$	$bd + \frac{d^2}{3}$	$\frac{b^3 + 3bd^2 + d^3}{6}$
k.	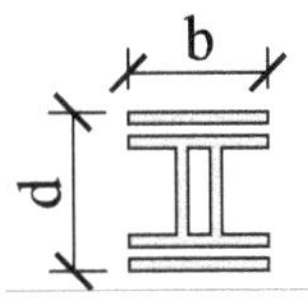	$\frac{b}{2}, \frac{d}{2}$	$2bd + \frac{d^2}{3}$	$\frac{2b^3 + 6bd^2 + d^3}{6}$
l.	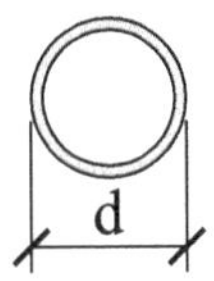	$\frac{d}{2}, \frac{d}{2}$	$\frac{\pi d^2}{4}$	$\frac{\pi d^3}{4}$

The polar moment of inertia, I_p, about the center of gravity of the weld group can be found by summing up the rectangular moments of inertia ($I_x + I_y$); however, Table 10-5 is provided as a reference for the speedy calculation of the polar moment of inertia of weld groups. In calculating the polar moment of inertia, the thickness of the weld is ignored, and the weld group is treated as line elements with a unit thickness.

Once all of the applied shear stress components are found (r_p, r_m), they are added together to determine the point at which the applied shear stress is highest, which is usually the point most remote or farthest from the centroid of the weld group, The resultant applied shear stress, r, in kips/inch is calculated as:

$$r = \sqrt{\left(r_{px} + r_{mx}\right)^2 + \left(r_{py} + r_{my}\right)^2}. \tag{10-28}$$

This resultant applied shear stress (in kips/inch) is then compared to the available strength of the weld, $\phi R_n = 1.392(D)(1\,\text{in.})$ in kips/inch, and the goal is to ensure that the strength is greater than or equal to the resultant applied shear stress; that is,

ϕR_n in kips per linear inch $\geq r$.

EXAMPLE 10-7

Eccentrically Loaded Weld: Inplane Shear Only

Determine whether the weld for the bracket connection shown in Figure 10-21 is adequate to support a factored load of P_u = 40 kips. Electrodes are E70XX. Use the Elastic Method.

Solution

From Table 10-5, the weld group is identified as type "e."

Center of gravity (x, y):

The location of the centroid (CG) of the Type "e" weld group is as shown in Table 10-5:
The weld dimensions for Type "e" weld group are:
b = 5 in.; and d = 10 in.

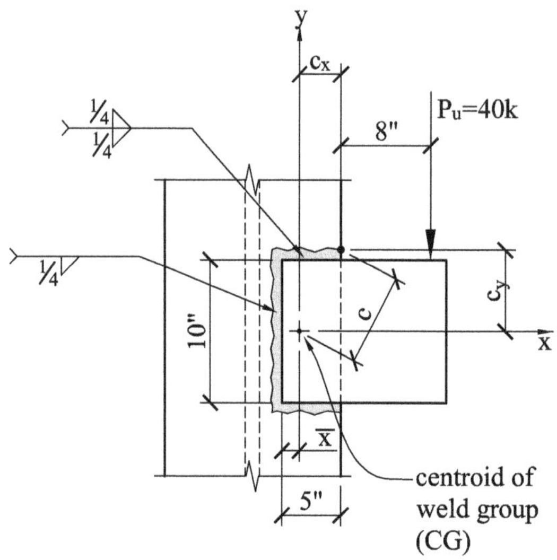

FIGURE 10-21 Connection detail for Examples 10-7 and 10-8

Centroid (CG):

$$\bar{x} = \frac{b^2}{2b+d}, \ \bar{y} = \frac{d}{2}$$

$$\text{That is, } \bar{x} = \frac{5^2}{2(5)+(10)} = 1.25\text{in}; \ \textit{and} \ \bar{y} = \frac{10}{2} = 5\text{in}.$$

$$e = 8 \text{ in.} + 5 \text{ in.} - \bar{x} = 8 \text{ in.} + 5 \text{ in.} - 1.25 \text{ in.} = 11.75 \text{ in.}$$

Total length of welds in the group, L = 5 in. + 10 in. + 5 in. = 20 in.

Polar moment of inertia:

The polar moment of inertia for Type "e" weld group from Table 10-5 is

$$I_p = \frac{(2b+d)^3}{12} - \frac{b^2(b+d)^2}{(2b+d)}$$

$$= \frac{\left[(2)(5)+(10)\right]^3}{12} - \frac{\left[(5)^2\right](5+10)^2}{\left[(2)(5)+10)\right]} = 385 \text{ in.}^4$$

The location of the most remote weld element from the CG are determined as follows:

c_x =5 in. $- \bar{x}$ = 5 in. − 1.25 in. = 3.75 in.

$$c_y = \frac{d}{2} = \frac{10}{2} = 5 \text{ in.}$$

$$c = \sqrt{c_x^{\ 2} + c_y^{\ 2}} = \sqrt{3.75^2 + 5^2} = 6.25 \text{ in.}$$

Note that c, c_x and c_y are measured from the centroid (CG) of the weld group to the farthest weld element.

The eccentricity of the load from the centroid of the weld is,

e = 8 in. + c_x = 8 + 3.75 in. = 11.75 in.

Weld stresses due to direct shear:

From equations (10-23) and (10-24), the applied weld stresses due to direct shear are:

$$r_{px} = \frac{P_x}{L} = \frac{0}{L} = 0; \text{ since } P_x = 0.$$

$$r_{py} = \frac{P_y}{L} = \frac{40 \text{ kips}}{(20 \text{ in.})} = 2 \text{ kips / in.}$$

Weld stress due to the moment from eccentric load:

From equations (10-26) and (10-27), the applied weld stresses due to the moment caused by the eccentric load are:

$$r_{mx} = \frac{Pec_y}{I_p}$$

$$= \frac{(40)(11.75)(5)}{385} = 6.10 \text{ kips / in.}$$

$$r_{my} = \frac{Pec_x}{I_p}$$

$$= \frac{(40)(11.75)(3.75)}{385} = 4.57 \text{ kips / in.}$$

Total applied weld stress:

From equation (10-28), the total applied weld stress is

$$r = \sqrt{(r_{px} + r_{mx})^2 + (r_{py} + r_{my})^2}$$

$$= \sqrt{(0 + 6.10)^2 + (2 + 4.57)^2} = 8.97 \text{ kips/in.}$$

Available weld strength:

For ¼ in. weld, D = 4

From equation (10-5a), the weld strength per unit length is

$$\phi R_n = 1.392DL$$

$$= (1.392)(4)(1 \text{ in.}) = 5.57 \text{ kips/in.}$$

$$< r = 8.97 \text{ kips/in. Not Good.}$$

or

$$\phi R_n = 1.392DL \geq r = 8.97$$

$$1.392D(1in.) \geq 8.97$$

$$D \geq 6.44 = 7 \text{ (}i.e., round\ up\ to\ the\ next\ whole\ number\text{)}$$

That is, a 7/16-in. fillet weld is required using the Elastic Method.

The applied weld stress, *r*, is found to be greater than the available weld strength, ϕR_n, therefore, the weld size would have to be increased; by inspection, a $^7/_{16}$-in. weld would be required, and this indicates a multiple-pass weld since its size is greater than $^{5}/_{16}$-in., the maximum fillet weld size for a single pass. Multi-pass welds are more expensive than single-pass welds, and since the volume of weld material in a fillet weld is proportional to the square of the fillet weld size, w, (see Figure 10-10), a $^{7}/_{16}$-in. weld will use almost twice as much volume of weld material compared to a $^{5}/_{16}$-in. weld. Note that in the preceding calculation, the weld length, L, was taken as 1 in. since the applied stress was calculated per unit length of the weld.

EXAMPLE 10-8

Eccentrically Loaded Weld: Inplane Shear Only, Using the *AISCM* Design Aids (i.e., the IC method)

Determine, using the appropriate design aid from the *AISCM*, whether the weld in Example 10-7 is adequate.

Solution

Note: In the tables in Part 8 of the *AISCM*, the weld symbol used, ℓ, is not the same as the total weld length, L, in the weld group.

$D = 4$ (for $^{1}/_{4}$-in. weld)

Determine the location of the center of gravity (CG) of the weld group and then determine c_x and c_y.

$c_x = 3.75$" (as determined previously in Example 10-7). Also, see Figure 10-21.

Since a type "e" weld (see Table 10-5 and Figure 10-21) is specified for this example, *AISCM* Table 8-8 is applicable. Since the applied load, P, is vertical, the angle between the load direction and the vertical, $\theta = 0°$.

In *AISCM* Table 8-8, the design parameters to be used are as follows:

ℓ = Vertical length of the weld = d = 10 in. (see type "e" weld in Figure 10-21 and Table 10-5)

$k\ell$ = Horizontal length of the weld = b = 5 in. (see type "e" weld in Figure 10-21 and Table 10-5)

The eccentricity of the applied load measured from the CG of the weld group is

$$e_x = a\ell = 8\text{ in.} + c_x = 8\text{ in.} + 3.75\text{ in.} = 11.75\text{ in.} \qquad \text{(see Figure 10-21)}$$

$$e_x = a\ell$$

$$\text{Therefore, } a = \frac{e_x}{\ell} = \frac{11.75\text{ in.}}{10\text{ in.}} = 1.175$$

$$k\ell = b = 5\text{ in.}$$

$$\text{Therefore, } k = \frac{5\text{ in.}}{\ell} = \frac{5\text{ in.}}{10\text{ in.}} = 0.5$$

Entering the *AISCM* Table 8-8 ($\theta = 0$) with $a = 1.175$ and $k = 0.5$, we obtain the following parameters:

$$C = 1.51 \ (\text{by interpolation})$$

$$D_{min} = \frac{P_u}{\phi C C_1 \ell}$$

$(C_1 = 1.0$ for E70XX, see *AISCM* Table 8-3$)$

The required minimum number of sixteenths-of-an-inch fillet weld size is

$$D_{min} = \frac{40\text{ kips}}{(0.75)(1.51)(1.0)(10\text{ in.})} = \mathbf{3.53} < D = 4 \quad \text{OK.}$$

The summary of required weld sizes from Example 10-7 (Elastic method) and Example 10-8 (IC method) is as follows:

Elastic method: 7/16-in. fillet weld required

IC method: ¼-in. fillet weld required.

Using *AISCM* Table 8-8, derived from the more accurate IC method, it is found that a $^1/_4$-in. weld is, in fact, adequate, in contrast to Example 10-7, where the elastic method indicates that a ¼ in. weld was inadequate and a 7/16-in. weld was required. As discussed previously, the elastic method is more conservative and less accurate than the IC method because of the simplified assumptions that are made in the elastic method. Note that in the elastic method, the *center of gravity of the weld group*, which can easily be determined, is the point of reference, whereas in the IC method, the *instantaneous center* is the point of reference, and has to be determined by an iterative procedure.

10-9 ECCENTRICALLY LOADED WELDS: SHEAR PLUS TENSION

A load that is applied eccentrically to a weld group that is not in the plane of the weld group (see Figure 10-22) produces both shear and tension stress components in the weld group. The eccentric load is resolved into a direct in-plane shear component due to the applied load, *P*, and a tension stress component caused by the moment, *Pe*. Using a unit width for the weld thickness, the stress components are calculated as follows:

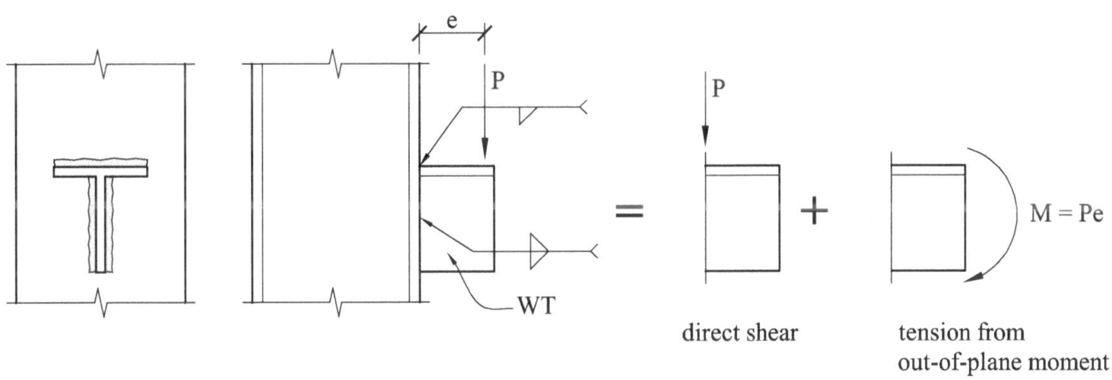

FIGURE 10-22 Eccentrically loaded weld in shear plus tension

Direct shear component:

$$f_p = \frac{P}{L} \qquad (10\text{-}29)$$

Tension component due to moments:

$$f_m = \frac{Pe}{S}, \qquad (10\text{-}30)$$

where

P = Applied load

e = Eccentricity of the load with respect to the plane of the weld

L = Total length of welds in the group

S = Section modulus of the weld group (see Table 10-5).

The resulting applied weld stress (in kips/inch) is found by adding the shear and tension components as follows:

$$f_r = \sqrt{f_p^{\,2} + f_m^{\,2}} \qquad (10\text{-}31)$$

This applied weld stress (in kips/in.) is compared to the design weld strength $\phi R_n = 1.392(D)(1\text{ in.})$ in kips/in., and the goal is to ensure that ϕR_n in kips per linear inch $\geq f_r$.

Alternatively, the required fillet weld dimension, D, can be determined by solving the equation

$$1.392D(1in.) \geq f_r$$

$$D_{reqd} \geq \frac{f_r}{1.392}$$

The fillet weld dimension, D, should be rounded to the next whole number, and the fillet weld size is $\frac{D}{16}in.$

EXAMPLE 10-9

Elastic Method for Eccentrically Loaded Weld: Shear Plus Tension (LRFD)

Determine the required fillet weld size for the bracket connection shown in Figure 10-23, considering the strength of the weld only. Compare the results with the appropriate design table from the *AISCM*. Electrodes are E70XX. Use the LRFD method.

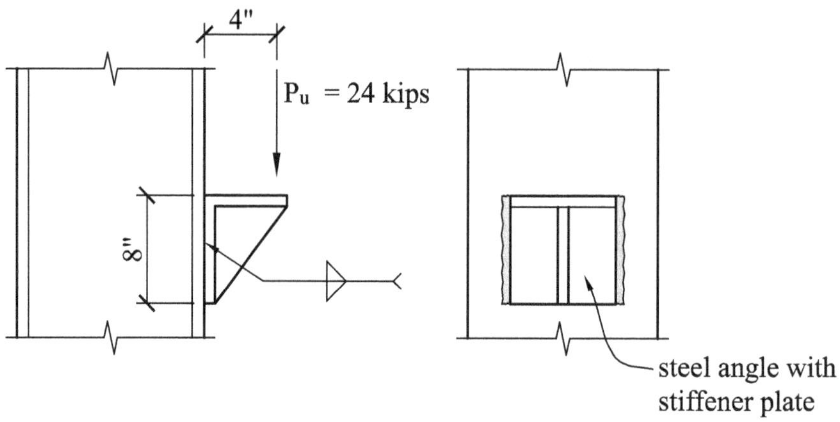

FIGURE 10-23 Connection detail for Example 10-9

Solution

The factored load, P_u = 24 kips

The load eccentricity, $e = 4$ in.

The total length of weld in the weld group is L = 8" + 8" = 16"

From Table 10-5, this indicates weld group type "b";

d = 8 in.

b is unknown but not needed to solve this problem.

From Table 10 - 5, the section modulus of the weld group, $S = \frac{d^2}{3}$

$$= \frac{(8)^2}{3} = 21.3 \text{ in.}^3/\text{in.}$$

Weld stress due to direct shear:

From equation (10-29), the factored applied direct shear stress is

$$f_{pu} = \frac{P_u}{L}$$

$$= \frac{24\text{kips}}{16\text{in.}} = 1.5 \text{ kips / in.}$$

Weld stress due to tension from the applied moment:

From equation (10-30), the factored applied tension stress is

$$f_{mu} = \frac{P_u e}{S}$$

$$= \frac{(24 \text{ kips})(4 \text{ in.})}{21.3 \text{ in.}^3/\text{in.}} = 4.5 \text{ kips / in.}$$

Combined applied weld stress:

From equation (10-31), the resultant applied weld stress is

$$f_{ru} = \sqrt{f_{pu}^{\ 2} + f_{mu}^{\ 2}}$$

$$f_{ru} = \sqrt{(1.5)^2 + (4.5)^2} = 4.74 \text{ kips / in.}$$

Required weld thickness (Elastic method):

$\phi R_n = 1.392DL$ (from equation 10-5a), and the limit states design (i.e., LRFD) equation requires that

$$\phi R_n / L = 1.392D \geq f_{ru}$$

Therefore, $1.392D \geq 4.74$ kips/in.

$D \geq 3.42 \approx 4$; That is, use four-sixteenths inch fillet weld or ¼ in. fillet weld.

∴ use ¼ in. weld

Note that the elastic method is more conservative.

We will now compare these results with the design aids in the *AISCM*, which are based on the more accurate but more cumbersome, instantaneous center of rotation (IC) method.

Required weld thickness (*AISCM* Table 8-4 or IC method)

Using Table 10-5, the weld group in Figure 10-23 corresponds to a Type "b" weld group.

Using Figure 10-23 and *AISCM* Table 8-4 (with Angle = 0°), since the applied load, P_u, is vertical, the following parameters can be determined:

$e = 4$ in.

In *AISCM* Table 8-4, $\ell =$ vertical length of weld = 8 in.

$e = a\ell = 4$ in.

$$a = \frac{e}{\ell} = \frac{4 \text{ in.}}{8 \text{ in.}} = 0.5,$$

$k = 0$ (This is a special case when the applied load is not in the plane of the weld group)

See *AISCM*, Table 8-4 when $\theta = 0$

$C = 2.30$,

The required fillet weld size in terms of the number of sixteenths-of-an-inch is

$$D_{min} = \frac{P_u}{\phi C C_1 \ell},$$

($C_1 = 1.0$ for E70XX, see *AISCM*, Table 8-3)

$$\text{Required } D_{min} = \frac{24 \text{ kips}}{(0.75)(2.30)(1.0)(8 \text{ in.})} = \mathbf{1.74 \approx two\text{-}sixteenths}$$ ∴ Use **$^1/_8$ in. weld.**

The required fillet weld size using *AISCM* Table 8-4 (i.e., the IC method) is a $^1/_8$ in. weld, whereas the required fillet weld size using the elastic method was previously calculated as ¼ in. Therefore, as noted previously, the elastic method is more conservative than the IC method - the analysis method that was used to develop the *AISCM* Table 8-4.

Note that while a ⅛-in. fillet weld is the minimum weld size required for strength from the IC method, the weld size may have to be increased to satisfy the minimum weld size required based on the thickness of the thinner connected element (see Table 10-2).

10-10 BOLTS USED IN COMBINATION WITH WELDS

Under certain conditions, bolts and fillet welds can be combined in the same connection to share the applied load, but the conditions that must be satisfied are as follows (see *AISCM* Section J1.8) [2]:

- Only longitudinal fillet welds can be used in combination with the bolts, and
- The bolts must be slip-critical.

For a connection subjected to a factored tension force, T_u, where the slip-critical bolts and the longitudinal fillet welds are subjected to shear, the strength of the combined weld and slip-critical bolt connection is given as follows [2]:

$$\phi R_n = \phi R_{n\,(slip-critical\,bolts)} + \phi R_{n\,(longitudinal\,welds)} = \phi \mu D_u h_f T_b n_s + 1.392DL \quad (10\text{-}32)$$

Where the following conditions must be satisfied:

(a) If high-strength bolts are pretensioned in accordance with *AISCM* Table J3.1 using the turn-of-nut method (see Chapter 9), $\phi R_{n\,(longitudinal\,welds)} = 1.392DL \geq 50\%T_u$

(b) If high-strength bolts are pretensioned using all other methods (see Chapter 9),

$$\phi R_{n\,(longitudinal\,welds)} = 1.392DL \geq 70\%T_u. \quad (10\text{-}33)$$

$$\phi R_{n\,(slip-critical)} = \phi \mu D_u h_f T_b n_s \geq 33\%T_u \quad (10\text{-}34)$$

T_u = factored tension load on the connection.

Transverse fillet welds are not allowed to be combined with bolts because of the lower ductility of transverse welds [2]. Note that welds and bolts do not deflect the same amount, and since welds are more rigid or stiffer than bolted connections, they will attract a greater share of the load compared to the bolts. The practical situations in which longitudinal fillet welds and slip-critical bolts may be combined include:

a. Retrofitting of existing structures where higher design loads are required.

b. Existing slip-critical connections where the slip coefficient assumed in the design has not been achieved and thus the connection needs to be reinforced.

References

[1] American Welding Society. 2020. *ANSI/AWS D1.1—Structural Welding Code*. 24th ed. Miami, FL: American Welding Society.

[2] American Institute of Steel Construction. 2023. *Steel Construction Manual*. 16th ed. Chicago, IL: AISC.

[3] American Institute of Steel Construction. 2006. *Steel Design Guide Series 21: Welded Connections—A Primer for Structural Engineers*. Chicago, IL: AISC.

[4] International Code Council. 2024. *International Building Code*. Falls Church, VA: ICC.

[5] Blodgett, Omer. 1966. *Design of Welded Structures*. Cleveland, OH: The James F. Lincoln Arc Welding Foundation.

[6] American Society of Civil Engineers. 2022. *ASCE-7: Minimum Design Loads for Buildings and Other Structures*. Reston, VA: ASCE.

[7] Lay, M. G. 1982. "Structural Steel Fundamentals – An Engineering and Metallurgical Primer." Australian Road Research Board, Australia.

[8] Drake, R. M., J. A. Memmott, and M. Bala. 2022. "Welded Connection Basics." *Modern Steel Construction*, April: 16–19.

Exercises

10-1. Determine the type of weld required for the connection shown in Figure 10-24. The steel is ASTM A36 and the weld electrodes are E70XX.

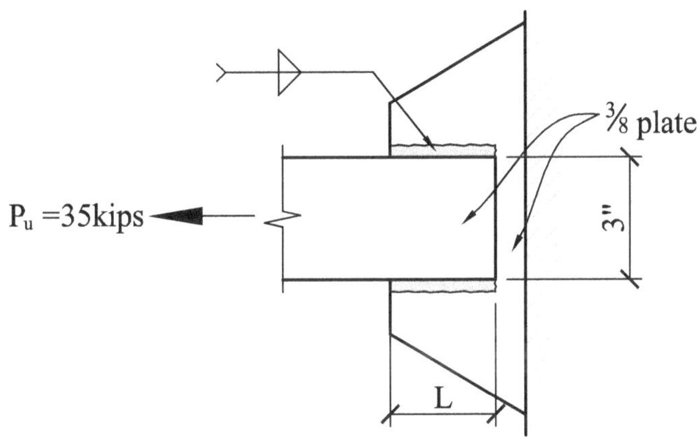

FIGURE 10-24 Detail for Exercise 10-1

10-2. Determine the maximum tensile load that may be applied to the connection shown in Figure 10-25 based on the weld strength only. The steel is ASTM A36 and the weld electrodes are E70XX.

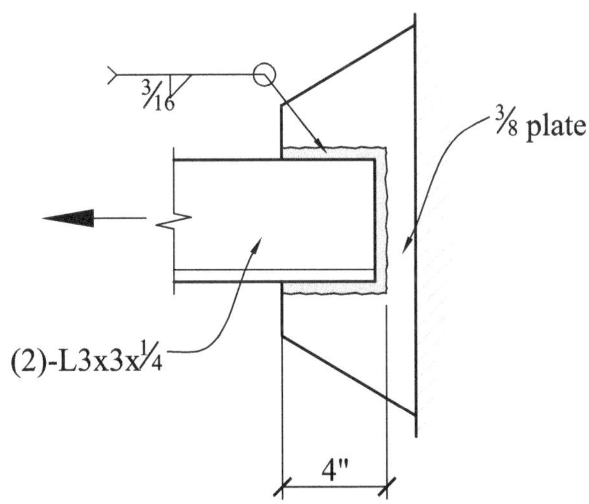

FIGURE 10-25 Detail for Exercise 10-2

10-3. Determine the fillet weld required for the lap splice connection shown in Figure 10-26. The steel is ASTM A36 and the weld electrodes are E70XX.

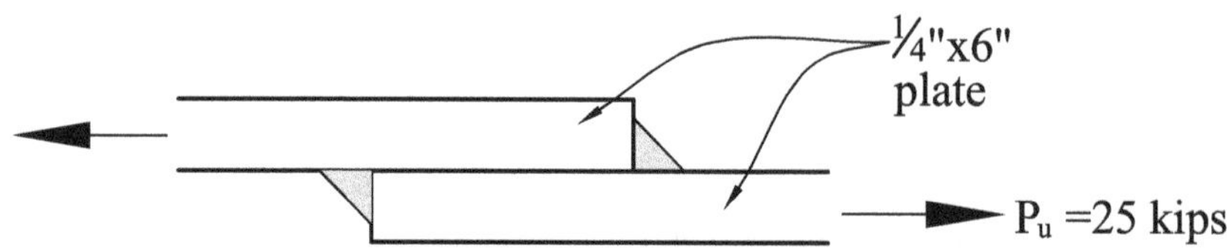

FIGURE 10-26 Detail for Exercise 10-3

10-4. Determine the length of the plates required for the moment connection shown in Figure 10-27. Determine the minimum and maximum weld sizes from the weld dimensional requirements, and select the most economical weld size. The steel is ASTM A36 and the weld electrodes are E70XX. Ignore the strength of the W12 × 26; consider the strength of the plate in tension.

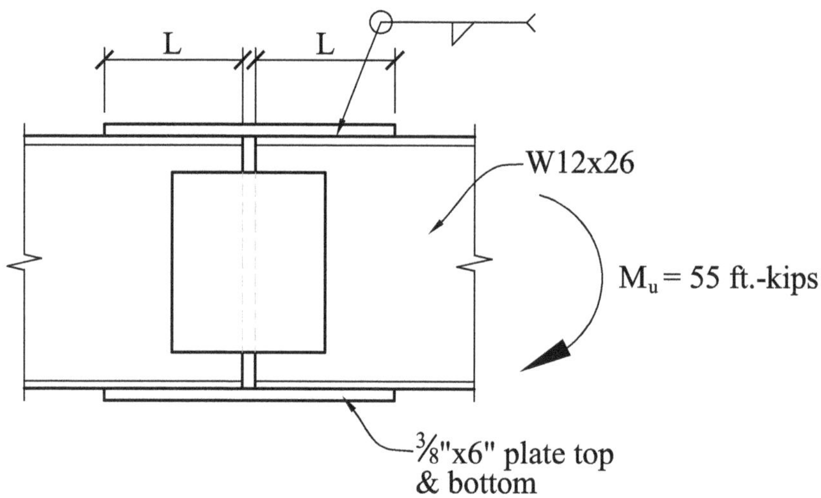

FIGURE 10-27 Detail for Exercise 10-4

10-5. Determine the capacity of the lap splice connection shown in Figure 10-28. The steel is ASTM A36 and the weld electrodes are E70XX.

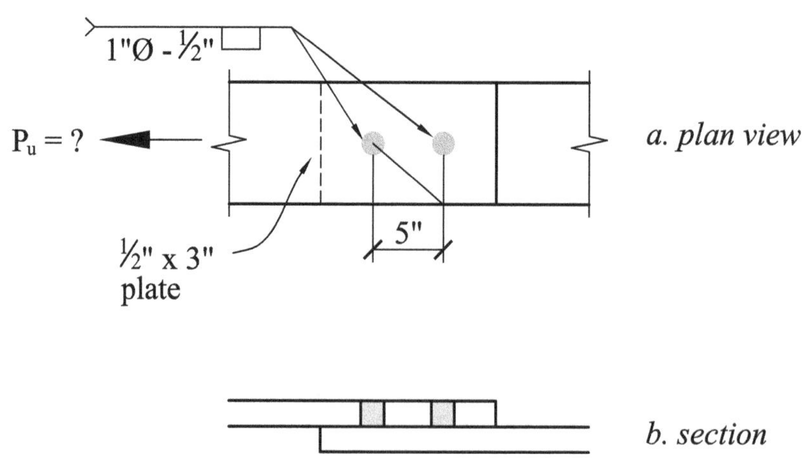

FIGURE 10-28 Detail for Exercise 10-5

10-6. Determine the required length of the slot weld shown in the connection in Figure 10-29. The steel is ASTM A992 Grade 50 steel, and the weld electrodes are E70XX.

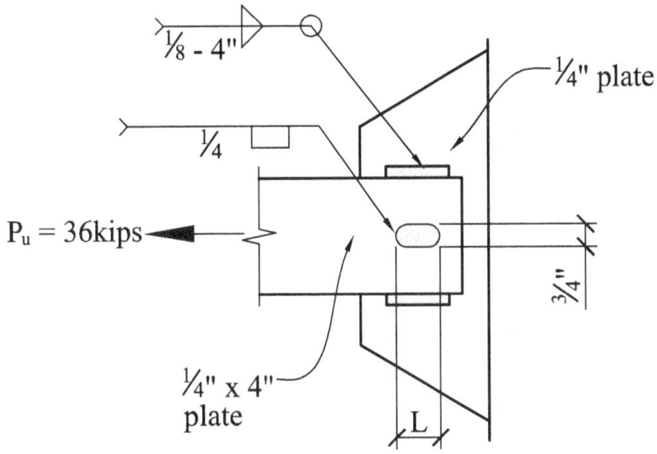

FIGURE 10-29 Detail for Exercise 10-6

10-7. Determine the capacity of the bracket connection shown in Figure 10-30 using the IC and the elastic methods. Compare the results. Weld electrodes are E70XX.

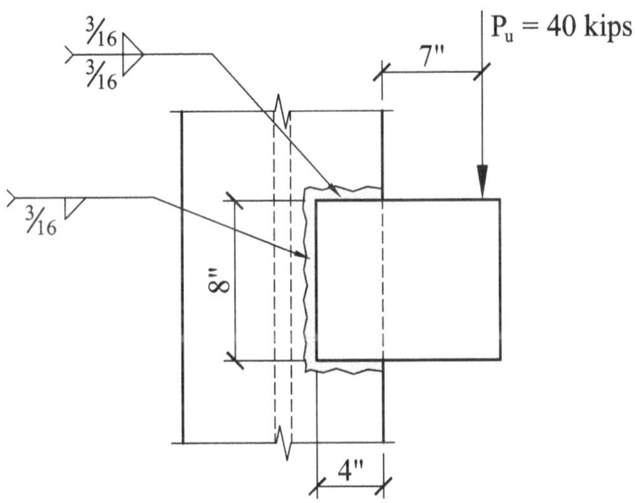

FIGURE 10-30 Detail for Exercise 10-7

10-8. Determine the required weld size for the bracket connection shown in Figure 10-31. Weld electrodes are E70XX.

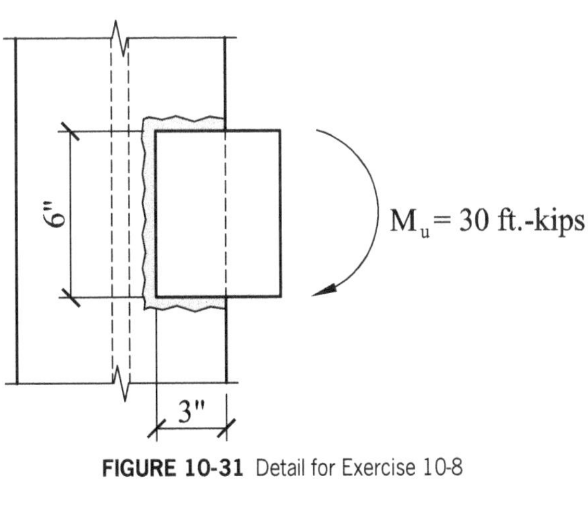

FIGURE 10-31 Detail for Exercise 10-8

10-9. Determine the required weld length for the seat connection shown in Figure 10-32. Weld electrodes are E70XX. Ignore the strength of the bracket. Use the elastic method, and compare the answer to that from the instantaneous center method using *AISCM* Table 8-4.

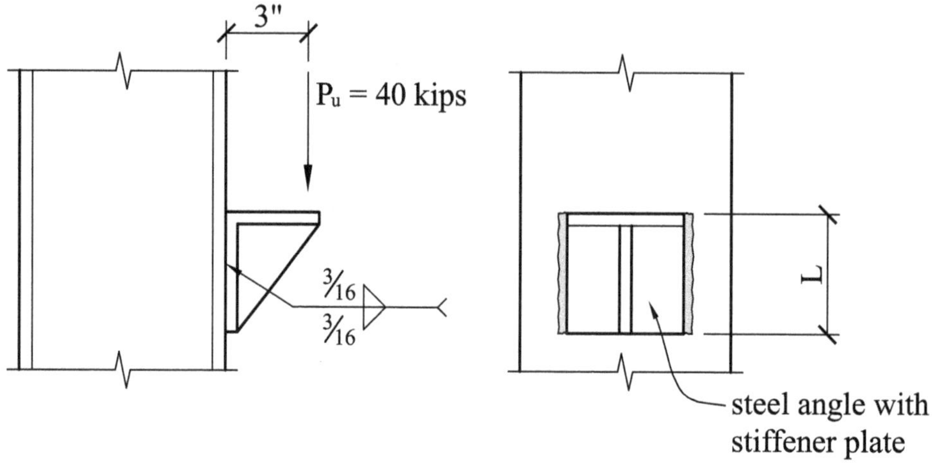

FIGURE 10-32 Detail for Exercise 10-9

10-10. Determine the maximum eccentricity for the connection shown in Figure 10-33. Weld electrodes are E70XX. Use the elastic method.

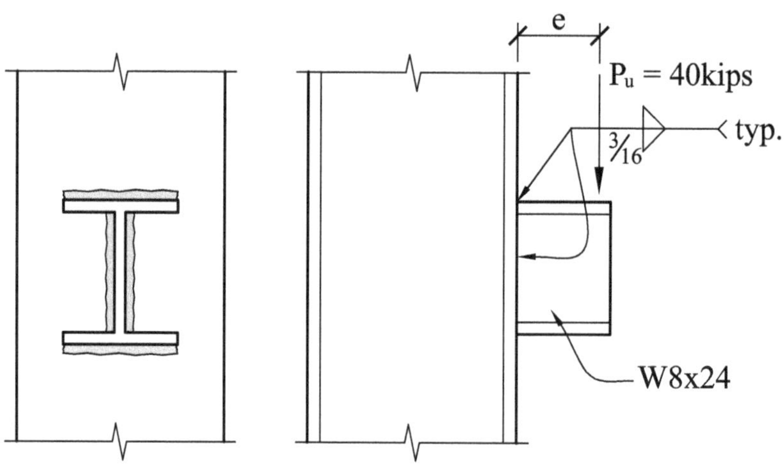

FIGURE 10-33 Detail for Exercise 10-10

10-11. Develop a spreadsheet to design a welded connection similar to that shown for Problem 10-2 and submit the output for Problem 10-2 with the following variables:

- Angle size
- Plate thickness
- Weld length along the three sides of the angle
- Weld size
- Applied tension load
- Check the design strength of the angle (see Chapter 4)

10-12. Develop a spreadsheet to design a welded moment connection like that shown for Problem 10-4. Submit the output for Problem 10-4 that could be used with the following variables:

- Applied moment
- Beam size (flange width, beam depth)
- Plate length, *L*, and plate width
- Weld size

10-13. Develop a spreadsheet to design a welded seat connection as shown in Figure 10-34 and submit the output for Problem 10-9 with the following variables:

- Weld length along the three sides of the angle
- Weld size
- Applied vertical load
- Load eccentricity

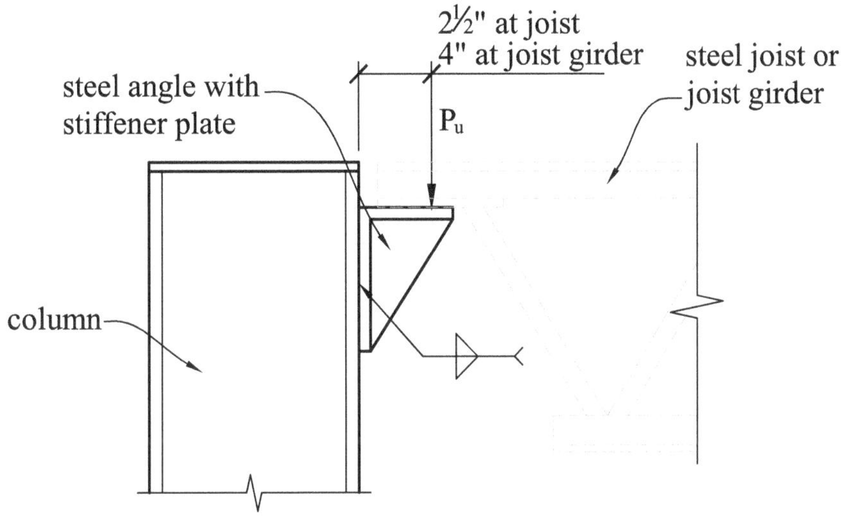

FIGURE 10-34 Detail for Exercise 10-13

10-14. Determine the strength of the splice connection shown in Figure 10-35. Welds are SMAW, E70XX, and the steel ASTM A36.

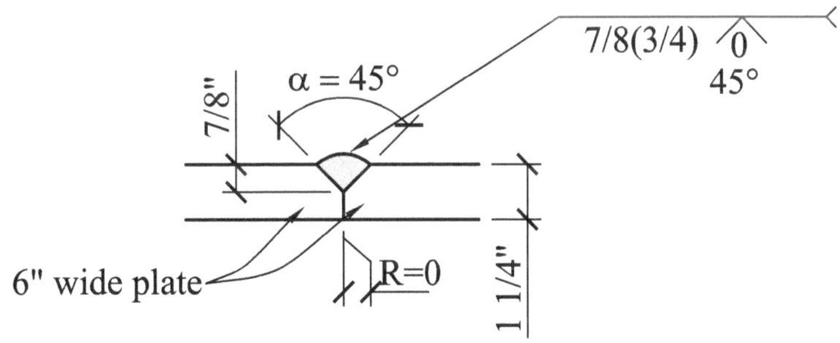

FIGURE 10-35 Details for Exercise 10-14

10-15. Determine the minimum PJP groove weld depth between the beam flange and the column to develop the moment shown in Figure 10-36. Provide a complete weld symbol for the required PJP weld and select a weld procedure.

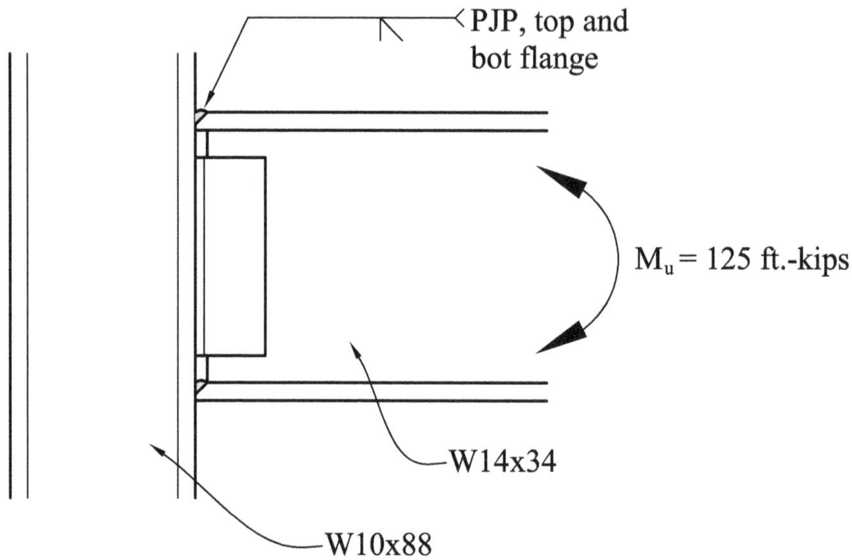

FIGURE 10-36 Details for Exercise 10-15

10-16. Determine the minimum weld size for the base of the handrail post shown in Figure 10-37. Welds are E70XX.

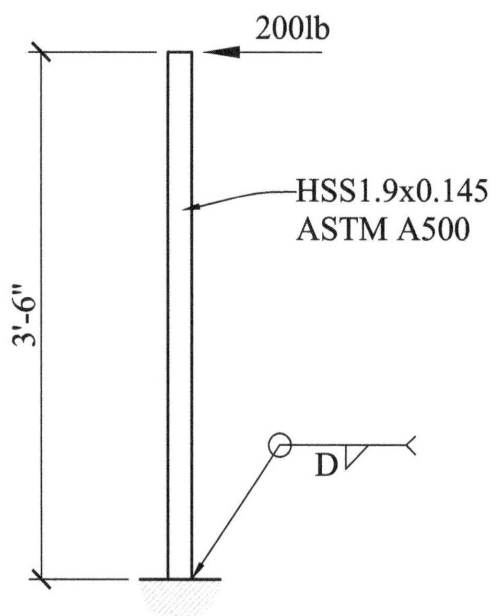

FIGURE 10-37 Details for Exercise 10-16

10-17. Design the weld size and length for the base of the handrail post shown in Figure 10-38. Welds are E70XX.

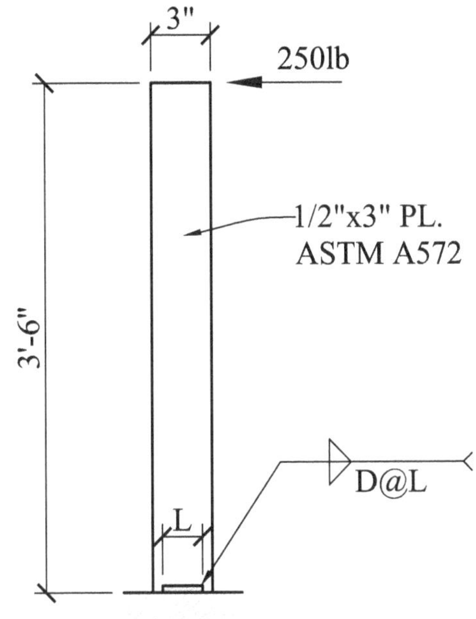

FIGURE 10-38 Details for Exercise 10-17

10-18. Determine the minimum weld size for the base of the tie-off post shown in Figure 10-39. The load can be applied in any direction. Welds are E70XX.

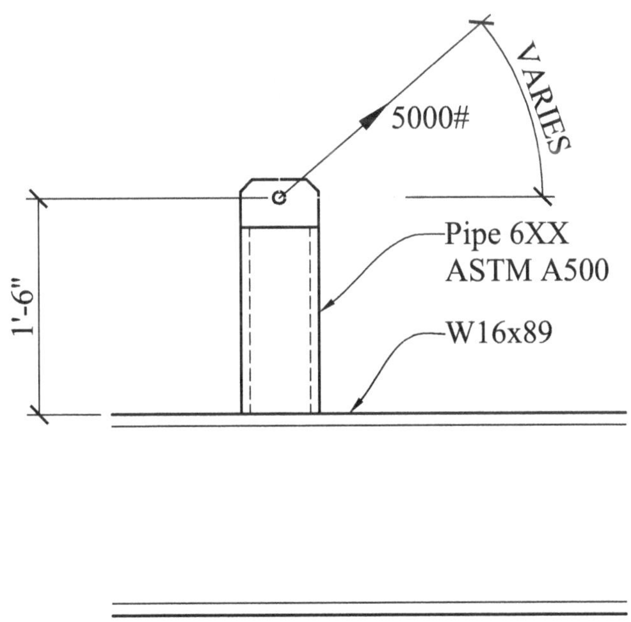

FIGURE 10-39 Details for Exercise 10-18

10-19. Design a welded connection for an HSS brace to gusset plate connection, assuming the following:

- HSS 5 × 5 × 3/16 brace centered on the gusset plate (ASTM A500)
- 5/16" × 8" × 8" thick gusset plate (ASTM A992 Grade 50 steel)
- Service load: W = 70 kips
- Design the weld from the brace to the plate and the plate to its support.

CHAPTER

11

Moment Connections, Bracing and Truss Connections, and Miscellaneous Details

11-1 INTRODUCTION

In Chapters 9 and 10, we covered the analysis and design of basic bolted and welded connections, including an introduction to the use of the *AISCM* design tables [1] for standard beam-to-girder and beam-to-column or girder-to-column connections. In this chapter, we will cover the design of moment connections that are used in unbraced or sway frames in buildings and other structures (see Chapter 8), the analysis and design of vertical bracing and truss connections, and other details that are not specifically covered in the *AISCM* design tables but are commonly used in design practice. The connections that we will consider in this chapter utilize a combination of bolts and welds, and many of these connections will have design components that are covered in the *AISCM* design tables. For the sake of brevity, only the LRFD method will be used throughout this chapter. Below is a summary of the connections and miscellaneous details that will be considered in this chapter:

11-2 Moment Connections: Introduction
11-3 Partially Restrained and Flexible Moment Connections
11-4 Fully Restrained Moment Connections
11-5 Moment Connections in Beams and Beam Splices
11-6 Column Stiffeners in Moment Connections
11-7 Column Splices
11-8 Design of Gusset Plates in Vertical Bracing and Truss Connections
11-9 Coped Beams
11-10 Holes in Beams

11-2 MOMENT CONNECTIONS: INTRODUCTION

A moment connection is capable of transferring across a joint or onto a structure a vertical shear force plus axial tension and compression couples caused by the moment. Most steel moment connections that occur in practice do not have absolute fixity because some rotation will occur at the connection. Recall from structural analysis that a perfectly fixed end condition is one in which the end rotation is zero, while the joint in a pinned connection are completely free to rotate. Virtually all steel connections have some degree of fixity, so that they are neither perfectly fixed nor perfectly pinned; they have a degree of fixity somewhere between these two extremes. In Ref. [4], a fixed connection is defined as one where the ratio of the applied moment to the fixed-end moment is at least 90%. A pinned connection is defined as one where the ratio of the applied moment to the fixed end moment is no greater than 20%.

For simplicity, connections that are designed to transfer shear only (e.g., the framed beam connections in Section 9-8) are considered to be pinned even though some degree of restraint, however small, may exist. Connections that are designed to resist some moment are considered to be moment connections, and they have a relatively smaller degree of rotation occurring at the joint (see Figure 11-1).

The AISC specification in Section B3.4 [2] identifies three types of steel connections (see Figures 11-1 and 11-3):

1. *simple shear connections*,
2. *fully restrained moment connections* (FR), and
3. *partially restrained moment connections* (PR).

Simple shear connections are assumed to allow complete rotation at a joint and will therefore transmit a negligible moment across the connection. Standard shear connections that connect the webs of the supported beams to the supporting girder or column with shear tabs, plates, or double angles are the most common types of simple shear-framed beam connections (see Section 9-8). These simple connections must have adequate rotation capacity in order to achieve the calculated joint rotations from the analysis of the structure.

Fully restrained (FR) moment connections transfer a moment across a joint with a negligible rotation. FR connections are designed to maintain the angle between the connected members at the factored load level. That is, the angle between the members remains virtually unchanged before and after the load is applied to the structure. FR moment connections form the lateral bracing system for many structures; the advantage of moment frames over other lateral force resisting systems is the architectural flexibility and unobstructed view that they provide compared to, say, shear walls or braced frames. Practical and economical frame spans for moment frames range from 20 ft. to a maximum of 40 ft. [7].

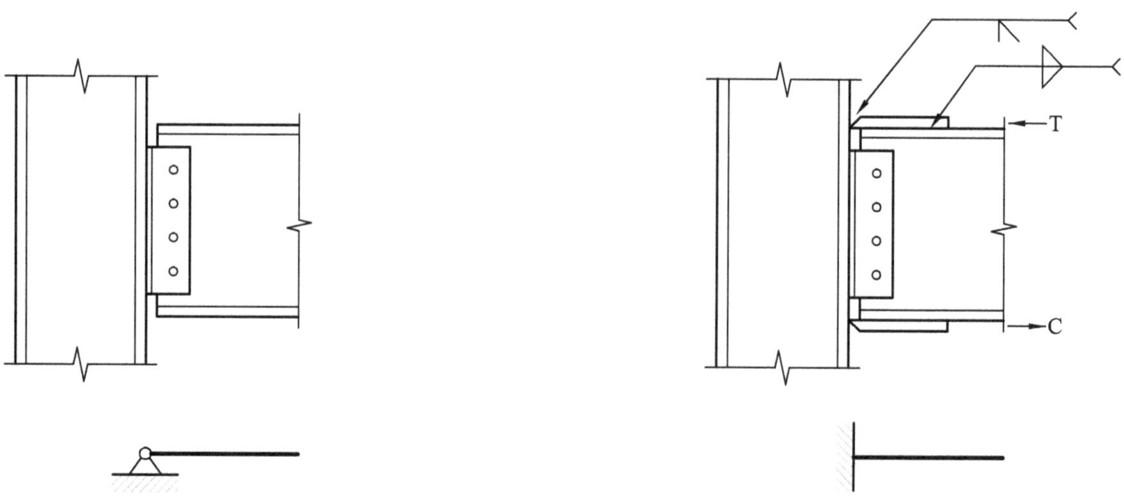

a) pinned (Simple shear connection) *b) Rigid beam-to-column connection*

FIGURE 11-1 Pinned and moment connections

Partially restrained (PR) moment connections will transfer some moment across a connection, but there will also exist a corresponding rotation at the connection. The use of PR connections is only permitted when the force versus deformation characteristics of the connection (i.e., the moment-rotation or M-θ curve, which are very nonlinear in nature) are known either by documented research or analysis. However, the deformation characteristics of a PR connection cannot be easily determined because such behavior is a function of the sequence in which the loads are applied. Since the actual load sequence for any structure cannot be known with certainty, engineering judgment must be used to identify the possible load sequences in order to properly design such a connection in accordance with the AISC Specifications [2, 3].

A more simplified and conservative approach to PR moment connection design is to use a **flexible moment connection** (FMC). In the past, FMCs have been referred to as "Type 2 with wind," "semirigid," or "flexible wind" connections [2, 3].

In the design of FMC, the gravity loads are resisted by the shear connection only, and any end restraint provided by the moment connection is neglected for gravity loads. For lateral loads, the FMC is assumed to have the same behavior as a FR moment connection, and all of the lateral loads are resisted by the moment connection. That is, the FMC is considered to act as a simple shear connection under gravity loads and a fully restrained (FR) moment connection under lateral loads. It has been shown that this type of connection is adequate for resisting lateral loads, provided that the plastic moment capacity of the connection is not exceeded [2, 3]. Therefore, the full plastic moment capacity of the beam is available to resist lateral loads.

The general moment-rotation curve for each of the above-mentioned connection types is shown in Figure 11-2. Each of the three curves (1, 2, and 3) represents a different connection type. In Figure 11-2, fully restrained or FR moment connection behavior would resemble curve 3 (i.e., the beam or girder framing into the column has a stiffness $K \geq \frac{20EI}{L}$), and partially restrained or PR moment connection behavior would resemble curve 2 in Figure 11-2 (i.e., the beam or girder has a stiffness, K, such that $\frac{2EI}{L} < K < \frac{20EI}{L}$).

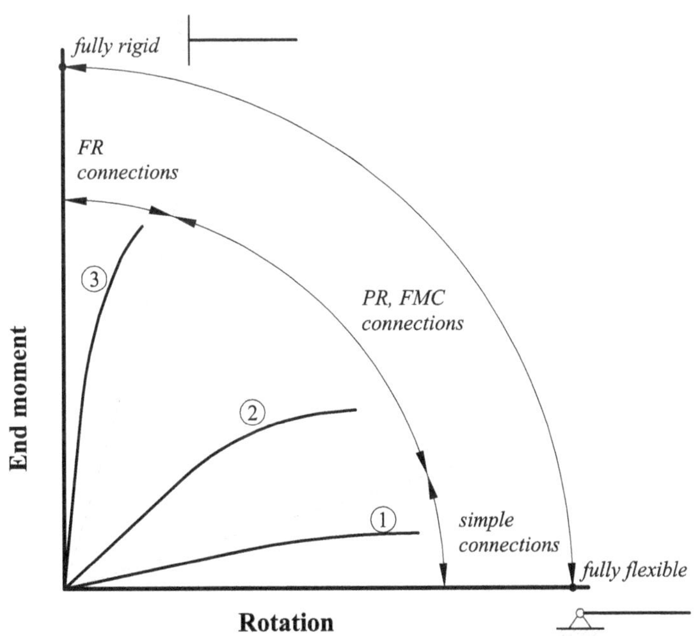

Adapted from AISC 360-22 Figure C-B3.3

FIGURE 11-2 Moment versus rotation curve for various connection types

Connections with beams having a stiffness $K \leq \frac{2EI}{L}$ are modeled as simple shear connections (see Figure 11-1a) and would resemble curve 1 in Figure 11-2. Fully restrained moment connections are typically achieved by connecting wide flange beams to wide flange columns. When a wide flange beam is directly welded to the face of an HSS column, the resulting moment connection is typically partially restrained (PR) if the width of the beam flange is less than 95% of the width of the HSS column, and the connection is assumed to be fully restrained (FR) if the minimum flange width of the beam is at least 95% of the width of the HSS column.

It is also important to note that some moment connections might provide little or no rotation even though an assumption has been made that a flexible moment connection (FMC) has been used when, in fact, a more rigid connection exists. For this reason, the designer should select a known flexible moment connection when one is required.

Figure 11-3a shows the common types of fully restrained (FR) moment connections—bolted flange plate moment connection, welded flange plate moment connection, and bolted end-plate moment connection; Figure 11-3b shows the common types of partially restrained (PR) or flexible moment connections. Figures 11-3c and 11-3d also shows fully restrained moment connections with the beam/girder top and bottom steel plates welded to the column.

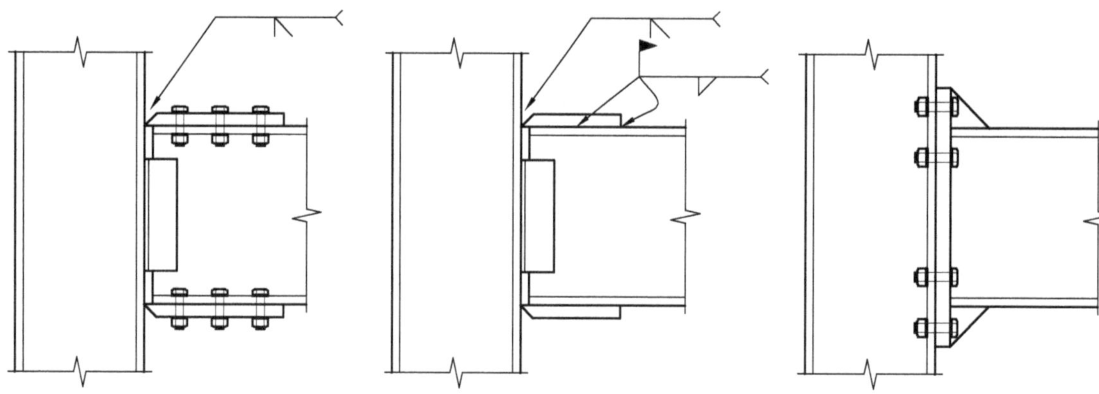

a) FR (fully-restrained) connections

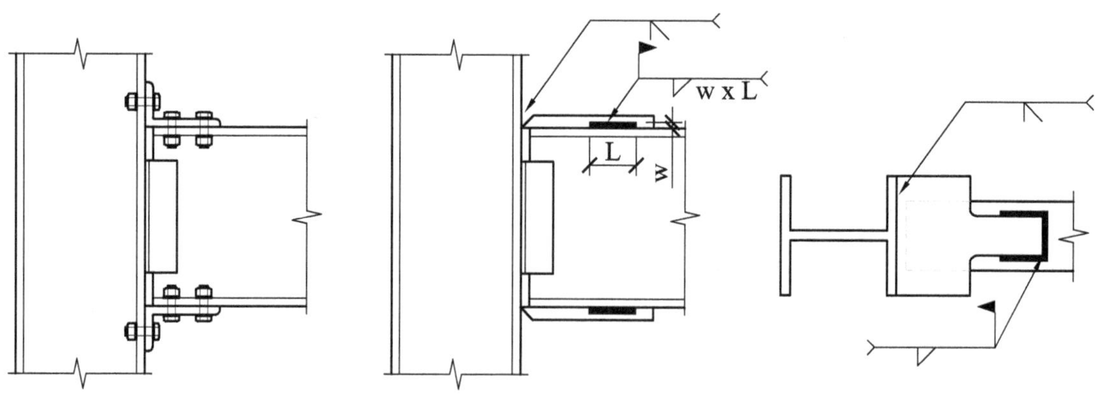

b) PR (partially-restrained) connections

FIGURE 11-3 FR and PR connection examples

FIGURE 11-3c Fully restrained moment connections about two orthogonal axes.

FIGURE 11-3d Moment connection top and bottom plates.

Figure 11-3a(iii) shows an 8-bolt end-plate moment connection which is very efficient with regards to the steel erection process as it obviates the need for any field welding, and the building does not need to be temporarily laterally braced during construction by guy cables because the building is instantly laterally braced once the end-plates are bolted to the column flanges, so no additional lateral bracing is needed [28]. For bolted end-plate moment connections, the beam flanges are typically shop welded to the end plates with complete joint penetration (CJP) groove welds, though partial joint penetration (PJP) groove welds or fillet welds are allowed in some situations [30]. The columns are usually sized with adequate flange and web thicknesses to obviate the need for column flange stiffener plates or column web doubler plates. In addition, no special surface preparation of the end plates is required. End-plate moment connections can be designed using *AISC Design Guide* 39 [29, 30] and are classified as fully restrained or FR moment connections. The behavior of the end plate can be classified as follows [29, 30]:

1. "Thin end-plate behavior," where failure of the end plate is the limit state that controls the design and prying action of the bolts assumed and considered.
2. "Thick end-plate behavior," where the bolts are assumed to fail in tensile rupture and no prying action of the bolts is considered; or
3. "Plastic hinging behavior" of the beam is assumed.

Design approach No. 3 is used in the design of end-plate moment connections in high seismic regions, while design approaches Nos. 1 and 2 are used for end-plate moment connections subject to gravity loads, wind, and low seismic loads.

There are also proprietary moment connections, such as the *SidePlate* special moment resisting connections. SidePlate moment connections uses "a shop-welded plate at the girder's top flange and shop-welded angles at the bottom flange. In the field, the girder slides between the plates that are welded to both sides of the column. The plate and angles on the girder are bolted to the angles at the top and bottom of the column plates to transfer the tension and compression components of the moment out of the girder and into the column plates, and then into the column" [31]. Other proprietary moment connections can be found in AISC's Prequalified Connections for Special and Intermediate Steel Moment Frames for Seismic Applications (ANS/AISC 358-22). Intermediate and special steel moment frames are typically used as lateral force resisting systems in high seismic zones.

11-3 PARTIALLY RESTRAINED (PR) MOMENT CONNECTIONS AND FLEXIBLE MOMENT CONNECTIONS (FMC)

As discussed in the previous section, the use of PR moment connections requires knowledge of the moment-rotation curve as well as the load sequence [2, 3]. Since very little data are available for PR moment connections, most designers will use a flexible moment connection (FMC), which allows conservative and simplifying assumptions to be made. As indicated in the previous section, in the analysis of moment frames with PR moment connections, or FMC, the gravity load is neglected in the lateral load analysis, and the moment connections resist the lateral load moments only; the gravity loads are resisted by the shear connections only.

It is important to note that the use of a PR moment connection or FMC requires that the seismic response modification factor, R, be taken as less than or equal to 3.0 (i.e., System "H" in ASCE 7 Table 12.2-1). The only PR connections with a specific R value from ASCE 7 Section 12.2-1 are partially restrained composite moment connections [8, 9]. When R is greater than 3.0, the moment connections must be designed as a fully restrained or FR moment connection and must include the gravity load effects. ASCE 7 Section 12.2.1.2 permits special moment frames (SMF) with partially restrained connections, provided it is justified with testing and analysis in addition to peer review. For special moment frames (SMF) and intermediate moment frames (IMF), the moment frame connections that meet the testing and performance requirements specified in ANSI/AISC 341 (Seismic Provisions for Structural Steel Buildings) can be found in AISC 358 (Prequalified Connections for Special and Intermediate Moment Frames for Seismic Applications) [5]. Currently, only a few proprietary moment connections have been prequalified for use in buildings as SMF [5, 7, 9].

Common flexible moment connections (FMC) are shown in Figure 11-3b, and the reader is referred to Example 9-8 for an analysis of a flange-angle FMC. For flange-plated FMCs, the flange plate has an unwelded length equal to 1.5 times the width of the plate in order to allow for the elongation of the plate, thus creating flexible behavior (see Figure 11-4).

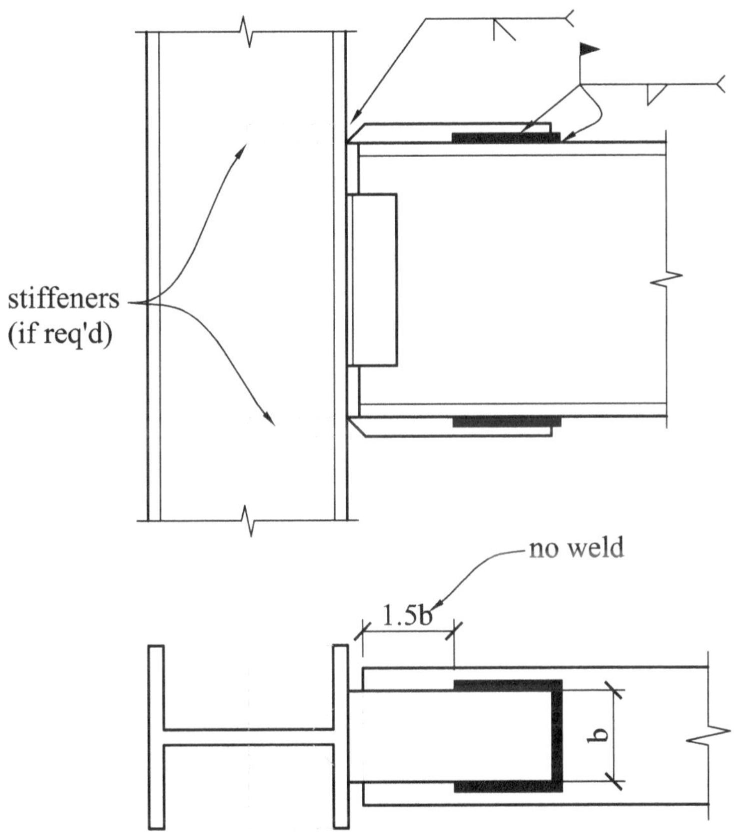

FIGURE 11-4 Flange-plated FMC

EXAMPLE 11-1

Flexible Moment Connection (FMC)

Determine whether the FMC shown in Figure 11-5 is adequate to support the factored moment due to wind loads. The beam and column are ASTM A992, the steel plate is ASTM A36, and the weld electrodes are E70XX.

Solution

From *AISCM* Table 1-1, the section properties for W12 × 26 are

$d = 12.2$ in., and

$t_f = 0.38$ in.

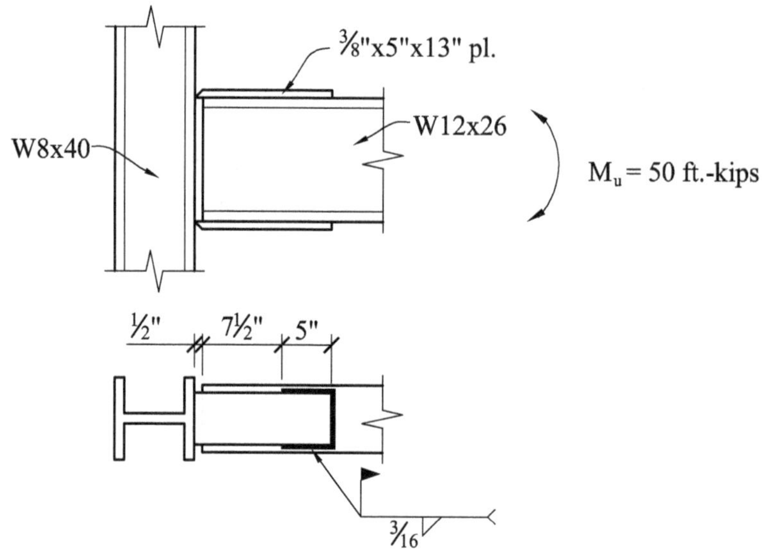

FIGURE 11-5 Detail for Example 1-1

Top and bottom plate: A_g = (3/8)(5) = 1.875 in.^{2}

Flange force:

$$P_{uf} = \frac{M}{d} = \frac{50}{12.2/12} = 49.2 \text{ kips.}$$

Tension on gross plate area:

$$\phi P_n = \phi A_g F_y = (0.9)(1.875)(36) = 60.8 \text{ kips} > 49.2 \text{ kips OK.}$$

Compression:

$K = 0.65$ (Assume fixity at both ends of the flange plate; see Figure 5-4)

The length of the plate from Figure 11-5 is,

$L = 0.5 \text{ in.} + 7.5 \text{ in.} = 8.0 \text{ in.}$

$$r_y = \frac{I_y}{\sqrt{A}} = \frac{t}{\sqrt{12}} = \frac{0.375}{\sqrt{12}} = 0.108 \text{ in.}$$

$$\frac{KL}{r_y} = \frac{(0.65)(8.0)}{0.108} = 48.0$$

From *AISCM* Table 4-14, $\phi_c F_{cr}$ = 28.7 ksi.

$$\phi_c P_n = \phi_c F_{cr} A_g = (28.7)(1.875) = 53.8 \text{ kips} > 49.2 \text{ kips OK.}$$

Weld strength:

$b = 5$ in.(⅜-in. by 5-in. flange plate); ($L = 8$ in. $\geq b = 5$ in.) OK.

From Table 10-2,

Minimum weld size = $\frac{3}{16}$ in.; $\therefore D_{min} = 3$

Maximum weld size = $t - \frac{1}{16} = \frac{5}{16}$ in.; $\therefore D_{max} = 5$ (hence, a $\frac{3}{16}$-in. weld is OK)

The nominal weld strength is obtained from Equation (10-7) as follows:

$$\begin{aligned}\phi R_n &= 0.85\phi R_{wl} + 1.5\phi R_{wt}\\ &= (0.85)1.392DL_l + (1.5)1.392DL_t\\ &= \left[(0.85)(1.392)(3)(5+5)\right] + \left[(1.5)(1.392)(3)(5)\right]\\ &= 66.8\text{ kips}\end{aligned}$$

Therefore, $\phi R_n = \mathbf{66.8\ kips}$

The design weld strength is $\phi R_n = 66.8\text{ kips} > P_{uf} = 49.2\text{ kips}$. OK.

11-4 FULLY RESTRAINED (FR) MOMENT CONNECTIONS

As discussed in Section 11-2, FR moment connections are sufficiently rigid to maintain the angle between the connected members, and they are designed to support both gravity and lateral loads. For seismic loads, when the seismic response modification factor, R, is equal to 3.0 (i.e., Structural Steel System "H" in ASCE 7 Table 12.2-1), the design approach is similar to that for flexible moment connections (FMC), and the connection type must be one that provides adequate rigidity (see curve 3 in Figure 11-2 and the details in Figure 11-3a).

When the seismic response modification factor is greater than 3.0, additional design and detailing requirements must be satisfied for resisting seismic loads [5]. These additional requirements are a combination of strength and stability design parameters, depending on the type of moment connections used. Additional material-specific requirements for structural systems with $R > 3.0$ are specified in Chapter 14 of ASCE 7 [6]. The AISC seismic provisions identify three basic types of moment frames with $R > 3.0$:

- Ordinary moment frames (OMF),
- Intermediate moment frames (IMF), and
- Special moment frames (SMF).

The seismic response modification factors for these three basic types of moment frames are 3.5, 4.5, and 8.0, respectively. Each of these moment frames requires varying degrees of additional strength, stability, stiffness, and detailing requirements; the OMF has the least stringent requirements, but they cannot be used in high seismic regions, and the SMF has the most stringent requirements. SMF's are not subject to any building height restrictions, even in high seismic zones. Intermediate moment frames (IMF) and special moment frames (SMF) are often used as lateral force resisting systems in high seismic zones. In general, each of these connection types is designed for a certain moment and rotation. These connections are generally designed around the concept of creating a plastic hinge in the beam away from the beam-column joint, thus creating a strong column-weak beam system, which is the preferred or desired failure mode [10]. Figure 11-6 illustrates this concept.

There are several published analyses and testing data that can be used for OMF, IMF, and SMF connections (see Refs. [5, 7] and [10]). Using these prequalified connections is generally preferred since a rigorous analysis would be required for other connections that have not been tested. Figure 11-7 illustrates some of the basic types of connections that can be used for OMF, IMF, and SMF frames. Note that only Figures 11-7d and 11-7e are recognized in the AISC seismic provisions (see Ref. [5]); the other connection types are found in Ref. [10]. For the reduced beam section moment connection (see Figure 11-7d), care should be taken to prevent any unauthorized welding or connection of structural or non-structural members to the beam at or near

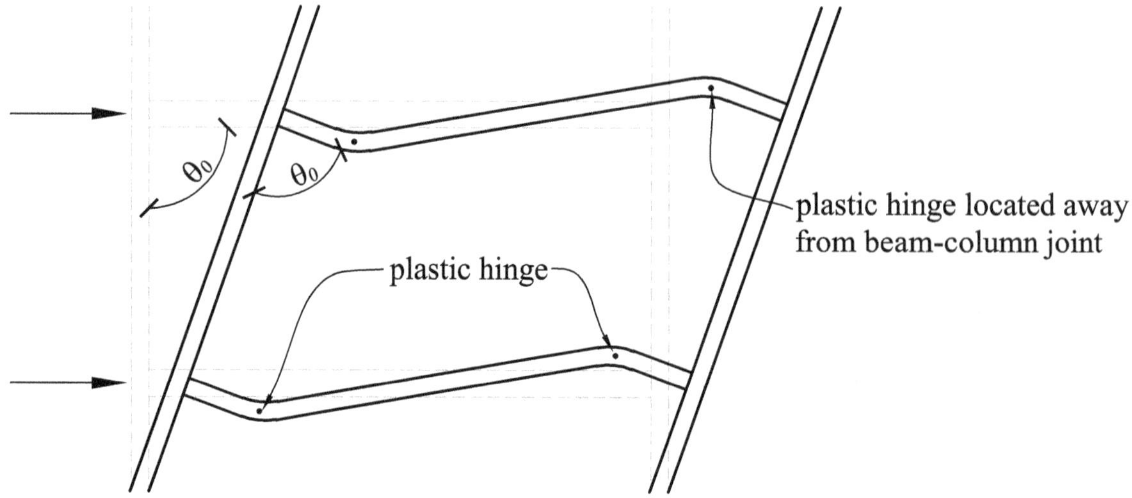

FIGURE 11-6 Plastic hinge formation in FR connections

the reduced section so as not to inadvertently increase the stiffness of the beam at the reduced section [10].

The prescriptive requirements for the connection types in Figure 11-7 are found in their respective standards (see Refs. [5] and [10]) and are beyond the scope of this text. It can be observed that using a seismic response modification factor, R, equal to or less than 3.0 is highly desirable because the analysis and design procedure is more simplified compared to the procedure for OMF, IMF, or SMF, and no material-specific detailing requirements have to be satisfied. In general, buildings with a Seismic Design Category of A, B, or C can usually be economically designed with $R \leq 3.0$. This approach is recommended where possible.

Weld Access Holes

Weld access holes are used to create adequate clearance for welding the beam or girder flange fully to the column flange in moment connections. Weld access hole dimensions should conform to AISC Specifications J1.6 and M2.2, and for heavy shapes (i.e., member thickness greater than 2 inches), the surface of the holes should be ground smooth to bright metal to eliminate any cracks that may form during the hole cutting or welding processes and to prevent the propagation of any microcracks. Without the weld access hole in the girder web, the bottom of the top flange and the top of the bottom flange of the girder at the intersection with the girder web location cannot be welded fully to the column flange because of the interference that exists due to the presence of the girder web. The weld access holes also allow space for a backing bar to be placed underneath the girder top flange and allow flat welding of the girder bottom flange to the column flange; thus, the weld access holes enable continuous full penetration welds between the top and bottom flanges of the girder and the column flange, as depicted in Figures 11-7a and 11-7d. In addition, the weld access holes help to minimize the effect of weld shrinkage and avoid orthogonal intersecting welds (i.e., at the joint where the weld connecting the girder flange to the column flange would intersect the orthogonal weld connecting the girder web to the girder flange). Weld access holes should not be closed in or filled in with weld metal in order to prevent a triaxial state of stress in the weld when loaded; instead, if aesthetics or corrosion protection is an issue, the weld holes can be filled in or sealed with mastic material [1]. The moment connections in Figures 11-7a and 11-7d have weld access holes (also sometimes called "rat holes").

Some examples of fully restrained moment connections in real life structures are shown in the following figures: Figures 11-7g and 11-7h show fully restrained moment connections at a cantilevered beam and girder, respectively; Figure 11-7i shows a fully restrained moment connection of a wide flange girder to a HSS column in a podium building. Figures 11-7j and 11-7k show structural details for a steel beam moment connection to a wide-flange column web.

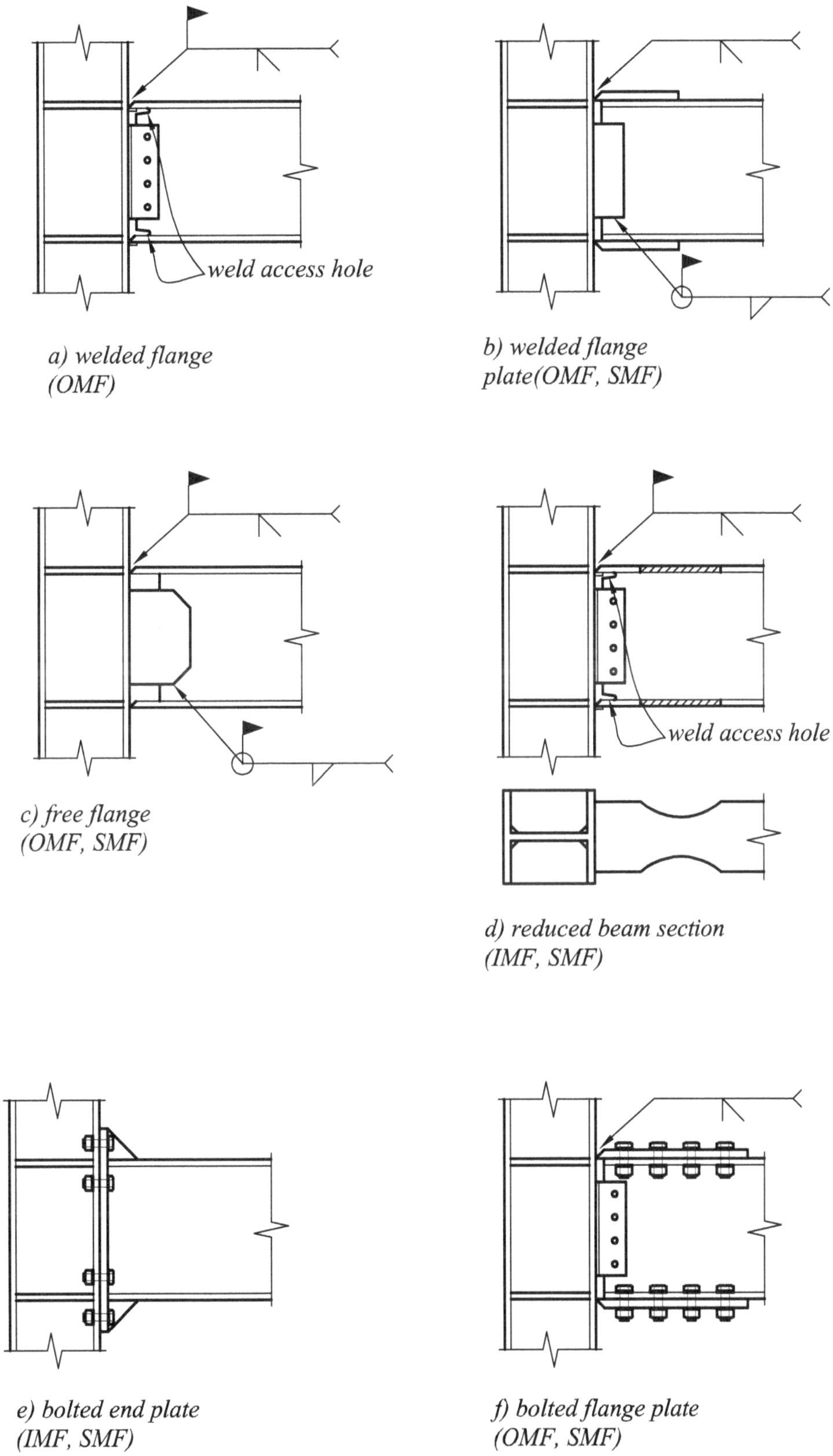

FIGURE 11-7 (a-f) Pre-qualified FR moment connections

FIGURE 11-7g Fully Restrained (FR) moment connection at a cantilevered beam (connection to column web)

FIGURE 11-7h Fully Restrained (FR) moment connection at a cantilevered girder (connection to column flange)

FIGURE 11-7i Fully Restrained (FR) moment connection of a wide-flange girder to HSS columns in a podium building

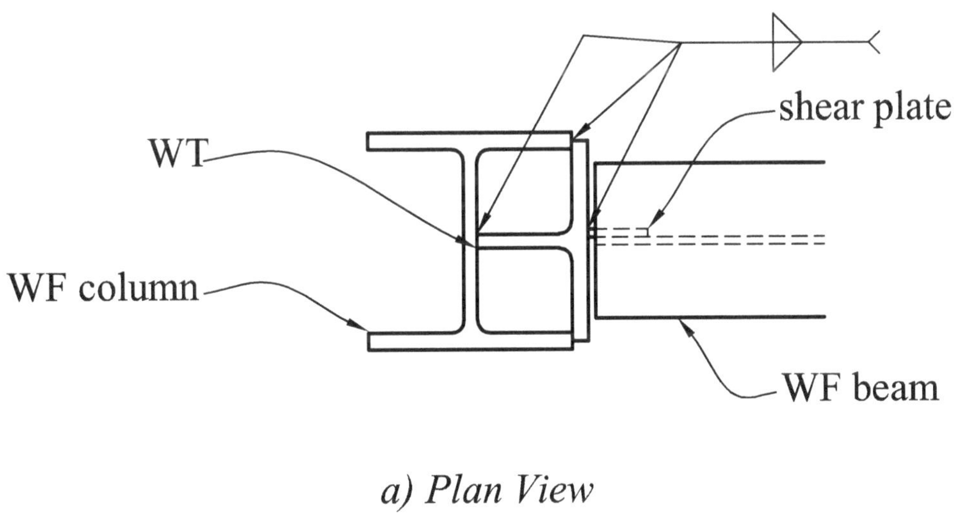

FIGURE 11-7j Plan detail of beam-to-column web moment connection (use with Figure 11-7k)

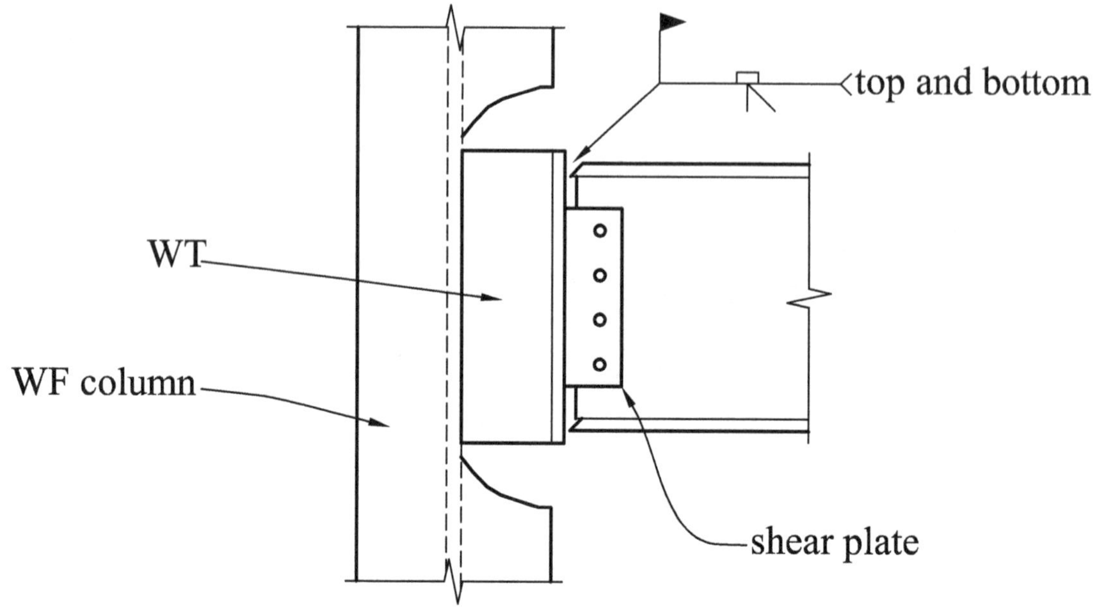

b) Elevation View

FIGURE 11-7k Elevation detail of beam-to-column web moment connection

EXAMPLE 11-2

Bolted Fully Restrained (FR) Moment Connection

Determine whether the fully restrained moment connection shown in Figure 11-8 is adequate to support the factored moment due to wind loads. The steel plate is ASTM A36, the beam is ASTM A992, and the bolts are $^3/_4$-in. A325N in standard (STD) holes.

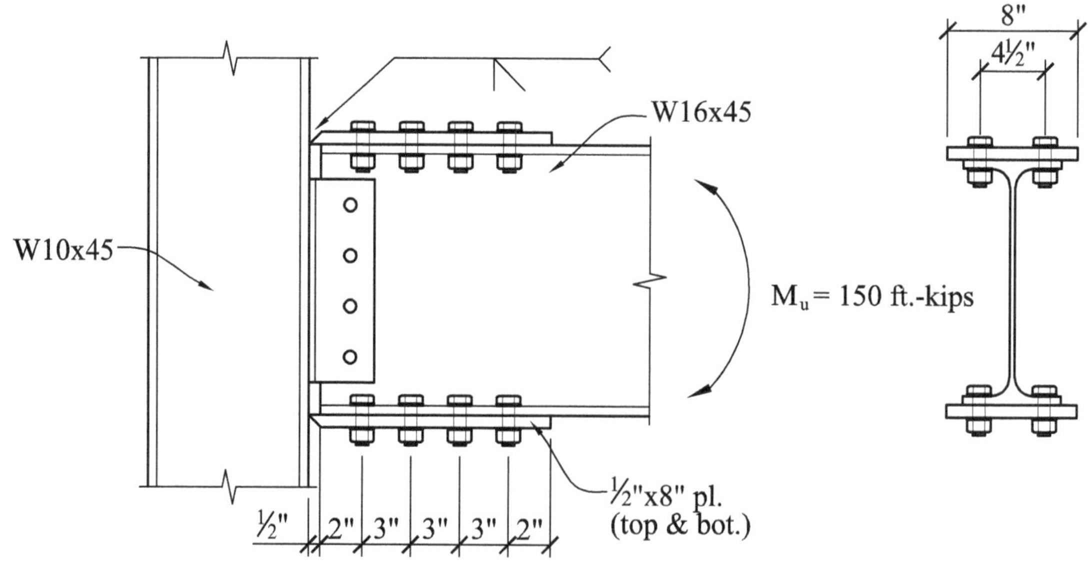

FIGURE 11-8 Detail for Example 11-2

Solution

From *AISCM* Table 1-1, the section properties for $W16 \times 45$ are

$d = 16.1$ in.

$b_f = 7.04$ in.

$t_f = 0.565$ in.

$S_x = 72.7$ in.^{3}

For ½ in. x 8 in. top and bottom flange plates, A_g = (½)(8) = 4 in.^{2}

Flange force:

$$P_{uf} = \frac{M}{d} = \frac{150}{16.1/12} = 112 \text{ kips.}$$

When the flanges of the beams and girders in moment connections are bolted, the flexural strength of the beam is reduced due to the presence of the bolt holes at the connection. The following provisions apply (AISC Specification, Section F13):

When $F_u A_{fn} \geq Y_t F_y A_{fg}$, (11-1)

the reduced flexural strength does not need to be checked. When $F_u A_{fn} < Y_t F_y A_{fg}$, the design flexural strength of the beam or girder at the moment connection is

$$\phi_b M_n = \phi_b F_u S_x \left(\frac{A_{fn}}{A_{fg}} \right), \quad (11\text{-}2)$$

Where,

ϕ_b = 0.9,

F_u = Tensile rupture strength,

S_x = Section modulus,

A_{fg} = Gross area of the tension flange of the beam,

A_{fn} = Net area of the tension flange of the beam,

Y_t = 1.0 for $F_y/F_u \leq 0.8$

= 1.1 for all other cases.

Check the reduced flexural strength of the W16 × 45 due to the bolt holes in the flanges:

$$F_y/F_u = 50/65 = 0.77 < 0.80 \therefore Y_t = 1.0$$

$$A_{fg} = (7.04)(0.565) = 3.98 \text{ in.}^2$$

$$A_{fn} = A_{fg} - \Sigma A_{holes}$$

$$= 3.98 - \left[(2\,\text{holes}) \left(\frac{3}{4} + \frac{1}{8} \right) (0.565) \right] = 2.99 \text{ in.}^2$$

$$F_u A_{fn} = (65)(2.99) = 194 \text{ kips}$$

$$Y_t F_y A_{fg} = (1.0)(50)(3.98) = 199 \text{ kips}$$

194 kips < 199 kips;

(i.e., $F_u A_{fn} < Y_t F_y A_{fg}$), Therefore, the reduced flexural strength must be checked.

$$\phi_b M_n = \phi_b F_u S_x \left(\frac{A_{fn}}{A_{fg}} \right) = (0.9)(65)(72.7) \left(\frac{2.99}{3.98} \right) = 3195 \text{ in.-kips}$$

$$= \mathbf{266\ ft.\text{-}kips} > 150 \text{ ft.-kips} \qquad \text{OK.}$$

Check direct shear in the bolts:

From *AISCM* Table 7-1, the single shear capacity for one ¾ in. diameter A325N bolt is $\phi_v r_n = 17.9$ kips/bolt.

Therefore, the number of bolts required is

$$N_{b,required} = \frac{P_{uf}}{\varphi_v r_n} = \frac{112}{17.9} = 6.26 < 8 \text{ bolts provided, OK.}$$

Check bolt bearing and tearout:

Since the thickness of the plate ($t_{pl} = 0.5$ in.) is less than the flange thickness ($t_f = 0.565$ in.), and since F_u for the plate is less than F_u for the beam, the bolt bearing and tearout on the plate will control. Figure 11-9 illustrates the failure modes for bearing, tension, and block shear.

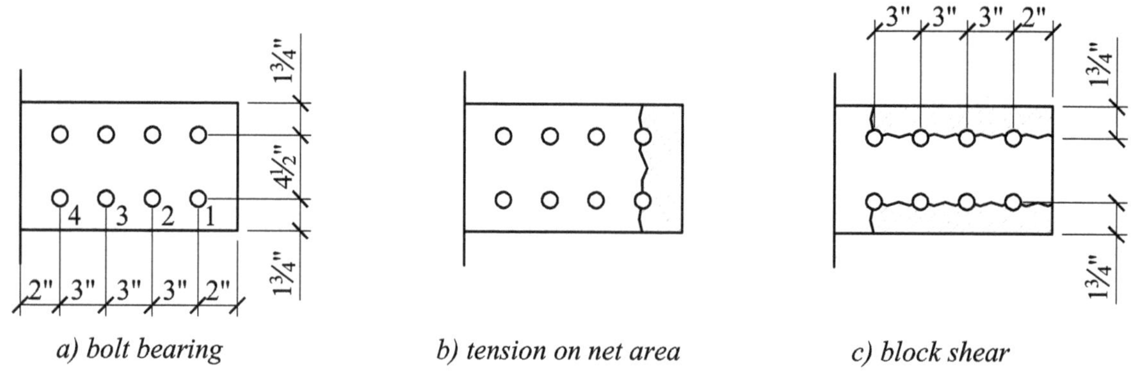

FIGURE 11-9 Bolt bearing and block shear

The clear distance between the edge of the plate and the edge of hole No. 1 in the direction of the applied force is,

$$L_{c1} = 2 \text{ in.} - 0.5\left(\frac{3}{4} + \frac{1}{8}\right) = 1.56 \text{ in.}$$

The clear distance between the edge of adjacent bolt holes in the direction of the applied force is,

$$L_{c2} = L_{c3} = L_{c4} = 3 \text{ in.} - \left(\frac{3}{4} + \frac{1}{8}\right) = 2.13 \text{ in.}$$

The bearing strength for each bolt will now be determined. It is assumed that the deformation at a bolt hole under service loads is a design consideration indicating that this deformation is limited; the smaller of the bearing capacity and the tearout capacity of the bolts will control:

$$\begin{aligned}\text{For bolt 1, } \phi R_n &= \phi 1.2 L_c t F_u \le \phi 2.4 d t F_u \\ &= (0.75)(1.2)(1.56 \text{ in.})(0.5 \text{ in.})(58 \text{ ksi}) \le (0.75)(2.4)(0.75)(0.5 \text{ in.})(58 \text{ ksi}) \\ &= 40.7 \text{ kips} \le 39.2 \text{ kips}\end{aligned}$$

Therefore, ϕR_n (bolt 1) = 39.2 kips.

For bolts 2, 3, and 4,

$$\phi R_n = (0.75)(1.2)(2.13 \text{ in.})(0.5 \text{ in.})(58 \text{ ksi}) \le (0.75)(2.4)(0.75)(0.5 \text{ in.})(58 \text{ ksi})$$

$= 55.4 \text{ kips} \leq 39.2 \text{ kips}$

Therefore, ϕR_n (bolt 2, 3, or 4) $= 39.2$ kips.

The total strength of the connection for the bolt bearing and tearout failure modes is

$\phi R_n = [(2)(39.2 \text{ kips})] + [(6)(39.2 \text{ kips})] = \mathbf{313 \text{ kips}} > P_{uf} = \mathbf{112 \text{ kips}}$ **OK.**

Check tension on the gross and net area of the flange plate:

$$A_g = (0.5 \text{ in.})(8 \text{ in.}) = 4 \text{ in.}^2$$

$$A_n = A_g - \Sigma A_{holes} = 4 - \left[(2\,\text{holes})\left(\frac{3}{4} + \frac{1}{8}\right)(0.5)\right] = 3.13 \text{ in.}^2$$

$$A_e = A_n U = (3.13 \text{ in.}^2)(1.0) = 3.13 \text{ in.}^2$$

Strength based on gross area is

$\phi P_n = \phi F_y A_g = (0.9)(36)(4 \text{ in.}^2) = \mathbf{129 \text{ kips}} > P_{uf} = 112$ kips OK.

Strength based on effective area is

$\phi P_n = \phi F_u A_e = (0.75)(58)(3.13 \text{ in.}^2) = 136 \text{ kips} > P_{uf} = 112$ kips OK.

Block shear:

Since the flange plate thickness of 0.5 in. is less than the beam flange thickness of 0.565 in., block shear will be more critical in the flange plate. The block shear failure mode in Figure 11-9c will now be checked as follows:

Number of holes in each shear plane = 3.5 holes
Number of holes in each tension plane = 0.5 hole
Number of shear planes in each flange plate = 2
Number of tension planes in each flange plate = 2

$$A_{gv} = (2)(3 \text{ in.} + 3 \text{ in.} + 3 \text{ in.} + 2 \text{ in.})(0.5 \text{ in.}) = 11 \text{ in.}^2$$

$$A_{gt} = (1.75 \text{ in.} + 1.75 \text{ in.})(0.5 \text{ in.}) = 1.75 \text{ in.}^2$$

$$A_{nv} = A_{gv} - \Sigma A_{holes} = 11 - \left[(3.5\,\text{holes})(2)\left(\frac{3}{4} + \frac{1}{8}\right)(0.5)\right] = 7.93 \text{ in.}^2$$

$$A_{nt} = A_{gt} - \Sigma A_{holes} = 1.75 - \left[\left(\tfrac{1}{2}\,\text{hole}\right)(2)\left(\frac{3}{4} + \frac{1}{8}\right)(0.5)\right] = 1.31 \text{ in.}^2$$

The available block shear strength is found from Equation (4-11a) as

$$\phi P_n = \phi(0.60 F_u A_{nv} + U_{bs} F_u A_{nt}) \leq \phi(0.60 F_y A_{gv} + U_{bs} F_u A_{nt})$$
$$= 0.75[(0.60)(58)(7.93) + (1.0)(58)(1.31)] \leq 0.75[(0.60)(36)(11) + (1.0)(58)(1.31)]$$

$$= 263 \text{ kips} \leq \mathbf{235 \text{ kips}}$$

Therefore, $\phi P_n = 235 \text{ kips} > P_{uf} = 112$ kips OK.

Compression buckling of flange plate:

$K = 0.65$ (Assume fixity for the flange plate at the bolts; see Figure 5-4)

$L = 3$ in. (i.e., distance between adjacent bolts, or distance between the last hole and the fixed supports for the flange plates)

$$r_y = \sqrt{\frac{I_y}{A}} = \frac{t}{\sqrt{12}} = \frac{0.5}{\sqrt{12}} = 0.144 \text{ in.}$$

$$\frac{KL}{r} = \frac{(0.65)(3)}{0.144} = 13.5$$

Since $KL/r < 25$, $F_{cr} = F_y$ (AISC specification, Section J4.4).

$$\phi P_n = \phi F_{cr} A_g = (0.9)(36)(4.0) = \mathbf{129\ kips} > P_{uf} = 112 \text{ kips OK.}$$

The connection is adequate for the applied moment. Note that the column should also be checked for the need for stiffeners to resist and transfer the concentrated flange force (see Example 11-5).

11-5 MOMENT CONNECTIONS IN BEAMS AND BEAM SPLICES

In the previous sections, we considered moment connections at the beam-to-column joints in moment frames that are used to resist lateral loads. In this section, we will consider moment connections between beam elements supporting mainly gravity loads.

The simplest type of moment connection is one in which welds are used to connect one beam to another beam or other element. Figure 11-10a shows a small beam cantilevered from the face of a column. The beam flanges and web are connected to the column with groove welds. Note that the flanges are welded to develop the flexural strength of the beam, and the web is welded to develop the shear capacity of the beam. Figure 11-10b shows the welds used at a beam splice. In each of these cases, the welds could be partial or full penetration welds, or fillet welds, depending on the magnitude of the applied shear and moment at the connection. Beams may be spliced because of restrictions in the maximum length of a member that can be transported by road.

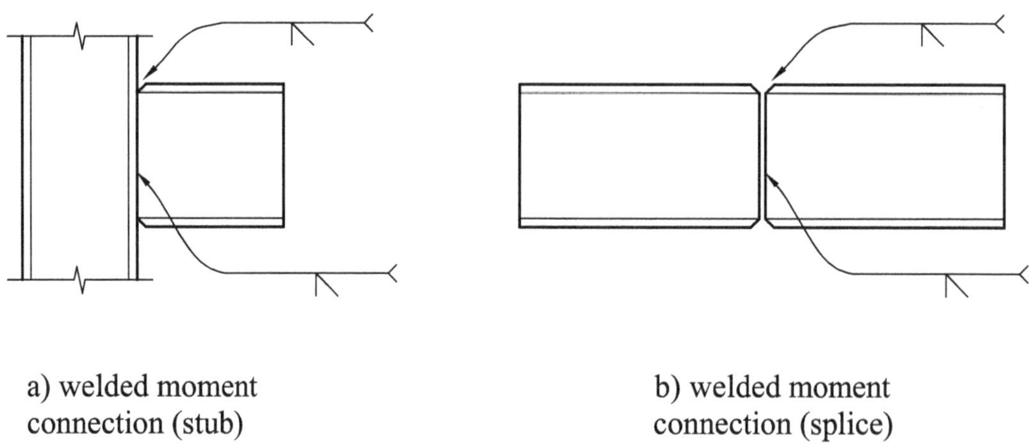

FIGURE 11-10 Basic moment connections

There are various types of connections where moment transfer occurs between beam segments. Figure 11-11 shows two types of moment connections that occur in beams. Figure 11-11a shows a new beam added to an existing floor framing plan. The existing conditions would usually make it impossible to place this beam in one section, and therefore one solution is to cut the beam at midspan and place the beam in two sections. The splice that is created would then be designed for the shear and moment that are required at the splice location. The top and bottom plates would be designed for the force-couple resulting from the moment, and the web plate would be designed for the shear (see Figures 11-11b and 11-11c). Figure 11-11d shows a cantilevered condition where a cantilevered beam frames into one side of an orthogonal girder and continues on the opposite side of the girder. To ensure continuity, the back-span and the cantilever segments of the beam are connected to each other with moment connections through the orthogonal girder. Thus, the connection is designed to transfer shear and moment across the intermediate supporting girder (see Figures 11-11e and 11-11f). In moment connections where bolted beam or girder flange plates are used, the beams/girders should have minimum flange widths, b_f, of 6 inches to accommodate two rows of bolts in each flange and ensure adequate edge distance for the bolts.

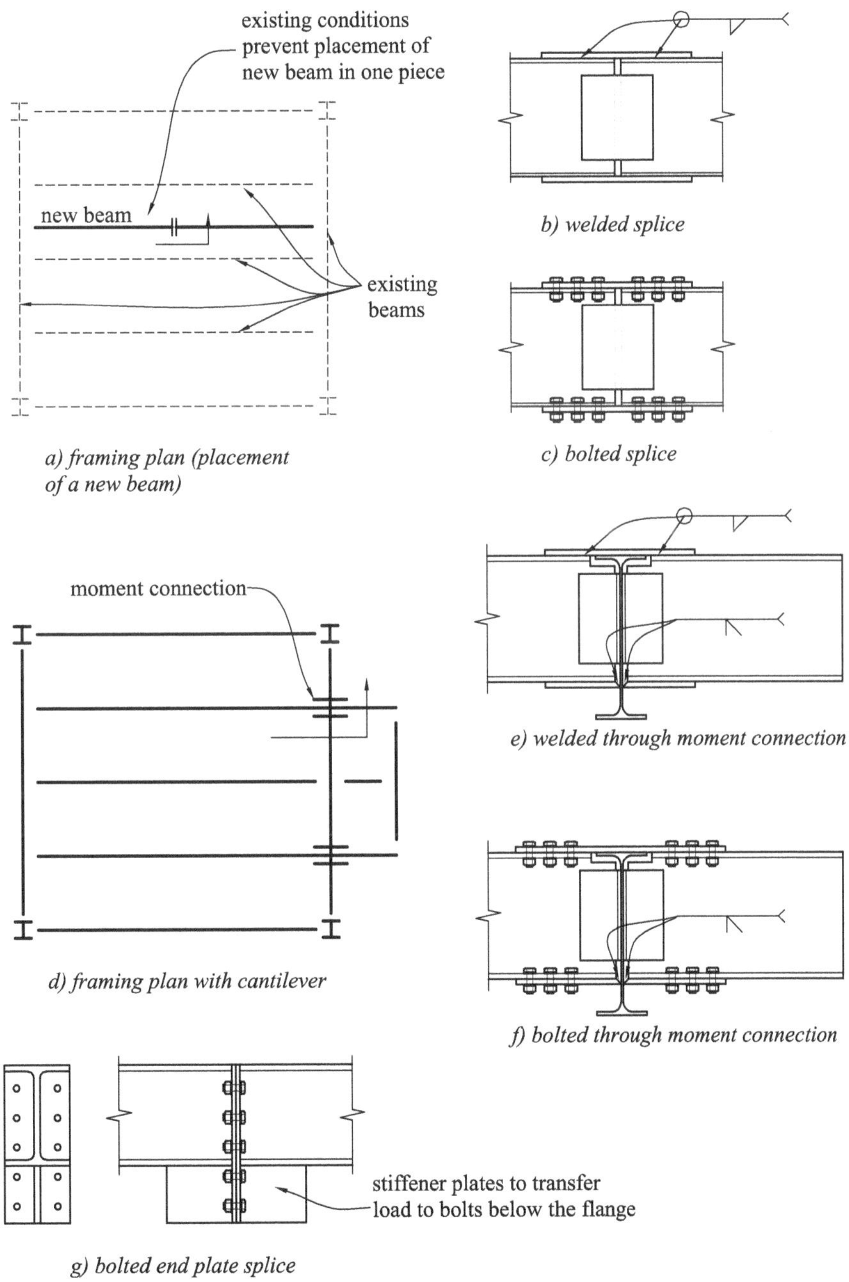

FIGURE 11-11 Beam splices

In each of these cases, the connection can be made with either bolts or welds, but not with bolts and welds in the same plane of loading unless slip-critical bolts and longitudinal welds are used (see AISC Specification J1.8). In many cases, the splice or moment connection is designed for the full moment capacity and full shear capacity of the beam for simplicity in the design. The location of the beam splice is also a design consideration. When a splice is located at the mid-span of a simply supported beam, the design moment is generally at a maximum, but the shear is generally at a minimum. The location of the splice may also have a practical significance. Figure 11-12 shows a two-span continuous beam, which is common in bridge construction. The basic moment diagram for this beam is such that the moment reaches zero at approximately the one-quarter point of each span measured from the center support. It is ideal to place a beam splice at this location since localized bending stresses are minimized.

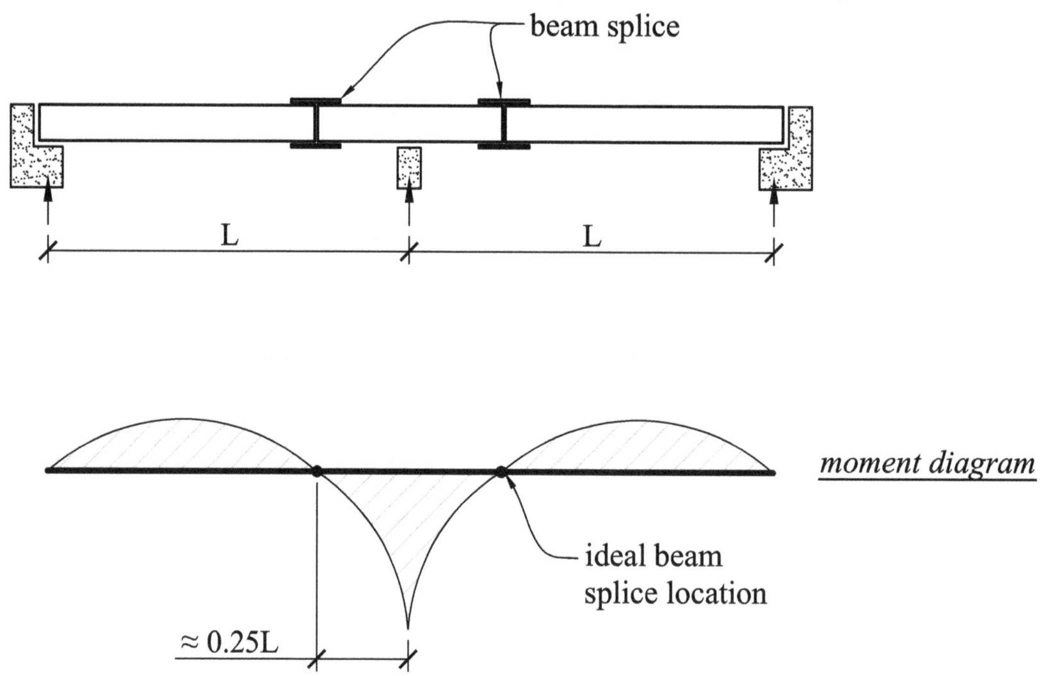

FIGURE 11-12 Beam splice location

EXAMPLE 11-3

Welded Beam Splice

A simply supported W16 × 36 beam requires a beam splice at midspan. Design the splice for the full moment capacity of the beam using welded plates. The beam is ASTM A992, the plates are ASTM A572, and the welds are E70XX.

Solution

From *AISCM* Table 1-1, the section properties for W16x36 are

$d = 15.9$ in. $\quad T = 13\text{-}5/8$ in.

$t_f = 0.43$ in. $\quad b_f = 6.99$ in.

$t_w = 0.295$ in.

From *AISCM* Table 3-2, $\phi_b M_p = 240$ ft. kips (i.e., the design moment capacity of a fully braced W16x36 beam).

Flange force:

$$P_{uf} = \frac{M}{d} = \frac{240}{15.9/12} = 182 \text{ kips.}$$

Tension on gross plate area:

$$\phi P_n = \phi A_g F_y = (0.9)(A_g)(50) = 182 \text{ kips}$$

Solving for A_g yields 4.04 in.^{2}

The flange plate size will now be selected and should be at least 1 in wider than the beam flange:

Size	A_g, in.^{2}	
$\frac{1}{2}$" × $11\frac{1}{4}$"	4.25	
$\frac{5}{8}$" × 8"	5	← Select
$\frac{3}{4}$" × 8"	6	

Weld strength:

The width and thickness of the beam flange are:

$b = 6.99$ in. $\quad$ (i.e., $b_f = 6.99$ in.)

Therefore, $L > b$

$t_f = 0.43$ in.

The 5/8 in. × 8 in. plate is selected, therefore, $t_{plate} = 5/8$ in.

From Table 10-2,

With thickness of the thinner part = 0.43 in.,

Minimum weld size = $\frac{3}{16}$ in. $\therefore D_{min} = 3$, but ¼ in. weld was specified; therefore, use D = 4

Maximum weld size $= t - \frac{1}{16} = 0.43 - \frac{1}{16} = 0.367$ in. = 5.9/16

$\therefore D_{max} = 5$; Therefore, the specified $\frac{1}{4}$-in. weld is O.K.

From Equation (10-7),

$$\phi R_n = 0.85\phi R_{wl} + 1.5\phi R_{wt}$$

$$182 = 0.85(1.392DL_l) + 1.5(1.392DL_t)$$

$$= (0.85)(1.392)(4)(2L_l) + (1.5)(1.392)(4)(6.99); \text{ Therefore, } L_l = 13.05 \text{ in.}$$

Therefore, Use $L_l = 13$ in. on each side of the beam splice.

Shear strength:

The amount of shear at the midspan of a simply supported beam is usually close to zero; however, heavy concentrated loads could be present on any given beam in a building, so engineering judgment will be needed to select the necessary amount of shear capacity. In this case, we will assume that 50 kips is the required shear capacity.

Since the throat length, T, is 13.625 in. for the W16 × 36 beams, this is the maximum depth of the shear plate that will be welded to the web of the beam. Solving for the web plate thickness, assuming that the height of the plate is 13 in.,

$$\phi_v V_n = \phi_v 0.6 F_y A_w C_{v1}$$

$50 = (1.0)(0.6)(50)(13)(t_p)(1.0);$ $\quad\quad \therefore t_p = 0.128$ in.

For practical reasons, a web plate that has a thickness equal to or greater than the web thickness of the W16 × 36 beam (t_w = 0.295 in.) should be used. Therefore, use $t_p = {}^3\!/_8$ in.

Thickness of the thinner part connected = 0.295 in.

From Table 10-2,

Minimum weld size = $\frac{3}{16}$ in. Therefore, D_{min} = 3 (therefore, the 3/16 in. weld specified is OK.)

Maximum weld size $= t - \frac{1}{16} = 0.295 - \frac{1}{16} = 0.232$ in. $= 3.7/16$. Therefore, $D_{max} = 3$

Assume that the length of the web plate parallel to the beam on each side of the splice = 3 in. Therefore, the horizontal weld dimension, b = 3 in. (see Figure 11-13):

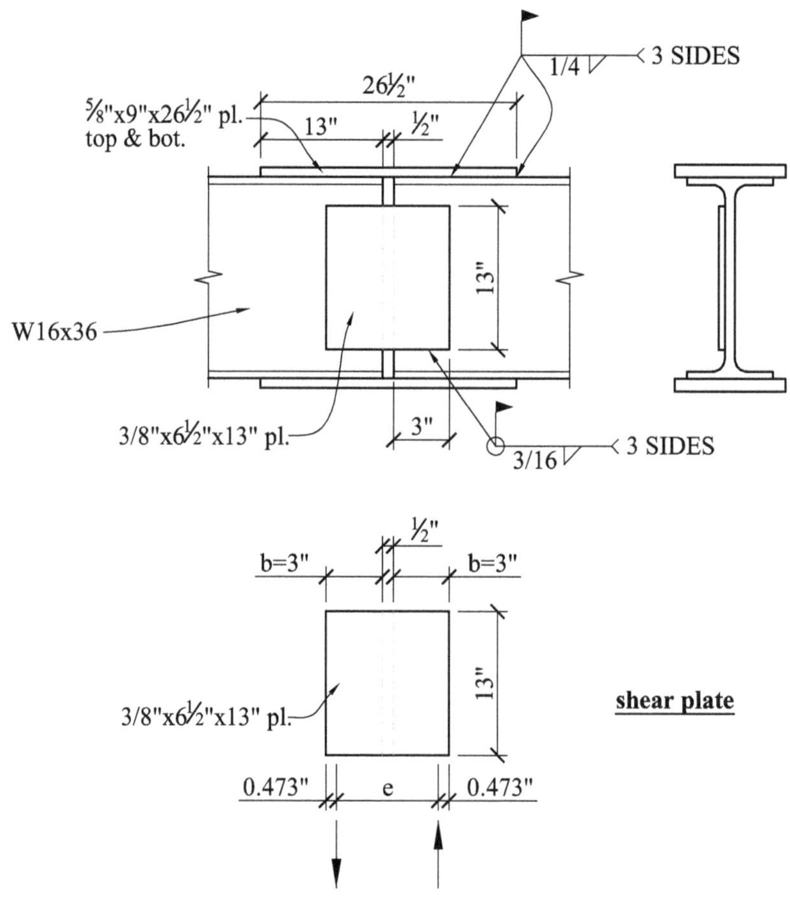

FIGURE 11-13 Detail for Example 11-3

The x – coordinate of the center of gravity of the weld group (Type "e") is found from Table 10-5 as follows:

$$\bar{x} = \frac{b^2}{2b + d} \quad \text{(see Table 10-5)}$$

$$= \frac{3^2}{(2)(3) + 13} = 0.473 \text{ in..}$$

The total horizontal dimension of the web plate = 3 in. + ½in. + 3 in. = 6.5 in.

The distance between the centroids of the web plate weld on both sides of the beam splice is e = 6.5 - $\bar{x}$ - $\bar{x}$ = 6.5 in. – 0.473 in. – 0.473 in. = 5.55 in

From *AISCM* Table 8-8,

ℓ = vertical length of weld in the web plate = d = 13 in.

$b = 3$ in. (see Figure 11-13)

$$a = \frac{e}{\ell} = \frac{5.55}{13} = 0.427$$

$$k = \frac{b}{\ell} = \frac{3}{13} = 0.230$$

$C = 2.10$ (obtained by linear interpolation)

$C_1 = 1.0$ for E70XX weld electrodes.

Recall that we assumed a required shear capacity of 50 kips in the splice.

$$\text{Therefore, } D_{min} = \frac{P_u}{\phi C C_1 \ell} = \frac{50\text{ kips}}{(0.75)(2.10)(1.0)(13)} = 2.46 < 3 \quad \text{OK}$$

Note that the final design detail includes a ½-in. gap between the ends of the beam that is added for construction tolerances.

EXAMPLE 11-4

Bolted Beam Splice

Repeat Example 11-3 for a bolted moment connection, assuming M_u = 175 ft.-kips and V_u = 50 kips. Use ⅞-in.-diameter (Group 150) A490 SC bolts in STD holes and a Class A faying surface.

Solution

From *AISCM* Table 1-1, the section properties for W16 × 36 are

$d = 15.9$ in. $\quad T = 13\text{-}5/8$ in.

$t_f = 0.43$ in. $\quad b_f = 6.99$ in.

$t_w = 0.295$ in. $\quad S_x = 56.5\text{ in.}^3$

From *AISCM* Table 3-2, $\phi_b M_p$ = 240 ft.-kips.

Flange force:

$$P_{uf} = \frac{M}{d} = \frac{175}{15.9/12} = 132\text{ kips.}$$

Bolts in Top and Bottom Flanges:

In this case, slip is a strength limit state since slip at the joint would cause additional beam deflection (see discussion in Chapter 9).

From *AISCM* Table 7-3, $\phi_v R_n$ = 16.6 kips/bolt; this is the single shear capacity for one ⅞ in. diameter (Group 150) A490 SC bolts in STD holes and Class A faying surface.

$$N_{b,required} = \frac{P_{uf}}{\varphi_v r_n} = \frac{132}{16.6} = 7.96.\text{ Therefore, use } 8 \text{ bolts.}$$

Check the reduced flexural strength of the W16 × 36 beam due to the presence of bolt holes:

$$F_y/F_u = 50/65 = 0.77 < 0.80;\ \text{therefore, } Y_t = 1.0$$

$$A_{fg} = (b_f)(t_f) = (6.99)(0.43) = 3.00 \text{ in.}^2$$

$$A_{fn} = A_{fg} - \Sigma A_{holes}$$

$$= 3.00 - \left[(2\,\text{holes})\left(\frac{7}{8}+\frac{1}{8}\right)(0.43)\right] = 2.14 \text{ in.}^2$$

$$F_u A_{fn} = (65)(2.14) = 139 \text{ kips}$$

$$Y_t F_y A_{fg} = (1.0)(50)(3.00) = 150 \text{ kips}$$

$$\phi_b M_n = \phi_b F_u S_x\left(\frac{A_{fn}}{A_{fg}}\right) = (0.9)(65)(56.5)\left(\frac{2.14}{3.0}\right) = 2357 \text{ in.-kips}$$

$$= 196 \text{ ft.-kips} > 175 \text{ ft.-kips,} \qquad \text{OK}$$

Flange Plates:

Assume that the top and bottom plates are still ⅝ in. × 8 in. Tension on the gross area was checked in the previous example; now we will check tension on the net area (see Figure 11-14).

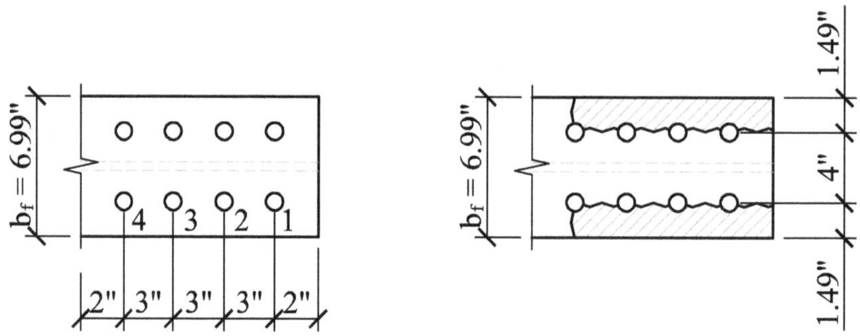

FIGURE 11-14 Bolt bearing and block shear for Example 11-4

Tension on the net plate area:

$$A_g = (0.625 \text{ in.})(8 \text{ in.}) = 5.0 \text{ in.}^2$$

$$A_n = A_g - \Sigma A_{holes} = 5.0 - \left[(2\,\text{holes})\left(\frac{7}{8}+\frac{1}{8}\right)(0.625)\right] = 3.75 \text{ in.}^2$$

$$A_e = A_n U = (3.75 \text{ in.}^2)(1.0) = 3.75 \text{ in.}^2$$

Strength based on the effective area is

$$\phi P_n = \phi F_u A_e = (0.75)(65)(3.75 \text{ in.}^2) = 182 \text{ kips} > P_{uf} = 132 \text{ kips OK.}$$

Check bolt bearing (see Figure 11-14):

Since $t_f F_{ub}$ = (0.43)(65) = 27.9 kips/in. < $t_p F_{up}$ = (0.625)(65) = 40.6 kips/in.; therefore, bearing and tearout on the beam flange will control.

The clear distance between the edge of bolt hole 1 and the plate edge is

$$L_{c1} = 2 \text{ in.} - \left[0.5\left(\frac{7}{8}+\frac{1}{8}\right)\right] = 1.50 \text{ in.}$$

The clear distance between the edge of adjacent bolt holes is

$$L_{c2} = L_{c3} = L_{c4} = 3 \text{ in.} - \left[\left(\frac{7}{8} + \frac{1}{8}\right)\right] = 2.0 \text{ in.}$$

The bolt bearing capacities are calculated assuming that deformation at the bolt hole at service load is limited and is therefore a design consideration. The smaller of the bearing capacity and the tearout capacity will control:

$$\begin{aligned} \text{For bolt 1, } \phi R_n &= \phi 1.2 L_c t F_u \leq \phi 2.4 d t F_u \\ &= (0.75)(1.2)(1.50 \text{ in.})(0.43 \text{ in.})(65 \text{ ksi}) \\ &\leq (0.75)(2.4)(0.875)(0.43 \text{ in.})(65 \text{ ksi}) \\ &= 37.7 \text{ kips} \leq 44.0 \text{ kips} \end{aligned}$$

Therefore, ϕR_n (bolt 1) = 37.7 kips.

$$\begin{aligned} \text{For bolts 2, 3, and 4, } \phi R_n &= (0.75)(1.2)(2.0 \text{ in.})(0.43 \text{ in.})(65 \text{ ksi}) \\ &\leq (0.75)(2.4)(0.875)(0.43 \text{ in.})(65 \text{ ksi}) \\ &= 50.3 \text{ kips} \leq 44.0 \text{ kips} \end{aligned}$$

Therefore, ϕR_n (bolt 2, 3, and 4) = 44.0 kips.

The total bearing and tearout strength, ϕR_n = (2)(37.7 kips) + (6)(44.0 kips)
= 339 kips > P_{uf} = 132 kips,
O.K. for bearing and tearout.

Block shear:

Since $t_f F_{ub}$ (for the beam flange) = 27.9 kips
$t_p F_{up}$ (for the plate) = 40.6 kips
Since $t_f F_{ub}$ (for the beam flange) < $t_p F_{up}$ (for the plate), therefore, the beam flange, (t_f = 0.43 in.) will control the capacity for block shear and will be checked (see Figure 11-14).
Number of holes in each shear plane = 3.5 holes
Number of holes in each tension plane = 0.5 hole
Number of shear planes in each plate = 2
Number of tension planes in each plate = 2

$$A_{gv} = (2)(3 \text{ in.} + 3 \text{ in.} + 3 \text{ in.} + 2 \text{ in.})(0.43 \text{ in.}) = 9.46 \text{ in.}^2$$

$$A_{gt} = (1.49 \text{ in.} + 1.49 \text{ in.})(0.43 \text{ in.}) = 1.28 \text{ in.}^2$$

$$A_{nv} = A_{gv} - \Sigma A_{holes} = 9.46 - \left[(3.5 \text{ holes})(2)\left(\frac{7}{8} + \frac{1}{8}\right)(0.43)\right] = 6.45 \text{ in.}^2$$

$$A_{nt} = A_{gt} - \Sigma A_{holes} = 1.28 - \left[\left(\tfrac{1}{2} \text{ hole}\right)(2)\left(\frac{7}{8} + \frac{1}{8}\right)(0.43)\right] = 0.85 \text{ in.}^2$$

The available block shear strength is found from Chapter 4 using U_{bs} = 1.0:

$$\begin{aligned} \phi P_n &= \phi(0.60 F_u A_{nv} + U_{bs} F_u A_{nt}) \leq \phi(0.60 F_y A_{gv} + U_{bs} F_u A_{nt}) \\ &= 0.75[(0.60)(65)(6.45) + (1.0)(65)(0.85)] \\ &\leq 0.75[(0.60)(50)(9.46) + (1.0)(65)(0.85)] \\ &= 230 \text{ kips} \leq 254 \text{ kips.} \end{aligned}$$

Therefore, ϕP_n = 230 kips > P_{uf} = 132 kips, OK.

Compression buckling of the flange plates:

Check the buckling of the flange plate; the plate is assumed to be fixed at both ends. Recall that A_g (flange plate) = 5.62 in.^{2}

$K = 0.65 \quad$ (see Figure 5-4).

Maximum unbraced length of the flange plate, $L = 2$ in. + 0.5 in. + 2 in. = 4.5 in. This is the distance between the bolts on each side of the splice.

Note: The unbraced length of the flange plate includes the 2 in. edge distance at each beam segment on each side of the splice, and the ½ in. gap at the splice.

$$r_y = \sqrt{\frac{I_y}{A}} = \frac{t}{\sqrt{12}} = \frac{0.625}{\sqrt{12}} = 0.180 \text{ in.}$$

$$\frac{KL}{r} = \frac{(0.65)(4.5)}{0.180} = 16.2$$

Since $KL/r < 25$, $F_{cr} = F_y$ (*AISCM* Section J4.4)

$\phi P_n = \phi F_{cr} A_g$ = (0.9)(50)(5.0) = 225 kips > P_{uf} = 132 kips, OK.

Shear strength:

A $\frac{3}{8}$-in. × 13-in. plate will be used as the web plate (the same size used in the previous example). Assuming that a shear strength of 50 kips is needed, *AISCM* Table 7-6 can be used to check the capacity of the bolt group (see Figure 11-15). The eccentricity, e_x, is the distance between the vertical row of bolts on both sides of the beam splice.

Therefore, e = 1.5 in. + 0.5 in. + 1.5 in. = 3.5 in.

Using the number of bolts in one vertical row, $n = 4$ and the bolt spacing, $s = 3$ in.,

By interpolation from *AISCM* Table 7-6, obtain $C = \frac{2.81 + 2.36}{2} = 2.58$

Assume (4)- $\frac{7}{8}$-in.-diameter (Group 150) A490N bolts; the single shear capacity of one $\frac{7}{8}$ in.diameter (Group 150) A490N bolt from AISCM Table 7-1) is $\phi_v R_n$ = 30.7 kips/bolt.

Therefore, the strength of the bolt group is

$(C)(\phi_v R_n)$ = (2.58)(30.7) = 79.2 kips > 50 kips. OK

Check bearing and tearout failure modes for the beam web (see Figure 11-15):

$$L_{c1} = 2 \text{ in.} - \left[0.5\left(\frac{7}{8} + \frac{1}{8}\right)\right] = 1.50 \text{ in.}$$

$$L_c(\text{bolts } 2, 3 \text{ and } 4) = 3 \text{ in.} - \left(\frac{7}{8} + \frac{1}{8}\right) = 2.0 \text{ in.}$$

$$\phi R_n \text{ (bolt 1)} = (0.75)(1.2)(1.5 \text{ in.})(0.295 \text{ in.})(65 \text{ ksi})$$

$$\leq (0.75)(2.4)(0.875)(0.295 \text{ in.})(65 \text{ ksi})$$

$$= 25.9 \text{ kips} \leq 30.2 \text{ kips}$$

Therefore, ϕR_n per bolt for bolt 1 (i.e. the top most bolts) = 25.9 kips

$$\phi R_n \text{ (bolts 2, 3, and 4)} = (0.75)(1.2)(2.0 \text{ in.})(0.295 \text{ in.})(65 \text{ ksi})$$

$$\leq (0.75)(2.4)(0.875)(0.295 \text{ in.})(65 \text{ ksi})$$

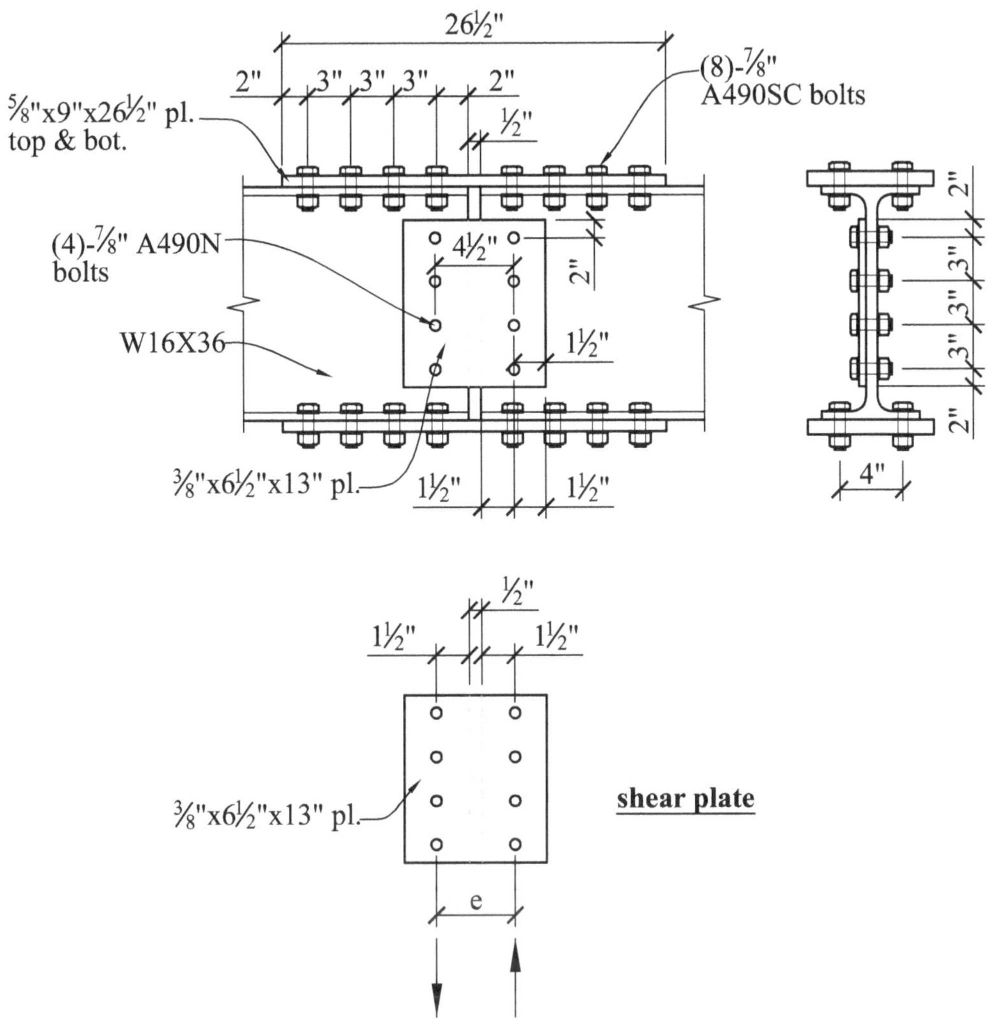

FIGURE 11-15 Detail for Example 11-4

$$= 34.5 \text{ kips} \leq 30.2 \text{ kips}$$

Therefore, ϕR_n per bolt = 30.2 kips

ϕR_n for the bolt group $= 25.9 + (3)(30.2) = \mathbf{117\ kips} > 50$ kips OK for bearing.

Block shear in ASTM A572 plate (3/8 in. x 6½ in. x 13 in. plate):

Use *AISCM* Tables 9-3a through 9-3c to determine the block shear strength per inch of thickness;

The distance from the top of the web plate to the center of the top bolt hole (see Figure 11-15), is L_{ev} = 2 in.;

L_{eh} = 1.5 in. (horizontal distance between the vertical bolt line in the web to the splice);

For (4) 7/8 in. diameter bolts in one vertical row, N = 4;

F_u = 65 ksi (ASTM A572 Grade 50 steel plate)

The available strength in block shear is calculated as follows:

Block shear – **Tension Rupture Component** (*AISCM* Table 9-3a):

For 3/8 in. thick ASTM A572 steel plate with 7/8 in. diameter bolt and L_{eh} = 1½ in., we obtain from *AISCM* Table 9-3a

$$\frac{\phi F_u A_{nt}}{t} = 48.8 \text{ kips / in.}$$

Therefore, $\phi F_u A_{nt} = 48.8t = 48.8(3/8) = 18.3 \text{ kips}$

Block shear – **Shear Yielding Component** (*AISCM* Table 9-3b):

For 3/8 in. thick ASTM A572 steel plate with 7/8 in. diameter bolt; n = N = 4; and L_{ev} = 2 in., we obtain from *AISCM* Table 9-3b

$$\frac{\phi 0.6F_y A_{gv}}{t} = 248\,\text{kips/in.}$$

Therefore, $\phi 0.6F_y A_{gv} = 248t = 248(3/8) = 93\,\text{kips}$

Block shear – **Shear Rupture Component** (*AISCM* Table 9-3c):

For ASTM A572 3/8 in. thick steel plate with 7/8 in. diameter bolt; n = N = 4; and L_{ev} = 2 in., we obtain from *AISCM* Table 9-3c

$$\frac{\phi 0.6F_u A_{nv}}{t} = 219\,\text{kips/in.}$$

Therefore, $\phi 0.6F_u A_{nv} = 219t = 219(3/8) = 82.1\,\text{kips}$

The total block shear strength for the 3/8 in. thick ASTM A572 steel plate is

$$\phi R_n = \phi\left(0.60F_u A_{nv} + U_{bs}F_u A_{nt}\right) \le \phi\left(0.60F_y A_{gv} + U_{bs}F_u A_{nt}\right)$$

$$= 82.1 + 18.3 \le 93 + 18.3$$

$$= 100.4\,\text{kips} \le 111.3\,\text{kips}$$

Therefore, ϕR_n = **100.4 kips** > 50 kips, OK.

11-6 COLUMN STIFFENERS IN MOMENT CONNECTIONS

In Chapter 6, we considered the design of beams for concentrated forces. In this section, we will expand on this topic as it applies to concentrated forces in columns. In the previous sections, we considered the design of moment connections at the ends of beams. When these connections are made to column flanges, there are several localized failure modes that need to be investigated. When an end moment from a beam is applied to the flanges of a column as shown in Figure 11-16, the flange force due to the moment is transferred to the column through its flanges and to the web. This force could be either tension or compression and could cause localized bending of the column flange and localized buckling of the column web. To prevent such behavior, stiffener plates can be added to the column flange and to the column web, or choose a column size with enough flange and web thickness that obviates the need for stiffener or doubler plates. In some cases, these stiffeners might be added even if the column were adequate to support the concentrated forces as a means of providing redundancy to the connection. However, the addition of these plates can create constructability issues because these stiffeners might conflict with the orthogonal beam framing into the web of the column. The stiffeners could be field applied, but this adds cost to the connection since field welds are more expensive than welds applied in the shop. Figure 11-17 shows several possible column stiffening details.

Another possible solution is to use an extended shear plate connection for the beam framing into the web of the column, as shown in Figure 11-18. This connection increases the bending moment in the column due to the increased eccentricity of the shear connection of the beam to the column web, and therefore the column would have to be designed accordingly. A simpler solution might be to increase the size of the column such that it has adequate capacity to support these localized concentrated forces without the use of stiffener plates.

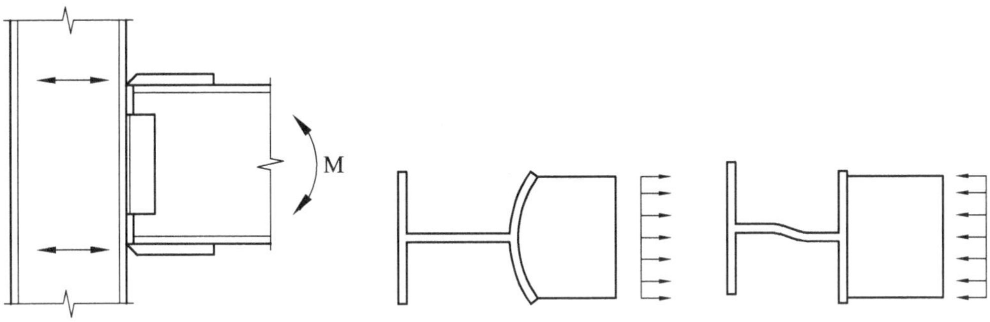

a) forces on flanges *b) local flange bending* *c) local web buckling*

FIGURE 11-16 Concentrated forces on columns

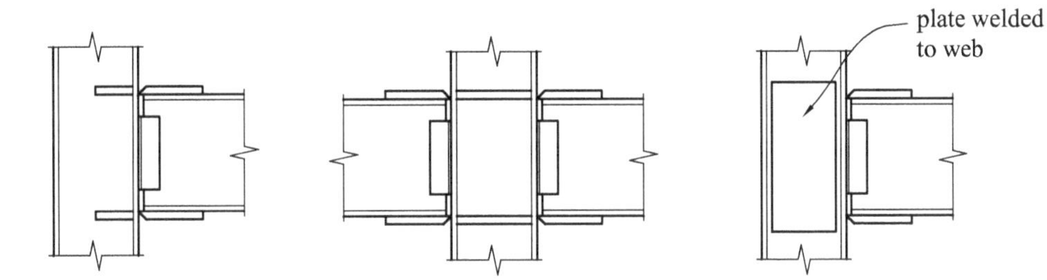

a) flange stiffener (one side) *b) flange stiffener (full-depth)* *c) web doubler plate*

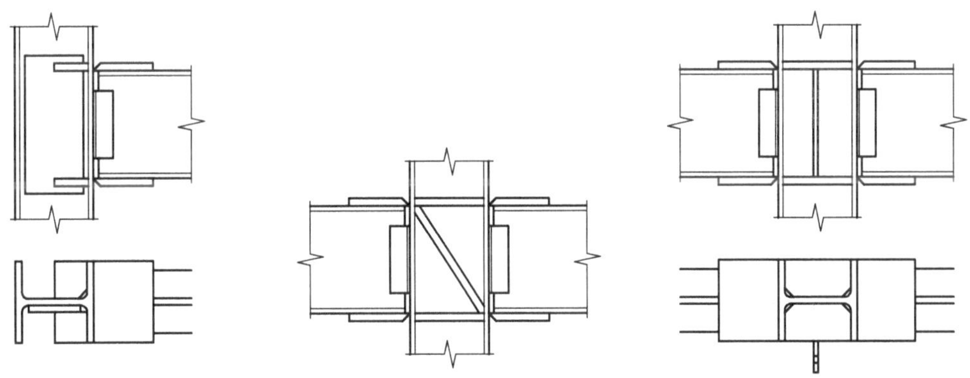

d) web doubler plate and flange stiffeners *e) diagonal stiffeners* *f) extended shear plate*

FIGURE 11-17 Column stiffeners for concentrated forces

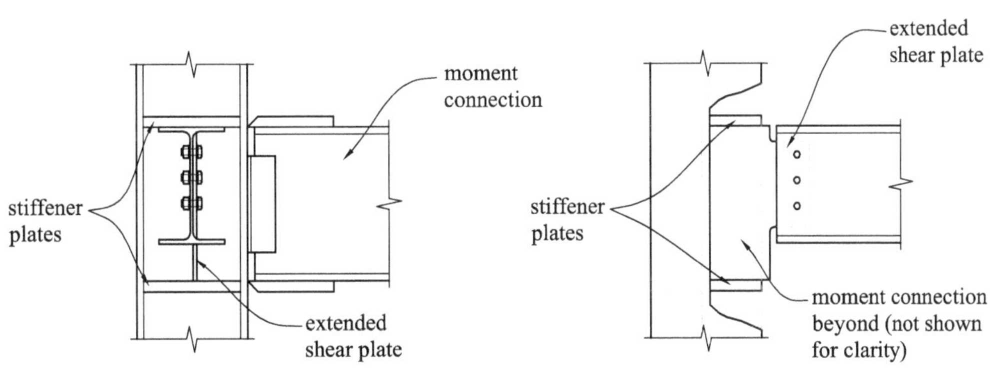

a. end view *b. side view*

FIGURE 11-18 Extended shear plate connection

The various failure modes arising from the concentrated horizontal forces acting on the column flanges due to the moment at the girder-to-column connections will now be described in greater detail (see *AISCM* Section J10) [1].

Flange local bending

Flange local bending (*AISCM* Section J10.1) occurs when a concentrated tension force is applied to the flange. The design strength for flange local bending when the concentrated force is applied at a distance greater than $10t_f$ from the end of the member is

$$\phi R_n = \phi 6.25 t_f^{\,2} F_y \qquad \text{for } y \geq 10b_f. \tag{11-3}$$

When the concentrated horizontal force occurs at a distance less than $10t_f$ from the end of the member (e.g., moment connection at a roof level), the design strength is

$$\phi_{fb} R_n = \phi_{fb} 3.125 t_f^{\,2} F_y \qquad \text{for } y < 10b_f, \tag{11-4}$$

Where,
ϕ_{fb} = 0.9 for flange local bending,
R_n = Nominal design strength,
t_f = Flange thickness,
b_f = Flange width, and
F_y = Yield strength of flange.

Note: For the ASD method, divide the nominal strength, R_n, which can be obtained from Equations (11-3) and (11-4) by the flange local buckling safety factor, Ω = 1.67.

When the concentrated force exceeds the design strength for flange local bending, transverse stiffeners are required (stiffener design will be covered later). When the extent of the loading across the flange is less than $0.15b_f$, then flange local bending does not need to be checked.

Web local yielding

Web local yielding (*AISCM* Section J10.2) applies to concentrated compression forces. The design equations for this case were discussed in Chapter 6 (beam bearing) but will be repeated here for clarity.

The design strength for web local yielding is

$$\phi_{wy} R_n = \phi_{wy} (5k + \ell_b) F_y t_w \qquad \text{for } y > d, \tag{11-5}$$

$$\phi_{wy} R_n = \phi_{wy} (2.5k + \ell_b) F_y t_w \qquad \text{for } y \leq d, \tag{11-6}$$

Where,
ϕ_{wy} = 1.0 for web local yielding,
R_n = Nominal design strength,
t_w = Column web thickness,
ℓ_b = Bearing length,
d = Column depth,
y = Distance from the concentrated applied force to the end of the column member (see Figure 11-19),
k = Section property from part 1 of *AISCM*
F_y = Yield strength, ksi.

Note: For the ASD method, divide the nominal strength, R_n, which can be obtained from Equations (11-5) and (11-6) by the web local yielding safety factor, Ω = 1.50.

When the concentrated force exceeds the design strength for web local yielding, either transverse stiffeners or a web doubler plate is required.

Web local crippling

Web local crippling (*AISCM* Section J10.3) occurs when the concentrated load causes a local buckling of the web. The design strength for web local crippling is obtained as follows:

$$\phi_{wc}R_n = \phi_{wc}0.8t_w^{\,2}\left[1+3\left(\frac{\ell_b}{d}\right)\left(\frac{t_w}{t_f}\right)^{1.5}\right]\sqrt{\frac{EF_{yf}t_f}{t_w}}\,Q_f \qquad \text{for } y \ge \frac{d}{2}, \tag{11-7}$$

$$\phi_{wc}R_n = \phi_{wc}0.4t_w^{\,2}\left[1+3\left(\frac{\ell_b}{d}\right)\left(\frac{t_w}{t_f}\right)^{1.5}\right]\sqrt{\frac{EF_{yf}t_f}{t_w}}\,Q_f \qquad \text{for } y < \frac{d}{2} \text{ and } \frac{\ell_b}{d} \le 0.2, \tag{11-8}$$

$$\phi_{wc}R_n = \phi_{wc}0.4t_w^{\,2}\left[1+\left(\frac{4\ell_b}{d}-0.2\right)\left(\frac{t_w}{t_f}\right)^{1.5}\right]\sqrt{\frac{EF_{yf}t_f}{t_w}}\,Q_f \qquad \text{for } y < \frac{d}{2} \text{ and } \frac{\ell_b}{d} > 0.2, \tag{11-9}$$

Where

Q_f = chord-stress interaction parameter

= 1.0 for wide flange sections, channels, box sections, and the connecting surface of HSS in tension.

ϕ_{wc} = 0.75 for web local crippling,

E = 29×10^6 psi.

Note: For the ASD method, divide the nominal strength, R_n, which can be obtained from Equations (11-7) through (11-9) by the web local crippling safety factor, $\Omega = 2.0$.

When the concentrated force exceeds the design strength for web crippling, either transverse stiffeners or a web doubler plate extending a distance of at least $d/2$ above and below the location of the horizontal concentrated force is required.

Web compression buckling

Web compression buckling (*AISCM* Section J10.5) can occur only when a pair of concentrated compression forces are applied to both sides of a member at the same location. This occurs at a column with moment connections on both column flanges, or where a pair of vertical compression forces (from the column below and the column above) act on the top and bottom flanges of a cantilevered beam. The design strength for web compression buckling is

$$\phi_{wb}R_n = \frac{\phi_{wb}24t_w^{\,3}\sqrt{EF_y}}{h}Q_f \qquad \text{for } y \ge \frac{d}{2}, \tag{11-10}$$

$$\phi_{wb}R_n = \frac{\phi_{wb}12t_w^{\,3}\sqrt{EF_y}}{h}Q_f \qquad \text{for } y < \frac{d}{2}, \tag{11-11}$$

Where,

ϕ_{wb} = 0.90 for web compression buckling, and

h = Clear distance between the flanges, excluding the fillets (see Figure 11-19) = d – 2k.

y = distance of the concentrated compression force from the end of the member whose web is subject to compression buckling

Q_f = chord-stress interaction parameter

= 1.0 for wide flange sections, channels, box sections, and for the connecting surface of HSS in tension

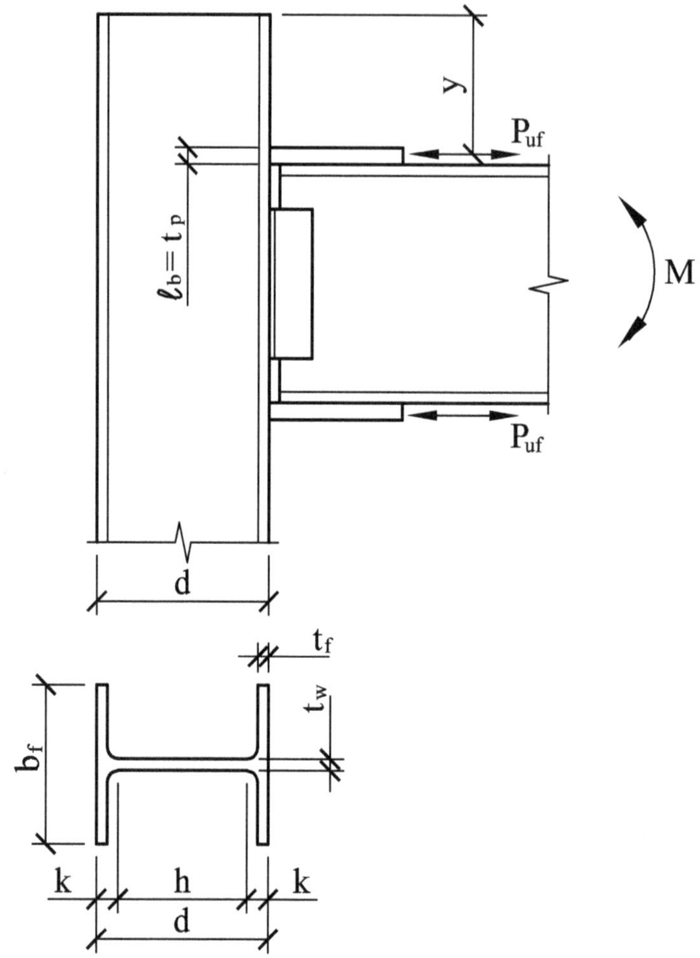

FIGURE 11-19 Dimensional parameters for concentrated forces

Note: For the ASD method, divide the nominal strength, R_n, which can be obtained from Equations (11-10) and (11-11) by the web compression buckling safety factor, $\Omega = 1.67$.

When the concentrated force exceeds the design strength for web buckling, a single transverse stiffener, a pair of transverse stiffeners, or a full-depth web doubler plate is required.

Web sidesway buckling

Web sidesway buckling occurs when a single compression load (e.g., a column supported on a transfer girder) is applied to the web of a wide flange girder through the top flange, and the relative lateral movement between the loaded compression or top flange and the tension or bottom flange is *not* restrained at the point of application of the concentrated load (e.g., the top flange is laterally restrained, but the bottom flange is not restrained and can therefore buckle sideways).

To determine the available girder web strength for this limit state, use Equations (11-12a) and (11-12b); see *AISC* Specification Section J10.4.

The design strength for web sidesway buckling *if the compression flange is restrained against rotation* is obtained as follows:

1. When $\left(\dfrac{h}{t_w}\right) \Big/ \left(\dfrac{L_b}{b_f}\right) \le 2.3$,

$$\phi_{wsb} R_n = \frac{\phi_{wsb} C_r t_w^{\,3} t_f}{h^2}\left[1 + 0.4\left(\frac{h}{t_w} \Big/ \frac{L_b}{b_f}\right)^3\right] \tag{11-12a}$$

2. When $\left(\frac{h}{t_w}\right) \Big/ \left(\frac{L_b}{b_f}\right) > 2.3$, the limit state of web sidesway buckling does ***not*** occur.

When the available strength in Equation (11-12a) is less than the required strength or factored force, local lateral bracing to the *tension flange*, a pair of transverse stiffeners, or a doubler plate must be provided.

The design strength for web sidesway buckling *if the compression flange is not restrained against rotation* is

1. When $\left(\frac{h}{t_w}\right) \Big/ \left(\frac{L_b}{b_f}\right) \leq 1.7$

$$\phi_{wsb} R_n = \frac{\phi_{wsb} C_r t_w^{\,3} t_f}{h^2}\left[0.4\left(\frac{h}{t_w}\Big/\frac{L_b}{b_f}\right)^3\right] \qquad (11\text{-}12b)$$

2. When $\left(\frac{h}{t_w}\right) \Big/ \left(\frac{L_b}{b_f}\right) > 1.7$, the limit state of web sidesway buckling does ***not*** occur.

When the available strength in Equation (11-12b) is less than the required strength or factored force, local lateral bracing to the *tension and compression flanges* must be provided at the point of application of the concentrated forces. Note that in this case where flange rotation is not restrained, neither web stiffeners nor doubler plates are effective and therefore should not be used.

Note: For the ASD method, divide the nominal strength, R_n, which can be obtained from Equations (11-12a) and (11-12b) by the web sidesway buckling safety factor, $\Omega = 1.76$.

Where,

ϕ_{wsb} = 0.85 for web sidesway buckling, and

h = Clear distance between the flanges, excluding the fillets (see Figure 11-19)
= d – 2k.
(The value of h above applies to doubly symmetric sections. For singly symmetric sections, h = twice the depth of the web in compression. See Commentary to AISC Specification Section C-J10.4)

L_b = largest lateral unbraced length along either flange at the point of the applied load, inches

C_r = 960,000 ksi when $M_u < M_y$ for LRFD or $1.5M_s < M_y$ for ASD
= 480,000 ksi when $M_u > M_y$ for LRFD or $1.5M_s \geq M_y$ for ASD,

$M_y = F_y S_x$,

F_y = Yield stress, and

S_x = Elastic section modulus,

M_u = Factored moment in kips-inch from the LRFD load combination in Section 2-3,

M_s = Service or unfactored total moment in kips-inch from the ASD load combination in Section 2-3.

Web panel-zone shear

The concentrated force on a column flange could also cause large shear forces across the column web. The region in which these forces occur is called the panel zone. The shear in the panel zone is the sum of the shear in the web and the shear due to the flange force, P_{uf}.

The design strength for *web panel-zone shear* (*AISCM* Section J10.6) assumes that the effects of panel zone deformation on frame stability are not considered:

When $\alpha P_r \leq 0.4P_y$,

$$\phi_{cw} R_n = \phi_{cw} 0.6 F_y d t_w \quad (11\text{-}13a)$$

when $\alpha P_r > 0.4P_y$,

$$\phi_{cw} R_n = \phi_{cw} 0.6 F_y \, d t_w \left(1.4 - \frac{\alpha P_r}{P_y}\right) \quad (11\text{-}13b)$$

Where,

α = 1.0 for LRFD and 1.6 for ASD

ϕ_{cw} = 0.9,

P_r = Required axial strength; for LRFD = factored axial load in the column, P_u,

P_y = Axial yield strength of the column = $F_y A$, and

A = Cross-sectional area of the column.

Note: For the ASD method, divide the nominal strength, R_n, which can be obtained from Equations (11-13a) and (11-13b) by the web panel-zone shear safety factor, $\Omega = 1.67$.

When the effects of panel zone deformation on frame stability are considered, the reader is referred to the AISC specification, Section J10.6 [1].

When the shear strength in the web panel zone is exceeded, a full-depth web doubler plate or a pair of diagonal stiffeners is required. It is typically more economical to avoid the use of doubler plates and instead increase the size of the column.

For all of the previous design checks for concentrated forces, stiffener plates are required when the applied forces are greater than the design strength for each failure mode. When transverse stiffeners are required, the force is distributed to the web or flange and the stiffener plates based on their relative stiffnesses. However, Section J10 of the AISC specification [1] allows a more simplified approach where the size of the plate is based on the difference between the required strength and the available strength of the failure mode in question, which is expressed as follows:

$$R_{u,\,st} = P_{uf} - \phi R_{n,\,min} \quad (11\text{-}14)$$

Where,

$R_{u,\,st}$ = Required strength of the stiffener (tension or compression),

P_{uf} = Factored applied flange force,

$\phi R_{n,\,min}$ = Lesser of the design strength for flange local bending, web local yielding, web local crippling, and web compression buckling failure modes.

The transverse stiffeners are then designed to provide adequate cross-sectional area as follows:

$$R_{u,\,st} \leq \phi F_{y,\,st} A_{st}. \quad (11\text{-}15)$$

Solving for the plate area gives

$$A_{st} \geq \frac{R_{u,\,st}}{\phi F_{y\,st}}, \quad (11\text{-}16)$$

Where,

ϕ = 0.9 (yielding),

A_{st} = Area of the transverse stiffeners, and

$F_{y,\,st}$ = Yield stress of the transverse stiffeners.

Web doubler plates are required when the shear in the column exceeds the web panel zone shear strength. The required design strength of the web doubler plate or plates is expressed as follows:

$$V_{u,\,dp} = V_u - \phi_{cw} R_n \quad (11\text{-}17)$$

Where,

$V_{u, dp}$ = Required strength of the web doubler plate or plates,

V_u = Factored shear in the column web at the concentrated force, and

$\phi_{cw}R_n$ = Design shear strength of the web panel zone (Equation (11-13a) or (11-13b)). With reference to Figure 11-20, column stiffeners are proportioned to meet the following requirements in accordance with Section J10.8 of the AISC Specification.

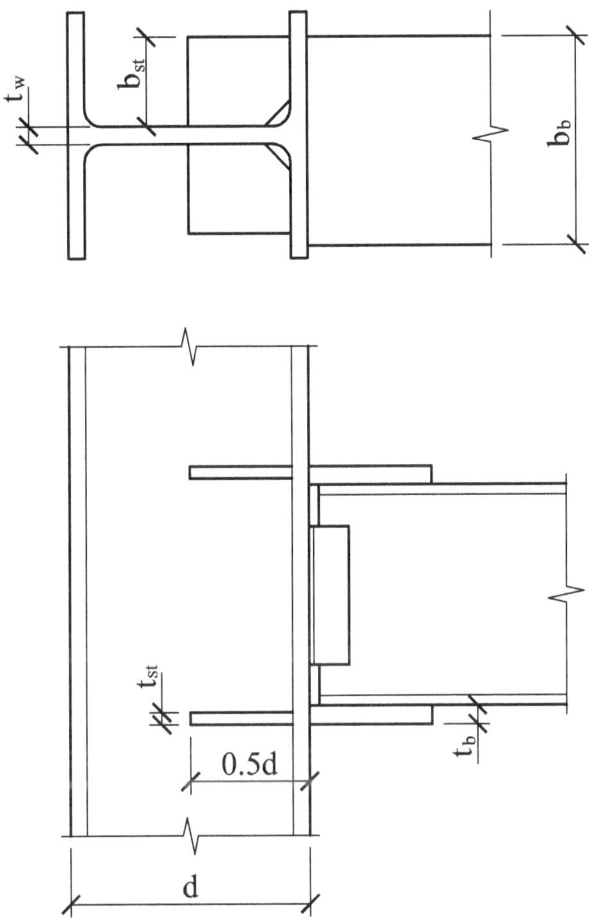

FIGURE 11-20 Dimensional requirements column stiffeners

The minimum width of a transverse stiffener is

$$b_{st} \geq \frac{b_b}{3} - \frac{t_w}{2}, \tag{11-18}$$

Where,

b_{st} = Width of the transverse stiffener,

b_b = Width of the beam flange or moment connection flange plate,

t_w = Column web thickness.

The minimum thickness of the stiffener is the larger of the following:

$$t_{st} \geq \frac{t_b}{2} \tag{11-19}$$

and

$$t_{st} \geq \frac{b_{st}}{15}, \tag{11-20}$$

Where,

b_{st} = Width of the transverse stiffener,

t_{st} = Thickness of the transverse stiffener, and

t_b = Thickness of the beam flange or moment connection flange plate.

Transverse stiffeners are required to extend the full depth of the column when there are applied concentrated horizontal forces on both sides of the column. For concentrated horizontal forces on one side of the column, the length of the transverse stiffener should extend to at least one-half of the column depth.

Transverse stiffeners are welded to both the web and the loaded flange. The weld to the flange is designed for the difference between the required strength and the design strength of the controlling limit state (Equation (11-14)).

When web doubler plates are required, they are designed for the shear in the column that exceeds the web panel zone shear strength (Equation (11-16)). This force in the doubler plate could be compression, tension, or shear, and therefore the doubler plate must be designed for these limit states (see *AISCM* Section J10.9). The web doubler plate is welded to the column web based on the force in the doubler plate (Equation (11-16)).

EXAMPLE 11-5

Column with Concentrated Flange Forces

Determine whether the W8 × 40 column in Example 11-1 is adequate for the applied concentrated flange forces. Assume that the beam connection occurs at a location remote from the ends (i.e., $y > d$ and $y > 10b_f$) and that $P_r \leq 0.4P_c$.

Solution

From Example 11-1,

$$P_{uf} = 49.2 \text{ kips}$$

The bearing length, $\ell_b = \frac{3}{8}$ in. = (flange plate thickness, t_b).

From *AISCM* Table 1-1, the section properties for W8 × 40 are,

$d = 8.25$ in. $\quad t_w = 0.36$ in.

$t_f = 0.56$ in. $\quad k = 0.954$ in.

$A_g = 11.7\text{ in,}^2$

Flange local bending (see Equation (11-3)):

$$\phi R_n = \phi 6.25 t_f^{\,2} F_y = (0.9)(6.25)(0.56)^2(50) = 88.2 \text{ kips} > P_{uf} = 49.2 \text{ kips} \quad \text{OK.}$$

Web local yielding (see Equation (11-5)):

$$\begin{aligned}\phi_{wy} R_n &= \phi_{wy}(5k + \ell_b)F_y t_w = 1.0\left[(5)(0.954) + 0.375\right](50)(0.36)\\ &= 92.6 \text{ kips} > P_{uf} = 49.2 \text{ kips} \quad \text{OK.}\end{aligned}$$

Web local crippling (see Equation (11-7)):

$$\phi_{wc} R_n = \phi_{wc} 0.8 t_w^2 \left[1 + 3\left(\frac{\ell_b}{d}\right)\left(\frac{t_w}{t_f}\right)^{1.5}\right]\sqrt{\frac{EF_y t_f}{t_w}}(Q_f)$$

$$= (0.75)(0.8)(0.36)^2\left[1 + 3\left(\frac{0.375}{8.25}\right)\left(\frac{0.36}{0.56}\right)^{1.5}\right]\sqrt{\frac{(29{,}000)(50)(0.56)}{0.36}}(1.0)$$

$$= 125 \text{ kips} > P_{uf} = 49.2 \text{ kips OK.}$$

Web panel zone shear:

$$P_y \text{ for W}8\times40 = F_y A_g = (50)(11.7) = 585 \text{kips}$$

$$\frac{P_{uf}}{P_y} = \frac{49.2}{585} = 0.084 < 0.4$$

Therefore, use Equation (11-12) to calculate the web panel zone shear strength,

$$\phi_{cw} R_n = \phi_{cw} 0.6 F_y d t_w = (0.9)(0.6)(50)(8.25)(0.36)$$
$$= 80.2 \text{ kips} > P_{uf} = 49.2 \text{ kips OK.}$$

The W8 × 40 column is adequate for the applied concentrated forces. Note that web compression buckling does not need to be checked since the concentrated forces are applied to **only one side** of the column.

11-7 COLUMN SPLICES

In buildings with less than four stories, it may be advantageous with regard to constructability to use the same column size and one continuous column for all of the stories instead of using smaller column sizes for the upper levels, even though the total weight of the column will be smaller when smaller columns are used for the upper levels. However, the addition of column splices also adds weight and complexity in terms of detailing to the structure. In this four-story case, the use of one continuous column without splices is usually the preferred option both from a constructability and cost perspective. In multistory buildings with four or more stories, the columns could be spliced every two, three, or four floor levels depending on the design and construction parameters, as previously discussed in Section 8-21 of Chapter 8.

The OSHA (Occupational Safety and Health Administration) safety regulations for the erection of perimeter columns and columns at the edges of floor openings require that a steel erector have adequate protection from fall hazards of not more than 30 ft. height above a lower level [6]. An additional safety requirement related to column splices is that perimeter columns should have holes or other attachment devices sufficient to support a safety cable or other similar rail systems. The holes or attachment devices are located 42 in. to 45 in. above the finished floor and at the midpoint between the top cable or rail and the finished floor. The column splice is therefore required to be a minimum of 48 in. (some engineers specify 60 in.) above the finished floor or at a higher distance in order to avoid interference with the safety attachments. The safety regulations do allow for exceptions to the above requirement where constructability does not allow such a distance, but the overall safety compliance would be left to the steel erector to resolve [12].

The simplest column splice is one in which only compression forces are transferred between two column segments that have the same nominal depth (see Figure 11-21). The design strength in bearing between the areas of contact is

$$\phi R_n = \phi 1.8 F_y A_{pb}, \tag{11-21}$$

Where,

$\phi = 0.75$,
F_y = Yield strength, and
A_{pb} = Contact area.

By inspection, it is evident that the bearing strength for a column with full-contact area will not be more critical than the design strength of the column in compression.

Column splice details for several framing conditions are presented in *AISCM* Table 14-3. This table, which consists of a series of tables for various column splice configurations, is used mainly for splices that support compression load only. Column splices with load conditions that result in tension, shear, bending, or some combination of these three forces in the splice would have to be designed accordingly using the procedures previously discussed in Chapter 4 for the design of tension members and in this chapter for the design of members for combined axial load, shear, and bending.

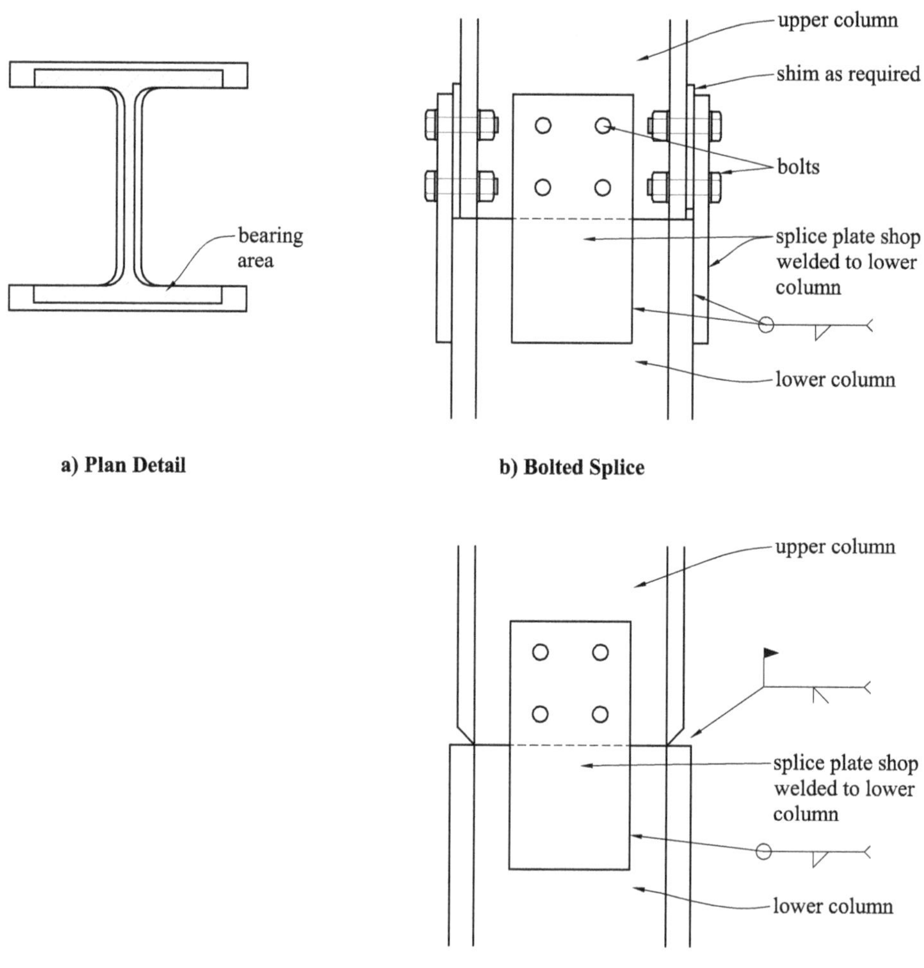

FIGURE 11-21 Column splice details

FIGURE 11-21d Column splice detail before field welding of the flanges.

EXAMPLE 11-6

Column Splice

Analyze the column splice between a W12 × 65 lower level column and a W12 × 53 upper level column. The steel is ASTM A992, Grade 50, and the floor-to-floor height is 15 ft.

Solution

From part 1 of the *AISCM*, the section properties for W12 × 65 and W12 × 53 are as follows:

W12 × 53:

A_g = 15.6 in.^{2}

Z_y = 29.1 in.^{3}

S_y = 19.2 in.^{3}

W12 × 65:

A_g = 19.1 in.^{2}

Z_y = 44.1 in.^{3}

S_y = 29.1 in.^{3}

A_{pb} = the smaller area of the two column segments at the splice location = 15.6 in.^{2}

Compression:

The depth of the W12 × 65 is equal to the depth of the W12 × 53, so there will be full contact for transferring the compression load. The design strength in bearing is

$$\phi R_n = \phi 1.8 F_y A_{pb} = (0.75)(1.8)(50)(15.6) = 1053 \text{ kips.}$$

This value far exceeds the design strength of the columns in compression, even at a very small unbraced length, so steel column-to-steel column bearing is not a concern.

Tension:

The design strength in tension for a W12 × 53 is

$$\phi_t P_n = \phi_t F_y A_g = (0.9)(50)(15.6) = 702 \text{ kips.}$$

While it is likely that the column does not experience that magnitude of tension under an actual design condition, the design strength in tension would likely have to be developed with a welded splice plate instead of bolts. Assuming that a plate on either side of the column web is provided with a cross-sectional area equal to the area of the column, a $\frac{7}{8}$-in.-thick plate would be required on each side with a width less than or equal to the *T* distance of the column ($T = 9\frac{1}{8}$ in. for the W12 × 65 and $T = 9\frac{1}{4}$ in. for the W12 × 53)

Assume $\frac{7}{8}$ in. x 9 in. plate on each side of the column web:

The area of the two steel plates is

$$A_p = (2\text{plates})\left(\tfrac{7}{8}\text{ in.}\right)(9\text{in.}) = 15.75 \text{ in.}^2 > 15.6 \text{ in.}^2 \text{, therefore OK.}$$

Since the cross-sectional area of the two plates is greater than the smallest column cross-sectional area, the plates will be adequate in tension.

Assuming a $\frac{5}{16}$-in. fillet weld (i.e., D = 5) and a 42-in. long x 9 in. wide plate (i.e., 21 in. long above and below the splice), the total fillet weld length on one side of the column web above or below the splice (see Figure 11-21c) is 21 in. + 9 in. + 21 in. = 51 in. Therefore, the weld strength above or below the splice (considering the plate on both sides of the column web) is

$$\begin{aligned}\phi R_n &= 1.392DL = (1.392 \text{ kips/in.})(5)(2\text{plates})(51 \text{ in.})\\ &= 709 \text{ kips} > \phi P_n = 702 \text{ kips.} \quad \text{OK.}\end{aligned}$$

Therefore, the welds are adequate since their capacity is greater than the tension capacity of the smaller column.

Shear and Bending in the W12 × 53:

From *AISCM* Table 3-6, ϕV_n = 125 kips.

From *AISCM* Table 3-10, for a W12x53 with unbraced length, L_b = 15 ft., the design moment capacity, $\phi_b M_{nx}$ = 258 ft.-kips;

The design moment capacity can also be obtained from *AISCM* Table 3-2 as follows:

$$\phi_b M_{nx} = C_b\left[\phi_b M_p - \phi_b BF(L_b - L_p)\right] \le \phi_b M_p$$

$$= (1.0)\left[292 - 5.48(15 - 8.75)\right] = 258 \text{ ft.-kips} \le 292 \text{ ft.-kips}$$

The moment capacity about the weak (y-y) axis is obtained as follows:

$$M_{ny} = M_{py} = F_y Z_y \leq 1.6 F_y S_y$$
$$= (50)(29.1) \leq (1.6)(50)(19.2) = 1455 \text{ in.-kips} \leq 1536 \text{ in.-kips}$$
$$\phi_b M_{ny} = \frac{(0.9)(1455)}{12} = 109 \text{ ft.-kips.}$$

The splice detail for shear and for the bending moments about the strong and weak axes will not be developed here; however, Examples 11-3 and 11-4 provide the procedure for designing these splice plates (welded or bolted).

11-8 DESIGN OF GUSSET PLATES IN VERTICAL BRACING AND TRUSS CONNECTIONS

Gusset plates are flat structural elements that are used to connect adjacent members meeting at vertical brace connections and at truss panel points, as shown in Figures 11-22a, 11-22b, 11-22c, and Figures 11-23a through 11-23d, with the gusset plate dimensional requirements. Gusset plates help to transmit loads from one member to another at a joint. The importance of gusset plates as connecting elements at structural joints, even though they are a small part of the overall structure, cannot be overemphasized. According to the National Transportation Safety Board (NTSB), the primary cause of the August 1, 2007, collapse of the I-35W bridge in Minneapolis, MN, that killed 13 people and injured hundreds of people, were its under-designed gusset plates [13, 14]. The NTSB in their report on the probable cause of the collapse of the I-35 bridge concluded the following [13]:

"*...Contributing to the design error was the failure of Sverdrup & Parcel's quality control procedures to ensure that the appropriate main truss gusset plate calculations were performed for the I-35W bridge and the inadequate design review by Federal and State transportation officials. Contributing to the accident was the generally accepted practice among Federal and State transportation officials of giving inadequate attention to gusset plates during inspections for conditions of distortion, such as bowing, and of excluding gusset plates in load rating analyses.*"

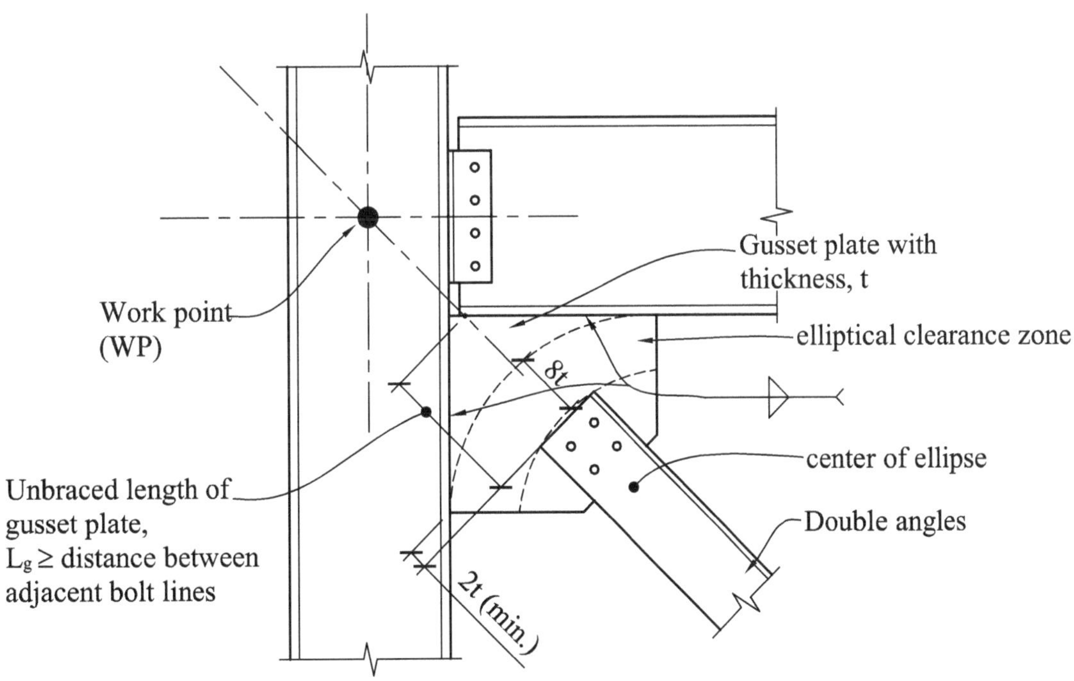

FIGURE 11-22a Gusset plate at diagonal brace

FIGURE 11-22b Chevron brace with gusset plate connections.

FIGURE 11-22c Diagonal braces with gusset plate connections.

The design of gusset plates is covered in Part 9 of the *AISCM*. Where the gusset plate is part of a seismic force-resisting system with a seismic response modification factor, *R*, greater than 3, the requirements of the AISC seismic provisions for steel buildings must also be satisfied [5]. The gusset plates may be bolted or welded to the members meeting at the joint, and the practical minimum thickness of gusset plates used in design practice is usually $^3/_8$ in. For diagonal brace connections with gusset plates, the gusset plate helps to transfer the diaphragm lateral forces from the beams or girders to the diagonal brace and the adjoining column. Several connection interfaces must be designed:

- Diagonal brace-to-gusset connection,
- Gusset-to-beam connection,
- Gusset-to-column connection,
- Beam-to-column connection.

At truss joints, the gusset plates connect the web members to the chord members. The gusset plates are connected to the adjoining members with welds or bolts. At truss panel joints or at diagonal brace connections, it is common practice to choose the geometry of the joint such that the centroidal axes of all the members meeting at the joint coincide at one point, called the work point (WP), in order to minimize bending moments in the gusset plate and in the adjoining members. Where it is not feasible to have a common work point for all the members meeting at a joint, there will be moments induced in the gusset plate and the connecting members, in addition to other stresses. These additional moments must be considered in the design of the gusset plates and the members meeting at the joint. In addition to bending moments, gusset plates are usually also subjected to shear and axial tension or compression stresses. For direct axial stresses (i.e., tension and compression), the effective cross-section - over which the axial load is spread - is defined by the Whitmore effective width, *W* (typical throughout), at the end of the connection, and this is obtained by projecting lines at an angle of spread of 30° on both sides of the connection, starting from the first row of bolts to the last row of bolts in the connection. For welded connections, the lines are projected on both sides of the longitudinal weld and at the spread-out angle of 30° starting from the edge of the longitudinal weld to the end of the weld (see Figures 11-23a, 11-23c, and 11-23d). The effective gross area of the plate is the Whitmore effective width, *W*, multiplied by the plate thickness, *t*.

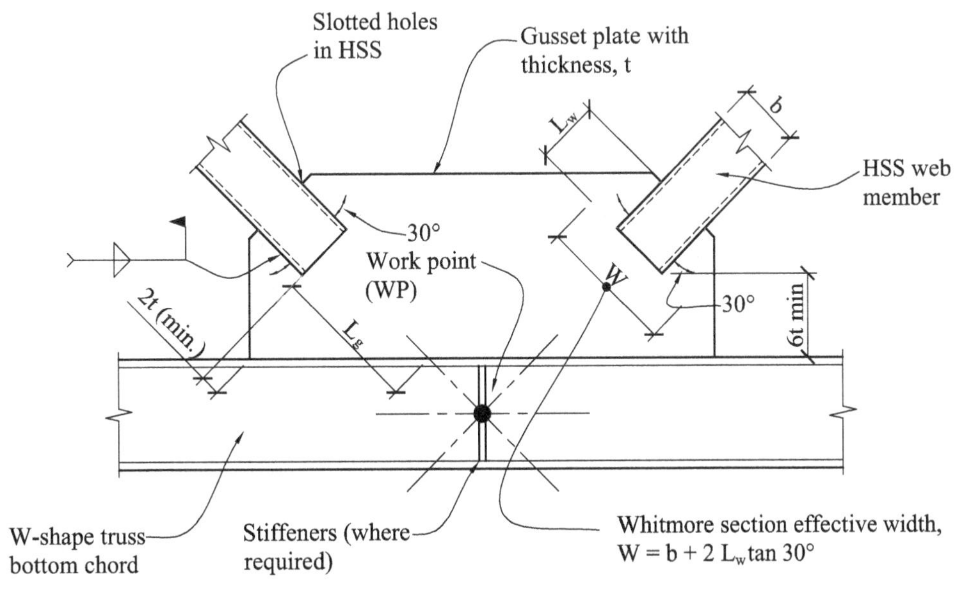

FIGURE 11-23a Gusset plate at a welded truss panel point

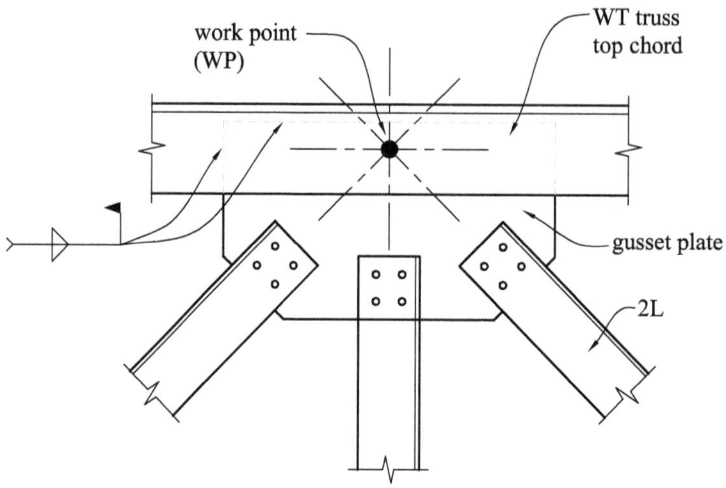

FIGURE 11-23b Gusset plate at truss panel point

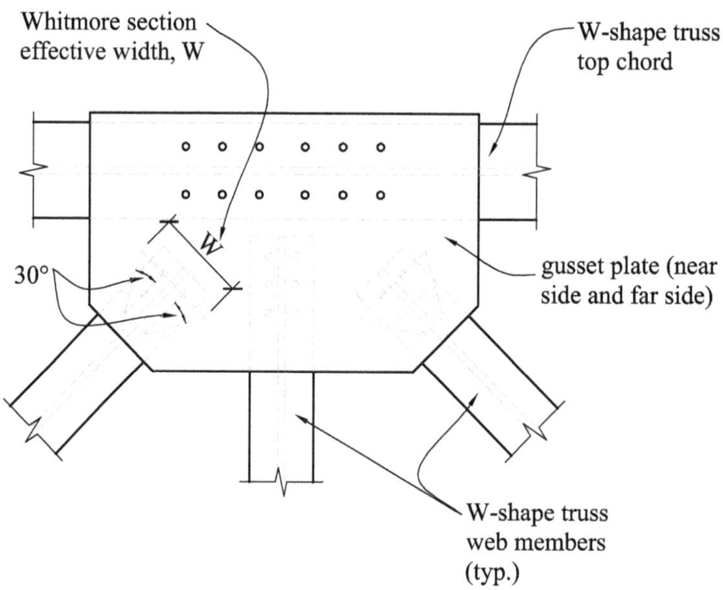

FIGURE 11-23c Gusset plate at a bolted truss panel point

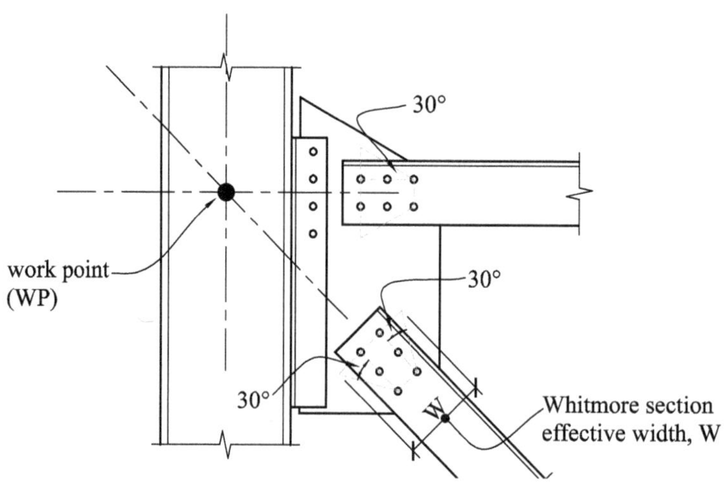

FIGURE 11-23d Gusset plate at truss support

Sometimes, the Whitmore section may be limited by the geometry of the connection, such as when the Whitmore effective width crosses a connected edge, or materials of differing strengths or thicknesses are used in the connection, or the geometry results in an unsymmetrical Whitmore section. For such cases, the recommendation in Ref. [15] can be followed to address these situations.

To ensure adequate out-of-plane rotation of the gusset plate when the bracing or truss web member is subjected to out-of-plane buckling under cyclic loading (e.g., seismic loads), Astaneh-Asl recommends that the end of the bracing member or truss web member be terminated at least a distance of $2t$ away from the re-entrant corner of the gusset plate at the gusset-to-beam and gusset-to-column interfaces [16] (see Figures 11-22 and 11-23a). This requirement can be relaxed for connections subject to monotonic or static loading. Although gusset plates may appear to be small and insignificant structural connecting elements, it is important that they be adequately designed, detailed, and protected against corrosion to avoid connection failures that could lead to the collapse of an entire structure.

In this section, only the design of the gusset plate is covered. The determination of the forces acting on gusset plate interface connections is discussed, but the design of the welds and bolts at these connection interfaces is not covered in this section because this matter has already been covered in previous chapters. The observed failure modes of gusset plates include the following [16]:

1. **Out-of-plane buckling** of the gusset plate due to the axial compression force:

   The unbraced length, L_g, of a gusset plate is taken as the larger of the length of the plate between adjacent lines of bolts parallel to the direction of the axial compression force and the length of the plate along the centroidal axis of the diagonal brace or the truss web member between the end of the diagonal brace or truss web member and the connected edge of the gusset plate (see Figures 11-22 and 11-23a). The gusset plate will buckle out of plane about its weaker axis, and the buckling is assumed to occur over a plate width equal to the Whitmore effective width, W. To determine the design compression load, determine the slenderness ratio, KL_g/r, where r is approximately $0.3t$ and the effective length factor, K, is conservatively taken as 1.2 to account for some end fixity of the gusset plate (see Figure 5-4). The design compression stress, ϕF_n, is obtained from *AISCM* Table 4-14, and the design compressive strength of the gusset plate is calculated as

   $$\phi P_n = \phi F_n W t, \tag{11-23}$$

   Where
   F_n = nominal compression stress (the design compression stress, ϕF_n can be obtained from *AISCM* Table 4-14),
   W = Whitmore effective width,
   t = Thickness of the gusset plate,
   ϕ = 0.9.

2. **Buckling of the free or unsupported edge** of the gusset plate:

   To prevent buckling at the unsupported edges of gusset plates, the minimum gusset plate thickness (without edge stiffeners) required in the American Association of State and Highway Transportation Officials (AASHTO) Code [17] and commonly used in design practice for monotonic or static loading is

   $$t \geq 0.5 L_{fg} \sqrt{\frac{F_y}{E}} \tag{11-24a}$$

for $F_y = 36$ ksi, $t \geq \frac{L_{fg}}{56}$, and

for $F_y = 50$ ksi, $t \geq \frac{L_{fg}}{48}$,

Where,

L_{fg} = Length of the free or unsupported edge of the gusset plate (see Figure 11-24),
F_y = Yield strength of the gusset plate, and
E = Modulus of elasticity of the gusset plate.

For gusset plates subjected to cyclic (or seismic) loading, the minimum required gusset plate thickness is

$$t \geq 1.33 L_{fg} \sqrt{\frac{F_y}{E}}. \quad (11\text{-}24b)$$

Note that in the 2022 AISC Seismic Provisions for Structural Steel Buildings ANSI/AISC 341-22 [32], the limitations above on the free edge length of gusset plates are no longer required, though some references recommend these limitations.

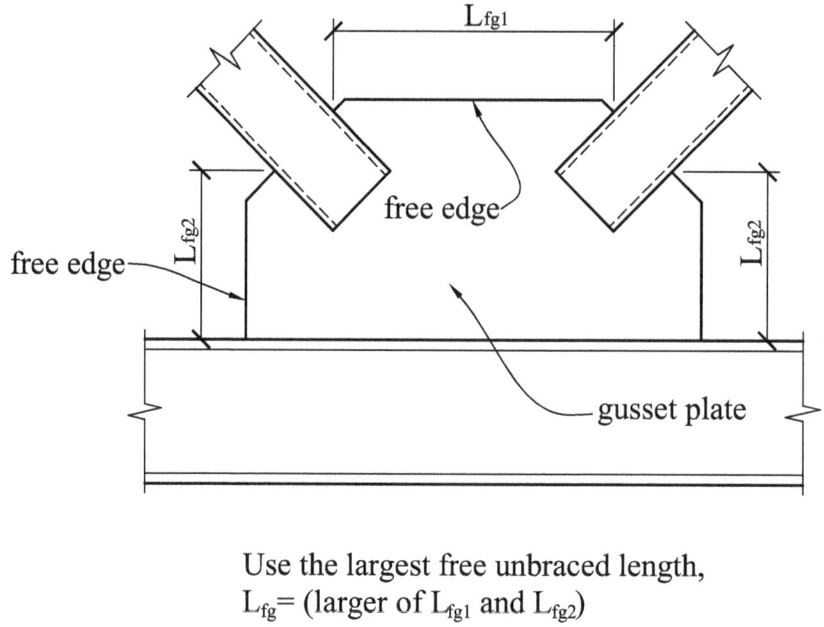

FIGURE 11-24 Free edge length of gusset plate

3. **Tension or compression failure of the gusset plate due to yielding within the Whitmore effective area**:

Tension yielding is the most desirable form of failure because of the ductility associated with this failure mode.

The design tension or compression yield strength is

$$\phi P_n = \text{Gross area of the Whitmore section} \times \phi F_y$$

$$= \phi F_y W t, \quad (11\text{-}25)$$

where

ϕ = 0.9,
W = Whitmore effective width (see Figures 11-23a, 11-23c, and 11-23d), and
t = Thickness of the gusset plate.

4. **Tension failure of the gusset plate due to *fracture* at a bolt line within the Whitmore effective area**:

   This is the least desirable form of failure and should be avoided for structures subjected to cyclic loading (e.g., seismic loads, traffic loads, and crane loads) because of the sudden and brittle nature of this failure mode.
   The design tension strength, ϕP_n = Net area of the Whitmore section × ϕF_u

$$= \phi F_u (W - nd_{hole})t, \qquad (11\text{-}26)$$

   Where,
   F_u = Tensile strength of the gusset plate,
   ϕ = 0.75,
   n = Number of bolt holes perpendicular to the applied axial force for each line of bolt, and
   d_{hole} = Diameter of the bolt hole = $d_{bolt} + \frac{1}{8}$ in. (see Chapter 4)

5. **Tension failure of the gusset plate due to block shear**:
   The calculation of the tensile strength of a plate for this failure mode has been discussed in Chapter 4.

6. **Fracture of the connecting welds and bolts**:
   The design of bolts and welds is covered in Chapters 9 and 10.

7. **Yielding failure of the gusset plate from combined axial tension or compression load, bending moment, and shear**:

   The applied loads on a diagonal brace or truss connection may result in a combination of shear, V_u, bending moment, M_u, and tension or compression force, P_u, acting on a critical section of the gusset plate. These forces are determined from a free-body diagram of the gusset plate using equilibrium and statics principles. The following interaction equation from plasticity theory (see *AISCM* Equation (9-1)) [1] is recommended for the design of gusset plates under combined loads [16, 18, 19]:

$$\frac{M_u}{\phi M_p} + \left(\frac{P_u}{\phi P_y}\right)^2 + \left(\frac{V_u}{\phi V_y}\right)^4 \leq 1.0, \qquad (11\text{-}27)$$

   Where,
   P_u = factored axial load on gusset plate,
   V_u = factored shear on gusset plate,
   M_u = factored in-plane bending moment on gusset plate,
   ϕ = 0.9,
   ϕM_p = In-plane plastic moment capacity of the gusset plate at the critical section = $\phi \frac{tb^2}{4} F_y$,
   t = Thickness of gusset plate,
   b = Length of the gusset plate at the critical section at the connection to the beam, column, or truss member; the length is taken parallel to the beam, column, or the truss web member,
   F_y = Yield strength of the gusset plate,
   ϕP_y = Axial yielding capacity = $A_{g,cr} F_y$,
   $A_{g,cr}$ = tb,
   ϕV_y = Shear yielding capacity of the gusset plate = $\phi(0.6 A_{g,cr} F_y) C_{v1}$.

   Note that Equation (11-27) accounts for the plastic strength of the gusset plate but does not account for any stability or torsional effects.

EXAMPLE 11-7

Gusset Plate at a Symmetrical Truss Joint

For the truss joint shown in Figure 11-25, determine the following, assuming a $^5/_8$-in. gusset plate, $^3/_4$-in.-diameter bolts, and ASTM A992 Grade 50 steel:

a. Whitmore effective width for the gusset plate on diagonal web members A and B,
b. Compression buckling capacity of the gusset plate on diagonal member A, and
c. Tension capacity of the gusset plate on diagonal member B (assume that block shear does not govern).

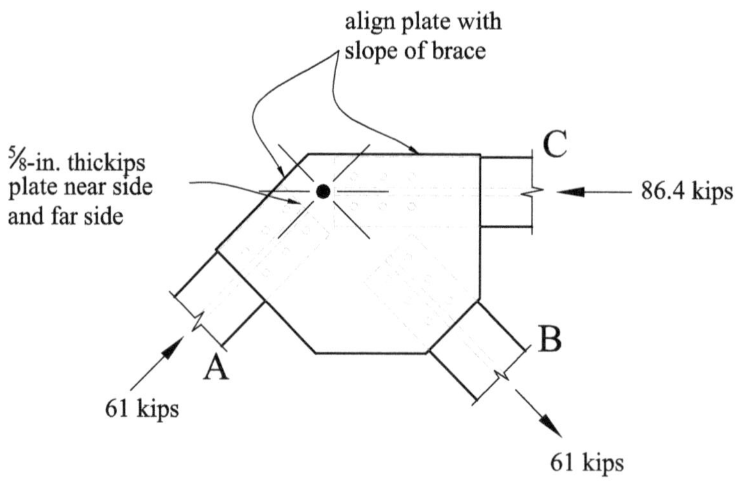

FIGURE 11-25 Gusset plate detail for Example 11-11

Solution

1. From Figure 11-26, the Whitmore effective width of the gusset plate on diagonal web members A and B is calculated as follows:

$$\begin{aligned} W_A &= 1.5 \text{ in.} + b_A + L_{wA} \tan\theta \\ &= 1.5 \text{ in.} + 3.08 \text{ in.} + (6)(\tan 30) \\ &= 8.04 \text{ in. (the part of the Whitemore section that falls outside of the gusset plate is neglected)} \end{aligned}$$

$$\begin{aligned} W_B &= b_B + 2L_{wB} \tan\theta = 3.08 \text{ in.} + (2)(6)(\tan 30) \\ &= 10 \text{ in.} \end{aligned}$$

Gusset plate thickness, $t = \frac{5}{8}$ in. = 0.625 in.

2. The maximum unbraced length of the gusset plate is the larger of the distance between adjacent bolt lines (3 in.), the maximum unsupported distance of the gusset plate measured along the centroid of the brace or diagonal member from the end of the diagonal member to the connected edge of the gusset plate (assume 0 in. in this case since the W6 diagonal members have been extended to the point where they abut other W6 members at the joint; see Figure 11-26).
Therefore, the maximum unbraced length of 5/8 in. thick gusset plate, L_g = 3 in.

$$\frac{KL_g}{r} = \frac{1.2(3 \text{ in.})}{0.3(0.625 \text{ in.})} = 19.2 < 25$$

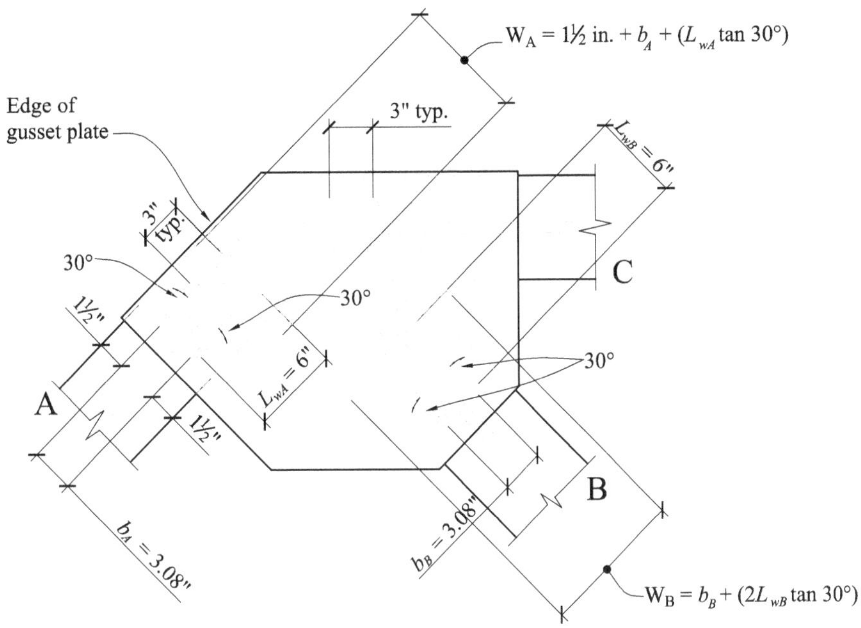

FIGURE 11-26 Whitmore effective width of gusset plates

Therefore, according to *AISCM* Section J4.4, buckling can be neglected, and $F_n = F_y$. Note that for cases where $KL/r > 25$, the buckling capacity, ϕP_n, is determined from $\phi F_n A_g$, where ϕF_n is obtained from *AISCM* Table 4-14.

a. The tension yielding capacity of the 5/8 in. thick gusset plate on diagonal member B is

$$\phi P_n = \phi F_y W_B t = (0.9)(50 \text{ ksi})(10 \text{ in.})(0.625 \text{ in.}) = 281 \text{ kips.}$$

b. The tension capacity of the 5/8 in. gusset plate on diagonal member B due to fracture of the 5/8 in. thick gusset plate is

$$\begin{aligned}\phi_t P_n &= \phi_t F_u (W_B - n d_{hole}) t \\ &= (0.75)(65 \text{ ksi})[10 \text{ in.} - (2)(\tfrac{7}{8} \text{in.})](0.625 \text{ in.}) \\ &= 251 \text{ kips,}\end{aligned}$$

Where,

$$\begin{aligned}W_B &= 10 \text{ in.,} \\ n &= 2 \text{ bolts per line,} \\ d_{bolt} &= \tfrac{3}{4} \text{ in.,} \\ d_{hole} &= (\tfrac{3}{4} + \tfrac{1}{8}) = \tfrac{7}{8} \text{ in., and}\end{aligned}$$

The smaller of the above two values will govern for the tension capacity of the gusset plate on diagonal member B. Therefore, the design tension strength, $\phi P_n = 251$ kips.

EXAMPLE 11-8

Gusset Plate at a Nonsymmetrical Truss Joint

The factored forces on each gusset plate at a joint of a truss bridge are shown in Figure 11-27. Assuming ASTM A992 Grade 50 steel,

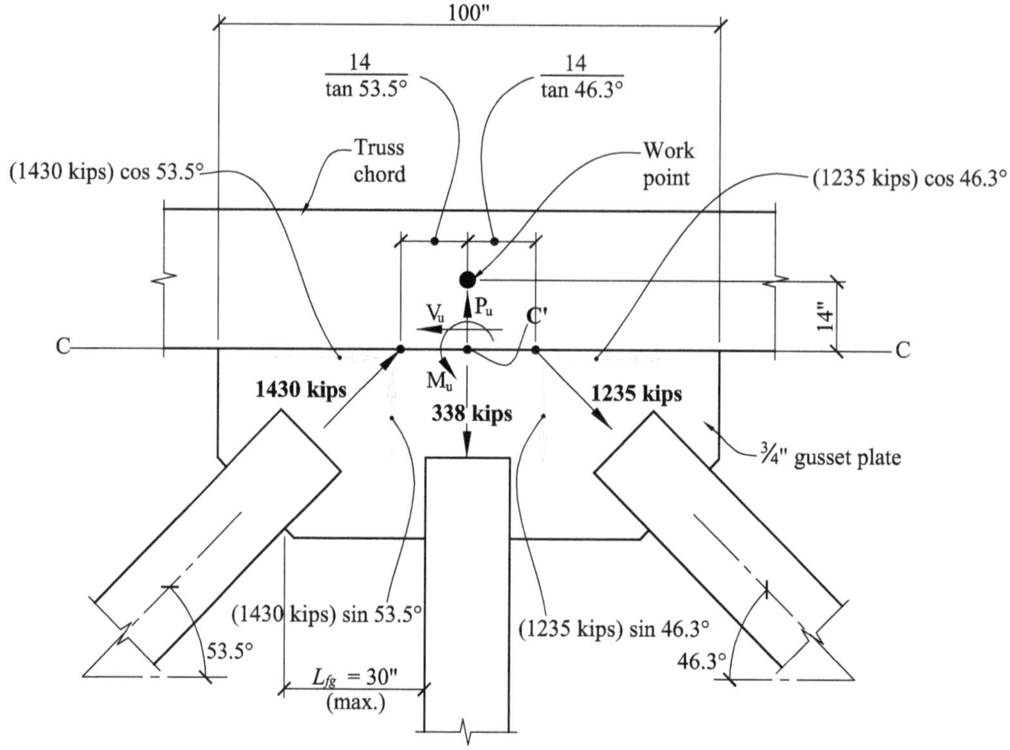

FIGURE 11-27 Gusset plate detail for Example 11-8

1. Determine the combined moment, M_u, shear, V_u, and axial load, P_u, acting on the gusset plate at point C' along the critical shear section C-C just below the work point.
2. Using the plasticity theory interaction formula, determine whether the gusset plate is adequate to resist the combined loads.
3. Determine whether the gusset plate is adequate for free-edge buckling.

Solution

The area of the ¾ in. thick gusset plate = tb = (¾ in.)(100 in.) = 75 in.^{2}

The depth of the gusset plate for in-plane bending (perpendicular to the axis of bending) is b = 100 in.

1. Summing the forces in the horizontal direction (i.e., $\Sigma F_x = 0$) yields

$$1430 \cos 53.5° + 1235 \cos 46.3° - V_u = 0.$$

Therefore, V_u = 1704 kips.

$$\phi V_y = \phi(0.6A_{g,cr}F_y) = (0.9)(0.6)(75\text{in.}^2)(50 \text{ ksi}) = 2025 \text{ kips} > V_u \text{ OK.}$$

Summing the forces in the vertical direction (i.e., $\Sigma F_y = 0$) yields

$$-338 + 1430 \sin 53.5° - 1235 \sin 46.3° + P_u = 0.$$

Therefore, P_u = 81 kips.

$$\phi P_y = \phi A_{g,cr}F_y = (0.9)(75\text{in.}^2)(50 \text{ ksi})$$
$$= 3375 \text{ kips} > P_u \text{ OK.}$$

Summing the moments about a point C' on the critical section C-C just below the work point (i.e., $\Sigma M = 0$) yields

$$[(1430 \sin 53.5°)(14/\tan 53.5°)] + [(1235 \sin 46.3°)(14/\tan 46.3°)] - M_u = 0.$$

Therefore, M_u = 23,854 in.-kips.

$$\phi M_p = \phi \frac{tb^2}{4} F_y = (0.9)\left[\frac{(0.75 \text{ in.})(100 \text{ in.})^2}{4}\right](50 \text{ ksi})$$

$$= 84{,}375 \text{ in.-kips} > M_u \text{ OK.}$$

2. From Equation (11-27), the plasticity theory interaction equation (see *AISCM* Equation (9-1)) for a gusset plate under combined loading is

$$\frac{M_u}{\phi M_p} + \left(\frac{P_u}{\phi P_y}\right)^2 + \left(\frac{V_u}{\phi V_y}\right)^4 = \frac{23{,}854}{84{,}375} + \left(\frac{81}{3375}\right)^2 + \left(\frac{1704}{2025}\right)^4 = 0.78 < 1.0. \text{ OK.}$$

3. The maximum unbraced length of the free (unsupported) edge of the gusset plate, L_{fg} = 30 in. From Equation (11-24a), the required minimum gusset plate thickness to avoid unsupported edge buckling is

$$t \geq 0.5 L_{fg} \sqrt{\frac{F_y}{E}} = (0.5)(30 \text{ in.})\left(\sqrt{\frac{50 \text{ ksi}}{29{,}000 \text{ ksi}}}\right)$$

$= 0.62$ in. $< \frac{3}{4}$ in. provided. OK

If this gusset plate were subjected to cyclic (i.e., seismic) loading, the minimum required thickness of the gusset plate, without edge stiffeners, would be 1.66 in. from Equation 11-24b. Also note that if a gusset plate of A36 steel were used in this connection, we would find that the gusset plate thickness of ¾ in. would be grossly inadequate to resist the loads. The reader should confirm that a 1 in. thick gusset plate would be required if an A36 steel gusset plate was used in this connection, and the value of the interaction equation from Step 2 would be 0.88 for that case.

Gusset Plate Connection Interface Forces

The analysis of the gusset plate connection interface forces is a complex indeterminate problem, and the most effective method for calculating the shear, axial tension or compression forces, and moments acting on gusset plate connections is the so-called **uniform force method** (UFM), illustrated in Figure 11-28 [1, 16, 20-23]. It results in the most economical design and the best prediction of the critical limit state and design strength for gusset plate connections [1]. For economic reasons, it is desirable to select a connection geometry that will avoid, or at least minimize, the in-plane moments acting on the gusset-to-beam, gusset-to-column, and beam-to-column connection interfaces, but in order for the joint to be free of moments leaving only axial and shear forces acting on the gusset connection interfaces, the UFM requires that the gusset plate connection satisfy the following conditions (see *AISCM* Part 13) [1]:

$$\alpha - \beta \tan\theta = 0.5 d_b \tan\theta - 0.5 d_c, \qquad (11\text{-}28)$$

$$r = \sqrt{(\alpha + 0.5 d_c)^2 + (\beta + 0.5 d_b)^2}, \qquad (11\text{-}29)$$

$$V_B = \frac{0.5 d_b}{r} P_{brace}, \qquad (11\text{-}30)$$

$$H_B = \frac{\alpha}{r} P_{brace}, \qquad (11\text{-}31)$$

$$V_C = \frac{\beta}{r} P_{brace}, \text{ and} \qquad (11\text{-}32)$$

$$H_C = \frac{0.5d_c}{r} P_{brace}, \tag{11-33}$$

where

θ = Angle between the diagonal brace and the vertical plane,

α = *Ideal* distance from the face of the column flange or web to the centroid of the gusset-to-beam connection (Note: The setback between the gusset and the face of the column flange can typically be assumed to be approximately 0.5 in.),

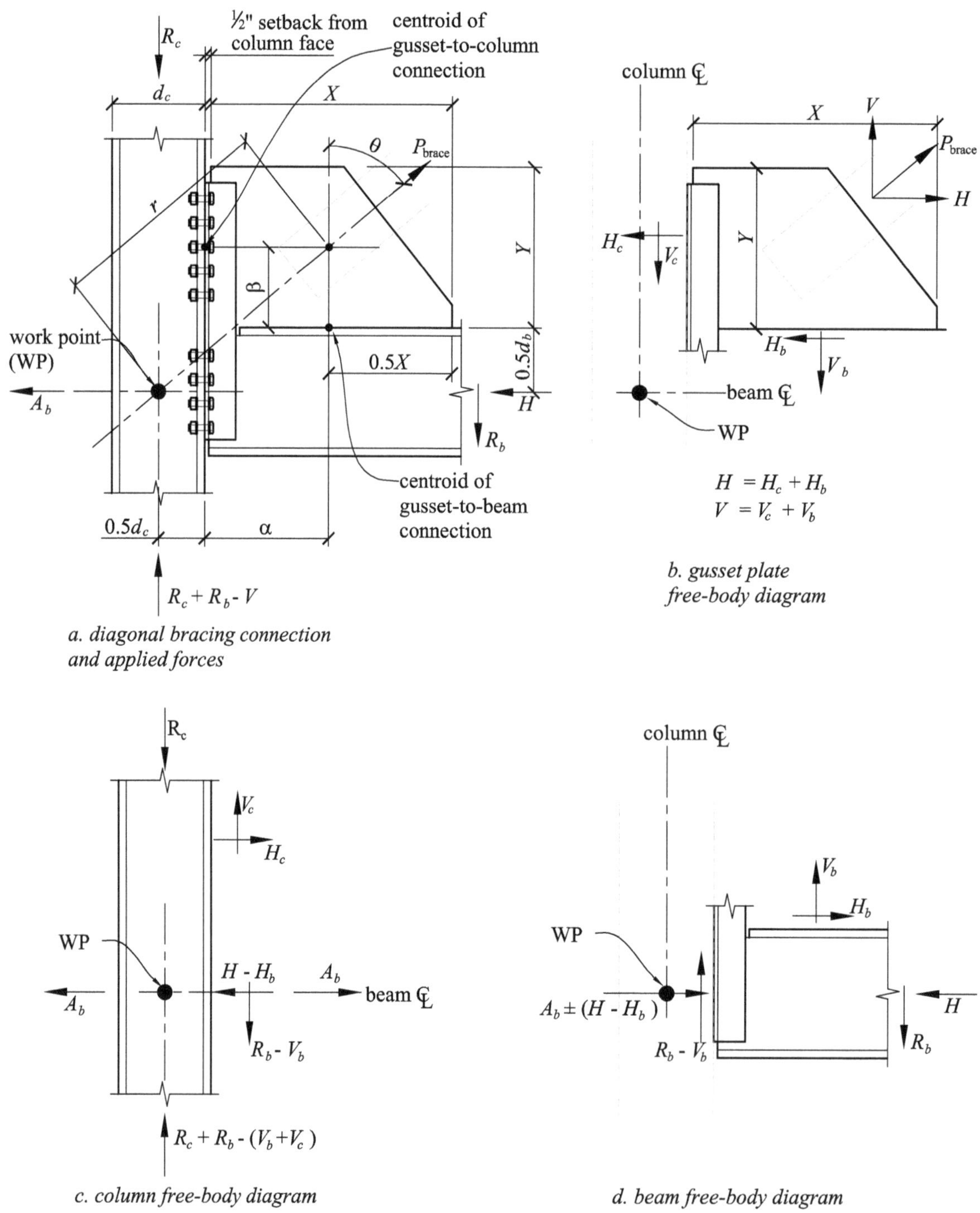

FIGURE 11-28 Gusset connection interface forces using the uniform force method
Adapted from Ref. [1]

β = *Ideal* distance from the face of the beam flange to the centroid of the gusset-to-column connection (where the beam is connected to the column web, set $\beta = 0$),

d_b = Depth of beam,

d_c = Depth of column (where the diagonal brace is not connected to a column flange, set $0.5d_c \approx 0$ and $H_C = 0$),

V_B, H_B = Vertical and horizontal forces on the gusset-beam connection interface,

V_C, H_C = Vertical and horizontal forces on the gusset-column connection interface,

X = Horizontal length of the gusset plate = $2(\alpha - 0.5\text{-in. setback})$, and

Y = Vertical length of the gusset plate = 2β.

The preceding development assumes that the work point, *WP*, is located at the intersection of the centroidal axes of the connecting members. For work points occurring at other locations, the reader should refer to the special cases presented in Part 13 of the *AISCM* [1].

TABLE 11-1 Gusset Connection Interface Design Forces

Connection	Shear force	Axial force	Moment	Remarks
Gusset-to-beam	H_B	V_B	0	Assuming that α and β satisfy equation (11-28)
Gusset-to-column	V_C	H_C	0	
Beam-to-column	$R - V_B$	$A_B \pm (H - H_B)$	0	

R = Factored or required end reaction in the beam,
A_B = Factored or required horizontal axial force from the adjacent bay (due to drag strut action),
H = Horizontal component of the factored or required diagonal brace force = Pbrace sin θ.

The *design* procedure for a ***new*** **gusset plate connection**, with the beam and column sizes already determined and assuming the work point, *WP*, lies at the intersection of the centroidal axes of the connecting members, is as follows:

1. Knowing the beam and column sizes and the brace geometry, determine $0.5d_b$, $0.5d_c$, and θ.
   Note: If the brace is connected to a column web, set $d_c = 0$ and $H_C = 0$. If the gusset is connected only to the beam and not to a column, set $d_C = 0$, $\beta = 0$, $V_C = 0$, and $H_C = 0$.
2. Select a value for Y, the vertical dimension of the gusset plate, and determine $\beta = 0.5Y$.
3. Substitute β, θ, $0.5d_b$, and $0.5d_c$ into Equation (11-28) to determine α.
4. Knowing α, with the centroid of the gusset-to-beam connection interface assumed to be at the midpoint of this interface, the horizontal length of the gusset plate can be determined from the relationship $\alpha = \frac{X}{2} + 0.5$ in., where 0.5 in. is the gusset setback from the column face.
5. Determine the horizontal and vertical forces on the gusset plate connection interfaces, V_B, H_B, V_C, and H_C, using Equations (11-29) to (11-33).

The forces on the gusset connection interfaces are given in Table 11-1.

For new connection designs, it is relatively easy to select a gusset geometry with values of X and Y or α and β that satisfy Equation (11-28) and thus ensure that there are no moments on the gusset connection interfaces. However, for existing gusset plate connections or where constraints have been placed on the gusset plate dimensions, it may not be possible to satisfy Equation (11-28), and therefore, moments may exist on one or both gusset connection interfaces or on the beam-to-column connection interface. It is usual in design practice to assume that the more rigid gusset connection interface will resist all of the moment required to satisfy equilibrium [24].

The procedure for the *analysis* of ***existing*** **gusset plate connections** is as follows:

1. Determine $0.5d_b$, $0.5d_c$, X, Y, and θ from the geometry of the connection. Note that where the brace is not connected to a column flange, set $0.5d_c \approx 0$ and $H_C = 0$.
2. Using the known horizontal and vertical lengths of the gusset plate, determine the actual α and β values as follows:

   $\bar{\alpha} = \dfrac{X}{2} + 0.5$ in. (0.5 in. is the assumed setback between the gusset plate and the face of the column flange),

   $\bar{\beta} = \dfrac{Y}{2}$,

   $\bar{\alpha}$ = *Actual* distance between the face of the column flange or web and the centroid of the gusset-to-beam connection, and

   $\bar{\beta}$ = *Actual* distance between the face of the beam flange and the centroid of the gusset-to-column connection.
3. If the gusset-to-beam connection is more rigid than the gusset-to-column connection (e.g., if a welded connection is used for the gusset-to-beam connection and a bolted connection is used for the gusset-to-column connection), set the ideal β equal to the actual β (i.e., $\bar{\beta}$ from Step 2) and calculate the ideal α using Equation (11-28). These ideal values of α and β are used in Equation (11-29) to calculate the parameter, r.

   If the ideal α calculated in Step 3 equals the actual α (i.e., $\bar{\alpha}$ from Step 2), then no moment exists on the *gusset-to-beam* connection interface. If not, the forces and moment on the *gusset-to-beam* connection interface are calculated as follows:

$$H_B = \frac{\alpha_{step3}}{r_{step3}} P_{brace}, \tag{11-34}$$

$$V_B = \frac{0.5d_b}{r_{step3}} P_{brace}, \text{ and} \tag{11-35}$$

$$M_{u,g-b} = V_B \left|\alpha - \bar{\alpha}\right|, \tag{11-36}$$

where

$$r = \sqrt{\left(\alpha_{step3} + 0.5d_c\right)^2 + \left(\bar{\beta}_{step2} + 0.5d_b\right)^2}. \tag{11-37}$$

4. If the gusset-to-column connection is more rigid than the gusset-to-beam connection (e.g., if a welded connection is used for the gusset-to-column connection and a bolted connection is used for the gusset-to-beam connection), set the ideal α equal to the actual α (i.e., $\bar{\alpha}$ from Step 2) and calculate the ideal β using Equation (11-28). These ideal values of α and β are used in Equation (11-29) to calculate the parameter r.

   If the ideal β, calculated in Step 4, equals the actual β (i.e., $\bar{\beta}$ from Step 2), then no moment exists on the *gusset-to-column* connection interface. If not, the forces and moment on the *gusset-to-column* connection interface are calculated as follows:

$$H_C = \frac{0.5d_c}{r_{step4}} P_{brace}, \tag{11-38}$$

$$V_C = \frac{\beta_{step4}}{r_{step4}} P_{brace}, \text{ and} \tag{11-39}$$

$$M_{u,g-c} = H_C \left|\beta - \bar{\beta}\right|, \tag{11-40}$$

where

$$r = \sqrt{\left(\bar{\alpha}_{step2} + 0.5d_c\right)^2 + \left(\beta_{step4} + 0.5d_b\right)^2}. \tag{11-41}$$

EXAMPLE 11-9

Design of gusset plate connection

Determine the dimensions of the gusset plate and the gusset connection interface forces for the diagonal brace connection shown in Figure 11-29.

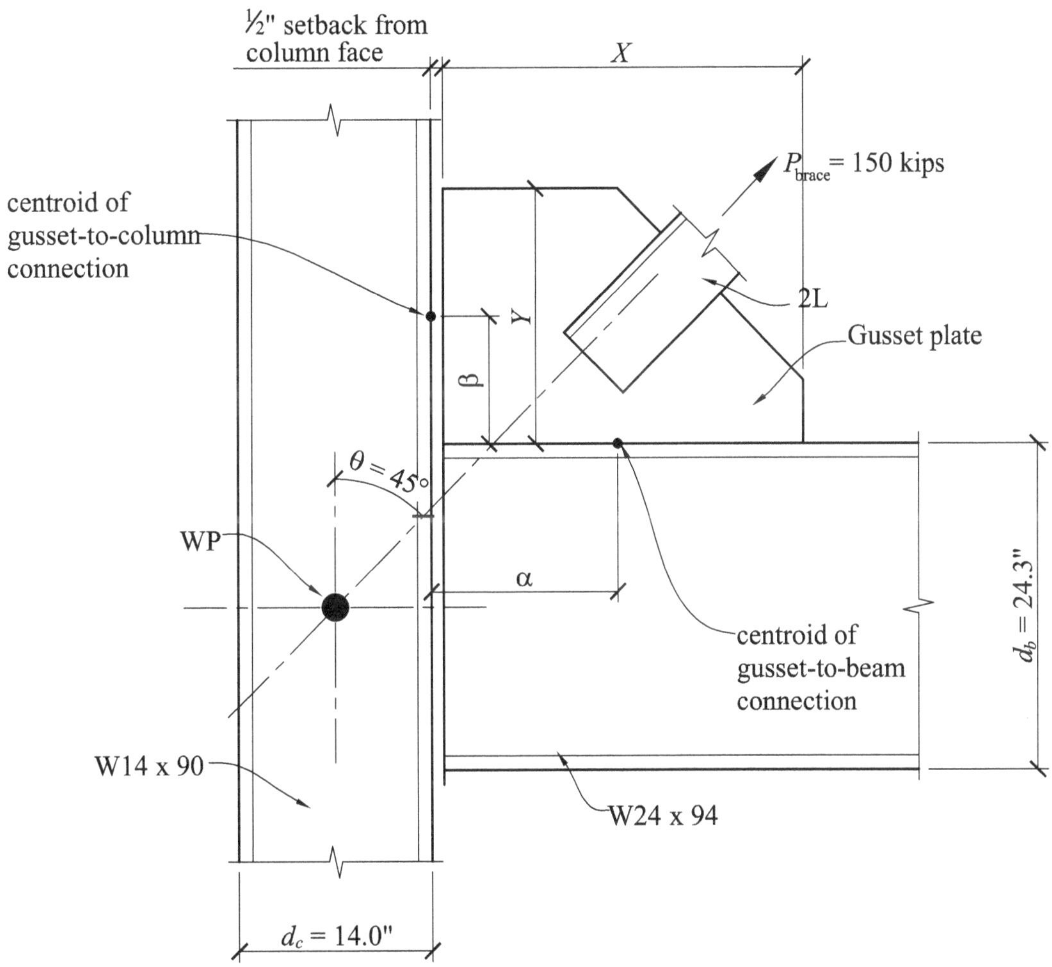

FIGURE 11-29 Diagonal brace connection design for Example 11-9

Solution

1. $d_c\,(W14 \times 90) = 14.0$ in.; $d_b\,(W24 \times 94) = 24.30$ in.

$0.5d_c\,(W14 \times 90) = 7.0$ in.; $\quad 0.5d_b\,(W24 \times 94) = 12.15$ in.

$\theta = 45°$ from the geometry of the connection.

2. Assume $Y = 18$ in.

$$\beta = \frac{Y}{2} = \frac{18}{2} = 9 \text{ in.}$$

3. Using Equation (11-28), we have

$\alpha - \beta \tan\theta = e_b \tan\theta - e_c$. Then,

$$\alpha - (9 \text{ in.})(\tan 45°) = (12.15 \text{ in.})(\tan 45°) - 7.0 \text{ in.}$$

Therefore, $\alpha = 14.15$ in.

4. Hence the horizontal length of the gusset plate, X, is obtained from

$$\alpha = \frac{X}{2} + 0.5 \text{ in.}$$

That is, $14.15 \text{ in.} = \frac{X}{2} + 0.5 \text{ in.}$

Thus, $X = (14.15 - 0.5)(2) = 27.3 \text{ in.} = 2 \text{ ft. } 4 \text{ in.}$

From Step 2, Y = 18 in. = **1 ft. 6 in.**

5. Use Equation (11-29) to determine

$$r = \sqrt{(\alpha + 0.5d_c)^2 + (\beta + 0.5d_b)^2} = \sqrt{(14.15 \text{ in.} + 7.0 \text{ in.})^2 + (9 \text{ in.} + 12.15 \text{ in.})^2}$$
$$= 29.92 \text{ in.}$$

Using Equations (11-30) to (11-33), the gusset connection interface forces are determined as follows:

$$V_B = \frac{0.5d_b}{r} P_{brace} = \frac{12.15 \text{ in.}}{29.92 \text{ in.}}(150 \text{ kips}) = 61.0 \text{ kips}$$

$$H_B = \frac{\alpha}{r} P_{brace} = \frac{14.15 \text{ in.}}{29.92 \text{ in.}}(150 \text{ kips}) = 71.0 \text{ kips}$$

$$V_C = \frac{\beta}{r} P_{brace} = \frac{9 \text{ in.}}{29.92 \text{ in.}}(150 \text{ kips}) = 45.0 \text{ kips}$$

$$H_C = \frac{0.5d_c}{r} P_{brace} = \frac{7.0 \text{ in.}}{29.92 \text{ in.}}(150 \text{ kips}) = 35.1 \text{ kips.}$$

EXAMPLE 11-10

Analysis of Existing Diagonal Brace Connection

Determine the force distribution at the gusset plate connection interfaces for the existing diagonal brace connection shown in Figure 11-30, assuming that

a. The gusset-to-beam connection is more rigid than the gusset-to-column connection, and
b. The gusset-to-column connection is more rigid than the gusset-to-beam connection.

Assume a ½-in. setback between the gusset plate and the face of the column.

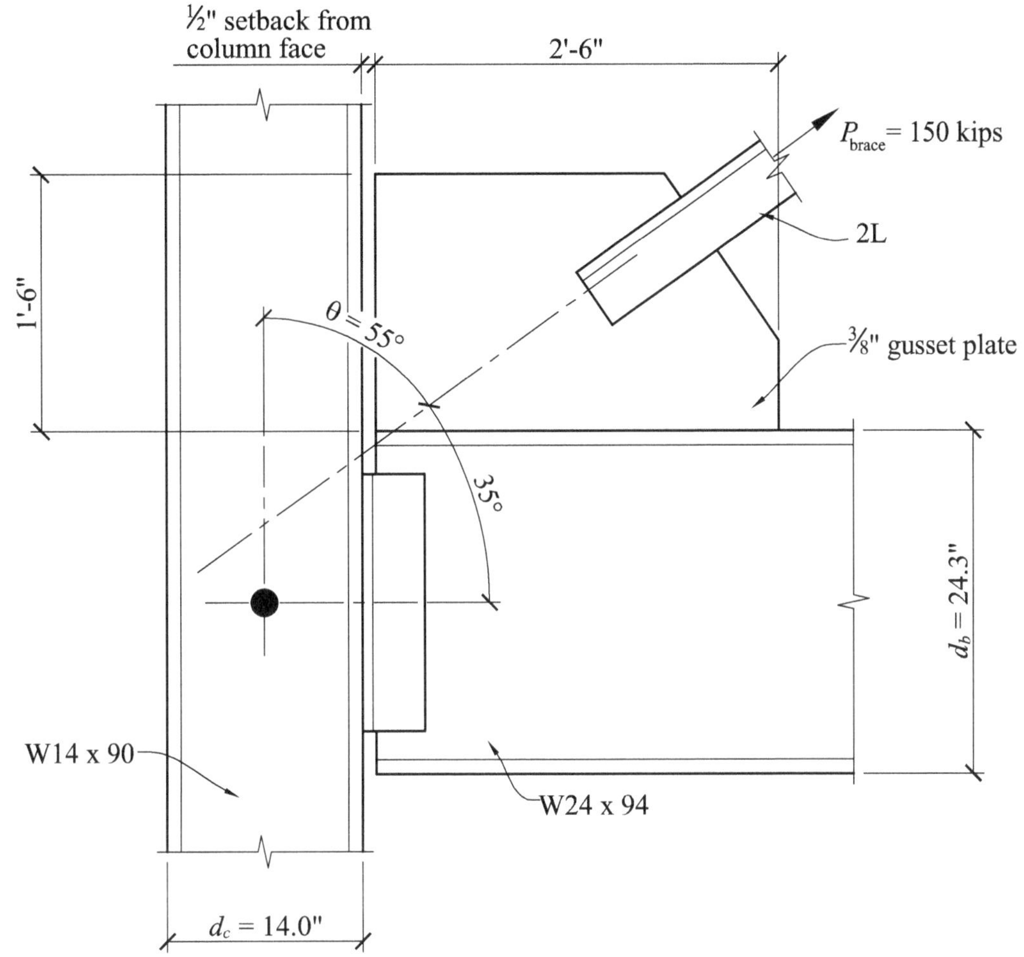

FIGURE 11-30 Diagonal brace connection alalysis for Example 11-10

Solution

1. $d_c\,(\mathrm{W14}\times 90) = 14.0$ in.; $d_b\,(\mathrm{W24}\times 94) = 24.3$ in.

   $0.5d_c\,(\mathrm{W14}\times 90) = 7.0$ in.; $0.5d_b\,(\mathrm{W24}\times 94) = 12.15$ in.

   $\theta = 55°$ from the geometry of the connection

   $X = 30$ in. and $Y = 18$ in. (see Figure 11-30)

2. The actual centroidal distances are calculated as follows:

$$\bar{\alpha} = \frac{X}{2} + 0.5\text{ in.} = \frac{30\text{ in.}}{2} + 0.5\text{ in.} = 15.5\text{ in.}$$

$$\bar{\beta} = \frac{Y}{2} = \frac{18\text{ in.}}{2} = 9\text{ in.}$$

3. If the gusset-to-beam connection is more rigid than the gusset-to-column connection (e.g., if a welded connection is used for the gusset-to-beam connection and a bolted connection is used for the gusset-to-column connection),

   $\beta = \bar{\beta}$ (from step 2) = 9 in. Substituting this in equation (11-28) yields

$$\alpha - \beta\tan\theta = e_b\tan\theta - e_c$$

$$\alpha - (9\text{ in.})(\tan 55°) = (12.15\text{ in.})(\tan 55°) - 7.0\text{ in.}$$

Therefore, set $\alpha = 23.2$ in. $= \bar{\alpha}$ from Step 2. Thus, we find that a moment exists on the gusset-to-beam connection interface; the moment and forces on this interface are calculated as follows:

$$r = \sqrt{\left(\alpha_{step3} + 0.5d_c\right)^2 + \left(\bar{\beta}_{step2} + 0.5d_b\right)^2}$$

$$= \sqrt{(23.2 \text{ in.} + 7 \text{ in.})^2 + (9 \text{ in.} + 12.15 \text{ in.})^2} = 36.9 \text{ in.}$$

The forces on the gusset-to-beam interface are

$$H_B = \frac{\alpha_{step3}}{r_{step3}} P_{brace} = \frac{23.2 \text{ in.}}{36.9 \text{ in.}}(150 \text{ kips}) = 94.3 \text{ kips},$$

$$V_B = \frac{0.5d_b}{r_{step3}} P_{brace} = \frac{12.15 \text{ in.}}{36.9 \text{ in.}}(150 \text{ kips}) = 49.4 \text{ kips},$$

$$M_{u,g\text{-}b} = V_B \left|\alpha - \bar{\alpha}\right| = (49.4 \text{ k})(23.2 \text{ in.} - 15.5 \text{ in.})$$

$$= 380.4 \text{ in.-kips} = 31.7 \text{ ft.-kips.}$$

The forces on the gusset-to-column interface are

$$H_C = \frac{0.5d_c}{r} P_{brace} = \frac{7.0 \text{ in.}}{36.9 \text{ in.}}(150 \text{ kips}) = 28.5 \text{ kips}$$

$$V_C = \frac{\beta}{r} P_{brace} = \frac{9 \text{ in.}}{36.9 \text{ in.}}(150 \text{ kips}) = 36.6 \text{ kips}$$

$$M_{u,g-c} = 0 \text{ ft.-kips.}$$

4. If the gusset-to-column connection is more rigid than the gusset-to-beam connection (e.g., if a welded connection is used for the gusset-to-column connection and a bolted connection is used for the gusset-to-beam connection), set $\alpha = \bar{\alpha}$ (from Step 2) = 15.5 in. Substituting this in Equation (11-28) yields

   $$15.5 \text{ in.} - \beta(\tan 55°) = (12.15 \text{ in.})(\tan 55°) - 7.0 \text{ in.}$$

   Therefore, $\beta = 3.6$ in. $\neq \bar{\beta}$ from Step 2. Thus, we find that a moment exists on the gusset plate-to-column connection interface, and the moment and forces on this interface are calculated as follows:

$$r = \sqrt{\left(\bar{\alpha}_{step2} + 0.5d_c\right)^2 + \left(\beta_{step4} + 0.5d_b\right)^2}$$

$$= \sqrt{(15.5 \text{ in.} + 7.0 \text{ in.})^2 + (3.6 \text{ in.} + 12.15 \text{ in.})^2} = 27.5 \text{ in.}$$

The forces on the gusset plate-to-column interface are:

$$H_C = \frac{0.5d_c}{r_{step4}} P_{brace} = \frac{7.0 \text{ in.}}{27.5 \text{ in.}}(150 \text{ kips}) = 38.2 \text{ kips},$$

$$V_C = \frac{\beta_{step4}}{r_{step4}} P_{brace} = \frac{3.6 \text{ in.}}{27.5 \text{ in.}}(150 \text{ kips}) = 19.6 \text{ kips},$$

$$M_{u,g-c} = H_C\left|\beta - \bar{\beta}\right| = (38.2\text{ k})\left|(3.6\text{ in.} - 9\text{ in.})\right|$$
$$= 206.3\text{ in.-kips} = \mathbf{17.2\ ft.\text{-}kips}.$$

The forces on the gusset plate-to-beam interface are:

$$H_B = \frac{\alpha}{r}P_{brace} = \frac{15.5\text{ in.}}{27.5\text{ in.}}(150\text{ kips}) = 84.5\text{ kips},$$

$$V_B = \frac{0.5d_b}{r}P_{brace} = \frac{12.15\text{ in.}}{27.5\text{ in.}}(150\text{ kips}) = 66.3\text{ kips},$$

$$M_{u,g-b} = 0\text{ ft.-kips}.$$

The available strength of the gusset plate connection is determined using the methods presented in Chapters 9 and 10 for calculating bolt and weld design capacities.

For more information and detailed design examples of gusset-plated connections, the reader should refer to Chapter 2 of Ref. [25].

11-9 COPED BEAMS

The analysis and design of coped beams, which occur on virtually every steel-framed project, is covered in this section. The geometry of a beam cope will generally be a function of the shape of the connected member. In some cases, field conditions may dictate the requirement of a cope. For example, Figure 11-31 shows a beam framing into the web of a column. When this beam is erected into its final position, it is dropped down in between the column flanges. This particular beam would require a coped bottom flange to enable it to be placed without any interference or obstruction. Other common coped beam connections are shown in Figure 11-32.

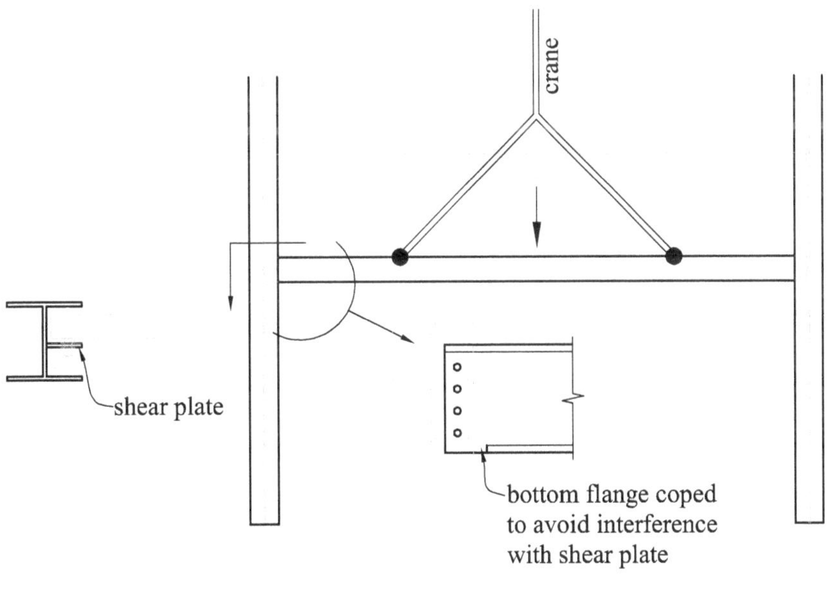

FIGURE 11-31 Coped bottom flange due to construction

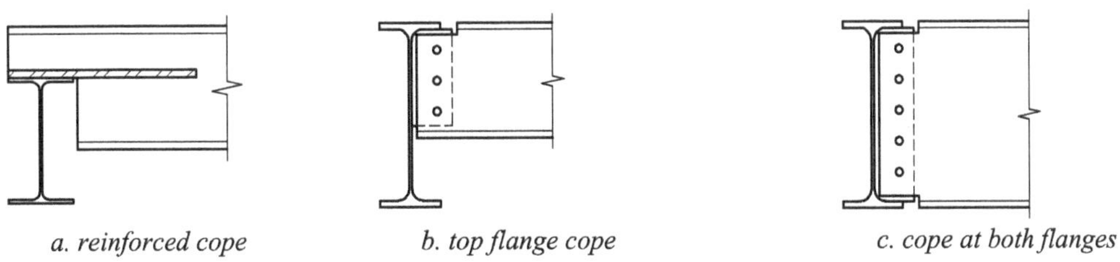

FIGURE 11-32 Types of beam copes

Additional beam modifications for connections are shown in Figure 11-33. A beam *cope* is defined as the removal of part of the beam web and flange. A *block* is the removal of the flange only, and a *cut* is the removal of one side of a flange. In each of these cases, it is common to refer to any of them as copes since the analysis and design procedures are similar.

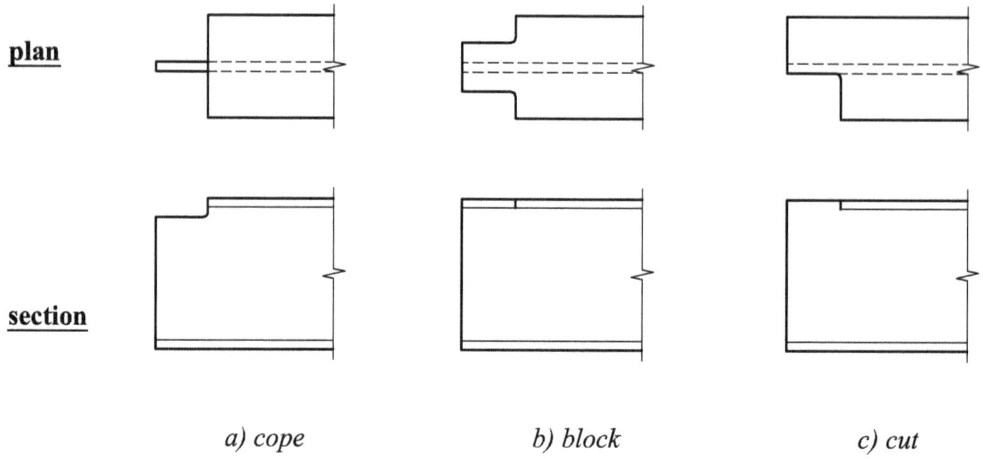

FIGURE 11-33 Cope, block, cut

When beams are coped, the load path of the end reaction must pass through a reduced section of the connected beam (see Figure 11-34). The strength of beams in block shear was covered in Chapter 9, and we will now consider the effect of the bending stresses at the critical (reduced) section of a coped beam due to the eccentricity of the beam end reaction, R_u, from the face of the cope. The coped section acts like a short cantilevered beam and is subjected to shear from the beam end reaction, plus bending stresses due to the moment, $R_u e$, of the beam end reaction about the face of the cope; the eccentricity, e, is equal to the distance from the point of inflection to the face of the cope, and due to the rotational rigidity of the beam end connection, the point of inflection will not always occur at the face of the supporting member; it will occur closer to the face of the cope, thus reducing the eccentricity. However, it is not practical to determine the inflection point for coped beams; therefore in this text, the eccentricity, e, is defined as the distance from the face of the supporting member to the face of the cope, or the distance from the vertical line of bolts in the coped member to the face of the cope (see Figure 11-34); in the former case, the line of action of the beam end reaction is assumed to occur at the face of the supporting member while in the latter case, the line of action of the beam end reaction is assumed to occur through the line of action of the bolt group [1, 26]. A lower value of the eccentricity, *e*, can be used if justified by analysis.

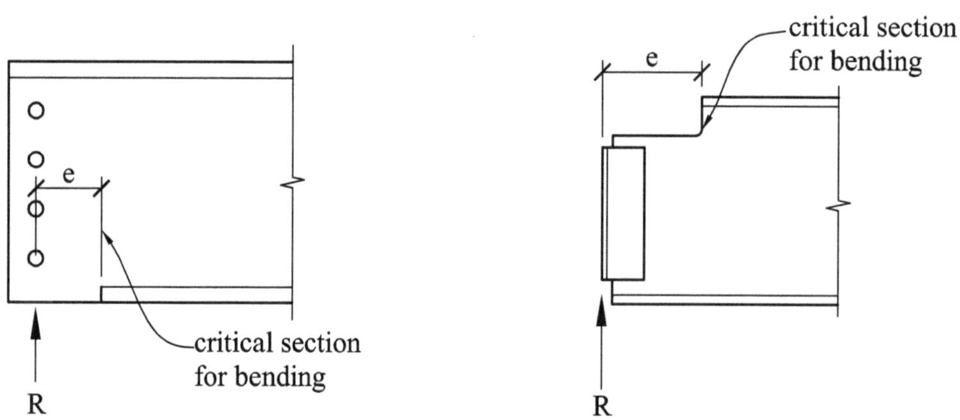

FIGURE 11-34 Reaction location and eccentricity at a coped beam

When the geometry of the beam cope is such that the localized stresses exceed the design strength, the beam can be reinforced at the cope. In Figure 11-35a, a web doubler plate is shown; in Figure 11-35b, a longitudinal stiffener is shown. In these two cases, the reinforcement should be extended a distance of d_c past the critical section, where d_c is the depth of the beam cope. In Figure 11-35c, both transverse and longitudinal stiffeners are shown. The given geometry of a connection will dictate the type of reinforcement to use where reinforcement is required.

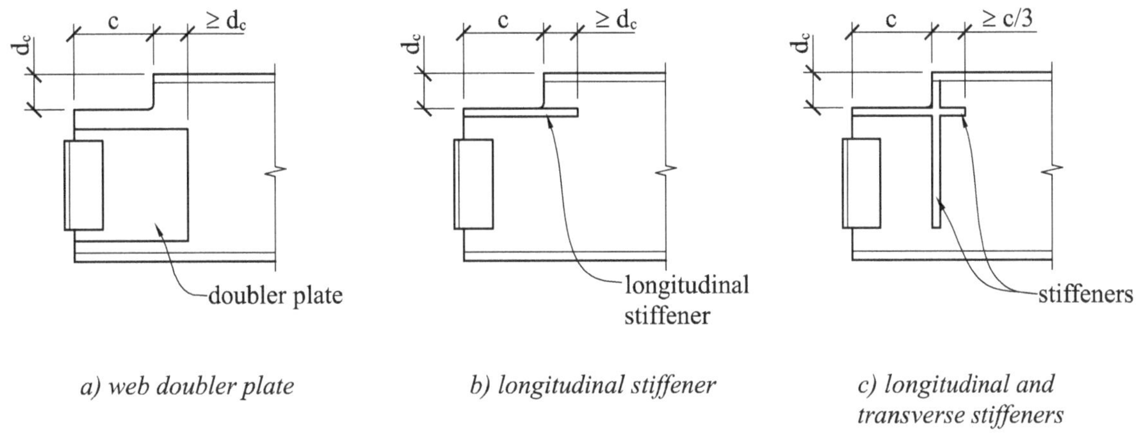

FIGURE 11-35 Reinforcement of coped beams

The failure modes to be considered for coped beams include bolt bearing, block shear, flexural rupture strength, and flexural local buckling strength, in addition to checking the strength of the beam web in shear yielding.

Flexural Rupture Strength:

The nominal flexural rupture strength is obtained from *AISCM* Equation (9-1) as

$$M_n = F_u Z_{net} \tag{11-42}$$

Where,

F_u is the ultimate tensile strength of the steel

Z_{net} = plastic section modulus of the reduced beam section at the cope

For LRFD, the design *flexural rupture* strength (see *AISCM* equation (9-4)) is given as

$$\phi_{br} M_n = \phi_{br} F_u Z_{net} \tag{11-43a}$$

The allowable maximum moment is,

$$M_a = \frac{M_n}{\Omega_{br}} = \frac{F_u Z_{net}}{\Omega_{br}} \tag{11-43b}$$

Where,

ϕ_{br} = 0.75 (LRFD strength reduction factor for the flexural rupture limit state)

Ω_{br} = 2.0 (ASD safety factor for the flexural rupture limit state)

As indicated at the beginning of this chapter, the ASD method is not used for the examples in this chapter, but the interested reader can easily and readily obtain the ASD solutions to the examples using the ASD equations.

Flexural Local Buckling Strength—Beams coped at the top flange only:

For beams coped at the top flange only, the reduced cross-section at the cope will be an inverted T-section, and the web will be subjected to local buckling from the compression stresses due to flexure. Thus, for this case, the flexural local buckling limit state will be applicable. The minimum length of the shear connector (i.e., the double angles or shear tab) should be no less than 50 percent of the depth of the beam at the cope, h_o. The nominal moment capacity at the coped section due to flexural local buckling will depend on the slenderness of the coped section and are obtained from Part 9 of the *AISCM* [1] as follows:

$$\text{For } \lambda \le \lambda_p \qquad M_n = M_p \tag{11-44}$$

$$\text{For } \lambda_p < \lambda \le 2\lambda_p \qquad M_n = M_p - (M_p - M_y)\left(\frac{\lambda}{\lambda_p} - 1\right) \le M_p \tag{11-45}$$

$$\text{For } \lambda > 2\lambda_p \qquad M_n = F_{cr} S_{net} \le M_p \tag{11-46}$$

$$\text{Critical stress}, F_{cr} = \frac{0.903Ek_1}{\lambda^2} = \frac{0.903Efk}{\left(\frac{h_c}{t_w}\right)^2} = 26{,}187 f k \left(\frac{t_w}{h_c}\right)^2 \tag{11-47}$$

$$\lambda_p = 0.475\sqrt{\frac{k_1 E}{F_y}} \tag{11-48}$$

The design moment capacity (LRFD) = $\phi_b M_n$ (11-49a)

The allowable maximum moment, $M_a = \dfrac{M_n}{\Omega_b}$ (11-49b)

Where,

ϕ_b = 0.9 (LRFD strength reduction factor for the flexural local buckling limit state)

Ω_b = 1.67 (ASD safety factor for the flexural local buckling limit state)

M_n = nominal moment capacity

M_p = plastic moment capacity of the reduced beam section at the cope = $F_y Z_{net}$

M_y = nominal yield moment capacity of the reduced beam section at the cope = $F_y S_{net}$

E = modulus of elasticity = 29,000 ksi

F_y = yield stress of the steel in ksi

S_{net} = elastic section modulus of the reduced beam section at the cope, in.^{3}

The net elastic section modulus, S_{net}, for various beams and cope depths is provided in *AISCM* Table 9-2. In this table, the cope depth is limited to the smaller of 0.5*d* and 10 in.

Z_{net} = plastic section modulus of the reduced beam section at the cope, in.^{3}

The net plastic section modulus, Z_{net}, is not tabulated in the *AISCM* but is tabulated in the *AISC* Design Examples, which can be downloaded from aisc.org. Z_{net} can be calculated for coped sections using the method presented in Chapter 1 for calculating plastic section modulus.

$$k_1 = f k \ge 1.61 \tag{11-50}$$

f = buckling adjustment factor

$$= 2\left(\frac{c}{d}\right) \qquad \text{for } \frac{c}{d} \le 1.0 \tag{11-51a}$$

$$= 1 + \left(\frac{c}{d}\right) \le 3 \qquad \text{for } \frac{c}{d} > 1.0 \tag{11-51b}$$

λ = slenderness of the web = $\dfrac{h_c}{t_w}$

c = cope length (see Figure 11-36)

d_c = cope depth (see Figure 11-36)

d = beam depth

h_c = depth of the coped section (see Figure 11-36)
t_w = beam web thickness

The plate buckling coefficient, k, is given by the following equations:

$$k = 2.2\left(\frac{h_c}{c}\right)^{1.65} \quad \text{for } \frac{c}{d} \leq 1.0 \tag{11-52a}$$

$$= 2.2\left(\frac{h_c}{c}\right) \quad \text{for } \frac{c}{d} > 1.0 \tag{11-52b}$$

Lateral Torsional Buckling Strength—Beams coped at both the *Top* and *Bottom* flanges:

For beams coped at both the top and bottom flanges, the coped section behaves in bending like a rectangular plate section [1, 26, 27], and AISC Specification Section F11 is applicable in calculating the flexural strength of the coped section, and the lateral torsional buckling limit state should be checked. The moment capacity and allowable moments for the lateral torsional buckling limit state should be modified by the lateral-torsional buckling modification factor, C_b, calculated as follows (see Equations (9-21) and (9-22) in Part 9 of the *AISCM* [1]):

When the **bottom cope is equal in length or longer than the top cope** (i.e., $c_b \geq c_t$), the lateral torsional buckling modification factor [1] will be

$$C_b = \left[3+\ln\left(\frac{L_b}{d}\right)\right]\left(1-\frac{d_{ct}}{d}\right) = \left[3+\ln\left(\frac{c_t}{d}\right)\right]\left(1-\frac{d_{ct}}{d}\right) \geq 1.84 \tag{11-53}$$

When $c_b \geq c_t$, the unbraced length of the cope, $L_b = c_t$, in.

When the **bottom cope is shorter than the top cope** (i.e., $c_b < c_t$), the lateral torsional buckling modification factor [1] will be

$$C_b = \left(\frac{c_b}{c_t}\right)\left[3+\ln\left(\frac{L_b}{d}\right)\right]\left(1-\frac{d_{ct}}{d}\right) = \left(\frac{c_b}{c_t}\right)\left[3+\ln\left(\frac{c_t+c_b}{2d}\right)\right]\left(1-\frac{d_{ct}}{d}\right) \geq 1.84 \tag{11-54}$$

When $c_b < c_t$, the unbraced length of the cope, $L_b = \dfrac{c_t + c_b}{2}$, in.

All the other parameters in Equations (11-53) and (11-54) above are defined as follows:

c_t = length of cope in top flange, in. (see Figure 11-36b)
c_b = length of cope in bottom flange, in. (see Figure 11-36b)
d_{ct} = depth of cope at the top flange, in. (see Figure 11-36b)
d_{cb} = depth of cope at the bottom flange, in. (see Figure 11-36b)

Note that buckling is checked at the top flange (which is usually in compression); hence, the cope depth at the top flange, d_{ct}, affects the lateral torsional stability factor in Equations (11-53) and (11-54). The larger the value of d_{ct}, the smaller the lateral torsional buckling modification factor, thus reducing the nominal moment strength.

For beams that are coped at both the top and bottom flanges, the rectangular coped section when subjected to flexure is susceptible to lateral torsional buckling, and the design moment capacity and allowable moment from AISC Specifications F11 for bending about the strong (or x-x) axis are given as follows as a function of the slenderness of the net section:

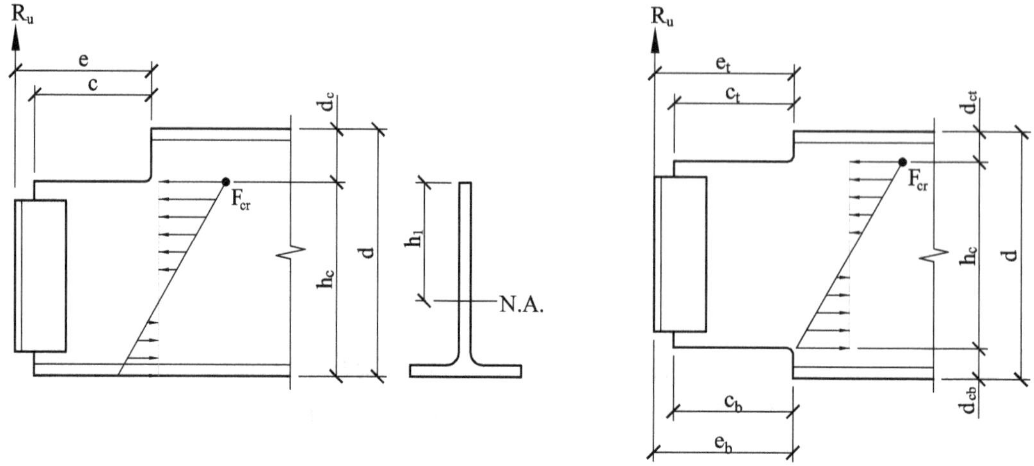

a) local buckling for one coped flange

b) local buckling for both flanges coped

FIGURE 11-36 Dimensional parameters for local buckling in a coped beam

1. For $\frac{L_b d}{t^2} \leq \frac{0.08E}{F_y}$, the design moment capacity and the allowable moment, respectively, are calculated as follows:

$$\phi_b M_n = \phi_b M_p = \phi_b F_y Z_x \leq \phi_b 1.6 F_y S_x \quad \text{for LRFD} \quad (11\text{-}55a)$$

$$\frac{M_n}{\Omega_b} = \frac{M_p}{\Omega_b} = \frac{F_y Z_x}{\Omega_b} \leq \frac{1.6 F_y S_x}{\Omega_b} \quad \text{for ASD} \quad (11\text{-}55b)$$

2. For $\frac{0.08E}{F_y} < \frac{L_b d}{t^2} \leq \frac{1.9E}{F_y}$, the design moment capacity and the allowable moment, respectively, are calculated as follows:

$$\phi_b M_n = \phi_b C_b \left[1.52 - 0.274 \left(\frac{L_b d}{t^2}\right)\left(\frac{F_y}{E}\right)\right] M_y \leq \phi_b M_p \quad \text{for LRFD} \quad (11\text{-}56a)$$

$$\frac{M_n}{\Omega_b} = \frac{C_b \left[1.52 - 0.274 \left(\frac{L_b d}{t^2}\right)\left(\frac{F_y}{E}\right)\right] M_y}{\Omega_b} \leq \frac{M_p}{\Omega_b} \quad \text{for ASD} \quad (11\text{-}56b)$$

3. For $\frac{L_b d}{t^2} > \frac{1.9E}{F_y}$, the design moment capacity and the allowable moment, respectively, are calculated as follows:

$$\phi_b M_n = \phi_b F_{cr} S_x \leq \phi_b M_p \quad \text{for LRFD} \quad (11\text{-}57a)$$

$$\frac{M_n}{\Omega_b} = \frac{F_{cr} S_x}{\Omega_b} \leq \frac{M_p}{\Omega_b} \quad \text{for ASD} \quad (11\text{-}57b)$$

For this limit state,

ϕ_b = 0.9
Ω_b = 1.67
L_b = unbraced length
C_b = lateral torsional buckling modification factor (see Equations (11-53) and (11-54))

d = bar depth (i.e., plate or bar dimension perpendicular to the axis of bending) = h_c for the beam coped at the top and bottom flange

t = bar thickness (i.e., plate or bar dimension parallel to the axis of bending) = t_w for the coped beam

S_x = coped section elastic modulus $= S_{net} = \frac{t_w h_c^2}{6}$ (for beam coped at the top and bottom flange).

Z_x = coped section plastic modulus $= Z_{net} = \frac{t_w h_c^2}{4}$ (for beam coped at the top and bottom flange).

M_y = yield moment for bending about the strong (i.e., x-x) axis = $F_y S_x$

M_p = yield moment for bending about the strong (i.e., x-x) axis = $F_y Z_x$

$$F_{cr} = \frac{1.9EC_b}{\frac{L_b d}{t^2}} \tag{11-58}$$

EXAMPLE 11-11

Beam Coped at the Top Flange

For the beam shown in Figure 11-37, determine whether the beam is adequate for flexural rupture and flexural yielding at the beam cope. The steel is ASTM A992, grade 50.

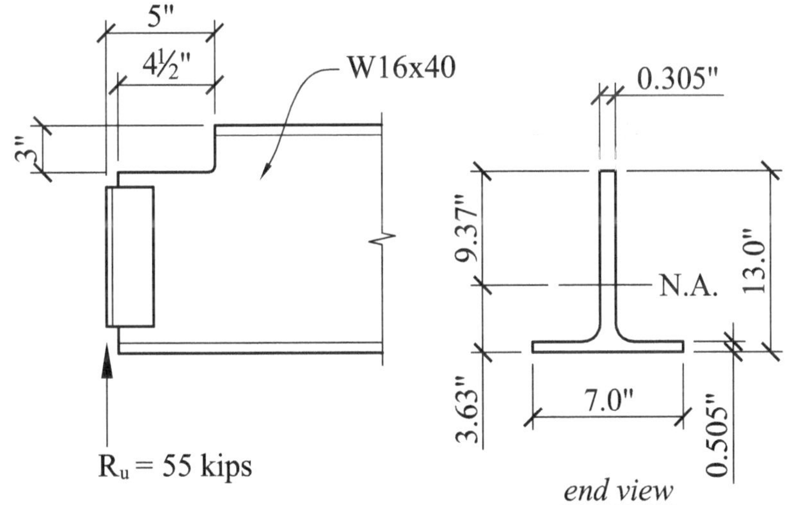

FIGURE 11-37 Detail for Example 11-11

Solution

From Figures 11-35 and 11-37, we obtain the following design parameters for the coped beam:

c = 4.5 in.

d_c = 3.0 in.

From Figures 11-34 and 11-37, we obtain the following design parameters:

R_u = 55 kips.

e = 5.0 in.

From *AISCM* Table 1-1, the section properties for W16 × 40 are

$d = 16.0 \text{ in.} \quad t_w = 0.305 \text{ in.}$

$b_f = 7.0 \text{ in.} \quad t_f = 0.505 \text{ in.}$

$h_c = d - d_c = 16.0 - 3.0 = 13.0 \text{ in.}$

$$\lambda = \frac{h_c}{t_w} = \frac{13.0}{0.305} = 42.6$$

Section properties at the critical section (see Table 11-2):

Web component is $(13.0 \text{ in.} - 0.505 \text{ in.}) \times 0.305 \text{ in.}$

Flange component is $7.0 \text{ in.} \times 0.505 \text{ in.}$

TABLE 11-2 Section Properties at the Beam Cope

Element	A	y	Ay	I	$d = y - \bar{y}$	$I + Ad^2$
Web	3.81	6.75	25.73	49.6	3.13	86.9
Flange	3.53	0.252	0.89	0.0751	–3.37	40.2
Σ =	**7.34**		**26.63**			I_{net} = **127.1**

$$\bar{y} = \frac{\Sigma Ay}{\Sigma A} = \frac{26.63}{7.34} = 3.63 \text{ in.}$$

$$h_1 = h_c - \bar{y} = 13.0 - 3.63 = 9.37 \text{ in.}$$

$$S_{net} = \frac{I_{net}}{h_1} = \frac{127.1}{9.37} = 13.6 \text{ in.}^3$$

Alternatively, using *AISCM* Table 9-2a for W16 × 40 with $d_c = 3$ in., confirm that $S_{net} = 13.6$ in.[3] Also, using the equal area principle for homogeneous sections (see Chapter 1), the plastic section modulus of the reduced section, Z_{net}, can also be calculated to be 24.64 in.^{3}. Alternatively, *AISCM* Table 9-2b can be used to determine the plastic section modulus for coped W-shapes. The factored applied moment due to the eccentricity of the beam end reaction is

$$M_u = R_u e = (55 \text{ kips})(5 \text{ in.}) = 275 \text{ in.-kips.}$$

Flexural rupture strength:

$\phi_{br} M_n = \phi_{br} F_u Z_{net}$ (see equation (11-43a))

$= (0.75)(65)(24.64) = 1201 \text{ in.-kips} > 275 \text{ in.-kips}$ OK.

Flexural local buckling strength:

$$\frac{c}{h_c} = \frac{4.5 \text{in.}}{13.0 \text{in.}} = 0.34 < 1.0,$$

$$\frac{c}{d} = \frac{4.5 \text{in.}}{16.0 \text{in.}} = 0.28 < 1.0,$$

Therefore, the plate buckling coefficient, $k = 2.2\left(\frac{h_c}{c}\right)^{1.65} = 2.2\left(\frac{13.0}{4.5}\right)^{1.65} = 12.67$

and, $$f = 2\left(\frac{c}{d}\right) = 2\left(\frac{4.5}{16}\right) = 0.56$$

$k_1 = fk \geq 1.61$
$= (0.56)(12.67) = 7.1 > 1.61, \text{ Use } k_1 = 7.1$

$$\lambda = \frac{h_c}{t_w} = \frac{13.0}{0.305} = 42.6$$

$$\lambda_p = 0.475\sqrt{\frac{k_1 E}{F_y}} = 0.475\sqrt{\frac{(7.1)(29{,}000)}{50}} = 30.5; \quad 2\lambda_p = 61$$

Since $\lambda_p < \lambda < 2\lambda_p$, therefore, use equation (11 - 45):

$M_p = F_y Z_{net} = (50\,\text{ksi})(24.64\,\text{in.}^3) = 1232\,\text{in.-kips}$

$M_y = F_y S_{net} = (50\,\text{ksi})(13.6\,\text{in.}^3) = 680\,\text{in.-kips}$

Therefore, $M_n = M_p - (M_p - M_y)\left(\frac{\lambda}{\lambda_p} - 1\right) \leq M_p$

$$= 1232 - (1232 - 680)\left(\frac{42.6}{30.5} - 1\right) \leq 1232\text{ in.-kips.}$$
$$= \mathbf{1013\,in.\text{-}kips} \leq 1232\text{ in.-kips}$$

Therefore, $M_n = 1013\,\text{in.-kips}$

Therefore, the flexural local buckling strength is

$\phi_b M_n = (0.9)(1013) = 912\text{ in.-kips} > R_u e = 275\text{ in.-kips.}$

Since the flexural local buckling strength (912 in.-kips) is less than the flexural rupture strength (1201 in.-kips), therefore, the flexural local buckling strength controls the flexural strength for this coped beam. In most cases, the flexural strength of the coped beam will be controlled by the flexural local buckling limit state.

Therefore, $\phi_b M_n = 912\,\text{in.-kips} = \mathbf{76\,ft.\text{-}kips}$

Shear strength of the coped section (from Chapter 6):

$$\phi_v V_n = \phi_v 0.6 F_y A_w C_{v1} = \phi_v 0.6 F_y \left[(h_c)(t_w)\right] C_{v1}$$
$$= (0.9)(0.6)(50)\left[(13.0)(0.305)\right](1.0) = 107\text{ kips} > V_u = 55\text{ kips} \qquad \text{OK.}$$

The coped beam is adequate in shear and flexure. The preceding discussion and solution assume that the coped beam is laterally braced in the compression zone (i.e., at the top flange) at the end of the uncoped region of the beam.

EXAMPLE 11-12

Beam Cope with Reinforcing

For the beam shown in Figure 11-38, determine whether the beam is adequate for flexural rupture and flexural yielding at the beam cope. The steel has a yield strength of 50 ksi.

Solution

This is a common connection condition when W-shape beams are used in the same framing plan as open-web steel joists that have a seat depth of 2.5 in. at the bearing ends. For the connection shown, the depth of the cope, d_c, is 5.49 in. (greater than $0.5d$ = 4.0 in.), and therefore it must be reinforced as shown with the $\frac{3}{8}$-in. × 2-in. plates. Note that the plates must extend beyond the face of the cope for a distance equal to at least the depth of the cope (i.e., at least 5.49 in.); see Figures 11-35 and 11-38.

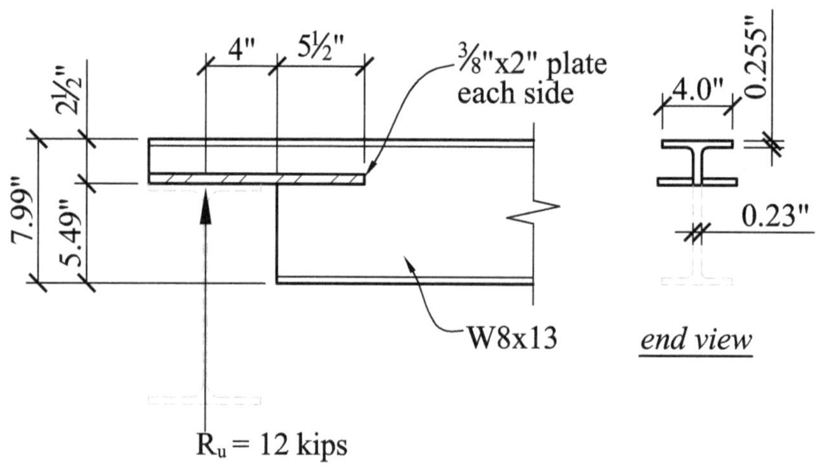

FIGURE 11-38 Detail for Example 11-12

The capacity of this connection must be checked for flexural rupture since flexural local buckling will not occur because the coped section is reinforced. The section properties of the critical section of the reinforced coped beam are summarized as follows:

From *AISCM* Table 1-1, the section properties of an uncoped W8 × 13 are

d = 7.99 in. t_w = 0.23 in.

b_f = 4.0 in. t_f = 0.255 in.

Section properties at the critical section (see Table 11-3 and Figure 11-38):

Web component is $(2.5 \text{ in.} - 0.255 \text{ in.}) \times 0.23$ in.

Flange component is 4.0 in. × 0.255 in.

Plate component is 4.0 in. × 0.375 in.

TABLE 11-3 Section Properties of Reinforced Beam Cope

Element	A	y	Ay	I	$d = y - \bar{y}$	$I + Ad^2$
Top flange	1.02	2.373	2.42	0.006	1.292	1.706
Web	0.516	1.123	0.58	0.217	0.042	0.218
Plates	1.5	0.188	0.281	0.018	−0.893	1.22
Σ =	**3.036**		**3.281**			I_{net} = **3.14**

$$\bar{y} = \frac{\Sigma Ay}{\Sigma A} = \frac{3.281}{3.036} = 1.08 \text{ in. (measured from the bottom of the } \tfrac{3}{8} \text{ in. plates)}$$

$$S_{net} = \frac{I_{net}}{2.5 - \bar{y}} = \frac{3.14}{2.5 - 1.08} = 2.21 \text{ in.}^3 \text{ (use the smaller section modulus)}$$

Using the equal area principle for homogeneous sections (see Chapter 1), the plastic neutral axis or equal area axis will be located at a distance of 0.359 in. from the bottom of the reduced 2.5 in. deep beam (i.e., from the bottom of the reinforcing plates). The plastic section modulus is found to be

$$Z_{net} = 2.735 \text{ in.}^3$$

Flexural rupture strength:

$$\phi_{br} M_n = \phi_{br} F_u Z_{net} = (0.75)(65)(2.735) = 133 \text{ in.-kips}$$

With the eccentricity, *e*, from the support reaction to the face of the cope equal to 4 inches., the factored moment at the face of the cope is

$$M_u = R_u e = (12 \text{ kips})(4 \text{ in.}) = 48 \text{ in.-kips} < 133 \text{ in.-kips} \quad \text{OK.}$$

Shear strength (from Chapter 6):

$$\phi_v V_n = \phi_v 0.6 F_y A_w C_{v1}$$
$$= (0.9)(0.6)(50)[(2.5)(0.23)](1.0) = 15.5 \text{ kips} > V_u = 12 \text{ kips} \quad \text{OK.}$$

EXAMPLE 11-13

Beam Cope at Both Flanges

For the beam shown in Figure 11-39, determine the maximum reaction that could occur based on flexural rupture and flexural yielding at the beam cope. The steel is ASTM A992, Grade 50.

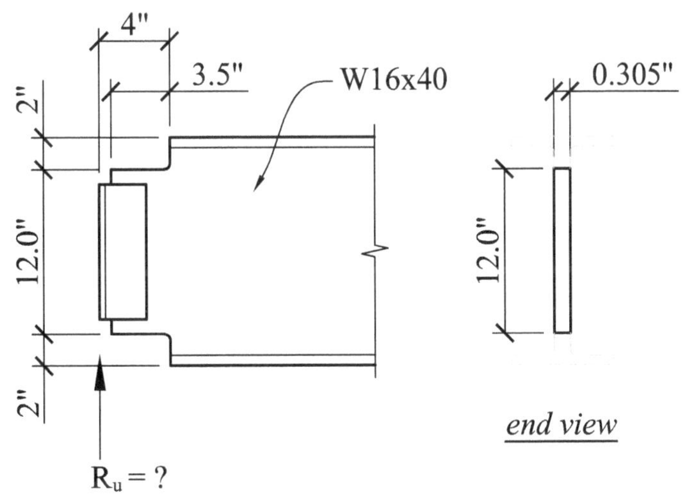

FIGURE 11-39 Detail for Example 11-13

Solution

From *AISCM* Table 1-1, the section properties for a W16 × 40 beam are

$d = 16.0$ in. $\quad t_w = 0.305$ in.

$b_f = 7.0$ in. $\quad t_f = 0.505$ in.

$$h_c = d - d_{ct} - d_{cb} = 16.0 - 2 \text{ in.} - 2 \text{ in.} = 12.0 \text{ in.}$$

$c_t = c_b = c = 3.5$ in. (beam is coped at both the top and bottom flanges)

$$S_{net} = \frac{t_w h_c^2}{6} = \frac{(0.305)(12.0)^2}{6} = 7.32 \text{ in.}^3$$

$$Z_{net} = \frac{t_w h_c^2}{4} = \frac{(0.305)(12.0)^2}{4} = 10.98 \text{ in.}^3$$

Flexural rupture strength:

$$\begin{aligned}\phi_{br} M_n &= \phi_{br} F_u Z_{net} \\ &= (0.75)(65)(10.98) = 535 \text{ in.-kips} \geq M_u = R_{u,\max} e\end{aligned}$$

The eccentricity, e, which is the distance between the support reaction and the face of the cope, is 4 inches; therefore, the required maximum reaction is

$$R_{u,max} = \frac{\phi_{br} M_n}{e} = \frac{535}{4} = 133.8 \text{ kips}$$

Flexural local buckling strength—beam coped at both the top and bottom flanges: Since the beam is coped at both the top and bottom flanges, the coped section behaves like a rectangular section bending about its strong axis. Since the length of the bottom (tension) cope is equal to that of the top cope (i.e., $c_t = c_b$), therefore, the unbraced length, $L_b = c_t = 3.5$ in. The lateral torsional buckling modification factor is calculated from Equation (11-53) as follows:

$$C_b = \left[3 + \ln\left(\frac{L_b}{d}\right)\right]\left(1 - \frac{d_{ct}}{d}\right) = \left[3 + \ln\left(\frac{c_t}{d}\right)\right]\left(1 - \frac{d_{ct}}{d}\right) \geq 1.84$$

$$= \left[3 + \ln\left(\frac{3.5}{16.0}\right)\right]\left(1 - \frac{2.0}{16.0}\right) \geq 1.84$$

$$= 1.295 \geq 1.84$$

Therefore, $C_b = 1.84$

$$M_y = F_y S_{net} = (50)(7.32) = 366 \text{ in.-kips}$$
$$M_p = F_y Z_{net} = (50)(10.98) = 549 \text{ in.-kips}$$

$$\frac{L_b d}{t^2} = \frac{L_b h_c}{t_w^2} = \frac{(3.5)(12.0)}{(0.305)^2} = 452$$

$$\frac{0.08E}{F_y} = \frac{(0.08)(29{,}000)}{50} = 46.4$$

$$\frac{1.9E}{F_y} = \frac{(1.9)(29{,}000)}{50} = 1102$$

$$\text{Since } \frac{0.08E}{F_y} < \frac{L_b d}{t^2} \leq \frac{1.9E}{F_y}$$

Use Equation (11-56a); therefore, the flexural capacity is

$$\phi_b M_n = \phi_b C_b \left[1.52 - 0.274\left(\frac{L_b d}{t^2}\right)\left(\frac{F_y}{E}\right)\right] M_y \leq \phi_b M_p$$

$$= (0.9)(1.84)\left[1.52 - 0.274(452)\left(\frac{50}{29{,}000}\right)\right](366) \leq (0.9)(549\text{ in.-kips}) = 494\text{ in.-kips}$$

$$= 792\text{ in.-kips} \leq \mathbf{494\text{ in.-kips}}$$

Therefore, the flexural lateral torsional buckling strength is $\phi_b M_n = 494$ in.-kips

The maximum reaction due to the flexural lateral torsional buckling limit state is

$$R_{u,max} = \frac{\phi_b M_n}{e} = \frac{494}{4} = 123.5\text{ kips}$$

Shear strength of coped beam (see Chapter 6):

$$\phi_v V_n = \phi_v 0.6 F_y A_w C_{v1} = (0.9)(0.6)(50)\left[(12.0)(0.305)\right](1.0) = 98.8\text{ kips}$$

The maximum reaction due to the shear yielding limit state is $R_u = 98.8$ kips.

Comparing the maximum reactions from the three limit states (133.8 kips, 123.5 kips, and 98.8 kips), we find that the shear yielding limit state controls. Therefore, for this coped beam, the maximum reaction that can be supported is $\boldsymbol{R_{u,\,max}}$ **= 98.8 kips**.

11-10 HOLES IN BEAMS

Holes in any steel section are generally not desirable, especially if the holes are made after construction and were not part of the original design. In many connections, bolt holes are intentional and necessary and are always considered in the original design. In other cases, field conditions might dictate the need for an opening in a steel beam, usually for the passage of mechanical, electrical, or plumbing ducts that conflict with a steel member (e.g., see Figures 3-8c, 11-40c and 11-40d). Careful attention to coordination among all of the construction trades could avoid such a conflict, but sometimes it is unavoidable. The use of open-web steel joists or castellated beams would provide a framing system that allows for the passage of moderately sized ducts through the framing members (see Figures 11-40a and 11-40b).

Other references provide more detailed coverage of beams with web openings [9], but the approach taken here will be similar to a common approach taken in practice. When an opening has to be made through a steel member, the following general guidelines apply:

1. Provide reinforcement above and below the opening so that all of the original section properties are maintained. For example, if an opening is required in a W-shaped beam, provide a steel angle or plate so that the addition of the angles or plates will yield a composite steel section that has a cross-sectional area, section modulus, and moment of inertia that is equal to or greater than the original section. This will help to ensure the same behavior of the beam with respect to the applied loads. For beams that are to be left unreinforced, refer to Ref. [9].
2. Concentrated loads should not be permitted above the opening, nor should the opening be within a distance d from the bearing end of the beam.

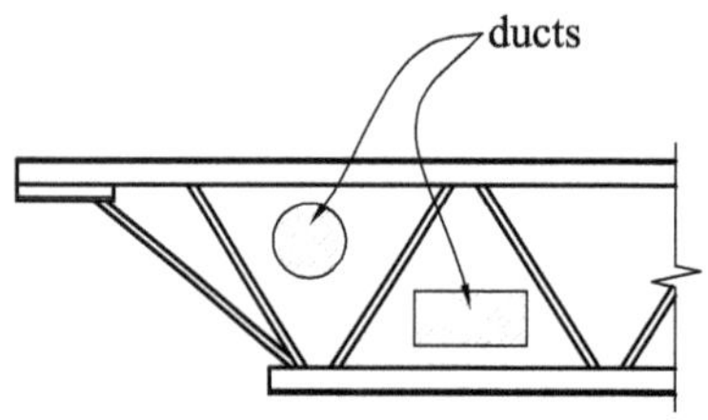

a) open web steel joist

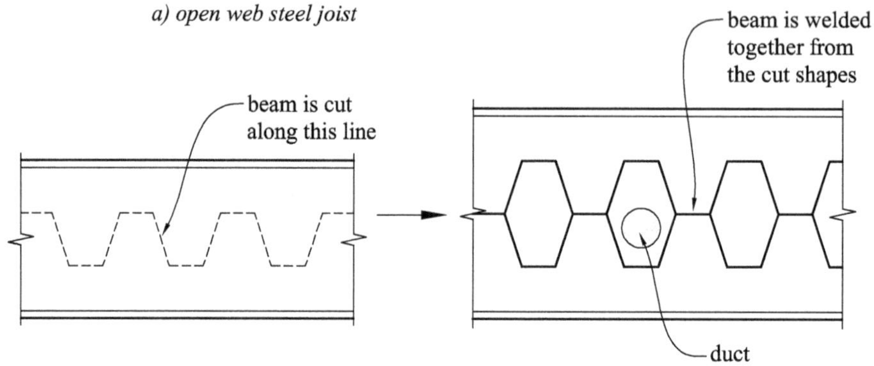

b) castellated beam

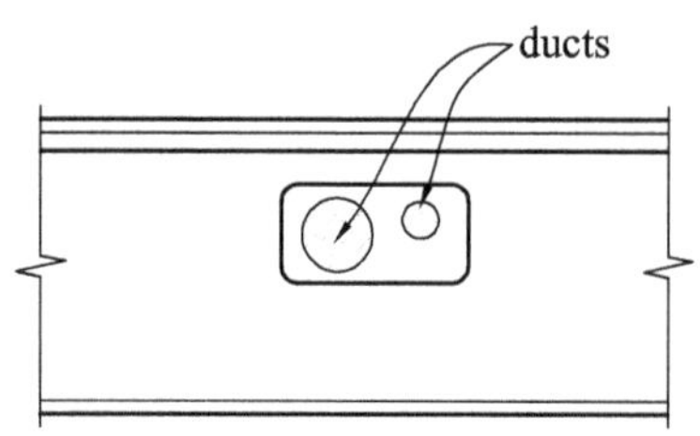

c) wide flange beam

FIGURE 11-40 Beams with openings

FIGURE 11-40d Holes in girder webs

3. Circular openings are preferred because they are less susceptible to stress concentrations. Square openings should have a minimum radius at the corners of $2t_w$ or $\frac{5}{8}$ in., whichever is greater.
4. Openings should be spaced as follows:

   a. *Rectangular openings:*

$$S \geq h_o \tag{11-59}$$

$$S \geq a_o\left(\frac{V_u/\phi V_p}{1-V_u/\phi V_p}\right) \tag{11-60}$$

   b. *Circular openings:*

$$S \geq 1.5D_o \tag{11-61}$$

$$S \geq D_o\left(\frac{V_u/\phi V_p}{1-V_u/\phi V_p}\right) \tag{11-62}$$

   c. *Openings in composite beams:*

$$S \geq a_o \tag{11-63}$$

$$S \geq 2d, \tag{11-64}$$

Where,
S = Clear space between openings,
h_o = Opening depth,
a_o = Opening width (parallel to the longitudinal axis of the beam),
D_o = Opening diameter,
d = Beam depth,
V_u = Factored shear,
ϕ = 0.9 for noncomposite beams
= 0.85 for composite beams,
V_p = Plastic shear capacity

$$= \frac{F_y t_w s}{\sqrt{3}}, \tag{11-65}$$

F_y = Yield stress,
t_w = Web thickness, and
s = Depth of the remaining section.

5. The weld strength within the length of the opening should be as follows:

$$R_{wr} = \phi 2P_r, \tag{11-66}$$

Where,
R_{wr} = Required weld strength,
ϕ = 0.9 for noncomposite beams,
= 0.85 for composite beams,

$$P_r = F_y A_r \leq \frac{F_y t_w a_o}{2\sqrt{3}}, \text{ and} \tag{11-67}$$

A_r = Cross-sectional area of the reinforcement above or below the opening.

6. The length of the extension beyond the opening should be as follows:

$$L_1 = \frac{a_o}{4} \geq \frac{A_r\sqrt{3}}{2t_w}. \tag{11-68}$$

7. The weld strength within the length of the extension should be as follows:

$$R_{wr} = \phi F_y A_r. \tag{11-69}$$

The design parameters indicated above are illustrated in Figure 11-41.

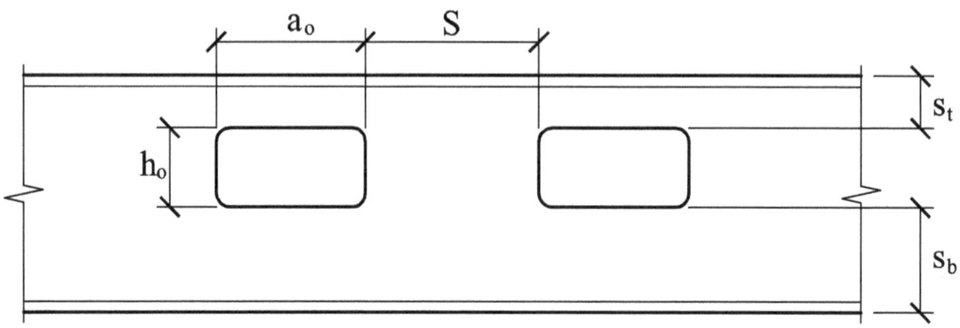

a) rectangular opening

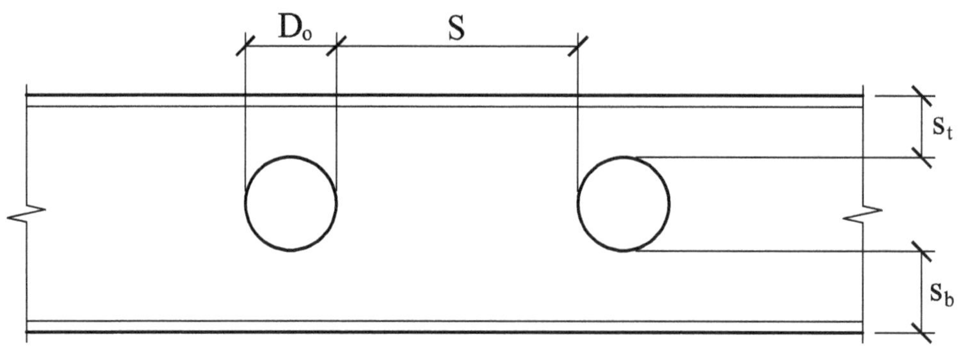

b) circular opening

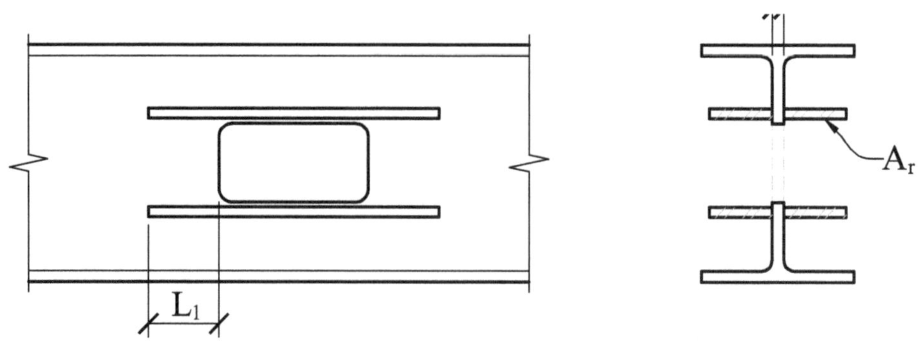

c) reinforcement

FIGURE 11-41 Web opening design parameters

EXAMPLE 11-14

Beam with Web Opening

Determine the required reinforcement for the noncomposite beam shown in Figure 11-42. The 6-in. opening is centered in the beam depth and is at midspan of a simple-span beam away from any significant concentrated loads. The steel is ASTM A992, Grade 50.

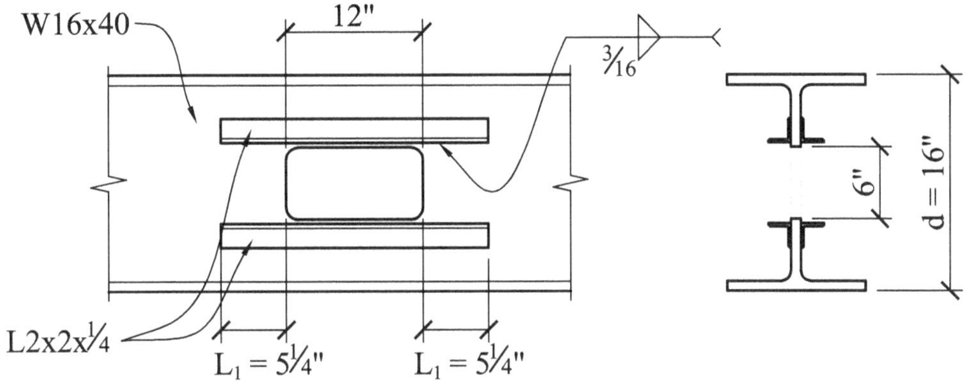

FIGURE 11-42 Detail for Example 11-14

Solution

From *AISCM* Table 1-1, the section properties for W16 × 40 are

$$d = 16.0 \text{ in.} \qquad t_w = 0.305 \text{ in.}$$
$$A = 11.8 \text{ in.}^2 \qquad I_x = 518 \text{ in.}^4$$
$$S_x = 64.7 \text{ in.}^3$$

The cross-sectional area of the web that has been removed is (6 in.)(0.305 in.) = 1.83 in.^{2}

Assuming that a pair of angles is provided on the top and bottom of the opening, the required area for each vertical leg of the angle reinforcement is

$$1.83 \text{ in.}^2/4 = 0.46 \text{ in.}^2$$

An $L2 \times 2 \times \frac{1}{4}$ angle is selected that has an area of (2 in.)(¼ in.) = 0.50 in.^{2} per vertical leg. This ensures that an equivalent amount of shear strength is provided to replace the portion of the web that has been removed. By inspection, the area provided is greater than the area removed, and the location of the angles is such that the composite section properties will be greater than the original section properties.

Required weld length within the opening:

Cross-sectional area of L2 × 2x¼ reinforcement = 0.938 in.^{2}

The area of the reinforcement for the web above or below the web opening is

$$A_r = \left(\frac{1}{2}\right)(0.938)(2\,\text{sides}) = 0.938\,\text{in.}^2$$

Note: Only the vertical leg of the angle reinforcement that lies on the same plane as the beam web with the opening is considered in calculating A_r, hence the factor, "½" , in the previous calculation for the area of the reinforcement.

The length of the web opening, a_o = 12 in.

$$P_r = F_y A_r \le \frac{F_y t_w a_o}{2\sqrt{3}} = (50)(0.938) \le \frac{(50)(0.305)(12)}{2\sqrt{3}}$$
$$= 46.9 \text{ kips } \le 52.8 \text{ kips}$$

Therefore, $P_r = 46.9$ kips

$$R_{wr} = \phi 2P_r = (0.9)(2)(46.9) = 84.4 \text{ kips.}$$

Assuming a $^{3}\!/_{16}$-in. weld (i.e., D = 3), the weld strength (given the length of the opening of 12 in.) is

$$1.392DL = (1.392)(3)(2\,\text{sides})(12\,\text{in.}) = 100 \text{ kips} > 84.4 \text{ kips. OK.}$$

Required extension length of reinforcement:

$$L_1 = \frac{a_o}{4} \ge \frac{A_r\sqrt{3}}{2t_w}$$
$$= \frac{12}{4} \ge \frac{0.938\sqrt{3}}{(2)(0.305)}$$
$$= 3 \text{ in.} \ge 2.66 \text{ in.}$$

Therefore, $L_1 = 3$ in.

Required weld strength within the length of the extension:

$$R_{wr} = \phi F_y A_r = (0.9)(50)(0.938) = 42.2 \text{ kips.}$$

Assuming a $^{3}\!/_{16}$-in. weld (i.e., D = 3), the weld strength (given the extension length L_1 = 3 in.) is

$$1.392DL = (1.392)(3)(2\,\text{sides})(3\,\text{in.}) = 25.1 \text{ kips} < 42.2 \text{ kips. Not good.}$$

The weld size could be increased, but a simpler solution would be to increase the extension length so that a $^{3}\!/_{16}$-in.-long weld could be used throughout. The minimum required length of the extension beyond the opening is:

$$L_1\,(\text{minimum}) = \frac{42.2 \text{ kips}}{(1.392)(3)(2 \text{ sides})} = 5.05; \quad \text{Therefore use } L_1 = 5.25 \text{ in. (see Figure 11-42)}$$

References

[1] American Institute of Steel Construction. 2023. *Steel Construction Manual*. 16th ed. Chicago, IL: AISC.

[2] Geschwinder, L. F., and R. O. Disque. 2005. "Flexible Moment Connections for Unbraced Frames—A Return to Simplicity." *Engineering Journal* 42, no. 2 (2nd quarter): 99–112.

[3] Carter, C. J., and R. O. Disque. 2005. "Flexible Moment Connections – A New Look at an Old Favorite." *Modern Steel Construction*, September.

[4] Blodgett, O. 1966. "Design of Welded Structures." Cleveland: The James F. Lincoln Arc Welding Foundation.

[5] American Institute of Steel Construction. 2022. *ANSI/AISC 358 - Prequalified Connections for Special and Intermediate Steel Moment Frames for Seismic Applications*. Chicago, IL: AISC.

[6] American Society of Civil Engineers. 2022. *Minimum Design Loads for Buildings and Other Structures (ASCE 7-22)*. Reston, VA: ASCE.

[7] Hamburger, R. O., and J. O. Malley. 2016. *Seismic Design of Special Moment Frames*. 2nd ed. NEHRP Seismic Design Technical Brief No. 2. Washington, DC: National Institute of Standards, U.S. Department of Commerce, June.

[8] Leon, R. T., J. J. Hoffman, and T. Staeger. 2003. *Partially Restrained Composite Connections*. AISC Design Guide 8. Chicago, IL: AISC.

[9] Darwin, D. 1990. *Steel Design Guide Series 2: Steel and Composite Beams with Web Openings*. Chicago, IL: AISC.

[10] Federal Emergency Management Agency. 2000. *FEMA-350: Recommended Seismic Design Criteria for New Steel Moment Frame Buildings*. Washington, DC: FEMA.

[11] American Welding Society. 2015. *ANSI/AWS D1.1 – Structural Welding Code*. Miami, FL: American Welding Society.

[12] Occupational Safety and Health Administration. 1998. *Part 1926: Safety and Health Regulations for Construction*. Washington, DC: OSHA.

[13] National Transportation Safety Board (NTSB). 2008. *Highway Accident Report—Interstate 35W Collapse Over the Mississippi River, Minneapolis, Minnesota. August 1, 2007.* NTSB/HAR—08/03. Washington, DC: NTSB, November 14.

[14] Holt, R., and J. Hartmann. 2008. *Adequacy of the U10 and L11 Gusset Plate Designs for Minnesota Bridge No. 9340 (I-35W over the Mississippi River)*. Interim Report. Washington, DC: U.S. Government Printing Office.

[15] Thornton, W. A., and C. Lim. 2011. "The Whitmore Section." *Modern Steel Construction*, July: 52–56.

[16] Astaneh-A. A. 1998. *Seismic Behavior and Design of Gusset Plates*. Moraga, CA: Structural Steel Educational Council.

[17] American Association of State Highway and Transportation Officials. 2004. *Standard Specification for Highway Bridges*. Washington, DC: AASHTO.

[18] Thornton, W. A. 2000. *Combined Stresses in Gusset Plates*. Roswell, GA: CIVES Engineering Corporation.

[19] Neal, B. G. 1977. *The Plastic Methods of Structural Analysis*. New York: Halsted Press, Wiley.

[20] Richard, R. M. 1986. "Analysis of Large Bracing Connection Designs for Heavy Construction." In *Proceedings of the AISC National Engineering Conference*, Nashville, TN: AISC.

[21] Thornton, W. A. 1991. "On the Analysis and Design of Bracing Connections." In *Proceedings of the AISC National Steel Construction Conference*, Washington, DC: AISC.

[22] Thornton, W. A. 1995. "Connections—Art, Science, and Information in the Quest for Economy and Safety." *AISC Engineering Journal* 32, no. 4 (4th quarter): 132–144.

[23] Muir, L. S. 2008. "Designing Compact Gussets with the Uniform Force Method." *AISC Engineering Journal* (1st quarter): 13–19.

[24] Carter, C. 1999. *Steel Design Guide Series 13: Wide-Flange Column Stiffening at Moment Connections*. Chicago, IL: AISC.

[25] Tamboli, A. R. 1999. *Handbook of Structural Steel Connection Design* and *Details*. New York: McGraw-Hill.

[26] Dowswell, B. 2018. "Design of Beam Copes." *Modern Steel Construction*, February.

[27a] Dowswell, B. 2016. "Stability of Rectangular Connection Elements." *Engineering Journal* (4th quarter): 171–201.

[27b] Dowswell, B. 2018. "Errata: Stability of Rectangular Connection Elements." *Engineering Journal* (2nd quarter): 133–142.

[28] Reynolds, K., and A. Font. 2023. "Shifting Gears." *Modern Steel Construction*, October: 28–35.

[29] Eatherton, M. R., and T. M. Murray. 2023. "The End-All, Be-All for End-Plate Moment Connections." *Modern Steel Construction*, December: 16–1.

[30] Eatherton, M. R., and T. M. Murray. 2023. *End-Plate Moment Connections – AISC Design Guide 39*. Chicago, IL: American Institute of Steel Construction.

[31] Savage, J. 2020. "Inside Job." *Modern Steel Construction*, March: 24–31.

[32] American Institute of Steel Construction. 2022. *Seismic Provisions for Structural Steel Buildings (ANSI/AISC 341-22)*. Chicago, IL: AISC.

Exercises

11-1. Determine whether the W14 × 34 beam shown in Figure 11-43 is adequate for flexural rupture and flexural yielding at the cope. The steel is ASTM A992, Grade 50.

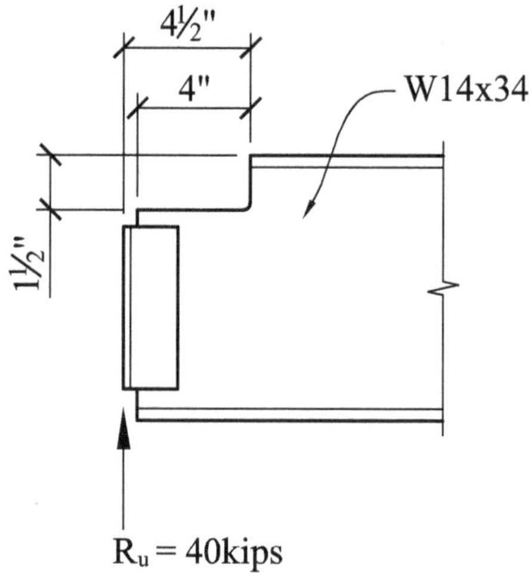

FIGURE 11-43 Detail for Exercise 11-1

11-2. Determine the maximum reaction, R_u, that could occur, considering flexural rupture and flexural yielding at the cope for the beam shown in Figure 11-44. The steel is ASTM A992, Grade 50. What is the required plate length for the transverse stiffener?

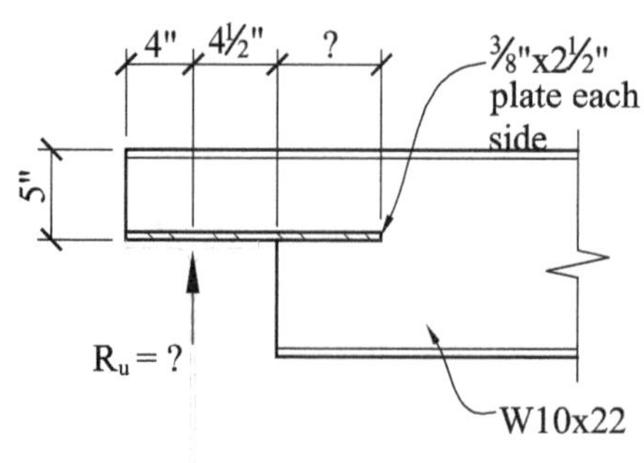

FIGURE 11-44 Detail for Exercise 11-2

11-3. Determine whether the W21 × 44 beam shown in Figure 11-45 is adequate for flexural rupture and flexural yielding at the cope. The steel is ASTM A36.

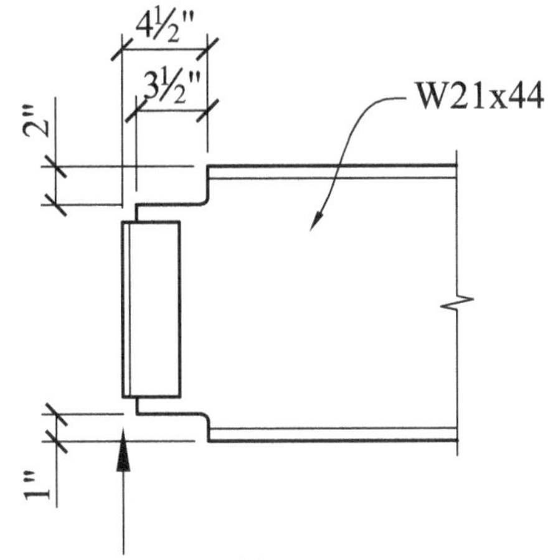

FIGURE 11-45 Detail for Exercise 11-3

11-4. Design the top and bottom plates for the moment connection shown in Figure 11-46, including the welds to the beam. The beam is ASTM A992 Grade 50 steel, and the welds are E70XX.

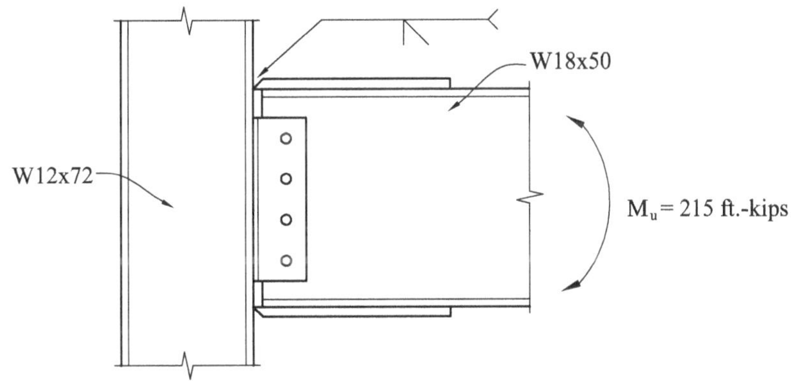

FIGURE 11-46 Detail for Exercises 11-4 and 11-5

11-5. Repeat Exercise 11-4, assuming that the plates are bolted to the beam flange. Use ¾-in.-diameter ASTM A490N bolts.

11-6. Figure 11-47 shows details for a column splice connection. The required strength in bending and shear across the splice is stated in terms of a percentage of the capacity of the column. All steel is ASTM A992, Grade 50. Determine the following:

a. The maximum moment that can be transferred across the splice is based only on the shear strength of the bolts. Does it meet the required moment?
b. The maximum moment that can be transferred across the splice is based only on the tensile rupture and tensile yielding strength of the plate. Does it meet the required moment?
c. Are the bolts in the end plate adequate to transfer the shear?

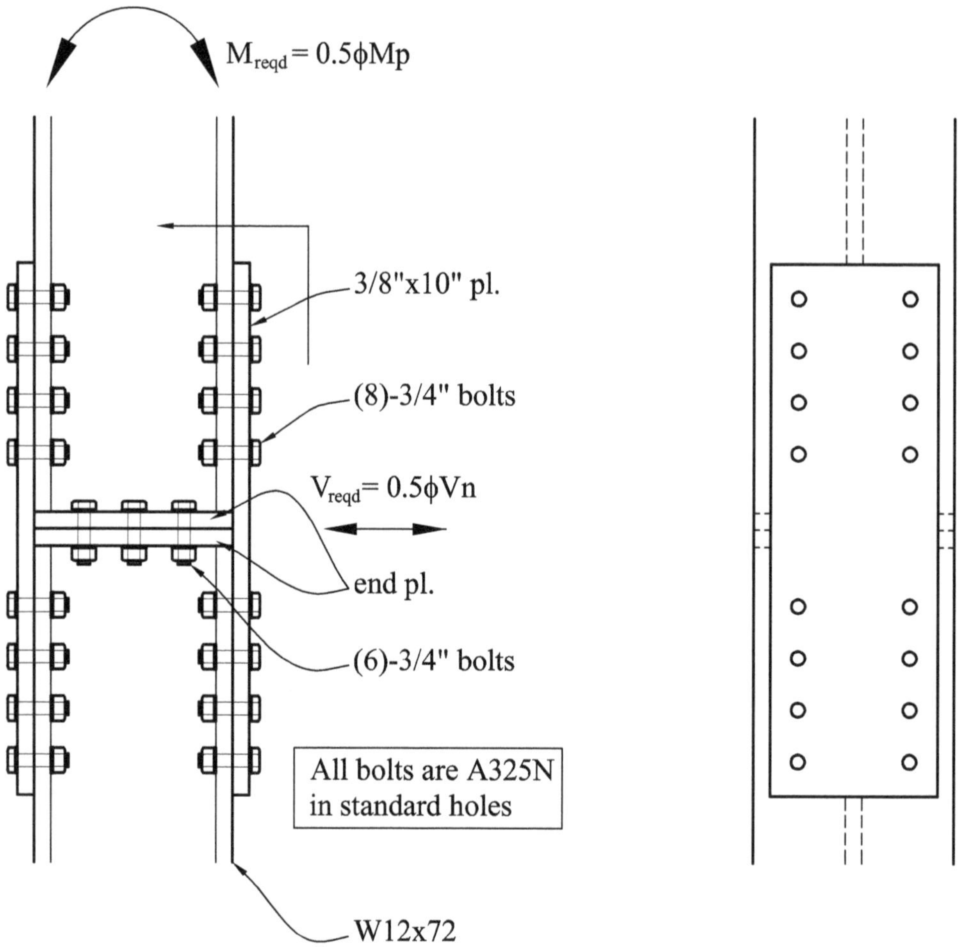

FIGURE 11-47 Details for Exercise 11-6

11-7. Determine the maximum reaction, R_u, that could occur in Figure 11-48. The details below show a W8 ×15 framing into another beam through a single plate shear connection. Bolts are ¾" in STD holes. All steel is ASTM A36.

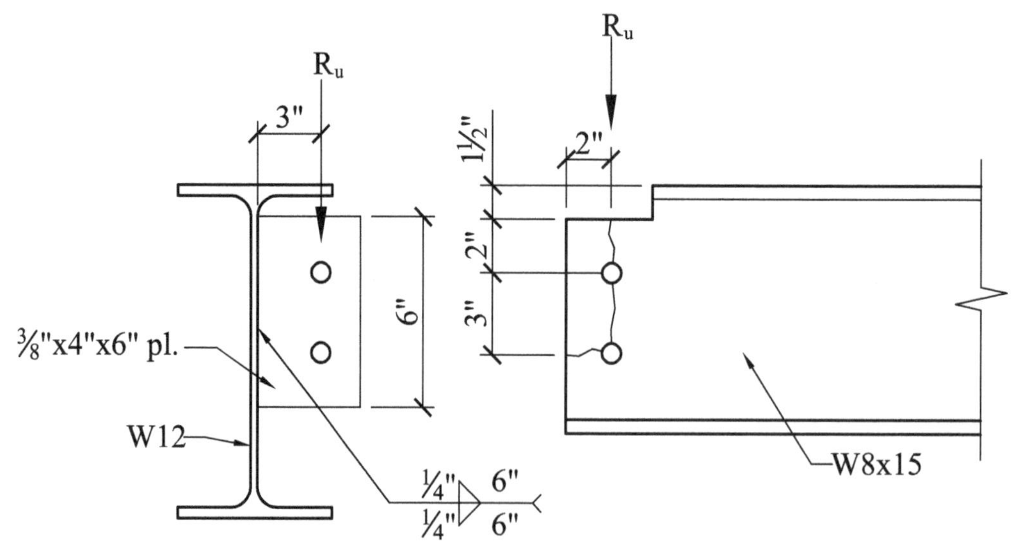

FIGURE 11-48 Details for Exercise 11-7

a. Find R_u based on shear on the bolts and bolt bearing. The bolts are ¾" A325 SC, and slip is a serviceability limit state. Faying surface is Class "C."
b. Find R_u based on block shear in the web of the W8 × 15.
c. Find R_u based on the weld strength of the single plate to the supporting beam. Use the *AISCM* design aids.
d. Based on the above, what is the maximum load (R_u) that could occur?

11-8. Design a beam splice for a W21 × 44 beam, assuming that the top and bottom plates and the shear plate are welded. The steel is ASTM A36, and the welds are E70XX. Assume that the required design bending strength is the full plastic moment capacity ($\phi_b M_p$) and that the required shear strength is one-third of the design shear strength ($\phi_v V_n$).

11-9. Determine whether the column in Exercise 11-4 is adequate for the concentrated forces due to the applied moment. Assume that the beam connection occurs at a location remote from the ends of the column.

11-10. For a W16 × 50 beam with the top and bottom flanges fully welded to the flanges of a W10 × 54 supporting column, determine the maximum end moment that could be applied without the use of stiffeners, taking into consideration the strength of the column in supporting concentrated forces. Assume that the beam connection occurs at a location remote from the ends. The steel is ASTM A992 Grade 50.

11-11. A diagonal brace in Chevron vertical bracing is connected to a W12 × 72 column and a W21 × 44 beam with a gusset plate in a connection similar to that shown in Figure 11-28. The factored brace force is 100 kips, and the diagonal brace is inclined 60° to the horizontal. Determine the dimensions of the gusset plate and the forces on the gusset column and gusset beam interfaces. Assume a ½-in. setback between the gusset plate and the face of the column.

11-12. A diagonal brace in Chevron vertical bracing is connected to a W12 × 72 column and a W21 × 44 beam with a gusset plate in a connection similar to that shown in Figure 11-29, but with a gusset horizontal length, *X* of 2 ft. 4 in., and a vertical length, *Y* of 1 ft. 2 in. The factored brace force is 90 kips, and the diagonal brace is inclined 35° to the horizontal. Determine the force distribution in the gusset plate connection interfaces, assuming that the gusset-to-beam connection is more rigid than the gusset-to-column connection. Assume a ½-in. setback between the gusset plate and the face of the column.

11-13. Repeat Exercise 11-12, assuming that the gusset-to-column connection is more rigid than the gusset-to-beam connection.

CHAPTER

12

Floor Vibrations due to Human Activity

12-1 INTRODUCTION

Steel-framed floors are generally designed to satisfy strength and deflection requirements, as discussed in previous chapters. However, deflection is not the only serviceability criterion that has to be satisfied. Floor vibrations due to human activity such as walking, dancing, running, aerobics, or jumping exercises are important serviceability criteria that should also be checked for steel-framed floors and other floor systems to ensure occupant comfort and proper functioning of sensitive equipment [1, 7, 8]. Floor vibration problems caused by resonance are difficult and expensive to fix after the fact; one famous example that illustrates this point is London's Millennium Bridge, a pedestrian bridge that cost several million dollars to retrofit with dampers after it developed vibration problems during its opening and had to be closed for two years for repairs. Long-span floor systems that require large column-free areas (e.g., members framing over exhibition halls, gymnasiums, auditoriums, ballrooms, or roadways) are particularly susceptible to floor vibration problems, even though the frequency of the forcing function of the source of vibration may be low, such as walking vibrations. Other vibration-sensitive structures include suspended running tracks and floor systems supporting aerobic or other rhythmic activities. The regular activity of human occupancy can be annoying to other occupants in the same space or in adjoining areas, and thus a vibration analysis should be part of the design process. Steel-framed floors have traditionally been designed to satisfy a live load deflection limit of $L/360$ (i.e., span/360), as well as a span-to-depth ratio, L/d, not to exceed 24. These rules of thumb are still in use today and are usually sufficient for shorter spans. Floors that are framed with longer span beams, girders, or trusses, or with open-web steel joists, are the most susceptible to vibration problems. In this chapter, we introduce the floor vibration analysis procedures presented in Ref. [1] to acquaint the reader with this important topic.

There are two basic types of floor vibrations: *steady-state* and *transient* vibrations. Steady-state vibrations are due to continuous harmonic dynamic forces, such as vibration due to equipment or rotating machinery. These vibrations are best controlled by isolating the equipment from the structure.

Transient vibrations are due to low-frequency and low-impact activities such as walking and dancing and high-impact activities such as aerobics, jumping, concerts, and athletics. This type of vibration decays or fades out eventually due to damping (i.e., friction or viscous forces). Transient vibrations are best controlled by relocating the activity, increasing the damping, increasing the mass of the structure, stiffening the structure, or a combination of these options. In this chapter, we will discuss the basic process for evaluating and mitigating transient vibrations in steel-framed floor systems. In general, the evaluation process involves the following steps:

1. Determine the natural frequencies and the various modes of vibration of the floor system.
2. Determine the characteristics of the vibration source, i.e., the frequency and force magnitude.
3. Calculate the maximum acceleration of the floor system caused by the vibration source. Alternatively, dynamic testing of the floor system can be carried out to determine the modal properties and the acceleration of the floor due to a specified dynamic force caused by walking, dancing, running, aerobics, or jumping exercises. Finite element analysis can also be used to model the dynamic behavior of the floor system.
4. Compare the calculated maximum or peak acceleration to the acceleration tolerance limits for the floor system; the acceleration tolerance limit is dependent on the floor damping available, the type of activity causing the vibration, as well as the type of human activity in the adjoining areas (e.g., in the case of a space used for aerobics adjacent to a floor area used for dining and dancing, the acceleration tolerance limits for dining and dancing will control).
5. Apply mitigation measures to the floor system if the acceleration tolerance limits are exceeded (see Section 12-10).
6. Re-evaluate the floor system with the mitigation measures in place, and re-calculate the maximum accelerations to ensure that the acceleration tolerance limits are not exceeded [13].

A detailed coverage of the evaluation and analysis of the vibrations of steel-framed floor systems can be found in Ref. [1].

12-2 VIBRATION TERMINOLOGY

In this section, we introduce and define some basic vibration terminology that will be referred to later in this chapter:

Damping ratio **(β)**: Damping in a structural system is usually expressed as a percentage of critical damp- ing. Critical damping is the amount of damping required to bring the system to rest in one-half cycle. See Table 12-2 for typical values of damping ratios.

Period **(T)**: Time, in seconds, for one complete cycle of oscillation.

Frequency **(f)**: Number of oscillations per second in hertz (Hz) or cycles per second, $f = 1/T$. A person walking at a pace of two steps per second is said to walk at a frequency of 2 Hz or 2 cycles per second.

Forcing Frequency **(f_f)**: The frequency, in hertz, of an applied dynamic force.

Harmonic: An integer multiple of a forcing frequency, f_f. Any forcing frequency can have an infinite number of harmonics, but human activities are generally limited to a maximum of three harmonics. For example, for an applied forcing frequency of 2 Hz, the first harmonic is 2 Hz, the second harmonic is 4 Hz, and the third harmonic is 6 Hz.

Natural Frequency **(f_n)**: The frequency of the first mode of vibration of a structure when it is displaced and then suddenly released from an at-rest position, without any applied external forces (i.e., free vibration). The natural frequency of a structure is proportional to its stiffness. A structure typically has modal frequencies that correspond to its modes of vibration. The first three modes of vibration of a structure are usually the most important, and the frequency of the first mode of vibration is also often referred to as the fundamental natural frequency, f_n. The natural frequency and mode shapes of a structure are independent of the dynamic force causing vibration in the structure.

Resonance: A phenomenon where the forcing frequency, f_f (or a harmonic multiple of the forcing frequency), of the dynamic activity coincides with one of the natural frequencies, f_n, of the structure. This causes very large displacements, velocity, acceleration, and stresses. For a person walking at a pace of 2 steps per second (i.e., $f_f = 2$ Hz), the floor will have to be checked for the first three harmonics of this forcing frequency (i.e., 2 Hz, 4 Hz, and 6 Hz).

Mode Shape: The deflected shape of a structural system that is subjected to free vibration. Each natural frequency of a structure has a corresponding deflected or mode shape.

Modal Analysis: Analytical method for calculating the natural frequencies, mode shapes, and responses of individual modes of a structure to a given dynamic force. A structure has multiple mode shapes, and the total structural system response to a dynamic force is a function of all the individual modal responses.

12-3 NATURAL FREQUENCY OF FLOOR SYSTEMS

The most critical parameter in the vibration analysis of a floor system is its natural frequency. There are several factors that affect the natural frequency of a floor system, such as the floor stiffness and its mass. The natural frequency of a complex floor system can be determined from a finite element analysis, but for regular floor systems, the simplified approach presented in Ref. [1] will be used in this text. The floor system is assumed to be a concrete slab with or without a metal deck supported by steel beams or joists that are supported by some combination of steel girders, walls, or columns. Even non-composite floor systems with a metal deck and a concrete slab (without shear studs) are assumed to have full composite action under vibration loads from human activity; this is because the interface shear forces between the steel beam and the steel deck/concrete slab caused by the vibration or dynamic loading are low enough so that the shear capacity of the puddle welds connecting the metal deck to the steel beams is sufficient to transfer the horizontal interface shear [5]. Therefore, the section properties for the floor members are calculated assuming full composite action between the beams and slab even when noncomposite action is assumed for ultimate strength design.

The natural frequency of a simply supported, uniformly loaded beam or joist can be estimated as follows [1]:

$$f_n = \frac{\pi}{2}\sqrt{\frac{gEI}{wL^4}}, \tag{12-1}$$

where

f_n = Natural frequency, Hz,

g = Acceleration due to gravity (386 in./s^2),

E = Modulus of elasticity of steel (29×10^6 psi),

I = Moment of inertia, in.^{4},

= Transformed moment of inertia, I_t, for composite floors,

w = Uniformly distributed load, lb./in. (see discussion below), and

L = Beam, joist, or girder span, in.

Equation (12-1) can be simplified in terms of the maximum deflection as follows:

$$f_n = 0.18\sqrt{\frac{g}{\Delta}},\ \text{Hz} \tag{12-2}$$

where

Δ = Maximum deflection at the midspan of the flexural member

$= \dfrac{5wL^4}{384EI}$, in.

For a combined beam (or joist or hollow core slab) and girder system, the natural frequency of the floor system can be estimated as follows [1, 7, 8]:

$$\frac{1}{f_n^2} = \frac{1}{f_j^2} + \frac{1}{f_g^2}, \tag{12-3}$$

where

f_j = Frequency of the beam or joist, and

f_g = Frequency of the girder.

Since the beams (or joist or hollow core slab) in a floor bay or panel will be supported on two girders (one on each end), use the properties of the girder with the smaller natural frequency (i.e., the girder with the smaller stiffness).

Combining equations (12-2) and (12-3) yields the natural frequency for the combined floor system:

$$f_n = 0.18\sqrt{\frac{g}{\Delta_j + \Delta_g}}, \text{ Hz} \tag{12-4}$$

Where,

Δ_j = Beam or joist maximum deflection

Δ_g = Girder maximum deflection (in a floor panel, use the girder with the larger maximum deflection)

The effect of column axial deformation must be considered for taller buildings with rhythmic activities since the axial deformation of a column increases as the length increases. For this case, the natural frequency of the floor system becomes

$$f_n = 0.18\sqrt{\frac{g}{\Delta_j + \Delta_g + \Delta_c}} \tag{12-5}$$

where

Δ_c = Column axial deformation

= PL/EA,

P = axial load on the column,

L = total length of the column below the floor level under consideration,

A = cross-sectional area of the column,

E = modulus of elasticity of the column.

The uniform load, w, in the natural frequency equations above represents the actual dead load plus the sustained live load (Ib/in.) supported by the beam or girder. Table 12-1 lists recommended values for the sustained live load, but the actual value should be used if it is known. The above equations are also based on the assumption that the beams and girders are uniformly loaded, which is typically the case. The exception to this assumption are the girders that support a single concentrated load from a beam or joist at midspan, in which case the deflection should be increased by a factor of $4/\pi$.

In practice, it is recommended that the vertical natural frequency of floor systems in buildings be greater than or equal to 3 Hz to avoid significant floor vibration problems caused by resonance because the floor natural frequency matches the first harmonic frequency of walking vibrations (1.6 Hz to 2.2 Hz). For pedestrian bridges, the recommended minimum *lateral* natural frequency is 1.3 Hz [1].

TABLE 12-1 Recommended sustained live load values

Occupancy	Sustained live load
Paper office Electronic office	11 psf 6-8 psf
Residential floors Assembly area	6 psf 0 psf
Footbridges, gymnasiums, and shopping center floors	0 psf

^Adapted from Table 3-1 of Ref. [1].

EXAMPLE 12-1

Frequency of a Floor System

For the floor system shown in Figure 12-1, calculate the natural frequency of the floor system. The dead load of the floor is 40 psf. Assume there are no sustained live loads. The moment of inertia of the beam is $I_b = 371\text{ in.}^4$, and the moment of inertia of the girder is $I_g = 500\text{ in.}^4$. Neglect the column axial deformation.

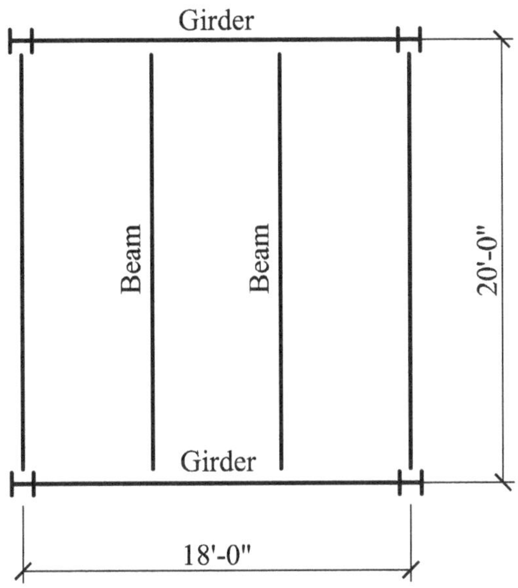

FIGURE 12-1 Frequency of a floor system

Solution

The free-body diagram of the beam is shown in Figure 12-2.

The dead load plus sustained live load on the beam with a tributary width of 6 ft. is

$w_j = (40\text{ psf} + 0\text{psf})(6\text{ ft.}) = 240\text{ lb./ft.} = 20\text{ lb./in.}$

The beam reaction is

$$R = (240\text{ lb./ft.})\left(\frac{20\text{ ft.}}{2}\right) = 2400\text{ lb.}$$

The beam maximum deflection is

$$\Delta_j = \frac{5w_jL^4}{384EI}$$

$$= \frac{(5)(20)(240)^4}{(384)(29\times10^6)(371)} = 0.0803\text{ in.}$$

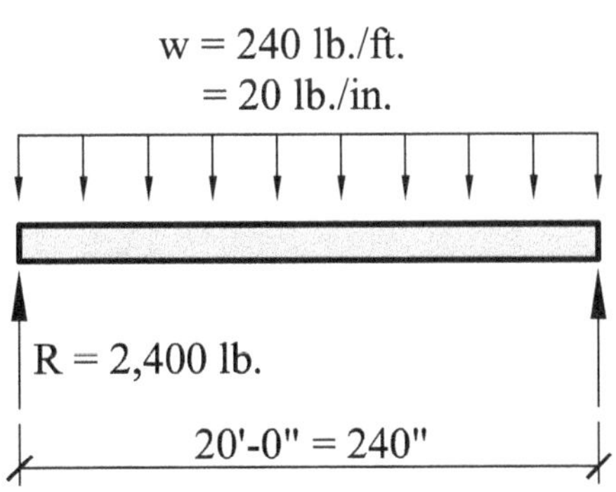

FIGURE 12-2 Free-body diagram of the beam

The natural frequency of the beam panel mode is

$$f_j = 0.18\sqrt{\frac{g}{\Delta_j}} \qquad \text{where, g = 386 in./s}^2$$

$$= 0.18\sqrt{\frac{386}{0.0803}} = 12.48 \text{ Hz.}$$

Using the beam reactions calculated previously, the free-body diagram of the girder is as shown in Figure 12-3. From *AISCM* Table 3-22 [3], the maximum deflection in the girder due to the concentrated reactions from the beams under dead load plus sustained live load is

$$\Delta_g = \frac{Pa}{24EI}\left(3L^2 - 4a^2\right)$$

$$= \frac{(2400)(72)}{(24)(29\times10^6)(500)}\left[(3)(216)^2 - (4)(72)^2\right] = 0.0592 \text{ in.}$$

The natural frequency of the girder panel mode is

$$f_g = 0.18\sqrt{\frac{g}{\Delta_g}}$$

$$= 0.18\sqrt{\frac{386}{0.0592}} = 14.53 \text{ Hz.}$$

The natural frequency of the combined system [1] is

$$\frac{1}{f_n^2} = \frac{1}{f_j^2} + \frac{1}{f_g^2}$$

$$= \frac{1}{(12.48)^2} + \frac{1}{(14.53)^2}$$

$$f_n = 9.47 \text{ Hz.}$$

Alternatively, the natural frequency of the combined system can be calculated as follows:

$$f_n = 0.18\sqrt{\frac{g}{\Delta_j + \Delta_g}} = 0.18\sqrt{\frac{386}{0.0803 + 0.0592}} = 9.47 \text{ Hz.}$$

The preceding calculations illustrate the application of the natural frequency equations to an isolated floor, with the beams and girders considered as simply supported members. The above calculations assumed that the weight of the steel framing (i.e., the beams and the girders) was already included in the given uniform dead load of 40 psf.

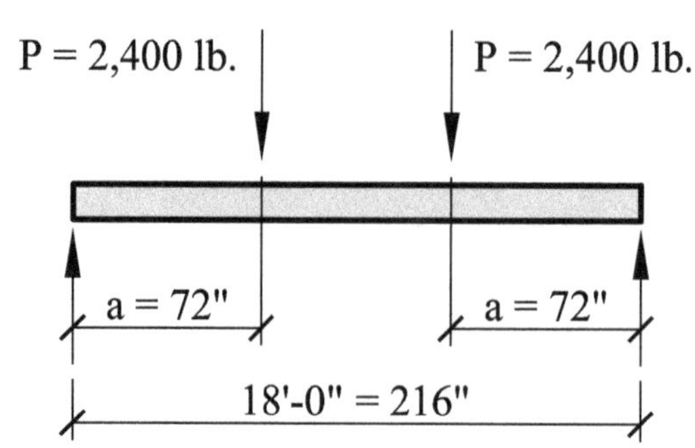

FIGURE 12-3 Free-body diagram of the girder

12-4 FLOOR SYSTEMS WITH OPEN-WEB STEEL JOISTS

For wide-flange beams with solid webs, the shear deformation is usually small enough to be neglected. For open-web steel joists, shear deformation in the chords occurs due to the eccentricity at the joints, which occurs as a result of the fabrication process (see Figure 12-4). For this reason, shear deformation must be considered for open-web steel joists [6] when designing for serviceability [6].

For the purposes of vibration analysis, the effective moment of inertia of a simply supported joist is calculated as follows [1]:

$$I_{jeff} = \frac{I_{comp}}{1 + \frac{0.15 I_{comp}}{I_{chords}}}, \tag{12-6}$$

where

I_{jeff} = Effective moment of inertia of the joist, in.^{4} to account for shear deformation

I_{comp} = Composite moment of inertia of the joist, in.^{4}

I_{chords} = Moment of inertia of the joist chords, in.^{4} (see Ref. [6]).

Equation (12-6) is valid for joist span-to-depth ratios greater than or equal to 12.

For the case when open-web steel joists are supported by wide-flange girders or joist girders, the girders do not act as a fully composite section with the concrete slab because the slabs are not directly supported by the girders but on the joist shoes or seats, so the girders are physically separated from the floor slab by the joist seats (see Figure 12-5).

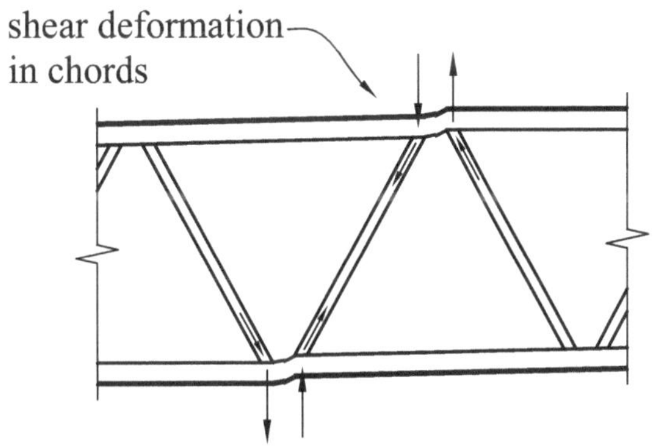

FIGURE 12-4 Shear deformation of a bar joist

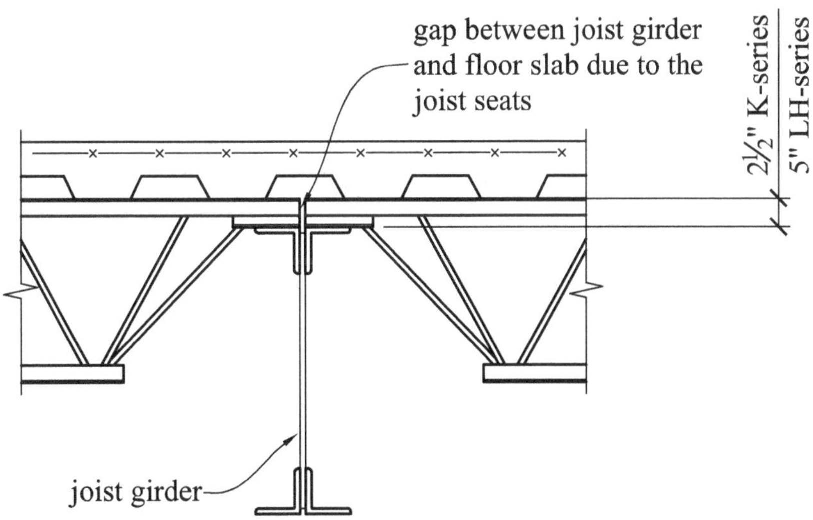

FIGURE 12-5 Floor section at a joist girder

For the purposes of vibration analysis, the effective moment of inertia of hot-rolled or built-up girders supporting open-web steel joists is calculated as follows:

$$I_{geff} = I_{nc} + \frac{(I_c - I_{nc})}{4}, \tag{12-7}$$

where

I_{geff} = Effective moment of inertia of the girder, in.^{4} to account for shear deformation,

I_{nc} = Noncomposite moment of inertia of the girder, in.^{4},

I_c = Composite moment of inertia of the girder in combination with the concrete slab, in.^{4}.

To account for shear deformations and the eccentricity at the joist panel points which affects the stiffness of the joist but not its strength [1], the effective moment of inertia of the joists is modified as follows for this case:

$$I_{jeff} = \frac{1}{\dfrac{\gamma}{I_{chords}} + \dfrac{1}{I_{comp}}}, \tag{12-8}$$

where

I_{jeff} = Effective moment of inertia of the joist, in.^{4},

I_{comp} = Composite moment of inertia of the joist, in.^{4},

I_{chords} = Moment of inertia of the joist chords in combination with the concrete slab, in.^{4}

$$\gamma = \frac{1}{C_r} - 1, \tag{12-9}$$

where C_r is a parameter that is determined as follows [1]:

For open-web steel joists with single- or double-angle web members,

$$C_r = 0.9\left(1 - e^{-0.28\left(\frac{L_j}{D}\right)}\right)^{2.8} \leq 0.9 \qquad \text{for } 6 \leq \frac{L_j}{D} \leq 24 \tag{12-10}$$

For open-web steel joists with round web members,

$$C_r = 0.721 + 0.00725\left(\frac{L_j}{D}\right) \leq 0.9 \quad \text{for } 10 \leq \frac{L_j}{D} \leq 24 \tag{12-11}$$

Where,

D = the nominal depth of the joist or joist girder, in.

L_j = span of the joist or joist girder, in.

12-5 EVALUATION OF FLOOR SYSTEMS FOR WALKING VIBRATIONS

Most common floor systems support live loads where the dynamic force is caused by walking. This motion could be annoying to other stationary occupants if the structure in question does not have adequate stiffness or damping. The perception of floor vibrations can be subjective and is a function of the individual sensitivities of the building occupants. Some common occupancy types that should be evaluated for walking vibrations include offices, residential buildings, retail spaces, pedestrian bridges, and hospitals. The floor systems in hospitals typically support sensitive equipment—such as MRI, CAT scan, and X-ray equipment—that have more stringent vibration criteria, and this topic is covered in Ref. [1]. The first harmonic frequency for walking vibrations is 1.6 to 2.2 Hz, and the maximum or 4th harmonic frequency for walking vibrations is 8.8 Hz [1].

The AISC Design Guide 11 [1] divides the evaluation of floor systems for walking vibrations into two categories depending on the combined natural frequency of the floor system:

low frequency floors (LFF) that have a natural frequency less than or equal to 9 Hz, and high frequency floors (HFF) that have a combined natural frequency greater than 9 Hz. High frequency floor (HFF) systems are not susceptible to resonance due to walking vibrations since their natural frequency is higher than the maximum harmonic frequency for walking vibrations which is 8.8 Hz.

The following evaluation procedure for walking vibrations is for low-frequency floor (LFF) systems. To satisfy the walking vibration criteria for a low-frequency floor system, the peak floor acceleration, a_p/g, due to walking vibrations should not exceed a specified limit, a_o/g, given in Table 12-2 [1]. The peak acceleration, a_p/g, due to walking vibration is calculated as

$$\frac{a_p}{g} = \frac{P_o e^{(-0.35 f_n)}}{\beta W} \leq \frac{a_o}{g} \tag{12-12}$$

Where,

$\frac{a_p}{g}$ = Peak floor acceleration as a fraction of gravity,

$\frac{a_o}{g}$ = Human acceleration tolerance limit as a fraction of gravity (see Table 12-2),

P_o = Constant force (see Table 12-2),

f_n = Natural frequency of the floor system (see equation (12-5)),

β = Modal damping ratio (varies between 0.01 and 0.05; see Table 12-2),

W = Weighted average mass of the floor system (see equation (12-13)).

Note that P_o is a constant force obtained by multiplying the body weight causing the walking vibration by the product of a dynamic load factor and a reduction factor that accounts for one-way or two-way mode shapes [1]. For vibration analysis of a framed or suspended track, the peak acceleration for running activity can be calculated using a modified form of equation (12-12), where P_o. varies from 133 Ib for recreational runners to 198 Ib for athletes (assuming the weight of an athlete is 250 Ibs), and the exponent in equation (12-12) changes from $-0.35f_n$ to $-0.173f_n$ [1].

TABLE 12-2 Recommended values of the dynamic parameters for walking vibrations

Occupancy	**Constant force, P_o**	**Damping ratio, β**	**Acceleration tolerance limit, $\frac{a_o}{g}$**
Offices, residences, and churches and quiet areas	65 lb.	0.02–0.05*	0.5%
Shopping malls	65 lb.	0.02	1.5%
Footbridges—indoor ^	92 lb. ^ ^	0.01**	1.5%
Footbridges—outdoor ^	92 lb. ^ ^	0.01**	5.0%

Source: Adapted from Tables 4-1, 4-2, and 4-4 of Ref. [1].

*β = 0.025 for an electronic office, including framing, ceiling, ductwork, and office fitout
= 0.03 for a paper office, including framing, ceiling, ductwork, and office fitout
= 0.02 for churches, schools, and malls
= 0.05 to 0.08 for a paper office with full-height partitions, including framing, ceiling, ductwork, and paper office fitout (i.e., 0.03 for paper office + 0.02 to 0.05 for partitions)

^ It is recommended that the lateral natural frequency of pedestrian bridges be ≥ 1.3 Hz.

^ ^ This force is for a single walker; for a group of random walkers, $P_o = \sqrt{n}\ (92\ kips)$; for a marching group of walkers, $P_o = n\ (92\ kips)$, where n is the number of walkers

**Assumes bare structural framing

The effective weight of the floor system, or the weighted average mass of the floor system, is a function of the weight of the beam or joist mode panel combined with the weight of the girder mode panel as follows:

$$W = \left(\frac{\Delta_j}{\Delta_j + \Delta_g}\right)W_j + \left(\frac{\Delta_g}{\Delta_j + \Delta_g}\right)W_g, \tag{12-13}$$

where

Δ_j = Maximum deflection of the beam or joist

$= \dfrac{5w_j L_j^4}{384EI_j}$, and

Δ_g = Maximum deflection of the girder

$= \dfrac{5w_g L_g^4}{384EI_g}$

Where,

$w_j = w_t \times$ tributary width of the beam or joist, Ib/ft.

$w_g = w_t \times$ tributary width of the girder (*if the girder self-weight is included in* w_t), Ib/ft.

$= w_t \times$ tributary width of girder + self-weight of the girder (*if the girder self-weight is not included in* w_t), Ib/ft.

w_t = dead load (includes weight of steel framing) plus sustained live load per unit area, psf.

Note that the sustained live load (see Table 12-1), which is the portion of the occupancy live load that remains permanently on the floor, should not be confused with the occupancy live load that is used in the ultimate strength design of the floor system.

Beam Panel:

The beam panel is an effective area of the floor system that is influenced by the vertical vibration of the floor beams; it is equal to the product of the joist or beam span, L_j, and the *effective floor width*, B_j, which is a function of the structural properties of the floor system but with a maximum value of two-thirds of the width of the entire floor perpendicular to the beams or joists. The weighted average mass of the beam panel, W_b (or W_j for a joist panel), is calculated as follows [1]:

$$W_j = \gamma w_t B_j L_j, \text{ lb.} \tag{12-14}$$

Where,

γ = 1.0 for joists or beams, except

= 1.5 for rolled steel beams shear-connected to the supports and with the adjacent beam span greater than 70% of the beam span under consideration

= 1.3 for joists with extended bottom chords

w_t = Dead load (includes weight of steel framing) plus sustained live load per unit area, psf,

L_j = Length or span of the beam or joist,

B_j = Effective width of the joist panel

$$= C_j\left(\frac{D_s}{D_j}\right)^{0.25} L_j \le \left(\frac{2}{3}\right)\text{floor width parallel to the girders (see Figure 12-7)}, \tag{12-15}$$

C_j = 2.0 for most beams and joists

= 1.0 for beams or joists at and parallel to a free edge (because of the reduced effective mass)

D_s = Transformed moment of inertia of the slab per unit width

$$= \frac{12d_e^3}{12n} = \frac{d_e^3}{n}, \text{ in.}^4/\text{ft.} \tag{12-16}$$

d_e = Average depth of the concrete slab on the metal deck (see Figure 12-6)

$$= t_c + \frac{h_r}{2}, \qquad (12\text{-}17)$$

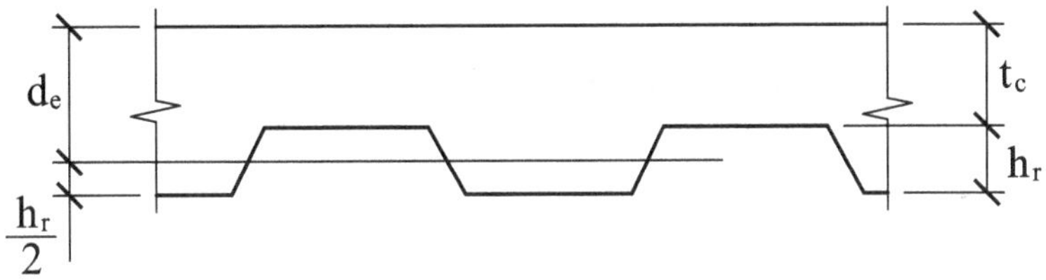

FIGURE 12-6 Floor slab section

Where,

t_c = Concrete slab thickness above the deck ribs,

n = Dynamic modular ratio,

$$= \frac{E_s}{1.35E_c}, \qquad (12\text{-}18)$$

E_s = Modulus of elasticity of steel

= 29,000,000 psi,

E_c = Modulus of elasticity of the concrete from Ref. [4]

$$= 33w_c^{1.5}\sqrt{f'_c}, \text{ psi}, \qquad (12\text{-}19)$$

w_c = Unit weight of concrete, lb./ft.^3,

f'_c = 28-day compressive strength of the concrete, psi,

D_j = Transformed moment of inertia of the beam or joist per unit width

$$= \frac{I_j}{S}, \text{ in.}^4/\text{ft.}, \qquad (12\text{-}20)$$

I_j = Transformed moment of inertia of the beam or joist,

S = Beam or joist spacing, ft.

Girder Panel:

The girder panel is an effective area of the floor system that is influenced by the vertical vibration of the floor girders; it is equal to the product of the girder span, L_g, and the *effective width of the girder panel*, B_g, where B_g is a function of the structural properties of the floor system but with a maximum value equal to two-thirds of the length of the entire floor perpendicular to the girders. The weighted average mass of the girder panel, W_g, is calculated as follows [1]:

$$W_g = w_t B_g L_g + \text{self-weight of girder*} \qquad (12\text{-}21)$$

*This is equal to zero if the self-weight of the girder is included in w_t.

Where,

B_g = Effective width of the girder panel

$$= C_g L_g \left(\frac{D_j}{D_g}\right)^{0.25} \le \left(\frac{2}{3}\right) \text{floor length parallel to the beams (see Figure 12-7)}, \qquad (12\text{-}22)$$

C_g = 1.8 for girders supporting rolled steel beams connected to the girder web

= 1.6 for girders supporting joists with joist seats on the girder flange.

D_g = Transformed moment of inertia of girder per unit width

$$= \frac{I_g}{L_j} \text{ for all but edge girders} \qquad (12\text{-}23)$$

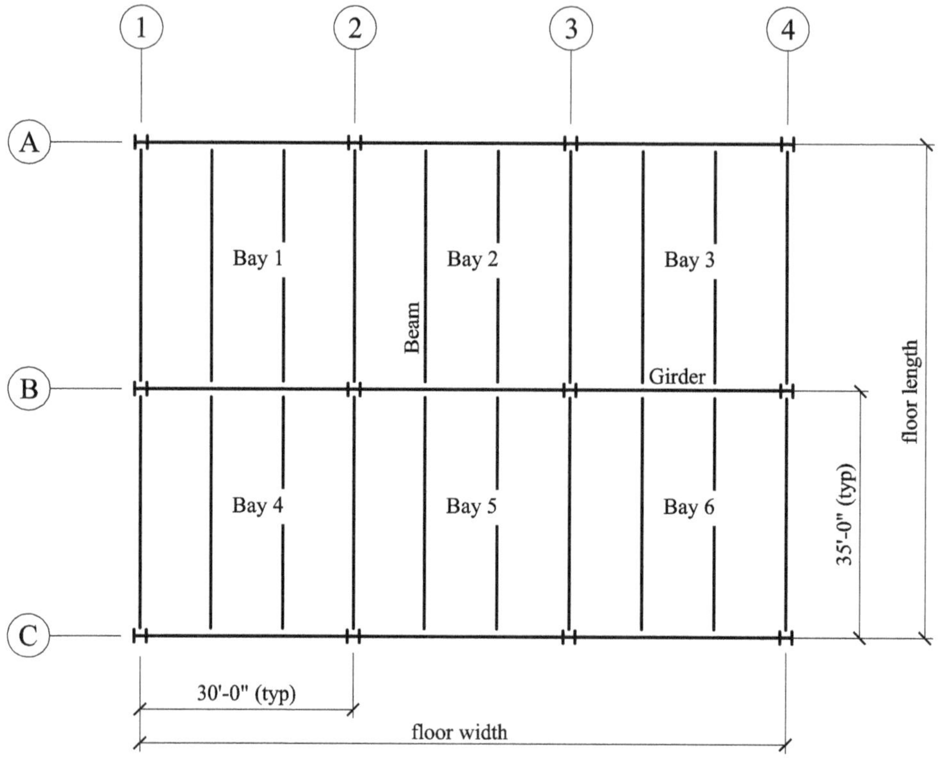

FIGURE 12-7 Floor width and floor length

$$= \frac{2I_g}{L_j} \text{ for edge girders} \tag{12-24}$$

L_g = Girder span

Note that the floor length (see Figure 12-7) is parallel to the beams or joists, and the floor width is parallel to the girders or perpendicular to the beams (or joists). The floor length and floor width will vary depending on the floor bay under consideration and will also be affected by any floor openings. As an example, assume that Bay 1 (see Figure 12-7) was completely open, then the floor length for Bays 2, 3, 5, and 6 will be 70 ft., while the floor length for Bay 4 will be 35 ft. On the other hand, the floor width for Bays 2 and 3 will be 60 ft., while the floor width for Bays 4, 5, and 6 will be 90 ft.

With reference to Figure 12-7, the plan aspect ratio of the floor needs to be checked. If the girder span is more than twice the beam or joist span (i.e., $L_g > 2L_j$), then the beam or joist panel mode, as well as the combined mode, should be checked separately. It should be noted that this is not likely to be the case in most structures since it is generally more economical to have beam or joist spans that are longer than the girder span in a typical steel-framed floor system.

When the girder span is less than the beam or joist panel width, $L_g < B_j$, the combined mode exhibits greater stiffness. To account for this, the girder deflection, Δ_g, used in equation (12-13) is modified as follows:

$$\Delta_g' = \frac{L_g}{B_j}\Delta_g \geq 0.5\Delta_g \tag{12-25}$$

The damping ratio, β, in a floor system is a function of the number of nonstructural components present, such as walls, furniture, and occupants (for rhythmic vibrations, see Section 12-6). Table 12-2 lists the recommended damping values for different types of building occupancy [1]. The floor vibration evaluation procedures presented thus far pertain to floor systems with vertical natural frequency less than or equal to 9 Hz, i.e., low-frequency floor (LFF) systems.

Walking Vibrations in High Frequency Floor (HFF) Systems (i.e., f_n > 9 Hz):

As noted previously, high frequency floor systems or HFF (i.e., floor systems with vertical natural frequency greater than 9 Hz) rarely exhibit problems with walking vibrations. For high frequency floors, a semi-empirical equation for the peak acceleration from Ref. [1] is given as follows:

$$\frac{a_p}{g} = \frac{154}{W}\left(\frac{f_{step}^{1.43}}{f_n^{0.3}}\right)\sqrt{\frac{1-e^{-4\pi h\beta}}{h\pi\beta}} \leq \frac{a_o}{g} \tag{12-26}$$

Where,

W = effective weight of the floor system (equation (12-13))

β = the damping ratio.

f_{step} = step frequency for walking from Table 12-3

h = the harmonic that corresponds to the natural frequency, f_n, from Table 12-4

Table 12-3 Forcing Frequencies for Walking Vibration and the Corresponding Harmonics

Harmonic, i	*Step Frequency if$_{step}$,* (Hz)
1	1.6 – 2.2
2	3.2 – 4.4
3	4.8 – 6.6
4	6.4 – 8.8

Adopted from Table 2-1 of Ref. [1].

Table 12-4. Matching harmonics corresponding to the natural frequency for f_n > 9 Hz.

Natural frequency, f_n (Hz)	Matching harmonic, *h*
9–11	5
11–13.2	6
13.2–15.4	7

Adopted from Table 2-2 of Ref. [1].

12-6 ANALYSIS PROCEDURE FOR WALKING VIBRATIONS

For a typical floor bay, use the interior beams and the girder with the lower natural frequency at both ends of the beam in the vibration analysis. Spandrel beams and girders tend to be stiffer than the typical interior flexural members. The spandrel girders can be restrained by the exterior cladding that they support, therefore, it is conservative to use the interior beam and interior girder in the vibration analysis because these members are more flexible and therefore have lower natural frequencies. The analysis procedure for evaluating walking vibrations for floor systems is outlined as follows:

1. Determine the effective slab width, b_{eff}, and use this to calculate the composite moment of inertia of the joists/beams and girders. Use the composite moment of inertia unless the upper flange of the beam, joist, or girder is separated from the concrete slab or where the deck frames into the web of the beam or girder (instead of the flange), in which case the noncomposite moment of inertia of the beam or girder is used. (see Example 12-2).

For a typical interior flexural member, the effective width is

$b_{eff} \le 0.4 \times \text{length of the flexural member}$

$\le$ Tributary width of the flexural member.

For a spandrel flexural member, the effective width is

$b_{eff} \le 0.2 \times \text{length of the member} + \text{the edge distance}$

$\le$ Tributary width of the flexural member

Note: For footbridges that are supported by beams only, with no girders, calculate only the beam properties and the beam panel mode.

2. Calculate the dead loads and sustained live loads supported by the joist, or beam, and the girder.
3. Calculate the joist or beam and girder deflections and the natural frequency, f_n, of the floor system using equation (12-5). For walking vibrations, ensure that the natural frequency of the floor system is greater than 3 Hz (i.e., $f_n > 3Hz$) to avoid resonance with the first harmonic of the forcing frequency for walking vibrations; otherwise, stiffen the joist or beams and girders.
4. Calculate the effective weight, W, of the floor system using equation (12-13).
5. Obtain the occupancy-dependent constant force, P_o, and the damping ratio, β, from Table 12-2. Use equation (12-12) to calculate the peak acceleration of the floor system due to walking vibration, a_p/g, for low frequency floor (LFF) systems (i.e., $f_n \le 9$ Hz).
6. For high frequency floor (HFF) systems (i.e., $f_n > 9$ Hz), use equation (12-26) to calculate the peak acceleration, a_p/g, of the floor system due to walking vibrations.
7. Ensure that the peak acceleration of the floor system, a_p/g, is less than or equal to the human acceleration limit, a_0/g, given in Table 12-2; otherwise, increase the floor stiffness and/or the floor dead load, and/or increase the floor damping (see Section 12-10 for floor vibration control measures).

EXAMPLE 12-2

Walking Vibrations with Wide-Flange Beam Framing

The floor framing system shown in Figure 12-8 is to be used in an office building (paper office). Check if the floor is adequate for walking vibrations assuming a walking frequency of 1.6 Hz. Assume normal-weight concrete with a density of 145 pcf and a 28-day compressive strength of 4000 psi. The floor loads are as follows:

Floor dead load = 55 psf (includes weight of slab, deck, finishes, mechanical and electrical fixtures, and partitions).

Sustained live load = 11 psf (see Table 12-1).

Solution

Beam Section Properties and Deflection

From part 1 of the *AISCM*, the section properties for W16 × 26 are as follows [3]:

$d = 15.7$ in.

$A = 7.68$ in.^{2}

$I = 301$ in.^{4}

Effective slab width, $b_{eff} \le 0.4 \times \text{beam span} = (0.4)(24 \text{ ft.}) = 9.6$ ft.

$\le$ Beam tributary width = 6 ft. = **72 in.** → controls

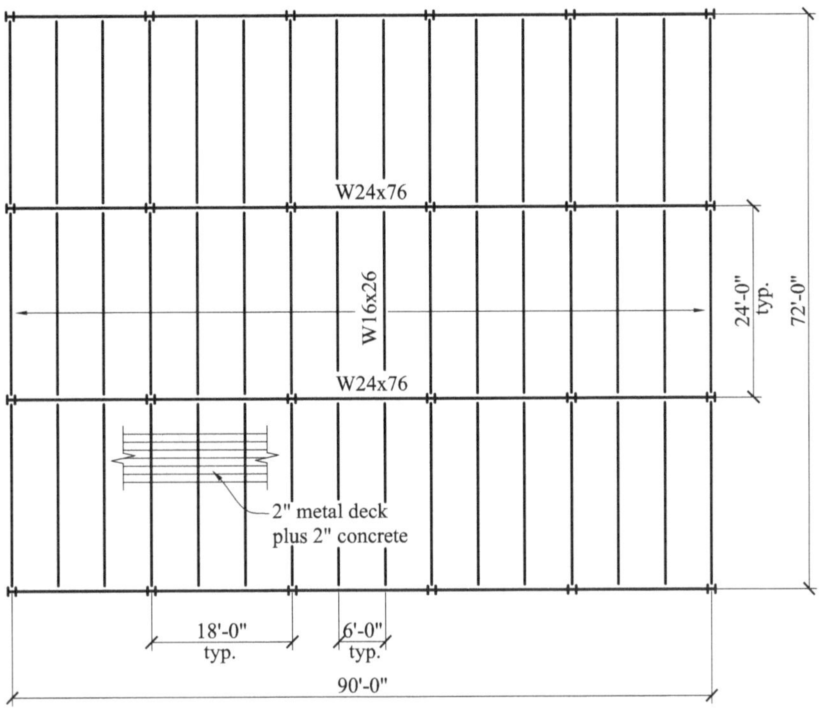

FIGURE 12-8 Floor framing plan for Example 12-2

t_c = Concrete slab thickness above deck ribs = 2 in.
h_r = Depth of metal deck ribs = 2 in.
Note: Only concrete above the deck ribs is used to compute the I_{comp} for the beam.
$E_s = 29{,}000$ ksi

$$E_c = 33 w_c^{1.5} \sqrt{f'_c}\text{, psi}$$

$$= (33)(145)^{1.5}\sqrt{4000} = 3{,}644{,}147\text{ psi} = 3644\text{ ksi,}$$

$$n = \frac{E_s}{1.35E_c}$$

$$= \frac{29{,}000}{(1.35)(3644)} = 5.9$$

b_e = Transformed effective width of the concrete slab

$$= \frac{b_{eff}}{n}$$

$$= \frac{72\text{ in.}}{5.9} = 12.2\text{ in.}$$

The centroid of the composite beam (see Figure 12-9) is found by summing the moments of areas about the top of the slab. Therefore,

$$\bar{y} = \frac{\Sigma Ay}{\Sigma A}$$

$$= \frac{\left[(2\text{ in.})(12.2\text{ in.})\left(\frac{2\text{ in.}}{2}\right)\right] + \left[(7.68\text{ in.}^2)\left(\frac{15.7\text{ in.}}{2} + 4\text{ in.}\right)\right]}{[(2\text{ in.})](12.2\text{ in.}) + 7.68\text{ in.}^2} = 3.6\text{ in.}$$

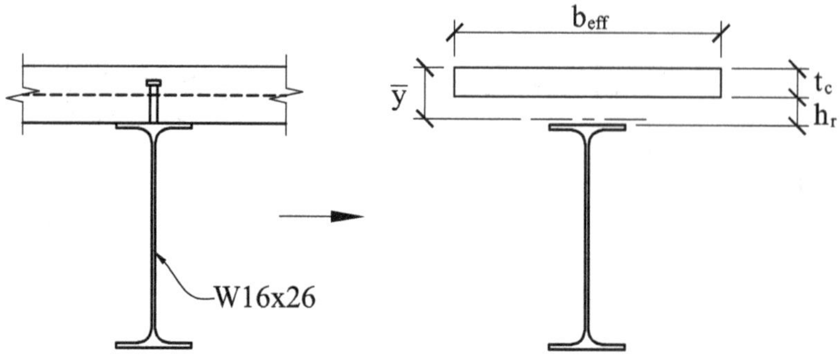

FIGURE 12-9 Section through composite beam

The composite moment of inertia of the beam is

$$I_{comp} = \Sigma I + Ad^2$$

$$= (12.2 \text{ in.})\left(\frac{(2)^3}{12}\right) + (12.2 \text{ in.})(2 \text{ in.})\left(3.6 \text{ in.} - \frac{2 \text{ in.}}{2}\right)^2$$

$$+ 301 \text{ in.}^4 + (7.68 \text{ in.}^2)\left(\frac{15.7 \text{ in.}}{2} + 4 \text{ in.} - 3.6 \text{ in.}\right)^2$$

$$= 996 \text{ in.}^4$$

The dead load plus sustained live load on the beam with a tributary width of 6 ft is

$$w_j = (55 \text{ psf} + 11 \text{ psf})(6 \text{ ft.}) = 396 \text{ lb./ft.}$$

$$= 33 \text{ lb./in.}$$

The maximum deflection of the beam at midspan is

$$\Delta_j = \frac{5w_jL_j^4}{384EI_j}$$

$$= \frac{(5)(33)(24 \text{ ft.} \times 12)^4}{(384)(29 \times 10^6)(996)} = 0.102 \text{ in.}$$

Girder Section Properties and Deflection

From part 1 of the *AISCM*, the section properties for W24 × 76 are as follows [3]:

$d = 23.9$ in.

$A = 22.4$ in.^{2}

$I = 2100$ in.^{4}

Effective slab width, $b_{eff} \leq 0.4 \times$ Girder span $= 0.4 \times 18$ ft. $= 7.2$ ft. $=$ **86.4 in. → controls**

$\leq$ Girder tributary width $= 24$ ft.

$t_c =$ Concrete slab thickness above deck ribs $= 2$ in.

$h_r =$ Depth of metal deck ribs $= 2$ in.

$$d_e = t_c + \frac{1}{2}h_r$$

$$= (2 \text{ in.}) + \frac{1}{2}(2 \text{ in.}) = 3 \text{ in.}$$

Note: Concrete within and above the deck ribs is used to compute the I_{comp} for the girder.

$b_e =$ Transformed effective width of the slab

$$= \frac{b_{eff}}{n}$$

$$= \frac{86.4 \text{ in.}}{5.9} = 14.6 \text{ in.}$$

The centroid of the composite girder (see Figure 12-10) is found by summing the moments of the areas about the top of the slab. Therefore,

$$\bar{y} = \frac{\Sigma Ay}{\Sigma A}$$

$$= \frac{\left[(3\text{ in.})(14.6\text{ in.})\left(\frac{3\text{ in.}}{2}\right)\right] + \left[(22.4\text{ in.}^2)\left(\frac{23.9\text{ in.}}{2} + 4\text{ in.}\right)\right]}{\left[(3\text{ in.})(14.6\text{ in.})\right] + 22.4\text{ in.}^2} = 6.4\text{ in.}$$

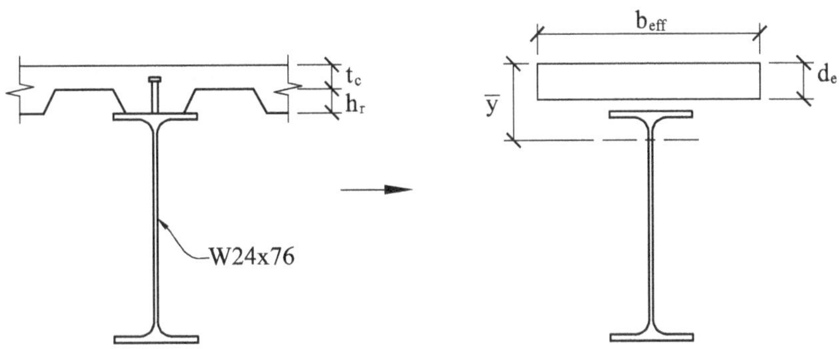

FIGURE 12-10 Section through composite girder

The composite moment of inertia of the girder is

$$I_{comp} = \Sigma I + Ad^2$$

$$= (14.6\text{ in.})\left(\frac{(3)^3}{12}\right) + (14.6\text{ in.})(3\text{ in.})\left(6.4\text{ in.} - \frac{3\text{ in.}}{2}\right)^2$$

$$+ 2100\text{ in.}^4 + (22.4\text{ in.}^2)\left(\frac{23.9\text{ in.}}{2} + 4\text{ in.} - 6.4\text{ in.}\right)^2$$

$$= 5230\text{ in.}^4$$

Assuming the girder self-weight is already included in the 55 psf dead load, the dead load plus sustained live load on the girder with a tributary width of 24 ft is

$$w_g = (55\text{ psf} + 11\text{ psf})(24\text{ ft.}) = 1584\text{ plf}$$

$$= 132\text{ lb./in.}$$

The maximum deflection of the girder is

$$\Delta_g = \frac{5w_g L_g^{\,4}}{384EI_g}$$

$$= \frac{(5)(132)(18\text{ ft.} \times 12)^4}{(384)(29 \times 10^6)(5230)} = 0.025\text{ in.}$$

Beam Effective Weight, W_j:

$W_j = \gamma w_t B_j L_j$ (see equation (12-14))

The beams are assumed to be connected to the girders at both ends with simple shear connectors (e.g., clip angles or shear plates/tabs), and the adjacent beam span is greater than 70% of the span of the beam being considered (since both spans are equal). Therefore, $\gamma = 1.5$.

$$D_s = \frac{d_e^{\,3}}{n}$$

$$= \frac{(3\text{ in.})^3}{5.9} = 4.6\text{ in.}^4/\text{ft.}$$

$$D_j = \frac{I_j}{S}$$

$$= \frac{996}{6 \text{ ft.}} = 166 \text{ in.}^4 / \text{ft.}$$

C_j = 2.0 (interior beam is being considered)

$$B_j = C_j L_j \left(\frac{D_s}{D_j}\right)^{0.25} \leq \left(\frac{2}{3}\right) \text{floor width parallel to the girders (see equation (12-15))}$$

$$= 2.0(24 \text{ ft.})\left(\frac{4.6}{166}\right)^{0.25} \leq \left(\frac{2}{3}\right)(90 \text{ ft.})$$

$$= 19.6 \text{ ft.} \leq 60 \text{ ft.}$$

Therefore, $B_j = 19.6$ ft.

From equation (12-14), the weighted average mass of the beam panel is

$W_j = (1.5)(55 \text{ psf} + 11 \text{ psf})(19.6 \text{ ft.})(24 \text{ ft.}) = \mathbf{46{,}570\ lb.}$

Girder Effective Weight, W_g:

$W_g = w_t B_g L_g$ (see equation (12-21))

Since the span of the girder is less than the width of the beam panel (L_g = 18 ft. < B_j = 19.6 ft.), the girder deflection has to be modified using equation (12-25):

$$\Delta_g' = \frac{L_g}{B_j}\Delta_g \geq 0.5\Delta_g$$

$$= \left(\frac{18 \text{ ft.}}{19.6 \text{ ft.}}\right)(0.025 \text{ in.}) \geq (0.5)(0.025 \text{ in.})$$

$$= 0.023 \text{ in.} \geq 0.013 \text{ in.}$$

Therefore, use $\Delta_g' = 0.023$ in.

$$D_g = \frac{I_g}{L_j}$$

$$= \frac{5230}{24 \text{ ft.}} = 218 \text{ in.}^4/\text{ft.}$$

C_g = 1.8 (girders supporting rolled steel beams)

$$B_g = C_g L_g \left(\frac{D_j}{D_g}\right)^{0.25} \leq \left(\frac{2}{3}\right) \text{floor length parallel to the beams (see equation (12-22))}$$

$$= 1.8(18 \text{ ft.})\left(\frac{166}{218}\right)^{0.25} \leq \left(\frac{2}{3}\right)(72 \text{ ft.})$$

$$= 30.3 \text{ ft.} < 48 \text{ ft.}$$

Therefore, B_g = 30.3 ft.

From equation (12-21), the weighted average mass of the girder panel is

$W_g = (55 \text{ psf} + 11 \text{ psf})(30.3 \text{ ft.})(18 \text{ ft.}) = \mathbf{36{,}000\ lb.}$

Effective Weight of the Floor System, W:

From equation (12-13), the weighted average mass of the floor system is calculated as

$$W = \left(\frac{\Delta_j}{\Delta_j + \Delta_g}\right)W_j + \left(\frac{\Delta_g}{\Delta_j + \Delta_g}\right)W_g$$

$$= \left(\frac{0.102 \text{ in.}}{0.102 \text{ in.} + 0.023 \text{ in.}}\right)(46{,}570 \text{ lb.}) + \left(\frac{0.023 \text{ in.}}{0.102 \text{ in.} + 0.023 \text{ in.}}\right)(36{,}000 \text{ lb.})$$

$$= 44{,}625 \text{ lb.}$$

The natural frequency of the floor system is determined from equation (12-4):

$$f_n = 0.18\sqrt{\frac{g}{\Delta_j + \Delta_g}}$$

$$= 0.18\sqrt{\frac{386}{0.102 \text{ in.} + 0.023 \text{ in.}}} = 10.0 \text{ Hz}$$

$$> 3Hz$$

$$> 9Hz$$

Since the combined natural frequency of the floor system is greater than 9 Hz, this is a high frequency floor (HFF) system, and it is unlikely that walking vibrations will cause any problems [1]. In any case, the peak acceleration for high frequency floors can be calculated using equation (12-26) as follows:

β = 0.03 (for a paper office)

W = 44,625 Ib. (calculated previously)

For f_n = 10 Hz, the matching harmonic, h = 5 (see Table 12-4)

$$f_{step} = \frac{if_{step}}{i} = \frac{1.6}{1} = 1.6 \text{ (see Table 12-3)}$$

The peak acceleration of the floor is

$$\frac{a_p}{g} = \frac{154}{W}\left(\frac{f_{step}^{1.43}}{f_n^{0.3}}\right)\sqrt{\frac{1-e^{-4\pi h\beta}}{h\pi\beta}} \le \frac{a_o}{g}$$

$$= \frac{154}{44{,}625}\left(\frac{1.6^{1.43}}{10^{0.3}}\right)\sqrt{\frac{1-e^{-4\pi(5)(0.03)}}{(5)\pi(0.03)}}$$

$$= 0.0045 = 0.45\% \le \frac{a_o}{g}$$

From Table 12-2, the recommended acceleration limit for offices is a_o/g = 0.5%, so the floor is adequate for walking vibrations.

EXAMPLE 12-3

Walking Vibrations with Open-Web Steel Joist Framing

The floor framing system with the partial plan shown in Figure 12-11 is to be used for an office building. The building has full-height partitions. Determine whether the floor is adequate for walking vibrations. Assume lightweight concrete with a density of 110 pcf and a 28-day compressive strength of 3500 psi. The floor has at least three bays in each orthogonal direction, and the floor loads are as follows:

Floor dead load = 55 psf (includes weight of slab, deck, finishes, mechanical and electrical fixtures, and partitions)

Sustained live load = 11 psf (see Table 12-1).

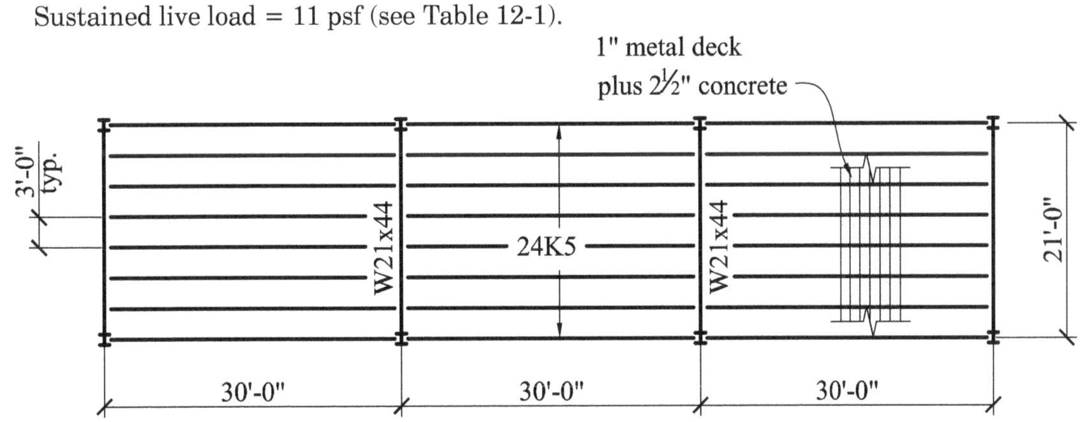

FIGURE 12-11 Floor framing plan for Example 12-3

Solution

Joist Section Properties and Deflection

For a 24K5 open-web steel joist (see Ref. [6]),

$d = 24$ in.

$A = 2.0 \text{ in.}^2 \left(1.0 \text{ in.}^2 \text{ per chord}\right)$

$I = 210 \text{ in.}^4$ (Note: Exact section properties for the joist would be provided by the joist supplier.)

Effective slab width, $b_{eff} \le 0.4 \times \text{joist span} = (0.4)(30 \text{ ft.}) = 12 \text{ ft.}$

$\le \text{Joist tributary width} = 3 \text{ ft.} = 36 \text{ in.} \rightarrow \text{controls}$

t_c = Concrete slab thickness above deck ribs = 2.5 in.

h_r = Depth of metal deck ribs = 1 in.

Note: Only concrete above the deck ribs is used to compute the I_{comp} for the joist (see Figure 12-12)

$E_s = 29{,}000$ ksi

$E_c = 33 w_c^{1.5} \sqrt{f'_c}$, ksi

$= (33)(110)^{1.5}\sqrt{3500} = 2{,}252{,}355 \text{ psi} = 2252 \text{ ksi},$

$$n = \frac{E_s}{1.35 E_c}$$

$$= \frac{29{,}000}{(1.35)(2252)} = 9.5,$$

b_e = Transformed effective width of the slab

$$= \frac{b_{eff}}{n}$$

$$= \frac{36 \text{ in.}}{9.5} = 3.8 \text{ in.}$$

The centroid of the composite joist (see Figure 12-9) is found by summing the moments of areas about the top of the slab. Therefore,

$$\bar{y} = \frac{\Sigma A y}{\Sigma A}$$

$$= \frac{\left[(2.5 \text{ in.})(3.8 \text{ in.})\left(\frac{2.5 \text{ in.}}{2}\right)\right] + \left[\left(2.0 \text{ in.}^2\right)\left(\frac{24 \text{ in.}}{2} + 3.5 \text{ in.}\right)\right]}{\left[(2.5 \text{ in.})(3.8 \text{ in.})\right] + 2.0 \text{ in.}^2} = 3.73 \text{ in.}$$

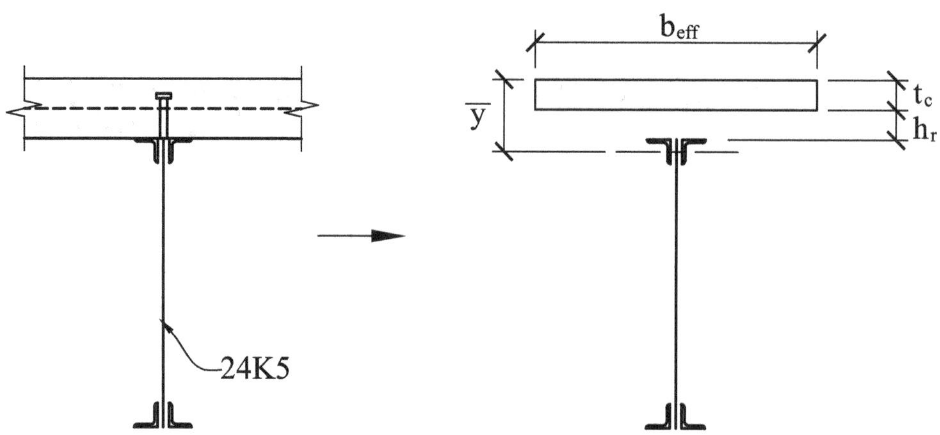

FIGURE 12-12 Section through composite joist

The composite moment of inertia of the joist is

$$I_{comp} = \sum\left(I + Ad^2\right)$$

$$= \left[(3.8\text{ in.})\left(\frac{(2.5)^3}{12}\right)\right] + \left[(3.8\text{ in.})(2.5\text{ in.})\left(3.73\text{ in.} - \frac{2.5\text{ in.}}{2}\right)^2\right]$$

$$+ 210\text{ in.}^4 + \left[\left(2.0\text{ in.}^2\right)\left(\frac{24\text{ in.}}{2} + 3.5\text{ in.} - 3.73\text{ in.}\right)^2\right]$$

$$= 550\text{ in.}^4$$

For joists with double-angle members with $6 \leq L_j/D = (30 \times 12)/24\text{ in.} = 15 \leq 24$, equation (12-10) is used to calculate C_r:

$$C_r = 0.9\left(1 - e^{-0.28\left(\frac{L_j}{D}\right)}\right)^{2.8} \leq 0.9$$

$$= 0.9\left(1 - e^{-0.28\left(\frac{30\times 12}{24}\right)}\right)^{2.8} = 0.862 < 0.9$$

Therefore, $C_r = 0.862$

$$\gamma = \frac{1}{C_r} - 1$$

$$= \frac{1}{0.862} - 1 = 0.159$$

The effective moment of inertia of the joists is modified as follows for this case:

$$I_{jeff} = \frac{1}{\dfrac{\gamma}{I_{chords}} + \dfrac{1}{I_{comp}}}$$

$$= \frac{1}{\dfrac{0.159}{210} + \dfrac{1}{550}} = 388\text{ in.}^4$$

The uniformly distributed dead load plus sustained load on each joist is,

$$w_j = (55\text{ psf} + 11\text{ psf})(3\text{ ft.}) = 198\text{ lb./ft.}$$

$$= 16.5\text{ lb./in.}$$

The joist maximum deflection is

$$\Delta_j = \frac{5w_jL_j^4}{384EI_j}$$

$$= \frac{(5)(16.5)(30\text{ ft.} \times 12)^4}{(384)\left(29 \times 10^6\right)(388)} = 0.321\text{ in.}$$

Girder Section Properties and Deflection

From part 1 of the *AISCM*, the section properties for W21 × 44 are as follows [3]:

$d = 20.7\text{ in.}$

$A = 13.0\text{ in.}^2$

$I = 843\text{ in.}^4$

For a typical interior bay in the floor, the effective width of the girder is calculated as follows [1, 3]:

Effective slab width, $b_{eff} \leq 0.4 \times \text{girder span}$

$$= 0.4 \times 21 \text{ ft.} = 8.4 \text{ ft.} = 100.8 \text{ in.} \rightarrow \text{controls}$$

$$\leq \text{Girder tributary width} = 30 \text{ ft.}$$

t_c = Concrete slab thickness above deck ribs = 2.5 in.

h_r = Depth of metal deck ribs = 1 in.

$$d_e = t_c + \frac{h_r}{2}$$

$$= 2.5 \text{ in.} + \frac{1 \text{ in.}}{2} = 3 \text{ in.}$$

Note: Concrete within and above the deck ribs is used to compute the I_{comp} for the girder.

b_e = Transformed effective width of the slab

$$= \frac{b_{eff}}{n}$$

$$= \frac{100.8 \text{ in.}}{9.5} = 10.6 \text{ in. (see Figure 12-13).}$$

The centroid of the composite girder (see Figure 12-13) is found by summing the moments of areas about the top of the slab. Therefore,

$$\bar{y} = \frac{\Sigma Ay}{\Sigma A}$$

$$= \frac{\left[(3 \text{ in.})(10.6 \text{ in.})\left(\frac{3 \text{ in.}}{2}\right)\right] + \left[(13.0 \text{ in.}^2)\left(\frac{20.7 \text{ in.}}{2} + 2.5 \text{ in.} + 3.5 \text{ in.}\right)\right]}{\left[(3 \text{ in.})(10.6 \text{ in.})\right] + 13.0 \text{ in.}^2} = 5.78 \text{ in.}$$

The composite moment of inertia of the girder is

$$I_{comp} = \Sigma\left(I + Ad^2\right)$$

$$= \left[(10.6 \text{ in.})\left(\frac{(3)^3}{12}\right)\right] + \left[(10.6 \text{ in.})(3 \text{ in.})\left(5.78 \text{ in.} - \frac{3 \text{ in.}}{2}\right)^2\right]$$

$$+ 843 \text{ in.}^4 + \left[(13.0 \text{ in.}^2)\left(\frac{20.7 \text{ in.}}{2} + 2.5 \text{ in.} + 3.5 \text{ in.} - 5.78 \text{ in.}\right)^2\right]$$

$$= 2901 \text{ in.}^4$$

The composite moment of inertia is reduced since the joist shoes (i.e., the 2½-in joist seat) separate the deck and the concrete from the top of the girder and thus prevent the slab from

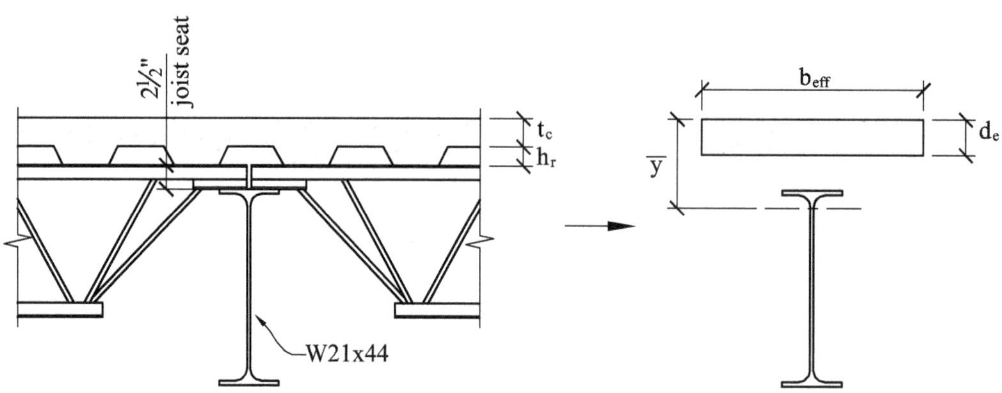

FIGURE 12-13 Section through composite girder

acting fully compositely with the girder. Therefore, the effective moment of inertia of the girder is:

$$I_{g,eff} = I_{nc} + \frac{(I_c - I_{nc})}{4}$$

$$= 843 + \frac{(2901 - 843)}{4} = 1357 \text{ in.}^4$$

Girder Deflection:

The uniformly distributed dead load plus sustained live load on the girder with a tributary width of 30 ft., assuming the girder self-weight is already included in the 55 psf dead load, is calculated as

$$w_g = (55 \text{ psf} + 11 \text{ psf})(30 \text{ ft.}) = 1980 \text{ lb./ft.}$$

$$= 165 \text{ lb./in.}$$

The girder maximum deflection is,

$$\Delta_g = \frac{5w_g L_g^4}{384EI_g}$$

$$= \frac{(5)(165)(21 \text{ ft.} \times 12)^4}{(384)(29 \times 10^6)(1356)} = 0.22 \text{ in.}$$

Joist Effective Weight, W_j:

$W_j = \gamma w_t B_j L_j$ (see equation (12-14))

Since joists are used, $\gamma = 1.0$.

$$D_s = \frac{d_e^3}{n}$$

$$= \frac{(3 \text{ in.})^3}{9.5} = 2.84 \text{ in.}^4/\text{ft.}$$

$$D_j = \frac{I_j}{S}$$

$$= \frac{388}{3 \text{ ft.}} = 129 \text{ in.}^4/\text{ft.}$$

$C_j = 2.0$ (interior joist is being considered)

$$B_j = C_j L_j \left(\frac{D_s}{D_j}\right)^{0.25} \leq \left(\frac{2}{3}\right) \text{floor width parallel to the girders}$$

$$= (2.0)(30 \text{ ft.})\left(\frac{2.84}{129}\right)^{0.25} \leq \left(\frac{2}{3}\right)(3 \text{ bays})(21 \text{ ft.}) = 42 \text{ ft.}$$

$$= 23.1 \text{ ft.} \leq 42 \text{ ft.}$$

Therefore, $B_j = 23.1$ ft.

From equation (12-14), the weighted average mass of the beam (or joist) panel is,

$$W_j = (1.0)(55 \text{ psf} + 11 \text{ psf})(23.1 \text{ ft.})(30 \text{ ft.}) = 45{,}738 \text{ lb.}$$

Girder Effective Weight, W_g:

$W_g = w_t B_g L_g$ (see equation (12-21))

Since the span of the girder is less than the width of the beam panel (L_g = 21 ft. < B_j = 23.1 ft.), the girder deflection has to be modified using equation (12-25) as follows:

$$\Delta_g' = \frac{L_g}{B_j}\Delta_g \geq 0.5\Delta_g$$

$$= \left(\frac{21 \text{ ft.}}{23.1 \text{ ft.}}\right)(0.22 \text{ in.}) \geq (0.5)(0.22 \text{ in.})$$

$$= 0.20 \text{ in.} \geq 0.11 \text{ in.}$$

Therefore, $\Delta_g' = 0.20$ in.

$$D_g = \frac{I_g}{L_j}$$

$$= \frac{1357}{30 \text{ ft.}} = 45.2 \text{ in.}^4/\text{ft.}$$

$C_g = 1.8$ (girders supporting rolled steel beams)

$$B_g = C_g L_g \left(\frac{D_j}{D_g}\right)^{0.25} \leq \left(\frac{2}{3}\right) \text{ floor length parallel to the beams/joist}$$

$$= (1.8)(21 \text{ ft})\left(\frac{129}{45.2}\right)^{0.25} \leq \left(\frac{2}{3}\right)(90 \text{ ft})$$

$$= 49.1 \text{ ft} \leq 60 \text{ ft}$$

From equation (12-21), the weighted average mass of the girder panel is,

$$W_g = (55 \text{ psf} + 11 \text{ psf})(49.1 \text{ ft.})(21 \text{ ft.}) = 68{,}095 \text{ lb.}$$

Effective Weight of the Floor System, *W*:

From equation (12-13), the weighted average mass of the floor system is calculated as

$$W = \left(\frac{\Delta_j}{\Delta_j + \Delta_g}\right) W_j + \left(\frac{\Delta_g}{\Delta_j + \Delta_g}\right) W_g$$

$$= \left(\frac{0.321 \text{ in.}}{0.321 \text{ in.} + 0.20 \text{ in.}}\right)(45{,}738 \text{ lb.}) + \left(\frac{0.20 \text{ in.}}{0.321 \text{ in.} + 0.20 \text{ in.}}\right)(68{,}095 \text{ lb.})$$

$$= 54{,}320 \text{ lb.}$$

The natural frequency of the floor system is determined from equation (12-4):

$$f_n = 0.18\sqrt{\frac{g}{\Delta_j + \Delta_g}}$$

$$= 0.18\sqrt{\frac{386}{0.321 \text{ in.} + 0.20 \text{ in.}}} = 4.90 \text{ Hz} \geq 3 \text{ Hz., OK.}$$

$$\leq 9 \text{Hz}$$

Since the combined natural frequency of the floor system is less than 9 Hz, this is a low-frequency floor (LFF) system, and the peak acceleration of the floor system can be calculated using equation (12-12) as follows:

From Table 12-2,

$P_o = 65$ lb., and

$\beta = 0.05$ (damping ratio for a floor with full-height partitions)

The peak acceleration of the floor system due to walking vibrations is calculated as follows from equation (12-12):

$$\frac{a_p}{g} = \frac{P_o e^{(-0.35 f_n)}}{\beta W} \leq \frac{a_o}{g}$$

$$= \frac{(65)\left(e^{(-0.35)(4.90)}\right)}{(0.05)(54{,}320)} = 0.0043, \text{ or } \mathbf{0.43\%}.$$

From Table 12-2, the recommended floor acceleration limit is a_o/g = 0.5%, which is greater than the peak acceleration of a_p/g = 0.43%, so the floor is adequate for walking vibrations. However, if this were an electronic office (which has a damping ratio of 0.025), the peak acceleration would be a_p/g = 0.86%; therefore, the floor would not be adequate for walking vibrations in that case, and therefore, remedial measures to control walking vibrations would be necessary.

12-7 RHYTHMIC VIBRATION CRITERIA

Rhythmic activities that cause floor vibrations include dancing, jumping exercises, aerobics, jogging, concerts, and sporting events. To avoid resonance during rhythmic activities, it is recommended that the natural frequency of the floor system should be greater than 7 Hz [14]. For any given rhythmic activity, there are several possible harmonics of vibration. For example, the participants in a lively concert might induce a forcing frequency of 2.0 Hz for the first harmonic and 4.0 Hz for the second harmonic. The floor system would need to be checked for both harmonics. For most rhythmic activities, not more than three harmonics would need to be checked.

The peak floor acceleration for each harmonic of the specific rhythmic activity is given as [1]

$$\frac{a_{pi}}{g} = \frac{1.3\alpha_i w_p}{w_t \sqrt{\left[\left(\frac{f_n}{i f_{step}}\right)^2 - 1\right]^2 + \left[\frac{2\beta f_n}{i f_{step}}\right]^2}}, \tag{12-27}$$

Where,

$\frac{a_{pi}}{g}$ = Peak floor acceleration for each harmonic, and

β = Damping ratio = 0.06 for rhythmic activity

Note: The damping ratio, β, used for rhythmic vibrations in equation (12-27) is higher than the values indicated in Table 12-2 because the rhythmic activity participants also contribute to the damping of the floor system [1].

f_{step} = step forcing frequency of the rhythmic activity (see Table 12-5),

f_n = natural frequency of the floor system using equation (12-4) or (12-5),

i = the harmonic number 1, 2, 3 (see Table 12-5),

α_i = Dynamic coefficient (see Table 12-5),

a_o/g = Acceleration limit for rhythmic activity (see Table 12-6),

w_p = Weight of the participants in the rhythmic activity, psf (i.e., sustained live load; see Table 12-5). Note that w_p can be reduced if the area used for the rhythmic activity is less than the total floor area, in which case:

$$w_{pj} = w_p \left(\frac{\textit{Area used for rhythmic acivity}}{\textit{Total floor area of the entire bay}} \right)$$

$$w_{pg} = w_p \left(\frac{\textit{Area used for rhythmic acivity}}{\textit{Total floor area of the entire bay}} \right)$$

TABLE 12-5 Recommended dynamic properties for rhythmic activities

Rhythmic Activity	Forcing frequency, if_{step}, Hz	Weight of participants, w_p, psf	Dynamic coefficient, α_i
Dancing:			
First harmonic (i = 1)	1.5–2.7	12.5*	0.5
Second harmonic (i = 2)	3.0–5.4	12.5*	0.05
Lively concert, sporting event:			
First harmonic (i = 1)	1.5–2.7	31.0^	0.25
Second harmonic (i = 2)	3.0–5.4	31.0^	0.05
Aerobics or jumping exercises:			
First harmonic (i = 1)	2.0–2.75	4.2^ ^	1.5
Second harmonic (i = 2)	4.0–5.5	4.2^ ^	0.6
Third harmonic (i = 3)	6.0–8.25	4.2^ ^	0.1

Source: Adapted from Table 5-2 of Ref. [1].

*Area of floor space per couple = 25 ft^2

^Area of floor space per person = 5 ft^2

^ ^Area of floor space per person = 35 ft^2

TABLE 12-6 Recommended floor acceleration limits for rhythmic activity

Affected occupancies (Rhythmic activities adjacent to ...)	*Acceleration Limit* a_o/g
Office or residential	0.5%
Dining or weightlifting	1.5%–2.5%
Dining and dancing	2%
Dining and aerobics	2%
Rhythmic activity only	4%–7%

Source: Adapted from Table 5-1, Ref. [1] and Ref. [9].

A typical bay is a rectangular floor area bounded by columns at the four corners.

w_t = Dead plus sustained live load per unit area, psf.

The effective maximum acceleration due to rhythmic activity, considering all the harmonics is given as [1]:

$$\frac{a_{pm}}{g} = \frac{\left(\Sigma a_{pi}^{\ 1.5}\right)^{\frac{2}{3}}}{g} \le \frac{a_o}{g} \tag{12-28}$$

$$= \left[\frac{\left(a_{p1}^{\ 1.5}\right)}{g} + \frac{\left(a_{p2}^{\ 1.5}\right)}{g} + \frac{\left(a_{p3}^{\ 1.5}\right)}{g}\right]^{\frac{2}{3}} \le \frac{a_o}{g}$$

For rhythmic vibrations, the natural frequency is calculated from either equation (12-4) or (12-5). The decision—whether or not—to include the effects of column axial deformation is left to the designer. Buildings with fewer than five stories generally do not have a significant contribution from column axial deformation. To avoid resonance, it is recommended that the vertical natural frequency of floor systems subjected to rhythmic vibrations (e.g., concerts, jogging, aerobics, and jumping) be greater than 7 Hz; if it is not feasible to achieve this, other vibration mitigation measures such as mechanical dampers or a structural vibration control device may need to be used [14].

EXAMPLE 12-4

Rhythmic Vibrations

Determine the adequacy of the floor framing shown in Figures 12-14 and 12-15 for jumping exercises. The precast hollow core slab properties are as follows:

Dead load = 120 psf

$I = 1400\ \text{in.}^4/\text{ft.}$

$f'_c = 5000$ psi

$\gamma_c = 145$ pcf

Solution

Slab Deflection:

From Table 12-5, the weight of participants in the rhythmic activity (i.e., jumping exercises), $w_p = 4.2$ psf

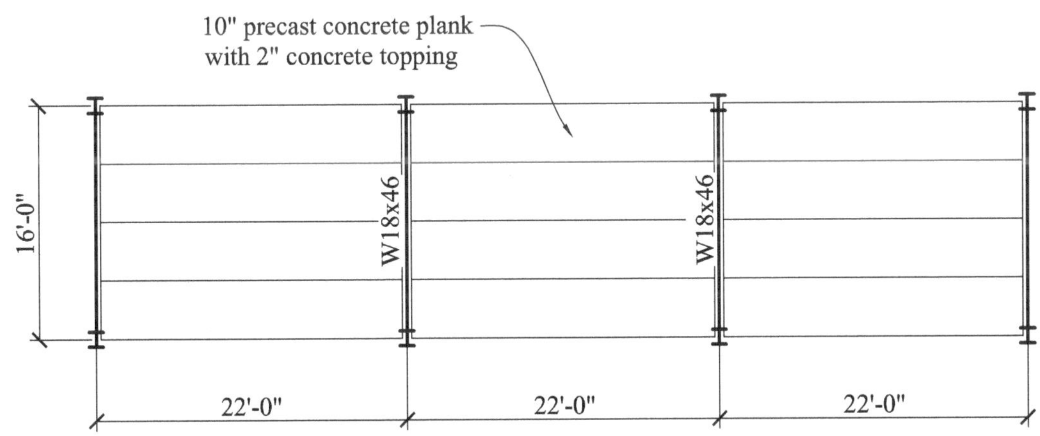

FIGURE 12-14 Floor framing plan for Example 12-4

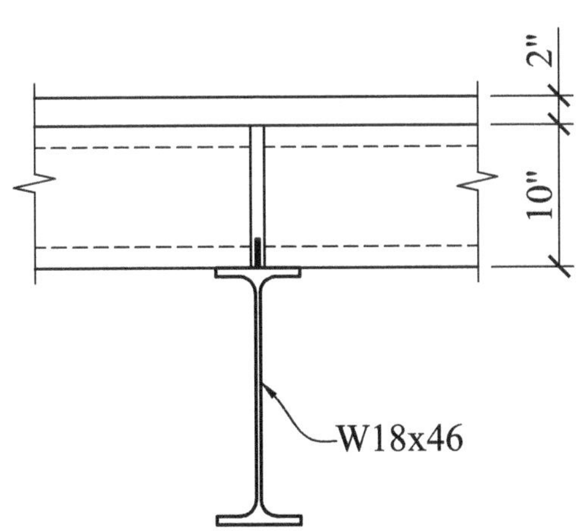

FIGURE 12-15 Floor section for Example 12-4

$$E_c = 33w_c^{1.5}\sqrt{f'_c}\text{, ksi}$$

$$= 33(145)^{1.5}\sqrt{5000} = 4,074,280\,\text{psi} = 4074\text{ ksi}$$

The total dead load plus weight of participants (see Table 12-5) on the slab (with a tributary width of 1 ft) is

$$w_t = (120\text{ psf} + 4.2\text{ psf})(1\text{ ft.}) = 124.2\text{ lb./ft.}$$

Therefore, $w_s = 10.35$ lb./in.

The precast hollow core slab maximum (i.e., mid-span) deflection is

$$\Delta_s = \frac{5w_sL^4}{384EI}$$

$$= \frac{(5)(10.35)(22\times12)^4}{(384)(4,074,000)(1400)} = 0.111\text{ in.}$$

Girder Deflection:

Assuming the girder self-weight is already included in the 120 psf dead load, the dead plus weight of participants (see Table 12-5) on the girder (which has a tributary width of 22 ft.) is

$$w_g = (120\text{ psf} + 4.2\text{ psf})(22\text{ ft.}) = 2732\text{ lb./ft.}$$

$$= 228\text{ lb./in.}$$

From part 1 of the *AISCM*, the moment of inertia, I, of W18 × 46 = 712 in.4
The girder maximum (i.e., midspan) deflection is

$$\Delta_g = \frac{5w_gL^4}{384EI}$$

$$= \frac{(5)(228)(16\times12)^4}{(384)(29,000,000)(712)} = 0.195\text{ in.}$$

Natural Frequency of the Floor System:

$$f_n = 0.18\sqrt{\frac{g}{\Delta_s + \Delta_g}}$$

$$= 0.18\sqrt{\frac{386}{0.111+0.195}} = 6.39\text{ Hz} \le 9\,\text{Hz}\text{; therefore, this is an LFF system.}$$

The peak acceleration for i^{th} harmonic of the rhythmic activity is computed using equation (12-27) as follows:

$$\frac{a_{pi}}{g} = \frac{1.3\alpha_i w_p}{w_t\sqrt{\left[\left(\frac{f_n}{if_{step}}\right)^2 - 1\right]^2 + \left[\frac{2\beta f_n}{if_{step}}\right]^2}},$$

Since the if_{step} for each harmonic is a range, it is not always readily apparent the step frequency within each harmonic that will produce the peak acceleration. It is recommended that a spreadsheet be used to compute the accelerations for each step frequency for each harmonic (i.e., $1f_{step}$, $2f_{step}$, and $3f_{step}$). The peak acceleration for jumping exercises for each step frequency (i.e., 2.0–2.75 Hz for the first harmonic; 4.0–5.5 Hz for the second harmonic; and 6.0–8.25 Hz for the third harmonic from Table 12-5) can be obtained using a spreadsheet, and they are presented in Table 12-7.

TABLE 12-7 Peak accelerations for example 12-4

α_i	$1f_{step}$	a_{p1}/g	α_i	$2f_{step}$	a_{p2}/g	α_i	$3f_{step}$	a_{p3}/g	a(peak)
1.5	2	0.007155	0.6	4	0.016867	0.1	6	0.02372	0.034647
1.5	2.05	0.007559	0.6	4.1	0.018302	0.1	6.15	0.029722	0.040852
1.5	2.1	0.007976	0.6	4.2	0.019872	0.1	6.3	0.03515	0.046736
1.5	2.15	0.008409	0.6	4.3	0.021595	0.1	6.45	0.036538	0.049247
1.5	2.2	0.008858	0.6	4.4	0.023494	0.1	6.6	0.033308	0.047992
1.5	2.25	0.009322	0.6	4.5	0.025594	0.1	6.75	0.028566	0.045865
1.5	2.3	0.009803	0.6	4.6	0.027926	0.1	6.9	0.024342	0.044651
1.5	2.35	0.0103	0.6	4.7	0.030529	0.1	7.05	0.021034	0.044635
1.5	2.4	0.010815	0.6	4.8	0.033449	0.1	7.2	0.018506	0.045669
1.5	2.45	0.011348	0.6	4.9	0.036742	0.1	7.35	0.016557	0.047591
1.5	2.5	0.011899	0.6	5	0.040481	0.1	7.5	0.015027	0.050311
1.5	2.55	0.01247	0.6	5.1	0.044755	0.1	7.65	0.013804	0.05381
1.5	2.6	0.013061	0.6	5.2	0.049678	0.1	7.8	0.012808	0.058129
1.5	2.65	0.013672	0.6	5.3	0.055398	0.1	7.95	0.011983	0.063362
1.5	2.7	0.014305	0.6	5.4	0.062104	0.1	8.1	0.011291	0.069665
1.5	2.75	0.014959	0.6	5.5	0.070043	0.1	8.25	0.010703	0.077258

The following are sample calculations for the peak acceleration for jumping exercises for step frequency, if_{step}, of 2.75 Hz, 5.5 Hz, and 8.25 Hz, respectively. Note that the peak values obtained are similar to those indicated in Table 12-7.

First harmonic (i.e., i = 1):

$$\frac{a_{p1}}{g} = \frac{(1.3)(1.5)(4.2)}{124.2\sqrt{\left[\left(\frac{6.39}{2.75}\right)^2 - 1\right]^2 + \left[\frac{(2)(0.06)(6.39)}{2.75}\right]^2}} = 0.015$$

Second harmonic (i.e., i = 2):

$$\frac{a_{p2}}{g} = \frac{(1.3)(0.6)(4.2)}{124.2\sqrt{\left[\left(\frac{6.39}{5.5}\right)^2 - 1\right]^2 + \left[\frac{(2)(0.06)(6.39)}{5.5}\right]^2}} = 0.070$$

Third harmonic (i.e., i = 3):

$$\frac{a_{p3}}{g} = \frac{(1.3)(0.1)(4.2)}{124.2\sqrt{\left[\left(\frac{6.39}{8.25}\right)^2 - 1\right]^2 + \left[\frac{(2)(0.06)(6.39)}{8.25}\right]^2}} = 0.0107$$

The effective maximum or peak acceleration, considering all the harmonics, is

$$\frac{a_{pm}}{g} = \frac{\left(\sum a_{pi}^{\ 1.5}\right)^{2/3}}{g} \le \frac{a_o}{g}$$

$$= \left(0.015^{1.5} + 0.07^{1.5} + 0.0107^{1.5}\right)^{\frac{2}{3}} = 0.077 = 7.7\%, \text{but this acceleration is greater than } \frac{a_o}{g}$$

From Table 12-6, the recommended acceleration limit is between 4% and 7% for rhythmic activities only, with the average of these two values being 5.5%.

Therefore, the peak floor acceleration obtained above is greater than the average of the specified limits and even greater than the upper limit of 7%, so the floor is not adequate for jumping exercises. If this floor were, say, to be used for jumping exercises adjacent to a dining area, the average acceleration limit from Table 12-6 would be 2% (i.e., for dining in the same area as aerobics). Therefore, this floor would also not be adequate in that case. One way to reduce the peak acceleration would be to increase the size of the girder, which results in increased stiffness and a smaller deflection.

12-8 SENSITIVE EQUIPMENT VIBRATION CRITERIA

In the previous sections, we considered transient floor vibrations in terms of peak floor acceleration (i.e., a_p/g) that was compared to the floor acceleration tolerance limit, a_o/g. . For floor structures that support sensitive equipment, the floor vibration also needs to be evaluated to ensure proper functioning of the equipment. Examples of floors supporting sensitive equipment include:

a. Operating rooms (OR) with surgical lasers and other sensitive equipment.
b. Microchip manufacturing equipment
c. Hospital floors supporting sensitive equipment like MRI machines, etc.

There are several methods available in Chapter 6 of Ref. [1] for evaluating the impact of floor vibrations on sensitive equipment, and these methods yield the peak accelerations due to slow, moderate, or fast walking; these peak accelerations are then compared to the peak acceleration limits for the sensitive equipment.

An alternate method involves calculating the one-third octave velocity due to slow, moderate, or fast walking, and the predicted peak velocity (in micro-inch per second or mips or μ-in./sec) is then compared to the velocity limit for the equipment [2, 10].

Of the two methods, the velocity method is more commonly used, and vibrational velocity limits (in mips or μ-in./sec) for sensitive equipment are readily available. For example, in hospital settings, the maximum vibrational velocity limits for MRI equipment can be as low as 500 micro-inch/sec or 500 μ-in./sec; laboratory floors may have vibrational velocity limits of 2000 μ-in./sec, while floors supporting computer equipment have a less stringent vibrational velocity limit of 8000 μ-in./sec [1].

In both methods, the predicted velocity or acceleration is a function of the natural frequency of the floor system (f_n), the damping ratio (β), the effective weight of the floor system (W) from equation (12-13), and the walking speed (i.e., whether very slow, slow, moderate, or fast walking). One common way to mitigate the impact of floor vibration on sensitive equipment is to isolate the equipment from the floor. More information on the evaluation of floor systems supporting sensitive equipment for the impact of walking vibrations can be found in Ref. [1].

12-9 VIBRATION OF MONUMENTAL STEEL STAIRS

Slender monumental steel stairs (e.g., see Figure 12-15b) are often used by architects as a showcase in the atriums of buildings. The architectural constraints usually require these stairs to have long spans with slender members for aesthetic appeal. The combination of long spans and heavy treads results in a structure with low stiffness and therefore low natural frequencies, which makes them more susceptible to perceptible vibrations from the dynamic forces induced by people descending and ascending the stairs.

FIGURE 12-15b Monumental steel stair

Typically, HSS, or built-up box sections, are used for the stair stringers or spine of monumental stairs because of large induced torsional moments and rotation. The stair may consist of two stair stringers or a single central stringer or spine used with the stair treads cantilevered on both sides from the central spine or stringer. Vibration is a critical criterion that must be checked, and the connections of the stair stringers to the framing at the floor levels are also critical. The design of monumental steel stairs for the ultimate limit state is usually straightforward using the methods discussed in previous chapters, but it is the design for vibration serviceability that is sometimes overlooked.

Davis and Murray [11] carried out a test and analyses of existing slender monumental steel stairs, and they found that stairs experience higher accelerations when people are descending the stairs compared to when they are ascending the stairs. The AISC Design Guide 11 [1] provides design guidance for evaluating steel stairs for floor vibration and recommends that the minimum vertical and lateral natural frequencies of monumental stairs should be 5 Hz and 2.5 Hz, respectively. Monumental steel stairs should be analyzed using 3-D finite element analysis to obtain the natural vertical and lateral frequencies, and designed for vibration serviceability. The structural designer should very early on in the design process make the architect and the owner aware of the implications of the acceleration tolerances on the stair member sizes. The floor beams or girders supporting the steel stair stringers should also be designed for the transmitted vibrations from the steel stairs, and these transmitted vibrations, if not mitigated, can lead to perceptible vibrations at the floor levels.

12-10 VIBRATION CONTROL MEASURES

For small dynamic forces such as those caused by walking vibrations, the vibration effects can be more effectively controlled by either increasing the mass of the structure, increasing the stiffness of the structure, increasing the damping, or a combination of the preceding measures.

For large dynamic forces such as those caused by aerobics, the vibration effects are most effectively controlled by keeping the natural frequency of any mode of vibration most affected by the dynamic force away from the forcing frequency causing the vibrations. To achieve this, the natural frequency of the structure must be much greater than the forcing frequency of the highest harmonic dynamic force causing the vibration. This can be achieved by stiffening the structure (e.g., by adding beam depth, columns, or posts).

Floors with a natural frequency greater than 10 Hz and a stiffness greater than 5.7 kips/in. do not generally have vibration problems due to human activities. Occupants may experience discomfort in floor systems with natural frequencies in the range of 5 Hz to 8 Hz because this frequency range coincides with the natural frequencies of many internal human organs.

The use of an isolated "floating floor" completely separated from the surrounding slabs is effective mostly for controlling vibrations due to aerobics and other rhythmic activities (see Figure 12-16 for examples). The floating floor concept is similar to that used in vibration isolation of equipment. The floor is supported on very soft springs (e.g., neoprene pads) attached to the structural floor. The combined natural frequency of the floating floor and the springs should be very small—less than 2 Hz to 3 Hz. This can be achieved by using a thick slab (4 in. to 8 in. thick). The space between the floating floor and the structural slab must be properly vented to prevent the change in pressure due to the movement of the floating floor from causing the structural floor to move. An isolated column base similar to that shown in Figure 12-16b can be used to isolate a building from ground vibrations induced by nearby vehicular or train traffic.

Rhythmic activities, such as aerobics, can be located on the ground floor of a building, or an isolated framing system can be used. Weightlifting activity can be accommodated on framed

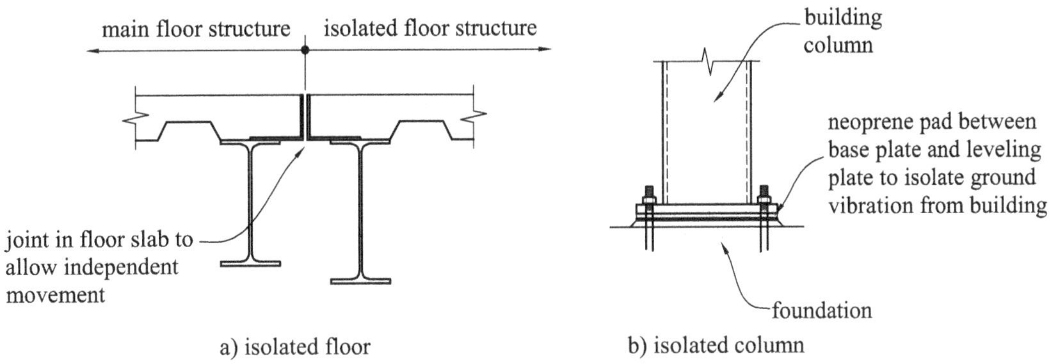

FIGURE 12-16 Floating or isolated structures

floors; however, one must consider the dynamic effect of weights being dropped suddenly on the floor, which is a common occurrence with weightlifting activity. This dynamic force can be mitigated by using an appropriate mat to absorb some of the energy of the falling weight, but it is generally advisable to locate weightlifting and aerobics activities on the ground floor of a building.

Floors with a natural frequency of less than 3 Hz should be avoided. Walking speed in an office is usually between 1.25 and 1.5 steps per second (or 1.25 Hz to 1.5 Hz). Floors with a natural frequency of 3 Hz may experience resonance at the second harmonic frequency of walking (i.e., between 2.5 Hz and 3 Hz).

Damping is a critical parameter in vibration analysis and has a significant impact on the calculated accelerations, but the designer typically has very little control over the amount of damping that is present in a floor system. For example, full-height partitions provide the best form of damping, but they are usually specified by someone other than the structural engineer. Furthermore, such partitions may not be present for the entire life of the structure. The same holds true for other damping components, such as furniture and ductwork. The point is that increasing damping to reduce vibration is generally not feasible, since damping is a parameter that is difficult to quantify and control, so the designer has to make reasonable assumptions as to the amount of damping that is present in a structure.

Sensitive equipment should ideally be located on the ground floor of a building, but when such equipment needs to be on elevated floors, it should be located close to columns and away from heavily traveled corridors. In buildings with sensitive equipment, such as laboratories or microchip fabrication facilities, avoid locating the equipment near long corridors since long corridors encourage fast walking and running, which cause vibrations that are detrimental to the operation and proper functioning of sensitive equipment. Where long corridors cannot be avoided, provide visually pleasing space along the corridor to encourage slower walking. Where possible, a building layout with short corridors and possibly with several turns should be used to discourage fast walking and thus minimize any adverse impact on sensitive equipment located close to the corridor.

To mitigate vibrations in existing floors, one possible solution is to add support posts. In some cases, this may be the only reasonable solution, but it is intrusive because the columns interrupt the amount of available open floor space. When steel posts are used to mitigate floor vibrations at an upper floor, the posts need to extend to the slab on grade in order to be effective. As mentioned earlier, fixing floor vibration problems after the fact can be quite expensive. For the Millennium Pedestrian Bridge, which had a low natural frequency of less than 1.3 Hz [12], dampers had to be installed to stop the excessive lateral swaying of the bridge that was first noticed as thousands of pedestrians walked across the bridge during its initial opening.

References

[1] Murray, T. M., Allen, D. E., E. E. Ungar, and D. B. Davis. 2016. *Steel Design Guide Series 11: Vibrations of Steel-Framed Structural Systems Due to Human Activity*. Second Edition. Chicago: AISC.

[2] Charlton, N. 1997. "Framing Systems for Microelectronic Facilities." *Modern Steel Construction*, May.

[3] American Institute of Steel Construction. 2023. *Steel Construction Manual*. 16th ed. Chicago: AISC.

[4] American Concrete Institute. 2019. *ACI 318: Building Code Requirements for Structural Concrete and Commentary*. Farmington Hills, MI: ACI.

[5] Vulcraft. 2018. *Steel Roof and Floor Deck*. Florence, SC: Vulcraft/Nucor.

[6] Steel Joist Institute. 2018. *Standard Specifications—Load Tables and Weight Tables for Steel Joists and Joist Girders*. 44th ed. Myrtle Beach, SC: SJI.

[7] Spancrete Manufacturers Association. 2005. *Span Limitations: Floor Vibrations—Rhythmic Activity*. Research Notes #1021. Waukesha, WI: SMA.

[8] Spancrete Manufacturers Association. 2005. *Span Limitations: Floor Vibrations—Flexible Supports*. Research Notes #1023. Waukesha, WI: SMA.

[9] National Research Council of Canada. 1990. *National Building Code of Canada, Supplement—Commentary A, Serviceability Criteria for Deflection and Vibration*. Ottawa, Canada.

[10] Lancaster, F. 2007. "Subduing Vibration in Laboratory Buildings." *Structure Magazine*, November.

[11] Davis, D.B. and Murray, T. M. 2009. "Slender Monumental Stair Vibration Serviceability." *Journal of Architectural Engineering*, ASCE, December: 111–121.

[12] Reina, P. 2002. "Dampers Stop Swaying, Enabling Crowds to Step Lively." *Engineering News Record* (ENR), February 14.

[13] Graham, J., S. Love, and S. Beaulieu. 2018. "When Humans Make Structures Shake." *Structure Magazine*, November: 16–19.

[14] DeVore, C. 2018. "Bad Vibrations." *Structure Magazine*, March: 20–23.

Exercises

12-1. Calculate the natural frequency of a W18 × 35 beam spanning 30 ft. with an applied load (dead plus sustained live loads) of 500 plf.

12-2. Given the floor plan and floor section shown in Figure 12-17 with the following design parameters:

- Dead load = 60 psf,
- Concrete is normal weight with f'_c = 3500 psi, and
- Assume loads to the joist and girder are uniformly distributed,

Determine the following:

1. Natural frequency of the floor system, and
2. Adequacy for walking vibrations if the floor is part of a shopping mall.

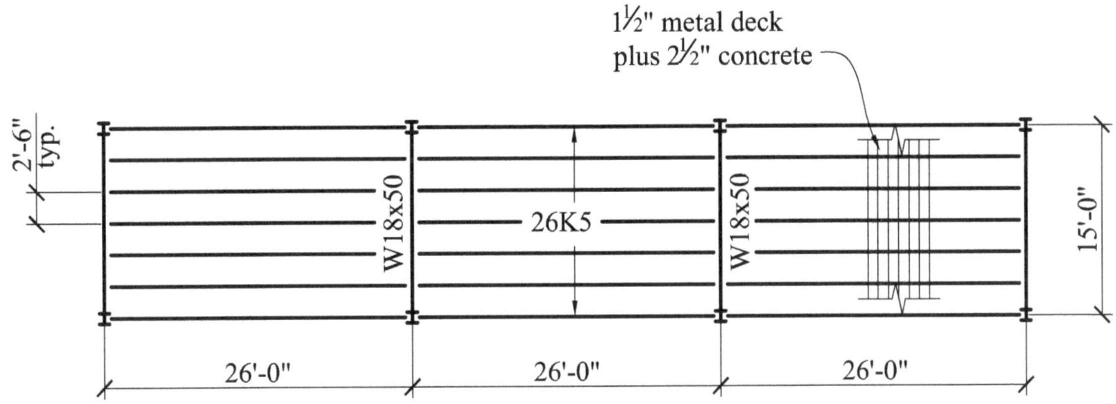

FIGURE 12-17 Floor framing plan for Exercise 12-2

12-3. Given the floor plan shown in Figure 12-18 with the following design parameters:

- Dead load = 70 psf; sustained live load = 11 psf,
- The floor is used as an open office area with few nonstructural components,
- Transformed moment of inertia, I_{BEAM} = 2100 in.^{4}, I_{GIRDER} = 3100 in.^{4}, and
- Assume loads to the beam and girder are uniformly distributed,
- Assume the weighted average mass of the floor system is 70,000 lb.,

Determine the following:

1. Natural frequency of the floor system, and
2. Adequacy for walking vibrations in an office.

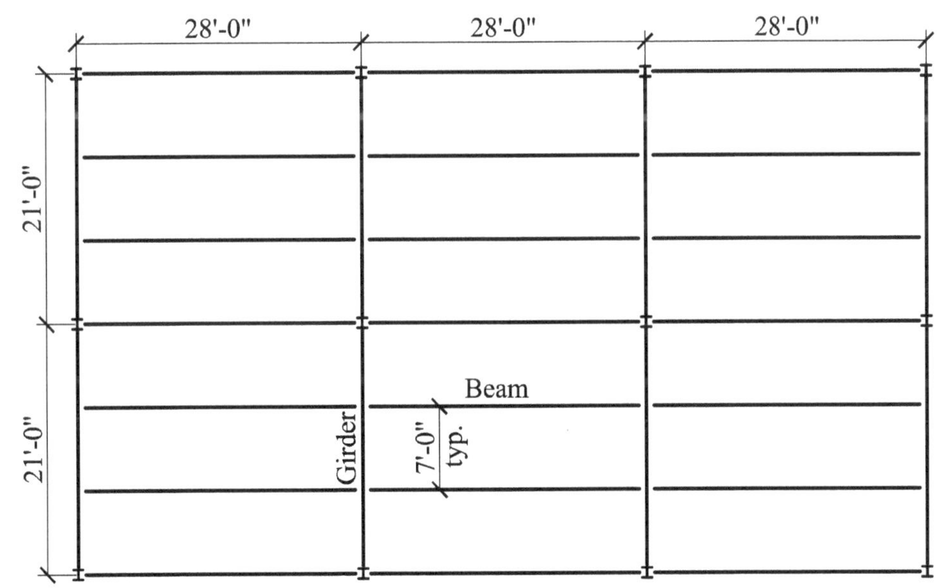

FIGURE 12-18 Floor framing for Exercises 12-3 and 12-4

12-4. Given the floor plan shown in Figure 12-19 with the following design parameters:

- Dead load = 75 psf, sustained live load = 11 psf, and
- Transformed moment of inertia, I_{BEAM} = 2100 in.^{4}, I_{GIRDER} = 3400 in.^{4},

Determine the following:

1. Natural frequency of the floor system, and
2. Adequacy for jumping exercises in a facility with rhythmic activities.

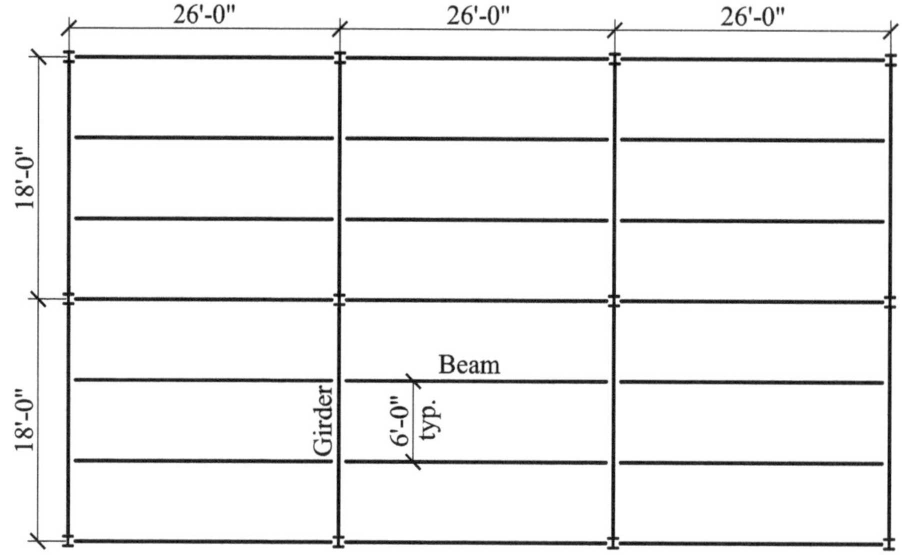

FIGURE 12-19 Floor framing for Exercises 12-5 and 12-6

12-5. Given the floor plan shown in Figure 12-20 with a dead load of 55 psf and a sustained live load of 11 psf, determine whether the floor system is adequate for walking vibrations in an office area with some nonstructural components. The concrete is normal weight with f'_c = 4000 psi.

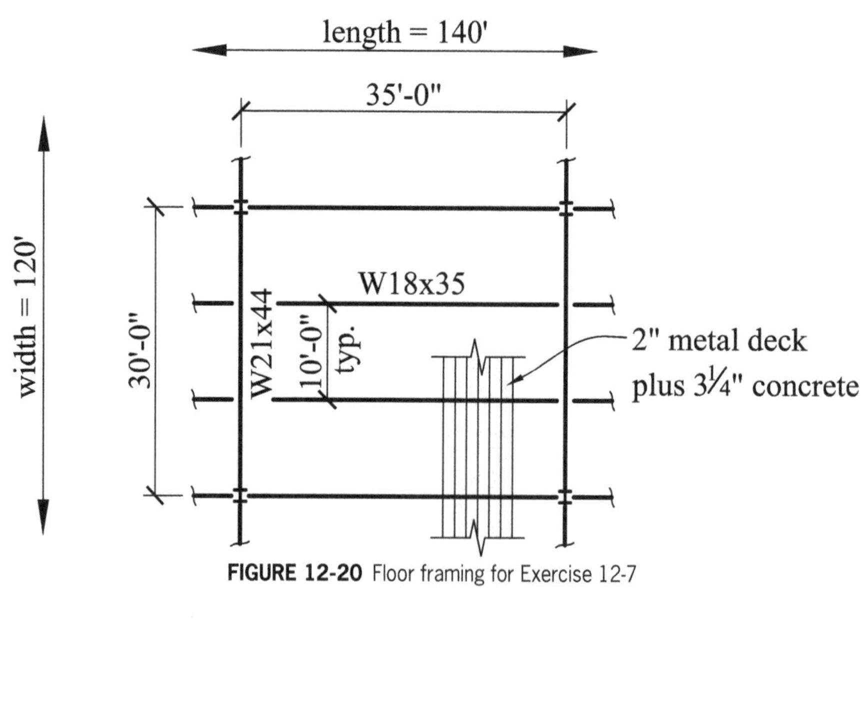

FIGURE 12-20 Floor framing for Exercise 12-7

12-6. Develop a spreadsheet to analyze a floor for walking vibrations. Include the following as variables:

- Beam and girder section properties
- Beam and girder lengths
- Deck and slab thickness

Use the spreadsheet to redo Exercise 12-5 and submit your results.

12-7. Calculate the natural frequency of the isolated floor system shown in Figure 12-21.

a. The total floor dead load is 85 psf (ignore beam self-weight.)
b. Assuming the girder size does not change, select the most economical beam to achieve a natural frequency of the floor system greater than 7 Hz.

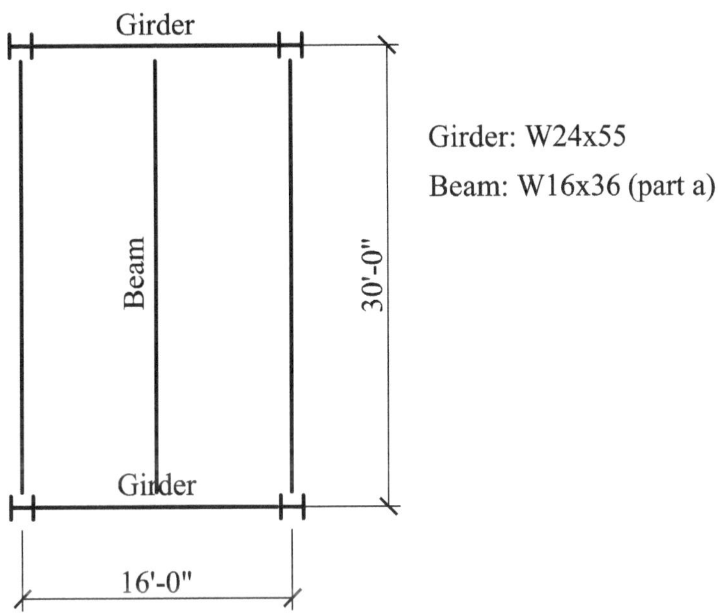

FIGURE 12-21 Floor framing for Exercise 12-9

CHAPTER

13

Built-up Sections—Welded Plate Girders

13-1 INTRODUCTION TO WELDED PLATE GIRDERS

Currently, the *heaviest* rolled section in the United States is W36 × 925 with a depth of 43.1 inches, a moment of inertia about the strong axis of 73,000 in.^{4}, and a plastic moment capacity about the strong axis, ϕM_{px}, of 15,500 ft.-kips, while the *deepest* rolled section in the United States is W44 × 408 with a depth of 44.8 inches, a moment of inertia about the strong axis of 38,700 in.^{4}, and a plastic moment capacity about the strong axis, ϕM_{px}, of 7,500 ft.-kips. For large spans in bridges and buildings or where large, concentrated column loads need to be transferred, and where no rolled wide flange shapes are available that can resist the moment and shear demand and satisfy the deflection and other serviceability criteria, welded plate girders or steel trusses are typically used as the framing members.

Welded plate girders (see Figure 13-1a), which are the most common form of plate girders, are built-up structural steel members that consist of flange plates welded to a web plate with fillet welds [1]. They are used to support large loads over long spans (60 ft. to 200 ft.) [2] and to support structural loads that are too large for which no rolled steel shapes can be found in the American Institute of Steel Construction Manual (*AISCM*). Plate girders are rarely used in building structures, but they are commonly used in bridge structures [3] and in oil and gas facilities [5]. Long-span tapered cantilevered plate girders are also used for stadium roof canopy framing. In building structures, they are used as transfer girders to support discontinued columns above large column-free areas such as atriums, auditoriums, ballrooms, exhibition halls, and other assembly areas, as shown in Figure 13-1a. An example of a welded plate girder is shown in Figure 13-1b.

Plate girders may also be used in the retrofitting of existing building structures where column-free areas are required and existing columns have to be cut off or removed below a certain floor level. An example of such detail is shown in Figure 13-2.

Plate girders are also used as crane support girders in heavy industrial structures. One disadvantage of using plate girders in building structures is that mechanical ducts may have to be placed below the girder to avoid cutting holes in the web of the girder which will reduce the strength of the transfer girder; locating mechanical ducts below a deep transfer plate girder increases the floor-to-floor height of the building, with a resulting increase in construction costs. In any case, it is more common to use transfer trusses in place of transfer girders in building structures because it allows for passage of mechanical

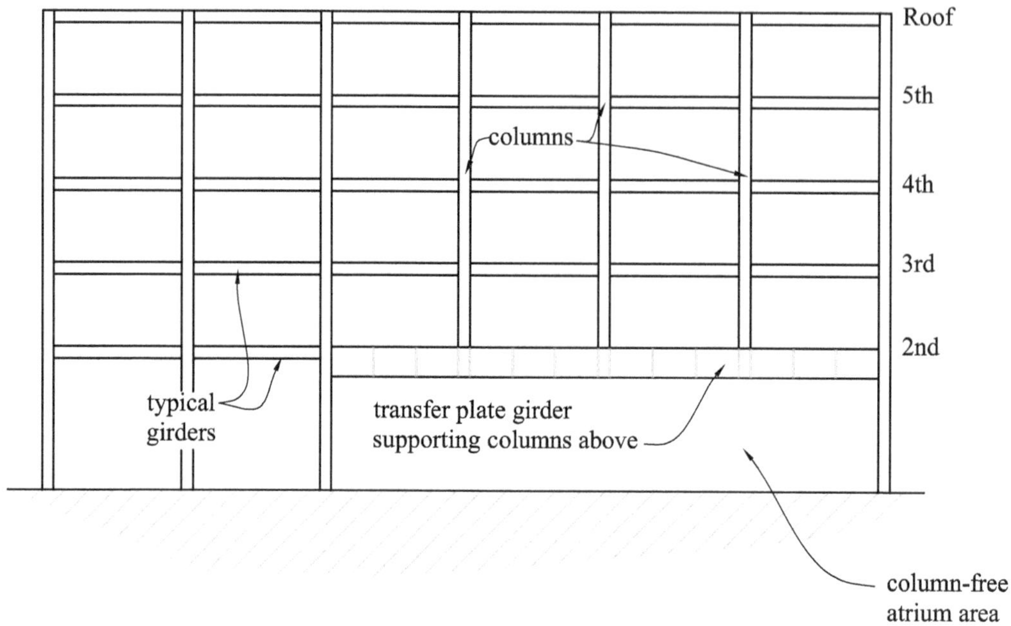

FIGURE 13-1a Transfer plate girder

FIGURE 13-1b Welded Plate Girder (Photo courtesy of JGM, Inc.)

ducts between the web members of the truss without adversely affecting the strength of the truss, and without increasing the required headroom.

The term *"plate girder"* no longer exists in the most recent AISC Specification [4]. Instead, this term has been replaced by the term "built-up sections," and the design requirements for flexure and shear for these sections are found in Sections F5 and G of the AISC Specification, respectively. In this text, we classify plate girders as built-up sections with noncompact or slender webs, and the design for flexure is carried out in accordance with Section F5 of the AISC Specification. The web depth-to-thickness ratios that define the limits of noncompactness or slenderness of the web of doubly symmetric and singly symmetric I-shaped built-up sections are given in Table 13-1.

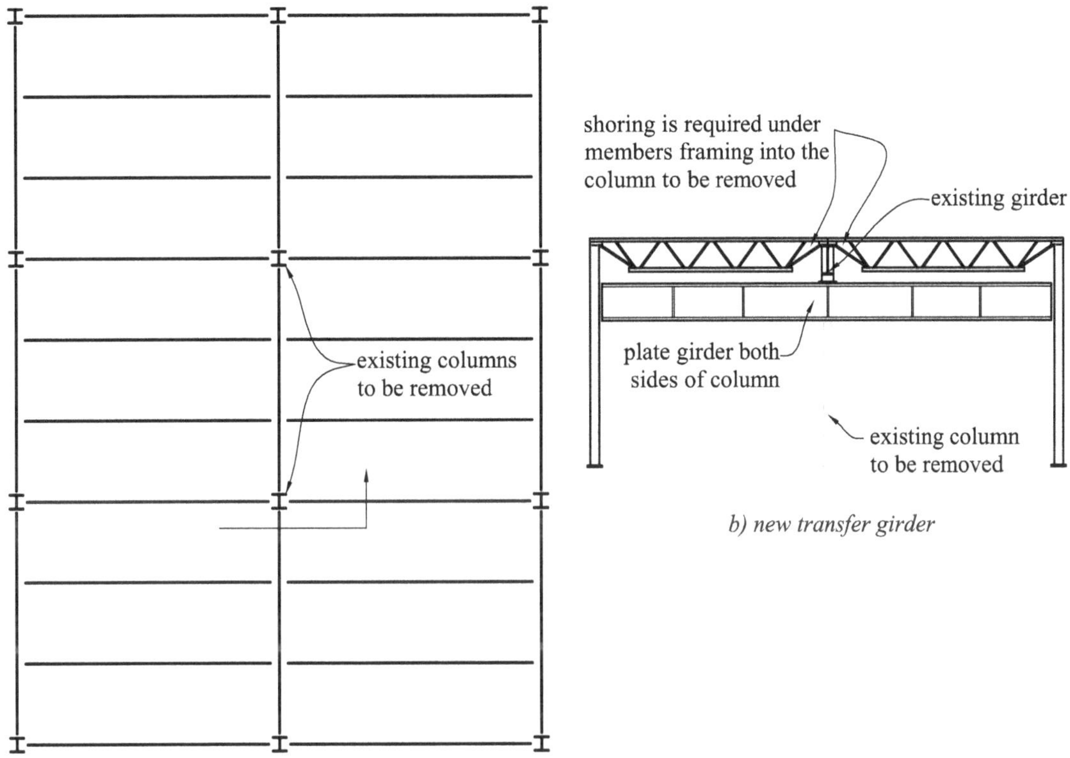

b) new transfer girder

a) existing roof plan

FIGURE 13-2 Transfer plate girder detail in retrofitting of an existing building

TABLE 13-1 Web Depth-to-Thickness Ratios for Noncompact and Slender webs (*AISCM* Table B4.1b, Cases 15 and 16)

	Webs of doubly symmetric I-shaped section and channels (Case 15)	**Webs of singly symmetric I-shaped section (Case 16)**
Noncompact web	$3.76\sqrt{\frac{E}{F_y}} < \frac{h}{t_w} \le 5.70\sqrt{\frac{E}{F_y}}$	$\dfrac{\frac{h_c}{h_p}\sqrt{\frac{E}{F_y}}}{\left(0.54\frac{M_p}{M_y} - 0.09\right)^2} < \frac{h_c}{t_w} \le 5.70\sqrt{\frac{E}{F_y}}$
Slender web	$\frac{h}{t_w} > 5.70\sqrt{\frac{E}{F_y}}$	$\frac{h_c}{t_w} > 5.70\sqrt{\frac{E}{F_y}}$

To prevent web buckling, web stiffeners can be added for stability (see Figure 13-3c). Section G2.4(b) of the AISC Specification permits the designer to either have the web stiffeners stopped short of the tension flange or welded to the tension flange. Where the stiffeners are subjected to bearing where they are required to transmit concentrated loads (e.g., at transfer columns) or concentrated reactions (e.g., at beam and girder supports on walls and tops of girders), the stiffeners must be welded to the tension flange. In past editions of the AISC specification, the stiffeners were cut short of the tension flange to prevent transverse welds in the flange that are perpendicular to the direction of the tensile stress in the bottom flange. It is now known that the fatigue strength of the stiffener detail is independent of its connection or lack thereof to the tension flange. For cut-short stiffeners that are connected to cross beams or frames, fatigue cracking in the web portion below the bottom of the transverse stiffener due to out-of-plane movement or out-of-plane bending in the web can occur.

The AISC Specification prescribes mandatory limits for the web depth-to-thickness ratios as a function of the clear spacing between the transverse stiffeners as presented in Table 13-2 [4]. The parameters in Tables 13-1 and 13-2 and Figure 13-3 are defined as follows:

a = Clear horizontal distance between transverse stiffeners, if any,
h = Depth of the web = Clear distance between the flanges of a plate girder minus the fillet or corner radius for rolled sections; for welded built-up sections, this is the clear distance between the flanges (see Figure 13-3),
h_c = Twice the distance between the *elastic* neutral axis (ENA) and the inside face of the compression flange for ***nonsymmetric*** welded built-up sections,
= Clear distance, h, between inside faces of the flanges for welded built-up sections with equal flange areas,
h_p = Twice the distance between the *plastic* neutral axis (PNA) and the inside face of the compression flange for ***nonsymmetric*** welded built-up sections,
= Clear distance, h, between inside faces of the flanges for welded built-up sections with equal flange areas,
t_w = Web thickness,
d = Overall depth of the plate girder,
M_p = Plastic moment of the section = $Z_x F_y$,
M_y = Yield moment of the section = $S_x F_y$,
A_w = Area of the web = ht_w,
A_{fc} = Area of the compression flange = $b_{fc} t_{fc}$,
I_y = Moment of inertia of the built-up cross section about the weak axis (y–y) or the vertical axis in the plane of the web, and
I_{yc} = Moment of inertia of the compression flange of the built-up cross section about the weak axis (y–y) or the vertical axis in the plane of the web.

In a typical welded plate girder (see Figure 13-3), the bending moment is resisted through a tension-compression couple in the top and bottom flanges, and the web plate primarily resists the vertical shear. Transverse vertical stiffeners are used to increase the web buckling capacity of the built-up section. Longitudinal web stiffener plates are rarely used except for very deep web [2]. The web plate is fillet welded to the top and bottom flange plates, and these welds resist the horizontal interface shear between the flanges and the web. The width and thickness of the compression and tension flanges may be varied along the span in proportion to the bending moment, but the web thickness is generally kept constant along the span of the girder. The depth of the plate girder may also be varied for long-span girders. When higher-yield steel is used for the flange plates and conventional steel is used for the web plates, the built-up section is known as a ***hybrid*** girder.

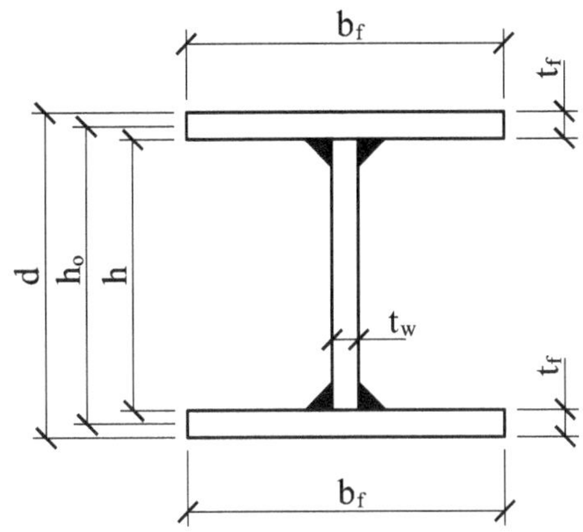

a. doubly symmetric plate girder section

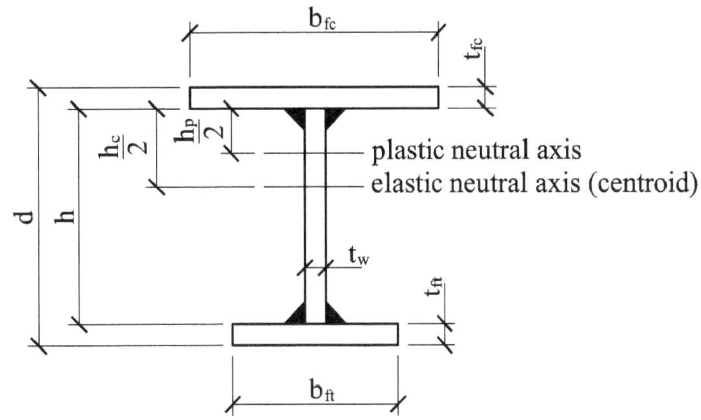

b. singly symmetric plate girder section

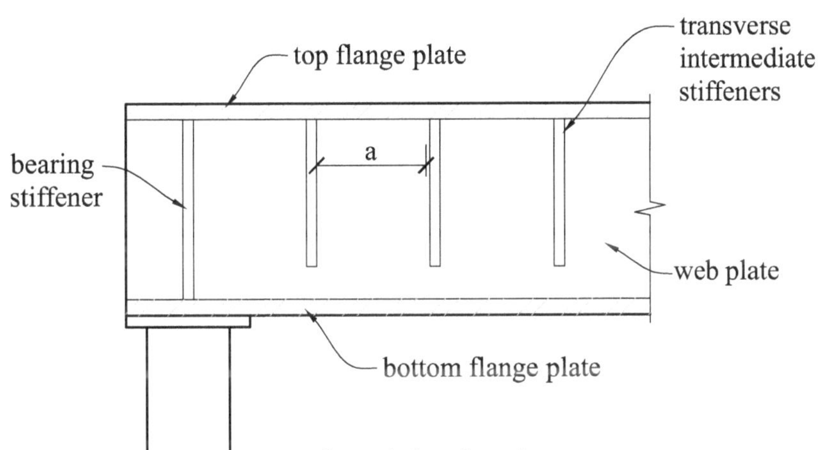

c. plate girder elevation

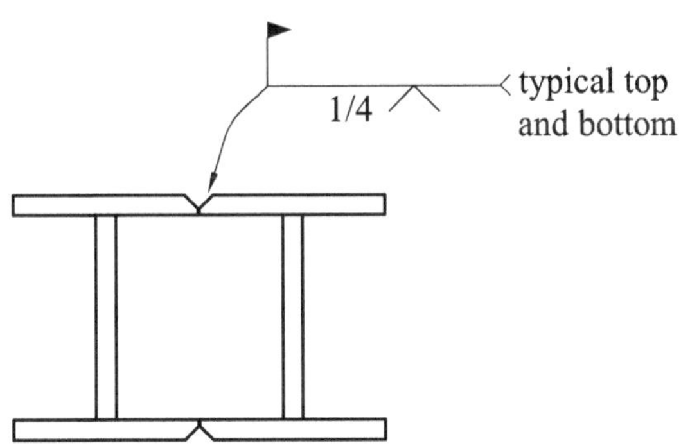

d. double web plate girder

FIGURE 13-3 Welded plate girder

TABLE 13-2 Size Limitations for I-shaped Built-up Sections (*AISCM*, Section F13.2)

Without transverse web stiffeners (i.e., unstiffened girders)	With transverse web stiffeners	
$\frac{h}{t_w} \le 260$ and $\frac{A_w}{A_{fc}} \le 10$	$\frac{a}{h} \le 1.5$	$\frac{a}{h} > 1.5$
	$\frac{h}{t_w} \le 12.0\sqrt{\frac{E}{F_y}}$	$\frac{h}{t_w} \le \frac{0.40E}{F_y}$
For **singly** symmetric sections, $0.1 \le \frac{I_{yc}}{I_y} \le 0.9$		

13-2 DESIGN OF PLATE GIRDERS

Whereas vertical shear is not typically a major consideration in the design of rolled I-shaped sections (see Chapter 6), however, shear is a major design consideration in the design of plate girders because of the relatively high web depth-to-thickness ratio. There are two possible approaches to the design of plate girders for shear:

a. Considering diagonal tension field action,
b. Not considering diagonal tension field action.

The condition under which diagonal tension field action can occur in a plate girder will depend on whether or not web stiffeners with an adequately small enough spacing to enable the development of a diagonal tension field are present. Therefore, in the design of plate girders, the two possible design and detailing scenarios to consider that will impact the development of the diagonal tension field are as follows:

1. ***Unstiffened Plate Girder:*** In this case, the plate girder is proportioned with adequate flange and web thicknesses to avoid the need for web stiffeners (see Figure 13-3a). The unstiffened plate girder option results in a thicker web and thicker flanges, therefore resulting in a heavier plate girder self-weight, but the complexity of fabrication is minimized. Since fabrication and erection costs make up more than 60% of the construction costs of structural steel buildings, this may be a more economical option in some cases. Note that the unstiffened web shear strength includes web shear buckling strength and web post-buckling strength, but diagonal tension field action cannot be developed because of the absence of stiffeners with adequately small spacing (i.e., stiffener spacing is greater than 3x the plate girder web depth or $a > 3h$).
2. ***Stiffened Plate Girder:*** In this case, web stiffeners are provided and spaced close enough to ensure the development of diagonal tension fields (i.e., stiffener spacing is less than or equal to 3x the plate girder depth or $a \le 3h$). This stiffened plate girder option results in a lighter-weight plate girder, but the fabrication costs are higher. It is possible to find optimum thicknesses for the web and flanges and an optimum size and spacing of stiffeners that will yield a stiffened plate girder with self-weight and fabrication and erection costs that are less costly than an equivalent unstiffened plate girder. The presence of vertical web stiffeners improves the buckling resistance of the web. After the initial shear buckling of the web, the plate girder is still capable of resisting additional loads because of the diagonal tension fields that develop in the web of the plate girder

between the stiffeners in the web post-buckling range [3]. A plate girder with diagonal tension field action behaves like a Pratt truss, as shown in Figure 13-4.

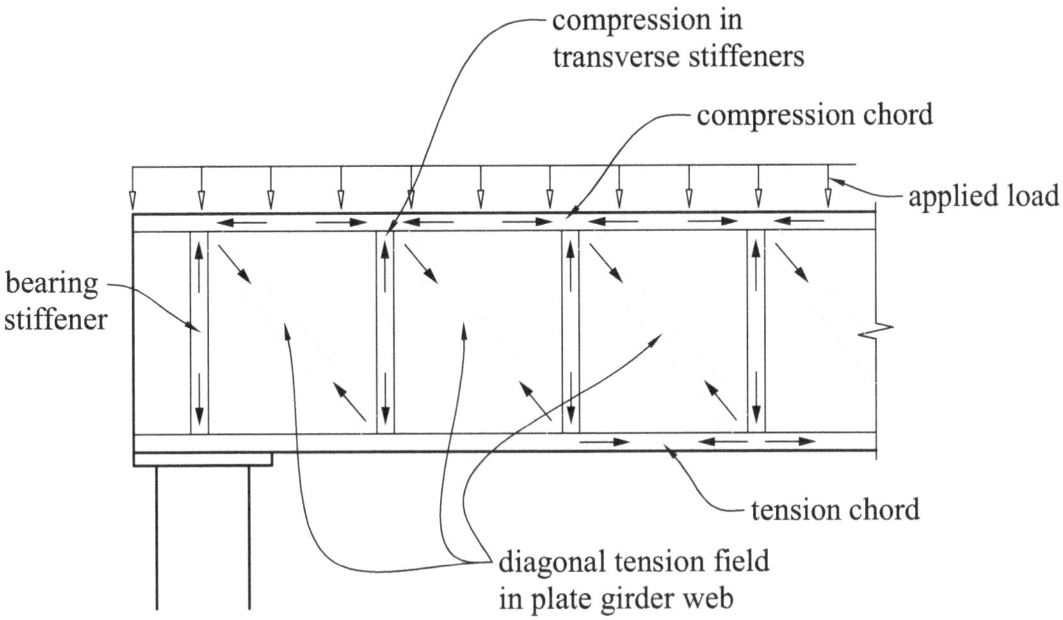

FIGURE 13-4 Diagonal tension field action in plate girders (Similar to a Pratt truss under load)

13-3 BENDING STRENGTH OF WELDED PLATE GIRDERS WITH SLENDER WEBS

When a plate girder section has a noncompact or slender web, AISC Specification Section F5 [4] is applicable; the nominal moment capacity, M_n, will be less than the plastic moment capacity, M_p, of the section because of several limit states that are attained before the section can reach its full plastic moment capacity. The following are the four possible limit states that may occur in built-up sections in bending:

- Compression flange yielding,
- Lateral torsional buckling,
- Compression flange local buckling, and
- Tension flange yielding.

The design moment capacity or allowable moment for a built-up section with slender webs depends on the compactness, noncompactness, or slenderness of the flanges and will be the smallest moment strength obtained for the following four possible limit states:

1. ***Compression flange yielding limit state***

Design moment capacity for LRFD, $\phi_b M_n = \phi_b R_{pg} F_y S_{xc}$, (13-1a)

The allowable moment for ASD, $M_a = \dfrac{M_n}{\Omega_b} = \dfrac{R_{pg} F_y S_{xc}}{1.67}$, (13-1b)

Where,

$\phi_b = 0.9$,

$\Omega_b = 1.67$

R_{pg} is as defined in equation (13-7),

F_y = Yield strength of the compression flange,

S_{xc} = Elastic section modulus about the strong axis $(x-x)$ relative to the outermost face of the compression flange of the built-up section $= I_{x-x}/y_c$, and

y_c = Distance from the neutral axis to the outer most face of the compression flange.

$Ix\text{–}x$ = moment of inertia of the plate girder about the strong axis $(x\text{–}x)$

M_n = nominal moment capacity $= R_{pg}F_yS_{xc}$

2. ***Lateral torsional buckling limit state***

Lateral torsional buckling is a function of the lateral unbraced length, L_b, of the compression flange of the plate girder. This limit state only applies when the unbraced length of the compression flange, L_b, is greater than L_p. The lateral torsional buckling limit state does not apply when $L_b \le L_p$. From AISC Specification Section F5,

The design bending strength for LRFD is $\phi_b M_n = \phi_b R_{pg} F_{cr} S_{xc}$, (13-2a)

The allowable moment for ASD, $M_a = \dfrac{M_n}{\Omega_b} = \dfrac{R_{pg}F_{cr}S_{xc}}{1.67}$, (13-2b)

Where,

The critical bending stress, F_{cr}, is obtained from Table 13-3.

$\phi_b = 0.9$,

$\Omega_b = 1.67$

R_{pg} is as defined in equation (13-7),

S_{xc} is as defined previously for the compression flange yielding limit state.

M_n = nominal moment capacity $= R_{pg}F_{cr}S_{xc}$

The lateral support length parameters, L_p and L_r, are defined in equations (13-5) and (13-6). L_p is the unbraced length at the transition point between plastic behavior and inelastic lateral-torsional buckling, while L_r is the unbraced length at the transition point between inelastic lateral-torsional buckling and elastic lateral-torsional buckling.

TABLE 13-3 Critical Bending Stress as a Function of the Unbraced Length

Unbraced length, L_b	Critical bending stress, F_{cr}	
$L_b \le L_p$	Lateral torsional buckling limit states does not apply.	
$L_p \le L_b \le L_r$	$F_{cr} = C_b\left[F_y - (0.3F_y)\left(\dfrac{L_b - L_p}{L_r - L_p}\right)\right] \le F_y$	(13-3)
$L_b > L_r$	$F_{cr} = \dfrac{C_b\pi^2 E}{\left(\dfrac{L_b}{r_t}\right)^2} \le F_y$	(13-4)

$$L_p = 1.1 r_t \sqrt{\frac{E}{F_y}}, \tag{13-5}$$

$$L_r = \pi r_t \sqrt{\frac{E}{0.7F_y}}, \text{ and} \tag{13-6}$$

C_b = Bending moment gradient factor (see Chapter 6). It is conservative to assume a C_b value of 1.0.

The bending strength reduction factor, R_{pg}, is given as

$$R_{pg} = 1 - \left(\frac{a_w}{1200 + 300a_w}\right)\left(\frac{h_c}{t_w} - 5.7\sqrt{\frac{E}{F_y}}\right) \leq 1.0, \tag{13-7}$$

where

a_w = Ratio of two times the web area in compression due to the application of major axis bending moment alone to the area of the compression flange components. Mathematically,

$$a_w = \frac{h_c t_w}{b_{fc} t_{fc}} \leq 10, \tag{13-8}$$

b_{fc} = Width of the compression flange,

t_{fc} = Thickness of the compression flange,

t_w = Thickness of the web,

h_c = Twice the distance between the elastic neutral axis (ENA) and the inside face of the compression flange for ***nonsymmetric*** welded built-up sections, and

= Clear distance, h, between inside faces of the flanges for welded built-up sections with equal flange areas.

The parameter r_t in equations (13-4) through (13-6) is the radius of gyration of the flange components in flexural compression plus one-third of the web area in compression due to the application of major axis bending moment. Mathematically, r_t can be approximated for I-shaped sections as

$$r_t \approx \frac{b_{fc}}{\sqrt{12\left(1 + \frac{1}{6}a_w\right)}} \text{ in.} \tag{13-9}$$

3. ***Compression flange local buckling limit state***

This limit state does not apply to built-up sections with ***compact*** flanges. For all other sections, the design moment capacity for LRFD is given as

$$\phi_b M_n = \phi_b R_{pg} F_{cr} S_{xc}, \tag{13-10a}$$

$$\text{The allowable moment for ASD, } M_a = \frac{M_n}{\Omega_b} = \frac{R_{pg} F_{cr} S_{xc}}{1.67}, \tag{13-10b}$$

where the critical bending stress, F_{cr}, is obtained from Table 13-4 and R_{pg} is as defined in equation (13-7).

S_{xc} is as defined previously for the compression flange yielding limit state, and

$\phi_b = 0.9$,

$\Omega_b = 1.67$

M_n = nominal moment capacity = $R_{pg} F_{cr} S_{xc}$

F_l, the nominal compressive flange stress for major axis bending above which inelastic buckling limit states apply, is obtained from Table 13-5 and used to calculate the critical stress, F_{cr}, in Table 13-4.

TABLE 13-4 Critical Bending Stress, F_{cr} due to Flange Local Buckling

Flange compactness	Controlling failure mode	Critical bending stress, F_{cr}
Compact flange $\frac{b_{fc}}{2t_{fc}} \leq 0.38\sqrt{\frac{E}{F_y}}$	Flange yielding	F_y
Noncompact flanges $0.38\sqrt{\frac{E}{F_y}} < \frac{b_{fc}}{2t_{fc}} \leq 0.95\sqrt{\frac{k_c E}{F_L}}$	Inelastic flange local buckling	$F_{cr} = \left[F_y - (0.3F_y)\left(\frac{\lambda - \lambda_{pf}}{\lambda_{rf} - \lambda_{pf}}\right)\right]$
Slender flanges $\frac{b_{fc}}{2t_{fc}} > 0.95\sqrt{\frac{k_c E}{F_L}}$	Elastic flange local buckling	$F_{cr} = \frac{0.9Ek_c}{\left(\frac{b_f}{2t_f}\right)^2}$

F_y = Yield strength of the compression flange,
For F_L, see Table 13-5.

$\lambda = \frac{b_{fc}}{2t_{fc}}$,

λ_{pf} = Limiting slenderness for the ***compact*** compression flange obtained from Table 6-2 (see *AISCM*, Table B4.1b (case 11))

$= 0.38\sqrt{\frac{E}{F_y}}$ for the compression flange,

λ_{rf} = Limiting slenderness for ***noncompact*** compression flange obtained from Table 6-2 (see *AISCM*, Table B4.1b (case 11))

$= 0.95\sqrt{\frac{k_c E}{F_L}}$ for the compression flange,

b_{fc} = Width of the compression flange,
t_{fc} = Thickness of the compression flange,

$k_c = \frac{4}{\sqrt{\frac{h}{t_w}}}$, *and* $0.35 \leq k_c \leq 0.76$

t_w = Web thickness, and
h = Clear distance between the inside faces of the flanges for welded built-up sections.

4. *Tension flange yielding limit state*

This limit state ***does not*** apply to built-up sections when the section modulus of the built-up section with respect to the tension face, S_{xt}, is greater than or equal to the section modulus with respect to the compression face, S_{xc} (i.e., when $S_{xt} \geq S_{xc}$).
For all other sections, the design moment capacity for LRFD is given as

$$\phi_b M_n = \phi_b F_y S_{xt} \quad (13\text{-}11a)$$

The allowable moment for ASD, $M_a = \frac{M_n}{\Omega_b} = \frac{F_y S_{xt}}{1.67}$, (13-11b)

TABLE 13-5 F_L for Major Axis Bending (use in Table 13-4)

Description	F_L
I-shaped built-up section with **noncompact** web ***and*** $\frac{S_{xt}}{S_{xc}} \geq 0.7$	$0.7F_y$
I-shaped built-up section with **noncompact** web ***and*** $\frac{S_{xt}}{S_{xc}} < 0.7$	$F_y \frac{S_{xt}}{S_{xc}} \geq 0.5F_y$

h_c = Twice the distance between the elastic neutral axis and the inside face of the compression flange for ***nonsymmetric*** welded built-up sections,
= Clear distance, h, between the inside faces of the flanges for ***symmetric*** welded built-up sections,
S_{xc} = Elastic section modulus about the strong axis (x–x) with respect to the compression flange of the built-up section, and
S_{xt} = Elastic section modulus about the strong axis (x–x) with respect to the tension flange of the built-up section.
Source: Adapted from *AISCM* Section F4.3 [4].

13-4 DESIGN FOR SHEAR IN PLATE GIRDERS *WITHOUT* DIAGONAL TENSION FIELD ACTION (*AISCM* SECTION G2.1)

Due to the relatively higher web-depth-to-thickness ratios used in plate girders compared to hot-rolled I-shaped sections, the design for shear in plate girders is more complicated than for rolled sections. The failure mode in shear for plate girders with low web-depth-to-thickness ratio is shear yielding, but for plate girders with moderate to high web depth-to-thickness ratio, failure occurs by shear buckling with a further increase in strength due to post-buckling in the web with or without tension field action [6].

There are two approaches for determining the shear strength of plate girders with moderate to high web-depth-to-thickness ratios when subjected to shear and bending in the plane of the web. In the first method, which is conservative, the shear strength takes into account the post-buckling strength of the web that occurs after shear buckling, but the development of the classical tension field action is not considered or included. In the second approach, which will be covered in another section of this chapter, the shear strength of the web includes the shear buckling strength and additional post-buckling strength due to the diagonal tension field action. Diagonal tension field action is only possible and can only be considered in plate girders with closely spaced transverse stiffeners (i.e., stiffener spacing-to-web depth ratio, $a/h \leq 3$). For ease of construction and to reduce the cost associated with providing stiffeners in plate girders, some engineers provide sufficient web thickness in plate girders to avoid the need for transverse stiffeners.

For *unstiffened* and *stiffened* webs of doubly symmetric and singly symmetric shapes subject to shear in the plane of the web, the design shear strength and the allowable shear, respectively, are given as follows:

The design shear strength, $\phi_v V_n = \phi_v 0.6 F_{yw} A_w C_{v1}$ for LRFD (13-12a)

The allowable shear, $V_a = \frac{V_n}{\Omega_v} = \frac{0.6 F_{yw} A_w C_{v1}}{\Omega_v}$, for ASD (13-12b)

where

$\phi_v =$ 0.9 for LRFD $(\Omega_v = 1.67 \textit{ for ASD})$, except

$\phi_v =$ 1.0 for LRFD $(\Omega_v = 1.5 \textit{ for ASD})$ for webs of rolled I-shaped members with $\frac{h}{t_w} \leq 2.24\sqrt{\frac{E}{F_y}}$

C_{v1} = Web shear strength coefficient for postbuckling with no tension field action (see Table 13-6)

h = *clear distance between the inside faces of the top and bottom flanges less the fillet at each flange, in inches*

a = *clear distance between transverse stiffeners, in inches*

A_w = Overall depth of built-up section times the web thickness = dt_w,

F_{yw} = Yield strength of the web material,

The web shear strength coefficient, C_{v1}, and the web plate shear buckling coefficient, k_v, are obtained from Table 13-6 and Table 13-7, respectively, as a function of the web depth-to-thickness ratio.

TABLE 13-6 Web Shear Strength Coefficient, C_{v1}

Web depth-to-thickness ratio, h/t_w	Web shear coefficient, C_{v1}
$\frac{h}{t_w} \le 1.10\sqrt{\frac{k_v E}{F_{yw}}}$	1.0
$\frac{h}{t_w} > 1.10\sqrt{\frac{k_v E}{F_{yw}}}$	$\frac{1.10\sqrt{\frac{k_v E}{F_{yw}}}}{h/t_w}$
^For the web plate shear buckling coefficient, k_v, see Table 13-7	

TABLE 13-7 Web Plate Shear Buckling Coefficient, k_v

Web depth-to-thickness ratio, h/t_w	Web plate shear buckling coefficient, k_v
Webs without transverse stiffeners	5.34
Webs with transverse stiffeners with $\frac{a}{h} \le 3$	$5 + \frac{5}{(a/h)^2}$
Webs with transverse stiffeners with $\frac{a}{h} > 3$	5.34

a = Horizontal clear distance between transverse stiffeners, and
h = Clear distance between the flanges for welded plate girders.

Stiffened Webs *without* Diagonal Tension Field Action (*AISCM* Section G2)

If web stiffeners are provided, but with a stiffener spacing-to-web depth ratio, a/h, greater than 3, then the diagonal tension field cannot develop and should not be considered.

Without diagonal tension field action, the transverse stiffeners, if provided, are there only to prevent web buckling. For transverse stiffener requirements with or without tension field action, see Section 13-5. Transverse stiffeners are ***not*** required if one of the following conditions is satisfied:

- $\frac{h}{t_w} \le 2.54\sqrt{\frac{E}{F_{yw}}}$

or

- $V_u \le \phi V_n$, *where* ϕV_n *is obtained from equation* $(13-12a)$ for $k_v = 5.34$.

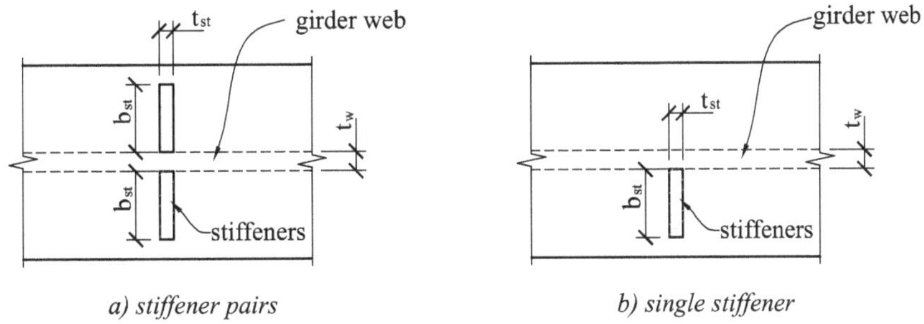

FIGURE 13-5 Plan view of web stiffeners

13-5 DIAGONAL TENSION FIELD ACTION IN PLATE GIRDERS (*AISCM* SECTION G2.2)

When a stiffened plate girder is loaded, it continues to support loads even after the initial shear buckling of the web panels between the transverse web stiffeners [3]. After this initial web buckling, diagonal tension fields develop in the webs of the plate girder between the transverse stiffeners during the post-buckling phase, leading to an increase in the strength of the plate girder. The formation of these diagonal tension bands, or fields, during the post-buckling phase, is balanced by vertical compression bands in the transverse stiffeners (which are assumed to support no loads prior to web buckling), thus creating a Pratt truss-like load-carrying mechanism [3] as shown in Figure 13-4. The consideration of diagonal tension field action, which is only permitted if certain conditions are met (i.e., $a/h \leq 3$) and if the web plate is supported on all four sides by the flanges and the transverse stiffeners, can increase the load-carrying capacity of a plate girder by up to two to three times the initial web-buckling capacity [3]. The AISC Specification does permit diagonal tension field action to be considered for end web panels in stiffened plate girders when $a/h \leq 3$. It should be noted, however, that in the AISC Specification, it is not mandatory to consider tension field action in the design of plate girders [4].

The AISC Specification does *not* permit diagonal tension field action to be considered for the following situations:

- Built-up sections with web members for which $\frac{a}{h} > 3$ or $\frac{a}{h} > \left(\frac{260}{h/t_w}\right)^2$ because of these a/h ratios, the diagonal tension bands become too flat to function effectively as truss members.
- Built-up sections with $\frac{2A_w}{A_{fc} + A_{ft}} > 2.5$ or $\frac{h}{b_{fc}}$ or $\frac{h}{b_{ft}} > 6.0$.

The design shear strength, when the diagonal tension field is considered and the tension field is assumed to yield, is given in Table 13-8 (design aids are presented in *AISCM* Table 3-16 (a through c) for 50 ksi steel to facilitate the solution of the equations in Table 13-8).

where

$\phi = 0.9$,

k_v has been defined previously and is obtained from Table 13-7.

C_{v2} = the web shear buckling coefficient obtained from Table 13-9 (for postbuckling with tension field action).

A_{fc} = Compression flange area = $b_{fc} t_{fc}$, in.^{2},

A_{ft} = Tension flange area = $b_{ft} t_{ft}$, in.^{2},

b_{fc}, t_{fc} = Compression flange width and thickness, respectively, in.,

b_{ft}, t_{ft} = Tension flange width and thickness, respectively, in.,

F_{yw} = Yield strength of the web material,

A_w = Overall depth of the built-up section times the web thickness = dt_w, and

a and h are as previously defined in Section 13-4.

Note that the *web shear strength coefficient*, C_{v1}, applies to the post-buckling strength of webs *without* tension field action considered (see AISC Specification Section G2.1). In contrast, the *web shear buckling coefficient*, C_{v2}, applies to the post-buckling strength of stiffened interior web panels with stiffeners that are spaced such that a/h ≤ 3, and *with* tension field action considered (AISC Specification Section G2.2).

Stiffened Webs with Diagonal Tension Field Action Considered

The design shear strength, $\phi_v V_n$, with the diagonal tension field considered, is given in Table 13-8. When diagonal tension field action is considered, which is applicable to interior web panels of plate girders with closely spaced stiffeners (i.e., stiffener spacing-to-web depth ratio, a/h ≤ 3), the transverse stiffeners must satisfy the requirements indicated below (AISC Specification Section G2.2).

Transverse Stiffener Requirements

Transverse stiffeners *are not required* if one of the following two conditions is satisfied:

- $\dfrac{h}{t_w} \le 2.46\sqrt{\dfrac{E}{F_{yw}}}$

  or

- $V_u \le \phi_v V_n$ for $k_v = 5.34$ (see equation (13-12a))

  Where, V_u = factored shear or required shear strength

  For all other conditions, transverse stiffeners *must* be provided, and the spacing, a, between the transverse stiffeners, the width of the stiffener, b_{st}, and the thickness, t_{st}, of the stiffener must be selected to satisfy the following two conditions [4]:

- $\dfrac{b_{st}}{t_{st}} \le 0.56\sqrt{\dfrac{E}{F_{y,st}}}$,

- $I_{st} \ge I_{st2} + (I_{st1} - I_{st2})\rho_w$

  Where,

  $F_{y,st}$ = Yield strength of the stiffener material,

  F_{yw} = Yield strength of the web material,

  I_{st} = Moment of inertia of the transverse stiffeners:

- For **a pair of transverse stiffeners** (i.e., stiffeners on both sides of the web), the moment of inertia of the transverse stiffeners **about a horizontal axis at the centerline of the web** (see Figure 13-5a) is,

  $$I_{st} = \frac{t_{st}(2b_{st} + t_w)^3}{12} \text{ in.}^4\text{, for a pair of stiffeners}$$

- For a **single** transverse stiffener (see Figure 13-5b), the **moment of inertia of the transverse stiffener about a horizontal axis at the face of the web in contact with the stiffener** is,

  $$I_{st} = \frac{t_{st}(b_{st})^3}{12} + t_{st}b_{st}\left(\frac{b_{st}}{2}\right)^2$$

$$I_{st1} = \frac{h^4 \rho_{st}^{1.3}}{40}\left(\frac{F_{yw}}{E}\right)^{1.5}$$ = minimum moment of inertia of the transverse stiffener required for development of the full shear post-buckling resistance of the stiffened web panels. $V_r = V_{c1}$, in^4.

$$I_{st2} = \left[\frac{2.5}{\left(\frac{a}{h}\right)^2} - 2\right] b_p t_w^3 \geq 0.5 b_p t_w^3$$ = minimum moment of inertia of the transverse stiffener required for the development of the web shear buckling resistance, $V_r = V_{c2}$, in^4.

V_{c1} = available shear strength, in kips = V_n (see equation (13-12a) or Table 13-8)
V_{c2} = available shear strength, in kips = $0.6F_y A_w C_{v2}$
(See Table 13-9 for C_{v2})
ρ_w = maximum shear ratio

$$= \left[\frac{V_r - V_{c2}}{V_{c1} - V_{c2}}\right] \geq 0$$, within the web panels on each side of the transverse stiffener

ρ_{st} = larger of $\frac{F_{y_w}}{F_{y,st}}$ *or* 1.0

V_r = required shear strength in the panel under consideration, in kips = V_u
b_{st} = width of stiffener perpendicular to the longitudinal axis of the plate girder, in inches (see Figure 13-5b)
t_{st} = thickness of stiffener, in inches (see Figure 13-5b)
$\frac{b_{st}}{t_{st}}$ = Stiffener width-to-thickness ratio
b_p = smaller of the dimensions a and h, in inches.
Note that I_{st} may be conservatively taken as I_{st1}.
A = Horizontal clear distance between transverse stiffeners, and
h = Clear distance between the flanges for welded plate girders.
V_u = Factored shear force or required shear strength, and
ϕV_n = Design shear strength.

TABLE 13-8 Design Shear Strength, $\phi_v V_n$, with Diagonal Tension Field Action Considered^

Web depth-to-thickness ratio, h/t_w	Design shear capacity, $\phi_v V_n$
$\frac{h}{t_w} \leq 1.10\sqrt{\frac{k_v E}{F_{yw}}}$	$\phi_v 0.6 F_{yw} A_w$
$\frac{h}{t_w} > 1.10\sqrt{\frac{k_v E}{F_{yw}}}$ When $2A_w/(A_{fc} + A_{ft}) \leq 2.5$, $h/b_{fc} \leq 6.0$ and $h/b_{ft} \leq 6.0$	$\phi_v 0.6 F_{yw} A_w \left(C_{v2} + \frac{1 - C_{v2}}{1.15\sqrt{1 + (a/h)^2}}\right)$
Otherwise =>	$\phi_v 0.6 F_{yw} A_w \left(C_{v2} + \frac{1 - C_{v2}}{1.15\left[a/h + \sqrt{1 + (a/h)^2}\right]}\right)$

^ (i.e., post-buckling strength considered and $a/h \leq 3$)
(See Table 13-9 for the web shear buckling coefficient, C_{v2} and Table 13-7 for the web plate shear buckling coefficient, k_v); $\phi_v = 0.9$

TABLE 13-9 Web Shear Buckling Coefficient, C_{v2}

Web depth-to-thickness ratio, h/t_w	Web shear buckling coefficient, C_{v2}
$\frac{h}{t_w} \leq 1.10\sqrt{\frac{k_v E}{F_{yw}}}$	1.0
$1.10\sqrt{\frac{k_v E}{F_{yw}}} < \frac{h}{t_w} \leq 1.37\sqrt{\frac{k_v E}{F_{yw}}}$	$\frac{1.10\sqrt{\frac{k_v E}{F_{yw}}}}{h/t_w}$
$\frac{h}{t_w} > 1.37\sqrt{\frac{k_v E}{F_{yw}}}$	$\frac{1.51 E k_v}{(h/t_w)^2 F_{yw}}$

To determine the design Shear Strength, $\phi_v V_n$ for **end web panels** with diagonal tension field action considered, the design shear strength for end web panels with a/h ≤ 3 from the AISC Specification Section G2.3, is given for built-up I-shaped members with equal flange areas as follows:

$$\phi_v V_n = \phi_v 0.6 F_{yw} A_w \left[C_{v2} + \beta_v \left(\frac{1 - C_{v2}}{1.15\sqrt{1 + (a/h)^2}} \right) \right]$$

Where

$$\beta_v = \frac{2.8\left(\sqrt{M_{pf} + M_{pm}} + \sqrt{M_{pst} + M_{pm}}\right)}{h\sqrt{F_{yw} t_w}\,(1 - C_{v2})} \leq 1.0$$

ϕ_v = 0.9

F_{yw} = Yield strength of the web material,

M_{pf} = plastic moment of a section consisting of the flange of the I-shape member and a segment of the web with a depth, d_e, kip-in (See Figure 13-5c, adapted from *AISCM* Figure C-G2.3.)

M_{pst} = plastic moment of a section consisting of the end stiffener plus the length of the web equal to d_e plus the distance from the inside face of the stiffener to the end of the beam, except that the distance from the inside face of the stiffener to the end of the beam shall not exceed $0.84 t_w \sqrt{E/F_y}$ for calculation purposes, kip-in. (See Figure 13-5b, adapted from *AISCM* Figure C-G2.3).

M_{pm} = smaller of M_{pf} and M_{pst}, kip-in.

When $C_{v2} \leq 0.8$, $d_e = 35 t_w (0.8 - C_{v2})^2$

When $C_{v2} > 0.8$, $d_e = 0$

The flexural stress in the tension flange in the end panel must not be greater than $0.35F_y$. That is,

For LRFD, $\frac{(1.0)M_u}{S_{xt}} \leq 0.35F_y$

$$\text{For ASD, } \frac{(1.6)M_s}{S_{xt}} \leq 0.35F_y$$

M_u = total factored moment for LRFD,
M_s = total service or unfactored moment for ASD,

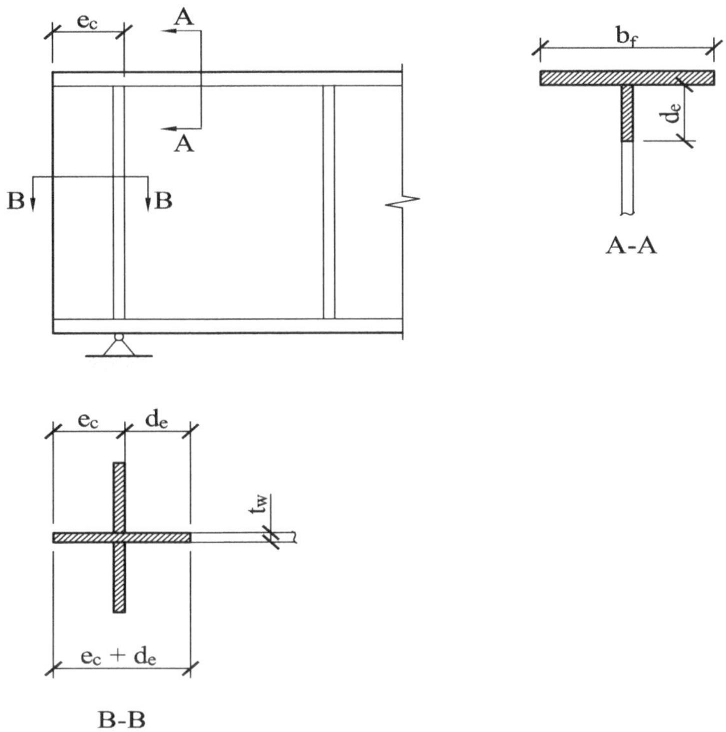

FIGURE 13-5c Effective cross sections for computing M_{pf} and M_{pst} (AISC Specification Section C-G2.3)
Source: Adapted from *AISCM* Figure C-G2.3 [1]. Copyright © American Institute of Steel Construction.

13-6 CONNECTION OF WELDED PLATE GIRDER COMPONENTS

The connections of the components of a welded plate girder include the fillet welds at the web-to-flange interface, the fillet welds at the transverse stiffener-to-web interface, and the fillet welds at the bearing stiffener-to-web interface. For very long-span girders, it might also include the welds at the flange and web splice locations. For some projects, the connections at the web-to-flange interface of the plate girders have been achieved using full penetration groove welds instead of fillet welds.

Connection of Plate Girder Flanges and Web

The connection between the flange plates and the web plate of the built-up section can be achieved by using intermittent or continuous fillet welds, with the latter more commonly preferred in design practice. On the other hand, care should be taken with continuous fillet welds, especially for thin plates, as the heat generated from the welding process could lead to distortion of these plates, creating undesirable residual stresses. The factored horizontal shear force per unit length or the shear flow, v_h, at the interface between the web plate and the flange plate is calculated as follows [1]:

$$v_h = \frac{V_u Q_f}{I_{x-x}}, \tag{13-13}$$

where

V_u = Factored vertical shear force for LRFD, kips,

Q_f = Statical moment of the flange area about the neutral axis of the built-up section, in.^{3}, and

I_{x-x} = Moment of inertia of the built-up section about the strong axis (x–x), in.^{4}.

Although the vertical shear force, V_u, may vary along the length of the girder depending on the loading, and the required weld size will also vary, it is practical to use the maximum factored shear force in the girder in equation (13-13) and to provide a constant size of fillet weld to resist the horizontal shear flow, v_h.

Connection of Intermediate Stiffeners to Girder Web (*AISCM* Section G2.2)

The intermediate stiffeners do not have to be connected to the tension flange, but they are connected to the web with intermittent fillet welds and the end distance between the intermittent stiffener-to-web fillet weld and the nearest toe of the web-to-flange fillet weld shall not be less than $4t_w$ nor more than $6t_w$ (see Figure 13-6). The clear spacing between the web-to-stiffener fillet welds shall not be greater than $16t_w$, or 10 in. Pairs of stiffeners do not have to be connected to the compression flange but to the web only. Single stiffeners should be connected to both the web and the compression flange to resist any uplift arising from torsion in the flange of the plate girder [4].

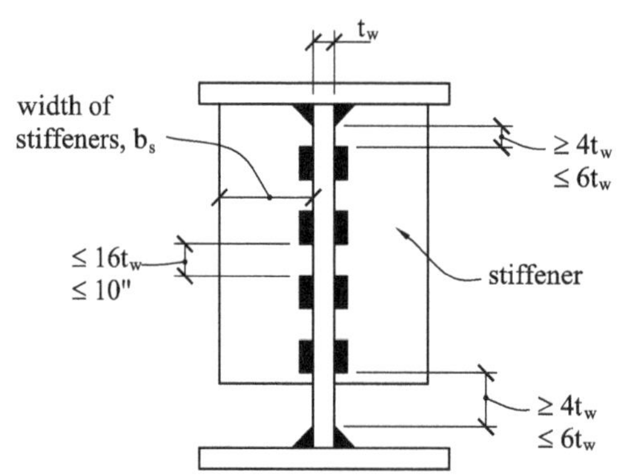

FIGURE 13-6 Spacing requirements for the stiffener-to-web fillet welds

Connection of Bearing Stiffeners to Girder Web

The bearing stiffeners are connected to the girder web with *continuous* fillet welds on both sides of the stiffener plate because bearing stiffeners are direct load-carrying elements that have to transfer the support reactions into the girder web. The design of bearing stiffeners has already been discussed in Chapter 6.

13-7 PLATE GIRDER PRELIMINARY DESIGN (LRFD)

In the preliminary design of plate girders, it can be assumed that the web resists all of the vertical shear force and the flanges resist all of the bending moment. The design procedure for the LRFD method is as follows:

1. Calculate the maximum factored moment, M_u, and maximum factored shear, V_u, initially neglecting the self-weight of the plate girder.
2. Assume that the plate girder's overall depth, d, is between $L/8$ and $L/12$, where L is the span of the girder [1].
   - For $d \leq 48$ in., it is recommended that a sufficient web thickness be provided to avoid the use of transverse stiffeners.
   - Even for $d > 48$ in., as few stiffeners as possible should be used to minimize the cost of fabrication [2].
3. Assume a flange width, $b_f > d/6$, but preferably greater than 12 in. [2].
4. Assume that the couple of forces in the top and bottom flanges, $T_u = C_u \approx M_u/0.95d$.
5. Assume that the initial flange thickness, $t_f = T_u/0.9F_y\, b_f \geq {}^3/_4$ in.
6. The depth of the web, $h = d - 2t_f$.
7. Select a web thickness, t_w, that meets the proportion limits given in Table 13-2, but it should be greater than or equal to ${}^1/_2$ in. [2].

13-8 PLATE GIRDER FINAL DESIGN (LRFD)

Using the size obtained from the preliminary design above, the procedure for the final design using the LRFD method is as follows:

1. Calculate the self-weight of the plate girder and recalculate the total factored moment, M_u, and the total factored shear, V_u, on the plate girder.
2. Decide whether to include diagonal tension field action. Note that tension field action can only be considered and included in the shear strength of the plate girder when stiffeners are used and only if their spacing is such that the stiffener spacing-to-web depth ratio, $a/h \leq 3.0$. It is ***not*** mandatory to account for the post-buckling strength of stiffened webs resulting from diagonal tension field action.
   - **a.** If the diagonal tension field action ***is not*** considered, stiffeners should not be used, and a sufficient web thickness for the plate girder should be provided.
   - **b.** If diagonal tension field action ***is*** to be considered as contributing to the shear strength of the plate girder, then select a stiffener spacing such that the a/h ratio is between 1.0 and 2.0. Ensure that the tension stiffeners meet the requirements in Section 13-5.
3. With the trial section selected in the preliminary design phase, determine the design moment capacity, ϕM_n, considering all four possible limit states (see Section 13-3). The smallest ϕM_n is the design moment capacity of the built-up section.
4. Determine k_v from Table 13-7, C_{v1} from Table 13-6, and C_{v2} from Table 13-9.
5. Based on the choice made in step 2 regarding diagonal tension field action, determine the design shear strength, $\phi_v V_n$.
6. Check deflections and vibrations as discussed previously in Chapters 6 and 12, respectively.

7. Design the bearing stiffeners at the support reactions or at concentrated load locations as discussed in Chapter 6.
8. Design the fillet welds required at the flange-to-web interface and the stiffener-to-web interface (see Section 13-6).

EXAMPLE 13-1

Design of a Welded Transfer Plate Girder

A five-story building has the elevation shown in Figure 13-7. The building has typical column grids at 30 ft. in the center in both orthogonal directions. The floor dead load is 100 psf and the live load is 100 psf (50 psf at the 2nd floor). The roof dead load is 30 psf with a flat roof snow load of 67 psf. In order to span over a large column-free atrium area at the ground floor, a series of transfer plate girders spanning 60 ft. and spaced at 30 ft. on the center are to be used at the second-floor level. Design a typical interior second-floor transfer plate girder to support the column loads from above as well as the second-floor load. Neglect any live load reduction and ignore diagonal tension field action. Assume ASTM A992 Grade 50 steel and the compression flange of the plate girder is fully braced.

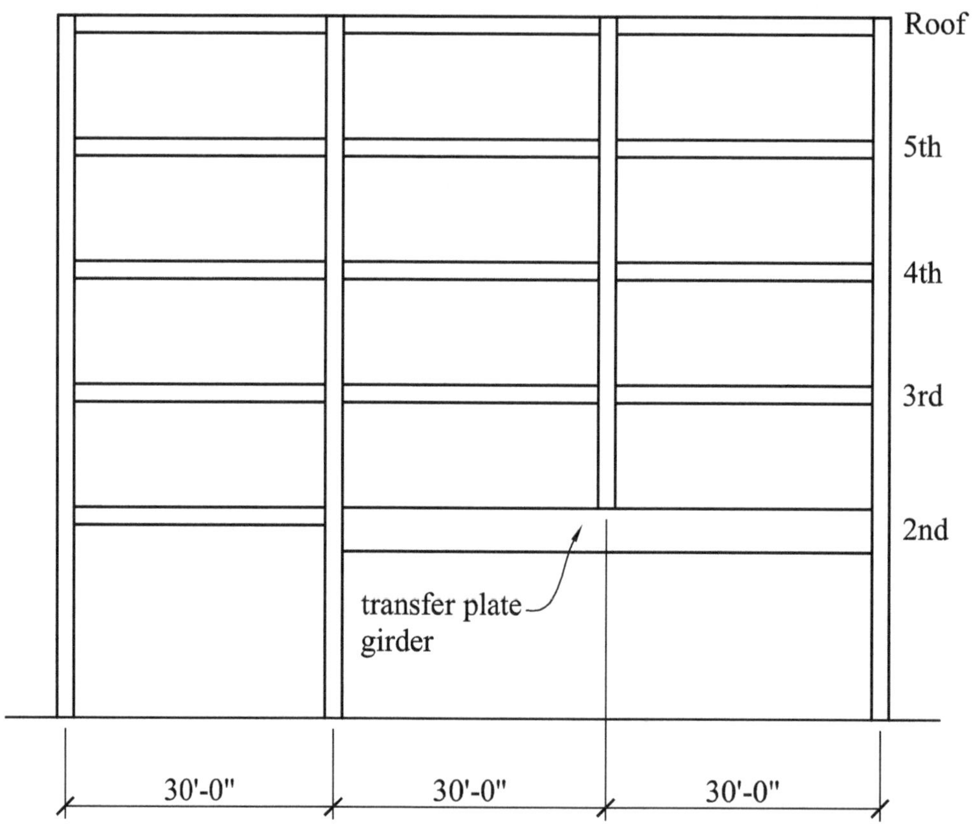

FIGURE 13-7 Interior building section for Example 13-1

Solution

Second-Floor Load on Transfer Girder

At this preliminary stage, we will initially neglect the self-weight of the girder, but this will be calculated later in the final design stage.

Tributary width of girder = 30 ft.

Assume that the infill beams are to be spaced such that a uniform floor load can be assumed on the girder. Therefore, the uniformly distributed loads on the girder from the 2nd floor are

$$w_D = (100 \text{ psf})(30 \text{ ft.}) = 3000 \text{ lb./ft.} = 3 \text{ kips/ft., and}$$

$$w_L = (50 \text{ psf})(30 \text{ ft.}) = 1500 \text{ lb./ft.} = 1.5 \text{ kips/ft.}$$

The column loads from the 3rd, 4th, and 5th floors, and the roof level are calculated as

$$P_D = [(30 \text{ psf} + (3 \text{ floors})\ (100 \text{ psf})](30 \text{ ft.} \times 30 \text{ ft.})$$

$$= 297{,}000 \text{ lb.} = 297 \text{ kips,}$$

$$P_S = (67 \text{ psf})(30 \text{ ft.})(30 \text{ ft.}) = 60{,}000 \text{ lb.} = 60 \text{ kips, and}$$

$$P_L = (3 \text{ floors})(100 \text{ psf})(30 \text{ ft.})(30 \text{ ft.}) = 270{,}000 \text{ lb.} = 270 \text{ kips.}$$

Note that the self-weight of the column has been neglected since this is usually negligible compared to the total dead load on the girder. Using the LRFD load combination No. 2 from Section 2-3, the factored uniformly distributed load and concentrated loads on the transfer girder are calculated as

$$w_u = 1.2(3 \text{ kips/ft.}) + 1.6(1.5 \text{ kips/ft.}) = 6 \text{ kips/ft., and}$$

$$P_u = 1.2(297 \text{ kips}) + 1.6(270 \text{ kips}) + 0.3(60 \text{ kips}) = 806.4 \text{ kips.}$$

A free-body diagram of the transfer girder is shown in Figure 13-8.

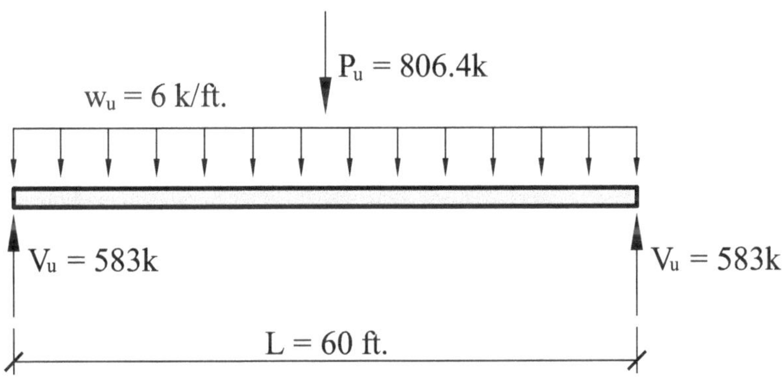

FIGURE 13-8 Freebody diagram of simply supported transfer girder

The factored or required shear and the bending moment are calculated as

$$V_u = 6 \text{ kips/ft.}\left(\frac{60 \text{ ft.}}{2}\right) + \frac{806.4 \text{ kips}}{2} = 583 \text{ kips, and}$$

$$M_u = 6 \text{ kips/ft.}\frac{(60 \text{ ft.})^2}{8} + \frac{806.4 \text{ kips}(60 \text{ ft.})}{4}$$

$$= 14{,}796 \text{ ft.-kips, respectively.}$$

Preliminary Design (LRFD)

1. M_u = 14,796 ft.-kips and V_u = 583 kips.
2. Assume a plate girder overall depth, $d = L/10 = (60 \text{ ft.})\ (12)/10 = 72$ in.
3. Assume a flange width, $b_f = d/6$ ($\geq$ 12 in.) = 72 in./6 = 12 in. Try b_f = 15 in.

4. The couple of forces in the bottom and tension flange is

$$T_u = C_u \approx \frac{M_u}{0.95d} = \frac{14{,}796 \text{ ft.-kips}}{0.95(72 \text{ in.}/12)} = 2596 \text{ kips.}$$

5. Assume an initial flange thickness, $t_f = \dfrac{T_u}{0.9F_y b_f} = \dfrac{2596 \text{ kips}}{0.9(50 \text{ ksi})(15 \text{ in.})}$

$$= 3.84 \text{ in.} \geq {}^3/_4 \text{ in.}$$

Try $t_f = 4$ in.

(**Note**: This is a heavy shape in accordance with AISC Specification A3.1d, and therefore, the requirements AISC Specification Sections A3.1d, J1.5, J1.6, J2.6, and M2.2 must be satisfied.)

6. Depth of web, $h = d - 2t_f = 72$ in. – (2)(4 in.) = 64 in.
7. From Table 13-2, assuming an unstiffened web, the minimum web thickness is

$$t_w = h/260 \geq \frac{1}{2} \text{ in.}$$

$$= 64/260 = 0.25 \text{ in.} < \frac{1}{2} \text{ in.}$$

Based on shear strength and web compactness, which will be checked later in this example, a $^1/_2$-in. web will not be adequate.
Therefore, try $t_w = 0.75$ in.
The trial plate girder cross-section resulting from the preliminary considerations above is shown in Figure 13-9.

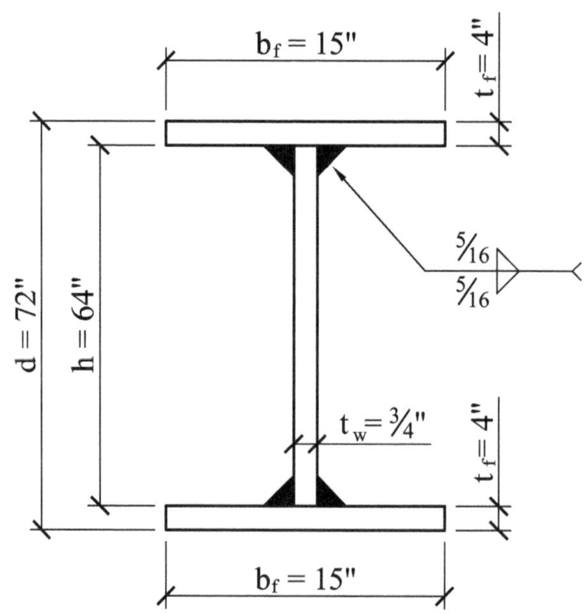

FIGURE 13-9 Trial plate girder for Example 13-1

After the preliminary sizing of the plate girder, the final design using the LRFD method is next carried out as follows:

1. Using the preliminary size selected, the self-weight of the plate girder is now calculated. Total area of the trial plate girder = (15 in.)(4 in.)(2 flanges) + (64 in.)(0.75 in.) = 168 in.^{2}

With a density of steel = 490 lb./ft.^{3}, the self-weight of the plate girder is

$$\left(\frac{168 \text{ in.}^2}{144}\right) 490 \text{ lb./ft.}^3 = 572 \text{ lb./ft.} \approx 0.6 \text{ kips/ft.}$$

The factored self-weight of the girder = 1.2(0.6 kips/ft.) = 0.72 kips/ft.
The revised total factored uniform load = 6 + 0.72 = 6.72 kips/ft. Therefore, the revised factored shear and bending moments are

$$V_u = 6.72 \text{ kips/ft.}\left(\frac{60 \text{ ft.}}{2}\right) + \frac{806.4 \text{ kips}}{2} = 605 \text{ kips, and}$$

$$M_u = 6.72 \text{ kips/ft.}\frac{(60 \text{ ft.})^2}{8} + \frac{806.4 \text{ kips}(60 \text{ ft.})}{4}$$

$$= 15{,}120 \text{ ft.-kips, respectively.}$$

2. To avoid using stiffeners, we will ignore diagonal tension field action and its contribution to the shear strength of the plate girder. Consequently, sufficient web thickness will be provided to avoid the use of intermediate stiffeners. The upper limit of the web depth-to-thickness ratio for a compact web of a doubly symmetric I-shaped built-up section is given in Table 13-1 as

$$3.76\sqrt{\frac{E}{F_y}} = 3.76\sqrt{\frac{29{,}000 \text{ ksi}}{50 \text{ ksi}}} = 90.55.$$

The web thickness from step 7 of the preliminary design, t_w = 0.75 in.
Therefore, the web depth-to-thickness ratio is
h/t_w = 64 in./0.75 in. = 85.3 < 90.55 OK.
Therefore, using Table 13-1, we conclude that this built-up section has a compact web, and therefore the plastic moment capacity of the section, M_p, is attainable. On the other hand, for illustrative purposes, we will still check the four limit states of flexure in Step 3.

$$A_w = (64 \text{ in.})(0.75 \text{ in.}) = 48 \text{ in.}^2$$

$$A_{fc} = (15 \text{ in.})(4 \text{ in.}) = 60 \text{ in.}^2$$

$$\frac{A_w}{A_{fc}} = \frac{48}{60} = 0.80 < 10, \text{ as required in Table 13-2 for unstiffened webs, OK.}$$

3. The four possible limit states for bending will now be checked (see Section 13-3).
The unbraced length of the compression flange, L_b = 0 (i.e., fully braced).

a. Compression flange yielding limit state

$$F_y = 50 \text{ ksi}$$

$$h_c = h = 64 \text{ in. (welded plate girder with equal flange areas)}$$

$$\phi_b = 0.9$$

From equation (13-8),

$$a_w = \frac{h_c t_w}{b_{fc} t_{fc}} = \frac{(64)(0.75)}{(15)(4)} = 0.8 < 10. \text{ OK.}$$

From equation (13-7),

$$R_{pg} = 1 - \left(\frac{a_w}{1200 + 300a_w}\right)\left(\frac{h_c}{t_w} - 5.7\sqrt{\frac{E}{F_y}}\right) \leq 1.0$$

$$= 1 - \left(\frac{0.8}{1200 + 300(0.8)}\right)\left(\frac{64}{0.75} - 5.7\sqrt{\frac{29,000}{50}}\right)$$

$$= 1.029 \leq 1.0.$$

Therefore, use $R_{pg} = 1.0$.

The moment of inertia of the plate girder about the strong axis is

$$I_{x-x} = (2\text{ flanges})\left[\frac{(15\text{ in.})(4\text{ in.})^3}{12} + (15\text{ in.})(4\text{ in.})\left(\frac{64\text{ in.}}{2} + \frac{4\text{ in.}}{2}\right)^2\right] + \frac{(0.75\text{ in.})(64\text{ in.})^3}{12}$$

$$= 155,264\text{ in.}^4$$

The distance from the elastic neutral axis to the outermost face of the compression flange is

$$y_c = \frac{72\text{ in.}}{2} = 36\text{ in.}$$

The elastic section modulus relative to the compression face is

$$S_{xc} = \frac{I_{x-x}}{y_c} = \frac{155,264}{36} = 4312\text{ in.}^3$$

The design moment capacity for this limit state is

$$\phi_b M_n = \phi_b R_{pg} F_y S_{xc} = \frac{(0.9)(1.0)(50\text{ ksi})(4312\text{ in.}^3)}{12}$$

$$= 16,173\text{ ft.-kips.}$$

b. Lateral torsional buckling limit state

From equation (13-9),

$$r_t \approx \frac{b_{fc}}{\sqrt{12\left(1 + \frac{1}{6}a_w\right)}} = \frac{15\text{ in.}}{\sqrt{12\left(1 + \frac{1}{6}(0.8)\right)}} = 4.07\text{ in.}$$

From equation (13-5),

$$L_p = 1.1 r_t \sqrt{\frac{E}{F_y}} = 1.1(4.07\text{ in.})\sqrt{\frac{29,000}{50}} = 107.8\text{ in.} = 9\text{ ft.}$$

The floor deck is assumed to be welded to the top of the plate girder; therefore, the lateral unbraced length, $L_b = 0$.

Since $L_b = 0 < L_p$, the lateral torsional buckling limit state is not applicable.

c. Compression flange buckling limit state

$$\frac{b_{fc}}{2t_{fc}} = \frac{15\text{ in.}}{2(4\text{ in.})} = 1.88 < 0.38\sqrt{\frac{29,000\text{ ksi}}{50\text{ ksi}}} = 9.15 \text{ (see Table 13-4)}$$

Therefore, from Table 13-4, this is a compact flange, and $F_{cr} = F_y$.

From equation (13-10), the design moment capacity for the LRFD method is

$$\phi_b M_n = \phi_b R_{pg} F_y S_{xc} = \frac{(0.9)(1.0)(50\text{ ksi})(4312\text{ in.}^3)}{12}$$

$$= 16,173\text{ ft.-kips.}$$

d. Tension flange yielding

Since the plate girder section selected is doubly symmetric, $S_{xt} = S_{xc}$; therefore, the tension flange limit state is not critical. In fact, since this is a compact section, the design moment capacity for tension flange yield is exactly the same as the design moment capacity for compression flange yielding.

Thus, the design moment capacity for the plate girder, which is the smallest moment strength from all of the four limit states, is

$\phi_b M_n = \mathbf{16{,}173\ ft.\text{-}kips} > M_u = 15{,}120$ ft.-kips. OK.

4. From Table 13-7, considering an unstiffened web (i.e., webs without transverse stiffeners); therefore, $k_v = 5.34$

The web depth-to-thickness ratio limits, from Table 13-6, are

$$1.10\sqrt{\frac{k_v E}{F_y}} = 1.10\sqrt{\frac{(5.34)(29{,}000 \text{ ksi})}{50 \text{ ksi}}} = 61.2$$

$$\frac{h}{t_w} = \frac{64 \text{ in.}}{0.75 \text{ in.}} = 85.33 > 61.2$$

Therefore, the web shear strength coefficient is obtained from Table 13-6 as

$$C_{v1} = \frac{1.10\sqrt{\frac{k_v E}{F_{yw}}}}{h/t_w} = \frac{1.10\sqrt{\frac{(5.34)(29{,}000 \text{ ksi})}{50 \text{ ksi}}}}{85.33} = 0.91$$

5. Based on the choice made in step 2 not to consider diagonal tension field action, determine the design shear strength, $\phi_v V_n$.

From equation (13-12a), the design shear strength for LRFD is

$$\begin{aligned}\phi_v V_n &= \phi_v 0.6 F_{yw} A_w C_{v1} \\ &= (0.9)(0.6)(50 \text{ ksi})(64 \text{ in.})(0.75 \text{ in.})(0.91) \\ &= 1179 \text{ kips} > V_u = 605 \text{ kips.} \quad \text{OK.}\end{aligned}$$

6. Check the deflections and vibrations as discussed previously in Chapters 6 and 12, respectively.

7. Design the bearing stiffeners at the support reactions or at concentrated load locations as discussed in Chapter 6.

8. Design the fillet welds required at the flange-to-web interface and the stiffener-to-web interface.

Since there are no intermediate stiffeners, only the weld at the web-flange interface will be designed for this plate girder.

The moment of inertia of the plate girder about the strong axis is

$$I_{x-x} = (2 \text{ flanges})\left[\frac{(15 \text{ in.})(4 \text{ in.})^3}{12} + (15 \text{ in.})(4 \text{ in.})\left(\frac{64 \text{ in.}}{2} + \frac{4 \text{ in.}}{2}\right)^2\right] + \frac{(0.75 \text{ in.})(64 \text{ in.})^3}{12}$$

$$= 155{,}264 \text{ in.}^4.$$

The statical moment of the flange area about the elastic neutral axis of the built-up section is

$$Q_f = (15 \text{ in.})(4 \text{ in.})\left(\frac{64 \text{ in.}}{2} + \frac{4 \text{ in.}}{2}\right) = 2040 \text{ in.}^3$$

Maximum factored vertical shear, V_u = 605 kips.

From equation (13-13), the maximum horizontal shear force per unit length at the web-to-flange fillet weld is

$$v_h = \frac{V_u Q_f}{I_{x-x}} = \frac{(605 \text{ kips})(2040 \text{ in.}^3)}{155{,}264 \text{ in.}^4}$$

$$= 7.95 \text{ kips/in. for welds on both sides of the web.}$$

Since the web is welded to the flange of the girder with fillet welds on two sides, therefore, the shear flow on the fillet weld on each side of the web is 7.95/2 = 3.97 kips/in.

A $^3/_{16}$-in. continuous fillet weld (i.e., D = 3) that has a design shear strength of (1.392 kips/in.) × (3) = 4.18 kips/in. (see Chapter 10) is adequate to resist the shear flow of 3.97 kips/in. Nevertheless, the minimum weld size from *AISCM* Table J2.4, is $^5/_{16}$ in. for a material (the girder web) that is $^3/_4$ in. thick and greater.

With respect to bending and shear, the 72-in.-deep unstiffened plate girder consists of two 15-in. × 4-in.-thick flange plates and a 64-in. × $^3/_4$ -in.-thick web plate as shown in Figure 13-9 is adequate to resist the applied loads and will suffice as the final girder selection. The reader should check deflections and bearings for the girder as was previously done in Chapter 6 to ensure that they are also within allowable limits.

References

[1] Spiegel, L., and G. F. Limbrunner. 2002. "*Applied Structural Steel Design.*" 4th ed. Upper Saddle River, NJ: Prentice Hall.

[2] New York State Department of Transportation. 2006. "*Bridge Mannual.*" Albany, NY: NYSDOT.

[3] Taly, N. 1998. "*Design of Modern Highway Bridges.*" New York: McGraw-Hill.

[4] American Institute of Steel Construction. 2023. *Steel Construction Manual.* 16th ed. Chicago, IL: AISC.

[5] Bedair, O. 2011. "Stress Analyses of Deep Plate Girders Used in Oil and Gas Facilities with Rectangular web Penetrations." *ASCE Practice Periodical on Structural Design and Construction,* American Society of Civil Engineers, New York: 112–120.

[6] Davis, B. 2017. "Sheer Improvement to Shear Design." *Modern Steel Construction,* August.

Exercises

13-1. Plate girders are frequently used in steel bridges. List some conditions that might warrant their use in building structures.

13-2. List some of the disadvantages of using plate girders in building structures and list one alternative to plate girders in building structures.

13-3. Determine the moment capacity of the welded built-up section shown in Figure 13-10, which has the following dimensions:

d = 82 in., t_f = 2 in., b_f = 13 in., t_w = $^3/_4$ in.

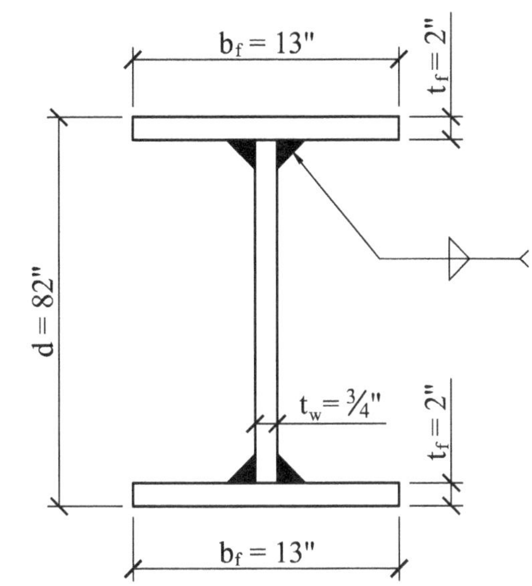

FIGURE 13-10 Plate girder for Exercise 13-3

13-4. What is diagonal tension field action, and when does it occur in plate girders? List the conditions under which a diagonal tension field may be considered in the design for shear in plate girders.

13-5. Repeat the design exercise in Example 13-1, taking into account diagonal tension field action.

13-6. Using the LRFD method, design a transfer-built-up plate girder to support a column carrying 18-floor levels and spanning 50 ft. across a ballroom. The ceiling space over the ballroom is limited to a maximum depth of 60 inches, and the discontinued column is supported at the midspan of the girder. The total cumulative loads supported by the column are D = 1000 kip and L = 500 kip. The girder also supports a floor dead load of 85 psf and a live load of 50 psf with a tributary width of 30 ft. Assume ASTM A992 Grade 70 steel. The girder is simply supported on wide flange columns at both ends and is continuously braced by the floor above the ballroom.

13-7. Using the LRFD method, design a transfer built-up plate girder that spans 55 ft. over a gallery with a 15 ft. cantilever or overhang at one end. The spacing of the girders is 30 ft., and the floor dead load (excluding the weight of the girder) is 110 psf, with a floor live load of 50 psf. The maximum depth of the plate girder is limited to 90 inches. Check strength, deflection, and floor vibration requirements. Assume ASTM A992 Grade 50 steel.

13-8. The typical floor plan of an office building consists of a 60 ft. by 30 ft. grid, with plate girders spanning 60 ft. between columns and spaced at 30 ft. on centers. The floor beams and the roof beams span 30 ft. between the plate girders and are spaced at 10 ft. on centers. The 5th floor deck system consists of 3-inch normal-weight concrete on a 2-inch 18-gauge composite metal deck. Assume a superimposed dead load on the 5th floor of 35 psf (includes 5 psf for the self-weight of the floor beams but does not include the self-weight of the plate girder). The roof dead load is 35 psf (includes all roof framing self-weight) with a 40 psf snow load. The exterior cladding supported by the 5th floor spandrel beams weighs 20 psf of the vertical surface area.

The welded plate girders consist of 23-inch wide by 2.5-inch-thick flanges and a web that is 21 inches deep by 1-inch thick. The typical columns in the building have a story height of 15 ft., but because of the set-back at the 2nd floor through the 4th floor levels, the perimeter columns on the west side of the building are super-tall built-up I-shaped columns that only support the 5th floor plate girders, the 5th floor spandrel beams, and the roof plate girders and roof spandrel beams. The supertall column has an unbraced height of 63 ft. between the ground floor and the 5th floor, and 15 ft. between the 5th floor and the roof level. Assuming ASTM A992 Grade 50 steel, determine the following design parameters:

a. The live load capacity of the floor deck and compare to the occupancy live load.
b. The elastic section modulus and the plastic section modulus of the built-up welded plate girder.
c. The moment of inertia of the plate girder.
d. The maximum factored moment and shear in the plate girder.
e. The maximum moment and shear strength of the plate girder using the LRFD method.
f. Is the girder adequate for strength?
g. Is the plate girder adequate for deflection, assuming the deflection limit is L/240 under total dead plus live load and L/360 under live load?
h. If the deflection criteria in part (g) are not satisfied and the plate girder size or properties cannot be changed, what else can be done to ensure that the deflection criteria are met? Support your answer with calculations.

13-9 A 60 ft span plate girder is used to transfer out a multi-story column supporting a factored load of 2250 kips at the midspan of the girder. Assume ASTM A992 Grade 50 steel. Design a 68 in. deep plate girder, assuming a flange thickness of 3.5 in. by selecting the appropriate flange width and web thickness. Ignore the deflection check.

13-10 A welded transfer plate girder has a total depth of 72 inches with top and bottom flange widths of 48 inches, flange thickness of 4 inches, and web thickness of 1 inch. The girder is simply supported over a grand ballroom for a span of 54 ft. The components of the plate girders are spliced with continuous joint penetration (CJP) welds conforming to the requirements of thick member welding (i.e., members with thickness greater than 2 inches) in *AISCM* Section M. Assume ASTM A992 Grade 50.

a. Calculate the moment capacity, ϕM_n, of the plate girder. Assume the girder compression flange is fully braced.
b. What, if any, is the size of a rolled steel wide flange section that will provide the same ϕM_n as this plate girder? Compare the weight of the rolled steel beam to that of the plate girder and discuss your observations.
c. If the plate girder supports transferred column loads at 3rd points (i.e., at 18 ft. and 36 ft.) in addition to the uniform load from the girder self-weight, determine the maximum factored column load, P_u, that can be supported by the girder considering bending moment alone.
d. Repeat part (c) if the girder deflection limit is span/480 and assuming the unfactored concentrated dead plus live load, P_{D+L}, is approximately equal to $P_u/1.5$. Assume the girder is cambered 1.0 inch.

13-11 A series of 72-inch deep welded built-up box girders with 3-inch flanges and web thicknesses is simply supported over a span of 143 ft over a column-free atrium area. The components of the box girders are spliced with continuous joint penetration (CJP) welds conforming to the requirements of thick member welding (i.e., members with thickness greater than 2 inches) in *AISCM* Section M. Assume ASTM A992 Grade 50.

a. Calculate the moment capacity, ϕM_n, of the plate girder. Assume the girder compression flange is fully braced.

b. What, if any, is the size of a rolled steel wide flange section that will provide the same ϕM_n as this box girder? Compare the weight of the rolled steel beam to that of the plate girder and discuss your observations.

c. Determine the maximum factored uniform load in kip/ft above and beyond the factored uniform load from the girder self-weight that can be supported by the box girder considering the bending moment alone.

d. Repeat part (c) if the girder deflection limit is span/480 and assuming the unfactored concentrated dead plus live load, P_{D+L}, is approximately equal to $P_u/1.5$. Assume the girder is cambered 1.0 inch.

13-12 Design a series of 60-inch-deep fully composite welded plate girders spaced at 27 ft. on centers to span 126 ft. over an ice rink. The plate girder supports fieldhouse occupancy, and the floor deck is 3.25-inch normal-weight concrete on a 3-inch composite metal deck. The plate girder is to be cambered for the deflection from the self-weight of the floor slab and the steel framing. Assume the self-weight of the girders and the secondary steel framing is 45 psf. Assume ASTM A992 Grade 50. Use a dead plus live load deflection limit of span/480. Neglect the floor vibration check.

13-13 Design a 60-in. deep double-web wide flange transfer plate girder that spans 60 ft over a ballroom and supports a factored concentrated of 1600 kips at its midspan from 18-story columns. In addition, the girder supports a floor load of 90 psf and a live load of 100 psf with a tributary width of 30 ft. The two identical wide flange plate girders are welded to each other intermittently with partial joint penetration (PJP) groove welds; the intermittent welds along the length of the beam are 4 inches long at 12 inches on centers at the top and bottom flanges. Assume ASTM A992 Grade 50 and use a dead plus live load deflection limit of span/480. Assume the plate girder is fully braced by the floor deck. Neglect floor vibration check.

13-14 Design a series of 60-inch-deep non-composite welded plate girders spaced at 25 ft. on centers to span 75 ft over an atrium area. In addition to a transfer column load of 160 kips dead load and 60 kips live load at 3rd points (i.e., at 25 ft and 50 ft), the girder also supports a uniform floor dead load of 100 psf and a live load of 40 psf. Assume ASTM A992 Grade 50 and use a dead plus live load deflection limit of span/480. Assume the plate girder is fully braced by the floor deck. Neglect floor vibration check.

CHAPTER

14

Practical Considerations in the Design of Steel Buildings

14-1 INTRODUCTION AND RULES OF THUMB FOR STRUCTURAL STEEL DESIGN

In this chapter, we present rules of thumb used for the quick selection of structural steel members and systems, and several topics of importance in the practical design of steel structures are covered.

Two structural steel design project exercises are also introduced in this chapter, with subsequent end-of-chapter design project questions. The goal is that by the time the student works through all the chapters of this text and the corresponding design project questions in this chapter pertaining to the topics covered in earlier chapters, the student will have completed the design of an entire structure, thus reinforcing the connection between the design of individual structural steel components and the design of an entire structural system.

Structural design is typically an iterative process where the member sizes are initially unknown but the loads on the structure are known. Design rules of thumb aid the designer in selecting preliminary member sizes that could be useful in laying out the framing for a structure. Such preliminary sizes are also useful to the engineer in getting a "feel" for the structure and the potential sizes of the members. Some design rules of thumb available for the preliminary sizing of structural steel members are presented in Table 14-1. In selecting steel members, the designer should strive for simplicity as much as possible and endeavor to use as few different details, member sizes, and connections as possible to keep the labor cost down since *fabrication* and *erection* account for approximately 60% of the total installed cost of structural steel buildings, while material cost accounts for only approximately 30% [1]. The approximate breakdown of the total installed cost of structural steel [56] is as follows:

- Material cost – 27%
- Erection cost – 27%
- Fabrication cost – 33%
- Other costs – 13%

Therefore, specifying the least weight for structural steel members does not necessarily lead to the most economical structure.

TABLE 14-1 Design Rules of Thumb [1], [3], [4], [54]

Structural steel components	Design rules of thumb
Joists	Joist depth $\geq \frac{L}{24}$ (use $L/20$ for floor joists; deeper joists may be needed to control floor vibrations)
Base plate thickness, t	t_p = larger of $\frac{m}{4}$ and $\frac{n}{4}$ m or n = Base plate extension beyond the critical column area (see Figures 8-33a and 8-33b)
Beams and girders	Depth $\geq \frac{L}{24}$ (use $L/20$ for floor beams); girder depth ≥ beam depth $I_{\text{required}} \geq \frac{wL^3}{64}$ (for uniform loads) or $I_{\text{required}} \geq \frac{ML}{8}$ (for nonuniform loads), Where, w = Unfactored uniform load, kips/ft., L = Beam or girder span, ft. M = Unfactored total maximum moment, ft.-kips I_{required} = Required moment of inertia, in.^{4}
Trusses	• Preliminary overall depth $\geq \frac{L}{10}$ *to* $\frac{L}{12}$ for buildings and $\geq \frac{L}{8}$ *to* $\frac{L}{5}$ for bridges, where L = span of truss • Optimum slope of diagonal web members within each panel relative to the horizontal = 40° to 60° • Preferably, choose a truss type so that the diagonals with the largest forces are in tension. Longer diagonals in compression are susceptible to buckling. • Check for net wind uplift forces on roof trusses that may result in tension forces in the top chord and compression forces in the bottom chord. • Ensure that the compression chord (and the tension chord) are laterally braced adequately.
Plate girders	Depth $\geq \frac{L}{8}$ to $\frac{L}{12}$, where L = plate girder span
Continuous girders	Locate hinge or splice locations between $0.15L$ and $0.25L$ *from the nearest support*, where L = Length of the span in which the hinge(s) are located.
Cantilevered roof system and cantilevered floor beams and girders	Maximum cantilever length should be between $0.15L$ and $0.25L$, where L = Length of the back span.
Braced frames and moment frames subject to wind loads	Limit total lateral drift under service wind loads to between $H/400$ and $H/500$, where H = Total height of building. The service wind loads could be calculated using the unfactored 10-year MRI wind speed (see the Commentary Section of Ref. [25] or the ASD wind load from Chapter 2 (i.e., 0.6W).
Diagonal braces in braced frames	Slope of the diagonal brace from the horizontal should preferably be between 35° and 55° to facilitate proper connections.

TABLE 14-1 (Continued)

Columns	The practical minimum size to accommodate beam-to-column and girder-to-column connections is a W8 column, but W10, W12, and W14 column sizes are more commonly specified by design engineers. The *approximate* axial compression load capacity for W-shapes (Grade 50 steel) not listed in the column load tables is $\phi P_n \approx 1.5\left(30 - 0.15\frac{KL}{r}\right)A_g,$ where L = Unbraced length of the column, r = Minimum radius of gyration, and A_g = Gross area of the column.
Optimum bay sizes (roof and floor framing)	Most economical bay size: • Rectangular bay size should have a plan length-to-width ratio between 1.25 and 1.50. • Bay area = 1000 ft.^{2} approximately 30 ft. × 30 ft. is a common bay size, that is, the plan dimension between adjacent columns in both orthogonal directions = 30 ft. • For a 10-ft. infill beam spacing, the 3-in. deep composite metal floor deck is commonly specified.

In designing the vertical lateral force resisting system (LFRS) for steel structures in seismic design category (SDC) A, B, or C, a structural systems response modification factor, R, of 3.0, corresponding to "*Structural steel systems not specifically detailed for seismic resistance*," from ASCE 7 Table 12.2-1 is recommended, in order to reduce the complexity of the member connection details and hence avoid the increased detailing, construction inspection, and construction costs associated with using higher R-values. In addition, the use of braced frames for the lateral force-resisting system should be given first consideration in the absence of other requirements because of its relative economy when compared to moment frames. For more ductile buildings or structures in SDC D, E, or F, where higher R-values are required, the lateral force resisting systems and other building components must be detailed to satisfy the special seismic detailing requirements in the *AISC Seismic Provisions for Structural Steel Buildings* (AISC 341-22) [2].

If moment frames cannot be avoided, wide-flange column sections bending about their strong axis should be used in these frames, and it is recommended that no column in the building be part of any two orthogonal moment frames to avoid wide-flange columns bending about their weak (y–y) axis, unless flanged cruciform columns (see Chapters 5, 6, and 8) are used for the columns in the orthogonal moment frames.

With flanged cruciform columns (see Figure 5-15) used in orthogonal moment frames, the moment frames in both orthogonal directions will consist of identical columns bending about their strong (x-x) axis. In this case, the moment capacity or strength of each column in the orthogonal moment frame will be the bending moment strength of the column about its strong (x-x) axis, taking into account its unbraced height (see Chapter 6).

14-2 LAYOUT OF STRUCTURAL SYSTEMS IN STEEL BUILDINGS

The practical and economic considerations necessary to achieve optimum and functional layout for the horizontal (roof and floor framing) systems and the vertical lateral force-resisting systems (LFRS) in steel buildings are discussed in this section.

Layout of Floor and Roof Framing Systems

The design process for a steel building begins with a review of the architectural drawings and laying out the roof and floor framing systems to support the gravity loads (i.e., providing an adequate gravity load path), as well as laying out the lateral force resisting systems (i.e., providing an adequate lateral load path), considering architectural, mechanical, and electrical constraints. Issues such as headroom limitations, roof or floor openings, and door and window openings must be considered when laying out the gravity load-resisting system and the lateral force-resisting system. The different types of floor and roof framing schemes commonly used in design practice to support gravity loads are presented in this section. The following design considerations should be factored into the selection of the floor and roof framing systems:

1. For rectangular bays, the beams and joists should span in a longer direction, while the girders should span in a shorter direction to produce optimum sizes for the structural framing.
2. The roof and floor deck should be framed in a direction perpendicular to the beams and joists. It is more economical to maximize the span and spacing of the roof and floor beams, thereby using fewer in-fill or filler beams, which results in fewer beam-to-girder connections. The fewer the connections in a steel structure, the lower the construction cost. The maximum depth of the floor girders and beams should be selected to fit within the ceiling/floor plenum space. This space, which is determined by the architect, consists of the steel girders and beams, the ventilation ducts and sprinkler system, and the ceiling and lighting. The ceiling/floor plenum space is typically approximately 4 ft. deep.
3. The framing around stair and floor or roof openings should consist of wide-flange sections, except for very short spans and lightly loaded members where channel or angle sections may suffice. It should be noted that at stair openings, the edge framing also supports the stair stringer reactions and partition wall loads around the stair. For floor beams or girders supporting stair stringers (and escalator support beams), the resulting torsion in the supporting floor beams due to the eccentricity of the reaction from the stair stringer (or the escalator beam) must be adequately considered (see Chapter 6).
4. Use wide-flange steel beams and girders along the column grid lines and along the grid lines or vertical planes containing the vertical lateral force resisting systems (LFRS) (e.g., shear walls, moment frames, or braced frames) and along the perimeter of the building. Open-web steel joists should not be used along the perimeter of a building since they cannot provide adequate lateral support for exterior cladding perpendicular to their span; open-web steel joists should also not be used along the grid lines or vertical planes where the lateral force resisting systems are located, unless the joists are specified to be part of the lateral force resisting system, something that is rarely done in design practice.
5. Heavy partition walls should be supported directly by a wide flange (W-shape) floor beam or girder located directly underneath the wall, with the longitudinal axis of the beam or girder spanning parallel to the partition wall. Similarly, skewed exterior walls should be directly supported by a skewed beam underneath the wall.
6. Roof or floor openings should be located offset from the column grid lines to minimize interruption of the beam and girder framing.
7. For industrial buildings where concentrated loads hanging from the roof or floor framing are common, it is more structurally efficient to use wide-flange beams instead of open-web steel joists for the roof or floor framing since the hanging loads can be located

anywhere along the beam bottom flange, whereas for open-web steel joists, the hanging loads would have to be located at the joist panel points; otherwise, the joist bottom chords would need to be reinforced for the resulting bending loads.

Several examples of roof and floor framing layouts for supporting gravity loads are shown in Figures 14-1 through Figure 14-5 [5].

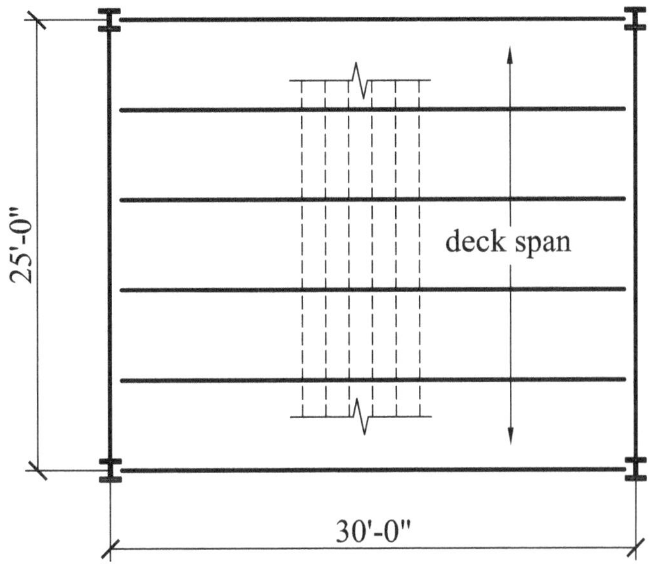

FIGURE 14-1 Simple roof/floor framing layout

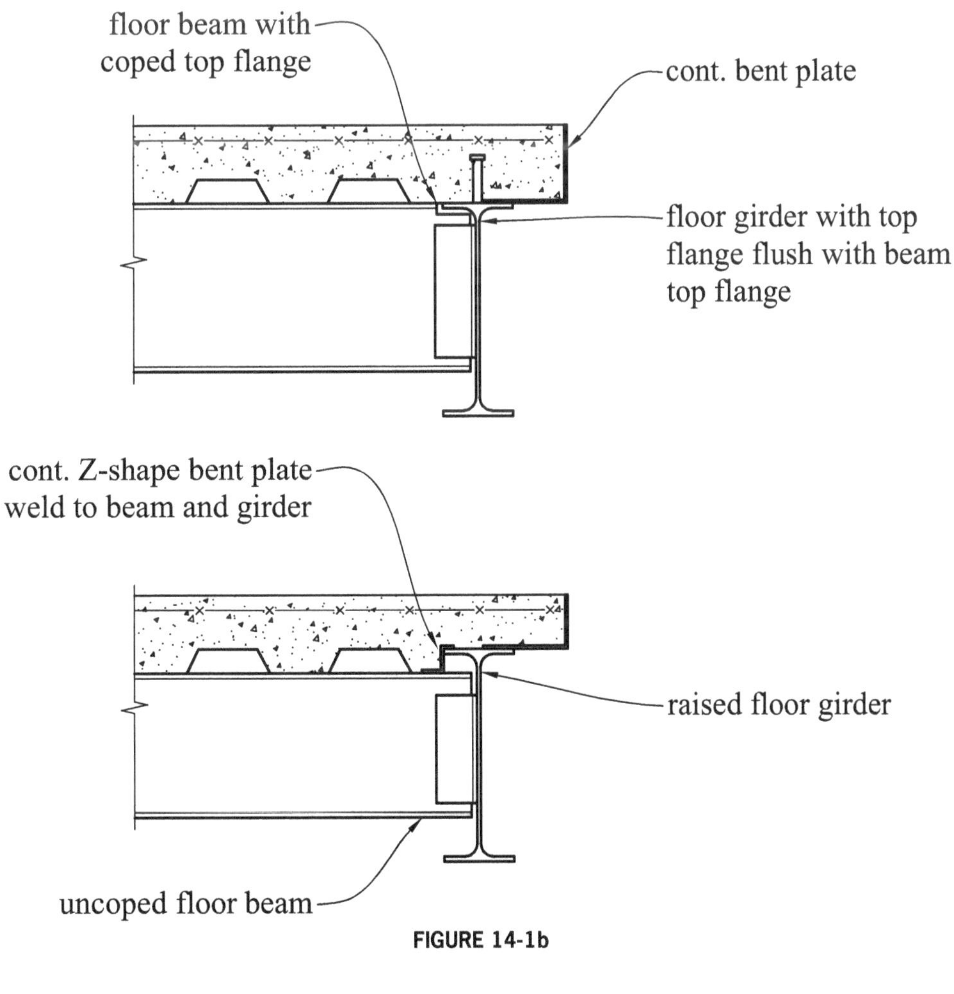

FIGURE 14-1b

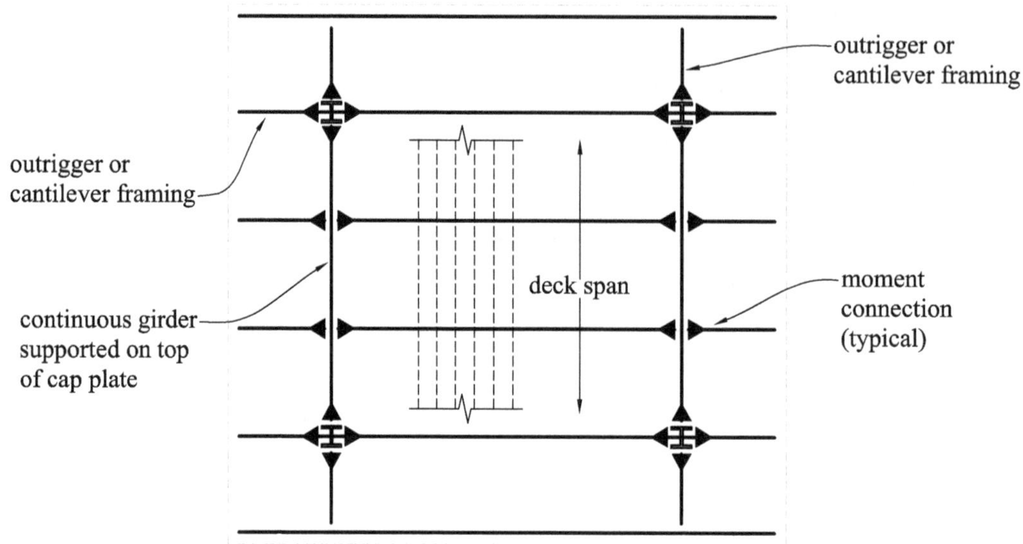

FIGURE 14-2 Roof/floor framing layout with outrigger framing schemes

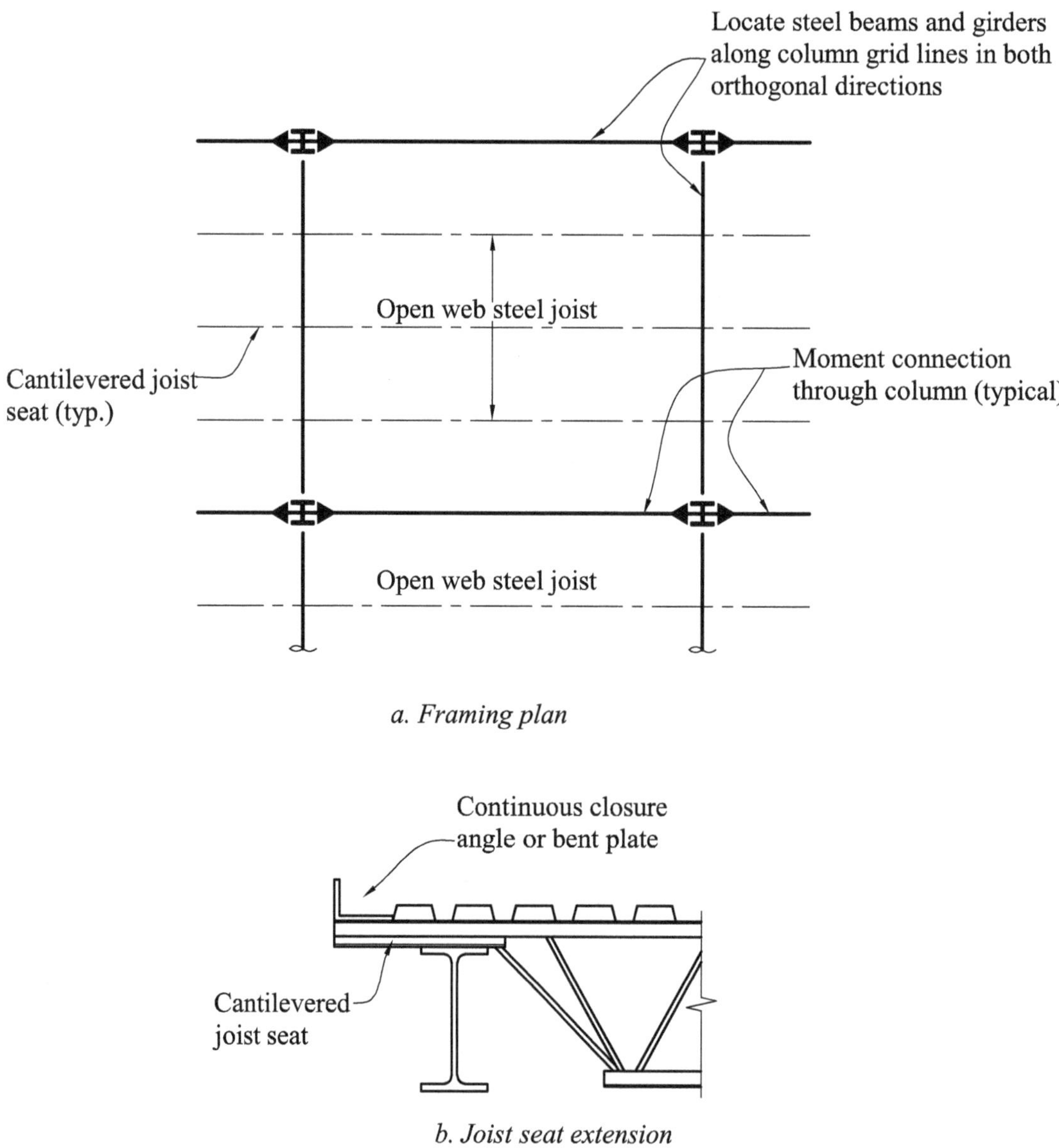

FIGURE 14-3 Roof framing with cantilevered joist shoes

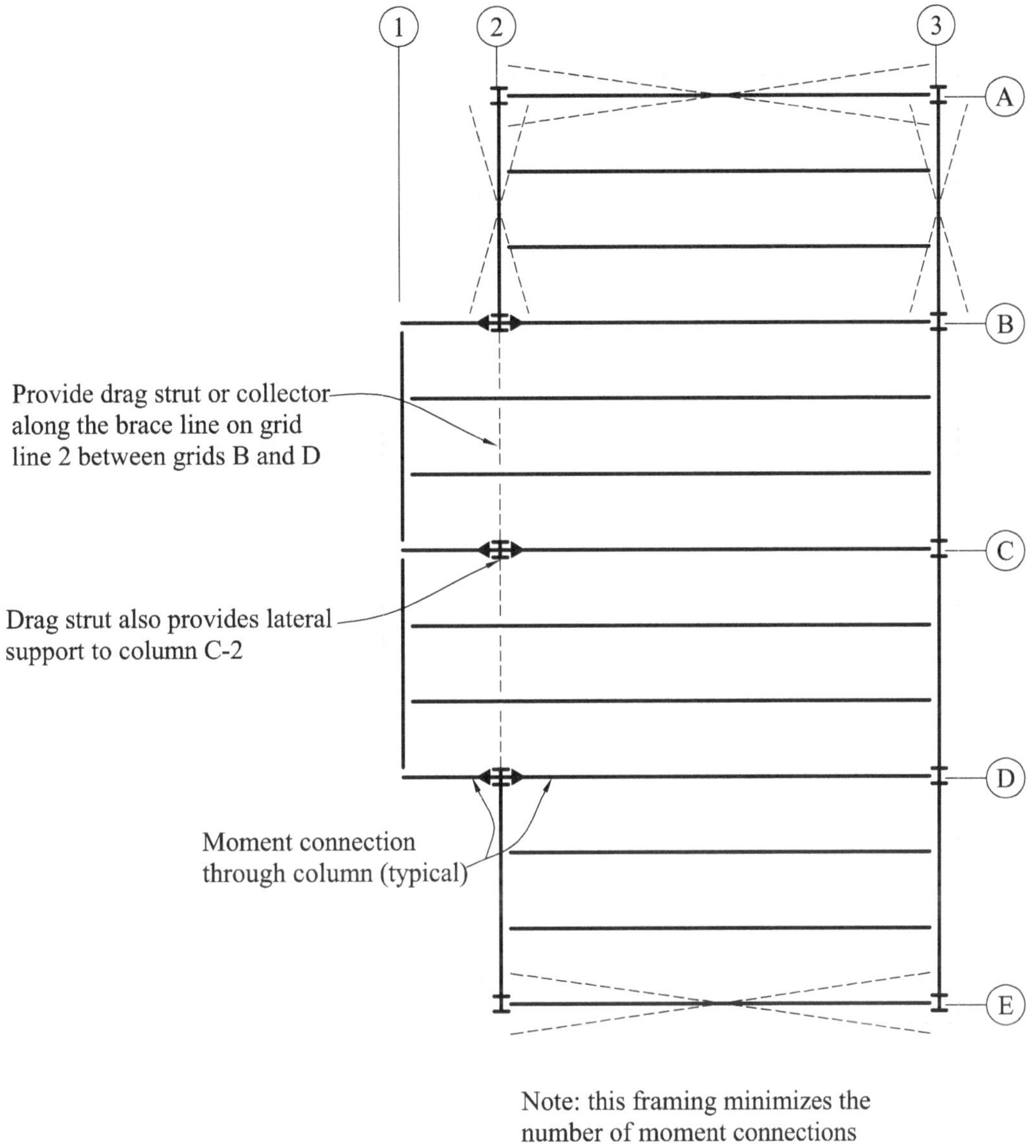

FIGURE 14-4 Roof/floor framing layout with interruption of drag strut and chords

Layout of Vertical Lateral-Force-Resisting Systems (LFRS)

The simple shear connections between the horizontal members (i.e., beams or girders) and the vertical linear members (i.e., columns) in steel buildings behave as a hinge, which provides little or no restraint to the moments caused by the lateral loads; therefore, these shear connections, which are the default connections specified in steel buildings, do not provide any inherent lateral stability to the structure. Consequently, steel buildings must be adequately braced for lateral loads by using either braced frames, moment frames, or shear walls, or a combination of these vertical lateral force resisting systems (LFRS). Moment frames provide the most flexibility regarding architectural considerations since the door or window layouts do not have any adverse effect or impact on the layout of moment frames. However, since the main lateral force-resisting members in braced frames are the diagonal brace members, these diagonal members must be located such that they do not conflict with the locations of the doors or windows. Careful coordination with the architect is necessary at the preliminary design stage to avoid the unpleasant situation of having to cut or modify the orientation of the diagonal members in a braced frame because of unforeseen conflict with a door or window opening.

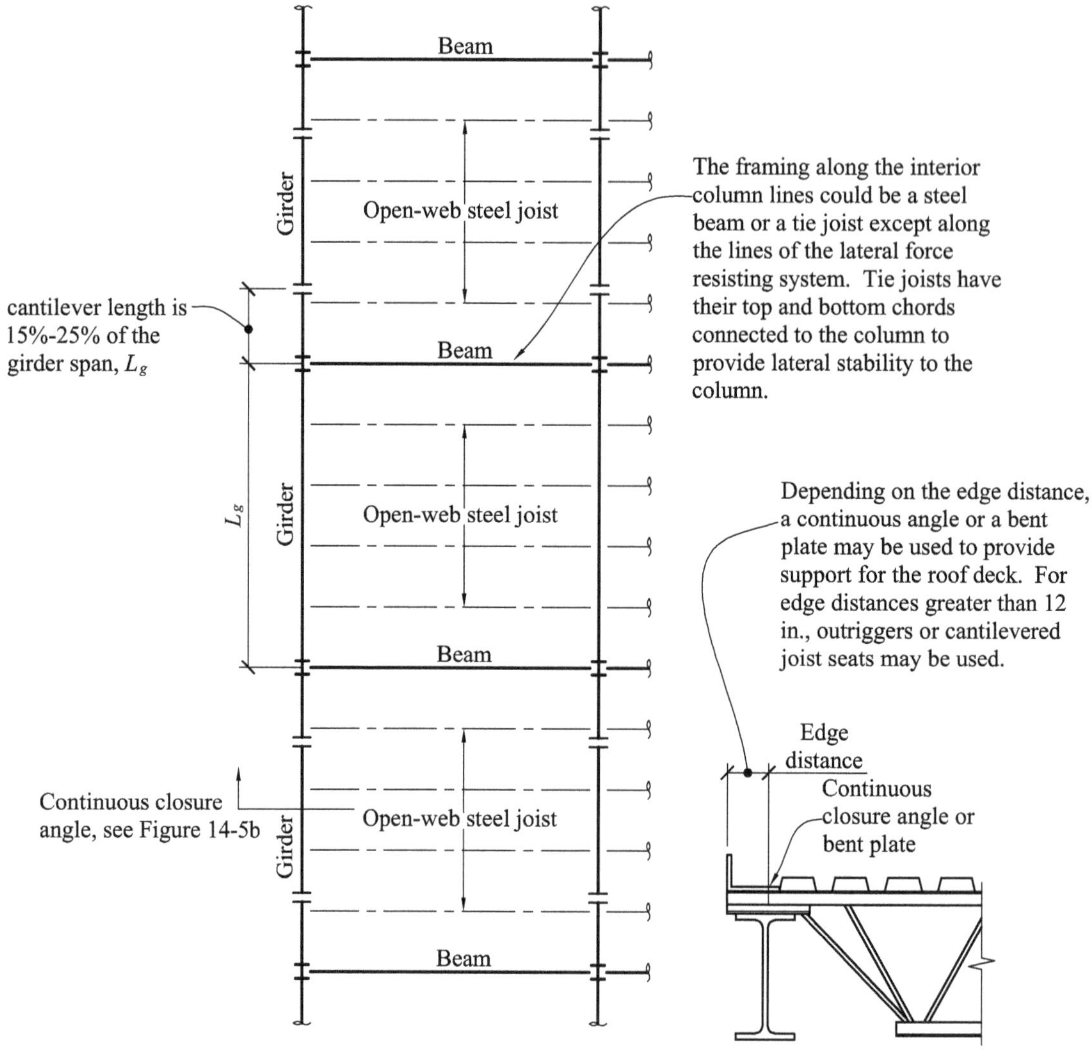

a. Framing plan *b. Roof edge detail*

FIGURE 14-5 Cantilevered roof framing

Braced frames should preferably be located along column lines at partition wall locations to help conceal the diagonal brace within the partition wall, unless the architect intends to expose the diagonal braces for aesthetic purposes. The size of the diagonal brace member should be selected to fit within the partition wall thickness. X-braced frames with single- or double-angle diagonal braces may require the use of gusset plates at the intersection of the diagonal members to ensure that the intersecting diagonal X-braced members lie on the same vertical plane, thus minimizing the required wall thickness necessary to accommodate the diagonal brace members. An example layout of vertical bracing in a floor framing plan is shown in Figure 14-6. The vertical bracing may be located on the interior of the building as shown on grid lines 2 and 4 or along the perimeter of the building as shown on grid lines A and D of Figure 14-6.

Masonry shear walls are sometimes used to resist lateral forces in low-rise steel buildings, in addition to serving as the exterior cladding to resist wind loads acting perpendicular to the face of the wall. In this system, the steel framing resists only gravity loads, and the connection between the steel beams or girders and the masonry shear wall is detailed in such a way as to prevent the gravity load from being transferred from the steel roof or floor framing into the masonry shear wall below. This is achieved by using a connection detail with vertically slotted bolt holes as shown in Figure 14-7a. This detail allows the lateral load (in the direction parallel

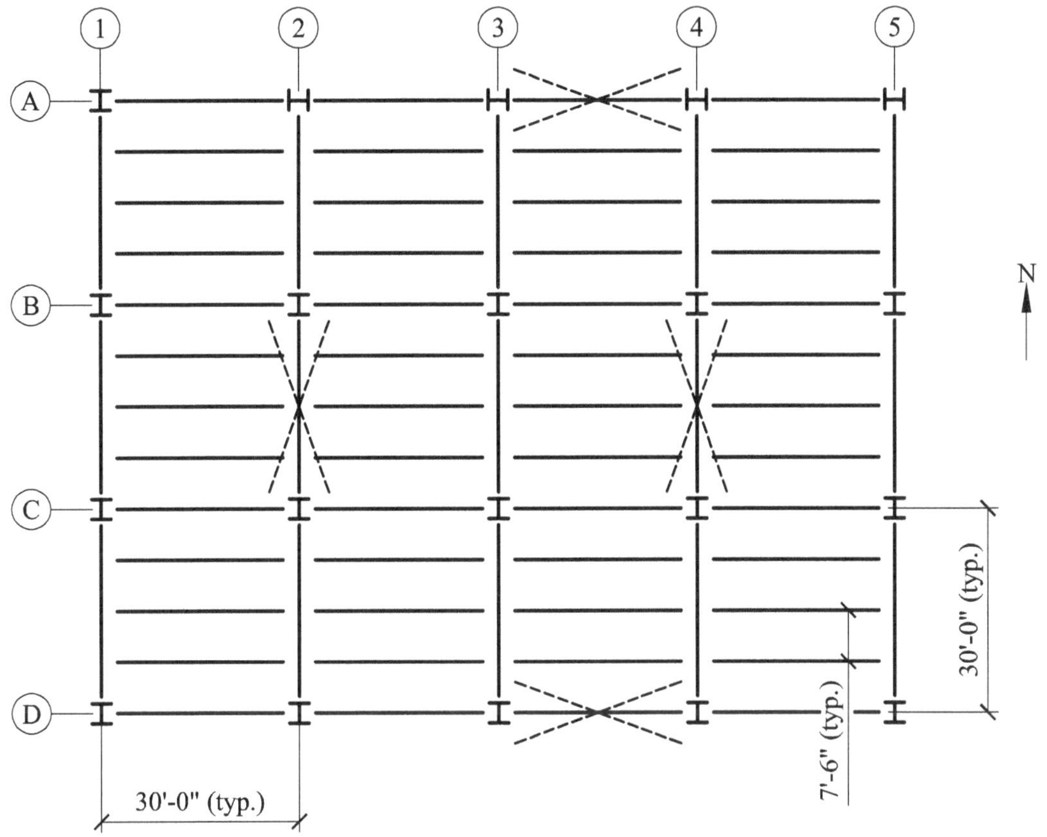

FIGURE 14-6 Diagonal bracing layout

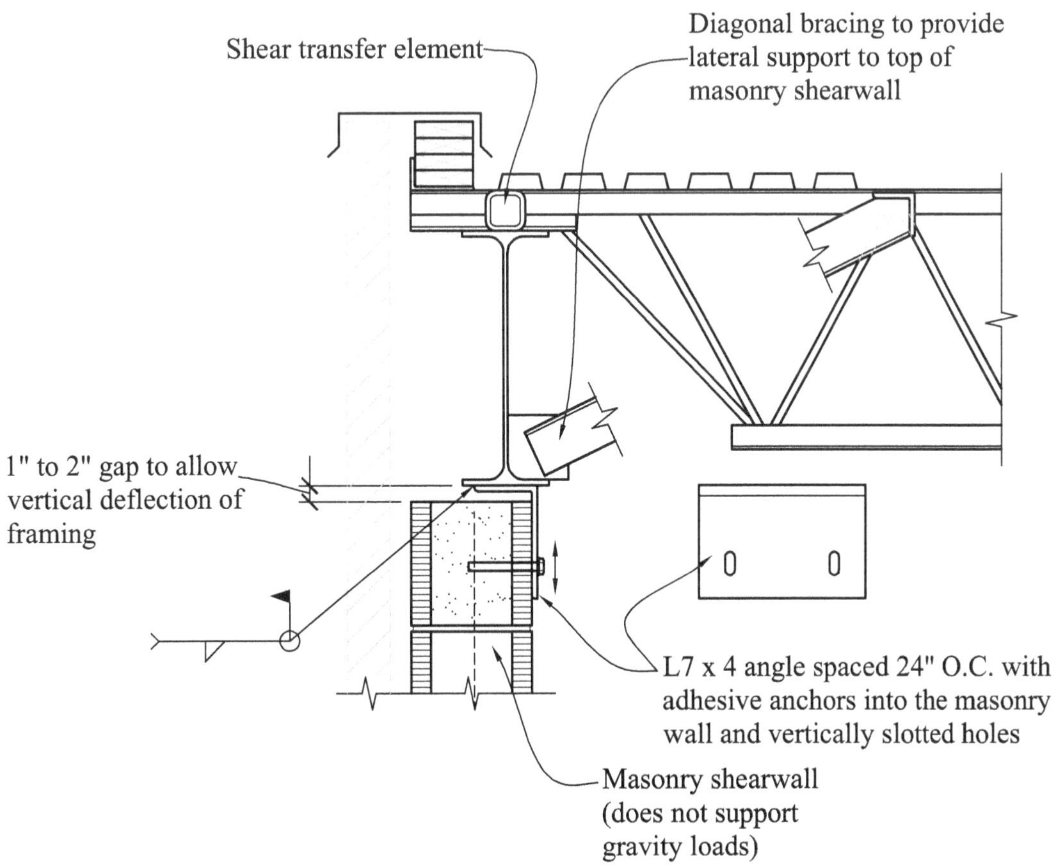

FIGURE 14-7a Lateral force transfer detail

to the plane of the wall) to be transferred from the roof or floor diaphragm into the masonry shear wall through the drag struts or collectors (i.e., the steel beams or girders that lie on the same vertical plane as the masonry shear wall) without transmitting the gravity loads onto the shear wall. The structural steel is usually erected before the masonry shear walls are built; therefore, the building must be temporarily braced (usually with guy wires) during construction until the masonry shear walls are built.

Masonry shear walls may also be used in infill frames where the block walls are built tightly against the columns and tightly to the underside of the steel beams and girders. In this system, both the gravity load and the lateral loads are resisted by the combined steel beam/girder/column and masonry wall system.

14-3 LATERAL LOAD DISTRIBUTION IN HORIZONTAL DIAPHRAGMS; DRAG STRUTS (OR COLLECTORS) AND CHORDS

The roof and floor deck or slabs in a steel building resist and transmit in-plane lateral loads to the vertical lateral force resisting systems (LFRS) parallel to the lateral loads by acting as a horizontal diaphragm or a deep in-plane horizontal beam spanning between adjacent vertical lateral force resisting systems (LFRS) (see AISC Specification, Section B3.5 [20]).

The component elements of the entire lateral force resisting systems (LFRS) in any structure—including the shear walls, braced frames, moment frames, horizontal diaphragms, and the chords and drag struts (or collectors) — can be modeled using 3-D finite element analyses, where the horizontal diaphragms and shear walls are modeled using planar shell elements and the moment resisting frames and braced frames are modeled using 3-D beam-column elements. The result of the 3-D finite element analysis, which implicitly accounts for in-plane torsional effects in the horizontal diaphragms, will give the forces distributed to each of the component elements in the lateral-force-resisting-system as well as the lateral displacements at each diaphragm level. The 3-D analysis just described is cumbersome and not commonly used in a typical design office, especially for low- to mid-rise building structures.

In lieu of such complicated 3-D finite element analyses, a simplified 2-D analysis is commonly used in practice to determine the lateral force distribution to the component elements of the lateral-force-resisting-system. In this simplified method, the diaphragms are modeled as simply supported horizontal beams spanning between the vertical lateral force-resisting system (LFRS), i.e., the shear walls, moment frames, or braced frames, or as a continuous horizontal beam spanning between several vertical LFRS. The horizontal diaphragm is subjected to in-plane bending moments, shear, and axial forces, and the chords and drag struts are subjected to in-plane axial tension and compression forces. The connections between all the component elements of the vertical LFRS along the lateral load path must be continuous and of sufficient strength to ensure the uninterrupted flow and transfer of the lateral load. The designer must identify a distinct and continuous lateral load path in both orthogonal directions through which the lateral load is transmitted from the diaphragm to the vertical LFRS, and subsequently to the foundations and finally to the soil or bedrock.

In summary, the horizontal diaphragm (i.e., roof, floor deck, or slab) performs the following functions, in addition to supporting the gravity loads and distributing them to the supporting beams, girders, and columns [6, 38, 39]:

- Distribute and transfer in-plane lateral or horizontal shear forces (from wind, earthquake, unbalanced soil pressure, or fluid loads) to the vertical lateral force resisting systems (LFRS) such as shear walls, braced frames, or moment frames.

- Act as a stabilizing tie for the structural elements at each level and provides lateral support for all the gravity-load supporting elements in the building, such as columns, walls, beams, and girders.
- They also help to transmit any horizontal thrusts from inclined or sloping columns into the vertical lateral force-resisting systems.
- Podium slabs in tall buildings are special diaphragms that help to transfer lateral forces from the more flexible shear walls or braced frames or moment frames into the much longer and therefore stiffer perimeter basement walls below the podium or ground floor level (see Figure 14-7b). Therefore, in addition to designing the podium slab level for the large gravity loads they support, the horizontal force transfer at the podium slab level is critical and must be properly designed and detailed for; thus, the required additional slab reinforcement (i.e., the drag strut or collector reinforcement) to ensure adequate connection of the LFRS to the podium level diaphragm and adequate transfer of the lateral forces must be provided.

A typical horizontal diaphragm in a steel-framed structure consists of the following structural elements:

- The **horizontal diaphragm web** which typically consists of the roof deck and floor deck/slab, which resist the in-plane shear (like the web of a beam or girder). In some cases where the roof or floor deck do not provide adequate in-plane shear resistance, in-plane or horizontal diagonal bracings or horizontal X-bracing elements may be required (see Figure 14-7c); examples of such cases include the following:

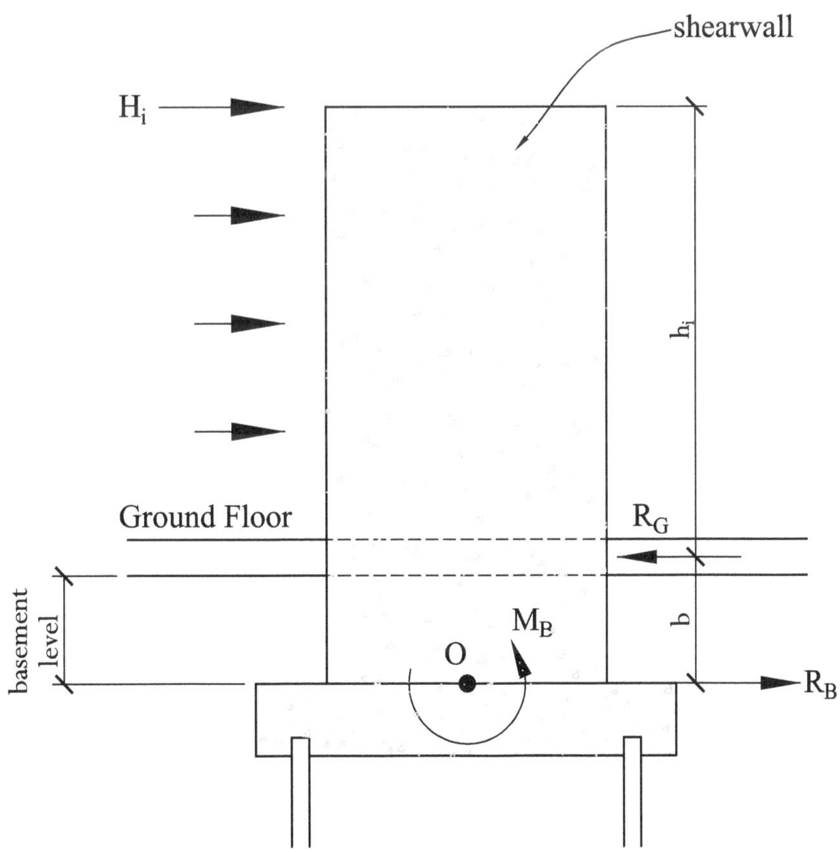

FIGURE 14-7b Shear wall forces due to lateral restraint at the podium (ground floor) slab level (Adapted from Figure 8.35 in Ref. [60])

- Where there is no roof or floor deck such as roof framing supporting large areas of glazing or skylight, or floors framed with steel bar grating, or elevated rooftop equipment platforms, or
- Where the in-plane shear resistance of the roof or floor deck is inadequate, such as diaphragms where the ratio of the diaphragm width (i.e., the dimension perpendicular to the lateral force) to the diaphragm length is large.

FIGURE 14-7c Horizontal rod X-bracing required for steel bar grating diaphragm in-plane shear resistance

The horizontal in-plane diagonal bracings or trusses may consist of rods, angles, HSS, or wide flange members. Rooftop equipment platforms typically do not have regular floor decks. These platform floors typically consist of steel bar grating which has inadequate in-plane diaphragm shear resistance, therefore, horizontal diagonal bracings or horizontal X-bracings are required to resist the in-plane shear. The vertical lateral bracing for these equipment platforms usually consists of diagonal knee braced frames with bolted angle or double angle diagonals or bolted or welded moment connection frames; continuous beams and girders framing over the tops of steel columns are also sometimes used as moment frames for equipment platforms. Note that since these platforms are exposed to the elements, the structural members are usually hot-dipped galvanized or coated with coats of zinc-rich primer to protect against corrosion [53].

- The **chords** at the perimeter of the roof or floor diaphragm are perpendicular to the lateral load, which resist the tension-compression couple from the in-plane bending of the diaphragm (they act like the flanges of a beam or girder). The chords are typically the beams at the edge of the floor or roof diaphragm; for roofs framed with open-web steel joists, the chords would be the continuous perimeter angle or bent plate at the edges of the roof to which the roof deck is welded (see Figures 14-3b, 14-5b, and 14-7a).
- The **collectors or drag struts**, which are situated on the same vertical plane or grid lines as the vertical LFRS, help to drag the in-plane lateral shear force from the diaphragm — beyond the location of the vertical LFRS — into the vertical LFRS. The drag struts or collectors are usually the beams or girders that lie on the same grid line or the same vertical plane as the vertical LFRS. These members would be subjected to axial tension and compression forces for which they must be designed. In addition, the connections of the chords and drag struts to the vertical LFRS and to the columns would have to be designed for the axial tension/compression forces resulting from the lateral loads.

In steel buildings, the vertical LFRS typically consists of one or more of the following systems:

- **Shear walls**: reinforced concrete, steel, steel-concrete composite, or masonry walls.
- **Braced frames**: X-braced frames or Chevron or inverted V-braced frames or Knee braces.
- **Multi-tiered braced frames (MTBF):** these consist of two or more levels of braced frames within each story with each braced frame panel connected to the ones above and below with horizontal struts, can be used for structures with tall story heights. The braced frame within each vertical panel could be X-braced frames, Chevron, or inverted V-braces. The columns in MTBF are oriented such that the weak (y-y) axis bending of the column occurs in the plane of the braced frame, and any out-of-plane bending of the columns is about their strong (x-x) axis.
- **Moment frames** (see Chapter 11 for the different types of moment frames).

Sometimes, braced frames or moment frames may extend over multiple bays of a building to help "spread out" the overturning moment and thus minimize the net uplift force at the bracing frame or moment frame columns. For any structure, there must be well-defined and continuous gravity load path and lateral load path, and they should be clearly identified on the structural drawings, indicating how the lateral load is transferred from the horizontal diaphragm (i.e., the floors and roof) into the vertical LFRS and safely to the foundation/ground.

Horizontal diaphragms can be classified into three categories as follows, depending on the in-plane stiffness of the diaphragm relative to the in-plane lateral stiffness of the lateral force-resisting systems: ***rigid***, ***semi-rigid***, *and* ***flexible*** diaphragms. The lateral stiffness of the vertical lateral force-resisting system (i.e., the vertical LFRS), expressed in kips/inch, is defined as the lateral force that causes a lateral deflection of 1-inch at the top of the vertical LFRS.

For cantilevered concrete shear walls, the lateral stiffness, K, is proportional to the cube of the length of the shear wall parallel to the lateral force. The lateral stiffness of a cantilevered shear wall, fixed at its base and including shear deformations, is given as

$$K = \frac{1}{\frac{H^3}{3E_c I_g} + \frac{H}{A_c G}} = \frac{1}{\frac{H^3}{3E_c\left(\frac{t\ell_w^3}{12}\right)} + \frac{H}{(t\ell_w)G_c}}$$

Where,

H = total height of the cantilevered concrete shear wall

t = thickness of shear wall

ℓ_w = length of shear wall at the base of the wall, parallel to the lateral force

E_c = modulus of elasticity of the concrete

G_c = shear modulus of the concrete

Rigid Diaphragms:

In rigid diaphragms, the in-plane stiffness of the roof and floor plate or slab is much greater than the lateral stiffness of the vertical LFRS; thus, the roof or floor slabs behave as rigid or continuous horizontal members that transmit the lateral seismic and wind loads to the vertical LFRS in proportion to the lateral stiffness of the vertical LFRS.

If the vertical LFRS in a building with rigid diaphragms are not symmetrically located, or the vertical LFRS do not have equal stiffnesses, or if the center of mass (CM) of the horizontal diaphragm or center of area of the vertical wall surface receiving the wind pressure does not coincide with the center of rigidity (CR) of all the vertical LFRS in the building, the building will be subjected to in-plane twisting (i.e., twisting on the horizontal plane about a vertical axis of rotation) due to the torsional moment from the lateral seismic or wind forces (see Figure 14-8).

Therefore, in addition to the direct lateral forces transmitted to the vertical LFRS that are parallel to the direction of the lateral wind or seismic load, a rigid diaphragm is also capable of transmitting the in-plane torsional or rotational moments, M_T, that result in additional lateral forces to **all** the vertical LFRS below that diaphragm level.

For ***seismic loads***, the in-plane torsional moment, M_T, at any floor level is a function of both the *floor eccentricity* and the *accidental eccentricity*. The floor eccentricity is the distance between the center of mass of the floor level (CM) and the center of rigidity (CR) of all the vertical LFRS just below that floor level under consideration (i.e., the floor eccentricities are X_{CR}-X_{CM} and Y_{CR}-Y_{CM}); this floor eccentricity is as a result of the non-symmetrical placement of the vertical LFRS in a building relative to the center of mass, and the in-plane torsion caused by this eccentricity is called the *natural or inherent torsion*. The in-plane torsion from seismic loads due to *accidental eccentricity*, which accounts for the uncertainties in calculating the actual location of the center of mass, X_{CM} and Y_{CM}, must also be included in calculating the in-plane torsional moment, M_T. The accidental eccentricity specified for seismic load calculations is defined as 5% of the width of the building perpendicular to the direction of the seismic force (see ASCE 7, Section 12.8.4.2.2). This accidental eccentricity must be considered separately to the right and to the left of the center of mass, and the largest forces in the LFRS from both scenarios are used for the seismic design. The inherent and accidental eccentricities, and the corresponding in-plane torsional moments are calculated for both orthogonal directions of the building.

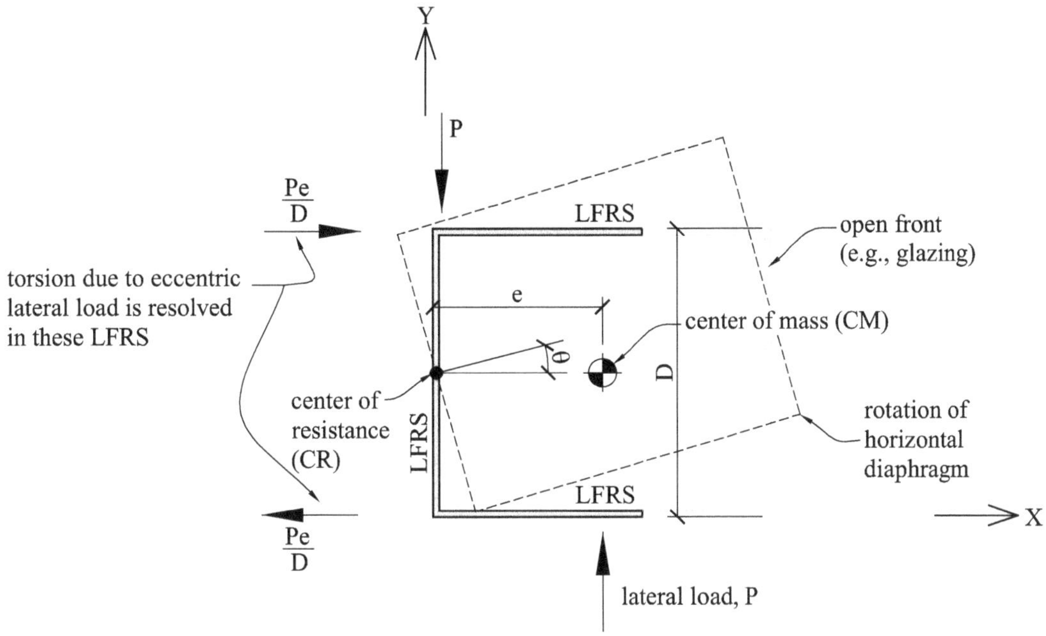

a) Open front building with LFRS on 3 sides

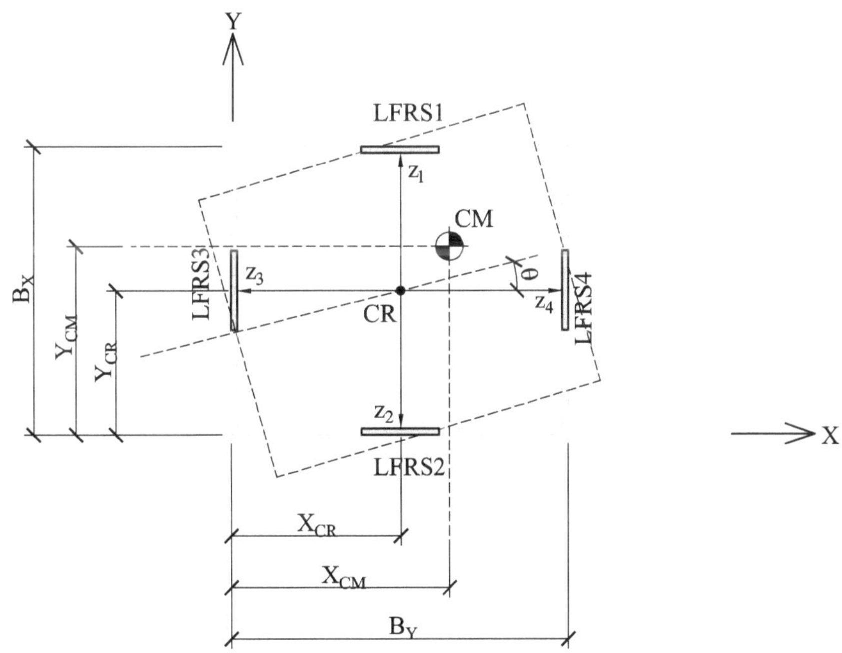

b) Building with LFRS on all sides

FIGURE 14-8 In-plane torsion of horizontal diaphragms

For ***wind* loads** on the main wind force resisting system (MWFRS), the in-plane torsional moments, M_T, are caused by the following factors [25]:

- The non-uniform wind pressures on the face of the building
- Dynamic effects of the wind
- The effects of adjacent or nearby buildings interfering with the wind flow (see ASCE 7, Section C27.3-5)

The eccentricity of the lateral wind forces is defined in the design Wind Load Cases 2 and 4 that are given in ASCE 7, Figure 27.3.8 [25]. Note that Load Cases 1 and 2 indicate only direct lateral wind forces on a building in the two orthogonal directions, without any in-plane torsion. The eccentricity is measured in a direction perpendicular to the direction of the lateral wind force. For wind force parallel to the X-direction, B_X is the width of the face of the building perpendicular to the X-direction, and the eccentricity, $e_X = 0.15B_X$. For wind force parallel to the Y-direction, B_Y is the width of the face of the building perpendicular to the Y-direction and the eccentricity, $e_Y = 0.15B_Y$ (see Figure 14-8b). Note that these eccentricities given in ASCE 7 apply to "rigid buildings" or buildings with fundamental or natural lateral frequency of vibration, $f \geq$ 1 Hz. For "flexible buildings" (i.e., $f < 1$ Hz), the dynamics of the wind loading can increase the torsional loading, and the eccentricity is calculated using ASCE 7 equation 27.3-4 (see ASCE 7, Section 27.3-5). Recall that approximate natural lateral frequencies of buildings for wind load calculations are obtained from Chapter 3 of this text or from ASCE 7, Section 26.11.3. The following floor or roof systems may be classified as ***rigid* diaphragms** according to ASCE 7, Section 12.3.1.2 [25]:

- Diaphragms of concrete on composite metal deck with *span-to-depth* ratios less than or equal to 3.0 in structures with no horizontal irregularities. The "*span*" is the horizontal distance between adjacent vertical LFRS, and the "*depth*" is the horizontal in-plane dimension of the diaphragm parallel to the lateral force.
- Diaphragms of concrete slabs with *span-to-depth* ratios less than or equal to 3.0 in structures with no horizontal irregularities (see ASCE 7 Table 12.3.1 and 12.3.2 for the definitions of horizontal and vertical structural irregularities).
- Precast hollow core concrete slabs with concrete topping having *span-to-depth* ratios of 3.0 or less.
- Metal roof decks can be classified as rigid diaphragms if the maximum *span-to-de*pth ratio is 2.0 or less [8].

In general, metal roof decks (except as indicated above) and plywood sheathing without any concrete topping are usually classified as *flexible* diaphragms. Since it is often difficult to ascertain whether a horizontal diaphragm is actually rigid or flexible, some engineers use the "envelope" approach, whereby the lateral forces are calculated assuming a rigid diaphragm and then recalculated assuming a flexible diaphragm; the higher forces from the two analyses are then used for the design of the diaphragm, the drag struts or collectors, the chords, and the vertical LFRS.

In analyzing rigid roof and floor diaphragms for in-plane torsion, the in-plane torsional moments cause an increase in the direct lateral force in some of the vertical LFRS and a decrease or offset in others. Nevertheless, in design practice, only the lateral force increases from in-plane torsion are considered; any lateral force reduction from in-plane torsional moments is usually ignored. This means that the lateral force for which each vertical LFRS is designed must not be less than the direct lateral force calculated for each vertical LFRS when in-plane torsion is neglected.

Flexible Diaphragms:

In flexible horizontal diaphragms, the in-plane stiffness of the roof and floor plate or slab is much smaller than the lateral stiffness of the vertical LFRS in the building. Therefore, the lateral load path consists of the roof and floor slabs acting as in-plane horizontal beams simply supported between adjacent vertical LFRS. Thus, the distribution of the lateral *seismic* load in a given direction is proportional to the in-plane tributary width (perpendicular to the lateral force direction) of each vertical LFRS that is parallel to the direction of the seismic force on the building.

For *wind* loads, the lateral load at each diaphragm level on each vertical LFRS that is parallel to the wind direction is proportional to the tributary vertical surface area of the exterior wall that is perpendicular to, and receiving, the wind pressure. Where there are more than one vertical LFRS along the same grid line or vertical plane, the tributary lateral force for that grid line is distributed to each vertical LFRS on that line in proportion to the relative lateral stiffnesses of the vertical LFRS along that grid line.

As previously discussed, only buildings with rigid or semi-rigid diaphragms can transmit in-plane torsional or rotational forces (see Figure 14-8); buildings with flexible diaphragms cannot resist or transmit in-plane rotational or torsional forces; therefore, the vertical LFRS in buildings with flexible diaphragms can only resist direct lateral forces. It is best, if possible, to avoid layouts of the vertical LFRS that result in in-plane torsion or twisting of the roof or floor diaphragms. For example, a rigid diaphragm building with an open front (e.g., the right hand side of the building in Figure 14-8a where there is no LFRS) would be subjected to large in-plane torsional moments; however, if the building in Figure 14-8a has a flexible diaphragm, it will be unstable and will collapse because there will be no lateral load path available to transfer the lateral load, *P*, from the center of mass (CM) of the horizontal diaphragm to the vertical LFRS on the left hand side exterior wall, since flexible diaphragms are incapable of transmitting in-plane rotation. Only rigid diaphragms are capable of transferring in-plane lateral forces through in-plane rotation of the diaphragm, but even then, large in-plane rotations of the horizontal diaphragms could lead to collapse of the structure. Some buildings have collapsed during earthquakes due to the irregular layout of the vertical LFRS and the subsequent in-plane twisting of the horizontal diaphragms, and the inadequacy of the structure to resist the in-plane rotation. One notable example is the JC Penney building in Anchorage that collapsed during the 1964 Alaskan earthquake [43].

Chords and Drag Struts or Collectors

Chords are structural elements located along the perimeter of the horizontal diaphragms perpendicular to the direction of the lateral load; they resist the tension-compression force couple resulting from the in-plane bending of the diaphragm due to the lateral seismic or wind forces on the building (see Figure 14-9).

Drag struts or collectors are structural elements in the plane of the horizontal diaphragms located parallel to the lateral load, and they are located on the same lines or the same vertical plane as the vertical LFRS. The drag strut or collector transfers the lateral wind or seismic forces from the roof and floor diaphragms into the vertical LFRS through the steel floor and roof framing and their connections. The floor and roof beams or girders and their connections would thus be subjected to axial tension or compression forces. Also, the eccentricity of the in-plane lateral force in the drag strut members must be considered since the diaphragm transfers the in-plane lateral load to the drag strut or collector beams at their top flange. This will result in additional vertical bending moment in the drag strut or girder. The drag strut beams will be designed for the combined axial force and the resulting moment from the eccentricity of the axial force, in addition to the moment from gravity and/or lateral loads; thus, the drag struts are designed as beam-columns (see AISC Commentary Section I7 [20]).

Where concrete shear walls are used to resist lateral loads in building structures, the drag strut (i.e., the roof and floor beams/girders) must be connected to the concrete walls using embedded steel plates (see Section 1 in Figure 14-10. An example of an embedded plate with headed studs before the concrete is poured is shown in Figure 14-10b). In addition, the roof and floor

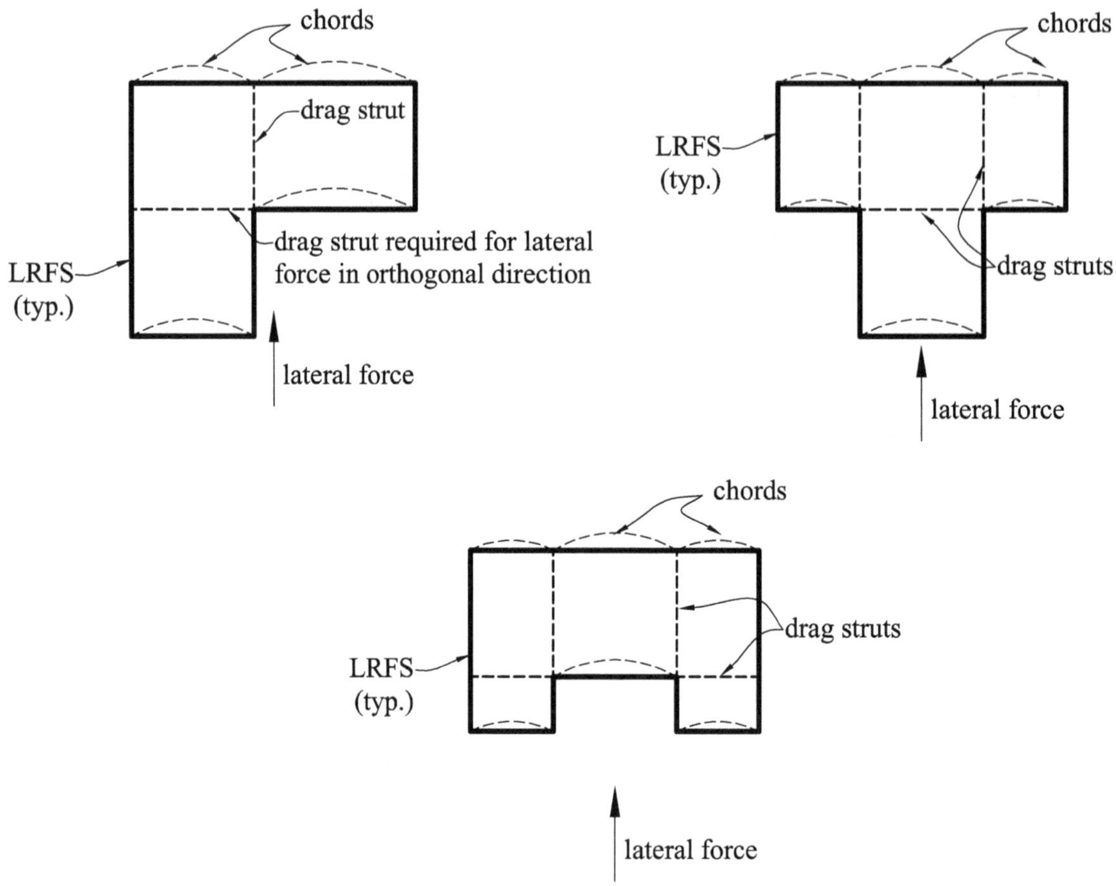

FIGURE 14-9 Diaphragm chords and drag struts

decks must be connected to the concrete shear walls by welding the deck to a continuous angle that is connected to the wall at regular intervals with embedded plates. The continuous angle acts as a drag strut that drags the diaphragm in-plane shear forces from the deck into the concrete shear walls (see Sections 2 and 3 in Figure 14-10). The beams and girders framing into the corners of the shear wall in Figure 14-10 act as drag struts or collectors that help to drag the diaphragm lateral forces into the shear walls that are parallel to the drag strut. The roof and floor diaphragms in buildings must be adequately connected to the vertical lateral force-resisting systems to ensure the transfer of the lateral loads (wind, seismic, or lateral earth pressure) from the diaphragm to the braced frames, shear walls, or moment frames.

When structural members are field welded to steel plates embedded in concrete shear walls, such as shown in Figure 14-10, it is recommended that a low-heat welding rod and a smaller weld thickness in multiple passes—instead of a thicker weld with a single pass—be specified. This will help to minimize the heat generated from the welding processes and thus minimize the potential loss of bond between the concrete and the steel plates and also minimize any cracking or spalling of the concrete that may arise from the expansion of the embedded plate [49]. In design practice, a minimum embedded plate thickness of 3/8 inch. is commonly specified, and the thicker the embedded steel plate, the smaller the distortion from the welding processes.

When drag struts and horizontal diaphragms are not properly connected to the lateral-force-resisting-systems, this results in an incomplete lateral load path, and the lateral stability of the structure will be compromised.

Drag struts also help prevent differential or incompatible horizontal or lateral displacements of the horizontal diaphragms in buildings with irregular shapes. Without drag struts or collectors, tearing forces would develop at the interface between the various diaphragm

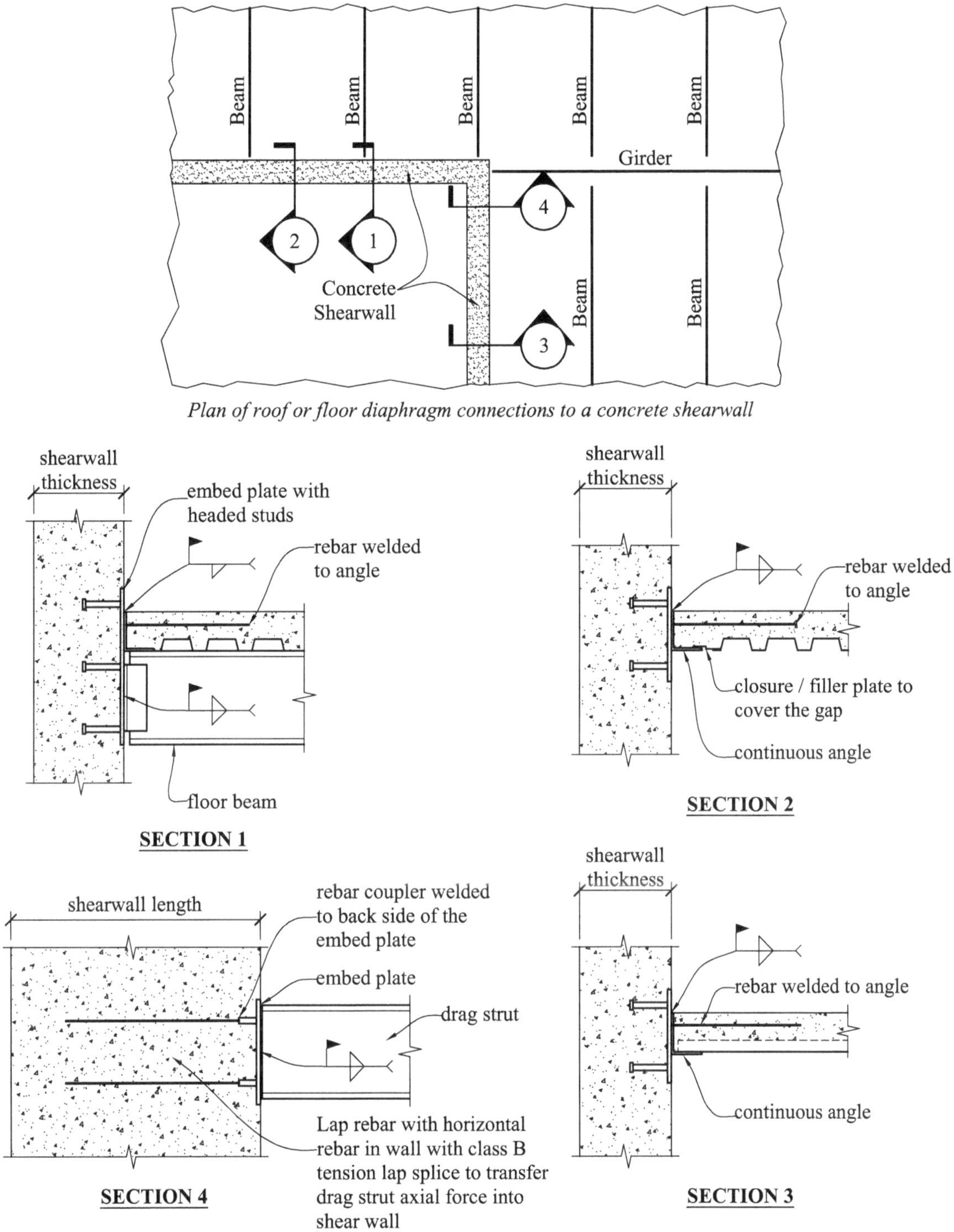

FIGURE 14-10 Horizontal diaphragm connections to concrete shearwalls

segments (see Figure 14-9). The ASCE 7 Load Standard specifies that drag struts be designed for the special seismic force, E_m, which includes amplification of seismic forces by the overstrength factor, Ω_o.

Section 7.10.1 of the AISC Code of Standard Practice [20] requires that "*the lateral force resisting system and connecting diaphragm elements that provide for lateral strength and stability in the completed structure*" must be identified. In the author's experience, this required information is not universally presented on structural drawings in the United States. Ruby [38, 39] has found several instances of roof and floor diaphragms on structural drawings that have no

FIGURE 14-10b Embedded plates with headed studs

collectors or drag struts; in addition, some design and analysis software do not supply the drag strut and chord forces. Some important conclusions from Ruby's articles regarding horizontal diaphragms and their component elements are as follows [38, 39]:

1. The beams and girders supporting the horizontal diaphragms should be framed to maximize the gravity loads that they deliver to the vertical LFRS so as to minimize the impact of uplift on the columns, shear walls, and foundations.
2. All structural drawings should indicate a well-defined continuous lateral load path showing the chords and drag struts and the axial forces, as well as their connections to the vertical LFRS.
3. Along every line of braced frames, moment frames, or shear walls, the collectors or drag struts (i.e., beams or girders) parallel to the vertical LFRS that are needed to drag the diaphragm in-plane lateral shear into the vertical LFRS must be identified. The required length of the drag struts or collectors must be selected so that the applied in-plane shear per unit length in the roof or floor metal deck diaphragm does not exceed the allowable (or design) in-plane shear strength per unit length presented in Ref. [9].

In-plane Shear Strengths of Diaphragms

For floor construction with a concrete slab on a composite metal deck supported by composite beams and girders (i.e., beams and girders with shear studs), the in-plane lateral shear force can be transferred from the composite floor diaphragm into the drag strut beams or chord beam members through the shear studs. A typical ¾ in. diameter shear stud has a maximum horizontal shear capacity of 19 kips. The in-plane shear strength of steel-framed diaphragms can be determined using the Steel Deck Institute (SDI) lateral in-plane shear strength tables [9], or the in-plane shear strength can be conservatively assumed to be equal to the in-plane shear capacity of the concrete slab above the deck ribs or flutes, ignoring the beneficial effect of the concrete within the deck flutes and any reinforcement within the slab (see *AISCM* Commentary, Section I7 [20]). Therefore, the approximate in-plane shear strength of the diaphragm is calculated using ACI 318 as

$$\phi V_c = 2\phi\sqrt{f_c'}\,bd$$
$$\geq V_u$$

where,

ϕ = 0.75
V_u = In-plane shear force due to lateral wind, seismic loads, or lateral soil pressures
f_c' = 28-day compressive strength of the concrete
b = thickness of the concrete slab above the deck ribs, inches
d = effective depth of the diaphragm ≈ 0.9 × plan dimension of the concrete diaphragm parallel to the lateral load, in inches.

For example, the shear capacity for a composite floor deck with 2.5 in. of concrete above the deck flutes (i.e., b = 2.5 in.) and a 28-day concrete strength of 3000 psi can be calculated approximately and conservatively using the ACI 318 one-way shear equation for slabs as: $2(0.75)\sqrt{3000}\,(2.5\text{ in.})(0.9)(12\text{ in./ft.}) = 2218$ Ib/ft. length of deck parallel to the lateral force.

The above value is conservative when compared to the lowest design in-plane shear strength for the 2.5 in. normal weight concrete slab on 1.5 in. × 20-gage composite floor deck of (0.5)(5499 Ib/ft.) or 2750 Ib/ft. (for wind or seismic loads); this is tabulated in the Steel Deck Institute (SDI) Diaphragm in-plane shear strength tables for a 20-gage composite floor deck with 2.5 in. normal weight concrete above the deck ribs and a 36/4 support fastener pattern plus 3 - #10 screws side lap connections per deck span [9].

Where the shear strength of the diaphragm is inadequate to resist the lateral in-plane shear from the lateral loads, horizontal steel truss bracing at the underside of the metal deck can be provided to resist the additional in-plane shear forces in the floor or roof diaphragms (e.g., see Figure 14-7c).

Analysis of Rigid Diaphragms for In-Plane Torsional Moments

As noted previously, only rigid and semi-rigid diaphragms can undergo in-plane torsional rotation or transmit in-plane torsional moments; for flexible diaphragms, in-plane torsion is not a consideration. As previously discussed, for seismic forces, there are two types of in-plane torsion that must be considered to obtain the in-plane torsional moment, M_T:

- **Accidental Torsion**: This torsion is caused by an eccentricity of the center of mass equal to 5% of the plan dimension of the building perpendicular to the direction of the seismic lateral force. See ASCE 7-22 Section 12.8.4.2.2.2.
- **Inherent or natural Torsion**: This torsion is caused by the difference in the locations of the center of mass, CM, and the center of rigidity, CR. See ASCE 7-22 Section 12.8.4.2.2.1.

For structures subjected to *wind loads*, the in-plane torsional moments, M_T, are defined by the equations given for Wind Load Cases 2 and 4 in ASCE 7 Figure 27.3.8 [25].

For *seismic loads*, the in-plane torsional moment at level x is the sum of the accidental in-plane torsion and the natural or inherent in-plane torsion at that level. The torsional moment at level x causes additional lateral forces in all the vertical lateral-force-resisting-systems (LFRS) in the building immediately below level x.

The eccentricity due to natural torsion, e, is equal to the distance between the center of rigidity (CR) and the center of mass, CM, at the level being considered.

Center of Rigidity

There is no clear consensus on the definition of the centers of rigidity (CR) of the floor levels of multistory buildings, and calculating the centers of rigidity of the floors for other than a one-story building is fraught with difficulties except for the so-called "proportional buildings," where the CR at each level lies on the same vertical axis. The actual centers of rigidity of the floors in a multistory building depend on the building geometry and the stiffness of the vertical LFRS above and below the floor level under consideration, and it also depends on the applied lateral load and any inelastic behavior [25, 44, 45]. The method presented in this text, which was originally developed for single-story buildings with rigid diaphragms, is applicable to "proportional buildings," but it has also been applied for the design of multistory buildings. This same approach for calculating the CR is also used by Wight and MacGregor for the calculation of the centers of rigidity for concrete floors in multistory reinforced concrete buildings [46].

The center of rigidity (CR) at a floor or roof level is the center of lateral stiffness of the vertical LFRS immediately below that floor level; that is, the point through which a lateral load applied at that floor level causes no in-plane rotation of that floor diaphragm. The CR, which is computed separately for each orthogonal direction and for each diaphragm level, is defined in this text as the sum of the first moment of the lateral stiffnesses of the vertical LFRS about a chosen origin divided by the total lateral stiffness of all the vertical LFRS in that direction [41, 45]. Thus, it is assumed in this text that the center of rigidity at any level is dependent only on the vertical LFRS below the floor level under consideration and independent of the lateral load distribution. Only the lateral stiffness of the vertical LFRS (i.e., its in-plane stiffness) is considered; the out-of-plane stiffness of the vertical LFRS is negligible and therefore neglected. Thus, the calculation of CR only includes the vertical LFRS that are parallel to the orthogonal direction being considered. The X- and Y-coordinate location of the center of rigidity of the diaphragm at any level measured from an arbitrarily chosen origin can be determined by taking statical moments of the stiffnesses of the vertical LFRS—that are immediately below the floor level under consideration—about the chosen origin, resulting in the following equations (see Figure 14-8b) [46]:

$$X_{CR,N-S} = \frac{\sum_{i=1}^{n} K_{i,N-S} X_i}{\sum_{i=1}^{n} K_{i,N-S}} \quad (14\text{-}1)$$

$$Y_{CR,E-W} = \frac{\sum_{i=1}^{N} K_{i,E-W} Y_i}{\sum_{i=1}^{N} K_{i,E-W}} \quad (14\text{-}2)$$

Where,

$K_{i,\,E\text{-}W}$ = in-plane lateral stiffness of the vertical LFRS that are parallel to the E-W direction (assumed as the X-axis) **with its centroid located at a perpendicular distance** Y_i from the chosen origin.

$K_{i,\,N\text{-}S}$ = in-plane lateral stiffness of the vertical LFRS that are parallel to the N-S direction (assumed as the Y-axis) **with its centroid located at a perpendicular distance** X_i from the chosen origin.

- The in-plane lateral stiffness (in kips/inch) of a vertical LFRS is the lateral force per unit lateral displacement at the top of the LFRS. That is, the force in kips that produces a lateral displacement of 1-in. at the top of the vertical LFRS.
- For isolated reinforced concrete shear walls, the lateral stiffness is proportional to the moment of inertia of the walls bending about their strong axis (i.e., proportional to $\left(\frac{t\ell_w^3}{12}\right)$, where t is the wall thickness and ℓ_w is the wall length parallel to the lateral load). This assumes that only flexural deformations of the cantilvered wall are considered and shear deformations are neglected.
- For compound shear walls with flanges (e.g., elevator or stair core walls), the moment of inertia of the compound shear walls can be calculated using the parallel axis theorem from mechanics of materials.

X_i = Location of the N-S vertical LFRS on the X-axis (i.e., perpendicular to the N-S direction) relative to the assumed origin

Y_i = Location of the E-W vertical LFRS on the Y-axis (i.e., perpendicular to the E-W direction) relative to the assumed origin

n = Total number of N-S vertical LFRS (i.e., parallel to the Y-axis)

N = Total number of E-W (i.e., parallel to the X-axis)

Center of Mass

The center of mass (CM) is the center of gravity of the floor; the resultant *seismic* lateral forces act through the CM, while the resultant *wind* load acts through the *center of the vertical wall surface area* receiving the wind, which for symmetrical buildings coincides with the center of mass. The X- and Y-coordinate location of the center of mass at any level of a multi-story building measured from an arbitrarily chosen_origin on the plan can be determined by taking the statical moments of the weights of the floors (including the walls and exterior cladding that are tributary to that floor level) about the chosen origin using the following equations [46]:

$$X_{CM} = \frac{\sum_{i=2nd\,flr}^{Roof} W_i\, X_{CM,i}}{\sum_{i=2nd\,flr}^{Roof} W_i} \qquad (14\text{-}3)$$

$$Y_{CM} = \frac{\sum_{i=2nd\,flr}^{Roof} W_i\, Y_{CM,i}}{\sum_{i=2nd\,flr}^{Roof} W_i} \qquad (14\text{-}4)$$

When the floor or roof weights are uniformly and symmetrically distributed, the CM will be located at the geometric centroid of the floor and roof diaphragm. Therefore, in that case,

$$X_{CM} = \frac{B_Y}{2}$$

$$Y_{CM} = \frac{B_X}{2}$$

B_X = plan dimension of the building perpendicular to the X-axis (or parallel to the Y-axis)

B_Y = plan dimension of the building perpendicular to the Y-axis (or parallel to the X-axis)

W_i = weight of floors or roof and walls tributary to level i.

$X_{CM,i}$ = Location on the X-axis (i.e., perpendicular to the N-S direction) of the center of mass of the floor/roof diaphragm element relative to the assumed origin.

$Y_{CM,i}$ = Location on the Y-axis (i.e., perpendicular to the E-W direction) of the center of mass of the floor/roof diaphragm element relative to the assumed origin.

Lateral Forces in LFRS due to In-Plane Torsion of Diaphragm

The lateral force in each vertical LFRS due to in-plane torsional moment, M_T, on a rigid diaphragm is derived below using the floor eccentricity concept [41], which is based on the following simplifying assumptions:

- Small in-plane displacements and small in-plane rotations of the horizontal diaphragm
- Linear elastic material behavior of the diaphragm and the vertical LFRS
- Rigid diaphragm
- The out-of-plane stiffness of the vertical LFRS is negligible and therefore neglected. Only the in-plane stiffness of the vertical LFRS is considered.
- For multi-story buildings, the CR is assumed to lie on the same vertical axis for every floor and roof level in the building. Similarly, the CM is also assumed to lie on one vertical axis in every floor in the building; thus, the eccentricity is assumed to be constant throughout the height of the building. This assumption may not always be the case, especially for structures with vertical and horizontal structural irregularities or asymmetric structures where the location of the CR and CM may vary from floor to floor and may not be at a consistent location throughout the height of the building, in which case the eccentricity will be different from floor to floor [41]. See ASCE 7 Tables 12.3.1 and 12.3.2 [25] for the definitions of horizontal and vertical irregularities.

The lateral force due to the in-plane torsion in each vertical LFRS below a floor or diaphragm level is assumed to be proportional to the lateral stiffness of the vertical LFRS using the following linear elastic equation (see Figure 14-8b):

$$F_{Ti} = K_i \delta_i = K_i \left(Z_i \theta\right) \tag{14-5}$$

In-plane torsional equilibrium requires that the applied cumulative in-plane torsional moment at any floor level be equal to or less than the total in-plane torsional resistance of the vertical LFRS just below that level. Using equation (14-5), the equilibrium equation relating the applied torsional moment to the torsional resistance can be written as

$$M_{Tx} = \sum F_{Ti} Z_i = \sum K_i (Z_i\theta) Z_i = \sum K_i (Z_i)^2 \theta \tag{14-6}$$

Therefore, the in-plane torsional rotation of the rigid diaphragm,

$$\theta = \frac{M_{Tx}}{\sum K_i (Z_i)^2} \tag{14-7}$$

The additional lateral force in each vertical LFRS just below the floor level under consideration due to the in-plane torsional moment is calculated as

$$F_{Ti} = K_i Z_i \theta = \frac{K_i Z_i M_{Tx}}{\sum K_i (Z_i)^2} \tag{14-8}$$

This additional lateral force due to the in-plane torsional moment is added to the direct lateral force for each vertical LFRS where the direct lateral force and the torsion-induced lateral force are additive. For some vertical LFRS where the torsion-induced lateral force opposes the direct lateral force, it is conservative to assume that the net lateral force on those vertical LFRS is the direct lateral force previously obtained with the in-plane torsional moment neglected. In summary, in no instance should the lateral force at each level in each vertical LFRS be smaller than the direct lateral force calculated when the rotation of the diaphragm or the in-plane torsion is neglected.

Therefore, the total lateral force on vertical LFRS i is

$$F_{total,i} = F_i + F_{Ti} \geq F_i \tag{14-9}$$

Where,

$\boldsymbol{Z_i}$ = Perpendicular distance from the center of rigidity (CR) to the longitudinal axes of vertical LFRS i, Thus, $Z_i = X_i$ for the North-South vertical LFRS and $Z_i = Y_i$ for the East-West vertical LFRS.

$\boldsymbol{K_i}$ = In-plane lateral stiffness of vertical LFRS i.

δ_i = Lateral deflection of the vertical LFRS due to in-plane torsion

$\boldsymbol{\theta}$ = In-plane rotation of the horizontal diaphragm about the center of resistance (CR) due to The in-plane torsional moment, M_T.

M_{Tx} = In-plane torsional moment just below the floor level under consideration (i.e., level x), caused by the product of the total lateral force at that floor or diaphragm level and the torsional eccentricity.

- For **wind loads**, see ASCE 7, Figure 27.3.8 [25] for the prescribed in-plane torsional eccentricities that are used to calculate the in-plane torsional moments.
- For **seismic loads**, the in-plane torsional eccentricities are obtained as follows:

For the lateral force in the **N-S direction** (i.e., parallel to the Y-direction) at any floor level (see Figure 14-8b), the in-plane torsional eccentricity is the sum of the natural or inherent torsional eccentricity and the accidental eccentricity of 5% B_Y (see ASCE 7, Section 12.8.4.2):

$$e_{N-S} = |X_{CR} - X_{CM}| \pm 0.05B_Y \tag{14-10}$$

and the resulting in-plane torsional moment due to the N-S lateral force is

$$M_{Tx\,(N-S)} = F_{N-S}\, e_{N-S} = F_{N-S}\left(|X_{CR} - X_{CM}| \pm 0.05B_Y\right) \tag{14-11}$$

For the lateral force in the **E-W direction** (i.e., parallel to the X-direction) at any floor level, the in-plane torsional eccentricity is the sum of the natural eccentricity and the accidental eccentricity of 5% B_X (see ASCE 7, Section 12.8.4.2):

$$e_{E-W} = |Y_{CR} - Y_{CM}| \pm 0.05B_X \tag{14-12}$$

and the resulting in-plane torsional moment due to the E-W lateral force is

$$M_{Tx\,(E-W)} = F_{E-W}\, e_{E-W} = F_{E-W}\left(|Y_{CR} - Y_{CM}| \pm 0.05B_X\right) \tag{14-13}$$

Where,

B_X = plan dimension of the building perpendicular to the X-axis (i.e., the Y-axis plan dimension of the building as shown in Figure 14-8b).

B_Y = plan dimension of the building perpendicular to the Y-axis (i.e., the X-axis plan dimension of the building as shown in Figure 14-8b).

Note that the accidental eccentricity can be neglected for structures not sensitive to torsion or for which there are no horizontal irregularities (i.e., Types 1a and 1b). See ASCE 7 Sections 12.8.4.1 and 12.8.4.2, and see ASCE 7 Table 12.3.1 for definitions of horizontal structural irregularities and ASCE 7 Table 12.3.2 for vertical structural irregularities.

Amplification of the 5% Accidental Torsional Eccentricity

For structures in Seismic Design Categories 'C' and higher with torsional irregularities 1a and 1b as defined in ASCE 7 Section 12.3.1, and for structures where the maximum or peak lateral displacement (i.e., the maximum lateral drift of the diaphragm when $A_x = 1$), δ_{max}, exceeds 1.2 times the average lateral drift of the diaphragm (i.e., the average of lateral displacement of the diaphragm at the extreme point of the diaphragm at level x when $A_x = 1$), δ_{avg}, (i.e., $\delta_{max} > 1.2\,\delta_{avg}$), the dynamic torsional instability and the resulting torsional irregularity caused by the earthquake are accounted for by multiplying the 5% accidental torsional eccentricity (see ASCE 7-22 Section 12.8.4.2) computed at each floor level by a torsional amplification factor, A_x, calculated using ASCE 7 Equation (12.8-15) from ASCE 7 Section 12.8.4.3 [25] as follows:

$$A_x = \left(\frac{\delta_{max}}{1.20\delta_{avg}}\right)^2 \geq 1.0 \quad \leq 3.0 \tag{14-14}$$

Note that for $\delta_{max} \leq 1.2\,\delta_{avg}$, $A_x = 1.0$, indicating no torsional irregularity. In the computation of δ_{max} and δ_{avg}, the diaphragm is assumed to be rigid.

δ_{max} = maximum in-plane displacement at level x with $A_x = 1$

δ_{avg} = average in-plane displacement at level x of the extreme corners of the structure with $A_x =$

The revised in-plane torsional moments due to the torsional amplification factor, A_x, and the redundancy factor, ρ, (see Section 2-3), are given as follows for structures in SDC C through F:

$$M_{Tx\,(N\text{-}S)} = (\rho)F_{N\text{-}S}\,e_{N\text{-}S}$$

$$e_{N\text{-}S} = \left(|X_{CR} - X_{CM}| \pm A_X 0.05 B_Y\right)$$

$$M_{Tx\,(E\text{-}W)} = (\rho)F_{E\text{-}W}\,e_{E\text{-}W}$$

$$e_{E\text{-}W} = \left(|Y_{CR} - Y_{CM}| \pm A_X 0.05 B_X\right)$$

Recall from Chapter 2 that $\rho = 1.0$ for SDC B or C, and $\rho = 1.3$ for SDC D through F.

Forces in the Chords and Drag Struts or Collectors

After the lateral shear forces in each vertical LFRS have been calculated at each diaphragm level, the in-plane shear force diagram for each diaphragm can be drawn, from which the in-plane bending moment diagram for the diaphragm can be obtained using the principle — from structural analysis — that the change in bending moment between two points is equal to the area under the shear force diagram between those two points. Knowing the maximum moment in the diaphragm, $M_{u,\max diaphragm}$, the maximum factored tension or compression forces in the diaphragm chords can be obtained from

$$T_{u,chord} = \left(\frac{M_{u,\max diaphragm}}{D}\right) \tag{14-15a}$$

Where,

D = the perpendicular distance between the chords (i.e., the plan dimension of the diaphragm parallel to the lateral load direction). See Figure 14-8b.

= B_X for the N-S direction lateral load,

= B_Y for the E-W direction lateral load

For *seismic* loads, ASCE 7 Section 12.10.2.1 requires that the forces be amplified in the *chords* by the overstrength factor, Ω_o, thus the factored tension or compression force in the chords becomes

$$T_{u,chord} = \Omega_o \left(\frac{M_{u,\max\ diaphragm}}{D} \right) \tag{14-15b}$$

As previously presented in Chapter 4, the required area of steel needed to resist the chord force in equation (14-15b) will be the larger of equations (14-16a) and (14-16b).

$$A_g = \frac{T_{u,chord}}{0.9F_y} \tag{14-16a}$$

$$A_g = \frac{T_{u,chord}}{0.75F_u} + \sum A_{holes} \tag{14-16b}$$

The maximum tension and compression forces in a *drag strut* or *collector* will be equal to the maximum diaphragm in-plane shear per unit length along the brace or shear wall line multiplied by the length of the drag strut. That is, the maximum tension force in the drag strut is given as

$$T_{u,drag\ strut} = \left(\frac{V_{u,diaphragm}}{D} \right) \left(\text{maximum } \ell_{\text{drag strut in tension}} \right) \tag{14-17}$$

Note that the chords and drag struts should be continuous where feasible, but where there is horizontal or vertical discontinuity in the drag struts or chords (e.g., caused by a vertical or horizontal offset in the diaphragm), an adequate structural detail that ensures an uninterrupted lateral load path—that is, the continuous transfer or flow of the chord or drag strut forces—must be ensured. Such details might include horizontal in-plane or vertical bracing. Recall that for the N-S lateral force (see Figure 14-8b), $D = B_X$, and for the E-W lateral force, $D = B_Y$.

For **seismic loads**, the drag strut and chord forces are amplified by the overstrength factor, Ω_o, to ensure that they and their connections will not collapse before the LFRS, and the factored tension force in the drag strut becomes

$$T_{u,drag\ strut} = \Omega_o \left(\frac{V_{u,diaphragm}}{D} \right) \left(\text{maximum } \ell_{\text{drag strut in tension}} \right) \tag{14-18}$$

The applied in-plane shear in the diaphragm should be such that

$$\Omega_o \frac{V_{u,diaphragm}}{D} \leq \text{allowable or design in-plane shear strength obtained from Ref. [9].}$$

Check the connections of the roof and floor diaphragms to the structural steel framing along the braced frame, moment frame, or shear wall lines. The required area of the drag strut is calculated as was done previously for the chords. Similarly, the connections of the chords and drag struts are also designed to resist these tension forces.

EXAMPLE 14-1

Lateral Load Distribution to Vertical LFRS – Simplified Method

The plan of a 3-story steel building with a 10 ft. floor-to-floor height and the locations of the vertical LFRS are shown in Figure 14-11a. The total lateral seismic loads on the building have been determined as follows: $F_{roof} = 10$ kips, $F_{3rd\ floor} = 20$ kips, and $F_{2nd\ floor} = 24$ kips. Assume that the relative lateral stiffnesses of the vertical LFRS have been determined as follows, and they are assumed to be constant over the full height of the building:

$K_{XB1} = K_{XB2} = K$

$K_{W1} = K_{W2} = 2K$

$K_{XB3} = K_{XB4} = 1.5K$

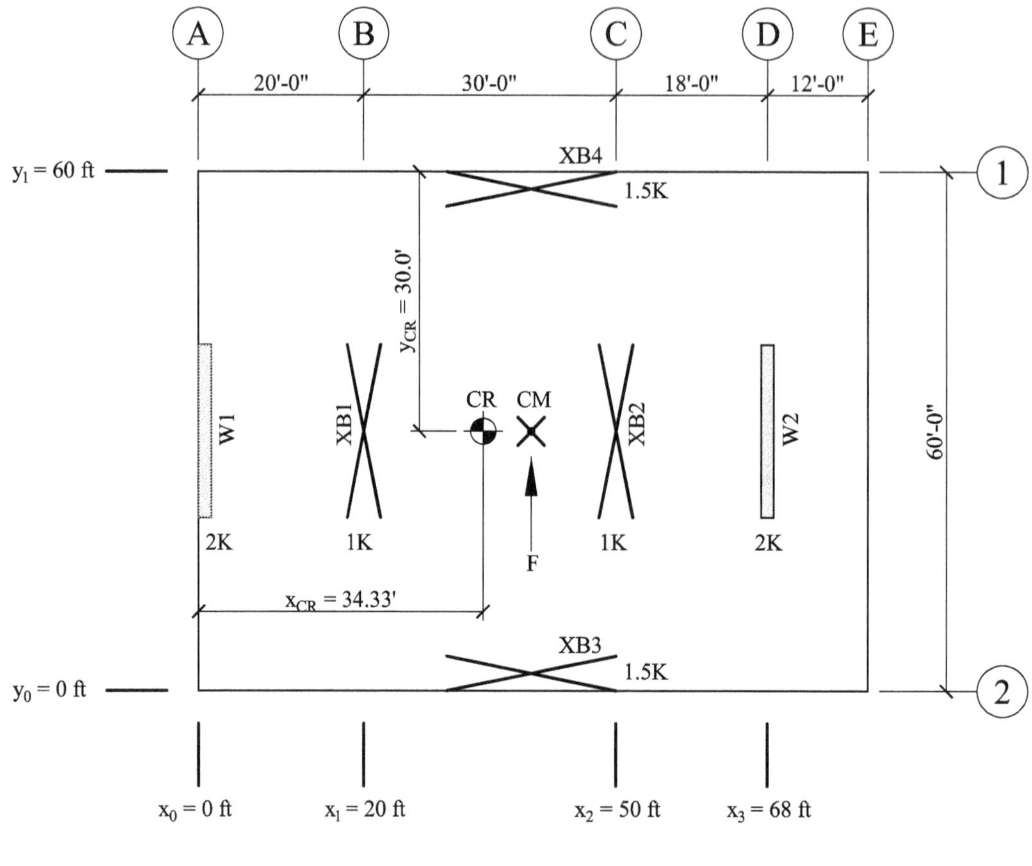

FIGURE 14-11a Diaphragm and LFRS layout for Example 14-1

In lieu of performing a 3-D finite element analysis, use the simplified lateral load distribution method previously presented to determine the following parameters:

a. The lateral force at each level for each vertical LFRS for the N-S lateral load, assuming a flexible diaphragm.

b. The lateral force at each level for each vertical LFRS for the E-W lateral load, assuming a flexible diaphragm.

c. The lateral force at each level for each vertical LFRS for the N-S lateral load, assuming a rigid diaphragm.

d. The lateral force at each level for each vertical LFRS for the E-W lateral load, assuming a rigid diaphragm.

e. The maximum base shear and overturning moment for shear wall *W2* and X-brace *XB3*.

Solution

Flexible **diaphragm:**

For the lateral load distribution for flexible diaphragms, the seismic lateral force on the building is distributed at each diaphragm level to each vertical LFRS in proportion to the width of the floor/roof that is tributary to each vertical LFRS parallel to the direction of the lateral force. In this case, no in-plane rotation of the diaphragm is possible. The vertical LFRS that are

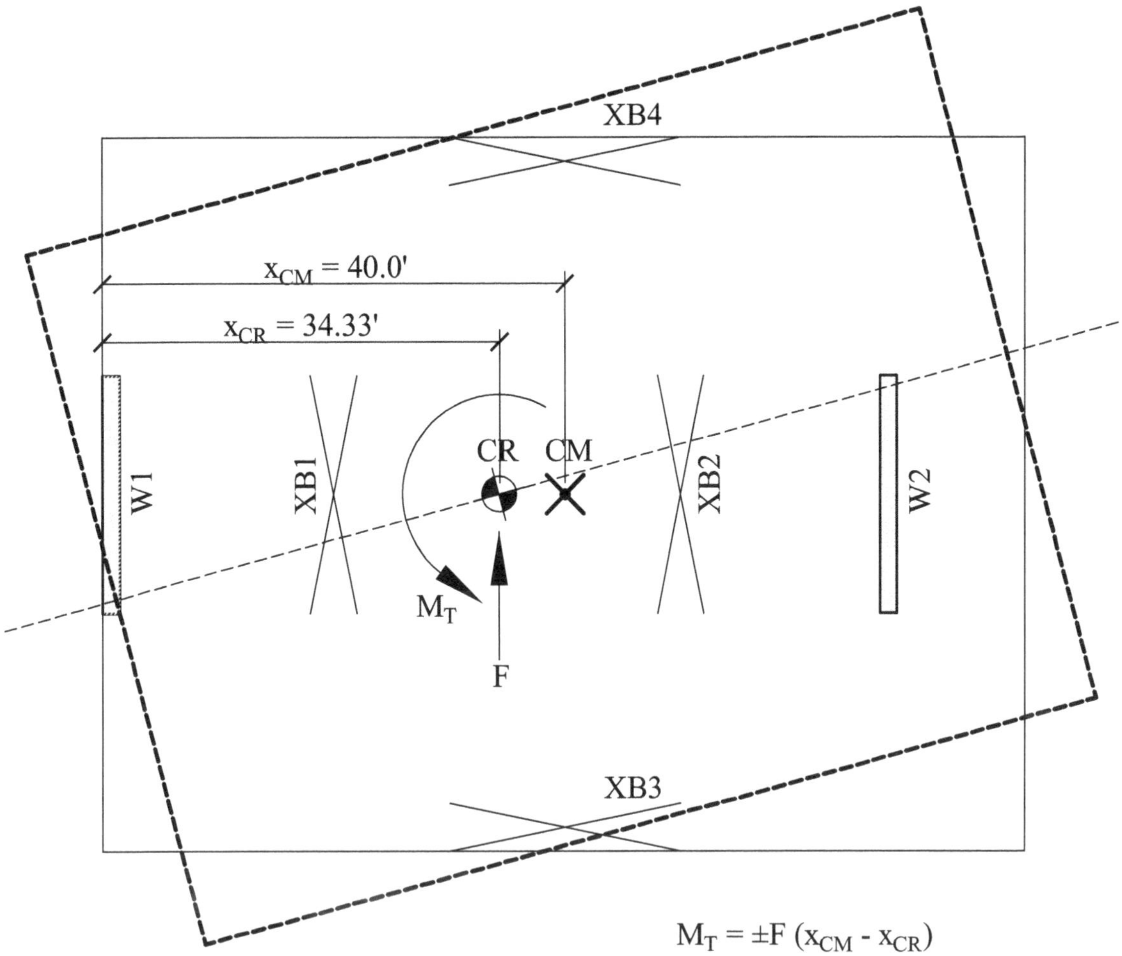

FIGURE 14-11b Rotation of rigid diaphragm about center of rotation

perpendicular to the direction of the lateral force have negligible out-of-plane lateral stiffness and therefore do not receive any direct lateral force.

North-South Lateral Force (flexible diaphragm):

The tributary width for each N-S vertical LFRS is calculated as follows:

N-S LFRS	Tributary width (TW)
Shear wall W1	20 ft./2 = 10 ft
X-Brace XB1	½ (20 ft. + 30 ft) = 25 ft
X-Brace XB2	½ (30 ft. + 18 ft) = 24 ft
Shear wall W2	½ (18 ft.) + 12 ft. = 21 ft
	$\sum TW_i = 80\text{ft.}$ OK.

In the North-South direction, the distribution of the lateral force to each vertical LFRS that are parallel to the lateral force is calculated using the following equation:

$$F_i = \left(\frac{TW_{i,N\text{-}S}}{\sum TW_{N\text{-}S}}\right) F_{N\text{-}S}$$

Using this equation, the North-South lateral force at each floor or diaphragm level for each N-S vertical LFRS (assuming a flexible diaphragm) is computed in the following table:

Diaphragm Level	N-S LFRS lateral Force at each floor level (flexible diaphragm)			
	W1	XB1	XB2	W2
Roof level	(10 ft/80 ft) (10 kips) = **1.25 kips**	(25 ft/80 ft) (10 kips) = **3.13 kips**	(24 ft/80 ft) (10 kips) = **3 kips**	(21 ft/80 ft) (10 kips) = **2.63 kips**
3rd floor level	(10 ft/80 ft) (20 kips) = **2.5 kips**	(25 ft/80 ft) (20 kips) = **6.25 kips**	(24 ft/80 ft) (20 kips) = **6 kips**	(21 ft/80 ft) (20 kips) = **5.25 kips**
2nd floor level	(10 ft/80 ft) (24 kips) = **3 kips**	(25 ft/80 ft) (24 kips) = **7.5 kips**	(24 ft/80 ft) (24 kips) = **7.2 kips**	(21 ft/80 ft) (24 kips) = **6.3 kips**

East-West Lateral Force (flexible diaphragm):

The tributary width for each E-W vertical LFRS is calculated as follows:

E-W LFRS	Tributary width (TW)
X-Brace XB3	½ (60 ft.) = 30 ft.
X-Brace XB4	½ (60 ft.) = 30 ft.
	$\sum TW_i = 60\text{ft. OK.}$

The East-West lateral force at each floor level for the flexible diaphragm is calculated using the following equation:

$$F_i = \left(\frac{TW_{i,E-W}}{\sum TW_{E-W}} \right) F_{E-W}$$

Using this equation, the East-West lateral force at each floor level for each E-W vertical LFRS (assuming a flexible diaphragm) is computed in the following table:

Diaphragm Level	E-W LFRS, Lateral forces at each level (flexible diaphragm)	
	XB3	XB4
Roof level	(30 ft/60 ft)(10 kips) = **5 kips**	(30 ft/60 ft)(10 kips) = **5 kips**
3rd floor level	(30 ft/60 ft)(20 kips) = **10 kips**	(30 ft/60 ft)(20 kips) = **10 kips**
2nd floor level	(30 ft/60 ft)(24 kips) = **12 kips**	(30 ft/60 ft)(24 kips) = **12 kips**

Rigid diaphragm:

The lateral force distribution for a rigid diaphragm consists of the direct lateral force, which is distributed to the vertical LFRS proportional to the relative stiffnesses of the vertical LFRS that are parallel to the direction of the applied lateral force, plus the lateral force due to the in-plane torsional moment in the diaphragm. The in-plane torsional moment at any level is resisted by all the vertical LFRS in the building below that level, whereas the direct lateral force is resisted only by the vertical LFRS that are parallel to the applied lateral force.

Direct Lateral Force:

North-South Direct Lateral Force (rigid diaphragm):

The relative stiffness for the N-S vertical LFRS is tabulated in the following table:

LFRS	Relative stiffness, K
Shear wall W1	2K
X-Brace XB1	K
X-Brace XB2	K
Shear wall W2	2K
	$\sum K_{i,N-S} = 6K$

The N-S lateral force at each floor level for each N-S vertical LFRS is calculated using the following equation and tabulated in the following table.

$$F_i = \left(\frac{K_{i,N-S}}{\sum K_{N-S}} \right) F_{N-S}$$

Diaphragm Level	N-S LFRS, Lateral Force at Each Floor Level (Rigid Diaphragm)			
	W1	**XB1**	**XB2**	**W1**
Roof level	(2K/6K)(10 kips) = 0.333(10) = **3.33 kips**	(1K/6K)(10 kips) = 0.167(10) = **1.67 kips**	(1K/6K)(10 kips) = 0.167(10) = **1.67 kips**	(2K/6K)(10 kips) = 0.333(10) = **3.33 kips**
3rd floor level	(2K/6K)(20 kips) = 0.333(20) = **6.66 kips**	(1K/6K)(20 kips) = 0.167(20) = **3.34 kips**	(1K/6K)(20 kips) = 0.167(20) = **3.34 kips**	(2K/6K)(20 kips) = 0.333(20) = **6.66 kips**
2nd floor level	(2K/6K)(24 kips) = 0.333(24) = **8 kips**	(1K/6K)(24 kips) = 0.167(24) = **4 kips**	(1K/6K)(24 kips) = 0.167(24) = **4 kips**	(2K/6K)(24 kips) = 0.333(24) = **8 kips**

East-West Direct Lateral Force (rigid diaphragm):

The relative stiffness for the E-W vertical LFRS is tabulated in the following table:

LFRS	Relative stiffness, K
X-Brace XB3	1.5K
X-Brace XB4	1.5K
	$\sum K_{i,E-W} = 3K$

The East-West direct lateral force at each floor level in each E-W vertical LFRS is calculated for the rigid diaphragm using the following equation and computed in the following table:

$$F_i = \left(\frac{K_{i,\,E-W}}{\sum K_{E-W}} \right) F_{E-W}$$

Diaphragm Level	E-W LFRS, Lateral force at each floor level (rigid diaphragm)	
	XB3	XB4
Roof level	(1.5K/3K)(10 kips) = 0.5(10 kip) = **5 kips**	(1.5K/3K)(10 kips) = 0.5(10 kip) = **5 kips**
3rd floor level	(1.5K/3K)(20 kips) = 0.5(20 kip) = **10 kips**	(1.5K/3K)(20 kips) = 0.5(20 kip) = **10 kips**
2nd floor level	(1.5K/3K)(24 kips) = 0.5(24) = **12 kips**	(1.5K/3K)(24 kips) = 0.5(24) = **12 kips**

In-plane Torsional Moments (*Rigid* diaphragms only):

The in-plane torsional moments from the North-South lateral force and the resulting lateral force at each floor level in all the vertical LFRS are determined next. Note that only rigid diaphragms can be subject to and able to tolerate in-plane rotation. Flexible diaphragms are incapable of resisting in-plane torsional rotation and would collapse if subject to such torsional forces. Calculate the coordinates of the center of mass (X_{CM} and Y_{CM}) and the coordinates of the center of rigidity (X_{CR} and Y_{CR}) at each floor level. These will be measured from an assumed origin located at the intersection of Grid Line A and Grid Line 2 (see Figures 14-11a and 14-11b).

N-S Lateral Force:

Since the vertical LFRS have no vertical discontinuity (see ASCE 7 Tables 12.3.1 and 12.3.2), the center of rigidity (CR) at **each** floor level will be assumed to occur at the same location and will be situated on the same vertical axis, and the coordinates can be calculated from equations (14-1) and (14-2) as follows:

$$X_{CR,N-S} = \frac{\sum_{i=1}^{n} K_{i,N-S} X_i}{\sum_{i=1}^{n} K_{i,N-S}} = \frac{(2K)(0\,\text{ft.})+(K)(20\,\text{ft.})+(K)(20\,\text{ft.}+30\,\text{ft.})+(2K)(20\,\text{ft.}+30\,\text{ft.}+18\text{ft.}))}{2K+K+K+2K}$$

$$= 34.33\,\text{ft. from the origin (i.e. measured from Grid Line A in Figure 14-11a)}$$

$$Y_{CR,E-W} = \frac{\sum_{i=1}^{N} K_{i,E-W} Y_i}{\sum_{i=1}^{N} K_{i,E-W}} = \frac{(1.5K)(0\,\text{ft.})+(1.5K)(60\,\text{ft.})}{1.5K+1.5K}$$

$$= 30\,\text{ft. from the origin (i.e., measured from Grid Line 2 in Figure 14-11a)}$$

For the North-South direction, the location of the center of mass is calculated from equations (14-3) and (14-4) as follows, assuming a uniform distribution of the weights at each floor level:

$$X_{CM,N-S} = \frac{\sum_{i=2nd\,flr}^{Roof} W_i X_{CM,i}}{\sum_{i=2nd\,flr}^{Roof} W_i} = \frac{80\,\text{ft.}}{2}$$

$$= 40\,\text{ft. measured from Grid Line A in Figure 14-11a}$$

$$Y_{CM,E-W} = \frac{\sum_{i=2nd\,flr}^{Roof} W_i\, Y_{CM,i}}{\sum_{i=2nd\,flr}^{Roof} W_i} = \frac{60\text{ ft.}}{2}$$

$= 30$ ft. measured from Grid Line 2 in Figure 14-11a

The torsional eccentricities, e_i, are calculated at **each** floor level using equations (14-10) and (14-12) as follows:

$$N-S\,direction:\; e_{i,N-S} = \left|X_{CR,N-S} - X_{CM,N-S}\right| + 0.05B_Y$$
$$= |34.33\text{ ft.} - 40\text{ ft.}| \pm (0.05)(80\text{ ft.}) = 9.67\text{ ft.}\;(\text{ or } 1.67\text{ ft.})$$

$$E-W\,direction:\; e_{i,E-W} = \left|Y_{CR,E-W} - Y_{CM,E-W}\right| + 0.05B_X$$
$$= |30\text{ ft.} - 30\text{ ft.}| \pm (0.05)(60\text{ ft.}) = 3\text{ ft. } or\ -3\text{ ft.}$$

For the N-S direction, only the case with eccentricity, e = 9.67 ft., will be checked since that eccentricity will control. For the E-W direction, the building will be checked for both e = 3 ft and e = - 3 ft. Note that because the vertical LFRS do not have any vertical discontinuity and the building is symmetrical, the torsional eccentricity is assumed to be the same at every floor level. The in-plane torsional moments at **each** floor level are calculated using equations (14-11) and (14-13) as follows:

$$M_{Tx,N-S} = F_{i,N-S}\, e_{i,N-S} = F_{i,N-S}\,(9.67\text{ ft.}) = 9.67\,F_{i,N-S}$$

$$M_{Tx,E-W} = F_{i,E-W}\, e_{i,E-W} = F_{i,E-W}\,(\pm 3\text{ ft.}) = \pm 3F_{i,E-W}$$

The lateral forces in each vertical LFRS at **each** floor level due to the North-South direct lateral force and the lateral force caused by the in-plane torsional moments from the North-South lateral force are computed in the following table (see equation (14-8)). Note that constant eccentricity is assumed throughout the height of the building.

LFRS	**Direct N-S lateral force on each LFRS at each floor level,** $F_i = \left(\frac{K_{i,N-S}}{\sum K_{N-S}}\right) F_{N-S}$	K_i	$\hat{Z_i}$ **Measured from the CR (ft.)**	**Lateral force on each LFRS at each floor level due to in-plane torsional moment from the N-S lateral force,** $F_{Ti} = \frac{K_i Z_i M_{Tx}}{\sum K_i (Z_i)^2}$ $(M_{Tx} = M_{T,N-S} = 9.67\,F_{N-S})$	**Maximum total lateral force on each LFRS at each floor level due to direct lateral force plus torsion (See equation (14-9))**
W1	$0.333F_{N\text{-}S}$	2K	−34.33	$-0.085\,F_{N\text{-}S}$	$\mathbf{0.333\,F_{N\text{-}S}}$
W2	$0.333F_{N\text{-}S}$	2K	33.67	$0.084\,F_{N\text{-}S}$	$\mathbf{0.417\,F_{N\text{-}S}}$
XB1	$0.167F_{N\text{-}S}$	K	−14.33	$-0.018\,F_{N\text{-}S}$	$\mathbf{0.167\,F_{N\text{-}S}}$
XB2	$0.167F_{N\text{-}S}$	K	15.67	$0.019\,F_{N\text{-}S}$	$\mathbf{0.186\,F_{N\text{-}S}}$
XB3	0	0* and 1.5K	−30	$-0.056\,F_{N\text{-}S}$	$-0.056\,F_{N\text{-}S}$

LFRS	Direct N-S lateral force on each LFRS at each floor level, $F_i = \left(\frac{K_{i,N-S}}{\sum K_{N-S}}\right) F_{N-S}$	K_i	Z_i^ Measured from the CR (ft.)	Lateral force on each LFRS at each floor level due to in-plane torsional moment from the N-S lateral force, $F_{Ti} = \frac{K_i Z_i M_{Tx}}{\sum K_i (Z_i)^2}$ $(M_{Tx} = M_{T,N-S} = 9.67 F_{N-S})$	Maximum total lateral force on each LFRS at each floor level due to direct lateral force plus torsion (See equation (14-9))
XB4	0	0* and 1.5K	30	0.056 $F_{N\text{-}S}$	0.056 $F_{N\text{-}S}$
			$\sum K_i (Z_i)^2 =$ 7775K		

*Out-of-plane lateral stiffness of XB3 and XB4 is zero for calculating N-S vertical LFRS direct lateral forces due to F_i, but for the in-plane torsional analysis, the in-plane stiffness for XB3 and XB4 of 1.5K is used in the calculation.
^ Note that $\mathbf{Z_i}$ = Perpendicular distance from the center of rigidity (CR) to the longitudinal axes of vertical LFRS i, thus, $Z_i = X_i$ for the North-South LFRS, and $Z_i = Y_i$ for the East-West vertical LFRS.

The lateral forces in the vertical LFRS due to **East-West** Seismic Force, F_i, and the in-plane torsional moment $M_{T,E\text{-}W}$ are computed in the following table:

$M_{T,E-W} = M_{Tx} = \pm 3 F_{E-W}$.

LFRS	Direct E-W lateral force on each LFRS at each floor level, $F_i = \left(\frac{K'_{i,E-W}}{\sum K_{E-W}}\right) F_{E-W}$	K_i	Z_i^ Measured from the CR (ft.)	Lateral force on each LFRS at each floor level due to in-plane torsional moment from the E-W lateral force, $F_{Ti} = \frac{K_i Z_i M_{Tx}}{\sum K_i (Z_i)^2}$ $(M_{Tx} = M_{T,E-W} = \pm 3 F_{E-W})$	Maximum total lateral force at each level due to direct lateral force plus torsion (See equation (14-9))
W1	0	2K*	−34.33	±0.026 $F_{E\text{-}W}$	± 0.026 $F_{E\text{-}W}$
W2	0	2K*	33.67	±0.026 $F_{E\text{-}W}$	±0.026 $F_{E\text{-}W}$
XB1	0	K*	−14.33	±0.0055 $F_{E\text{-}W}$	± 0.0055 $F_{E\text{-}W}$
XB2	0	K*	15.67	±0.006 $F_{E\text{-}W}$	±0.006 $F_{E\text{-}W}$
XB3	0.5$F_{E\text{-}W}$	1.5K	−30	±0.017 $F_{E\text{-}W}$	**0.517 $F_{E\text{-}W}$**
XB4	0.5$F_{E\text{-}W}$	1.5K	30	±0.017 $F_{E\text{-}W}$	**0.517 $F_{E\text{-}W}$**
			$\sum K_i (Z_i)^2 =$ 7775K		

*Out-of-plane lateral stiffness of W1, W2, XB1, and XB2 is zero for calculating the direct lateral force in these vertical LFRS due to the E-W F_i, but for torsional analysis, the in-plane stiffness for W1, W2, XB1, and XB2 of 2K, 2K, 1K, and 1K, respectively, is used in the calculation.
^ Note that $\mathbf{Z_i}$ = Perpendicular distance from the center of rigidity (CR) to the longitudinal axes of vertical LFRS i, thus, $Z_i = X_i$ for the North-South LFRS, and $Z_i = Y_i$ for the East-West LFRS.

The maximum total lateral forces for each of the vertical LFRS at each level due to the direct lateral force plus the lateral force from the in-plane torsional moment are shown in bold font in the preceding tables. Given the seismic lateral forces for this example, the lateral forces at each diaphragm level of all the vertical LFRS in the building are tabulated in the following table:

			Maximum lateral force on each vertical LFRS at each floor level due to the direct lateral force plus in-plane torsional force (rigid diaphragm)					
Level	**Height, ft.**	**Total lateral force on the building**	**W1**	**W2**	**XB1**	**XB2**	**XB3**	**XB4**
			^0.333 $F_{N\text{-}S}$	**^0.417 $F_{N\text{-}S}$**	**^0.167 $F_{N\text{-}S}$**	**^0.186 $F_{N\text{-}S}$**	**^0.517 $F_{E\text{-}W}$**	**^0.517 $F_{E\text{-}W}$**
Roof	30	10 kips	3.33 kips	4.17 kips	1.67 kips	1.86 kips	5.17 kips	5.17 kips
3rd floor	20	20 kips	6.66 kips	8.34 kips	3.34 kips	3.72 kips	10.34 kips	10.34 kips
2nd floor	10	24 kips	8 kips	10 kips	4 kips	4.46 kips	12.41 kips	12.41 kips

^ Note that for this problem, these values are constant at each floor level because the building is symmetrical from floor to floor, and there is no vertical discontinuity of the vertical LFRS over the height of the building (see ASCE 7 Tables 12.3.1 and 12.3.2).
^ For this type of building, the floor eccentricity at each level is assumed in this text to be located on the same vertical axis.
^ Also, note that the in-plane torsional moment at any level from both the N-S and E-W lateral forces are resisted by all of the vertical LFRS in the building immediately below that level; therefore, for each vertical LFRS, the larger of the two in-plane torsional moment-induced lateral forces (see the numbers in bold font in the preceding tables) is used, which is then added to the larger of the direct lateral forces to obtain the total maximum lateral force in each vertical LFRS at each floor level.
^ Note that from equation (14-9), the minimum lateral force in each vertical LFRS is the direct lateral force obtained with in-plane torsion neglected.

Using the maximum forces from the preceding table, the maximum base shear and overturning moment for wall W2 and X-brace XB3 are calculated as follows:

Wall W2:

Base shear = 4.17 kips + 8.34 kips + 10 kips = 22.5 kips

Overturning moment at base of wall W2 = (4.17 kips)(30 ft.) + (8.34 kips)(20 ft.) + (10 kips)(10 ft.) = 392 ft.-kips

XB3:

Base shear = 5.17 kips + 10.34 kips + 12.41 kips = 28 kips

Overturning moment at the base of X-Brace frame, XB3 = (5.17 kips)(30 ft.) + (10.34 kips)(20 ft.) + (12.41 kips)(10 ft.) = 486 ft.-kips

14-4 DESIGN OF ROOF AND FLOOR DIAPHRAGMS IN STEEL-FRAMED BUILDINGS

As previously noted, in addition to supporting gravity loads, the roof and floors of steel buildings also act as horizontal diaphragms in resisting lateral loads such as wind or earthquake loads and help to transfer the lateral load to the lateral force resisting system. The seismic force used for the design of diaphragms, chords, and drag struts and their connections to the vertical lateral force resisting systems (e.g., braced frames, shear walls, or moment frames) is determined using the larger of the F_x seismic lateral force (see equation (3-20)) and the F_{px} diaphragm seismic lateral (see equation (3-22)) [25].

The horizontal diaphragm (e.g., see Figures 14-4, 14-9, and 14-12) consists of the following elements [6]:

- Metal deck at roof level supported on steel beams and girders
- Concrete slab on composite metal deck at the floor levels supported on steel beams and girders (with or without shear studs)
- The diaphragm chord members which are perpendicular to the lateral force direction
- The drag struts or collectors which are parallel to, and lie along the same vertical plane as, the vertical lateral force resisting system (LFRS)

The drag struts or collectors transfer or "collect" (or drag) the lateral force from the roof or floor diaphragm into the lateral force resisting system (LFRS) while ensuring that the applied in-plane shear in the horizontal diaphragm is uniform and does not exceed the in-plane shear strength of the diaphragm. The drag struts, which resist axial tension and compression forces due to the reversal of wind and seismic lateral forces, are provided along the same lines as the lateral force-resisting systems (i.e., at the diaphragm levels in the same vertical plane as the vertical LFRS) for the entire length of the roof and floor diaphragms in both orthogonal directions [7]. The drag strut beam members and their connections to the columns must be designed for these axial tension and compression forces (plus any resulting moments from the eccentricity of the axial forces), in combination with the vertical shear and moments from the gravity loads. The beams and girders along the drag lines should be designed as noncomposite members for gravity loads, even for floors with shear studs, because of the possibility of stress reversals that may occur due to upward bending caused by the axial compression forces in the drag beams, thus resulting in tension stress (and hence tension cracks) in the concrete and therefore rendering the studs ineffective. In designing the drag strut beams for the axial compression forces, the unbraced length should be taken as the smaller of the length of the beam between column supports and the distance between adjacent infill framing members that are supported by the drag strut beam. Thus, the drag struts are designed as beam-columns under combined loads using the methods previously discussed in Chapters 4, 5, 6, and 8. The chord and drag strut forces must be shown on the structural plans and on the elevation views of the lateral force-resisting systems. Typical drag strut-to-steel column connections are illustrated in Figure 14-15 and will be discussed later in this Chapter, and Section 1 in Figure 14-10 shows typical drag strut connections to concrete shear walls.

The use of drag struts reduces the applied in-plane shear in the diaphragm compared to the case where the lateral force is assumed to be transferred from the deck directly to the vertical LFRS *only* within the length of the vertical LFRS. This can be illustrated using Figure 14-14a in Example 14-3: If there were no drag struts between points B and F in Figure 14-14a, the applied in-plane shear in the roof diaphragm would be (100 kips/2)/20 ft. = 2.5 kips/ft. since the applied lateral shear would be transferred only between points A and B (i.e., distributed over the length of the lateral force resisting system only, which in this example is 20 ft.), instead of the uniform in-plane shear of 0.5 kip/ft. (i.e., (100 kips/2)/100 ft.) in the diaphragm obtained when drag struts are provided between points B and F, thus utilizing the full 100-ft. length of the diaphragm (parallel to the lateral force) in resisting the total in-plane shear of 100 kips/2 or 50 kips. It is recommended that no roof or floor openings should be located adjacent to the drag struts or near the vertical plane that contains the drag struts, as this increases the shear in the diaphragm and the forces in the drag struts.

The common standard width for metal, roof, or floor decks is usually 36 in. center to center of the end flutes; they are usually available in lengths up to 40 ft. The common practice is for the deck to be fastened to the steel framing members using 5/8-in.-diameter puddle welds in the flutes. Sometimes, mechanical fasteners such as screws or power-actuated (i.e., power-driven) fasteners are also used. These fasteners may be more efficient in areas where the metal deck roofs may be subjected to high uplift loads, such as in tornado-prone regions [33]. The sidelap connections, or the attachment of adjacent deck panels to each other, and the attachment of deck panels to the diaphragm chords, are made using self-drilling screws, button punching, or welds with a spacing not exceeding 2 ft.-6 in. [9]. The number of puddle welds for each standard 36 in. width of deck can vary from a minimum of three (i.e., 36/3 pattern) to a maximum of nine (i.e., 36/9 pattern) (see Figure 14-12). Figure 14-12 is an example of a deck with a 36/5 fastener pattern. The diaphragm shear strength of a roof or floor deck is a function of the number of puddle welds used within a standard deck width and the type and spacing of sidelap connections. An excerpt from a typical diaphragm lateral in-plane shear strength table from the Steel Deck Institute (SDI) *Diaphragm Design Manual* (SDI *DDM03*) [9] is shown in Table 14-2.

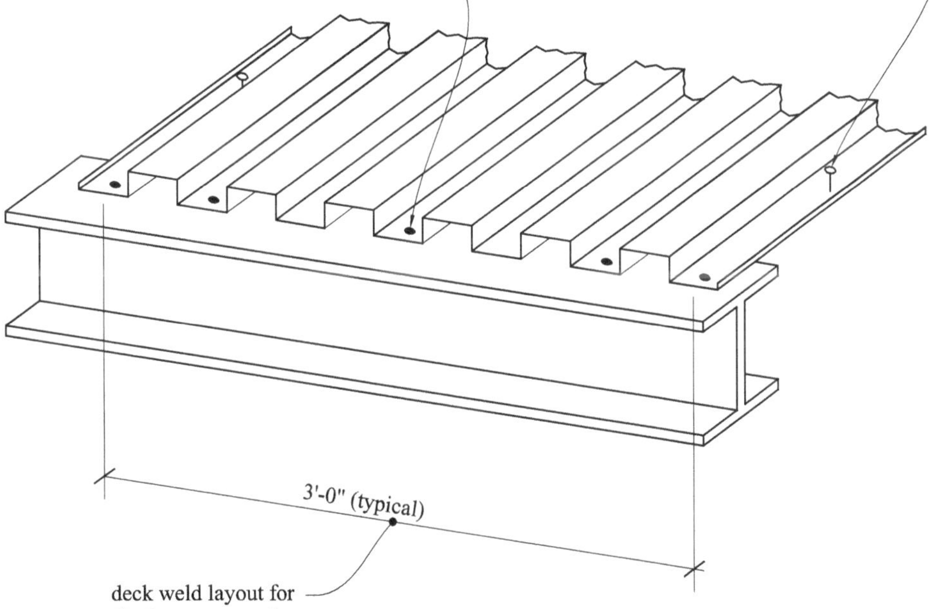

FIGURE 14-12 Deck diaphragm weld layout detail

When the required lateral shear strength of the deck cannot be achieved using puddle welds and sidelap connections, different and stronger connection materials may be used, or a horizontal in-plane bracing system consisting of angle members in the plane of the roof or floor may be provided at the underside of the deck; or, alternatively, more lateral force resisting systems could be added to reduce the span of the diaphragm and hence reduce the applied diaphragm shear. If a diaphragm system is not capable of transferring lateral loads, such as a roof deck or concrete slab on composite floor deck, or diaphragms with glass skylights or open floor areas in industrial structures, a horizontal in-plane bracing system (i.e., a horizontal truss diaphragm) in the plane

TABLE 14-2 Nominal Shear Strength in lb/ft. for Selected Diaphragm Assemblies (Roof Deck)

Strength factors: ϕ (EQ) = 0.55; Ω (EQ) = 3.00
ϕ (Wind) = 0.70; Ω (Wind) = 2.35
ϕ (Other) = 0.60; Ω (Other) = 2.65

1.5 in. (WR, IR, NR) t = 0.0358 in. (20 gage)		Support fastening: 5/8-in. puddle welds (or equivalent) Sidelap fastening: #10 screws									
Fastener layout	Sidelap connection per span	Nominal shear strength, plf — Span, ft.									
		4.0	4.5	5.0	5.5	6.0	6.5	7.0	7.5	8.0	K1
36/7	0	1035	915	820	740	675	620	575	530	495	0.535
	1	1220	1090	975	880	805	—	—	—	—	0.415
	2	1385	1245	1130	1020	930	855	795	735	690	0.340
	3	1540	1390	1265	1160	1060	975	900	840	785	0.287
	4	1695	1530	1395	1280	1185	1095	1010	940	880	0.249
	5	1835	1665	1525	1400	1295	1205	1120	1045	975*	0.219
	6	1970	1795	1645	1515	1405	1305	1220*	1145*	1070*	0.196
36/5	0	945	845	760	685	625	575	530	490	455	0.642
	1	1100	995	905	825	750	—	—	—	—	0.477
	2	1245	1130	1030	950	880	810	750	695	650	0.380
	3	1375	1255	1150	1065	985	920	860	800	745	0.315
	4	1495	1370	1265	1170	1090	1015	950	895	840	0.270
	5	1605	1480	1370	1270	1185	1110	1040	980	925	0.236
	6	1705	1580	1465	1365	1280	1200	1130	1065*	1005*	0.209
36/4	0	725	640	575	515	470	430	395	365	340	0.803
	1	875	795	725	655	600	—	—	—	—	0.561
	2	1015	925	845	780	725	665	615	570	535	0.431
	3	1135	1040	960	890	825	770	725	675	630	0.350
	4	1240	1145	1060	990	925	865	810	765	725	0.294
	5	1335	1240	1155	1080	1010	950	895	845	800	0.254
	6	1415	1325	1240	1165	1095	1030	975	920	875	0.224

Notes: Courtesy of Steel Deck Institute, Glenshaw, PA.
1. Diaphragm shear and stiffness values are based on the Steel Deck Institute (SDI) Diaphragm Design Manual, Third Edition (DDM03) [9].
2. Diaphragm shear and stiffness values are based on a minimum 3-span condition.
3. An asterisk (*) denotes condition may be limited by shear buckling of the deck panel. See the complete diaphragm design tables in the SDI Diaphragm Design Manual, Third Edition (DDM03) [9].
4. Shear strength values shown shaded do not comply with minimum SDI sidelap connection requirements and shall not be used except with properly spaced sidelap connections.

of the roof or the floor diaphragm must be provided to transfer the lateral loads to the vertical lateral force resisting system as illustrated in Figures 14-7c and 14-13.

The horizontal in-plane bracing should be oriented such that horizontal brace angle members have their legs turned downward, and tee shape or wide-flange members should have their webs oriented vertically. In these orientations, the horizontal brace members will be less prone to collecting and retaining dust, debris, or rainwater — thus reducing the likelihood of corrosion [26].

Deck fasteners should also be checked for combined in-plane shear and uplift loading; the in-plane shear is due to the deck acting as a diaphragm under lateral wind loads, and the uplift forces are due to wind loads acting perpendicular to the surface of the roof of the building. This combined in-plane shear plus tension would require the use of interaction equations that can be found in the SDI Roof Deck Design Manual.

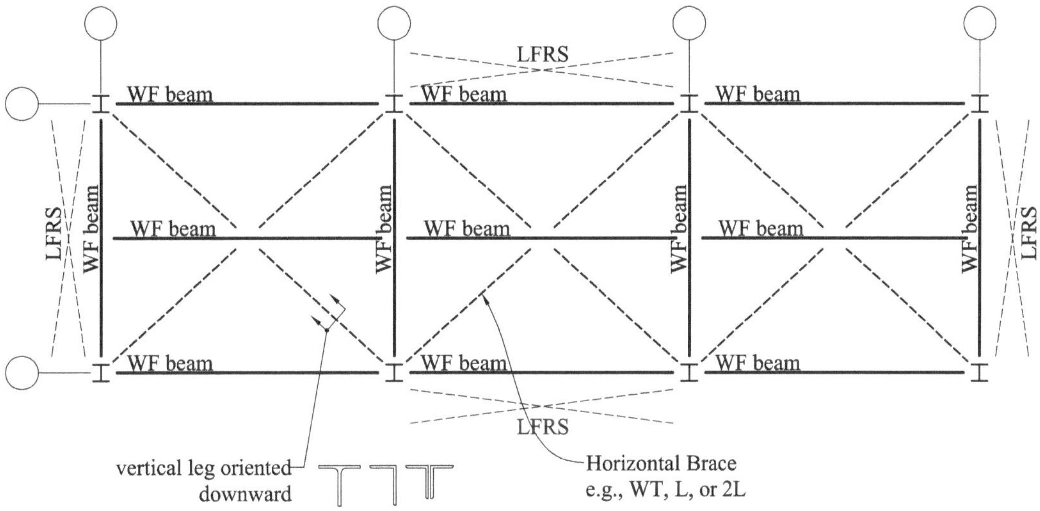

FIGURE 14-13 Plan view – Horizontal brace systems

EXAMPLE 14-2

Design of Horizontal Diaphragms

The roof framing for a five-story building with a 15-ft. floor-to-floor height is similar to that shown in Figure 14-6. The roof deck is $1^1/_2$-in. × 20-gage metal deck, and the floor deck consists of $2^1/_2$-in. concrete on 3-in. × 20-gage galvanized composite metal deck. The building is subjected to a uniform design wind load of 40 psf in the East–West direction. Determine the required size and spacing of the puddle weld and sidelap connection, the chord forces along grid lines 1 and 5, and the drag strut forces along lines A and D at the roof level. Assume a standard 36-in. deck width.

Solution

The horizontal diaphragm is modeled as a simply supported horizontal beam, where the supports are the vertical LFRS at both ends of the diaphragm.

For the East–West wind, the diaphragm spans between the X-brace frames along grid lines A and D in Figure 14-6. Therefore,

Diaphragm span, $l = 90$ ft. (for East–West wind), and

Diaphragm width (dimension parallel to the lateral force), $D = 120$ ft.

The SDI deck diaphragm design load tables in Table 14-2 [9] will be used to determine the diaphragm shear strength for the roof deck that will be compared to the applied factored unit shear.

Factored uniform horizontal load on the roof diaphragm $= (1.0)(40 \text{ psf})\left(\frac{15 \text{ ft.}}{2}\right)$

$= 300$ lb./ft.

Maximum factored shear in the roof diaphragm, $V_u = 300 \text{ lb./ft.}\left(\frac{90 \text{ ft.}}{2}\right)$

$= 13{,}500$ lb.

Maximum factored moment in the roof diaphragm, $M_{du} = 300 \text{ lb./ft.}\frac{(90 \text{ ft.})^2}{8}$

$= 303{,}750$ ft.-lb.

Applied factored unit shear in the roof diaphragm,

$$v_{d_u} = \left(\frac{\text{Lateral shear, } V}{\begin{matrix}\text{Length of diaphragm} \\ \text{parallel to LFRS}\end{matrix} - \begin{matrix}\text{Cumulative length of diaphragm openings} \\ \text{adjacent to drag struts}\end{matrix}}\right)$$

$$= \frac{13{,}500 \text{ lb.}}{(120 \text{ ft.} - 0 \text{ ft.})}$$

$$= 112.5 \text{ lb./ft.}$$

Determine the diaphragm unit shear strength from the deck diaphragm load tables (see Table 14-2):

Thickness of deck (20 gage) = 0.0358 in.

Span of deck (between adjacent steel beams or joists) = 7.5 ft.

Assume four 5/8-in.-diameter puddle welds on a 36-in. deck section (i.e., **36/4 fastener pattern**; this is the minimum weld configuration presented in the SDI *Diaphragm Design Manual* for the 36-in. standard deck width). Assume #10 TEK screws sidelap connections spaced at no more than 30 in. on center (o.c.) (7.5 ft./2.5 ft. = 3 spaces), which implies **two sidelap connections** within the span (i.e., 3 spaces minus 1 = 2).

From Table 14-2 [9]), the nominal diaphragm shear strength is $v_n = 570$ lb./ft.

The design diaphragm shear strength for a 20-gage deck with a span of 7.5 ft. is $\phi v_n = (0.70)(570 \text{ lb./ft.}) = 399 \text{ lb./ft.} > 112.5 \text{ lb./ft.}$,
(where $\phi = 0.70$ for wind, from Table 14-2)

Chord Forces:

Maximum factored chord force, $T_u = C_u = \frac{M_{du}}{D}$

$$= \frac{303{,}750 \text{ ft.-lb.}}{120 \text{ ft.}} = 2531 \text{ lb.} = 2.53 \text{ kips.}$$

The chord member for this roof framing would typically be a continuous deck closure angle. Assuming a continuous $L3 \times 3 \times {}^1/_4$ (ASTM A36 steel), the tensile capacity of the angle is $\phi P_n = \phi A_g F_y = (0.9)(1.44 \text{ in.}^2)(36 \text{ ksi}) = 46.7 \text{ kips} > 2.53 \text{ kips.}$ OK.

EXAMPLE 14-3

Drag Strut Forces

The building in Figures 14.14a through 14.14f is subjected to a factored 100-kips lateral force at the roof level in the North-South direction. For each X-brace configuration and layout, calculate the drag strut forces in the roof diaphragm and determine which X-brace configuration results in the lowest drag strut forces.

Solution

The lateral shear force to be transferred from the diaphragm to the drag struts along the vertical brace lines is

$$V = \frac{100 \text{ kips}}{2} = 50 \text{ kips}.$$

Assuming the drag struts extend through the full length of the building, and since there are no openings in the roof diaphragm adjacent to the drag struts, the uniform unit factored shear in the diaphragm.

$$= \frac{\text{Lateral shear, } V}{\begin{matrix}\text{Length of diaphragm} \\ \text{parallel to LFRS}\end{matrix} - \begin{matrix}\text{Cumulative length of diaphragm openings} \\ \text{adjacent to drag struts}\end{matrix}}$$

$$= \frac{50 \text{ kips}(1000)}{100 \text{ ft.} - 0 \text{ ft.}} = 500 \text{ lb./ft.} = 0.5 \text{ kips/ft.}$$

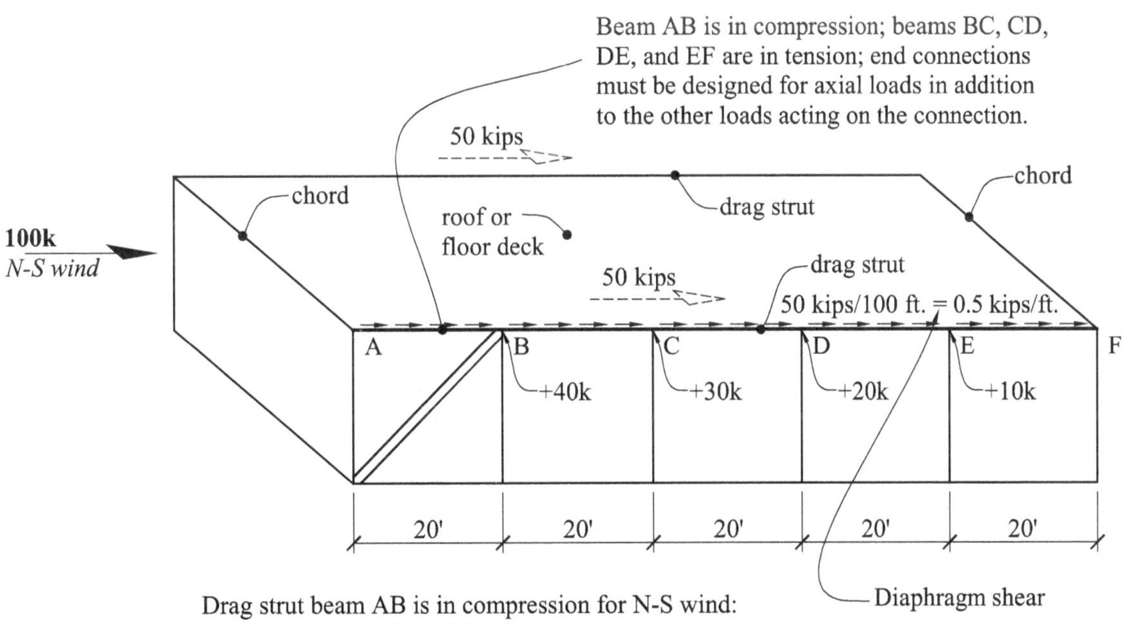

Drag strut beam AB is in compression for N-S wind:

$P_{uAB} = (0.5 \text{ kips/ft.})(20 \text{ ft.}) = 10 \text{ kips}$

Drag strut beams BC, CD, DE, and EF are in tension:

$T_{uBC} = (0.5 \text{ kips/ft.})(20 \text{ ft.} + 20 \text{ ft.} + 20 \text{ ft.} + 20 \text{ ft.}) = 40 \text{ kips}$
$T_{uCD} = (0.5 \text{ kips/ft.})(20 \text{ ft.} + 20 \text{ ft.} + 20 \text{ ft.}) = 30 \text{ kips}$
$T_{uDE} = (0.5 \text{ kips/ft.})(20 \text{ ft.} + 20 \text{ ft.}) = 20 \text{ kips}$
$T_{uEF} = (0.5 \text{ kips/ft.})(20 \text{ ft.}) = 10 \text{ kips}$

FIGURE 14-14a Drag strut detail for Example 14-3

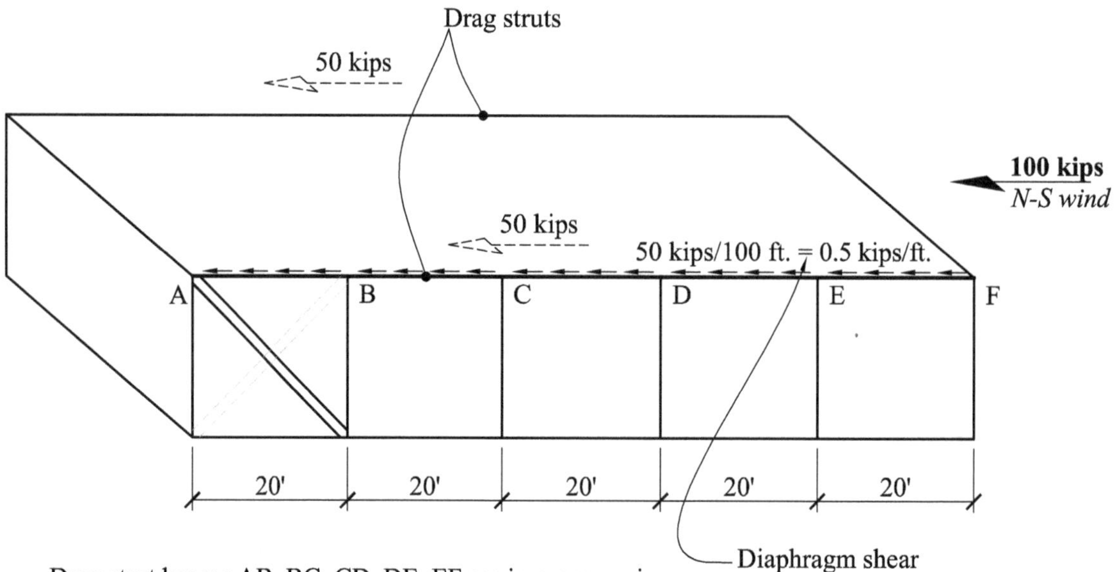

Drag strut beams AB, BC, CD, DE, EF are in compression:

P_{uAB} = (0.5 kips/ft.)(20 ft. + 20 ft. + 20 ft. + 20 ft. + 20 ft.) = 50 kips
P_{uBC} = (0.5 kips/ft.)(20 ft. + 20 ft. + 20 ft. + 20 ft.) = 40 kips
P_{uCD} = (0.5 kips/ft.)(20 ft. + 20 ft. + 20 ft.) = 30 kips
P_{uDE} = (0.5 kips/ft.)(20 ft. + 20 ft.) = 20 kips
P_{uEF} = (0.5 kips/ft.)(20 ft.) = 10 kips

FIGURE 14-14b Drag strut detail for Example 14-3

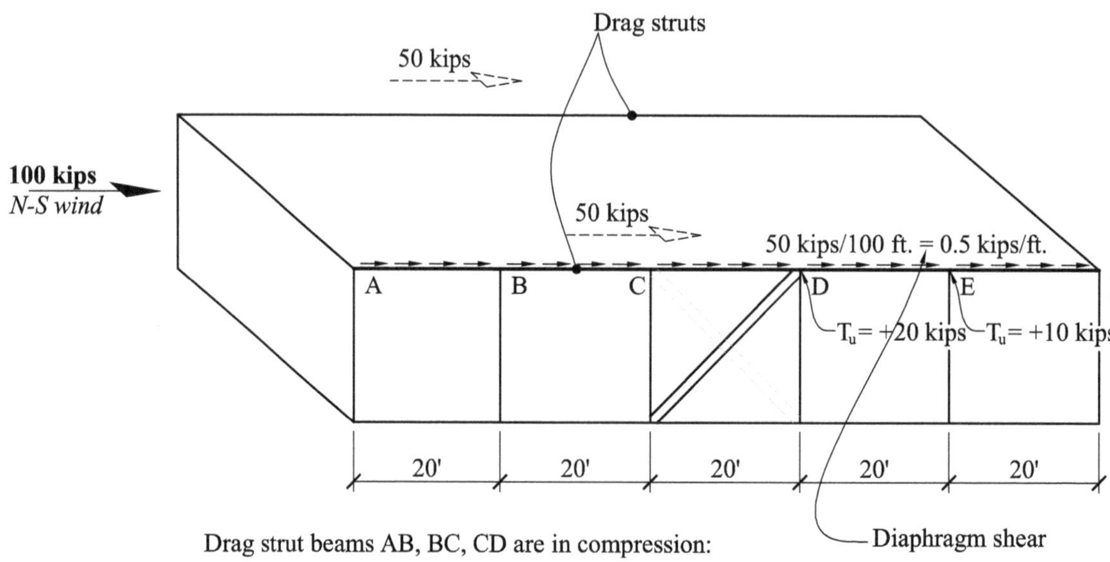

Drag strut beams AB, BC, CD are in compression:

P_{uAB} = (0.5 kips/ft.)(20 ft.) = 10 kips
P_{uBC} = (0.5 kips/ft.)(20 ft. + 20 ft.) = 20 kips
P_{uCD} = (0.5 kips/ft.)(20 ft. + 20 ft. + 20 ft.) = 30 kips

Drag strut beams DE, EF are in tension:

T_{uDE} = (0.5 kips/ft.)(20 ft. + 20 ft.) = 20 kips
T_{uEF} = (0.5 kips/ft.)(20 ft.) = 10 kips

FIGURE 14-14c Drag strut detail for Example 14-3

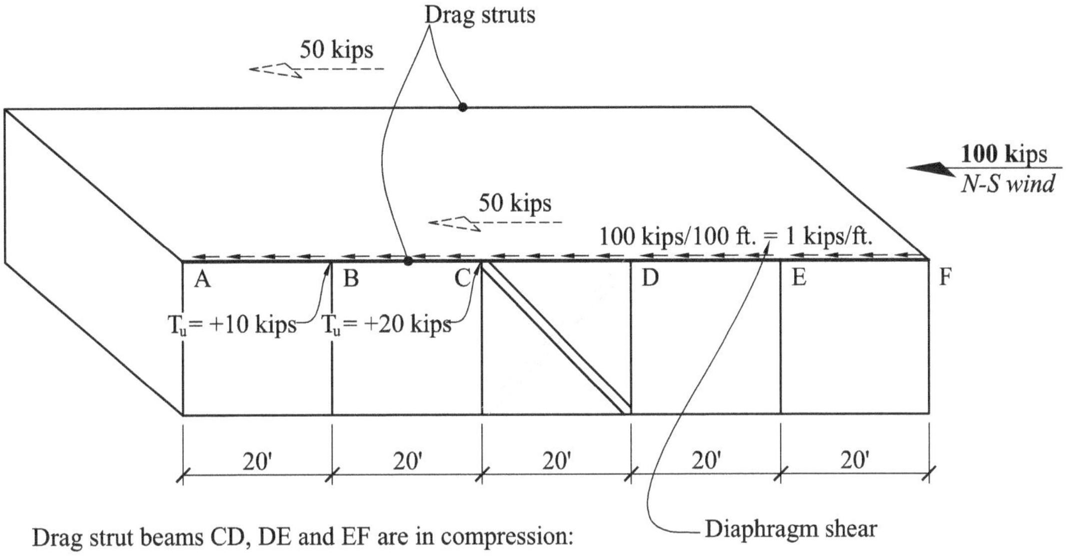

Drag strut beams CD, DE and EF are in compression:

$$P_{uCD} = (0.5\ \text{kips/ft.})(20\ \text{ft.} + 20\ \text{ft.} + 20\ \text{ft.}) = 30\ \text{kips}$$
$$P_{uDE} = (0.5\ \text{kips/ft.})(20\ \text{ft.} + 20\ \text{ft.}) = 20\ \text{kips}$$
$$P_{uEF} = (0.5\ \text{kips/ft.})(20\ \text{ft.}) = 10\ \text{kips}$$

Drag strut beams AB and BC are in tension:

$$T_{uAB} = (0.5\ \text{kips/ft.})(20\ \text{ft.}) = 10\ \text{kips}$$
$$T_{uBC} = (0.5\ \text{kips/ft.})(20\ \text{ft.} + 20\ \text{ft.}) = 20\ \text{kips}$$

FIGURE 14-14d Drag strut detail for Example 14-3

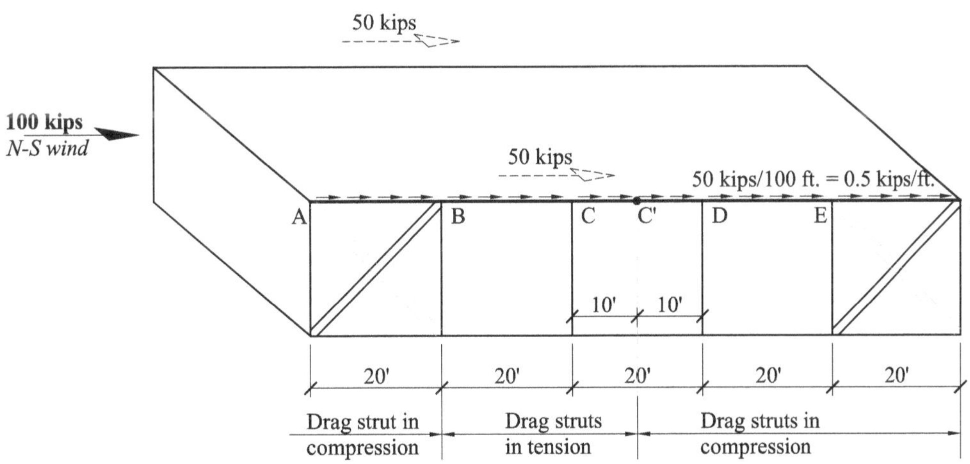

Drag strut beam AB is in compression:

$$P_{uAB} = (0.5\ \text{kips/ft.})(20\ \text{ft.}) = 10\ \text{kips}$$

Drag strut beams DE, EF, and ½C'D are in compression:

$$P_{uC'D} = (0.5\ \text{kips/ft.})(0.5)(20\ \text{ft.}) = 5\ \text{kips}$$
$$P_{uDE} = (0.5\ \text{kips/ft.})[(0.5)(20\ \text{ft.}) + 20\ \text{ft.}] = 15\ \text{kips}$$
$$P_{uEF} = (0.5\ \text{kips/ft.})[(0.5)(20\ \text{ft.}) + 20\ \text{ft.} + 20\ \text{ft.}] = 25\ \text{kips}$$

Drag strut beams BC and ½CC' are in tension:

$$T_{uC'C} = (0.5\ \text{kips/ft.})(0.5)(20\ \text{ft.}) = 5\ \text{kips}$$
$$T_{uBC} = (0.5\ \text{kips/ft.})[(0.5)(20\ \text{ft.}) + 20\ \text{ft.}] = 15\ \text{kips}$$

FIGURE 14-14e Drag strut detail for Example 14-3

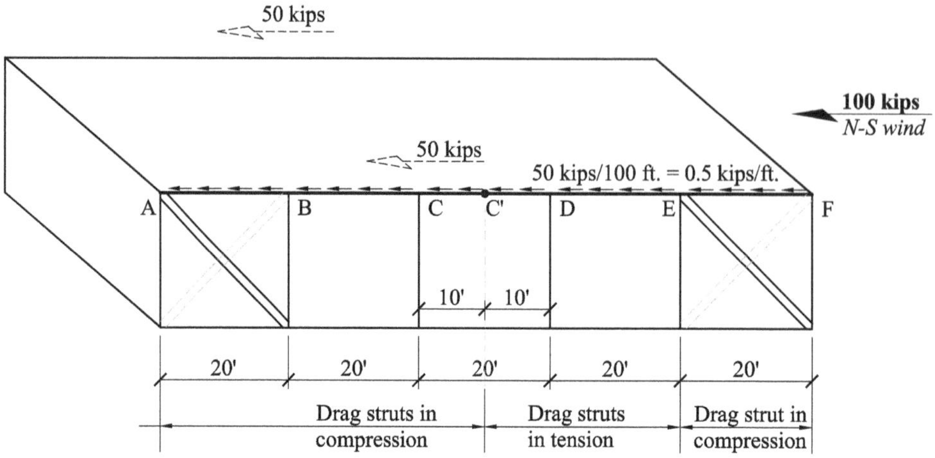

Drag strut beam EF is in compression:

$P_{uEF} = (0.5 \text{ kips/ft.})(20 \text{ ft.}) = 10 \text{ kips}$

Drag strut beams AB, BC, and ½C'C are in compression:

$P_{uC'C} = (0.5 \text{ kips/ft.})(0.5)(20 \text{ ft.}) = 5 \text{ kips}$
$P_{uBC} = (0.5 \text{ kips/ft.})[(0.5)(20 \text{ ft.}) + 20 \text{ ft.}] = 15 \text{ kips}$
$P_{uAB} = (0.5 \text{ kips/ft.})[(0.5)(20 \text{ ft.}) + 20 \text{ ft.} + 20 \text{ ft.}] = 25 \text{ kips}$

Drag strut beams DE and ½C'D are in tension:

$T_{uC'D} = (0.5 \text{ kips/ft.})(0.5)(20 \text{ ft.}) = 5 \text{ kips}$
$T_{uDE} = (0.5 \text{ kips/ft.})[(0.5)(20 \text{ ft.}) + 20 \text{ ft.}] = 15 \text{ kips}$

FIGURE 14-14f Drag strut detail for Example 14-3

The reader should refer to Figures 14.14a to 14.14f for the drag strut forces. For the single X-brace frame, the configuration with a symmetrically located X-braced frame (i.e., X-braced frame located within bay CD) yields the smallest drag strut force. If the lateral load on the building is caused by wind, the drag strut force calculated in this example is substituted, as W, directly into the load combination equations in Section 2-3 to determine the design loads for the drag beams. On the other hand, if the lateral load is caused by an earthquake or seismic effects, a special seismic force must be calculated for the drag struts as discussed in Section 2-3 before substituting into the load combination equations to determine the design loads. The special seismic force is

$$E_m = \Omega_0 Q_E \pm 0.2\, S_{DS} D,$$

where

Ω_0 = Overstrength factor from ASCE-7 Table 12.2.1 (ensures that this structural member—the drag strut in this case—remains elastic during a seismic event),
Q_E = Drag strut force due to horizontal earthquake
= Drag strut forces calculated in Figures 14.14a through 14.14f,
D = Dead load supported by the drag beams,
S_{DS} = Design spectral response acceleration at short period (see Section 2-3 and Chapter 3),
$\Omega_0 Q_E$ = Horizontal component of the earthquake force on the drag strut, and
$0.2\, S_{DS} D$ = Vertical component of the earthquake force on the drag strut.

To illustrate the calculation of this special seismic force for the drag beam BC in Figures 14.14a and 14.14b, assume $S_{DS} = 0.27$, and that the building is located in SDC C and can be classified as "Steel systems not specifically detailed for seismic resistance" (i.e., System H in ASCE-7, Table 12.2.1); therefore, the overstrength factor is $\Omega_0 = 3.0$.

The horizontal seismic forces in drag beam BC are $\Omega_0 Q_E = (3.0)(\pm 40 \text{ kips}) = 120$ kips (tension and compression). The vertical seismic force in drag beam BC $= 0.2\, S_{DS}D = (0.2)(0.27)D = 0.054D$. Therefore, the special seismic force in the drag beam, $E_m = (120 \pm 0.054D)$ kips. To complete the design load calculations for the drag beams, this special seismic force is substituted, as the seismic force, E, into the load combination equations from Section 2-3.

Thus, the drag beam-to-column connections will be subjected to combined vertical shear and axial tension, or compression loads. The design of simple connections for vertical shear and axial loads was covered in Chapters 9 and 10. There are several possible connection details that could be used to transfer the drag strut axial force and the gravity loads, but the most economical method is to use a simple shear connection. In Figures 14.15a and 14-15b, the connection will be subjected to vertical and horizontal loads. For the single plate connection (Figure 14-15a), the bolts will be subjected to shear only, and the weld to the column will be subjected to both shear and tension. In Figure 14-15b, the weld to the supported beam (or drag strut) will be subjected to shear and tension, and the bolts connecting the double angles to the column will also be subjected to both shear and tension. The double angles will be subjected as well to prying action (see Chapter 9). An alternative detail would be to provide a shear connection to resist only the

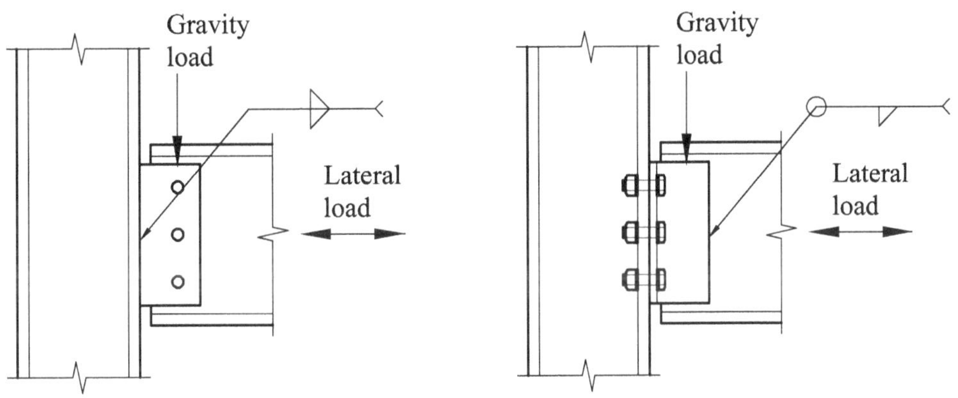

a. single plate

b. double angle

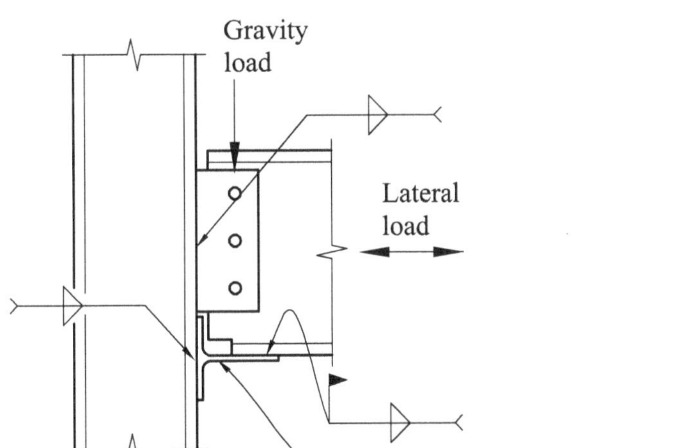

c. drag strut connector

FIGURE 14-15 Drag strut-to-column connection

gravity loads and then provide an additional member to transfer the drag strut force from the beam to the column (see Figure 14-15c) without introducing any unnecessary moment restraint at the beam-to-column connections.

14-5 TRANSFER OF LATERAL LOADS IN ROOFS WITH OPEN-WEB STEEL JOISTS

It is important that the diaphragm shear be adequately transferred from the roof or floor diaphragm to the lateral force resisting system; as previously discussed, the drag struts or collectors help make this transfer of lateral shear forces possible. Where the roof or floor diaphragm is attached directly to the infill beams and girders, the transfer of the lateral shear is direct, since the perimeter beams and girders act as the chords and drag struts, and the tops of these beams and girders are at the same elevation as the underside of the deck. However, one structural framing configuration in which proper lateral shear transfer is commonly overlooked in practice is shown in Figure 14-16.

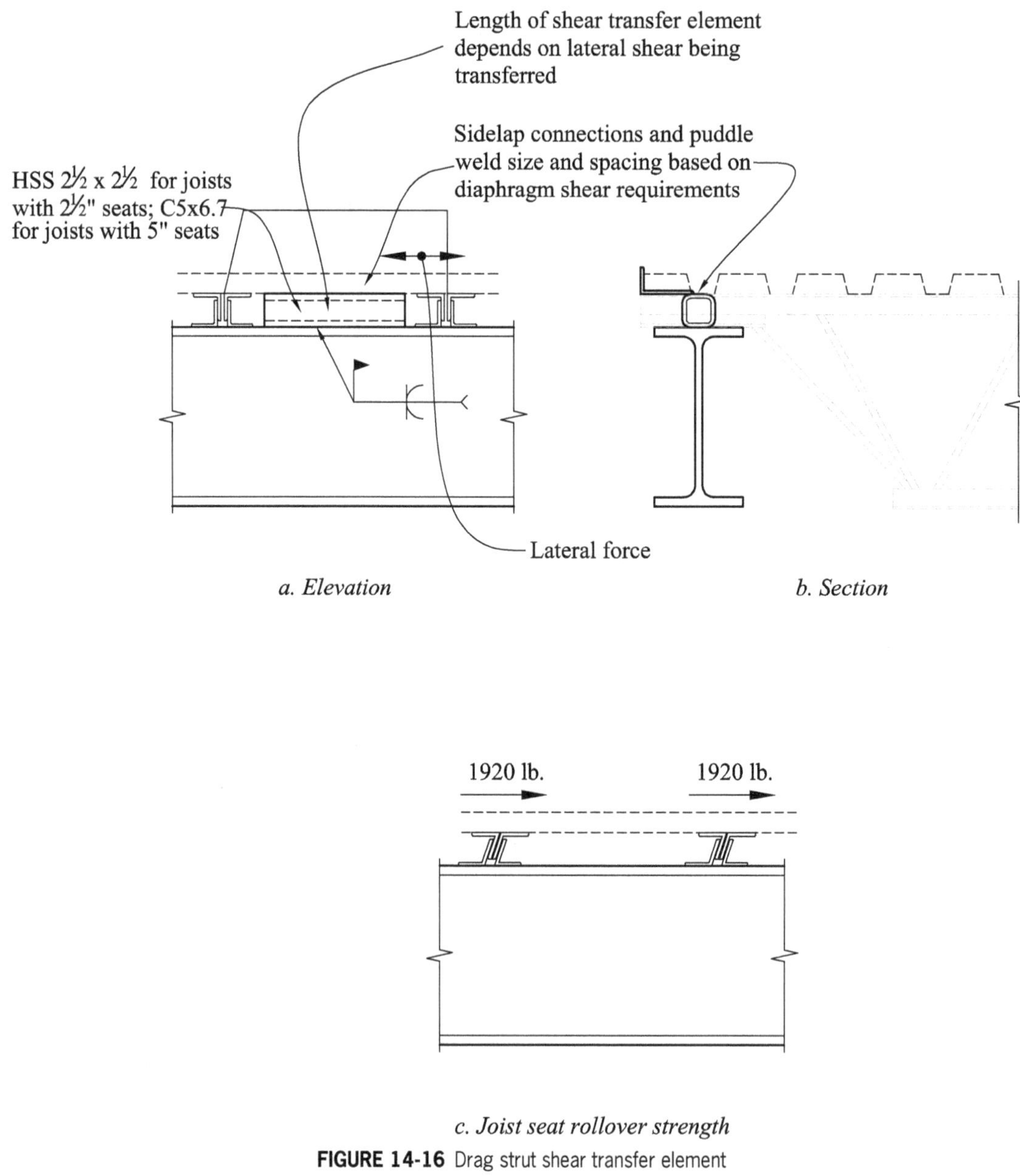

FIGURE 14-16 Drag strut shear transfer element

In this figure, the joists are framed in a direction perpendicular to the beam or girder framing, and without a drag strut or collector element located between the joist seats and welded to the top flange of the roof beams along the vertical brace lines, the web of the joist seats will be subjected to roll-over or racking about its weak axis when the lateral force in the roof diaphragm (i.e., the roof deck) is acting in a direction perpendicular to the roof joists. The racking or roll-over capacity of the joist shoe can be enhanced by welding stiffeners to both sides of the joist shoes at the joist bearing locations. Testing has indicated that a standard $2^1/_2$ in. joist seat has a nominal roll-over strength of more than 8000 lb. Using an appropriate factor of safety yields a safe working load of 1920 lb. at each joist seat, as shown in Figure 14-16c. For joist girders with $7^1/_2$ in. seats, the safe working roll-over load is 4000 lb. [10].

An alternate method for transferring the lateral force from the deck diaphragm to the drag struts and chords while preventing roll-over or racking of the steel joist shoes is to provide a drag strut shear transfer element (or a shear collector) as shown in Figures 14-16a and 14-16b. The shear transfer element could be a channel stub beam or, preferably, an HSS stub beam located between the joist shoes and welded to the drag strut or chord member. The size and spacing of the puddle weld connection between the steel roof deck and the drag strut shear transfer element are dependent on the applied lateral shear transferred from the deck diaphragm to the shear transfer element.

14-6 GIRTS AND WIND COLUMNS

For exterior wall systems in steel buildings, the maximum vertical span of the exterior metal wall cladding is usually limited, and for large floor-to-floor heights, girts or horizontal bending members may be required to reduce the vertical span of the cladding (see Figure 14-17a, b, c). The girts, which are usually C-shaped rolled sections or light-gauge C- or Z-shaped sections, or HSS members (see Figure 14-17 c), span between the building columns and the wind columns and are usually oriented with the channel toes pointing downward to prevent moisture or debris from accumulating on the channel girt. The channels support their self-weight through bending about their weak axis, and for this reason, hanger or sag rods are typically used in these situations to limit the vertical deflection of the channel girt. The sag rods also provide lateral bracing for the girts as they bend about their strong axis in resisting wind pressures acting perpendicular to the face of the exterior wall cladding (see Figure 14-17a).

Wind columns are vertical structural members that resist lateral wind loads, but they may or may not support any axial compression loads from the roof or floor framing. They span vertically from floor to floor and resist bending from the wind load applied perpendicular to the face of the exterior cladding, in addition to any gravity loads to which they may or may not be subjected. Wind columns are used when the exterior wall cladding system is unable to span vertically from floor to floor because of the large story height (see Figures 14-17a through 14-17c); therefore, horizontal girts, which span between the wind columns and the building columns, must be used. When the wind columns are not gravity-load-supporting elements, the connection details of the wind columns to the underside of the beams or girders at the top of the wind column must allow for a vertical deflection gap to enable the floor or roof framing above to deflect without imposing any gravity load on the wind column. The deflection gap is achieved by using a vertically slotted connection, as shown in Figure 14-17b. Note that the wind columns are laterally braced by the girts.

Figure 14-17c shows HSS girts in a steel building under construction; HSS girts at two levels are used because of the large floor-to-floor height (between the ground and the second floor) and the large distance between columns (i.e., the span of the girts).

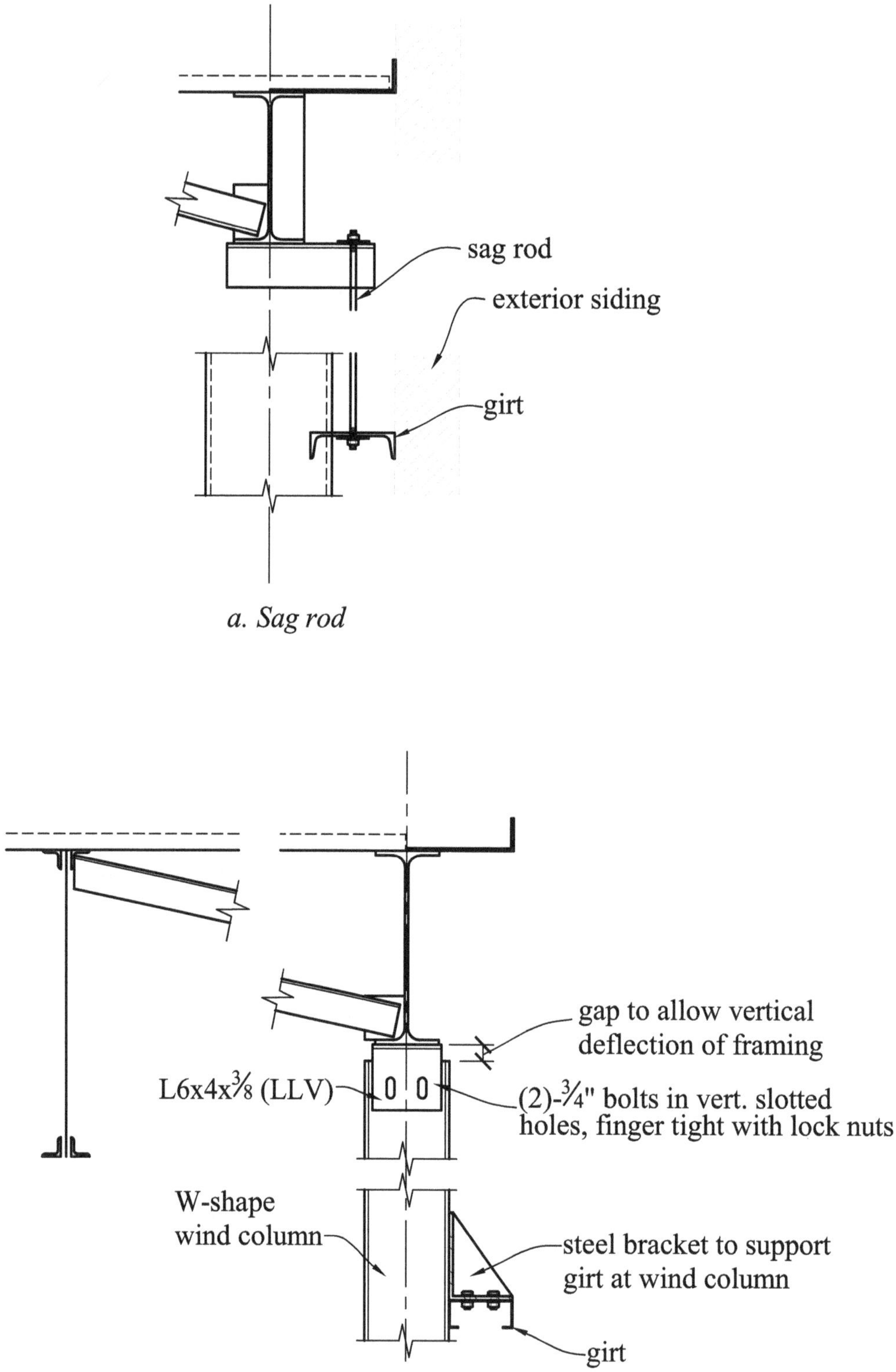

a. Sag rod

b. Wind column connection

FIGURE 14-17a-b Sag rod and wind column details

FIGURE 14-17c HSS girts and wind columns (Photo by Abi Aghayere)

4-7 RELIEF OR SHELF ANGLES FOR SUPPORTING BRICK VENEER, AND PARAPET DETAILS

In multistory buildings, horizontal angles supported by the floor framing are normally used to support the brick cladding and to provide horizontal expansion joints for the brick veneer at each floor level [11]. On the other hand, for buildings with three stories or less, and with the height of the brick cladding from the top of the foundation wall to the top of the brick cladding not exceeding 30 ft., the brick cladding can span the full 30-ft. height without any relief angle. In this case, the entire height of the brick wall will be supported at the ground-floor level on a concrete foundation wall; however, adequate provisions should be made in the lateral tie system to account for the differential vertical movement between the brick veneer and the building frame. Typical relief or shelf angle details used in multistory buildings are shown in Figure 14-18 [27].

Roof parapets in buildings are subjected to lateral wind and seismic loads, and they require adequate lateral support to prevent collapse. Figure 14-19 shows some typical parapet details that provide the needed lateral support.

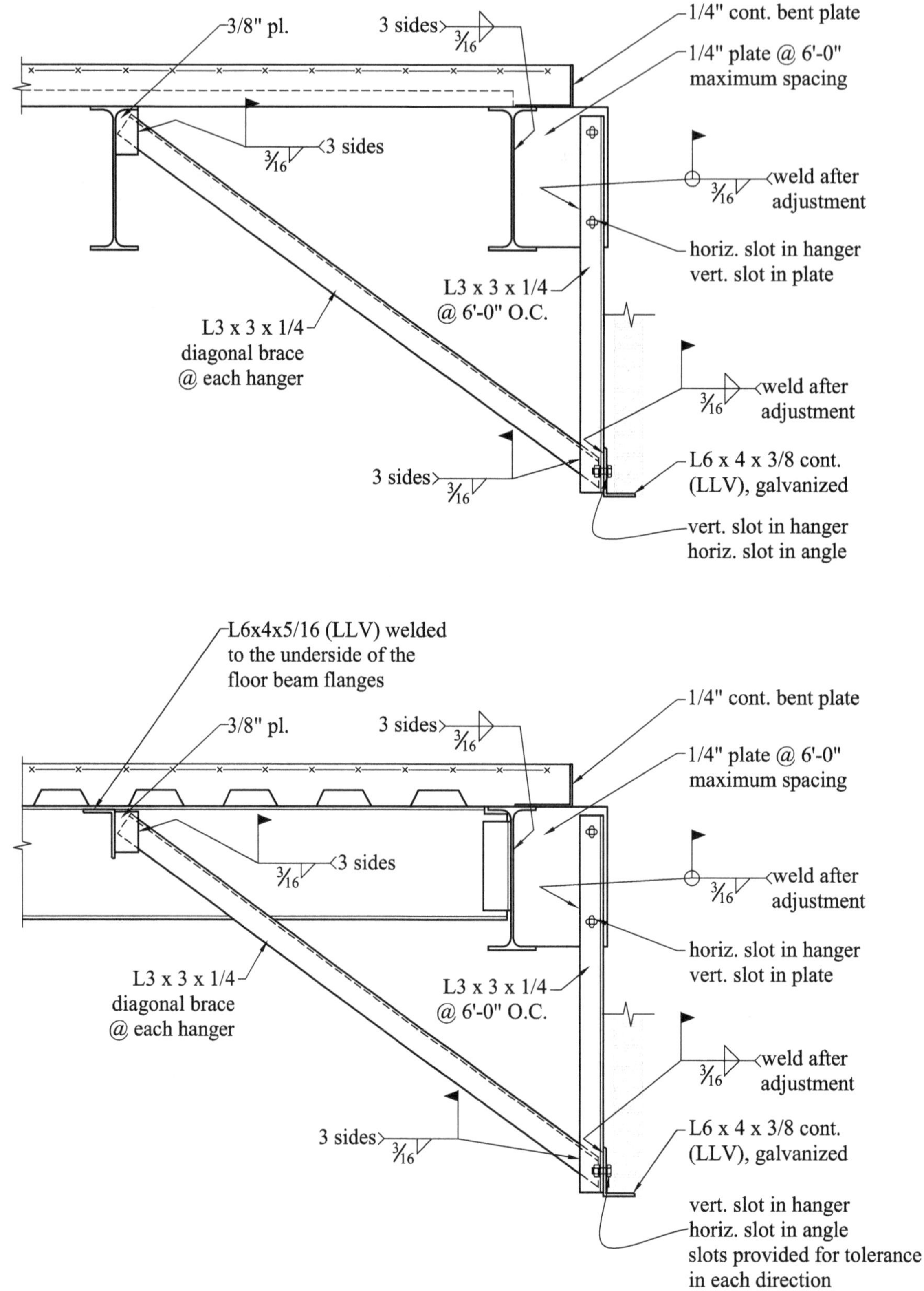

FIGURE 14-18 Relief angle details

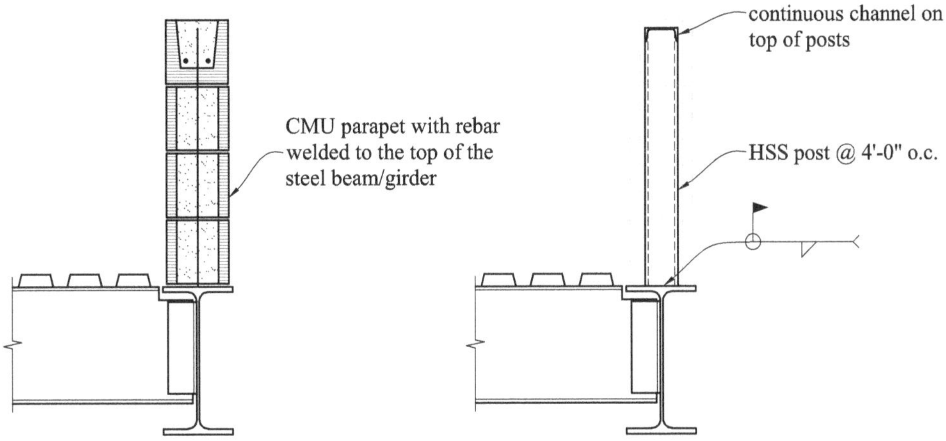

a. CMU parapet with vertical reinforcement

b. CMU parapet with HSS posts

FIGURE 14-19 Roof parapet details

14-8 ACHIEVING ADEQUATE DRAINAGE IN STEEL-FRAMED ROOFS

Most steel buildings have relatively flat roofs. These types of roofs may be highly susceptible to ponding—the tendency for flat roofs to be subjected to increased loads due to additional accumulation of rain or melted snow in the deflected areas of a flat roof. To avoid drainage problems, the roof slope in steel-framed roofs is achieved in one of two ways:

1. Maintain an essentially flat structural roof framing and use tapered insulation that is placed underneath the roofing membrane and on top of the metal roof deck to achieve the roof slope. A commonly used rule of thumb to ensure adequate drainage is to use a minimum roof slope of $^{1}/_{4}$-in. per 1 ft. [50], and this could result in large insulation thicknesses, especially at the high points on the roof. Note that a roof slope of $1/4$ in. per ft. may not always ensure proper drainage because the deflection of the sloped beams will result in an average roof slope of less than $^{1}/_{4}$-in. per 1 ft. between the midspan of the deflected beam and its lower end, creating a "bowl" effect. This effect, which is particularly more pronounced for roof beams that are flexible (i.e., long-span beams), can prevent proper drainage or even result in negative drainage. One way to mitigate this effect is to specify a minimum "average roof slope" of $^{1}/_{4}$-in. per ft. [50].
2. The more frequently used method for achieving proper roof drainage is to slope the structural roof framing with the high and low points (i.e., drain location) located near the columns. A typical roof framing plan layout showing the top of steel elevations (TOS) with allowance for proper drainage (i.e., minimum slope of $^{1}/_{4}$-in. rise per 1-ft. horizontal length) is shown in Figure 14-20. As previously discussed, the steel elevations specified to achieve proper drainage should take into account the maximum deflection of the roof beams [50].

In Figure 14-20, the tops of steel elevations at columns B-2 and C-2 (i.e., the low points) to accommodate a minimum slope of $^{1}/_{4}$-in. rise per 1 ft. horizontal length are the larger of $8^{3}/_{4}$ in. (i.e., 35 ft. × $^{1}/_{4}$ in./ft.) and $7^{1}/_{2}$ in. (i.e., 30 ft. × $^{1}/_{4}$ in./ft.); therefore, we will assume that the top of steel elevation at the low points (i.e., drain locations) is $-8^{3}/_{4}$ in. relative to the top of steel

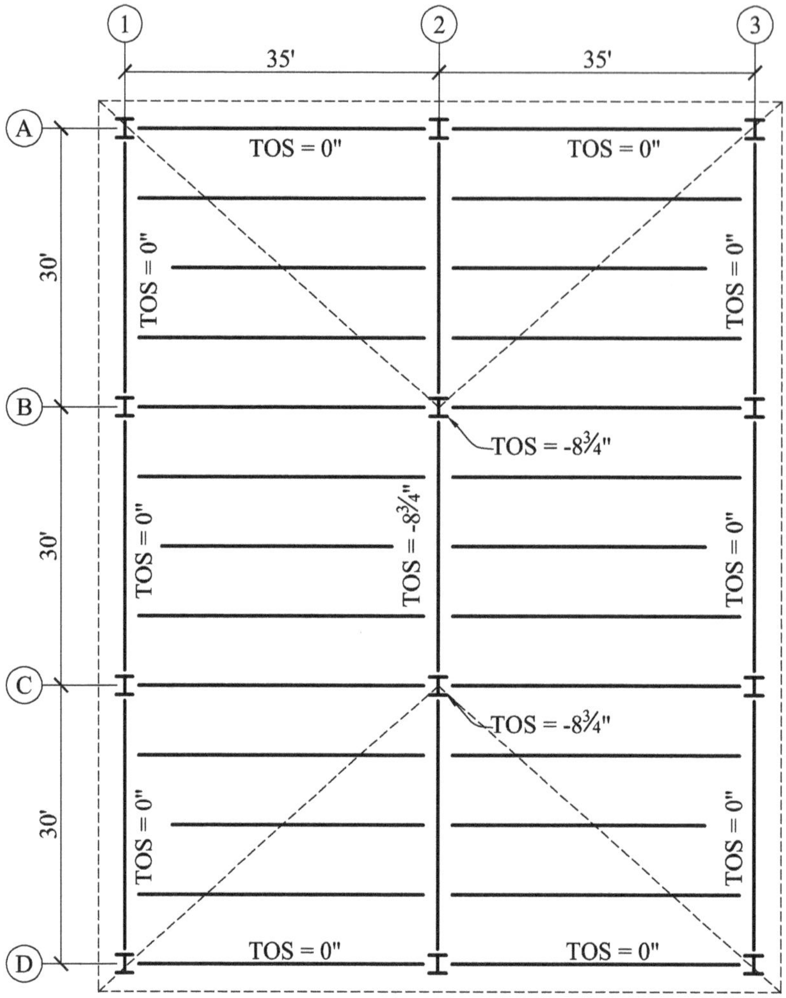

TOS = top of steel

FIGURE 14-20 Typical roof framing plan with top of steel elevations for proper drainage

elevation at the high points of the roof. The roof drains will be located near columns B-2 and C-2. The roof framing around the roof drain opening usually consists of steel angles spanning between the roof joists or roof beams.

If the infill members in the roof framing in Figure 14-20 were open-web steel joists instead of steel beams, the top of steel elevations for the girders along grid lines 1, 2, and 3 would be lowered further by an amount equal to the depth of the joist seat. Typical joist seat depths are $2^1/_2$ in. and 5 in., depending on the span of the steel joist. The joist seat depth that is appropriate for a given span is usually tabulated in joist manufacturers' catalogs. For example, if a joist with a seat depth of $2^1/_2$ in. were used, the top of the steel elevation for the girders along grid lines 1 and 3 would be $-2^1/_2$ in.

14-9 PONDING IN STEEL-FRAMED ROOF SYSTEMS

Ponding occurs when rainwater accumulates on a roof due to progressive deflection of the roof framing. In general, many roofs are constructed to be flat, but they could have some initial deflection due to the dead weight of the roofing and structure. When rainwater accumulates on a roof (see Chapter 2 for rain load calculations), the roof framing will deflect under the weight of the rainwater; this deflected shape of the roof beams and girders will result in more room for the rainwater to accumulate, thus creating a situation in which instability and eventual collapse of the roof framing could occur due to "unbonded incremental deflections" [20].

Section 1611 of the International Building Code [12] (IBC) requires that, irrespective of roof slopes, roofs should be investigated for adequate stiffness to ensure that instability does not occur. Unless the roof surface is laid out such that accumulation of rainwater cannot occur, ponding analysis of the roof framing must be carried out. (See also *AISCM* Commentary Section B3.10) [20]. The IBC also requires that roofs have a primary and secondary drainage system to accommodate the possibility that the primary drainage system may become blocked. The height of the rainwater used for design is the distance from the roof to the inlet elevation of the secondary drainage system. Careful attention to the detailing of the roof drains will allow rainwater to properly drain from the roof and mitigate the effects of ponding. This detailing, however, is usually the responsibility of the project architect. See Ref. [61] for the analysis of steel-framed roofs for ponding loads.

14-10 STABILITY BRACING FOR BEAMS, COLUMNS, AND BEAM–COLUMNS

Individual beam, column, and beam-column members require bracing for lateral stability at their support points and at the points along the member where lateral stability is required for the limit state of buckling. We will consider bracing requirements for individual beams, columns, and beam-columns in this section.

When beams are supported by direct bearing on their bottom flange, their top flange needs to be restrained against rotation. The AISC Specification, Section J10-7, requires that a full-depth stiffener plate be added when no other restraint is provided (see Figure 14-21a). This is often the case when a steel beam is bearing on a concrete wall. When the bearing occurs at a steel column, for example, a stabilizing angle could be added at the top of the beam (see Figure 14-21b).

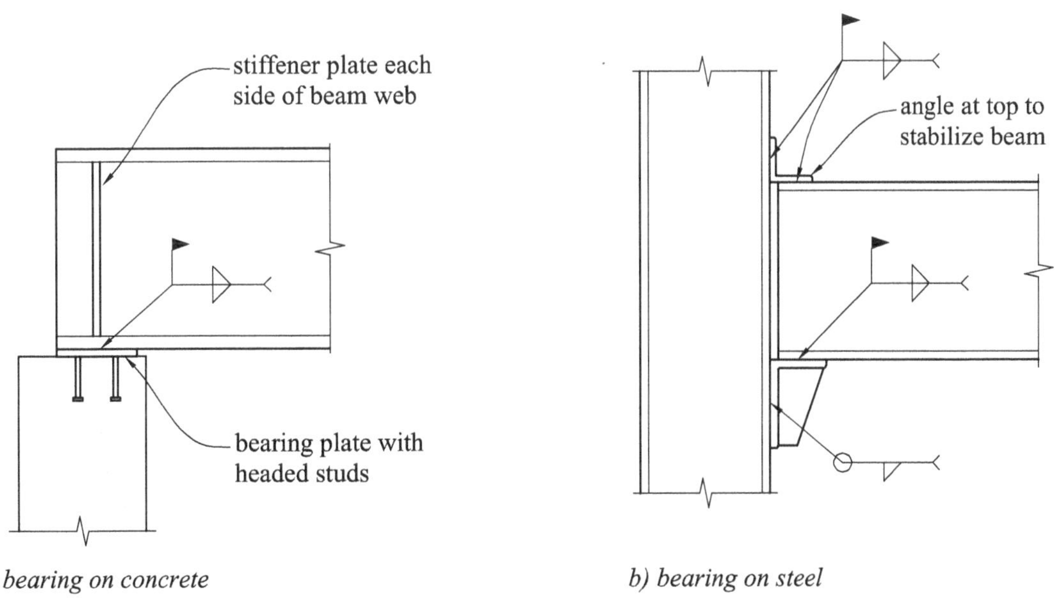

FIGURE 14-21 Stiffener plate for end bearing

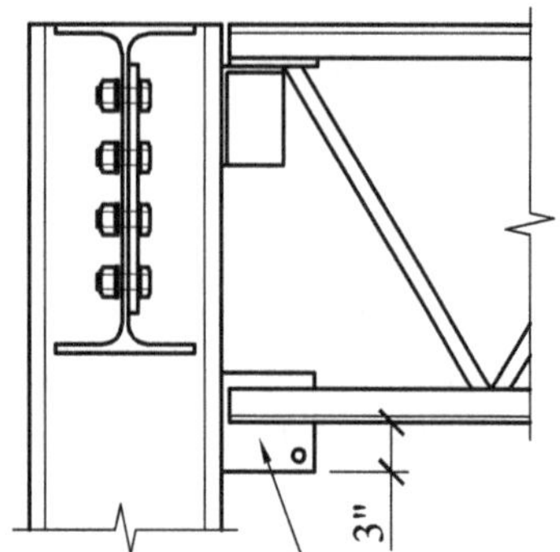

a) girder frames into column

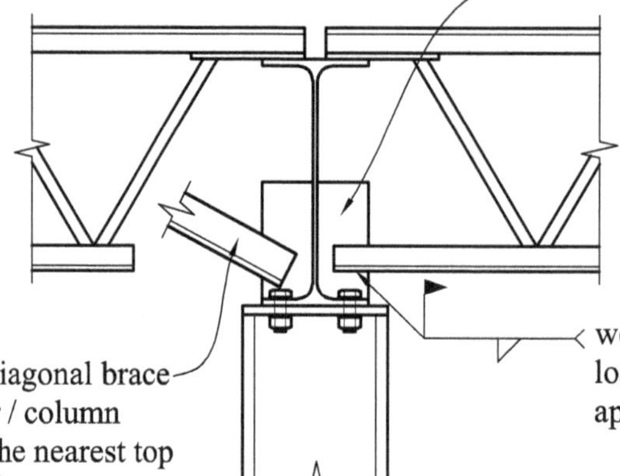

b) girder frames over column

FIGURE 14-22 Column bracing with steel joists

In roof framing with open-web steel joists, the two basic framing conditions at the columns that are commonly used are girders framing *into* the columns and girders framing *over* the top of the columns. When the girders and beams frame *into* the column in both orthogonal directions, the column is adequately stabilized, and no further detailing needs to be considered. The steel joists, however, will require a stabilizer plate at the bottom chord to meet OSHA regulations. This plate must be at least 6 in. × 6 in., and it should extend at least 3 in. below the bottom chord of the joist and have a $^{13}/_{16}$-in. hole for attaching guy wires (see Figure 14-22a).

When the girder frames *over* the column, the top of the column needs lateral support perpendicular to the longitudinal axis of the girder to prevent lateral instability since the bottom flange of the girder will be in compression and therefore subject to out-of-plane buckling. This can be accomplished by either adding a diagonal brace or by extending the bottom chords of the joists to brace the column (see Figure 14-22b); such joists are known as "tie joists."

When individual beams, columns, or beam-columns require lateral bracing to prevent the limited state of buckling, the strength and stiffness of the brace must conform to the minimum values given in Appendix 6 of the AISC Specifications. When bracing is provided, the axial force in the bracing is assumed to act perpendicular to the member being braced. When bracing is oriented diagonally, the axial force must be adjusted in accordance with the geometry of the bracing. Additionally, the strength and stiffness of the bracing end connections should also be considered. Stability bracing should always be located between the compression flange and the neutral or centroidal axis of the beam or column (see discussion on lateral torsional buckling in Chapter 6). For the bracing of beams or girders, the goal is to prevent twisting of the beam or girder (i.e., preventing the relative lateral displacement of the top and bottom flanges of the beam or girder). For beams loaded in the web, the lateral brace on the other side of the web should be located at the load level where the framing connects to the beam web or higher.

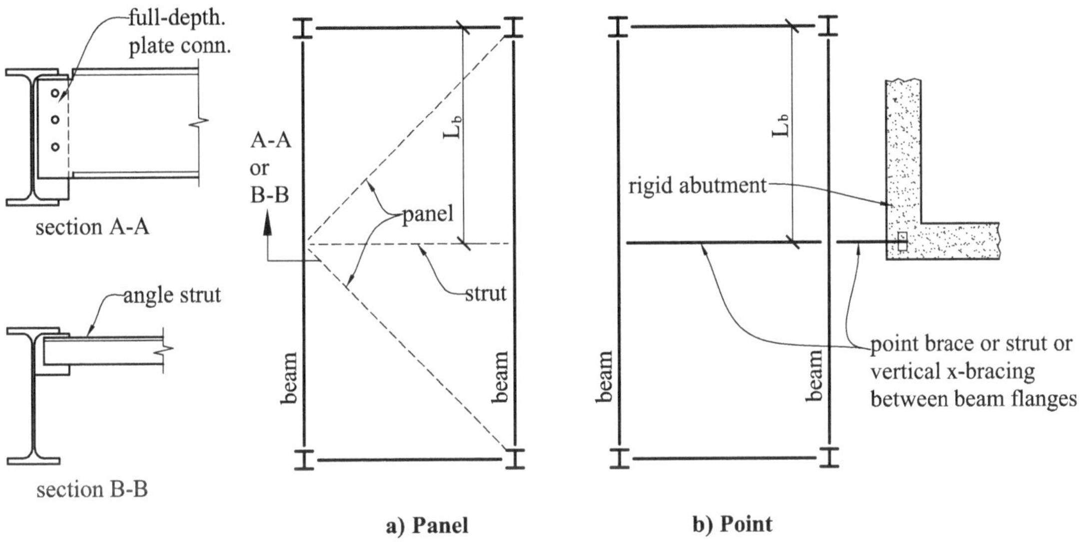

FIGURE 14-23a, b Beam bracing

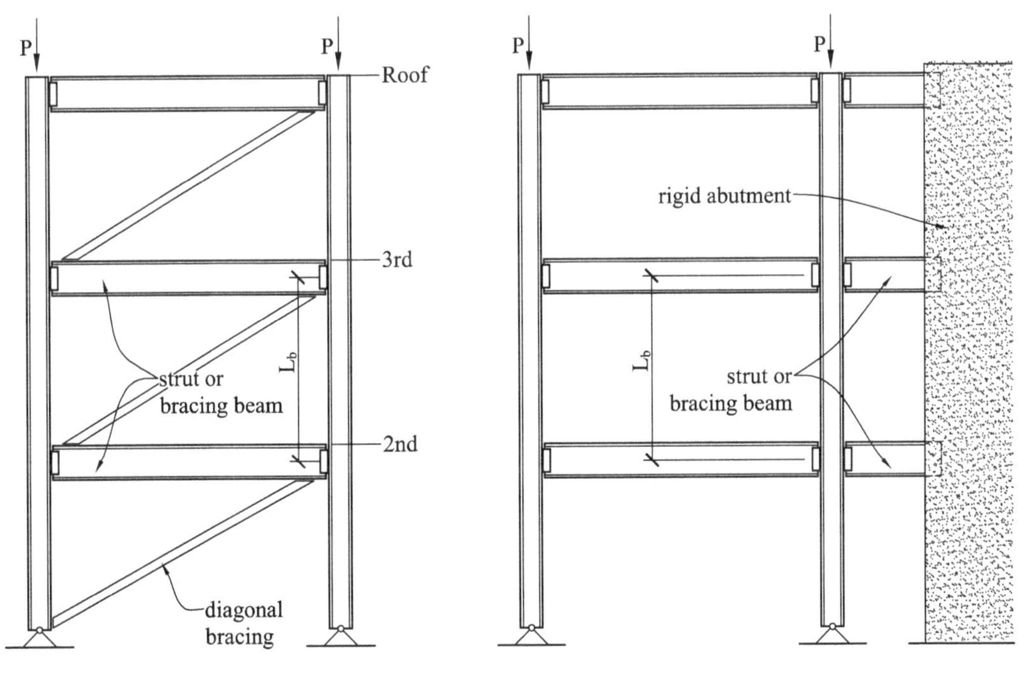

FIGURE 14-24a, b Column bracing

The two types of bracing systems used to stabilize individual columns, beam-columns, and beams recognized in Appendix 6 of the AISC Specifications are *Panel* and *Point* bracings (see Figures 14-23 and 14-24). These were previously referred to as *Relative* and *Nodal* bracings in earlier editions of the AISC Specifications. *Panel* bracing consists of diagonals and struts that are connected to more than one brace point on the braced member (e.g., the diagonal braces for the columns and beams in Figures 14-23a and 14-24a). For a *Panel* brace, the lateral displacements at adjacent brace points on the braced member are related and interdependent on each other. Thus, the displacement at one-panel brace point affects the displacement at other brace points on the member. A *Point* brace—which is axially loaded members—connects the braced member to a rigid or immovable support or abutment, and the displacement at one *Point* brace location is independent of the displacement at any other *Point* brace locations on the member (see Figures 14-23b and 14-24b). *Point* braces can also be identified as follows: if the second-floor *Point* brace for the columns in Figure 14-24b were removed, there would be lateral displacement of the columns at the second-floor level, whereas no lateral displacement would occur in the columns at the third floor and roof level *Point* brace. Thus, we see that the lateral displacements at the *Point* brace locations are independent of one another. Where there is any doubt as to how to classify the bracing system for a column, beam-column, or beam, the use of a *Point* brace solution will result in a conservative and safe design [30].

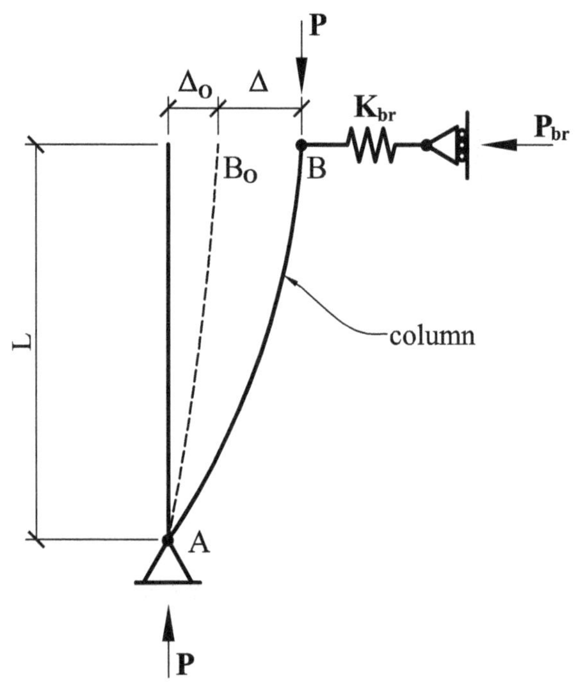

FIGURE 14-25a Column bracing - strength and stiffness requirements

The force and stiffness required to brace a beam, column, or beam-column can be derived from statics. Consider the column in Figure 14-25a with an initial out of plumbness, Δ_o, at the top of the column, which is braced by a spring with stiffness k_{br} at the top. If the column is assumed to deflect laterally to point B, a distance Δ at the top, under the axial load P, we can determine the required brace force by summing moments about the bottom support of the column at A:

For equilibrium, $\sum$ Moments about $A = P(\Delta+\Delta_o) - F_{br}(L) = 0$,

which yields $P(\Delta+\Delta_o) = F_{br}(L)$.

Therefore, the brace force, $F_{br} = \dfrac{P(\Delta+\Delta_o)}{L}$

Since $F_{br} = k_{br}\Delta$,

Therefore, the brace stiffness, $k_{br} = \dfrac{F_{br}}{\Delta} = \dfrac{P(\Delta+\Delta_o)}{(\Delta)(L)} = \dfrac{P}{L}\left(1+\dfrac{\Delta_o}{\Delta}\right)$

The maximum allowable out of plumbness in the AISC Specification is $L/500$ [20]. If we assume that $\Delta_o = L/500$ and also that the column deflects laterally a further distance $\Delta = L/500$ just before it buckles, we can calculate the required brace force and stiffness as follows:

Assuming $\Delta = \Delta_o = L/500$

The required brace force, $F_{br} = \dfrac{P(\Delta+\Delta_o)}{L} = \dfrac{P(\frac{L}{500}+\frac{L}{500})}{L} = \dfrac{P}{250} = 0.004P$

The required brace stiffness, $k_{br} = \dfrac{P}{L}\left(1+\dfrac{\Delta_o}{\Delta}\right) = \dfrac{2P}{L}$,

Where,

P = the applied axial compression force in the braced member.

Note: For W-shaped bending, the compression force for which bracing would be provided will be M_u/h_o, where M_u is the maximum factored moment in the braced member, and h_o is the distance between the centroids of the top and bottom flanges of the W-shape.

Δ_o = Initial out of plumbness of the column (i.e., the column in its undeflected state),

Δ = The maximum lateral deflection of the top of the column = the compression of the lateral brace spring,

k_{br} = Stiffness of the lateral brace,

F_{br} = Design strength of the lateral brace.

The following equations (only LRFD considered) for the minimum design strength and stiffness of lateral bracing of individual beams, columns, and beam-columns from Appendix 6 of the AISC Specifications were derived using a similar approach to the preceding equations. For column or beam-column bracing, the lateral bracing is assumed to be located at the shear center of the column. Beams shall be laterally braced to prevent twisting or rotation about their longitudinal axis at the brace points.

Recall that *Panel* bracing typically consists of diagonals and struts, while *Point* bracing is axially loaded members.

Columns—Panel Bracing

Required shear strength of *panel* bracing system:

$$V_{br} = 0.005P_u \tag{14-23}$$

Required shear stiffness of *panel* bracing system:

$$\beta_{br} = \frac{2P_u}{\phi L_b} \tag{14-24}$$

Columns—Point Bracing

Required axial strength of *point* bracing system:

$$P_{br} = 0.01P_u \tag{14-25}$$

Required axial stiffness of *point* bracing system:

$$\beta_{br} = \frac{8P_u}{\phi L_b} \tag{14-26}$$

Beams—Panel Bracing

Required shear strength of *panel* bracing system:

$$V_{br} = \frac{0.01M_uC_d}{h_o} \tag{14-27}$$

Required shear stiffness of *panel* bracing system:

$$\beta_{br} = \frac{4M_uC_d}{\phi L_b h_o} \tag{14-28}$$

Beams—Point Bracing

Required axial strength of *point* bracing system:

$$P_{br} = \frac{0.02M_uC_d}{h_o} \tag{14-29}$$

Required axial stiffness of *point* bracing system:

$$\beta_{br} = \frac{10M_uC_d}{\phi L_b h_o} \tag{14-30}$$

Beam-Columns—Panel Bracing

Required shear strength of *panel* bracing system:

$$V_{br} = 0.005P_u + \frac{0.01M_uC_d}{h_o} \tag{14-31}$$

Required shear stiffness of *panel* bracing system:

$$\beta_{br} = \frac{2P_u}{\phi L_b} + \frac{4M_uC_d}{\phi L_b h_o} \tag{14-32}$$

Beam-Columns—Point Bracing

Required axial strength of *point* bracing system:

$$P_{br} = 0.01P_u + \frac{0.02M_uC_d}{h_o} \tag{14-33}$$

Required axial stiffness of *point* bracing system:

$$\beta_{br} = \frac{8P_u}{\phi L_b} + \frac{10M_uC_d}{\phi L_b h_o}, \tag{14-34}$$

where

P_u = Factored axial load on the braced member,

M_u = Factored bending moment in the braced member within the panel under consideration,

L_b = Unbraced length of the beam, column, or beam-column (i.e., distance between lateral braces),

C_d = 1.0 for single curvature bending

= 2.0 for double-curvature bending,

h_o = Distance between flange centroids, and

ϕ = 0.75.

Torsional Bracing

We first encountered the use of torsional bracings in Section 6-9 of Chapter 6. The details in Figures 6-18(a-d) show methods for controlling torsion or for bracing members against torsion. Torsional bracings are typically provided *as point bracing*, and examples include the following (see Appendix 6.3.2 of the AISC Specifications [20]):

- Cross or diagonal bracings or frames attached to the flanges of the beam being braced (see Figure 6-18c)
- Moment-connected beams connected to the beam being braced (see Figure 6-18d)
- A horizontal diaphragm element connected to the beam being braced, and with the diaphragm element approximately the same depth as the beam being braced.

Torsional Bracing—Point Bracing

Required flexural strength of *point* bracing system about the longitudinal axis of the beam that is being braced:

$$M_{br} = \frac{3.6L}{nEI_{y,eff}}\left(\frac{M_u}{C_b}\right)^2\left(\frac{L_{br}}{500h_o}\right) \geq 0.02M_u \qquad (14\text{-}35)$$

The required flexural stiffness of the *point* bracing system is:

$$\beta_{br} = \frac{\beta_T}{\left(1-\frac{\beta_T}{\beta_{sec}}\right)} \qquad (14\text{-}36)$$

$$\beta_T = \frac{1}{\phi}\frac{3.6L}{nEI_{y,eff}}\left(\frac{M_u}{C_b}\right)^2 \qquad (14\text{-}37)$$

$$\beta_{sec} = \frac{3.3E}{h_o}\left(\frac{1.5h_o t_w^3}{12} + \frac{t_{st} b_s^3}{12}\right) \qquad (14\text{-}38)$$

where

M_u = Factored largest bending moment in the braced member within the unbraced length adjacent to the point brace, using the LRFD load combinations

n = number of brace points within the span of the braced beam

E = 29,000 ksi

ϕ = 0.75

$I_{y,eff}$ = Effective out-of-plane moment of inertia of the beam being braced, in.^{4}

$= I_{yc} + \left(t/c\right) I_{yt}$

I_{yc} = moment of inertia of the compression flange about the y-axis, in.^{4}
I_{yt} = moment of inertia of the tension flange about the y-axis, in.^{4}
L = span length of the beam being braced, in.
L_{br} = Unbraced length of the beam being braced adjacent to the *point brace* (i.e., distance between lateral braces)
C_b = Moment gradient or lateral torsional buckling factor with the unbraced length adjacent to the point brace
b_s = stiffener width for one-sided stiffeners, in.
= twice the individual stiffener width for pairs of stiffeners, in.
c = distance from the neutral axis to the extreme compression fibers, in.
t = distance from the neutral axis to the extreme tension fibers, in.
t_{st} = thickness of web stiffener width for one-sided stiffeners, in.
t_w = thickness of beam web, in.
β_T = overall brace stiffness required stiffness, kip-in//rad
β_{sec} = Web distortional stiffness, including the effect of web transverse stiffeners, if any,| kip-in./rad
h_o = Distance between flange centroids, and
ϕ = 0.75.

Notes:

- For doubly symmetric members (e.g. wide flange beams), c = t and $I_{y,eff}$ = out-of-plane or weak axis moment of inertia, I_y.
- $\beta_{sec} = \infty$ and $\beta_{br} = \beta_T$ when a cross-frame is connected near both flanges of the beam being braced or a horizontal diaphragm element is connected to the beam being braced, and with the diaphragm element approximately the same depth as the beam being braced.
- If $\beta_{sec} < \beta_T$, this implies that Equation (14-36) will be negative, indicating that the torsional beam bracing will not be effective due to a lack of adequate web distortional stiffness. Using web stiffeners increases the web distortional stiffness of the beam being braced and improves the effectiveness of the torsional brace.
- If the torsional brace is attached to the flanges of the beam being braced, web stiffeners, if required, must be in full depth and be connected to both flanges of the beam.

EXAMPLE 14-4

Lateral Bracing for a Column

A 32-ft.-long W8 × 31 column requires a lateral brace for weak axis bending at the mid-height. Determine whether an *L*3 × 3 × $^1/_4$ *point* brace provides adequate lateral bracing if the factored axial load in the column is P_u = 200 kips. The *L*3 × 3 × $^1/_4$ brace is ASTM A36 steel and has a length of 6 ft. from the column web to an isolated support.

Solution

The unbraced length of the column, L_b = 32 ft./2 = 16 ft. (midspan brace).

From equation (14-25), the required strength of the point brace is
$P_{br} = 0.01P_u = (0.01)(200) = 2$ kips.

From *AISCM* Table 4-11, the design strength of the *L*3 × 3 × $^1/_4$ at an unbraced length of 6 ft. is ϕP_n = 20.9 kips > 2 kips, OK.

From equation (14-26), the required stiffness of the point brace is

$$\beta_{br} = \frac{8P_u}{\phi L_b} = \frac{(8)(200)}{(0.75)(16)} = 134 \text{ kips/ft.} = 11.1 \text{ kips/in.},$$

From *AISCM* Table 1-7, the gross cross-sectional area for an L3 × 3 × $^1/_4$ is,

$A = 1.44 \text{ in.}^2$

Length of brace, L = 6 ft. = 72 in.

The stiffness of the brace is

$$\beta_{br} = \frac{EA}{L} = \frac{(29{,}000)(1.44)}{(72)} = 580 \text{ kips/in.} > 11.1 \text{ kips/in.}$$

The *L*3 × 3 × $^1/_4$ lateral point bracing provides adequate strength and stiffness.

EXAMPLE 14-5

Lateral Bracing for a Beam

A 32-ft.-long W18 × 50 beam requires a lateral brace at midspan of a simple span. Determine whether an *L*3 × 3 × $^1/_4$ *point brace* provides adequate lateral bracing if the factored bending moment in the beam is M_u = 240 ft.-kips. The *L*3 × 3 × $^1/_4$ brace is ASTM A36 steel and has a length of 8 ft. from the compression flange of the beam to an isolated support.

Solution

Unbraced length of the beam, L_b = 32 ft./2 = 16 ft. (midspan brace)

Length of the *L*3 × 3 × $^1/_4$ brace, *L* = 8 ft. = 96 in.

From *AISCM* Table 1-1,

$h_o = 17.4$ in. for a W18 × 50

From equation (14-29), the required strength of the point brace is

$$P_{br} = \frac{0.02M_u C_d}{h_o} = \frac{(0.02)[(200)(12)](1.0)}{(17.4)} = 2.76 \text{ kips.}$$

From *AISCM* Table 4-11, the design strength of the *L*3 × 3 × $^1/_4$ at an unbraced length of 8 ft. is ϕP_n = 12.0 kips > 2.76 kips, OK.

From equation (14-30), the required stiffness of the point brace is

$$\beta_{br} = \frac{10M_u C_d}{\phi L_b h_o} = \frac{(10)[(200)(12)](1.0)}{(0.75)[(16)(12)](17.4)} = 9.58 \text{ kips/in.},$$

where L_b = 32 ft./2 = 16 ft. (midspan brace) and C_d = 1.0 (simple span, therefore single curvature bending in the beam).

From *AISCM* Table 1-7, the gross cross-sectional area for an L3 × 3 × $^1/_4$ is,

$A = 1.44 \text{ in.}^2$.

The stiffness of the point brace is

$$\beta_{br} = \frac{EA}{L} = \frac{(29{,}000)(1.44)}{(96)} = 435 \text{ kips/in.} > 9.58 \text{ kips/in.}$$

The *L*3 × 3 × $^1/_4$ lateral point bracing provides adequate strength and stiffness.

14-11 STEEL PREPARATIONS, FINISHES, AND FIREPROOFING

The preparation and finishing of structural steel is generally beyond the scope of responsibilities of the structural engineer; however, it is important for any designer to have a basic understanding of steel coatings because they often impact the detailing and selection of steel members. This section will provide a general overview of steel coatings and fireproofing, but given the wide variety of paint and fireproofing products available, it is recommended that the designer review the required steel preparation and coating options for any given project.

We will first consider the preparation of steel for the application of coatings. The Society for Protective Coatings identifies several preparation methods based on the final coating to be applied [13]. When a coating system is selected, the required surface preparation is usually dictated

Figure 14-25b Continued

FIGURE 14-25b Sprayed-on cementitious fireproofing of beams, girders, columns, and diagonal braces

by the manufacturer of the coating. Table 14-3 summarizes the various surface preparations that can be specified for structural steel sections.

A more detailed description of each method is given in Ref. [13], but most are self-explanatory. When structural steel is intended to be left uncoated, the AISC Code of Standard Practice states that SSPC-SP 1 (Solvent Cleaning) is required. Solvent cleaning uses a commercial cleaner to remove visible grease and soil. When shop painting is required, the minimum surface preparation is SSPC-SP 2 (Hand Tool Cleaning). Hand tool cleaning removes loose mill scale, rust, and other deleterious matter using nonpowered hand tools. The other surface preparations indicated are more labor-intensive and should be specified in accordance with the coating manufacturer's recommendations.

The use of coatings is a function of several factors, such as aesthetics and the intended use of the structure. For most structural steel applications, the use of a shop coat of paint is sufficient in that the shop coat provides moderate resistance to environmental conditions during the construction process when the steel is directly exposed to the weather. Once the steel is in place, adequate protection from weather is usually assured for covered structures. It should be noted that structural steel can be left uncoated for this typical condition, and it is recommended that the steel be left uncoated when spray-applied fireproofing is specified because the presence of a primer often reduces the adhesion quality of the fireproofing. When the term "shop coat" is used to describe a coating, it is assumed that the paint has a minimum dry film thickness of 1 mil, which is $1/1000^{th}$ of an inch.

TABLE 14-3 Description of Various Steel Surface Preparations [13]

SSPC-SP 1: Solvent Cleaning
SSPC-SP 2: Hand Tool Cleaning
SSPC-SP 3: Power Tool Cleaning
SSPC-SP 5: White Metal Blast Cleaning
SSPC-SP 6: Commercial Blast Cleaning
SSPC-SP 7: Brush-Off Blast Cleaning
SSPC-SP 8: Pickling
SSPC-SP 10: Near-White Blast Cleaning
SSPC-SP 11: Power Tool Cleaning to Bare Metal
SSPC-SP 12: Surface Preparation and Cleaning of Metals by Water jetting Prior to Recoating

When structural steel is intended for use in an exterior environment or in a corrosive environment such as an indoor swimming pool, corrosion protection must be provided; the coating system should be carefully selected and should be applied in consultation with the coating manufacturer. The coating system for these types of service conditions usually consists of a more extensive surface preparation (often SSPC-SP 6, Commercial Blast Cleaning) in conjunction with an epoxy primer and a finish coat. In addition to the preparation and painting of the structural steel members, there are often field welds at the connections that also need a proper coating system. The surface preparation for field welds is typically SSPC-SP 11, Power Tool Cleaning to Bare Metal, and the coating system would match the system for the other members. The variety of coating products available requires that the coating system for any project be examined on a case-by-case basis to determine the best system for a given exposure condition.

In lieu of the coating systems mentioned above for exterior exposed or corrosive environments, the process of hot-dip galvanizing could be specified. Galvanizing can provide a service life of up to 100 years to structural steel [55]. Hot-dip galvanizing is the process of coating structural steel with a layer of zinc by dipping it into a tank of molten zinc. The zinc layer will react with oxygen to create zinc oxide while in service; therefore, the layer will decrease in thickness over time. The minimum thickness of this zinc layer is given in ASTM A123; it varies, depending on the thickness of the steel section, from 1.4 mils to 3.9 mils. These minimum values of zinc layer thickness have proven to provide a service life that varies from 20 years to 50 years, depending on the zinc thickness and the exposure conditions.

When hot-dip galvanizing is specified with bolted connections, the fabricator may prefer to use larger bolt holes to accommodate the added layers of zinc. The standard hole size is $\frac{1}{16}$ in. or 62.5 mils larger than the bolt, but the addition of zinc layers on the bolts and at least two connected parts (clip angles or shear tabs) could reduce the ease of installing the bolts given that the hole diameter is 62.5 mils larger than the bolt in each connected part and that a hole size reduction of as much as 3.9 mils could be present in each connected part. For field welds, cold galvanizing repair paint in accordance with ASTM A780 is used to repair previously galvanized areas affected by welding.

1. AISC Specification Appendix 4 presents the provisions for structural design of steel members for fire. Structural steel loses strength at high temperatures (e.g., in a fire), so steel members are often required to be protected from fire depending on the occupancy category of the building and the member type. Steel can be left exposed in some buildings, but in cases where the aesthetics require an exposed steel member, as well as adequate protection from fire, *intumescent coating* could be applied. During a fire event, the intumescent coating will expand up to 50 times its original thickness, creating a foam-like layer of insulation that will help reduce the rate of heat transfer and therefore reduce the loss of strength in the structural steel. Intumescent coating is usually applied in several layers, varying from $\frac{1}{8}$ in. to $\frac{5}{8}$ in., depending on the product and the required fire rating for the member.
2. Another method used for meeting the fire protection requirements for a building with exposed steel is the use of a sprinkler system. The use and applicability of this system would be coordinated between the architect and the plumbing engineer.
3. HSS columns filled with reinforced concrete are also used to provide fire resistance in structures with exposed structural steel. These reinforced concrete-filled columns resist fire by gradually transferring the load from the structural steel to the in-fill fire-resistant reinforced concrete. Ventilation or weep holes at the bottom of the HSS or steel pipe allow for the release of steam as the concrete is heated under fire. These reinforced concrete-filled steel members, which are used in architecturally exposed structural steel (AESS), can be coated, and they provide an aesthetic expression of the structure. Some proprietary fire-resisting systems are also available [28].
4. One of the more common methods for protecting steel from fire is to use spray-applied fire-resistant materials (SFRM). SFRM can be either fibrous or cementitious, both of which have noncombustible properties. The use of a primer prior to the application of SFRM is not recommended because the primer could reduce the adhesion quality of the SFRM. See Figure 14-25b for examples of sprayed-on cementitious fireproofing of beams, girders, columns, and diagonal braces.
5. Another common method used for protecting steel members from fire is to encapsulate or enclose the steel member with gypsum wallboard (GWB). This type of protection can be specified regardless of the coating used on the structural steel. The steel could also be encased in concrete or masonry, which was a common method of fire projection in the past but is not as economical today compared to the other fire protection methods.

The selection of a fire protection system is often the responsibility of someone other than the structural engineer, and it rarely has a major impact on the final design of the main structural system. For more information and details on fire-rated assemblies, the reader should refer to Ref. [14], [15], and [28].

Architecturally Exposed Structural Steel (AESS)

Sometimes, the architect and structural engineer have the opportunity to showcase their structure by exposing the structural steel elements. The five different categories of architecturally exposed structural steel (**AESS**) are as follows (see AISC Code of Standard Practice Section 10.1.1) [51, 52]:

1. **AESS Category 1**: These are basic elements that are far from view but have requirements that are more stringent than non-AESS construction [20]. This involves a three-layer coating system, which might include zinc coating as the first coat, followed by an epoxy coating, which can then be followed by a polycarbamide (i.e., Tnemec) coating that provides the finish color and gloss to the steel member [51].
2. **AESS Category 2**: Feature elements that are not in close view (i.e., viewing distance greater than 20 ft.)
3. **AESS Category 3**: Feature elements in close view (i.e., viewing distance of not more than 20 ft). For example, an AESS building might involve AESS Category 3 for the first 15 ft height of the column, Category 2 for the remaining column height, and Category 1 for the exposed roof steel framing.
4. **AESS Category 4**: Showcase elements and all welds are usually ground smooth.
5. **AESS Category C** (Custom): Custom elements. For AESS exposed to the elements, it is recommended to use galvanized steel with AESS Category C.
   The surface preparation for all the categories should conform to SSPC-SP 6, and all sharp edges shall be ground smooth. Mockups or visual samples are required for AESS categories 3, 4, and C.

Exposed steel must be specifically designated in the contract documents for the AESS provisions to be applicable according to Section 10.1 of the AISC Standard Code of Practice.

14-12 CORROSION PROTECTION OF STRUCTURAL STEEL

Corrosion protection is not necessary or required for steel members that are enclosed within a building and are not exposed to the elements (i.e., the presence of water and oxygen) or to a chemically corrosive environment. However, exposed structural steel (e.g., steel used in a parking garage structure), or structural steel in an interior corrosive environment (e.g., steel framing over a swimming pool area or in a chemical plant), must be protected from corrosion, and extra care should be taken to protect structural steel connections—bolted and welded—from corrosion because structural failure due to corrosion of connections can occur suddenly and without warning. The gusset plates in connections should be oriented in such a way that they do not become traps for dirt, water, and debris that could lead to corrosion. Therefore, the use of vertical gusset plates is preferred because it sheds water and debris compared to flat horizontal gusset plates, which can collect and retain dirt and debris that can lead to corrosion [37]. The corrosion of the steel H-piles supporting the piers of the Leo Frigo Bridge on Interstate 43 in Green Bay, Wisconsin, caused them to buckle and led to a major sag of up to 2.5 ft. in the bridge in September 2013 [29]. A building failure that led to the loss of two lives in Elliot Lake, Ontario, Canada, was apparently caused by the corrosion of the steel beam-to-column connections that occurred from long-term leakage of road salt and water into these welded connections from the vehicle parking lot on top of the steel-framed roof structure [32].

Corrosion protection should also be provided to prevent galvanic corrosion when dissimilar metals are connected and in contact with one another in a humid environment. Examples of situations where galvanic corrosion can occur are when carbon steel members come in contact with fasteners or other members made from stainless steel, copper, brass, or aluminum; Table 2.8 of the *AISCM* provides guidance on "galvanic corrosion potential between steel and common construction materials" [20, 40]. One solution for preventing galvanic corrosion is by inserting a nonconductive material such as a neoprene pad (or Mylar or Teflon) to separate or insulate the two dissimilar metal material surfaces from each other.

Protection of exposed steel members from corrosion usually involves the use of zinc-rich primers or galvanizing the steel surfaces. These coatings and the required minimum thickness are typically prescribed by the structural engineer of record in the material specifications or in the general notes on the structural drawings. To remediate the effects of corrosion in existing structures, the structure and the connections should be inspected, and temporary shoring may need to be provided to support the loads on the structure before a permanent fix is designed and constructed. The permanent fix for a corroded member or structure may involve replacing the corroded member or reinforcing the corroded member in situ. Irrespective of the approach, this remedial work should be done after appropriately shoring the structure and relieving the load on the corroded member by jacking up the structure, and then installing a new replacement member or reinforcing the existing member in place. The replacement member and reinforcement should have appropriate corrosion protection, such as a zinc-rich primer, coal tar epoxy, or galvanizing.

FIGURE 14-26a Canopy collapse due to exterior HSS column corrosion (Courtesy of Darren A. Brooks, P.E.)

FIGURE 14-26b Corroded steel beam (Courtesy of Jason, P.E., S.E.)

FIGURE 14-26c Corroded steel column base (Courtesy of Jason, P.E., S.E.)

FIGURE 14-26d Corroded steel column base (Courtesy of Jason, P.E., S.E.)

FIGURE 14-26e Corroded HSS column base (Courtesy of Jason, P.E., S.E.)

FIGURE 14-26f Corroded steel beams and floor deck (Courtesy of Jason, P.E., S.E.)

FIGURE 14-26g Corroded steel beams and floor deck (Courtesy of Jason, P.E., S.E.)

FIGURE 14-26h Corroded steel column in exterior wall (Courtesy of Jason, P.E., S.E.)

FIGURE 14-26i Corroded exterior steel column (Courtesy of Jason, P.E., S.E.)

FIGURE 14-26j Corroded exterior steel framing (Courtesy of Jason, P.E., S.E.)

The pictures in Figures 14-26a through 14-26j are examples of corroded steel structures. When investigating structures that have deteriorated, it is important to bring any concerns to the attention of the building owner, building official, or other interested parties if the safe functioning of the structure is in question. Note that the Existing Building Code of New York State, Section 202 [31], classifies a *Dangerous Condition* as one in which either

a. The stress in any member or portion thereof due to all factored dead and live loads is more than one and one-third (i.e., 133%) the nominal strength, or
b. Any portion of the structure is likely to fail or collapse.

It may be necessary in some cases that the structure be shut down from regular use until proper repairs have been made.

Once corrosion has occurred and the amount of section loss has been ascertained by "scraping off the rust scale and measuring the remaining section with calipers or ultrasonic testing instruments," the possible solutions required to bring the structure back to a state of proper functioning usually consist of some combination of the following:

1. For steel with minimal section loss such that the remaining section is adequate to support the structure, the steel should be cleaned and painted with an appropriate coating to prevent further corrosion. The process of cleaning the steel usually involves sandblasting and/or use of a wire brush. The coating could be a coal tar epoxy or a zinc-rich paint product. In most of the cases shown in Figures 14-26a through 14-26j, proper coatings were not provided, but rather a standard primer was used. A standard primer will initially look adequate during the construction phase, but it is not sufficient for the long-term protection of the steel in a corrosive environment.
2. For steel with significant section loss, the steel can be removed and replaced in kind by field welding. This process usually involves shoring the structure to allow the complete removal and replacement of the corroded steel sections. Once the corroded steel is removed, the new steel should be coated as described earlier. For steel columns, elevating the base above the ground level is recommended, where possible, since moisture has a natural tendency to always be present at column bases in an exposed environment.
3. For steel with significant section loss, the steel could be cleaned and reinforced with steel plates (i.e., plating the member) to replace the steel sections that have been lost to corrosion. This process would likely require minimal or no shoring but will involve field welding or field bolting the reinforcing steel section to the corroded member outside of the corroded regions of the member. Once the steel reinforcement is in place, the new and existing steel should be coated to prevent corrosion, as described earlier.

While corrosion can often be considered a normal maintenance item, there are several design solutions that can help minimize and even prevent the type of corrosion seen in Figures 14-26a through 14-26j. For each picture in Figures 14-26a through 14-26j, a description of the problem and possible remedial measures that could have been taken during the design phase are presented in the table below:

Figures	Problem description	Possible design solution
	HSS columns corroded at the base due to exposure.	Allow galvanizing to coat the interior wall surfaces; elevate the column base above grade.
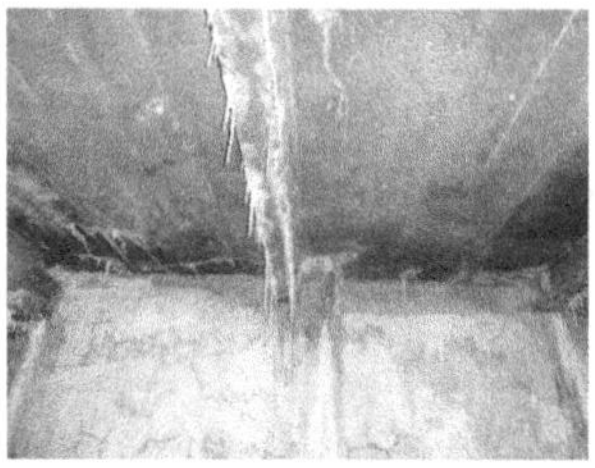	Steel beam supporting a concrete slab structure used as a sidewalk subject to water and salt infiltration.	Provide proper coating on the steel; provide a layered slab system with waterproofing that allows water to channel away from the slab.
  	Steel column corroded at the base in an interior condition but in an environment with chemicals.	Provide proper coating on the steel; elevate the column base above the ground level.
	Steel beams and decking support an interior concrete slab structure used as a garage with vehicle traffic subject to water and salt infiltration.	Provide proper coating on the steel; provide a layered slab system with waterproofing that allows water to channel away from the slab. It is recommended to periodically inspect the beam/girder-to-column connections for corrosion.
	Steel beams and decking support an interior concrete slab structure with foot traffic subject to water and salt infiltration.	Provide proper coating on the steel; provide a layered slab system with waterproofing that allows water to channel away from the slab. It is recommended to periodically inspect the beam/girder-to-column connections for corrosion.

	Steel column within an exterior wall encased by brick veneer.	Provide proper coating on the steel; provide a moisture barrier between the veneer and steel.
	Steel column corroded at the base in an exterior condition.	Provide proper coating on the steel; elevate the column base above the ground level.
	Steel framing corrosion in an exterior condition.	Provide proper coating on the steel.

14-13 INTRODUCTION TO STRENGTHENING AND RETROFITTING OF EXISTING STEEL STRUCTURES

Structural rehabilitation (i.e., additions and renovations) of existing buildings are common during the life of a structure. These modifications are often made to accommodate the changing use or occupancy category of a building. For example, a building addition in which the new building is taller than an adjacent existing building would create a condition in which snow can accumulate on the lower roof of the existing building. The original roof of the existing lower building would not likely have been designed for this additional snow load and would have to be investigated for structural adequacy. Another common change in the use of a roof structure occurs due to the addition or replacement of new equipment that is supported by the roof framing. This would also require an investigation into the existing roof framing for structural adequacy. Occupancy changes to floor structures are common as well, such as the addition of a high-density file storage system to a portion of a floor. In any occupancy change, both strength and serviceability must be checked, and in some cases, serviceability can be the primary concern (e.g., when a structure must support deflection-sensitive elements or if the structure is intended to support dynamic loads from equipment or human occupants).

Chapter 34 of the IBC [12] provides direction for repairs, additions, and occupancy changes to existing structures. In general, modifications to existing structures must comply with the design requirements for new construction, and any modification should not cause any part of the existing structure to be in violation of the requirements for new construction and should not cause distress in the existing structure. Unaffected portions of the existing structure are not required to comply with the current code requirements unless an existing member is found to be structurally inadequate, in which case it must be modified or replaced to conform to the current code. Another key provision is that modifications to an existing structure shall not increase the stress on an existing member by more than 5% unless the member can adequately support the increased stress, and the strength of any existing member shall not be decreased by more than what is required for new construction. An existing structure does not have to be investigated for seismic loads if the addition to the structure is seismically independent of the existing structure. If this is not the case, Section 806 of the International Existing Building Code (IEBC) requires that the existing vertical lateral force resisting systems (LFRS) and the addition must conform to the requirements for new construction, unless the following conditions are met:

- The gravity demand-to-capacity ratio due to the addition does not exceed 5% (i.e., $\frac{Q_u}{\phi R_n} \leq 1.05$).
- The lateral demand-to-capacity ratio for the vertical LFRS due to the addition does not exceed 10% (i.e., $\frac{Q_u}{\phi R_n} \leq 1.10$).

Once a new occupancy or proposed structural modification is defined, an investigation of the existing structural condition is required. If the existing construction drawings are available, the investigation is simpler; however, drawings for older structures are often not available, and a more extensive field investigation is required. For any structural modification, a field investigation is usually necessary to help assess practical issues, such as the constructability of an addition or alteration. There are often physical barriers to structural modifications, such as mechanical equipment, lights, ducts, and other similar elements.

Once a structure is examined for adequacy, the nature of the required reinforcement can be selected. For floor or roof systems that require a relatively large increase in load-carrying capacity, the addition of new intermediate beams between existing beams may be required, as shown in Figure 14-27. This scheme also decreases the span of the deck system and therefore increases the load-carrying capacity of the slab system, which may be desirable in some cases. Note that the constructability of this scheme must be carefully considered. There will likely be lights, ducts, and other physical obstructions that might make the placing of new beams difficult or even impossible. Additionally, it may not be possible to install the new beams in one piece; a beam splice may be needed (see Figure 11-11).

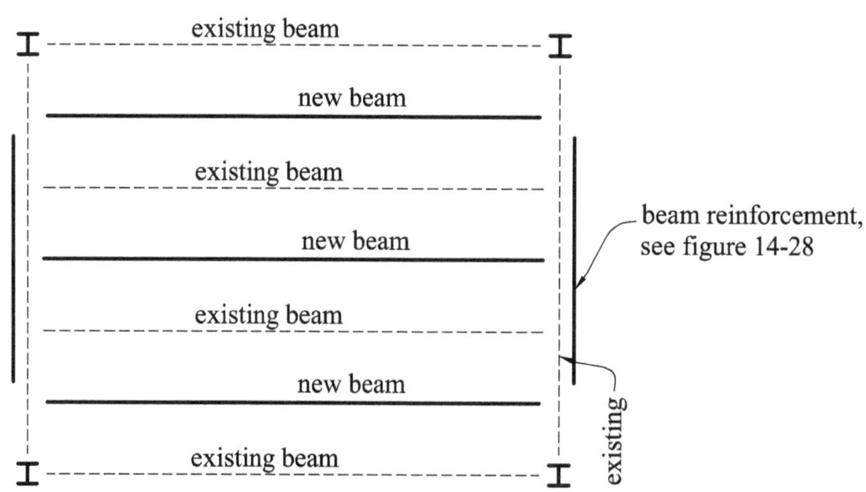

FIGURE 14-27 Additional framing to increase load capacity

The girders in Figure 14-27 would probably require reinforcement to support the new design loads. Figure 14-28 shows several possible beam reinforcement details to increase the flexural strength of a beam section. In each case, the intent is to increase the section properties by adding more steel near the flanges or by increasing the depth of the section. Noncomposite beams usually need reinforcement at both flanges; however, it is not usually possible to add reinforcement at the top flange of a beam or girder in an existing roof or floor system due to the presence of the existing steel deck. Composite beams usually only require reinforcement at the bottom flange since additional compression stresses can often be resisted by the concrete slab. Field welds should be installed by certified welders, and the position of the field weld should be considered when selecting a reinforcement detail. Note that in the details in Figures 14-28a, 14-28c, and 14-28d, the reinforcing steel is wider than the flange of the existing beam, which allows for the desirable downward field weld. To avoid overhead welding, the use of a wider steel section as reinforcement in these cases would likely be preferred regardless of the increase in material cost.

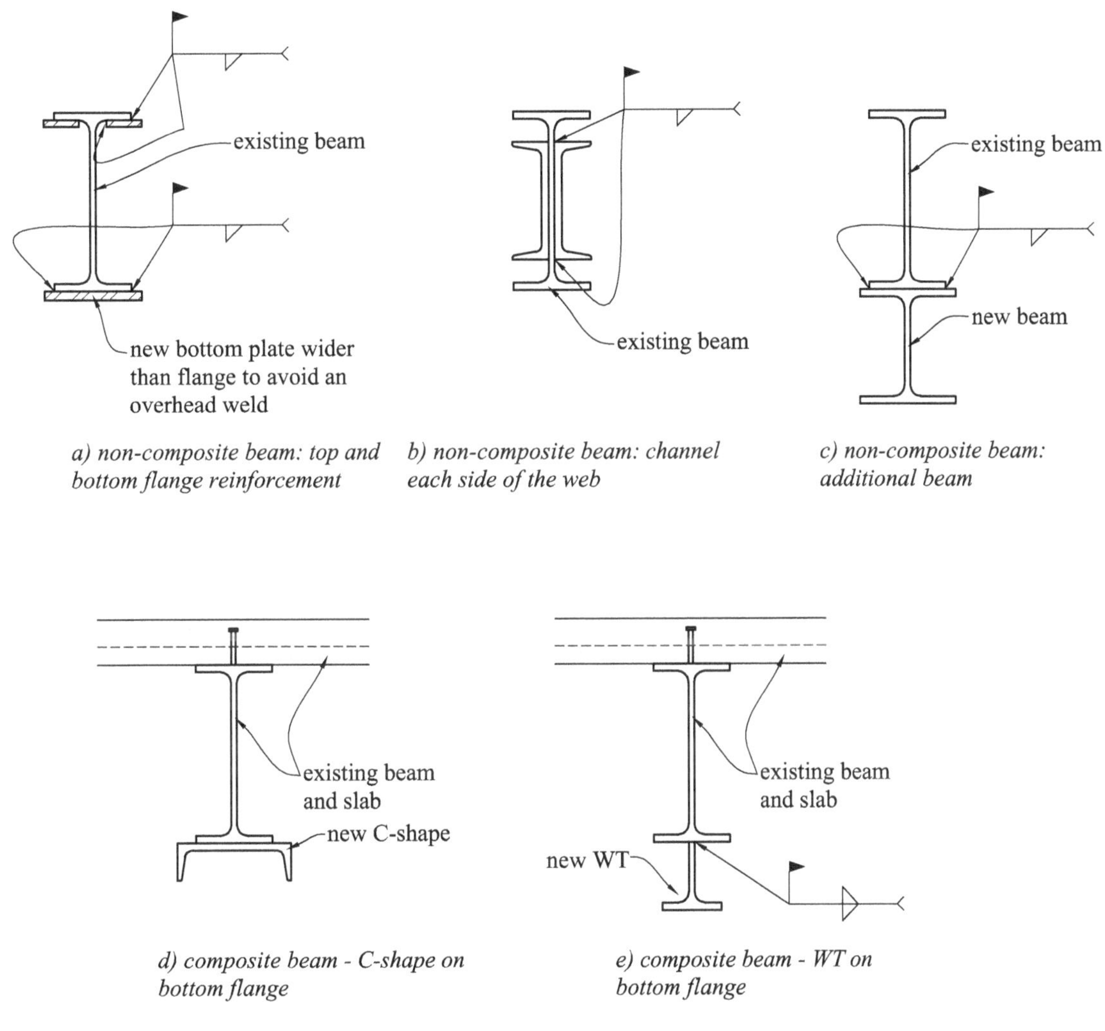

FIGURE 14-28 a to e Flexural reinforcement of beams

Noncomposite and partially composite floor beams can also be reinforced to increase their degree of composite action and therefore their load-carrying capacity, as shown in Figure 14-29. To accomplish this for noncomposite beams and girders, round holes are core-drilled into the slab at intervals along the length of the beam to allow for the placement of shear studs. The size of each hole must be such that the code-required clearances are provided around the shear stud (see Chapter 7). A shear stud is then placed in each hole, and the holes are filled with shrinkage-compensating cementitious (SCC) grout. It is possible to increase the degree of composite action of an existing partially composite beam by adding new shear studs, but it is difficult to accomplish in this case because the location of the existing shear studs may not be easily ascertained.

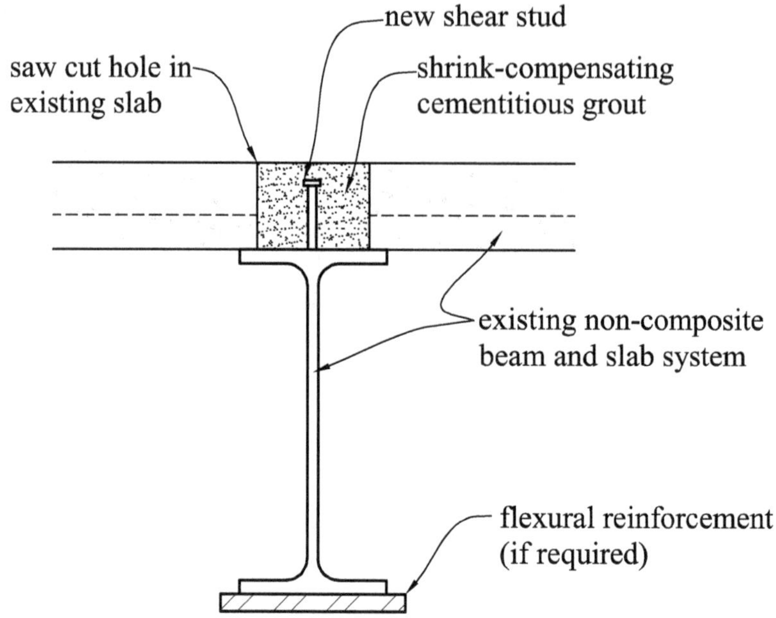

FIGURE 14-29 Reinforcement of a non-composite beam

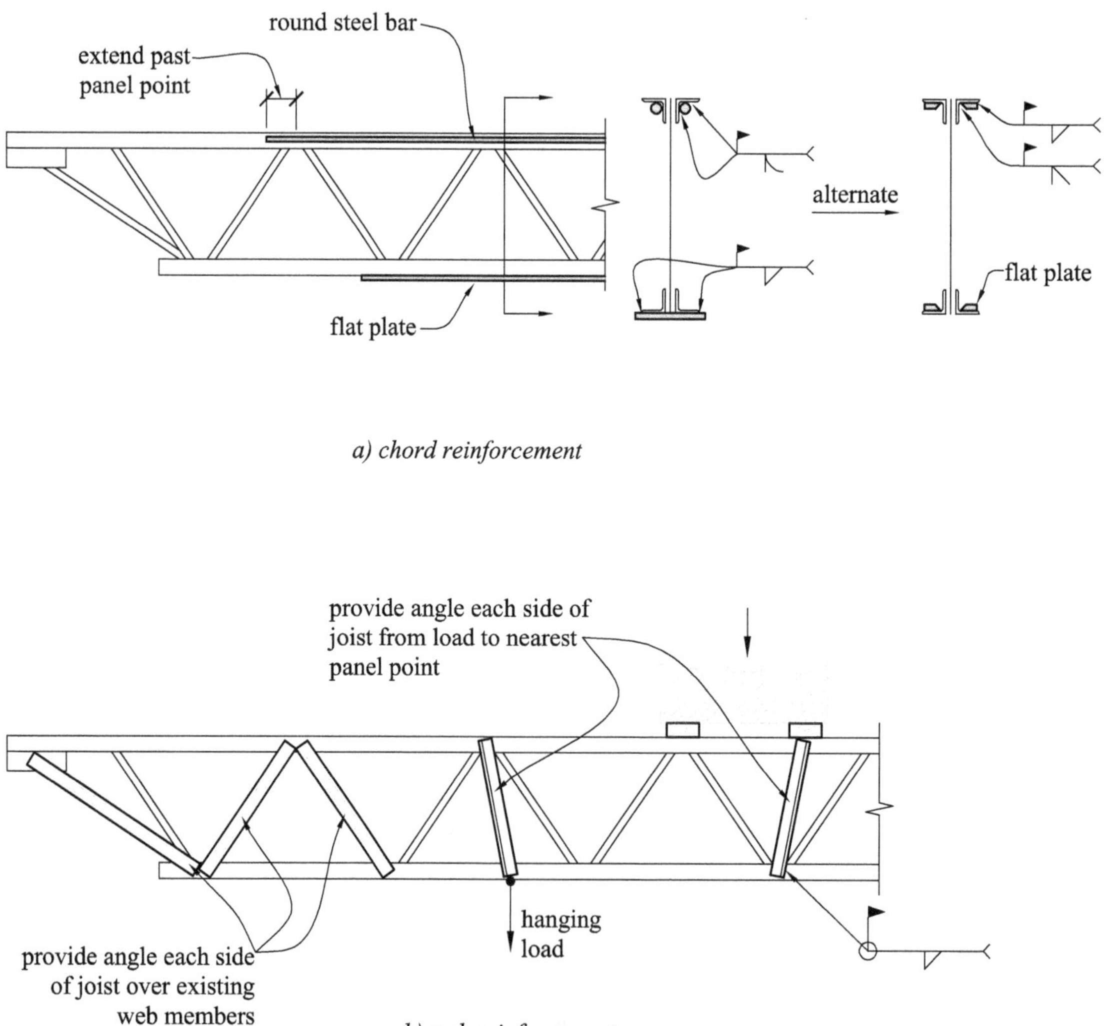

FIGURE 14-30 a, b Flexural reinforcement of steel joists

Open-web steel joists can also be reinforced to increase their load-carrying capacity. The investigation of these members is usually more difficult because the individual web and chord shapes can usually only be determined by field investigation. Furthermore, the geometry of the truss members likely has joint eccentricities that must be accounted for in the analysis.

Figure 14-30 shows common joist reinforcement methods. In Figure 14-30a, round bars are added near the top and bottom chords to increase the flexural strength and stiffness of the overall open-web steel joist. The web members may also require reinforcement to support additional loads. In Figure 14-30b, new web members are added adjacent to existing web members to increase the axial load capacity of the existing 4eb member. Figure 14-30b also shows the reinforcement of a joist when a concentrated load is applied away from a panel point. When loads are applied away from panel points, the joist chords will be subjected to bending, but the section properties and therefore the flexural strength of the chords are such that they cannot support any significant concentrated loads (200 lb. is a common upper limit on the concentrated loads that can be supported away from a panel point). In Figure 14-30b, a new web member is added from the location of the concentrated load to the nearest panel point at the top chord of the joist to ensure that the concentrated load is resisted by direct tension in a web member rather than by bending of the joist chord member. Note that some engineers prefer not to reinforce joists by welding new members to the joist because joists are pre-engineered structural members. Other retrofitting methods, such as adding additional joists or additional wide flange beams and columns, might be preferred to avoid welding to pre-engineered members.

When beam or girder members are reinforced to support additional loads, it is also important to consider the strength of their connections to the supporting members and the strength of any other elements along the load path. Two possible shear reinforcement details are shown in Figures 14-31a and 14-31b. In these details, welds are added to existing bolted connections, which is a common method of reinforcing existing connections. The AISC Specification J1.8 provides guidance on the design of connections that combine both welds and bolts in the same shear plane. Unless the bolts are pretensioned, the connection strength should only be based on the larger of either the longitudinal weld length alone or the bolt shear strength alone. Ref. [16] also provides some background on the use of combined bolt and weld details. One important parameter for combined welded and bolted connections is the use of welds that are oriented longitudinally with respect to the load direction. Test results have shown that when bolts are loaded in combination with transverse welds, the contribution of bolt shear strength is negligible, but when bolts are loaded in combination with longitudinal welds (i.e., welds that are parallel to the load direction), the contribution of bolt shear is approximately 75% of the ultimate strength of the bolts [16, 20].

In lieu of adding additional steel components to an existing connection, it may be possible to completely remove and replace the existing connection or some selected components of the existing connection, such as the replacement of A325 bolts with A490 bolts. An existing welded connection could also be reinforced by the placement of additional weld length or by increasing the weld size.

The end connections of open-web steel joists can be reinforced by the addition of vertical stiffeners at the bearing points, as shown in Figure 14-31c.

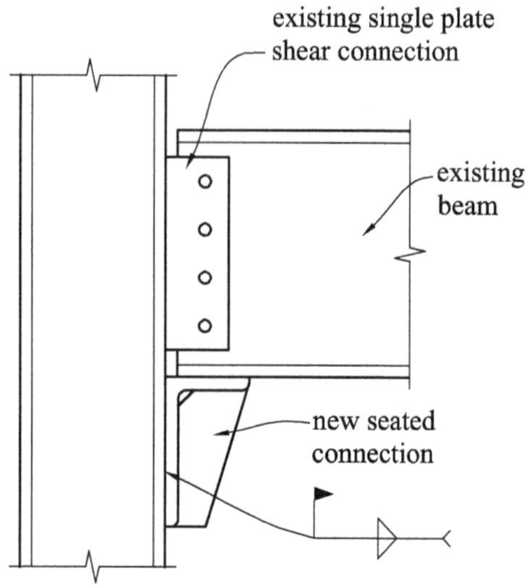

a) seat connection reinforcement

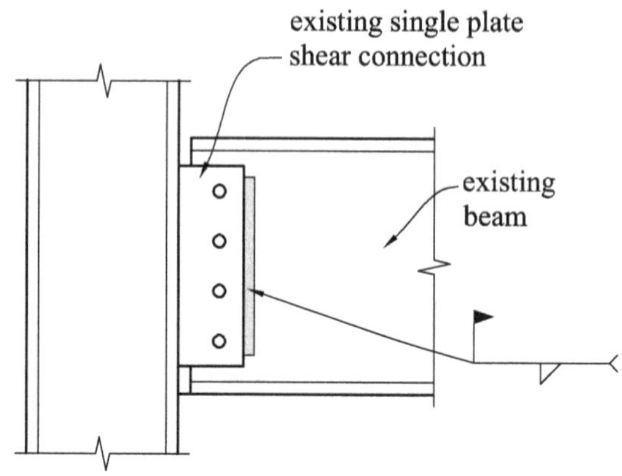

b) welded reinforcement or existing bolted connection

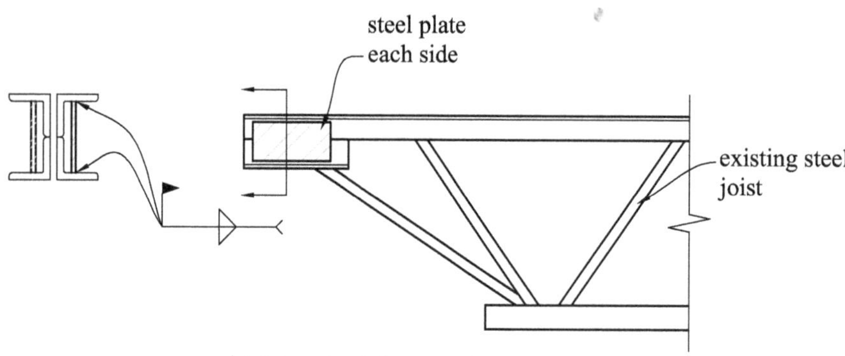

c) joist seat reinforcement

FIGURE 14-31a to c Reinforcement of connections

The load-carrying capacity of columns can be increased by the addition of steel sections, similar to the reinforcement for beams. Figures 14-32a, 14-32b, and 14-32c show several possible column reinforcement details. The addition of steel sections to an existing column will increase the column design strength because of the additional area of steel and because the location of the steel sections could increase the radius of gyration, thus decreasing the column slenderness ratio, leading to an increase in the design compressive strength. When existing steel columns are reinforced by welding additional steel members to the column, the heat generated from the welding process can lead to a reduction in the load-carrying capacity of the existing column, which may necessitate shoring of the floor or roof framing supported by the column while the remedial work is in progress. The effect of the welding process on the column capacity can be accounted for by using a reduced cross-section that considers the heated parts of the column as inactive in resisting loads. Some ways to mitigate the effect of the heat from the welding process include using low-heat welding processes, using multiple-pass welds and allowing time for the welds to cool between passes, and using intermittent fillet welds instead of continuous welds [42].

One common issue with older steel structures is the weldability of the existing material. Steel replaced wrought iron as a building material after about 1900, but the chemical and

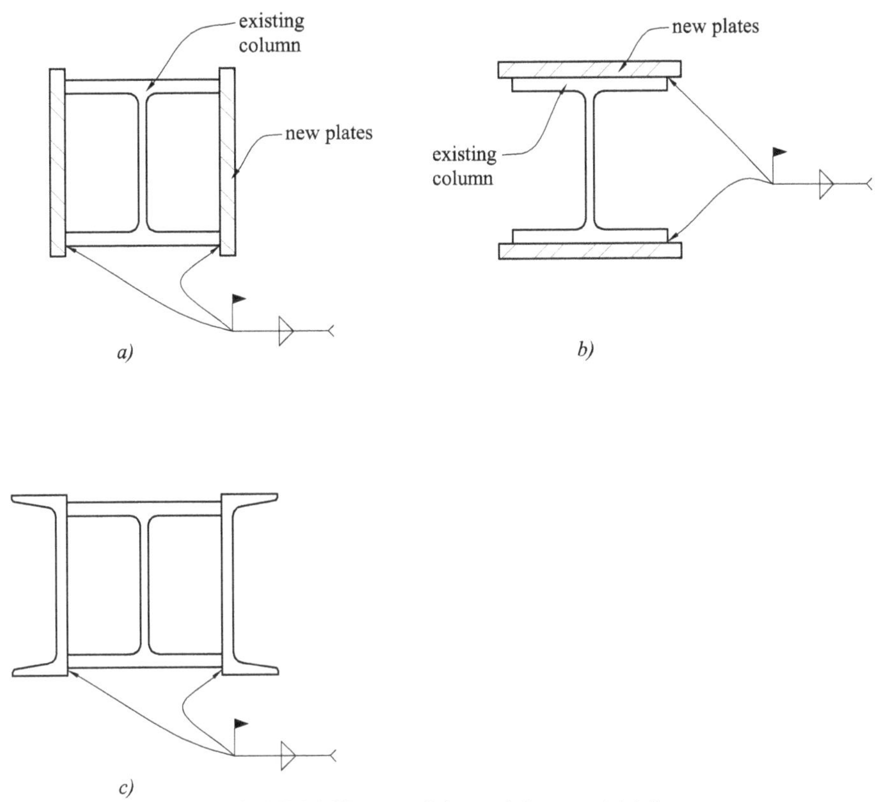

FIGURE 14-32a to c Column reinforcement details

mechanical properties were not initially standardized, thus some of the steel grades were weldable while others were not. One could quickly determine the weldability of an unknown steel section by looking for existing welds on the member, which would indicate that the steel was weldable. From about 1910 to 1960, most steel used in buildings was ASTM A7 or A9, which is a medium-carbon steel and is generally weldable; however, it has a relatively wide range of acceptable carbon content and should therefore be tested prior to welding (see Ref. [17] for the test method). After about 1960, ASTM A36 and A572 were primarily used in building construction, and these steel grades are weldable.

Sometimes, the retrofitting or structural renovation of an existing building might lead to an increase in the net uplift tension force transferred at the base of the existing columns of braced frames or moment frames. The detail in Figure 14-32d shows an existing steel column that is reinforced to transfer an increased uplift force to the substructure.

Another important consideration in the reinforcement of existing members is the use of shoring to relieve existing stresses. When existing members are shored, the strength of the reinforced section is maximized, but it may not be possible in some cases to use shoring. Columns are usually not shored prior to reinforcement because it has been shown that the addition of shoring does not significantly increase the load-carrying capacity of a reinforced section [18].

The design of the shoring and reinforcement for existing composite beams must consider several factors. If the existing structure was constructed with shoring prior to the placement of the concrete slab, then shoring can be used to reinforce the existing member since the dead and live loads on the member were applied after the concrete hardened. If the existing structure was not constructed without shoring prior to the placement of the concrete slab, then it is not recommended to use shoring while adding the reinforcement. When reinforcement is required for an unshored floor system, then any unnecessary dead and live loads should be removed from the floor area while the reinforcement is being installed.

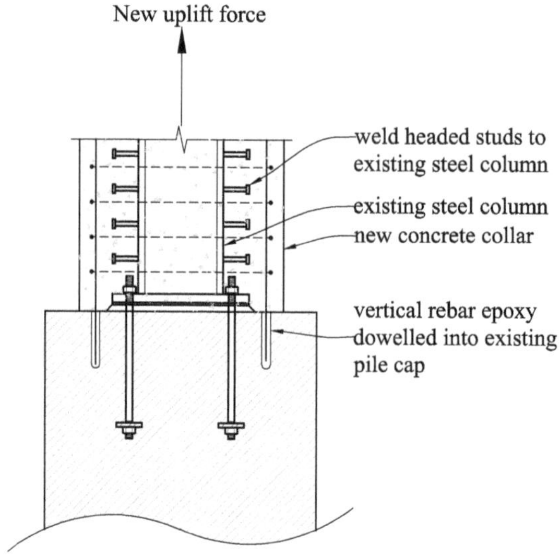

FIGURE 14-32d Transfer of increased uplift force at existing column

14-14 ELASTIC SHORTENING OF COLUMNS IN TALL STEEL BUILDINGS

The axial deformation of steel columns in tall buildings can result in differential vertical deformation between the exterior steel columns and the interior concrete core walls because the core walls usually have a lower axial stress and are used predominantly to resist lateral wind and seismic loads. Therefore, the axial deformation of the interior concrete walls will be much smaller than the axial deformation in the exterior steel columns. In fact, if all the vertical elements in a building (i.e., the columns and the core walls) had the same axial deformation, then the axial deformation would cause little or no problem; it is the *differential* axial deformation of the exterior steel columns relative to the interior concrete core walls that causes problems. This becomes even more acute with the use of high-strength steels (e.g., ASTM A913 Grade 65) for columns in tall buildings with large bay sizes (e.g., 40 ft × 50 ft bay size) to resist higher axial loads using standard rolled column sections—a more economical option than using built-up column sections [35]. The use of higher-strength steels results in higher axial stress demand in the perimeter steel columns and thus higher column axial shortening, which results in higher differential axial shortening. Similarly, the taller the building, the greater the effects of axial shortening, as is evident from the equation for linear elastic shortening of columns, PL/AE, where L is the length of the column. Note that linear elastic behavior is assumed in the steel columns under service loads. Nevertheless, for concrete columns or walls, inelastic behavior due to the stress levels and the effect of creep and shrinkage complicates the calculation of their vertical shortening. Differential vertical shortening in tall structures could induce unintended stresses in the beam and girder members and their connections, which could also result in the misfit of the exterior cladding. Differential axial shortening could also lead to a leaning building as the exterior columns shorten more than the interior concrete core walls, creating serviceability problems such as uneven or sloping floors [36]. To mitigate the effects of elastic shortening of columns in tall buildings, the lengths of the exterior steel columns in tall buildings could be fabricated longer than the theoretical floor-to-floor height. That is, the exterior columns are overbuilt at every floor level by the expected amount of axial shortening per floor. In one

18-story office building, the lengths of the perimeter steel columns were increased by 1/16 inch per floor [34], and in a 450-ft tall structural steel office building with 31 stories, the lengths of the perimeter steel columns needed to be increased by 1/8 inch per floor, resulting in an increase in column length of 1-in. for every eight floors, because of the higher column axial stress and to mitigate the effect of differential axial shortening [35]. Note that the actual elastic shortening of a vertical steel member in building structures has been found to vary between 70% and 100% of the predicted elastic shortening because the simple elastic models do not account for all the elements in the structure; thus, the structure may be stiffer than indicated by the model [36]. The axial deformation of a column in a multi-story building can be calculated as follows from the well-known formula from the mechanics of materials:

$$\Delta_{axial} = \frac{PL}{AE} = \frac{P}{A}\left(\frac{L}{E}\right) = f_{axial}\left(\frac{L}{E}\right)$$

The cumulative axial deformation in a column is

$$\Delta_m = \Delta_{m-1} + \frac{\left(\sum_{i=m}^{n} P_i\right) L_m}{A_m E}$$

Where,

$\sum_i^n P_i$ = cumulative axial load on the column between level i and level $i-1$ from floor level i to the roof level, n.

L_m = length of column between level m and level $m-1$ A_m = Cross-sectional area of the column between level m and level $m-1$

At the first framed floor level of the building, that is, level 2 or the 2nd floor level, m = 2; level 1 is assumed to be the slab-on-grade level or the lowest unframed floor level, while n is the roof level. Note that if level 1 is a slab-on-grade, there is no axial deformation at that level, thus $\Delta_1 = 0$. Therefore, the axial deformation of the column between level 2 and level 1 is

$$\Delta_2 = \Delta_1 + \frac{\left(\sum_{i=2}^{n} P_i\right) L_2}{A_2 E} = \frac{\left(\sum_{i=2}^{n} P_i\right) L_2}{A_2 E}$$

Where, L_2 is the length of the column between level 2 and level 1, and A_2 is the cross-sectional area of the column between level 2 and level 1; E is the modulus of elasticity. The cumulative axial deformation of the upper-level columns is as follows:

$$\Delta_3 = \Delta_2 + \frac{\left(\sum_{i=3}^{n} P_i\right) L_3}{A_3 E}$$

$$\Delta_4 = \Delta_3 + \frac{\left(\sum_{i=4}^{n} P_i\right) L_4}{A_4 E}$$

$$\Delta_n = \Delta_{n-1} + \frac{\left(\sum_{i=n}^{n} P_i\right) L_n}{A_n E}$$

Where,

$\sum_{i=n}^{n} P_i$ = Axial load on the column between level n and level $n-1$

L_n = length of column between level n and level $n-1$

A_n = Cross-sectional area of the column between level n and level $n-1$

EXAMPLE 14-6

Elastic Axial Shortening in Steel Columns

A 10-story steel-framed building has a typical bay size of 30 ft × 30 ft, with a roof dead load of 30 psf and a typical floor dead load of 100 psf. The story height is 12 ft, and the lowest floor level (level 1) is an unframed slab-on-grade. Calculate the elastic axial shortening in a typical interior column at each level.

Solution:

The solution is presented in a tabular format below using the formula for axial deformation presented previously:

Level	Cumulative service axial dead load on column, kips $\sum_{i=m}^{n} P_i$	Column size	Column cross-sectional area, in.^{2}	Story height, in.	Column axial deformation or shortening, in.	Required correction to column length, in.
n (i.e., roof)	27	W14 × 43	12.6	144	0.714	0.011
10	117	W14 × 43	12.6	144	0.703	0.046
9	207	W14 × 61	17.9	144	0.657	0.057
8	297	W14 × 61	17.9	144	0.600	0.082
7	387	W14 × 90	26.5	144	0.518	0.073
6	477	W14 × 90	26.5	144	0.445	0.089
5	567	W14 × 120	35.3	144	0.356	0.080
4	657	W14 × 120	35.3	144	0.276	0.092
3	747	W14 × 145	42.7	144	0.184	0.087
2	837	W14 × 145	42.7	144	0.097	0.097

$$\Delta_2 = \Delta_1 + \frac{\left(\sum_{i=2}^{n} P_i\right) L_2}{A_2 E} = 0 + \frac{(837)(144)}{(42.7)(29{,}000)} = 0.097\,\text{in.}$$

$$\Delta_3 = \Delta_2 + \frac{\left(\sum_{i=3}^{n} P_i\right) L_3}{A_3 E} = 0.097 + \frac{(747)(144)}{(42.7)(29{,}000)} = 0.184\,\text{in.}$$

$$\Delta_4 = \Delta_3 + \frac{\left(\sum_{i=4}^{n} P_i\right) L_4}{A_4 E} = 0.184 + \frac{(657)(144)}{(35.3)(29{,}000)} = 0.276\text{ in.}$$

$$\Delta_5 = \Delta_4 + \frac{\left(\sum_{i=5}^{n} P_i\right) L_5}{A_5 E} = 0.276 + \frac{(567)(144)}{(35.3)(29{,}000)} = 0.356\text{ in.}$$

$$\Delta_6 = \Delta_5 + \frac{\left(\sum_{i=6}^{n} P_i\right) L_6}{A_6 E} = 0.356 + \frac{(477)(144)}{(26.5)(29{,}000)} = 0.445 \text{ in.}$$

$$\Delta_7 = \Delta_6 + \frac{\left(\sum_{i=7}^{n} P_i\right) L_7}{A_7 E} = 0.445 + \frac{(387)(144)}{(26.5)(29{,}000)} = 0.518 \text{ in.}$$

$$\Delta_8 = \Delta_7 + \frac{\left(\sum_{i=8}^{n} P_i\right) L_8}{A_8 E} = 0.518 + \frac{(297)(144)}{(17.9)(29{,}000)} = 0.6 \text{ in.}$$

$$\Delta_9 = \Delta_8 + \frac{\left(\sum_{i=9}^{n} P_i\right) L_9}{A_9 E} = 0.6 + \frac{(207)(144)}{(17.9)(29{,}000)} = 0.657 \text{ in.}$$

$$\Delta_{10} = \Delta_9 + \frac{\left(\sum_{i=10}^{n} P_i\right) L_{10}}{A_{10} E} = 0.657 + \frac{(117)(144)}{(12.6)(29{,}000)} = 0.703 \text{ in.}$$

$$\Delta_n = \Delta_{n-1} + \frac{\left(\sum_{i=n}^{n} P_i\right) L_n}{A_n E} = \Delta_{10} + \frac{\left(P_{roof}\right) L_{roof}}{A_{roof} E} = 0.703 + \frac{(27)(144)}{(12.6)(29{,}000)} = 0.714 \text{ in.}$$

-15 STRUCTURAL INTEGRITY REQUIREMENTS

The structural integrity provisions from *AISCM* Section B3.9 are minimum connection design requirements that pertain to column splices, beam and girder end connections, and the bracing connections for columns. The goal is to prevent connection failure and disproportionate or progressive collapse when a structure is subjected to abnormal or unanticipated loads that cause tension forces at these connections [20]. These provisions ensure a continuous load path and improved ductility under extreme load events. The forces needed to ensure structural integrity are applied only to the connections and not to the structural members, so the structural integrity provisions are connection design requirements only, and they should be considered separately and distinct from all other load cases. The unanticipated tension forces may be caused by blast loading or the failure of a portion of the structure that may result in overload on other parts of the structure, or from cooling of a structure after a fire. These structural integrity provisions should only be considered when required by the applicable local building code, such as required by the IBC for highrise buildings in Risk Categories III and IV, or if they are requested by the owner of the building who desires better or improved structural performance under abnormal or unanticipated loads.

1. **Column splices**: The nominal tensile strength of column splices should be greater than or equal to $P_D + P_L$, where P_D is the service dead load and P_L is the service live load on the column based on the tributary area of the column between the splice under consideration and the splice or base immediately below. If, as is usually the case, there is a splice every two floors, then the tributary area will be the tributary area for two floors supported by the column (see Figure 14-32e).

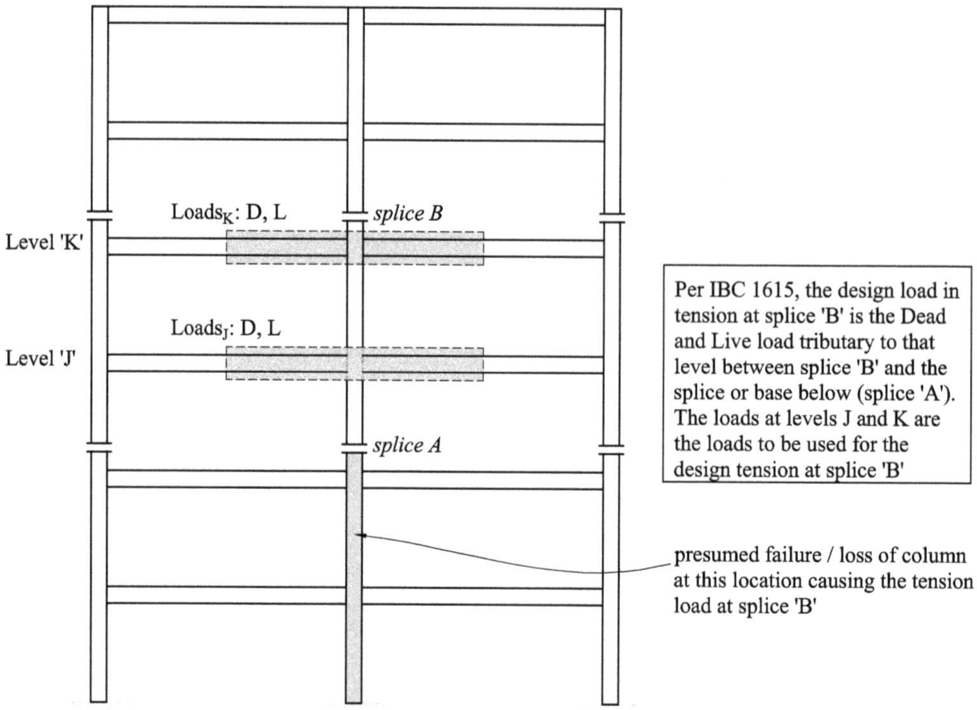

FIGURE 14-32e Column splice design load

2. **Beam and girder end connections:**

   LRFD: Nominal axial tension strength $\geq \frac{2}{3} V_u \geq 10\,\text{kips}$,

   where, V_u = required vertical shear strength or factored vertical shear at the beam or girder connection.

   ASD: Nominal axial tension strength $\geq V_{D+L} \geq 10\,\text{kips}$.

3. **End connections of lateral braces for columns:**

   LRFD: Nominal axial tension strength $\geq \dfrac{\frac{2}{3}P_u}{100}$,

   ASD: Nominal axial tension strength $\geq \dfrac{P_{D+L}}{100}$,

   where,

   P_u is the factored load, and

   P_{D+L} is the unfactored load on the column at the level where it is laterally braced.

   Note that the above structural integrity requirements must be evaluated independently of all other strength requirements. That is, these requirements are additional to and separate from other strength requirements [20].

In addition to the structural integrity provisions in the *AISC* Specifications, ASCE 7-22 Section 1.4 also prescribes general structural integrity requirements applicable to any structure with an emphasis on the continuity of the lateral and gravity load paths and thus ensuring adequate connections along the load paths.

Load path connections

A continuous load path for the lateral force resisting system is required in any structure so that the connections along the load path are adequate to transfer the lateral forces. The connection of any smaller part of the structure to the remaining parts of the structure along the lateral load path should be designed to resist a minimum force of 5% of the weight of the smaller part (see ASCE 7 Section 1.4.1).

Roof and Floor Diaphragm connections to beams/girders

The connection between the supported roof and floor diaphragms and the supporting structural members such as beams, girders, or trusses must be designed to resist a minimum horizontal force parallel to the supporting member of 5% of the unfactored dead plus live load reaction from the supported element (i.e., roof or floor slab) on the supporting beam, girder, or truss (ASCE 7 Section 1.4.3).

Anchorage of structural walls to roof and floors

The connection of structural walls to roofs and floors shall be designed to resist a minimum factored horizontal force perpendicular to the surface of the wall equal to "0.2 times the weight of the wall tributary to the connection, but not less than 5 lb/ft^2" applied on the surface of the wall (see ASCE 7 Section 1.4.4).

Minimum lateral force

The minimum lateral force, F_x, at each level of a structure shall not be less than $0.01W_x$, where W_x is the total dead load of the structure assigned to level x (see ASCE 7 Section 1.4.2).

Progressive or disproportionate collapse prevention

Progressive or disproportionate collapse is the collapse of an entire structure precipitated by the failure of a single critical element or component. For example, what would happen to a structure if a building column or bridge pier were taken out? Will the entire structure collapse? ASCE 7 does not define precisely what the resistance to progressive collapse is, but some federal, state, and local agencies like the General Services Administration and the Department of Defense (DOD) have developed policies that ensure a robust structure—that is, a structure that has the ability to survive the removal of a key load-bearing structural element (e.g., the corner column in a multi-story building such as that shown in Figures 14-32f and 14-32g) due to an abnormal hazard event (e.g., vehicular impact) by load redistribution through alternate load paths, such that only local failure occurs and a global failure of the entire structure is prevented or avoided (see ASCE 7 Section 2.5).

Progressive or disproportionate collapse prevention should be considered for signature structures or structures with high occupancy whose collapse will be catastrophic to the community. More information on progressive or disproportionate collapse prevention can be found in Ref. [25, 57, 58, 59].

FIGURE 14-32f A corner column in a multi-story steel-framed building under construction, located close to a major road

FIGURE 14-32g A corner column in a completed multi-story steel-framed building located close to a major road

14-16 STRUCTURAL DRAWING NOTES, AND SAMPLE ROOF/ FLOOR PLANS AND DETAILS

In this section, some typical plans and details, and sample general notes that are included on the structural drawings are presented; in addition to the plans and details, the general notes are used to convey information to the contractor regarding the materials and other specifications for the structural members. Figure 14-33 shows sample general notes pertaining to structural steel, steel deck, and open-web steel joist members. These general notes, which are usually presented on the first structural drawing together with the typical details, provide an additional opportunity for the engineer of record (EOR) to specify additional information such as the following:

- The code used in the design of the structure.
- The required material properties.
- The coating for the structural steel.
- The requirements for the design of the steel connections.
- The design requirements and the loads for steel joists and steel connections that will be designed by the contractor.
- The delegated designs are to be carried out by the contractor, such as the simple shear connections, the open-web steel joists, the diagonal bracing connections, and sometimes the moment connections.

To aid the reader in completing the student design projects presented in the next section, some sample roof and floor framing plans together with a set of plan notes and lateral bracing elevations and moment connection details are provided in Figures 14-34 through 14-37. The plan notes are used by the EOR to specify requirements that may be specific to a floor or roof, such as the top of the steel elevation, the slab or deck specification, loading for the open-web steel joists, and so on. The roof and floor plans illustrate how structural information obtained from the structural design is presented in a form that the contractor can use to erect the structure. The reader is reminded that while quality engineering calculations are necessary, the design contained within those calculations needs to be properly conveyed to the contractor in the form of drawings and notes as well as the project specifications. It is imperative that the structural drawings and structural notes, together with the material specifications, convey to the contractor and steel fabricator all the information necessary to build the structure—in a way that allows for efficiency in the construction process and the safe functioning of the structure. The drawings also need to have ample dimensions and details to avoid delays and confusion during construction. When certain designs are delegated to the contractor, such as the steel connections, it is also important that the drawings and notes provide the design loading and other design information—such as beam and girder end reactions, dimensional constraints and material properties—to avoid delays during construction due to the time it takes to generate and process the request for information (RFI's) from the contractor.

The drawings, plan notes, lateral bracing elevations, and moment connection details shown in Figures 14-33 through 14-37 are provided as examples to expose the reader to some of the basic types of information needed on structural drawings for typical structural steel design projects in engineering practice.

STRUCTURAL STEEL NOTES:

1.) STRUCTURAL STEEL SHALL CONFORM TO THE AISC CODE OF STANDARD PRACTICE - 2022

2.) STRUCTURAL STEEL ARE TO BE AS FOLLOWS GRADES:
- A) STRUCTURAL STEEL (W-, C-): ASTM A572 OR A992, Fy=50ksi
- B) STRUCTURAL STEEL (M-, S, L-, PLATES): ASTM A36, Fy=36ksi
- C) STRUCTURAL TUBING (RECT.): ASTM A500, Fy=46ksi
- D) STRUCTURAL TUBING (RND.): ASTM A500, Fy=42ksi
- E) STRUCTURAL STEEL PIPE: ASTM A53 GRADE B, Fy=35ksi
- F) BOLTS: ASTM A325N OR F1852, PRETENSIONED
  - i.) BOLTS IN BRACED FRAME CONNECTIONS AND MOMENT CONNECTIONS ARE TO BE SLIP CRITICAL
- G) ANCHOR RODS: F1554, GRADE 36
- H) SHEAR CONNECTORS: ASTM A108, Fy = 50ksi
- I) CLEVISES AND TURNBUCKLES: AISI C-1035
- J) RAISED FLOOR PLATE: ASTM A786
- K) WELDS: E70xx
- L) GALVANIZING:
  - i.) STRUCTURAL STEEL: ASTM A123
  - ii.) BOLTS, FASTENERS, HARDWARE: ASTM A153

3.) ALL STRUCTURAL STEEL SHALL BE COATED AS INDICATED BELOW. APPLY COATINGS IN ACCORDANCE WITH MANUFACTURERS RECOMMENDATIONS, INCLUDING SURFACE PREPARATIONS AND COMPATIBILITY REQUIREMENTS BETWEEN ALL COATINGS. AFTER ERECTION TOUCH UP ALL AREAS WHERE PAINT OR GALVANIZING IS MISSING OR DAMAGED INCLUDING FIELD WELDS.

A.)EXTERIOR EXPOSED STEEL AND LINTELS WITHIN EXTERIOR WALLS TO BE HOT-DIPPED GALVANIZED.

B.)ALL OTHER STRUCTURAL STEEL SHALL BE SHOP PAINTED WITH A MODIFIED ALKYD RUST INHIBITIVE PRIMER.

C.)EXTERIOR EXPOSED STEEL (WHERE INDICATED) TO BE PRIMED & PAINTED WITH A COATING SYSTEM RATED FOR EXTERIOR USE.

D.) THE FOLLOWING SURFACES ARE NOT TO BE COATED:
- i) THE TOP FLANGE OF BEAMS THAT RECEIVE DECKING.
- ii) CONNECTION MATERIAL AT SLIP CRITICAL JOINTS; SURFACES SHALL BE PREPARED WITH A WIRE BRUSH (SSPC-SP2).
- iii) SURFACES THAT ARE TO RECEIVE SPRAY-APPLIED FIRE-RESITIVE MATERIAL.

4.) BEAM CONNECTIONS SHALL BE DESIGNED FOR THE LARGEST OF THE FOLLOWING:
- i) END REACTIONS COMPUTED FROM THE TABLE "UNIFORM LOAD CONSTANTS" IN THE AISC MANUAL FOR BEAMS LONGER THAN 20ft
- ii) THE REACTIONS SHOWN ON THE PLANS
- iii) FACTORED LOAD OF 15k.

5.) SUBMIT SHOP DRAWINGS FOR STRUCTURAL STEEL FOR REVIEW PRIOR TO CONSTRUCTION.

FIGURE 14-33 Sample drawing notes

STEEL JOIST NOTES:

1.) STEEL JOISTS AND JOIST GIRDERS SHALL CONFORM TO THE STANDARD SPECIFICATIONS FOR JOISTS AND JOIST GIRDERS PUBLISHED BY THE STEEL JOIST INSTITUTE - 2020

2.) STEEL JOISTS AND JOIST GIRDERS SHALL BE DESIGNED BY THE MANUFACTURER. JOIST SUPPLIER TO PROVIDE CALCULATIONS AND DETAILS FOR EACH JOIST AND JOIST GIRDER. ALL CALCULATIONS SHALL BE SIGNED AND SEALED BY THE MANUFACTURERS ENGINEER AND SHALL BE SUBMITTED WITH THE SHOP DRAWINGS.

3.) THE MANUFACTURER SHALL DESIGN AND PROVIDE BRIDGING AND ATTACHMENTS IN ACCORDANCE WITH THE STEEL JOIST INSTITUTE SPECIFICATIONS, INCLUDING BRIDGING REQUIRED FOR UPLIFT.

4.) MINIMUM BEARING SHALL BE AS FOLLOWS:

K-SERIES:	2 1/2" ON STRUCTURAL STEEL 4" ON CONCRETE OR MASONRY
LH-SERIES:	4" ON STRUCTURAL STEEL 6" ON CONCRETE OR MASONRY
JOIST GIRDERS:	4" ON STRUCTURAL STEEL 6" ON CONCRETE OR MASONRY

5.) MINIMUM END ATTACHMENT SHALL BE AS FOLLOWS:

K-SERIES:	(2)-1/8"x1" FILLET WELDS OR (2)-1/2" BOLTS
LH-SERIES:	(2)-¼"x2" FILLET WELDS OR (2)-3/4" BOLTS
JOIST GIRDERS:	(2)-¼"x2" FILLET WELDS OR (2)-3/4" BOLTS

6.) TOP CHORD EXTENSIONS SHALL BE DESIGNED FOR THE UNIFORM LOAD CAPACITY OF THE MAIN JOIST.

7.) STEEL JOISTS AND JOISTS GIRDERS TO BE DESIGNED FOR A NET UPLIFT OF 10 psf.

8.) SEE STRUCTURAL STEEL NOTES FOR COATING REQUIREMENTS.

9.) SUBMIT JOIST CALCULATIONS AND SHOP DRAWINGS FOR REVIEW PRIOR TO CONSTRUCTION.

FIGURE 14-33 (Continued)

STEEL DECK NOTES:

1.) STEEL DECK CONSTRUCTION SHALL CONFORM TO THE MANUAL OF CONSTRUCTION AND CODE OF STANDARD PRACTICE FOR STEEL DECK PUBLISHED BY THE STEEL DECK INSTITUTE - 2023.

2.) STEEL DECK SHALL BE CONTINUOUS OVER 3 OR MORE SPANS UNLESS NOTED OTHERWISE.

3.) STEEL DECK SHALL BE GALVANIZED, G60.

4.) PERMANENT STEEL FORM DECK AND FLOOR DECK FASTENING SHALL CONFORM TO THE SDI RECOMMENDATIONS FOR FASTENING TO SUPPORTING MEMBERS.

5.) STEEL ROOF DECK SHALL BE FASTENED TO SUPPORTS AT 12" O.C. MAXIMUM AND 6" O.C. AT THE PERIMETER WITH 5/8" PUDDLE WELDS. SIDE LAPS SHALL BE FASTENED WITH #10 TEK SCREWS AT 18" O.C. OR 3 PER SPAN, WHICHEVER IS GREATER.

6.) ALTERNATE DECK FASTENING MAY BE SUBMITTED FOR REVIEW. CONTRACTOR TO SUBMIT ENGINEERING DATA INDICATING EQUIVALENCE OF THE ALTERNATE FASTENING TO THAT WHICH IS SPECIFIED.

7.) SUBMIT SHOP DRAWINGS FOR STEEL DECK PRODUCT DATA, DECK LAYOUT, AND DECK FASTENERS FOR REVIEW PRIOR TO CONSTRUCTION.

FIGURE 14-33 (Continued)

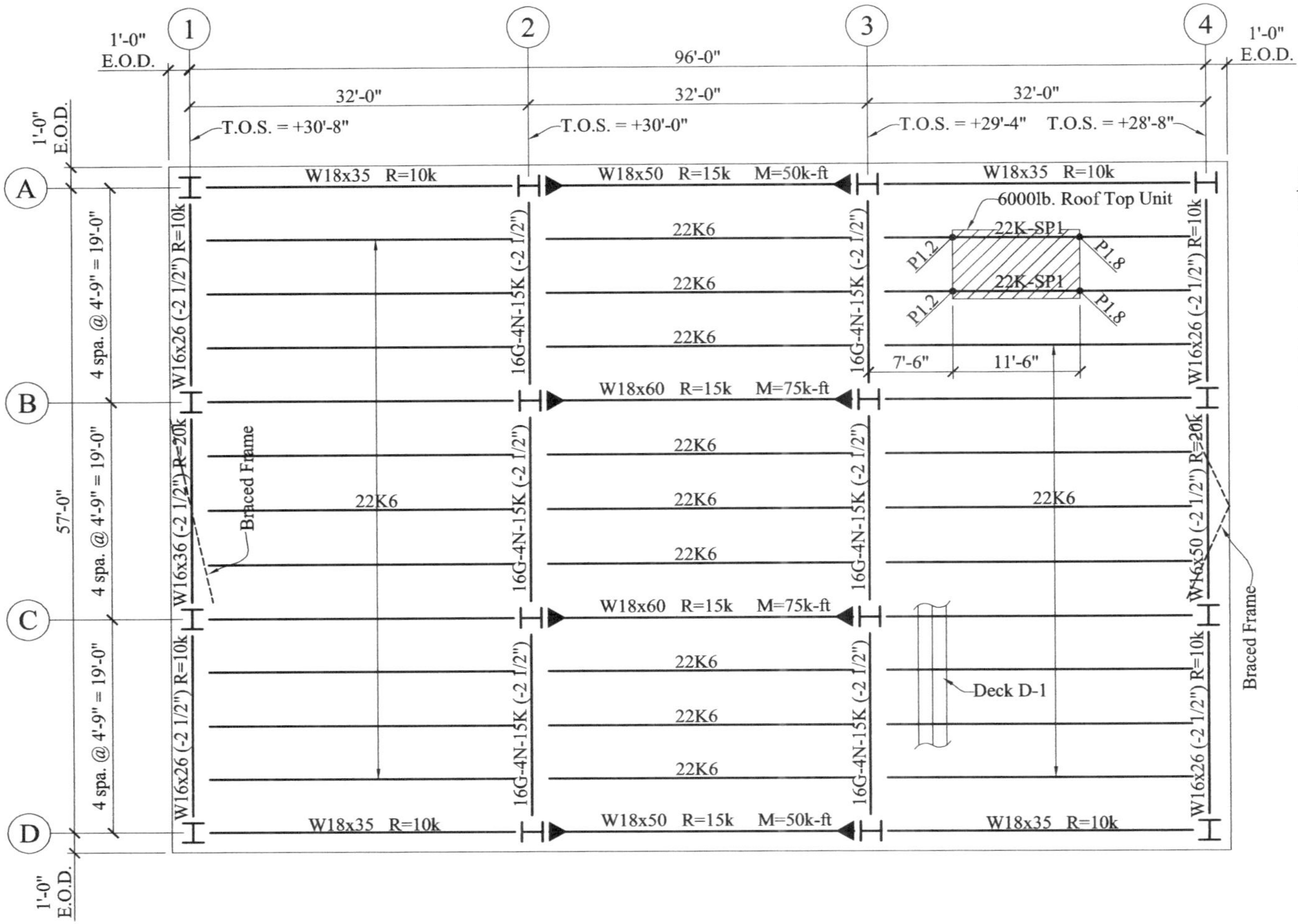

Roof Framing Plan

1/8" = 1'-0"

1. Top of steel (deck bearing) elevation is referenced from ground floor elevation of +0'-0". T.O.S. = top of steel.

2. Deck D-1: 1 1/2"-22ga, type 'B' roof deck.

3. Provide a 12ga bent plate at the perimeter of all deck edges. E.O.D. = edge of deck.

4. Special joist loading, 22K-SP1 is as follows:
   a) uniform load based on specified loads for 22K6
   b) concentrated loads on the top chord where indicated
      P1.8: service dead load of 1800lb
      P1.2: service dead load of 1200lb

4. Framing legend:

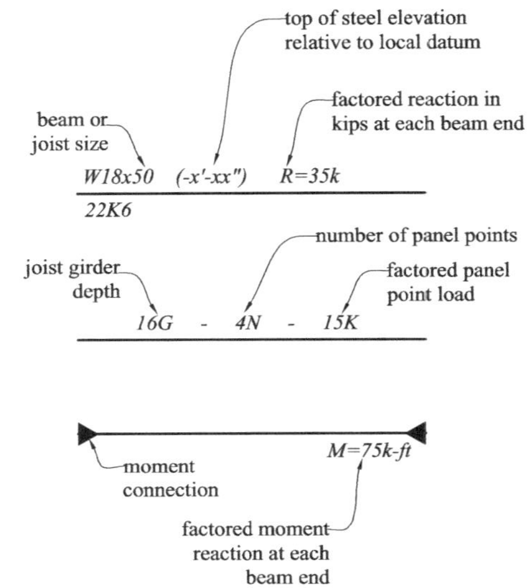

FIGURE 14-34 Sample roof framing plan

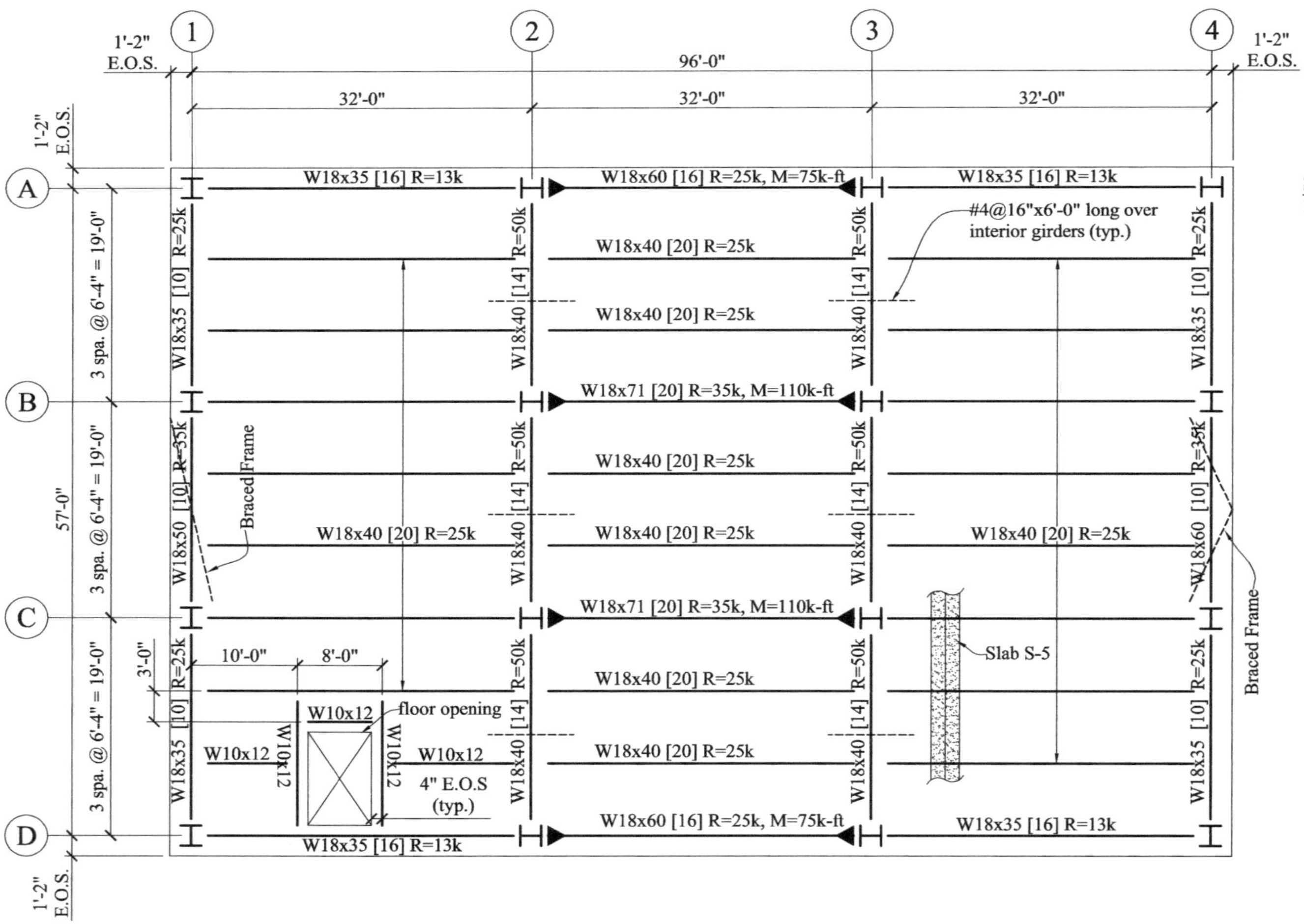

Second Floor Framing Plan

1/8" = 1'-0"

1. Top of slab elevation is +13'-6". Top of steel is +13'-0 3/4".

2. Slab S-5 indicates 2"-20ga composite deck with 3 1/4" lightweight concrete reinforced with 6x6-W2.9xW2.9 WWF. Total slab thickness is 5 1/4".

3. Provide a 10ga bent plate at the perimeter of all slab edges. E.O.S. = edge of slab.

4. Framing legend:

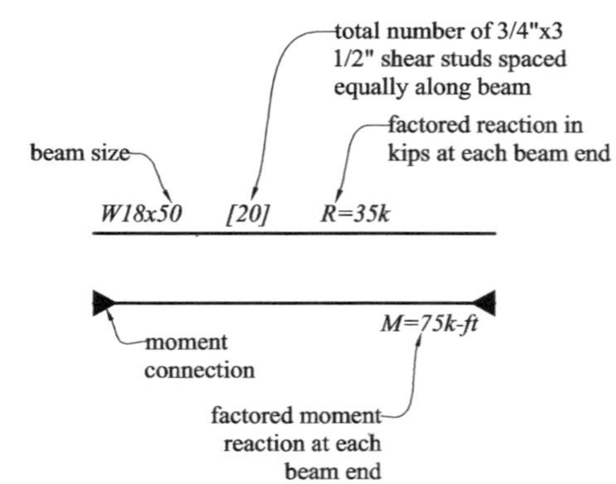

FIGURE 14-35 Sample floor framing plan

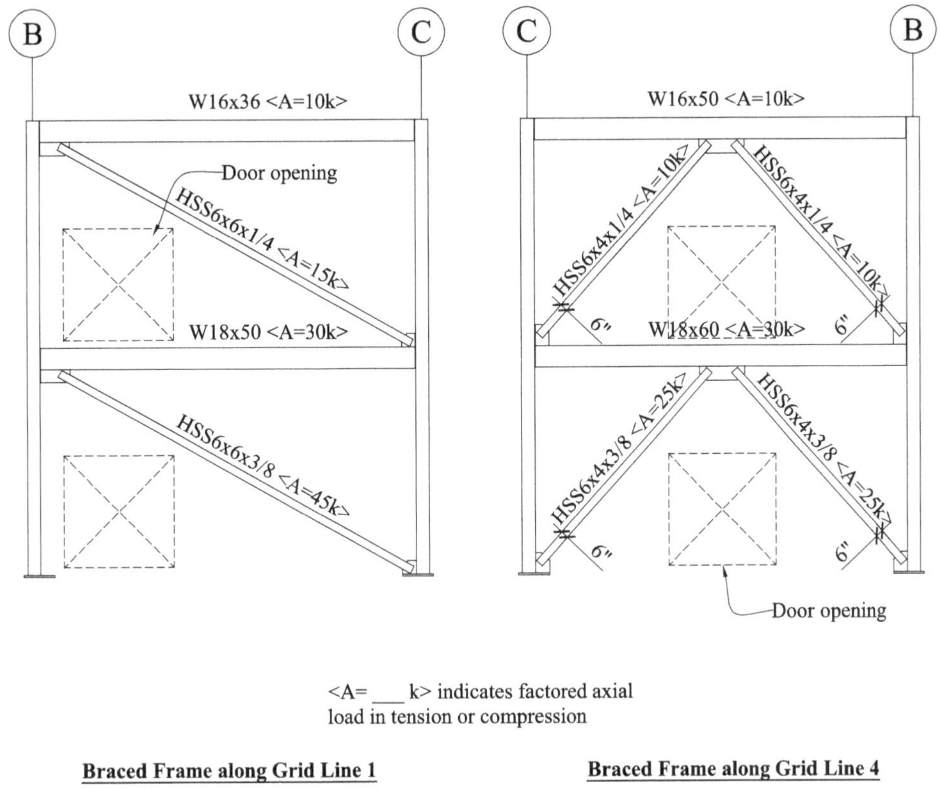

FIGURE 14-36 Braced frame elevations

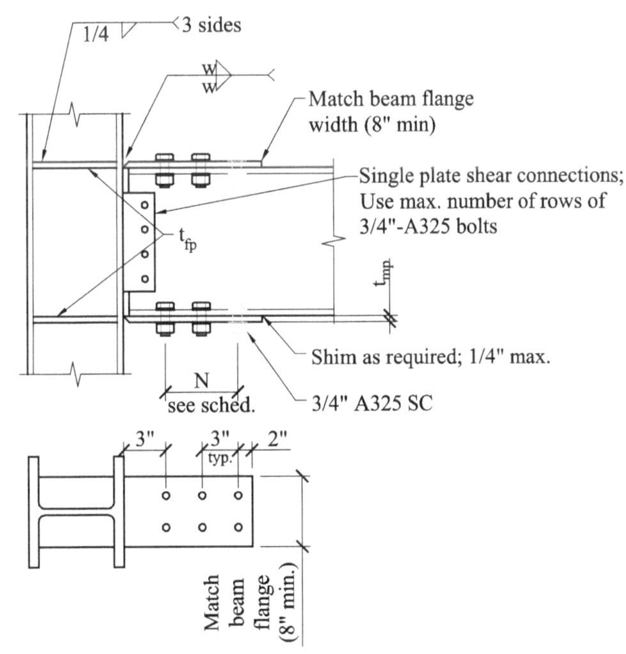

	BEAM	COLUMN	MOMENT, k-ft.	N	t_{mp}	t_{fp}	w
MC1	W18	W10	110	5	1/2"	3/8"	5/16
MC2	W18	W10	75	4	3/8"	3/8"	1/4
MC3	W18	W10	50	3	3/8"	3/8"	1/4

Legend:
N: number of bolts along each line
t_{mp}: thickness of top and bottom plates
t_{fp}: thickness of column stiffener plate
w: weld size

FIGURE 14-37 Moment connection schedule

14-17 STUDENT DESIGN PROJECTS

In this section, two structural steel design projects that can be assigned as part of a typical structural steel design course are presented (see Figures 14-38 through 14-41). The tasks needed to complete this project are intended to coincide with the work done on the end-of-chapter exercises throughout the text. Several analysis and design options are presented. The intent is that by the time the student has worked through all the chapters of the text and the corresponding design project tasks related to each chapter, they will have completed the design of an entire steel-framed structure. All hand calculations must be presented, and all computer-aided designs (e.g., using Excel spreadsheets) must be verified with hand calculations.

Student Design Project Exercise Option 1

1. **Office Building:** A steel-framed, two-story office building with plan dimensions of **72 ft. by 108 ft.** The floor-to-floor height is **15 ft.** (see Figures 14-38 and 14-39).
    - The building is in Buffalo, New York, or an alternate location selected by the instructor.
    - Main structural members are of structural steel.
    - The AISC LRFD Specification (or ASD depending on the instructor's preference) and the ASCE 7 load standard should be used.
    - The floors can be designed as either noncomposite or composite construction, with a concrete slab on a metal deck supported on steel infill beams and girders.
    - Roofing is 5-ply plus gravel supported on a metal deck on noncomposite steel framing.
    - Roof framing can be steel beams or open-web steel joists.
    - Assume that the perimeter cladding is supported on the foundation wall at the ground floor level and bypasses the floor and the roof.
    - Refer to building section for dead load components.
    - Assume a 1-ft. edge distance at each floor level around the perimeter of the building.
    - Assume a 1-ft. high parapet at the roof.
    - Assume that the stair and elevator shafts are located outside of the 72-ft. by 108-ft. footprint on the east and west sides of the building, and they will be designed by others.
    - Use Grade 50 steel for all wide flange beams and columns, and use Grade 36 steel for all plates, angles, and other connection elements.
    - The final submission should include all calculations and drawings with framing plans and details.
2. **Drawings:** These may be any appropriate size. Plans should be drawn to a scale of ⅛" =1'-0", and details should mainly be drawn to a scale of ¾" = 1'-0". Drawings should include the following:

    Roof Plan that shows framing members and roof deck pattern. Indicate camber and end reactions.

    Floor Plan that shows framing members and floor deck structure. Indicate number of studs, camber, and end reactions.

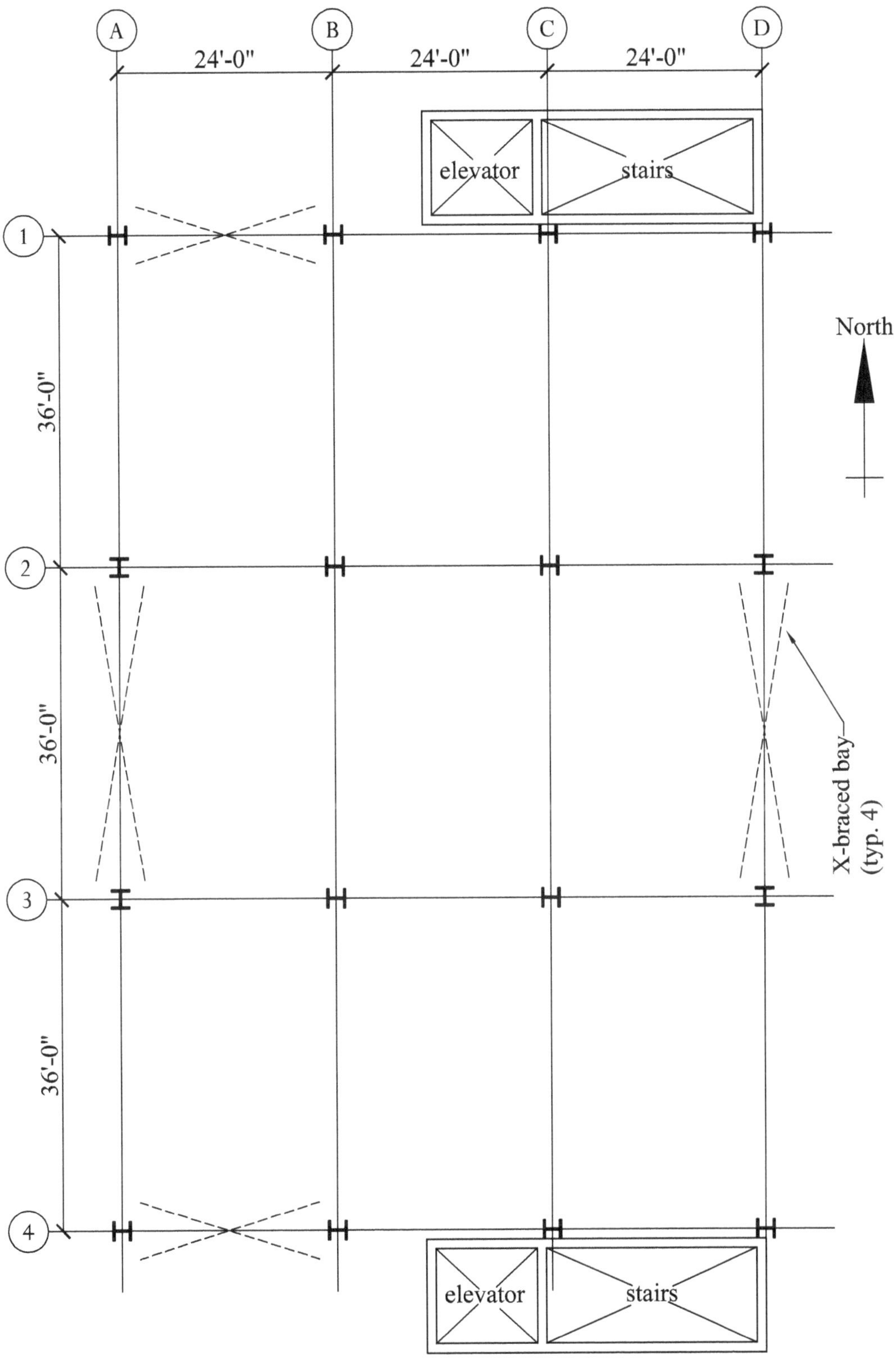

FIGURE 14-38 Grid layout for the student design project – option 1

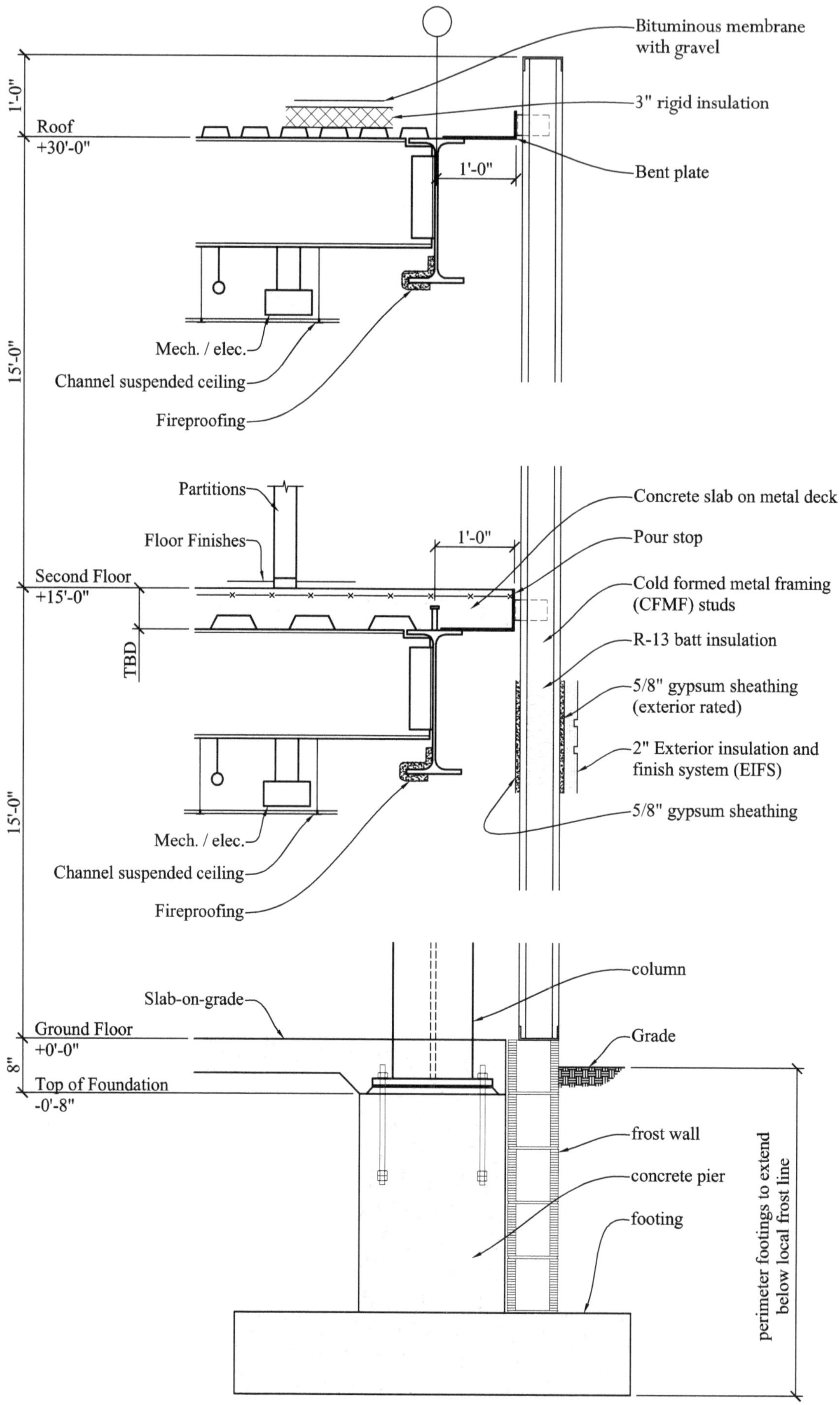

FIGURE 14-39 Building section for the student design project – option 1

Foundation Plan that shows column sizes or marks. Sketch an approximate footing size by using a net allowable soil bearing pressure of 3000 psf.

Elevations that show the X-braces or moment frames and connection configurations.

Column Schedule showing column sizes and reference elevations.

Connection Details that show the following:

- Typical beam and girder connections to other beams and columns
- Typical slab section showing the metal deck profile and reinforcing
- Typical slab section showing the reinforcing over the girders, if required by design
- For open-web steel joists, details at joists bearing on girders and seated connections at columns
- X-braced connections or moment frame connections
- Base plate and anchor rod details

3. **Loads:**

Calculate the dead load of the floor and roof and the live loads on the floor and roof.
Calculate the snow load using ASCE 7.
Follow the ASCE load standard to calculate the wind loads and seismic loads.
Use the maps in ASCE 7 to determine the basic wind speed, ground snow load, and seismic data for the selected building location.
Assume an exposure category of C for wind and use an importance factor of 1.0 for snow loads.
Beam deflection criteria: Use a limit of $L/360$ for live loads and $L/240$ for total loads.
Lateral deflection: Assume a lateral deflection limit of $H/400$ or $h/400$ under the unfactored 10-year MRI wind loads, where H is the total height of the building and h is the story height. See Figure CC.2.1 in Appendix CC of the Commentary Volume of Ref. [25] for the 10-year MRI 3-second gust wind speed map.

4. **Checklist of Design Items:**

- *Gravity and Lateral Loads:* Provide a load summation table for the floor and roof loads (gravity) and the wind and seismic loads (lateral). Provide a column load summation table.
- *Roof Framing Members:* Use wide flange beams or open-web steel joists. Select open-web steel joists from a manufacturer's catalog. Provide the manufacturer's load data to confirm the joist selection. Design the roof girders as wide-flange sections.
- *Floor Framing:* Use wide-flange beams that will be designed as composite or noncomposite.
- *Columns:* Design for maximum loads using the appropriate load combinations. Design columns for axial load and moments due to girder/beam connection eccentricities. Design the column base plates and anchor rods assuming a concrete strength (f_c') of 3000 psi and a 1-in. grout thickness.
- *Floor and Roof Deck:* Determine the exact gage and material weight from a deck manufacturer's catalog. Provide manufacturers load data to confirm the deck selection. Use normal-weight concrete and a concrete strength of f_c' = 3000 psi.

Lateral Force Resisting System

Vertical LFRS Option 1:

X-braces: Analyze the X-brace frames manually or using a structural analysis software program. Design the diagonal brace connections and check all connection failure modes.

Vertical LFRS Option 2:

Moment frames: Assume that moment frames are used along grid lines 1 and 4 in lieu of the X-braces. Reanalyze the frame and design the moment frames. Design the beam and girder connections and check all connection failure modes.

Design Project Tasks - Structural Steel Building (Option 1):

The following end-of-chapter tasks related to the design project (Option 1) are intended to help provide direction to the student in completing the design project since it serves as a checklist of the tasks that need to be completed after completing each chapter in the text.

P2[1]-1. Load calculations:

- **a.** For each elevated level, use a legend to show the different building areas and occupancy, and the corresponding live loads and dead loads.
- **b.** Estimate the floor and roof dead loads and provide a summation table showing each item.
- **c.** Determine the basic floor and roof live loads using ASCE 7.
- **d.** Calculate the flat roof snow load using ASCE 7.

P2-2. Lay out framing options for the floor and roof, showing the beam and girder locations and beam spacing. Show beam spacing to be 3, 4, and 5 equal spaces with the beams spanning the 36'-0" direction. Select one of these beam spacing schemes for both the floor and the roof to be used in the deck selection.

P2-3. Floor deck:

- **a.** Composite deck option: Select an appropriate deck and slab thickness using the deck span determined in the previous exercise.
- **b.** Noncomposite deck option: Select an appropriate deck and slab thickness using the deck span determined in the previous exercise.
- **c.** Revise dead loads from P2-1 as necessary based on actual slab and deck selection.

P2-4. Roof deck: Select an appropriate deck using the deck span determined in the previous exercise.

P3. Lateral Loads: For each lateral force resisting system (i.e., braced frame or moment frame), show the lateral forces acting at each level for wind and seismic. Note that for braced frames, the diagonal braces are assumed to resist lateral loads only and are not considered as supports for gravity loads.

P3-1. Wind loads:

- **a.** Calculate the wind loads in each direction of the building for the main wind force resisting system (MWFRS); Ignore the Components and Cladding.
- **b.** Determine the wind force at each level in each direction and provide a sketch showing these loads.

---

[1] The number immediately after "P" (e.g., P2-) denotes the chapter number in the text from which the information needed to complete the listed tasks for the design project can be found.

P3-2. Seismic loads:

a. Calculate the seismic base shear in each direction of the building. It is recommended that a value of $R = 3.0$ be used in both directions to simplify the connection details for the lateral force-resisting system. An alternate value of "R" can be used if appropriate.

b. Determine the seismic lateral force at each level in each direction and provide a sketch showing these loads.

c. Determine the diaphragm seismic lateral force at each level.

P4-1. X-bracing option:

a. Determine the controlling brace forces (wind or seismic) in each level and in each braced frame.

b. Plate option: Design tension-only plate X-brace frames assuming a single line of $^3\!/_4$-in.-diameter bolts at the ends. Check the limit states for tension on gross and net areas, and block shear. Use ASTM A572 Grade 50 steel.

c. Double angle option: Design tension-only braces using double angles and assume a single line of $^3\!/_4$-in.-diameter bolts at the ends. Check the limit states for tension on gross and net areas, and block shear. Use ASTM A572 Grade 50 steel.

d. Tension rod option: Design tension-only braces using tension rods. Check all limit states for tension rods. Use ASTM A36 steel.

e. Provide a final sketch of the designs in (b) or (c) showing all dimensions and material properties.

P5-1. Preliminary column selection: Neglecting the beam-to-column and girder-to-column connection eccentricities (and, therefore, neglecting the moments caused by those eccentricities), calculate the cumulative factored axial loads on the typical interior, exterior, and corner columns for each level of the building (see Example 2-13). Determine the required column size for each level of the building, assuming an effective length factor, K, of 1.0.

P6-1. Floor beams, noncomposite option:

a. Design the interior floor beams and girders as noncomposite.

b. Design the spandrel floor beams and girders as noncomposite.

c. Provide a sketch of the floor framing plan showing member sizes and end reactions.

P6-2. Roof framing:

a. Beam option: Design the interior beams using wide-flange steel.

b. Joist option: Design the interior members using open-web steel joists.

c. Design the interior girders using wide-flange steel.

d. Design all spandrel members using wide-flange steel.

e. Provide a sketch of the roof framing plan showing member sizes and end reactions.

P7-1. Floor beams, composite option:

a. Design the interior floor beams and girders as composite.

b. Design the interior floor beams and girders as composite.

c. Design reinforcing over the girders. Provide a sketch of this design.

d. Provide a sketch of the floor framing plan showing member sizes, stud layout, and end reactions.

P8-1. Column design:

a. Determine the factored axial loads and the no-translation moments (at each level of the building for the typical interior, exterior, and corner columns. Sketch each condition showing column marks and loads.

b. Using the column design template, select the most economical W-shape columns. Assume the columns are the full height of the building (i.e., the same size for the lower and upper levels).

c. Provide final designs in a column schedule.

P9-1. Shear connections: Select the following connections from the appropriate *AISCM* table; check block shear and bolt bearing as required.

a. Beam-to-girder connections at the second floor and roof using single-plate shear connections.

b. Beam-to-girder connections at the second floor and roof using all-bolted, double-angle shear connections.

c. Beam-to-column and girder-to-column connections at the second floor and roof using all-bolted, double-angle shear connections.

d. Provide a sketch of each design showing all appropriate dimensions and material properties. Show copes where needed.

P10-1. Seat angle, joist option: Design a welded seat angle connection for the typical roof joist and joist girders, using the detail shown in Figure 10-23 as a guide. Provide a sketch of this design.

P10-2. X-brace option: For the double-angle X-bracing, connections are to be welded. Check the appropriate limit states for block shear on the gusset plates. Provide a sketch of this design.

P11-1. Moment frame option:

a. Perform a lateral load analysis of the building assuming that the X-braced frames are replaced by partially restrained moment frames with a seismic response modification coefficient, $R = 3.0$. Provide a summary of each controlling end moment.

b. Design bolted moment connections to the column, assuming a single top and bottom plate for each critical moment. Check the appropriate limit states for the beam and moment plates.

c. Check the column for concentrated loads on the flange and web. Design stiffeners as required.

d. Provide a sketch of each connection detail.

P11-2. For the shear connections previously designed, determine which of the connections require a coped connection and determine whether the beam is adequate for the appropriate limit states for coped beams. Revise the connection detail, if needed.

P11-3. For the vertical X-brace option in the design project exercise, design the gusset. plate connections between the diagonal brace, the ground floor column, and the second-floor beam. Assume a $\frac{1}{2}$-in. setback between the gusset plate and the face of the column. Provide a final sketch of this design.

P12-1. Floor vibrations:

a. Analyze the floor structure for walking vibrations in an office, assuming a damping ratio, $\beta = 0.03$.

b. Redesign option: If the floor structure is not adequate, provide a new design that is adequate.

c. In lieu of a redesign, provide a discussion of how the structure could be improved to be adequate for floor vibrations. Include an assessment of how the structure would perform as is.

P13-1. Assuming it is desired to create a column-free atrium area at the ground floor by terminating columns B-3 and C-3 at the second floor. Design a transfer plate girder spanning between column A-3 and column D-3 to support the second floor and the loads from columns B-3 and C-3. Neglect diagonal tension field action.

P14-1 Cost estimate:

a. Determine the building square footage at each level.
Roof:______________
Second floor:______________
Total______________

b. Contact local steel fabricators and ask for estimates on the current material cost of steel and the installed cost for steel.

c. Develop a table similar to Table 14-4 indicating the total weight of steel and estimated cost.

d. Determine the cost per square foot for each level.

TABLE 14-4 Total Weight of Steel and Estimated Cost for Student Design Project—Option 1

Member	Weight (pounds)	Weight (tons)	Material cost	Installed cost
Roof beams (or joists)				
Roof girders (or joists)				
Roof columns				
Roof connections				
Roof X-braces				
Metal deck				
Total roof:				
Second floor beams				
Second floor girders				
Second floor columns				
Second floor connections				
Second floor X-braces				
Metal Deck				
Total floor:				
Total building:				

Student Design Project Exercise Option 2

In lieu of the structural steel building design project presented in the previous section, a steel-framed pedestrian bridge design project is introduced here as Option 2. The project requirements are as follows:

1. **Pedestrian Bridge:** A steel-framed bridge with plan dimensions of 80 ft. long and 6 ft. clear width inside the railings (see Figures 14-40, 14-41a, and 14-41b).
   - The structure is in Buffalo, New York, or an alternate location selected by the instructor.
   - Main structural members are of structural steel.
   - Use the AASHTO LRFD Guide Specification for the Design of Pedestrian Bridges for loading and for provisions related to the stability of the top chords [48].
   - The AISC LRFD Specification and the ASCE 7 load standard should be used for all other loading and design information not covered in Ref [48].
   - The floors can be designed as either a noncomposite slab on deck or steel bar grating.
   - Railings are designed for handrail loading.
   - Use Grade 50 steel for all wide flange beams and columns, and use Grade 36 steel for all plates, angles, and other connection elements. HSS is ASTM A500.
   - The final submission should include all calculations and drawings with framing plans and details.
2. **Drawings:** May be any appropriate size. Plans should be drawn to a scale of $\frac{1}{8}$"= 1'-0", and details should mainly be drawn to a scale of $\frac{3}{4}$" = 1'-0". Drawings should include the following:

   Floor and roof Plan that shows framing members and floor deck structure. Indicate end reactions. For the floor plan, show the slab or bar grating design.

   Elevations that show framing members with axial forces for the trusses along the sides, top, and bottom of the pedestrian bridge.

   Elevation view of the railings.

   Elevations with loads for the sides, top, and bottom trusses are shown as a line diagram with axial loads in each member.

   Foundation Plan that shows bearing plates and end reactions. Sketch an approximate footing size by using a net allowable bearing pressure of 3000 psf.

   Connection details that show the following:
   - Typical floor beams,
   - Truss member end connections,
   - Typical slab section (for concrete slab scheme),
   - Typical bar grating section,
   - Railings components,
   - Bearing plate and anchor rod details.
3. **Loads:**
   - Calculate the dead load of the floor and structure.
   - Live loads are 100 psf or an H-5 or H-10 single maintenance truck live load per AASHTO Code if the clear bridge width is greater than or equal to 7 ft.
   - Winds loads are 50 psf on the projected area.
   - Calculate the snow load using ASCE 7.

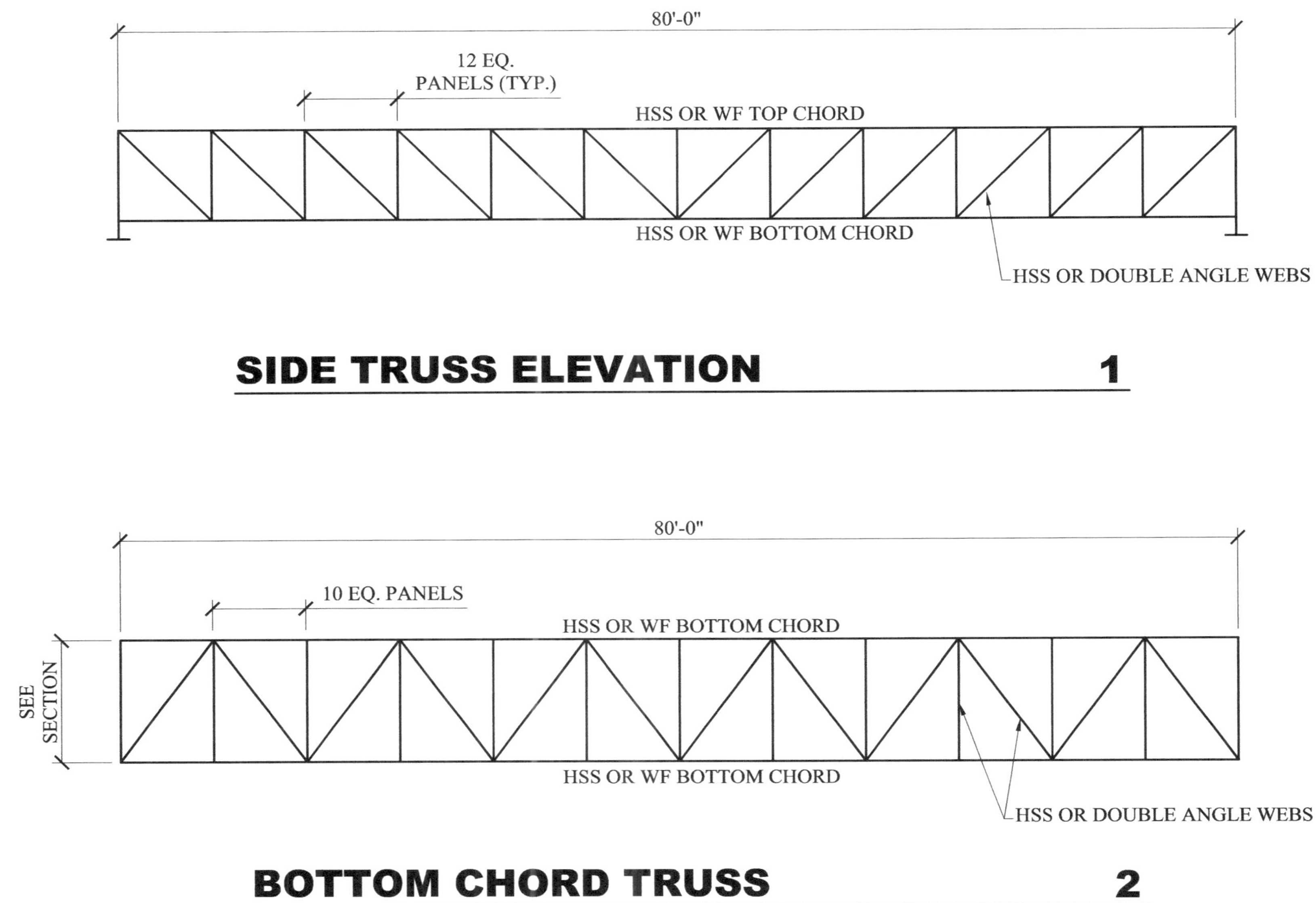

FIGURE 14-40 Pedestrian bridge truss

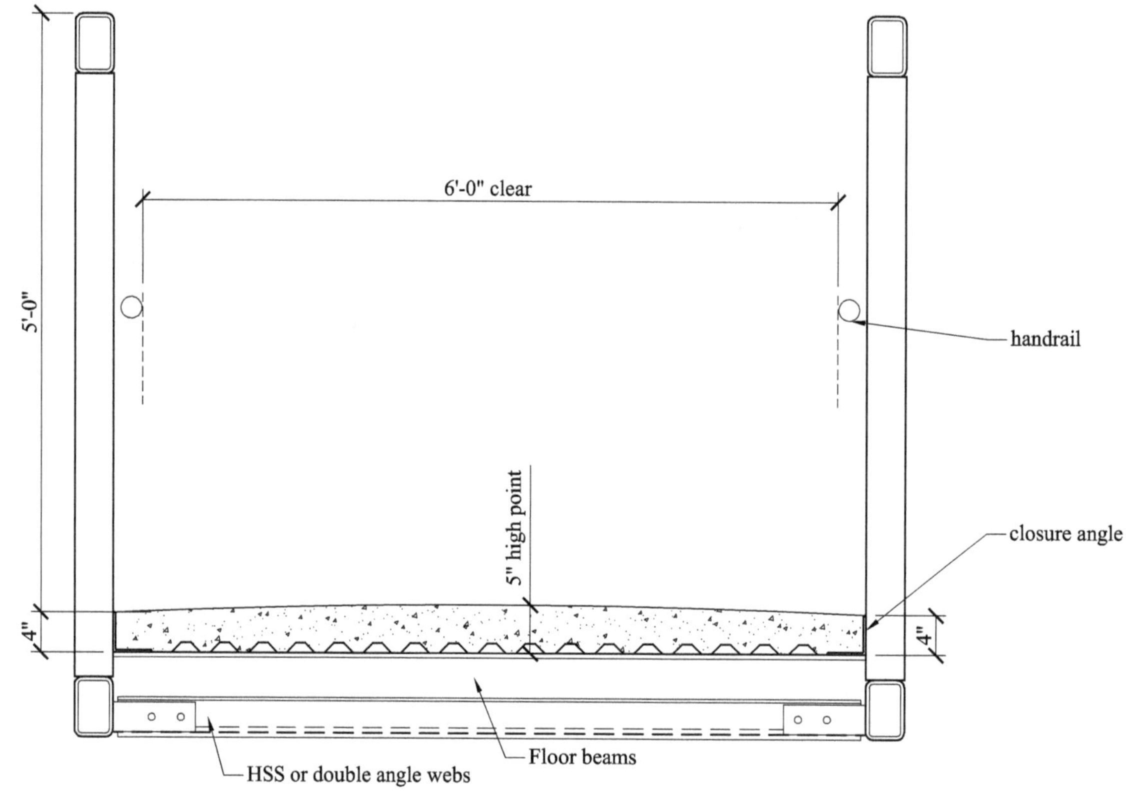

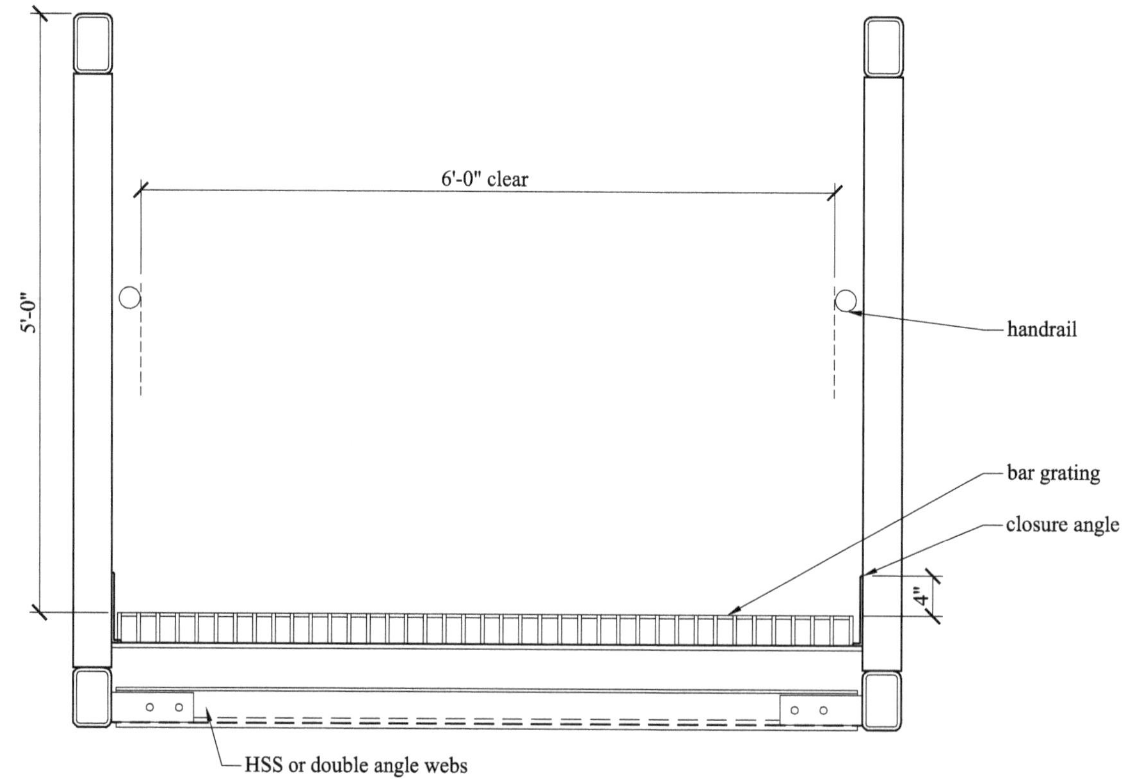

FIGURE 14-41a Pedestrian bridge truss sections

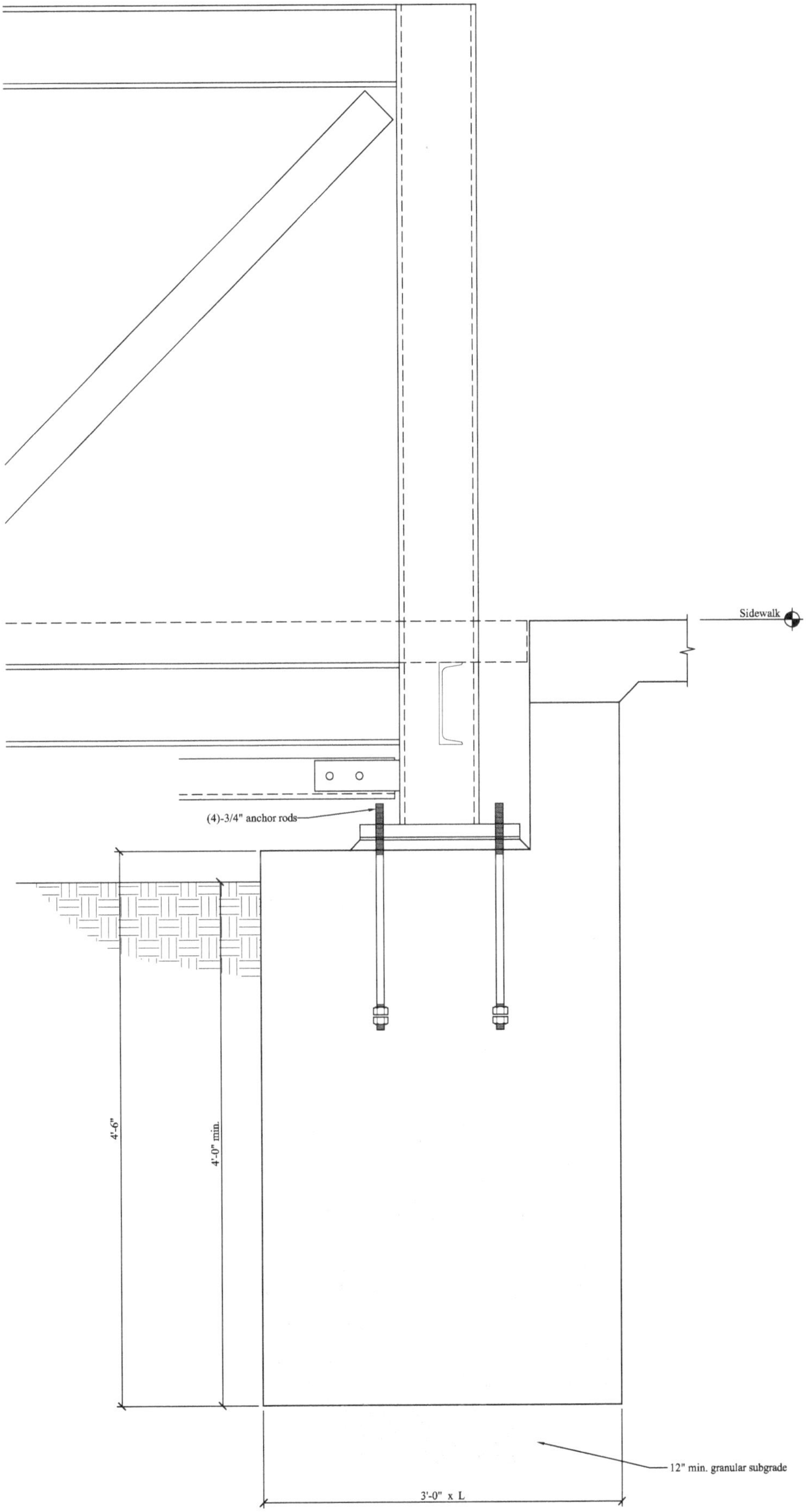

FIGURE 14-41b Pedestrian bridge section at truss support

- Use an importance factor of 1.0 and assume an exposure category of C for wind and snow loads.
- Ignore seismic loading.
- Beam deflection criteria: Use a limit of L/480 for live loads and L/360 for total loads. Use L/600 for lateral wind load.

4. **Checklist of Design Items:**
   - *Gravity and Lateral Loads:* Provide a load summation table for the floor and roof loads (gravity) and the wind loads (lateral).
   - *Floor Framing:* Use Channels or wide flange sections or double angles for the floor beams.
   - *Truss framing: Use HSS members for the chords*; use double angles or HSS members for the webs.
   - *Floor Deck:*
   - Slab on deck option: Determine the exact gage and material weight from a deck manufacturer's catalog. Provide manufacturers load data to confirm the deck selection. Use normal-weight concrete and a concrete strength of f_c' = 3000 psi.
   - Bar grating option: Select from a manufacturer's catalog and verify the selection with calculations.
   - Railings: Design for code-required loading for handrails. Post spacing should be close to 48 inches.

Lateral Force Resisting System

Design for wind loads only on the vertical projected area of the pedestrian bridge; assume All lateral load is transferred to the bottom horizontal truss below the bridge floor.

Design Project Tasks – Structural Steel Pedestrian Bridge (Option 2):

The following end-of-chapter tasks related to the design project (Option 2) are intended to help provide direction to the student in completing the design project since it serves as a checklist of the tasks that need to be completed after completing each chapter in the text.

P2-1. Load calculations:

a. Estimate the floor and roof dead loads and provide a summation table showing each item.

b. Determine pedestrian floor live load from ASCE 7 and AASHTO (see Ref. [48]).

c. Calculate the snow load using ASCE 7 on the deck surface

P2-2. Floor deck:

a. Noncomposite deck option: Select an appropriate deck and slab thickness using the deck span determined in the previous Exercise.

b. Bar grating option: select an appropriate grating size for the uniform and concentrated loads.

c. Revise dead loads from P2-1 as necessary based on actual slab and deck selection.

P2-3. Lay out framing options for the floor showing the beam locations and beam spacing.

P2-4. Lay out the side trusses and top and bottom trusses, showing the actual dimensions of the member centerlines that will be used in the structural analysis.

P3-1. Wind loads:

a. Calculate the wind loads in each direction of the bridge

b. Determine the wind force along the truss bottom chord and provide a sketch showing these loads.

P4-1. Bottom chord truss

a. Analyze the truss bottom chord for the lateral wind load

b. Double angle option: Design web members in tension using double angles and assume a single line of two or three 3/4-in.-diameter bolts at the ends. Check the limit states for tension on gross and net areas, and block shear. Use ASTM A36 steel.

c. HSS option: Design web members in tension, assuming 3/8" gusset plates. Check the limit states for tension on gross and net areas, and block shear. Use ASTM A36 steel.

d. Provide a final sketch of the design showing all dimensions and material properties.

P4-2. Gravity Trusses

a. Analyze the side trusses for gravity loads.

b. Double angle option: Design web members in tension using double angles and assume a single line of two or three 3/4-in.-diameter bolts at the ends. Check the limit states for tension on gross and net areas, and block shear. Use ASTM A36 steel.

c. HSS option: Design web members in tension, assuming 3/8" gusset plates. Check the limit states for tension on gross and net areas, and block shear. Use ASTM A36 steel.

d. Provide a final sketch of the design showing all dimensions and material properties.

P5-1. Bottom chord truss

a. Double angle option: Using the previous analysis, design web members in compression using double angles and assume a single line of two or three 3/4-in.-diameter bolts at the ends. Check the limit states for tension on gross and net areas, and block shear. Use ASTM A36 steel.

b. HSS option: Design web members in tension, assuming 3/8" gusset plates. Check the limit states for tension on gross and net areas, and block shear. Use ASTM A36 steel.

c. Provide a final sketch of the design showing all dimensions and material properties.

P5-2. Gravity Trusses

a. Double angle option: Using the previous analysis, design web members in compression using double angles and assume a single line of two or three 3/4-in.-diameter bolts at the ends. Check the limit states for tension on gross and net areas, and block shear. Use ASTM A36 steel.

b. HSS option: Design web members in compression, assuming 3/8" gusset plates. Check the limit states for tension on gross and net areas, and block shear. Use ASTM A36 steel.

c. Provide a final sketch of the design showing all dimensions and material properties.

P6-1. Floor beams, noncomposite deck option:

a. Design the interior floor beams as noncomposite.

b. Revise the analysis model for any changes to the loading or configuration.

c. Provide a sketch of the floor framing plan showing member sizes and end reactions.

P6-2. Floor beams, bar grating option:

a. Design the interior floor beams.

b. Revise the analysis model for any changes to the loading or configuration.

c. Provide a sketch of the floor framing plan showing member sizes and end reactions.

P8-1. Truss chord member design:

a. Determine the factored axial loads and moments in the truss chords. Sketch each condition showing loads and locations.

b. Using the column design template, select the most economical HSS member for the compression chord.

c. Select the most economical HSS member for the tension chord.

d. Update the drawings for the final sizes and loads.

P9-1. Shear connections: Select the following connections from the appropriate *AISCM* table; check block shear and bolt bearing as required.

a. Beam-to-truss chord connections at the floor using single-plate shear connections.

b. Beam-to-truss chord connections at the floor using all-bolted, double-angle shear connections.

c. Provide a sketch of each design showing all appropriate dimensions and material properties. Show copes where needed.

P10-1. Truss connections—double angle option: design all welds for the gusset plates welded to the supports.

P10-2. Truss connections—HSS option: design all welds for the gusset plates welded to the supports and HSS members welded to the plates.

P10-3. Railings: design all connections of the railing-to-railing and railing to support connections.

P11-1. For the shear connections previously designed, determine which of the connections require a coped connection and determine whether the beam is adequate for the appropriate limit states for coped beams. Revise the connection detail, if needed.

P14-1. Cost estimate:

a. Contact local steel fabricators and ask for estimates on the current material cost of steel and the installed cost for steel.

b. Develop a table (see Table 14-5) indicating the total weight of steel and estimated cost.

TABLE 14-5 Total Weight of Steel and Estimated Cost for Student Design Project—Option 2.

Member	Weight (pounds)	Weight (tons)	Material cost	Installed cost
Floor beams				
Horizontal railing				
Vertical handrail posts				
Connections				
Bar Grating				
Metal deck				
Total Floor:				
Side Trusses				
Bottom truss				
Side Truss connections				
Bottom Truss connections				
Total floor:				
Total bridge:				

References

[1] Ricker, D. T. 2000. "Value Engineering for Steel Construction." *Modern Steel Construction* April.

[2] American Institute of Steel Construction. 2022. "*Seismic Provisions for Structural Steel Building.*" Chicago, IL: AISC.

[3] Ioannides, S. A., and J. L. Ruddy. 2000. "Rules of Thumb for Steel Design." *Modern Steel Construction*, February.

[4] Carter, C. J., and M. A. Steven. 2004. "Look Before You Leap: Practical Advice for Laying out Steel Framing." *Modern Steel Construction*, March.

[5] Ambrose, J. 1997. *Simplified Design of Steel Structures.* 7th ed. Hoboken, NJ: John Wiley and Sons.

[6] Naeim, F. 2001. *The Seismic Design Handbook.* Country Club Hills, IL: International Conference of Building Officials (ICBO).

[7] Lochrane, M. L., and W. C. Honeck. 2004. "Design of Special Concentric Braced Frames (with Comments on Ordinary Concentric Braced Frames)," *Structure Steel Education Council*, Moraga, CA.

[8] Matthes, J., ed. 1993. *Masonry Designers' Guide.* The Masonry Society, Boulder, CO.

[9] Luttrel, L. 2004. *Diaphragm Design Mannual.* 3rd ed. Fox River Grove, IL: Steel Deck Institute (SDI).

[10] Fisher, J. M., M. A. West, and J. P. Van de Pas. 2002. "*Designing with Vulcraft Steel Joists, Joist Girders and Steel Deck.*" Milwaukee, WI: Nucor Co.

[11] Krogstad, N. V. 2003. "Using Steel Relief Angles." *Masonry Construction*, May.

[12] International Codes Council. 2024. *2024 International Building Code*. Falls Church, VA: ICC.

[13] The Society for Protective Coatings. 2006. *Surface Preparation Specification and Practices*. Chicago, IL: SPC.

[14] American Institute of Steel Construction. 2003. *Fire Resistance of Structural Steel Framing*. Chicago, IL: AISC.

[15] Underwriters Laboratories, Inc. 2003. *Fire Resistance Directory*. Vol. I. Northbrook, IL: Underwriters Laboratories, Inc.

[16] Kulak, G. L., and Y. G. Gilbert. 2003. "Strength of Joints that Combine Bolts and Welds." *Engineering Journal*, (2nd Quarter).

[17] Ricker, D. T. 1988. "Field Welding to Existing Structures." *Engineering Journal*, (1st Quarter).

[18] Tall, L. 1989. "The Reinforcement of Steel Columns." *Engineering Journal*, (1st Quarter).

[19] Al Nageim, H. and T. G. MacGinley. 2005. *Steel Structures—Practical Design Studies*. 3rd ed. Taylor and Francis.

[20] American Institute of Steel Construction. 2023. *Steel Construction Manual*. 16th ed. *AISC*, Chicago, IL: AISC.

[21] American Institute of Steel Construction. 2003. *Steel Design Guide Series 3: Serviceability design considerations for steel buildings*, AISC, Chicago, IL, 2003.

[22] ASTM International. 2013. *Standard specification for Zinc (hot-dip galvanized) coatings on iron and steel products—ASTM A123*. West Conshohocken, PA: ASTM International.

[23] American Institute of *Steel Construction*. 2002. *Steel design guide series 15: AISC rehabilitation and retrofit guide*. Chicago, IL: AISC.

[24] Miller, J. P. 1996. "Strengthening of existing composite beams using LRFD procedures." *Engineering Journal* (2nd Quarter).

[25] American Society of Civil Engineers. 2016. *ASCE-7, Minimum Design Loads for Buildings and* ASCE Reston, VA: ASCE.

[26] Dowswell, B., A. Brice, *and B.* Blain. 2010. "Horizontal Bracing." *Modern Steel Construction*, July.

[27] Williams, E. 2009. "Different Angles." *Modern Steel Construction*, February.

[28] Stockman, M. 2009. "Fire Protection Basics." *Modern Steel Construction*, September.

[29] McCampbell, C. 2013. "Fix Under Way for Sagging Span." *Engineering News Record*, December 29: 12.

[30] Geschwindner, L. F., and A. Lepage. 2013. "Notes on Nodal and Relative Lateral Stability Bracing Requirements of AISC 360." *Engineering Journal*, (3rd Quarter).

[31] International Code Council. 2010. *Existing Building Code of New York State*. Washington, DC: ICC.

[32] Friscolanti, M. 2013. "Elliot Lake: How Could So Many Engineers Be So Wrong?" *Maclean's*, (March 20). *https://www.macleans.ca/news/canada/elliot-lake-how-could-so-many-engineers-be-so-wrong/* (Accessed August 15, 2024).

[33] Geise, C. 2012. "Lessons Learned from the Joplin Tornado." *STRUCTURE Magazine*, July.

[34] Eckmann, D. E. 1999. "Column-Free Office Space Rises in Chicago's Loop." *Modern Steel Construction*, April.

[35] Whitely, M. 2018. "Elevated Efficiency." *Modern Steel Construction*, September.

[36] Martar, S. S. and W. J. Faschan. 2017. "A Structural Engineer's Approach to Differential Vertical Shortening in Tall Buildings." *International Journal of High-Rise Buildings*, Vol. 6, (1) March: 73-82.

[37] Sebastian, S. A. 2015. "Considering Corrosion." *Modern Steel Construction*, April.

[38] Ruby, D. 2018. "Construction Meets Constructability: Part I." *Modern Steel Construction*, May.

[39] Ruby, D. 2018. "Construction Meets Constructability: Part II." *Modern Steel Construction*, September.

[40] Hewitt, C., A. Humphreys, and E. Twomey. 2019. "Opposites Attract – A Primer on Galvanic Corrosion of Dissimilar Metals." *Modern Steel Construction*, July: 16-19.

[41] Tso, W. K. 1990. "Static Eccentricity Concept for Torsional Moment Estimations." *ASCE Journal of Structural Engineering*, Vol. 116, (5) May.

[42] Dowswell, B. 2014. "Reinforcing the Point." *Modern Steel Construction*. January.

[43] Green, N. 1981. *Earthquake Resistant Building Design and Construction*. New York: Van Nostrand Reinhold.

[44] Goel, R. A., and A. K. Chopra. 1993. "Seismic Code Analysis of Buildings Without Locating Centers of Rigidity." *ASCE Journal of Structural Engineering*, Vol. 119, (10) October.

[45] Rafezy, B. and W. P. Howson. 2009. "A New and Practical Approach for Calculating the Static Eccentricity of Doubly Asymmetric, Non-Proportional, Multi-story Buildings." *Proceedings of the World Congress on Engineering,* Vol II, July 1-3, London U.K.

[46] Wight, J. K. and J. G. MacGregor. 2012. "Reinforced Concrete: Mechanics and Design." 6th ed. Boston, MA: Pearson.

[47] American Society of Civil Engineers. 2016. "*ASCE-7, Minimum Design Loads for Buildings and Other Structures.*" Reston, VA: ASCE.

[48] AASHTO. 2015. "LRFD Guide Specification for the Design of Pedestrian Bridges." 2nd ed. Washington D.C: AASHTO.

[49] The Precast/Prestressed Concrete Institute. 1999. "PCI Design Handbook." 5th ed., 6-16 – 6-17.

[50] Coffman, S. D. 2017. "¼ in 12 Design Slope and Water Drainage." *STRUCTURE Magazine*, August: 24-27.

[51] Curtis, Ryan. 2020. "AESS Success." Modern Steel Construction, September: 40-45

[52] Collins, Jacinda and Pazdon, A Jennifer. 2020. "AESS Answers." *Modern Steel Construction*, September: 46-53.

[53] Phelan, J. P. and M. F. Hughes. 2023. "The Devil in the Details." *STRUCTURE Magazine*, August: 64-66.

[54] Schwinger, C. 2024. "Builder Friendly." *Modern Steel Construction,* February: 55-57.

[55] Patel, V. and H. Patel. 2024. "Flying By." *Modern Steel Construction*, February: 32-39.

[56] Barg, S., F. Flager, and M. Fischer. 2017. "An Analytical Method to Estimate the Total Installed Cost of Steel Frames During Early Design." *Center for Integrated Facility Engineering, Civil and Environmental Engineering Dept, Stanford University. https://stacks.stanford.edu/file/druid:yb503ws4475/TR220.pdf* (Last accessed August 31, 2024)

[57] GSA (General Services Administration). 2003. *Progressive collapse analysis and design guidelines for new federal office buildings and major modernization projects.* Washington, DC: GSA.

[58] DOD (US Department of Defense). 2009. *Design of Buildings to Resist Progressive Collapse.* UFC 4-023-03, Washington, DC: DOD.

[59] American Society of Civil Engineers. 2023. *Standard for Mitigation of Disproportionate Collapse Potential in Buildings and Other Structures (76-23).* ASCE, 62.

[60] Aghayere A. and J. Vigil, 2024. *Reinforced Concrete Design.* 10th ed. Boston, MA: Mercury Learning, 841 pp.

[61] American Institute of Steel Construction. 2024. *Design Guide 40: Rain Loads and Ponding.* Chicago, IL: AISC.

APPENDIX

Plastic Analysis and Design of Continuous Beams and Girders

In Chapter 6, we discussed the design of statically determinate (i.e., simply supported single span) beams and girders using the ASD and LRFD methods. The load effects, such as moment and shear demands, were calculated using elastic analysis. This chapter focuses on the design of statically indeterminate or continuous beams and girders using the plastic design (PD) method introduced in Chapter 2. Plastic design is an economical method for the design of statically indeterminate beams, girders, and frames, and it is covered in Appendix 1 of the AISC Specification [1]. In plastic design, the structure is assumed to undergo inelastic deformations under the factored loads from the formation of plastic hinges [2]. The only difference between the plastic design (PD) method and the LRFD method is the analysis procedure used to determine the load effects. In fact, the LRFD load combinations introduced in Chapter 2 apply to both methods. In plastic design, the load effect (e.g., required plastic moment strength) is determined using *plastic* structural analysis, whereas the LRFD method uses *elastic* structural analysis to determine the required moment strength or factored moment, M_u (see Chapter 6). For both design methods, the moment capacity or design moment strength, ϕM_n, of the member is determined in exactly the same way. In this text, the focus will be on the plastic design of continuous beams and girders. More extensive coverage of the plastic design of statically indeterminate frames can be found in Reference [3].

The process of plastic hinge formation is as follows: when a steel member is subjected to a bending moment, M, the stress in the critical section increases from zero as the load is increased. It continues to rise until the yield stress is reached at the extreme fibers at the critical section when the moment reaches M_y, the yield moment. As the load is further increased, the extreme fiber stress at this critical section can no longer increase beyond the yield stress, F_y, but the yielding spreads further into the section between the neutral axis and the extreme fibers (see Figure A-1). As the load or moment is increased further, the yielding at the critical section continues to spread until the whole section reaches the yield stress, F_y. At this stage, the critical section is in the plastic state, and a plastic hinge forms. Beyond this stage, the section where plastification has occurred can no longer participate in resisting any increase in applied load or moment. Any further increase in load now has to be resisted, through a process of redistribution of moments and stresses, to other sections that have not yet reached plastification. though there will be increased strain or rotations at the sections that have already achieved plastification.

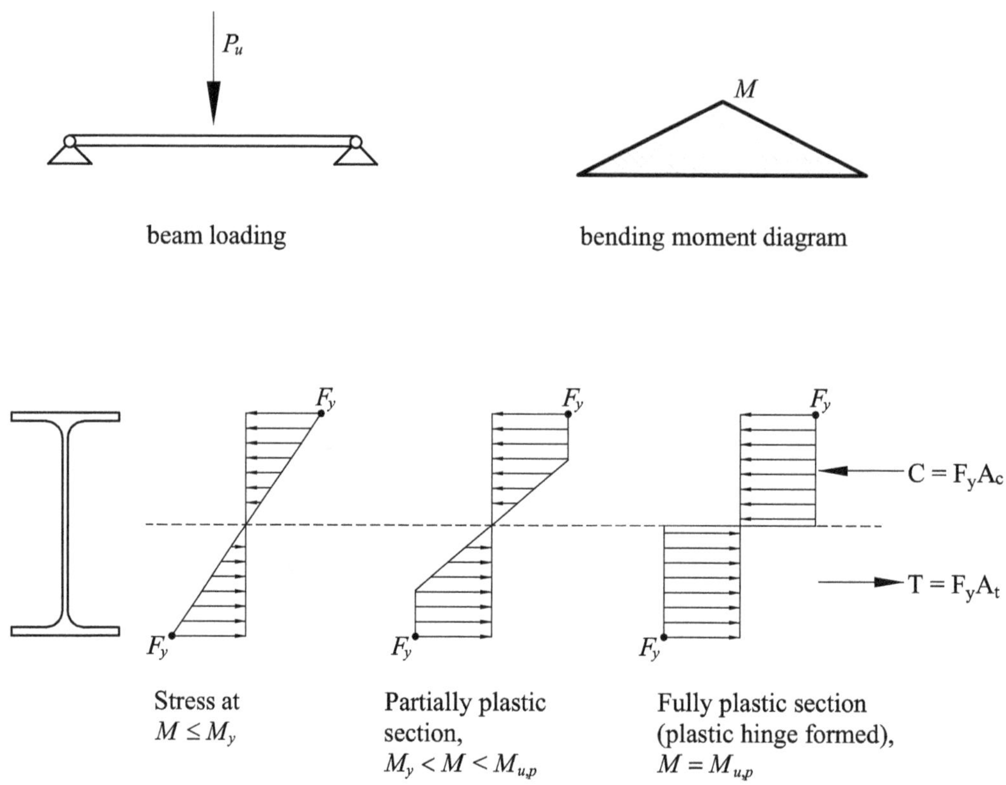

FIGURE A-1 Process of plastic hinge formation

This process of plastification is continued until a sufficient number of plastic hinges are formed at all the critical sections within the beam span to create a collapse mechanism. The principle of equilibrium or the principle of virtual work can then be used to determine the plastic moment strength or moment capacity of the structural member. It should be noted that in order for the plastic moment capacity to be reached, the ductility of the beam must be ensured, and lateral-torsional buckling must be prevented by adequately bracing the compression zone of the beam. If plastic design is used, the maximum permitted unbraced length for the compression flange of I-shaped members to avoid lateral torsional buckling is obtained from equation (A-1-5) in Section 1.2.3(2c) of Appendix 1 of the AISC specification as follows:

$$L_{pd} = \left[0.12 - 0.076\left(\frac{M_1'}{M_2}\right)\right]\left(\frac{E}{F_y}\right)r_y, \tag{A-1}$$

where

M_1 = Smaller moment at the end of the unbraced length of the beam, kip-in.,
M_2 = Larger moment at the end of the unbraced length of the beam, kip-in. = $M_{u,p}$,
r_y = Radius of gyration about the weak axis,
M_1' = effective moment at end of unbraced length

When the moment within the unbraced length $\geq M_2$, use $\dfrac{M_1'}{M_2} = +1$

When $M_{mid} \leq \dfrac{M_1 + M_2}{2}$, *use* $M_1' = M_1$

When $M_{mid} > \dfrac{M_1 + M_2}{2}$, *use* $M_1' = (2M_{mid} - M_2) < M_2$

M_1 and M_{mid} are positive (+) when they cause compression in the same flange as M_2, and they are negative otherwise.
E = modulus of elasticity of steel = 29,000 ksi.

Additionally, the section has to be a compact shape (i.e., it must not be susceptible to local buckling). Table 6-2 (i.e., AISC Specification Table B4.1b (cases 10 and 15)) provides the limits required for compact I-shaped sections as follows:

$$\frac{b_f}{2t_f} \leq \frac{65}{\sqrt{F_y}} \quad and$$
,
$$\frac{h}{t_w} \leq \frac{640}{\sqrt{F_y}}$$

where F_y is the yield stress of the steel in ksi. Other requirements include the following:

- F_y should be less than or equal to 65 ksi, and
- The design axial strength in compression should be no greater than $0.75\phi A_g F_y$ (AISC Specification Section 1.3.2d).

If the maximum permitted unbraced length in equation (A-1) cannot be satisfied, the load effects on the member must be obtained using elastic analysis and not plastic analysis. The principle of virtual work is used in this text for plastic analysis and involves the following steps:

1. Based on the elastic moment distribution, determine where the plastic hinges (i.e., apart from the naturally occurring hinges at rollers or pinned supports) need to form within a span to create a mechanism in that span. Note that each beam span is considered separately and independently of the other spans.
2. Assume a virtual vertical displacement at the plastic hinge within the span; also assume that rotations can only occur at the plastic hinges and at roller or pinned supports. Therefore, the member segments between the plastic hinges and roller or pinned supports deflect rigidly (i.e., linearly between the hinges).
3. Determine the virtual rotations at the hinges resulting from the applied virtual displacement.
4. At each formed hinge, excluding pinned or roller supports at the ends of a member, calculate the internal work done by the plastic moment, $M_{u,p}$.
5. Determine the external work done by the applied loads acting through the applied virtual displacement. For a concentrated load, the external work done is the product of the factored concentrated load and the virtual vertical displacement at that load. For a uniform load, the external work done is the product of the factored uniform load and the area of the virtual displacement diagram within the extent of the uniform load.
6. Sum up the external work done by all of the applied factored loads within the span.
7. Sum up all of the internal work done by the plastic moment, $M_{u,p}$, at the formed hinges (i.e., excluding pinned or roller supports at the exterior end of the span).
8. Using the principle of virtual work, equate the total internal work done to the total external work done and solve for the required plastic moment strength, $M_{u,p}$, as a function of the applied factored loads. $M_{u,p}$ is calculated for each span separately and independently of the other spans, and the largest value is the required plastic moment strength for the continuous beam or girder.
9. Determine the required plastic section modulus, Z_x, using the limit states design relationship that the required plastic moment strength, $M_{u,p}$, shall not be greater than the design strength, ϕM_n; that is, $M_{u,p} \leq \phi M_n = \phi Z_x F_y$.

The plastic analysis and design of continuous beams and girders will now be illustrated with several examples.

EXAMPLE A-1

Plastic Load Capacity for Statically Indeterminate Beams

Determine the plastic load capacity for the statically indeterminate beams shown in Figure A-2.

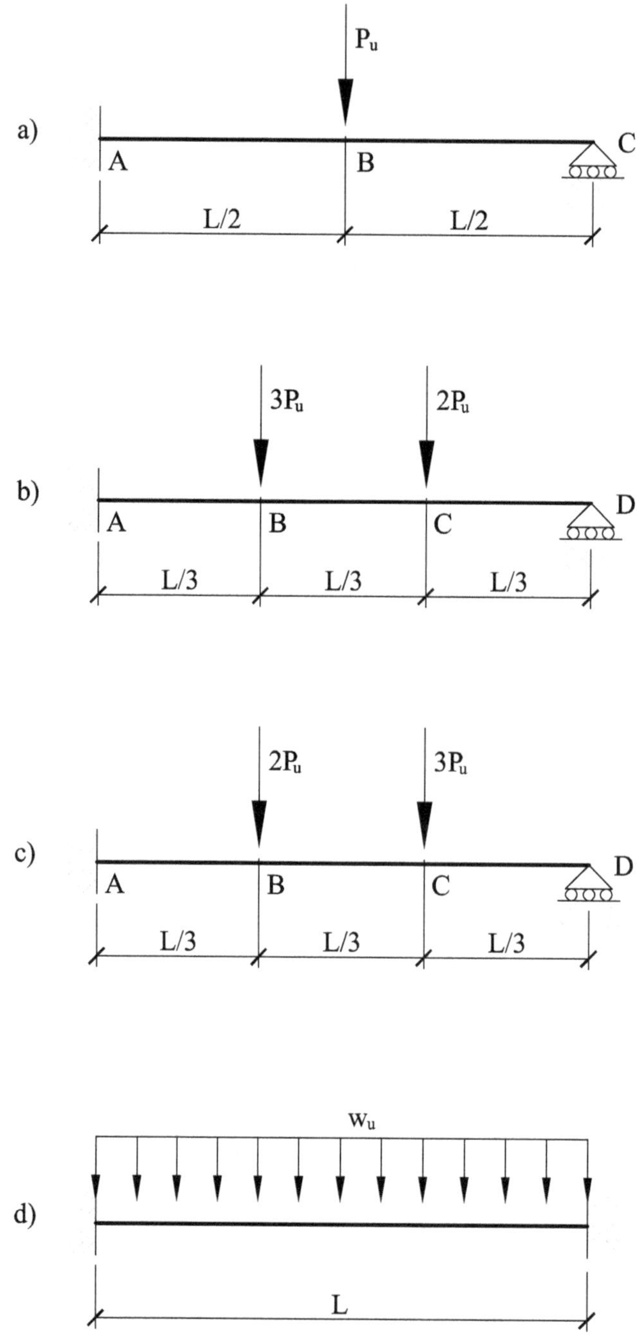

FIGURE A-2 Statically indeterminate beams for Example A-1

Solution

a. For the beam in Figure A-2a, two plastic hinges are necessary to create a collapse mechanism (i.e., an unstable structure that will continue to undergo deflection without any increase in load). The plastic hinges will form at points *A* and *B*. Because point *C* is already a hinge, these hinges will result in a mechanism or an unstable structure.

From the elastic moment distribution shown in Figure A-3, the plastic hinge will first form at point A because point A has the highest elastic moment compared with point B or any other point within the beam.Figure A-4 shows the beam in its deflected shape with plastic hinges formed at points A and B. The member segments between the hinges are assumed to deflect as rigid bodies.

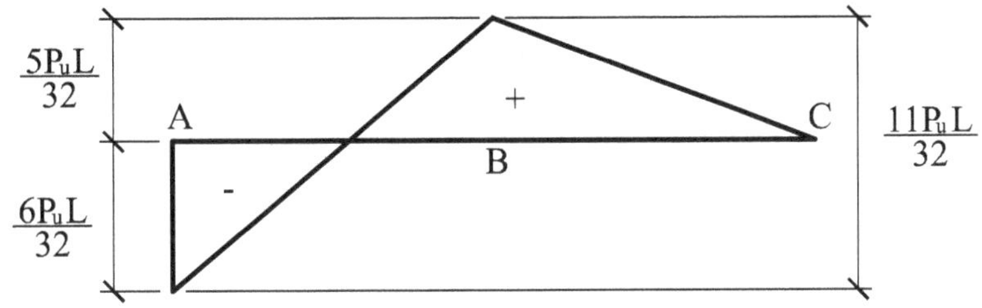

FIGURE A-3 Elastic moment distribution for figure A-2a

To determine whether there are sufficient hinges to form a mechanism or an unstable structure, the reader should recall from structural analysis that Figure A-4 is an unstable structure and thus a mechanism. Assuming that point A rotates through an angle, q, as shown, the rotation at point C, by geometry, will also be q, and the rotation at point B will therefore be 2θ (i.e., θ + θ).

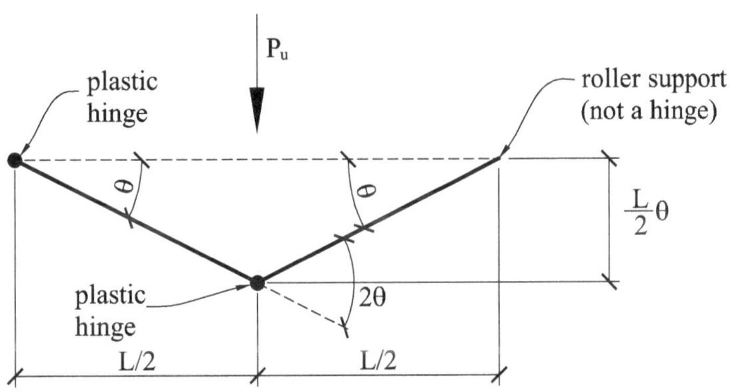

FIGURE A-4 Virtual displacement of the beam in Figure A-2a due to plastic hinges

The total internal work done by the plastic moments at the plastic hinges is the sum of the product of the plastic moments and the corresponding rotations at the hinges. Thus, the total internal work done is $M_{u,p}\,\theta + M_{u,p}\,(2\theta) = 3M_{u,p}\,\theta$.

The total external work done by the applied factored loads is S(Factored concentrated load × Vertical displacement of the mechanism at the concentrated load) = $P_u\,(L/2)\theta = P_u L\theta/2$.

Using the principle of virtual work, the internal work done must be equal to the external work done; thus,

$$3M_{u,p}\theta = P_u L\theta / 2.$$

Therefore, the required plastic moment strength, $M_{u,p} = P_u L/6$.

A steel section can be selected that will provide this required plastic moment.

b. For the beam in Figure A-2b, the plastic hinges will form at point A and at point B or C. Reviewing the applied loads, it is more likely that the plastic hinge will form at point B rather than at point C.

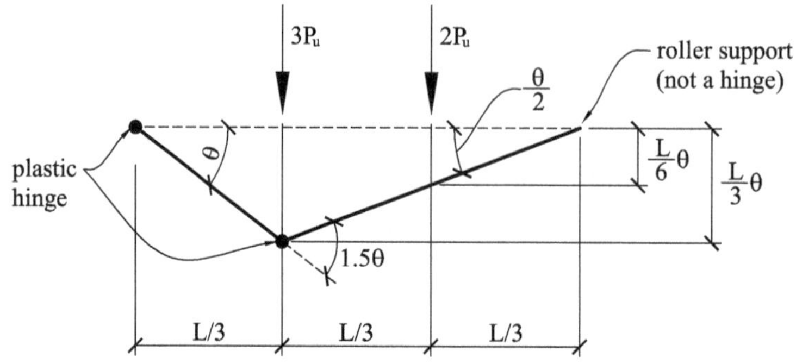

FIGURE A-5 Virtual displacement of the beam in Figure A-2b due to plastic hinges

From Figure A-5, the total internal work done at the plastic hinges is $M_{u,p}\,\theta + M_{u,p}(1.5\theta) = 2.5\,M_{u,p}\,\theta$.

The total external work done by the factored loads is S(Factored concentrated load × Vertical displacement of the mechanism at the load) = $3P_u(L/3)\theta + 2P_u(L/6)\theta = 1.33P_uL\theta$.

Using the principle of virtual work, the internal work done must be equal to the external work done; thus,

$$2.5M_{u,p}\theta = 1.33P_uL.$$

Therefore, the required plastic moment capacity is $M_{u,p} = 0.533P_uL$.

A section can be selected that will provide this required plastic moment.

c. For the beam in Figure A-2c, plastic hinges will form at point *A* and at point *B* or *C*, depending on the elastic moment distribution. Given the higher load applied at point *C*, it is more likely that a plastic hinge will form there first before forming at point *B*.

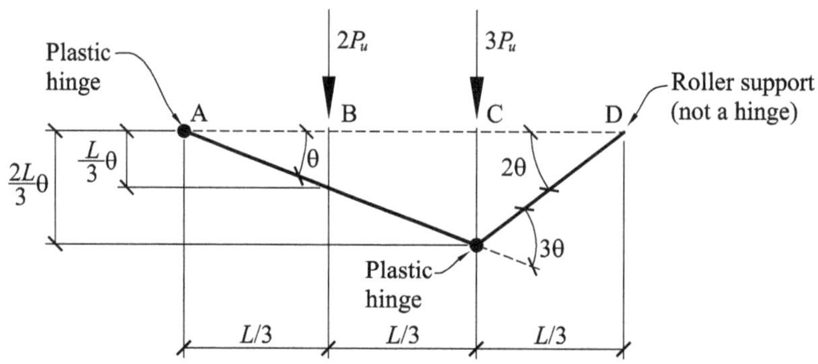

FIGURE A-6a Virtual displacement of the beam in Figure A-2c due to plastic hinges at points A and C

From Figure A-6a, the total internal work done at the plastic hinges is $M_{u,p}\theta + M_{u,p}(3\theta) = 4M_{u,p}\theta$.

The total external work done by the factored loads is S(Factored concentrated load × Vertical displacement of the mechanism at the load) = $2P_u(L/3)\theta + 3P_u(2/3L\theta) = 2.67P_uL\theta$.

Using the principle of virtual work, the internal work done must be equal to the external work done; thus,

$$4M_{u,p}\theta = 2.67P_uL\theta.$$

Therefore, the required plastic moment capacity, $M_{u,p} = 0.667P_uL$.

As a check, it can be verified that the plastic hinge does occur at point C and *not* at point B. If a plastic hinge were assumed to form at point B instead of point C, the deflected shape of the mechanism would be as shown in Figure A-6b.

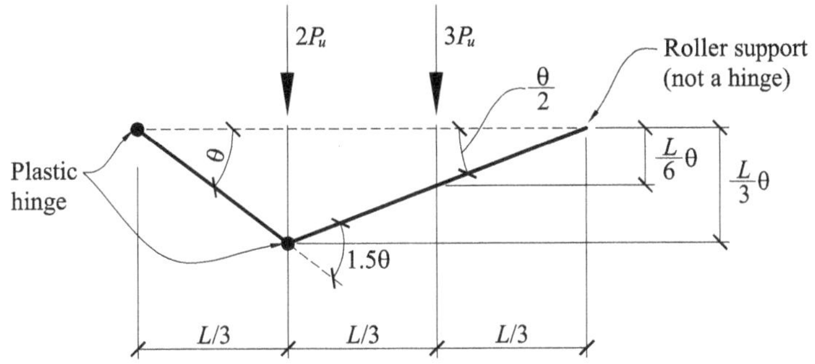

FIGURE A-6b Virtual displacement of the beam in Figure A-2c due to plastic hinges at points A and B

From Figure A-6b, the total internal work done at the plastic hinges is $M_{u,p}\theta$ + $M_{u,p}(1.5\theta) = 2.5M_{u,p}\theta$.
The total external work done by the factored loads is S(Factored concentrated load × Vertical displacement at the load) = $2P_u(L/3)\theta + 3P_u(L/6)\theta = 1.17P_uL\theta$.
Using the principle of virtual work, the internal work done must be equal to the external work done; thus,

$$2.5M_{u,p}\theta = 1.17P_uL\theta.$$

Therefore, the required plastic moment capacity, $M_p = 0.467\,P_uL$, is less than the plastic moment capacity of $0.667P_uL$ required for the plastic hinge forming at point C. Therefore, the condition with the plastic hinge at point C requires a higher capacity and thus governs the design, as expected.

d. For the beam in Figure A-2d, based on the elastic moment distribution (see Figure A-7), the plastic hinges will form at points A, B, and C to create a mechanism.
From Figure A-8, the total internal work done at the plastic hinges is $M_{u,p}\theta + M_{u,p}(2\theta) + M_{u,p}\theta = 4M_p\theta$.

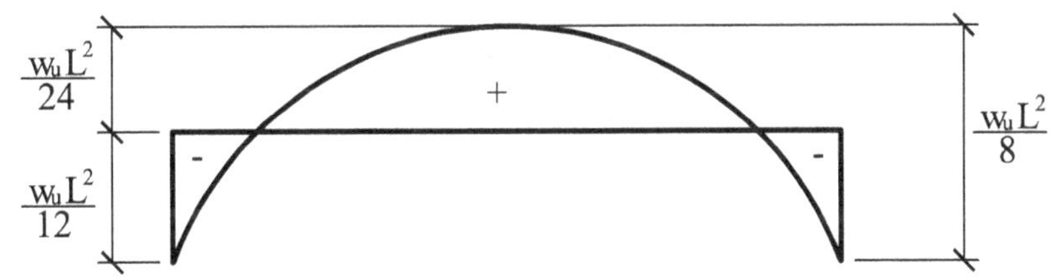

FIGURE A-7 Elastic moment distribution for Figure A-2d

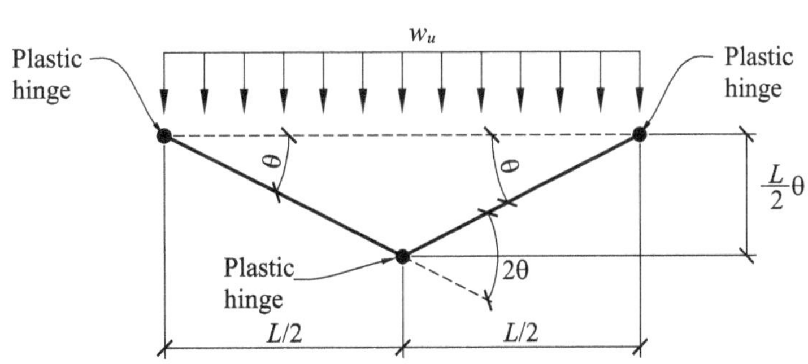

FIGURE A-8 Virtual displacement of the beam in Figure A-2d due to plastic hinges

The total external work done by the factored uniformly distributed loads is S(Factored uniform load × Area of the deflected shape of the mechanism under the load) = $w_u\,(^1/_2L)(L/2)\theta = w_uL^2\theta/4$.

Using the principle of virtual work, the internal work done must be equal to the external work done; thus,

$$4M_{u,p}\theta = w_uL^2\theta/4.$$

Therefore, the required plastic moment capacity, $M_{u,p} = w_uL^2/16$.

EXAMPLE A-2

Plastic Moment Capacity for a Beam with Uniform Loads

Determine the required plastic moment, $M_{u,p}$, for the statically indeterminate three-span beam shown in Figure A-9 with a typical span of 40 ft. The beam is part of a floor framing with a uniformly distributed dead load of 3.0 kips/ft. and a uniformly distributed live load of 3.0 kips/ft. Select a W-shaped section to support this load. Assume ASTM A992 Grade 50 steel.

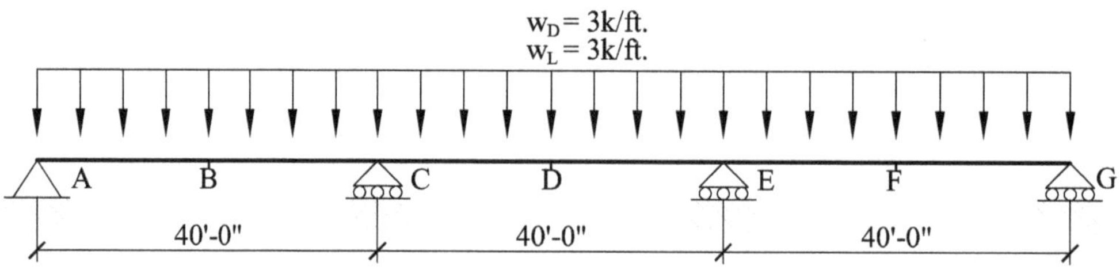

FIGURE A-9 Statically indeterminate beam for Example A-2

Solution

Based on the elastic moment distribution (see Figure A-10), the collapse mechanism will form first in the end spans with plastic hinges at points *B* (not at midspan) and *C*, or at points *E* and *F* (not at midspan). If plastic hinges were to form in the middle span, based on the symmetrical loading, they would form at points *C* and *E* and at the midspan between points *C* and *E*. Considering span *EFG*, assume that the distance from the plastic hinge at point *F* to the first interior support is x_p.

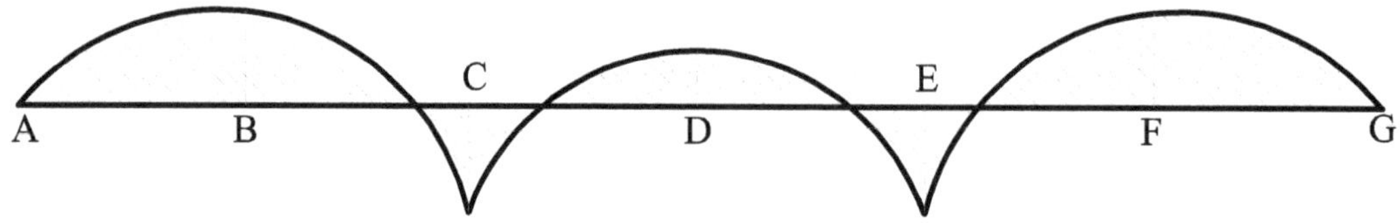

FIGURE A-10 Elastic moment distribution for Figure A-9

From Figure A-11, the total internal work done at the plastic hinges is $M_{u,p}\theta + M_{u,p}\left(\dfrac{L}{L-x_p}\right)\theta$.

The total external work done by the factored uniformly distributed loads is S(Factored uniform load × Area of the deflected shape of the mechanism under the load) $= \dfrac{1}{2}(L)(x_p\theta)w_u$.

Using the principle of virtual work, the internal work done must be equal to the external work done; therefore,

$$\frac{1}{2}Lx_pw_u\theta = M_{u,p}\left(\frac{2L-x_p}{L-x_p}\right)\theta.$$

Thus, the plastic uniform load capacity, $w_u = \dfrac{2M_{u,p}}{L}\left(\dfrac{2L-x_p}{Lx_p - x_p^2}\right)$.

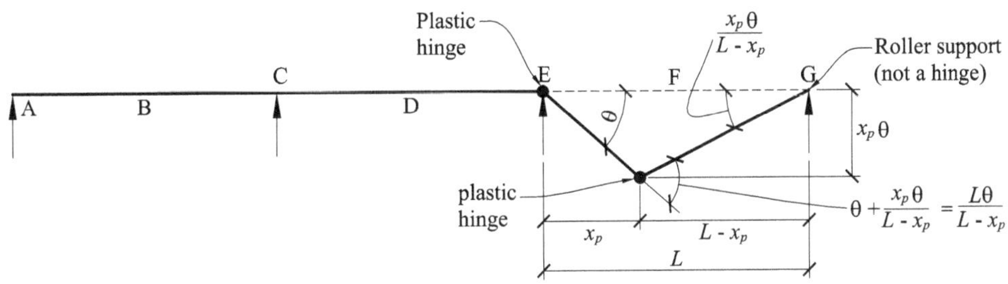

FIGURE A-11 Virtual displacement of the beam in Figure A-9 due to plastic hinges in span EFG

The distance x_p required to obtain the maximum load capacity can be determined using the principles of calculus by requiring that $\frac{dw_u}{dx_p} = 0$, which yields $x_p = 0.586L$.

Therefore, the required plastic moment capacity is obtained by solving the virtual work equation above:

$$\frac{1}{2}Lx_p w_u \theta = M_{u,p}\left(\frac{2L - x_p}{L - x_p}\right)\theta$$

$$\frac{1}{2}L(0.586L)w_u = M_{u,p}\left(\frac{2L - 0.586L}{L - 0.586L}\right).$$

Thus, $M_{u,p} = 0.0858w_u L^2$

The factored load, $w_u = 1.2w_D + 1.6w_L = 1.2(3.0 \text{ kips/ft.}) + 1.6(3.0 \text{ kips/ft.}) = 8.4 \text{ kips/ft.}$

Therefore, $M_{u,p} = 0.0858(8.4 \text{ kips/ft.})(40 \text{ ft.})^2 = 1153 \text{ ft.-kips} \le \phi M_n = \phi Z_x F_y$.

Hence, the section modulus required, $Z_x = \frac{M_{u,p}}{\phi F_y} = \frac{1153 \text{ ft.-kips}(12 \text{ in./ft.})}{(0.9)(50 \text{ ksi})} = 307.5 \text{ in}^3$.

Therefore, use W30 × 99 ($Z_x = 312 \text{ in.}^3$; see *AISCM* Table 3-2).

EXAMPLE A-3

Plastic Moment Capacity for a Girder with Concentrated Loads

Determine the required plastic moment capacity M_p for the statically indeterminate three-span girder shown in Figure A-12 with a typical span of 35 ft. and the service loads shown. Select a W-shaped section to support this load. Assume ASTM A992 Grade 50 steel.

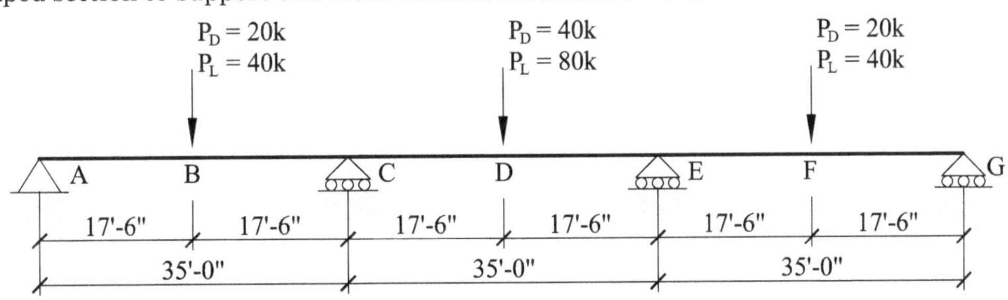

FIGURE A-12 Statically indeterminate beam for Example A-3

Solution

We will investigate the maximum load required for mechanisms to develop in spans *ABC*, *CDE*, and *EFG*. The mechanism that yields the highest required plastic moment will be the governing mechanism.

Span *ABC*:

For this span, the factored concentrated load at midspan, $P_u = 1.2P_D + 1.6P_L = 1.2(20 \text{ kips}) + 1.6(40 \text{ kips}) = 88$ kips.

The location of the hinges in this span will be at the concentrated load at point *B* and at point *C*, as shown in Figure A-13.

From Figure A-13, the total internal work done at the plastic hinges is $M_{u,p}\theta + M_{u,p}(2\theta) = 3M_{u,p}\theta$.

The total external work done by the factored concentrated loads is S(Factored concentrated load × Vertical displacement of the mechanism at the load) $= P_{u,ABC}\left(\frac{L}{2}\theta\right)$.

Using the principle of virtual work, the internal work done must be equal to the external work done; thus,

$$3M_{u,p}\theta = P_{u,ABC}\left(\frac{L}{2}\theta\right) = (88 \text{ kips})\left(\frac{35 \text{ ft.}}{2}\right)\theta.$$

Therefore, $M_{u,p} = 513.3$ ft.-kips.

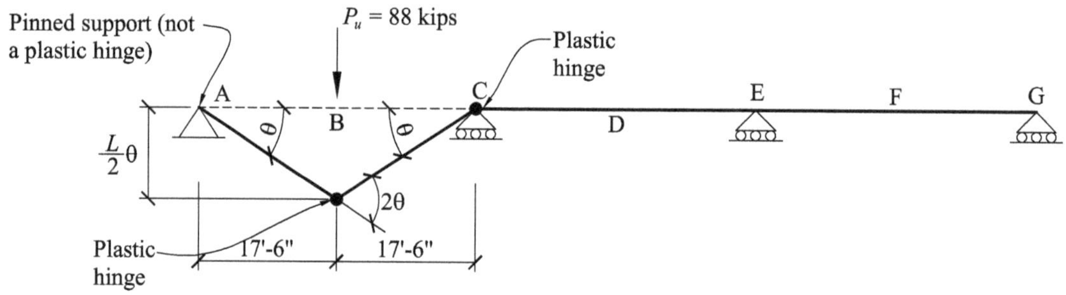

FIGURE A-13 Virtual displacement in span ABC

Span *CDE*:

For this span, the total factored concentrated load at midspan, $P_u = 1.2P_D + 1.6P_L = 1.2(40 \text{ kips}) + 1.6(80 \text{ kips}) = 176$ kips.

The location of the hinges in this span will be at the concentrated load at point *D* and at points *C* and *E*, as shown in Figure A-14.

From Figure A-14, the total internal work done at the plastic hinges is $M_{u,p}\theta + M_{u,p}(2\theta) + M_{u,p}\theta = 4M_{u,p}\theta$.

The total external work done by the factored concentrated loads is S(Factored concentrated load × Vertical displacement of the mechanism at the load) $= P_{u,CDE}\left(\frac{L}{2}\theta\right)$.

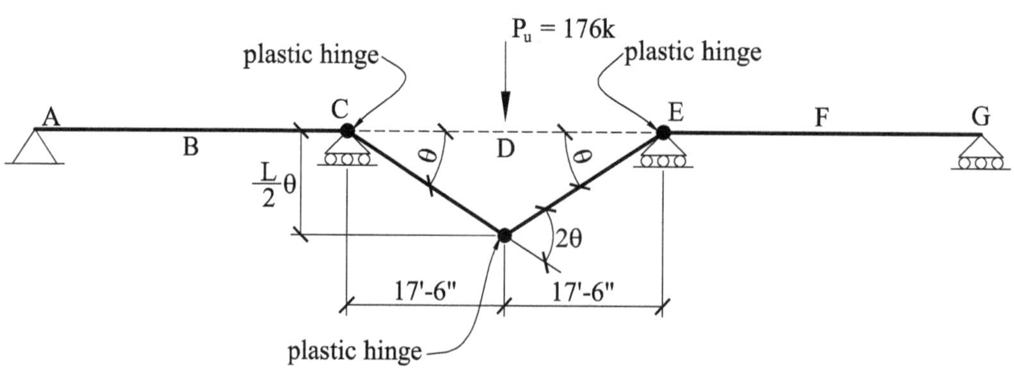

FIGURE A-14 Deflected shape of beam after formation of plastic hinges in span CDE

Using the principle of virtual work, the internal work done must be equal to the external work done; thus,

$$4M_{u,p}\theta = P_{u,CDE}\left(\frac{L}{2}\theta\right) = (176\text{ kips})\left(\frac{35\text{ ft.}}{2}\right)\theta.$$

Therefore, $M_{u,p}$ = 770 ft.-kips. Governs

Span *EFG*:

The location of hinges for this span will be similar to that for span *ABC*, and thus the required plastic moment will also be $M_{u,p}$ = 513.3 ft.-kips.

The highest required plastic moment for all three spans will govern the capacity of the girder; therefore, $\mathbf{M_{u,p}}$ **= 770 ft.-kips.**

Using the limit states design equation ($M_{u,p} \le \phi M_n$) yields

$770\text{ ft-kips} \le \phi M_n = \phi Z_x F_y$.

The required section modulus, $Z_x = \dfrac{M_{u,p}}{\phi F_y} = \dfrac{770\text{ ft.-kips}(12\text{ in./ft.})}{0.9(50\text{ ksi})} = 205.3\text{ in.}^3$.

Therefore, use W24 × 84 (Z_x = 224 in.^{3}).

References

[1] American Institute of Steel Construction. *Steel Construction Manual*. 16th ed. Chicago: AISC, 2023.

[2] Bhatt, P., and H. M. Nelson. *Marshall and Nelson's Structures*. 3rd ed. London: Longman, 1990.

[3] Disque, R. O. *Applied Design in Steel*. New York: Van Nostrand Reinhold, 1971.

Exercises

A-1. Determine the required plastic moment capacity, $M_{u,p}$, for the statically indeterminate two-span girder shown in Figure A-15. Select a W-shaped section to support this load. Use ASTM A992 Grade 50 steel.

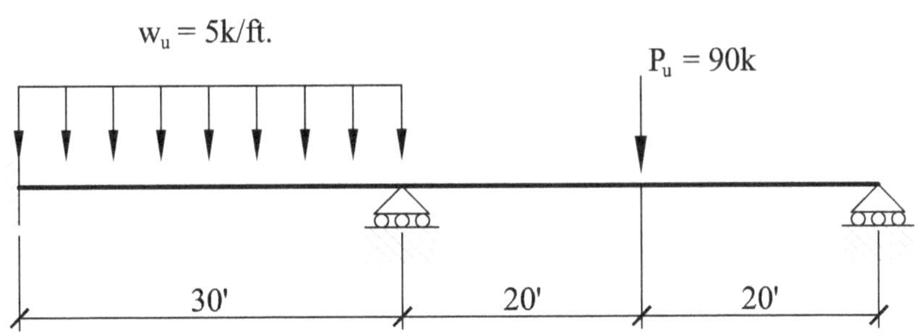

FIGURE A-15 Statically indeterminate beam for Exercise A-1

Index

C

D

E

F

G

H

I

J

K

L

M

N

O

P

R

S

T

U

V

W

X

Y

Z

www.ingramcontent.com/pod-product-compliance
Lightning Source LLC
LaVergne TN
LVHW060315240926
848814LV00005B/34